Climate Change Management

Series Editor

Walter Leal Filho, International Climate Change Information and Research Programme, Hamburg University of Applied Sciences, Hamburg, Germany

The aim of this book series is to provide an authoritative source of information on climate change management, with an emphasis on projects, case studies and practical initiatives – all of which may help to address a problem with a global scope, but the impacts of which are mostly local. As the world actively seeks ways to cope with the effects of climate change and global warming, such as floods, droughts, rising sea levels and landscape changes, there is a vital need for reliable information and data to support the efforts pursued by local governments, NGOs and other organizations to address the problems associated with climate change.

This series welcomes monographs and contributed volumes written for an academic and professional audience, as well as peer-reviewed conference proceedings. Relevant topics include but are not limited to water conservation, disaster prevention and management, and agriculture, as well as regional studies and documentation of trends. Thanks to its interdisciplinary focus, the series aims to concretely contribute to a better understanding of the state-of-the-art of climate change adaptation, and of the tools with which it can be implemented on the ground.

More information about this series at http://www.springer.com/series/8740

Walter Leal Filho · Gustavo J. Nagy ·
Marco Borga · Pastor David Chávez Muñoz ·
Artur Magnuszewski
Editors

Climate Change, Hazards and Adaptation Options

Handling the Impacts of a Changing Climate

Springer

Editors
Walter Leal Filho
HAW Hamburg
Hamburg, Germany

Marco Borga
Department of Land, Environment,
Agriculture and Forestry
University of Padova
Legnaro, Italy

Artur Magnuszewski
Department of Hydrology
University of Warsaw
Warsaw, Poland

Gustavo J. Nagy
Oceanografia y Ecologia Marina.
Coastal Climate Science
Universidad de la República de Uruguay
Montevideo, Uruguay

Pastor David Chávez Muñoz
Department of Engineering
Pontifical Catholic University of Peru
Lima, Peru

ISSN 1610-2002 ISSN 1610-2010 (electronic)
Climate Change Management
ISBN 978-3-030-37424-2 ISBN 978-3-030-37425-9 (eBook)
https://doi.org/10.1007/978-3-030-37425-9

This Springer imprint is published by the registered company Springer Nature Switzerland AG
The registered company address is: Gewerbestrasse 11, 6330 Cham, Switzerland

Preface

Hazards may be defined as the potential occurrence of a natural or human-induced physical event, trend or physical impact, which may cause loss of life, injury or other health impacts, as well as damages and/or losses to property, infrastructure, livelihoods, service provision and environmental resources.

Due to climate change, the frequency and intensity of hazard events such as extreme weather events (e.g. floods, droughts and heatwaves) on the one hand, but also of forest fires and damages to agricultural productivity (especially crop yield) on the other, are expected to increase in the future. Hazards may change exposure patterns, lead to substantial breakdowns of infrastructure, damages to property and ultimately decrease the resilience of households and communities. Moreover, the links between climate hazards and health are strong, with mortality and morbidity rates rising as a result of them. Hazards pose additional pressures to both human and natural systems. The Intergovernmental Panel on Climate Change (IPCC) outlined on its 4th (2007) and 5th (2014) Assessment Reports the need to manage the risks of extreme events and the hazards they bring about, to advance climate change adaptation.

It is important to better understand what climate change hazards are, how the different components of climate change including frequency, intensity, variability and uncertainty play together and relate to ecosystems, and how a better understanding of vulnerability of people to climate change in terms of sensitivity can be achieved. Also, it is equally important to identify adaptation options based on the adaptive capacity—which does vary—between sectors, populations and ecosystems.

This book is an attempt to address the need for interdisciplinary publications, which look at the subject matter of climate change and hazards on the one hand, but which outline adaptation options on the other. It contains articles presented and discussed at a series of events organised by the International Climate Change Information and Research Programme (ICCIRP), held in Lima (Peru), Padova (Italy) and Warsaw (Poland) in 2019. The events gathered a set of specialists from different backgrounds and disciplines, whose know-how is amassed and documented on this publication. It is structured in two main parts:

Understanding Climate Change Hazards: In this part, a variety of studies and technical assessments are showcased, aimed at leading to a better understanding of what climate hazards are and their implications.
Handling the Impacts of Climate Hazards via Adaptation Methods and Options: In this part, a wide range of experiences and case studies is presented, which illustrate the many ways via which climate hazards are being dealt with, along with the role of education and communication.

Thanks to its scope, the book *Climate Change, Hazards and Adaptation Options* not only provides essential scientific information, but also describes facts, trends and case studies from various geographical regions.

We hope this book will foster a broader understanding of the subject matter of climate hazards and risks and will support the search for suitable adaptation measures so as to ensure that their impacts may be minimised.

Hamburg, Germany — Walter Leal Filho
Montevideo, Uruguay — Gustavo J. Nagy
Legnaro, Italy — Marco Borga
Lima, Peru — Pastor David Chávez Muñoz
Warsaw, Poland — Artur Magnuszewski

Contents

Understanding Climate Change Hazards

Hydrometeorological Analysis of an Extreme Flash-Flood: The 28 September 2012 Event in Murcia, South-Eastern Spain

A. Amengual and Marco Borga

Abstract Heavy precipitation following a prolonged summer drought led to widespread flash flooding across Andalucía, Murcia and Valencia in south-eastern Spain on 27, 28 and 29 September 2012. On September 28, an extreme flash-flood developed when 214 mm of rain fell in 8 h over the semi-arid and medium-sized Guadalentín River basin up to Paretón (~2800 km^2). Six fatalities were reported, hundreds of homes were evacuated and a bridge spanning an ephemeral channel was undermined as the flood bore routed through normally dry river beds. Current estimates of flood damage are of €64 million, including extensive losses in livestock and agriculture. The last event of this magnitude over the Guadalentín occurred on 19 October 1973. Availability of high-resolution rainfall estimates from dense rain-gauge networks and radar observations, together with flood response observations derived from stream-gauge data and post-event surveys, provides the opportunity to study the hydrometeorological mechanisms associated with the responsible convective systems and the associated flash-flood. Results show that the basin faced a very rare rainfall event with extreme intensities and accumulations which, in combination with the catchment properties, led to extreme runoff. The distinct soil substrates and basin morphology led to varied runoff responses that required a multisite calibration of a hydrological model so as to successfully reproduce this flash-flood. Heavy precipitations resulted from deep convection triggered by local orography as well as the subsequent passage of a slow-moving mesoscale convective system (MCS). The motion of the MCS was crucial for exacerbating peak discharges, whereas times to peaks were modulated by the river network geometry and the temporal distribution of the rainfall rates. The roles of the different anthropogenic activities on the mitigation or intensification of the hazardous effects of this extreme event are also highlighted. Finally, some recommendations are proposed in order to mitigate future impacts of such catastrophic floods in a changing climate era.

A. Amengual (✉)
Grup de Meteorologia, Departament de Física, Universitat de Les Illes Balears, Palma, Mallorca, Spain
e-mail: arnau.amengual@uib.es

M. Borga
Department of Land, Environment, Agriculture and Forestry, University of Padova, Legnaro, Italy
e-mail: marco.borga@unipd.it

W. Leal Filho et al. (eds.), *Climate Change, Hazards and Adaptation Options*, Climate Change Management, https://doi.org/10.1007/978-3-030-37425-9_1

Introduction

Climate change poses major challenges for current societies given that it will impact mean and extreme regimes. A warming of 0.85 °C has been observed in the average global temperature over the period 1880–2012. Associated with this warming, the number of heavy precipitation episodes has likely increased worldwide since the mid-twentieth century. Confidence in a likely increase in either the frequency or intensity of heavy precipitation is highest over Europe (Stocker et al. 2013). In particular, the Mediterranean emerges as an especially responsive region to climate change and extremes. Basin-wide, annual mean temperature has increased by 1.4 °C since the late nineteenth century (Cramer et al. 2018).

Future warming in the Mediterranean region is projected to be associated with a prevailing increase in the annual number of precipitation extremes (Beniston et al. 2007). In summer, precipitations are projected to substantially decrease, but the frequency of extreme rainfalls is expected to increase over large regions (Giorgi and Lionello 2008). Heavy rainfall episodes are likely to intensify by 10–20% in all seasons, except summer (Toreti et al. 2013; Toreti and Naveau 2015). Furthermore, high levels of exposure and vulnerability to flash-flooding are currently present across this area.

The Western Mediterranean region is especially prone to heavy precipitation and flash-flooding during late summer and early autumn. The relatively high sea surface temperature increases the convective available potential energy of the overlying moist air masses through sensible and latent heat flux exchanges. Together with the intrusion of polar cold air masses aloft, the complex orography and land-sea contrasts promote the lifting of low-level conditionally unstable air, favouring the triggering of deep moist convection.

High precipitation rates can remain during several hours over individual catchments. This persistence is often associated with prominent orography that anchors quasi-stationary mesoscale convective systems (MCSs; Doswell et al. 1996; Kolios and Feidas 2010). In the Spanish Mediterranean semi-arid area, many small- to medium-sized catchments are steep, densely urbanized and close to the coastline. Many of these rivers are ephemeral, with short hydrological response times, and dominated by extreme events of low frequency but high magnitude. All these factors enhance flood risk and further exacerbate unexpected and extensive flood damage (Camarasa-Belmonte and Segura-Beltrán 2001; Amengual et al. 2007, 2015, 2017).

Within this framework, the extreme flash-flood over the Guadalentín River basin in Murcia, south-eastern Spain, on 28 September 2012 is examined in detail (Fig. 1). Six fatalities were reported, hundreds of homes were evacuated and a bridge spanning an ephemeral stream was undermined as the flood bore routed through normally dry river beds. Current estimates of flood damage are of €64 million, including extensive losses in livestock and agriculture. This catchment is a paradigmatic example of the semi-arid and medium-sized watersheds in Mediterranean Spain that suffer sudden hydrological responses to heavy precipitation.

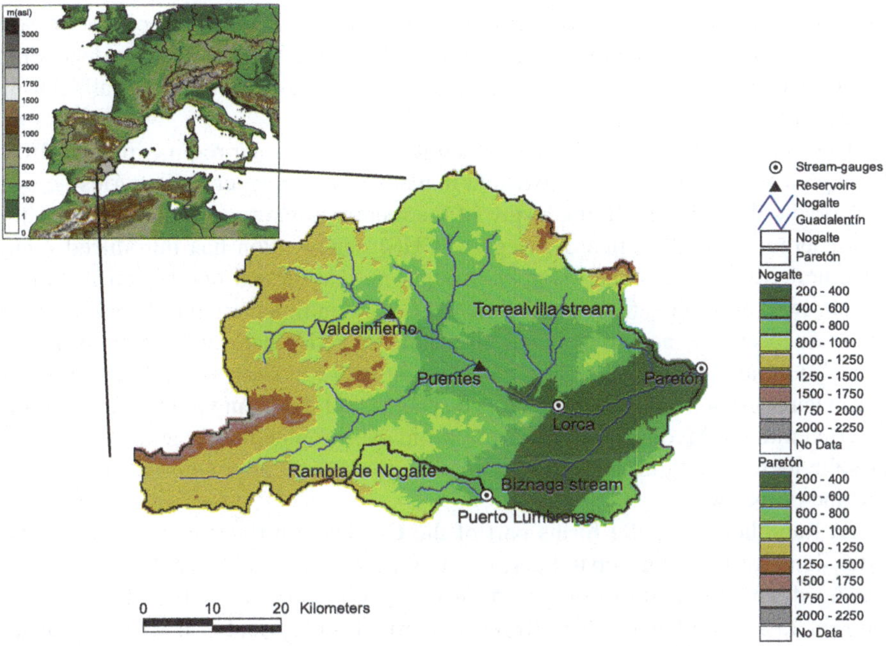

Fig. 1 Top left figure: overview of the Western Mediterranean region. The grey shaded area displays the CHS demarcation where the Guadalentín basin is located. Bottom centre figure: The Guadalentín River basin up to Paretón. The affected populations, available stream-gauges and reservoirs are indicated as well as tributaries mentioned in the text

The availability of high-resolution rainfall estimates from radar observations and dense automatic rain-gauge networks, together with flow measurements from stream-gauge data and maximum discharge estimates from post-event surveys, has provided the opportunity to highlight the most relevant hydrometeorological mechanisms associated with this flash-flood. The consequences on the hydrological response of the different anthropogenic activities carried out inside the Guadalentín basin have been assessed as well.

The Study Region

The Guadalentín River is the most important affluent of the Segura River, ending in the vicinity of Murcia City. This catchment spans over a drainage area of 2848.1 km^2 up to Paretón (Fig. 1). The height transition is from above 2000 m in its mountainous headwaters to 210 m at this hydrometric section. The Guadalentín basin is located in one of the most arid regions of Spain. High mountainous ranges shelter this area from the passage of the rainfall-bearing Atlantic cold fronts. Thus, precipitation mainly comes from easterly moist flows associated with sub-synoptic scale, less frequent Mediterranean disturbances. The rainfall regime is typical of a semi-arid climate: annual precipitations range from barely 300 to above 500 mm, depending on altitude. Most of the annual amounts are collected as torrential rainfalls during the extended warm season.

The Guadalentín River is characterized by a very irregular regime, passing from large periods of very low—or inexistent in many of its ephemeral tributaries—flows to sporadic flash-floods. Being aware of the recurrent nature of these episodes, the Confederación Hidrográfica del Segura (CHS) demarcation has introduced many structural elements for water supply, drought mitigation and flood control. In particular, two reservoirs are located along the Guadalentín River up to Paretón. These dams are located at the outlets of the Valdeinfierno and Puentes mountainous basins. In addition, an artificial diversion connects directly the river to the Mediterranean Sea at Paretón (Fig. 1). Therefore, large discharge and sediment volumes are partially diverted into the Mediterranean so as to avoid catastrophic flooding in Murcia City. Despite these structural measures, extreme episodes still pose significant threats to life and property.

La Rambla de Nogalte forms part of the Guadalentín river network. The basin extension is of 124.3 km^2 up to Puerto Lumbreras town, where this ephemeral stream is gauged. The height transition is from close to 1100–360 m in 30 km and with a mean slope of 25% (Gil-Ocina 2016). However, livestock and agricultural activities carried out downstream of Puerto Lumbreras have produced the stream bed disconnection with the Guadalentín River. That is, La Rambla de Nogalte ends in an alluvial fan over the Guadalentín valley (Fig. 1). Therefore, the Biznaga basin collects most part of the water contribution of La Rambla de Nogalte when it is active. Biznaga has an extension of 404.8 km^2, spanning over the flattest portion of the Guadalentín basin, the river valley.

The Hydrometeorological Episode: A Brief Description

A strong closed low formed southwest of Portugal on 27 September 2012 to later slowly move eastward along southern and eastern Spain (Fig. 1). Deep convection was promoted in its forward flank due to the lifting of low-level warm and moist conditionally unstable air and its later destabilization by the intrusion of cold air aloft. Subsequent convection organized in bands that were frequently anchored by the complex orography of the region. Extraordinary torrential precipitations took place over Murcia on 28 September 2012 due to a V-shaped MCS that developed along a low-level convergence line in front of the coastline (Ducrocq et al. 2014).

According to the rain-gauge networks, daily precipitations were up to 240 mm over the Guadalentín River basin. Most of heavy precipitation occurred between 06 and 14 UTC on 28 September 2012, with recorded accumulations up to 37 and 119 mm in 5 min and 1 h, respectively. A maximum cumulative of 214 mm in 8 h was registered in Puerto Lumbreras. In Lorca town, the observed maximum discharge was of 616.3 $m^3\ s^{-1}$ at 13:15 UTC. At Paretón, two almost consecutive peak discharges of 1067.9 and 1081.2 $m^3 s^{-1}$ at 16:00 and 17:20 UTC were registered. Peak flows up to 939.7 $m^3 s^{-1}$ were diverted into the Mediterranean Sea by the artificial diversion.

This extreme flash-flood produced 6 fatalities and the evacuation of many inhabitants in Puerto Lumbreras and Lorca towns. Material losses were of 64 M€, as several infrastructures were destroyed and extensive livestock and agricultural areas were flooded. This natural hazard is known as the San Wenceslao flash-flood. In Spain, it is customary to name these extremes with the Christian Saint of the occurrence day. It is important to note that the Valdeinfierno and Puentes reservoirs were both closed during the entire event. Therefore, no discharge contributed downstream for these basins. Further details about the study region, Guadalentín River basin and hydrometeorological episode can be found in Amengual et al. (2015).

Analysis of the San Wenceslao Flash-Flood

The San Wenceslao flash-flood exemplifies the impact of torrential precipitations over a basin with persistent dry soils. The lithology of the Guadalentín features large spatial heterogeneities: most parts of the basin are settled on karstic and dolomitic fractured bedrocks, but others areas lay over relative impermeable metamorphic substrates (e.g. phyllites, schists, quartzites and micachists). A clear example of the spatial heterogeneities is found in the neighbouring La Rambla de Nogalte and Biznaga basins. The former is highly impermeable, whereas the latter is extremely permeable, despite being located a few kilometres away (Fig. 1).

Figure 2 shows the climatological soil moisture of the Guadalentín River basin, based on the computation of the daily water balance over the 2000–2014 period.

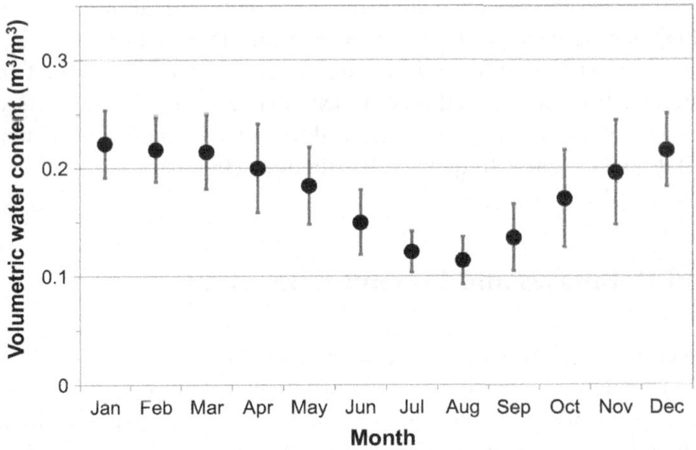

Fig. 2 Mean monthly climatological values of the soil moisture over the Guadalentín basin for the 2000–2014 period. Vertical bars denote the standard deviation

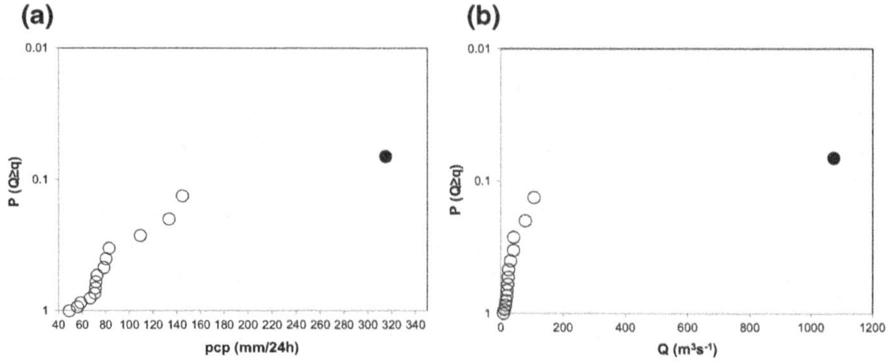

Fig. 3 Annual observed **a** maximum daily precipitation amounts and **b** hourly peak flows represented as cumulative distribution functions on a Gumbel chart. Black solid dots denote the 28 September 2012 episode. Recall that the 15-year time-series span over the 2000–2014 period

Monthly mean climatic volumetric water content is less than—or close to—0.25 throughout the entire hydrological year. Thus, persistent low antecedent water contents and high soil moisture capacities are inherent to this semi-arid catchment. Sparse vegetation, thin soils and convective precipitations—which easily exceeds the high initial soil infiltration capacity—lead to the generation of fast Hortonian surface flows and rapid flow velocities in the river streams.

Intense precipitations on 28 September 2012 ended a long summery drought period with very warm conditions. This episode can be considered as a paradigm of the organized convective systems that are likely responsible for much of flash-floods in this region. Two previous similar extreme events were recorded on 19 October 1973 and 2 September 1989.

An annual observed frequency analysis of the daily rainfall amounts and hourly peak discharges have been performed over the same 15-year period. At this aim, 28 automatic pluviometric stations and two stream-gauges deployed in the Guadalentín River basin have been considered (Fig. 1; see next section). The frequency analysis shows the rarity of the San Wenceslao flash-flood, which dominates the upper tail of the precipitation and runoff frequency distributions (Fig. 3).

Observed Databases and Precipitation Analysis

The region is monitored by the radar network of the Spanish Agency of Meteorology (AEMET). This network consists of a set of Doppler dual-polarized C-band radars located close to the most important cities of the country. Reflectivity scans are carried out with spatial and temporal resolutions of 1 km and 10 min, respectively. Three different radars cover the Guadalentín catchment and they are located in Murcia, Almería and Valencia cities. Out of these, only the Almería and Valencia radars were entirely operative during the 28 September 2012 episode. The closest weather radar to the Guadalentín (the Murcia radar, 50 km away) was severely affected by heavy precipitations, being off-line almost all the day. The Almería and Valencia radars were operative, being approximately located 100 and 200 km south and north of the basin, respectively.

Raw precipitation is observed with a temporal resolution of 5 min by a dense rain-gauge network of 108 automatic stations that are distributed over the CHS demarcation (Amengual et al. 2015). Out of these, 28 rain-gauges were selected to evaluate the hourly radar-derived rainfalls against the pluviometric values, because being sited within or very close to the Guadalentín basin. The rain-gauge spatial density is of ~125 km^2 (Fig. 2). 5-min runoff data are available at three different flow-gauges along the basin. These stream-gauges are located in Puerto Lumbreras, Lorca and Paretón (Fig. 1). 5-min stage data are also available at Valdeinfierno and Puentes dams. In addition, valuable post-flood field information was gathered at different hydrometric sections after the San Wenceslao flash-flood as well (Table 1).

Quantitative rainfall estimations have been derived from the Almería radar reflectivity scans from 27 to 29 September 2012 00 UTC. These observations were taken over a highly rouged terrain, which partially shielded the radar beams. Partial beam blocking was amended by using the procedure developed by Pellarin et al. (2002). Next, the WSR-88D convective relationship was used to convert the 10-min radar reflectivity to precipitation estimations (Hunter 1996). Finally, the hourly radar rainfall estimations were compared against observations from the selected rain-gauges. Biases in the precise hourly rainfall spatial distributions and amounts where corrected by applying a dynamical adjustment based on the pluviometric data and the mean fields (Cole and Moore 2008). Once the errors were removed, the statistical comparisons between the hourly rain-gauge and radar-derived precipitation fields depicted a good agreement, with a squared-correlation over 0.85.

Table 1 Available hydrological information over the Guadalentín basin for the San Wenceslao flash-flood

Hydrometric section	Basin area (km^2)	Available information
Valdeinfierno	430.6	Real-time reservoir stage time-series. Stage-discharge relationship not available
Puentes	1441.6	Real-time reservoir stage time-series. Stage-discharge relationship not available
Torrealvilla	236.4	Post-event survey (Hooke 2016)
Lorca	1831.4	Real-time river stage time-series; stage-discharge relationship Post-event survey (Benito et al. 2012)
Paretón	2848.1	Real-time river stage time-series; stage-discharge relationship
Nogalte	124.3	Real-time river stage time-series; stage-discharge relationship (malfunctioning) Post-event survey (Benito et al. 2012; Hooke 2016)

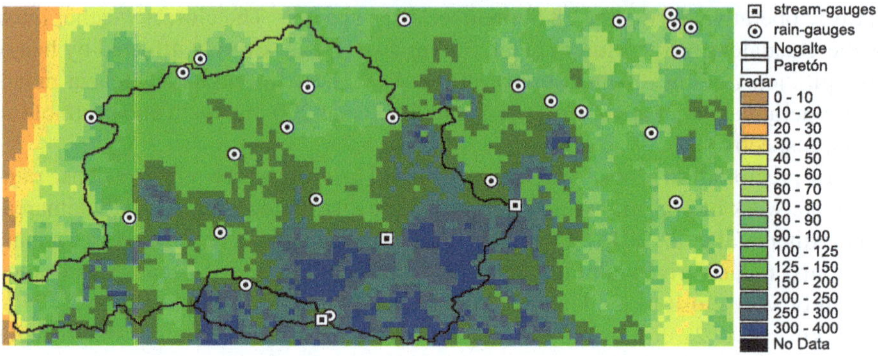

Fig. 4 Spatial distribution of the 48-h accumulated radar-observed precipitation from 27 to 29 September 2012 00 UTC. Rain-gauges used for the bias correction of the hourly radar rainfall estimations are shown as circles. Also displayed stream-gauges as squares

The spatial track of the MCS passage along the southernmost part of the basin is clearly visible from the distribution of the 48-h accumulated radar-derived precipitation (Fig. 4). The MCS followed a south-west to north-east direction, affecting the basin from 09 to 13 UTC on September 28, with an estimated speed ranging from 3.0 to 3.5 ms^{-1}. Initially, the MCS crossed transversally La Rambla de Nogalte catchment, to later move almost parallel over the southernmost tributary of the Guadalentín basin, the Biznaga (Figs. 1 and 4).

Basin Response and Hydrological Modelling

Stream-gauge and post-event survey data have been used to examine the main features of the basin response to the torrential precipitations. As shown in Table 2, initially very dry soils and acute spatial heterogeneities enhance the nonlinear hydrological response of the Guadalentín basin to intense precipitations and large accumulated rainfall amounts. Areas of high infiltration capacities coexist with zones relatively impermeable: while the Guadalentín sub-basins feature very low runoff coefficients, Nogalte exhibits an impressive unit peak discharge (Table 2). The former catchment noticeably mitigates the magnitude of the runoff discharges. These high infiltration losses and large soil moisture storage capacities are linked to the presence of calcareous and dolomitic fractured bedrocks, favouring the recharge of the deep aquifers. On the other hand, La Rambla de Nogalte basin is mostly formed by highly impermeable terrain (i.e., phyllites, quartzites, micadites and gypsum on schist bedrock).

Note that the contributing catchment and basin areas differ due to the fact that Valdeinfierno and Puentes did not contribute to flow discharge downstream (Tables 1 and 2). Accordingly, the computation of the main hydrometeorological features over

Table 2 Main hydrometeorological features of the 28 September 2012 flash flood for the different hydrometric sections of the Guadalentín basin

Basin	Contributing basin area (km^2)	Total rainfall (mm)	Total runoff (mm)	Peak discharge (m^3 s^{-1})	Unit peak discharge (m^3 s^{-1} km^{-2})	Runoff ratio (−)
Valdeinfierno	430.6	123.8	13.6	NA	NA	0.11
Puentes	1011.0	130.7	9.3	NA	NA	0.07
Torrealvilla	236.4	136.6	NA	300–450[a]	1.27–1.90	NA
Lorca	389.8	154.8	18.7	607.6	1.56	0.12
Paretón	1016.7	210.4	28.6	1081.2	1.06	0.14
Nogalte	124.3	211.7	NA	1050[a]–1500[b]	8.45–12.07	NA

The total rainfall amounts are expressed as area-averaged values. *NA* denotes not available information. Data indicated with [a] denote field estimations by Hooke (2016). Data indicated with [b] stand for field estimations by Benito et al. (2012)

the Lorca and Paretón basins has just accounted for the contributing areas downstream of Puentes. Analogously, the derivation of the hydrometeorological features in Puentes has not considered the Valdeinfierno contribution. Also note that Hooke (2016) estimated the peak flows close to the outlet in Torrealvilla (Table 2; Fig. 1). In La Rambla de Nogalte, the peak estimations correspond to the middle of the basin, where overflow embankment was reported (Hooke 2016), and at Puerto Lumbreras (Benito et al. 2012).

The Flood Event–Based Spatially Distributed Rainfall–Runoff Transformation–Water Balance (FEST-WB) model has been implemented to further assess the hydrological response of the Guadalentín catchment to the intense precipitations. FEST-WB is a fully-distributed and physically-based hydrological model that accounts for evapotranspiration, infiltration, surface runoff, subsurface flow and flow routing (Rabuffetti et al. 2008). The model computes soil moisture fluxes by solving the water balance equation at each grid point. In particular, the evolution of the soil moisture, θ_{ij}, for the generic mesh point at (i, j) is given by:

$$\frac{\partial \theta_{ij}}{\partial t} = \frac{1}{Z_{ij}}\left(P_{ij} - R_{ij} - D_{ij} - ET_{ij}\right) \tag{1}$$

where P is the precipitation rate, R and D are the runoff and drainage fluxes, ET is the evapotranspiration rate, and Z is the soil depth. Runoff is calculated according to a modified Soil Conservation Service–Curve Number (SCS-CN; USDA 1986) method extended for continuous simulation (Ravazzani et al. 2016). At this aim, the maximum potential retention S is updated at the beginning of a storm as a linear function of the degree of saturation. That is,

$$S = S_1(1 - \varepsilon) + S_3\varepsilon \tag{2}$$

where S_1 and S_3 are the values of S when the soil is dry and wet (i.e., antecedent moisture conditions I and III, respectively). The actual evapotranspiration is calculated as a fraction of the potential rate tuned by the beta function that, in turn, depends on soil moisture content (Montaldo et al. 2003). Potential evapotranspiration is computed according to a modified version of the Hargreaves–Samani equation (Ravazzani et al. 2012). The surface and subsurface flow routing is based on the Muskingum method (Chow et al. 1988).

Regarding the calibration of the infiltration parameters, the procedure presented by Borga et al. (2007) has been adopted so as to better encompass the strong nonlinearities of the runoff generation on the Guadalentín basin. Accordingly, the infiltration storativity (S_0) and the initial abstraction ratio (λ) have been selected as calibration parameters. S_0 is a site storage index—with a default value of 254 mm—related to the SCS-CN scheme by Ponce and Hawkins (1996):

$$S = S_0\left(\frac{100}{CN} - 1\right) \tag{3}$$

As pointed out by Borga et al. (2007), the infiltration storativity is considered as a calibration parameter so as to properly simulate the observed flood water balance when using the CN spatial distribution. The initial abstraction is defined to be proportional to the maximum potential soil retention. In the original SCS-CN formulation, the constant of proportionality (λ) is set to 0.2 as the standard value (Ponce and Hawkins 1996):

$$I_a = \lambda S \tag{4}$$

Calibration of λ copes with the specific lithological features of the Guadalentín basin. Model calibration tasks have focused on the different flow information available at each hydrometric section (Table 1). In Valdeinfierno and Puentes dams, the real-time observed stages have been used to calibrate the modelled discharge volumes. In Lorca and Paretón, calibration has focussed on peak discharges, timings and runoff volumes. In Torrealvilla and Puerto Lumbreras, efforts have centred in reproducing the peak discharge estimations from the available post-flood field measurements.

Heterogeneities in the hydraulics of the basin response to flash-floods must be also taken into account. These heterogeneities emerge as result of the gradual decrease of the catchment response with increasing precipitation amounts. The main factor regulating these heterogeneities is the expansion of the stream network to not previously channelled topographic elements during flash-flooding (Borga et al. 2007). Consequently, the hillslope and channel velocities have been calibrated by means of the Strickler coefficients. FEST-WB has been driven by the radar-derived precipitation fields from 27 to 29 September 2012 00 UTC. Note that the FEST-WB simulation started on 1 August 2012 00 UTC. This warm-up period permits a good initialization of the soil moisture content for the 28 September 2012 flash-flood.

Analogously to the procedure presented by Amengual el at. (2007), a multi-site calibration of the FEST-WB parameters has been carried out to account for the high nonlinearities in the infiltration and rainfall-runoff transformation mechanisms (Table 3). The infiltration storativity and initial abstraction ratio have been accordingly adjusted to the subjacent substrates in the Guadalentín catchment. No further calibration of the infiltration parameters has been performed in La Rambla de Nogalte, although the model parameters associated with the dynamical formulation have been adjusted in all the sub-basins.

After calibration, FEST-WB succeeds in simulating the highly nonlinear runoff production, the Hortonian infiltration excess mechanism, and the fast times to peaks and flood wave celerities in the river channels (Table 4 and Fig. 5). As extensive overflowing was reported in the Biznaga, Lorca and Nogalte basins (Benito et al.

Table 3 Calibrated parameters according to FEST-WB

Basin	Contributing basin area (km^2)	Total rainfall (mm)	CN (II)	So (mm)	λ	V_h (ms^{-1})	V_c (ms^{-1})
Valdeinfierno	430.6	123.8	70.1 (11.3)	304.8	0.30	0.28	2.9
Puentes	1011.0	130.7	68.1 (12.1)	508.0	0.40	0.32	3.3
Torrealvilla	236.4	136.6	67.7 (11.3)	558.8	0.35	0.34	4.0
Lorca	389.8	154.8	70.2 (11.0)	508.0	0.30	0.31	2.2
Biznaga	404.8	255.8	72.6 (11.1)	355.6	0.30	0.06	1.1
Paretón	1016.7	210.4	70.8 (11.1)	304.8	0.30	0.24	2.1
Nogalte	124.3	211.7	81.0 (9.8)	254.0	0.20	0.38	5.0

Curve numbers are expressed as area-averaged values, while their standard deviations are shown between brackets. Note that curve numbers correspond to normal antecedent conditions

Table 4 Observed and radar-driven simulated flow volumes and peak discharges for the 28 September flash-flood at the different hydrometric sections of the Guadalentín basin

Basin	Flow volumes			Flow peaks			
	OBS (mm)	FEST (mm)	Rel. Error	OBS (m^3 s^{-1})	FEST (m^3s^{-1})	Rel. error	NSE
Valdeinfierno	13.6	13.0	−0.04	NA	595.8	–	–
Puentes	9.3	9.4	0.01	NA	1172.0	–	–
Torrealvilla	NA	11.9	–	300–450[a]	369.8	–	–
Lorca	18.7	18.8	0.01	607.6	646.2	0.06	0.89
Biznaga	NA	33.3	–	NA	495.6	–	–
Paretón	28.6	31.2	0.09	1081.2	1182.6	0.09	0.84
Nogalte	NA	124.8	–	1050[a]–1500[b]	1544.3	–	–

Data indicated with [a] and [b] denote field estimations by Hooke (2016) and Benito et al. (2012), respectively

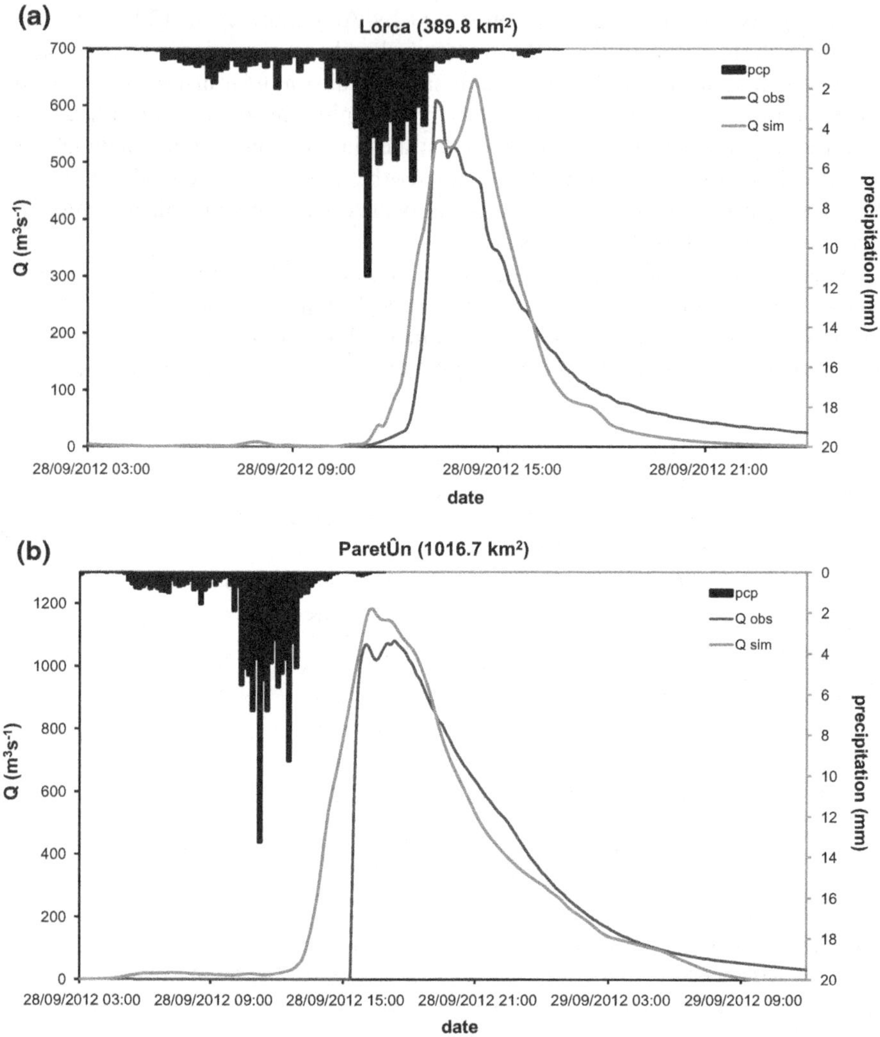

Fig. 5 Observed and radar-driven discharges for the 28 Sep 2012 episode at **a** Lorca and **b** Paretón

2012; Hooke 2016), the radar-driven runoff experiment has intended to account for these water losses by simulating peak discharges higher than observed at the Lorca and Paretón hydrometric sections (Table 4). Even after applying this proxy, uncertainties in estimating the actual peak discharges and runoff volumes still remain. The spilled water volumes were not estimated during the different post-flood field campaigns.

The times to peak are satisfactorily reproduced at Paretón, denoting an effective calibration of the model parameters for the dynamical routing (Fig. 5). At Lorca, although these are accurate enough, the experiment fails in simulating the magnitude of the first peak discharge. Note that the relative errors have been computed where observations were available in Table 4. Negative errors denote model underestimation.

Anthropogenic Impacts on the Basin Response

The effects of the human activities on the hydrological response have been quantified by modelling the Guadalentín River basin in its natural regime. That is, this catchment has been modelled without reservoirs. Table 5 and Fig. 6 show the intercomparison between the actual and natural basins. In Lorca, increases in runoff volume and peak discharge would have been of more than 200% of the actual basin simulations. In Paretón, increments would have been of 130% in peak discharge and 50% in total runoff volume. The total water amount carried by the Guadalentín River during this episode would have been of 22.5 and 46.8 Hm^3 at Lorca and Paretón, respectively. In addition, the maximum discharges would have been above 2200 and 2700 $m^3\ s^{-1}$, respectively.

Considering that the river bed was very close to its maximum water stage level in Lorca, the outcome of a river overflowing of this magnitude would have been catastrophic for the city. Therefore, the reservoirs acted as suitable structural measures so as to avoid catastrophic flood impacts in Lorca. In addition, the river diversion at Paretón would have been overwhelmed by the peak discharges and flow volumes under a natural regime, as its maximum evacuation capacity is of approximately 1000 $m^3\ s^{-1}$. Even with the diversion, a considerable bore would have propagated downstream towards Murcia City.

On the other hand, La Rambla de Nogalte carried a total water volume of 15.5 Hm^3, according to the radar-derived rainfall driven runoff simulation. Most of this amount was poured into the Biznaga basin (Fig. 1 and Table 4). To this incoming runoff volume, the intense precipitations added additional 13.9 Hm^3 over the latter catchment and the subsequent flood bore destroyed a bridge. The simulated peak discharge has been of 495.6 $m^3\ s^{-1}$ at the junction of the Biznaga stream with the Guadalentín River. These water volumes converted the Guadalentín valley in

Table 5 Comparison of the peak discharges and flow volumes between the actual and natural basins according to the radar-derived rainfall driven runoff simulation for the 28 September flash-flood

Sub-basin	Runoff volumes			Peak discharges		
	Actual (Hm^3)	Natural (Hm^3)	Rel. Diff.	Actual ($m^3\ s^{-1}$)	Natural ($m^3\ s^{-1}$)	Rel. Diff.
Valdeinfierno	5.6	5.6	0.0	595.8	595.8	0.0
Puentes	9.5	15.1	0.6	1172.0	1708.7	0.5
Lorca	2.8	22.5	2.1	646.2	2280.8	2.5
Paretón	31.7	46.8	0.5	1182.6	2708.3	1.3

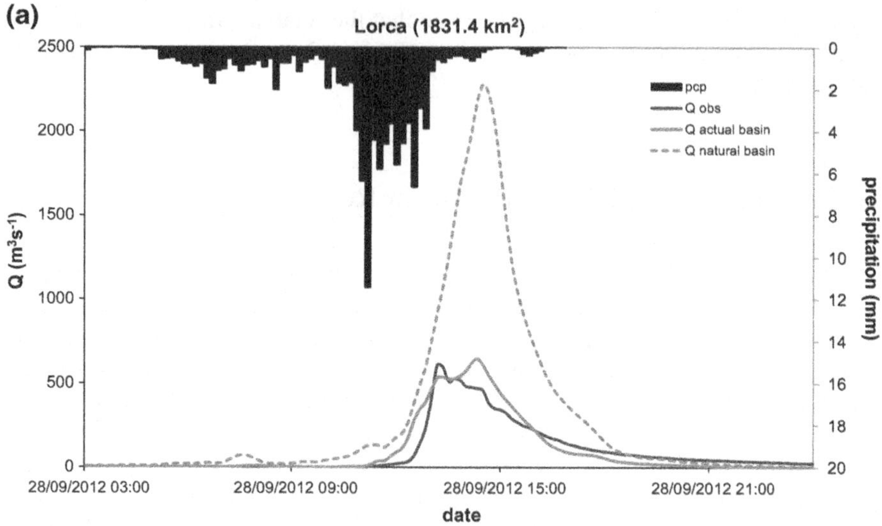

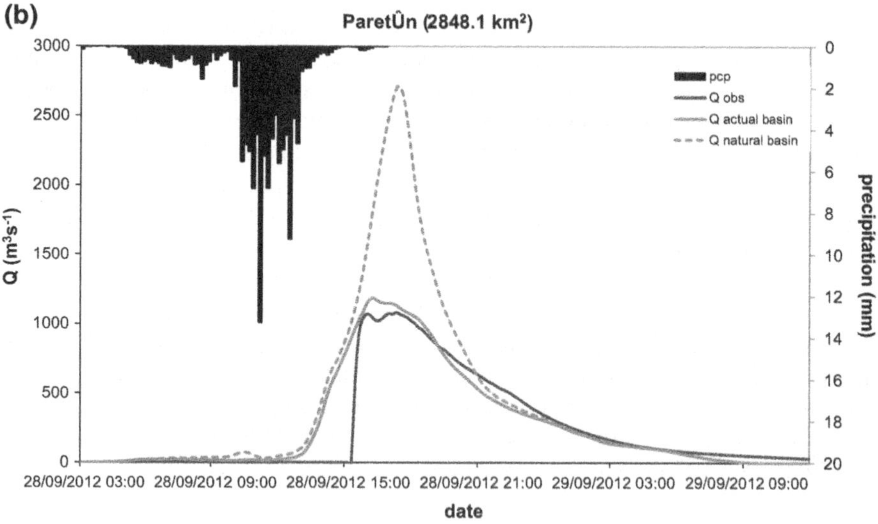

Fig. 6 Observed (black solid line) and radar-driven discharges for the 28 September 2012 episode and the actual (grey solid line) and natural (grey dashed line) basins at **a** Lorca and **b** Paretón

an extensive flood-plain. Note that the slow recession limbs of the observed and simulated hydrographs at Paretón indicate that substantial amounts of water remained for a long time in the Biznaga hillslopes (Fig. 6b).

Kinematics of the San Wenceslao Flash-Flood

The kinematic features of this extreme flash-flood have been examined through the spatial moments of the catchment rainfall and the catchment-scale storm velocity. The underlying idea is to relate the rainfall spatial organization to the flow distance, which is a basic descriptor of the structure of the drainage network. Zoccatelli et al. (2011) showed that n-order rainfall spatial moments ($p_n(t)$) can be defined based on the precipitation at a certain point of the basin and at a specific time ($r(x, y, t)$) and flow distance through the flow path ($d(x, y)$) between this point and the catchment outlet as:

$$p_n(t) = |A|^{-1} \int_A r(x, y, t) \cdot d(x, y)^n dA \tag{5}$$

The zeroth-order spatial moment of the rainfall field, $p_0(t)$, corresponds to the basin-averaged precipitation rate at time t, where A denotes the basin total area. Analogously, n-order moments of the flow distance (g_n) can be introduced as:

$$g_n = |A|^{-1} \int_A d(x, y)^n dA \tag{6}$$

The first-order moment of the flow distance, g_1, is equal to the basin-averaged flow distance. Next, the spatial moments of the catchment rainfall ($\delta_n(t)$) are defined as the ratio between different orders of the rainfall spatial and flow distance moments. Specifically, the first two moments are given by:

$$\delta_1(t) = \frac{1}{g_1}\left[\frac{p_1(t)}{p_0(t)}\right] \tag{7}$$

$$\delta_2(t) = \frac{1}{g_2 - g_1^2}\left[\frac{p_2(t)}{p_0(t)} - \left(\frac{p_1(t)}{p_0(t)}\right)^2\right] \tag{8}$$

$\delta_1(t)$ describes the flow distance between the centroids of the catchment rainfall and the basin. $\delta_2(t)$ is related to the dispersions of the rainfall fields—with respect to its mean position—and the flow distances. Values of $\delta_1(t)$ equal to one mean that the rainfall distribution is either concentrated on the basin centroid position or it is uniform over the catchment. Values of $\delta_1(t)$ less (greater) than one indicate that rainfall is distributed close to the outlet (headwaters). Values of $\delta_2(t)$ close to 1 reflect a uniform-like rainfall distribution, while values less (greater) than the unity indicate that precipitation is characterized by a unimodal (multimodal) distribution along the flow distance.

From Eq. (7), it is possible to derive an expression for the effective storm velocity (v_{eff}) as it is filtered by the catchment drainage properties (Zocattelli et al. 2011). V_{eff} is obtained as the temporal derivative of the first-order spatial moment of the catchment rainfall:

$$v_{eff} = g_1 \frac{d}{dt} \delta_1(t) \quad (9)$$

Also note that p_n and δ_n can be used to describe the organization of the precipitation over a certain time period (i.e., typically the total storm duration, T_s) by defining:

$$P_n = \frac{1}{T_s} \int_{T_s} p_n(t)dt \quad (10)$$

$$\Delta_1 = \frac{1}{g_1}\left[\frac{P_1}{P_0}\right] \quad (11)$$

$$\Delta_2 = \frac{1}{g_2 - g_1^2}\left[\frac{P_2}{P_0} - \left(\frac{P_1}{P_0}\right)^2\right] \quad (12)$$

Δ_1 and Δ_2 are the first and second time-integrated scaled moments, respectively. These concepts describe the overall rainfall organization at catchment scale, controlling the shape of the flood hydrograph. That is, Δ_1 has an influence on the runoff timing, while Δ_2 affects the hydrograph shape and the flood peak value. The first time-integrated scaled moment represents the ratio between the routing time corresponding to the rainfall centre of mass with respect to the basin response time. Values of Δ_1 less (larger) than one indicate that rainfall is located towards the basin outlet (headwaters). The second time integrated scaled moment expresses the ratio between the differential variance in runoff timing generated by the rainfall spatial distribution and the variance of the basin response time. Values of Δ_2 smaller than unity suggest that the rainfall field is spatially concentrated across the basin. On the contrary, values of Δ_2 larger than 1 point out that the rainfall field has a bimodal spatial distribution: concentrations are found both at catchment's headwaters and outlet.

Finally, a catchment-scale storm velocity (v_s) is defined from the first and second time-integrated scaled moments as:

$$v_s = g_1\left[\frac{cov_t[T, \delta_1(t) \cdot w(t)]}{var[T]} - \frac{cov_t[T, w(t)]}{var[T]}\Delta_1\right] \quad (13)$$

where T is the time, var[ ] is the variance, and $cov_t[\ ]$ denotes the temporal covariance. For a time duration equal to the total storm extent, T_s, $var[T] = \frac{1}{12}T_s^2$ (Zoccatelli et al. 2011). In order to consider the variation over time of the first-order moment and the mean areal rainfall, a rainfall weight (w(t)) must be introduced in Eq. (13). This weight is prescribed as:

$$w(t) = \frac{p_0(t)}{P_0} \quad (14)$$

Briefly, this method quantifies the dependency among the spatial and temporal organization of precipitation, catchment morphology and basin response in terms of surface runoff. The velocities of the convective cells are also appraised at basin scale. v_{eff} and v_s take into account the spatial orientation of the basin and its morphology with respect to the kinematics of the convective systems. In this way, it is possible to determine the impact of these ingredients into the hydrograph shape. It is worth noting that the kinematic variables have been computed by considering the natural basin so as to avoid the possible disturbances on the hydrological response introduced by the Valdeinfierno and Puentes reservoirs.

During the last hours of 27 September 2012, the Guadalentín basin was affected by small-scale, scattered and intense convective cores mechanically forced by the local orography. At the beginning of the next day, a first organized convective band mainly impacted the headwaters from south-west to north-east in approximately 3 h (01-03 UTC; Fig. 7a). From 05 UTC, successive organized convective bands emerged from the littoral and precipitated following the same direction, but affecting the central and

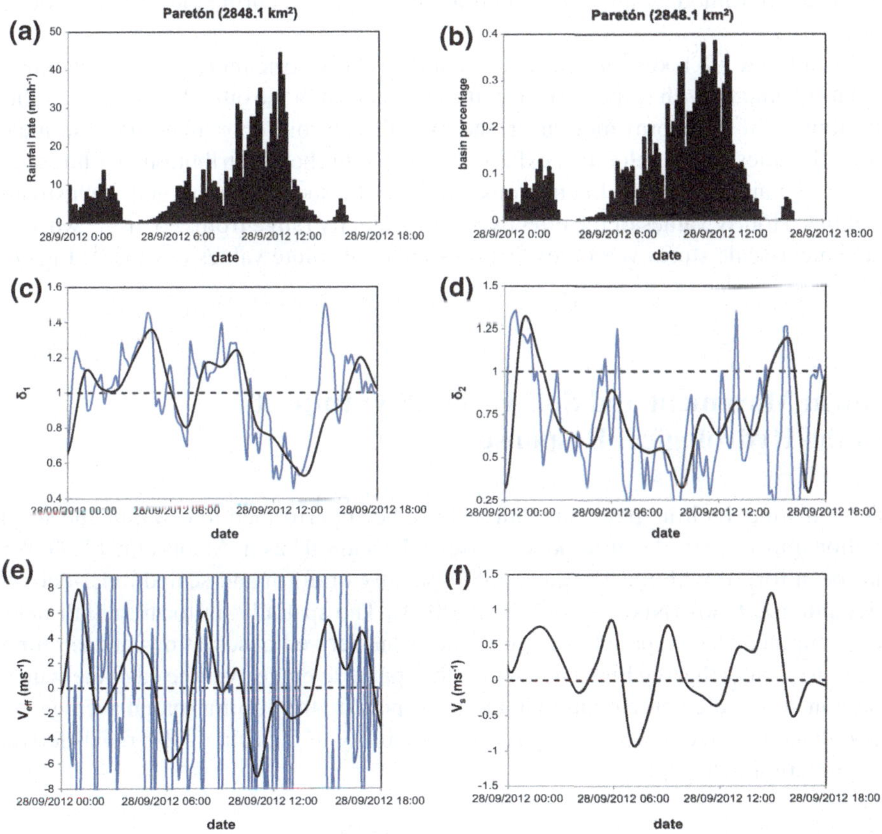

Fig. 7 Time series showing: **a** 10-min basin-averaged rainfall, **b** 10-min fractional area of the basin covered by rain >20 mmh^{-1}, **c** 10-min (blue line) and hourly (black line) δ_1, **d** 10-min (blue line) and hourly (black line) δ_2, **e** 10-min (blue line) and hourly (black line) effective storm velocity (v_{eff}) and, **f** hourly catchment-scale storm velocity (v_s)

southern part of the catchment. These bands were formed over the Mediterranean Sea and were instable enough to affect the flattest areas without any external triggering mechanism. Nonetheless, some of the convective nuclei embedded in these bands reinforced when impinging the basin topography. Finally, the MCS impacted the south-westernmost part of the basin from 09 UTC on September 28th (Fig. 7a). This MCS went across the basin with an approximate advective velocity of 3.0–3.5 ms^{-1}. Its passage lasted 4 h, leaving the Guadalentín catchment by Paretón (Fig. 4).

Maxima fractional coverages of the torrential precipitation were close to 0.4 during the passage of the MCS, affecting an area slightly greater than 1000 km^2 (Fig. 7b). The different episode kinematics are accordingly described by the oscillations of δ_1 (Fig. 7c). That is, δ_1 reflects the movement of the different convective rainfall bands through the basin and it clearly quantifies the MCS movement across the Guadalentín valley. The temporal variation of the rainfall dispersion (δ_2) shows a majority of values less than 1, thus indicating a spatial concentration over the catchment centroid. However, multimodal rainfall distributions also appeared during the first hours of the day.

Recall that v_{eff} takes into consideration the relative orientation of the catchment and its geometry with respect to the convective cells. On the other hand, v_s quantifies the impact of the storm motion on the rainfall centroid coordinate and the mean rainfall values. Both velocities exhibit symmetry in their distributions with respect to the sign and noticeable fluctuations that describe successive up- and downstream motions. Hourly values of the effective storm velocity range from ± 6 ms^{-1}, whereas catchment-scale storm velocities feature smaller absolute values (~1 ms^{-1}; Fig. 7e, f).

Storm Movement and Soil Variability: Impact on the Hydrological Response

Three distinct hydrological simulations have been performed to examine the storm motion impact on the flash-flood response of the natural basin. At this aim, FEST-WB has been forced with the 10-min actual, spatially uniform and scaled radar-derived precipitation fields (Nikolopoulos et al. 2014). The spatially uniform rainfall fields are computed as the spatial average of the radar-derived observations at each time-step, $\overline{p(t)}$. Note that the high bar denotes the spatial average over the catchment up to Paretón. The scaled precipitation fields correspond to the 10-min spatial average, but modulated by a factor accounting for the variability of the actual 48-h radar-derived precipitation field, P_T:

$$p_{scaled}(t) = \overline{p(t)}\frac{P_T}{\overline{P_T}} \tag{15}$$

These rainfall scenarios encompass different levels of spatial variability. On the one hand, the spatially uniform rainfall experiment neglects the spatial variability of the 10-min actual radar fields. On the other hand, the scaled precipitation scenario maintains a constant spatial variability equal to the total accumulated rainfall pattern. Note that the synthetic rainfall fields conserve the actual basin-averaged precipitation volume at each time-step.

The inclusion of these additional precipitation scenarios arises from the concept of the aforementioned spatial moments (Nikolopoulos et al. 2014). That is, the shape of the flood hydrograph is controlled by the following factors: (i) the catchment drainage structure; (ii) the temporal pattern of the basin-averaged rainfall rates; (iii) the first and second time integrated scaled moments (i.e. Δ_1 and Δ_2); and (iv) the catchment-scale storm velocity. The actual discharge is the outcome of combining all these ingredients. The hydrograph resulting from the spatially uniform precipitation is modulated by components (i) to (iii), whereas the flow associated with the scaled rainfall is just controlled by (i) and (ii). The comparison of the actual and spatially uniform radar driven runoff simulations isolates the combined effect of Δ_1 and Δ_2, and v_s on the flood hydrograph. The contrasting of the actual and scaled radar driven runoff experiments unravels the effect of v_s on the flood hydrograph. That is, it neglects the storm movement. Note that the overall spatial rainfall organization is preserved in the scaled experiment, yielding identical Δ_1 and Δ_2 scores than the actual radar observed fields.

This strategy disentangles the effects of the event kinematics on the runoff timing. Zoccatelli et al. (2011) showed that Δ_1 is related to the normalised mean time difference between the hydrographs obtained by considering the actual and spatially uniform rainfall patterns. The normalisation is given by the catchment response time. Δ_1 smaller (larger) than 1 indicates an advancement (a delay) of the mean hydrograph time with respect to the spatially uniform precipitation field. In this case study, $\Delta_1 = 0.89$ and $\Delta_2 = 1.03$ illustrate that the location of the cumulative rainfall amounts are found close to the Paretón outlet, but featuring a slightly multimodal distribution (Fig. 4). This Δ_1 score also points out that the mean time difference between the two hydrographs corresponds to approximately a 10% of the basin response time, indicating an advancement of the hydrograph obtained from the actual with respect to the uniformly distributed rainfalls (Fig. 8a).

The value of Δ_1 confirms that the San Wenceslao flash-flood was only partially triggered by convective precipitation anchored by local topography. Undoubtedly, condensation and deep convection formed as consequence of the mechanical forcing of the low-level warm and moist air impinging against the local orography. As aforementioned, the orographic forcing originated and sustained most of the convective cells that affected the Guadalentín headwaters during the first hours of the episode on 27 September. But what mostly overwhelmed the catchment was the subsequent entrance of the convective bands and, very especially, the V-shaped MCS. This mesoscale convective system formed along a convergence line over the Mediterranean Sea due to the high atmospheric instability associated with the strong moisture and thermal gradients. The south-westerly low-level maritime flow advected the MCS inland, traversing the basin north-eastwards. The slightly multimodal spatial

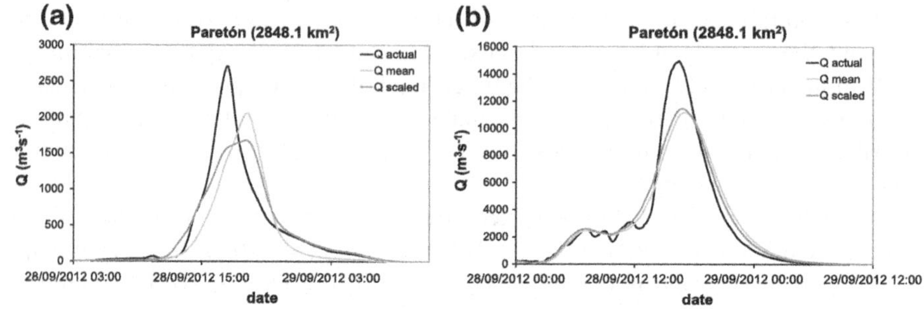

Fig. 8 Actual (Q actual), spatially-uniform (Q mean) and scaled (Q scaled) radar-derived rainfall driven runoff experiments for the **a** natural and **b** impervious basin

distribution of the precipitation pattern is the outcome of the conjunction of these different rainfall-bearing mechanisms.

The consequence of the MCS motion on the flood hydrograph is to reduce its base time, focussing and exacerbating the peak discharge. Due to the storm kinematics, the peak discharge is enhanced up to 161.3%. The structure of the basin drainage, the temporal pattern of the rainfall rates and the rainfall location modulate the time to peak, resulting in a remarkable advancement of the San Wenceslao flash-flood (~2 h; Fig. 8a).

Catchments act as spatial and temporal filters with specific dampening characteristics to the rainfall inputs (Skøien and Blöschl 2006). These filtering properties can be enough acute to systematically smooth out some features of the spatial and temporal rainfall variability (Amengual et al. 2007). And just some specific characteristics of space and time rainfall organisation could eventually emerge as spatial and temporal variability on the runoff generation (Skøien et al. 2003). Therefore, another important issue is to assess the impact of the soil heterogeneities on the catchment response and its timing.

As the rainfall-runoff transformation acts as a highly non-linear filter, a completely impervious natural basin has been considered. This strategy avoids the filtering effects on the kinematic features of the hydrograph via the runoff generation mechanism. That is, an impermeable basin erases the effects of the soil heterogeneities and highlights the role of the river network geometry and its impact on the structure of the dampening properties. Thus, a clearer picture emerges of the effect of the precipitation and basin geomorphic properties on the catchment response.

Figure 8b depicts the impact of the different spatial rainfall variabilities—or its lack—on the time to peak and peak discharge for the impervious basin. Again, time to peak is modulated by the river network geometry, the temporal pattern of the rainfall rates and the precipitation location. The storm motion shapes the hydrograph and enhances the peak discharge. These features appear to be little sensitive to soil heterogeneities, but highly dependent on the hillslope and channel velocities of the different sub-basins.

Effectively, extreme floods in the semi-arid Mediterranean Spain are consequence of fast surface flows as soil is quickly overwhelmed by high rainfall intensities. On the other hand, the stream network spans to previously unchannelized topographic elements, increasing the drainage density and the flow response rate of hillslopes. The morphologic features of the river watersheds play a central role in the runoff response of extreme floods. The hydrograph response is mainly determined by the distribution of travel times. That is, runoff routing through branched channel networks imposes an effective averaging of the spatial rainfall excess across locations with equal routing times, in spite of the inherent spatial variability (Naden 1992; Woods and Sivapalan 1999; Nicótina et al. 2008; Sangati et al. 2009).

Conclusions

Climate change results in a redistribution of the hydrometeorological mean and extreme regimes. This rearrangement affects both the lower and upper tails of the precipitation frequency distribution over the Mediterranean region. Longer-lasting droughts and more frequent and intense torrential precipitations will especially impact this responsive area to extremes. The Mediterranean is also characterized by a high level of exposure to natural hazards, because of its highly populated coastal areas. Rough reliefs close to the sea, steep and highly urbanized river basins, sparse vegetation, thin soils and intense but irregularly distributed precipitations make this region prone to flash-floods.

The Spanish Mediterranean region is persistently affected by hazardous flash-floods. Many of its small- to medium-sized semi-arid basins are ephemeral gravel-bed streams, hydrologically dependent on rainfall. As return periods for these natural hazards over single catchments go well beyond the decade, the associated risks are often forgotten. Sudden increases in discharge rates, rapid flow velocities, and high peaks can cause unexpected, substantial and extensive damage to human life and property. Furthermore, strong nonlinearities arise from the hydrological response of these catchments to intense rainfalls. Predominantly low initial soil moisture contents, high soil moisture storages and the expansion of the river networks result in differential spatial and temporal runoff responses.

The San Wenceslao episode provides a paradigm of an extreme flash-flood over the Mediterranean Spain and it has been examined in detail. First, the geomorphologic and lithological features of the Guadalentín River catchment have been assessed. Next, high spatial and temporal resolution precipitation fields derived from radar measurements have been used to force a fully-distributed and physically-based hydrological model. Subsequent runoff simulations have permitted a thorough analysis of the hydrological response and the interplay between the hydrometeorological factors, the morphologic features and the anthropogenic activities. The main conclusions are:

- Heavy precipitations and flash-flooding had a multifaceted origin. Large rainfall amounts were the result of deep convection triggered by local orography and the passage of successive organized convective bands and a slow-moving MCS.
- Distinct soil substrates led to myriad soil responses and runoff generation. Persistent low soil moisture contents and fractured bedrocks played an important role in flood response, resulting in very low runoff ratios.
- A multisite calibration of a hydrological model has been necessary to satisfactorily reproduce the basin response. However, uncertainties in the actual peak discharges and runoff volumes still remain owing to the extensive river overflows reported along the Guadalentín and La Rambla de Nogalte watersheds.
- Anthropogenic modifications resulted in contrasting outcomes. On one side, the Valdeinfierno and Puentes dams acted as effective flood-control structural measures. On the other side, the disconnection of the river bed between the Nogalte and Biznaga sub-basins resulted in large flood damages.
- The entrance of the MCS was the main factor modulating this extreme flood, overwhelming the catchment. Its downstream motion exacerbated the peak discharge. On the other hand, time to peak was mainly modulated by the river network geometry, the temporal pattern of the rainfall rates and the rainfall position. These features appear to be little sensitive to soil heterogeneities.

In summary, the Guadalentín River basin faced a very rare rainfall episode with high intensities which, in combination with the catchment properties, led to extreme runoff. The deployed structural measures undoubtedly mitigated partially the impacts of the San Wenceslao flash-flood. However, no structural measure can fully guarantee complete safety before such hazardous events. These preventive infrastructures should be accompanied by diverse and complementary non-structural measures.

Awareness, information and training campaigns addressed to local populations living in areas with a high flood risk arise as fundamental to minimize casualties and material damages. Even though the precise forecasting of small-scale convective cores and MCSs is a challenging task—as many factors concur in determining such events—early warning systems based on advanced hydrometeorological ensemble prediction systems can expand the lead times associated to flood alerts many hours ahead of its occurrence.

Finally, a changing climate exacerbates the risk of suffering flash-floods of similar magnitude in the Mediterranean more frequently. Policymakers and water management agencies should also devote their efforts in reducing the high level of exposure to flash-flooding currently present in this region. Building and agricultural activities should be carried out well beyond the floodplains of the Mediterranean ephemeral rivers and always respecting their morphology. Hopefully, the implementation of new structural and non-structural measures over the Mediterranean region will avoid future catastrophic impacts as those occurred during the widespread flash-flooding across south-eastern Spain on 27, 28 and 29 September 2012.

Acknowledgements The Hydrographic Confederation of the Segura river (CHS) and, in special, Mr. Fernando Toledano Sánchez, former head of the management of the observational networks of the CHS, are acknowledged for providing the SAIH rain and flow data. The Spanish Agency of

Meteorology (AEMET) is also acknowledged for providing the radar and automatic weather stations precipitation and temperature data. This work has been sponsored by the CAS18/00009 research project, granted to Dr. Amengual by El Ministerio de Ciencia, Innovación y Universidades. This work has been also sponsored by the CGL2014-52199-R (EXTREMO), PCIN-2015-221 (METE-OforSIM) and CGL2017-82868-R (COASTEPS) Spanish research projects, which are partially supported with FEDER funds.

References

Amengual A, Romero R, Gómez M, Martín A, Alonso S (2007) A hydrometeorological modeling study of a flash flood event over Catalonia, Spain. J Hydrometeor 8:282–303. https://doi.org/10.1175/JHM577.1

Amengual A, Homar V, Jaume O (2015) Potential of a probabilistic hydrometeorological forecasting approach for the 28 September 2012 extreme flash flood in Murcia, Spain. Atmos Res 166:10–23. https://doi.org/10.1016/j.atmosres.2015.06.012

Amengual A, Carrió DS, Ravazzani G, Homar V (2017) A comparison of ensemble strategies for flash flood forecasting: the 12 October 2007 case study in Valencia, Spain. J Hydrometeor 18(4):1143–1166

Beniston M et al (2007) Current and future extreme climatic events in Europe: observations and modeling studies conducted within the EU PRUDENCE project. Clim Change 81:71–95

Benito G, Machado MJ, Lloveras XR (2012) Cómo un evento" moderado" puede causar un impacto extremo. Enseñanza de las Ciencias de la Tierra 20(3):301

Borga M, Boscolo P, Zanon F, Sangati M (2007) Hydrometeorological analysis of the 29 August 2003 flash flood in the Eastern Italian Alps. J Hydrometeor 8(5):1049–1067

Camarasa-Belmonte AM, Segura-Beltrán F (2001) Flood events in Mediterranean ephemeral streams (ramblas) in Valencia region, Spain. CATENA 45:229–249. https://doi.org/10.1016/S0341-8162(01)00146-1

Chow VT, Maidment DR, Mays LW (1988) Applied hydrology. In: Civil engineering series. McGraw-Hill International Editions, 572 pp

Cole SJ, Moore RJ (2008) Hydrological modelling using raingauge- and radar-based estimators of areal rainfall. J Hydrol 358(3):159–181

Cramer W et al (2018) Climate change and interconnected risks to sustainable development in the Mediterranean. Nat Clim Change 1

Doswell CA, Brooks HE, Maddox RA (1996) Flash flood forecasting: an ingredients-based methodology. Weather Forecast 11:560–581

Ducrocq V et al (2014) HyMeX-SOP1, the field campaign dedicated heavy precipitation and flash flooding in the northwestern Mediterranean. Bull Am Meteorol Soc . https://doi.org/10.1175/BAMS-D-12-00244.1

Gil-Olcina A (2016) Causas atmosféricas de dos descomunales llenas en la Rambla de Nogalte (Cuenca del Segura). Libro Homenaje al Profesor Alfredo Morales Gil, Universidad de Alicante, pp 935–956. https://doi.org/10.14198/librohomenajealfredomorales2016-42

Giorgi F, Lionello P (2008) Climate change projections for the Mediterranean region. Glob Planet Change 63(2–3):90–104

Hooke JM (2016) Geomorphological impacts of an extreme flood in SE Spain. Geomorphology 263:19–38

Hunter SM (1996) WSR-88D radar rainfall estimation: capabilities, limitations and potential improvements. Natl Wea Dig 20:26–38

Kolios S, Feidas H (2010) Warm season climatology of mesoscale convective systems in the Mediterranean basin using satellite data. Theor Appl Climatol 102(1–2):29–42

Montaldo N, Toninelli V, Albertson JD, Mancini M, Troch PA (2003) The effect of background hydrometeorological conditions on the sensitivity of evapotranspiration to model parameters: analysis with measurements from an Italian alpine catchment. Hydrol Earth Syst Sci 7:848–861. https://doi.org/10.5194/hess-7-848-2003

Naden PS (1992) Spatial variability in flood estimation for large catchments: the exploitation of channel network structure. Hydrol Sci J 37:53–71

Nicótina L, Alessi Celegon E, Rinaldo A, Marani M (2008) On the impact of rainfall patterns on the hydrologic response. Water Resour Res 44:W12401. https://doi.org/10.1029/2007WR006654

Nikolopoulos EI, Borga M, Zoccatelli D, Anagnostou EN (2014) Catchment-scale storm velocity: quantification, scale dependence and effect on flood response. Hydrol Sci J 59(7):1363–1376

Pellarin T et al (2002) Hydrologic visibility of weather radar systems operating in mountainous regions: case study for the Ardèche catchment (France). J Hydrometeor 3(5):539–555. https://doi.org/10.1175/1525-7541

Ponce VM, Hawkins ERH (1996) Runoff curve number: has it reached maturity? J Hydrol Eng 1:11–19

Rabuffetti D, Ravazzani G, Corbari C, Mancini M (2008) Verification of operational quantitative discharge forecast (QDF) for a regional warning system—the AMPHORE case studies in the upper Po River. Nat Hazards Earth Syst Sci 8:161–173

Ravazzani G, Corbari C, Morella S, Gianoli P, Mancini M (2012) Modified Hargreaves-Samani equation for the assessment of reference evapotranspiration in Alpine river basins. J Irrig Drain Eng 138:592–599. https://doi.org/10.1061/(ASCE)IR.1943-4774.0000453

Ravazzani G, Amengual A, Ceppi A, Homar V, Romero R, Lombardia G, Mancini M (2016) Potentialities of ensemble strategies for flood forecasting over the Milano urban area. J Hydrol 539:237–253. https://doi.org/10.1016/j.jhydrol.2016.05.023

Sangati M, Borga M, Rabuffetti D, Bechini R (2009) Influence of rainfall and soil properties spatial aggregation on extreme flash flood response modelling: an evaluation based on the Sesia river basin, North Western Italy. Adv Water Res 32:1090–1106

Skøien JO, Blöschl G (2006) Catchments as space-time filters—a joint spatio-temporal geostatistical analysis of runoff and precipitation. Hydrol Earth Syst Sci 10:645–662. https://doi.org/10.5194/hess-10-645-2006

Skøien JO, Blöschl G, Western A (2003) Characteristic space scales and timescales in hydrology. Water Resour Res 39:1304. https://doi.org/10.1029/2002WR001736

Stocker T, Qin D, Plattner G, Tignor M, Allen S, Boschung J, Nauels A, Xia Y, Bex V, Midgley P (2013) IPCC 2013: climate change 2013: the physical science basis. Contribution of working group I to the fifth assessment report of the intergovernmental panel on climate change. IPCC, Geneva, 1535 pp

Toreti A, Naveau P (2015) On the evaluation of climate model simulated precipitation extremes. Environ Res Lett 10:014012

Toreti A et al (2013) Projections of global changes in precipitation extremes from coupled model intercomparison project phase 5 models. Geophys Res Lett 40:4887–4892

USDA (1986) Urban hydrology for small watersheds. USDA TR-55, 164 pp

Woods RA, Sivapalan M (1999) A synthesis of space-time variability in storm response: rainfall, runoff generation and routing. Water Resour Res 35:2469–2485

Zoccatelli D, Borga M, Viglione A, Chirico GB, Blöschl G (2011) Spatial moments of catchment rainfall: rainfall spatial organisation, basin morphology, and flood response. Hydrol Earth Sys Sci 15(12):3767–3783

A Multiple Linear Regression-Based Approach for Storm Surge Prediction Along South Brazil

Arthur Ohz, Antonio H. F. Klein and Davide Franco

Abstract Knowledge of the characteristics of storm surges is of paramount importance for an effective coastal hazard risk assessment. The vast Majority of storm surges in southern Brazil are generated by extratropical cyclones. In this context, this study focuses on spatial and temporal analysis of storm surges caused by the cyclones within a 30-year period. The wind and pressure from CFSv2 data were analyzed from 1979 to 2010 time span, with a time lag of 9 h identified for pressure and 26 h for wind stress. A multiple linear regression model was then employed to estimate the expected intensity of the storm surge. The results suggest that the model represented 82% of the variations in the storm surge levels. The accuracy of the surge predictions was validated by the recorded storm surges using tide gauge data for Itajaí port. The agreement between the modelled and the observed values were considered good by explaining 74% of the variations in the storm surge. Finally, Analysis of future scenarios of 1-, 10-, 50- and 100-return period revealed that positive storm surge events may be higher, increasing coastal risks in the coastal zone of southern Brazil.

Introduction

The State of Santa Catarina has a coastline with approximately 564 km of extension, considering the headland, bays and the more expressive islands can reach 1874 km

A. Ohz (✉) · A. H. F. Klein
Coastal Oceanography Laboratory, Department of Geosciences, Federal University of Santa Catarina, Rua dos Coroas, S/N, Barra da lagoa, 88061-600 Florianópolis, Santa Catarina, Brazil
e-mail: ohz.arthurn@gmail.com

A. H. F. Klein
e-mail: ahfklein@gmail.com

D. Franco
Maritime Hydraulic Laboratory, Department of Sanitary and Environmental Engineering, Federal University of Santa Catarina, Campus Reitor João David Ferreira Lima, 88040-900 Florianópolis, Santa Catarina, Brazil
e-mail: franco@ens.ufsc.br

W. Leal Filho et al. (eds.), *Climate Change, Hazards and Adaptation Options*, Climate Change Management, https://doi.org/10.1007/978-3-030-37425-9_2

(Santa Catarina 2006). Most of the municipalities of this coastal segment are in an advanced stage of urbanization, being involved with environmental problems arising from the lack of regional and urban planning policies (Klein et al. 2006). Tourism represents 12.5% of the Gross Domestic Product of the state, reaching up to 20% in some coastal municipalities. Every year storm waves cause coastal erosion in the state, which increasing the erosive power when they occur concomitantly during the spring tide and storm surges.

Woth et al. (2005) has observed in their studies that one of the consequences of climate change is that it becomes more likely to increase the probability of the occurrence of surges due to storms, which boost the waves produced by the storms and often lead to the destruction of property. Storm surges are occurring with greater intensity, thus becoming a problem which tends to increase with the passage of time if we consider as part of this scenario the rise in average sea-level which has already attracted attention in various regions of the globe (IPCC 2014). According to Church and White (2006) and Baart et al. (2012) this rise will result in the retreat or loss of the beach zone where there is an insufficient supply of sediments to off-set the loss.

Hurricanes are one of the processes which can generate extreme positive storm surges. From the South of Brazil, we may quote the example of the hurricane Catarina in 2004, an unprecedented event (Mctaggart-Cowan et al. 2006) which resulted in great environmental and socio-economic damage even though the storm surge levels attained were not extreme.

Besides having an extensive, completely urbanized coastline, the state of Santa Catarina possesses five maritime ports: Itapoá, São Francisco do Sul, Navegantes, Itajaí and Imbituba. Together, they are responsible for approximately 18 million tons of freight. The Port of Itajaí (Fig. 1) is the second largest in Brazil and the 13th in Latin America and the Caribbean. Eventually the complex is closed because of the high waves and strong winds due to the storms. The estimated losses may vary between US$25,000 and US$50,000 for each 24 idle hours. During the last six years the port interrupted its activities 76 times due to strong winds or big waves. Such storms lead to great economic loss and environmental damage in Santa Catarina.

Studies conducted by Gan and Rao (1991) and confirmed by Sinclair (1994), Reboita (2008) and Parise et al. (2009) registered the existence of two centers of maximum cyclonic activity in South America—one in Argentina and the other one above Uruguay. Due to the proximity of these two cyclones, Saraiva et al. (2003) declared that these two cyclones caused storm surges in Southern Brazil with a time lag of 24 h between them in 45% of cases. This result was confirmed by Parise et al. (2009) in a study undertaken for another period, in which the authors identified the same time lag in 39% of the cases. By time lag, we mean the period of time between the wind action parallel to the coast of Argentina and south of Brazil and the beginning of the storm surge signal at the study area.

Based on past studies, it should be emphasized that studies on storm surges which affect Brazil, normally used local data for the forecasting models or focused on the storm surge analysis on the atmospheric events which generated them. Further, because of those studies it was concluded that the surges are generated by the presence of storms (cyclones) which occur between Argentina and the South of Brazil,

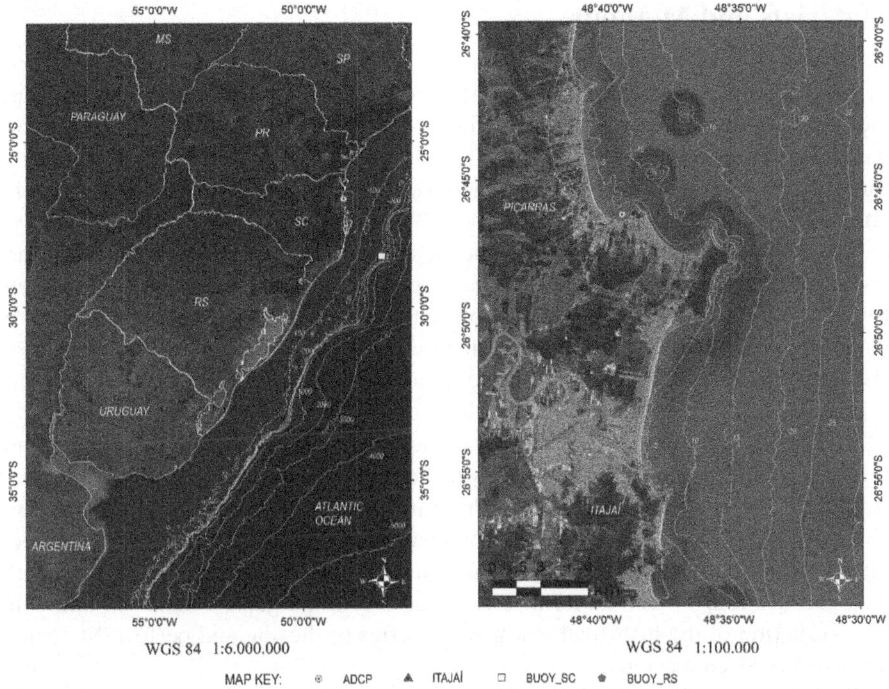

Fig. 1 Position of the buoys of the Brazilian Navy and of the ADCP installed in the region of Itapocorói Bay, Santa Catarina. *Images source* Esri, Digital Globe, Geo Eye, Earthstar Geographics, CNES/Airbus DS, USDA, USGS, AEX, Get mapping, Aerogrid, IGN, IGP, Swiss topo, and the GIS User Community

presenting a periodicity of between 6.5 and 11 days, the winds parallel to the coast being the most important force and the variation in atmospheric pressure being a secondary, local component. The storm surge is propagated northward from the South over the continental shelf and are of the same order of magnitude as the astronomical tides.

Understanding the atmospheric conditions responsible for generating storm surges, identifying the periods in which each event occurred and the wind fields and pressure for each event, is of paramount importance for the productive sector and society of Santa Catarina in general. In most regions of Brazil there is no long historical series of storm surge that can be used for statistical analyzes of future scenarios. For these reasons the main question that this research will try to answer is: what will happen to embayed beaches in south of Brazil subject to the action of extreme events of storm surge in different future scenarios?

Materials and Methods

In the south of Brazil the main atmospheric forcing that generate the storm surge are the wind and the pressure, so in this work we are limited to working with the wind parallel to the coast and the pressure. This study thus proposes: Create a statistical forecasting model of storm surges, which uses the wind and pressure data, from the buoys of the National Buoy Program (PNBOIA) of the GOOS-Brazil and CFSv2 data, extracted at hourly intervals at the closest coordinates at which the buoys are located. To verify whether the proposed model can be used as a monitoring system for storm surges, thus making it possible satisfactorily to represent storm surges in the north of Santa Catarina state, specifically in the studies of the case of Itapocorói Bay and Itajaí Port (Fig. 1). Assess the duration of the preceding period necessary for the forecast of the storm surge for these case studies.

The steps taken for the development of the forecasting model for storm surges are presented in the flowchart in Fig. 2. They make use of wind and pressure data collected from the buoys of the National Buoy Program (PNBOIA) of the GOOS-Brazil, and data from the CFSv2 extracted at hourly intervals at the closest coordinates at which the buoys are located. The model was based on the CFSv2 series of wind and pressure data because it is a much larger historical series with no gaps, which will allow the reconstruction of the historical storm surge series of the site and perform statistical analysis of extreme events.

With the data of pressure it will be traced the trajectory of the cyclones. The analysis of partial serial correlation between the storm surge and the wind stress and the pressure will serve to find the time lag between the storm and the beginning of the storm surge in the Itapocorói Bay to develop a storm surge forecast model in the Bay.

Astronomical Tide and Storm Surge

In order to identify and analyze the rise in the free sea-level surface in the Itapocorói Bay (26.73° S, 48.65° W), an ADCP was anchored at a depth of 3.5 m (26° 46′ 5.79″ S, 48° 39′ 17.79″ W), for a period of 33 days (from 19 August 2011 to 22 September 2011) (Fig. 1). A temporal, hourly, water series was obtained, of which the water-level was extracted from the data of the height of the water column measured by the ADCP (at each half hour) and the occurrence and the duration of each storm surge were recorded. Further, the atmospheric conditions responsible for generating each of them were registered.

The storm surge was obtained based on the difference between the data of the variation of the tide measured at the site and the data calculated from the forecast of the harmonic components of the tide. The t-tide program developed by Pawlowicz et al. (2002) was used to verify the storm surges occurring during the events analyzed while the experiment was in progress. This program analyzes the harmonic components

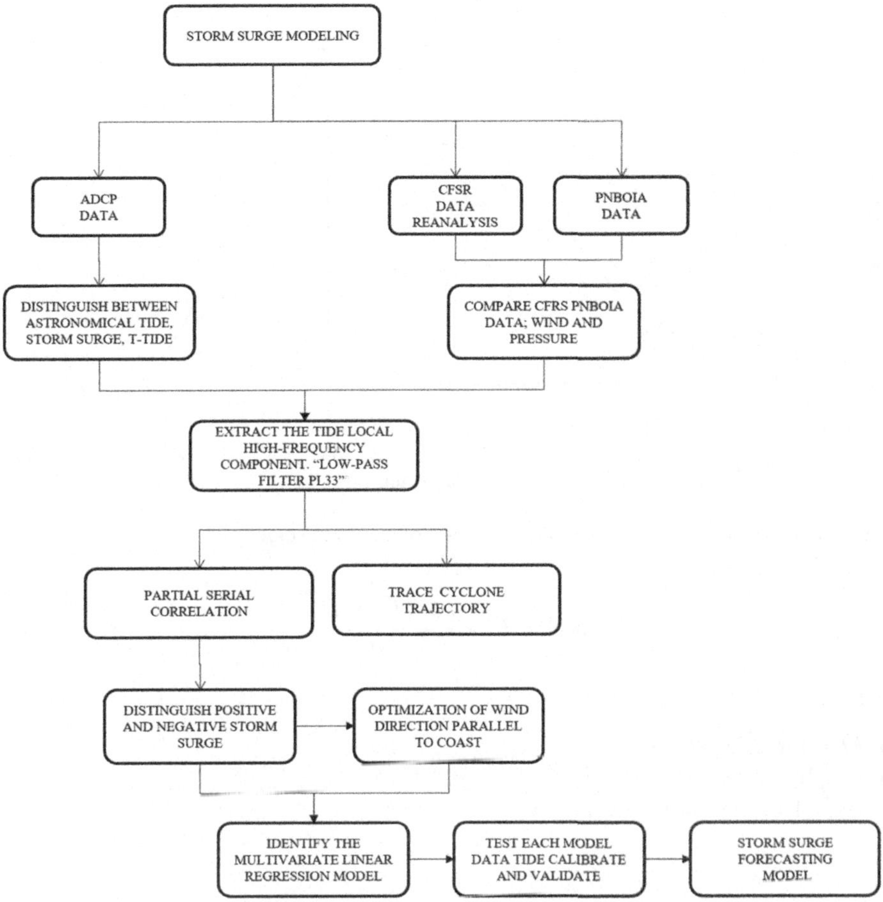

Fig. 2 Flowchart of the methodology for the development of the statistical model for storm surges

of the tide over time and distinguishes the astronomical tide from residual factors. These residual factors, whether negative or positive, are the difference between the levels of the sea observed and the astronomical tide (Pugh and Woodworth 2014).

The determination of each event was undertaken after distinguishing the storm surge from the astronomical tide. The positive storm surge was identified as from the moment when the level began to increase and terminated when the level began to fall (Fig. 3), that is to say, the event was defined from trough to crest of the wave of the storm surge, inverting the analysis in the case of the negative storm surge. The risk factor (RF) of each event was calculated from Eq. 1, values close to zero offer a lower risk, positive values indicate a positive storm surge that offers erosion and flood risk in coastal areas and negative values indicate a negative storm surge that pose risk to navigation.

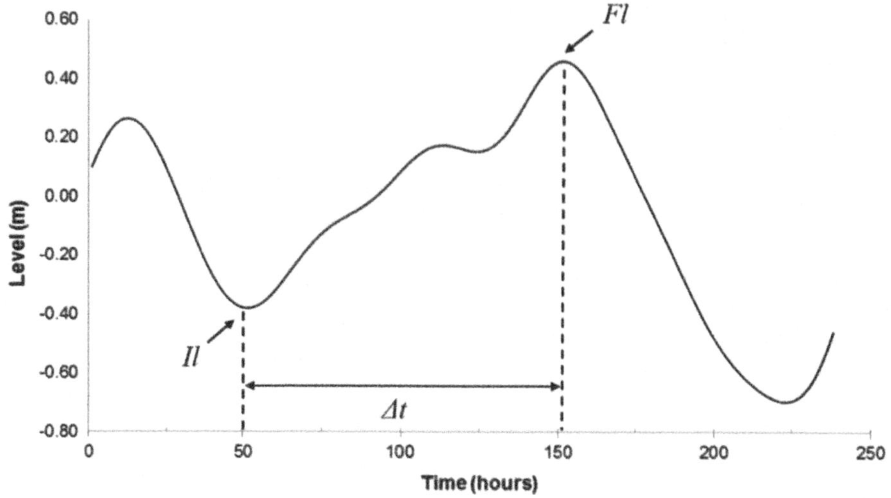

Fig. 3 Scheme of the calculation of the risk factor of the storm surge

$$RF = Fl * \frac{abs(Fl - Il)}{\Delta t}, \quad (1)$$

where:

- Fl = Final level;
- Il = Initial level;
- Δt = Duration of the event.

Meteorological Data and the Mapping of the Cyclones' Paths

The wind data of the re-analysis of the NCEP/NOAA of the CFSv2 with a spatial resolution of 0.2045° × 0.2045° and a temporal hourly resolution at 10-m height, extracted at the closest sites as the buoys of the National Buoy Program (PNBOIA) of the GOOS-Brazil, were used. The buoys used in the analysis were that of no. 69152, situated in front of the cape of Santa Marta (SC Buoy), and no. 69253, located to the north-east of the city of Rio Grande (RS Buoy). The wind stress was calculated on that basis.

To calculate the wind stress generated by the wind two different equations were used. Van Dorn's (1953) equation which uses the density of the water and Wu's (1969) which makes use of the air density. Also, six different drag coefficients were tested with the objective of discovering which of them was most applicable to the climatic conditions of Southern Brazil. It was thus decided which of the equations best described the partial correlation with time lag between the wind stress generated by the action of the wind and the storm surge.

Equation 2 presents the calculation of the wind stress according to Van Dorn (1953):

$$\tau = \rho_w C_d U_{10} |U_{10}|, \tag{2}$$

where:

- ρ_w = Water density (ρ = 1025 kg/m^3);
- C_d = Drag coefficient varying according to the different velocities of the wind.

Equation 3 presents the calculation of the wind stress according to Wu (1969):

$$\tau = \rho_{air} C_d U_{10} |U_{10}|, \tag{3}$$

where:

- ρ_{air} = Air density (ρ = 1.22 kg/m^3);
- C_d = Drag coefficient;
- U_{10} = Wind speed (m/s) at 10 m height.

The drag coefficients tested are presented in Tables 1 and 2. The first four use one equation for all wind speeds, whereas the drag coefficient proposed by Van Dorn (1953) and Wu (1982) are divided into two wind speed bands, all of them using a wind speed at a height of 10 m.

Table 1 Drag coefficient for wind stress in Van Dorn's (1953) equation

Author	Drag coefficient
Van Dorn (1953)	$C_d = \begin{cases} 1.2 \times 10^{-6} & U_{10} < U_c \\ 1.2 \times 10^{-6} + 2.25 \times 10^{-6}\left(1 - \frac{U_c}{U_{10}}\right) & U_{10} \geq U_c \\ where\ U_c = 5.6\ \text{m/s} \end{cases}$

Table 2 Drag coefficients for wind stress for Wu's (1969) equation

Authors	Drag coefficients
Smith and Banke (1975)	$C_d = (0.63 + 0.066 \times \|U_{10}\|) \times 10^{-3}$
Wang and Connor (1975)	$C_d = (1.10 + 0.053 \times \|U_{10}\|) \times 10^{-3}$
Garratt (1977)	$C_d = (0.75 + 0.067 \times \|U_{10}\|) \times 10^{-3}$
Large and Pond (1981)	$C_d = (0.49 + 0.065 \times \|U_{10}\|) \times 10^{-3}$
Wu (1982)	$C_d = \begin{cases} 1.2875 \times 10^{-3} & U_{10} < U_c \\ (0.8 + 0.65 \times \|U_{10}\|) \times 10^{-3} & U_{10} \geq U_c \\ where\ U_c = 7.5\ \text{m/s} \end{cases}$

The paths of the meteorological systems were traced using data of the minimum pressure at average sea-level which relate to the CFSv2 Reanalysis of the NCEP/NOAA, with spatial resolution of 0.5° × 0.5° and hourly data. For the identification of the cyclogenetic areas which generated the storm surges during the period of the experiment, a program was developed to generate synoptic wind and pressure charts. The same program seeks in a specific grid (latitude and longitude) to discover where the pressure minima occurred at each hour, thus extracting from the coordinate grid the coordinates and the value of the hourly pressure (Eq. 4) where the extremes occurred, later storing them for future analysis of the cyclones' paths.

By virtue of the fact that the average annual velocity of the cyclones in the southern hemisphere at intermediate latitudes (45°) attains 54 km/h (Sinclair 1995), the program after undertaking the quest for the initial grid and finding the latitude and longitude of the cyclone's minimum pressure, reduces the grid of the search for the following period taking as the central point the coordinates of the previous point and diminishing the search radius by two degrees in the latitude and longitude of the new movable grid.

$$[[Lat(t), Long(t), Pmin(t)]] = minimum\left[P_{i,j}(t);\ P_{m,n}(t)\right]. \tag{4}$$

Statistical Analysis and Data Filtering

Analysis of the correlations between temporal series were used in this study, and initially the Low-Pass filter PL 33 (Signell 1996) was applied to analyze the low frequency components and remove the local high frequency components of the tide, considering the local effect of the wind and pressure, in accordance with the proposal of Jenkins and Watts (1968), Chatfield (2004) and Truccolo et al. (2006). The Low-Pass filter PL 33 uses The Fourier transform to withdraw the high frequencies from the series leaving the signal only with the elements with the greatest period.

The partial correlation between wind stress (Variable x), the atmospheric pressure at the sea-level (Variable y) and the storm surge (Variable z) was calculated in this study by the use of Eqs. 5 and 6.

$$\rho_{zx|y} = \frac{\rho_{xz} - \rho_{yz} * \rho_{xy}}{\sqrt{\left(1 - \rho_{xy}^2\right)\left(1 - \rho_{yz}^2\right)}}, \tag{5}$$

$$\rho_{zy|x} = \frac{\rho_{yz} - \rho_{xz} * \rho_{xy}}{\sqrt{\left(1 - \rho_{xy}^2\right)\left(1 - \rho_{xz}^2\right)}}, \tag{6}$$

where:

- Pxy = Correlation between wind stress and Pressure;
- Pxz = Correlation between wind stress and storm surge;

- Pyz = Correlation between Pressure and storm surge.

For the analysis of the Partial Serial Correlation (PSC) which exists between the local storm surge, the Mean sea-level pressure and the wind stress of the wind parallel to the coast at the place where the buoys are anchored, an algorithm was developed. This shifts the whole storm surge series at each hour in the past and analyzes the partial correlation between the wind stress and the pressure without losing the data of the temporal series.

Later multiple linear regression analysis was undertaken. The dependent variable was the storm surge level, the pressure and the wind, in this case, the two independent variables. The multiple linear regression with time lag uses the optimum lag calculated on the PSC and shifts the pressure and wind series with regarding the storm surge (Eq. 7).

$$SS_t = \beta_0 + \left(\beta_p * P_{t-lagP}\right) + \left(\beta_\tau * \tau p_{t-lag\tau}\right) \quad (7)$$

where:

- SS_t = Storm Surge
- $\beta 0$ = intercept on the y axis;
- βp = angular pressure coefficient;
- $\beta \tau$ = angular wind stress coefficient;
- P = Pressure at average sea-level;
- τ_p = Wind stress of the wind parallel to the coast at 10 m height;
- $lag\ P$ = Optimum time lag between pressure and SS;
- $lag\ \tau$ = Optimum time lag between wind stress and SS.

The wind stress parallel to the coast may be written, as an example, with the Wu (1969) equation and the Wang and Connor (1975) drag coefficient as presented in Eq. 8. The component of the wind parallel to the coast can be calculated with Eqs. 9–12 as a function of the zonal and southern component of the wind. In this one, one tests among different directions of the wind which of them best accounts for the storm surge (variation in level).

$$\tau_p = \rho_{ar}\left(1.10 + 0.053 * \left|\vec{U}_{pc}\right|\right)10^{-3} * \vec{U}_{pc}\left|\vec{U}_{pc}\right| \quad (8)$$

$$\vec{V}_{10} = \sqrt{v_{10}^2 + u_{10}^2} \quad (9)$$

$$\beta = \tan^{-1}\left(\frac{u_{10}}{v_{10}}\right) \quad (10)$$

$$\alpha = abs(\theta - \beta) \quad (11)$$

$$\vec{U}_{pc} = \vec{V}_{10}\cos(\alpha) \quad (12)$$

where:

- τ_p = Wind stress parallel to the coast;
- U_{pc} = Component of the wind parallel to the coast;
- V_{10} = Wind speed (m/s) at 10 m height;
- u_{10} = Zonal component of the wind at 10 m height;
- v_{10} = Southern component of the wind at 10 m height;
- θ = Coast line azimuth;
- β = Course wind angle;
- α = Angle between the wind direction and the coastline direction.

Extreme Data Analysis

The methodology for the analysis of extreme EVA stationary values was developed by Mentaschi et al. (2016), the methodology consists of three steps (Eq. 13). The first is to transform a non-stationary time series $y(t)$ into a stationary series $x(t)$, then perform the analysis of EVA stationary extreme values and finally transform back the distribution of extreme values to the time domain

$$x(t) = f(y, t) = \frac{y(t) - T_y(t)}{S_y(t)} \tag{13}$$

where:

- $y(t)$ = Non-stationary series;
- $x(t)$ = Stationary series;
- $T_y(t)$ = Series trend;
- $S_y(t)$ = Long-term standard deviation.

This approach consists of transforming a non-stationary time series into a stationary one, to which the stationary EVA theory can be applied, and transform the result into a distribution of non-stationary extreme values. With the transformed series, a simple normalization of the time signal is performed and a more comprehensive formulation of generalized non-stationary (GEV) and generalized Pareto generalized extreme-value models (GPD) is applied with a constant parameter

Results

Analysis of the ADCP Data—Itapocorói Bay

The astronomical tide was distinguished from the level read on the ADCP by making the analysis of the harmonic components. The principle components of the astronomical tide were identified and are presented in Table 3. In Fig. 4, one may see the datum observed, the Tide and the storm surge. In this latter, it is possible to identify the periods during which the storm surge occurred, as it is also possible to separate the series by events.

It may be observed that in the period studied the main components of the amplitude of the tide are the principal lunar semi-diurnal (M2) with an amplitude of 28.53 cm, the principal solar semi-diurnal (S2) with an amplitude of 24.2 cm. Also the principal lunar diurnal (O1) with an amplitude of 10.57 cm and the principal lunar shallow water component (M4) with an amplitude of 8.84 cm and the shallow water quarter-diurnal component (MS4) with an amplitude of 5.08 cm and the larger lunar elliptic semidiurnal (N2) with an amplitude of 5.37 cm. The astronomical tide receives a

Table 3 Amplitude of the tides and phase with a 95% confidence interval

Component of the tide	Period (h)	Amplitude (cm)	Phase (°)	SNR
ALP1	29.1	1.46	291.26	11.00
2Q1	28.0	1.49	83.25	11.00
Q1	26.9	2.82	318.95	38.00
O1	25.8	10.57	286	470.00
NO1	24.8	2.02	70.14	15.00
K1	23.9	4.97	8.14	85.00
OO1	22.3	0.62	177.78	1.80
MU2	12.9	2.81	328.18	4.70
N2	12.7	5.37	274.92	17.00
M2	12.4	28.53	125.7	370.00
S2	12.0	24.2	301.45	390.00
MO3	8.4	2.62	271.96	9.20
M3	8.3	4.86	270	24.00
MK3	8.2	0.79	337.44	1.10
SK3	8.0	2.16	51.74	6.30
MN4	6.3	3.78	255.1	7.90
M4	6.2	8.84	237.56	51.00
MS4	6.1	5.08	138.95	18.00
M6	4.1	0.29	120.26	0.29
2MS6	4.1	0.78	292.73	1.20

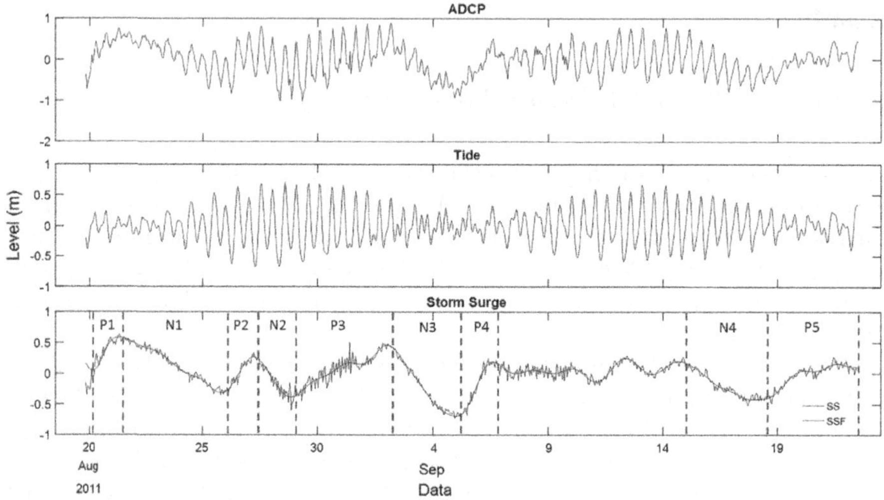

Fig. 4 Water level observed, storm surge level (SS) and filtered storm surge level (SSF) for the ADCP data from Itapocorói Bay

contribution from the other components, but those mentioned representing more than 70% of the tide.

During the experiment the astronomical tide presented three quadrature phases with an amplitude of 0.63 m and two spring tides ones with amplitude of 1.39 m, while during this period there occurred five positive storm surges and four negative ones (Fig. 4). In the graph of the level observed in Fig. 4 the influence of the storm surge on the astronomical tide may be observed. When the level is positive the level observed is 80% greater than that forecast and when the level is negative the storm surge can reduce the level observed by 75%.

Table 4 gives the dates of the beginning and end of each event defined, as well as its amplitude, the storm surge speed and the risk factor, which is calculated by multiplying the final level by the speed of the surge. It may be observed that the positive surge levels were of shorter duration and bigger speed than the negative surge levels. The response time of the positive levels is normally of the same duration as the passage of the cyclones while the negative levels correspond to the periods between the passage of the cyclones, when the action of the semi-permanent anticyclone of the South Atlantic occurs (Fig. 5), blowing north and northeast winds.

It may be observed in the first event that the storm surge affected the sea level positively (Fig. 4) and that whereas the neap tide presented a maximum level of 0.24 m in the period, this coincided with the surge level of 0.30 m resulting in an observed level of 0.54 m. During the same event, the greatest surge level was of 0.59 m, which coincided with the astronomical level of 0.03 m thus resulting in an observed level of 0.62 m, not presenting, a variation of more than 4%. However, in this period there occurred a maximum observed level of 0.69 m, when the astronomical level was of

Table 4 Amplitude and risk factor of the surge level

Event	Date of beginning	Date of end	Duration (h)	Initial level (cm)	Final level (cm)	Amplitude (cm)	Speed (cm/h)	Risk Factor (cm*cm/h)
P1	19/Aug. 20.00	21/Aug. 13.00	41	−17	59	76	1.85	109.37
N1	21/Aug. 14.00	25/Aug. 22.00	104	59	−30	89	0.86	−25.67
P2	25/Aug. 23.00	27/Aug. 7.00	32	−30	26	56	1.75	45.50
N2	27/Aug. 8.00	29/Aug. 2.00	42	26	−35	61	1.45	−50.83
P3	29/Aug. 3.00	2/Sep. 4.00	97	−34	45	79	0.81	36.65
N3	2/Sep. 5.00	5/Sep. 3.00	70	44	−69	113	1.61	−111.37
P4	5/Sep. 4.00	6/Sep. 15.00	35	−68	19	87	2.50	47.23
N4	14/Sep. 18.00	18/Sep. 4.00	82	20	−41	61	0.75	−30.50
P5	18/Sep. 5.00	21/Sep. 3.00	70	−41	10	51	0.73	7.29

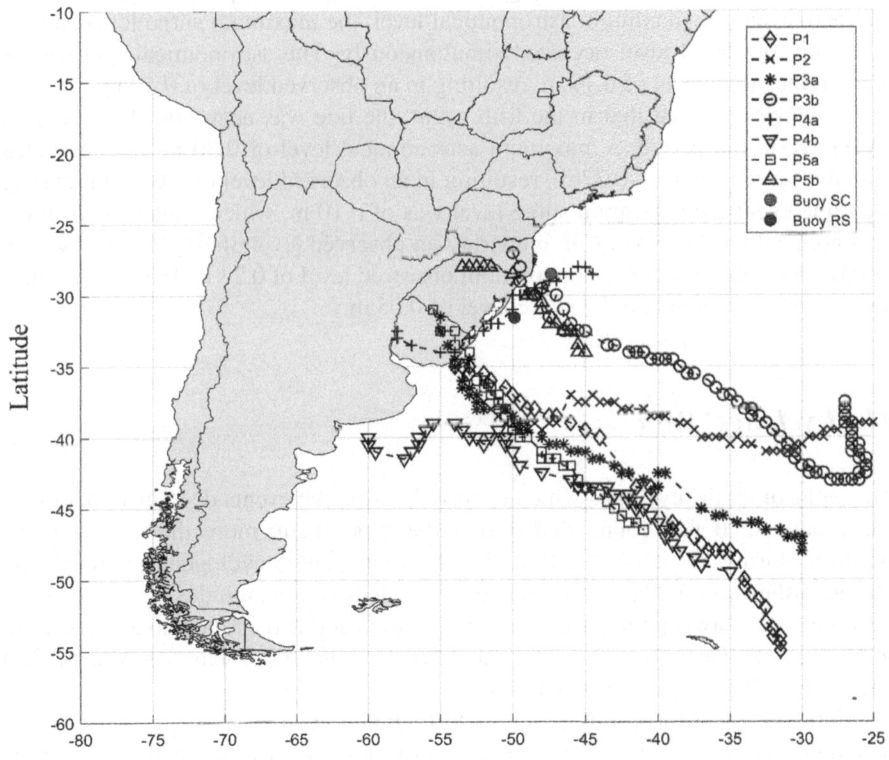

Fig. 5 The paths followed by the cyclones during the experiment (from 19 August 2011 to 22 September 2011) (EV = Event)

0.14 m and the surge level was of 0.55 m, resulting in a measured level much higher than that forecast astronomically.

In the second event, the storm surge occurred slightly before the spring tide (Fig. 4), presenting a maximum astronomical level in the period of 0.64 m, which coincided with the surge level of −0.03 m, resulting in an observed level of 0.61 m, with no significant variation. During the same event, the maximum surge level was of 0.27 m, which coincided with the astronomical level of −0.67 m, resulting in an observed level of −0.40 m. However, in this period there occurred an observed maximum level of 0.7 m coinciding with an astronomical level of 0.46 m and a surge level of 0.24 m.

In the third event the storm surge began during the spring tide and finished before the neap tide (Fig. 4). It presented in the period a maximum astronomical level of 0.69 m, which coincided with the surge level of −0.18 m, resulting in an observed level of 0.51 m, reducing the astronomical level foreseen. During the same event the surge level coincided with the maximum observed level, when the surge level was of 0.46 m and the astronomical level 0.45 m, resulting in an observed level of 0.91 m.

In the fourth event the storm surge occurred during the neap tide (Fig. 4). During the same event the maximum astronomical level, the maximum surge level and the maximum observed level occurred simultaneously. The astronomical tide was of 0.34 m, the surge level of 0.19 m, resulting in an observed level of 0.53 m.

It may be observed that in the fifth event, the tide was neap tide (Fig. 4). This presented, in the period, a maximum astronomical level of 0.30 m and coincided with the surge level of −0.37 m, resulting in an observed level of −0.07 m. During the same event the maximum surge level was of 0.10 m, which coincided with the astronomical level of −0.17 m, rendering an observed level of −0.07 m. However, in this period there occurred a maximum observed level of 0.23 m, the astronomical level being of 0.19 m and the surge level of 0.05 m.

The Cyclones' Path

The paths of all the cyclones which occurred during the events described in Table 4 were mapped. It was found that during the experiment, more than one cyclone occurred during the third, fourth and fifth events. Three cyclogenesis areas were also identified, as may be observed in Fig. 5. One between latitudes 29° S and 31° S near the coast between the states of Santa Catarina and Rio Grande do Sul; a second in Uruguay between latitudes 33° S and 35° S and a third near San Matias Gulf (Argentina), between 40° S and 42° S.

All the cyclones moved in a southeasterly direction and were generally followed by a high-pressure cell which moved eastwards, creating a wind fetch coming from the South-Southwest, where the high and low pressure cells. This began in southern Argentina and as the days passed the cyclone moved and the wind fetch came to extend as far as southern Brazil.

Usually the cyclones closest to Brazil occur separately, but sometimes another low pressure cell located in the south of Argentina, during the same period of the event P1, is generating southwest winds, parallel to the coast, increased the wind fetch and generating a storm surge greater than that resulting from the action of a single cyclone.

Partial Serial Correlation (PSC) and Time Lag Multiple Linear Regression (TLMLR)

When one undertakes an analysis of the wind stress for different wind speed scales, one observes that the results are very similar; up to a speed of 25 m/s, all the equations behaved similarly. Above that limit only Van Dorn (1953) equation behaved slightly differently, but the wind distribution is concentrated below 20 m/s, due to this the time lag multiple linear regression, can use any of the equations under study.

In the analysis undertaken of buoy 69,152 (Buoy SC) situated closest to Santa Catarina, the PSC was analyzed with variable pressure and constant wind stress (Fig. 6a). An average partial correlation, with $r = 0.5$, is seen for a time lag of -3 h. Undertaking an analysis of PSC with variable wind stress and constant pressure (Fig. 6a), a partial correlation of about $r = 0.81$ is found, with a time lag of -19 h.

To optimize the model, the time lag multiple linear regression (TLMLR) was undertaken varying the inclination of the wind parallel to the coastline (Fig. 6b). It was found that a 23° inclination of the wind was that which best represented the action of the wind on the surge level, with a coefficient of determination of $R^2 = 0.69$, using the wind and pressure data extracted at the site of the SC Buoy. The coefficient of determination presented values above 0.69 as from 18° and up to 28° of inclination, showing that the wind parallel to the coast has a level of action, which can influence the surge level.

The analysis of the PSC showed a better correlation when the series was analyzed with variable tension and constant pressure and with a greater time lag. The basic equation found considering the time lag of -3 h for pressure and of -19 h is presented as Eq. 14.

$$SS_t = \beta_0 + \left(\beta_p * P_{t-3}\right) + \left(\beta_\tau * \tau p_{t-19}\right) \tag{14}$$

Six different equations were used for the drag coefficient and the Time Lag Multiple Linear Regression (TLMLR) was calculated for each one, to purpose of assessing which of these equations best represents the storm surge for the different atmospheric conditions of Southern Brazil.

The coefficients for each equation are presented in Table 5. These are closely similar in their orders of greatness for all the equations. The storm surge is described in 69% with the drag coefficient proposed by Wang and Connor (1975) (Fig. 7).

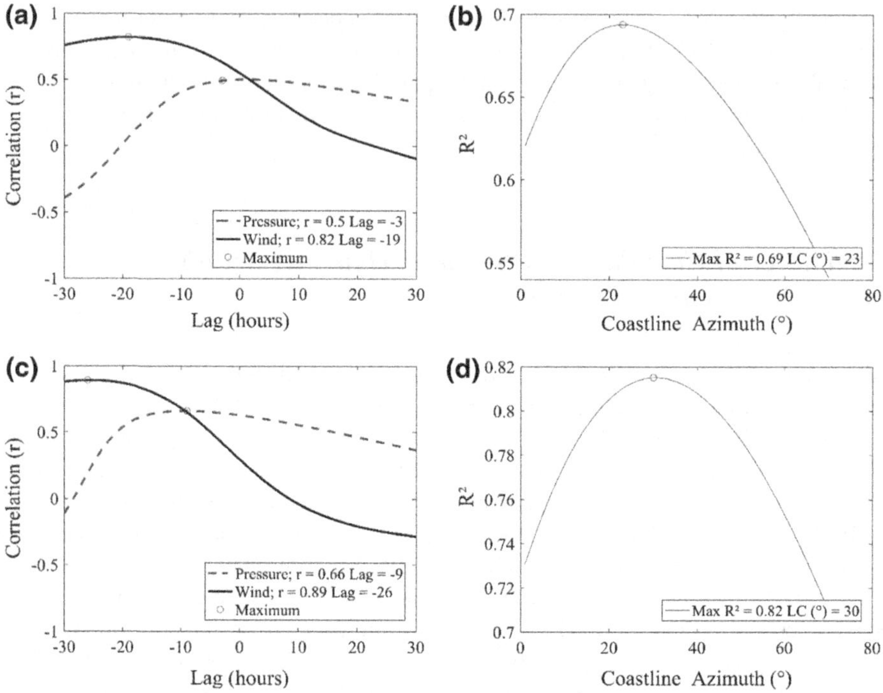

Fig. 6 Partial serial correlation at: **a** SC Buoy, **c** RS Buoy and the wind direction analysis at: **b** SC Buoy, **d** RS Buoy

Table 5 Coefficients of the model at the Santa Catarina buoy

	S and B	Garrat	W and C	L and P	Wu	Van D
β_0	−13.536	−13.506	−13.394	−14.636	−14.057	−14.002
β_p	0.0132	0.0132	0.0131	0.0143	0.0137	0.0137
β_τ	0.760	0.703	0.655	0.045	0.641	0.658

However, all the models show practically the same behavior for the variations of wind and pressure.

In the analysis of the PSC undertaken at buoy 69,153 (RS Buoy) (Fig. 6c), with variable pressure and constant tension, a partial correlation of r = 0.66 and a time lag of the SSt of −9 h were found for all the equations. Making the wind stress of the wind parallel to the coast variable and the pressure constant, a partial correlation of r = 0.89 was found for the first three coefficients and a correlation r = 0.88 for the last three, and all of them with a time lag of the SSt of −26 h.

The optimization of the azimuth angle of the wind parallel to the coast (Fig. 6d) was undertaken using the wind and pressure data extracted at the site of the RS Buoy. It was found that a 30° inclination of the wind with respect to the North is what best

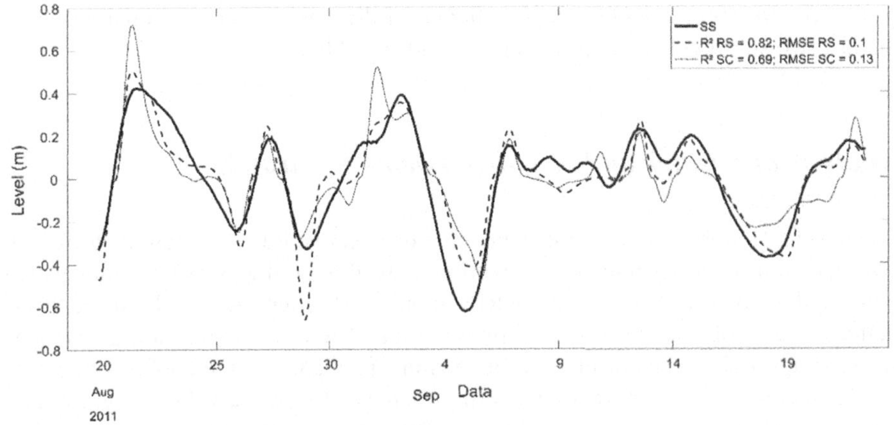

Fig. 7 TLMLR in the Itapocorói Bay

represents the variation of the surge level, with a coefficient of determination $R^2 = 0.82$. The coefficient of determination had values above 0.82 as from an inclination of 29° up to 32°, showing that the wind parallel to the coast at the site of the RS buoy has a smaller band of action than the wind parallel to the coast at the SC buoy.

For the analysis of the RS Buoy model, the basic equation is the same as the previous one, what changes is the time lag of −26 h found in the partial correlation with time lag for the wind stress and −9 for the pressure, presented as Eq. 15.

$$SS_t = \beta_0 + (\beta_1 * P_{t-9}) + (\beta_2 * \tau p_{t-26}) \tag{15}$$

One observes in Fig. 7 that the behavior of the model in relation to the variations of wind and pressure is similar, with a coefficient of determination $R^2 = 0.82$ using the equation proposed by Wang and Connor (1975). The models presented a delayed response to the changes in the direction of the storm surge, overestimating or underestimating the level of the storm surge.

The coefficients calculated for the TLMLR for the RS buoy are given in Table 6. The orders of greatness for all the coefficients of drag were closely similar, even though the first five equations used that of Wu (1969), which calculates the wind stress using the density of the air and Van Dorn (1953) which uses the density of the

Table 6 Coefficients of the model for the Rio Grande buoy

	S and B	Garrat	W and C	L and P	Wu	Van D
β_0	−3.089	−3.060	−2.953	−4.340	−3.783	−3.712
β_1	0.0030	0.0030	0.0028	0.0042	0.0037	0.0036
β_2	0.876	0.809	0.754	0.051	0.738	0.757

water. The first four calculate the coefficient of drag only with one equation while the last two sub-divide the equation into wind speed bands.

Analysis of Positive and Negative Events Separately

Each event of storm surge is either positive or negative due to different storms in: size, location, intensity and wind direction as well as in the pressure gradient that generated them. The buoy located in Rio Grande do Sul represented the surge levels better. As a result, we performed a time lag multiple linear regression analysis as well as the serial partial correlation for each independent event at the buoy location.

The methodology consists of analyzing each event separately, both negative and positive, and calculating the regression coefficients as well as the response time of each one. Then, it is used each set of coefficients to model the whole series with the different time lags of each event. Thus, to evaluate how much each event can represent the others, in Table 7 are the results of the regression.

In the analysis it was observed that positive events had better results, with determination coefficient $R^2 = 0.78$ in the fourth positive event (P4), while the highest negative coefficient of determination was $R^2 = 0.64$ in the third negative event (N3). The mean of the coefficients and time lags for the positive and negative events were calculated as well as the average of the event P4 with the event N3, in order to evaluate how much this new coefficient represents the observed surge level.

The average of the coefficients of P4 and N3 (A P4 and N3) presented a coefficient of determination $R^2 = 0.72$. An analysis was performed using the coefficients of P4 when the wind component parallel to the coast was positive and the coefficients of

Table 7 Coefficients of the model in the Buoy of Rio Grande do Sul separated by events

Event	B	Bp	Bt	lagP (h)	lagW (h)	R^2
P1	−26.944	0.027	0.181	−12	−30	0.52
P2	−5.986	0.006	0.506	−12	−24	0.63
P3	−15.529	0.015	0.385	−32	−39	0.37
P4	−23.807	0.023	0.536	−3	−26	0.78
P5	26.203	−0.026	0.582	−31	−23	0.53
Average P	−9.213	0.009	0.438	−18	−28	0.56
N1	−28.851	0.028	0.525	−33	−32	0.50
N2	−20.509	0.020	0.248	−15	−37	0.48
N3	−27.760	0.027	0.622	−3	−39	0.64
N4	−61.473	0.060	0.115	−1	−35	0.56
Average N	−34.648	0.034	0.377	−13	−36	0.60
Average P4 and N3	−25.783	0.025	0.579	−3	−32	0.72

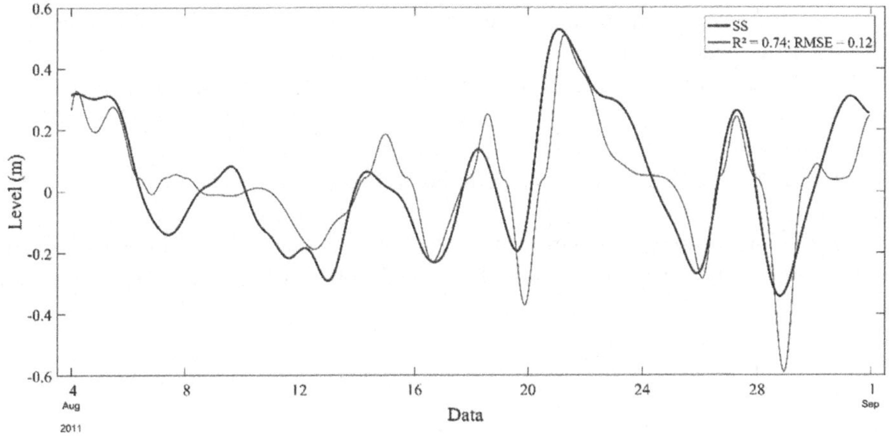

Fig. 8 Validation of the TLMLR in Itajaí for the month of August

N3 when the component was negative (P4 & N3), using the time lag calculated in the average of the two events, presenting a reestablishment coefficient $R^2 = 0.82$.

Validation of the Methodology

For the validation of the methodology a series measured in the river Itajaí was used the same treatment as that for the series measured in the Itapocorói Bay being applied. This covered the months of August and September the same period as the experiment. The monitoring in the Itapocorói bay lasted only 31 days, and we did not have a larger data series for the beach, the model was tested in the port of Itajaí, but as the tide is influenced by the river flow, these data were used only to verify if the methodology is applicable everywhere.

The validation of the TLMLR model was undertaken with the pressure and wind data extracted at the closest site of the RS Buoy, the tide series being divided into the months of August and September and analyzed separately. The inclination of the wind parallel to the coast was considered as being of 30° in relation to the North. The model presented behavior similar to that of the storm surge (Fig. 8), with a maximum coefficient of determination of $R^2 = 0.74$, as the negative storm surge was overestimated on days 19 and 29 by more than 20 cm.

Historical Analysis of Storm Surge

With the storm surge model developed for the Itapocorói bay, a historical study was carried out with pressure and wind data CFSv2 from 1979 to 2010. When the meteorological levels calculated by the model are observed in Fig. 9 from 1995 to 2000 to better observe the seasonality in the storm surge.

The EVA extreme events analysis (Fig. 10) presents the positive storm surge levels and their return periods. The period of 1 year, we have an extreme level of 0.59 m; for the period of 10 years, we have 0.65 m; for the period of 50 years, we find an extreme level of 0.76 m and for the period of 100 years, we find an extreme level of 0.80 m.

Positive storm surge have a maximum of 0.89 m, in the analysis of negative storm surge, historically it was a minimum of −68 m, For both positive and negative storm

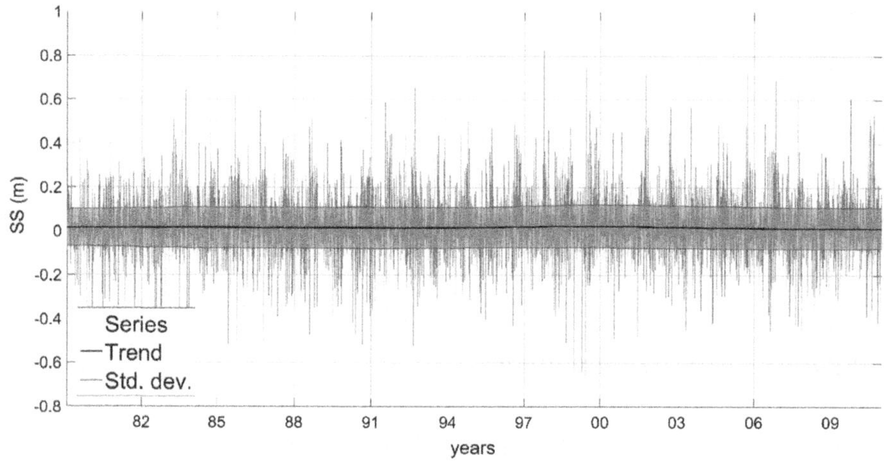

Fig. 9 Storm surge modeled from 1979 to 2010

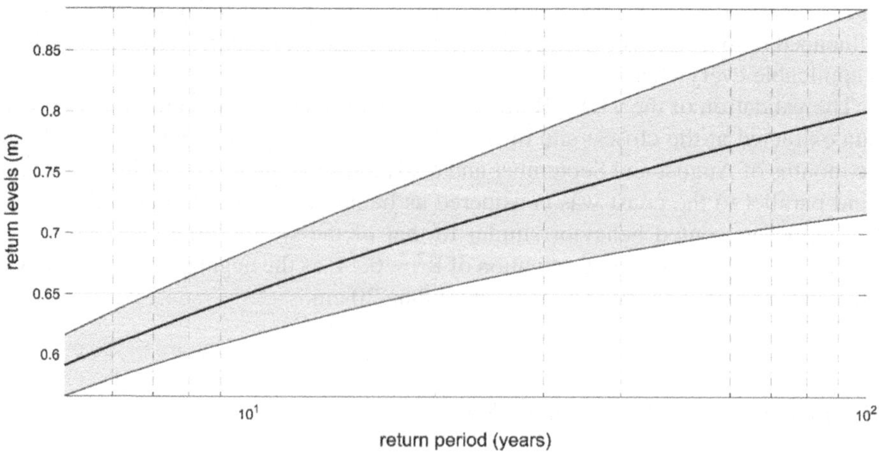

Fig. 10 Storm surge extreme analysis EVA from 1979 to 2010

surge there is bimodal distribution, with average 0.02 m and the quartile of 50% very close to zero.

Discussion

The comparison between the buoy and CFSv2 data is not shown in this paper because the buoy data presented many gaps and temporal incompatibility of sampling, the idea in this work is to test the wind and pressure at the nearest location of the buoys with the CFSv2 data and verify if they can represent the storm surge in the study area, so that in the future when the buoys work correctly can be used the data to mount a system of continuous monitoring of storm surge.

The results of the variation of the astronomical tide were similar to those found by Schettini (2002), Mesquita (2003) and Truccolo et al. (2006). In the period of the study maximum amplitude of the astronomical tide in the quadrature of 0.63 m was found, with amplitude in the spring tide of 1.39 m. It was also observed that when the surge level is positive, the level observed may be up to 80% higher than the level forecast and when the level is negative the storm surge can make the forecast level as much as 75% lower. Truccolo et al. (2006) in a study conducted in the north of Santa Catarina state found similar results. The level observed presented a variation 55 cm above the forecast level, whereas for Truccolo et al. (2006) it was of 90 cm.

In the study period three areas in which the cyclones arose were identified, one between latitudes 29° S and 31° S, between the states of Santa Catarina and Rio Grande do Sul, previously identified by Saraiva et al. (2003). The second cyclogenetic area was in Uruguay and the third near San Matias Gulf (Argentina): the same areas had been identified in previous results by Gan and Rao (1991), Sinclair (1994) and Parise et al. (2009). In some cases the formation of a second cyclone was identified during the same event, thus increasing the duration and intensity of the storm surge; according to Saraiva et al. (2003) this phenomenon occurs in 15% of the frontal systems which reach the coast of southern Brazil.

One of the meteorological phenomena which generate positive storm surge in Itapocorói Bay is a low-pressure system lying to the East of the continent, circulating clockwise, and a high-pressure one to the West, turning anti-clockwise, increasing the action of the winds and creating a wind fetch coming from the South-Southeast, a phenomenon identified in studies conducted by Saraiva et al. (2003). It was often found that the cyclone remained stationary between Argentina and Uruguay, and that the southerly winds do not get as far as the Santa Catarina coast, though they do produce positive surge levels in the study area.

The differences in the results of the PSC study show that the variations in storm surge are more susceptible to variations in the wind than to variations in pressure, as has already been commented by Truccolo et al. (2006). The time differences between the two buoys indicate that by virtue of the RS buoy's being located closer to the areas where the cyclones arise, it succeeds, with greater precision and with a greater time lag, in describing the storm surges in southern Brazil than does the SC buoy.

The variations in pressure with a time lag, both at the site of the SC buoy as at that of the RS buoy, reflected the behavior of the storm surges.

The storm surge in the south of Brazil is due to the coriolis effect, that causes that the masses of water from greater latitudes, that move towards the north tend to bend towards the west due to the increase of the circular speed in smaller latitudes, that generates the stacking of water on the coast and consequently the positive storm surge. The opposite occurs when the water bodies move from lower latitudes to higher latitudes, causing negative storm surge.

The intensification of the heights of the surge level coincides with the change in the direction of the wind fetch, relating the positive response of the storm surge to the winds coming from the South and Southwest, and a negative response to winds coming from the North and Northeast. With an excellent inclination of the coastline in relation to the North of 30° for the RS buoy and of 23° for the SC buoy.

The results show that all the equations of the wind stress represent the storm surge closely, but that Wu's (1969) equation, which uses the density of the air instead of that of the water, presents a slight difference. The drag coefficient which best describes the conditions of the wind for the South of Brazil, in this study, is that proposed by Wang and Connor (1975). However, the differences between all the equations are within the margin of error, without being possible to establish one single equation which represents the atmospheric conditions of the South of Brazil, in this study, better than any of the others.

The validation of the model showed that it reproduces the level of the water faithfully by virtue of the wind and pressure forces, R^2 being smaller due to the tide meters being in the Itajaí River. The discharge of the river exercises its influence. At least 30 days of data are necessary for a forecast of the level of the site desired to be made and to gain knowledge of the zone of origin of the disturbance of the storm surges.

Analyzing the historical results obtained with the storm surge model from the CFRv2 data series, it can be inferred that the meteorological levels follow a seasonal trend, varying according to the season and other atmospheric factors not analyzed in this work. The analysis of external EVA events to evaluate the external levels for recurrence times of 1, 10, 50 and 100 years resulted in already recorded external values indicating that the time series used to make the model must be larger in order to obtain better results.

Conclusion

The final inundation level on the coast due to storms depends on several factors, the proposed model is only intended to predict the additional level resulting from the storm surge, predicting 81% of the surge level at best. However, the proposed methodology is in the initial stage, which can be improved in future works, but the work achieved the objective of proposing a model that can be applied to any beach in southern Brazil, requires little computational infrastructure and that in a few seconds can if you have a forecast of the local storm surge for a few days. As well as carrying out studies of future scenarios that can provide information for coastal urban planning and elaboration of coastal risk prevention plans.

Although the time series used in this study is small and does not cover the various meteorological conditions, which can give rise to extreme surge levels on the coast of Santa Catarina. It is understood that the methodology proposed in this study has shown itself to be valid and indicates that it would be possible to use the data of the National Buoy Program (PNBOIA) of the GOOS-Brazil to forecast and monitor the storm surges in the South of Brazil.

On the basis of the model proposed it is further concluded that to prepare a monitoring system for any desired location (whether port or city) this methodology has to be applied with the tide data measured in situ and the coefficients calculated for each specific place. For a system of short-term forecasting one must calculate new coefficients with more than 30 days data in advance for the storm surge, wind and pressure so as to be able to build the model and foresee the next days of storm surge. All equations were shown to be compatible with the study and there is no one set of coefficients for the equation proposed.

Acknowledgements The authors express their gratitude to UNIVALI for making the data on Itajaí port available and to Chicago Bridge and Iron (CBI) for their participation in the project. The research was supported by CAPES (Rede Riscos Costeiros 09/2009), CNPQ (Universal/2008 [Proc. No. 471068/2008-0] and CTTrans/44-2008 [Proc. No. 575008/2008-3] and Scholarship for Research Productivity (Bolsa de Produtividade em Pesquisa – Nível 2) [Proc. No. 303550/2012-0]), PFRH-PB240 (PETROBRAS) and the Postgraduate Geography Program of the Federal University of Santa Catarina.

References

Baart F, Van Gelder PHAJM, De Ronde J, Van Koningsveld M, Wouters B (2012) The effect of the 18,6-year lunar nodal cycle on regional sea-level rise estimates. J Coast Res 28(2):511–516

Chatfield C (2004) The analysis of time series, an introduction, 6th edn. Chapman & Hall/CRC, New York, USA, p 333

Church JA, White NJ (2006) A 20th century acceleration in global sea level rise. Geophys Res Lett L01602:33

Gan MA, Rao BV (1991) Surface cyclogenesis over South America. Mon Weather Rev 119(5):1293–1302

Garratt JR (1977) Review of drag coefficients over oceans and continent. Mon Weather Rev 104:418–442

IPCC (2014) Climate changes 2014: impacts, adaptation, and vulnerability, part A: global and sectoral aspects. In: Field CB, Barros VR, Dokken DJ, Mach KJ, Mastrandrea MD, Bilir TE, Chatterjee M, Ebi KL, Estrada YO, Genova RC, Girma B, Kissel ES, Levy AN, MacCracken S, Mastrandrea PR, White LL (eds) Contribution of working group II to the fifth assessment report of the intergovernmental panel on climate change. Cambridge University Press, Cambridge, United Kingdom and New York, NY, USA, 1132 p

Jenkins GM, Watts DG (1968) Spectral analysis and its applications, Holden-Day series in time series analysis. Holden-Day, San Francisco, USA

Klein AHF, Menezes JT, Diehl FL, Abreu JGN, Polette M, Sperb RM, Sperb RC, Horn N (2006) Santa Catarina. In: Muehe D (ed) Erosão e progradação no litoral brasileiro. Brasília, DF: MMA, pp 402–412. ISBN 85-7738-028-9

Large WG, Pond S (1981) Open ocean momentum flux measurements in moderate to strong winds. J Phys Ocean 11:324–336

Mctaggart-Cowan R, Bosart LF, Davis CA, Atallah EH, Gyakum JR, Emanuel KA (2006) Analysis of hurricane Catarina (2004). Mon Weather Rev 134:3029–3053

Mentaschi L, Vousdoukas M, Vousdoukas E, Sartini L, Feyen L, Besio G, Alfieri L (2016) The transformed-stationary approach: a generic and simplified methodology. Hydrol Earth Syst Sci 20:3527–3547

Mesquita AR (2003) Sea level variations along the Brazilian coast: a short review. J Coast Res 35:21–31

Parise CK, Calliari LJ, Krusche N (2009) Extreme storm surges in the South of Brazil: atmospheric conditions and shore erosion. Braz J Oceanogr 57(3):175–188

Pawlowicz R, Beardsley B, Lentz S (2002) Classical tidal harmonic analysis including error estimates in MATLAB using T_TIDE. Comput Geosci 28(8):929–937

Pugh DT, Woodworth P (2014) Sea-level science. In: Understanding tides, surges, tsunamis and mean sea level, 2nd edn. Cambridge University Press, UK

Reboita MS (2008) Ciclones Extratropicais sobre o Atlântico Sul: Simulação Climática e Experimentos Numéricos de Sensibilidade. Universidade de São Paulo, São Paulo, SP, Brazil

Santa Catarina (2006) Decreto nº 5.010, de 22 de dezembro de 2006. Regulamenta a Lei n. 13.553, de 16 de novembro de 2005. Lex, coletânea de legislação e jurisprudência, Florianópolis, SC, Brazil

Saraiva JMB, Bedran C, Carneiro C (2003) Monitoring of storm surges on Cassino Beach. J Coast Res 35:323–331

Schettini CAF (2002) Caracterização física do estuário do rio Itajaí-Açu, Revista Brasileira de Recursos Hídricos, Porto Alegre, RS. Brazil 7(1):123–142

Signell R (1996) Low-pass filters hourly data using the pl33 filter. USGS, Reston, Virgínia, USA

Sinclair MR (1994) An objective cyclone climatology for the Southern Hemisphere. Mon Weather Rev 122(10):2239–2256

Sinclair MR (1995) A climatology of cyclogenesis for the southern Hemisphere. Mon Weather Rev 123:1601–1619

Smith S, Banke E (1975) Variation of the sea surface drag coefficient with wind speed. Q J Royal Meteorol Soc 101:665–673

Truccolo EC, Franco D, Schettini CAF (2006) The low frequency sea level oscillations in the northern coast of Santa Catarina, Brazil. J Coast Res 39:547–552

Van Dorn WC (1953) Wind stress on an artificial pond. J Mar Res 12:249–276

Wang JD, Connor JJ (1975) Mathematical modelling of near coastal circulation. R.M. Parsons Laboratory for Water Resources and Hydrodynamics, Report No. 200. Cambridge, Massachusetts, USA

Woth K, Weisse R, Von Storch H (2005) Climate change and North Sea storm surge extremes: an ensemble study of storm surge extremes expected in a changed climate projected by four different regional climate models. Ocean Dyn 56:3–15

Wu J (1969) Wind stress and surface roughness at sea interface. J Geophys Res 74:444–453

Wu J (1982) Wind-stress coefficients over sea surface from breeze to hurricane. J Geophys Res 87(C12):9704–9706

Hydrodynamic Study of Free Standing Drilling Riser Under Hurricane Conditions

Hong Yaw Yong, M. S. Liew, Mark Ovinis, Kamaluddeen U. Danyaro and Eu Shawn Lim

Abstract Offshore structures are susceptible to damages due to hurricane. The main subject of this research, free standing drilling riser (FSDR), is an innovation to the drilling riser that enables rapid disconnection prior to hurricane conditions. This allows the offshore drilling rig and operator to move away from hurricane path prior to hurricane landing. Yet, currently, the emergency disconnection decision of FSDR is decided by a rig manager. Past incidents have shown that human error can cause catastrophic damage. Studies on FSDR operations under hurricane condition are scarcely available. The most relevant studies, free standing riser (FSR) behaves differently. With the increase in hurricane intensity over the past 50 years, this increases the risk in FSDR operations. In this work, the hydrodynamic response of FSDR under hurricane conditions was investigated. Four simulation studies were conducted in Orcaflex finite element modeling (FEM) software. These studies were conducted to determine the effect of emergency disconnection location, the optimum emergency disconnection location, the effect of various Saffir Simpson maximum significant wave height and hurricane generated swell on the hydrodynamic response of FSDR. Hurricane generated wave spectrum was used to replicate hurricane conditions. The simulation study shows that the lateral displacement of FSDR decreases in a 3rd order polynomial function with the emergency disconnection location, $LD = -6e^{-6}d^3 + 0.0024d^2 - 0.3348d + 19.179$. In addition, the optimum emergency disconnection location along drilling riser line is at 87.50% of the original water depth. The lateral displacement of FSDR increases exponentially with Saffir Simpson scale storm intensity, $y = 1.7537e^{0.0386x}$. Lastly, the stronger the swell component of hurricane generated swell, the higher the hydrodynamic responses. Wind wave component of hurricane generated swell did not influence FSDR as the riser was in the free standing

H. Y. Yong · M. S. Liew (✉) · K. U. Danyaro · E. S. Lim
Offshore Engineering Centre, Universiti Teknologi PETRONAS, 32610 Bandar Seri Iskandar, Malaysia
e-mail: shahir_liew@utp.edu.my

M. Ovinis
Department of Mechanical Engineering, Universiti Teknologi PETRONAS, 32610 Bandar Seri Iskandar, Malaysia

W. Leal Filho et al. (eds.), *Climate Change, Hazards and Adaptation Options*, Climate Change Management, https://doi.org/10.1007/978-3-030-37425-9_3

mode. The limitation of this research is the findings are applicable for this simulation setup only. Further validations are required with experimental setup and various types of offshore structure.

Introduction

Hurricanes are one of the frequent threats to offshore facilities. They occur annually, generating strong winds and high waves, causing damage to the offshore facilities (Kaiser 2007; Reed 2005; Thies et al. 2015; Veldman et al. 2011; Wisch 2006; Zhu et al. 2009). Past damages have shown that the offshore industry is susceptible to damage by hurricanes. In 1979, dynamically positioned (DP) vessels HAVDRILL, PELICAN and SEDCO 445 were all damaged by hurricanes. These DP vessels were all caught in various stages of the riser and Blow Out Preventer (BOP) stack pulling with multiple damages reported on the BOP stack, risers and moonpool structure. In 2005, Hurricane Katrina and Rita both of which category 5 hurricane had destroyed at least 113 platforms, severely damaged 53 offshore platforms and 457 pipelines (Cruz and Krausmann 2008). These two hurricanes damaged a third of drilling rig (Deluca 2005). Certainly, these damages have shown us that the offshore industry is still vulnerable to the impact of hurricane (Cruz and Krausmann 2009; Duhaney et al. 2004; Zoethout and Brugts 2006).

With the increase in hurricane intensity due to global warming (Bushnell and Falvey 2017; Knutson et al. 2010), it is important to study the hydrodynamic response of FSDR under hurricane conditions. This will guide the operator for the safe usage of FSDR. In 2017, with preventive measures been put in place, Hurricane Harvey did not bring as much damage to the US oil and gas industry when compared to the 2005 hurricane season (Lempiere 2017).

The subject of this research, FSDR is designed to enable rapid disconnection prior to storm landing. With FSDR mechanism, a total time of 12 h is required to securely prepare for emergency disconnection procedure (Kogure et al. 1998). Guo et al. (2015) stated a similar time required that FSDR mechanism can enable emergency disconnection within 8 h instead of 36 h. Despite numerous past studies available, (i) the functional relationship of emergency disconnection location along drilling riser line and lateral displacement of FSDR, (ii) the optimum emergency disconnection location along drilling riser line in technical terms, (iii) the functional relationship of hurricane maximum significant wave height and lateral displacement of FSDR and (iv) the effect of hurricane generated swell on hydrodynamic response of FSDR were rarely seen in past work. The decision to disconnect in an emergency and the location of disconnection were commonly decided by the Rig Manager. Uncertainty in weather forecast system and human error effect can lead to catastrophic damage (Corona et al. 1996; Kogure et al. 1998; Kumar 2005). In normal conditions, weather forecasts are only available around 48 h prior to the storm landing. This short time frame is insufficient for the whole drilling riser to be pulled back (Kogure et al. 1998; Saruhashi et al. 2014; Sheng et al. 2016).

Problem Statement

Failure to emergency disconnect in time or wrong decision in the determining the disconnection location can lead to structural failure. The lesson learnt from the delay in response of Fukushima Daiichi Nuclear Meltdown incident is to be studied and remembered. Delay in decision making in emergency response led to a catastrophic event (Yang 2014). Apart from that, historically, there were two free standing riser (FSR) failure accidents been reported. In the first incident, a free-standing production riser that was installed in the GoM used chain connection instead of flex joints below the buoyancy tanks and between the riser and the sea bed. Due to human error in designing, the riser failed and was damaged (Millheim and Aune 2011). On the other hand, the second FSR failure accident happened at offshore Indonesia where wells were being drilled by surface blow out preventer (SBOP). When the drilling depths have reached the limits of the rig tensioner, buoyancy device was added just below the water to support the riser. However, due to no stress joints were installed at the critical bending stress areas, the riser pipe failed just below the buoyancy device, causing damage to the rig structure (Millheim and Aune 2011). These two incidents have shown that human error is a risk in designing of the FSR system. Proactive measures must be taken to prevent similar accidents from repeating itself. Current recommended practice for drilling risers was defined by the American Petroleum Institute (API), API-RP 16Q (APIb 2010). The relationship of hydrodynamic response of drilling riser under extreme weather conditions should be studied in terms of hurricane intensity and emergency disconnection location. This study can serve as a recommendation for the operators and legislative body to develop guidelines for safe operations of FSDR.

The guideline for safe operation of FSDR is scarcely available in the public domain. Guidelines such as API-RP 95 J should be put in place. API-RP 95 J address guidance on the minimum required air gaps for jack up rigs under hurricane conditions and evacuation checklist for jack up platform during hurricane season. Several improvements were also made to the recommended practice after hurricane Katrina and Rita (Defu et al. 2017). The main objective of this research is to study the hydrodynamic response of FSDR under various hurricane conditions. Besides, this research develops the following:

(1) To develop the functional relationship between the lateral displacement of FSDR and disconnection location along the riser line.
(2) To identify the optimum emergency disconnection location of FSDR along riser line in terms of technical aspects.
(3) To develop the functional relationship between the lateral displacement of FSDR and Saffir Simpson scale maximum significant wave height.
(4) To identify the effect of hurricane generated swell on the hydrodynamic response of FSDR.

Free Standing Drilling Riser

The motivation of FSDR system (Fig. 1) can be dated back to 1973 where Burke (1973) stated that in adverse condition, riser design and operating procedures should be developed to maintain the integrity of the drilling risers. Under survival conditions, disconnection of the risers from the wellhead should be initiated to assure its survival (Burke 1973). Kwan et al. (1979) stated that the most comprehensive solution for protecting the risers in a storm event is to disconnect it at a point somewhere below the wave zone. Kwan et al. (1979) also mentioned that with the riser being disconnected several meters below mean sea level (MSL), the drilling rig can sail away to safety. However, Kwan et al's. (1979) research work did not quantify the emergency

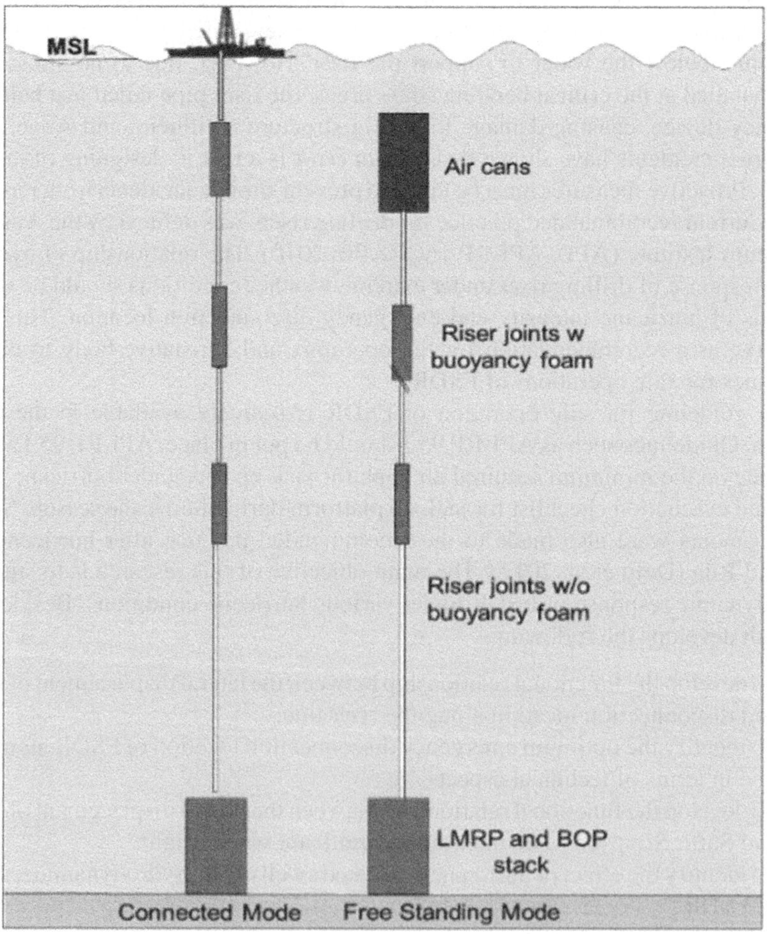

Fig. 1 Drilling riser under connected mode and free-standing mode (Yong et al. 2018)

disconnection location. With the knowledge of the functional relationship between the emergency disconnection location of FSDR and the lateral displacement, the rig manager can better decide on the emergency disconnection location with lesser time required.

Studies Conducted on Free Standing Drilling Riser (FSDR)

A time domain analysis was performed by Kogure et al. (1998) on the FSDR mechanism. The analysis was conducted to compare FSDR mechanism and conventional drilling riser system. The results show that FSDR mechanism is feasible and its hydrodynamic response is close to conventional drilling riser system. For their study, it was noted that there was a rapid fluctuation of stresses at the upper part of the FSDR flex joint and just below the air cans. This was due to its increase in diameter that increased the hydrodynamic drag. The emergency disconnection location for this case is located at 100 m from MSL. Yet, there is no supporting reason for the selection of the location. The disconnection location is 95% of the original water depth (WD). Despite the preliminary study on the FSDR mechanism, the relationship as per the objectives of this research were not explored.

API Recommended Practice for Design, Selection, Operation and Maintenance of Marine Drilling Riser System (API-RP 16Q)

According to API-RP 16Q (APIb 2010), there were three operating modes for marine drilling riser operation. Namely, drilling mode, connected non-drilling mode and disconnected mode. The disconnected mode was only initiated when the environmental conditions exceeded the limits of safe operations under connected non-drilling mode. This mode was used as the study conditions for the FSDR. Table 1 shows the maximum operating and design guidelines for marine drilling riser (APIb 2010). The maximum allowable stress under disconnected mode will be used to evaluate the integrity of the FSDR mechanism.

Table 1 Maximum operating and design guidelines for marine drilling riser (APIb 2010)

Design parameter	Drilling	Non-drilling	Disconnected
Max flex joint angle	4.0°	90% available	90% available
Maximum allowable stress	0.67*yield stress	0.67*yield stress	0.67*yield stress

Table 2 Saffir-Simpson hurricane severity scale (Zahibo and Didenkulova 2008; Ochi 2003)

Category	Wind speed (m/s)	Estimated damage
1	33–42	Minimal
2	43–49	Moderate
3	50–58	Extensive
4	59–69	Extreme
5	≥70	Catastrophic

Hurricane

With the term “hurricane” being the commonly accepted terminology for a tropical cyclone, “hurricane” was used throughout this research to describe tropical cyclone. A tropical cyclone is a non-frontal synoptic scale low-pressure system over tropical or subtropical waters with organized convection and definite cyclonic surface wind circulation. A tropical cyclone with maximum sustained surface wind speed of less than 17 m/s is called tropical depression. Once the tropical cyclone reaches wind speed of at least 17 m/s, they are called a tropical storm. When the cyclone’s wind speed reaches 33 m/s, it is classified as a “hurricane” for the North Atlantic Ocean and East of the dateline in Northeast Pacific Ocean; “typhoon” for the West of the dateline in Northwest Pacific Ocean (Collins 2014; Kumar 2005; Zahibo and Didenkulova 2008; Ochi 2003; Thies et al. 2015). Saffir-Simpson’s scale was adapted to describe the hurricane intensity. Table 2 shows the Saffir-Simpson’s Hurricane Severity Scale and the estimated damage (Zahibo and Didenkulova 2008; Ochi 2003).

Hurricane Generated Sea State

Hurricane generated sea state was replicated to represent the sea state under hurricane conditions. Several authors had classified and represented the sea state using mathematical equations. Young (2003) provided information on the spatial distribution of a wave field. The wind generates a spectrum of the wave with different values of group velocity, Cg (velocity where the waves propagate). For all but the fastest moving storm, spectral components with the higher value of Cg can outrun the hurricane and becomes swell traversing in front of the hurricane. The swell radiates out from the intense wind generation area to the right of the storm center (Young 2003). Wave field generated by a hurricane is asymmetrical. The maximum waves are located slightly ahead of the storm center. When the velocity of the forward movement increase, the region gradually moves to a point behind and to the right of the storm center (Young 2003). Additionally, when the wave spectra were recorded close to the storm center, the spectra were uni-modal.

Yang et al. and Young (2003) stated that JONSWAP wave spectra fitted well with the actual hurricane generated sea state. The only difference was at high frequency section the measured spectra were larger than the JONSWAP spectrum (Yang et al. 2015). Due to lack of actual hurricane wind field parameters, vector speed of forwarding movement, V_{fm}, maximum wind velocity in a storm, V_{max} and radius to the maximum wind, R; modified JONSWAP wave spectrum was used to represent the hurricane generated sea state.

Ochi (2003) wrote in a book that for hurricane generated seas, the input energy that generated the waves was usually in the same direction as the wave energy was progressing. The duration taken for given wind speed to increase in speed was faster than normal storms. The sea states normally increase rapidly when the eye of the hurricane reach within 150 km of the site (Ochi 2003). Ochi (2003) added that the wave spectrum to best represent the hurricane generated sea state is by modifying the JONSWAP wave spectrum with the scale parameter, α and peak shape parameter, γ. Equation 1 shows the original form of the JONSWAP wave spectrum.

$$S(f) = \alpha \frac{g^2}{(2\pi)^4} \frac{1}{f^5} exp\{-1.25(f_m/f)^4\} \gamma^{exp\{-(\omega-\omega_m)^2/2(\sigma f_m)^2\}} \quad (1)$$

where

γ = peak shape parameter, 3.3 being the average,
f_m = modal frequency $= 3.5\left(\frac{g}{U}\right)x^{-0.33}$
$\omega = 2\pi f$
$\alpha = 0.076x^{-0.22}$
$\sigma = 0.07$ for $f \leq f_m$, 0.09 for $f > f_m$.
X = dimensionless fetch $= \frac{gX}{U^2}$
U = mean wind speed
g = gravity constant.

The modified JONSWAP spectrum to represent the hurricane generated sea state is as per Eq. 2. The spectrum is represented by significant wave height, H_s, modal frequency, f_m and scale parameter, σ.

$$S(f_m) = \frac{4.5}{(2\pi)^4} g^2 H_s^2 \frac{f_m^4}{f^5} e^{-1.25(f_m/f)^4} \left(9.5 f_m H_s^{0.34}\right)^{\exp\{-(f-f_m)^2/2(\sigma f_m)^2\}} \quad (2)$$

where H_s is in m and f_m is in Hz.

Significant wave height (SWH), H_s and modal frequency, f_m were identified using Foster (1982) method as shown in Eqs. 3 and 4.

$$\alpha = 4.5 f_m^4 H_s^2 \quad (3)$$

$$\gamma = 9.5 f_m H_s^{0.34} \quad (4)$$

Hurricane Generated Waves

Hurricane generated waves were affected by various factors. Several studies were done to capture the wave parameters. For example, a study conducted on Hurricane Juan in 2003 reported several measurements of significant wave height (SWH) at locations with different water depths (Xu et al. 2007). When the measuring station was located on the storm track, the SWH increased up to 12.1 m when compared to 6.9 m when the hurricane was 150 km away from the hurricane. The closer the measuring station was located to the storm track, the higher the SWH measurement.

Ochi (2003) produced the functional relationship between wind speed and the sea severity with regards to the sea conditions 7–10 days prior to the hurricane arrival. Equation 5 shows the relationship between significant wave height and the mean wind speed when the sea conditions are relatively calm for 7–10 days prior to the hurricane arrival. Equation 6 shows the relationship between significant wave height and the mean wind speed when the sea conditions are severe for 7–10 days prior to the hurricane arrival.

$$H_s = 0.24 * U_{10} \tag{5}$$

$$H_s = 0.078 * U_{10}^{1.57} \tag{6}$$

Some of the studies stated that hurricanes with wind speed of 32.7–36.9 m/s can generate wave height up to 14.0 m (Fang and Duan 2014). Besides, another hindcast study performed on Hurricane Katrina and Rita that landed in the GoM reaffirmed that the significant wave height was increased to the range of 15–16 m during hurricane Katrina and 14–15 m range during hurricane Rita where these two hurricanes were category 5 hurricane (Cardone et al. 2007). These findings were used as reference wave height in the current research.

Hurricane Generated Swell

Hurricane being a frequent event in the SCS, on rare occasions, will occur in the Sulu Sea; offshore Sabah, Malaysia and moving towards the Gulf of Thailand (Bernama 2019; Raghupathi et al. 2016). According to Chan, Yeh et al. (2006) and Hong Kong Observatory (2005), a distant hurricane can generate swell that are able to travel several hundred kilometers from the hurricane origin. The swell can travel at a faster speed when compared to the hurricane itself. Hence, there were occasions of huge waves crossing during clear weather (Yeh et al. 2006).

Pei et al. (2016) stated that hurricane generated swell originating from Northwest Pacific region can be observed in the East China Sea. Tao et al. (2017) stated that swells in the East China Sea produce wave heights of about 0.1–2.5 m with the highest frequency of occurrence at 0.1–0.5 m. The swell period recorded to range between

4 and 15 s with the highest frequency recorded in the range of 9–15 s. The study also concluded that 43% of swells experienced in the East China Sea originated from nearby seas due to hurricanes (Tao et al. 2017). These swell parameters were used as input to study the hydrodynamic response of FSDR under hurricane generated swell.

In terms of hurricane swell spectrum, Collins (2014) stated that the wave spectrum was bi-modal when it was far away from the hurricane core. The bi-modal spectrum consists of both the low frequency swell components (0.04–0.12 Hz) and the high frequency wind wave components (0.12–0.50 Hz). In the year of 1995, DiMarco et al. (1995), made an observation that when Hurricane Andrew passed by 30 km away from a measuring gauge, a strong bi-modal wave spectrum was observed. Xie et al. (2013) added that there was commonly three types of sea states namely, swell dominated sea state, wind sea dominated sea state and mixed sea state. The mathematical representation of sea state spectrum is as per Eq. 7.

$$S(f) = S_{swell}(f) + S_w(f) \tag{7}$$

Research Methodology

The research work starts with creating both the FSDR model and the hurricane generated sea states. FSDR model was reconstructed in Orcaflex 10.1c FEM software and the hurricane generated sea states were generated using a modified JONSWAP wave spectrum. The scale, α and peak shape parameter, γ used were derived from maximum significant wave height based on Saffir Simpson scale wind speed. Past hurricane data from Hurricane Eloise (Lee 1980) were used to replicate hurricane conditions as well. The FSDR model was then validated with published work to ensure the model created was correct. Then, the FEM analyses were conducted. The FEM analyses were conducted to study the four objectives to understand the hydrodynamic response of FSDR under hurricane conditions. Figure 2 shows the overall process flow chart of this research.

Procedures to Develop the Functional Relationship Between Emergency Disconnection Location and Lateral Displacement of FSDR

The study aimed to study the mechanical behavior of FSDR under 100 years extreme weather conditions as stated in API-RP 16Q standards. Lateral displacement, bending moment and von Mises stress of various disconnection location were studied in relation with the percentage of original WD. However, in offshore operations, drilling risers are normally installed in joints of 15.24 m length. Therefore, this research work

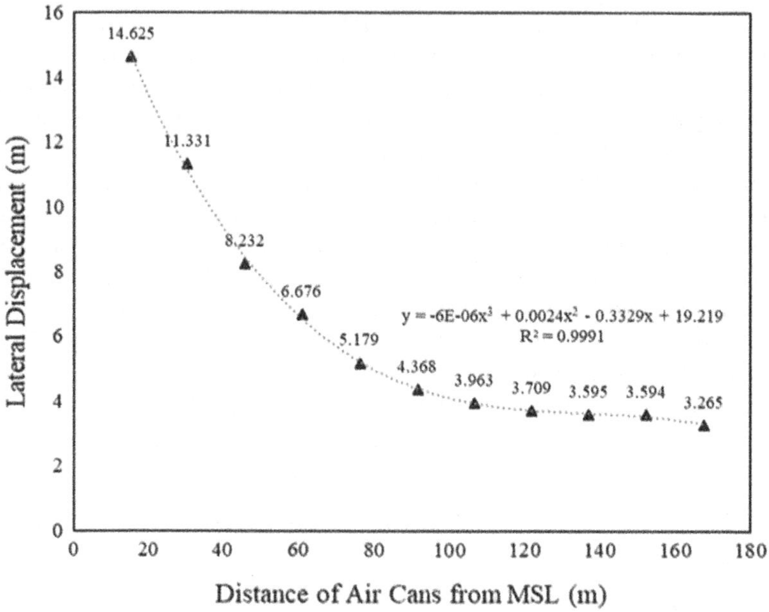

Fig. 2 Process flow chart

included the study of the functional relationship in terms of removing the drilling riser joints. Table 3 shows the 100 years extreme weather conditions at the Spiced Jar semi-submersible drilling location (APIb 2010).

The simulation study was conducted under two conditions. (i) under 100 years extreme weather conditions with JONSWAP wave spectrum, (ii) under hurricane generated sea states. Data for hurricane generated sea state was based on the data

Table 3 100 years extreme weather condition for spiced jar semi-submersible (APIb 2010)

Parameters	Values	
Significant wave height (m)	8.2	
Zero crossing period (s)	9.7	
Current speed (m/s)	WD from MSL (m)	Values
	0	1.158
	10.06	1.158
	114.91	0.386
	266.09	0.345
	459.94	0.232
	609.60	0.144

recorded during Hurricane Eloise by Withee and Johnson (1975). Hurricane Eloise's scale parameter, α and peak shape parameter, γ were calculated based on Eqs. 3 and 4.

$$\begin{aligned}\alpha &= 4.5 f_m^4 H_s^2 \\ &= 4.5 * (0.09)^4 * 8.8^2 \\ &= 0.023\end{aligned}$$

$$\begin{aligned}\gamma &= 9.5 f_m H_s^{0.34} \\ &= 9.5 * (0.09) * \left(8.8^{0.34}\right) \\ &= 1.7910\end{aligned}$$

The simulation was conducted by removing the drilling risers from the sea surface by one riser joint per run. The functional relationship between the emergency disconnection location and the lateral displacement of FSDR was then developed.

Procedures to Identify the Optimum Emergency Disconnection Location Along Drilling Riser Line in Technical Terms

The hydrodynamic responses of the previous sub-section were used to identify the optimum emergency disconnection location along drilling riser line. The reduction of von Mises stress from the previous run was evaluated. When the reduction of von Mises stress from the previous run is less than 30%, the emergency disconnection location was then identified. The von Mises stress at the optimum emergency disconnection location was further validated with the guidelines determined in API-RP-16Q.

Procedures to Develop the Functional Relationship Between Saffir Simpson Scale Storm Maximum Significant Wave Height and Lateral Displacement of FSDR

For this sub-section, hurricane sea state with various Saffir Simpson scale storm wind speed was used to identify its relationship with the lateral displacement of FSDR. The FSDR was simulated under free standing mode with the buoy located at the MSL. The purpose of this study is to identify at which hurricane intensity, the drilling rig will need to be disconnected. SWH and sea states were estimated using Saffir Simpson's scale hurricane intensity, Eqs. 5 and 6. The peak shape parameter, γ and

Table 4 Maximum significant wave height for various Saffir-Simpsons scale storm intensity

Storm intensity	Maximum sustained wind speed (m/s)	Maximum wave height (m)—calm sea prior to storm landing	Maximum wave height (m)—moderate sea prior to storm landing
Tropical depression	<17	4.08	6.67
Tropical storm	17–32	4.08–7.68	6.67–18.00
Category 1 hurricane	33–42	7.68–10.08	18.00–27.58
Category 2 hurricane	43–49	10.08–11.76	28.62–35.13
Category 3 hurricane	50–58	11.76–13.92	36.26–45.78
Category 4 hurricane	59–69	13.92–16.56	47.03–60.13
Category 5 hurricane	$\geq$70	16.80	61.50

scale parameter, α were identified using Eqs. 3 and 4. These parameters formulated the hurricane generated sea state with the assumption of the modal frequency being 0.0817 Hz.

Table 4 shows the maximum significant wave height developed for various Saffir Simpsons scale storm intensity. The maximum SWH were used as input to replicate the hurricane generated sea state. The maximum significant wave height developed was conducted using top down approach to predict what will be the possible maximum significant wave height based on the functional relationship identified by Ochi (2003) from various historical data. Table 5 shows the scale parameter, α and peak shape parameter, γ for each storm intensity. The SWH developed agrees with the observation stated by Fang and Duan (2014).

Table 5 The α and γ value for each storm intensity

Storm intensity	Maximum SWH (m)		Scale parameter (α)		Peak shape parameter (γ)	
	Calm	Rough	Calm	Rough	Calm	Rough
Tropical depression	4.08	6.67	0.005	0.013	0.653	0.772
Tropical storm	7.68	18.00	0.017	0.096	0.810	1.082
Category 1 hurricane	10.08	27.58	0.030	0.225	0.888	1.251
Category 2 hurricane	11.76	35.13	0.041	0.364	0.936	1.358
Category 3 hurricane	13.92	45.78	0.057	0.619	0.991	1.486
Category 4 hurricane	16.56	60.13	0.081	1.067	1.052	1.631
Category 5 hurricane	16.80	61.50	0.083	1.117	1.057	1.643

Procedure to Determine the Effect of Hurricane Generated Swell on Hydrodynamic Response of FSDR

The sensitivity analysis was conducted to isolate the effect of each swell and wind wave component to better understand its effect on FSDR mechanism. With that, it can clearly identify which the parameters that were influencing the hydrodynamic response of FSDR. Table 6 shows the parameter used for the sensitivity analysis. Sensitivity analysis 1 and 3 varied the swell height and period respectively. These two analyses were used to study the effect of hurricane generated swell (swell dominated sea state and mix sea state). On the contrary, sensitivity analysis 2 and 4 varied the wind wave height and period respectively. These two analyses were used to study the effect of wind wave components (wind wave sea state). These phenomena normally occur when hurricane was near to the study location.

Table 6 Swell height, period, wind wave height, period and current parameters used in the sensitivity analysis

Run	Swell height (m)	Swell period (s)	Wave height (m)	Wave period (s)	Current speed (m/s)
Sensitivity analysis 1					
1	1	15	7	5	1.158
2	2				
3	3				
4	4				
Sensitivity analysis 2					
5	4	15	3	5	1.158
6			4		
7			5		
8			6		
Sensitivity analysis 3					
9	4	11	7	5	1.158
10		12			
11		13			
12		14			
Sensitivity analysis 4					
13	4	15	7	1	1.158
14				2	
15				3	
16				4	

Results for Determination of Functional Relationship Between Emergency Disconnection Location and Lateral Displacement of FSDR

The functional relationship between emergency disconnection location and lateral displacement of FSDR have been successfully identified. Based on Figs. 3 and 4, it can be observed that with the increase in the number of riser joints being removed from the FSDR mechanism, the lateral displacement reduces.

Besides, the results obtained between modified JONSWAP sea state and hurricane sea state—data from Hurricane ELOISE (Withee and Johson 1975) agrees with each other. The responses were quite similar. This proved that modified JONSWAP wave spectrum can replicate the hurricane generated sea state. Equation 8 shows the functional relationship under the JONSWAP sea state. Equation 9 shows the functional relationship under Hurricane sea state.

$$LD = -6E^{-6}d^3 + 0.0024d^2 - 0.3329d + 19.219 \tag{8}$$

$$LD = -6E^{-6}d^3 + 0.0024d^2 - 0.3348d + 19.179 \tag{9}$$

The relationship between the lateral displacement of FSDR with respect to emergency disconnection location was 3rd order polynomial relationship. The graph

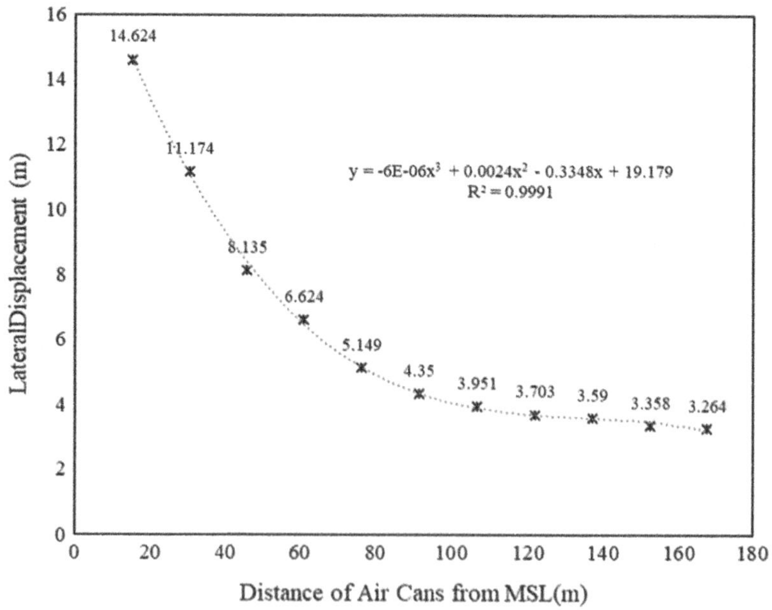

Fig. 3 Lateral displacement (m) versus distance of air cans from MSL—JONSWAP

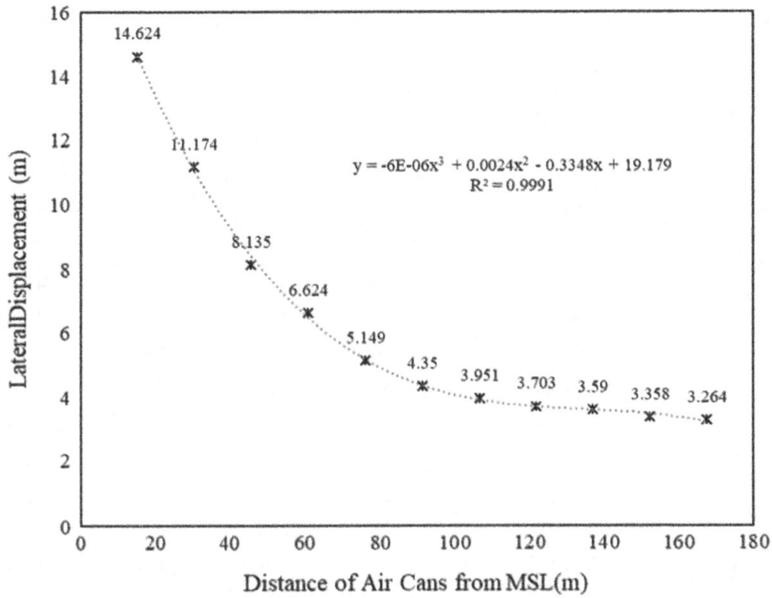

Fig. 4 Lateral displacement (m) versus distance of air cans from MSL—JONSWAP

decreased as the distance of air cans from MSL increased. The lateral displacement of FSDR reduced lesser when the emergency disconnection location moves away further from the MSL. This result agreed with the findings of Qin et al.'s (2011) sensitivity analysis of varying buoyancy tank top design water depth location. Besides, this is in line with the deep water current profile where the wave and current effect reduces as the water depth increases. With the removal of the riser by joints instead of predefined WD, the functional relationship of lateral displacement of FSDR with respect to emergency disconnection location was more precise and provide a more sensitive functional relationship.

Results for Determination of Optimum Emergency Disconnection Location Along Riser Line in Technical Terms

Table 7 shows the von Mises stresses and the reduction of stress for each simulation run with relation to numbers of riser joints removed. Based on Table 7, it can be noted at the beginning of the study, the von Mises stress reduction from the previous run reduces as more riser joints were removed. There was a clear segregation point where beyond riser joint number 5 being removed, the von Mises stress reduction fluctuated with no clear trend. The optimum emergency disconnection location was with 5th riser joints being removed, 76.20 m from MSL (87.5% of original WD). The

Table 7 von Mises stresses and the reduction of stress for each run with relation to numbers of riser joints removed

Run	Number of riser joints removed	Distance from MSL (m)	von Mises stress (MPa)		von Mises stress reduction from previous run (%)	
			JONSWAP	Hurricane	JONSWAP	Hurricane
1	1	15.24	2349.63	2314.42	N/A	N/A
2	2	30.48	1053.64	856.96	55.16	62.97
3	3	45.72	593.79	463.08	43.64	45.96
4	4	60.96	353.49	276.74	40.47	40.23
5	5	76.20	248.01	216.48	29.84	21.77
6	6	91.44	217.82	209.33	12.17	3.30
7	7	106.68	189.48	183.75	13.01	12.22
8	8	121.92	185.33	181.86	2.19	1.03
9	9	137.16	160.85	158.61	13.21	12.78
10	10	152.40	159.85	159.54	0.62	0.58
11	11	167.64	137.56	136.69	13.94	14.32

study has shown that the optimum emergency disconnection location along riser line is at 87.5% of original water depth. Further reduction of the drilling riser location did not reduce the hydrodynamic response of FSDR > than 30%, thus making it not effective. It will only increase the time and effort required to emergency disconnect the drilling riser as more riser joints need to be pulled back to the drilling rig.

In addition, the study also showed that the optimum emergency disconnection location is in accordance with API-RP 16Q standards. The drilling risers were safe for service when the von Mises stress does not exceed the 67% of the material's yield strength which in this case, 369.84 MPa. In an event of an emergency, with these proper guidelines on where the optimum emergency disconnection location for FSDR is, the operations team will safe time to decide where to emergency disconnect.

Results for Determination of Functional Relationship Between Saffir Simpson Scale Maximum Significant Wave Height and Lateral Displacement of FSDR

The hydrodynamic study of FSDR under Saffir Simpson scale storm was performed under 2 conditions. (i) calm sea conditions prior to hurricane landing and (ii) rough sea conditions prior to hurricane landing. Table 8 shows the hydrodynamic response of FSDR under various Saffir Simpson scale storm intensity.

Based on Table 8, it can be noted that the hydrodynamic responses of FSDR increased when the storm intensity increased. With the simulations being conducted

Table 8 Hydrodynamic response of FSDR under various Saffir Simpson scale storm intensity

Storm intensity	Lateral displacement (m)		Bending moment (kN m)		von Mises stress (MPa)	
	Calm	Rough	Calm	Rough	Calm	Rough
Tropical depression	22.39	30.44	287.56	511.19	5716.48	14,379.31
Tropical storm	37.58	58.22	638.07	1311.47	19,846.90	45,378.04
Category 1	44.71	75.86	782.07	1580.77	25,777.24	52,097.23
Category 2	50.07	83.59	937.98	1874.84	31,996.50	62,483.79
Category 3	54.35	87.00	1145.84	2074.95	38,847.15	67,714.22
Category 4	55.96	93.88	1260.38	2179.28	43,222.84	71,944.78
Category 5	59.85	102.27	1312.58	2321.51	44,766.08	77,772.87

at the MSL, none of the simulation runs was within the acceptable limit as per API-RP 16Q. This shows that the drilling riser should be disconnected in the event of a hurricane or apply dynamic positioning to safeguard the drilling riser as even a tropical depression can damage the FSDR mechanism. Figures 5, 6 and 7 show the graph of lateral displacement (m), bending moment (kN m) and von Mises stress (MPa) versus riser length from sea bed (m)-FSDR under category 5 hurricane, severe sea condition prior to the storm.

Figure 6 shows that the maximum lateral displacement is at the top section of the drilling riser.

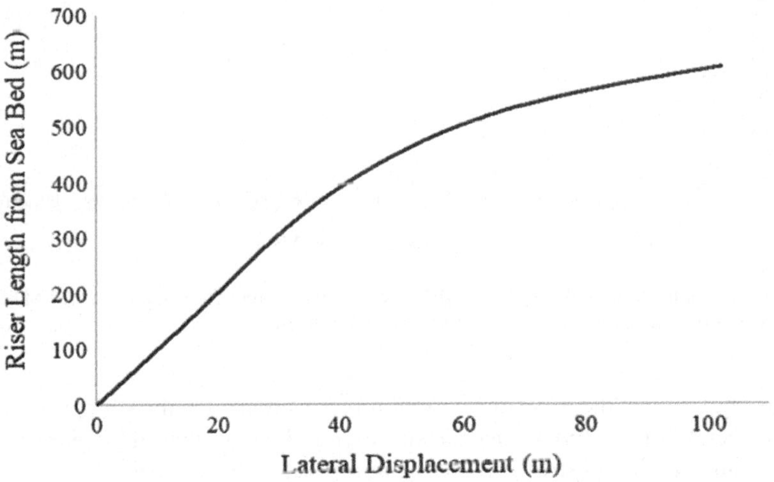

Fig. 5 The graphs of lateral displacement (m) versus riser length from sea bed (m)-FSDR under category 5 hurricane, severe sea conditions prior to the storm

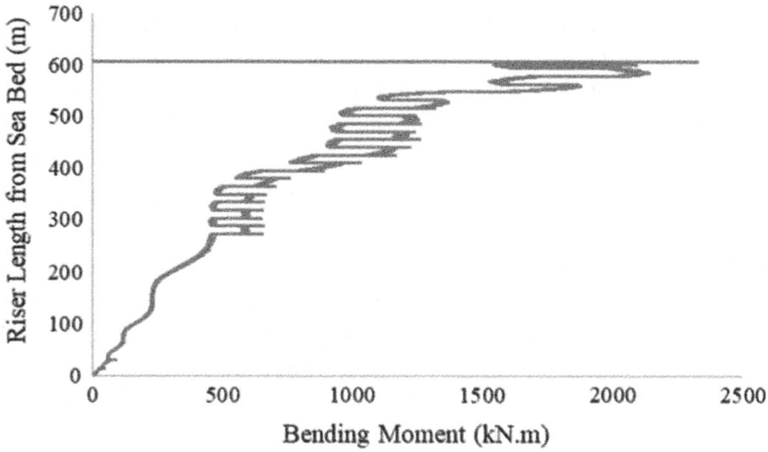

Fig. 6 The graphs of bending moment (kN m) versus riser length from sea bed (m)-FSDR under category 5 hurricane, severe sea conditions prior to the storm

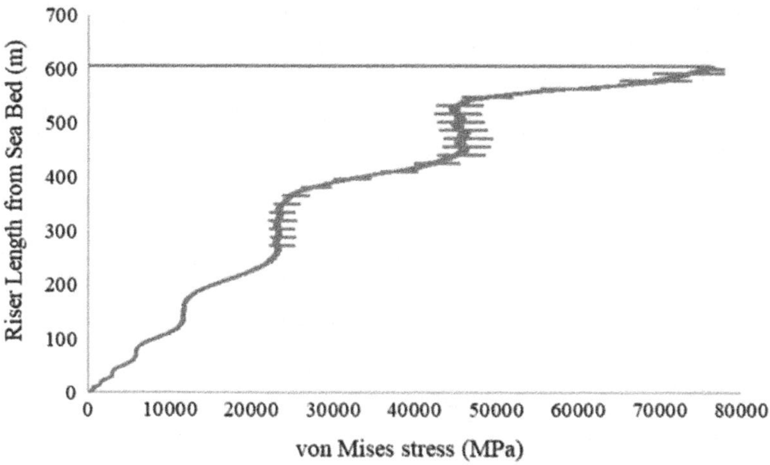

Fig. 7 The graphs of von Mises stress (MPa) versus riser length from sea bed (m)-FSDR under category 5 hurricane, severe sea conditions prior to the storm

This showed that the wave and current forces were having the strongest effect on the top section of the FSDR mechanism. Figure 7 shows that the highest bending moment and von Mises stress was acting at just below the air cans.

This was due to the installation of air cans increased the hydrodynamic drag and thus result in the increase in bending moment and von Mises stress. This is similar to the past findings by Kogure et al. (1998); Morooka and Shiguemoto (2012); Pereira et al. (2006). There were some fluctuations in terms for bending moment and von Mises stresses, the fluctuations were caused by various wall thickness used and

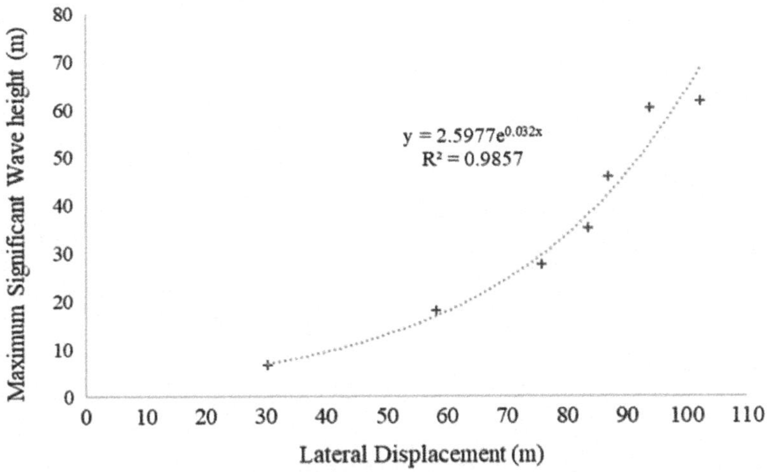

Fig. 8 Maximum significant wave height (m) versus lateral displacement (m)—calm sea conditions prior to storm landing

the installation of buoyancy foams. With the installation of buoyancy foams, the hydrodynamic drag diameter increased, causing a higher drag force.

With all these hydrodynamic responses, the functional relationship of the lateral displacement of FSDR and maximum significant wave height was plotted. The lateral displacement was used to represent the hydrodynamic responses. Figure 8 shows the functional relationship of maximum significant wave height (m) versus lateral displacement (m)-calm sea conditions prior to storm landing. Figure 9 shows the functional relationship of maximum significant wave height (m) versus lateral displacement (m)-severe sea conditions prior to storm landing. Based on Figs. 8 and 9, it can be noted that the lateral displacement increases exponentially with the increase of maximum significant wave height.

Equation 10 shows the functional relationship of maximum significant wave height and lateral displacement under calm sea conditions prior to storm landing. Equation 11 shows the functional relationship of maximum significant wave height and lateral displacement of FSDR under severe sea conditions prior to storm landing.

$$LD = 1.7537e^{0.0386H_{smax}} \tag{10}$$

$$LD = 2.5977e^{0.032H_{smax}} \tag{11}$$

where LD is lateral displacement in (m) and Hs_{max} is maximum significant wave height (m).

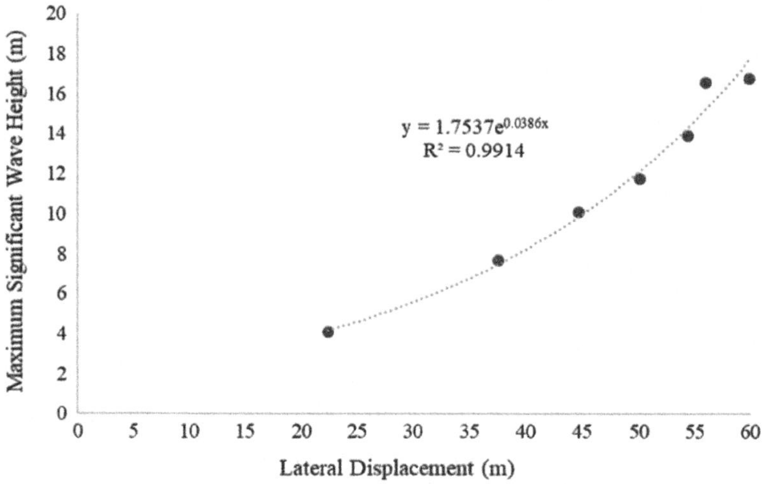

Fig. 9 Maximum significant wave height (m) versus lateral displacement (m)—severe sea conditions prior to storm landing

Results for Determination of Hydrodynamic Response of FSDR Under Hurricane Generated Swell

Four sensitivity analyses were performed by varying the swell height, swell period, wind wave height and wind wave period to study the effect of hurricane generated swell on FSDR mechanism. Table 9 shows the results for sensitivity analysis 1, 2, 3 and 4.

Sensitivity Analysis 1 (Varying Swell Height): Sensitivity analysis 1 was used to replicate the drilling location located a distance away from the hurricane location. Past studies have shown that even though the hurricane did not happen at the location, it still generated swells that can travel to distance location (Hong Kong Observatory 2005; Liang 2003; Yeh et al. 2006). Hence, this analysis was conducted to study the effect of hurricane generated swell on FSDR after disconnection. Varying swell height of 1–4 m and swell period of 15 s was selected for this analysis. Based on Table 9, it can be deduced that with the increase in the swell height, the lateral deflection and bending moment increases. Despite the increasing trend, the differences between each swell height were small. This was due to the FSDR mechanism was under the free-standing mode (90% of WD). The risers were located out of the wave zone. Hence, the increase of swell impact on the riser was lesser. In relation to the von Mises stress, when the FSDR mechanism was in free standing mode, the von Mises stress is 216.729 MPa, safe for operation under the API-RP-16Q standards.

Table 9 Results for sensitivity analysis 1, 2, 3 and 4

Sensitivity analysis 1			
Swell height (m)	Maximum lateral displacement (m)	Maximum bending moment (kN m)	Maximum von Mises stress (MPa)
1	7.84	15.67	216.728
2	7.86	16.11	216.728
3	7.89	18.04	216.729
4	7.94	20.32	216.729
Sensitivity analysis 2			
Wind wave height (m)	Maximum lateral displacement (m)	Maximum bending moment (kN m)	Maximum von Mises stress (MPa)
3	7.97	20.87	216.726
4	7.98	20.37	216.726
5	7.97	21.04	216.727
6	7.95	20.73	216.728
Sensitivity analysis 3			
Swell period (s)	Maximum lateral displacement (m)	Maximum bending moment (kN m)	Maximum von Mises stress (MPa)
11	7.89	20.16	216.727
12	7.91	20.91	216.728
13	7.93	20.86	216.726
14	7.94	20.96	216.728
Sensitivity analysis 4			
Wind wave period (s)	Maximum lateral displacement (m)	Maximum bending moment (kN m)	Maximum von Mises stress (MPa)
3	7.95	20.47	216.396
4	7.95	20.55	216.731
5	7.94	20.72	216.729
6	7.94	20.75	216.729

Sensitivity Analysis 2 (Varying Wind Wave Height): Sensitivity analysis 2 was used to investigate the effect of varying wind wave height. When a hurricane was close to FSDR location, it can generate wind wave up to 7 m. This was to study the reliability of FSDR systems after disconnection near to hurricane. 3–6 m wind wave with 5 s period and swell of 4 m with a period of 15 s were selected for this sensitivity analysis. Based on the results obtained above, it can be noted that when the wind wave height was increasing, there were only minor changes on the lateral displacement, bending moment and von Mises stress. This is due to FSDR mechanism being disconnected out of the wave zone (90% WD), hence, reducing the wind wave effect.

Sensitivity Analysis 3 (Varying Swell Period): Sensitivity analysis 3 was used to study the effect of varying swell period on the hydrodynamics of FSDR mechanism. The longer the swell period the higher the energy content. Past studies have shown that hurricane generated swell was approximately 4 m (Yeh et al. 2006). The swell height of 4 m and period of 11–14 s was therefore selected for this analysis. It can be observed that the lateral displacement and bending moment increased with the increase of the swell period. This finding is in line with the findings of sensitivity analysis 1.

Sensitivity Analysis 4 (Varying Wind Wave Period): Sensitivity analysis 4 was used to study the effect of the varying wave period. Wave period of 3–6 s and wave height of 7 m was selected for this analysis.

Conclusion

In conclusion, the research work was conducted successfully with the findings of the four objectives. Namely, (i) Determination of functional relationship between emergency disconnection location and lateral displacement of FSDR, $LD = -6E^{-6}d^3 + 0.0024d^2 - 0.3348d + 19.17$. The lateral displacement of FSDR increases in 3rd order polynomial relationship with the emergency disconnection location. (ii) Determination of optimum emergency disconnection location along riser line in technical terms. The location identified was at 87.5% of the original WD. (iii) Determination of functional relationship of Saffir Simpson scale maximum significant wave height and lateral displacement of FSDR, $LD = 1.7537e^{0.0386H_{smax}}$ for calm sea conditions prior to storm landing and $LD = 2.5977e^{0.032Hs_{max}}$ for severe sea conditions prior to storm landing. (iv) Determination of hydrodynamic response of FSDR under hurricane generated swell. The results shown that with the increase in swell components, the hydrodynamic response of FSDR increases. With these findings and relationship, this will add to the body of knowledge to develop guidelines on safe operation of offshore drilling equipment under hurricane conditions. Future recommendations are to replicate the current work with other offshore structures, in various geographical area and to conduct experimental testing under hurricane conditions to validate and strengthen the findings.

References

APIb (2010) Recommended practice for design, selection, operation and maintenance of marine drilling riser systems. In: Operating procedures. American Petroleum Institute, pp 21–25

Bernama (2019) MetMalaysia: Storm Pabuk expected to cause strong winds, choppy seas in northern Melaka Straits. TheStar

Burke BG (1973) An analysis of marine risers for deep water. Paper presented at the OTC Houston, Texas, USA

Bushnell JM, Falvey RJ (2017) Annual tropical cyclone report-2017. Retrieved from http://www.metoc.navy.mil/jtwc/products/atcr/2017atcr.pdf

Cardone VJ, Cox AT, Forristall GZ (2007) Hindcast of winds, waves and currents in Northern Gulf of Mexico in Hurricanes Katrina (2005) and Rita (2005). Paper presented at the OTC, Houston, Texas, USA

Collins III CO (2014) Typhoon generated surface gravity waves measured by NOMAD-type Buoys. Ph.D. in Applied Marine Physics Dissertation, University of Miami. http://scholarlyrepository.miami.edu/oa_dissertations/1286

Corona EN, Lynch RD, Riffe D, Cardone VJ, Cox A, Chen H (1996) Typhoon emergency response planning for the South China Sea. Paper presented at the OTC, Houston, Texas, USA

Cruz AM, Krausmann E (2008) Damage to offshore oil and gas facilities following hurricanes Katrina and Rita: an overview. J Loss Prev Process Ind 21(6):620–626. https://doi.org/10.1016/j.jlp.2008.04.008

Cruz AM, Krausmann E (2009) Hazardous-materials releases from offshore oil and gas facilities and emergency response following Hurricanes Katrina and Rita. J Loss Prev Process Ind 22(1):59–65. https://doi.org/10.1016/j.jlp.2008.08.007

Defu L, Guilin L, Fengqing W, Longzhi H (2017) Typhoon/hurricane disaster prediction and prevention for coastal, offshore, and nuclear power plant infrastructure. In: Collins JM, Walsh K (eds) Hurricanes and climate change, vol 3. Springer International Publishing, Cham, pp 135–165

Deluca M (2005) Offshore escapes Katrina's Wrath—Louisiana lookout after Katrina. Offshore Engineer

DiMarco SF, Kelly F, Zhang J, Guinasso Jr NL (1995) Directional wave spectra on the Louisiana—Texas shelf during Hurricane Andrew. J Coast Res 217–223

Duhaney H, Keys H, Daigle P, Laird J (2004) Case study of a 300-ft water depth jackup in a class IV Hurricane. Paper presented at the IADC/SPE drilling conference, Dallas, Texas

Fang H, Duan M (2014) The environment and environmental load of offshore oil engineering. In: Offshore operation facilities. Elsevier Inc, pp 1196

Foster ER (1982) JONSWAP spectral formulation applied to Hurricane generated seas

Guo Y, Song L, Zeng H, Qiu Z (2015) A field trial of a new system using both free-standing drilling riser and mid-depth BOP in the South China Sea. Paper presented at the 25th international ocean and polar engineering conference, Hawaii, USA

Hong Kong Observatory (2005) Dangers of the seas

Kaiser MJ (2007) World offshore energy loss statistics. Energy Policy 35(6):3496–3525. https://doi.org/10.1016/j.enpol.2006.12.020

Knutson TR, McBride JL, Chan J, Emanuel K, Holland G, Landsea C, Sugi M (2010) Tropical cyclones and climate change. Nat Geosci 3:157. https://doi.org/10.1038/ngeo779; https://www.nature.com/articles/ngeo779#supplementary-information

Kogure E, Ohashi M, Urabe S, Tanabe A (1998) Applications of a near surface disconnectable drilling riser in deepwater. Paper presented at the IADC/ SPE Asia Pacific drilling technology conference, Jakarta, Indonesia

Kumar A (2005) Oceanic disaster. Daya Publishing House, Delhi, India

Kwan CT, Marion TL, Gardner TN (1979) Storm disconnect of deepwater drilling risers. Paper presented at the OTC, Houston, Texas

Lee YK (1980) Hurricane ELOISE wave spectra. Coast Eng 4:151–156

Lempiere M (2017). Offshore oil: recovery from Harvey. Retrieved from Offshore Technology website

Liang NK (2003) The typhoon swell Doppler effect. Ocean Eng 30(9):1107–1115. https://doi.org/10.1016/S0029-8018(02)00099-9

Millheim KK, Aune G (2011) Self standing riser offers alternative to conventional deep water approach for developing marginal deep water oil and gas resources. Paper presented at the international petroleum technology conference, Bangkok, Thailand

Morooka CK, Shiguemoto DA (2012) Dynamic behavior analysis of a deepwater self standing hybrid riser system. Paper presented at the 31st international conference on ocean, offshore and arctic engineering, Rio de Jenario, Brazil
Ochi M (2003) Wave spectra of hurricane generated seas in deepwater. In: Hurricane generated seas, 1 edn. Elsevier Ltd, Oxford, UK, pp 25–61
Pei Y, Tao AF, Zhang YF, Li H, Yan J (2016) The generation mechanisms of low frequency swell in the E3 sea area of the East China Sea. Trans Oceanol Limnol 1(1):17–24
Pereira PSD, Morooka CK, Champi DF (2006) Dynamics of a vertical riser with a subsurface Buoy. Paper presented at the sixteenth international offshore and polar engineering conference, San Francisco, California
Qin W, Kang Z, Kang Y (2011) Free standing hybrid riser global parametric sensitivity analysis and optimum design. Paper presented at the international conference on offshore mechanics and arctic engineering, Rotterdam, The Netherlands. http://dx.doi.org/10.1115/OMAE2011-49606
Raghupathi L, Randell D, Jonathan P, Ewans KC (2016) Consistent design criteria for South China sea with a large-scale extreme value model. Paper presented at the OTC Asia, Kuala Lumpur, Malaysia
Reed RM (2005). In 1077691 (ed), vol 3008X1960. https://www.dvidshub.net/image/1077691/hurricane-dennis, US Coast Guard District 8
Saruhashi T, Sawada I, Kyo M, Miyazaki E, Yamazaki Y, Yokoyama T, Gong L (2014) Planning and feedback for deepwater drilling riser operation in high currents, typhoons and cold front. Paper presented at the OTC, Houston, Texas, USA
Sheng L, Xu L, Zhou J, Jiang S, Li X (2016) Experience on deepwater drilling riser about hang-off modes in typhoon condition of South China sea. Paper presented at the OTC Asia, Kuala Lumpur, Malaysia
Tao A, Yan J, Pei Y, Zheng J, Mori N (2017) Swells of the East China sea. J Ocean Univ China 16(4):674–682. https://doi.org/10.1007/s11802-017-3406-5
Thies PR, Crowley S, Johanning L, Micklethwai W, Ye H, Tang D, Li X (2015) Novel mooring design options for high-intensity typhoon conditions-an investigation for wave energy in China. Paper presented at the structural load and fatigue on floating structures conference, London, UK
Veldman AEP, Luppes R, Bunnik T, Huijsmans RHM, Duz B, Iwanowski B, van der Plas P (2011) Extreme wave impact on offshore platforms and coastal constructions. Paper presented at the 30th international conference of ocean, offshore and arctic engineering, Rotterdam, Netherlands
Wisch DJ (2006) Observations of hurricane impacts on deepwater facilities. Paper presented at the OYC, Houston, Texas, USA. https://doi.org/10.4043/18414-MS
Withee GW, Johson A (1975) Buoy observation during hurricane ELOISE
Xie B, Lei F, Wang J (2013) The statistical characteristics of sea waves with double-peaked spectrum in deep area of South China Sea. Paper presented at the 23rd international offshore and polar engineering, Anchorage, Alaska
Xu F, Perrie W, Toulany B, Smith PC (2007) Wind-generated waves in Hurricane Juan. Ocean Model 16(3):188–205. https://doi.org/10.1016/j.ocemod.2006.09.001
Yang JE (2014) Fukushima Dai-Ichi accident: lessons learnt and future actions from risk perspective. Nucl Eng Technol 46(1):27–38
Yeh SP, Tseung CM, Lee BC, Doong DJ (2006) Typhoon swell observed by data Buoy. Paper presented at the 16th international offshore and polar engineering conference, San Francisco, USA
Yong HY, Liew MS, Ovinis M, Danyaro KU, Lim ES (2018) Hydrodynamic study of free standing drilling risers under typhoon generated swell. Paper presented at the OTC Asia, Kuala Lumpur, Malaysia
Young IR (2003) A review of the sea state generated by hurricanes. Mar Struct 16(3):201–218. https://doi.org/10.1016/S0951-8339(02)00054-0
Zahibo IN, Didenkulova Ira (2008) Extreme waves generated by cyclone in Guadeloupe. In: Pelinovsky E, Kharif C (eds) Extreme ocean waves. Springer Ltd., Milton Keynes, UK, pp 159–177

Zhu S, Xizhao J, Rong P, Gang L (2009) The development of deepwater riser recovery technology in Liuhua oil field. Paper presented at the OTC, Houston, Texas, USA

Zoethout M, Brugts H (2006) Experience with a dynamically positioned FPSO in a Hurricane area. Paper presented at the OTC, Houston, Texas, USA

Assessing Contribution of Climate Change on Wetlands by Using Multi-temporal Satellite Data

Nebiye Musaoglu, Adalet Dervisoglu, Nur Yagmur, Baha Bilgilioglu, Aylin Tuzcu and Aysegul Tanik

Abstract Numerous natural lentic systems of the world are under the threat of drying due to either natural or anthropogenic activities. Natural circumstances are basically due to climate change effects remarkably sensed within the past decades. However, human-induced activities in some cases dominate over natural events due to demand of water particularly for irrigation and domestic needs. This study investigates the changes that have occurred on the water surfaces of 3 significant water bodies located in the Konya Closed Basin of Turkey. Geographically inner Anatolia region of Turkey receives half of the annual average precipitation of the overall country; on the contrary, the region is known to be an important agricultural area. The majority of fertile arable land is spared to wet agriculture necessitating vast amounts of water. As expected, all the 3 water bodies have incredibly lost their water within years. In the multi-temporal analysis of the study area, Landsat 5 TM, 7 ETM+ and 8 OLI satellite images belonging to every 10-years interval between 1987 and 2017, and corresponding meteorological data were used. Water losses in the 3 ecosystems, one is a salt production lake, Tuz (Salt) Lake, the other is a Ramsar Site (Meke Maar),

N. Musaoglu (✉) · A. Dervisoglu · N. Yagmur · A. Tuzcu
Civil Engineering Faculty, Geomatic Engineering Department, Istanbul Technical University, 34469 Maslak, Istanbul, Turkey
e-mail: musaoglune@itu.edu.tr

A. Dervisoglu
e-mail: adervisoglu@itu.edu.tr

N. Yagmur
e-mail: yagmurn@itu.edu.tr

A. Tuzcu
e-mail: tuzcuay@itu.edu.tr

B. Bilgilioglu
Faculty of Engineering and Natural Sciences, Geomatic Engineering Department, Gumushane University, 29000 Gumushane, Turkey
e-mail: bahabilgilioglu@gumushane.edu.tr

A. Tanik
Civil Engineering Faculty, Environmental Engineering Department, Istanbul Technical University, 34469 Maslak, Istanbul, Turkey
e-mail: tanika@itu.edu.tr

W. Leal Filho et al. (eds.), *Climate Change, Hazards and Adaptation Options*,
Climate Change Management, https://doi.org/10.1007/978-3-030-37425-9_4

and the last one is an internationally recognized wetland (Akgol), are detected. Resultantly, it is observed that only Tuz Lake has been artificially fed in the last decade as its economic value is higher than the other two systems. Meke Maar and Akgol Wetland have high ecological value; however, their original functions vanished as they have almost lost their water. It is important to state that humans' need of water is essential; but, unconscious utilization and wrong irrigation applications lead to water scarcity. On the other hand, climate change accelerates the reduction of water resources. Satellite data coupled with ground truth measurements are accepted as modern technological tools to detect short and long term temporal changes in the water surfaces of such vulnerable water systems especially for attracting the attention of decision-makers responsible of sustainable management of water resources.

Introduction

Wetlands are often described as transitional habitats situated between dry land and water body such as estuary, lake, river, or stream (Mitsch and Gosselink 2000). They have different functions like water storage, nutrient retention and transformation, carbon sequestration, sediment and particulate retention, stream flow maintenance and provision of substrate for plant colonization. Advances in remote sensing (RS) science and technology is highly utilized in inventing detailed qualifications of wetlands; and thus, RS has greatly supported wetlands by producing information on their status (Tiner et al. 2015). Closely linked with the well-being of such vulnerable water bodies' climate change has started to affect both water quality and quantity in wetlands causing land use/cover deterioration with time. Naturally occurring disasters like droughts, heat waves, floods and salinization of groundwater especially in coastal areas associated with several human-induced activities are also among the signs of climate change (Ozcan et al. 2018). Climate change has intense impact on the hydrological cycles as it causes temperature increase. Radiation, surface temperature, wind, water vapour, gases, aerosols, clouds and identification of properties, precipitation and sea ice are important meteorological parameters that can be measured by RS. Among all, precipitation is addressed as the most significant parameter for the water cycle (Ilcev 2019). Wetlands, as sensitive water bodies, are among those that are affected more than the other water bodies from both climate change and anthropogenic factors as they cover large areas with shallow water depths. The detection of temporal and spatial changes of wetlands via field surveys and tests are time and money consuming practices. However; RS data with their high resolution capacity provide valuable information on the changes of wetlands in shorter period of times. Effects of climate change can be observed by coupling multiple satellites and in situ measurements.

There are many case studies all over the world that investigate climate change with RS data. In a recently conducted case study in Lake Kyoga in Uganda, Landsat images were used to compare water surface area and precipitation, and it was seen that there is a correlation between the water surface area and rainfall (Nsubuga et al.

2017). In another study, lakes in Nam Co Basin located in Tibet were examined by Landsat TM/ETM+, SRTM 3 DEM and ICESat elevation data. The study area that was not much affected by human activities indicated that the surface area and quantity of glacial lakes increased in parallel to temperature and precipitation rise (Wang et al. 2016). The purpose of another study was to calculate the lake area increment and glaciers decrease in the region between Himalayas and Karakoram by using meteorological and satellite data to observe the impact of climate change (Li et al. 2015). Similarly, Li and Zha (2019) conducted long term detection of global warming in China by satellite data during years 2001 and 2015. On the other hand, Yang et al. (2013) searched the advantages of RS on the climate system, and it was seen that integrating active and passive RS provided better information for understanding the complex climate system. Moreover, components of the atmosphere and a fraction of cloud cover were evaluated with climate analysis and RS by using MODIS-MAIAC data and it was concluded that cloud cover was 80% higher in the North Western region in one-year period and 20% lower in the Southern Amazon Region in the dry season (Martins et al. 2018).

It is not always the climate change that causes the degradation of ecosystems; some regions of the world experience the ruin of the environment due to human-induced activities. There are several case studies conducted in areas where anthropogenic activities become the major cause of land and water degradation within years. Bai et al. (2011) investigated the areal change of inland lakes located in Central Asia via RS and tried to associate the findings with precipitation. It was concluded that the impact of irrigation schedule and applications dominated over the climate change effects in the plain lakes. Changes of evapotranspiration in Heihe Region of China that is known to be an agricultural region was studied by utilizing RS and evapotranspiration data by Zou et al. (2017), and it was seen that impact of agriculture was higher than the effect of climate change. In another study, areal change in the Honghu Lake in China was determined by RS between 1990 and 2013, and impacts of human activities and climate change were analysed. It was determined that the influence of climate change and human activities differentiated according to different periods (Chang et al. 2015). Moreover, Ji et al. (2015) searched both the human effect and the climatic conditions on the dynamics of wetlands for Kansas in United States (USA) by using RS. Supervised and knowledge-based classification algorithms were set in this research, and it was stated that knowledge-based classification approach was better than the traditional supervised classification on the detection of changes.

In Turkey, similar studies on the observation of both the climatic changes and human activities have been carried out in various regions of the country basically focusing on the Central Anatolia Region where water is comparatively scarcer and agricultural activities are more profound. Ekercin and Ormeci (2010) tried to determine climate change effects of Tuz Lake located in the Konya Closed Basin (KCB) of Turkey with the aid of RS, and it was seen that both the salt and water amount of the lake decreased between years 1987 and 2005. Dervisoglu et al. (2017) monitored long term temporal change of Akgol Wetland located in KCB by RS data based on annual, seasonal and even monthly data with the aim of determining the main cause of water decrease in the wetland. By utilizing the water surface areas and Digital Elevation Models (DEM) generated from 1:25,000 scaled topographical maps, water volumes were calculated. The adverse impact of dam constructions on the water ecosystems was clearly observed.

It is seen from the recent case studies conducted in different regions of the world that degradation of ecosystems is either due to climate change affects or to human-induced activities. In some cases, climate change dominates and in others, it is the anthropogenic activities that cause the negative impact on the well-being of water systems. Therefore, the aim of this study was to investigate and assess the contribution of natural occurrences and man-made practices on the 3 vulnerable water ecosystems located in the KCB of Turkey.

Meke Maar, Akgol Wetland and Tuz Lake located in the KCB were monitored by RS coupled with meteorological data within years 1984–2017 with the aim of inspecting the surface water changes with time, and to evaluate the findings regarding to climate change and human-induced activities. The main objective of this study was to generate long term temporal data and perform data analyses to be further utilized by the decision- makers responsible of the sustainable management of water resources. Figure 1a shows the geographical location of the KCB whereas Fig. 1b displays the 3 water bodies and the location of the meteorological stations from which long-term data are gathered. Tuz Lake is a well-known water body in the country from where salt production with 99% purity is realized. It is the 2nd significant salt lake of the world. Geologically, it bears a tectonic character and is rich in biodiversity. According to Ramsar criteria, it is of "Class A". Meke Maar was generated in a volcanic eruption and has acted as a crater lake for many years. However, it is entirely opposed to danger of dryness. As it houses a variety of endemic species and acts as a shelter for many migrating birds, it has been declared as a Ramsar site in 2005. Akgol Wetland is an internationally recognized wetland with rich biodiversity; hosting many migrating birds especially, flamingos. The land of the basin is highly fertile and the majority of the economic income is based on agricultural activities.

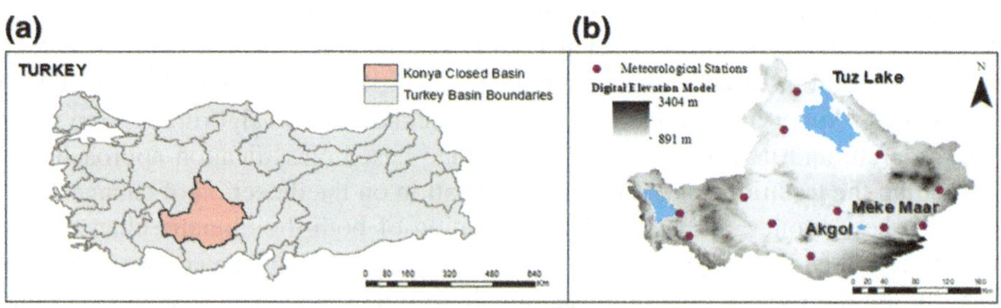

Fig. 1 **a** Geographical location of the Konya Closed Basin, **b** Location of the 3 water bodies together with the meteorological stations established in the basin

Data and Methodology Used

In the multi-temporal analysis of the study area, Landsat 5 TM, 7 ETM+ and 8 OLI satellite images belonging to every 10-years interval between 1987 and 2017, and the corresponding meteorological data were used. Landsat satellite images have 30 m spatial resolution with 16 days temporal resolution, and associated spectral bands had wavelength changes between 0.433 and 2.35 nm. Meteorological data covering precipitation and temperature values representing the 3 water bodies were considered in the evaluation of the results obtained from the satellite images. For this purpose, data gathered from 12 existing meteorological stations located in the study area were used (Fig. 1b). Long term data were provided from the Turkish State Meteorological Service.

Satellite images were analyzed on Google Earth Engine (GEE) platform. GEE is one of the most available platforms which stores the satellite imagery and provides visualization and analyses of the geographical data. It also delivers opportunity for making further analysis by utilizing different methods via algorithms, and compiles RS data which is taken in a time interval. It is one of the most wide-spread geographical data browsers and provides access to RS data which allows users to produce numerical results. It has recently been gaining popularity due to its large global imagery database, fast cloud computing time, and easy-to-use interface. Water body extraction from satellite imagery is an efficient way to monitor and investigate water resources. Normalized Difference Water Index (NDWI) was applied to images by using Google Earth Engine platform. NDWI is one of the most commonly used water indices initially developed by McFeeters in 1996 to detect surface waters in wetland environments, and to allow measurement of surface water extent (McFeeters 1996; McFeeters 2013). It uses the green and infrared regions of the electromagnetic spectrum that bears maximum and minimum water reflectance, respectively, and the achieved results vary between -1 and $+1$; where the water covered areas always take positive values. Its expression may be given as follows:

$$\text{NDWI} = (\text{Green} - \text{NIR})/(\text{Green} + \text{NIR}) \quad (1)$$

As a ratio combining the two different bands, NDWI index enhances water spectral signals by contrasting the reflectance between different wavelengths and removing a large portion of noise components in different wavelength regions (Gao 1996; Gao et al. 2016). Temporal changes of Tuz Lake, Akgol and Meke Maar specifically were analysed by using RS data in a period of about 30 years (1987–2017) by using GEE platform. Initially, boundaries of the study area were formed from satellite images at GEE platform. NDWI was then used to obtain water surface areas from the satellite images. The flow diagram of the methodology used in the study is given in Fig. 2. This study was performed during the past 2 years (2017–2018) within the context of two national projects.

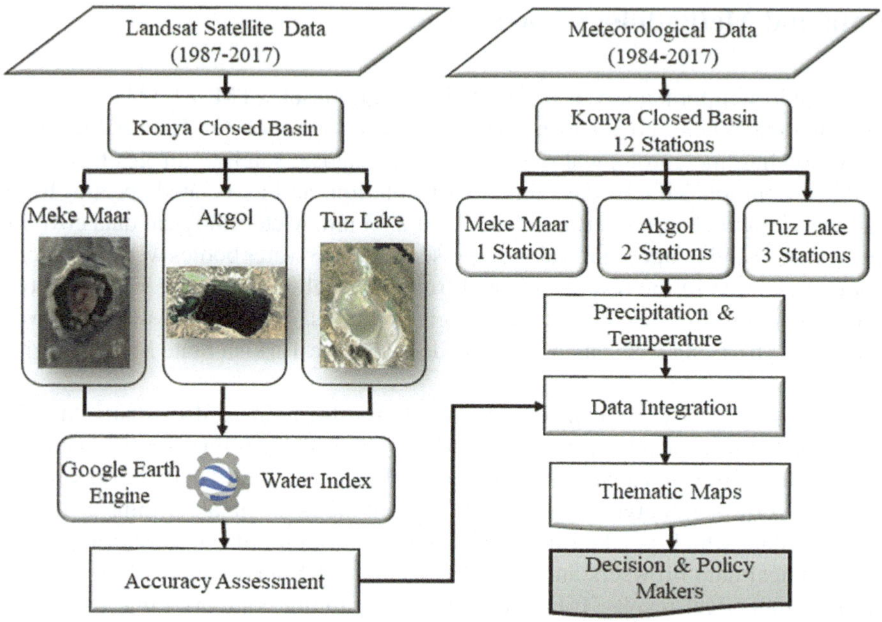

Fig. 2 Flow diagram of the methodology used

Results and Discussion

The study mainly focused on the temporal changes of water surfaces of 3 different water bodies located in KCB over a period of 30 years. Natural characteristics of the water systems were different and investigations were conducted via RS technology. Water covered areas in these 3 vulnerable systems have considerably decreased in time mainly due to construction of dams. As KCB is known as one of the most important agricultural areas of the country; dams have been built and put into operation for supplying the irrigation demand of the basin. KCB is one of the 25 watersheds of Turkey and is the largest one with a surface area of 5.3 million ha. In general, the basin bears an arid climatic character; and thus, suffers from water resources on the contrary to its fertile soil characteristics. Most of the water streams/rivers get dry during summer which makes the situation even more difficult as water does not reach to the wetlands. Dry and hot summers accelerate evaporation leading to more water losses. For example, excess evaporation causes the formation of salt layers of almost 30 cm thickness during summer in the Tuz Lake. A management plan for the Tuz Lake covering years 2014–2018 has been declared by the Ministry of Environment and Urbanization (Tuz Lake Management Plan 2014) which stated that water loss of the lake was compensated by diverting the excess water of another nearby water body and by discharging the treated domestic wastewaters of Konya Province to the Tuz Lake via a main drainage channel of about 122 km in length. There exists 3 pumping stations along the route; each having a water head of 6 m. As such, water volume increases and/or stays stagnant. However, the situation in the 2 other systems, Meke Maar and Akgol Wetland, is quite different. Water feeding and/or diversions from other water resources has not been experienced so far leading to remarkable water losses.

Areal values of the inspected water bodies were calculated from the satellite images at every 10-years interval between 1987 and 2017, whereas the meteorological data like temperature and precipitation were assessed starting from 1984. In the general evaluation of the overall basin data of the 12 meteorological stations were considered. However, only the nearby and representative ones were taken into consideration in the status argument of the individual water bodies. Meteorological data were gathered from the General Directorate of Meteorological Service (DMS). Figure 3a, b shows the long term annual average temperature and precipitation values between 1984 and 2017, respectively.

Long term annual average temperature within 1984–2017 of the overall basin is 11.5 °C; whereas this value is 11.7 °C for Meke Maar and Akgol Wetland and 11.6 °C for Tuz Lake. On the other hand, the similar long term annual average temperature for Turkey is calculated as 13.1 °C covering years 1970–2017. As KCB is located in the Central Anatolia Region of Turkey, the average temperature values fall below the country's average. From 1994 onwards, it is clearly observed that the average temperature values slightly lied above the long term average value calculated for KCB. This increasing trend has become even more noticeable after 2012 reflecting

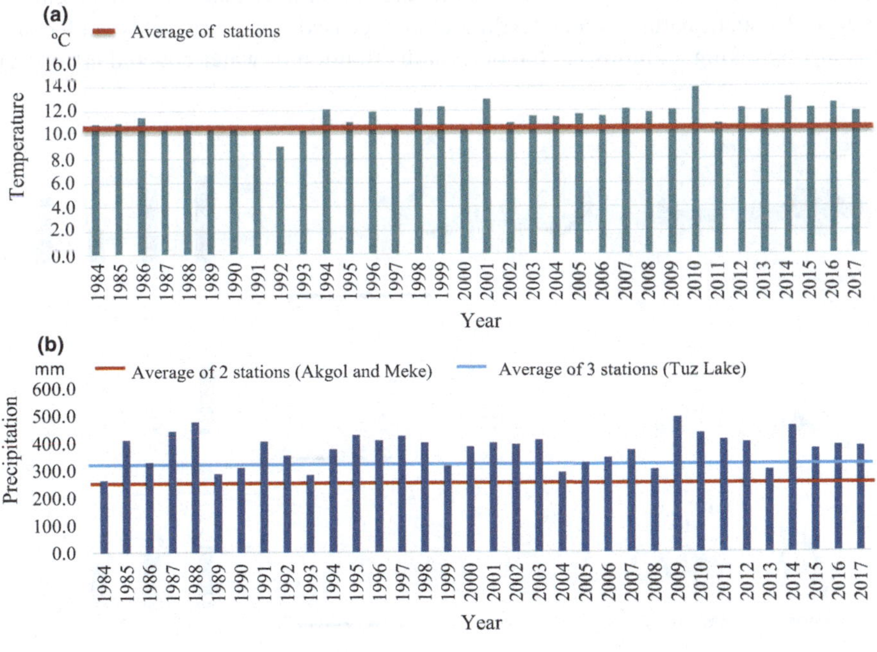

Fig. 3 Long term variation of **a** temperature and **b** precipitation values in KCB

the climate change. Regarding precipitation, the long term annual average for Tuz Lake is around 350 mm and for Meke and Akgol approximately 300 mm. In the graphical presentation of the precipitation profile (Fig. 3b), it is hard to refer to an increasing or decreasing trend as is the case in many other regions of the country. In some years like 1989, 1993, 2004 and 2013 comparatively less rainfall is realized. However, the highest wet years were during 2009–2012; similarly 2014–2017 can also be regarded as comparatively wet years. In general, this basin is receiving rainfall at almost half of the country's average value which is 625 mm. Thus, it can be named as one of the drier regions of the country. On the contrary, it is one of the important agricultural areas of the country, and usually wet agriculture is practised. Figure 4 designates the temporal change of the water covered surfaces of each of the inspected water bodies at every 10-years interval. These values are derived by means of image processing.

As seen from the temporal change of the 3 water bodies, it can easily be stated that the majority of the original water areas have already been lost prior to the inspection years. The dramatic decrease in the water covered areas of Meke Maar and Akgol have been ongoing during the past 30 years, and almost both of these significant water bodies run out of water. However, the situation at Tuz Lake does not reflect this remarkable decrease, and even in the last decade, water surface has doubled up due to water feeding activities to the lake. The common reason of this general decrease in the water areas of the 3 water systems is the diversion of the feeding streams and creeks, etc. to these lakes for the purpose of collecting water for irrigation by the construction of dams and reservoirs rather than the climate change effects. As such, natural water feeding almost ceased. One can easily understand this fact by taking a glimpse to the percent distribution of water covered areas even

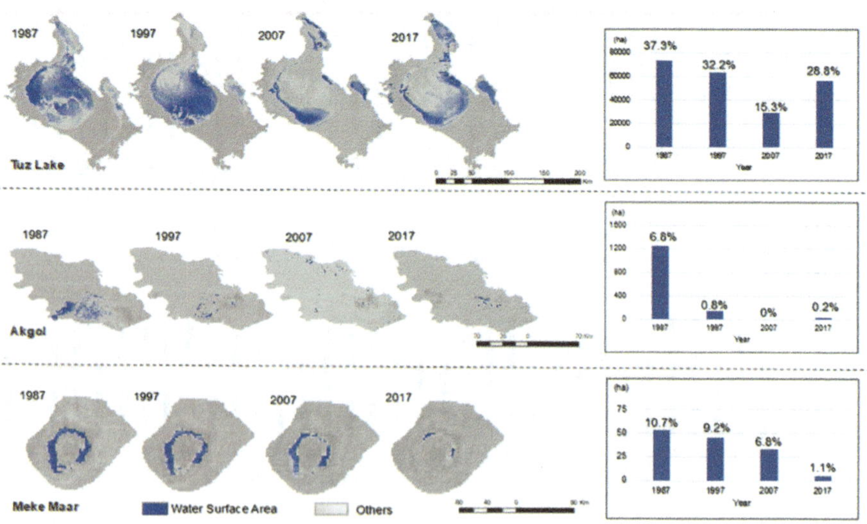

Fig. 4 Temporal change of the water covered surfaces in the inspected water bodies

before the inspection years in which the original water areas have already been lost significantly. KCB is an example of a closed system with almost no important streams on which water flows. Therefore, both the domestic and irrigation water needs are supplied from groundwater resources. This is the main reason of observing many wells in the basin.

Tuz Lake Management Plan (2014) clearly states that groundwater level around the lake decreased approximately 14 m within the past 33 years due to excess withdrawal of irrigation water. It is seen that many external factors like climate change, geological and hydrogeological conditions, and more importantly agricultural structure and crop pattern affected this remarkable water loss pointing out the need for urgent plans to diminish this unconceivable decrease. The increase in water amount within the past decade is due to an urgently taken measure to save the lake as still salt production is ongoing with an important economic value. Treated effluent of Konya Province is discharged to the lake as well as some water being fed from the nearby Beysehir Lake.

Meke Maar unfortunately almost dissapeared. Even in 1987, the majority of its water surface had been lost. This is basically due to water diversions that once fed the lake. On the other hand, human-induced activities dominate over the climate change effects on Akgol Wetland. Water cuts are the important reasons of water decrease. 3 important dams have been in operation since 1958, 1984 and 1988, respectively. A dike with a height of 1.65 m and length of 6,700 m has been built in 2013 for rehabilitation purposes where the slight positive impact of the dike was observed by Landsat images. However, this increase is just less than 1%. Treated effluent of a nearby residential district together with the treated discharge of an organized industrial district are sent to the wetland partly by a closed channel. However, the farmers withdraw this nutrient-rich water on the way to the wetland; and hardly, a limited amount of effluent finds its way to the wetland.

To sum up, it can be stated that KCB is an example of an area where human-induced activities dominate over climate change effects. Huge volumes of water being decreased in the past 30 years cannot be correlated and explained by the climatic conditions. Even there appears a temperature increase, precipitation also performs an above average trend basically in the last decade. This trend overcomes the climate change effects.

No significant drawbacks has been experienced while performing the desk work (processing of satellite data, obtaining and analysing of meteorological data). It is important to note that satellite data must be coupled with groundtruth measurements to obtain reliable results. Groundtruth measurements were conducted only in 2017 simultaneously with the corresponding satellite image. As such measurements could not be done for the previous inspection years as expected, accuracy of the images could only be performed by obtaining information from the local people and authorities. This problem can be addressed as the only limitation faced while conducting the study.

Conclusions and Recommendations

Water losses in the natural lakes and wetlands in the long term may partly be due to climate change effects and to wrong and unconsious use of water resources by the humans. KCB is an example of an enclosed basin with limited natural streams feeding the water systems. At the start of the study, the authors were almost sure that the climatic changes affected more on the lessening of water volumes in the stagnant water bodies in the basin; however, the detailed research on the temporal change analysis put forth the reality that human-induced withdrawals, and especially the construction of dams have significantly cut the water feeding to these natural water bodies. As such, human-induced activities dominated over natural climatic changes. This is one of the important lessons learnt from the study. Determining and analysing temporal changes on land use/cover within a short period can only be realised through RS with today's technology. Accurate areal values achieved in detail can be addressed the main success of the study. Hydraulic detention time, and inlet and outlet flows of the lakes may change extending from days to years based on precipitation, evaporation, seasons and on the water quantity. At this point, satellite images become useful tools to detect the long term (10–30 years interval) and even the short term (seasonal and monthly) variations. Temporal changes may also be due to artificial environmental factors like effluent discharges besides the natural cycle.

It is recommended to systematically monitor these vulnerable water bodies in future for the decision-makers who are responsible of their sustainable management. As uncertainities in dynamic ecosystems are the main natural drawbacks commonly experienced, continuos analysis and assessments can only be achieved by monitoring efforts that can be realised by using different satellite images of various resolutions.

Acknowledgements The authors would like to express their thanks to TUBITAK for their financial support to Project number 116Y142, and to Istanbul Technical University, Scientific Research Project Funding for their financial support to ITU BAP Project number MGA-2017-40803.

References

Bai J, Chen X, Li J, Yang L, Fang H (2011) Changes in the area of inland lakes in arid regions of central Asia during the past 30 years. Environ Monit Assess 178(1–4):247–256

Chang B, Li R, Zhu C, Liu K (2015) Quantitative impacts of climate change and human activities on water-surface area variations from the 1990s to 2013 in Honghu Lake. China. Water 7(6):2881–2899

Dervisoglu A, Musaoglu N, Tanik A, Seker DZ, Kaya S (2017) Satellite-based temporal assessment of a dried lake: case study of Akgol Wetland. Fresenius Environ Bull 26(1):352–359

Ekercin S, Ormeci C (2010) Evaluating climate change effects on water and salt resources in Salt Lake, Turkey using multitemporal SPOT imagery. Environ Monit Assess 163(1–4):361–368

Gao BC (1996) NDWI—a normalized difference water index for remote sensing of vegetation liquid water from space. Remote Sens Environ 58(3):257–266

Gao H, Wang L, Jing L, Xu J (2016) An effective modified water extraction method for Landsat-8 OLI imagery of mountainous plateau regions. IOP Conf Ser Earth Environ Sci 34, Article 012010
Ilcev SD (2019) Satellite remote sensing in meteorology. In global satellite meteorological observation (GSMO) applications, pp 129–182. Springer, Cham
Ji W, Xu X, Murambadoro D (2015) Understanding urban wetland dynamics: cross-scale detection and analysis of remote sensing. Int J Remote Sens 36(7):1763–1788
Li L, Zha Y (2019) Satellite-based regional warming hiatus in China and its implication. Sci Total Environ 648:1394–1402
Li Z, Fan K, Tian L, Shi B, Zhang S, Zhang J (2015) Response of glacier and lake dynamics in four inland basins to climate change at the transition zone between the Karakorum and Himalayas. PLoS ONE 10(12):e0144696
Martins VS, Novo EM, Lyapustin A, Aragão L, Freitas SR, Barbosa CC (2018) Seasonal and interannual assessment of cloud cover and atmospheric constituents across the Amazon (2000–2015): Insights for remote sensing and climate analysis. ISPRS J Photogramm Remote Sens 145 (Part B):309–327
McFeeters SK (1996) The use of the Normalized Difference Water Index (NDWI) in the delineation of open water features. Int J Remote Sens 17(7):1425–1432
McFeeters SK (2013) Using the Normalized Difference Water Index (NDWI) within a geographic information system to detect swimming pools for Mosquito abatement: a practical approach. Remote Sens 5(7):3544–3561
Mitsch W, Gosselink J (2000) Wetlands, 3rd edn. Wiley, New York
Nsubuga FWN, Botaib JO, Olwoch JM, de Rautenbach WCJ, Kalumba AM, Tselab P, Adeolab AM, Ssentongob AA, Mearns KF (2017) Detecting changes in surface water area of Lake Kyoga sub-basin using remotely sensed imagery in a changing climate. Theoret Appl Climatol 127(1–2):327–337
Ozcan O, Musaoglu N, Turkes M (2018) Assessing vulnerability of a forest ecosystem to climate change and variability in the western Mediterranean sub-region of Turkey. J For Res 29(3):709–725
Tiner RW, Lang MW, Klemas VV (2015) Remote sensing of wetlands applications and advances CRC Press
Tuz Lake Management Plan (2014) Management plan of the Tuz Lake special protection area 2014–2018. Turkish Ministry of Environment and Urbanization, General Directorate on the Protection of Natural Assets Ankara
Wang X, Zhou A, Sun Z (2016) Spatial and temporal dynamics of lakes in Nam Co Basin, 1991–2011. J Earth Sci 27(1):130–138
Yang J, Gong P, Fu R, Zhan M, Chen J, Liang S, Xu B, Shi J, Dickinson R (2013) The role of satellite remote sensing in climate change studies. Nat Clim Chang 3(10):875–883
Zou M, Niu J, Kang S, Li X, Lu H (2017) The contribution of human agricultural activities to increasing evapotranspiration is significantly greater than climate change effect over Heihe agricultural region. Sci Rep 7(1) 8805:1–13

City-scale Modeling of Urban Heat Islands for Kolkata

Ansar Khan, Soumendu Chatterjee, Walter Leal Filho, Rupali Khatun, Apurba Dinda and Aprajita Minhas

Abstract The Kolkata metropolitan area (KMA) is one of the sub-tropical urban environments experiencing swift and incessant urbanization process that has resulted in significant energetic differentiation in rural-urban domain as advection of heat wave and urban heat island (UHI) effect, it has to persuade on urban climate, biological environment and socio-economic atmosphere of urban society. The preparation of UHI susceptibility zonation is the preliminary measure for UHI risk assessment and hazard mitigation. The present study has been adopted the city-scale modeling of UHI by means of geographic information system (GIS) based statistical models for building the UHI susceptibility zonation using remote sensing (RS) data and other ancillary data. Initially, the UHI inventory map with 350 random pixels were extracted from mono-window algorithm (MWA) derived land surface temperature (LST) map using e-cognition approach. As such, UHI locations in LST map were then split into a ratio of 70/30 for building the UHI models and model validations. Finally, a spatial database of socio-economic, structural and radiative drivers with nine UHI conditioning factors has been prepared, including population distribution, land use and land cover (LULC), building material, building height, building roof

A. Khan (✉)
Department of Geography, Lalbaba College, Howrah, India
e-mail: khanansargeo@gmail.com

S. Chatterjee · A. Dinda
Department of Geography, Presidency University, Kolkata, India
e-mail: soumendu.geog@presiuniv.ac.in

A. Dinda
e-mail: apurbadinda0@gmail.com

W. L. Filho · A. Minhas
Research and Transfer Center "Sustainability and Climate Impact Management", Hamburg University of Applied Sciences, Hamburg, Germany
e-mail: walter.leal2@haw-hamburg.de

A. Minhas
e-mail: aprajita.minhas@haw-hamburg.de

R. Khatun
School of Oceanographic Studies, Jadavpur University, Kolkata, India
e-mail: rupaliocean@gmail.com

W. Leal Filho et al. (eds.), *Climate Change, Hazards and Adaptation Options*, Climate Change Management, https://doi.org/10.1007/978-3-030-37425-9_5

type, building roof reflectance, building age, building association and road pavement and these database were extracted from multi-spectral scanning (MSS), thematic mapper-5 (TM-5) of Landsat images, Google Earth (GE) historical images and OpenStreetMap (OSM) with intensive rapid visual field survey (RVFS). The geospatial relationships between UHI inventory's pixels locations and nine conditioning thematic factors were recognized by using four GIS-based statistical models i.e. analytical hierarchy process (AHP), two-class kernel logistic regression (KLR), support vectors machines (SVM) and spatial multi-criteria evaluation (SMCE) model. These models were constructed based on of training dataset and model-derived results have been validated and compared with the area under receiver operating characteristic (ROC) curve, kappa index and five different statistical evaluation measures to corroborate the noteworthy differences on the overall performance. The results of goodness of fit are of 86, 87, 85 and 89% and corresponding prediction capabilities are of 81, 85, 82 and 87% for AHP, KLR, SVM and SMCE models respectively. The statistical measures show that the SMCE model gives up overall better performance and precise results than the AHP, KLR and SVM models. The KLR and AHP models have produced to some extent better results than the SVM model in provisions of positive spatial prediction values. Hence, the study revealed that SMCE and KLR are the promising physical data mining approach to be considered to map the spatiality of UHI susceptibility zonation.

Introduction

The pattern of urban growth and resultant haphazard urbanization process is the most evident aspects of human development on the urban system (DeFries 2008). It is proverbial that the city centres experience higher thermal heterogeneity than contagious rural surrounds in terms of temperature differences, a phenomenon known as the urban heat island (UHI) effects (Oke 1988; Kalnay and Cai 2003; Zhou et al. 2004; Stone 2007; Jones et al. 2008; Fujibe 2009; Parker 2010).

As the process of incessant urbanization and subsequent UHI effects are hasten across the urban world, especially in cities of developing country like India, this has gained in interesting to urban climatologist and geographers (Kalnay and Cai 2003; Zhou et al. 2004; Grimm et al. 2008; DeFries and Pandey 2010). In the context of green house warming it has been argued that the increasing warming trends observed at urban weather observatories could be somewhat persuaded by local UHI, although this connotation has been disagreement by other studies (Jones et al. 2008; Parker 2010). On the other hand, the adverse energy and environmental effects of UHI phenomena, and searching a novel approach to assuage them, has become an important research topic in urban sustainability and environmental engineering (Wang and Akbari 2014). An urbanized area with lower albedo offset absorbs more than 80% of the incoming solar energy that increases their in situ surface air temperature as compared to rural surrounds (Touchaei and Wang 2015).

In neighbourhood scale, the UHI effects are closed to the urban surface; surface temperature has an indirect effect, but significant influence on air temperature and thermal comfort of urban environment (Bretz and Akbari 1997). The absorption of day-time incoming solar energy is the primary source of UHI phenomena during summer seasons. In generally, pavements and roofs comprise over 60% of urban surfaces (Akbari et al. 1997). The absorption capacity of dark materials, dark pavements and roofs are 80–90% and light materials, white roofs and light-coloured pavements are 30–65% of the incoming solar energy. Hence, the use of reflective building surface materials is a critical solution for UHI mitigation (Taha et al. 1988; Berdahl and Bretz 1994; Akbari et al. 1997; Bretz and Akbari 1997; Bretz et al. 1998; Konopacki and Akbari 2001; Synnefa et al. 2006, 2007). Therefore, it is need to understand the main drivers of UHI intensity is of critical importance for the study of urban climate, which has a strong implication for urban planning (Arnfield 2003; Rizwan et al. 2008). The UHI intensity is often measured by the difference in a weather station of urban area and one in a less urbanized outskirt, which is generally named as surface air UHI phenomena (Arnfield 2003; Stone 2007; Fujibe 2009). Since the spatial density of in situ weather stations are sparse and usually influenced by local environment. So, it is very difficult to rely on these data alone for obtaining information about UHI at city-scale.

The researches in large-scale UHI variation for the cities are mostly adopted in the fields of urban climate and meteorology. Therefore, the studies on urban climate are being carried out world-wide to know the nature of temperature change in urban world during the era of green house warming at different geographical scales (Chatterjee et al. 2013; Khan et al. 2014; Bisai et al. 2014; Khan and Chatterjee 2016). To mitigate the UHI effect on city-scale policies, the urban ventilation, pollution dispersion and greening are mostly examined using meso-scale models (MMs) simulations (Mirzaei 2015). These models are based on the governing equations of fluid dynamics whilst equally important models like radiation, cloud cover and soils are integrated into the computation process (Mirzaei 2015).

City-scale modeling of UHI is a challenging task to urban climatologists and geographers using satellite remote sensing (SRS) data. The SRS data are important devices to spatial correlate surface air temperatures and land use-land cover (LULC) of a city area. For this, SRS is now offers the opportunity to characterize spatio-temporal structures of land surface temperature (LST) with available resolution to discriminate between urban and rural thermal anomalies (Voogt and Oke 2003; Imhoff et al. 2010; Wan 2008). During recent past, a number of studies are available using SRS derived LST to map UHI effects, which mainly focused on a single or several big cities across the globe (Tran et al. 2006; Imhoff et al. 2010). However, a systematic and robust technique of UHI for city-scale using GIS-based statistical approach is still insufficient. It is urgent need to develop a novel approach for UHI mapping in tropics where the lower atmospheric temperature variability is very complex and persistence. The mounting surface air temperature of an urban area, UHI phenomenon amplifies the summer heat wave and pollution related human mortality, decreasing the human physiological comfort and promoting the average and peak energy demand of buildings (Mirzaei 2015). To countermeasures this superfluous phenomenon, a string of approaches and policies were recommended and accustomed to the cities (Akbari et al. 2003). Also to assess the impact of light-coloured surfaces (roofs and pavements) and urban green coverage (trees, grasses and shrubs) on meteorology and air quality of the city, it is indispensable to accurately characterize the various unreceptive urban surfaces (Akbari et al. 2003).

A number of models have been developed to appraise and credential effectiveness of such strategies to predict the UHI effects. It is, therefore, obvious that the summer time heat waves may occur anytime in the forthcoming decades unless any anticipatory measures are taken towards mitigation purposes. Thus, the analysis of UHI susceptibility in KMA as carried out in the study involves multi-criteria-based state-of-art evaluation through the integration of thematic components of UHI drives including, demography, LULC, building structure and radiative properties. The regional level classification approach has been adopted to map the population distribution, LULC, building materials, building age, building height, building association, building roof type, building roof reflectance and road pavements using SRS data (except population distribution) and geographic information system (GIS). The classification system comprises an integral component which is used to generate the socio-economic, structural and radiative themes as vector and raster layers in GIS environment. The spatial distribution of different susceptibility themes is generated on subsequent GIS processing platform and finally integrated to multi-criteria techniques including (a) analytical hierarchical process (AHP), (b) two-class kernel logistic regression (KLR), (c) support vector machines (SVM) and (d) spatial multi-criteria evaluation (SMCE) for UHI mapping. The AHP is a multi-criterion decision making approach which is permits a mutual agreement through subjective rational (Saaty 1980). The KLR is a kernel version of logit model which is used for modeling the binary responses of predictive variable using a posterior probability function of class membership (Cawley and Talbot 2008). The SVM is a premier margin classifier approach based on structural jeopardy minimization rule from the theory of statistical learning (Vapnik and Vapnik 1998). The SMCE is a spatial approach than can be combined the spatial factors and multi-criteria evaluation (MCE) to renovate the spatial and non-spatial input which generates output decision (Malczewski 1999). These GIS-based multi-criteria and machine learning techniques have resulted in new promising techniques to have first time tested on city-scale UHI modeling. The objectives of study is to develop the elements that can be used towards UHI monitoring and mitigation. Thus, UHI susceptibility zonations of cities enable to characterize potential UHI susceptibility, which is needs to be considered into account while developing new planning, sustenance facilities or retrofitting the existing urban surface. The UHI susceptibility assessments, aimed at endorsing reasonable surface temperature mitigation, are generally based on susceptibility aspects such as socio-economic characteristics of LULC and demographic distribution and the structural feature of building typology.

Health Impacts of UHI

The increases in global surface temperature have been paralleled with the frequency and intensity of temperature extremes (Agarwal et al. 2018). Due to the increased temperature, and as a result of the release of heat from buildings, vehicles and due to human activities, people in urban areas in India and elsewhere have been experiencing very hot weather in the summer months. Low wind speeds in urban areas also hinders the entry of fresh air (Sharma and Bharat 2009). The intensity of UHI is especially felt in such areas, which may cause impacts to human health (Leal Filho et al. 2017, Sharma et al. 2019, Shahmohamadi et al. 2011).

Various adverse effects are associated with exposure to UHI, such as general discomfort, heat stroke, heat cramps, heat exhaustion, heat syncope, and respiratory problems. Prolonged exposure to heat can lead to aggravation of minor conditions, to hospitalization and death. In future, there are high chances of increased heat-related mortality due to climate change (Heaviside et al. 2016; US EPA 2014). The consequences of long heat waves have more serious health implications on vulnerable people such as the elderly, children and people with pre-existing illnesses (Agarwal et al. 2018; Mavrogianni et al. 2011).

When body temperature exceeds over 41 °C, heat stroke and even death can occur. This disturbs the thermo-regulation process of the body causing fever, increased pulse, and hot and dry skin. Heat-related illnesses occur as a result of exposure to high temperature over a period of time which includes heat syncope, heat cramps, heat exhaustion, heat stroke and eventually death (Luber and McGeehin 2008). Table 1 shows some of the health impacts of UHI.

It is thus important to foster a better understanding of IHU and they affect cities, especially large ones such as Kolkatta.

Methods and Materials

Study Area

The present study of city-scale UHI modeling was conducted on Kolkata metropolitan area (KMA), delimited by the latitudes of 22.01°N and 23.08°N, and the longitudes of 88.03°E and 88.45°E, developed mainly along the banks of the river Hugli about 150 km to the north of the Bay of Bengal (BoB), right over the Gangetic delta plains (Nath et al 2015). Figure 1 depicts the study region, which are urban and suburban Kolkata. The KMA is one of the leading metropolis in the urban world and third largest urban agglomeration in India (census of India, 2011). It provides major financial hub of east and northeast India and also supports crucial industrial and transportational facilities to hinterland regions. The population of Kolkata is increasing trend as 1.5 million in 1901, 11 million in 1991 and a phenomenal 14.2 million in 2011 (census of India, 2011; Nath et al. 2015). The concentration of population is

Table 1 Some of the health impacts of UHI

Impacts of UHI	Signs and symptoms	Effect
Heat stroke	• Body temperature suddenly rises to more than 40 °C and is associated with central nervous system abnormalities, such as stupor, confusion or coma • Hot, dry skin, hypotension, nausea, tachycardia and tachypnoea are frequently present	Cardiovascular system, central nervous system abnormality, decreased kidneys function, and muscle cramps
Heat exhaustion	• Severe thirst, weakness, discomfort, anxiety, dizziness, fainting and headache • Pulse is rapid, with postural hypotension and rapid shallow breathing • Core body temperature less than 40 °C	Cardiovascular system, kidney damage, muscle cramps, CNS abnormality
Heat cramps	• Painful muscular spasms occur, in the legs, arms or abdomen, usually at the end of continuous exercise • This can lead to dehydration, loss of electrolytes through heavy sweating and muscle fatigue	Muscle spasms, cardiovascular system, metabolic system
Heat syncope	• Loss of consciousness or postural hypotension. More commonly seen in patients with cardiovascular diseases or taking diuretics before acclimatization takes place • It is attributed to dehydration, peripheral vasodilatation and decreased venous return resulting in reduced cardiac output	Cardiovascular system, central nervous system, decreased kidney function
Heat oedema	• Oedema of the lower limbs, typically ankles • This is attributed to heat induced peripheral vasodilatation and retention of water and salt	Kidney damage

(continued)

Table 1 (continued)

Impacts of UHI	Signs and symptoms	Effect
Heat rash	• Small, red, itchy papules on the face, neck, upper chest, under breast, groin and scrotum areas • More prevalent in young children	Affects skin
Depression, aggressive behaviour, anxiety, mental stress	• Reduced emotional well-bring, irritable, feeling agitated, restlessness, fatigue, increased emergency room visits	Psychological illness
Heat related mortality	• Death	Increase mortality risk due to exposure to extreme heat

WHO—Heatwave and Health: Guidance on warning system development (2015)

univocal, in a few parts of city comprises population density >100,000 persons/km^2 (Nath et al. 2015). The aesthetic facility in city are also increase the huge pressure of population and subsequent haphazard urban development, it has now continuous intruded into surrounding back swamp and marshy land to east, filling up large areas, particularly in Salt Lake City, Newtown-Rajarhat regions, in an unplanned and swift manner (Nath et al. 2015). Such types of haphazard growth pattern have a direct influence on urbanization process as city-scale as well as neighbourhood levels (Bhatta 2009). More than 80% of Kolkata has haphazard built-up areas with ill-conditioned high-rise residential buildings, congested business district, hospital, administrative sites and schools etc., a few of these are historically old structure in dilapidated physical condition with unplanned construction remaining to blocking effect in process of urban climate circulation (Nandy 2007; Nath et al. 2015). Kolkata's business district shows unmistakable signs of UHI effects mirroring the reality in other big cities in the country.

The UHI of KMA are affected by a series of conditioning factors such as socio-economic, structural and radiative drivers due to it emblematic urban attributes (Fig. 2). The increasing summer temperatures not only undoubtedly increase the cooling power consumption but also might increase the risk of summer heat wave over the study area; this makes the worse climate in city in terms of local and regional circulations of winds. Now, a big challenge is faced by urban climatologist and geographers to establish a protocol between process of urbanization and city climate by means of different climate forcing extreme events. It would be understand how to adapt and mitigate summer heat wave induced by climate variability and UHI phenomena. Therefore, a better understanding is needed to know the nature of UHI pattern, which is a critical important for the study of urban climate and subsequent sustainable urban planning.

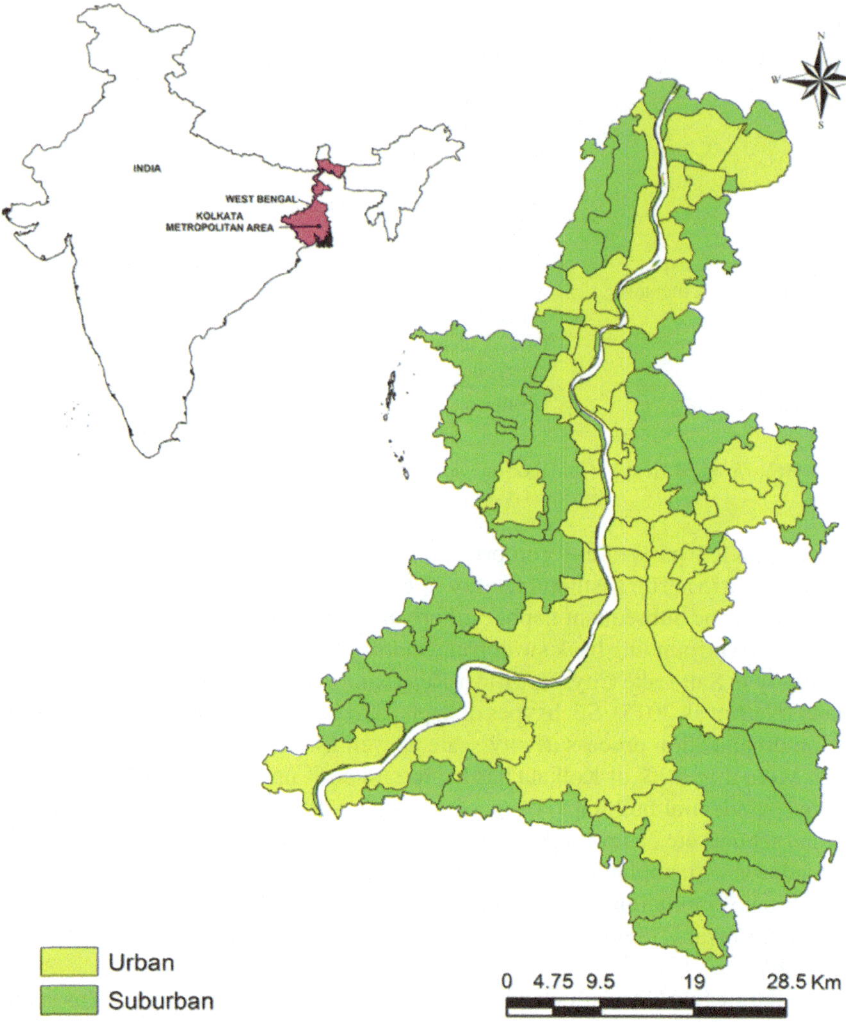

Fig. 1 Location map of Kolkata metropolitan area (KMA) and suburban-urban structure

Preparation of Spatial Database

The trend of haphazard development in unplanned urban growth process confront building codes is constantly increasing the UHI susceptibility in the developing countries like India, necessitating the systematic evaluation of UHI susceptibility by using those conditioning factors to create risks in terms of socio-economic, structural and radiative drivers (Table 2). To better understand the impacts of the built-up environment and economic infrastructure of city layout, a geo-spatial database of socio-economic (population distribution, and LULC), structural (building materials,

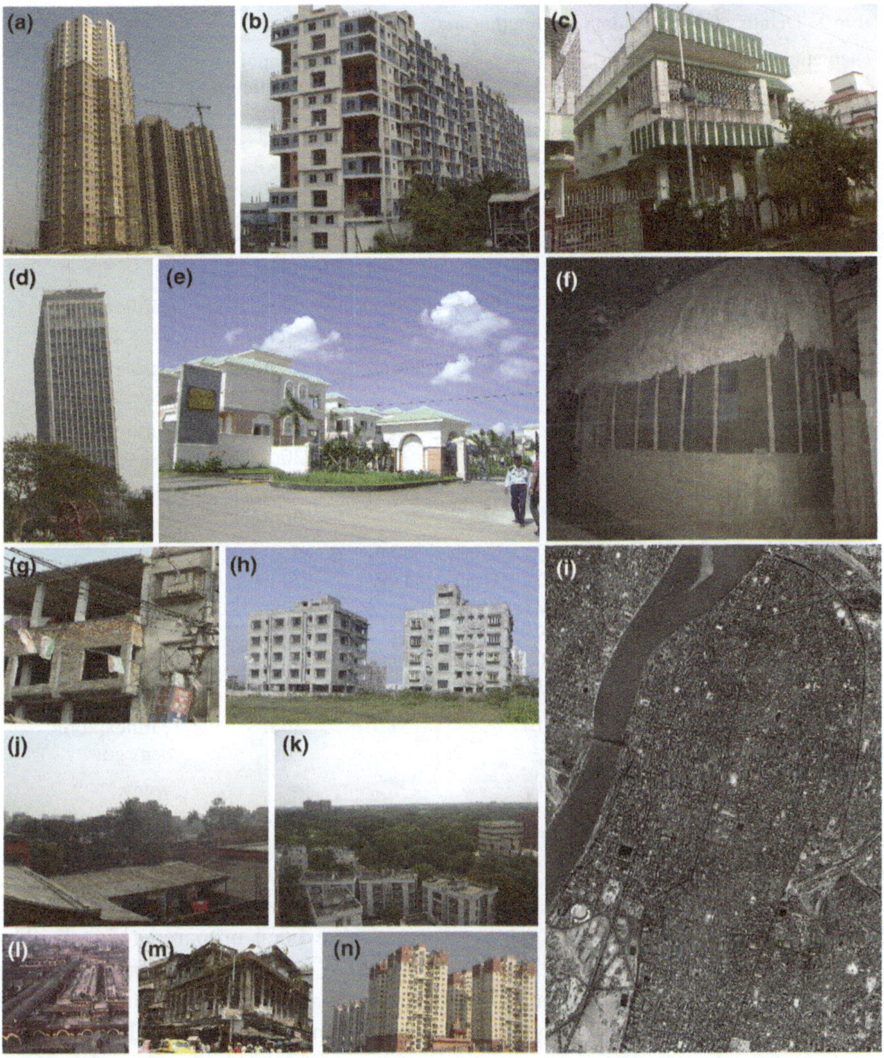

Fig. 2 Typical urban-suburban attributes in KMA: **a** A high rise building (South City); **b** A medium rise building (Bidhannagar Road); **c** A low rise building (AK Block, Saltlake) **d** Steel frame buildings (Tata Centre) **e** Newly built-up concrete buildings (Rajarhat); **f** Mud and non-brick wall (Kuleshwar Village, Diamond Harbour, South 24 Parganas); **g** Burnt brick wall (Paikpara, Kolkata) **h** Concrete building (New Town); **i** Urban association in middle Kolkata; **j** Metal roofs (Bowbazar, Kolkata); **k** Concrete roofs (Royal Calcutta Golf Club, Kolkata); **l** Dark colour roofs (New Market); **m** Old building (Central Kolkata) and **n** Recent building (Newtown, Kolkata)

Table 2 Details of thematic data used in this study

Thematic layer	Satellite/data type	Time	Sensor	Spatial resolution	Source URL
Population distribution	Census enumeration	2011	–	–	censusindia.gov.in
Land use/land cover	IRS-P6	2010	LISS-IV	5.8 m × 5.8m	earth.esa.int
	Landsat 5		TM	30 m × 30 m	earthexplorer.usgs.gov
Building material	Landsat 5	2010	TM	30 m × 30 m	earthexplorer.usgs.gov
Building height	Landsat 5	2010	TM	30 m × 30m	earthexplorer.usgs.gov
Building roof type	Landsat 5	2010	TM	30m × 30m	earthexplorer.usgs.gov
Building roof reflectance	Landsat 5	2010	TM	30m × 30m	earthexplorer.usgs.gov
Building age	Landsat 2	1975,1980, 1985, 1990, 1995, 2000, 2005 and 2010	MSS	79 m × 79m	earthexplorer.usgs.gov
	Landsat 4		TM	30 m × 30 m	
	Landsat 5				
Building association	Landsat 5	2010	TM	30m × 30m	earthexplorer.usgs.gov
Road pavement	OSM	2010	–	–	www.openstreetmap.org

building age, building height and building association) and radiative (building roof, building roof reflectance and road pavements) have been prepared (Fig. 3). The susceptibility of selected nine conditioning factors is computed and ordered by the potential to create UHI effect in the study area. Finally, the susceptibility of UHI effect map has been prepared to represent the all nine factors in single entities by means of amalgamation of socio-economic, structural and radiative drivers.

The accuracy of efficiency and quality of each satellite images derived conditioning factors have been examined by using root mean square error (RMSE). The sample ground control points (GPCs) have been taken for checking the efficiency of each conditioning layers based on given formula (Gibson et al. 2000).

$$RMSE = \sqrt{\mu^2 + \nu^2} \quad (1)$$

where μ is the residual at x axis and ν is the residual at y axis. Then the total RMSE is derived by following formula.

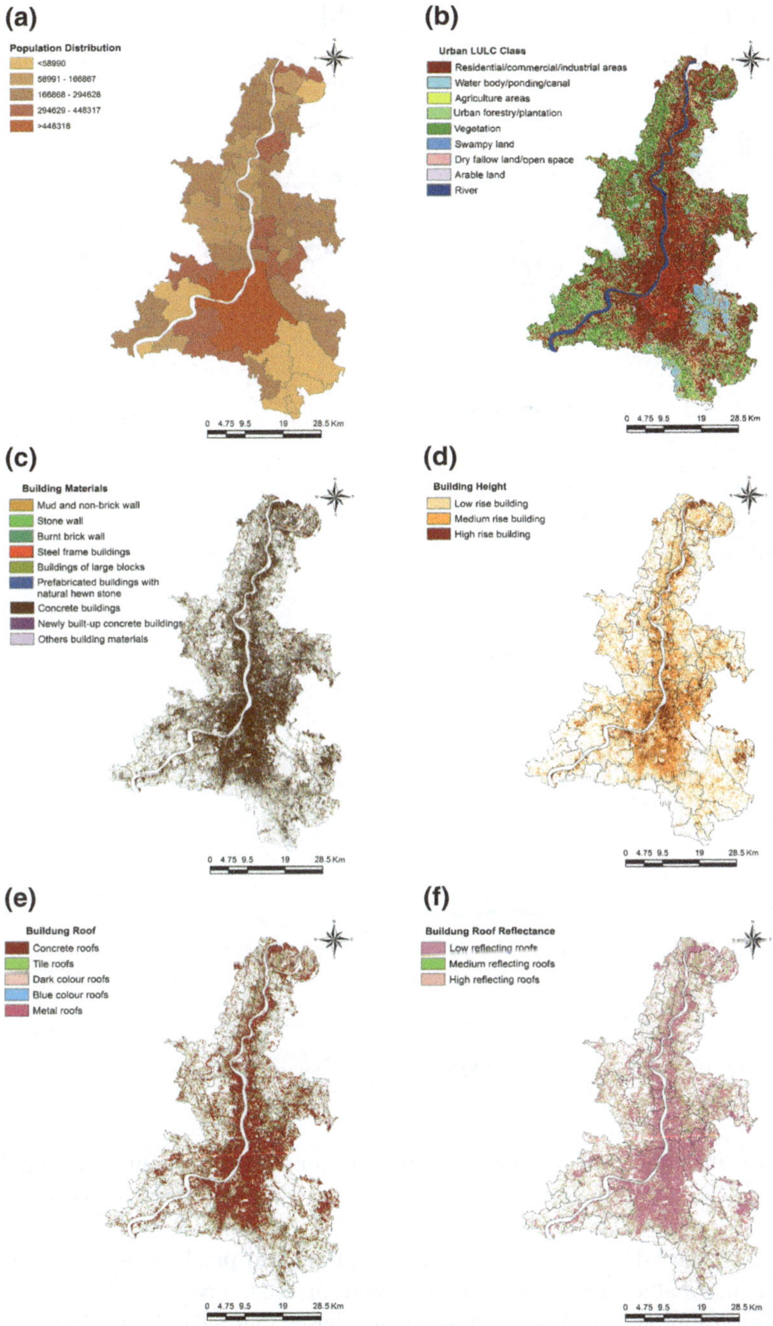

Fig. 3 Database of socio-economic, structural and radiative aspects: **a** population distribution **b** land use/land cover **c** building materials, **d** building age **e** building height **f** building association **g** building roof **h** building roof reflectance **i** road pavement

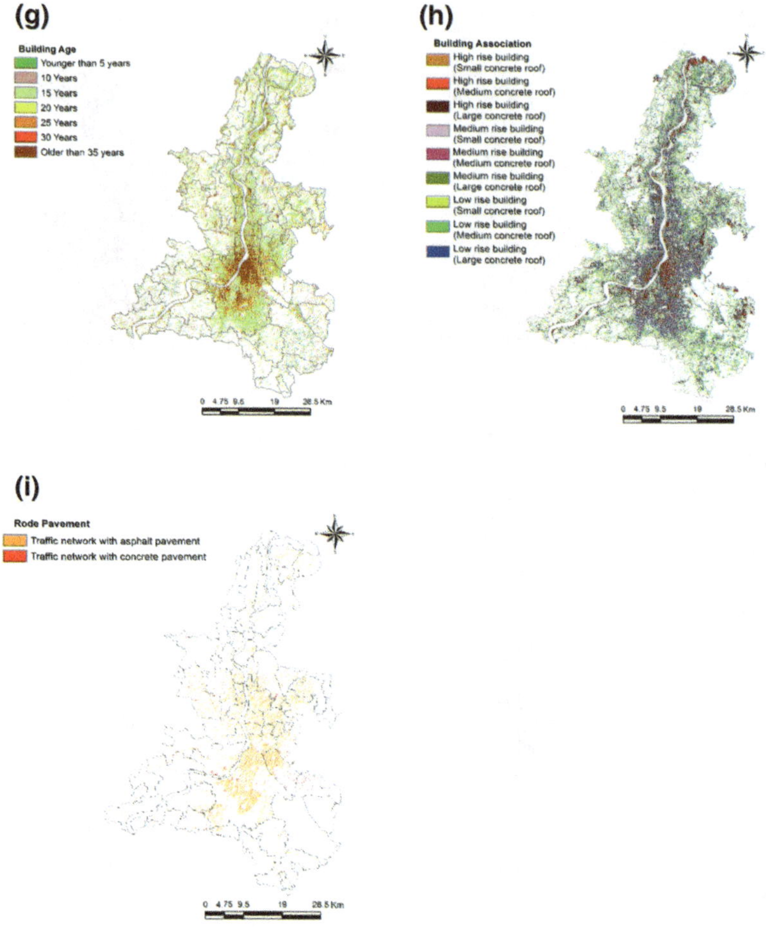

Fig. 3 (continued)

$$Total\ RMSE = \sqrt{\frac{1}{n}\sum_{i=n}^{n} \mu^2 + \nu^2} \tag{2}$$

where n is the total number of GCPs (100 GCPs for each layer). Thus the total RMSE was computed for each conditioning factors based on nearest neighbour resampling method.

Initially, the UHI inventory map with 350 random pixels were extracted from mono-window algorithm (MWA) derived land surface temperature (LST) map using e-cognition approach. Hence, a single channel based thermal radiation transfer equation has been employed to derive the LST by using MWA for Landsat 5 TM band 6. In this algorithm three parameters were used: surface emissivity, atmospheric transmittance and effective mean atmospheric temperature (Qin et al. 2001). The surface emissivity and atmospheric transmittance were estimated and assuming the land surface as Lambertia. The effective mean atmospheric temperature was taken from local metrological station (Alipore weather observation) because lack of in situ atmospheric data when the satellites pass. The upwelling and down welling atmospheric emittances were computed as an integrative function of atmospheric emittances with an effective mean atmospheric temperature (T_a). With help of these assumptions and approximations, the LST was solved for TM 5 band 6 data as (Qin et al. 2001):

$$T_s = [a_6(1 - C_6 - D_6) + (b_6(1 - C_6 - D_6) + C_6 + D_6)T_6 - D_6 T_a]/C_6 \quad (3)$$

where T_s is the LST in Kelvin degrees; T_a is the effective mean atmospheric temperature; T_6 is the brightness temperature of TM 5 Band 6; a_6 and b_6 are the coefficients used to approximate the derivative of the Planck radiance function for the thermal band; and C_6 and D_6 are the internal parameters for the algorithm based on the atmospheric parameters and ground emissivity.

In the study, socio-economic, structural and radiative aspects of UHI susceptibility exposures have been derived from satellite images and in case of building height, coupling of Landsat imageries and 3-D aspect of Google Earth (GE) have been employed to generate the building height map (Nath et al. 2015). All these thematic layers were used as classified data, while those derived from extensive rapid visual field survey (RVFS) using 1500 ground locations being considered as reference data used for accuracy assessment through error matrix of all spatial themes (Congalton 1991; Story and Congalton 1986; Jensen 1986; Nath et al. 2015). For RVFS a hand held global positioning system (GPS) is used for generating the coordinate system at each of the 1500 location points. The kappa statistic is also used to measure the differences in classified and reference data and allows to be determined the statistically significant results if one error matrix significantly differ from another matrix (Bishop et al. 1975; Jensen 1986; Congalton and Mead 1983; Nath et al. 2015). The kappa statistic is given by:

$$k = \frac{N\sum_{i=1}^{r} X_{ii} - \sum_{i=1}^{r} (X_{i+}X_{+i})}{N^2 - \sum_{i=1}^{r} (X_{i+}X_{+i})} \quad (4)$$

where N is the total number of sites in the error matrix, r is the number of rows in the error matrix, X_{ii} is the number in row i and column i, X_{i+} is the total for row i and X_{+i} is the total for column i.

A normalized level of accuracy is obtained by sum up the values on the major diagonal and dividing by the sum of the total values in the normalized error matrix (Nath et al. 2015). As a result, both the producer's and user's accuracy has been included in the normalized cell value, which is based on a balanced effect of the two accuracy measures (Congalton and Green 1999; Nath et al. 2015).

Population Distribution

The spatial distribution of human population has been estimated from census data, which provides the total number population in an administrative unit and its relation to household types. The trends in population growth of the Kolkata had been increased from 1.5 million in 1901 to 14.2 million in 2011. The demographic characteristics and urban population of KMA has been massively increased in the last four decades (Nath et al. 2015). This increasing rate of urban population in KMA is mainly concentrated due large-scale migration of persons from rural and small towns to bigger cities in search for better employment opportunities and lifestyle. Kolkata metropolitan development authority (KMDA) reports states the total area of KMA about 1886.67 km^2, making the population density 7840 persons/km^2. This makes the Kolkata is the highest density of population among other leading ones in the country and it is well exemplified by the questionnaire for the Indian census bureau (census of India, 2011). Due to huge demand of livelihood and aesthetic attraction, KMA reached third most populous metropolitan area in India after Delhi and Mumbai. It consists of many cities and towns around the city core of Kolkata. This region is developing very fast with new technologies, building block, commercial infrastructure, parks, sports clubs, malls and riverside region etc. Such high populous metropolis releases enormous anthropogenic heat which may cause the UHI effect at screen height temperature of the city. For mapping UHI susceptibility, the distribution of population in entire KMA has been classified into five demographic zone like <58,990; 58,991–166,867; 166,868–294,628; 2,946,289–448,317 and >448,318 using Jenks natural break classifier in Arc GIS 10.1 (Fig. 3a). It has observed that the population distribution is very high at Burabazar, Taltala, Kalidaha, Beniatola, Khidirpur, Metiaburuz, Shyambazar and Howrah regions of Kolkata municipal corporation (KMC) and Howrah municipal corporation (HMC) regions respectively and rest of areas population distribution are monotonic pattern (Nath et al. 2015). As population centres grow, these areas being changed the open lands which then undergo a corresponding increase in local average temperature.

Land Use/Land Cover

The spatial distribution of UHI pattern has been changed as of a mixed pattern, where open bare land, semi-bare land and other land going to be developed were warmer than other unreceptive urban surface to all-embracing UHI centres (Chen et al. 2006; Pielke et al. 2011). The LULC of urban centres provides the information concerning predominant urban surface types and other socio-economic attributes. The present LULC classification of KMA has been extracted by carrying out an object-based approach on national atlas and thematic mapping organization (NATMO) scheme of nomenclature (Nath et al. 2015). The regional scale LULC classes were designated by the association of building blocks, water bodies, agricultural fields, vegetation coverage, plantation areas, wet lands and open bare lands etc. (Nath et al. 2015).

In the study, Landsat-5 TM and GE images have been used to map LULC using maximum likelihood algorithm (MLA). The LULC map of KMA depicts nine major LULC units viz. residential/commercial/industrials areas; water body/ponding/canal; agricultural areas; vegetation; plantation; swampy land; dry fallow land/open spaces; arable land and river (Fig. 3b). The accuracy assessment was done by using the test samples collected from RFVS and GE image for LULC classification. The overall accuracy (OA) and kappa coefficient of LULC classifications were 85% and 0.903 respectively. Moreover, the LULC classification results have been further checked visually and refined manually, in order to have the most reliable product for the subsequent step.

Building Materials

The nature and types of material used in urban construction is one of the most imperative attributes in evaluating UHI susceptibility zonation (Oke 1982; Bretz et al. 1998). The thermal conductivity of urban unreceptive surface may significantly intensify the UHI at diurnal scale. However, the detection of urban building materials is a challenging task though the visual interpretation of image elements such as tone, texture, shape, size, shadow, pattern association and location were explicitly available on satellite image. The layout of building footprint map could be prepared using coarse resolution SRS imageries (Nath et al. 2015). Hence, Landsat-5 TM and MSS imagery have been used for its coarse resolution and better spatial enhancement of urban attributes using a bilinear 4-nearest neighbour algorithm. For extraction of building materials, principal component analysis (PCA), textural analysis and normalized difference built-up index (NDBI) have been performed to detect the building materials (Geneletti and Gortea 2003; Zhang et al. 2002; Zha et al. 2003; Nath et al. 2015). The materials were classified on the basis of building materials and technology promotion council (BMTPC 1997) nomenclature system. Thus the building materials of entire KMA have been regarded as nine classes like mud and non-brick wall, stone wall, burnt brick building, steel frame buildings, building of large blocks, prefabricated buildings with natural hewn stone, concrete buildings, newly built-up concrete building and other building materials (Fig. 3c). The confidence level of accuracy statistics has been established and OA and KC produces the agreement of 83% and 0.851 respectively. In addition, the classification result was auxiliary verified in order to enclose the more consistent artefact for further processing.

Building Age

The SRS imagery is immensely used for the extraction of built-up areas in urban environment for detecting and monitoring LULC changes that occur repeatedly in urban and periurban interface as a result of haphazard urban growth (Zha et al. 2003). The nature of LULC in urban areas tend to change considerably over a short period of time than elsewhere because of rapid cost-effective development and urban sprawl.

In the study, the built-up areas were extracted from the Landsat MSS (1975) and TM (1985, 1990, 1995, 2000, 2005 and 2010) classified images of seven different periods of time in order to scrutinize the dynamic changes of urban growth using NDBI approach (Small 2002; Zhang et al. 2002; Zha et al. 2003; Nath et al. 2015). For detection of built-up age, a rule of change detection has been adopted which portray the differences between the classified image of same scene at different periods of temporal framework. Thus building age (built-up age) has been break into seven classes in order to detect the urban growth process. These classes are included as younger than 5, 10, 15, 20, 25, 30 years and older than 35 years (Fig. 3d). The older building those are constructed before four decades ago, which has basically composed of concrete materials. And these types of concrete materials absorbed maximum amount of solar radiation in day-time and release least in night-time which may cause the higher temperature than those are pre-fabricated coated structures. The accuracy statistics between the RFVS-derived reference and the classified maps establish the confidence level of this building age classification and the OA and KC was 86% and 0.834 respectively. In addition, the classification result was further verified visually and refined manually, in order to have the most reliable outcome for further processing.

Building Height

The geometry of urban form is composed of urban canyons that are defined by the ratio of building height to width of street (H/W ratio) and the alignment of their long axis (Ali-Toudert and Mayer 2006). These two geometric descriptors are domineering the physical process of absorption and reflection of incoming solar radiation and emission of the thermal radiation that persuade the ambient screen height air temperature to be higher than the rural surrounds. The high rise building within Kolkata urban areas possessing multiple surface factors for reflecting and absorbing of incoming solar radiation which increasing the efficiency of heating capacity of urban areas are called urban canyon effect (UCE). The blocking effect of buildings to wind circulation and inhibits cooling by process of advection and prevents the pollution from dissipation (Bottyán and Unger 2003). The extraction of urban building information using coarse resolution SRS data reflects the plentiful challenges, such as spatially intermixed spectral response of built-up and other land class. Considering that the SRS image texture is one of the important geo-spatial information for detecting the urban LULC. This approach holding two basic steps, first multivariate texture is computed through multivariate variogram and then spectral bands and multivariate texture are then combined in image classification process for extracting built-up areas. One-class support vector machine (OCSVM) classifier technique has been used in this classification process. In the study, the structural building heights have been generated from coupling of TM data with 3-D aspect of GE as classified data and which has been finally portrayed on Landsat-5 TM image. In addition, RFVS-derived coordinate points have been generated from hand held GPS survey of 1500 ground locations (low rise-500, medium rise-800 and high rise-200) and being considered as reference data for accuracy assessment of corresponding building height. The OA and KC value was 86% and 0.932 respectively. Furthermore, the classification was further tartan visually and refinement process has manually done in order to have the largely produces the trustworthy for the next step. Finally, the entire KMA buildings have classified map into three regional building height such as low rise (<3 floors), medium rise (4–8 floors) and high rise (>8 floors) in Fig. 3e.

Building Roof Type

The nature and persistence of UHI effects varies climate to cities. The mitigation of UHI effect can be proficient through the application of green roofs and light coloured surfaces in urban areas, which reflects more incoming solar radiation and absorb heat. The Landsat images and a suitable set of spectral image bands are basically used to extract general features and problems in details mapping of urban LULC (Herold et al. 2003). The spectral response of urban built-up materials have been identified and separated by wavelength specific electronic and vibrational absorption features related with concrete type material properties (Hepner et al. 1998; Heiden et al. 2001; Chen and Hepner 2001). Subsequently, the spectral resolutions are used in image classification to map the building roof. The types of roof incorporate a fair number of different concretes and tile materials that vary with materials in use. The nature of roof, age and roof geometry have been separated by within-class variability of spectral range and contributing to the complexity in spectral range of the urban built-up environment (Herold et al. 2003). The spectral response of gray-brown tile roof is similar to a fair number of other unreceptive urban surface types like asphalt roads and composite shingle roofs. This variability in different classes does not properly separate on materials scale (Herold et al. 2003). Generally, the inter-class spectral variability was extracted from objects brightness value. In this regard, the Bhattacharyya distance (B-distance) was applied for measuring the spectral separability to select a most apposite subset of image bands for mapping of building roof (Herold et al. 2003). The B-distance is a sum, represents the component of differences of mean and covariance (Heiden et al. 2001). The B-distance can be calculated by given formula.

$$B = \frac{1}{8}[\mu_1 - \mu_2]^T \left[\frac{\sum_1 + \sum_2}{2}\right]^{-1} [\mu_2 - \mu_1] + \frac{1}{2} \ln \frac{\left|\frac{1}{2}[\sum_1 + \sum_2]\right|}{\sqrt{|\sum_1 + \sum_2|}} \tag{5}$$

where μ_i and $\sum_i$ are the mean vector and the covariance matrix of class respectively. Thus, the building roof has been classified on the basis of spectral separability into five types like concrete roofs; tiles roofs; dark colour roofs; blue colour roofs and metal roofs (Fig. 3f). Finally, the performance of spectral setting with RFVS has been assessed for determining the accuracy level of different roof types. The OA and KC was 83% and 0.853 respectively. Moreover, the classification result was further corroborate visually and refined manually, in order to have the most upright classification for subsequent processing.

Building Roof Reflectance

The reflections from unreceptive surface of urban areas are determined the environmental condition of urban climate. The surface reflection is first order measurement of energy flux but reflection may depend on soil moisture and surface temperature. The actual correlation between surface temperature and reflection from objects provide the information of surface composition and emissivity and different physical process (heat flux). This entire system effects on the surface energy balance of urban environment. For this reason, an effort has been made to understand the environmental factors that persuade the relationship between surface temperature and optical reflectance (Small 2002). The RFVS and highly correlated among spectral bands have been selected to handle the general problems in the analysis of multispectral data. Most of image classification techniques require a condensed spectral dimension to prioritize the set of spectral bands that are enough for real application (Price 1998; Chang et al. 1999; Jimenez and Landgrebe 1999). In this regard, different approaches are available for band selection, including PCA, class separability measures and band correlation measures (Price 1997; Jimenez and Landgrebe 1999). Recently, B-distance was proposed for a useful tool to measure the class separability and band correlation of an image (Jimenez and Landgrebe 1999; Landgrebe 2000). However, this measure has been employed to detect the separability of roof reflectance of urban environment and to prioritize bands that indicate most to spectral discrimination among the region of interest (Chang et al. 1999). In this study, Landsat TM data have been processed to detect apparent surface reflectance using modified Modtran radiative transfer codes and subsequent refinement are done using a ground reflectance target (Clark et al. 1993; Green et al. 1998; Roberts et al. 1997). The spectral response of TM-5 bands in normalized transmittance values convolved with 10-nm increments (Herold et al. 2003). The roof reflectance mapping has been done on the basis of three different spatial sampling schemes like random sampling, neighbourhood sampling and case-specific sampling techniques. These sampling techniques allow a comprehensive database to be acquired that was digitally processed to prop up the image classification and mapping of roof reflectance. For accuracy assessment, spectral random sampling techniques within map polygon (referred as stratified cluster sampling) are applied to extract 300 individual and statistically independent pixels for each reflectance type to examine the classification performance. Afterwards, spectral value is then considered for mapping of building roof types into three basic categories on the basis of reflectance criteria like high reflecting roofs, medium reflecting roofs and low reflecting roofs (Fig. 3g). The accuracy statistics between the RFVS-derived reference points and the Landsat TM derived classified maps establish the confidence level of this building roof reflectance classification and the OA and KC was 84% and 0.823 respectively. In addition, the classification result was further established visually and refined manually, in order to have the most dependable artefact for further processing.

Building Association

Building association is the agglomerated urban scene of different structural building blocks. In metropolis Kolkata, most of building roofs are categorised as concrete roof. For assessing UHI effects, the study has coupled with building height and their roof types. Further on a single thematic map building height and roof types were reclassified into nine classes including, high rise building (small concrete roof), high rise building (medium concrete roof), high rise building (large concrete roof), medium rise building (small concrete roof), medium rise building (medium concrete roof), medium rise building (large concrete roof), low rise building (small concrete roof), low rise building (medium concrete roof), low rise building (large concrete roof) are shown in Fig. 3h. A high solar reflectance or albedo is the most important characteristic of a roof as it helps to reflect sunlight and heat away from a building, reducing roof temperatures (Akbari et al. 1999; Levinson et al. 2007). On contrary, a high thermal admittance also plays a significant role, particularly in tropical climates that are warm and sunny. Together, these properties help roofs to absorb more heat during peak summer weather (Akbari and Konopacki 2005). On the other hand, the narrow arrangements of buildings along the city's streets form urban canyons that inhibit the less escape of the reflected radiation from most of the three-dimensional urban surface to space. This radiation is ultimately absorbed by the building walls due to low sky view factor (SVF) of closed building association, thus reducing the urban heat release and creates UHI effects. The accuracy statistics between the RVS-derived reference and the Landsat TM-5 derived classified maps establish the confidence level of this building association classification and the OA and KC was 86% and 0.863 respectively. In addition, the classification result was further verified visually and refined manually, in order to have the most reliable product for further step.

Road Pavement

The surface transport systems of KMA are a mixed of modern mass rapid transport and old modalities like the rickshaws. The entire city is connected with different modes of transport web. Since, the development of web likes transport systems; Kolkata suffers from traffic congestions akin to other Indian cities. A surface network of expressways including, Kona expressway, Belghoria expressway, flyovers, widening of southern stretch of eastern metropolitan bypass (EMB) and other cellular networks are being created to cease up road traffics. In this study, all the road networks have been directly derived from OpenStreetMap (OSM) by means of ArcGIS software (Fig. 3i). These transport system of KMA have constructed mostly by concrete and asphalt materials, as a results, the surface temperature of road pavement reached up to 50 °C during the summer seasons. Thus, the road pavement acts as additional conditioning factor to indemnify the UHI effects. The traditional road pavements with less solar reflectance usually augment the surface and subsurface temperature of pavements, as more heat available at the surface to absorb into the pavements. The dry and permeable road pavement surface temperatures could be higher than impermeable alike; but studies showed that the subsurface basically is analogous to or even cooler than the traditional equivalent, because the permeable road pavements trim down the heat transfer into subsurface. Hence, more information on road pavement heat transfer is needed to understand the potential UHI effects because the different storage capacity of heat in surface and subsurface layers of pavement. The accuracy statistics between the Google-derived reference and the OSM derived classified road maps establish the confidence level of this network association and the OA and KC was 90% and 0.950, respectively.

Modeling of UHI Susceptibility

Preparation of UHI Inventory Map and Training and Validation Dataset

For city-scale modeling, the study has incorporated three broad protocols. Firstly, a visual comparison was made using overlying UHI vector image (a preliminary UHI information map prepared through integration of different sources) on LST raster image (application of mono-window algorithm in thermal image) and whether results are well-matched or not, if matched then proceed next step. Secondly, classify the image derived from first step to separate the significant UHI thermal zones from other temperatures. Finally, a multi-resolution object based e-cognition approach was employed to generate UHI map for building training and validation dataset. The preparation of training and validation dataset is the crucial step in the evaluation of UHI modeling. The splitting of datasets could be used in validation of model derived results (Hong et al. 2015). In order to build dataset inventory, 350 vectorized pixel patches were split into training dataset and validation dataset. The training dataset and validation dataset are consisted of 70% (245) and 30% (105) of the UHI inventory respectively (Fig. 4). Although, most of UHI areas were found in compact built-up areas where SVF is very low with a distinct microclimate zones, therefore, all UHI were converted into point in ArcGIS 10.1 software (Hong et al. 2015). These points were located at the centres of vectorized pixels patches. For mapping of UHI susceptibility, a binary classification technique was adopted as non-UHI and UHI points on the basis of probability. Hence, value would be assigned as 1 for UHI points and 0 for non-UHI points in inventory map (Hong et al. 2015). Finally, the causative conditioning factors maps were transformed into raster format with a spatial resolution of 30 × 30 m.

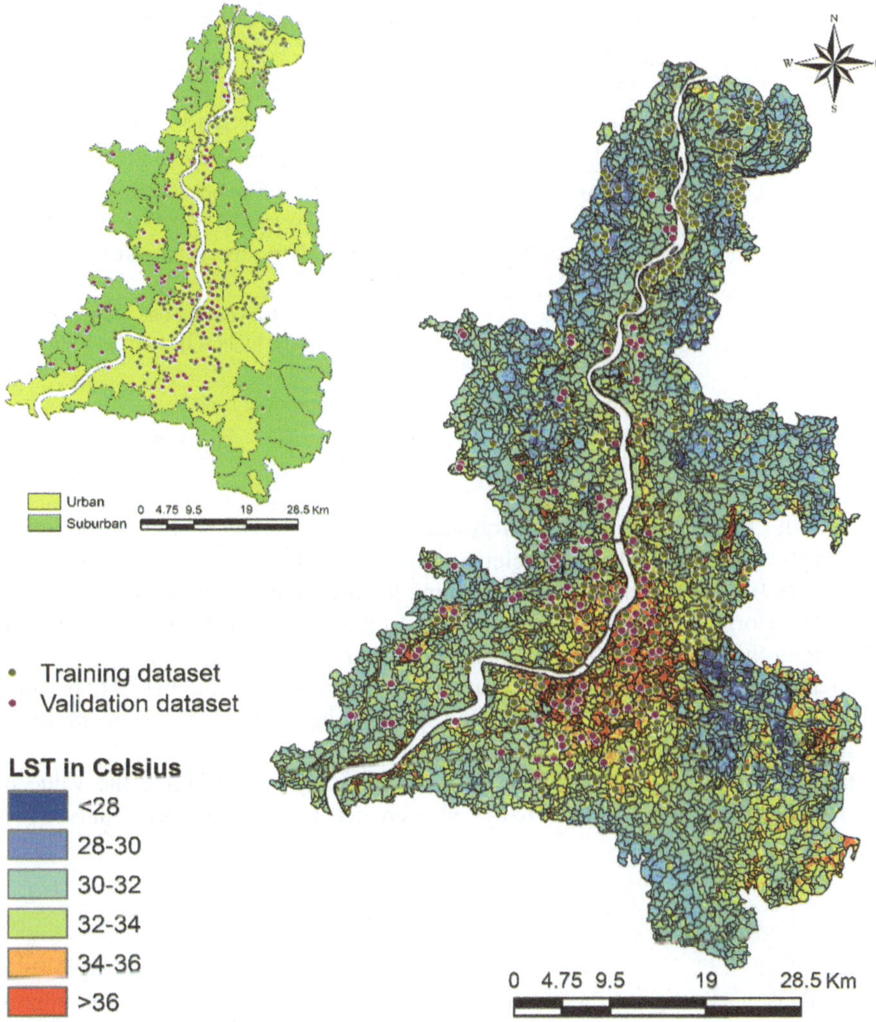

Fig. 4 Final urban heat island inventory map extracted from Landsat TM satellite images using e-Cognition approach

AHP Model

The AHP is semi-quantitative and flexible tool, which involves a matrix-based pair-wise comparison of the contribution of different conditioning factors and analyzing complicated problems focusing onsite selection, regional planning, routing modeling and modeling of environmental phenomena (Pourghasemi et al. 2012; Shahabi and Hashim 2015). It is multi-criteria decision-making multi-objective approach allows the users to infer a rational agreement on scale of preference (Saaty 1980). In the

Table 3 Fundamental scale of absolute numbers between two parameters in AHP model (Saaty 1980)

Intensity of importance	Degree of preference	Explanation
1	Equally	Equal importance to objective is equal
3	Moderately	Attribute is slightly favoured over another
5	Strongly	Experience and judgment strongly or essentially favour one activity over another
7	Very strongly	Attribute is very strongly favoured over another
9	Extremely	The evidence of favouring one activity over another is of the highest degree possible of an affirmation
2,4,6,8	Intermediate values	When compromise is needed
Reciprocals	Opposites	Used for inverse comparison

AHP model, the conditioning factors are arranged in a hierarchical order and assigned numerical values on the basis of relative importance of factors by user subjectivity (Saaty 1980). Subsequently, the conditioning factors are synthesized rating to determine the priorities to be assigned according to their importance (Saaty and Vargas 2012; Pourghasemi et al. 2012; Shahabi and Hashim 2015).

Apart from that, reciprocal pair-wise comparison matrix is established to perform AHP and each thematic factor based on 9-point rating scale (Table 3) is employed in the matrix (Saaty 1977).

In AHP model, the consistency ratio (CR) is used to determine the value of probability that the judgement matrix was created random manner (Saaty 1977, 1980)

$$CR = \left(\frac{CI}{RI}\right) \tag{6}$$

where RI is the average of the resulting consistency index depending on the order of the matrix given by Saaty (1980) and CI is the consistency index and can be expressed as:

$$CI = \left(\frac{\lambda\max - n}{n - 1}\right) \tag{7}$$

where $\lambda_{\max}$ is the largest or principal eigenvalue of the matrix and can be easily calculated from the matrix and n is the order of the matrix.

The CR is a result of the division of CI by RI, and in universal range varies from 0 to 1. A CR of 0.1 or less is a reasonable level of consistency (Malczewski 1999). A CR above 0.1 requires reconsideration of the decision in the matrix due to an inconsistent dealing for particular factor ratings (Pourghasemi et al. 2012).

Kernel Logistic Regression

The KLR is an improved version of kernel based logistic regression to estimate the posterior probability of class membership for binary classification (Cawley and Talbot 2008; Hong et al. 2015). With training dataset T with $T = (X_i, y_i), X_i \in R^n, y \in \{0, 1\}$, X is an input vector that includes nine UHI conditioning factors and y indicates for two classes i.e., non-UHI and UHI zones. The KLR is a high-dimensional feature space that could be separated the non-UHI and UHI zones from the training dataset. This can be expressed as a non-linear regression as (Hong et al. 2015).

$$\text{logit}\{y(X)\} = w.\varphi(X) + b \tag{8}$$

where, w is a vector of model parameters and $\varphi(X)$ indicates a non-linear transformation of the input vectors.

The logit transformation is expressed as:

$$y(X) = \frac{1}{1 + \exp\{-\omega.\varphi(X) - b\}} \tag{9}$$

Thus the logit function limits the values of UHI of the model in a range (0.1).

The non-linear transformation of input vectors called kernel function and it can be expressed as.

$$K\left(X, X'\right) = \phi(X).\phi\left(X'\right) \tag{10}$$

In the present study, most common GRBF kernel function has been used to modeling the UHI susceptibility (Cawley and Talbot 2008; Hong et al. 2015).

$$K(X_i, X_j) = \exp\left(\left(-\|X_i - X_j\|^2\right) / 2\delta^2\right) \tag{11}$$

where δ is a turning parameter.

The optimal vector of model parameters (W) have been derived from minimizing the cost function and written in linear combination as.

$$W = \sum_{i=1}^{n} \alpha_i.\phi(X_i) \tag{12}$$

Thus, the kernel logistic regression is expressed as.

$$\text{logit}\{y(X)\} = \sum_{i=1}^{n} \alpha_i K(X_i, X) + b \tag{13}$$

$$\alpha = (\alpha_1, \alpha_2, \ldots, \alpha_1) \tag{14}$$

To find out the optimal kernel parameters of KLR, a heuristic test was carried out by using the training dataset (Hong et al. 2015). And best tuning and regularize parameter are 0.03 and 0.045 respectively.

SVM Model

The SVM is a statistical maximum-margin classifier technique based on the principle of structural risk minimization (Vapnik and Vapnik 1998; Hong et al. 2015). In this model, T regard as the UHI training dataset, the inputs of SVM maps into high-dimensional feature space, then optimal hyper-plane with maximum margin to be determined to split two classes i.e., non-UHI and UHI zones. The optimal hyper-plane can be solved by using following expression (Hong et al. 2015).

$$\text{Min} \sum_{i=1}^{n} \alpha_i - \frac{1}{2} \sum_{i=1}^{n} \sum_{j=1}^{n} \alpha_i \alpha_j y_i y_j \left(X_i X_j \right) \tag{15}$$

$$\text{Subject to}\, i = \sum_{i=j}^{n} \alpha_i y_j = 0 \text{ and } 0 \leq \alpha_i \leq C \tag{16}$$

where α_i are the Lagrange multipliers, C is the penalty.

The decision function can be written as:

$$g(X) = sign\left(\sum_{i=1}^{n} y_i \alpha_i K\left(X_i, Y_j\right) + b \right) \tag{17}$$

where $K\left(X_i, Y_j\right)$ are the kernel functions. For UHI susceptibility modeling, the Gaussian radial basis function kernel (GRBF) has been used.

$$K\left(X_i, Y_j\right) = \exp\left(-\gamma \left\| X_i - Y_j \right\|^2\right) \tag{18}$$

where γ is a kernel parameter.

With training dataset T, the optimal C and γ have been extracting as 49 and 0.201 respectively. To employing the best pairs of the GRFB kernel parameters, the SVM model was build on the basis of training data and subsequently model was applied to UHI mapping.

SMCE Model

The SMCE allows performing the assessment of multi-criteria in a spatial approach to transform spatial and non-spatial input which generates output decision (Malczewski 1999; Shahabi and Hashim 2015). The output of SMCE is one or more composite map, which provides the extent of criteria are met or not in different aspects, and thereby support decision making (Rahman and Saha 2008). There are number of stages in pursuing the SMCE, including tree analysis, normalization, weighting and map integration (Shahabi and Hashim 2015). In order to normalize input maps in SMCE environment, one of the normalization techniques such as numerical, Boolean and qualitative methods were used (Shahabi and Hashim 2015). In the study, SMCE process comprised three sequence steps. Firstly, the normalization map values are to convert the actual map values to a range 0–1. Secondly, this step is to determine the each conditioning factor for intermediate or overall objectives using fuzzy operations (Gopal et al. 2016). Finally, the UHI conditioning factors are weighted by means of direct, pair-wise and order of ranking comparison and output is a composite index map.

The fuzzy algebraic sum and fuzzy algebraic product creates the resultant set larger than, or equal to the maximum value and smaller than, or equal to the minimum value among all fuzzy sets respectively (Bhaskaran et al. 2010). The selection of fuzzy operator depends on the types of spatial data to be integrated (Morano et al. 2014). The fuzzy gamma operator was selected to integrate conditioning factors using the formula (Gopal et al. 2016).

$$f_\gamma(x) = (Fuzzy\,algebric\,sum)^\gamma \times (Fuzzy\,algebric\,product)^{1-\gamma} \tag{19}$$

$$Fuzzy\,algebric\,product = \prod_{i=1}^{n} R_i \tag{20}$$

$$Fuzzy\ algebric\ sum = 1 - \prod_{i=1}^{n} (1 - R_i) \tag{21}$$

where x denotes the membership functions and R_i denotes fuzzy membership function of ith map, $i = 1, 2 \ldots n$. Using Eqs. (18), (20) and (21), UHI susceptibility maps were prepared. Further UHI maps were classified into six zones namely, very low, low, moderate, high and very high and severe zones using Jenks natural break classifier in Arc GIS 10.1.

Model Evaluation and Assessment

The model evaluation and assessment is a critical aspect of UHI modeling. In this regard, the overall performance of the UHI susceptibility models might be assessed and compared on the basis of training and validation dataset. The assessment of the model performance only on the training dataset and it provides the degree of fit of UHI model with dataset. The model performance on the validation dataset could be used to measure the prediction capability and accuracy of the trained UHI models. In the study, the ROC curve, Kappa index, and measures of different statistical evaluation have been adopted for the assessment of overall performance of the UHI models (Arsanjani et al. 2013; Hong et al. 2015).

The ROC Curve and Kappa Index of Agreement

In order to build the assertion of performance of UHI susceptibility is considered to be a decisive measure in choice of model and evaluation. In the study, UHI susceptibility is taken as a binary classification (0, 1) for splitting into non-UHI and UHI susceptibility zones. The receiver operating characteristic (ROC) curve is employed to assess the performance of model-based classification. The ROC curve is drawn by simply plotting the true positive rate (sensitivity) against the false positive rate (1-specificity) with the various thresholds (Hong et al. 2015). The area under ROC curve (AUC) has also been used for comparison evaluation of city-scale models. An AUC value of 1 and 0 indicates a perfect and non-informative model respectively. The agreement of kappa index is also used for measure the consistency of models, a value equals to zero indicates that agreement between the predicted and the observed pixels of the UHI modes has transpired by chance, while the value equals to one designates perfect agreement (Eq. 3).

Statistical Measures and Evaluation of City-scale Models

To measure the performance metrics for UHI and non-UHI classes, five statistical evaluation measures were used (Hong et al. 2015). The metrics based evaluation was computed on the basis of confusion matrixes resulting from four performed city-scale model. These evaluation measures can be expressed as.

$$\text{Accuracy} = \frac{TP + TN}{TP + TN + FP + FN} \tag{22}$$

$$\text{Sensitivity} = \frac{TP}{TP + FN} \tag{23}$$

$$\text{Specificity} = \frac{TN}{FP + TN} \tag{24}$$

$$\text{Positive predictive value} = \frac{TP}{FP + TP} \tag{25}$$

$$\text{Negative predictive value} = \frac{TN}{FN + TN} \tag{26}$$

where true positive (TP) is the number of UHI points correctly classified to the UHI class, true negative (TN) is to the total number of non-UHI points correctly classified to the non- UHI class. False positive (FP) is the number of UHI points classified to the non- UHI class; false negative (FN) is the non-UHI points classified to the UHI class.

Contribution measures for UHI Susceptibility

In order to find out the contribution of conditioning factors in UHI susceptibility is an important aspect of variations, evaluation criteria and modeling approach (Xu et al. 2014; Hong et al. 2015). However, the majority voting ensemble of AHP, KLR, SVM and SMCE are to have good performed for contribution measures (Hong et al. 2015). The majority voting ensembles represents that a thematic factor that has a high contribution in a particular model may have a little contribution for other and vice versa (Hong et al. 2015). Hence, the relative significance of a thematic layer may imply a large difference. The contribution of each thematic layer to UHI susceptibility model is assessed by using attributes evaluation and ranking method (Hong et al. 2015).

Results and Discussion

The mapping of thermal heterogeneity, extent and location of UHI phenomenon are largely depends on unreceptive surface coverage of urban areas. The sparse weather station based UHI measurement are inaccurate, therefore not suitable for continuous monitoring of UHI effects in city-scale or by means of any predetermined UHI inventory map. Hence, unremitting repetition of SRS TM data and other allied information have been used to detect the potential UHI zone by using segmentation and classification techniques. The pixel based spectral values are used to proper representing the themes of different conditioning factors map through GIS-based statistical manipulation and machine learning algorithm. In this study, only surface UHIs are mapped, because the occurrence of other UHI types (boundary and canopy layer) were scared and meteorologically very complex to determine through 2-D aspect of SRS data in use. To assess the level of confidence and quality of theme, the total RMSE was 0.136 which is satisfactory and shows high quality of efficiency to map the UHI susceptibility zonation. Through the GIS-based statistical environment, all themes are processed through four models with a precise window-filtering technique to map UHI susceptibility zonation. And subsequent model generated UHI maps were overlaid on UHI build inventory map then raster files of each model were reclassified into six relative susceptibility zones like very low, low, moderate, high, very high and severe using Jenks natural break classifier in ArcGIS 10.1. Figure 5a–d show

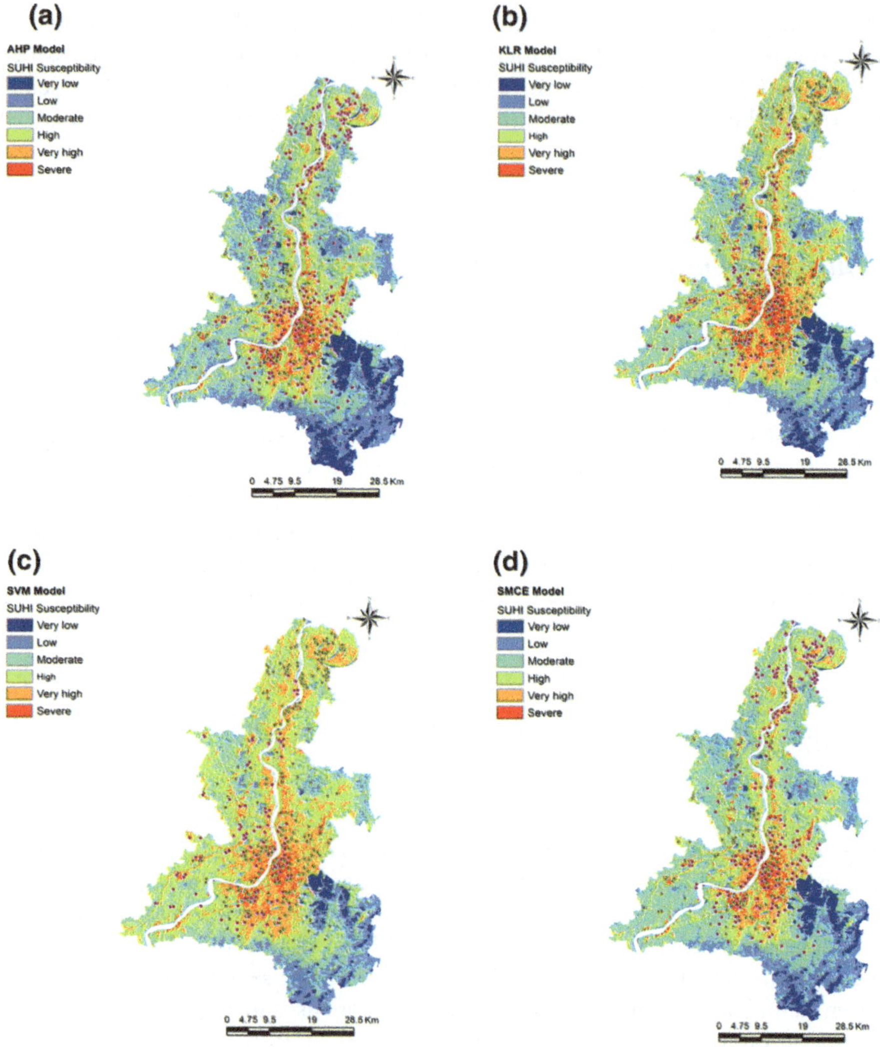

Fig. 5 Surface heat island susceptibility map derived from **a** AHP model **b** KLR model **c** SVM model and **d** SMCE model

UHI maps using the AHP, KLR, SVM and the SMCE models. The percentage area for each model generated UHI susceptibility classes were shown in Table 4.

Table 4 Comparison between susceptibility level and areal coverage

Susceptibility level	SUHI under statistical model (%)			
	AHP	KLR	SVM	SMCE
Very low	13	13	2	9
Low	24	17	11	22
Moderate	30	29	24	29
High	21	27	43	27
Very high	7	8	15	7
Severe	5	6	5	6

AHP Model and UHI Zonation

In this multi-criteria decision making situation AHP model mostly well suited where the contributing factors can be organized in a hierarchical structure (Saaty 1980). The top level of the hierarchy defines a general objective of the problems i.e. quantification of UHI susceptibility. The second level involves contributing factors and the third level considers sub classes pertaining to each of the factors in the second level. The pair wise comparison of an element at each level of the hierarchy allows the assessment of the relative significance of all elements in the hierarchy system and to assign a rating to conditioning factors to a level as opposed to a specific element in the next higher level (Saaty 1980). This rating was computed through Expert Choice software, which is reveals that the rating of each factors were prioritized subject to cognitive limitation with uncertainty and users subjectivity (Pourghasemi et al. 2012). The rating values of each conditioning factors and corresponding CR values shown in Table 5. The computed CR implies a reasonable level of consistency (<0.1), which is enough to recognize the factors rating in UHI susceptibility zonation. The weight ratings of AHP model in building material, building roof type and building roof reflectance are 0.200, 0.178 and 0.156 respectively, were found to be important conditioning factors in UHI susceptibility. On the other hand, population distribution (0.044) and LULC (0.022) received a lower degree of scale importance as compared to other factors. The results of class rating, factors rating and CR values of all thematic data layers are presented in Table 5. Afterwards, for mapping the UHI susceptibility, the study adopted the following equation (Pourghasemi et al. 2012).

$$SUHIS = \sum_{i=1}^{n} (R_i \times W_i) \tag{27}$$

where R_i is the normalized rating class of each thematic layer and W_i is the weights for each layer of UHI conditioning factors. The UHI index values were then divided into six susceptibility zones including very low, low, moderate, high, very high and severe based on the natural breaks method. Based on obtained results of UHI susceptibility map, 13% of the total area shows very low UHI susceptibility. Low, moderate, high,

Table 5 Normalized weights and ranks assigned to respective themes and the features of susceptibility class for thematic integration on GIS platform

Factor	Factor weight (W_i)	Class	Class rating to potential UHI phenomena	Normalized rating (R_i)
Population distribution	0.044	<58,990	1	0.000
		58,991–166,867	2	0.250
		166,868–294,628	3	0.500
		294,629–448,317	4	0.750
		>448,318	5	1.000
Land use/land cover	0.022	Residential/commercial/industrial areas	9	1.000
		Water body/ponding/river/canal	1	0.000
		Agriculture areas	8	0.875
		Vegetation	2	0.125
		Urban forestry/plantation	3	0.250
		Swampy land	4	0.375
		Dry fallow land	5	0.500
		Arable land	7	0.750
		Open space	6	0.625
Building material	0.200	Mud and non-brick wall	2	0.125
		Stone wall	5	0.500
		Burnt brick wall	4	0.375
		Steel frame buildings	9	1.000
		Buildings of large blocks	6	0.625
		Prefabricated buildings with natural hewn stone	7	0.750
		Concrete buildings	8	0.875
		Newly built-up concrete buildings	3	0.250
		Others building materials	1	0.000
Building height	0.089	High rise building	3	1.000
		Medium rise building	2	0.500
		Low rise building	1	0.000
Building roof type	0.178	Concrete roofs	5	1.000
		Tile roofs	2	0.250
		Dark colour roofs	3	0.500
		Blue colour roofs	1	0.000
		Metal roofs	4	0.750
Building roof reflectance	0.156	Low reflecting roofs	3	1.000

(continued)

Table 5 (continued)

Factor	Factor weight (W_i)	Class	Class rating to potential UHI phenomena	Normalized rating (R_i)
		Medium reflecting roofs	2	0.500
		High reflecting roofs	1	0.000
Building age	0.067	< 5 Years	1	0.000
		10 Years	2	0.167
		15 Years	3	0.333
		20 Years	4	0.500
		25 Years	5	0.667
		30 Years	6	0.833
		>35 Years	7	1.000
Building association	0.133	High rise building (small concrete roof)	7	0.750
		High rise building (medium concrete roof)	8	0.875
		High rise building (large concrete roof)	9	1.000
		Medium rise building (small concrete roof)	4	0.375
		Medium rise building (medium concrete roof)	5	0.500
		Medium rise building (large concrete roof)	6	0.625
		Low rise building (small concrete roof)	1	0.000
		Low rise building (medium concrete roof)	2	0.125
		Low rise building (large concrete roof)	3	0.250
Road pavement	0.111	Traffic Network with asphalt pavement	2	1.000
		Traffic Network with concrete pavement	1	0.000

very high and severe susceptible zones comprise 24, 30, 21, 7 and 5% of the total KMA area respectively (Fig. 5a; Table 5).

KLR Model and UHI Zonation

The KLR has an ordinary extension to multi-criteria classification system, while SVM, there are multi-ways to pull out it into multi-criteria classification. KLR is stimulated by the similarity in geometric shape between negative log likelihood (NLL) of a binomial distribution and the pivot loss of the SVM. KLR is a kernel version of logistic regression that usually used for classification of binary responses of predictive variable by estimating a posterior probability function of class membership (Cawley and Talbot 2008; Hong et al. 2015). A heuristic test was carried out to search the optimum kernel parameters for KLR using the training dataset T (Hong et al. 2015). The optimum and refined parameter and normalize parameter are 0.03 and 0.045 respectively. Thus, the model was implementing to map the UHI susceptibility. The derived values were categorized into six zones including very low, low, moderate, high, very high and severe based on the natural breaks technique. Based on the result of the obtained UHI susceptibility map, 13% of the total area shows very low UHI susceptibility. Low, moderate, high, very high and severe susceptible zones make up 17, 29, 27, 8 and 6% of the total area, respectively (Fig. 5b; Table 5).

SVM Model and UHI Zonation

The SVM is a conventional technique is well-known for its reasonable performance to binary classification and generates a non-linear classification boundary for input dataset by producing a linear classification boundary through transformation output dataset. The dimension and magnitude of space conversion is very large, even some cases it will infinite. The complex and prohibited computation is attained through a positive definite reproducing kernel *K*, which provides an inner product in the transformed space. In this study, the SVM fitting for UHI modeling pays well attention on overall spatial features, while neglecting the noise caused by the random factors and the strong thermal heterogeneity. Hence, this technique is a powerful tool for examining the spatial pattern of UHI pockets in the urban thermal heterogeneity. The UHI training dataset T with SVM was employed to recognize the different UHI zones. The SVM with a GRBF requires two basic parameters for prediction; C and γ. The C is a regularized parameter that controls the trade-off between maximum the margin and minimizing the training error, while γ illustrates the kernel width (Hong et al. 2015). The optimum values of C and γ parameters were obtained grid search techniques and these parameters were tuned using 10 fold cross-validation method, the C and γ was set up to 49 and 0.201 respectively (Hong et al. 2015). Thus, the final UHI susceptibility map was derived from SVM model and reclassified into six zones including, very low, low, moderate, high, very high and severe based on natural break classification method (Fig. 5c). According to UHI susceptibility map, 2% of the entire area is found to be of very low UHI susceptibility zone. Low, moderate high and very high susceptible zones showed 11, 24, 43, and 15% of the entire area respectively. The severe UHI susceptibility zone is 5% of the total study area (Fig. 5c).

SMCE Model and UHI Zonation

The SMCE technique permits the users to carry out multi-criteria appraisal in a spatial approach (Shahabi and Hashim 2015). In this technique, at first, normalized the input thematic layers based on their original thematic values (0–1). This normalization process was performed as a SMCE module is available in integrated land and water information system (ILWIS) software to derive composite map and subsequent final UHI map (Malczewski 1999; Shahabi and Hashim 2015). The SMCE was built on the basis of weight values by using bivariate statistical technique for the thematic layers of the conditioning factors (Table 6).

For normalizing the values of thematic conditioning layers, the fuzzy approach was employed. The pair-wise comparisons between factors were performed using scale of AHP model (Table 3). The fuzzy sets were incorporated to represent the linguistic expressions in classification process, which makes it easy to read efficiently (Shahabi and Hashim 2015). In order to input the linguistic expressions in classification process, the six-class fuzzy sets have been employed.

(a) Very low = (1/1, 0.75/2, 0.5/3, 0.25/4, 0/5, 0.2/6)

Table 6 The weight value of each group and weight value of factors using pair wise comparison for the SMCE model

Group factor	Group factor weight (W_{gi})	Factor	Factor weight (W_i)	Fuzzy membership
Socio-economic	0.167	Population distribution	0.091	0.189
		Land use/land cover	0.081	0.000
Structural	0.500	Building material	0.134	1.000
		Building height	0.126	0.849
		Building age	0.117	0.679
		Building association	0.111	0.566
Radiative	0.333	Building roof type	0.115	0.642
		Building roof reflectance	0.112	0.585
		Road pavement	0.113	0.604
CR	0.057	*CR*	0.059	

(b) Low = (0/1, 0.2/2, 0.75/3, 0.2/4, 0.2/5, 0.1/6)
(c) Moderate = (0/1, 0.25/2, 0.75/3, 0.5/4, 0.25/5, 0.2/6)
(d) High = (0/1, 1/2, 0.5/3, 0.75/4, 0.25/5, 0.2/6)
(e) Very high = (0/1, 0.2/2, 0.5/3, 0.25/4, 0.2/5, 0.5/6)
(f) Severe = (0/1, 0.25/2, 0.5/3, 0.75/4, 1/5, 0.2/6)

In the preceding fuzzy sets, the values of slash before after are degree of confidence and members of sets respectively (Pourghasemi et al. 2014; Shahabi and Hashim 2015). The fuzzy sets of prompting factors of UHI susceptibility are as follows.

(a) μ_s Population distribution = (0.09/1, 0.73/2, 0.82/3, 0.91/4, 1/5)
(b) μ_s Land use/land cover = (0/1, 0.87/2, 1/3, 0.75/4, 0.96/5, 0.39/6, 0.46/7, 0.65/8, 0.54/9)
(c) μ_s Building material = (1/1, 0.17/2, 0.13/3, 0.34/4, 0.22/5, 0.36/6, 0.38/7, 0.07/8, 0/9)
(d) μ_s Building height = (0/1, 1/2, 0.56/3)
(e) μ_s Building roof type = (1/1, 0.93/2, 0.72/3, 0.83/4, 0.49/5)
(f) μ_s Building roof reflectance = (0.88/1, 0.56/2, 0.68/3
(g) μ_s Building age = (0.33/1, 0.81/2, 0.48/3, 0.54/4, 0.68/5, 0.76/6, 0.95/7)
(h) μ_s Building association = (1/1, 0.95/2, 0.84/3, 0.79/4, 0.71/5, 0.79/6, 0/7, 0.74/8, 0.81/9)
(i) μ_s Road pavement = (0.86/1, 0.69/2)

Afterwards, the employing the fuzzy rules in classification system, the grid-based analysis were carried out to map the fuzzified index in ArcGIS environment (Shahabi and Hashim 2015). In broad aspect, the UHI mapping process was further divided into three sub-group like socio-economic (population distribution and LULC); structural (building material, building height, building age and building association); radiative (building roof type, building roof reflectance, road pavement) factors and these are input in SMCE analysis.

The influence of sub-factors group and respective weigh values of principal indicators have been derived by using AHP technique (Table 6). The results obtained from SMCE analysis, it has found that the structural driving factors largely promote the UHI occurrence (0.500). Whereas, population distribution (0.091) is significant susceptible to UHI but LULC is a lesser amount of prone to UHI as it weights value of 0.081 as compared to other factors. For structural conditioning factors, weight corresponding to building materials (0.134) is large, while building association is lowest (0.111). In case of radiative factors, it was found the building roof type; road pavements and building roof reflectance have a weight of 0.115, 0.113 and 0.112 respectively. In general, the building material is highly prone to UHI susceptibility, and on contrary, LULC has lowest influence on UHI susceptibility. For all cases of the obtained weights (sub-factors and thematic conditioning factors), the CR further <0.1 indicates the good agreement to recognize the weight values. The final UHI map is extracted from SMCE model and reclassified into six susceptibility zones like very low, low, moderate, high, very high and severe based on Jenk natural break classification method (Fig. 5d). According to the UHI susceptibility map obtained

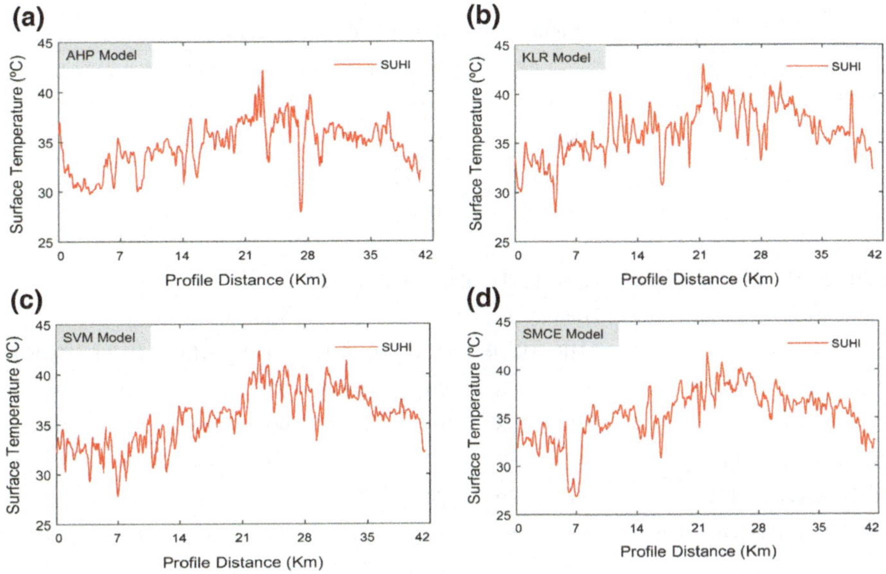

Fig. 6 Model derived SUHI over KMA area (0–42 km horizontal sample line) viz. **a** the AHP model, **b** the KLR model, **c** the SVM model and **d** the SMCE model

from the SMCE approach, 9% of the entire area is found to be of very low UHI susceptibility zone. Low, moderate high and very high susceptible zones showed 22, 29, 27, and 7% of the whole area respectively. The severe UHI susceptibility zone is 6% of the entire study area (Fig. 5d).

The UHI extension over the KMA area extending from 88.46°E, 22.74°N (0-km) to 88.38°E, 22.38°N (42-km) is shown in Fig. 6a–d for the AHP model, the KLR model, the SVM model and the SMCE model respectively. The urban and suburban temperatures are profile more prominent in SMCE model as compared to other models.

Performances of City-scale UHI Models

In order to assess the performance of city-scale UHI models, training dataset *T* training has been employed for constructing the AHP, KLR, SVM and SMCE model (Tables 7 and 8). The results of ROC curves and AUC for the each model are shown in Fig. 7a. The values of AUCs for SMCE, KLR, AHP and SVM are 89, 87, 86 and 85% respectively. This fitting of curve reveals that the SMCE model has the high degree of fit to the training dataset as compared to other models. The values of kappa index for SMCE, KLR, AHP and SVM model are 0.752, 0.711, 0.635 and 0.612 respectively and indicating that a substantial agreement between the models and certainty. Thus, the UHI models have high performance in regard to classification accuracy. In this

Table 7 Kappa index of agreement for the AHP, the KLR, the SVM, and the SMCE model

Model	Kappa index of agreement	
	Training dataset	Validation dataset
AHP	0.635	0.523
KLR	0.711	0.602
SVM	0.612	0.504
SMCE	0.752	0.673

Table 8 Model validation using statistical evaluation measures

Parameter	Training dataset				Validation dataset			
	AHP	KLR	SVM	SMCE	AHP	KLR	SVM	SMCE
True positive (TP)	201	206	201	210	72	75	70	76
True negative (TN)	204	208	202	201	90	89	90	88
False positive (FP)	41	37	43	44	15	16	15	17
False negative (FN)	44	39	44	35	33	30	35	29
Positive predictive value (%)	83	85	82	86	83	87	81	89
Negative predictive value (%)	82	84	82	85	73	75	72	75
Sensitivity (%)	82	84	82	86	69	71	67	72
Specificity (%)	83	85	82	82	86	85	86	83
Accuracy (%)	83	84	82	85	77	78	76	78

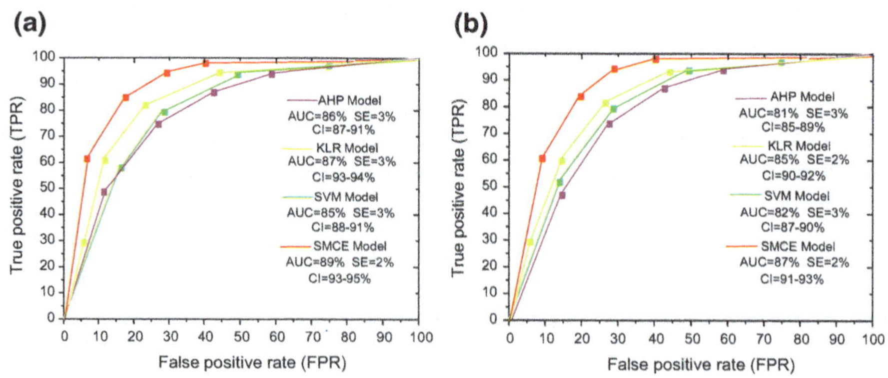

Fig. 7 The ROC curves and AUC for the AHP, the KLR, the SVM and the SMCE models using **a** the training dataset and **b** the validation dataset

regard, the highest classification accuracy is for the SMCE (85%) followed by KLR (84%), AHP (83%) and SVM (82%). The maximum positive predictive value has been found in SMCE model (86%) which specifies the probability that the model correctly detect the pixels in training dataset to UHI class is 86% and followed by KLR model (85%), AHP (83%) and SVM model (82%). For maximum negative predictive value, the model SMCE produces 85% of accuracy indicating that the probability to accurately classify the pixels in training dataset to non-UHI class is 85% as compared to KLR model (84%), AHP (82%) and SVM model (82%). Both the SMCE and KRL models have produced maximum values of sensitivity (86 and 84%), which indicates that 86 and 84% of the UHI areas are correctly classified to UHI class respectively, as compared to AHP (82%) and SVM model (82%). The highest specificity is for the KLR model (85%), which elucidating that 85% of non-UHI areas have been properly classified to the non-UHI class and it is followed by AHP (83%), SVM (82%) and SMCE model (82%). The highest level of accuracy is for the SMCE model (85%) providing that 85% of non-UHI areas were correctly classified to the non-UHI class as compared to KLR model (84%), AHP (83%) and SVM model (82%).

Assessment and Comparison of City-scale UHI Models

The model comparison and evaluation is critical step in the UHI susceptibility mapping. The model-derived results were validated using the validation dataset of UHI inventory through AUC, kappa index and statistical evaluation measures (Fig. 7b; Tables 7, 8). The results reveal that all the four UHI models were produced well prediction capability with maximum in SMCE model (AUC = 87%) and followed by KLR model (AUC = 85%), SVM model (82%) and AHP model (AUC = 81%). The values of kappa index are 0.673, 0.602, 0.523 and 0.504 for SMCE, KLR, AHP and SVM model respectively, which explain a well agreement between observed and predicted UHI areas. The better results of classification accuracy in validation dataset are for the SMCE model (78%) as compared to KLR model (78%), AHP (77%) the SVM model (76%). The SMCE model has the highest positive predictive value (89%), which infer the probability that to classify properly the pixels in respect to the UHI class at 89% and it followed by the KLR model (87%), AHP (83%) and the SVM model (81%). On the other hand the highest occurrence of probability to well classify the pixels to the non-UHI class is for the SMCE model (75%), to some extent followed by the KLR (75%), AHP (73%) and SVM model (72%). The SMCE and KLR models have the maximum sensitivity (72 and 71%), which may reveal that 72 and 71% of the UHI areas are well classified into UHI class respectively. In case of AHP and SVM model the sensitivity values are 69 and 67% respectively. The specificity is for the SVM model is 86% which explain the 86% non-UHI areas which are correctly classified the non-UHI class as compared to KLR model (85%), AHP model (86%) and SMCE (84%). In view of the fact that the AUC is a graphical measure of level of overall performance of UHI models. Hence, a statistical test was carried out to measure if there exit any statistical differences between models performance. It is imperative to note that when comparing the AUCs of susceptibility models with the purpose of an equal AUC value reveals that the two models are produced the similar overall performance, but it does not inevitably means that the two ROC curves are indistinguishable. Statistically, null hypothesis take for granted that there is no difference between the performances of UHI models, then *p*-value estimate. For this, pair-wise comparisons of the ROC curves between the SMCE, KLR, AHP and SVM have been carried out at 95% level of confidence that in prediction capability of the models ($p < 0.05$). In compare, the reliable prediction capability of the SMCE model is significantly higher than those in other UHI models.

Applicability of the Models and Priority Voting Ensemble of Contributing Factors

The UHI susceptibility modeling approach is an important tool for urban climatologists and geographers to predict the UHI effects and future urban planning. Therefore, it is urgent to need that accurate mapping of UHI effects with reliable statistical models. The reliability and objectivity of city-scale modeling is still deliberated. Hence, the exploration of new models for UHI modeling approach needs to be required as reasonable aspects. In this regard, coupling of GIS and machine learning modeling approach are new and emerging techniques that have been advanced in terms of overall performance. The state-of-the art machine learning urban modeling approaches such as AHP, KLR, SVM and SMCE were used in urban study with a better accuracy level. However, the searching for these advanced modeling approaches for UHI modeling has rarely been carried out. This study deals with the issues of investigation and comparison of AHP, KLR, SVM and SMCE for UHI susceptibility modeling. For model evaluation and comparison, the AHP and the SVM have been taken since approaches provide the best predictability in terms of overall performance. But, SMCE model give up significant better results as compared to KLR, AHP and SVM models in respect to overall performance on both training and validation datasets. In order to build the SMCE model, only the boosting iteration needs to be determined. Moreover, the SMCE depicts a simple transparent model structure with straightforward to interpret the rules of classification. But, the KLR and SVM is the complex mathematical model that has no transparent view. The process of building these models, the kernel parameters C and γ are required for GRBF function. In case of SVM model, data-driven grid search technique was used to get kernel parameters C and γ which are to be optimized values. On contrary, for the case of KLR model, a heuristic test was carried out to determine the optimum kernel parameters C and γ.

The goodness-of-fit of all susceptibility models on the basis of training dataset are produced better results but AUC varied significantly for each model. For this, AUC not only provides the exclusive for goodness-of-fit, because it does not bear out the results as high spatial accuracy of the models. Therefore, other statistical evaluation

Table 9 Contribution of nine urban heat inland conditioning factors for the AHP, the KLR, the SVM and the SMCE model

Minus conditioning factors	Ranking				Priority voting ensemble of AHP KLR, SVM and SMCE model
	AHP	KLR	SVM	SMCE	
Population distribution	74	75	78	78	76
Land use/land cover	68	68	67	68	68
Building material	80	82	81	84	83
Building height	80	82	79	82	81
Building roof	80	80	80	80	80
Building roof reflectance	80	82	80	81	81
Building age	82	83	81	83	81
Building association	81	81	79	80	80
Road pavement	81	83	80	83	82

measure needs to be required for further assessment (Table 8). The results show that the there is no significant differences in SMCE and KLR models in respect to prediction probability. Hence, the SMCE model is considered for better UHI susceptibility, because the positive predictive value for SMCE model is higher as compared to other three models. In addition, the SMCE model has produced steadiness results in terms of positive and negative predictive values. It is well-known fact that the overall performances of the model-derived results have been influenced by underlying conditioning factors. The evaluation of thematic conditioning factors to the models is of immense interest in UHI susceptibility modeling. The classifier attribute evaluation approach was employed to measure the contribution of socio-economic, structural and radiative drivers in UHI models, (Witten et al. 2011; Hong et al. 2015). The results reveal that the building materials have the highest contribution in UHI models and other thematic contributions are varied with the use of methods. To reduce the biasness caused by the varying preference of individual models. Hence, the quantification of the contributions of different conditioning factors has been carried out on the basis of majority voting combination rules and ranking (Kuncheva 2004; Witten et al. 2011; Hong et al. 2015). The technique of majority voting reveals that the building materials, road pavement, building age, building roof reflectance and building height have the highest contribution whereas LULC has the lowest contribution to the priority voting ensemble model (Table 9).

Conclusion

The mapping of UHI susceptibility in cities of tropical environment is a challenging task due to changing urban environment and climatic persistence. For the city-scale modeling, the GIS-based statistical modeling approaches for UHI susceptibility were performed on KMA as a tropical city. The study comprises three protocols including preparation of UHI inventory by using e-cognition approach, modeling of UHI susceptibility through four machine learning statistical techniques and validation of model derived results using different statistical measures. The inventory map of UHI with a total number of 350 vectorized LST pixel patches extracted from Landsat data to the distribution of UHI pattern of the urban surface. The total RMSE of all thematic layers are acceptable and efficient for subsequent analysis, which is qualified for UHI susceptibility mapping. A total of nine UHI driving factors have been used, including socio-economic (population distribution and LULC), structural (building materials, building age, building height and building association) and radiative (building roof, building roof reflectance and road pavements) were obtained from the SRS data and other secondary databases. These causative drivers were selected carefully on the basis of importance, availability and scale attributes. Moreover, robust a statistical approach needs to be implemented the causative drivers on certain properties with a conditional independence. Hence, the selected drivers were used as input for accurate modeling of UHI susceptibility on a city-scale, which reduced the dependency on conditional circumstances. The nature of UHI pattern is very sensitive and diverse, as results, the susceptibility modeling can only be achieved by employing the suitable statistical models. Hence, the selections of actual UHI zones and appropriate statistical models for UHI mapping are quite difficult and have to influence the final UHI susceptibility mapping. However, high capable four statistical models like AHP, KLR, SVM and SMCE are integrated with GIS and SRS data to map the UHI susceptibility in the study. The present analysis adopted the GIS-based statistical approach to improve the overall performance of UHI modeling. The KLR and SMCE is the state-of-art machine learning approaches, these techniques could not be applied in case of UHI susceptibility mapping in available conventional methods. The model based study proved that SMCE has the more reliable prediction ability as compared to KLR, AHP and SVM models. Hence, study inferred that the SMCE has to be considered as a qualifying and promising approach that could be used for UHI modeling. In order to ensure the overall performance of the SMCE and KLR approaches, these techniques need to be tested in other geographic to map the UHI effects. These UHI susceptibility approaches could be useful for urban planners and decision makers to UHI mitigate with a suitable prophylactic assessments and minimization procedures.

References

Agarwal AK, Dwivedi S, Ghanshyam A (2018) Summer heat: making a consistent health impact. Indian J Occup Environ Med 22:57–58. https://doi.org/10.4103/ijoem.IJOEM_94_17

Akbari H, Konopacki S (2005) Calculating energy-saving potentials of heat-island reduction strategies. Energy Policy 33(6):721–756

Akbari H, Bretz S, Kurn DM, Hanford J (1997) Peak power and cooling energy savings of high-albedo roofs. Energy Build 25(2):117–126

Akbari H, Konopacki S, Pomerantz M (1999) Cooling energy savings potential of reflective roofs for residential and commercial buildings in the United States. Energy 24(5):391–407

Akbari H, Rose LS, Taha H (2003) Analyzing the land cover of an urban environment using high-resolution orthophotos. Landsc Urban Plan 63(1):1–14

Ali-Toudert F, Mayer H (2006) Numerical study on the effects of aspect ratio and orientation of an urban street canyon on outdoor thermal comfort in hot and dry climate. Build Environ 41(2):94–108

Arnfield AJ (2003) Two decades of urban climate research: a review of turbulence, exchanges of energy and water, and the urban heat island. Int J Climatol 23(1):1–26

Arsanjani JJ, Helbich M, Kainz W, Boloorani AD (2013) Integration of logistic regression, Markov chain and cellular automata models to simulate urban expansion. Int J Appl Earth Obs Geoinf 21:265–275

Berdahl P, Bretz S (1994) Spectral solar reflectance of various roof materials. In: Cool building and paving materials workshop

Bhaskaran S, Paramananda S, Ramnarayan M (2010) Per-pixel and object-oriented classification methods for mapping urban features using Ikonos satellite data. Appl Geogr 30(4):650–665

Bhatta B (2009) Analysis of urban growth pattern using remote sensing and GIS: a case study of Kolkata, India. Int J Remote Sens 30(18):4733–4746

Bisai D, Chatterjee S, Khan A (2014) Detection of recognizing events in lower atmospheric temperature time series (1941-2010) of Midnapore Weather Observatory, West Bengal, India. J Environ Earth Sci 4(3):61–66

Bishop YMM, Fienberg SE, Holland PW (1975) Discrete multivariate analysis theory and practice. MIT Press, Carnbridge, Massachusetts, p 557

BMTPC (1997) Vulnerability atlas of India: earthquake, windstorm and flood hazard maps and damaged risk to housing, ministry of housing and urban poverty alleviation, Government of India, First Revision

Bottyán Z, Unger J (2003) A multiple linear statistical model for estimating the mean maximum urban heat island. Theoret Appl Climatol 75(3–4):233–243

Bretz SE, Akbari H (1997) Long-term performance of high-albedo roof coatings. Energy Build 25(2):159–167

Bretz S, Akbari H, Rosenfeld A (1998) Practical issues for using solar-reflective materials to mitigate urban heat islands. Atmos Environ 32(1):95–101

Cawley GC, Talbot NL (2008) Efficient approximate leave-one-out cross-validation for kernel logistic regression. Mach Learn 71(2–3):243–264

Chang CI, Du Q, Sun TL, Althouse ML (1999) A joint band prioritization and band-decorrelation approach to band selection for hyperspectral image classification. Geosci Remote Sens IEEE Trans On 37(6):2631–2641

Chatterjee S, Bisai D, Khan A (2013) Detection of approximate potential trend turning points in temperature time series (1941–2010) for Asansol weather observation station, West Bengal. Atmos Clim Sci, India, p 2014

Chen CM, Hepner GF (2001) Investigation of imaging spectroscopy for discriminating urban land covers and surface materials (Doctoral dissertation, Department of Geography, University of Utah)

Chen XL, Zhao HM, Li PX, Yin ZY (2006) Remote sensing image-based analysis of the relationship between urban heat island and land use/cover changes. Remote Sens Environ 104(2):133–146

Clark RN, Swayze G, Heidebrecht K, Goetz AF, Green RO (1993) Comparison of methods for calibrating AVIRIS data to ground reflectance. In: 5th annual airborne geoscience workshop. AVIRIS. Jet Propulsion Laboratory, Pasadena, Calif, pp 35–36

Congalton RG (1991) A review of assessing the accuracy of classifications of remotely sensed data. Remote Sens Environ 37(1):35–46

Congalton RG, Green K (1999) Assessing the accuracy of remotely sensed data: principles and applications. Lewis Publishers, Boca Raton, Fla

Congalton R, Mead RA (1983) A quantitative method to test for consistency and correctness in photointerpretation. Photogra Eng Remote Sens 49(1):69–74

DeFries R (2008) Terrestrial vegetation in the coupled human-earth system: contributions of remote sensing. Annu Rev Environ Resour 33:369–390

DeFries R, Pandey D (2010) Urbanization, the energy ladder and forest transitions in India's emerging economy. Land Use Policy 27(2):130–138

Fujibe F (2009) Detection of urban warming in recent temperature trends in Japan. Int J Climatol 29(12):1811–1822

Geneletti D, Gorte BGH (2003) A method for object-oriented land cover classification combining Landsat TM data and aerial photographs. Int J Remote Sens 24(6):1273–1286

Gibson PJ, Power CH, Goldin SE, Rudahl KT (2000) Introductory remote sensing: digital image processing and applications, vol 11. Routledge, London, UK

Gopal S, Tang X, Phillips N, Nomack M, Pasquarella V, Pitts J (2016) Characterizing urban landscapes using fuzzy sets. Comput Environ Urban Syst 57:212–223

Green RO, Eastwood ML, Sarture CM, Chrien TG, Aronsson M, Chippendale BJ, … Olah MR (1998) Imaging spectroscopy and the airborne visible/infrared imaging spectrometer (AVIRIS). Remote Sens Environ 65(3):227–248

Grimm NB, Faeth SH, Golubiewski NE, Redman CL, Wu J, Bai X, Briggs JM (2008) Global change and the ecology of cities. Science 319(5864):756–760

Heatwave and Health: Guidance on warning system development (2015) WHO and WMO Geneva. https://www.who.int/globalchange/publications/WMO_WHO_Heat_Health_Guidance_2015.pdf. Assessed on 17 July 19

Heaviside C, Vardoulakis S, Cai X-M (2016) Attribution of mortality to the urban heat island during heatwaves in the West Midlands. UK Environ Health 15(Suppl 1):27. https://doi.org/10.1186/s12940-016-0100-9

Heiden U, Roessner S, Segl K, Kaufmann H (2001) Analysis of spectral signatures of urban surfaces for their identification using hyperspectral HyMap data. In: Remote sensing and data fusion over urban areas, IEEE/ISPRS joint workshop 2001, IEEE, pp 173–177

Hepner GF, Houshmand B, Kulikov I, Bryant N (1998) Investigation of the integration of AVIRIS and IFSAR for urban analysis. Photogram Eng Remote Sens 64(8):813–820

Herold M, Gardner ME, Roberts DA (2003) Spectral resolution requirements for mapping urban areas. Geosci Remote Sens IEEE Trans On 41(9):1907–1919

Hong H, Pradhan B, Xu C, Bui DT (2015) Spatial prediction of landslide hazard at the Yihuang area (China) using two-class kernel logistic regression, alternating decision tree and support vector machines. CATENA 133:266–281

Imhoff ML, Zhang P, Wolfe RE, Bounoua L (2010) Remote sensing of the urban heat island effect across biomes in the continental USA. Remote Sens Environ 114(3):504–513

Jensen JR (1986) Introductory digital image processing: a remote sensing perspective. University of South Carolina, Columbus

Jimenez LO, Landgrebe DA (1999) Hyperspectral data analysis and supervised feature reduction via projection pursuit. Geosci Remote Sens IEEE Trans On 37(6):2653–2667

Jones PD, Lister DH, Li Q (2008) Urbanization effects in large-scale temperature records, with an emphasis on China. J Geophys Res Atmos 113(D16)

Kalnay E, Cai M (2003) Impact of urbanization and land-use change on climate. Nature 423(6939):528–531

Khan A, Chatterjee S (2016) Numerical simulation of urban heat island intensity under urban–suburban surface and reference site in Kolkata, India. Model Earth Syst Environ 2(2):1–11

Khan A, Chatterjee S, Bisai D, Barman NK (2014) Analysis of change point in surface temperature time series using cumulative sum chart and bootstrapping for Asansol weather observation station, West Bengal, India. Am J Clim Chang 3(1):83

Konopacki SJ, Akbari H (2001) Measured energy savings and demand reduction from a reflective roof membrane on a large retail store in Austin. Lawrence Berkeley National Laboratory

Kuncheva LI (2004) Combining pattern classifiers: methods and algorithms. Wiley, New Jersey

Landgrebe D (2000) On the relationship between class definition precision and classification accuracy in hyperspectral analysis. In: Geoscience and remote sensing symposium, 2000. Proceedings. IGARSS 2000. IEEE 2000 international, IEEE, vol 1, pp 147–149

Leal Filho W, Echevarria Icaza L, Emanche VO, Quasem Al-Amin A (2017) An evidence-based review of impacts, strategies and tools to mitigate urban heat islands. Int J Environ Res Public Health 14. https://doi.org/10.3390/ijerph14121600

Levinson R, Akbari H, Reilly JC (2007) Cooler tile-roofed buildings with near-infrared-reflective non-white coatings. Build Environ 42(7):2591–2605

Luber G, McGeehin M (2008) Climate change and extreme heat events. Am J Prev Med Theme Issue Clim Chang Health Public 35:429–435. https://doi.org/10.1016/j.amepre.2008.08.021

Malczewski J (1999) GIS and multicriteria decision analysis. Wiley, London

Mavrogianni A, Davies M, Batty M, Belcher S, Bohnenstengel S, Carruthers D, Chalabi Z, Croxford B, Demanuele C, Evans S, Giridharan R, Hacker J, Hamilton I, Hogg C, Hunt J, Kolokotroni M, Martin C, Milner J, Rajapaksha I, Ridley I, Steadman J, Stocker J, Wilkinson P, Ye Z (2011) The comfort, energy and health implications of London's urban heat island. Build Serv Eng Res Technol 32:35–52. https://doi.org/10.1177/0143624410394530

Mirzaei PA (2015) Recent challenges in modeling of urban heat island. Sustain Cities Soc 19:200–206

Morano P, Locurcio M, Tajani F, Guarini MR (2014) Urban redevelopment: a multi-criteria valuation model optimized through the fuzzy logic. In: Computational science and its applications–ICCSA 2014, pp 161–175. Springer, Cham

Nandy DR (2007) Need for seismic microzonation of Kolkata megacity. In: Proceedings of workshop on microzonation, Indian Institute of science, Bangalore, India, vol 2627

Nath SK, Adhikari MD, Devaraj N, Maiti SK (2015) Seismic vulnerability and risk assessment of Kolkata City, India. Nat Hazards Earth Syst Sci 15(6):1103–1121

Oke TR (1982) The energetic basis of the urban heat island. Q J R Meteorol Soc 108(455):1–24

Oke TR (1988) The urban energy balance. Prog Phys Geogr 12(4):471–508

Parker DE (2010) Urban heat island effects on estimates of observed climate change. Wiley Interdiscip Rev Clim Chang 1(1):123–133

Pielke RA, Pitman A, Niyogi D, Mahmood R, McAlpine C, Hossain F, …, Reichstein M (2011). Land use/land cover changes and climate: modeling analysis and observational evidence. Wiley Interdiscip Rev Clim Chang 2(6):828–850

Pourghasemi H, Moradi H, Aghda SF, Gokceoglu C, Pradhan B (2014) GIS-based landslide susceptibility mapping with probabilistic likelihood ratio and spatial multi-criteria evaluation models (North of Tehran, Iran). Arab J Geosci 7:1857–1878

Pourghasemi H, Pradhan B, Gokceoglu C (2012) Application of fuzzy logic and analytical hierarchy process (AHP) to landslide susceptibility mapping at Haraz watershed. Iran Nat Hazards 63:965–996

Price JC (1997) Spectral band selection for visible-near infrared remote sensing: spectral-spatial resolution tradeoffs. Geosci Remote Sens IEEE Trans On 35(5):1277–1285

Price JC (1998) An approach for analysis of reflectance spectra. Remote Sens Environ 64(3):316–330

Qin Z, Karnieli A, Berliner P (2001) A mono-window algorithm for retrieving land surface temperature from Landsat TM data and its application to the Israel-Egypt border region. Int J Remote Sens 22:3719–3746

Rahman R, Saha SK (2008) Remote sensing, spatial multi criteria evaluation (SMCE) and analytical hierarchy process (AHP) in optimal cropping pattern planning for a flood prone area. J Spat Sci 53(2):161–177

Rizwan AM, Dennis LY, Chunho LIU (2008) A review on the generation, determination and mitigation of Urban Heat Island. J Environ Sci 20(1):120–128

Roberts DA, Green RO, Adams JB (1997) Temporal and spatial patterns in vegetation and atmospheric properties from AVIRIS. Remote Sens Environ 62(3):223–240

Saaty TL (1977) A scaling method for priorities in hierarchical structures. J Math Psychol 15(3):234–281

Saaty TL (1980) The analytic hierarchy process: planning, priority setting, resources allocation. McGraw, New York

Saaty TL, Vargas LG (2012) Models, methods, concepts and applications of the analytic hierarchy process (vol 175). Springer, Cham

Shahabi H, Hashim M (2015) Landslide susceptibility mapping using GIS-based statistical models and Remote sensing data in tropical environment. Sci Rep 5(9899). https://doi.org/10.1038/srep09899

Shahmohamadi P, Che-Ani AI, Etessam I, Maulud KNA, Tawil NM (2011) Healthy environment: the need to mitigate urban heat island effects on human health. In: Procedia engineering, 2nd international building control conference vol 20, pp 61–70. https://doi.org/10.1016/j.proeng.2011.11.139

Sharma R, Hooyberghs H, Lauwaet D, De Ridder K (2019) Urban heat island and future climate change—implications for Delhi's heat. J Urban Health 96:235–251. https://doi.org/10.1007/s11524-018-0322-y

Sharma D, Bharat A (2009) Urban heat island effect—causes, impacts, methods of measurement and mitigation options. Inst Town PlanS, India J 6–2:69–77. https://www.researchgate.net/publication/265477727_Urban_Heat_Island_Effect_-_Causes_Impacts_Methods_of_Measurement_and_Mitigation_Options. Accessed 17 July 19

Small C (2002) Multitemporal analysis of urban reflectance. Remote Sens Environ 81(2):427–442

Stone B (2007) Urban and rural temperature trends in proximity to large US cities: 1951–2000. Int J Climatol 27(13):1801–1807

Story M, Congalton RG (1986) Accuracy assessment-A user\'s perspective. Photogramm Eng Remote Sens 52(3):397–399

Synnefa A, Santamouris M, Akbari H (2007) Estimating the effect of using cool coatings on energy loads and thermal comfort in residential buildings in various climatic conditions. Energy Build 39(11):1167–1174

Synnefa A, Santamouris M, Livada I (2006) A study of the thermal performance of reflective coatings for the urban environment. Sol Energy 80(8):968–981

Taha H, Akbari H, Rosenfeld A, Huang J (1988) Residential cooling loads and the urban heat island-the effects of albedo. Build Environ 23(4):271–283

Touchaei AG, Wang Y (2015) Characterizing urban heat island in Montreal (Canada)-Effect of urban morphology. Sustain Cities Soc 19:395–402

Tran H, Uchihama D, Ochi S, Yasuoka Y (2006) Assessment with satellite data of the urban heat island effects in Asian mega cities. Int J Appl Earth Obs Geoinf 8(1):34–48

US EPA, O (2014) Heat Island Impacts [WWW Document]. US EPA. https://www.epa.gov/heat-islands/heat-island-impacts Accessed 17 June 19

Vapnik VN, Vapnik V (1998) Statistical learning theory, vol 1. Wiley, New York

Voogt JA, Oke TR (2003) Thermal remote sensing of urban climates. Remote Sens Environ 86(3):370–384

Wan Z (2008) New refinements and validation of the MODIS land-surface temperature/emissivity products. Remote Sens Environ 112(1):59–74

Wang Y, Akbari H (2014) Effect of sky view factor on outdoor temperature and comfort in Montreal. Environ Eng Sci 31(6):272–287

Witten IH, Frank E, Mark AH (2011) Data mining: practical machine learning tools and techniques, 3rd edn. Morgan Kaufmann, Burlington, USA

Xu L, Li J, Brenning A (2014) A comparative study of different classification techniques for marine oil spill identification using RADARSAT-1 imagery. Remote Sens Environ 141:14–23

Zha Y, Gao J, Ni S (2003) Use of normalized difference built-up index in automatically mapping urban areas from TM imagery. Int J Remote Sens 24(3):583–594

Zhang Q, Wang J, Peng X, Gong P, Shi P (2002) Urban built-up land change detection with road density and spectral information from multi-temporal Landsat TM data. Int J Remote Sens 23(15):3057–3078
Zhou L, Dickinson RE, Tian Y, Fang J, Li Q, Kaufmann RK, … Myneni RB (2004) Evidence for a significant urbanization effect on climate in China. Proc Natl Acad Sci U S A 101(26):9540–9544

GIS Hazard Assessments as the First Step to Climate Change Adaptation

Nelson Rangel-Buitrago, Adriana Gracia C., Giorgio Anfuso and Jarbas Bonetti

Abstract A scientific consensus exists regarding the significant impacts of global climate change over coastal zones. These effects include sea level rise, variability in the patterns of rainfall and runoff and changes in frequency, intensity, and duration of extreme wave events. Natural disasters have substantial adverse impacts on human activities and structures as well as on social and political concerns of human life and ecological and conservation aspects. In this work, a detailed methodology for the evaluation and characterization of coastal hazard associated with extreme wave events was developed under a GIS environment and tested and applied in Cartagena city, located in the Caribbean coast of Colombia. The analysis was prepared utilizing a semi-quantitative approximation method, involving variables representative of intrinsic coastal zone properties and extreme wave related hazards using GIS analytical tools. Results obtained reveal that there are several areas affected by extremely high erosion, inundation, and flooding. Hazard maps generated with this methodology are the first step in any Climate Change Adaptation strategy to be used. In the same way is a guideline contributing to the determination of causes, processes, and consequences derived from hazards magnified due climate change.

N. Rangel-Buitrago (✉)
Departamento de Física, Facultad de Ciencias Básicas, Universidad del Atlántico, Antigua vía Puerto Colombia, Barranquilla, Atlántico, Colombia
e-mail: nelsonrangel@mail.uniatlantico.edu.co

N. Rangel-Buitrago · A. Gracia C.
Departamento de Biología, Facultad de Ciencias Básicas, Universidad del Atlántico, Antigua vía Puerto Colombia, Barranquilla, Atlántico, Colombia

G. Anfuso
Departamento de Ciencias de la Tierra, Facultad de Ciencias del Mar y Ambientales, Universidad de Cádiz, Polígono Río San Pedro s/n, 11510 Puerto Real, Cádiz, España

J. Bonetti
Coastal Oceanography Laboratory, Federal University of Santa Catarina, Florianopolis, SC 88040-900, Brazil

W. Leal Filho et al. (eds.), *Climate Change, Hazards and Adaptation Options*, Climate Change Management, https://doi.org/10.1007/978-3-030-37425-9_6

Introduction

Climatic change-related hazards in coastal areas became an important issue in past years and nowadays threaten many human settlements and activities worldwide. Coastal hazards are linked to erosion processes and flooding associated with sea level rise and the increased strength of extreme wave events. As suggested by Rangel-Buitrago and Anfuso (2013) and Pilkey and Cooper (2014), sea level, wave climate, occurrence and distribution of storms are significant issues in the occurrence and amount of coastal erosion, the deterioration and (at some places) complete loss of ecosystems. In the currents scenarios of rising sea levels and increasing wave heights, the coastline will suffer vast impacts in terms of erosion and flooding especially with respect to low-lying regions that may partly or entirely disappear (Pilkey and Cooper 2014; Rangel-Buitrago et al. 2018a).

The evaluation of coastal hazards is a key issue in the geosciences field and a huge literature detailing system responses to perturbation exists (Rangel-Buitrago and Neal 2018). Methodologies used to assess coastal hazards can be classified according to different characteristics but the establishment of a succinct classification results many times in a hard task where limits between classes are not exact (Di Paola et al. 2011). This intrinsic hazard is determined using different information such as physical and ecological coastal features, human occupation, present and future shoreline trends, etc. (Gornitz 1991, McLaughlin and Cooper 2010). First studies employed single approach methods (i.e. UNEP Methodology, Carter et al. 1994). They progressively evolved and were superseded by more recent techniques.

The improved consideration of physical and non-physical factors, as well as of the associated uncertainties, has given rise to more consistent methods (i.e. USGS-CVI, Gornitz et al. 1994; SURVAS, Benassai et al. 2009) some of which are very recent and have to be validated. Their associated maps have been obtained for several coastal sectors around the world by means of Geographical Information Systems (GIS), computer-assisted multivariate analysis and numerical models (Cooper and McLaughlin 1998). Specifically, recent works have been focused on the determination of coastal hazards related to the specific impacts of extreme wave events under a climate change context (Li and Li 2011; Raji et al. 2013).

Mapping is usually the basis for zoning, the standard management approach for appropriate land use planning, especially in susceptible-prone areas. Ideally, coastal hazard mapping should be the first step in climate change adaptation schemes, because this is the broadest approach to define those coastal areas which are at risk.

This work deals with a methodological approach to the hazard determination for coasts as a first step to climate change adaptation by the use of matrixes concerning physical parameters under a GIS environment.

The approach is based on the selection and evaluation of two types of variables:

- Forcing variables contributing to coastline erosion due extreme wave events.
- Dynamic variables that determine the coast resilience (Susceptibility).

These are combined into a separate index, the Hazard Index (combining forcing and susceptibility) a single numerical measure of the hazard for a given area.

Study Area

Cartagena is a large seaport at the north coast of Colombia. Founded in 1533, it was the major centre of early Spanish settlements in South America (Fig. 1). The coastal area is about 30 km long and the metropolitan area has a population of 1,240,000 inhabitants, which are particularly concentrated in the city of Cartagena. In fact the Cartagena Municipality occupies about 21% of the Caribbean territory and it is the fifth largest urban area in Colombia. It represents an important centre of economic activities in the region and a popular tourist destination.

Coastal morphology is characterized by a sediment deficit resulting in erosion problems. In detail, the erosion processes were induced in 1936 by the construction of the jetty of Bocas de Cenizas. This structure had favoured the transport of the Magdalena River supply into a submarine canyon, located in front of Barranquilla (north of the investigated area). The structure works also as a barrier to the natural sediment transport approaching from the northeast. A few decades ago the evident coastal erosion and interests to preserve the historical and cultural patrimony in the study area have motivated the local decision-makers to implement different protection measures.

Erosive events are also related to the impact of hurricanes and cold fronts (Rangel-Buitrago et al. 2015 and 2018b). Hurricanes, usually originated in the Caribbean area from June to November may affect the Colombia coast with strong winds, heavy rains and storm waves. Hurricane Joan reached Cartagena region in October 1988 producing great damages. Cold fronts, occurring during January, February and March, cause strong swell waves which impact may be increased by trade winds blowing from ENE. They usually hit the coast for about 48 h and have an average occurrence of six events per year (Ortiz et al. 2013).

Coastal occupation shows important tourist activities linked to both cultural resources and beach attractiveness. Good weather conditions make tourist activities very attractive during the entire year, converting Cartagena in the most popular destination along the Caribbean littoral of Colombia—147,280 of national and international arrivals were recorded in the last year.

Methods

This work was developed during July 2017 and the method proposed for the assessment of coast hazard, has been based on the combination of two components or sub-indices within a GIS environment:

Fig. 1 Location of Cartagena city, Colombia

(i) the Forcing variables contributing to extreme waves-induced erosion (Table 1)
(ii) the Susceptibility sub-index which described the coast resilience and susceptibility to erosion according to its specific morphological characteristics, within two main coast types (sandy and rocky, Tables 2 and 3).

Much of used variables ("an") for the determination of used sub-indices have been chosen according to previous studies focused on chronic coastal erosion or sea

Table 1 Forcing variables used

Parameter	Very low (1)	Low (2)	Medium (3)	High (4)	Very high (5)
Significant wave height at a specific coastal sector (% of initial H_s)	Less than 20%	20–40%	40–60%	60–80%	80–100%
Storm Surge at a specific coastal sector	Less than 20%	20–40%	40–60%	60–80%	80–100%
Degree of littoral exposition to wave fronts	10–45° Oblique	x	0–10° Sub-parallel	x	0° Parallel
Tidal range	Macrotidal	x	Mesotidal	x	Microtidal

Table 2 Susceptibility variables—sandy coast

Parameter	Very low (1)	Low (2)	Medium (3)	High (4)	Very high (5)
Dune height	≥6	≥3	≥2	≥1	<1
Percentage of washovers	0%	≤5%	≤25%	≤50%	≥50%
Dry beach	5 times ICZ	4 times ICZ	3 times ICZ	2 times ICZ	Equal to ICZ
Beach slope Morphodynamic state	Dissipative ($\tan\beta \leq 0.02$)	x	Intermediate ($0.02 < \tan\beta < 0.08$)	x	Reflective ($\tan\beta \geq 0.08$)
K Index	Extreme (K > 1)	Maximum (K = 0.51÷1)	Average (K = 0.11÷0.5)	Minimum (K = 0.0001÷0.1)	No structures (K = 0)

Table 3 Susceptibility variables—rocky coast

Parameter	Very low (1)	Low (2)	Medium (3)	High (4)	Very high (5)
Type	Cliff with horizontal shore platform	x	Cliff with sloping shore platform	x	Plunging cliff
Lithology	Granitic rocks, resistant metamorphic	limestone	Flysch, shale, tertiary sedimentary rocks	Quaternary deposits	Volcanic ejecta
Structures	Virtual absence of discontinuities cracks, joints, faults	x	Some evidence of discontinuities, cracks, faults	x	High density of discontinuities, cracks, faults
Slope	<30°	31°÷40°	41°÷50°	51°÷60°	>60°
Cliff edge	5 times ICZ	4 times ICZ	3 times ICZ	2 times ICZ	equal to ICZ
Weathering	Un weathered	Slightly un weathered	Moderately weathered	Highly weathered	Decomposed

level rise related hazards (Gornitz 1991, Gornitz et al. 1997; Cooper and McLaughlin 1998; McLaughlin et al. 2002; Coelho et al. 2009; McLaughlin and Cooper 2010, Özyurt and Ergin 2009, 2010; among others) and/or storm impacts (Burzel et al. 2010; Ceia et al. 2010; Carrasco et al. 2012; Di Paola et al. 2011; Li and Li 2011; Maio et al. 2012; Raji et al. 2013).

Gornitz (1991), Gornitz et al. (1997) and Hammer-Klose and Thieler (2001), used variables have been classified on a 1–5 scale; 1 indicated a low contribution to specific key variable for the studied sector, while 5 indicated a high contribution. Classes have been set on a numerical base and an ordinal scale approach was adopted in case of a semi-quantitative variable difficult to quantify (Cooper and McLaughlin 1998). Hence, different variables have been calculated for a coastal area segmented in a number of sectors in which dimensions were defined according to data availability and coastal uniformity. In this sense, the methodology can be applied at different spatial scales; in the presented study case investigated coastal sector presented a length of 30 km and were further divided into 500 m wide sectors.

Used variables were combined under a GIS environment into the Forcing, Susceptibility and Hazard indexes. The scores of each variable have been summed with the scope of obtaining an absolute value for each sub-index.

According to Gornitz (1991), Gornitz et al. (1997) and Hammer-Klose and Thieler (2001), used variables have been classified on a 1–5 scale; 1 indicated a low contribution to specific key variable for the studied sector, while 5 indicated a high contribution. Classes have been set on a numerical base and an ordinal scale approach was adopted in case of a semi-quantitative variable difficult to quantify (Cooper and McLaughlin 1998).

Dealing with the used variables, they were combined under a GIS environment into the Forcing, Susceptibility and Hazard indexes. The scores of each variable have been summed with the scope of obtaining an absolute value for each sub-index according to the follow equations:

$$\boldsymbol{Coastal Forcing Index} = \frac{\sum Cf\,an - n\,Cf}{nCf * 4} * 100 \tag{1}$$

$$\boldsymbol{Susceptibility Index} = \frac{\sum S\,an - n\,S}{n\,S * 4} * 100 \tag{2}$$

$$\boldsymbol{Hazard} = \frac{(Coastal\ Forcing\ Index * n\,Cf) + (Susceptibility\ Index * n\,Sb)}{n\,Cf + n\,S} \tag{3}$$

Being an each one of the variables used in each index (Coastal Forcing: Cf, Susceptibility: S) and n the number of variables in each index.

Index Distribution

The Coastal Forcing (CFI), Susceptibility (CSI) and Hazard Indexes are presented in Fig. 2 showing values ranging from high to very high. CFI essentially showed medium to high values respectively for the 38% and 21% of the littoral. Such values were recorded at Crespo, Marbella and Bocagrande sectors. The rest of the coast (41%) presented low CFI values.

Obtained data suggested the dominance of medium to very high CFI values that were related to the conjunction of several factors determining the way in which wave energy is distributed along the coastline:

(i) the absence of obstacles in front of them.
(ii) the degree of littoral exposition to waves.
(iii) the local bathymetric characteristics.

Specifically, coastal orientation determines the level of exposure to the main approaching wave direction and hence the prevalence of longshore or shore normal transport important in beach and dune erosion, meanwhile bathymetric conditions determine wave shoaling and dissipation. The rectilinear orientation of the coastline and the homogeneous characteristics of the nearshore area allow the arrival of high energetic waves with associated, elevated storm surge values.

Further, the higher values of CFI are linked to the greater exposition of this coast; it is parallel to the incoming wave fronts. Low CFI values were observed in sheltered areas, at the different bays located at the northern part of Cartagena coast.

The Cartagena area was mainly classified (62%) inside the medium class of CSI. Results showed how this was the case of Bocagrande—La Boquilla sector, a series of dissipative urban beaches with medium to high erosion rates (Rangel-Buitrago et al. 2015 and 2018) which score is improved because of the high degree of armouring (Rangel-Buitrago et al. 2013 and Stancheva et al. 2011).

High (28%) and very high (5%) CSI values were recorded at Tierrabomba and the Northern part of Cartagena area, where small cliffs of soft diapiric Tertiary materials, composed by faulted and weathered sandstones and clays, are directly exposed to wave action and experience important retreat rates (Rangel-Buitrago et al. 2015 and 2018b). Low CSI values do not exceed the 6% of the area.

Hazard is the probability of occurrence of a potential damaging phenomenon (in this case, storm event) within a specific period of time and within a given area. The HI shows the potential of different sectors to experience significative damages associated with the effect of storm wave events (Fig. 2 and Table 4) and it is the result of the crossing of Forcing and Susceptibility Indexes.

In Cartagena area, the CHI presented from low to high values. Medium values were clearly dominant, accounting for the 60% of the coastline—essentially the Crespo—Marbella sector. High values accounted for the 35% of the coastline, occurring in Bocagrande, El Laguito and Tierrabomba sectors. Low hazard values covered just over the 5% of the coastline in areas essentially located north of Cartagena.

The areas with high and very high CHI values presented important erosion rates in past decades. An elevated percentage of areas with high and very high values of CHI,

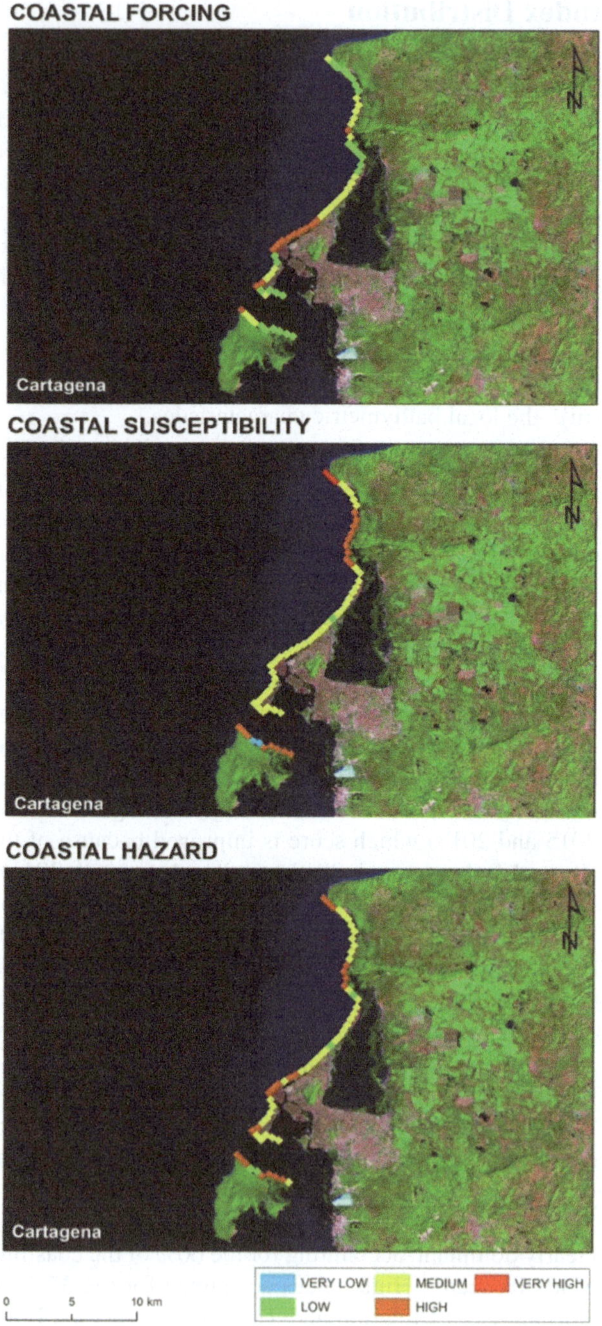

Fig. 2 Coastal forcing, susceptibility and hazard index calculated for Cartagena city area

Table 4 Distribution of forcing, susceptibility and hazard index calculated for Cartagena

Indexes			
Class	Forcing (%)	Susceptibility (%)	Hazard (%)
Very low	0	4	0
Low	41.0	1.6	6.56
Medium	37.7	62.3	59.02
High	21.3	27.9	34.42
Very high	0	4.2	0

presented erosion rates that exceeded 1.5 m yr-1, as observed by Rangel-Buitrago et al. (2015, 2018b). In this sense, such good correspondence confirms the validity of the applied methodology, e.g. the higher the Hazard Index the greater the recorded erosion, as observed in similar circumstances in Northern Ireland by McLaughlin et al. (2002).

Management Considerations

The use and implementation of different kinds of information into a GIS project, e.g. wave data propagation, geomorphologic units, etc., and the generation of new one by means of spatial analysis, allowed the creation of an initial data base containing the hazard, vulnerability and risk assessment for Cartagena areas.

The obtained information can be easily used in its entirety or partially, i.e. at different levels and scales, by local planning staff or other kind of end users according to their specific goals, for example environmental programs, tourism and coastal zone management plans, etc.

Further, according with Nunes et al (2009), the use of GIS tools allow the collection of a great amount of different data and, one of its main advantages, is the possibility of constantly update the initial information with new sets of data. In this sense variables as significant wave height, land uses, percentage of urbanised areas, population density, among others, can be easily integrated, spatially analysed and updated into a GIS environment in order to provide an Integrated Coastal Zone Management plan (Fischer and Arredondo 1999; Anfuso and Gracia 2005).

The used method gives an "instantaneous picture" analysis of coastal risk to storm wave related hazard for a given area. In fact, great advantages of the proposed method are the fact that it is based on recent coastal evolution and erosion processes effects as well as its flexibility and adaptability to the specific features of each area to be investigated. In this sense it represents a basic and appropriate evaluation tool previous to the elaboration of any kind of coastal management plan.

The present work deals with a topic of great importance and actuality. Natural disasters have huge negative impacts on human activities and structures as well as on social and political concerns of human life and on ecological and conservation aspects (Li and Li 2011). Further, even if disasters are due to natural processes, the associated impacts are many times increased by human interventions/actuations such as an inexistent or inappropriate coastal planning (Komar and Allan 2008; Jones and Phillips 2011).

The methodological approach development in this work has been designed with the objective of being a scientifically complete planning tool, easy to use, that takes into consideration the higher amount of factors involved into the coastal hazard. The proposed methodology consists in a relatively simple series of indexes applicable to different coastal areas, becoming a general and not site-specific method.

One of the points to be taken into account in the application of the indices is the zoning, e.g. the method used to define coastal sectors. For this work, in order to have an optimum zoning scheme, each one of the investigated variables was calculated along a constant and uniform coastal stretch of 500 × 500 m.

Many considered parameters and their division in the five-classes considered intervals, were not expressed by means of predefined absolute values, which can greatly range from one place to another, but using specific, calculated absolute values referred to used parameters at investigated places.

In most other cases, in order to qualitatively or quantitatively describe some specific parameter and to slow down subjectivity in defining its division into five classes, the parameter definition and variance were established according to existing previous studies carried out on considered specific subjects.

Of special attention in this kind of assessments is the scale of work due to the spatial resolution of the used zoning because obtained results will be strongly linked to the spatial scale. Given that hazard, vulnerability and risk assessments have to be evaluated for coastal management purposes, spatial resolution of the segmentation must be in agreement with the level at which it is projected to support stakeholders and management decision-making (i.e. local, regional and national).

Since the data can be easily collected the application of the methodology should not introduce significant limitations in the hazard assessment. Maybe additional enhancements in the development of the indices can be made, for example by including more variables in the assessment. An addition of different variables must avoid redundancy and ambiguity. In the same sense, increasing in the number of variables will require the increasing of the complexity of the index, so in any case a balance should be found among applicability, scientific validity and facility of use.

Conclusions

The present work deals with a topic of great importance and actuality. Natural disasters have huge negative impacts on human activities and structures as well as on social and political concerns of human life and on ecological and conservation aspects. Further, even if disasters are due to natural processes, the associated impacts are many times increased by human interventions/actuations such as an inexistent or inappropriate coastal planning.

Natural hazards and disasters produce important economic losses. In the United States of America, during the last years, flooding, severe storms and hurricanes have been responsible for economic loses estimated in more than $1.14 billion. In 2013, a total of 19 disasters affected around 552,000 peoples in Latin America and the Caribbean, and Philippines was strike by a huge typhoon which produced the displacement of 4.1 millions of people, 6069 fatalities and the damage and destruction of 1.1 million houses.

Exacerbation of existing coastal hazards by climate change will create new wider-scale risks not previously experienced in coastal areas. This has implications for making decisions today, especially for those activities or assets that have long lifetimes in coastal areas likely to be affected by climate change-related impacts.

In addition, responses to the more intense impacts can affect wider community values at the coast, such as public amenity, recreation, coastal habitats, and other environmental aspects. Adverse effects can thus arise from both the hazards themselves and the human responses to them. The capacity to adapt will be different for different groups in the community, according to their vulnerability.

Data presented in this work underline the importance and interest of further research on such topics in order to fully understand the natural process and evaluate coastal hazard to mentioned events, in this case, extreme wave events.

Since the goal of this study was to develop a general methodological approach easily and objectively applicable at different areas, the evaluation of the Coast hazard was based on an objective and quantitative methodology to remove uncertainty and subjectivity.

References

Anfuso G, Gracia FJ (2005) Morphodynamic characteristics and short-term evolution of a coastal sector in SW Spain: implications for coastal erosion management. J Coastal Res 21(6):1139–1153

Benassai G, Chirico F, Corsini S (2009) Una metodologia per la definizione del rischio da inondazione costiera. Studi Costieri 16:51–72

Burzel A, Dassanayake D, Naulin M, Kortenhaus A, Oumeraci H, Wahl T, Mudersbach C, Jensen J, Gönnert G, Sossidi K, Ujeyl G, Pasche E (2010) Integrated flood risk analysis for extreme storm surges (XTREMRISK). COPRI, Shangai

Carrasco AR, Ferreira Ó, Matias A, Pacheco A, Freire P (2012) Short-term sediment transport at a backbarrier beach. J Coast Res 27(6):1076–1084

Carter TR, Parry ML, Nishioka S, Harasawa H (1994) Technical guidelines for assessing climate change impacts and adaptation. University College and Centre for Global Environmental Research, Tskuba

Ceia F, Patrício J, Marquez J, Alveirinho Dias J (2010) Coastal vulnerability in barrier islands: the high-risk areas of the Ria Formosa (Portugal) system. Ocean Coast Manage 53:478–486

Coelho CR, Silva F, Veloso-Gomes F, Taveira-Pinto F (2009) A vulnerability analysis approach for the Portuguese west coast. In: Popov V, Brebbia CA (eds) Risk analysis V: simulation and Hazard Mitigation. Institute of Technology, Wessex

Cooper JAG, McLaughlin S (1998) Contemporary multidisciplinary approaches to coastal classification and environmental risk analysis. J Coast Res 14:512–524

Di Paola G, Iglesias J, Rodriguez G, Benassai G, Aucelli P, Pappone G (2011) Estimating coastal vulnerability in a meso-tidal beach by means of quantitative and semi-quantitative methodologies. J Coast Res SI 61:303–308

Fischer DW, Arredondo MC (1999) Municipal coastal hazard planning: Los Angeles and Orange County city responses, California. J Coast Res 15(4):974–984

Gornitz V (1991) Global coastal hazards from future sea level rise. Palaeogeogr Palaeocl 89:379–398

Gornitz VM, Daniels RC, White TW, Birdwell KR (1994) The development of a coastal risk assessment database: vulnerability to sea-level rise in the U.S. Southeast. J Coast Res 12:327–338

Gornitz VM, Beaty TW, Daniels RC (1997) A coastal hazards data base for the U.S. West coast. Oak Ridge National Laboratory, Oak Ridge

Hammer-Klose ES, Thieler ER (2001) Coastal vulnerability to sea-level rise, a preliminary database for the U.S. Atlantic, Pacific, and Gulf of Mexico coasts. U.S. Geological Survey, Charleston

Jones A, Phillips M (2011) Disappearing destinations. CABI, London

Komar PD, Allan JC (2008) Increasing hurricane-generated wave heights along the U.S. East Coast and their climate controls. J Coastal Res 24(2):479–488

Li K, Li GS (2011) Vulnerability assessment of storm surges in the coastal area of Guangdong Province. Nat Hazards Earth Syst Sci 11:2003–2011

Maio CV, Gontz AM, Tenenbaum DE, Berkland EP (2012) Coastal hazard vulnerability assessment of sensitive historical sites on Rainsford Island, Boston Harbor, Massachusetts. J Coastal Res 28(1A):20–33

McLaughlin S, Cooper JAG (2010) A multi-scale coastal vulnerability index: A tool for coastal managers? Environ Haz 9(3):233–248

McLaughlin S, McKenna J, Cooper JAG (2002) Socio-economic data in coastal vulnerability indices: constraints and opportunities. J Coast Res 36:487–497

Nunes MA, Ferreira O, Schaefer M, Clifton J, Baily B, Moura D, Loureiro C (2009) Hazard assessment in rock cliffs at Central Algarve (Portugal): a tool for coastal management. Ocean Coast Manage 52:506–515

Ortiz JC, Otero LJ, Restrepo JC, Ruiz J, Cadena M (2013) Cold fronts in the Colombian Caribbean Sea and their relationship to extreme wave events. Nat Haz 13:2797–2804

Özyurt G, Ergin A (2009) Application of sea level rise vulnerability assessment model to selected coastal areas of Turkey. J Coast Res 51:248–251

Özyurt G, Ergin A (2010) Improving coastal vulnerability assessments to sea-level rise: a new indicator-based methodology for decision makers. J Coast Res 26(2):265–273

Pilkey O, Cooper A (2014) The last beach. Duke University Press, Durham

Raji O, Niazi S, Snoussi M, Dezileau L, Khouakhi A (2013) Vulnerability assessment of a lagoon to sea level rise and storm events: Nador lagoon (NE Morocco). J Coast Res 65:802–807

Rangel-Buitrago N, Anfuso G (2013) Winter wave climate, storms and regional cycles: the SW Spanish Atlantic coast. Int J Climatol 33:2142–2156

Rangel-Buitrago N, Anfuso G, Williams AT (2015) Coastal erosion along the Caribbean coast of Colombia: magnitudes, causes and management. Ocean Coast Manage 114:129–144

Rangel-Buitrago N, Anfuso G, Correa I, Ergin A, Williams AT (2013) Assessing and managing scenery of the Caribbean coast of Colombia. Tour Manage 35:41–58

Rangel-Buitrago N, Neal WJ (2018) Coastal erosion management. In: Finkl C, Makowski C (eds) Encyclopedia of coastal science. Encyclopedia of earth sciences series. Springer, Cham

Rangel-Buitrago N, Williams AT, Anfuso G (2018a) Hard protection structures as a principal coastal erosion management strategy along the Caribbean coast of Colombia. A chronicle of pitfalls. Ocean Coast Manage 156:58–75

Rangel-Buitrago N, de Jonge V, Neal WJ (2018b) How to make integrated coastal erosion management a reality. Ocean Coast Manage 156:290–299

Stancheva M, Rangel N, Anfuso G, Palazov A, Stanchev H, Correa I (2011) Expanding level of coastal armouring: case studies from different countries. J Coast Res 64:1815–1819

Landscape Ecology and Conservation for Building Resilience and Adaptation to Global Change in Venezuela

Eulogio Chacón-Moreno, Isabel Olivares, Georgina Navarro, Anderson J. Albarrán, Yorman Paredes, Carla I. Aranguren and Gustavo J. Nagy

Abstract The elements and components of the landscape serve as a scale for the analysis, action and planning of the conservation and restoration strategies of biodiversity, and climate adaptation. This study presents an analytical framework focused on building resilience and adaptation measures to changing climate and

E. Chacón-Moreno (✉) · I. Olivares · G. Navarro
Instituto de Ciencias Ambientales y Ecológicas, Facultad de Ciencias, Universidad de Los Andes, Mérida 5101, Venezuela
e-mail: eulogio@ula.ve; eulogio.chacon@gmail.com
URL: http://www.ciencias.ula.ve/icae/

I. Olivares
e-mail: paleoecologia.cimar@gmail.com; isabelolivares@ula.ve
URL: http://www.ciencias.ula.ve/icae/

G. Navarro
e-mail: ginavarrog@gmail.com
URL: http://www.ciencias.ula.ve/icae/

I. Olivares
Programa de Ciencias Ambientales de la Universidad Nacional Experimental Francisco de Miranda Santa Ana de Coro, Estado Falcón 4101, Venezuela

A. J. Albarrán
Escuela de Geografía, Facultad de Ciencias Forestales y Ambientales, Universidad de Los Andes, Mérida 5101, Venezuela
e-mail: aalbarran9@gmail.com

Y. Paredes
Departamento de Medicina Preventiva y Social de la Escuela de Medicina, Facultad de Medicina, Universidad de Los Andes, Mérida 5101, Venezuela
e-mail: paredesy@gmail.com

C. I. Aranguren
Departamento de Biología, Facultad de Ciencias, Universidad de Los Andes, Mérida 5101, Venezuela
e-mail: arangurencarla@gmail.com

G. J. Nagy
Oceanografía y Ecología Marina, Instituto de Ciencias Ambientales y Ecología, Facultad de Ciencias, Universidad de la República (UdelaR), Iguá 4225, POB 11400 Montevideo, Uruguay
e-mail: gnagy@fcien.edu.uy

W. Leal Filho et al. (eds.), *Climate Change, Hazards and Adaptation Options*, Climate Change Management, https://doi.org/10.1007/978-3-030-37425-9_7

environmental conditions. Such a framework builds on the diagnosis of previous evaluations of landscapes of western Venezuela in which geomatics tools and Open Standards for the Practice of Conservation (OSpC) were used. The integration of information on ecosystems and landscapes in the design of adaptation proposals with a socio-ecological approach would make it possible to compensate for the lack of perception of the existing synergy between drivers of change, the degree of exposure to climatic disturbances. Such an approach would also promote the resilience of the livelihoods of the inhabitants by favouring the conservation of vulnerable areas and biodiversity, and the adaptation of local communities to develop an anticipatory climate adaptation.

Introduction

Venezuela is a country with a high biological diversity which offers numerous ecosystem services, which can be affected to a different degree by the impact of climate change. Distribution of disease vectors, farming systems, and local livelihoods may also be affected directly or indirectly (Stocker et al. 2013). The transformation of ecosystems is generating, in addition to climate change, a rapid loss of biodiversity, which could lead to a new process of mass extinction, affecting even the survival of the human species (McNeely and Mainka 2009). One of the main drivers of landscape change in Latin America is deforestation and forest degradation, which weak the ecosystem resilience and increase the promotion of a vicious cycle of ecosystem degradation with an increased greenhouse of gas emissions (SCBD 2017).

Such global, regional and local threats have been confronted in different ways and through different methodological approaches. Instruments and actions are developed to face these threats and the potential impacts of climate and environmental changes regionally, taking into account the adaptation capacities and the building of socio-ecological resilience using the ecology of the landscape as a tool. A recent evaluation of the research priorities for the conservation and sustainable governance of Andean forest landscapes to face the global changes, it is recognized the necessity of building-integrated climate management responses allowing understanding the dynamic and complex interactions between their ecological and social components (Mathez-Stiefel et al. 2017). Such priorities are in agreement with the United Nations recommendations "integrating adaptation, sustainable development and disaster risk reduction can contribute to strengthening resilience, reducing vulnerabilities and increasing the understanding and implementation of adaptation actions" (UNCCS 2017).

An analytical framework focused on resilience was developed to assess the adaptation to climate and environmental changes based on previous experiences on the evaluation of landscapes of western Venezuela (Fig. 1) in which geomatics tools and Open Standards for the Practice of Conservation (OSpC) were used. Most of the research has been carried out in the Andes (Fig. 1, area B), where the studies include geographic scales from 1:100.000 to local areas, and from a simple description of the

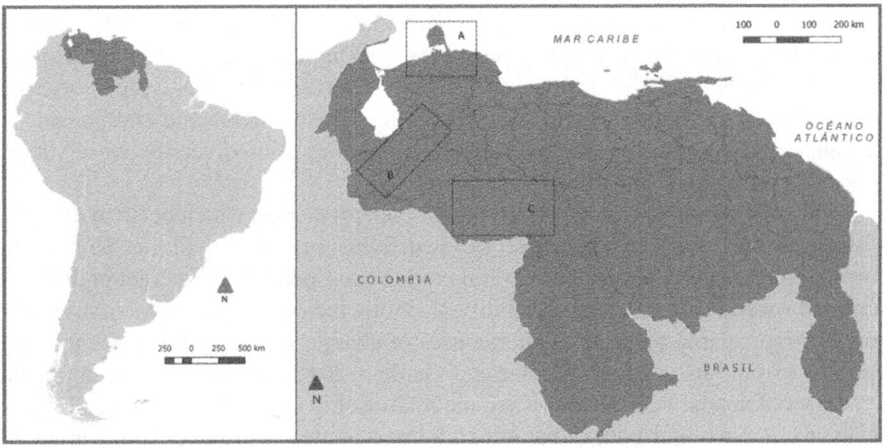

Fig. 1 Left South America. Right Venezuela. The rectangles of dotted lines indicate the study areas of western Venezuela. A. Coastal ecoregion. B. Eco-Region Los Andes. C. Eco-Region Los Llanos

landscapes to modelling ecosystem in climate change scenarios. Also, conservation planning strategies with Andean local communities were developed. Some aspects of landscape and conservation were studied in area C (eco-region Los Llanos), while area A (coastal eco-region) is currently being addressed, including geomatics, conservation, and climate adaptation. This chapter consists of two main parts; the first part contains a conceptual framework on fundamental aspects of landscape ecology, dynamics of changes, modelling, adaptation and conservation; the second part analyses the methodological contributions of these investigations for the building of resilience.

Conceptual Theoretical Framework About the Landscape Analysis

Landscape Ecology: Spatial Patterns and Dynamics of Changes

Historically, the development of landscape ecology has been marked by the influence of two schools of thought. The European school is characterised by a holistic and humanitarian approach, and the North American one, which is more analytical and biophysical. Nowadays, it is necessary to combine both approaches to propose solutions to the various problems that humanity faces. Such combination can be seen as the science and art of studying and influencing the relationship between spatial pattern and ecological processes through hierarchical levels of biological organisation at different scales in space and time (Wu and Hobbs 2007).

Characterising and understanding the relationships and configuration of elements or systems in space and ecological processes is an approach that landscape ecology has (Wiens 1999; Turner and Gardner 2015). Therefore, the landscape serves as a spatial scale for the analysis and action, where the elements and components of the landscape system are instruments for the planning of biodiversity conservation, restoration strategies, and climate adaptation.

Landscape dynamics are the result of complex processes, which occur at different scales and are of great importance to most living organisms (Farina 1998). The current landscape is the result of several causes that operate and interact in different time and space scales (e.g. the variability of abiotic factors, biotic interactions, which can generate spatial patterns). Other causes of change are natural disturbances and patterns of land-use history (Turner and Gardner 2015). The understanding of the dynamics of spatial heterogeneity and the relationship between pattern, process and scale are fundamental both in natural and domesticated landscapes to integrate the biophysical, socio-economic and cultural components of the system (Wu and Hobbs 2007).

In this sense, research at different scales allowed characterising the processes and functioning of ecosystems and determined their relationship with global, regional and local changes and transformations. However, research is needed to facilitate the diagnosis and evaluation of the state of the environmental elements and processes in the spatial context at different scales.

The ecological landscape approach serves to develop ecosystem maps for an ecological classification that considers a nested hierarchy where the country is subdivided into large eco-regions and landscapes. Ecosystems are the spatial units of analysis, integrating non-natural areas as secondary ecosystems (Chacón-Moreno et al. 2013) (Box 1). This map is based on the ecosystem map of the Northern Andes that includes the Andean regions of Bolivia, Colombia, Ecuador, Peru, and Venezuela. The NatureServe classification system is the conceptual basis used to incorporate the bioclimatic aspects (Josse et al. 2009).

Box 1 A conceptual and methodological approach for an ecological classification system and ecosystem map.

A system of ecological classification was developed for the elaboration of maps of ecosystems of medium-scale (1: 250.000–1: 100.000), under an environmental approach of the landscape. This classification system subdivides the country into large ecoregions and landscapes. The spatial units of analysis are the ecosystems, characterized by their structure, functioning, and assembly or spatial configuration (horizontal structure) within the landscape. The elaboration of ecosystem maps was done under this approach through ecological surveys, remote sensing tools and geographic information systems (GIS). Such an approach allows integrating natural and secondary ecosystems by facilitating the determination of the biodiversity conservation status and land coverage. This methodology was applied to the elaboration of the Map of Ecosystems of the Venezuelan Andes (Josse et al. 2009, Chacón-Moreno et al. 2013), where 16 ecosystems were defined. The Perijá area represents 14.2% of the total area (>5300 km^2); while the Cordillera de Mérida 85.8% (>32,000 km^2) for a total of more than 37,400 km^2 surface. In the area, montane and low montane rainforests predominate, with more than 22% of the total area. The ecosystems that represent the páramo (altimontane and high Andean ecosystems) cover an area of 328,108 ha, which represents 8.6% of the area of the Map of Ecosystems of the Venezuelan Andes, and 0.35% concerning the Venezuelan national area. An analysis of the state of conservation of these ecosystems was carried out concerning the percentage of the area of each ecosystem represented in the national parks (NP), of which there are 13 in the studied area, 12 of them in the Cordillera de Mérida. Of the total area of the ecosystem map, approximately 33% is under the figure of PN, which means that a large part of the ecosystems is under legal protection and that almost half of the surface still conserves their natural ecosystems. However, not all ecosystems are equally represented. The paramos ecosystems are included in a large percentage in the PN. Even the highest altitude ecosystems such as the altándino páramo are almost entirely protected; while nearly 80% of the frailejonales are in the PN. Both montane and low montane forests are represented in the PN (40–60%, respectively). Important to highlight, 20% of the areas intervened are within the PN.

In Venezuela, there are ecosystem maps developed using geomatic tools and ecological recognition following the proposed approach (Chacón-Moreno 2007; Oyola 2009; Tovar 2013; Chacón-Moreno et al. 2013; Betancourt 2013). This allows knowing the structure of the landscape at a specific moment as the primary basis for analysis and application in aspects related to planning for conservation and adaptation to global changes (Suárez 2005; Ruíz 2005; Ulloa 2005; Rodríguez 2005; Chacón-Moreno 2007; Santaella 2007; Barrios 2008; Sulbarán-Romero 2009; Roig 2011; Ulloa 2009; Montilla 2012; Betancourt and Chacón-Moreno 2015).

Motivated by the effects of anthropic activities on global ecosystems, in the last decade, studies that attempt to integrate the social and ecological drivers of landscape patterns with changes have increased significantly (Turner and Gardner 2015). It is difficult to establish the interrelation between agents of change because social and natural systems almost always operate on different spatial and temporal scales (Mora and Hurtado 2015). In principle, the interrelation is unknown due to the difficulty in establishing, which are the forces that determine the transformations of the landscape since they generally act together. It is for this reason that landscape transformations vary according to the social, economic, political and cultural context of the regions or places (Ojima et al. 1994).

The resilience approach is aimed at creating systems robust enough to persist and adapt in the long-term and has become a proper way of managing ecosystems in the face of an uncertain future (Beller et al. 2015). Hence the importance of incorporating landscape studies as useful tools in the construction and integration of two schools of thought (ecological systems and complex social systems), and addressing the issue of change and adaptation through the resilience approach.

Modelling of Climate Change Scenarios

The Intergovernmental Panel on Climate Change (IPCC) has reviewed the empirical evidence at the international level, showing that the global average temperature of the planet's surface has increased during the last century (Stocker et al. 2013). The possible global scenarios of climate change and their direct and indirect general impacts, as well as their consequences, have been projected and summarised in the Fifth Assessment Report of the IPCC (Stocker et al. 2013). Temperature plays a critical role in climate conditions by affecting the hydrological cycle and the radiation absorbed by the surface of the earth. In this way, changes in magnitude and the rate of change in temperature can have significant effects on climate conditions, ecosystems and, consequently, on socioeconomic activities.

For Venezuela, it is estimated that in the year 2100 the average temperatures will increase between 1.3 and 4.6 °C, which represents ranges of global warming between 0.1 and 0.4 °C per decade (Hulme and Sheard 1999; Stocker et al. 2013), and between 2 and 6 °C for the area of the Amazon (Salazar et al. 2007). The IPCC AR-5 report (IPCC 2013) projections for the Venezuelan Andes indicate that in the best scenario (RCP2.0) the temperature will increase between 0.4 and 2.5 °C, while the precipitation could decrease up to 17%.

Spatial models of ecosystem distribution are analysed and evaluated since 2012, which take into account previous works (Chacón-Moreno 2007; Arzac 2008; Suárez del Moral and Chacón-Moreno 2011; Arzac et al. 2011; Chacón-Moreno et al. 2013) on:

(i) The distribution of ecosystems in the Venezuelan Andes (Ataroff and Sarmiento 2004; Suárez del Moral and Chacón-Moreno 2013; Chacón-Moreno et al. 2013).
(ii) The climate dynamics (Monasterio and Reyes 1980; Sarmiento 1986; Andressen, et al. 1996; Suárez del Moral 2009, Suárez del Moral and Chacón Moreno 2013)

Currently, projection models for the distribution of the ecosystems of the arid coasts of the Falcón state are developed, subjected to multiple impacts, using representative concentration routes (CRP). These studies allow defining the scenarios under different climate change projections, on a spatial scale according to their use in regional and local planning (1:100,000) (Box 2). The example presented in Box 2 shows the importance of modelling the scenarios of the agricultural areas of the Andes for a better understanding of adaptation to climate change.

Conservation and Adaptation

The Convention on Biological Diversity generated the bases for the formulation of strategies for the conservation and sustainable use of biodiversity (Shepherd 2006). However, climate change is still considered one of the main threats to biodiversity, as it is estimated to be one of the most important forces of change in the coming decades (Board 2005). Among the strategies developed are the ecosystem approach and adaptive management, which allow the generation of strategies aimed at the conservation of ecosystem services and local livelihoods. Consequently, the activities and policies of local governance have as objectives the conservation of biodiversity and the improvement of human well-being (Andrade et al. 2011).

The Open Standards for Conservation Practice is an open-source adaptive planning framework used by governments and NGOs around the world to conserve flora and fauna collaboratively and systematically. This proposal was created by the Conservation Measures Partnership (CMP 2013), to learn from the deficiencies of previous models and promote innovation in conservation monitoring and evaluation. The Open Standards for the Practice of conservation focus on human welfare and ecological integrity goals; linking actions to impacts, incorporating adaptation and assessment from the beginning of the planning process. From 2010, the focus of the standards was included in the landscape research work in the ICAE landscape ecology laboratory. In the last three years, adaptation strategies were included, consolidating an integrated vision, generating proposals to adapt to climate change and creating resilience.

Box 2 Potential distribution of crops in climate change scenarios in the state of Merida, Venezuela (Paredes 2014)
The Merida state is located in the central part of the Venezuelan Andes. Its surface occupies approximately 11,300 km^2 and represents 1.24% of the national territory. The state is crossed by the Cordillera de Mérida formed by two mountain ranges: the Sierra Nevada that separates the Andes from the high western plains, and the Sierra La Culata that separates it from the lake plains of Maracaibo.

Sixteen different ecosystems have been recognized for the Venezuelan Andes (Josse et al. 2009), which represents a high diversity of environments with great potential for crop distribution. The Mérida state percentage of agrarian activities is divided as follows (Gómez and Azócar 2002): (a) Metropolitan area (32%); (b) South of the Lake (39%, (c) Valles del Mocotíes (41%), (d) villages in the South (15%), and Páramo (9%).

To understand the relationship between environmental variables and the effects that climate changes could have on the distribution of the leading crops, the potential distribution of potato, carrot, coffee, cocoa, and plantain crops was determined using Gaussian distribution models. According to the ecosystem map of the Andes by Josse et al. (2009), the potential areas of potato cultivation are distributed in 36% of areas currently intervened and 64% of ecosystems under natural conditions, concluding that the State of Merida has the climatic conditions for the development of potential regions of these crops.

The distribution of crops in climate change scenarios was modelled using the maximum probable temperature and precipitation averages generated by the Representative Concentration Pathways (RCP 2,6, RCP4.5, RCP6.0, and RCP8.5) for the years 2035, 2065 and 2100 extracted from the Fifth Assessment Report (AR5) of the IPCC (Stocker et al. 2013). However, in the scenarios leading to a low level of forcing (RCP2,6) and very high level of greenhouse gas emissions (RCP8,5), it was found that the potato crop will occupy higher altitudes over time. Likewise, a contrary tendency of the potential probability of occurrence (at 47 thousand hectares) was seen. Such a trend would increase the fragmentation and the degree of intervention of the potential areas with replacement patterns of the natural ecosystems, while the potential zones of the crops of cocoa, banana and shade coffee would tend to disappear.

These scenarios would force agricultural producers to adapt to a set of new climatic conditions that would generate the abandonment of current crops by varieties that best adapt to the new climatic conditions (Läderach et al. 2011).

On the other hand, the Cancun Adaptation Framework created in COP 16 of the United Nations Framework Convention on Climate Change, promote ecosystem-based adaptation (EBA) to strengthen the resilience of socioeconomic and ecological systems, through measures such as economic diversification and sustainable management of natural resources. Besides, it establishes that adaptation actions must promote a gender-sensitive, participatory and transparent approach that considers vulnerable groups, communities, and ecosystems (Lhumeau and Cordero 2012).

Methodological Contributions for Building Resilience

In this analysis, the building of resilience is handled based on "the perspective of adaptive change", perceived as the capacity of a socio-ecological system to adaptively self-organise to preserve its essential attributes after a disturbance (Levin et al. 1998; Holling 2001). Resilience emerges as a concept capable of helping decision-makers identify vulnerable socio-geographical areas and improve preparedness for extreme climate events, which affect livelihoods and induce migrations, leaving in evidence the underlying weaknesses of the communities (Alcayna et al. 2016).

Currently, multidisciplinary approaches on resilience have emerged to favour the ability of people to reduce their vulnerability through adaptation, which is seen as a critical element for the achievement of resilient societies, because it seeks to ensure the maintenance of the provision of ecosystem services that sustain human well-being in the face of change (Convention on Biological Diversity 2005). Because threats to species and ecosystems, such as climate change and habitat fragmentation occur at large scales, it is necessary to recognise that conservation, management, and adaptation actions must be applied at the landscape level (Andrade 2010).

Building Resilience

Resilience recognises several properties that promote the renewal and reorganisation of a system after disturbances (Holling 1973). Resilience depends on a system's ability to adapt, the biophysical and social legacies that contribute to diversity, the ability of people to long-term planning under uncertainty and change, the ability to adjust government structures to cope with changes and needs (Holling and Gunderson 2002; Walker and Salt 2006; Chapin et al. 2009).

The current pressure on different landscapes and their components in the face of global change poses a significant challenge for the management of natural areas and their conservation. Our research is based on the proposal of Beller et al. (2015), which highlight a scientific vision for the health and resilience of ecosystems, as follows:

1. *The configuration* allows understanding how the landscape responds to the drivers, what will be the distribution of infrastructure and other elements in the landscape and discover which elements have persisted over time and which are limited. Aspects related to this principle are being studied by Olivares and Chacón-Moreno (2016) from the analysis of cumulative impacts and resilience in marine-coastal ecosystems of Venezuela.
2. *Processes* provide information on how resources are distributed, how controllers interact with the environment, and how species can adapt to environmental change. Highlights the work of Chacón-Moreno (2007) that deals with the ecological and spatial modelling in the Llanos del Orinoco de Venezuela.
3. *Connectivity* facilitates the connections, space and biophysical gradients required for species to be mobilised in response to changing conditions, allowing the ability of species and communities to adapt to varying scenarios. Betancourt and Chacón-Moreno (2015) address connectivity with the approach of ecological corridors as a strategy for conservation in a forest reserve.
4. *Diversity and complexity*, in turn, help maintain the variability necessary for the adaptation and evolution of species by supporting a range of responses to a heterogeneous and dynamic environment. This process was evaluated by Chacón-Moreno et al. (2013) by developing an Ecological Classification System and Ecosystem Maps for Venezuela at the landscape scale (Box 2).
5. *The spatial and time scales* allow to know the capacity of recovery before disturbances and to consider the delay times between the management actions and the response of the ecosystem. Aspects related to this principle were evaluated

by Rodríguez-Morales et al. (2009) in his work on the transformation of the mountain forest landscape in the Rio Capaz basin.

6. *The human factor*, the construction of resilient landscapes depends on the deep understanding of the place and the political will in the management and conservation of resources. The proposal for the conservation of the sub-basin of the State of Mérida, Venezuela (Gutiérrez 2014), is an example of the application of the time and scale management principle.
7. Also relevant are the training programmes in planning for the conservation and climate adaptation developed by the Institute of Environmental and Ecological Sciences (ICAE) of the Universidad de Los Andes. These programmes are aimed at training of community leaders in rural areas, government technicians, young researchers and the community and in general, contributing to the linking of its field of research with participatory action in the communities, promoting the application of strategies in matters of conservation and climate change.

This work is an agenda to develop future ecosystem maps intended to facilitate the conservation of biodiversity and climate management of Venezuelan landscapes; however, it has some limitations. The study does not cover the totality of the landscapes of Venezuela, highlighting mainly the terrestrial landscapes. The "Conservation and Adaptation" section may be expanded in subsequent works, due to the limited availability of previous research under this approach. There is a need to expand research on climate change modelling for other regions of the country than the Andean region. Studies at the landscape level allow the integration of several approaches; however, this research requires further study in the new trends of socioecological systems. The socioeconomic and political situation of Venezuela during the last decade affected the scientific research and adaptation studies. As a consequence, this study is a small contribution, but it reflects a robust proposal to carry out in these conditions with very few budgets.

The integration of information on ecosystems and landscapes in the design of adaptation proposals promotes socio-ecological resilience. Such an integration favours the maintenance of biodiversity, the conservation of vulnerable areas and the adaptation of local communities to anticipate and adapt to changes. Similarly, the ecosystem approach allows building climate resilience when conservation activities, governance, and managers adopt better management practices by decreasing the exposure of local populations, their sensitivity to develop preventive plans and increase their adaptive capacity.

The building of socio-ecological resilience requires political willing and socio-economic development consistent with the United Nations Sustainable Development Goal 13 (Climate Action). Some examples are taking actions to increase adaptive capacity or reduce vulnerability (e.g. early warning systems, Ecosystem-based Adaptation and/or planning for the future) (Leal Filho et al. 2017; Nagy et al. 2018).

Conclusions

Biodiversity and its ecosystem services in Venezuela are threatened by climate change, including the determination of the possible survival of the most vulnerable communities. A conceptual-methodological approach is proposed to respond to these threats. The approach includes the assessment of the state of landscape conservation, the potential impacts of global changes at the regional level, and recommending strategies for the adaptation and building of socio-ecological resilience within the framework of landscape ecology. The use of geomatics has contributed to evaluating the processes of landscape transformation and model the impact of climate change through the generation of scenarios. This integrated approach is considered to help strengthen resilience, reduce vulnerabilities and encourage adaptation actions.

References

Alcayna T, Bollettino V, Dy P, Vinck P (2016) Resilience and disaster trends in the Philippines: opportunities for national and local capacity building. PLoS Currents 8

Andrade A (2010) Adaptation to climate change based on ecosystems. In: Naranjo G (ed) Climate change in a living landscape: vulnerability and adaptation in the Eastern Cordillera Real of Colombia, Ecuador and Peru. Cali, WWF-Natura Foundation, pp 65–74

Andrade A, Arguedas S, Vides R (2011) Guide for the application and monitoring of the ecosystem approach. CEM-UICN, CI-Colombia, ELAP-UCI, FCBC, UNESCO-MAB Program, Colombia

Andressen R, Robok A, Acevedo M (1996) Scenarios of climate change due to the greenhouse effect and deforestation for Venezuela. Rev Geográfica Venez 37:221–250

Arzac A (2008) Spatial distribution of plant life forms in an altothermic gradient in the Tropical Andes. Special Work of Bachelor Degree in Biology. University of the Andes, Venezuela

Arzac A, Chacón-Moreno E, Llambí LD, Dulhoste R (2011) Distribución de formas de vida de plantas en el límite superior del ecotono bosque paramo en los Andes Tropicales. Ecotrópicos 24(1):26–46

Ataroff M, Sarmiento L (2004) The ecological units of the andes of Venezuela. In: Marca EL, Soriano P (eds) Reptiles of the andes of Venezuela. BIOGEOS, Venezuela, pp 11–26

Barrios J G (2008) Transformation of the landscape of the micro-basin of the Quebrada Turmero, páramo de Turmero, Mérida state. Special Work of Bachelor Degree in Biology. University of the Andes, Venezuela

Beller E, Robinson A, Grossinger R, Grenier L (2015) Landscape resilience framework: operationalizing ecological resilience at the landscape scale. Richmond, San Francisco

Betancourt A (2013) Ecological corridors as a strategy for the conservation of semi-deciduous forests of the Caparo Forest Reserve, Edo. Barinas, Venezuela Special Work of Bachelor Degree in Biology. University of the Andes, Venezuela

Betancourt A, Chacón-Moreno E (2015) Ecological corridors as a strategy for the conservation of forest ecosystems of the Caparo Forest Reserve. Interciencia 40(4):275–281

Board MA (2005) Millennium ecosystem assessment. New Island, Washington

Chacón-Moreno E (2007) Ecological and spatial modeling: Mapping ecosystems, landscape changes, and plant species distribution in Llanos del Orinoco, Venezuela. Ph.D. Thesis. Enschede, The Netherlands: International Institute for Geo-Information Science and Earth Observation and Wageningen University

Chacón-Moreno E, Ulloa A, Tovar W, Márquez T, Sulbarán-Romero E, Rodríguez-Morales M (2013) Ecological classification system and ecosystem maps: conceptual approach - methodological for Venezuela. Ecotrópicos 26(1–2):1–27
Chapin FS, Folke C, Kofinas G P (2009) A framework for understanding change. In: Stuart Chapin III F, Carl F, Gary PK (eds) Principles of ecosystem stewardship. Springer, New York, pp 3–28
CMP, The Conservation Measures Partnership (2013) Open standards for the practice of conservation version 3.0. http://www.conservationmeasures.org/
Convention on Biological Diversity (CDB) (2005) Handbook of the convention on biological diversity, including its Cartagena protocol on biosafety, 3rd edn. Montreal, CDB
Farina A (1998) Principles and methods in landscape ecology. Springer, Dordrecht
Gómez A, Azócar A (2002) Potential areas for the development of the cocoa crop in the state of Merida. Agronomy Trop 52(4)
Gutiérrez Y (2014) Planning for the conservation of the La Toma Municipality Rangel sub-basin, Mérida state. Special Work of Bachelor Degree in Biology. University of the Andes, Venezuela
Holling CS (1973) Resilience and stability of ecological systems. Annu Rev Ecol Syst 4:1–23
Holling CS (2001) Understanding the complexity of economic, ecological, and social systems. Ecosystems 4:390–405
Holling CS, Gunderson LH (2002) Resilience and adaptive cycles. Panarchy: understanding transformations in human and natural systems. Island Press, Washington, pp 25–62
Hulme M, Sheard N (1999) Scenarios of climate change for countries of the northern andes. Climate Research Unit, Norwich, United Kingdom
Josse C, Cuesta F, Navarro G, Barrena V, Cabrera E, Chacón-Moreno E, Ferreira W, Peralvo M, Saito J, Tovar A (2009) Ecosystems of the andes of the north and center. Bolivia, Colombia, Ecuador, Peru and Venezuela. Lima: General Secretariat of the Andean Community
Läderach P, Haggar J, Lau C, Eitzinger A, Ovalle O, Baca M, Jarvis A, Lundy M (2011) Mesoamerican coffee: development of a strategy of adaptation to climate change. CIAT, Cali, Colombia
Leal Filho W, Modesto F, Nagy GJ, Saroar M, Nsani Y, Ha'apio MO (2017) Fostering coastal resilience to climate change vulnerability in Bangladesh, Brazil, Cameroon, and Uruguay: a cross-country comparison. J Mitig Adapt Strateg Global Change 23:579–602. https://doi.org/10.1007/s11027-017-9750-3
Levin SA, Barrett S, Aniyar S, Baumol, W, Bliss C, Bolin B, …, Holling CS (1998) Resilience in natural and socioeconomic systems. Environ Dev Eco 3(2):222–235
Lhumeau A, Cordero D (2012) Adaptation based on ecosystems: a response to climate change. IUCN, Quito
Mathez-Stiefel Peralvo M, Báez S, Rist S, Buytaert W, Cuesta F, Fadrique B, Feeley KJ, Groth AAP, Homeier J, Llambí LD, Locatelli B, López Sandoval MF, Malizia A, Young KR (2017) Research priorities for the conservation and sustainable governance of andean forest landscapes. Mt Res Dev 37(3):323–339
McNeely JA, Mainka SA (2009) Conservation in a new Era. IUCN, Gland, Switzerland
Monasterio M, Reyes S (1980) Environmental diversity and variation of vegetation in the páramos of the Venezuelan Andes. In: Monasterio M (ed) Ecological studies in the andean páramos. Merida, University of Los Andes, pp 47–91
Montilla A (2012) Ecological and social impact due to the transformation of the landscape in the Carache river basin (1966–2010). Master's Degree in Tropical Ecology. University of the Andes, Venezuela
Mora AM, Hurtado RJ (2015) The theory of change in the landscape from the change of land use and land cover (LUCC approach): its usefulness for environmental history. Reflections 94(2):25–33
Nagy GJ, Leal Filho W, Azeiteiro UM, Heimfarth J, Verocai JE, Li C (2018) An assessment of the relationships between extreme weather events, health vulnerability and the impacts on human Wellbeing in Latin American Countries. Int J Environ Res Public Health 15(9):1802. https://doi.org/10.3390/ijerph15091802

Ojima DS, Galvin KA, Turner BL (1994) The global impact of land-use change. Bioscience 44(5):300–304

Olivares I, Chacón-Moreno E (2016) Cumulative impacts and resilience in marine-coastal ecosystems. General considerations in the context of global climate change. In: 2nd symposium on climate change. Faculty of forestry and environmental sciences, Universidad de Los Andes, Venezuela

Oyola A (2009) Heterogeneity of the vegetation in a landscape of glacial origin in the páramos of Las Cruces, Sierra de La Culata, Mérida-Venezuela. Master's Degree in Ecology. University of the Andes, Venezuela

Paredes Y (2014) Potential distribution of the leading crops in climate change scenarios in the state of Merida, Venezuela. Special Work of Bachelor Degree in Biology. University of the Andes, Venezuela

Rodríguez M (2005) Changes in the landscape in the Capaz river basin. Special Work of Bachelor Degree in Biology. University of the Andes, Venezuela

Rodríguez-Morales M, Chacón-Moreno E, Ataroff M (2009) Transformation of the landscape of mountain forests in the Capaz river basin, Venezuelan Andes. Ecotrópicos 22(2):64–82

Roig E (2011) Is the tropical dry forest regenerated in the La Jarreta farm, Cuba? Analysis of the reconversion process through the characterization of the successional stages. Master's Degree in Tropical Ecology. University of the Andes, Venezuela

Ruíz R (2005) Piaroa agroecological system in the sipapo forest reserve. An ecological approach to the landscape. Special Work of Bachelor Degree in Biology. University of the Andes, Venezuela

Salazar LF, Nobre CA, Oyama MD (2007) Climate change consequences on the biome distribution in tropical South America. Geophys Res Lett 34(9)

Santaella W (2007) Change of landscape in the transitional páramo-cloud forest, La Aguada sector of the northern slope of the Sierra Nevada de Mérida. De Thesis at is Yacambú University, Venezuela

Sarmiento G (1986) Ecological features of climate in high tropical mountains. In: Vuilleumier F, Monasterio M (eds) High altitude tropical biogeography. Oxford: Oxford University Press, pp 11–45

Secretariat of the Convention on Biological Diversity (2017) The lima declaration on biodiversity and climate change: contributions from science to policy for sustainable development. Technical Series No. 89. In: Rodríguez L, Anderson I (eds) Secretariat of the convention on biological diversity. Montreal, 156p

Shepherd G (2006) The ecosystem approach: five steps for its implementation. IUCN, United Kingdom

Stocker TF, Qin D, Plattner GK, Tignor M, Allen SK (2013) Climate change 2013, the physical science basis. In: Working group I contribution to the fifth assessment report of the intergovernmental panel on climate change. Cambridge, UK: Cambridge University

Suárez DC (2005) Comparative analysis of the diversity of the ecological landscape of the semideciduous forest caused by human impact. Special Work of Bachelor Degree in Biology. University of the Andes, Venezuela

Suárez del Moral P (2009) Modeling the distribution of the forest-páramo contact line in the Venezuelan Andes. Special Work of Bachelor Degree in Biology. University of the Andes, Venezuela

Suárez del Moral P, Chacón-Moreno E (2011) Spatial model of distribution of the forest-páramo ecotone in the Venezuelan Andes. Potential location and climate change scenarios. Ecotrópicos, 24(1):3–25

Suárez del Moral P, Chacón-Moreno E (2013) Bioclimatic map of Venezuela. Technical report and working protocol, Mérida: ECOMAP_CC (2011000350)

Sulbarán-Romero JE (2009) Model the of spatial distribution of Triatomines habitat and risk zones of Chagas disease in the Andrés Bello and Morán municipalities of Lara state, Venezuela. Master's degree in tropical ecology. University of the Andes, Venezuela

Tovar W (2013) Proposal for classification of land cover/use in the Andes, case of the La Grita river basin, Venezuela. Master's Degree in tropical ecology. University of the Andes, Venezuela

Turner MG, Gardner RH (2015) Landscape ecology in theory and practice, 2nd edn. Springer, New York

Ulloa A (2005) Chigüire habitat distribution (Hydrochaeris hydrochaeris Linne 1766) in flooded savannas of El Frio Biological Station, Venezuela. Special Work of Bachelor Degree in Biology. University of the Andes, Venezuela

Ulloa A (2009) Conservation priorities in the floodplains of the Llanos del Orinoco. Master's degree in tropical ecology. University of the Andes, Venezuela

UNCCS (2017) Opportunities and options for integrating climate change adaptation with the sustainable development goals and the Sendai framework for disaster risk reduction 2015–2030. United Nations Climate Change Secretariat, Bonn, Germany

Walker BH, Salt D (2006) Resilience thinking: sustaining ecosystems and people in a changing world. Island Press, Washington

Wiens JA (1999) The science and practice of landscape ecology. In: Gardner RH, Klopatek JM (eds) Landscape ecological analysis: issues and applications. Springer, New York, 371–383

Wu J, Hobbs R (2007) Landscape ecology: the state of the science. In: Wu J, Hobbs R (eds) Key topics in landscape ecology. Cambridge: Cambridge University Press, pp 271–287

Influence of Vegetation in the Creation of Urban Microclimates

Bruna Giacomelli, Júlia Calvaitis Padilha, Paula Renata Albrecht Mantovani, Fabiane Benche and Natalia Hauenstein Eckert

Abstract Big part of the urban land occupation is composed of high density of buildings and large paved and impermeable areas, which supports the high increase of the air temperature in the urban centers, which creates microclimates in different zones of the city, causing the heat islands. The objective of the research is the identification of the thermal variability between the downtown of a medium-sized Brazilian city and the University campus, in order to evaluate the urban ambience in these different venues. The methodology used was the collection of climatic elements by means of specific devices in the winter and summer climatic season, taking into consideration forestation and built areas, and later the discussion and comparison of the data obtained in the surveys. As partial results, the climatic environment generated by the green spaces was highlighted, what propitiates the increase of the quality of life for the urban users. In addition, managers should introduce studies about the microclimates in municipal management plans, seeking to identify the influence of urban morphology and incorporating urban planning processes in order to improve conditions and minimize the effects caused by heat islands.

Investigation of Urban Heat Islands in Cruz Alta: Comparison between downtown and the University Campus.

B. Giacomelli (✉) · J. Calvaitis Padilha · P. Renata Albrecht Mantovani · F. Benche · N. Hauenstein Eckert
Center for Human and Social Sciences, Architecture and Urbanism Course, University of Cruz Alta, Municipal Highway Jacob Mella Mèa, Km 5.6 CEP 98005-972, Parada Benito, Cruz Alta, Rio Grande do Sul, Brazil
e-mail: brunagiacomelli1@gmail.com

J. Calvaitis Padilha
e-mail: juliacpadilha@outlook.com

P. Renata Albrecht Mantovani
e-mail: re.mantovani.21@hotmail.com

F. Benche
e-mail: fabinha-b@hotmail.com

N. Hauenstein Eckert
e-mail: eckert@unicruz.edu.br

W. Leal Filho et al. (eds.), *Climate Change, Hazards and Adaptation Options*, Climate Change Management, https://doi.org/10.1007/978-3-030-37425-9_8

Introduction

The steady increase in man-made impacts on climate change, such as the emission of greenhouse gases, deforestation, erosion, rising seas, rising temperatures, and so on, remain troubling challenges for the global community (United Nations New York 2015). Climate changes are defined by climate change, which persists over a long period of time. Climate change can trigger natural events, such as volcanic eruptions and variations of solar phases, or even anthropogenic changes affecting the soil and atmosphere (Burkett et al. 2014).

Another reason for climate change is the deforestation of forests. About 30% of the land area is composed of forests, which is responsible for playing the important role in adapting and mitigating climate change. Despite of the decrease of deforestation, forests are still under constant threat around the world, because it continues to compromise the lives of species and the sustenance of thousands of people. South America and Africa were the continents that had the most losses of forest areas until the year 2010. Oceania also recorded large losses, mainly due to severe drought and forest fires in Australia. Asia, meanwhile, recorded between 2000 and 2010 a net gain of 2.2 million hectares per year. This gain is due to the implementation of large-scale reforestation projects and programs, mainly in China (2015).

According to the Intergovernmental Panel on Climate Change (IPCC) in the last three decades, the earth's surface has gotten warmer. This period of almost 30 years (1983–2012) was certainly the one that presented higher temperatures, taking into consideration the last 1400 years. Information on the average temperature of the terrestrial and global oceanic surface, indicates that there was a heating of 0.85 °C in the period between 1880 and 2012 (IPCC 2014).

It is assumed that the increase in temperature is due to the frequency and intensity of meteorological and hydrological events, due to changes in the water cycle, the increase in the temperature of the terrestrial territory, changes in the parameters of the winds and especially the increase of precipitations (IPCC 2013).

The main impacts generated by climate change in urban areas are air pollution, storms, heat waves, water shortages caused by droughts, floods, landslides, among others. (Piazza et al. 2016). These effects can be felt more intensely mainly by developing countries as these regions are more vulnerable to climate change (United Nations Conference on Sustainable Development 2012).

Population growth in urban centers is another issue that interferes with climate change. These changes are the most pressing environmental, social and economic problem facing the entire planet. (Leite 2012). According to Fialho (2012), tropical countries recorded a more intense growth rate due to the process of urbanization and late industrialization, with developed countries accounting for 72.0% of the population, compared to 42.6% in developing countries.

As urban centers expand in population and size, the changes inherent in the process of landscape modifications affect the essential conditions for an urban life with quality, such as water, soil and air, affecting the climate (Fialho 2012). These climate changes are the most urgent environmental, social and economic problem that the planet faces, including temperature, precipitation, wind and meteorological patterns (Leite 2012).

The studies on the temperature encompass diverse subjects as well as the air, the soil, the buildings and even the perception of the inhabitants of a certain place, because it influences in the thermal comfort of the city. The temperature is one of the main climatic factors that causes alteration in the local climate. The action of precipitation and winds can cause the temperature to suffer great influence on the climate (de Neto 2007). O que determines the air temperature are the heating and cooling rates of the earth's surface (de Almeida 2009).

All the solar radiation that is emitted into our Earth's atmosphere is transmitted by the sun, in the form of magnetic waves, whose intensity is reduced as soon as it penetrates the atmosphere and its distribution is carried out through the absorption, reflection and diffusion of the solar rays through various air elements (Romero 2000). Mascaró and Mascaró (2009) suggests that one of the efficient resources against heat, be the implantation of vegetated areas, because they allow the passage of the wind, thus absorbing the solar radiation and collaborating with the shading of the built areas, protecting them against the insolation. However, the illuminance under the vegetative cover can vary according to the height, age and the type of foliage of the trees (Mascaró and Mascaró 2009).

The expansion of the urban spot, with large buildings and extensive waterproofed surfaces, intensifies intra-urban flows, and this differentiation in relation to rural areas entails a difference in the absorption time of solar energy available during the day and the transmission of energy into space at night. As a result, the uncontrolled increase in urban temperatures, resulting in lower air quality, and the slow cooling of the city after sunset, which constitute the heat islands (Fialho 2012). The heat islands usually have their most intense effect on clear and calm days, because on cloudy days and with wind the solar radiation is captured with less intensity, which makes the heat islands' effect weaker. On the other hand, in clearer days, the solar incidence becomes more accentuated and with the calmer winds the urban environment heats up, causing the formation of the heat islands to be more intense (Gartland 2010).

Romero (2011a) points out that the possible causes of the formation of heat islands are the capacity of the materials used in the buildings to store heat, the anthropogenic heating, the reduction of air humidity caused by surface paving and lack of vegetation, the reduction of air velocity caused by vertical elevations, increased absorption of solar radiation and albedo decrease.

The effects caused by heat islands favor air pollution, because when the air heats up, it tends to concentrate in the center of the city, carrying with it all pollutants, thus making it difficult to dissipate air pollution (Romero 2011b). The formation of the heat islands is only one of the consequences that the population faces with the neglect of the natural disregard of the urban centers. As these areas extend, they directly affect fundamental conditions to urban life, with the quality of water, soil and air, affecting the climate as a whole (Fialho 2012).

As a result, this study aims to analyze the formation of heat islands in the central area of the city of Cruz Alta in the state of Rio Grande do Sul, in comparison to its rural area, where the Campus of the University of Cruz Alta—UNICRUZ is located, also aiming the indication of strategies to mitigate the negative effects of this phenomenon.

Methodology

The object of study is the Brazilian municipality of Cruz Alta, which is characterized by being a medium sized city, belonging to the northwest of the state of Rio Grande do Sul, according to Fig. 1. The city presents a warm and temperate subtropical climate, with the four seasons well defined and the month of January presents the warmest temperatures (IBGE 2017).

Figure 2 represented below deals with the methodological framework on which the research structuring was based. The research was realized through the analysis of the thermal conditions of four urban points and four rural points of the municipality,

Source: Google Earth, adapted by authors, 2018.

Fig. 1 Location of the municipality under study

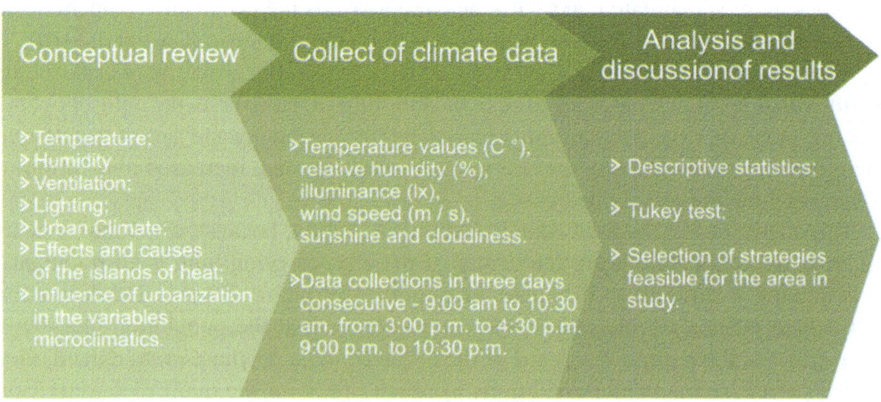

Source: Authors, 2018

Fig. 2 Methodological outline

so that it was possible to analyze the formation of heat islands and the influence of vegetation on the regulation of the urban ambience.

As shown in Fig. 2, the measurements were performed on 3 consecutive days in each season of the years 2017 and 2018 and describes in this article a relevant part of the results obtained, where the greatest thermal differences were presented in comparison of the environments. The data were collected on August 15, 16 and 17 of 2017 in the winter season and March 12, 13 and 14 of 2018 in the summer season, at three different times, from 9:00 a.m. to 10:30 a.m., from 3:00 p.m. to 4:30 p.m. and from 9:00 p.m. to 10:30 p.m. The points of downtown were chosen due to their territorial occupation and high flow of passers-by, since the points of the rural zone were delimited within the university campus of the University of Cruz Alta, which is located 6 km of the densely urbanistic zone, already being considered belonging to the rural area. The location of the points is represented in Fig. 3.

The criteria listed for the study object of the research, shown in Fig. 3, were the following: open spaces without buildings, with and without forestation (points 02, 03, 06 and 07); open spaces with buildings, with and without forestation (points 01, 04, 05 and 08). Data collection took place during three consecutive days, according to the methodology recommended by the World Meteorological Organization (2009).

Measurements were performed with a multi-parameter meter, HM-875 HIGH-MED portable equipment. For the measurements, the instrument was positioned at 1.10 meters above the ground as recommended by Mayer and Hope (1987), and the data of temperature and relative humidity of the air were collected, from three repetitions with a break of one minute each. All the data were obtained in clear sky time, which has less than 1/3 of clouds in its extension, according to methodology characterized by Mascaró (1991).

Legend: 01 and 05 without forestation and with edification; 02 and 06 without forestation and without building; 03and 07 with forestation and without building; 04 and 08 with forestation and with edification

Source: Google Earth, adapted by the authors, 2017

Fig. 3 Distribution of measurement points in the two chosen areas of Cruz Alta—RS

With the data obtained, the database was developed, containing all the necessary mathematical information for the discussions on the subject. To analyze the data, a descriptive statistical method was applied through the Tukey test, which evaluated the magnitude of the existing differences, with multiple comparisons, based on the minimum significant difference (UFPR 2009; Correa 2003). The statistical procedures sought results that presented statistical significance, according to the pre-established limit, with the level of ($p < 0.05$) and sample number ($n = 9$), since each of the data, in the collected points had three repetitions of measure.

After analysis and discussion of the results obtained, were listed green infrastructure strategies that would reduce the impact of temperature and the reduction of humidity that were generated by the heat island in the central area of the municipality. The selected infrastructures serve as a subterfuge of performance, faced with the aggravation of climate change generated by large urban centers, paved and overcrowded.

Results and Discussions

The climatic variations that affect the world are accompanied by man through researches, data collection or analysis of existing materials into the environment. This information shows that in the course of time the effects of man's action or of nature itself has been influencing human life directly (de Almeida Junior 2005). The studies developed over the years on the urban climate, has given greater prominence in large cities. Primarily, by compromising people's quality of life and aggravating environmental problems. Only on an international level, a broader portfolio of studies is found in small and medium-sized cities (Viana and Amorim 2009).

In the winter season of 2017 the main relevant statistical variations of temperature form in the night period with significance at all points, regardless of whether there is shading or not, as shown in the winter column of Table 1.

It is observed that points 1 and 5, of Table 1, that present a characteristic without forestation and with edification, the difference in temperature between the campus and downtown is of 2.66 °C in the afternoon and of 2.7 °C at night. In points 2 and 6, which is listed as without forestation and without building, downtown presented 4.01 °C higher than the campus in the evening, where the same happens with points 3 and 7, with forestation and without building, which registered a rise of temperature of 3.93 °C in the same period. It can also be observed that in points 4 and 8, points with forestation and with edification, that the highest temperatures are recorded in downtown with 2.1 °C in the afternoon and 2.6 °C in the evening compared to the university campus.

In the summer season, it was possible to observe the influence of vegetation in the more open areas, because in the three periods of measurements, the statistical difference between the temperature of the campus and downtown was significant, and in regions with surrounding buildings, the difference was found in the morning and evening shifts.

Table 1 Descriptive measures, Student's t test of the temperature in the central urban area and the university campus of UNICRUZ. Cruz Alta/RS, 2018

Collect point	Shift	Local	Winter			Summer		
			N	Average (°C)	*p*-value	N	Average (°C)	*p*-value
Without a forestation and with edification (1 and 5)	Morning	Campus	9	18.59	0.160	9	30.13	0.592
		Downtown	9	20.11		9	30.52	
	Afternoon	Campus	9	24.62	**0.003**	9	34.41	0.948
		Downtown	9	27.28		9	34.36	
	Night	Campus	9	16.66	**0.001**	9	25.30	*0.023*
		Downtown	9	19.36		9	26.19	
No forestation and no building (2 and 6)	Morning	Campus	9	20.59	0.980	9	28.09	*0.015*
		Downtown	9	20.62		9	30.19	
	Afternoon	Campus	9	27.72	0.929	9	35.39	0.651
		Downtown	9	27.66		9	34.74	
	Night	Campus	9	14.78	**0.000**	9	24.30	0.055
		Downtown	9	18.79		9	25.60	
With forestation and without building (3 and 7)	Morning	Campus	9	18.33	0.627	9	26.00	*0.011*
		Downtown	9	18.80		9	30.41	
	Afternoon	Campus	9	26.04	0.794	9	33.47	*0.041*
		Downtown	9	25.89		9	34.14	
	Night	Campus	9	14.34	**0.000**	9	23.87	*0.007*
		Downtown	9	18.27		9	25.78	
With forestation and with edification (4 and 8)	Morning	Campus	9	18.39	0.786	9	25.73	*0.006*
		Downtown	9	18.72		9	29.70	
	Afternoon	Campus	9	23.51	**0.047**	9	33.39	0.712
		Downtown	9	25.61		9	33.00	
	Night	Campus	9	15.54	**0.001**	9	23.64	*0.023*
		Downtown	9	18.14		9	25.47	

Source Authors (2018)

It can be observed in points 1 and 5, without forestation and with edification, and in points 2 and 6, without forestation and without building, a small difference in temperature compared to the forested spots, being 0.89 °C in the night at points 1 and 5 and at points 2 and 6 a difference of 2.1 °C in the morning and 1.3 °C in the night period. On points 3 and 7, with forestation and without building, the temperature difference between downtown and the campus is 4.41 °C in the morning, 0.67 °C in the afternoon and 1.91 °C at night, and in points 4 and 8 listed as points with forestation and building, the differences vary from 3.97 °C in the morning, from 0.39 °C in the afternoon and from 1.83 °C at night. The Graphic 1, shows the characteristics of the environment of each point listed for the study. This was elaborated

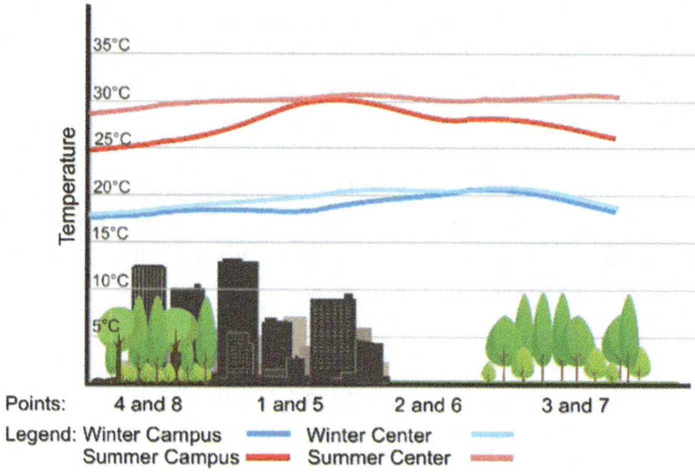

Source: Authors, 2018
Note: 1 ° Celsius equals 33.8 ° Fahrenheit

Graphic 1 Analysis of temperatures in the demarcated areas in the morning

through the averages of the obtained temperatures through the measurements made in these stations, as already mentioned, proving that in these periods the temperature in the urban center of the city is higher in relation to the university campus, thus demonstrating the formation of heat islands.

It is perceived, through the Graphic 1 a greater difference of the temperatures in the winter between the campus and downtown. On summer, this difference is smaller, but still significant for comparison. These differences show that downtown always have the highest temperatures in relation to the university campus at these points, which leads us to prove the formation of the heat islands and the performance of urban forestation in the control and regulation of downtown temperatures. However, Table 2 will be presented below, which will show the performance of the humidity on campus and in the city at the points analyzed for a better understanding of the study.

In the winter, according to Table 2, season the humidity had statistically significant percentages in the period of the night in three points in the night period. The highest mean moisture difference was found at points 2 and 6 in the evening, where the difference in humidity between the campus and downtown was 9.51%. According to the World Health Organization (2018), relative air humidity considered ideal ranges from 50 to 80% and in the study, the morning period presents a percentage superior to 50% in all points and in the night above 60%, only in the afternoon that the percentages fall in all points, being between 38 and 46% of humidity. Downtown, the average humidity remained at 64.21% and campus 70.26%, which shows the benefits of rural areas, which normally have more vegetation.

Table 2 Descriptive measures, Student's t-test of humidity in the central urban area and the university campus of UNICRUZ. Cruz Alta/RS, 2018

			Winter			Summer		
Collect point	Shift	Local	N	Average (%)	*p*-value	N	Average (%)	*p*-value
Without forestation and with edification (1 and 5)	Morning	Campus	9	62.02	0.272	9	47.62	0.638
		Downtown	9	57.87		9	44.92	
	Afternoon	Campus	9	43.31	**0.002**	9	36.88	0.817
		Downtown	9	38.41		9	35.12	
	Night	Campus	9	64.80	0.060	9	58.48	0.616
		Downtown	9	62.17		9	55.82	
No forestation and no building (2 and 6)	Morning	Campus	9	57.54	0.732	9	52.44	0.054
		Downtown	9	58.69		9	42.34	
	Afternoon	Campus	9	38.02	0.311	9	37.68	0.924
		Downtown	9	39.57		9	38.40	
	Night	Campus	9	72.59	**0.000**	9	58.13	0.584
		Downtown	9	63.08		9	54.52	
With forestation and without building (3 and 7)	Morning	Campus	9	63.79	0.230	9	55.37	*0.070*
		Downtown	9	60.20		9	45.96	
	Afternoon	Campus	9	39.79	0.193	9	43.22	0.783
		Downtown	9	41.47		9	41.48	
	Night	Campus	9	74.47	**0.023**	9	62.38	0.206
		Downtown	9	67.89		9	55.20	
With forestation and with edification (4 and 8)	Morning	Campus	9	63.78	0.297	9	52.46	*0.051*
		Downtown	9	60.87		9	45.19	
	Afternoon	Campus	9	45.69	0.287	9	44.73	0.802
		Downtown	9	43.20		9	46.29	
	Night	Campus	9	69.18	**0.001**	9	62.37	0.261
		Downtown	9	63.71		9	55.30	

Source Authors (2018)

It is observed, from the table, that the points that had the lower humidity was the 1 and 5, without forestation and with building, in the afternoon registering 38.41% of average and with a difference of 4.9% in relation to the campus.

In the summer season the relative humidity of the air in the afternoon shift in downtown was below 40% in points 5 and 6, areas that do not have any influence of forestation around the points, which may justify the percentage of up to 46, 29% in the forested area in the same period in points 7 and 8. It is generally observed the afternoon period in both the campus and downtown, that the points where there is forestation, the relative humidity of the air is greater in relation to the non-forested spots.

In the morning the humidity downtown remained below 46%, while on campus it can be observed that in this same period the humidity reached to register about 56% in average. It is noticed that in the period of the night the humidity increases again in all points, but only on campus in the forested spaces that humidity goes from 62%, in other points is less than this percentage. It can be noticed that the forest spots in the morning presented a statistic more significant than in the non-vegetated points, which characterizes the importance of the surrounding vegetation to assist in the decrease of the temperature and increase of the relative humidity of the air.

Graphic 2 shows the comparisons between the analyzed points in relation to the relative humidity of the air. It is noted that in the campus area the humidity is higher in relation to downtown in the two seasons and that in summer, downtown proved to be less humid compared to the campus, one of the reasons is due to the lack of vegetation that help in the cooling and humidification of the air.

The points that reached a higher humidity were the, shown in Graphic 2, 3 and 7 that are characterized as vegetated points and that do not suffer with the influence of buildings and the points that suffered with the decrease of the humidity were the 2 and 6 (without forestation and buildings) and the points 1 and 5 (without forestation and with buildings).

When comparing the data collected, it can be observed that the performance of the heat islands increases during the course of the day, appearing more intense after

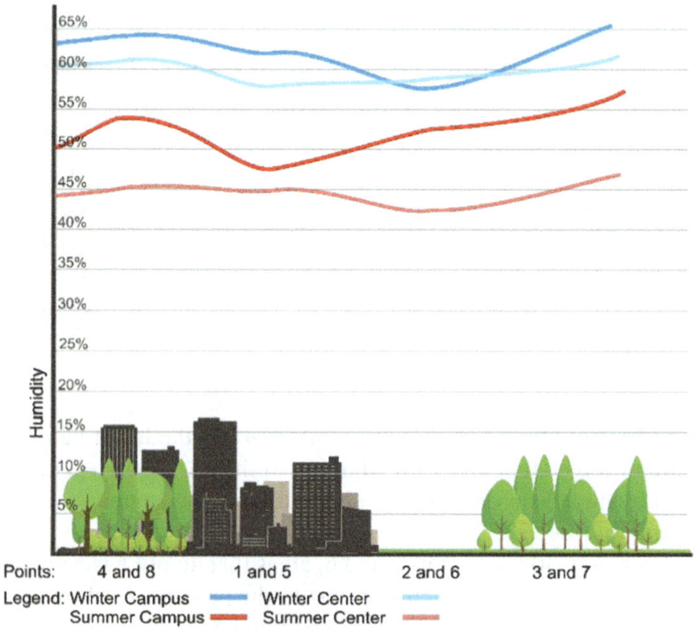

Source: Authors, 2018

Graphic 2 Analysis of humidity in the demarcated areas in the morning sunset. Gartland (2010) already reported that the intensity of the effects varies during the day and night, because in the morning shift temperatures are constantly lower, and over the day temperatures tend to increase, along with heating of urban centers. The points that have buildings in the surroundings, show an increase in temperature even in the afternoon period and lasts until night, proving that the built areas influence the temperatures of the urban areas.

Due to the information presented and analyzed, the formation of the heat island in the city of Cruz Alta, which contributes to an urban area with less quality for its inhabitants. However, through this research it is recommended some mitigation strategies of the heat islands that provide a better quality of life and an environmental comfort for the population of the municipality, bringing positive changes to the existing microclimate.

Strategies

Urban landscaping is one of the green infrastructure strategies indicated for the qualification and assistance in the improvement of the urban microclimate. It can be present in green corridors, parks and preserved natural landscape environments, but in addition to providing beauty to the locals, its use can be linked to the resolution of problems in the existing urban infrastructure, in particular on water quality and urban drainage, for example, the rainforest system and the stormwater planters that can be implanted in public roads (Cormier and Pellegrino 2008).

It is the responsibility of the rain gardens or rainforest to carry out the process of infiltration of rainwater into the soil, because the pollutants brought by the water are filtered, being through the air, the sun and micro-organisms present in these areas. The vegetation inserted in these places has extremely important functions, the imprisonment and decomposition of pollution, and the act as a humidifier of the enclosure (Cormier and Pellegrino 2008).

According to Herzog (2013) the green corridors have a characteristic of multi-functionality, because they act as an interconnected network of permeable vegetation residues, which are connected by forested or local areas with the presence of water resources, in which biodiversity aims to reorganize the urban landscape at different scales.

Many are the benefits that public green spaces offer to the hostile environment of urbanized cities. Areas of urban parks, green corridors, linear parks and gardens, contribute to the reduction of air pollutants, because they perform the photosynthesis, contribute with the permeability and humidity of the soil, assist in the control of temperature and humidity of the air, protect the slopes of rivers and lakes from soil erosion and they reduce urban noise by creating barriers. These benefits contribute in general to the quality of life in urban spaces and environmental comfort (Benini 2015).

Green corridors and linear parks are green areas of public character that are introduced from the urban mesh through the configuration of the environment, adding features and performing various functions (Herzog 2010). They can be implanted in either natural or reclaimed areas, can be projected along rivers and streams, in flooded or easily flooded areas, to protect the fragility of these ecosystems. For the elaboration of green corridors, it is necessary to propose tracks and cycle paths that interlude the neighborhoods that are part of the environment, so that these roads are an alternative form of transportation, where there is no influence of traffic or it is controlled in a way that avoids direct impacts on the corridors (Herzog 2008).

Green corridors are not solutions to all problems in urban centers, but it has been shown a possibility to improve them and that can be implanted in different situations and cities of the country, helping to prevent and mitigate the impacts caused by man's influence on the environment.

The planting of trees in urban areas has been highlighted in several ecological advantages, environmental and social aspects, as they are responsible for integrating urban spaces. For this, there are guidelines that define the diversity of species that can be implanted in urban centers, for example, it is not indicated that insert more than 10% of any kind of species, 20% less than any gender and no more than 30% of any family, so that the maintenance and preservation of the species can be protected (Sever Mutlu et al. 2017). One of the other factors is the high amount of native tree species, which are capable of enhancing ecological integrity, but this also does not exclude the possibility of insertion of non-native tree species, which also contribute to urban spaces (Almas and Conway 2016).

Its main advantage is to be able to reduce from 0.5 to 4 °C the urban temperature in the urban centers affected by the islands of heat (Qiu et al. 2013). In addition, vegetation in urban areas acts against noise pollution and maximizes pollution in the atmosphere, minimizing ambient heat and improving the quality of life (Pisello et al. 2018).

As the population likes to enjoy outdoor space in the presence of vegetation and also in the company of people, this would already be one of the reasons to encourage the creation of more green spaces, even if it is on the street or block itself. These green spaces only need to be part of people's social and recreational landscape, making these places comfortable and attractive so that users can appreciate it.

Conclusions

Studies on the consequences of heat islands in urban environments, mainly highlights the problem with thermal comfort, since the behavior of cold and heat influence the population health, on high energy consumption, urbanity, income at work, and finally, the quality of life of the communities (Souza and Nery 2012).

One of humanity's greatest challenges today is to transform urban spaces into places that are resistant to climatic problems that can lead to the aggravation of diseases or lead to the death of human beings (Monteiro et al. 2012). A worrying point is that a large part of human beings, particularly those with political and decision-making power in environmental management, lives in artificial urban environments and do not relate the importance of the natural environment as inexhaustible resources, being necessary, that we environmentally re-educate us so that we realize how important our natural resources are (Primavesi et al. 2007).

It is possible to verify that the heat islands are formed during the daytime period and dissipate in the night period, but the apex of the temperature difference between urban and rural areas happen at nightfall, because the rural environment cools faster than the urban one, due to the vegetal cover and existing forestation. The size of urban centers is also related to heat islands, not discarding that small and medium-sized cities also occur. The effects caused by the heat islands contribute in the air pollution, because when the air heats up, it tends to focus on the center of the city, transporting all the pollutants with them, making it difficult the dissemination of them (Romero 2011b).

Managers should further introduce the purpose of urban climate studies in the management plans, which would require incorporate into the process of urban planning of cities and to improve the conditions and quality of life of the population (de Moura and Zanella 2012). Climate should be considered as an essential resource for understanding, planning and managing urban space (de Souza and Nery 2011).

The strategies must go beyond the implementation of green areas, alternative infrastructure techniques should be inserted, with the incentive to use transport that does not pollute the air, the conscious consumption of available natural resources, the use of light colors in the pavements and coatings to increase the reflectance of the heat, careful with the heights of the buildings so that no canyons of wind are created, systems for the reuse of gray and rainwater, among other sustainable urban planning strategies.

References

Almas AD, Conway TM (2016) The role of antive species in urban forest planning and practice: a case study of Carolinian Canada. Urban Forestry and Urban Greening, pp 54–62

Benini SM (2015) Infraestrutura verde como prática sustentável para subsidiar a elaboração de planos de drenagem urbana: estudo de caso da cidade de Tupã/ SP. Tese de Doutorado pela Universidade Estadual Paulista. Presidente Prudente

Burkett et al (2014) Climate change 2014: impacts, adaptation, and vulnerability. Part A: global and sectoral aspects. Contribution of working group II to the fifth assessment report of the intergovernmental panel on climate change. Cambridge University Press, Cambridge, pp 169–194

Cormier NS, Pellegrino PRM (2008) Infra-estrutura verde: uma estratégia paisagística para a água urbana. São Paulo. Paisagem Ambiente: ensaios, no 25, pp 125–142

Correa SMBB (2003) Probabilidade e Estatística. - 2nd edn. PUC Minas Virtual, Belo Horizonte

de Almeida, AM (2009) Conforto térmico e eficiência energética em edifício multifamiliar na cidade de Maceió. Maceió: Dissertação (Mestrado). Universidade Federal de Alagoas, Centro de Tecnologia

de Almeida Junior NL (2005) "Estudo de clima urbano: uma proposta metodológica". Cuiabá: Dissertação (Mestrado). Departamento de Física, Instituto de Ciências Exatas e da Terra, Universidade Federal de Mato Grosso de Moura MO, Zanella ME (2012) Os estudos de clima urbano no nordeste do Brasil. (Dissertação de Doutorando). Concepções e Ensaios da Climatologia Geográfica. 1st edn, vol 1. Editora da UFGD, Dourados, pp 39–60

de Neto AAC (2007) Estudo Bioclimático no Campus da Universidade Federal de Mato Grosso. Cuiabá- MT: Dissertação (mestrado). Universidade Federal do Mato Grosso

de Souza DM, Nery JT (2011) Parâmetros climáticos de Ourinhos, Estado de São Paulo. Maringá: acta Scientiarum. Hum Soc Sci 33(1):107–115

de Souza DM, Nery JT (2012) O Conforto térmico na perspectiva da Climatologia Geográfica. Londrina: Revista Geografia 21(2):65–83

Fialho ES (2012) Ilha de calor: reflexões acerca de um conceito. Universidade Federal de Viçosa. ACTA Geográfica, Boa Vista, ed esp, pp 61–76

Gartland L (2010) Ilhas de Calor: como mitigar zonas de calor em áreas urbanas. Oficina de Textos, São Paulo

Herzog CP (2008) Corredores verdes: expansão urbana sustentável através da articulação entre espaços livres, conservação ambiental e aspectos históricos culturais. In: Terra CG. e de Andrade R. Coleção Paisagens-Culturais – Materialização da Paisagem através das Manifestações Sócio-Culturais. UFRJ-EBA. Available in: http://inverde.files.wordpress.com/2010/08/herzog_cecilia_corredores_verdes.pdf

Herzog CP (2010) Infra-estrutura verde para cidades mais sustentáveis. Produtos e sistemas relativos a infra-estrutura. In: Cadernos Virtuais de Construção Sustentável. Secretaria do Ambiente (SEA) do Estado do Rio de Janeiro. Available in: http://inverde.wordpress.com/artigos-eteses/. Acesso em 19 ago 2018

Herzog CP (2013) Cidades para Todos: (re) aprendendo a conviver com a natureza. 1. Ed. Rio de Janeiro: Mauad X: Inverde, p 312

IBGE (2017) Instituto Brasileiro de Geografia e Estatística. Available in: http://www.ibge.gov.br/home/

IPCC (2014) Intergovernmental panel on climate change. Impacts, adaptation and vulnerability

Leite C (2012) Cidades sustentáveis, cidades inteligentes: desenvolvimento sustentável num planeta urbano. Bookman, Porto Alegre

Mascaró L (1991) Energia na edificação: estratégias para minimizar seu consumo. Projeto Editores Associados, São Paulo

Mascaró L, Mascaró JJ (2009) Ambiência Urbana, 3rd edn. Masquatro, Porto Alegre

Mayer H, Hope P (1987) Thermal conforto f man in diferente urban enviromments. In: Theoretical and applied climatology, 38p

Monteiro A, Velho S, Góis J (2012) A importância da fragmentação das paisagens urbanas na Grande Área Metropolitana do Porto para a modelização das ilhas de calor urbano – uma abordagem metodológica. Revista da Faculdade de Letras- Geografia- Universidade do Porto, III série, vol I, pp 123–159

Piazza GA, Torres E, Moser P, Gotardo R, Kaufmann V, Pinheiro A (2016) Análise Espacial e Temporal dos Dados de Precipitação das Estações de Entorno da Bacia do Ribeirão Concórdia, Lontras (SC) Visando Sua Inserção no Contexto de Mudanças Climáticas. São Paulo: Rev. Brasileira de Meteorologia 31(4) supl 1

Primavesi O, Arzabe C, dos Pedreira MS (2007) Aquecimento global e mudanças climáticas: uma visão integrada tropical. 1st edn. São Carlos: Embrapa Pecuária Sudeste

Romero MAB (2000) Princípios bioclimáticos para o desenho urbano. CopyMarket.com

Romero MAB (2011) Correlação entre o microclima urbano e a configuração do espaço residencial de Brasília. Belo Horizonte: Fórum Patrimônio, Mudanças climáticas e o impacto das cidades 4(1):9–22

Romero MAB (2011) Arquitetura do lugar: uma visão bioclimática da sustentabilidade em Brasília. São Paulo: Nova Técnica

Seven Mutlu S, Selim C, Un, G (2017) Plant biodiversity of urban roadside trees in Antalya, Turkey. Kastamonu Uni. Orman Fakültesi Dergisi: 80–87

United Nations Conference on Sustainable Development RIO + 20 (2012) The future we want. Rio de Janeiro, Brazil. Outcome or the conference 20–22 June 2012

United Nations New York (2015) The millennium development goals report. Available in: http://www.un.org/millenniumgoals/2015_MDG_Report/pdf/MDG%202015%20rev%20(July%201).pdf

UNIVERSIDADE FEDERAL DO PARANÁ – Setor de ciências exatas, departamento de estatística. Apostila Estatística II. Curitiba, 2009. Disponível em: https://www.est.ufpr.br/ce003/material/apostilace003.pdf. Acessado em: 06/07/2018

Viana, SSM, de Amorim MCC (2009) Caracterização do clima urbano em Teodoro Sampaio/SP: Uma introdução. Uberlândia: Sociedade & Natureza, pp 19–42, dez. 2008. World Meteorological Organization. Handbook on Climat and Climat Temp Reporting. Available in: https://www.wmo.int/pages/prog/www/OSY/Publications/TD1188/HandbookCLIMAT-CLIMATTEMP_en.pdf

World Meteorological Organization (2009) Handbook on climat and climat temp reporting. Available in: https://www.wmo.int/pages/prog/www/OSY/Publications/TD1188/HandbookCLIMAT-CLIMATTEMP_en.pdf. Acesso em: 05 abr. 2018

Preliminary Evaluation of Emergency Shelters for Disasters Associated with Landslides at the Hydrographic Basin of Corrego D'Antas, Nova Friburgo, Rio de Janeiro, Brazil

Tomás Coelho Netto Duek, Leonardo Esteves de Freitas and Marcos Barreto de Mendonça

Abstract Socio-environmental disasters associated with landslide events are usually triggered by intense rainfall. In Southeastern Brazil, rains are becoming more severe, reaching more often extreme values due to climate change, making it increasingly relevant to implement risk management processes capable of reducing its consequences. In disaster emergency situations, shelter must be provided before, during and after the occurrence of the adverse event, in order to minimize its consequences. The emergency shelter's adequacy evaluation intends to establish common minimum requirements to protect public health, safety and general the well being for those affected by disasters. In 2011, a major disaster occurred in the mountainous region of Rio de Janeiro State in Brazil. Some deaths could have been avoided with proper planning on where people at risk should take shelter. The purpose of this study is to propose and apply a methodology of quantitative-qualitative assessment of emergency shelters suitability, putting forward, through a multi-criteria analysis, an Evaluation Coefficient for Emergency Shelters for the Córrego d'Antas basin (**ECES-CD**). The results ranked 5 shelters as 'suitable', 2 as 'moderate' and 4 as 'not suitable. It was also noted in this study that all eleven shelters presented accessibility problems. It is possible, from the different analyzed items, to identify what measures should be taken to improve the condition of shelters.

T. C. N. Duek (✉)
Defense and Civil Security, Fluminense Federal University, Rio de Janeiro, Brazil
e-mail: tomascoelhonetto@gmail.com

L. E. de Freitas
Laboratory of Geo-Hydroecology, Geography Department, Federal University of Rio de Janeiro (UFRJ) and Observatory of Sustainable and Healthy Territories of Fundação Oswaldo Cruz (Fiocruz), Rio de Janeiro, Brazil
e-mail: leonardofreitas73@gmail.com

M. B. de Mendonça
Department of Civil Construction, Polytechnic School, Federal University of Rio de Janeiro (UFRJ), Rio de Janeiro, Brazil
e-mail: mbm@poli.ufrj.br

W. Leal Filho et al. (eds.), *Climate Change, Hazards and Adaptation Options*,
Climate Change Management, https://doi.org/10.1007/978-3-030-37425-9_9

Introduction

Climate change will affect large parts of the world's population in the coming decades, and will impact, increasingly and above al else, the most vulnerable groups (Bowen and Friel 2012; Magrin et al. 2014). Climate change has various consequences in different parts of the world, including the rise in frequency and magnitude of disasters associated with natural phenomena (Hoegh-Guldberg 2018).

In the Amazon Basin, for instance, diverse changes in the local climate have been observed, such as droughts and excessive rainfall, profoundly impacting natural ecosystems and local populations: "*(...), the impacts of climate change in the Basin vary significantly and are vast; (...), increased severity of drought can greatly affect the Amazon's freshwater ecosystems and the people that rely on them, change in rainfall and temperature could impact the spread of disease, and sea level rise and storm surge will have substantial impacts on lowland areas of the Amazon delta*" (Usaid 2018 p. 1).

In Southern Brazil, the Brazilian Panel in Climate Change and the Intergovernmental Panel on Climate Change Reports indicated an increase in temperature of up to 2.5 °C between 1901 and 2012, as well as an increase in the number of days with more than 30 mm rainfall (MMA 2015). In this region, intense climate events impacts populations and economies generating high-risk situations (Bonatti 2016). Figueiró and Coelho Netto (2011) observed a change in rainfall regime in this region in the 20th century, associating that with climate change. The authors argue that, while the frequency of rainfall below 20 mm/day is dropping, reducing the total number of rainy days per year, extreme rainfall totals above 100 mm/day are increasingly more frequent compared to late 1990s. Other authors also observed the increase in frequency of extreme rains in Southeastern Brazil (PBMC 2012; Marengo et al. 2009, Groisman et al. 2005).

Considering the direct association between extreme rainfall events in the mountainous areas of Brazil's Southeastern region and the occurrence of landslides[1], this elevation in frequency of extreme rainfall—probably associated with the recent process of climate change—makes the landslides more likely to occur. Here, landslides often reach the dimension of disasters, since they generate losses and damages of social, economic and ecological nature. Therefore, climate change, by increasing the frequency of extreme rains, tends to increase the occurrence of disasters.

The data indicate an increase of more than 21% in landslide related disasters between 1991 and 2012 in Brazil and the occurrence of 80% of these phenomena in the Southeast Region, affecting almost 4 million people (Ceped UFSC 2013). Most of the landslide related deaths are concentrated in Rio de Janeiro State, amounting to 418 of the 525 total deaths in the period, when landslides directly or indirectly affected 991,393 people (Ceped UFSC 2013). It is important to note that it was in the last few years, precisely those for which several studies identify a more pronounced change

[1]In this paper the term landslide is used to refer to different types of mass movement, such as shallow or deep landslides, debris flow and rock mass movements

in precipitation regimes, that concentrated the largest number of people affected by these types of disaster.

During the same period, flash floods, including debris flow, were responsible for most of the deaths and losses Rio de Janeiro State, affecting a total of 1395 million people and causing 1153 deaths (Ceped UFSC 2013). Again, the frequency and magnitude increase in recent years is clear.

The Mountainous Region of Rio de Janeiro is the one that concentrates the greatest occurrence of landslides. These phenomena are part of the landscape evolution itself, recurring for at least 10,000 years (Coelho Netto et al. 2015). After human occupation, the occurrence of landslides began to generate disasters, with records that go back to at least the 19th century. Deaths related to landslides and floods in the Mountainous Region occurred in almost every year between 2000 and 2012 (Fundação Coppetec 2014). The disasters of January 11 and 12, 2011 in the municipalities of Petrópolis, Teresópolis and Nova Friburgo (Coelho Netto et al. 2011) alone claimed 964 lives, according to official data (Coppetec Foundation 2014).

The Socio-Environmental Disasters Associated with Landslide and the Reduction of Vulnerability Through Emergency Shelters (ES's)

According to Highland and Bobrowsky (2008), "landslide" is a general term used to describe the downslope movement of soil, rock, and organic materials under the effects of gravity, but also the landform that results from such movement. In the present study, water is the main trigger agent of this process, by increasing the pore pressure and thus reducing the shear strength of the slope and, consequently, detonating its rupture.

A socio-environmental disaster is considered to happen when natural events, such as landslides, affect a determined social system (Freitas and Coelho Netto 2016). In order to reduce the risks of socio-environmental disasters, it is of great importance to create public policies focused on the implementation of mitigation measures. In Brazil, the Civil Defense is responsible for taking risk reduction measures (Brasília 2007), including the implementation of alert and alarm systems and the establishment of emergency shelters.

When a disaster of great magnitude occurs, one of the greatest consequences, apart from death, is lack of shelter. There is, therefore, the need to give refuge and protection for those affected or at risk. Lack of shelter may be temporary or even occur before the event, due to a decision to evacuate homes that are in risk areas. Thus, shelter availability can be critical in disaster risk reduction processes, since shelter is a determining factor for survival in the early stages of a disaster (McConnan 2004).

The provision of emergency shelters, therefore, is one of the key measures that should be adopted by government officials concerned with climate change, especially in view of the increase in the occurrence of extreme precipitation events (Aminzadeh 2007; Melgarejo and Lakes 2014).

This is particularly relevant in a climate change context, when demand for emergency shelters will be far higher than their availability planned by the public authorities, as Dasgputa et al. (2010) pointed out in their assessment of cyclone protection shelters in Bangladesh.

However, it is common for residents at risk not to perceive or to underestimate degree of risk indicated by technicians, and not rely on existing warning and alarm systems, even in scenarios of climate change in which extreme rainfall is frequent. Carvalho and Coelho Netto (2014) verified that 94% of the residents at Córrego d'Antas, a risk area located in the Mountainous Region of Rio de Janeiro, refuse leave their houses when the siren of the alarm system sounds, and that 75% of them do not feel secure in moving to the Emergency Shelters previously set by the local government.

Emergency Shelters (ES)

Emergency shelters are classified according to the need of shelter and the time needed to house disaster victims in each one, designating each one as "emergency shelter", "temporary shelter", "temporary housing" and "permanent housing" (Chou et al. 2013, Quarantelli 1995).

The object of the present study are emergency shelters, such as defined by Quarantelli (1985, 1995), as immediate response shelters so that the victims of disaster can shelter for short periods of time before and after the impact. Emergency shelters are spontaneously used and in a timely manner by local inhabitants, as the need arises (Anders 2007 Duek et al. 2017), meeting needs ranging from pre-impact flight, the use of a place that is connected to the power grid or that is equipped with generators and where one can spend time until the threat of imminent danger is reduced or terminated. They can also take the form of emotional protection, a place to store one's belongings, a place exclusively for the use of services or a stopping point before future actions to reduce the consequences of the disaster. Therefore, emergency shelter offers considerations beyond "four walls and a roof" for the people affected before, during and after a disaster (Smyth and McInerney 2012).

However and is spite of that, it is common that the definition of shelter locations prove to be inadequate in times of disaster, disregarding local climatic conditions as well as local social and cultural issues and the shelter's own security (Bashawria et al. 2014; Zahari and Hashim 2018).

It is essential, therefore, to assess whether emergency shelters available in disaster-prone areas meet basic criteria, ensuring the safety and minimum comfort for those seeking refuge (Melgarejo and Lakes 2014, Zahari and Hashim 2018).

In ES, people are looking for short-term relief in a safer place. The duration of stay for unsheltered people at an ES ranges from 1 to 18 h, so no beds or substantial meals is required (Smith and Parsons 2014). The present study aims to develop a methodology of preliminary evaluation based on a multi-criteria analysis for ES established by the Civil Defense in the Córrego d'Antas basin in the municipality of Nova Friburgo, in Rio de Janeiro's Mountainous Region.

Study Area

The Córrego d'Antas watershed (53 km^2), located in the municipality of Nova Friburgo in the mountainous region of Rio de Janeiro (Fig. 1), was strongly affected by landslides in 2011. There is a prevalence of man-made vegetation, especially grasses in this drainage basin. Slopes and valley floors include agricultural areas and urban areas spread on valley floors, especially in the lower part of the drainage basin. The average annual rainfall in the region is about 2000 mm, and its geological composition consists of granite and gneiss with low cohesion soils (Silva et al. 2016). The area is a mountainous basin with an elevation of 1180 m, with a maximum altitude of 2016 m (Fig. 1), with a tendency to evolve geomorphological processes by incising canals and landslides, with few forest remnants and occupied by about 23,000 people (Coutinho 2014) (Reger-CD Database).

Methodology for Constructing a ES Evaluation Coefficient

The present study proposes the construction of a methodology offering technical support to the practical exercise of choosing ES in urban or rural environments, for eventual disasters caused by landslides. A multi-criteria Evaluation Coefficient for Emergency Shelters is proposed for the Córrego d'Antas basin (ECES-CD), based on qualitative and quantitative aspects. The ECES-CD aims to evaluate existing and future ES's, constituting a tool to support processes of adaptation to climatic changes with respect to disaster risks management.

The methodology's development was based on a bibliographical review on the subject, performed in 2017, especially in the studies of Sphere Project (2011), Anhorn and Khazai (2015) and Smith and Parsons (2014) and in consultations with professionals involved in risk management, realized in the same year.

This proposal, which was finalized in November 2017, was used to evaluate the 11 ES established by the Municipal Civil Defense of Nova Friburgo for the Córrego d'Antas basin. They are all located in public schools or churches.

Table 1 presents the indicators for assessing ES adequacy during landslides events in the Córrego d'Antas basin. The following describes the meaning of the indicators—Implementation items, Environmental items, Structure and Accessibility items and their respective sub-items—for the evaluation of these 11 shelters:

(a) Implementation Items

Implementation items are divided into types of establishment and types of use.

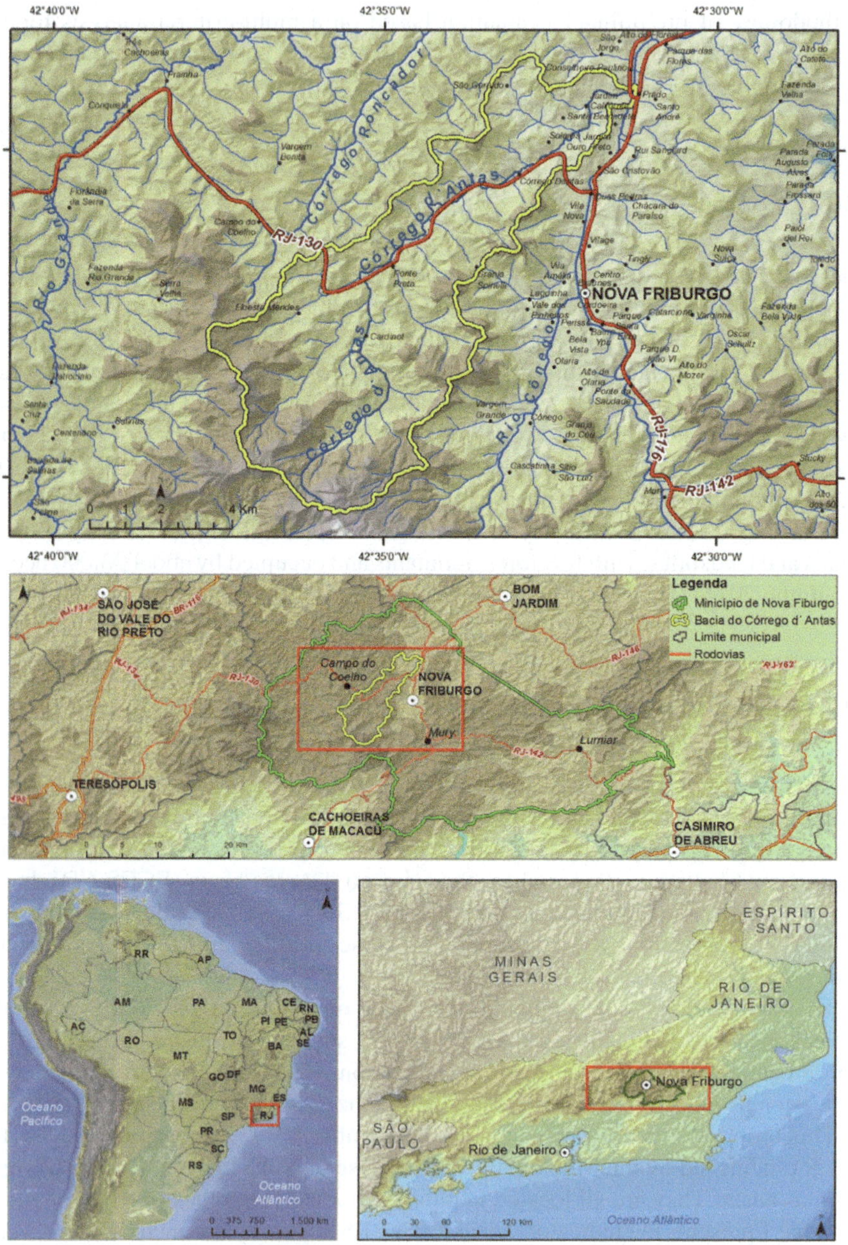

Fig. 1 Location of the Córrego D'Antas drainage basin. *Source* Reger-CD Database. Layout by Nunes, FSB)

Table 1 Categories of indicators, weights and scores for assessing emergency shelter locations adequacy

Category	Weight	Indicator	Score	Typification
Implementation	0.1	Types of establishment	1	Public (Governmental, community, religious, institutional, educational …)
			0.8	Private
	0.05	Types of use	0.9	School
			0.7	Church
			0.5	Community Center
			0.4	No use
			0.3	Companies in general
			0.2	Private parks and stadiums
Environmental	0.2	Susceptibility associated with landslides and floods[a]	1	No danger
			0.7	Flood danger
			0.5	Landslide danger (slides, detrital flow, and falling blocks)
			0.3	Flood and landslide danger
	0.05	Types of pollution	1	No pollution
			0.7	Exposed solid wastes
			0.5	Nearby raw sewage
			0.3	Solid waste and raw sewage nearby
Structure and accessibility	0.1	Electricity	1	Distribution line and generator or other alternative energy source
			0.9	Generator or alternative source
			0.7	Distribution line
			0.2	No electricity
	0.1	Water supply	1	Public water supply, reservoir and fountain
			0.8	Fountain and reservoir
			0.7	Fountain and running water

(continued)

Table 1 (continued)

Category	Weight	Indicator	Score	Typification
			0.6	Any kind of source (natural, water sheets or well)
			0.4	Reservoir
			0.2	Piped water
			0	No water
	0.05	Capacity (number of people)	1	100 or more
			0.9	50–100 people or more
			0.8	1–50 people
	0.2	Obstacle of any kind on the escape route	1	No obstacles
			0.2	Obstacles present
	0.15	Critical facilities	1	Hospital between 2 and 5 km
			0.8	Hospital between 5 and 10 km
			0.6	Hospital between 10 and 15 km
			0.4	Hospital more than 15 km
			0	Unknown distance

[a]Consider areas of medium or low susceptibility to landslides at the site of the emergency shelter

(a.1) Types of establishment: public or private. The highest score was awarded to public establishments, as they facilitate articulation and access to shelters when compared to private ones. Fema (2007) confirms this consideration that public spaces should be the preferred choice for shelters, since the management of private spaces depends on the decisions made by the owners of these places.

(a.2) Types of use: schools or churches. It was established that schools would have higher scores because they are public spaces with greater ease of articulation than churches and because they count with basic structure and characteristics that attend immediate and emergency use.

(b) Environmental Items:

These are items that influence the shelter regarding the environment in which it they are inserted. In order to evaluate this aspect, the work was subdivided into two analysis, the first one referring to the shelter's terrain susceptibility to landslides or floods and the second one referring to the types of pollution affecting the shelter's environment and surroundings.

(b.1) Susceptibility to landslides: In order to carry out this evaluation, the map of susceptibility to landslides or flood events in the Córrego d'Antas basin

(Coutinho 2014) was used and in situ analysis where the shelters were located were made.

(b.2) Types of pollution: This analysis was based on the data collected during fieldwork consisting of observation and interviews with those in charge of the shelters regarding the disposal of solid wastes, effluents and sound pollution, in order to evaluate the shelter's sanitary condition.

c) Structure and accessibility Items:

This item is divided into electricity supply, water supply and holding capacity of a given shelter (infrastructure) and the presence of obstacles barring the shelter's access (accessibility). Anhorn and Khazai (2015) corroborate this issue by stating that when a person leaves home looking for emergency shelter, he or she focuses on accessibility and space availability.

(c.1) Electricity: This topic assesses how many and which energy sources the establishment holds, giving a higher score according to the number of sources the shelter offers.

(c.2) Water supply: As in the previous topic, the larger the number of sources, the higher the score, thus, availability of a water reservoir, public water supply and water source give the shelter maximum score.

(c.3) Holding Capacity: The total capacity of a shelter was calculated by measuring the available covered area divided by the minimum space required for one person. The minimum space required per person considered was 1.2 m^2, as recommended for ES by Smith and Parsons (2014). The higher the shelter's absolute capacity, the higher the score. The ideal would be to compare the shelter area with the amount of people exposed to the threat of landslides that will access the shelter. However, it was not possible to obtain this information at this time and this will be the subject of further studies.

(c.4) Obstacles of any nature in the escape route: Escape routes are routes used by inhabitants from their houses to the emergency shelter during an emergency. This topic evaluated the existence of potential natural or social obstacles in the shelters accesses. The obstacle may exist due to the very event responsible for the disaster, such blocked streets, a flooded area or another existing obstacle. The more obstacles on the route, the lower the score.

(c.5) Critical facilities: In this study, the proximity to hospitals and health posts was considered a significant amenity for shelter adequacy; the closer to these locations, the higher scores.

Calculation of ECES-CD and Classification of Shelters

The ECES-CD to evaluate emergency shelter adequacy for risks associated with landslides, considering the items in Table 1, is defined according to the equation below:

Table 2 Criteria for classification of shelters adequacy according to the score obtained through the ECES-CD

Classification	Criterion	Observation
Not suitable	ECES-CD <0.5	Shelters that have obstacles in the routes to reach them (rivers and possible floods or landslides) are also considered inadequate
Moderately adequate	0.5< ECES-CD < 0.7	Shelters that have reached average scores but do not yet offer fully satisfactory conditions
Suitable	ECES-CD >0.7	Shelters that satisfactorily meet the requirements for an ES, according to the criteria presented

$$\mathrm{ECES-CD} = \Sigma(\mathrm{n}=1)^{\mathrm{n}}(\mathrm{Wi}\times\mathrm{Ii}+\cdots+\mathrm{Wn}\times\mathrm{In}); \quad (1)$$

In which:

n number of indicators, which in this case, is 9.
i indicator number, variable from 1 to 9.
Ii score of indicator i according to its typification, variable from 0 to 1.
Wi indicator weight i

Both the score (I) and the weight (W) are shown in Table 1.

The index results are presented in a qualitative way, establishing a relation between the value of the ECES-CD and shelter suitability, according to Table 2. Depending on the score obtained by a given shelter, it is considered Unfit, Moderate Adequate or Suitable. In addition, the impossibility of access to the shelter is a criterion that defines the shelter as inappropriate, regardless of the score reached.

ECES-CD Application Methodology

The schools and churches defined as ES in the study area were visited in December 2017 in order to collect the necessary information for methodology application. During these visits, interviews were conducted with officials (school directors or religious leaders) to collect information such as water and electricity supply and covered area size (measurements were taken with measuring tape).

The distances between the shelters and the nearest health service unit, the existence of obstacles in escape routes, the distribution of population clusters, water network, areas of susceptibility to landslides as well as the road network were established using ArcGis 10.3 and Google Earth software and with the support of Databases made available by Geoheco/UFRJ (Geohidroecology Laboratory of the Federal University of Rio de Janeiro) and by the Córrego d'Antas Risk Management Network (Reger-CD). A map was then drawn, indicating areas where new ES should be inserted. The definition of these areas was made from the visual identification of clusters of houses that did not have shelters nearby.

Results

To facilitate the understanding of the results, the shelters were numbered from 1 to 11, as shown in Fig. 2, which also shows the adequacy of each ES. There is a higher concentration of shelters in the basin's lower section, close to Córrego d'Antas' mouth. This is an expected result, since this is an urban portion of the basin, with a higher population density, whereas the middle and upper portions count with three shelters only, covering a very wide area, indicating a difficulty for the area's population to reach such shelters.

The following are the study area's ES main characteristics:

ES 1: Souza Cardinot Municipal School. ECES-CD: 0.60. Classification: not suitable. Positives: ample space and structure with two bathrooms. Negative aspects: it is located near the river and was flooded in the disaster of 2011, making it little suitable for ES function. Capacity: 116 people.
ES 2: Alberto Meyer Municipal School. ECES-CD: 0.62. Classification: Moderately suitable. Positive aspects: it is one of the few that counts with a water reservoir. Negative aspects: it sits near an area with high susceptibility to landslides and where it is not possible to hear the siren of the warning and alarm system. Capacity: 127 people.
ES 3: Etelvina Schottz State School. ECES-CD: 0.61. Classification: not suitable. Positive aspects: this is a large space and accommodates 125 people. Negative aspects: only serves part of the residents of the neighborhood, leaving another significant portion of residents without access, as they are on the other side of the river.
ES 4: Francisco Silveira Municipal School. ECES-CD: 0.71. Classification: suitable. Positives: ample space with several bathrooms, dining room and classrooms, capable of holding 140 people. Negative aspects: even if the structure is good, part of the population finds it difficult to access the shelter, due to obstacles along the way, such as the river itself which crosses this neighborhood.
ES 5: Umbelina Breder de Queiroz Municipal School. ECES-CD: 0.71. Classification: suitable. Positive aspects: ample space, with capacity for 142 people. Negative aspects: limited structure, counting with only one source of water, which makes it difficult to accommodate large numbers of people.
ES 6: Menino Jesus Municipal Infant Education Center. ECES-CD: 0.62. Classification: Moderately suitable. Positive aspects: reasonable space and two bathrooms. Negative aspects: being a structure designed to be a kindergarten, it is unable to provide shelter for longer than a few hours. Capacity: 83 people.
ES 7: Hope Presbyterian Church. ECES-CD: 0.65. Classification: not suitable. Positive aspects: good location with good accessibility by being on the main road with

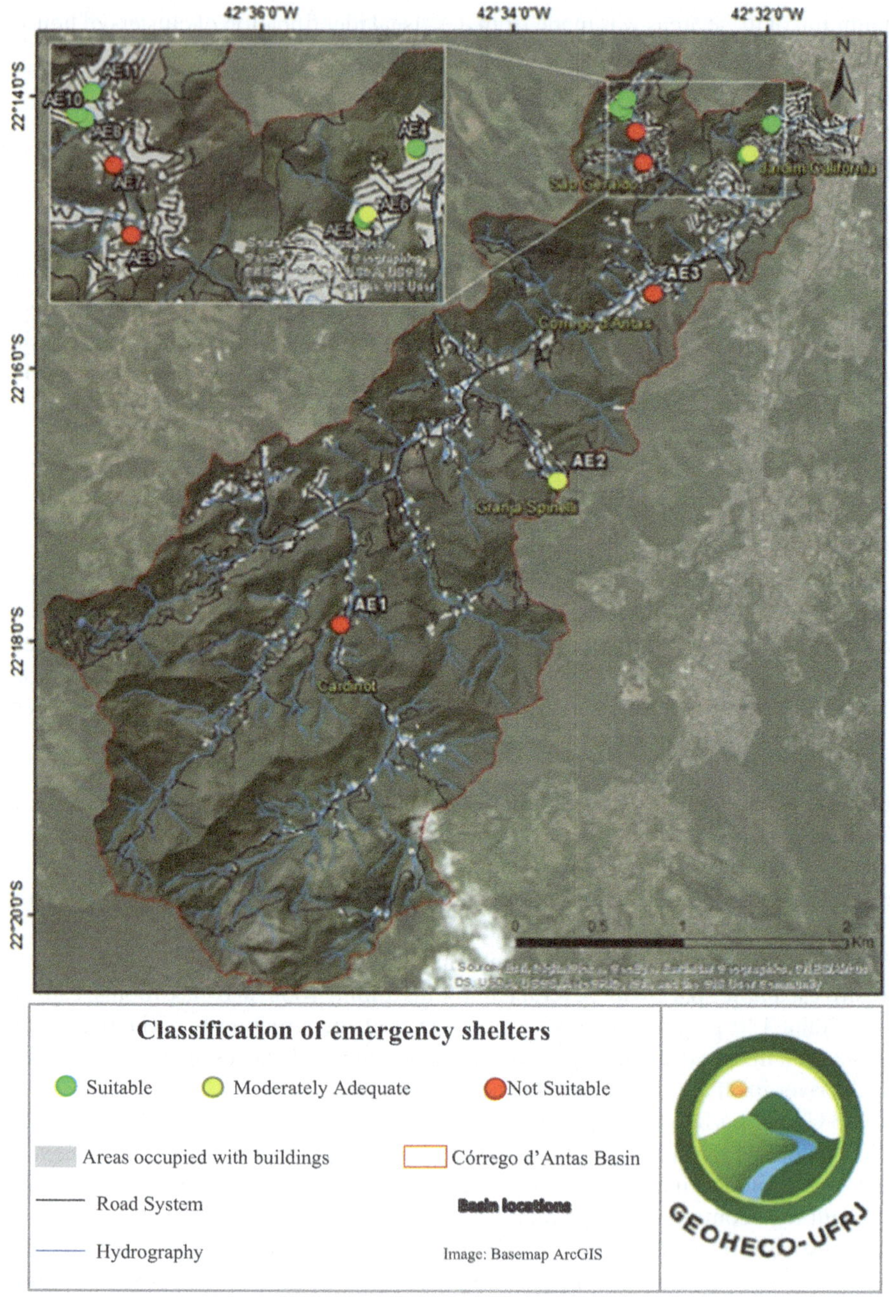

Fig. 2 Map showing the suitability of ES listed by Nova Friburgo's Civil Defense in the Córrego d'Antas basin. *Source* GeoHidroecologia Laboratory—UFRJ

good visibility. Negative aspects: steep slope just behind the shelter, which in rains can suffer slides. No one responsible to the premises was found to open and grant access to the church. Capacity: 98 people.

ES 8: 5th Baptist Church. ECES-CD: 0.71. Classification: adequate. Positive aspects: ample capacity for 250 people, kitchen and three bathrooms for women and three for men. Should it be necessary, this shelter could easily evolve into a temporary shelter. Negative aspects: part of the population need to cross areas of high susceptibility to landslides to reach this shelter.

ES 9: New Hope Catholic Church. ECES-CD: 0.59. Classification: not suitable; Positive aspects: the person in charge of the church keys showed concern on the issue of unsheltered people in disasters; Negative aspects: the access in this shelter is completely obstructed, since it sits in an elevated area and its surroundings flood completely in heavy rains. Capacity: 250 people.

ES 10: New Dawn National Baptist Church. ECES-CD: 0.71. Classification: suitable. Positive aspects: a large space, with capacity for 378 people, five bathrooms and a kitchen with two industrial stoves. This emergency shelter can evolve into a temporary shelter, housing people for days. The health service unit is nearby. Negative aspects: lack of a water reservoir and few accesses to the church's area.

ES 11: Assembly of God Church. B ECES-CD: 0.70. Classification: suitable. Positive aspects: the person in charge lives in an annex of the shelter, which facilitates the issue of shelter access. Negative aspects: There are only 98 places for unsheltered people and little or no structure beyond that required for an emergency shelter.

The results indicated that five shelters were considered adequate, two were moderately adequate and four inadequate shelters.

The highest coefficient (ECES-CD = 0.74) was attained by ES 5. One item that stood out in this shelter was the presence of a water reservoir. In comparison with analyzed other shelters, this item was relevant, since only three other had a water resrvoir. ES 5 holds 142 people, which was on average in comparison with the others, totaling 7 shelters (54.5%) that fall in the range of 100–199 people.

The lowest coefficient (ECES-CD = 0.59) was assigned to the ES 8. This shelter was considered inappropriate because it sits in a high area in relation to the community and therefore difficult to access and because its vicinity is often flooded in heavy rain. Another negative point was the presence of nearby raw sewage canal.

Based on the different indicators analyzed for the 11 shelters, it is possible to identify measures that should be taken to increase their adequacy, aiming at a better Córrego d'Antas basin community's adaptation to risk situations.

A relevant question is the placement of water reservoirs in all 8 analyzed shelters that do not count this item, to avoid water supply collapse during events, turning these shelters unviable. Another issue to be dealt with is the presence of raw sewage near ES 8. One of the most relevant items in the evaluation research of the ES is accessibility, which is closely related to access to the road network and to the presence or absence of obstacles at the time or after occurrence of the emergency.

All analyzed shelters had access to the road network, which is positive in terms of accessibility. However, most also presented some kind of obstacle for people in need of shelter to reach them, which caused the coefficients to drop considerably. This is an item that the Civil Defense and communities should be aware of as it may render a shelter unsuitable to significant numbers of people who are under threat during heavy rains. This discussion can be made from two general aspects. On the one hand, one could think of new routes to reach the shelters already established. On the other hand, more emergency shelters can be established, reducing the need for displacement of people at risk to the nearest shelter.

Defining new escape routes to reach existing shelters does not seem the best solution as there are few shelters in the basin and long routes are not an ideal solution as it increases the exposure of unsheltered people to risk. In addition, the 11 shelters are on the right bank of the Córrego d'Antas river (Fig. 2), meaning that this river will always be an obstacle for the residents on the left bank (where a significant population density is observed) to access these shelters, even if changes are made in the escape routes. Therefore, establishing new shelters is certainly the best solution, as it will reduce the need for resident displacement and the presence of obstacles in escape routes.

An analysis by means of GIS (Geographic Information System) also evidenced the absence of shelters in some population clusters in the basin, making their implementation to be necessary both to increase the available space, as well as to reduce the displacement needs.

Based on this analysis, the present study points out suitable sites to be used as ES. This analysis could serve as support to help the Nova Friburgo Civil Defense choose new shelters.

Figure 3 shows a map that indicates important places for the implementation of ES based on a visual analysis of population clusters that do not have ES inside or near them. There is a preeminent need on Córrego d'Antas' left bank, which is totally deprived of shelters, as well as in the middle and upper portions of the basin, where they are very sparse and do not meet the region's large population demand, living, in most cases, in localities far from the shelters.

It should be emphasized that installing more shelters would increase the suitability of existing shelters as it would alter the escape route used by people living in the basin. In this way, existing shelters would only be used by the people living around or near them, increasing their accessibility and, consequently, their suitability. The need to install new shelters occurs in every basin, but the Cardinot, Granja Spinelli and Córrego d'Antas neighborhoods are in most in need.

Based on a possible ES better distribution, as indicated in Fig. 3, one of the planned developments to improve the methodology forwarded in this study is the inclusion of information regarding shelter utilization demand, rather than the use of shelter capacity absolute values in the ECES-CD calculation. This information can be obtained through population census data from the study area or even counting dwellings from aerial photographs and/or satellite images.

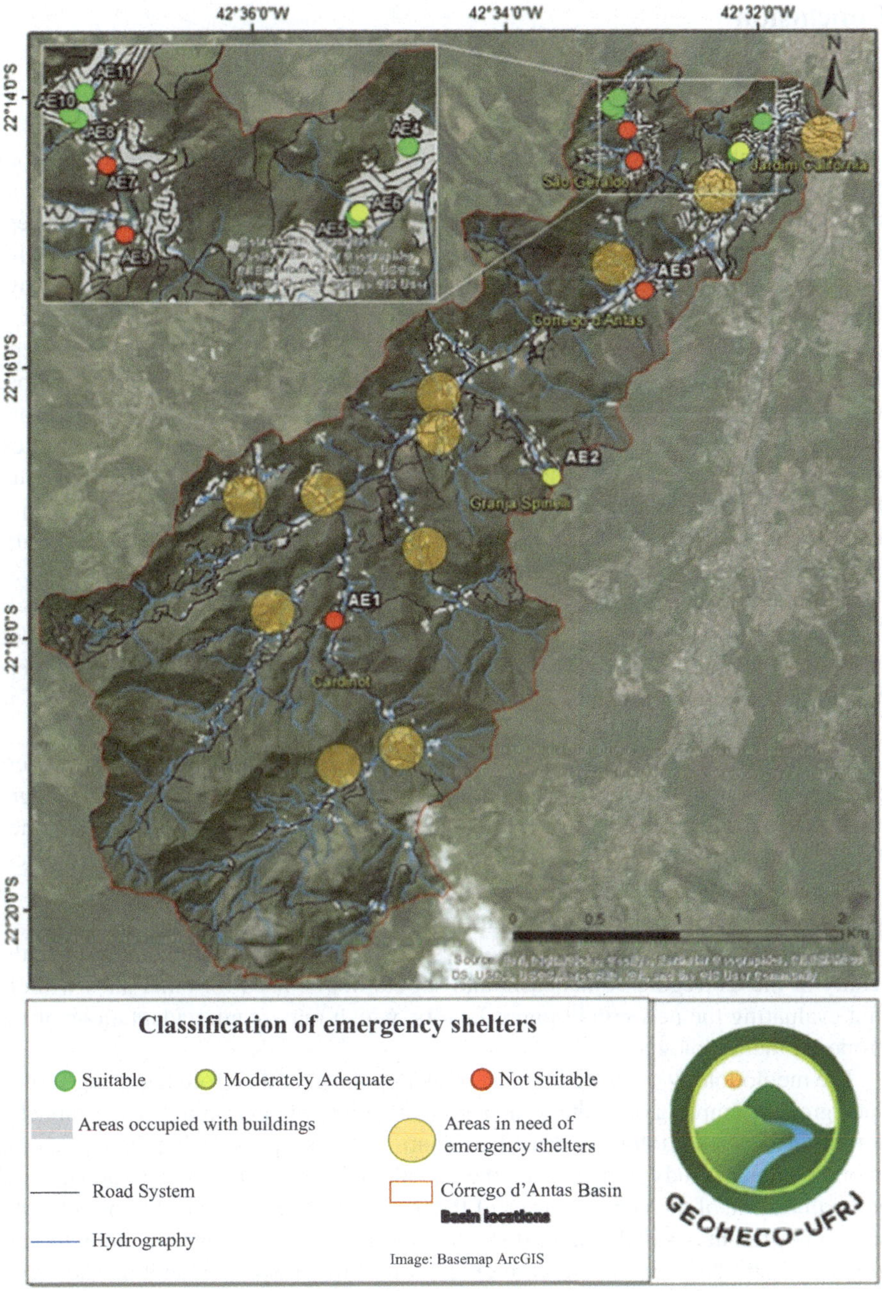

Fig. 3 Map with indication of areas in need of ES. *Source* GeoHidroecologia Laboratory, GEOHECO—UFRJ, 2017

Conclusion

In addition to the need to provide emergency shelters to the affected population, it is equally important to ensure their suitability for use.

The present paper presented an emergency shelter suitability assessment methodology. We believe that one of the main qualities of the proposed methodology is its simplicity and ease of application, requiring only the mapping of landslide susceptibility, simple analyzes in a GIS platform and the accomplishment of fieldwork for pre-defined indicator qualification. Thus, studies based on this methodology have the potential to integrate disaster prevention plans.

The evaluation by calculating the ECES-CD of the eleven ES in the Córrego dAntas basin indicated the need of at least another 10 shelters in addition to those already in existence. This would increase the supply of total capacity and improve the evaluation of the currently existing ES, since it would shorten escape routes, reducing their vulnerability. The visualization of the basin through GIS platforms, with their respective shelters, shows the areas in need and groups of residents which are without access to satisfactory shelters, especially in the central and upper parts of the basin.

Through this investigation it was possible to highlight some recommendations to improve existing ES's suitability in the study area. In general, they should be provided with multiple sources of electricity and water supply; they should be located near hospitals and fire departments, and be established in places of low susceptibility to landslides and free of obstacles in their access routes.

Although the distribution, availability, and infrastructure of existing shelters present problems, they are still important to mitigate the impacts and challenges involved. In the 2011 event, many residents did not leave their homes for emergency shelters and were victimized by the landslides, since there were no shelters in place. If the area had been evacuated and people had gone to some emergency shelter, at least some of that population could have been saved.

It is important to highlight that the present study was developed considering the reality of the Córrego d'Antas basin and should not be used for other sites without first evaluating the new conditioning factors, which may require adaptations of the proposed methodology.

The methodology presented here is in its initial stage, therefore requiring further data, greater sampling of shelters and a greater variety of environments to establish a definition of a methodology of general evaluation of this type of shelter. Its application in other areas of study to reevaluate the way the ECES is calculated is fundamental and will be one of the consequences of this research. There is a need to broaden the sampling of contexts and analyzed shelters in order for the proposed index to gain greater reliability, becoming a more efficient tool for shelter evaluation. Therefore, the application of this methodology in other contexts should be done in a judicious way, taking into account the evaluation of the specific characteristic of each analyzed shelter's locality. Evaluating the ECES-CD calculation methodology and its results from the consultation of different professionals involved in disaster risk management and the vulnerable population itself is also an important process to improve the methodology presented here.

Another shortcoming identified in this study that should be addressed in future work is the evaluation of which buildings located in areas close to those indicated in Fig. 3 may become future ES. Furthermore, it is important to reassess the ECES-CD of the shelters evaluated in this study, after the implementation of new ones.

References

Anders GC (2007) Abrigos temporários de caráter emergencial. 2007.119 f. Dissertação (Mestrado em Arquitetura e Urbanismo)—Universidade de São Paulo, São Paulo

Aminzadeh F (2007) Image is everything. Lead Edge 26(2):120–120

Anhorn J, Khazai B (2015) Open space suitability analysis for emergency shelter after an earthquake. Nat Hazards Earth Syst Sci 15:789–803. https://doi.org/10.5194/nhess-15-789-2015. Disponível em:http://www.nat-hazards-earth-syst-sci.net/15/789/2015/nhess-15-789-2015.html

Bashawri A, Garrity S, Moodley K (2014) An Overview of the Design of Disaster Relief Shelters. Procedia Econ Finance 924–931

Bowen KJ, Friel S (2012) Climate change adaptation: where does global health fit in the agenda? Globalization health 8(1):10

Bonatti M, Sieber S, Schlindwein SL, Lana MA, de Vasconcelos AC, Gentile E, Malheiros TF (2016) Climate vulnerability and contrasting climate perceptions as an element for the development of community adaptation strategies: case studies in Southern Brazil. Land Use Policy 58:114–122

Brasília (2007) Política Nacional de Defesa Civil—Ministério da Integração Nacional. Available at http://www.defesacivil.gov.br/publicacoes

Carvalho NL, Coelho-Netto AL (2014) A Gestão de Áreas de Risco a Deslizamentos e Estratégias de Resposta a Desastres: um estudo sobre a percepção de risco da população local através do sistema de alerta e alarme. In: VI Congreso Iberoamericano de Estudios Teeritoriales y Ambientales, 2014, São Paulo. Anais do VI Congreso Iberoamericano de Estudios Teeritoriales y Ambientales. São Paulo

Centro de Estudos e Pesquisas em Desastres Naturais/Universidade Federal de Santa Caratina. Atlas Brasileiro de Desastres Naturais: 1991 a 2012/Centro Universitário de Estudose Pesquisas sobre Desastres. 2. ed. rev. ampl. Florianópolis: CEPED UFSC (2013)

Chou J-S, Ou Y-C, Cheng M-Y, Cheng M-Y, Lee C-M (2013) Emergency shelter capacity estimation by earthquake damage analysis. Nat Hazards 65:2031–2061. https://doi.org/10.1007/s11069-012-0461-5

Coelho Netto, AL, Sato AM, Freitas LE (2015) Land use-vegetation-landslide interactions in the mountainous region of Rio de Janeiro State: scientific basis for risk assessment and management. In: Anais regional conference of international geographical union. Moscow

Coelho Netto AL, Sato AM, Avelar AS, Vianna LG, Araújo IS, Ferreira DLC, Lima PH, Silva APA, Silva RP (2011) The extreme landslide disaster in Brazil. In: Claudio Margottini; Paolo Canuti; Kyoji Sassa. (Org.). Landslide Science and Practice. 1ed. Springer, Berlin, Heidelberg

Coutinho BH (2014) Indicadores Geo-Hidroecológicos de Susceptibilidade das Encostas frente à erosão e movimentos de massa em Região Montanhosa Tropical Úmida: suporte metodológico para zoneamentos de susceptibilidade e risco em diferentes escalas de análise espacial. Relatório Final de Pós-Doc

McConnan I (2004) Humanitarian charter and minimum standards in disaster response. Sphere Project, Geneva

Duek TCN, Freitas LE, Mendonça MB (2017) Critérios e aplicabilidade de abrigos de emergência em caso de desastres por movimento de massa. Anais do XII Encontro Nacional da ANPEGE

Fundação Coppetec/Laboratório de Hidrologia e Estudos de Meio Ambiente. Elaboração do Plano Estadual de Recursos Hídricos do Estado do Rio de Janeiro R3-A (2014) Temas técnicos estratégicos RT-03 - Vulnerabilidade a Eventos Críticos, Volume 2 - Ocorrências de Desastres Naturais entre 2000 e 2012 por Região Hidrográfica. Rio de Janeiro 120 p

Fema (2007) Emergency temporary group housing site selection guidelines—Minimizing environmental/historic/safety problems, available at. http://www.fema.gov. Last Access May 2017

Figueiró AS, Coelho Netto AL (2011) Climatic variability and pluviometric trends in a humid tropical environment at Resende municipality- middle Paraíba do Sul river valley (Rio de Janeiro-Brazil). Brazilian Geographical Journal 2:256–273

Freitas LE, Netto ALC (2016) Reger Córrego Dantas: uma ação coletiva para enfrentamento de ameaças naturais e redução de desastres socioambientais. Ciência & Trópico, 40(1)

Groisman PY, Knight RW, Easterling DR, Karl TR, Hegerl GC, Razuvaev VN (2005). Trends in intense precipitation in the climate record. J Clim 18:1326–1350

Highland LM, Bobrowsky P (2008) The landslide handbook: a guide to understanding landslides. USGS Geologic Hazards. Circular 1325, p 129

Hoegh-Guldberg O, Jacob D, Taylor M, Bindi M, Brown S, Camilloni I, Guiot K (2018) Impacts of 1.5 °C global warming on natural and human systems

Marengo JA, Jones R, Alves LM, Valverde MC (2009) Future change of temperature and precipitation extremes in South America as derived from the PRECIS regional climate modeling system. Int J Climatol 15:2241–2255

Magrin GO, Marengo JA, Boulanger JP, Buckeridge MS, Castellanos E, Poveda G, Scarano FR, Vicuna S (2014) Central and South America. Climate change 2014: impacts, adaptation, and vulnerability. Part B: regional aspects. Contribution of Working Group II to the Fifth Assessment Report of the intergovernmental panel on climate change. In: Barros VR, Field CB, Dokken DJ, Mastrandrea MD, Mach KJ, Bilic TE, Chatterjee M, Ebi KI, Estrada YO, Genova RC, Birma B, Kissel ES, Levy AN, MacCracken S, Mastrandrea PR, White LL (eds) Cambridge University Press, Cambridge, pp 1499–1566

Melgarejo LF, Lakes T (2014) Urban adaptation planning and climate-related disasters: An integrated assessment of public infrastructure serving as temporary shelter during river floods in Colombia. Int J Disaster risk reduction 9:147–158

Ministério do Meio Ambiente—Grupo Executivo do Comitê Interministerial de Mudança do Clima (2015). Plano Nacional de Adaptação à Mudança do Clima, Volume I: Estratégia Geral, Versão Consulta Pública. Brasília 67 p

Painel Brasileiro de Mudanças Climáticas (2012). Sumário Executivo do Volume 1 - Base Científica das Mudanças Climáticas. Contribuição do Grupo de Trabalho 1 para o 1 o Relatório de Avaliação Nacional do Painel Brasileiro de Mudanças Climáticas. Volume Especial para a Rio +20. PBMC, Rio de Janeiro, Brasil, 34p. ISBN 978-85-285-0208-4

Quarantelli EL (1985) An assessment of conflicting views on mental health: the consequences of traumatic events. In: Figley C (ed) Trauma and its wake: the treatment of post-traumatic stress disorder. Brunner/Mazel, New York

Quarantelli EL (1995) Patterns of sheltering and housing in US disasters. Disaster Prev Manag 4:43–53. https://doi.org/10.1108/09653569510088069

Silva RP, Lima OHM, Facadio A, Coelho Netto AL (2016) Condicionantes geomorfológicos e geológicos relacionados à deflagração de movimentos gravitacionais de massa: bacia do Córrego Dantas, Nova Friburgo/RJ. In: XI Sinageo, 2016, Maringá. Anais do XI Sinageo

Smyth J, Mcinerney P (2012) From silent witnesses to active agents: student voice in reengaging with learning. Peter Lang Publishing, New York

Smith C, Parsons C (2014) Australian red cross, preferred sheltering practices for emergency sheltering in Australia. Australian Red Cross, Brisbane (Queensland)

Sheet KPAF USAID [online]. Washington DC: USAID 2018, 28 March 2018, Accessed on 22 Apr 2018

Sphere Project. The Sphere Project (2011) Humanitarian Charter and Minimum Standards in Humanitarian Response. Northampton: The Sphere Project

Zahari NZ, Hashim AM (2018) Adequacy of Flood Relief Shelters: a case study in perak, Malaysia. In E3S Web of Conferences, Vol 34. p 02016 EDP Sciences

Post-catastrophic Disaster Induced Laws for Climatic Change Adaptation: A Case Study in SE-Brazil

Leonardo Esteves de Freitas, Raiza Fernandes da Silva and Ana Luiza Coelho Netto

Abstract Climate change has increased the frequency and magnitude of socio-environmental disasters in Brazil's southeastern region, especially on State of Rio de Janeiro's Mountainous Region, where a major disaster occurred in 2011. This disaster had a great international repercussion and, in response, the public power at the three levels of government (Federal, State and Municipal) reformulated legal instruments seeking greater efficiency in risk management processes. In this paper we analyse two legal instruments, enacted before and after the disaster, related to risk management at each level of government (at the municipal level for Nova Friburgo, the municipality most affected by the disaster of 2011), in order to evaluate the impact of this event in these instruments. The results indicate a great advance at the federal and municipal level in the analyzed legal instruments, making them more robust and helping the design of risk management systems that integrate different sectors of society and public power. But none of these instruments incorporated the relationship between climate change and disaster occurrence consistently. Regarding the State of Rio de Janeiro, the impact of the 2011 disaster on the analyzed legal instruments was very small, and no relevant differences were found. In this case, the relationship between disasters and climate change has been totally neglected. The next steps of the study include the analysis of other legal and management

(*) Supported by grants from CNPq and FAPERJ.

L. E. de Freitas (✉)
Laboratório de Geo-Hidroecologia e Gestão de Riscos do Departamento de Geografia da UFRJ and Centro de Estudos e Pesquisas em Desastres e Emergências em Saúde and Observatório de Territórios Sustentáveis e Saudáveis da Bocaina da Fundação Oswaldo Cruz (Fiocruz), Universidade Federal do Rio de Janeiro (UFRJ) and Fundação Oswaldo Cruz (Fiocruz), Rio de Janeiro, Brazil
e-mail: leonardofreitas73@gmail.com

R. F. da Silva · A. L. Coelho Netto
Laboratório de Geo-Hidroecologia e Gestão de Riscos do Departamento de Geografia da UFRJ, Universidade Federal do Rio de Janeiro (UFRJ), Rio de Janeiro, Brazil
e-mail: raiza_fernandes@hotmail.com

A. L. Coelho Netto
e-mail: ananetto@acd.ufrj.br

W. Leal Filho et al. (eds.), *Climate Change, Hazards and Adaptation Options*, Climate Change Management, https://doi.org/10.1007/978-3-030-37425-9_10

instruments, including those specifically focused on climate change, to understand whether these instruments have incorporated the issue of disaster risk management.

Introduction

The rise in frequency and magnitude of disasters associated with natural phenomena is one of the drastic consequences of climate change, directly affecting thousands of people, leading to deaths and incalculable losses (Hoegh-guldberg et al. 2018).

Many efforts have been directed towards understanding the relationship between climate change and changes in the patterns of occurrence of natural processes that generate disasters, but local-level predictability is still fragile. This is even truer for the poorest regions of the world, where available data on precipitation is still scarce (Hartmann et al. 2013; Stocker et al. 2013). Even so, the Intergovernmental Panel on Climate Change states that "(…) it was assessed that there is medium confidence that anthropogenic forcing has contributed to a global-scale intensification of heavy precipitation over the second half of the 20th century (…)" (Hoegh-guldberg et al. 2018).

Since climate change is recognized as an importante elemento in the verified increase in disaster frequency, it is of paramount importance to integrate this issue into disaster reduction planning and measures (UN/ISDR 2005). Regarding the southeastern region of Brazil, the data indicate a process of intense rainfall concentration throughout the 20th century. Figueiró and Coelho Netto (2011), after analyzing long-term data, indicated that the frequency of rainfall below 20 mm/day is dropping and extreme rainfall totals above 100 mm/day are increasingly more frequent towards the early XXI century. Other authors also observed the increase of extreme rainfall in the southeastern Brazilian region during the same period (Groisman et al. 2005; Marengo et al. 2009; PBMC 2012; MMA 2015).

This concentration of extreme climatic events generates impacts on this region's population, inducing high-risk situations related to the incidence of disasters (Bonatti et al. 2016).

Since it is impossible to prevent the occurrence of disaster inducing natural phenomena, especially in a context of climate change in which these phenomena are becoming more frequent, the process of adaptation to climate change and related disasters becomes critical to reduce their consequences. However, the implementation of climate change adaptation processes that are capable of anticipating the consequences of disasters is difficult to achieve, as described by Kates et al. (2012):

> (…), anticipatory transformational adaptation may be difficult to implement because of uncertainties about climate change risks and adaptation benefits, the high costs of transformational actions, and institutional and behavioral actions that tend to maintain existing resource systems and policies. Implementing transformational adaptation requires effort to initiate it and then to sustain the effort over time. In initiating transformational adaptation focusing events and multiple stresses are important, combined with local leadership (Kates et al. 2012, p. 7156)

This process is even more difficult in developing countries such as Brazil, where land occupation processes are often performed in a disorganized way, making it difficult to implement effective climate change adaptation policies.

Such is the case in Southeastern Brazil's mountainous regions, an área where human occupation takes place in steep slopes and valley floors, where disasters are mostly related to landslides caused by extreme rainfall as reported by Carvalho et al. (2018). Chrono-stratigraphic evidences in Rio de Janeiro state indicate that landslides have been occurring throughout the Holocene in response to bio-climatic changes (Behling and Safford 2010, Coelho Netto et al. 2015). These authors point out that changes in the original forest cover brought about by urban expansion, extensive agriculture and cattle grazing on steep slopes throughout the XX century led to a progressive reduction of the recurrence time interval of landslides in the area.

The period between 1991 and 2012 saw an increase in frequency of extreme rainfall related disasters by 21% all over Brazil, with the southeastern region accounting for around 80% of the ocurrences, especially in Rio de Janeiro, where flash floods, including debris flows, affected 1.395 million and killed 1153 people in the same period (Ceped UFSC 2013).

The concentration of deaths from disasters in the mountainous region of the State of Rio de Janeiro is mainly associated with three factors: (1) It is a region of steep relief and with great topographical unevenness and narrow valleys, characteristics that favor the occurrence of landslides and debri flows, which are high energy phenomena and possessing great power of destruction; (2) It is a high rainfall region, even compared with other regions in the Southeast, the wettest in Rio de Janeiro state, with rainfall concentrated in the summer, when it exceeds 300 mm on average; (3) Human occupation takes place mainly in valley floors and at the foot of the slopes, where the movement of mass propagates, especially from the formation of debri flows, advancing to the steep slopes, where landslides are common. Thus, over the 21-year period described above, this region has concentrated most of the deaths from disasters in Brazil. Most of the deaths observed took place during the extreme rains of January 2011. Officially, this disaster led to the death of 964 people (Fundação Coppetec 2014).

This disaster had a major impact on disaster risk management in Brazil, which has been reflected in the formation of a conceptual, legal and institutional framework on risk management (Silva 2017).

An important conceptual shift perceived after this catastrophic event refers to the denaturalization of disasters. It is increasingly perceived that human actions interfere not only in the local but also on regional climate, while also altering ecosystems, influencing the increase of frequency and magnitude of natural phenomena that cause disasters such as landslides and flash floods. Social relations condition land use and the possibilities of responding to natural phenomena, influencing the magnitude of the consequences of these phenomena. Therefore, the disasters that are here analysed should not be considered as natural, as they result from the impact of a natural phenomenon induced by extreme rainfall events and are induced by a specific social system (Freitas and Coelho Netto 2016).

This conceptual standpoint gained strength after the catastrophic event of January 2011 in Rio de Janeiro, and has strengthened the notion that disaster risk management must be carried out in a horizontal structure capable of integrating the various segments of society, emphasizing a direct participation of communities at risk, meaning that decisions on risk management should be discussed with these communities (Saito 2018; Avila and Mattedi 2017; Freitas and Coelho Netto 2016).

Another shift refers to legal instruments dealing with risk management, especially those about land use planning. At municipal, state and federal levels, many of these instruments were reviewed after the 2011 disaster. New Federal National Defense and Civil Protection Policies were established, and at state level, a new Civil Defense System was drawn; within the municipality of Nova Friburgo, several legal instruments were prepared, including a proposal to revise the Municipal Master Plan, now focusing on risk management. The present work analyzes these legal instruments that resonate in Nova Friburgo's management, trying to understand how disaster risk management elements were incorporated by them.

All references, throughout this study, refers to the following territories: the Brazilian territory, where Federal legislation is analysed; the State of Rio de Janeiro, analyzing the state legislation; and the municipality of Nova Friburgo, where we analyse the municipality's legislation. Special attention is given to Nova Friburgo, a municipality located in Rio de Janeiro's mountainous region (Fig. 1), which was most strongly affected in the catastrophic event of January 2011.

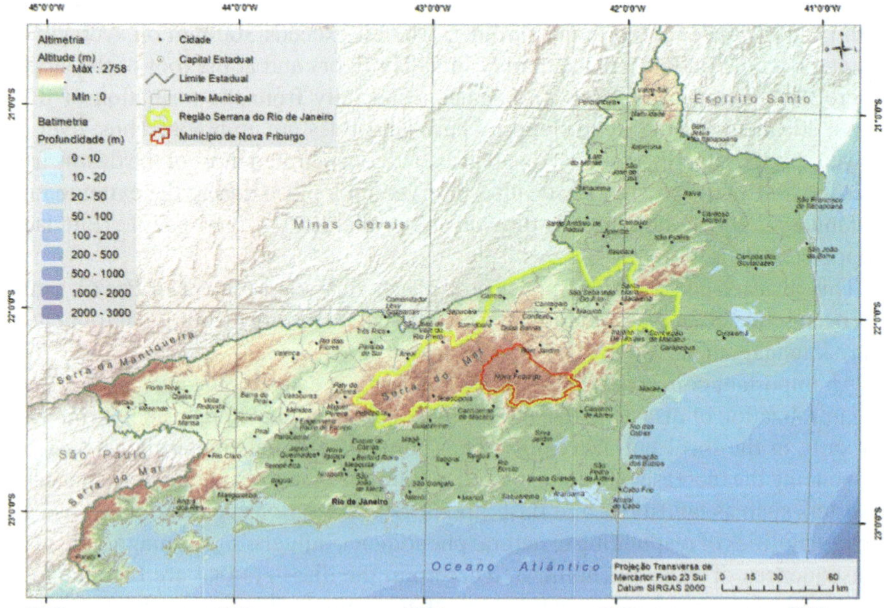

Fig. 1 Hypsometric map of the state of Rio de Janeiro, highlighting the Mountainous Region and Nova Friburgo

Methods

To understand the changes in legislation related to the impact caused by the event of January 2011, we sought to analyze the laws dealing with the issue that were written before as well as after this event, including:

(1) at the federal level: Law 12.340 of December 1st, 2010 (National System of Civil Defense Law) and Law 12.608 of April 10, 2012 (which establishes the National Policy of Protection and Civil Defense);
(2) at the state level: Decree No. 40.908 of August 17, 2007 (measures for the State System of Civil Defense of Rio de Janeiro) and Decree No. 43.599 of May 17, 2012 (measures for the reorganization of the State System of Civil Defense from Rio de Janeiro);
(3) at the municipal level (Nova Friburgo): Participative Master Plan of Nova Friburgo, promulgated through Law 24 of December 28, 2006, and the Draft Law of the Participative Master Plan of Nova Friburgo sent by the City Hall to the Council in 2015. It is worth mentioning that the Master Plan is the municipality's main risk management instrument and that the proposal prepared in 2015 sought to revise the Plan promulgated in 2006.

The comparisons between legislations were conducted based on a qualitative analysis, striving to understand how disaster risk management is dealt with in each of the laws; as well as a quantitative analysis, through the citation counting in each of the legal instruments of the following keywords: "disaster", "risk", "landslides", "mitigation", "prevention", "flood", "rain" and "climate change". The term "landslides" has been deployed in other terms of equal meaning, such as "mass movement", "gravitational mass movement," and "slipping". The term "flood" also included the terms flooding and inundation, and "rain" included the words "precipitation" and "rainfall". All selected terms are directly related to the management of socioenvironmental disasters (as defined by Freitas and Coelho Netto 2016) in Nova Friburgo.

Regarding the legal instruments in force at the time of the 2011 disaster, only the legal texts valid at the time of the disaster were analyzed. Changes that occurred later were disregarded, as they reflect, in most cases, the effect of instruments enacted after the disaster on previous legislation.

Results

The quantitative analysis of the differences between legislation at all levels (federal, state and municipal) indicates a significant shift after 2011 at the federal and municipal levels regarding risk management. The legal instruments that came into force after the 2011 disaster at these two levels were more detailed and more focused on risk management.

All words analyzed were present in the two legal instruments at these two management levels post 2011, but several of them were missing on legal pieces before the disaster, with the exception for the word "rain", which was not cited in Federal Law 12.680/2012. As for the searched words' number of citations, there was an increase, for all of them, in federal and municipal instruments (Table 1), again with the exception of "rain".

It is important to highlight the municipal level legal instruments. The proposed revision of the 2015 Master Plan is the only instrument analyzed that mentions all words, including climate change (twice), and a comparison with the 2006 Master Plan evidences a significant increase in the use of these keywords. In the 2006 document only the words "risk" and "mitigation" were cited, and in a much smaller frequency than the 2015 Proposal. These results indicate that the 2015 proposal incorporated the issue of risk management as a fundamental element of territorial planning and municipal development, whereas in the 2006 Master Plan, this concern was only peripheral.

Regarding state legislation, the comparison of the two decrees did not show important differences regarding the number of keyword citations. The words "risk" and "disaster" were cited in similar amounts in both documents. Therefore, there is an indication that the discussion of risk management at the state level did not lead to a significant change in the existing legislation.

A similar pattern was found searching for the term "climate change", but the number of citations in the analyzed legislation is small. Changes were detected in federal and municipal legislation, which did not mention "climate change" before 2011 and have been cited once (federal) or twice (municipal) after the 2011 event. The few citations of this term indicate that the relationship between disasters and climate changes is not something central in the main Federal Civil Defense and Municipal Land Management policies, although its citation could be indicative of concern with this issue. The little relevance given by the municipal legislation to these issues is an expected result, considering that Nova Friburgo does not have specific legislation on climate change. When it comes to the Federal legislation, however, this fact is surprising, given the existence of Federal Law No. 12,187, promulgated on December 29, 2009, which establishes the National Policy on Climate Change, its principles, objectives, guidelines and instruments (Brasil 2009). This Act addresses only once the issue of disasters as a consequence of climate change, and yet only to define the concept of vulnerability. However, as Law 12,680/2012 was enacted after the National Policy on Climate Change, it was expected that it would be more closely integrated with this policy, which certainly is not the case.

The analyzed state legislation does not cite this term once before or after the 2011 disaster, showing a disregard for the relationship between disasters and climate changes. This is also an unexpected result, as the state of Rio de Janeiro has specific legislation on climate change: Law 5690, dated April 14, 2010, which establishes the State Policy on Global Climate Change and Sustainable Development (Rio de Janeiro 2010); one of this law's guidelines is precisely the prevention of extreme weather events. Even so, the state Law analyzed in this paper ignores the issue of climate change and its relation to the increase in the frequency of disasters.

Table 1 Number of articles and citations of the keywords searched in the analyzed legal documents

Level	2011 disaster	Legal tool	Number of articles	Number of keyword citations							
				Disaster	Risk	Landslide	Mitigation	Prevention	Flood	Rain	Climate change
Federal	Before	Law 12.340/2010	17	2	13	–	–	–	–	–	–
	After	Law 12.608/2012	31	36	57	10	5	12	10	–	1
State	Before	Decree 40.908/2007	17	1	19	–	–	–	–	–	–
	After	Decree 43.599/2012	10	2	15	–	–	–	–	–	–
Municipal	Before	Law 24/2006	199	–	8	–	1	–	–	–	–
	After	Draft bill 2015	331	5	37	2	4	3	6	4	2

Federal Legislation: Comparison Between Law 12.340/2010 and Law 12.608/2012—National Policy on Protection and Civil Defense

Law 12.340 from December 1st, 2010:

> Provides for the National System of Civil Defense—SINDEC regarding resource transfers for relief actions, assistance to victims, restoration of essential services, reconstruction in areas affected by disaster and the Special Fund for Public Disasters and indicates other measures. (Brasil 2010)

And law 12.608 from April 10th, 2012:

> Establishes the National Policy of Protection and Civil Defense—PNPDEC; provides for the National System of Protection and Civil Defense—SINPDEC and the National Council for Protection and Civil Defense—CONPDEC; authorizes the creation of a disaster information and monitoring system; amends Laws No. 12.340, December 1, 2010; 10.257, July 10, 2001; 6.766, December 19, 1979; 8.239, October 4, 1991; and 9.394, December 20, 1996; and indicates other measures. (Brazil 2012)

From the definition of the objectives of the two laws presented above, it is clear that Law 12.608/2012 replaces the previous law enacted less than two years earlier. This is one of the key indicators that the 2011 disaster had a great impact on national legislation regarding disaster risk management. Law 12.340/2010, which was promulgated only one month before the disaster, was expected to become a legal reference. However, after the disaster, the same law was considered inadequate and only 19 months later, Law 12.608/2012 was promulgated.

Another difference that demonstrates a legal improvement towards a more efficient management of disaster risks is that even though both laws deal with the National Civil Defense System, the 2012 Law presents a much broader and more detailed model defining the competencies of each federative entity. The model advocated by the 2012 Law involves actions of prevention, mitigation, preparation, response and recovery in a perspective that includes all phases of risk management, whereas the 2010 Law does not even cover the term "mitigation", mentioning a system that addresses prevention, response, and recovery, but focuses solely on response. So much so that of the four chapters that comprise the Law, only one discusses the National System of Protection and Civil Defense, dealing with prevention, recovery and response. The other three chapters only discuss response. A chapter is devoted to emergency situations and public calamity (typical response situations); another chapter is specific to resource transfers for response actions and the other is focused on monitoring the resources made available for response.

Law 12.608/2012 also presents an integrated discussion about risk management, stating that issues such as territorial planning, urban development, health, environment, water resources management, geology, infrastructure, education, science and technology need to be integrated with a sustainable development perspective. This integrated view is absent in the 2010 Law.

Regarding climate change, it is not mentioned at all in Law 12.340/2010. Meanwhile, the 2012 Law cites it once, indicating that the National Policy on Protection and Civil Defense must be integrated with the policies of Climate Change, evidencing the recognition by the Brazilian government of the relationship between these changes and the management of disaster risks.

Therefore, Law 12.680/2012 when compared to Law 12.340/2010 deserves special attention in the search for an integrated management able to include the various sections in Brazilian society. This Law establishes the need for an institutional articulation between the spheres of public power and those of society, including universities, public managers, organized civil society and communities susceptible to disaster risk. This need for articulation is mentioned in the guidelines of the National Policy, and in the objectives of this policy.

A requirement of Law 12.608/2012, which did not exist in the 2010 Law and is essential for the management of risks at the local level, is the obligation that all municipalities included in the *National Register of Municipalities with Areas Susceptible to the Occurrence of Large Impact Landslides, Severe Flood or Related Geological or Hydrological Processes* review their Master Plans and include risk analyzes in this review.

This meant a legal breakthrough for the discussion of risks, since it ties urban development to risk management in municipalities where disasters are recurrent. Since the occurrence of socio-environmental disasters is closely related to urban development models, this shift is fundamental.

Therefore, at the federal level, the legislation has made progress in terms of disaster risk management, incorporating a perspective of integration among different governmental entities and between government and society, including vulnerable communities.

However, it is important to emphasize that this integration is not part of the day-to-day management processes and that there are difficulties of integration between the spheres of power and articulation of public policies with civil society, especially with vulnerable communities (Freitas and Coelho Netto 2018, 2016).

Rio de Janeiro's State Legislation—Detailed Comparison Between Decree 40.908, August 17, 2007 and Decree 43.599, May 17, 2012

Decree 40.908, August 17, 2007:

> Provides for the reorganization of the State System of Civil Defense—SIEDEC, without increase of expenses, and indicates other measures (Rio de Janeiro 2007).

Decree 43.599, May 17, 2012:

> Provides for the reorganization of the State System of Civil Defense—SIEDEC, without increase of expenses, and indicates other measures (Rio de Janeiro 2012).

Both decrees dispose on the same subject, without any modification in their initial statement. The main difference between them is that the legal instrument enacted in

2012 aims to adapt the State Civil Defense System to the National Civil Defense Policy, which was implemented following Federal Law 12.340, December 1st, 2010. This poses a problem, since in 2012 a new Law (Law 12,608 of 2012) was promulgated, restructuring the National Policy of Protection and Civil Defense on which the State Law based itself on, in 2010. Another important consideration that supported State Decree No. 43,599 of May 17, 2012 was the creation of the State Secretariat of Civil Defense (SEDEC) in June 2011, making it necessary to adapt the State Civil Defense System to this new structure of the State's Executive Branch.

Besides these administrative distinctions, there are conceptual differences that underlie the two decrees. The concepts of "Civil Defense" and of "disaster" have remained exactly the same, whereas the concepts of "emergency situation" and "state of calamity" have different understandings. According to the decree prior to the 2011 disaster, "emergency situation" and "state of public calamity" could only be enacted through the recognition of the public power of the existence of such situations. The 2012 Decree indicates that "emergency situation" and "state of public calamity" are situations consisting of the abnormal situation itself—not necessarily declared by the government—and are associated with the partial (state of emergency) or substantial (state of public calamity) capacity of response coming from the public power. This is an important shift, since it made possible in legal terms that other social actors, besides the public power, to require that a given situation be classified in one of these two concepts. Since the enactment of these two abnormal situations has administrative, political and juridical implications, this shift in legislation extends the possibility of different actors operating in abnormal situations.

The comparison between the two decrees indicates small differences between them, evidencing that the impact of the 2011 event on the main state legal instrument was minimal, lower than at the Federal level. The approach and conceptualization regarding risk management are very similar in both decrees. Part of this result can be explained by the fact that the 2012 Decree was drafted based on the National Policy on Protection and Civil Defense of 2010 rather than 2012, which generated serious problems in state legislation. In any case, it is indicative that, in terms of legal instruments, the state of Rio de Janeiro is not prepared for risk management as it should.

Nova Friburgo Municipal Legislation—Detailed Comparison Between Complementary Law no.28 from December 28, 2006 and 2015 Draft Law Which Presents the Proposal to Review the Municipal Master Plan

The Municipal Master Plan, according to Article 3 of Brazil's Federal Constitution, is each city's basic instrument for development policy and urban expansion. It has the competence to analyze the physical, economic and social characteristics, the predominant activities, the vocations, the problems and the potentialities of a municipality. Regarding Nova Friburgo, the Master Plan in force was established by Complementary Law 24 of December 28, 2006. A Proposed Revision of this Master Plan was filed in the City Council of Nova Friburgo on October 31, 2015.

This revision is subject to requirements established by Federal Law 12,608/2012, since Nova Friburgo is part of the National Register of Municipalities with Areas Susceptible to the Occurrence of Large Impact Slips, Sudden Flooding or Related Geological and Hydrological Processes. Therefore, the proposed revision itself arose from the direct discussion of risk management, which is reflected in the proposal's development.

As previously seen in Table 1, a discrepancy was verified between the current Master Plan (Nova Friburgo 2006) and its proposed revision (Nova Friburgo 2015) with respect to disaster risk management. No words analyzed cited in the Master Plan's proposed revision appear in the 2006 document, and those mentioned in the two Plans appear much more often in the revision proposal. The difference is significant, demonstrating that the issue of risk management is peripheral in the 2006 Master Plan and central in the revision proposal, reflecting the 2011 disaster's impact. That is, at the local level this event marked a change of culture in the population and in public power as previously indicated by Freitas and Coelho Netto (2018, 2016).

The fact that risk management is peripheral in the 2006 Master Plan demonstrates a lack of awareness among legislators of a matter of great relevance to the municipality's development and indicates the fragility in the formation of a risk management culture, even in a place where risks are evident and disasters recurrent (Freitas and Coelho Netto 2016). The Proposed Revision of the Master Plan seeks to overcome this fragility, through the internalization of a disaster prevention culture and the integrated management of vulnerabilities caused by climate change and its impacts on the physical environment.

To this end, the Draft Law's second chapter addresses the resettlement of populations occupying risk areas, stating that the priority of relocation of families living in vulnerable areas is intended for residents living in or near areas of geohydrological risk. This is a controversial topic and should be discussed very carefully and with the effective participation of these populations. This discussion became problematic in Nova Friburgo after the 2011 disaster, generating serious conflicts between the public power and the populations affected by the disaster (Freitas et al. 2017). Often, disaster-adaptive processes from structuring and non-structuring measures prove to be more effective and less traumatic. Even so, the priority given by the 2015 Preliminary Draft to the resettlement of people located in risk areas of geological and hydrological disasters shows that the issue of risk management has become a priority.

Another indication of the priority given by the Master Plan's proposed revision for the issue of risk management compared with the current Master Plan is the fact that the proposal contains an entire chapter dedicated to risk management: *Chap. 4—Geological and Hydrological Risk*. This chapter contains guidelines directly related to the management of socioenvironmental disasters that impact on the urban development process, such as the definition of geological and hydrological risk areas and the Geological and Geotechnical Areas of Interest (GGAI).

This chapter outlines thirteen GGAI guidelines directly related to risk management, including establishing the precise location of risk areas, participatory monitoring of risk areas, constant meteorological monitoring, the training of civil servants to act on risk management, the integration of civil defense actions, etc. In addition, the last article of this chapter has a direct impact on urban development, highlighting the relationship between land parceling and land occupation and risk management.

It should be noted that the mapping of the so-called GGAI poses great methodological challenges, since there are no relevant cartographic and thematic bases, at an appropriate level of detail, covering the whole municipality and mountain region (Freitas et al. 2017). Even so, the definition of specific areas where urban policies must have restrictions is a step forward.

From a conceptual point of view, there are many differences between the 2006 Master Plan and the proposed revision prepared in 2015. The concept of "risk" should be highlighted in the comparison. In the current Master Plan, "risk", as a rule, is associated with constructions or human impacts. There is mention to areas of risk, control of risks, risks to the population, risks to the environment and health risks. In the 2015 proposal, the concept of risk is broad and focused on vulnerability. It includes areas of risk, risk-generating activities, measures to reduce risks, risks to the environment, risks to the urban landscape, health risks and, especially, integrated risk management.

Despite these developments, the conceptual association between disasters and climate change in the 2015 proposal is still weak. The discussion of the need to adapt communities and ecosystems to climate change runs laterally in the text, exposing a fragility in these documents, as several authors are evidencing the trend of extreme rainfall concentration in southeastern Brazil, increasing the likelihood of socio-environmental disasters (Figueiró and Coelho Netto 2011; PBMC 2012; Marengo et al. 2009).

The differences between Nova Friburgo's 2006 Municipal Master Plan and its Proposed Revision show that risk management has become central in the municipality. However, this is only an initial step, since the construction of risk management processes requires not only changes in legal and management instruments, but also the effective construction of a risk culture that encompasses public power and civil society.

In this sense, one of the main problems of the 2015 Draft is the lack of attention given to the participation of vulnerable communities in the specific topics related to risk management. There is an entire chapter devoted to democratic and participatory management and several other references to participatory management. However, in the specific sections on risk management, participatory management appears only once, referring to the need for participatory risk monitoring. There is no discussion of participatory risk management in its various phases, including the definition of participatory risk reduction policies constructed together with the population. This is a structural problem that indicates that participatory risk management is still a challenge.

Nevertheless, the proposed revision of the Master Plan is undeniably a major advance in the construction of effective risk management policies in Nova Friburgo.

Conclusions

The disaster of January 2011 was a hallmark for the discussions on risk management in Brazil and specifically in the city of Nova Friburgo, increasing the perception among society and public authorities that risk management should be prioritized. This paper confirms this prioritization based on the analysis of legislation focused on risk management, which has become more robust at the federal level and in Nova Friburgo. At these two levels, legal instruments have been built or proposed that have risk management as a central aspect.

At the federal level, the law promulgating the National Protection and Civil Defense Policy has qualified the previous legislation, promulgated only 19 months earlier, regarding essential aspects, especially the construction of an integrated management model advocating the participation of different groups in society, including vulnerable communities and public authorities at all three levels of government. This result corroborates the hypothesis that the 2011 event was a hallmark for Brazilian legislation related to disaster risks.

At the municipal level, the proposed revision in Nova Friburgo's Master Plan completely reorganizes risk management, turning it into a central part of the proposal and linking the municipality's own development to the discussion of risk management, whereas this theme was only peripheral in the city's previous Master Plan of 2006.

Nevertheless, at both the federal and municipal levels, important problems remain, with emphasis on the near absent discussion of climatic changes and variabilities, neglecting their relation to the occurrence of disasters. This is one aspect that needs to be discussed, as climate change adaptation and mitigation policies need to be strongly integrated into disaster risk management policies, or the latter will become inefficient.

At the state level there was no significant change in the legislation covering the State System of Civil Defense. This was a surprising result, given that the 2011 disaster struck seven cities in that state. Therefore, a significant change in this policy was expected, but no major alterations were made, incorporating only administrative changes. Climate change and its impacts on disasters, which was peripheral in federal and municipal legislation, is not even mentioned in state legislation. These are worrying results, considering the state's fundamental role in the National System of Protection and Civil Defense. This seems to be related, among other issues, to the fact that the state decree of 2012 was drafted based on the outdated version of the National Policy on Protection and Civil Defense (2010) and not on that of 2012. It is important to note, however, that this result is not a definitive proof that the State of Rio de Janeiro was not impacted by the 2011 disaster and was not prepared in any sphere for risk management. But it is indicative that in the legal sphere this preparation still falls short of what is necessary. At the state level, it is important to evaluate legislation built after the National Protection and Civil Defense Policy promulgation to more consistently ascertain whether the changes observed at federal and municipal levels also occur at the federal level.

This is in fact one of the main limitations of this article, since it analyzed only two legal instruments at each level of government, one before and one after the events of 2011. Evaluating other legal instruments at all levels is important and may indicate the impact of this event on the Brazilian legislation as a whole.

In spite of these limitations, results show that Brazilian disaster legislation has undergone major changes following the disaster in the mountainous region in 2011. However, a shift in risk management culture is still ongoing and needs to be reinforced, given that disasters are recurrent in Brazil and in the mountainous region itself, and yet they have always been neglected. Risk management, in order to truly generate a change in culture, must be continuously discussed, especially with communities living in or near risk areas.

On this path, the next steps of research include the evaluation of other legislation relevant to disaster risk management in the three levels of government, to confirm or challenge the results presented here, as well as a specific discussion on both management and legal instruments that focus on climate change, seeking to understand how these instruments relate to disaster risk management.

Another important development of this research is the evaluation of the analyzed legislation implementation in the municipality of Nova Friburgo, especially with regard to the participation of vulnerable communities in disaster risk management.

After all, in a world permeated by climate change and strong human impacts, extreme events and great damage may occur in the coming years if the various levels of government and society do not pay enough attention to this problem—and a significant part of this attention is established through the Legislative power.

References

Avila MRR, Mattedi MA (2017) Desastre e território: a produção da vulnerabilidade a desastres na cidade de Blumenau/SC. Revista Brasileira de Gestão Urbana 9(2):187–202

Behling H, Safford HD (2010) Late-glacial and Holocene vegetation, climate and fire dynamics in the Serra dos Órgãos, Rio de Janeiro State, southeastern Brasil. Glob Change Biol 16:1661–1671

Bonatti M, Sieber S, Schlindwein SL, Lana MA, de Vasconcelos ACF, Gentile E, Boulanger JP, Plencovich MC, Malheiros TF (2016) Climate vulnerability and contrasting climate perceptions as an element for the development of community adaptation strategies: Case studies in Southern Brazil. Land Use Policy 58:114–122. Available at: http://dx.doi.org/10.1016/j.landusepol.2016.06.033 0264-8377/© 2016 Elsevier Ltd

Brasil (2010) Lei nº 12.340 de 1º de dezembro de 2010

Brasil (2009) Lei nº 12.187 de 29º de dezembro de 2009. Avaiable at http://www.planalto.gov.br/ccivil_03/_ato2007-2010/2009/lei/l12187.htm. Acessed in Jan 2019

Carvalho NL, Silva EV, Coelho Netto AL (2018) O sistema de alerta e alarme como estratégia de redução de riscos: participação popular e avaliação de políticas públicas. Desenvolvimento e Meio Ambiente (UFPR) 46:276–288

Centro Universitário de Estudos e Pesquisas sobre Desastres—Universidade Federal de Santa Catarina—Ceped (2013) "Atlas brasileiro de desastres naturais". Florianópolis

Coelho Netto AL, Sato AM, Freitas LE (2015) Land use-vegetation-landslide interactions in the mountainous region of Rio de Janeiro State: scientific basis for risk assessment and management. In: Anais regional conference of international geographical union. Moscow

Figueiró AS, Coelho Netto AL (2011) Climatic variability and pluviometric trends in a humid tropical environment at Resende municipality- middle Paraíba do Sul river valley (Rio de Janeiro-Brazil). Braz Geogr J 2:256–273

Freitas LE, Coelho Netto AL (2016) Reger Córrego Dantas: uma ação coletiva para enfrentamento de ameaças naturais e redução de desastres socioambientais. Ciência & Trópico, Ed. Massangana, Rio de Janeiro, pp 165–190

Freitas LE, Coelho Netto AL (2018) Gestão de riscos de desastres e participação popular: Lições aprendidas e a relevância da educação para a consolidação da Rede de Gestão de Riscos da Bacia Hidrográfica do Córrego d'Antas (Reger-CD), Nova Friburgo/RJ, vol 4, no 7

Freitas LE, Vicente A, Coelho Netto AL (2017) Conflicts after the tragedy in the mountains of the state of Rio de Janeiro in 2011: the relationship between residents of Córrego d'Antas and the zoning of evacuation areas for an adaptation to climate change. In: Leal Filho W, Freitas LE (Org) Climate change adaptation in Latinamerican: managing vulnerability, fostering and resilience. 1st edn, vol 1. Springer, Nova York, pp 225–241

Fundação Coppetec/Laboratório de Hidrologia e Estudos de Meio Ambiente. Elaboração do Plano Estadual de Recursos Hídricos do Estado do Rio de Janeiro (2014) R3-A - Temas técnicos estratégicos RT-03 - Vulnerabilidade a Eventos Críticos, Vol 2—Ocorrências de Desastres Naturais entre 2000 e 2012 por Região Hidrográfica. Rio de Janeiro, 120p

Groisman PY, Knight RW, Easterling DR, Karl TR, Hegerl GC, Razuvaev VN (2005) Trends in intense precipitation in the climate record. J. Clim. 18:1326–1350

Hartmann DL et al (2013) Observations: atmosphere and surface. In: Stocker TF, Qin D, Plattner GK, Tignor M, Allen SK, Boschung J, Nauels A, Xia Y, Bex V, Midgley PM (eds) Climate change 2013: the physical science basis. Contribution of working group I to the fifth assessment report of the intergovernmental panel on climate change. Cambridge University Press, Cambridge, United Kingdom and New York, USA, pp 159–254

Hoegh-guldberg O, Jacob D, Taylor M, Bindi M, Brown S, Camilloni I, Diedhiou A, Djalante R, Ebi K, Engelbrecht F, Guiot J, Hijioka Y, Mehrotra S, Payne A, Seneviratne SI, Thomas A. Warren R, Zhou G (2018) Impacts of 1.5 °C global warming on natural and human systems. In: Masson-Delmotte V, Zhai P, Pörtner HO, Roberts D, Skea J, Shukla PR, Pirani A, Moufouma-Okia W, Péan C, Pidcock R, Connors S, Matthews JBR, Chen Y, Zhou X, Gomis MI, Lonnoy E, Maycock T, Tignor M, Waterfield T(eds) Global warming of 1.5 °C. An IPCC Special Report on the impacts of global warming of 1.5 °C above pre-industrial levels and related global greenhouse gas emission pathways, in the context of strengthening the global response to the threat of climate change, sustainable development, and efforts to eradicate poverty, In Press

Kates RW, Travis WR, Wilbanks TJ (2012) Transformational adaptation when incremental adaptations to climate change are insufficient. Proc Natl Acad Sci USA 109(19):7156–7161

Marengo JA, Jones R, Alves LM, Valverde MC (2009) Future change of temperature and precipitation extremes in South America as derived from the PRECIS regional climate modeling system. Int J Climatol 15:2241–2255

Ministério do Meio Ambiente—Grupo Executivo do Comitê Interministerial de Mudança Do Clima (2015) Plano Nacional de Adaptação à Mudança do Clima, Volume I: Estratégia Geral, Versão Consulta Pública. Brasília, 67p

Painel Brasileiro de Mudanças Climáticas (2012) Sumário Executivo do Volume 1 - Base Científica das Mudanças Climáticas. Contribuição do Grupo de Trabalho 1 para o 1 o Relatório de Avaliação Nacional do Painel Brasileiro de Mudanças Climáticas. Volume Especial para a Rio + 20. PBMC, Rio de Janeiro, Brasil, 34p. ISBN 978-85-285-0208-4

Nova Friburgo (2015). Anteprojeto de Lei "Proposta de Revisão para o Plano Diretor Participativo de Nova Friburgo"

Nova Friburgo (2006) Lei Complementar n° 24, 28 de dezembro de 2006

Rio de Janeiro (2012) Decreto n° 43.599, de 17 de maio de 2012

Rio de Janeiro (2007) Decreto n° 40.908, de 17 de agosto de 2007

Rio de Janeiro (2010) Lei nº 4.690, de 14 de abril de 2010. Avaiable at http://alerjln1.alerj.rj.gov.br/contlei.nsf/f25571cac4a61011032564fe0052c89c/a9593961f9d00ab28325770a005bd6a4?OpenDocument. Acessed at Jan 2019

Saito SM (2018). Vulnerabilidades no contexto de sistemas de alerta de risco de desastres. Rev. gest. sust. ambient., Florianópolis 7(esp):618–630, jun. https://doi.org/10.19177/rgsa.v7e02018618-630

Silva RF (2017) Análise de legislação referente à gestão de desastres socioambientais: estudo de caso do município de Nova Friburgo (RJ), em um cenário pós-desastre. Dissertação (Mestrado em Geografia) Departamento de Geografia da Universidade Federal do Rio de Janeiro. Rio de Janeiro, p 42

Stocker TF et al (2013) Technical summary. In: Stocker TF, Qin D, Plattner GK, Tignor M, Allen SK, Boschung J, Nauels A, Xia Y, Bex V, Midgley PM (eds) Climate change 2013: the physical science basis. Contribution of working group I to the fifth assessment report of the intergovernmental panel on climate change. Cambridge University Press, Cambridge, United Kingdom and New York, USA, pp 33–115

UN/ISDR (2005) World conference on disaster reduction Kobe, 18–22 Jan 2005. http://www.unisdr.org/

Assessment of Carbon Sequestration Potential of a Disturbed Humid Tropical Ecosystem, Southeast Nigeria

Chris O. Nwoko, Samuel C. Anuna and Jonathan C. Anyanwu

Abstract Potentially, vegetation plays an important role in reducing the CO_2 emitted by anthropogenic activities in cities, especially by extensive green vegetation. Accurate estimation of C stock by urban vegetation is difficult due to the fragmented nature of urban ecosystems and inherent variability in the distribution of tree species. This research work was carried out to assess the carbon sequestration ability of selected woody species within the main campus of Federal University of Technology Owerri, Nigeria as a means of mitigating the effects of GHGs emissions. The carbon sequestration analysis was conducted on woody species (*Terminalia catapa, Azadirachta indica, Platanus occidentalis, Cassia sp., Magnifera indica, Milicia excels, Bouea macrophylla* etc.) in four land use categories viz high forest, secondary re-growth forest, bush fallow area and built up areas. All measurements were taken at DBH $\geq$5 cm. The total carbon stock sequestered by individual trees was highest in the high forest areas and lowest in the built up areas. Also, the highest diameter at breast height (DBH) of an individual tree was recorded in the built up areas while the lowest was recorded in the forests. The total C stock sequestered per hectare of trees ranged from 44.2 to 311.7 tC/ha. The highest was recorded in the forested areas while the lowest sequestration was recorded in the built up areas.

Introduction

Global warming is one of the environmental issues experienced worldwide. Ambient concentration of green house gases such as CO_2, N_2O, CH_4 and CFCs are on the increase, particularly occasioned by increased industrialization, deforestation and

C. O. Nwoko (✉) · S. C. Anuna · J. C. Anyanwu
Department of Environmental Technology, Federal University of Technology, Owerri 1526, PMB Imo State, Nigeria
e-mail: conwoko2002@yahoo.com

S. C. Anuna
e-mail: anunachigoziri@gmail.com

J. C. Anyanwu
e-mail: anyanwujonathan50@gmail.com

W. Leal Filho et al. (eds.), *Climate Change, Hazards and Adaptation Options*, Climate Change Management, https://doi.org/10.1007/978-3-030-37425-9_11

forest degradation (Pragasan 2015a). As a consequence, it has increased the average temperature of the atmosphere (IPCC 2013). It is on record that many developed economies have signed the Kyoto protocol to reduce their green house gas emissions (Cao et al. 2010). There has been global campaign to reduce CO_2 and other green house gas emissions, yet average concentrations are on the increase. National Oceanic and Atmospheric Administration (NOAA 2015) has posted that the current concentration of CO_2 gas in the atmosphere is 400 ppm and it is estimated that by 2070 the concentration will reach up to 500 ppm (Jackson et al. 2014). Researchers also have predicted significant global surface temperature increase from 1 to 3.5 °C by 2100 (IPCC 2013).

Obvious environmental consequences of this change cannot be ruled out. Global warming has since been implicated in the melting of glaciers and ice sheets, sea level rise, migration of fishes and animals (Pragasan 2015b), acidification of oceans and changes in physiological processes in plants (NASA 2015). It has also led to catastrophic outbreak of diseases (NASA 2015). There is urgent need to mitigate rising surface temperature by removing excess atmospheric carbon dioxide and other GHGs through afforestation programs, reduction in forest degradation and improvement in forest conservation.

Tropical forests play a leading role in the global carbon cycle. Trees act as sink for atmospheric carbon which capture the gas and store it in form of plant biomass through photosynthesis and aid in limiting global warming to below 2 °C (Mitchard 2018). Mitra et al. (2011) noted that about 30–40% of terrestrial net production of plant biomass is from tropical forest. Knowledge on the process of carbon storage in forest ecosystem is key to the management of carbon pools within the forest system. Many studies have quantified carbon stock in forest ecosystems and tree species in many parts of the world (Pragasan 2015b). The aim of this study was to determine the vegetation cover and carbon sequestration potential of a disturbed forest ecosystem.

Materials and Methods

The study was conducted at the Federal University of Technology, Owerri Imo state which lies between longitude 6°58′30″E to 7°1′0″E and between latitude 5°22′30″N to 5°24′0″N in the South-eastern part of Nigeria. Figure 1 shows a satellite imagery of the Federal University of Technology Owerri. The soil formation which is Pliocene to Miocene in age consists of coastal plain sands, which is about 0.05–2.0 mm in size, with minor clay beds. It contains some isolated gravels, conglomerates, and very coarse sandstone in some places. Benin Formation is overlain by alluvium deposits and underlain by Ogwashi-Asaba Formation which consists of lignite, sandstones, clays and shale. The area has humid tropical climate, with mean annual precipitation of 2200 mm and average annual temperature of 30°C. Rain fall pattern is bimodal with peaks in July and August.

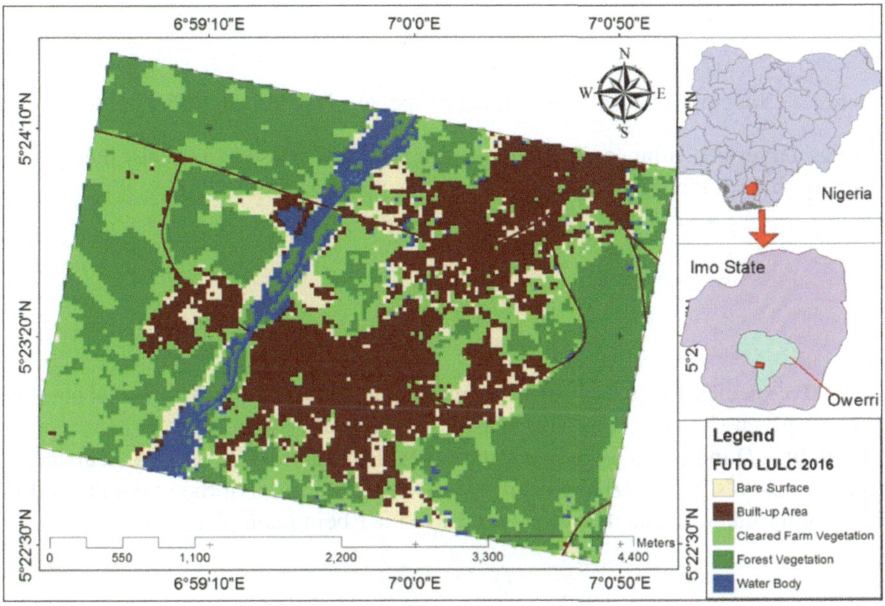

Fig. 1 A classified map of FUTO showing natural vegetation zones, built up zones and Otamiri River which traverses the institution (geographic information system centre, FUTO)

Sample Collection

The study was conducted between June and July 2018. The area was stratified into four land use categories viz high forest, secondary re-growth forest, bush fallow area and built up areas. In each strata a transect of 20 × 20 m were set up and trees ≥5 cm DBH (Diameter Breast Height) and 1.3 m in height were inventoried.

Three (3) randomly selected sub transects each of 4 m by 4 m were set up within each land use category and 100% enumeration of the ground cover including seedlings and saplings were carried out.

Vegetation Analysis

The following ecological indices were determined; species richness, species diversity, equitability

Species Richness—Species richness, also referred to as species abundance, Indices of species richness are based on the total number of individuals in a sample or habitat. Menhinick's (1964) method as modified by Nwoko et al, (2018) was used to calculate species richness index.

$$Y = \frac{S}{\surd N}$$

where Y = Species richness, S = Number of species, N = Total number of individuals

Species Diversity Simpson's diversity index (D), was used to obtain the species diversity in each land use category.

$$D = \frac{1}{\sum_{i=1}^{Y} p_i^2}$$

where D = species diversity, P_i proportion of each species relative to the total number.

For a given richness (Y), D increases as equitability increases, and for a given equitability, D increases as richness increases. Equitability (E_D) was calculated by using Simpson's index (D) and expressing it as a proportion of the maximum value. Equitability takes a value between 0 and 1, with 1 being complete evenness.

$$E_D = \frac{D}{D\mathrm{max}} = \frac{1}{\sum_{i=1}^{S} p_i^2} \times \frac{1}{S}$$

where: D = Simpson's diversity index, S = total number of species in the community (richness)

$P_{i=}$ proportion of S made up of the i th species, E_D = equitability (evenness).

Determination of Carbon Stock of the Woody Species

Non destructive sampling method was adopted which involved the use of allometric equations with measurable variables. This method has been widely used for the estimation of carbon stock of natural vegetations (Chavan and Rasal 2010; Pragasan 2015a). Carbon stock for each tree was determined by the following equations.

Above ground biomass (AGB) was estimated using allometric equation following Brown et al. 1989

$$W = 34.3703 - 8.0671D + 0.6589D^2 \quad (1)$$

where, 'W' is AGB in kg, 'D' is diameter at breast height in cm.

There is strong evidence that AGB is strongly correlated with tree diameter (Brown 1997; Nelson et al. 1999).

Below ground biomass (BGB) was calculated based on the assumption that 15% of the aboveground biomass gives a fair estimate of the BGB as proposed by previous authors (MacDicken 1997; Karthick and Pragasan 2014)

$$BGB = AGB \times 0.15 \tag{2}$$

Therefore the total carbon content is assumed to be 50% of the total biomass (Pragasan 2015a; Takimoto et al. (2008).

$$\text{The total carbon stock(TCS)} = (AGB + BGB) \times 0.5 \tag{3}$$

Statistical Analysis

The data collected were statistically analyzed using Minitab statistical package version 16. Analyses of variance (ANOVA) was performed on the data and significant means separated using Tukey least significant different at 5% probability level. Regression analysis was used to test the relationship between the carbon stock and the various ecological parameters.

Major Limitations of This Research

(a) Species-specific allometric equation, though, ideal were not used for the different plant species
(b) Data on plant species height were not used in the determination of plant Biomass.

Results and Discussion

Ecological Characteristics of the Study Area

The high forest zone recorded the highest species richness index (3.35 ± 0.00), species diversity (13.25 ± 0.005 and equitability (0.74 ± 0.001) and was followed by the forest regrowth zone and Bush fallow area (Table 1). All the ecological characteristics significantly affected the study area. The high forest zone significantly differed from the regrowth and bush fallow areas. Similarly, the bush fallow area recorded the least values of the ecological characteristics when compared to other land use categories. Viz: species richness (1.25 ± 0.00), species diversity (5.88 ± 0.00) and equitability (0.26 ± 0.001). The high species richness and species diversity of high forest zone is expected since the zone is less disturbed compared to others.

Table 1 Ecological characteristics of the study area

Land use categories	Species richness index	Species diversity index	Equitability index (E_d)
High Forest	3.35 ± 0.000^a	13.25 ± 0.005^a	0.74 ± 0.001^a
Secondary Re-growth Forest	2.21 ± 0.005^b	8.21 ± 0.005^b	0.48 ± 0.001^b
Bush Fallow Area	1.25 ± 0.000^c	5.88 ± 0.000^c	0.26 ± 0.001^c

Same letter within column is not significantly different at 5% probability level

Carbon Sequestration Across Land Use Categories

The values of the carbon stock across the land use categories are presented in Table 2. The total carbon sequestered by individual trees varied from 1.59 to 77.75 tC/tree/ha. The highest value of 77.75 tC/tree/ha was recorded in the high forest zone and lowest in the built up area 1. Also, the mean diameter at breast height (DBH) of each tree varied in all the sampled land use categories. It ranged from 20.9 to 46.5 cm. The highest DBH occurred in high forest zone followed by built up area 1. The total carbon stock recorded in each land use category ranged from 44.2 tC/ha (built up area 1) to 311.7 tC/ha (high forest). The total carbon stock per hectare recorded according to the land use category is in the order 311.7 > 213.7 > 188.1 > 67.0 > 44.2 tC/ha, (high forest, built up 2, bush fallow, regrowth forest and built up 1, respectively (Table 2). The fairly high value of carbon stock recorded at the built up area 2 may be attributed to the nature, age and type of trees found in the zone. The zone has exotic economic plant species e.g. *Magnifera indica.* Eneji et al. (2014)

Table 2 Tree Growth and Carbon stock across the Land Use Categories

Land use categories	Sample size(n)	DBH (cm)	AGB (kg)	BGB(kg)	Total carbon stock (kg/m^2)	Total C stock/ha (tC/ha)	Average C stock/tree/ha (tC/tree/ha)
High forest	28	46.5 ± 0.05a	1084	162.6	623.3	311.7	77.75
Re-growth Forest	19	24.53 ± 0.005^d	232.9	35.0	134.1	67.0	3.5
Bush fallow area	4	37.4 ± 0.005^c	654.3	98.1	376.2	188.1	47.0
Built up area 1	4	20.9 ± 0.005^e	153.0	23.1	88.45	44.2	1.59
Built up area 2	5	39.5 ± 0.005^b	743	111.5	427.3	213.7	42.7

Same letter within column is not significantly different at 5% probability level, AGB = above ground biomass, BGB = below ground biomass noted that the high level of CO_2 sequestered in the high traffic area (1,405,168.1 kg) in Benue State was attributed to the nature, age, location and species of the trees.

However, a low value of 44.2 tC/ha recorded in the built up area 1 may be as a result presence of trees with small girth size.

Relationship Between Ecological Characteristics and Carbon Sequestered by Tree Species

The relationship between ecological characteristics and the carbon sequestered by tree species is presented in Table 3. The species richness index significantly related with diameter at breast height (DBH) but not significant with average C stock ($p =$ 0.082). However, species richness index significantly ($p = 0.019$) correlated with total carbon stock. The species diversity index positively related with total carbon stock and negatively correlated with DBH and average carbon stock.

There is strong relationship ($R^2 = 88.6$, $p = 0.001$) between C sequestered and tree diameter at breast height (DBH) (Fig. 2). This finding corroborates the works of previous authors (Vargas-Larreta et al. 2017; that the size of the plant girth directly influences the Carbon sequestered in woody plant species.

There was linear relationship between species richness and carbon sequestered ($R^2 = 92.6\%$ $p = 0.002$) (Fig. 3). The higher the number of woody species present in an area, the more carbon sequestered.

The carbon stock by individual tree species ranged from 2.2 to 72.3 t C/ha (Table 4). The tree species with the highest carbon stock of 72.3 t C/ha: (*Platanus occidentalis*), followed by 68.2 t C/ha: (*Cassia sp*) and 65.7 t C/ha (*Azadirachta indica*). A tree species from the built up area had the largest girth size whereas tree species found in the forest area had the highest average diameter breast height

Table 3 Relationship between ecological diversity parameters and carbon sequestration

Ecological characteristics	DBH	Total C stock/ha	Average C stock/tree/ha
Species richness index	−0.879 0.002**	0.877 0.0194	−0.608 0.082 ns
Species diversity index	−0.791 0.011*	0.612 0.080 ns	−0.728 0.026*
Equitability index (E_d)	−0.879 0.002**	0.478 0.193 ns	−0.609 0.082 ns

* = 5%, ** 1%, ns = not significant

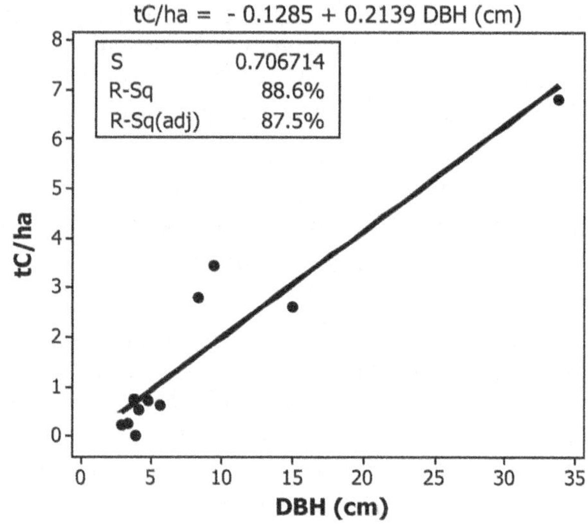

Fig. 2 Regression curve of DBH against carbon stock per hectare

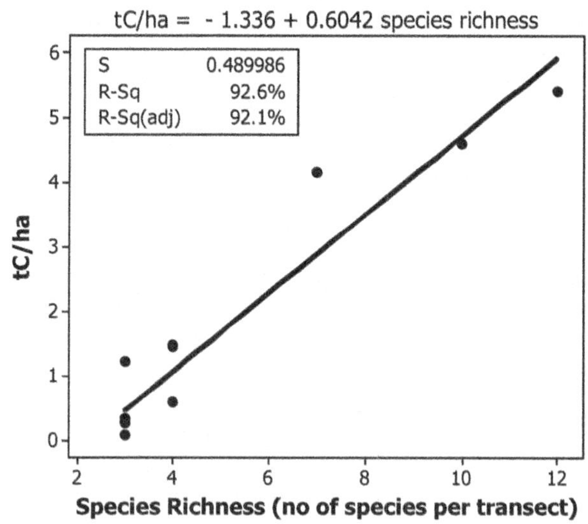

Fig. 3 Regression curve of species richness against carbon stock per hectare

Discussion

The Allometric Equations

Three model formulations were used in the estimation of carbon stock in this study (Eqs. (1)–(3)) which has been used by previous authors (Pragasan 2015b; Takimoto et al. 2008, Castedo-Dorado et al. 2012). Diameter at breast height was the main explanatory tree variable used to estimate the tree biomass components for

Table 4 Carbon stock of some selected woody species in the study area

s/n	Plant species	Land use	DBH(cm)	Carbon stock (C/kg)	Carbon stock (tC/ha)
1	*Terminalia catapa*	Forest	17.2	51.9	12.9
2	*Azadirachta indica*	Fallow	56.4	263.1	65.7
3	*Platanus occidentalis*	Built up	60.5	289.5	72.3
4	*Cassia sp.*	Forest	57.2	273	68.2
5	*Magnifera indica*	Built up	48.5	182.2	45.5
6	*Milicia excels*	Forest	51.3	222	55.5
7	*Bouea macrophylla*	Fallow	53.2	232.8	58.2
8	*Pterocarpus osun*	Forest	44.5	103	25.7
9	*Canthium arnoldianum*	Built up	9.8	10.9	2.7
10	*Strombosia pustalata*	Fallow	9.4	10.5	2.73
11	*Maesobotrya dusenic*	Forest	10.3	12.2	3.0
12	*Trichilia prieuriana*	Built up	9.4	10.2	2.55
13	*Millietia drastic*	Fallow	21.0	89.4	22.4
14	*Allanblackia floribunda*	Fallow	14.1	16.7	3.5
15	*Klainedoxa gabonensis*	Built up	8.25	9.34	2.3
16	*Holarrhena floribunda*	Forest	37.6	65.4	16.3
17	*Bosqueia angolensis*	Built up	7.2	8.8	2.2
18	*Tabernaemontana pachysiphon*	Built up	11.9	13.4	3.4

all plant species. Other authors have also noted that inclusion of total height does not usually lead to a significant increase in the predictive ability of diameter-based biomass regressions, and they conclude that diameter is sufficient to give a reliable tree biomass prediction (Jenkins et al. 2003; Johansson 1999).

However, other researchers have made significant improvement in model fits (Reed and Tomé 1998) or an increase in the accuracy of the tree biomass estimates (Menéndez-Miguélez et al. 2013) when plant height is also used in addition as a predictor. In this study, diameter alone was a good predictor of tree biomass.

Carbon Sequestration Across the Land Use Categories

The High forest area recorded the highest carbon stock per hectare of 311.7 tC/ha (Table 2), it serves as a better carbon sink compared to the built-up areas with low carbon stock per hectare. According to Luyssaert et al., (2008), old growth forests represent about one-third of global forests and store 1.3 ± 0.5 Pg of carbon every year which equates 10% of global ecosystem net production (GENP). In Italy, forest stands with old growth features amount to 93,100 hectares (Barbati et al. 2012) revealing that they are carbon sinks with 192–268 Mg of carbon/hectare. From the result, we noted significant linear relationship between the DBH and the total carbon stock per hectare. This finding is in-line with the work of previous authors for example, Lung and Espira (2015) observed that tree stems larger than 19. 68 inches DBH had the greatest effect on CO_2 sequestered and less than 16% of the species pool accounted for over 62% of the above ground carbon. Also Afzal and Akhtar (2013) made similar observation that *Eucalyptus botryoides* tree with DBH of 65.02 cm sequestered 3527 kg of CO_2 while *Robinia pseudoacacia* with a DBH of 4.57 cm sequestered 6 kg of CO_2. The following plant species viz: *Platanus occidentalis, Azadirachta indica, Cassia sp, Magnifera indica, Bouea macrophylla*, were very remarkable in sequestering C in their biomass.

Ecological Characteristics as Affected by Land Use Categories

There is positive relationship between the total carbon stock per hectare and the species richness, species diversity and equitability (Table 3). The positive relationship between species diversity and productivity in natural habitats has been reported by other authors (Sanchez-Martinez and Cabrales 2012). The high values of ecological characteristics obtained in the High Forest viz: species richness 3.35, species diversity 13.25 and equitability index 0.74 (Table 1) corroborated the work of Labriere et al. (2015), that species composition of natural forests is always high. Species diversity had significant negative relationship with carbon stock per tree per hectare ($p = 0.026$). This outcome may suggest that the effects of diversity on forest carbon may vary with other factors such as types and succession stages of the forests (Lasky et al. 2014; Wu et al. 2015) including the specific dimension of the diversity measurement used (Con et al. 2013; Lasky et al. 2014; Ouyang et al. 2016).

Conclusion

The assessment of carbon sequestration potential of tree species within different land use categories of the study area revealed that less C was sequestered per hectare of trees in the built up areas (University campus) compared to the forest region on the outskirts of the University where more C was stored by trees per hectare. The tree girth size was found to significantly influence the amount of carbon stored in the plant species evaluated. Also species diversity index did not significantly affect total carbon stock per ha. There is need to plant more exotic economic trees in the built up areas to aid in the sequestration of excess Carbon from the atmosphere. Further investigation on the incorporation of other tree variables like height and foliage area in the estimation of carbon stock is recommended.

References

Afzal M, Akhtar MA (2013) Factors affecting carbon sequestration in trees. J Agric Res 51(1):61–69

Barbati A, Salvati R, Ferrari B, Di santo D, Quatrini A, Portoghesi L, Travaglini D, Iovino F, Nocentini S (2012) Assessing and promoting old-growthness of forest stands: lessons from research in Italy. Plant Biosyst 146:167–174

Brown S (1997) Estimating biomass and biomass change of tropical forests: a primer. Rome, Italy: FAO Forestry Paper, p 134

Brown S, Gillespie AJR, Lugo AE (1989) Biomass estimation methods for tropical forests with applications to forest inventory data. Forest Sci 35:881–902

Cao T, Valsta L, Makela A (2010) A comparison of carbon assessment methods for optimizing timber production and carbon sequestration in Scots pine stands. For Ecol Manage 260:1726–1734

Castedo-Dorado F, Gómez-García E, Diéguez-Aranda U, Barrio-Anta M, Crecente-Campo F (2012) Aboveground stand-level biomass estimation: a comparison of two methods for major forest species in northwest Spain. Ann Forest Sci 69:735–746

Chavan BL, Rasal GB (2010) Sequestered standing carbon stock in selected tree species grown in university campus in Aurangabad, Maharashtra, India. IJEST 2(7):3003–3007

Con TV, Thang NT, Ha DTT, Khiem CC, Quy TH, Lam VT, … Sato T (2013) Relationship between above-ground biomass and measures of structure and species diversity in tropical forests in Vietnam. For Ecol Manage 310:213–218

Eneji IS, Ofoegbu O, Azua ET (2014) Sequestration and Carbon Storage Potential of Tropical Forest Reserve and Tree Species Located Within Benue State of Nigeria. J Geosci Environ Prot 2:157–166

IPCC 2013: Summary for Policymakers. In: Climate change 2013: the physical science basis. Contribution of working group I to the fifth assessment report of the intergovernmental panel on climate change. In: Stocker TF, Qin D, Plattner GK, Tignor M, Allen SK, Boschung J, Nauels A, Xia Y, Bex V, Midgley PM (eds) Cambridge University Press, Cambridge, United Kingdom and New York

Jackson M, Ford-Lloyd B, Parry M (2014) Plant genetic resources and climate change. CAB International

Jenkins JC, Chojnacky DC, Heath LS, Birdsey RA (2003) National-scale biomassestimation for United States tree species. Forest Sci 49:12–35

Johansson T (1999) Biomass equations for determining fractions of pendula and pubescent birches growing on abandoned farmland and some practical implications. Biomass Bioenergy 16:223–238

Karthick A, Pragasan LA (2014) Stand structure and above-ground biomass of two tree plantations at Bharathiar University, Coimbatore. Indian For 140:29–33

Labriere N, Laumonier Y, Locatelli B, Vieilledent G, Comptour M (2015) Ecosystem services and biodiversity in a rapidly transforming landscape in Northern Borneo

Lasky JR, Uriarte M, Boukili VK, Erickson DL, Kress JW, Chadon RL (2014) The relationship between tree biodiversity and biomass dynamics changes with tropical forest succession. Ecol Lett 17:1158–1167

Lung M, Espira A (2015) The influence of stand variables and human use on biomass and carbon stocks of a transitional African forest: implications for forest carbon projects. For Ecol Manage 351:36–46

Luyssaert S, Schulze ED, Borner A, Knohl A, Hessenmoller D, Law BE et al (2008) Old-growth forests as global carbon sinks. Nature 455:213–215
MacDicken KG (1997) A guide to monitoring carbon storage in forestry and agro forestry projects USA. Winrock Int Institute Agri Development pp 19–99
Menéndez-Miguélez M, Canga E, Barrio-Anta M, Majada J, Álvarez-Álvarez P (2013) A three level system for estimating the biomass of Castanea sativa Mill. Coppice stands in north-west Spain. Forest Ecol Manage 291:417–426
Menhinick EF (1964) A comparison of some species-individuals diversity indices applied to samples of field in-sects. Ecology 45:859–861
Mitchard ETA (2018) The tropical forest carbon cycle and climate change. Nature 559:527–534
Mitra A, Sengupta K, Banerjee K (2011) Standing biomass and carbon storage of above-ground structures in dominant mangrove trees in the Sundarbans. Forest Ecol Management 261:1325–1335
NASA (2015) Global climate change. https://www.who.int/globalchange/climate/summary/en/index5.html
Nelson BW, Mequita R, Pereira JLG, de Souza SGA, Batista GT et al (1999) Assessing tree and stand biomass: a review with examples and critical comparisons. Forest Sci 45:573–593
NOAA (2015) National oceanic and atmospheric administration, US
Nwoko CO, Ngumah CJ, Ihejirika CE, Ngumah C, Ojiaku AA, Mbuka-Nwosu IE (2018) Arbuscular mycorrhizal fungi diversity in selected heavy metal contaminated soils in Owerri, Nigeria. J Appl Environ Microbiol 6(2):30–36
Ouyang S, Xiang W, Wang X, Zeng Y, Lei P, Deng X, Peng C (2016) Significant effects of biodiversity on forest biomass during the succession of subtropical forest in south China. For Ecol Manage 372:291–302
Pragasan AL (2015a) Tree carbon stock assessment from the tropical forests of Bodamalai hills located In India. Jou Earth Sci Clim Change 6:314
Pragasan LA (2015b) Total carbon stock of tree vegetation and its relationship with altitudinal gradient from the Shervarayan hills located in India. Jou Earth Sci Climatic Change 6:273
Reed D, Tomé M (1998) Total aboveground biomass and net dry matter accumulation by plant component in young Eucalyptus globulus in response to irrigation. Forest Ecol Manage 103:21–32
Sanchez-Martinez JL, Cabrales LC (2012) Is there a relationship between floristic diversity and carbon stocks in tropical vegetation in mexico? Afr J Agric Res 7(17):2584–2591
Takimoto A, Nair PKR, Nair VD (2008) Carbon stock and sequestration potential of traditional and improved agroforestry systems in the West African Sahel. Agric Ecosyst Environ 125:159–166
Vargas-Larreta B, Carlos Antonio López-Sánchez ID, Corral-Rivas José Javier, López-Martínez Jorge Omar et al (2017) Allometric equations for estimating biomass and carbon stocks in the temperate forests of North-Western Mexico. Forests 8:269–280
Wu X, Wang X, Tang Z, Shen Z, Zheng C, Xia X, Fang J (2015) The relationship between species richness and biomass changes from boreal to subtropical forests in China. Ecography 38:602–613

Flash-Floods: More Often, More Severe, More Damaging? An Analysis of Hydro-geo-environmental Conditions and Anthropogenic Impacts

Axel Bronstert, Irene Crisologo, Maik Heistermann, Ugur Ozturk, Kristin Vogel and Dadiyorto Wendi

Abstract In recent years, urban and rural flash floods in Europe and abroad have gained considerable attention because of their sudden occurrence, severe material damages and even danger to life of inhabitants. This contribution addresses questions about possibly changing environmental conditions which might have altered the occurrence frequencies of such events and their consequences. We analyze the following major fields of environmental changes.

- Altered high intensity rain storm conditions, as a consequence of regional warming;
- Possibly altered runoff generation conditions in response to high intensity rainfall events;
- Possibly altered runoff concentration conditions in response to the usage and management of the landscape, such as agricultural, forest practices or rural roads;
- Effects of engineering measures in the catchment, such as retention basins, check dams, culverts, or river and geomorphological engineering measures.

We take the flash-flood in Braunsbach, SW-Germany, as an example, where a particularly concise flash flood event occurred at the end of May 2016. This extreme cascading natural event led to immense damage in this particular village. The event is retrospectively analyzed with regard to meteorology, hydrology, geomorphology and damage to obtain a quantitative assessment of the processes and their development.

The results show that it was a very rare rainfall event with extreme intensities, which in combination with catchment properties and altered environmental conditions led to extreme runoff, extreme debris flow and immense damages. Due to the complex and interacting processes, no single flood cause can be identified, since only the interplay of those led to such an event. We have shown that environmental changes are important, but—at least for this case study—even natural weather and hydrologic conditions would still have resulted in an extreme flash flood event.

A. Bronstert (✉) · I. Crisologo · M. Heistermann · U. Ozturk · K. Vogel · D. Wendi
University of Potsdam, Karl-Liebknecht-Str. 24-25, 14476 Potsdam-Golm, Germany
e-mail: Axel.Bronstert@uni-potsdam.de

W. Leal Filho et al. (eds.), *Climate Change, Hazards and Adaptation Options*,
Climate Change Management, https://doi.org/10.1007/978-3-030-37425-9_12

Introduction: Peculiarities of Flash Floods and the Question of Anthropogenic Effects

The question of the possible influence of global warming on flood conditions on a regional scale is being intensively and controversially discussed by both the public and by experts, especially during and shortly after major flood events. A scientifically sound flood analysis must be based on adequate scales, i.e. on scales in which the processes of flood generation occur and for which management measures are appropriate. This typical adequate spatial scale is the upper meso-scale of about 1000–100,000 km^2 for floods on the major rivers, i.e. not the continental or even global scale. For flash floods, the adequate spatial scale is the lower meso-scale of about 50–1000 km^2. The relevant time scale of flood generation for large river floods is usually several days to one or two weeks, with a temporal resolution of days. For flash floods, the relevant time scale is between hours and one day, with an hourly to about 5-min temporal resolution.

A profound knowledge synopsis applicable for these scales is all the more relevant, as there is little concrete information about regional conditions in global "assessment reports". In the second chapter of the first working group in Sect. 2.6.2.2 (Hartmann et al. 2013) on floods, the fifth progress report (AR5) of the IPCC only states: "… trends in floods are strongly influenced by changes in river management" and "…other studies for Europe and Asia show evidence for upward, downward or no trend in the magnitude and frequency of floods". The third chapter of the second working group mentions in Sect. 3.2.7 (Jiménez Cisneros 2014) on extreme hydrological events and their effects: "There is no strong evidence for trends in observed flooding in the USA, Europe, South America, and Africa. However, at smaller spatial scales, an increase in annual maximum discharge has been detected in parts of northwestern Europe, while a decrease was observed in southern France."

It is clear that those statements on floods in general and Europe in particular, derived from global assessment reports, are very fuzzy and uncertain and are not applicable for any management or adaptation measures. Moreover, the mentioned reports do not distinguish regarding their main physical flood formation mechanisms, e.g. large river flooding versus small scale flash flooding triggered through high intensity rain storms.

Therefore, in this contribution we follow two lines to gain insight into the question of potentially changing occurrence, magnitude and consequences of flash flood events and the possible relations to global changes. First we present a summary of state-of-the-art knowledge of global change effects on different processes of flood generation and flood propagation. In this regard we address changing environmental boundary conditions, such as climate change, land-use change and river construction measures, with a main focus on climate change. Secondly, we present results from a forensic hydrological disaster analysis (Bronstert et al. 2018), where a disastrous flash flood event was analyzed retrospectively as detailed as possible. The aim of this analysis was to identify and quantify the intertwining processes and consequences from meteorology, hydrology, hydraulics and geomorphology and to understand the different natural and anthropogenic influences.

In the beginning, because of the general importance of the peculiarities of flash flood events, we briefly summarize their particular features.

Flash floods are sudden flood events which are typically caused by a somewhat spatially small-scale high-intensity convective rainstorm, potentially amplified by orographic effects. Such events are characterized by a relatively short response time, i.e. the delay between the barycenter of the rainfall input and the flood peak is usually less than six hours (Borga et al. 2011, 2014). Because of the rather small area covered by a convective rainstorm, their spatial extent is usually below 500 km^2.

The runoff generated from such rainstorms may be amplified by soils with reduced infiltrability, due to e.g. high soil saturation, or hydrophobic crusting, or sealing effects at the soil surface, either natural or man-made. Steep slope gradients, efficient channel networks and low soil surface roughness support high runoff velocities in a catchment and thus the generation of a severe flash flood. Flash floods often create a particularly strong threat to the affected populated areas, with the following key features:

Short warning time: The time available for a flash flood warning is short by definition. However, a fast and timely warning is not only difficult due to the short response time from rainfall to runoff, but also because of the highly uncertain predictability regarding the location, time and magnitude of the rainfall event. This is why the options for counter measures are limited and the mortality rate (# of fatalities/# of affected people) from flash floods is significantly higher compared to large river floods Jonkman (2005).

High runoff velocities: In the case of mountainous or hilly relief, extreme runoff will develop high flow velocities, both in the channel system and the inundation areas. This results in strong momentum towards any obstacles and resulting in much damage to houses and other infrastructure. The momentum from the water flow can be amplified through rocks, bed load or woody debris transported with the water flow.

Singular (erratic) behavior: The dynamics of such runoff and inundation conditions are difficult to predict and might be heavily influenced by singularities occurring during the flood. Typical examples for such conditions are the sudden blockage of culverts or bridge openings through rocky or woody debris. This may lead to spontaneous backwatering, new flow paths or surge waves.

Geomorphological processes may be triggered through high runoff rates and discharge velocities. This is a result from the strong momentum of such water flows, both on the hillslopes of the catchment area and in the channel system. Such strong momentum may cause soil surface erosion, bank erosion, slope failures, transport of sediment debris and of woody debris in the channel system. In areas with gentler slopes, the sediments will be deposited and thus cause another type of hazard, including the mentioned blockages of river sections or bridges. Such events bear particular high individual risk of material damage or even loss of lives (Jonkman 2005; Kron 2016).

The chances to conduct observations and measurements during those events are very limited. Even a strategic pre-event installation of measurement equipment of, e.g. pluviometers or discharge gauges, is rarely possible because the limited spatial extent of such events in combination with the rather stochastic kind of occurrence will not allow prediction of the location and timing of such an event that would be required for such an on-site observation campaign. Having in mind such limited options for direct observations, Marchi et al. 2010 were among the first to adopt forensic analysis as a suitable method to analyze such events retrospectively. They showed that it is possible to derive a relatively comprehensive and multi-faceted picture of the triggering meteorological and catchment conditions, the hydrological, hydraulic and geomorphological processes, and of the subsequent damage of the affected area.

In the case of an extraordinary flash flood event, e.g. like the one in Braunsbach, SW Germany on 29th May 2016, questions have arisen in public debates such as if—and if yes, how much—anthropogenic activities may catalyze or intensify such events. On the one hand, it is obvious that almost any anthropogenic activities in the climate, landscapes or water resources systems *do influence hydrological processes and such subsequent processes* as sediment transport. On the other hand, the more tricky and relevant question is *to what extent* such processes and systems are influenced by human activities. There have been many investigations recently of influences of environmental changes on hydrological processes and hydro systems and the possible relevant effects regarding this particular event are elaborated in the following sections.

Effects of Climate Warming

When analyzing the effects of an altered warming climate on precipitation it is essential to focus on a regional scale: i.e. statements about global precipitation have neither any practical value nor can the precipitation results from global circulation models be given with an acceptable reliability. Kunz et al. (2016) reviewed the state of knowledge on precipitation changes as a consequence of climate change for Germany. They found that for average conditions, i.e. annual values, changes in rainfall conditions are not clearly identifiable up to now. However, they state that a seasonal rainfall shift from summer towards winter, i.e. wetter winters and drier summers, can be seen although with considerable uncertainty.

If climate change effects on flash floods are of concern, one crucial question is whether the intensity of short rainstorms is becoming higher, or—in other words—if the occurrence probability of high intensity rain storms is increasing. Therefore, the relevant question is not if total rainfall increasing, but if the events with high rain intensities are increasing. To answer this question, analysis of rainfall series with a very high temporal resolution, e.g. time steps below one hour down to a few minutes are required. Analyzing daily series—as conducted in most studies in the past—cannot deliver the required information.

Such detailed analyses have been conducted, e.g. for stations in the Ruhr area (Müller and Pfister 2011) and for regions in Austria, Switzerland and West Germany (Bürger et al. 2014; Bürger et al. 2018). Müller und Pfister (2011) analyzed precipitation series for eight stations in the Emscher-Lippe basins in North-Rhine-Westphalia, which were available in an extraordinarily high temporal resolution of one minute for the period of the 1940s until 2009. They analysed events above an intensity of 0.3 mm/min (which is 20 mm/h, respectively). The results showed, see Fig. 1, that for all the stations investigated, the occurrence of short but high-intensity rain storms has increased. These trends were more pronounced in the last 35 years as compared to earlier decades. The increasing trends were particularly strong during the summer months July, August and September. Such high-intensity rainstorms occur between 4 and 15 times a year, depending on the station. An increase of 0.5 events/year, or 5 events/decade, means that the occurrence rate of such events has been multiplying over a period of few decades.

Two other studies analyzed the same data set: the study by Fiener et al. (2013) for the same region confirms the key messages, i.e. the erosion-relevant heavy rain events have increased significantly since the mid-1970s, by about 21% per decade. On the other hand, the study by NRW (2010) did not find significant trends for extreme rain event occurrence in their analysis period 1950–2008. However, an analysis of transient changes was not conducted in this study.

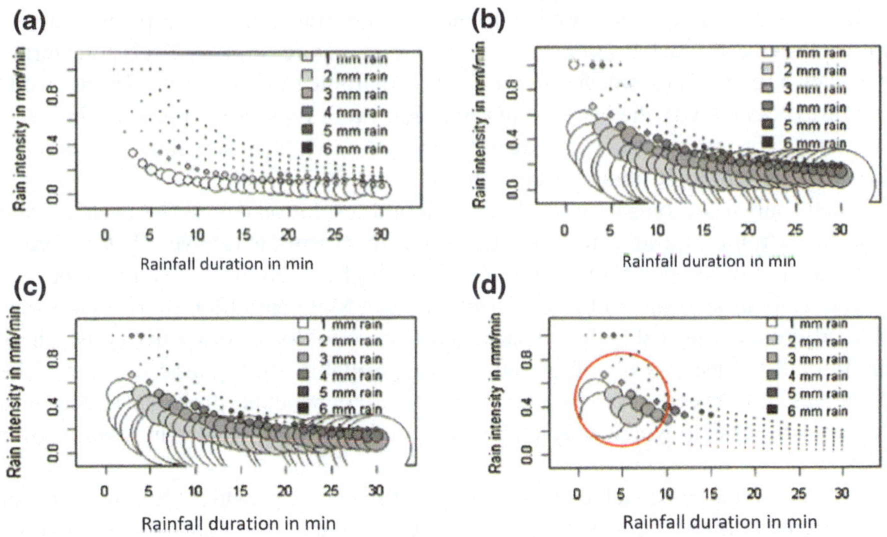

Fig. 1 Trend analysis of high-intensity rain storms at station Essen, Germany, in high temporal resolution. Event duration from 1 to 30 min. Bubbles are scaled uniformly: the largest bubble diameter corresponds to a trend increase of 0.5 events/year, and a pixel points to an increase of zero; **a** trends for 1940ths–2009; **b** trends for 1975–2009; **c** trends for 1975–2009, statistically significant; **d** trends for 1975–2009, statistically significant and erosion-relevant (from Müller and Pfister 2011)

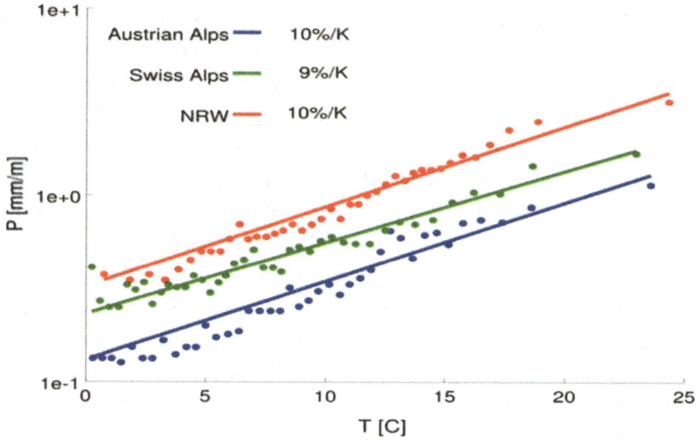

Fig. 2 Relationship of high-intensity rainstorm intensities (99.9% quantiles) and station temperature at the corresponding day of the events. Analysis based on 15 min-precipitation data for stations in the Austrian Alps, Swiss Alps, North-Rhine-Westphalia (from Bürger et al. 2014)

Bürger et al. (2014) investigated the relationship between high intensity rain storms (for 15 min time increments) and the day-time air temperature during the events. For three analyzed European Regions (Austrian Alps, Switzerland, North-Rhine-Westphalia in Germany) they found a strong, statistically significant relationship between extreme local rainstorms and air temperature, see Fig. 2. It is noteworthy that such a relationship was not found for non-extreme rainfall events. These results can be explained with the increase in saturation vapor pressure with increasing temperature according to the law of Clausius-Clapeyron (which means about 7% increase in water vapour per °C temperature increase).

Such analysis of long-term and high-temporal resolution data series even showed that the extreme rainfall intensities for relatively warm air (above 12 °C) were—as a function of temperature—up to two times higher than the Clausius–Clapeyron relationship, as also shown by Lenderink und van Meijgaard (2008). This considerable difference is probably due to particular features of local convective cells, where the rainfall intensity is restricted only by the available atmospheric moisture. The dynamics of convection cells is also temperature dependent, probably resulting in superposing effects which may lead to rates of rainfall intensity increase of over 10%/°C.

In general one can conclude that a warmer atmosphere provides the conditions for more frequent high-intensity rain storms. This is in agreement with general knowledge of higher rainfall intensities in warmer climates, e.g. tropical compared with mid-latitude climates.

The results of such studies, thus, support the assumption of an increased occurrence probability of flash flood triggering high intensity rainfall events as a consequence of global warming on a regional scale. This has been shown at least by the analyzed data for the mentioned regions in Central Europe. Similar studies for other regions would be rather interesting, but unfortunately such high-resolution, sub-hourly time series are not routinely available.

Projections for the future development of temperature are much more reliable than those for future rainfall. Therefore, such temperature projections could be used to derive temperature-dependent high-intensity rainfall projections, applying the relationship detected in these data series. The actual value of temperature dependency, however, is strongly influenced by the region under consideration and from the data resolution in time (e.g. hourly vs. minute time steps), e.g. as discussed by Loriaux et al. (2013).

Effects of Altered Land Use and Land Cover

Land use, land cover and land surface conditions have a particular influence on surface and near-surface hydrological processes. With respect to flood generation due to high-intensity rainfall, the governing processes are related to infiltration and surface runoff flow. Thus, overland flow generation is significantly generated due to infiltration excess, so-called Hortonian overland flow, i.e. where the actual rainfall intensity I_N exceeds the actual soil surface infiltration capacity I.

Figure 3 shows a general scheme of infiltration excess overland flow generation, accounting for a soil containing micro-pores (or "soil matrix") and larger voids (or "macro pores"). In the left panel, the actual net rainfall intensity I_{Nnet} is smaller than the infiltration capacity into the micro-pores I_{mic}. Thus all rainfall is entering the soil surface. In the case of very high rainfall intensity (right panel) I_{mic} might be exceeded and local microscopic surface flow will divert excess water into the macro-pores (infiltration into macro-pores I_{mac}). In the case when the rainfall intensity even

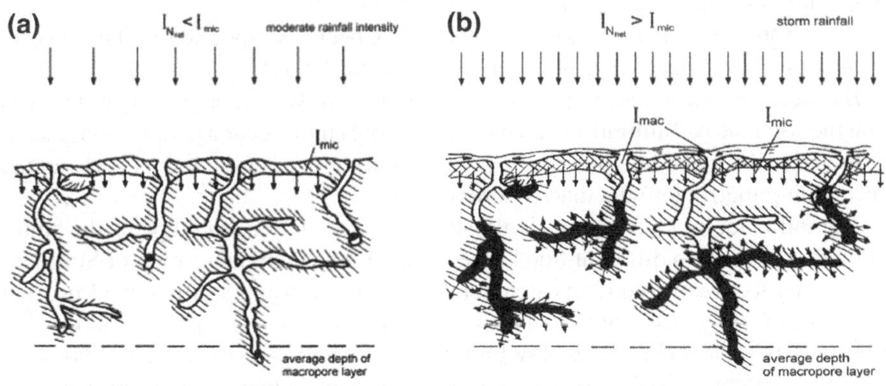

Fig. 3 Scheme of surface runoff generation as a response to rainfall excess: **a** moderate rainfall is fully absorbed by the soil matrix/micro-pores; **b** high rainfall intensity exceeds matrix infiltration capacity and possibly even macro-pore infitation capacity, leading to surface runoff (from Bronstert 1999)

exceeds the total infiltration capacity ($I_{mic} + I_{mac}$) surface runoff will be generated. It becomes clear that this process is derived directly from the balance between rainfall intensity and infiltration capacity. All conditions leading to high rainfall intensity, e.g. convective rain formation, increase the probability of surface runoff. All conditions favoring/hindering soil infiltrability, e.g. macro-porosity/soil compaction or crusting will decrease/increase potential surface runoff generation.

Besides the rainfall intensity, the formation of surface runoff also depends on the local infiltration conditions. This means surface runoff is significantly smaller if infiltration enhancing soil features, e.g. macro-porosity or other options for bypass-flow, are present. Those features are more likely in soil surfaces with many bioturbations, such as worm pipes or voids from rotten roots. But also surface features such as cracks in drying and thus shrinking clay-rich soils, or gravel layers close to and in connection with the soil surface may enhance bulk infiltration capacity by an order of magnitude or even more, see Beven and Germann (1982). On the other hand, soil infiltrability can also be reduced, if macro-pores are clogged, e.g., either by fine soil particles mobilized through high rainfall energy on un-covered bare soil, and then transported and finally deposited into such voids, or by soil compaction, e.g. caused by agricultural machinery, or grazing livestock. Other causes for hindered infiltration can be chemical or biological soil crusting as a result of e.g. salt on soils in semi-arid regions, or algal occurrence on the soil, respectively.

Based on the infiltration model sketched in Fig. 3, (Niehoff et al. 2002; Bronstert et al. 2002) conducted a series of simulation experiments. Those included case studies with catchment and field data in three different meso-scale catchments in Germany, i.e. two in Baden-Württemberg and one in North Rhine-Westphalia. They found that in the case of high-intensity rainfall events the land-surface features affect flood generation considerably and, as a consequence, the flood magnitude, shape and volume. The higher the intensity is, the greater is the share of overland flow generated on the land surface and this is why the real rainfall intensities—non-averaged over time—are thus important (Bronstert and Bárdossy 2003).

The infiltration scheme mentioned above has been incorporated in the process-oriented and distributed hydrological model WaSiM-ETH (Schulla 1997; Niehoff 2002). Here, macro-pores are treated as a distinct soil water storage which interacts with the soil matrix. Infiltration into the macro-pores may occur as infiltration-excess (from the soil surface into the macro-pore system) or as saturation-excess (from the surrounding matrix into the macro-pores).

The influence of macro-pores is illustrated in Fig. 4, which compares the simulation results for two different flood types in the Lein catchment, south of Stuttgart. The mainly loess-covered Lein catchment is dominated by agricultural land-use and covers an area of approximately 115 km^2. For this example catchment, spatially distributed data on actual and possibly future land-use and land-surface conditions are available. The hydrological response to the scenarios is being simulated using the mentioned modified version of WaSiM-ETH, including consideration for macro-pore infiltration, siltation effects and the retention capacity of urban sewer systems. In the following, an important feature of the land-surface characteristics is assessed—the

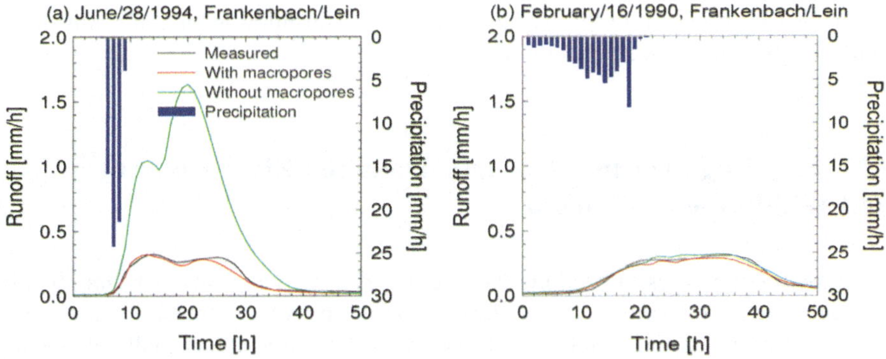

Fig. 4 Exemplary simulation of two flood events in the Lein catchment with a return period of approximately 3 years as response to **a** a convective and **b** an advective storm event, showing the impact of macro-pores on runoff generation

role of macro-pore infiltration—which is directly related to land-use and thus the formation of surface runoff.

The surface impacts are demonstrated for rather different rainfall intensities, e.g. for convective and advective rain events. As Fig. 4a demonstrates, generation of infiltration-excess overland flow and resultant direct runoff due to intensive precipitation is greatly dependent upon the way infiltration is accounted for i.e., if infiltration is only effective as matrix infiltration, controlled by soil hydraulic conductivity, the resulting surface runoff is about 5 times higher compared to simulations including macro-pore infiltration. For the advective event (Fig. 4b), because of lower precipitation intensities and higher initial soil moisture, the differences are less obvious. In this context it is important to notice that flash floods are almost always triggered by convective rain cells. In contrast, for advective rainfall events, covering wide areas for longer time periods, the event type which may generate floods in large river basis, such as the whole Rhine basin, infiltration-excess overland flow is of very minor importance. For such conditions, as shown in Fig. 5, the main runoff generation

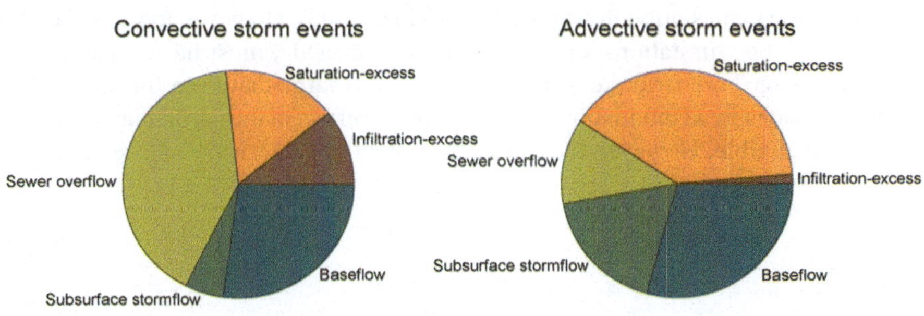

Fig. 5 Runoff components simulated with WaSiM-ETH for the Lein catchment for five convective (left panel) and six advective (right panel) heavy rainfall events with return periods between two and eight years (from Niehoff and Bronstert 2001)

mechanisms are saturation-excess and subsurface stormflow which are influenced much less by land-surface conditions.

Effects of Engineering Measures Along the River Course and Within the Catchment

Construction measures in or along the river course, such as dikes or weirs, mainly influence the runoff velocity, water levels, and the possible inundation extent, but hardly the runoff volume and peak discharge, at least for rather small catchments as in many flash-flood event cases. Hydraulic constructions in the catchment area, such as retention basins, may retain a—limited—volume of flood runoff. Thus, it becomes clear that river construction measures can have positive effects (inundation reduction), worsening effects (amplifying or even triggering inundation) or hardly any effect at all, depending on the individual event.

Particular problems for high flood discharges are channel contractions, caused by either natural narrows or by man-made reduction of the river cross section. Such effects can lead to local congestion (blockages) of the water, especially in connection with geomorphological processes (heavy stony or woody debris transport). The flood discharge then seeks either an alternative course (such as through the streets) or the backflow upstream of the conglomeration will lead soon to the breakage of this obstacle with a subsequent, often very dangerous surge wave in the channel downstream.

Such structural bottlenecks are often bridges that do not span the entire cross section of the water or passages under roads or railways embankments. In particular, a piping of a river section will definitely limit the maximum discharge. The analyses of many flash flood-induced damage, have shown that blockings resulting from a natural or man-made narrow were often a major cause of very hazardous inundations. Impressive examples are the conglomeration of the Saltina River with subsequent flooding of the city center of Brig in Valais, Switzerland in September 1993 (BUFA 1993) or the blockage of the Weisseritz in the urban area of Dresden with subsequent flooding of the main station in August 2002 (Kinze 2002). However, the actual effect of such discharge limitations, either natural or man-made, must be compared with a "non-blockage" scenario: i.e. if the discharge is so large that even for un-blocked conditions widespread inundations would occur, the effect of the particular bottleneck might be negligible. However, the risk of the potential surge-wave still remains.

Case Study: The Extreme Braunsbach Flash Flood, 29th May, 2016

On the evening of May 29, 2016, the municipality of Braunsbach in the state of Baden-Württemberg, SW-Germany, was hit by an extreme flash flood. This flash flood was triggered by one of the various rain storms occurring over Eastern France and Southern Germany in the early summer of 2016, caused by the low pressure system Elvira. This particular extreme runoff and debris flow event led to immense damage in this municipality of over 100 Mio Euro. Figure 6 shows three photographs enabling the reader to get an idea of the processes and consequences during and after this event.

In the following paragraphs, a brief summary of this event and an analysis of possible anthropogenic influences are given.

Description of the Affected Region

The Orlacher Bach drains a small 6 km^2, predominantly agricultural catchment in the southwest German stratified landscape and flows southwest through a steep valley ("Orlacher Klinge") until it flows into the regional main river Kocher after about 3 km. Shortly before the river mouth, where the "Klinge" opens to the Kocher-valley, is the village Braunsbach. The side-slopes of the valley, above Braunsbach, have a high inclination between 20° and 40° and are mainly covered with forest, as an agricultural use is not possible there.

Fig. 6 Photographs of the Braunsbach Flash flood event at 2016-05-29. Left: discharge of water and sediment through the town's center at 18.30 UTC; Center: Sediment debris deposited in the town center (picture taken on 2016-06-01); Right: damaged house in the town center (picture taken on 2016-06-03). *Photos* J. Kolb, left; A. Lucia, University of Tübingen, center; V. Rözer, GFZ, right

Rainfall Conditions

In order to estimate the fallen rainfall of this event, a retrospective meteorological analysis was conducted. In particular precipitation data of surrounding rainfall stations and radar data of the German Weather Service were analyzed. No direct rainfall station data were taken because no pluviometers did exist in the catchment area. A thorough multi-method rainfall analysis in the highest possible spatial and temporal resolution (i.e. 5 min time steps, 1 km^2 spatial resolution) gave the following results.

- It was a very intensive, very rare, i.e. “extreme” rain storm with a return period of >100 years. The total rainfall amounted to about 140 mm in two hours. The 5 min resolution reveals that the main rainfall over the Orlacher Creek catchment took place in about 70 min (from 16:50 to 18:00 UTC) and amounted to 130 mm.
- Both the temporal as well as the spatial variability of the rain were very high. Information about these variabilities is only partly detectable from the station data and lead to underestimations of local peak values.
- The inclusion of the radar data yields important information about the spatio-temporal patterns of the intensity. From that, it is recognizable that the catchment area of the Orlacher Creek was exactly in the center of the local thunderstorm. A correction of the radar signals’ reflectivity attenuation is essential to yield more reliable estimates of the rainfall rates. The attenuation correction applied here resulted in at least a doubling of the peak intensities.

Discharge Situation

The estimation of peak water levels and discharge rates is a key issue for a forensic analysis of an extreme hydrological event. In the Braunsbach case, four independent approaches were performed in order to obtain a rather reliable estimate of the extreme runoff.

- Derivation of maximum water levels based on information from local residents, flood marks along the valley slopes and at house walls and other markers in the terrain. A subsequent estimation of runoff cross sections of the “channel” in the village during the flood and estimates of flow velocities, based on analysis of video data yielded peak discharge estimates.
- Estimation of runoff volume and hydrograph shape, based on total rainfall, runoff coefficient estimates and duration of the flood hydrograph.
- Hydrological modelling of the event, based on the spatial rainfall data, runoff coefficient approach and flow-velocity estimates.
- Comparison and differencing of discharge rates in the main river Kocher upstream and downstream the inflow of the Orlacher Creek.

In summary, these independent methods yielded rather similar results, i.e. peak discharge in the range of 100–150 m^3/s. Considering that the long term mean discharge of the creek is about 0.2 m^3/s, then the peak discharge rates during the flash flood were at least 500–800 times larger. This very high difference demonstrates both the typical flash flood type and the extreme characteristics of this event. It is nonetheless obvious that, in the absence of direct discharge measurements during the event, the analysis of flow rates presented here is still subject to considerable uncertainty.

Geomorphological Analysis

The event caused very strong and interacting geomorphological processes. The most severe process was the mobilization, transport and subsequent deposit of large rocks/debris. This debris was mobilized and transported by the extreme creek discharge and finally deposited in the village. It amounted to 42,000 m^3 of gravel and boulder, according to information provided by a clearing company. Further geomorphological processes include surface soil erosion on the arable land in the highlands, river bank erosion, mobilization and transport of woody debris and damage to buildings and streets.

The analysis of these geomorphological processes was conducted in the catchment area and the affected municipality in several field studies shortly after the event:

- One could see strong surface erosion traces almost all over the agricultural fields. Assuming an average erosion rate of 2–3 mm on the approx. 5 km^2 large highlands of the catchment results in a removal of suspended sediment of approx. 10,000–15,000 m^3. Since this is fine, suspended sediment, it can be assumed that this was transferred almost completely into the main river Kocher and not stored before.
- Due to the high discharge rates during the flash flood, sediments from the banks and channel bed from the Orlacher Creek was removed and transported downwards. Almost throughout the valley section signs of bank and bed erosion were clearly visible. These surveys suggest that much of the 42,000 m^3 of material deposited in the village was originating from those erosion processes in the valley. Taking an average erosion depth of 2 m over a width of 5–10 m over all valley sections (about 2 km), results in a volume of 20,000–40,000 m^3 debris.
- Further debris were mobilized by several landslides in the Orlacher “Klinge” and transported by the extreme discharge into Braunsbach. Based on the mentioned field measurements, the total volume of such material mobilized by the landslides towards the stream is estimated to be about 8000 ± 1600 m^3. One important result refers to the fact that the creek had been partly equipped with some check dams about 100 years ago (see below). Those check dams collapsed during the event and significantly contributed and enlarged the amount of debris.

Damage Analysis

The high discharge rates and large amounts of entrained sediments caused enormous damage in the village of Braunsbach. In order to get a better impression of the damage as well as to better understand damaging processes in case of flash floods, on 7 and 8 June 2016, i.e. nine days after the event, externally visible damage to buildings and other information on the buildings and the flooding systematically were recorded, see Vogel et al. 2017. Some damage to infrastructure and vehicles, and the nature and extent of building contamination could not be analyzed at the time of the field investigations.

On the other hand, impact damage to buildings and identifiable contamination by oil or sludge were documented. The water level at the buildings was estimated on the basis of flow marks, mud contamination and moisture penetration into the outer walls. In addition, temperature differences in the house walls were mapped with a thermal imager. The thermal imaging also helped in the classification of building material. Further field data of the buildings included type of basement, number of floors, use, age, condition of the houses, recognizable precautionary measures and characterization of the immediate building environment, e.g. sealing and exposure of the building to the flow direction.

As a final result, Fig. 7 shows the estimated water levels for the areas affected by the flash flood in Braunsbach, where the water levels estimated on the buildings were taken as a basis.

The discharge rates and large amounts of entrained material caused enormous damage in the village of Braunsbach. The damage estimated by the district authorities, amount to 112 Mio Euro for the Braunsbach municipality, where the major part (104 Mio Euro) occurred in the village. The municipality includes, next to the main village Braunsbach, some more villages and hamlets. Although these figures are rough estimates, they illustrate the particular magnitude of destruction.

In the following paragraphs the most important anthropogenic effects which might have amplified this event are elaborated.

Intensified Rainfall as a Consequence of Global/Regional Warming

These studies mentioned in the earlier paragraphs of this article imply that the occurrence probability of high intensity rainfall events has increased over the last 70 years, with accelerated rates in the last 30 years, probably as a result of regional warming and, thus, as a consequence of climate change. This has been analyzed, however, not (yet) for all regions in Germany. For instance, it is still missing for the region of Northwester Baden-Württemberg, where Braunsbach is located. The biggest obstacle for such an analysis is the lack of long term rainfall data in an appropriate, i.e. highly temporal resolution. Otherwise this could be done rather straight forward.

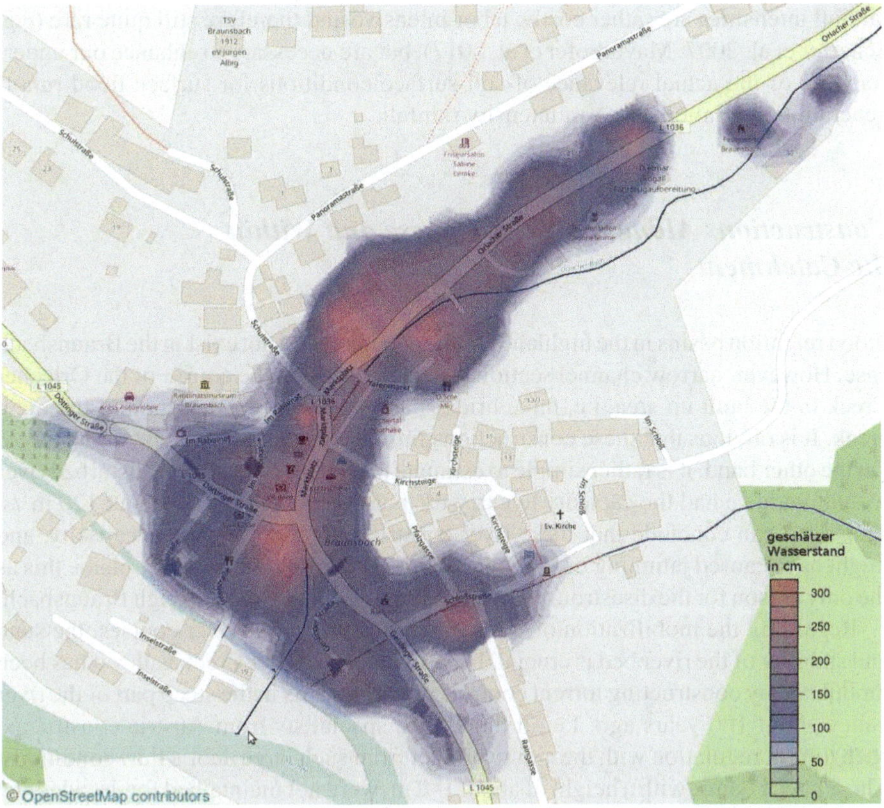

Fig. 7 Estimates of the water level at the buildings affected by the flash flood in Braunsbach in cm above the terrain surface. The Orlacher Bach (parallel to the Orlacher Str. and marketplace) and the Schlossbach (parallel to Schlossstr.) are shown in dark blue or the piped sections in dark red (from Vogel et al. 2017)

Land Use, Land Cover and Land Surface Conditions

In the case of Braunsbach, over 70% of the catchment area is used for arable land, with less favorable conditions for high infiltration capacities. In particular, the 1.4 km^2 used for corn farming (ca. 32% of total arable land), where the crops during the event were just in their initial development state. Those fields may have contained few bioturbations and additionally were prone to soil siltation due to the high kinetic rainfall energy impinging directly on the soil. Therefore, one may argue that the agricultural land use, in particular the corn field, might have enhanced surface runoff generation. On the other hand, Salazar et al. (2012) argued that the higher the rainfall amount, and thus the infiltration exceedance rate, the smaller the relative effect of different land cover and land uses on surface runoff production. Actual field observations and/or controlled experiments on runoff generation processes under such high rainfall intensities are rather costly, labor intensive and therefore still quite rare (e.g. Scherrer et al. 2007; Mayerhofer et al. 2017), but are necessary to enhance our understanding of the actual relevance of soil surface conditions for surface flood runoff generation in relation to storm intensity rainfall.

Constructions Along the River Course and Within the Catchment

Flood retention basins in the highlands of the catchment do not exist in the Braunsbach case. However, narrow channel sections are present along the course of the Orlacher Creek in the built up area, i.e. three bridges and a 150 m piped section of the main creek. It is obvious that these constructions limit the discharge capacity of the creek. On the other hand, it is rather simple to estimate that even an open, non-disturbed river would not have had the capacity to carry the peak discharge rate of about 120 m^3/s. Thus, one can conclude that these river constrictions limit discharge capacity, and might have caused jamming of sediment debris. However, one must not blame this as the only reason for the disastrous inundations and sediment flow through Braunsbach.

Regarding the mobilization of sediment and debris in the river courses, the state and stability of the river bed is crucial. In case of the Orlacher Creek valley it has been modulated by constructing torrent controls or check dams in the steep part of the river valley about 100 years ago. I.e. in the 1920s, specialists from Austria constructed such torrent regulation with the aim to protect from such large debris flow conditions. These check dams, with a height of about 1–2 m, were not maintained for decades and thus were, partly damaged and almost forgotten, before the flash flood occurred. As the flash flood happened, most of those check dams were finally destroyed and their remainders transported downstream with the flood water, together with sediment deposited earlier behind these dams. Therefore, the extraordinarily large amount of coarse sediment, mobilized, transported and deposited downstream might partly be explained as a consequence of the failure of the old check-dam cascade.

Discussion and Conclusions

This contribution has first summarized present knowledge about effects of environmental changes on different flood generation and propagation processes, with a main focus on climate change effects on flash floods. Secondly, we presented results from a forensic analysis of an extreme and disastrous flash flood event which happened end of May 2016 in SW-Germany. By combining a more general review of process knowledge with a very detailed analysis of a particular flash flood disaster we derived insights regarding the most important processes, the relevance and difference of various anthropogenic influences on such events.

We could show that both climate and hydrologic boundary conditions are highly relevant for the formation of a flash flood event. Keeping this in mind, it is obvious that changes of such boundary conditions will result in altered flash flood situation. However, the actual magnitude of such influences is strongly dependent on the local conditions and interaction of both rainfall intensity and soil surface infiltration capacity. As these governing variables—rainfall intensity and infiltration capacity—can vary over orders of magnitude, it is hardly possible to derive generic statements on the quantity of human influences on such flash floods. However, it was possible to identify the mentioned key processes, which have to be taken into account for studies about anthropogenic impacts on this kind of events.

The detailed, forensic hydro-disaster analysis resulted in a rather clear picture of the Braunsbach flash flood event. It was essential to combine data from different disciplines (here: meteorology, hydrology, geomorphology, damage and risk analysis) and from rather different types of data, e.g. rainfall station data from the surrounding region, rainfall radar data enabling necessary corrections and delineating spatial patterns, runoff volume and peak discharge estimates and assessment of sediment mobilization,

This particular event which occurred in this small catchment area of only 6 km^2 showed once again that in case of the formation of such events many different factors play an important role. Therefore, an interdisciplinary approach is indispensable for the analysis. The special features of the event investigated here are:

- the location of the storm cell directly above the catchment area of the Orlacher stream,
- the very high precipitation intensities over a period of less than two hours (recurrence interval clearly >100 years),
- high and rapid runoff formation in the area,
- mobilization of sediment in the area and of very large boulders (diameter of single blocks >1 m) and woody debris in the steep valley of the Orlacher Bach and neighboring valleys,
- overspilling and outbreak of the brook, resulting in the transport of both water and debris, just through the community, and
- sludge and water masses pushed against the house walls in the village, which ultimately caused the exceptionally high damage.

Besides these specific event characteristics, also typical, general features of flash floods can be seen:

- Only the interaction of different and multi-layered processes can cause such a particular strong event.
- The question of anthropogenic effects cannot be discussed without an overall view of all influencing factors. A singular cause or even "blame" is scientifically dubious and misleading.
- At the same time, it is also clear that anthropogenic activity and measures have an impact on the occurrence, frequency, magnitude and course of such flood events.

- For very heavy rain storms, however, the relative importance of land use is probably rather small. Still, there is not enough knowledge on how soil infiltration conditions develop during such events. The scientific challenge is to quantify the extent of the mentioned anthropogenic effects and to provide a comparison to flood events without human impacts.

A complete "protection" from such events in the sense of prevention is not possible. There will always remain a residual risk, as with all natural and technical hazards. However, the goal must be to better deal with such events and to establish suitable precautionary measures. Such precautionary measures are to be understood as:

- An improvement in the forecast of short-term extreme precipitation and resulting flash floods. Future efforts to improve the prediction should be directed towards the development of operational systems for the use of radar information in combination with ground measurements in real time and to use this information for short-term forecasting ("near real time forecasting") of convective events with associated hydrological prediction.
- Constructions at waters courses and in the catchment area to be designed in a way to prevent bottlenecks during flood discharges. This also includes the examination of existing bridges, road culvert and other measures along the channels regarding their capacity for extreme discharge.
- Also, fluvial-geomorphological assessments of the stability of the river bed and adjacent slopes may be important, see Vogel et al. 2017. There are proven experiences at the authorities dealing with torrential events in mountain regions.

In order to determine the future impact of climate change on the frequency and amplitude of flash flood events precipitation projections for short durations, e.g. less than one hour, and for small spatial scales are required. Such projections cannot be obtained from the current global climate models, even not from regional climate models (RCMs). As an alternative method, a new perspective has developed in recent years: the consideration of the dependency of extreme precipitation intensity on the air temperature during the time of the event. The correlation between extreme local precipitation and air temperature seems to be that much pronounced that empirical relations can be derived and thus, projections for rain storm intensity under a future warner climate can be established. Based on such estimates on future rain storm intensities, refined hydrological and hydraulic models may serve as the basis for projections of future flash flood conditions. Still, considerable uncertainties will remain for processes under rare hydro-meteorological conditions, such as infiltration conditions under extreme rain intensity, runoff concentration for wide-spread overland flow, or stability of channel banks, channel bed and adjacent hillslopes in case of extreme discharge rates. It will be worthwhile to focus future process research at least partly on such extreme conditions rather than on average or "normal" conditions.

References

Beven K, Germann P (1982) Macropores and water flow in soils. Water Resour Res 18(5):1311–1325

Borga M, Anagnoustou EN, Blöschl G, Creutin JD (2011) Flash flood forecasting, warning and risk management: the HYDRATE project. Environ Sci Policy 14(7):834–844

Borga M, Stoffel M, Marchi L, Marra F, Jakob M (2014) Hydrogeomorphic response to extreme rainfall in headwater systems: flash floods and debris flows. J Hydrol 518:194–205

Bronstert A (1999) Capabilities and limitations of detailed hillslope hydrological modelling. Hydrol Process 13:21–48

Bronstert A, Bárdossy A (2003) Uncertainty of runoff modeling at the hillslope scale due to temporal variations of rainfall intensity. Phys Chem Earth 28:283–288

Bronstert A, Niehoff D, Bürger G (2002) Effects of climate and land-use change on storm runoff generation: present knowledge and modelling capabilities. Hydrol Process 16(2):509–529

Bronstert A, Agarwal A, Boessenkool B, Crisologo I, Fischer M, Heistermann M, Köhn-Reich L, López-Tarazón JA, Moran T, Ozturk U, Reinhardt-Imjela C (2018) Forensic hydro-meteorological analysis of an extreme flash flood: the 2016-05-29 event in Braunsbach, SW-Germany. Sci Total Environ 630:977–991

Bundesamt für Umwelt (BAFU) (1993) Die Hochwasser 1993 im Wallis und Tessin. Messdaten und ausgesuchte Auswertungen. Mitteilungen des BAFU Nr. 19. Landeshydrologie und -geologie der Schweiz, Bern, 1993, 65 Seiten

Bürger G, Heistermann M, Bronstert A (2014) Towards sub-daily rainfall disaggregation via Clausius Clapeyron. J Hydrometeorol 15(3):1303–1311

Bürger G, Pfister A, Bronstert A (2018) Sub-hourly extreme rainfall scenarios, consistent with Clausius-Clapeyron. (Under review)

Fiener P, Neuhaus P, Botschek J (2013) Long-term trends in rainfall erosivity–analysis of high resolution precipitation time series (1937–2007) from Western Germany. Agric For Meteorol 171–172(2013):115–123

Hartmann DL, Klein Tank AMG, Rusticucci M, Alexander LV, Brönnimann S, Charabi Y, Dentener FJ, Dlugokencky EJ, Easterling DR, Kaplan A, Soden BJ, Thorne PW, Wild M, Zhai PM (2013) Observations: atmosphere and surface. In: Stocker TF, Qin D, Plattner GK, Tignor M, Allen SK, Boschung J, Nauels A, Xia Y, Bex V, Midgley PM (eds) Climate change 2013: the physical science basis. Contribution of working group I to the fifth assessment report of the intergovernmental panel on climate change. Cambridge University Press

Jiménez Cisneros BE, Oki T, Arnell NW, Benito G, Cogley JG, Döll P, Jiang T, Mwakalila SS (2014) Freshwater resources. In: Climate change 2014: impacts, adaptation, and vulnerability. Part A: global and sectoral aspects. In: Field CB, Barros VR, Dokken DJ, Mach KJ, Mastrandrea MD, Bilir TE, Chatterjee M, Ebi KL, Estrada YO, Genova RC, Girma B, Kissel ES, Levy AN, MacCracken S, Mastrandrea PR, White LL (eds) Contribution of working group II to the fifth assessment report of the intergovernmental panel on climate change. Cambridge University Press, pp 229–269

Jonkman SN (2005) Global perspectives on loss of human life caused by floods. Nat Hazards 34(2):151–175

Kinze M (2002) Das Augusthochwasser 2002. In: Akademie der Sächsischen Landesstiftung Natur und Umwelt (Hrsg): Aktuelle Hochwasserereignisse und ihre Folgen. Bericht zur wissenschaftlichen Arbeitstagung am 15. und 16. November 2002 in Dresden, pp 13–21

Kron W (2016) Überraschend, tödlich, zerstörerisch: Sturzfluten. Munich Re: Topics Schadenspiegel 2/2016, pp 36–39

Kunz M, Mohr S, Werner P (2016) Niederschlag. In: Brasseur G, Jacob D (Hrsg) Klimawandel in Deutschland, chap 7. Heidelberg, Springer/Spektrum, pp 57–66

Lenderink G, van Meijgaard E (2008) Increase in hourly precipitation extremes beyond expectations from temperature changes. Nat Geosci 1(8):511–514

Loriaux JM, Lenderink G, De Roode SR, Siebesma AP (2013) Understanding convective extreme precipitation scaling using observations and an entraining plume model. J Atmos Sci 130814132040006. https://doi.org/10.1175/jas-d-12-0317.1

Marchi L, Borga M, Preciso E, Gaume E (2010) Characterisation of selected extreme flash floods in Europe and implications for flood risk management. J Hydrol 394(1–2):118–133

Mayerhofer C, Meissl G, Klebinder K, Kohl B, Markart G (2017) Comparison of the results of a small-plot and a large-plot rainfall simulator—effects of land use and land cover on surface runoff in Alpine catchments. CATENA 156:184–196

Müller EN, Pfister A (2011) Increasing occurrence of high-intensity rainstorm events relevant for the generation of soil erosion in a temperate lowland region in Central Europe. J Hydrol 411:266–278

Niehoff D (2002) Modellierung des Einflusses der Landnutzung auf die Hochwasserentstehung in der Mesoskala. Brandenburgische Umweltberichte, 11. http://opus.kobv.de/ubp/volltexte/2005/398/pdf/vol11.pdf

Niehoff D, Bronstert A (2001) Influences of land-use and land-surface conditions on flood generation: a simulation study. In: Marsalek J et al (eds) Advances in urban stormwater and agricultural source controls. NATO science series IV. Earth and environmental sciences. Kluwer Academic Publishers, pp 267–278

Niehoff D, Fritsch U, Bronstert A (2002) Land-use impacts on storm-runoff generation: scenarios of land-use change and simulation of hydrological response in a meso-scale catchment in SW-Germany. J Hydrol 267(1–2):80–93

NRW (2010) ExUS Extremwertstatistische Untersuchungen von Starkregen in Nordrhein-Westfalen, aqua_plan GmbH, hydro & meteo GmbH & Co. KG und dr. papadakis GmbH

Salazar S, Francés F, Komma J, Blume T, Franke T, Bronstert A, Blöschl G (2012) A comparative analysis of the effectiveness of flood management measures based on the concept of retaining water in the landscape in different European hydro-climatic regions. Nat Hazards Earth Syst Sci 12:3287–3306

Scherrer S, Naef F, Faeh AO, Cordery I (2007) Formation of runoff at the hillslope scale during intense precipitation. Hydrol Earth Syst Sci 11(2):907–922. https://doi.org/10.5194/hess-11-907-2007

Schulla J (1997) Hydrologische Modellierung von Flußgebieten zur Abschätzung der Folgen von Klimaänderungen. Zürcher Geographische Schriften, Heft 69. Zürich

Vogel K, Öztürk U, Riemer A, Laudan J, Sieg T, Wendi D, Agarwal A, Rözer V, Korup O, Thieken A (2017) Die Sturzflut von Braunsbach am 29. Mai 2016 - Entstehung, Ablauf und Schäden eines „Jahrhundertereignisses". 2. Geomorphologische Prozesse und Schadensanalyse. Hydrologie und Wasserbewirtschaftung 61(3):163–175

Landslide Hazard Induced by Climate Changes in North-Eastern Romania

Mihai Niculiţă

Abstract Climate change driven by humans is a certitude, with uncertainties regarding the level of climate variability. This variability will influence natural hazard processes and the climate change uncertainty is transferred also to natural hazard modelling. The uncertainty is further multiplied by the fact that humans continue to modify the land surface, thus influencing the natural hazards, especially in the case of landslides. This mixture of factors can only be investigated through modelling under various scenarios, with results that can be validated in the near future and that can be used to establish measures and capabilities of adaptation to changing patterns of natural hazards induced by climate changes. We present the empirical evaluation of the landslide hazard in Northeastern Romania based on the current climate change scenarios, which show that this area will present an increase in mean annual and seasonal rainfall quantities. The historic rainfall data was empirically correlated with the historical landslide events. The correlation show that this increase in rainfall quantity will probably increase the frequency of the landslide events and of the reactivation of existing landslides, which cover almost 20% of the study area. The maximum 24 h rainfall during summer and the coupling of the autumn rainfall, with spring snowmelt and spring and summer rainfall which are the main triggers of landslides in the study area have the potential to reach a maximum similar with that from the 1960–1990 wet period, in 2030–2050, considering the most probable climate scenarios. This situation require actions in order to manage these natural hazards, because the study area has a high degree of vulnerability, being one of the poorest areas from Romania. This increase in landslide hazard is worrying since the study area exhibit an increase of anthropic pressure which was previously also related with landslides.

M. Niculiţă (✉)
Department of Geography, Faculty of Geography and Geology, Alexandru Ioan Cuza University of Iaşi, Carol I, 20A, 700505 Iaşi, Romania
e-mail: mihai.niculita@uaic.ro

W. Leal Filho et al. (eds.), *Climate Change, Hazards and Adaptation Options*, Climate Change Management, https://doi.org/10.1007/978-3-030-37425-9_13

Introduction

At the present state-of-the-art knowledge there is a general agreement between scientists (but not between politicians or the general public) that climate is changing toward a general increase in temperature, and the driving forces that influence the speed of this change are human actions, mainly through the increase of greenhouse gas concentrations (Brocker 1975; Eggleton 2012; IPCC 1990a, 1992a, 1996a, b, 2001a, d, 2007a, d, 2013, 2014c, 2018). The effect of greenhouse gases on the energy balance of the Earth-Atmosphere system was called radiative forcing (IPCC 1992a, 1994a) and the knowledge regarding this interaction evolved from a direct forcing model to a complicated interaction given either by the depletion of ozone (IPCC 1996a) or by the increase in aerosols (Myhre et al. 2013; Samset et al. 2018) which produced a negative radiative forcing. The need to predict these changes is important because of the uncertain effects of the climate change to other Earth systems (hydrology, landforms, biota, soils, humans and society) (IPCC 1990b, 1992b, 1992c, 1996b, 1996d, 1997, 2000a, b, c, 2001b, 2007b, 2014a, 2018). In order to mitigate the climate change effects, specific actions are needed (IPCC 1990c, 1992c, 1994b, 1996c, 2001c, 2007c) in order to decrease the emissions level until a no anthropogenic emission level or even storage are reached. Various measures and strategies were proposed (IPCC 1990c, 1994b, 2000d, 2005a, 2005b, 2011, 2014c, 2018) and applied in order to mitigate the climate change and its effects, while the future plans are still in discussion (Khan et al. 2018).

Currently the conclusions are that human activities caused an increase of 0.8–1.2 °C of the global mean temperature above the pre-industrial levels, value which will rise to 1.5 °C between 2030 and 2052, and to 3 °C in 2100, with a rate between 0.1 and 0.3 °C per decade at the current rate of the emissions (IPCC 2018).

While for the temperature forecast IPCC estimation was close, there are evidences that on many other linked aspects the reports forecasts are underestimated (Brysse et al. 2013; Ni and Groffman 2018). Also there it seems that many other factors, beside the radiative forcing is affecting the climate so the uncertainty increases (Steffen et al. 2018). Currently these aspects are being taken into account (IPCC 2001a, 2001b, 2014a, c, 2018). The change in precipitation is much more heterogeneous than the change in temperature, but in the temperate zone of the northern Hemisphere precipitation likely increased after 1901 (IPCC 2014c, 2018). Heavy precipitation in some areas and drought in other is to be expected, and the magnitude and return period of the extreme events will probably be bigger as the temperature rise (IPCC 2018).

In recent times, given the irreversible nature of climate changes, the single possibility to reduce the economic and social impact is to manage the risks and to adapt by decreasing the vulnerability and increasing the resilience (IPCC 2012, 2014b, 2018; Leal Filho 2015, 2016, 2017; Leal Filho et al. 2017; Leal Filho and Keenan 2017; Leal Filho et al. 2018; Leal Filho and Nalau 2018).

In North Eastern Romania (Fig. 1) Croitoru and Minea (2015) established a general positive trend of temperature and rainfall for the 1950–2006 period, as annual and

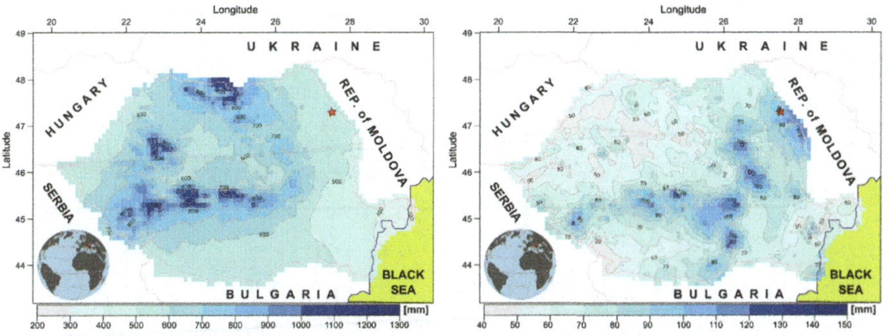

Fig. 1 The mean multiannual precipitation amount (left) and the maximum 24 h rainfall (right) in Romania for the period 1961–2013. *Source* ROCADA—Dumitrescu and Bîrsan (2015); the red star is the location of Iași municipality

seasonal values (with the exception of autumn for temperature and winter for precipitation which have a negative trend). Regarding the precipitation extremes Croitoru et al. (2016) established that for the 1961–2013 period in Moldavia the linear trends are generally positive, but for Iași there is a negative or no trend at all.

The Moldavian Plateau occupy an important part of the NE Romania, here the landslides covering almost 20% and representing an important geomorphological hazard (Mărgărint and Niculiţă 2017).

Landslides in Northeastern Romania

Gariano and Guzzetti (2016) review on the problem of landslides and climate change has given us the context of empirical analysis of the last century landslides: since the main trigger of the landslides in our study area is the rainfall (Pujină 2008; Niculiţă et al. 2017; Mărgărint and Niculiţă 2017; Niculiţă et al. 2018), an increase in rainfall amount and frequency will increase the frequency of landslides. Beside triggering, which is done by big rainfall events which happen very quickly, precipitation as annual (Fig. 1) or seasonal mean sum (Fig. 2) act as a preconditional factor for the landslides in the study area (Mărgărint and Niculiţă 2017).

Landslides in NE Romania appear mainly in the spring and summer, but previous rainy autumns will also favors landsliding in these periods (Niculiţă et al. 2017; Niculiţă et al. 2018). While the general increase in temperature and rainfall is related to an increase in landslide activity, the feedbacks are not well understood and are difficult to assess because of a multitude of factors involved (Sidle and Ochiai 2006). Anyway, study cases from around the world show that an increase in rainfall amount, either as yearly, seasonal or daily maximum, will generate a higher frequency of landslides, shallow or deep seated (Sidle and Dhakal 2002; Polemio and Petrucci 2010; Gariano and Guzzetti 2016).

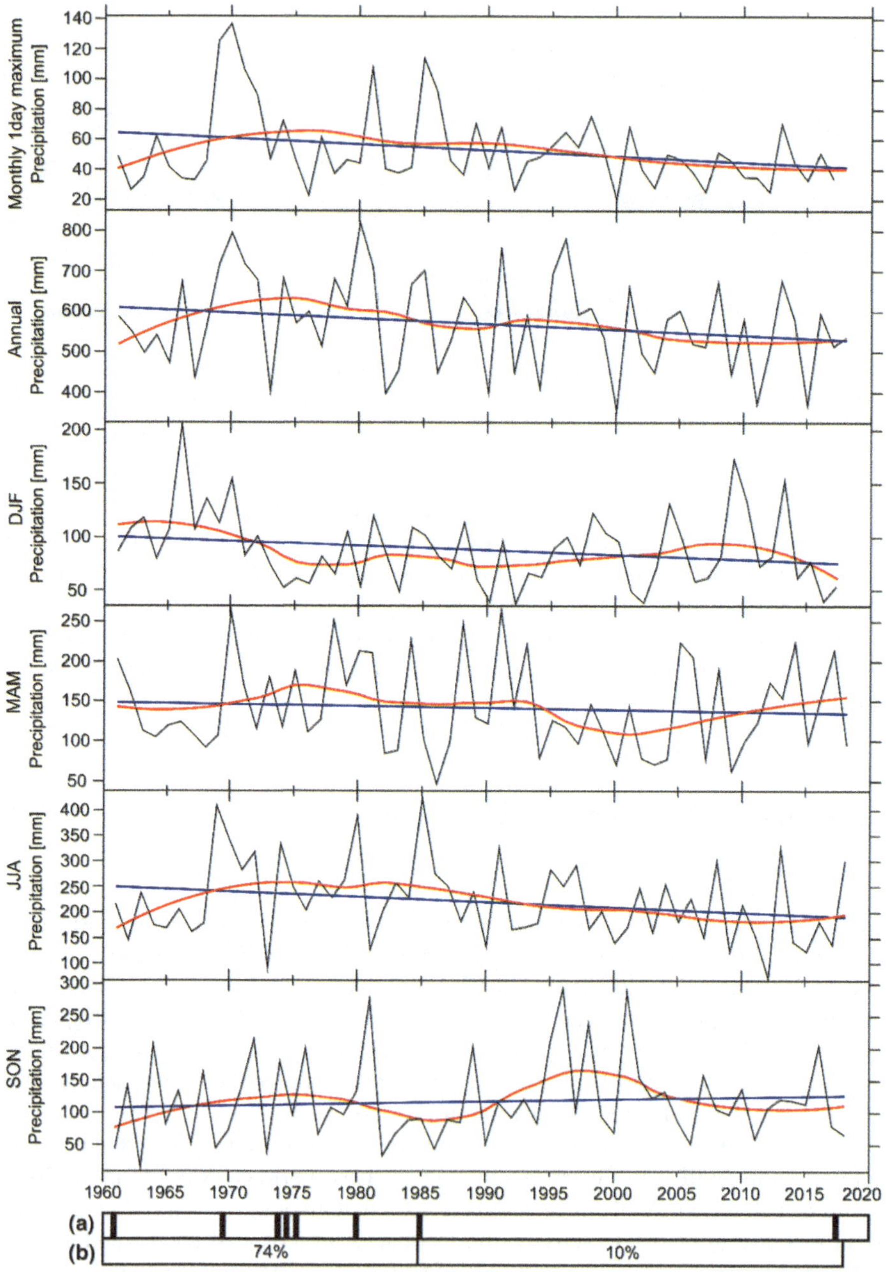

Fig. 2 The monthly 1 day maximum rainfall (position 1, uppermost), the annual rainfall and yearly seasonal rainfall (positions 2–6) for the Iaşi meteorological station (ECA&D data: 1961–2018) and the main landslide events in Iaşi municipality (**a**) and the percentage of the 518 landslides mapped by Niculiţă et al. (2018) in Iaşi municipality (**b**); the blue line represents the linear trend while the red line curve is the loess smooth with 0.5 span

Recently Niculiţă et al. (2018) have shown that for Iaşi Municipality (located in the study area) the wet period from between 1960 and 1990 generated the reactivations of the Holocene (Niculiţă et al. 2016) landslides (Fig. 2), situation that is similar with other areas around the world (Bovis and Jones 1992). From 518 mapped landslides, 74% were triggered between 1956 and 1984. In Fig. 2 the evolution of monthly maximum 1-day precipitation and the sum of the annual and seasonal precipitation amount for the Iași meteorological station (according to ECA&D data—Haylock et al. 2008, for the 1961–2018 period) is plotted against the main landslide events and the percentage of recent landslides identified by Niculiţă et al. (2018). This graph show clearly that the main landslide events from Iași Municipality are related to the wet cycle from between 1960 and 1990 and to the maximum 24 h rainfall amount. After 1990 both the annual rainfall and the monthly 1 day maximum rainfall have decreased, and after the 2010 lower trend, the quantities tend to increase gently. The seasonal trends show that during the 1960–1990 wet period all but the winter trends were raised, the tendency of general gentle increase, but not for the winter season, which is decreasing being found after 1990. The 2017 landslide event of Ursulea (Necula and Niculiţă 2017) pops up in this general and gently trend of precipitation increase, both in annual, maximum and seasonal values. The 1990-present cycle is a dry one, probably in the near future there will be shift to a new wet cycle.

Our analysis of the rainfall data is in opposition with the conclusion of Croitoru and Minea (2015) regarding the annual and seasonal linear trends, but supports the conclusion of Croitoru et al. (2016) regarding the linear trend of the maximum precipitation. It is clear that by increasing/decreasing the length of the historical dataset the linear trend might have different slope directions and that the analysis of the non-linear trends can better approximate the past and future changes of climate.

By analyzing other climatic indices related to precipitation (monthly precipitation sum—Fig. 3 and monthly maximum consecutive 5-day precipitation, annual total precipitation in wet days, annual total precipitation when daily precipitation amount is bigger than 1 mm, called RR >95p, annual total precipitation when RR >99p, annual count of days when precipitation is $\geq$1 mm—Fig. 4) we can see that the 1984–1985 landslide events can be associated with a maximum of monthly maximum

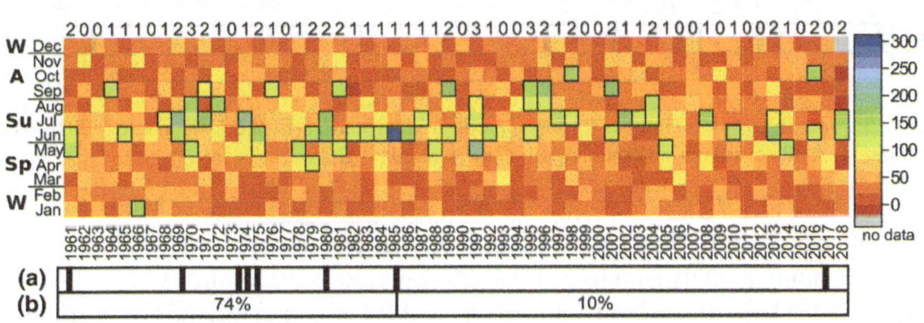

Fig. 3 Monthly precipitation at Iași meteorological station (ECA&D—1961–2018) and the main landslide events in Iași municipality (idem as for Fig. 2); framed months have sums over 100 mm; at the top of the graph is the number of months with over 100 mm rainfall

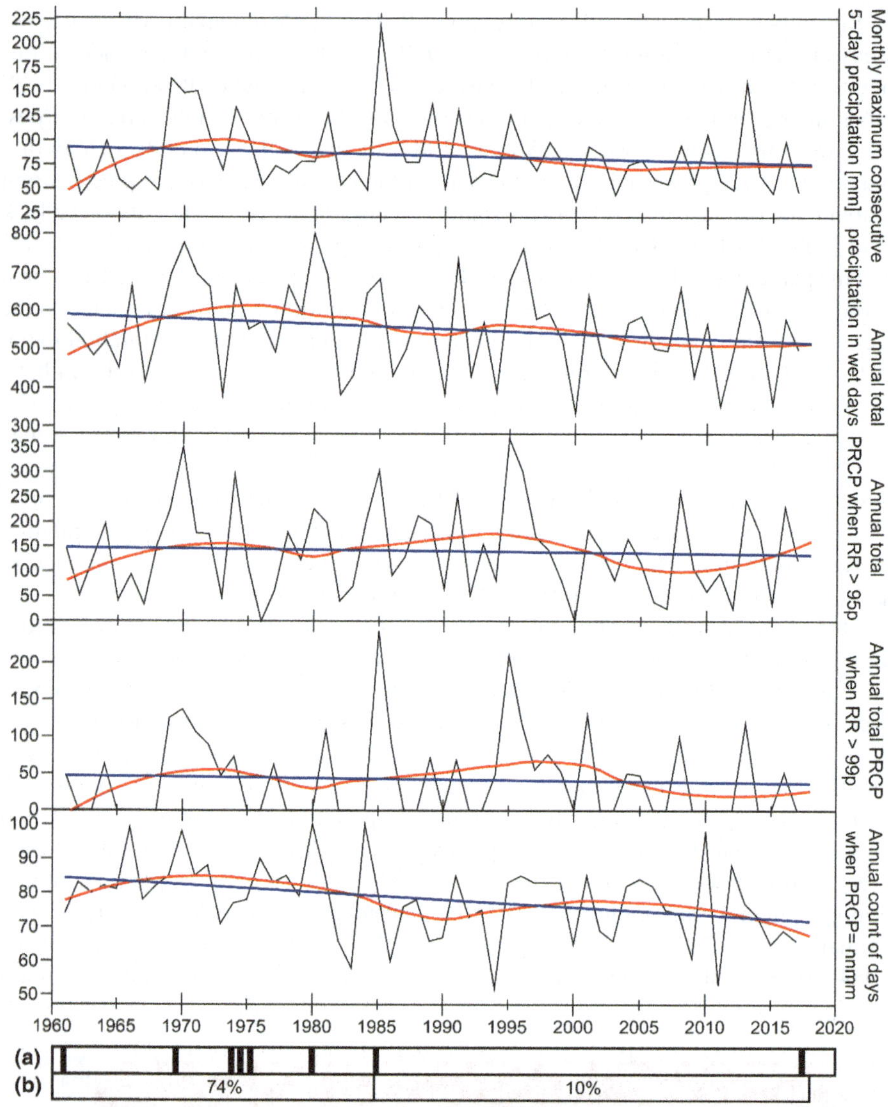

Fig. 4 Several precipitation indices for Iași meteorological station (ECA&D data: 1961–2018) and the main landslide events in Iași municipality (idem as for Fig. 2)

consecutive 5-day precipitation and the monthly precipitation from May 1985 (which is the maximum monthly sum from 1961–2018 period—291.9 mm). The 1960–1990 wet period can be depicted both in Fig. 3 where there is a bigger frequency of summer wet months having over 100 mm, a successive period with at least 1 month with over 100 mm rainfall and on Fig. 4 where all the indices have maximum values. This wet period seems to be characterized both by high maximum 24 h rainfall, maximum consecutive 5-day precipitation, annual total precipitation in wet days and annual count of days when precipitation is $\geq$1 mm. The 1990–2005 period is characterized by a higher precipitation sum fallen in days with a high quantity of rainfall, while the period after 2005 has the lowest trend regarding all the precipitation indices. Regarding the seasonal precipitation (Fig. 2), the 1960–1990 wet period is characterized by high precipitation in all seasons, the 1990–2005 period by normal quantities in summer and winter, by low quantities in spring and by high quantities in autumn, while the post 2005 period by an increase in precipitation quantities in all season but winter, which are decreasing.

The empirical correlation between climatic data and landslide events in Iași Metropolitan Area does reflect the preconditioning, preparatory and triggering role of the precipitation on landslides. The increase of the annual precipitation in the 1960–1990 wet cycle, favored the increase of the groundwater table level (Minea and Croitoru 2015, 2017; Minea et al. 2016) which has a direct effect for the reactivations of the relict and old landslides (Minea et al. 2015; Niculita et al. 2016; Margarint and Niculiță 2017; Necula et al. 2017), although there might be a delay due to the groundwater recharging. In the present dry cycle (after 1990) there is the inverse effect, although there it seems to be a trend of groundwater table increase (Minea et al. 2015; Minea and Croitoru 2015, 2017; Minea et al. 2016). The seasonal rainfall and monthly rainfall act as preparatory factors as we can see from the Ursulea 2017 landslide event (Fig. 3 and Necula and Niculiță 2017). The consecutive 5-day and 1-day maximum rainfall act as trigger factors as shown by the landslide events from between 1960 and 1985 (Fig. 3).

Beside rainfall triggering, urbanization also was an important factor that prepared the landslide events (Niculiță et al. 2018), since human impact generate disturbances in the drainage, permeability and hydrogeology. The increase of the urbanized area in the Iaşi Municipality, which nowadays hosts 50% of the population of Iaşi County will probably exacerbate the climate factor regarding the triggering of landslides.

Climate Change Scenarios

The climate change prediction is performed today using numerical models, which became more complex in terms of input data (IPCC 2013). These model simulations return atmospheric and land surface meteorological variables that can be used for assessing the influence on Earth systems, ecosystems and human society, and can incorporate biogeochemical models, becoming Earth System Models (ESMs). The usage of the climate models results must be done in accordance with their limits, assumptions and errors (Kreienkamp et al. 2012). Global Climate Models (GCM), also called Global Circulation Models are time-dependent numerical models of the global atmosphere, oceans, land surface and ice (Roberts et al. 2018a, b).

These GCMs are based on the reanalyses of the historical climate (1948 to present, but in general periods as 1979 to present are favored since there is a general agreement that after 1980 the climate show a human influenced non-natural variability—IPCC

2013, 2018). Regional Climate Models (RCMs) are one of the means of downscaling GCMs, in order to improve the size of the targeted phenomenon, but for specific areas around the globe (with a limited spatial domain). These models are dynamic, as the global ones are. Statistical methods (ESD) can also be used to downscale GCMs to regional level. Currently the Coupled Model Intercomparison Project Phase 5 (CMIP5) is used, the future CMIP6 being the most complex framework design for climate changes forecast (Eyring et al. 2016a, b).

For assessing the climatic changes, a model cascade approach is used, in the sense that currently greenhouse emissions are converted into atmospheric greenhouse gas (GHG) concentrations which are used to assess the radiative forcing of the atmosphere. The radiative forcing (RF) is used to estimate the climate change signal in terms of the difference between a climate variable at one time and at one later time. This climate variability can be natural, but also various scenarios attribute it to humans. These scenarios take into account various GHG levels and their impact to climate, through climate change signal. In this way the climate is projected through models and forecasts are obtained. Currently (CMIP5) four Representative Concentration Pathways (RCP) are used: RCP 2.6, RCP 4.5, RCP 6.0 and RCP 8.5, which correspond to RFs in 2100 of 2.2, 3.8, 4.8 and 7.6 W m^{-2} (IPCC 2013).

What climate changes scenarios can tell us is how the temperature and rainfall will change (increase/decrease), how much will change and how frequent some phenomenon will appear (Haarsma et al. 2013; Baatsen et al. 2015). The prediction power and the resolution of the climate models is increasing constantly (IPCC 2013). In 2002 Canadian Climate Center Model (CCM) was considered the most powerful in terms of prediction power (Cuculeanu et al. 2002) for Romania, but there was no regional resolution. This model predicted a 20% decrease of rainfall during the cold season and a 20% decrease in the summer season. Currently a general increase of precipitation is foreseen with the increase in temperature, but substantial spatial variation is to be expected (IPCC 2013 2018). In general for Romania nowadays the CORDEX projections are preferred (Jurchescu et al. 2017).

EURO-CORDEX (Coordinated Regional Climate Downscaling Experiment) regional climate model ensemble (Gobiet and Jacob 2012; Jacob et al. 2014) was initiated as a domain of the CORDEX project (Giorgi et al. 2009) and represent the best public available data regarding the climate change scenarios in Europe. The best resolution is 0.11 degree grid (EUR-11—~12.5 km cell size). EURO-CORDEX uses the E-OBS data for the historical period.

EURO-CORDEX models evaluations show that these RCMs are able to represent the basic patterns of the European climate both spatially and temporally, but the spatial resolution is regional (NE Romania, rather than the NE part of the Moldavian Plateau or Jijia Hills, for example), and biases are present both in terms of space and time. The same general bias to a hotter and wetter climate toward north as found in global forecasts is still present in this regional downscaling (Kotlarski et al. 2014; Van Haren et al. 2015a, b). Giot et al. (2016) shows that ALARO-0 has the potential to be used as a European model, having reasonable errors.

Methods

We used the CORDEX bias-adjusted precipitation data (prAdjust) at daily time resolution for the EUR-11 domain (Vautard et al. 2013; IPCC 2015; Dosio 2016), from which we selected 10 models (description of the models are in Table 1) for the rcp4.5 and rcp8.5. These datasets cover the 1971–2070 period. The data was downloaded from ESGF WRCP CORDEX nodes (https://esgf-data.dkrz.de/search/cordex-dkrz/).

In first step we used the Climate Data Operators (CDO—https://code.mpimet.mpg.de/projects/cdo/) command line software to interpolate, cut and reproject the CORDEX data to the ROCADA spatial grid. The remapbil command was used, which performs bilinear interpolation. The .nc formatted data was further imported and processed in R stat (Core Team 2014) using the eds package (Benestad et al. 2015, 2017).

For the analysis of the past rainfall variability we have used the ROCADA dataset (Fig. 1—Dumitrescu and Bârsan 2015) which cover the 1961–2013 period and the ECA&D dataset (Klein Tank et al. 2002) which cover the 1961–2018 period. The ROCADA dataset is gridded and is homogenized, that is why it appears more smoothed than ECA&D data for Iași meteorological station, but the trends are similar for both datasets (Fig. 5).

By plotting the main landslide events that were identified in Iași Municipality by Niculiţă et al. (2018) we have conceptually associated them with the trends of the climatic data regarding precipitation (Figs. 2–4). The trends were computed as local polynomial regression estimates in R stat with the loess function (locally estimated scatter plot smoothing), at a 0.5 span (50% of data is considered as the size of the neighborhood). Linear regression trends were also plotted, but regarding wet and dry cycles the loess trends were considered useful.

Forecasting the Influence of Climate Change for Landslide Hazard

Previous approaches for Romania (Jurchescu et al. 2017) have used the same climate change data as we used in the present approach, but have opted to use this data in a probabilistic statistical approach of landslide susceptibility. Their conclusions is that in north-eastern Romania in the most probable scenario the precipitation amount will increase which will also generate a higher landslide hazard.

While we must take into account that every climate model is a realization of the reality and while acknowledging the limitations of climate change models, we can observe that the cyclic nature of precipitations (which derive from climatic and hydrologic variability—Allan and Zveryaev 2011; Adler et al. 2008) is modeled by all the models (Fig. 5). In this context is it clear that future wet cycles will increase the landslide hazard, so the question is how will these cycles will be: stronger or weaker than the last one. To answer to this question we have chosen three models

Table 1 The EURO-CORDEX bias adjusted data used in this work

ID	Short name	Driving model ID	Driving model ensemble member	Institution	Bias model
M1	CNRM.CNRM-CERFACS-CNRM-CM5	CNRM-CERFACS-CNRM-CM5	r1i1p1	CNRM	CNRM-ARPEGE51
M2	DMI.ICHEC-EC-EARTH	ICHEC-EC-EARTH	r3i1p1	DMI	DMI-HIRHAM5
M3	IPSL-INERIS.IPSL-IPSL-CM5A-MR	IPSL-IPSL-CM5A-MR	r1i1p1	IPSL	IPSL-INERIS-WRF331F
M4	KNMI.ICHEC-EC-EARTH	ICHEC-EC-EARTH	r1i1p1	KNMI	KNMI-RACMO22E
M5	KNMI.MOHC-HadGEM2-ES	MOHC-HadGEM2-ES	r1i1p1	KNMI	KNMI-RACMO22E
M6	MPI-CSC.MPI-M-MPI-ESM-LR	MPI-M-MPI-ESM-LR	r1i1p1	MPI	MPI-CSC-REMO2009
M7	SMHI.CNRM-CERFACS-CNRM-CM5	CNRM-CERFACS-CNRM-CM5	r12i1p1	SMHI	SMHI-RCA4
M8	SMHI.ICHEC-EC-EARTH	ICHEC-EC-EARTH	r12i1p1	SMHI	IPSL-CDFT22
M9	SMHI.MOHC-HadGEM2-ES	MOHC-HadGEM2-ES	r1i1p1	SMHI	SMHI-RCA4
M10	SMHI.MPI-M-MPI-ESM-LR	MPI-M-MPI-ESM-LR	r1i1p1	SMHI	IPSL-CDFT22

IPSL—Institut Pierre Simon Laplace, CNRM—Centre National de Recherches Meteorologiques, DMI—Danish Meteorological Institute, KNMI—Royal Netherlands Meteorological Insitute, SMHI—Swedish Meteorological and Hydrological Institute Rossby Centre, MPI—Helmholtz-Zentrum Geesthacht Climate Service Center

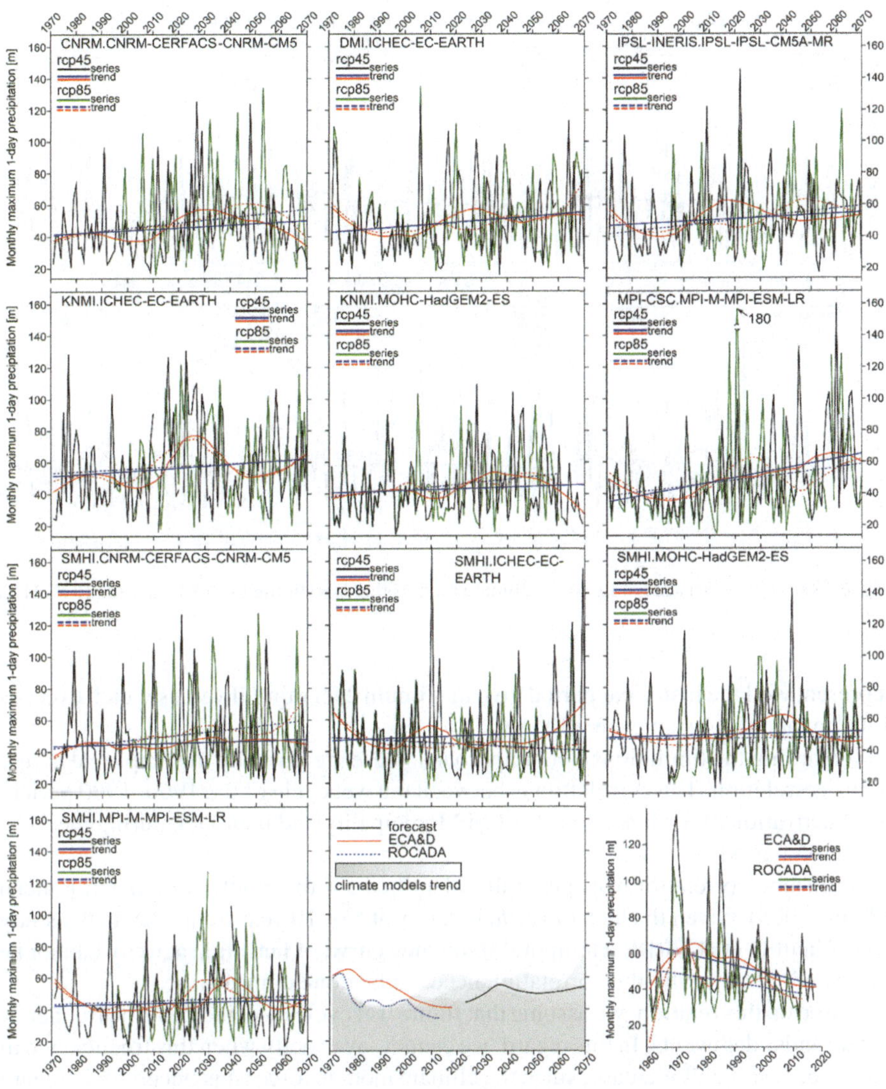

Fig. 5 Rx1 day at Iași according to various climate models; on the last line in the center are the trend lines overlaid and on the right are the ECA&D and the ROCADA dataset overlaid; the blue line represents the linear trend while the red line curve is the loess smooth with 0.5 span

which model the best the period for which the historical data (Fig. 6): SMHI.MPI-M-MPI-ESM-LR, DMI.ICHEC-EC-EARTH and IPSL-INERIS.IPSL-IPSL-CM5A-MR. These model show the return of a new wet cycle either between 2020 and 2030, either later, in 2030–2050. Some realizations have already missed the forecast, by indicating a wet period in the present time. The maximum 24 h rainfall amount does not exceeds the 140 mm historical maximum value. The other models, which do not

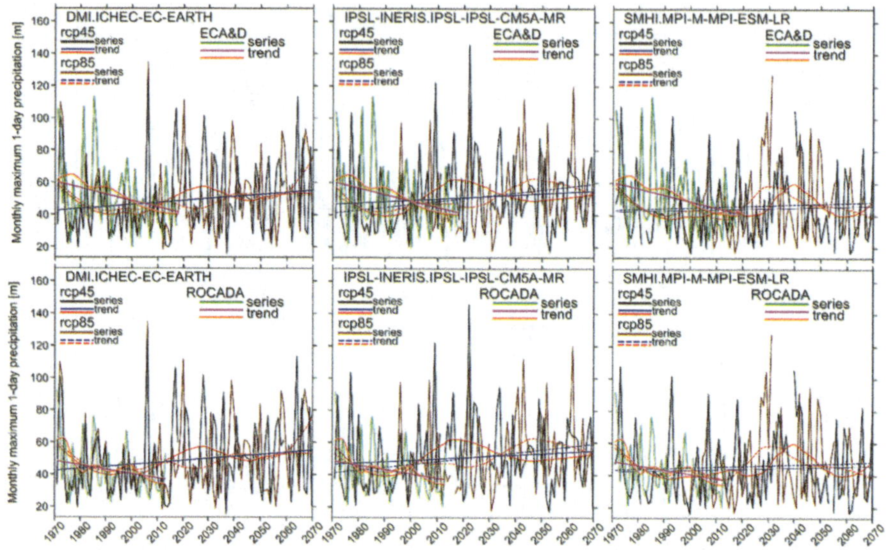

Fig. 6 Rx1 day at Iaşi according to the climate models that best fit the ECA&D and the ROCADA datasets

represent well the historical period give maximum 24 h rainfall values which exceeds 140 mm and can reach even 180 mm.

Regarding the limitations of our approach, these are given by the empirical nature of the correlation. The correlation is between the wet cycle of the 1960–1990 period, the reactivation of the landslides for Iaşi Municipality and their lack during the 1990 and 2005 period.

The same correlation was generally identified for the whole Moldavian Plateau (Pujină 2008). Given the lack of aerial imagery at 5 or 10 year frequency in the study area (Niculiţă et al. 2018) a temporal frequency curve of landslide activity cannot be estimated. That is why the correlation needs to be done empirically.

Based on this relation, we assume that future wet cycles will increase the frequency of the landslides events. In this regard, we cannot say exactly when this frequency will increase, or if it will increase, since the climate models forecast is uncertain in regard to rainfall data. Nevertheless, is of common sense to consider such an increase in the context of a future increase in rainfall amount. Because exact frequency data cannot be estimated, the best approach is to consider this potential increase and include it in planning regulations or zoning, especially in areas which already had reactivations.

In terms of 5-day consecutive maximum precipitation the SMHI.MPI-M-MPI-ESM-LR data predict values similar with historical 1985 maximum in the period between 2030 and 2050 (Fig. 7). In terms of monthly and annual values the same model predicts an increase in precipitation for all the seasons at values bigger than the historical 1960–1990 wet period (Fig. 7). This model is the single one that fits

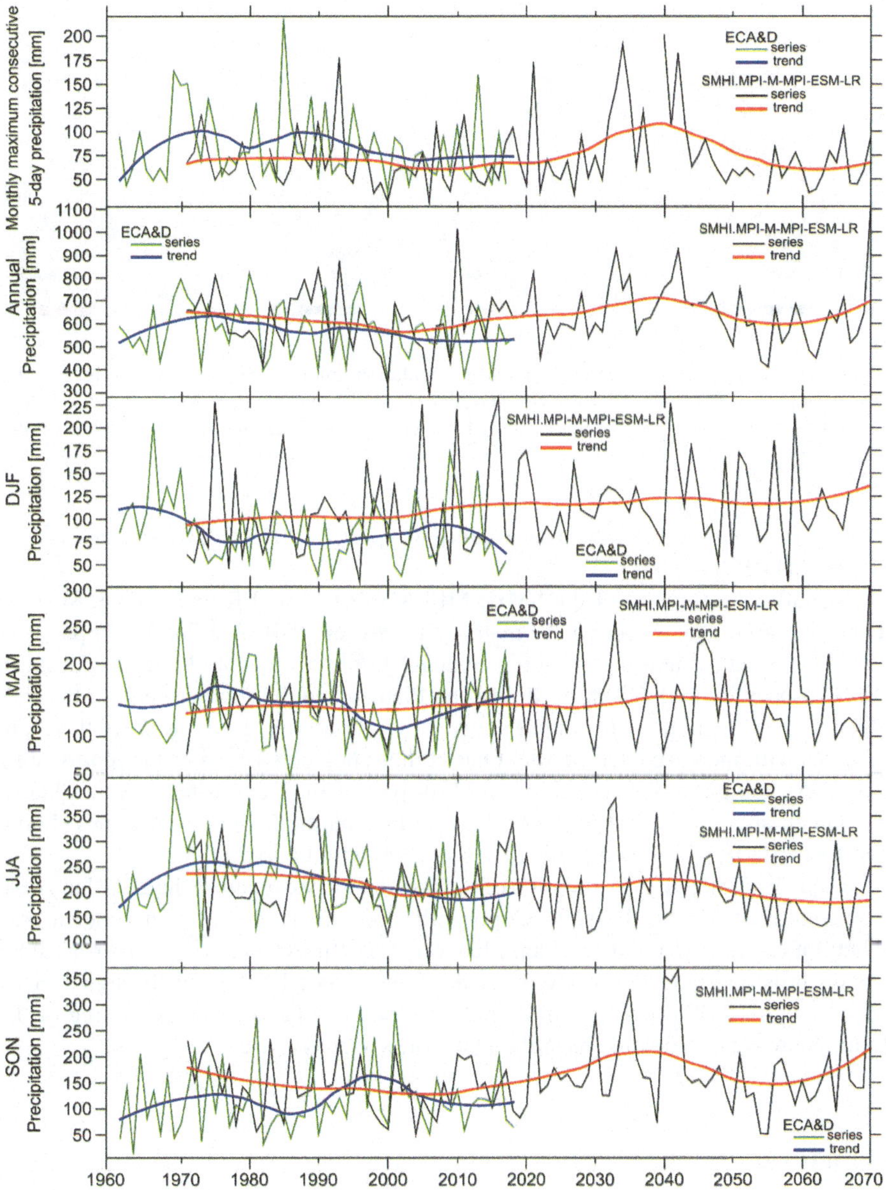

Fig. 7 Forecasts of the trends for the SMHI.MPI M MPI ESM LR model

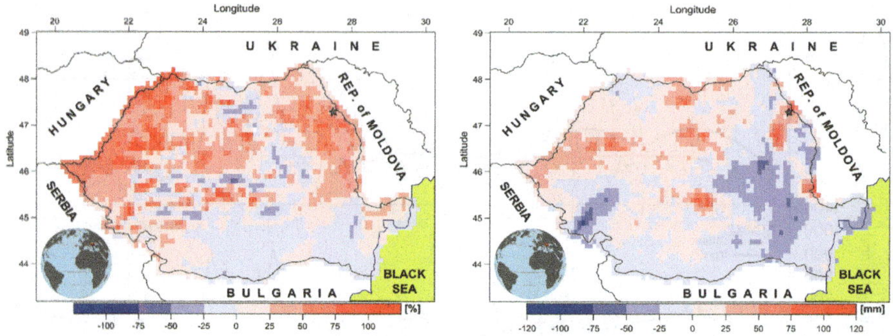

Fig. 8 The difference of mean multiannual precipitation amount (left) and the maximum 24 h rainfall (right) in Romania, between the period 1971–1990 (*Source* ROCADA) and 2030–2050 (according to the SMHI.MPI-M-MPI-ESM-LR model); the black star is the location of Iași municipality

the historical data and is indicating a future wet period (2030–2050), similar or more intense than 1960–1990.

Spatially, the forecast of the SMHI.MPI-M-MPI-ESM-LR model (Fig. 8) show a certain variability in the NE Romania. In between 2030 and 2050, compared to 1971–1990, the annual mean should increase in the Moldavian Plateau with 25 to 100%, especially in the Central Moldavian Plateau and the southern part of the Jijia Hills. The maximum 24 h rainfall show more variability. In three areas there are hotspots with increase over 50 mm. One of these hotspots is Iaşi Municipality area, the others being the SW Jijia Hills and NW part of the Central Moldavian Plateau and Galaţi area. In the rest of the Moldavian Plateau the 24 h rainfall values should decrease in 2030–2050 compared to 1971–1990 with up to 50 mm.

As an initial step in the validation of our conclusion we mention the 2017 landslide reactivation of the Northeastern Copou hill slope landslide scarp in Ursulea area (Niculiţă et al. 2018) and the identified very slow movement of the Northeastern Copou hillslope landslide body in Ţicău neighborhood, revealed through InSAR (Necula et al. 2017), which could represents events related to the beginning of the future increase of landslide hazard due to a new wet cycle.

Conclusions

The presented empirical approach is based on the assumption that future climate changes influenced by humans will generate a wet period, similar with the 1960–1990 period, which in NE Romania generated an increase in landslide frequency (Pujină 2008; Margarint and Niculiţă 2017). While the timing of this new wet period is uncertain, at least one model that fits well the historical rainfall data, show that this period will start in 2030 or later. Another uncertainty is the level of intensity of this new wet cycle. The climate changes models show that it will be at least of the same magnitude, or even bigger then the historical one. In the studied landslide dataset, around Iași Municipality (Niculiţă et al. 2018) a 24 h rainfall amount over 100 mm triggered landslides. The climate predictions show that this values will be reached and exceeded 3 to 5 times until 2070. The maximum value of 24 h rainfall in the Iași climatic ECA&D dataset was 136.7 mm. This value is similar with the maximum value returned by the most probable climate change scenario.

Assessing uncertainties in regard to both climate change and landslide hazard is an important aspect and has to be performed (Bonnard et al. 2008; Gariano and Guzzetti 2016). In this regard, performing landslide susceptibility modelling is needed. In order to mitigate the possible effects of the increase of landslide hazard due to climate changes, the susceptibility assessment need to be used by land planners (McInnes et al. 2007). Also, vulnerability needs to be assessed in order to plan the best practices for decreasing it.

Acknowledgements This work was supported by a grant of Ministry of Research and Innovation, CNCS—UEFISCDI, project number PN-III-P1-1.1-PD-2016-0154, within PNCDI III. We acknowledge the data providers in the ECA&D project (https://www.ecad.eu).

References

Adler RF, Gu G, Wang JJ, Huffman GJ, Curtis S, Bolvin D (2008) Relationships between global precipitation and surface temperature on interannual and longer timescales (1979–2006). Geophys J Res 113: Art. no. D22104. https://doi.org/10.1029/2008jd010536

Allan RP, Zveryaev II (2011) Variability in the summer season hydrological cycle over the Atlantic-Europe region 1979–2007. Int J Clim 31:337–348. https://doi.org/10.1002/joc.2070

Baatsen M, Haarsma RJ, Van Delden AJ, De Vries H (2015) Severe autumn storms in future western Europe with a warmer Atlantic Ocean. Clim Dyn 45(3–4):949–964. https://doi.org/10.1007/s00382-014-2329-8

Benestad RE, Mezghani A, Parding K (2015) An open source R package designed for climate and weather data analysis, empirical statistical downscaling, and visualization. PCC Work RegNal Clim Proj Their Use Impacts Risk. https://doi.org/10.13140/RG.2.1.1138.7764

Benestad R, Haensler A, Hennemuth B, Illy T, Jacob D, Keup-Thiel E, Kotlarski S, Nikulin G, Otto J, Rechid D, Sieck K, Sobolowski S, Szabó P, Szépszó G, Teichmann C, Vautard R, Weber T, Zsebehazi G (2017) Guidance for EURO-CORDEX climate projections data use. v1.0

Bonnard C, Tacher L, Beniston M (2008) Prediction of landslide movements caused by climate change: modelling the behaviour of a mean elevation large slide in the Alps and assessing its uncertainties. In: Chen Z, Zhang JM, Li ZK, Wu FQ, Ho K (eds) Landslides and engineering slopes: from the past to the future, pp 217–227. https://doi.org/10.1201/9780203885284-c13

Bovis MJ, Jones P (1992) Holocene history of earthflow mass movements in south-central British Columbia: the influence of hydroclimatic changes. Can J Earth Sci 29:1746–1755. https://doi.org/10.1139/e92-137

Brocker W (1975) Climatic change: are we on the brink of a pronounced global warming? Science 189(4201):460–463. https://doi.org/10.1126/science.189.4201.460

Brysse K, Oreskes N, O'reilly J, Oppenheimer M (2013) Climate change prediction: erring on the side of least drama? Glob Environ Change 23(1):327–337. https://doi.org/10.1016/j.gloenvcha.2012.10.008

Core Team R (2014) A language and environment for statistical computing. R Foundation for Statistical Computing, Vienna, Austria
Croitoru A-E, Minea I (2015) The impact of climate changes on rivers discharge in Eastern Romania. Theor Appl Clim 120:563–573. https://doi.org/10.1007/s00704-014-1194-z
Croitoru A-E, Piticar A, Burada DC (2016) Changes in precipitation extremes in Romania. Quaternary Int 415(10):325–335. https://doi.org/10.1016/j.quaint.2015.07.028
Cuculeanu V, Tuinea P, Bălteanu D (2002) Climate change impacts in Romania: vulnerability and adaptation options. GeoJournal 57:203–209. https://doi.org/10.1023/B:GEJO.0000003613.15101.d9
Dosio A (2016) Projections of climate change indices of temperature and precipitation from an ensemble of bias-adjusted high-resolution EURO-CORDEX regional climate models. J Geophys Res-Atmos 121:5488–5511. https://doi.org/10.1002/2015JD024411
Dumitrescu A, Bîrsan M (2015) ROCADA: a gridded daily climatic dataset over Romania (1961–2013) for nine meteorological variables. Nat Hazards 78(2):1045–1063. https://doi.org/10.1007/s11069-015-1757-z
Eggleton T (2012) A short introduction to climate change. Cambridge University Press, p 249
Eyring V, Bony S, Meehl GA, Senior CA, Stevens B, Stouffer RJ, Taylor KE (2016a) Overview of the coupled model intercomparison project phase 6 (CMIP6) experimental design and organization. Geosci Model Dev 9:1937–1958. https://doi.org/10.5194/gmd-9-1937-2016
Eyring V, Gleckler PJ, Heinze C, Stouffer RJ, Taylor KE, Balaji V, Guilyardi E, Joussaume S, Kindermann S, Lawrence BN, Meehl GA, Righi M, Williams DN (2016b) Towards improved and more routine Earth system model evaluation in CMIP. Earth Syst Dynam 7:813–830. https://doi.org/10.5194/esd-7-813-2016
Gariano SL, Guzzetti F (2016) Landslides in a changing climate. Earth-Sci R 162:227–252. https://doi.org/10.1016/j.earscirev.2016.08.011
Giorgi F, Jones C, Asrar GR (2009) Addressing climate information needs at the regional level: the CORDEX framework. WMO Bulletin 58(3):175–183
Giot O, Termonia P, Degrauwe D, De Troch R, Caluwaerts S, Smet G, Berckmans J, Deckmyn A, De Cruz L, De Meutter P, Duerinckx A, Gerard L, Hamdi R, Van den Bergh J, Van Ginderachter M, Van Schaeybroeck B (2016) Validation of the ALARO-0 model within the EURO-CORDEX framework. Geosci Model Dev 9:1143–1152. https://doi.org/10.5194/gmd-9-1143-2016
Gobiet A, Jacob D (2012) A new generation of regional climate simulations for Europe: the EURO-CORDEX initiative. Geophys Res Abstracts 14, EGU2012–8211
Haarsma RJ, Hazeleger W, Severijns C, De Vries H, Sterl A, Bintanja R, Van oldenborgh GJ, Van den Brink HW (2013) More hurricanes to hit western Europe due to global warming. Geophys Res Lett 40(9):1783–1788. https://doi.org/10.1002/grl.50360
Haylock MR, Hofstra N, Klein Tank AMG, Klok EJ, Jones PD, New M (2008) A European daily high-resolution gridded dataset of surface temperature and precipitation. J Geophys Res-Atmos 113:D20119. https://doi.org/10.1029/2008JD10201
Intergovernmental Panel on Climate Change (IPCC) (1990a) Climate change. In: Houghton JT, Jenkins GJ, Ephraums JJ (eds) The IPCC scientific assessment. Cambridge University Press, p 365
Intergovernmental Panel on Climate Change (IPCC) (1990b) Climate change. In: McG Tehart WJ, Sheldon GW, Griffiths DC (eds) The IPCC impacts assessment. Australian Government Publishing Service, Canberra, p 296
Intergovernmental Panel on Climate Change (IPCC) (1990b) Climate change. The IPCC Response Strategies, IPCC, Geneva, Switzerland, p 270
Intergovernmental Panel on Climate Change (IPCC) (1992a) Climate change: the 1990 and 1992 IPCC assessments. IPCC, Geneva, Switzerland, p 168
Intergovernmental Panel on Climate Change (IPCC) (1992b) Climate change 1992. In: Houghton JT, Callander BA, Varney SK (eds) The supplementary report to the IPCC scientific assessment. Cambridge University Press, p 200

Intergovernmental Panel on Climate Change (IPCC) (1992c) Climate change 1992. In: McG Tegart WJ, Sheldon GW (eds) The supplementary report to the IPCC impacts assessment. Australian Government Publishing Service, Canberra, p 115
Intergovernmental Panel on Climate Change (IPCC) (1994a) Climate change 1994. In: Houghton JT, Meira Filho LG, Bruce J, Lee H, Callander BA, Haites E, Harris N, Maskell K (eds) Radiative forcing of climate change and an evaluation of the IPCC IS92 emission scenarios. Cambridge University Press, p 339
Intergovernmental Panel on Climate Change (IPCC) (1994b) IPCC technical guidelines for assessing climate change impacts and adaptations. University College London and Center for Global Environmental Research, p 59
Intergovernmental Panel on Climate Change (IPCC) (1996a) Climate change 1995. In: Houghton JT, Meira Filho LG, Callander BA, Harris N, Kattenberg A, Maskell K (eds) The science of climate change. Cambridge University Press, p 572
Intergovernmental Panel on Climate Change (IPCC) (1996b) Climate change 1995. In: Watson RT, Zinyowera MC, Moss RH, Dokken DJ (eds) Impacts, adaptations and mitigation of climate change: scientific-technical analyses. Cambridge University Press, 878 p
Intergovernmental Panel on Climate Change (IPCC) (1996c) Climate change 1995. In: Bruce JP, Lee H, Haites EF (eds) Economic and social dimensions of climate change. Cambridge University Press, p 448
Intergovernmental Panel on Climate Change (IPCC) (1996d) IPCC second assessment. Climate change 1995. IPCC, Geneva, Switzerland, p 63
Intergovernmental Panel on Climate Change (IPCC) (1997) IPCC special report on the regional impacts of climate change: an assessment of vulnerability. In: Watson RT, Zinyowera MC, Moss RH, Dokken DJ (eds) Cambridge University Press, UK, p 517
Intergovernmental Panel on Climate Change (IPCC) (2000a) Aviation and the global atmosphere. In: Penner JE, Lister DH, Griggs DJ, Dokken DJ, Farland MMc (eds) Prepared in collaboration with the scientific assessment panel to the montreal protocol on substances that deplete the ozone layer. Cambridge University Press, UK, p 373
Intergovernmental Panel on Climate Change (IPCC) (2000b) Land use, land-use change, and forestry. In: Watson RT, Noble IR, Bolin B, Ravindranath NH, Verardo DJ, Dokken DJ (eds) Cambridge University Press, UK, p 375
Intergovernmental Panel on Climate Change (IPCC) (2000c) Methodological and technological issues in technology transfer. In: Metz B, Davidson O, Martens J-W, Van Rooijen S, Van Wie Mcgrory L (eds) Cambridge University Press, UK, p 432
Intergovernmental Panel on Climate Change (IPCC) (2000d) Special report on emissions scenarios. Cambridge University Press, p 599
Intergovernmental Panel on Climate Change (IPCC) (2001a) Climate change 2001: the scientific basis. In: Houghton JT, Ding Y, Griggs DJ, Noguer M, van der Linden PJ, Dai X, Maskell K, Johnson CA (eds) Cambridge University Press, p 881
Intergovernmental Panel on Climate Change (IPCC) (2001b) Climate change 2001: impacts, adaptation, and vulnerability. In: McCarthy JJ, Canziani OS, Leary NA, Dokken DJ, White KS (eds) Cambridge University Press, p 1032
Intergovernmental Panel on Climate Change (IPCC) (2001b) Climate change 2001: mitigation. IPCC, Geneva, Switzerland, p 753
Intergovernmental Panel on Climate Change (IPCC) (2001d) Climate change 2001: synthesis report. In: Watson RT, Core Writing Team (eds) IPCC, Geneva, Switzerland, p 151
Intergovernmental Panel on Climate Change (IPCC) (2005a) IPCC special report on carbon dioxide capture and storage. In: Metz B, Davidson O, de Conink H, Loos M, Meyer L (eds) Cambridge University Press, p 431
Intergovernmental Panel on Climate Change (IPCC) (2005b) Safeguarding the ozone layer and the global climate system. In: Metz B, Kuijpers L, Solomon S, Andersen SO, Davidson O, Pons J, de Jager D, Kestin T, Manning M, Meyer L (eds) Cambridge University Press, UK, p 478

Intergovernmental Panel on Climate Change (IPCC) (2007a) Climate change 2007—the physical science basis. In: Solomon S, Qin D, Manning M, Marquis M, Averyt K, Tignor MMB, Leroy Miller Jr H, Chen Z (eds) Cambridge University Press, p 996
Intergovernmental Panel on Climate Change (IPCC) (2007b) Climate change 2007—impacts, adaptation and vulnerability. In: Parry M, Canziani O, Palutikof J, van der Linden P, Hanson C (eds) Cambridge University Press, p 976
Intergovernmental Panel on Climate Change (IPCC) (2007c) Climate change 2007—mitigation of climate change. In: Metz B, Davidson O, Bosch P, Dave R, Meyer L (eds) Cambridge University Press, p 851. https://doi.org/10.1017/cbo9780511546013
Intergovernmental Panel on Climate Change (IPCC) (2007d) Climate change 2017: synthesis report. In: Core Writing team, Pachauri RK, Reisinger A (eds) IPCC, Geneva, Switzerland, p 104
Intergovernmental Panel on Climate Change (IPCC) (2011) Renewable energy sources and climate change mitigation. In: Edenhofer O, Pichs-Madruga R, Sokona Y, Seyboth K, Matschoss P, Kadner S, Zwickel T, Eickemeier P, Hansen G, Schloemer S, von Stechow C (eds) Special report of the intergovernmental panel on climate change. Cambridge University Press, Cambridge, United Kingdom and New York, NY, USA, p 1075
Intergovernmental Panel on Climate Change (IPCC) (2012) Managing the risks of extreme events and disasters to advance climate change adaptation. In: Field CB, Barros V, Stocker TF, Qin D, Dokken DJ, Ebi KL, Mastrandrea MD, Mach KJ, Plattner GK, Allen SK, Tignor M, Midgley PM (eds) Special report of the intergovernmental panel on climate change. Cambridge University Press, UK, p 582. https://doi.org/10.1017/cbo9781139177245
Intergovernmental Panel on Climate Change (IPCC) (2013) Climate change 2013: the physical science basis. In: Stocker TF, Qin D, Plattner G-K, Tignor M, Allen SK, Boschung J, Nauels A, Xia Y, Bex V, Midgley PM (eds) Contribution of working group I to the fifth assessment report of the intergovernmental panel on climate change. Cambridge University Press, Cambridge, United Kingdom and New York, NY, USA, p 1535. https://doi.org/10.1017/cbo9781107415324
Intergovernmental Panel on Climate Change (IPCC) (2014a) Climate change 2014: impacts, adaptation, and vulnerability. In: Field CB, Barros VR, Dokken DJ, Mach KJ, Mastrandrea MD, Bilir TE, Chatterjee M, Ebi KL, Estrada YO, Genova RC, Girma B, Kissel ES, Levy AN, MacCracken S, Mastrandrea PR, White LL (eds) Part A: global and sectoral aspects working group II contribution to the fifth assessment report of the intergovernmental panel on climate change. Cambridge University Press, p 1131. https://doi.org/10.1017/cbo9781107415379
Intergovernmental Panel on Climate Change (IPCC) (2014b) Climate change 2014: mitigation of climate change. In: Edenhofer O, Pichs-madruga R, Sokona Y, Minx JC, Farahani E, Kadner S, Seyboth K, Adler A, Baum I, Brunner S, Eickemeier P, Kriemann B, Savolainen J, Schlomer S, von Stechow C, Zwickel T (eds) Working group III contribution to the fifth assessment report of the intergovernmental panel on climate change. Cambridge University Press, p 1435. https://doi.org/10.1017/cbo9781107415416
Intergovernmental Panel on Climate Change (IPCC) (2014c) Climate change 2014: synthesis report. In: Core Writing Team, Pachauri RK, Meyer LA (eds) Contribution of working groups I, II and III to the fifth assessment report of the intergovernmental panel on climate change. IPCC, Geneva, Switzerland, p 151
Intergovernmental Panel on Climate Change (IPCC) (2015) Workshop report of the intergovernmental panel on climate change workshop on regional climate projections and their use in impacts and risk analysis studies. In: Stocker TF, Qin D, Plattner GK, Tignor M (eds) IPCC working group I technical support unit, University of Bern, Bern, Switzerland, p 171
Intergovernmental Panel on Climate Change (IPCC) (2018) Global warming of 1.5 °C. In: Masson-Delmotte V, Zhai P, Portner H-O, Roberts D, Skea J, Shukla PR, Pirani A, Moufouma-Okia W, Pean C, Pidcock R, Connors S, Matthews JBR, Chen Y, Zhou X, Gomis MI, Lonnoy E, Maycock T, Tignor M, Waterfield T (eds) An IPCC special report on the impacts of global warming of 1.5 °C above pre-industrial levels and related global greenhouse gas emission pathways, in the context of strengthening the global response to the threat of climate change, sustainable development, and efforts to eradicate poverty. IPCC, Geneva, Switzerland, p 562

Jacob D, Petersen J, Eggert B, Alias A, Christensen OB, Bouwer LM, Braun A, Colette A, Déqué M, Georgievski G, Georgopoulou E, Gobiet A, Menut L, Nikulin G, Haensler A, Hempelmann N, Jones C, Keuler K, Kovats S, Kröner N, Kotlarski S, Kriegsmann A, Martin E, van Meijgaard E, Moseley C, Pfeifer S, Preuschmann S, Radermacher C, Radtke K, Rechid D, Rounsevell M, Samuelsson P, Somot S, Soussana J-F, Teichmann C, Valentini R, Vautard R, Weber B, Yiou P (2014) EURO-CORDEX: new high resolution climate change projections for European impact research. Reg Environ Change 14:563–578

Jurchescu M, Micu D, Sima M, Bălteanu D, Dragotă C, Micu M, Bojariu R, Dumitrescu A (2017) An approach to investigate the effects of climate change on landslide hazard at a national scale (Romania). In: Niculiță M, Mărgărint MC (eds) Proceedings of Romanian geomorphology symposium, vol 1, pp 121–124. https://doi.org/10.15551/prgs.2017.121

Khan MR, Timmons Roberts J, Saleemul H, Hoffmeister V (2018) The paris framework for climate change capacity building. Routledge Advances in Climate Change Research, Routledge, p 244

Klein Tank AMG, Wijngaard JB, Können GP, Böhm R, Demarée G, Gocheva A, Mileta M, Pashiardis S, Hejkrlik L, Kern-Hansen C, Heino R, Bessemoulin P (2002) Daily dataset of 20th-century surface air temperature and precipitation series for the European climate assessment. Int J Clim 22:1441–1453. https://doi.org/10.1002/joc.773

Kotlarski S, Keuler K, Christensen OB, Colette A, Déqué M, Gobiet A, Goergen K, Jacob D, Lüthi D, van Meijgaard E, Nikulin G, Schär C, Teichmann C, Vautard R, Warrach-Sagi K, Wulfmeyer V (2014) Regional climate modeling on European scales: a joint standard evaluation of the EURO-CORDEX RCM ensemble. Geosci Model Dev 7:1297–1333. https://doi.org/10.5194/gmd-7-1297-2014

Kreienkamp F, Huebener H, Linke C, Spekat A (2012) Good practice for the usage of climate model simulation results—a discussion paper. Enviro Syst Res 2012:1–9. https://doi.org/10.1186/2193-2697-1-9

Leal Filho W (ed) (2015) Handbook of climate change adaptation. Springer-Verlag, Berlin Heidelberg, p 2198

Leal Filho W (ed) (2016) Innovation in climate change adaptation. Climate change management. Springer, Cham, p 388. https://doi.org/10.1007/978-3-319-25814-0

Leal Filho W (ed) (2017) Climate change adaptation in pacific countries fostering resilience and improving the quality of life. Climate change management. Springer, Cham, p 441. https://doi.org/10.1007/978-3-319-50094-2

Leal Filho W, Keenan J (eds) (2017) Climate change adaptation in North America. Fostering resilience and the regional capacity to adapt. climate change management. Springer, Cham, p 354. https://doi.org/10.1007/978-3-319-53742-9

Leal Filho W, Simane B, Kalangu J, Wuta M, Munishi P, Musiyiwa K (eds) (2017) Climate change adaptation in Africa. Fostering resilience and capacity to adapt. climate change management. Springer, Cham, p 757. https://doi.org/10.1007/978-3-319-49520-0

Leal Filho W, Esteves Freitas L (eds) (2018) Climate change adaptation in latin America. Managing vulnerability, fostering resilience. climate change management. Springer, Cham, p 537. https://doi.org/10.1007/978-3-319-56946-8

Leal Filho W, Nalau J (eds) (2018) Limits to climate change adaptation. Climate change management. Springer, Cham, p 410. https://doi.org/10.1007/978-3-319-64599-5

Mărgărint MC, Niculiţă M (2017) Landslide type and pattern in Moldavian Plateau, NE Romania. In: Rădoane M, Vespremeanu–Stroe A (eds) Landform Dynamics and evolution in Romania. Springer geography, springer cham, pp 271–304. https://doi.org/10.1007/978-3-319-32589-7_12

McInnes R, Jakeways J, Fairbank H, Mathie E (eds) (2007) Landslides and climate change: challenges and solutions. Proceedings of the International Conference on Landslides and Climate Change. Taylor and Francis, Ventnor. https://doi.org/10.1201/noe0415443180

Minea I, Croitoru AD (2015) Climate changes and their impact on the variation of the groundwater level in the Moldavian Plateau (Eastern Romania). 15th International Multidisciplinary Scientific GeoConference SGEM2015, pp 137–144. https://doi.org/10.5593/sgem2015/b31/s12.018

Minea I, Andrei A, Niculiţă M (2015) Interpolating phreatic level altitude around Mădârjac village using geomorphometric variables as covariates. 15th International Multidisciplinary Scientific GeoConference (SGEM2015), pp 403–410. https://doi.org/10.5593/sgem2015/b31/s12.052

Minea I, Hapciuc OE, Bănuc G, Jora I (2016) Trends and variations of the groundwater level in the North-eastern part of Romania. 16th International Multidisciplinary Scientific GeoConference (SGEM2016), pp 1053–1060. https://doi.org/10.5593/sgem2016/b11/s02.133

Minea I, Croitoru AD (2017) Groundwater response to changes in precipitations in North-Eastern Romania. Environ Eng Manag J 16(3):643–651. https://doi.org/10.30638/eemj.2017.066

Myhre G, Myhre CEL, Samset BH, Storelvmo T (2013) Aerosols and their relation to global climate and climate sensitivity. Nature Educ Know 4(5):7

Necula N, Niculiţă M (2017) Landslide reactivation susceptibility modelling in Iaşi municipality. Revista de Geomorfologie 19:101–117. https://doi.org/10.21094/rg.2017.021

Necula N, Niculiţă M, Tessari G, Floris M (2017) InSAR analysis of Sentinel-1 data for monitoring landslide displacement of the North–Eastern Copou hillslope, Iaşi city, Romania. In: Niculiţă M, Mărgărint MC (eds) Proceedings of Romanian geomorphology symposium, vol 1, pp 85–88. https://doi.org/10.15551/prgs.2017.85

Ni X, Groffman PM (2018) Declines in methane uptake in forest soils. PNAS. https://doi.org/10.1073/pnas.1807377115

Niculiţă M, Mărgărint C, Santangelo M (2016) Archaeological evidence for Holocene landslide activity in the Eastern carpathian lowland. Quaternary Int 415(10):175–189. https://doi.org/10.1016/j.quaint.2015.12.048

Niculiţă M, Andrei A, Lupu C (2017) The landslide database of the North–Eastern Romania. In: Niculiţă M, Mărgărint MC (eds) Proceedings of romanian geomorphology symposium, vol 1, pp 81–84. https://doi.org/10.15551/prgs.2017.81

Niculiţă M, Stoilov-Linu V, Necula N (2018) Recent landslides from Iaşi metropolitan area. Revista de Geomorfologie 20:90–101. https://doi.org/10.21094/rg.2018.030

Polemio M, Petrucci O (2010) Occurrence of landslide events and the role of climate in the twentieth century in Calabria, southern Italy. Quaterly J Eng Geol Hydrogeology 43:403–415. https://doi.org/10.1144/1470-9236/09-006

Pujină D (2008) Alunecările de teren din Podişul Moldovei. Editura Performantica, Iaşi (in Romanian)

Roberts MJ, Vidale PL, Senior C, Hewitt HT, Bates C, Berthou S, Chang P, Christensen HM, Danilov S, Demory M-E, Griffies SM, Haarsma R, Jung T, Martin G, Minobe S, Ringler T, Satoh M, Schiemann R, Scoccimarro E, Stephens G, Wehner MF (2018a) The benefits of global high-resolution for climate simulation: process-understanding and the enabling of stakeholder decisions at the regional scale. BAMS. https://doi.org/10.1175/BAMS-D-15-00320.1

Roberts CD, Senan R, Molteni F, Boussetta S, Mayer M, Keeley SPE (2018b) Climate model configurations of the ECMWF integrated forecasting system (ECMWF-IFS cycle 43r1) for HighResMIP. Geosci Model Dev 11:3681–3712. https://doi.org/10.5194/gmd-11-3681-2018

Samset BH, Sand M, Smith CJ, Bauer CJ, Forster PM, Fuglestvedt JS, Osprey S, Schleussner C-F (2018) Climate impacts from a removal of anthropogenic aerosol emissions. Geophys Res Lett 45(2):1020–1029. https://doi.org/10.1002/2017GL076079

Sidle RC, Dhakal AS (2002) Potential effects of environmental change on landslide hazards in forest environments. In: Sidle RC (ed), Environmental change and geomorphic hazards in forests, IUFRO research series 9. CABI Publishing, Wallingford, Oxen, UK, pp 123–165. https://doi.org/10.1079/9780851995984

Sidle RC, Ochiai H (2006) Landslides: processes, prediction, and land use. Water Resour Monogr Ser 18. AGU, Washington DC, p 312. https://doi.org/10.1029/wm018

Steffen W, Rockström J, Richardson K, Lenton TM, Folke C, Liverman D, Summerhayes CP, Barnosky AD, Cornell SE, Crucifix M, Donges JF, Fetzer I, Lade SJ, Scheffer M, Winkelmann R, Schellnhuber HS (2018) Trajectories of the Earth system in the anthropocene. PNAS 115(33):8252–8259. https://doi.org/10.1073/pnas.1810141115

Van Haren R, Haarsma RJ, Van Oldenborgh GJ, Hazeleger W (2015a) Resolution dependence of European precipitation in a state-of-the-art atmospheric general circulation model. J Climate 28(13):5134–5149. https://doi.org/10.1175/JCLI-D-14-00279.1

Van Haren R, Haarsma RJ, De Vries H, Van Oldenborgh GJ, Hazeleger W (2015b) Resolution dependence of circulation forced future central European summer drying. Enviro Res Lett 10(5):055002. https://doi.org/10.1088/1748-9326/10/5/055002

Vautard R, Gobiet A, Jacob D, Belda M, Colette A, Déqué M, Fernández J, García-Díez M, Goergen K, Güttler I, Halenka T, Karakostas T, Katragkou E, Keuler K, Kotlarski S, Mayer S, van Meijgaard E, Nikulin G, Patarcic M, Scinocca J, Sobolowski S, Suklitsch M, Teichmann C, Warrach-Sagi K, Wulfmeyer V, Yiou P (2013) The simulation of European heat waves from an ensemble of regional climate models within the EURO-CORDEX project. Clim Dynam 41:2555–2575. https://doi.org/10.1007/s00382-013-1714-z

Objective Analysis of Envelope Curves for Peak Floods of European and Mediterranean Flash Floods

William Amponsah, Francesco Marra, Lorenzo Marchi, Hélène Roux, Isabelle Braud and Marco Borga

Abstract Flash floods rank highly among natural disasters in terms of number of affected people and number of fatalities. This paper analyzes the scaling of the highest flash flood peaks at multiple spatial scales for different hydro-climatic regions in Europe and in the Mediterranean.The analysis is based on an integrated, high-resolution dataset of discharges concerning a number of high-intensity flash floods that occurred in these regions from 1991 to 2015. Quantile regression has permitted to define regional envelope curves of unit peak discharge versus drainage basin area, which summarize the current bound to extreme flash floods in a given region. Mean and standard error of the envelope curves' parameters are objectively derived, permitting to explore the similarities in the slopes of the regional envelope curves. Results indicate that the exponent of the envelope curves shows almost negligible variations among climatic region whereas the multiplier depends on the climatic regions.

W. Amponsah (✉)
Department of Agricultural and Biosystems Engineering, College of Engineering, KNUST, Kumasi, Ghana
e-mail: wamponsah@knust.edu.gh

F. Marra
Institute of Earth Sciences, Hebrew University of Jerusalem, Jerusalem, Israel

L. Marchi
CNR IRPI, Padua, Italy

H. Roux
Institut de Mécanique Des Fluides de Toulouse, Université de Toulouse, INPT, UPS, IMFT, Allée Camille Soula, Toulouse 31400, France

I. Braud
INRAE, RiverLy, Villeurbanne F-69625, France

M. Borga
Department of Land, Environment, Agriculture and Forestry, University of Padova, Legnaro, Italy

W. Leal Filho et al. (eds.), *Climate Change, Hazards and Adaptation Options*, Climate Change Management, https://doi.org/10.1007/978-3-030-37425-9_14

Introduction

Flash floods are usually the consequence of short, high-intensity rainfalls mainly of spatially confined convective origin and often orographically enhanced (Gaume et al. 2009). Due to the relatively small temporal scales, catchment scales impacted by flash floods are generally less than 2000–3000 km^2 in size (Marchi et al. 2010; Braud et al. 2016). Given the large rainfall rates and the rapid concentration of streamflow promoted by the topographic relief, flash floods often shape the upper tail of the flood frequency distribution of small- to medium-size catchments.

Understanding the hydro-meteorological processes that control flash flooding is therefore important from both scientific and societal perspectives. On one hand, the analysis of these events often shows unexpected features of flood response or displays expected but previously undocumented behaviour (Delrieu et al. 2005; Archer et al. 2007). On the other hand, flash floods rank highly among natural disasters as well as the most destructive process among weather-related hazards in many parts of the world in terms of the number of people affected and the number of fatalities (Barredo 2007; Borga et al. 2011; Terti et al. 2017). In the typical data-poor conditions which characterize flash flood forecasting and warning, the relationship between catchment size and flood peaks provides a regional context to evaluate the severity of flash flood events. Also, derivation of empirical envelope curves (upper bounds) for these relationships could be used as probabilistic surrogate for design flood estimation at ungauged sites (Castellarin 2007).

Flash floods are very difficult to monitor, predict and simulate, due to the mismatch between the space-time scales of occurrence and the features of the typical hydro-meteorological sampling networks (Borga et al. 2010). This has stimulated the development of an integrated post-event hydrologic observation strategy, which involves post-flood indirect estimation of peak discharges, use of weather radar rainfall estimates, and hydrological modelling for water budget analysis (Gaume and Borga 2008; Borga et al. 2008; Marchi et al. 2010; Amponsah et al. 2016; Borga et al. 2019). The systematic application of this observation strategy in specific regions permitted the development of a freely accessible European and Mediterranean flash flood database including hydrometeorological data and related observations concerning a wide number of flash floods that occurred between 1991 and 2015, referred to as the EuroMedeFF Database (Amponsah et al. 2018). The EuroMedeFF Database was used as the reference for the work proposed in this article.

This paper aims at exploring the extremal characteristics of flash flood regimes for different hydro-climatic regions in Europe in order to develop envelope curves for peak flows. Regional envelope curves are developed which summarize the current bound of extreme flash floods in a given region. In this study, the method of quantile regression is used to derive objective estimators for the regional envelope curves for a number of flash flood events included in the EuroMedeFF database. The method permitted the derivations of the mean and standard deviation for the parameters of the envelope curve, which allowed to explore the similarities in the intercept and slope of the envelope curves for different climatic regions in Europe and the Mediterranean.

Materials and Methods

The EuroMedeFF Database

The EuroMedeFF database consists of an integrated, high-resolution dataset of hydro-meteorological and geographical variables (rainfall, discharge, topography) concerning 49 high-intensity flash flood events that occurred in France (30), Israel (7), Germany (1), Slovenia (1), Romania (3), and Italy (7) between 1991 and 2015. Data comprises digital elevation models (DEM) of the impacted area (resolutions 5–90 m), high space-time resolution radar-rainfall estimates (60 min or less), flood hydrographs and flood peak estimates from post-flood surveys. The EuroMedeFF Database was created in the framework of the HyMeX international programme (Drobinski et al. 2014) and made available to the public (Amponsah et al. 2018).

Figure 1 presents the spatial location of the flash flood events included in the EuroMedeFF database, which covers from Continental Europe through the Mediterranean to Arid climates. A unit peak discharge of 0.5 $m^3\ s^{-1}\ km^{-2}$ is considered as the lowest value for defining a flash flood event. This means that, for an event to be included in the database, at least one measured flood peak should exceed the above value. The upper limit for a catchment impacted by the flood is 3000 km^2, whereas the upper limit for the duration of the flood-triggering storm is up to 48 h.

Discharge data in the EuroMedeFF database derive from both streamflow monitoring stations and post-flood indirect estimates of flow peak. Streamflow data permit recording of flood hydrographs, thus enabling assessment of not only discharge, but also time response and flood runoff volume estimation. Flood hydrographs were checked for the uncertainties affecting rating curves at high-flood stages by using

Fig. 1 Location of the flash floods included in the EuroMedeFF database

hydraulic models and topographic data. Discharge data from reservoir operations, water levels and use of the continuity equation, when available, were also included in the database after accurate quality control.

Different methods have been used for the indirect reconstruction of flow velocity and peak discharge from flood marks, such as slope-area, slope-conveyance, flow-through culvert, and lateral super-elevation in bends. Amongst these methods, the most commonly used for the implementation of the dataset presented in this paper is the slope-conveyance, which consists of the application of the Manning–Strickler equation, under assumption of uniform flow, and requires the topographic survey of cross-section geometry and flow energy gradient, computed from the elevation difference between the high water marks along the channel reach surveyed (Gaume and Borga 2008; Lumbroso and Gaume 2012).

Overall, 680 peak discharge data are included in the archive: 32% (219) were recorded by river gauging stations or based on data from reservoir operations, and 68% (461) from post-flood surveys. We followed the geomorphic impact-based linear error analysis of the slope conveyance discharge determination presented in Amponsah et al. (2016) for the uncertainty assessment of the post flood estimates. Almost 90% of the included discharge data are from the Mediterranean region, which is consistent with increasing collation and analysis of flash flood data in this region compared to other climatic regions in Europe (e.g. Gaume et al. 2009; Marchi et al. 2010).

Use of Quantile Regression for Derivation of Flash Flood Envelope Curves

We use here a power law equation for the flash flood envelope curve, in the form of:

$$Q_u = a \cdot A^b \tag{1}$$

where Q_u is the unit peak discharge in $m^3\ s^{-1}\ km^{-2}$, A is the watershed area in km^2, and a and b are the parameters describing the specific envelope curve. Equation (1) shares the same structure as typical equations relating watershed area and the annual peak discharge with a given return period. Empirical values of the exponent b, often called scaling exponent, have been found to lie in the range $-0.5 < b < 0$ (Benson 1962; Leopold et al. 1964; Alexander 1972). Hereinafter, the parameter 'a' is termed coefficient.

Linear regression represents the dependent variable, as a linear function of one or more independent variables, subject to an 'error' term. It estimates the expected value of the dependent variable for given levels of the independent variables. For this type of regression, where the central tendency in a dataset is the aim of the analysis, ordinary least squares (OLS) is an effective parameters estimation method. OLS loses its effectiveness when the analysis aims to examine the extremes of a data set. Koenker and Bassett (1978) introduced Quantile Regression (QR) as an extension of OLS for the estimation of models for conditional quantile functions of a data distribution. The central special case is the median regression estimator that minimizes a sum of absolute errors. The remaining conditional quantile functions are predicted by minimizing an asymmetrically weighted sum of absolute errors, weights being the function of quantile of interest. This makes QR a robust technique even in presence of outliers.

We propose a methodology based on QR to objectively estimate the parameters of the envelope curves. By definition envelope curves are mathematical relations that represent the upper limit (envelope) of a given dataset, therefore, if a sufficiently high quantile q is sought by the QR, the estimated curve will represent an envelope curve. We propose to define q as a quantile which is exceeded by less than 1 data point:

$$q = (N - 0.5)/N \tag{2}$$

where N is the number of data points in the dataset of interest. This permits objective identification of the parameters in presence of different sample sizes limiting the numbers of data points exceeding the threshold in the case of large data samples.

Uncertainty in the estimated parameters depends on the available sample from the population; a proper quantification should thus account for the characteristics of the dataset. Bootstrap methods (Efron and Tibshirani 1993) allow for a consistent quantification of such uncertainties and are used in different fields of climatology and hydrology in which data sample issues are dominant factors, such as frequency analyses (Overeem et al. 2008) and thermodynamic scaling of extreme precipitation with temperature (Peleg et al. 2018).

In this study, the QR was computed using the *quantreg* matlab function (*MathWorks*) and the uncertainty in the parameters was quantified as the standard error of 200 bootstrap iterations (with replacement) for each dataset.

Results and Discussion

The aim of the analysis is to assess the similarities between the slopes of the envelope curves estimated from a linear quantile regression model for the four categories of climatic regions in Europe (Mediterranean, Alpine-Mediterranean/Alpine, Inland Continental, and Arid-Mediterranean/Arid) as delineated by Amponsah et al. (2018). Figures 2, 3 and 4 show the envelope curves and their relative 95% confidence intervals for the full EuroMedeFF dataset (Fig. 2), for the four climatic regions (Fig. 3) and for the two discharge estimation methods (Fig. 4). Table 1 summarizes the envelope curve parameters and their relative standard error.

The adopted quantiles range between 99.93 (full dataset, 680 data points) and 95.0 (Arid-Mediterranean/Arid region, 10 data points). The envelope curve characterizing the full dataset (Fig. 2) is parametrized as $Q_u = 96.8 \cdot A^{-0.37}$ and is fully overlapping with the envelope curve characterizing Mediterranean climates (Fig. 3a). This clearly

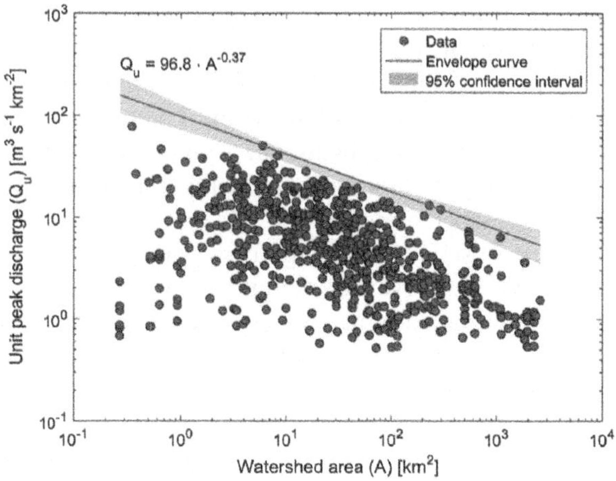

Fig. 2 Unit peak discharge versus watershed area for the EuroMedeFF database showing the envelope curve at 95% confidence interval

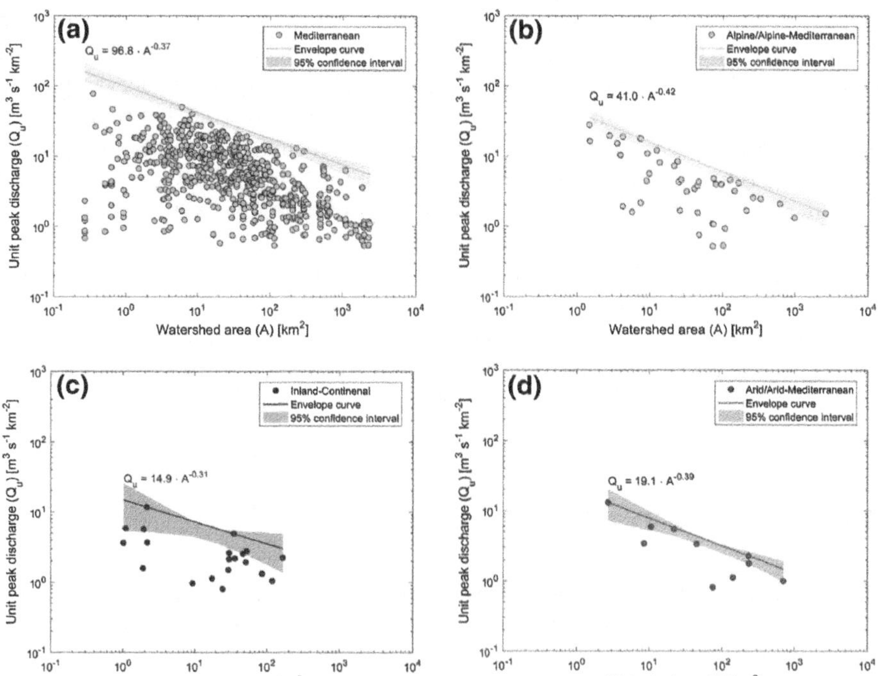

Fig. 3 Unit peak discharge versus watershed area based on climatic regions: **a** Mediterranean catchments, **b** Alpine-Mediterranean and Alpine catchments, **c** inland continental, and **d** Arid and Arid-Mediterranean catchments. The envelope curves at 95% confidence interval are reported

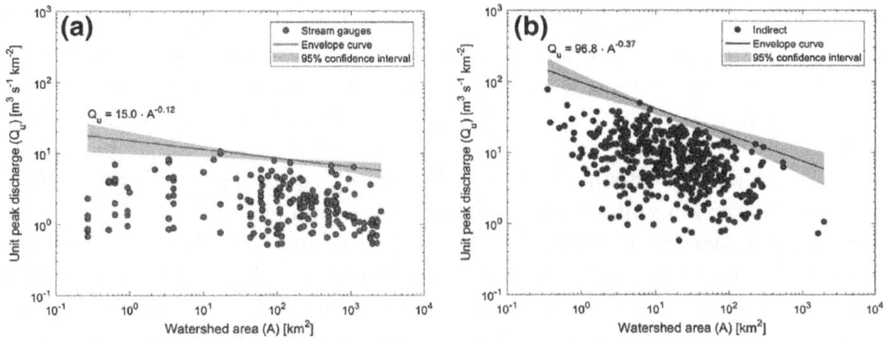

Fig. 4 Unit peak discharge versus watershed area based on discharge assessment methods: **a** stream gauges and **b** indirect methods. The envelope curves at 95% confidence interval are reported

Table 1 Estimated parameters of the envelope curves for the EuroMedeFF database and relative standard error. The used quantiles (q) are defined as in Eq. (2)

	Data sample (N)	Quantile used (q)	Parameters		Standard error	
			Coefficient	Exponent	Coefficient	Exponent
Data (EuroMedeFF database)	680	99.93	96.8	−0.37	0.12	0.030
Climatic regions						
Mediterranean	606	99.92	96.8	−0.37	0.12	0.032
Alpine-Mediterranean/Alpine	44	98.86	41.0	−0.42	0.15	0.042
Continental	20	97.50	14.9	−0.31	0.34	0.117
Arid-Mediterranean/Arid	10	95.00	19.1	−0.39	0.26	0.062
Discharge assessment method						
Stream gauges	219	99.77	15.0	−0.12	0.19	0.038
Indirect method	461	99.89	96.8	−0.37	0.15	0.052

comes from the fact that a large portion of the heaviest events belong to the Mediterranean region, but also indicates that other regions, despite all having comparable exponents within the data sample uncertainties, are characterized by smaller coefficients. This indicates that the scaling of the unit peak discharge with area is rather independent on the climate within the EuroMedeFF dataset. The Mediterranean climates are able to provide the largest unit peak discharges for a given catchment area. This can be associated both to meteorological conditions typical of the region, as well as to specific characteristics of the catchments. It should be however pointed out that limited samples size in the other regions is likely to cause underestimation of the envelope curves due to the insufficient sampling of extreme floods.

Coefficient values (parameter 'a' equal to 97) reported by Gaume et al. (2009) and Marchi et al. (2010) is consistent with the parameter range reported for the whole dataset (Fig. 2) and the Mediterranean climatic region (Fig. 3a). Coefficient value of 120 reported by Amponsah et al. (2018) falls within the uncertainty range of the 200 bootstrap iterations for the whole dataset and the Mediterranean climatic region. These authors also reported an exponent of −0.4, which is comparable to the −0.37 reported for these datasets and also fall within the uncertainty range of the 200 bootstrap iterations. The coefficients and exponents reported by Amponsah et al. (2018) for the other climatic regions (Alpine-Mediterranean/Alpine, Inland Continental, and Arid-Mediterranean/Arid) also fall within the uncertainty range of the 200 bootstrap iterations from the quantile regression method.

Figure 4 shows how the envelope curve characterizing the full dataset is mostly shaped by indirect discharge estimates. The exponent is slightly lower for the stream gauge data compared to the indirect methods. This may be ascribed to the fact that highest values of the peak discharge are often missed by the gauging stations because of insufficient density of streamgauge networks and/or damage to the stations during floods. Furthermore, since the envelope curve of the EuroMedeFF database (as well as the Mediterranean region) fully overlaps with that of the indirect discharge estimation method, it is worth noting that the chosen discharge estimation methods may have an impact on the shape of the envelope curve.

For basin areas less than 10 km^2, the distribution for the stream gauges (Fig. 4a) shows a largely smaller amount of extreme floods with respect to the large areas. On the contrary, smaller amount of extreme floods are observed at the large-area end of the distribution for the indirect methods (Fig. 4b). These points out the major role post-flood surveys play in the documentation and analysis of extreme floods in small- to medium-sized catchments due to the small space–time scales of flash flood occurrence compared to the typical hydrometeorological monitoring networks (Marchi et al. 2010; Amponsah et al. 2016; Borga et al. 2019).

Conclusions

The dependence of peak discharge on drainage area has been investigated for a wide number of floods in different climatic regions. Herschy and Fairbridge (1998) analysed the relationship between the drainage area and the unit peak discharge for 41 maximum floods impacting catchments greater than 100 km^2 from the world whereas Herschy (2002) investigated for the maximum floods in Europe. Gaume et al. (2009) and Marchi et al. (2010) also characterized for different climatic regions in Europe between 1994 and 2008 whereas Amponsah et al. (2018) compiled data for European and Mediterranean flash floods between 1991 and 2015. Whereas the envelope curves developed by these authors may be subjective, the quantile regression approach establishes a more objective way to estimate parameters of the envelope curves and supports their use for the estimation of maximum flood discharge in a given region.

From the quantile regression analysis presented in this work, the exponents of the envelope curves have been found to be quite stable for the different climatic regions, provided the data samples include indirect peak discharge estimates. On the other hand, the coefficients have been found to depend on the climatic regions. Although we have identified the small sample sizes in some of the climatic region as a major limitation of this work, our choice of quantile (Eq. 2) represents an objective upper bound for variable sample sizes. Overall, we can conclude that the proposed approach is a promising method for the objective estimation and comparison of the parameters of regional envelope curves of flash floods.

Acknowledgements This paper contributes to the HyMeX programme (www.hymex.org). Flash flood data were obtained from the EuroMedeFF database (https://doi.org/10.6096/mistrals-hymex.1493) Francesco Marra was funded by the Israel Science Foundation [grant no. 1069/18], by NSF-BSF grant [BSF 2016953], and by a Google gift grant.

References

Alexander GN (1972) Effect of catchment area on flood magnitude. J Hydrol 16:225–240

Amponsah W, Ayral PA, Boudevillain B, Bouvier C, Braud I, Brunet P, Delrieu G, Didon-Lescot JF, Gaume E, Lebouc L, Marchi L, Marra F, Morin E, Nord G, Payrastre O, Zoccatelli D, Borga M (2018) Integrated high-resolution dataset of high-intensity European and Mediterranean flash floods. Earth Syst Sci Data 10:1783–1794

Amponsah W, Marchi L, Zoccatelli D, Boni G, Cavalli M, Comiti F, Crema S, Lucía A, Marra F, Borga M (2016) Hydrometeorological characterisation of a flash flood associated with major geomorphic effects: assessment of peak discharge uncertainties and analysis of the runoff response. J Hydrometeor 17(12):3063–3077

Archer DR, Leesch F, Harwood K (2007) Learning from the extreme river Tyne flood in January 2005. Water Environ J 21(2):133–141

Barredo JI (2007) Major flood disasters in Europe: 1950–2005. Nat Hazards 42(1):125–148

Benson MA (1962) Factors influencing the occurrence of floods in a humid region of diverse terrain. U.S. Geol Surv Water Supply Pap 1580-B, p 64

Borga M, Comiti F, Ruin I, Marra F (2019) Forensic analysis of flash flood response. Wiley interdisciplinary reviews: water 6(2). https://doi.org/10.1002/wat2.1338

Borga M, Gaume E, Creutin JD, Marchi L (2008) Surveying flash flood response: gauging the ungauged extremes. Hydrol Processes 22:3883–3885

Borga M, Anagnostou E, Blöschl G, Creutin JD (2011) Flash flood forecasting, warning and risk management: the hydrate project. Environ Sci Policy 14:834–844

Borga M, Anagnostou EN, Blöschl G, Creutin JD (2010) Flash floods: observations and analysis of hydrometeorological controls. J Hydrol 394(1–2):1–3

Braud I, Borga M, Gourley J, Hürlimann M, Zappa M, Gallart F (2016) Flash floods, hydrogeomorphic response and risk management. J Hydrol 541:1–5

Castellarin A (2007) Probabilistic envelope curves for design flood estimation at ungauged sites. Water Resour Res 43:W04406

Delrieu G, Ducrocq V, Gaume E, Nicol J, Payrastre O, Yates E, Kirstetter PE, Andrieu H, Ayral PA, Bouvier C, Creutin JD, Livet M, Anquetin A, Lang M, Neppel L, Obled C, Parent-du-Chatelet J, Saulnier GM, Walpersdorf A, Wobrock W (2005) The catastrophic flash-flood event of 8–9 September 2002 in the Gard region, France: a first case study for the Cévennes-Vivarais Mediterranean hydro-meteorological observatory. J Hydrometeor 6:34–52

Drobinski P, Ducrocq V, Alpert P, Anagnostou E, Béranger K, Borga M, Braud I, Chanzy A, Davolio S, Delrieu G, Estournel C, Filali Boubrahmi N, Font J, Grubisic V, Gualdi S, Homar V, Ivancan-Picek B, Kottmeier C, Kotroni V, Lagouvardos K, Lionello P, Llasat MC, Ludwig W, Lutoff C, Mariotti A, Richard E, Romero R, Rotunno R, Roussot O, Ruin I, Somot S, Taupier-Letage I,

Tintore J, Uijlenhoet R, Wernli H (2014) HyMeX, a 10-year multidisciplinary program on the Mediterranean water cycle. Bull Am Meteor Soc 95(7):1063–1082
Efron B, Tibshirani RJ (1993) An introduction to the bootstrap. Chapman and Hall, New York
Gaume E, Borga M (2008) Post-flood field investigations in upland catchments after major flash floods: proposal of a methodology and illustrations. J Flood Risk Manag 1(4):175–189
Gaume E, Bain V, Bernardara P, Newinger O, Barbuc M, Bateman A, Blaškovicová L, Blöschl G, Borga M, Dumitrescu A, Daliakopoulos I, Garcia J, Irimescu A, Kohnová S, Koutroulis A, Marchi L, Matreata S, Medina V, Preciso E, Sempere-Torres D, Stancalie G, Szolgay J, Tsanis I, Velasco D, Viglione A (2009) A collation of data on European flash floods. J Hydrol 367:70–78
Herschy RW, Fairbridge RW (1998) Encyclopedia of hydrology and water resources. Kluwer, Dordrecht, The Netherlands
Herschy RW (2002) The world's maximum observed floods. Flow Meas Instrum 13(5–6):231–235
Koenker R, Bassett G (1978) Regression quantiles. Econometrica 46(1):33–50
Leopold LB, Wolman MG, Miller JP (1964) Fluvial processes in geomorphology. W. H. Freeman, New York
Lumbroso D, Gaume E (2012) Reducing the uncertainty in indirect estimates of extreme flash flood discharges. J Hydrol 414–415:16–30
Marchi L, Borga M, Preciso E, Gaume E (2010) Characterisation of selected extreme flash floods in Europe and implications for flood risk management. J Hydrol 394:118–133
Overeem A, Buishand TA, Holleman I (2008) Rainfall depth–duration–frequency curves and their uncertainties. J Hydrol 348:124–134
Peleg N, Marra F, Fatichi S, Molnar P, Morin E, Sharma A, Burlando P (2018) Intensification of convective rain cells at warmer temperatures observed from high-resolution weather radar data. J Hydrometeor 19:715–726
Terti G, Ruin I, Anquetin S, Gourley JJ (2017) A situation-based analysis of flash flood fatalities in the United States. Bull Am Meteorol Soc 98:333–345

Treatment of Natural Hazards Within Planning Documents in Serbia in Relation to Climate Change Issues

Tijana Crnčević, Jelena Živanović Miljković and Omiljena Dželebdžić

Abstract Corresponding to contemporary international frameworks that promote sustainable development following the adoption of the *Paris agreement* at the 21st Conference on the Parties of the United Nations Framework Convention on Climate Change (UNFCCC), 17 sustainable development goals are specifically promoted, and, in addition to sustainability, resilience and, accordingly, disaster risk reduction. In this respect, the harmonization of global, regional and local development is conditioned by climate change issues. The Republic of Serbia (RS), according to the given scenarios, expects a rise in temperature and long-lasting droughts and fires, as well as more intense precipitations that could result in the occurrence of natural hazards such as floods, escarpments and other similar events. Thus, this paper provides an overview of the current planning framework that addresses the problem of preparing areas for emergency situations and disaster risk reduction with a special reference to these climate change issues and their impact in terms of natural disasters. An overview of selected case studies shows that disaster risk reduction is present in the planning documents—the spatial and urban plans—and that the natural hazards that have appeared within the last decade in Serbia have resulted in the initiating and innovating of the legal and planning bases, as well as concrete adjustments and amendments within the planning documents.

T. Crnčević (✉) · J. Ž. Miljković · O. Dželebdžić
Institute of Architecture and Urban and Spatial Planning of Serbia, Bulevar Kralja Aleksandra 73/II, 11000 Belgrade, Serbia
e-mail: tijana@iaus.ac.rs

J. Ž. Miljković
e-mail: jelena@iaus.ac.rs

O. Dželebdžić
e-mail: omiljena@iaus.ac.rs

W. Leal Filho et al. (eds.), *Climate Change, Hazards and Adaptation Options*, Climate Change Management, https://doi.org/10.1007/978-3-030-37425-9_15

Introduction

As a result of global changes within climate in the last decades, increasingly frequent occurrences of natural hazards[1] are evident. As stated within the report of the Weather Meteorological Organization (WHO) titled "*Atlas of Mortality and Economic Losses from Weather, Climate and Weather Extremes (1970–2012)*" the natural disasters of repeated floods, droughts and temperature extremes "are nearly five times as frequent as they were in the 1970s", (UNFCCC 2018). When people's lives and livelihoods are damaged and destroyed natural hazards become disasters. Around 80% of the 3496 registries of disasters in the last decade were due to flooding and storms. Wildfires and heat waves, which were not recorded in registries during the 1970s were, by 2010, along with storms, one of "the leading causes of deaths in natural disasters". In Russia alone, during 2010, more than 55,000 people died (Ibid 2018). As a response to the climate change issue, within the Paris Agreement's[2] 17 adopted sustainable development goals, the issue of climate change and its impacts were addressed. As stated in Goal 13, "Climate Action", there is the need to "*Take urgent action to combat climate change and its impacts*" and one of the targets is to "*Strengthen resilience and adaptive capacity to climate/related hazards and natural disasters in all countries*" together with "*Integrating climate change measures into national policies, strategies and planning*" (17 GOALS 2018).

Within the issue of disasters, among the most important international strategic frameworks is, above all, the *International Strategy for Disaster Reduction* (ISDR) from 1989, which represents the basic international strategic framework for disaster risk reduction (DRR), formulated on the principles contained in the *Yokohama Strategy.*[3] The *Johannesburg Plan of Implementation of the World Summit on Sustainable Development* (2002) highlights the need for an integrated approach in terms of vulnerability, risk, assessment and disaster management, including prevention, mitigation, preparedness, response and recovery. Following the *Hyogo Framework for Action 2005–2015*, which was dedicated towards building the resilience of states and communities to disasters (caused by natural disasters and related environmental and technological hazards and risks) (UNISDR 2005), the *Sendai Framework for Disaster Risk Reduction 2015–2030* was adopted at the third UN World Conference on Disaster Risk Reduction. In relation to the Hyogo framework, the *Sendai*

---

[1]Within the context of climate change, natural hazards are "naturally occurring physical phenomena caused either by rapid or slow onset events which are climatological (extreme temperatures, drought and wildfires) or meteorological (cyclones and storms/wave surges)" (IFRC 2018).

[2]The Paris Agreement was adopted after the 21st Conference of the Parties of the United Nations Framework Convention on Climate Change (UNFCCC) in December 2015 (UN, Paris Agreement, 2018).

[3]The Yokohama Strategy and Plan of Action for a Safer World contains guidelines for preventing natural disasters, the mitigation of the effects of natural disasters and the action plan adopted in 1994. The Progress Report highlights the importance of reducing disaster risk based on a proactive approach to information and motivating and involving people in all aspects of disaster risk reduction in their local communities (UNISDR 2018a, b).

Framework puts emphasis on disaster risk management in relation to disaster management, prevention of emergence of new risks, as well as reduction of existing risk, the strengthening of resistance, and the engagement of the whole society and all state institutions (UNISDR 2015). Taking into consideration the global framework mentioned above, it is evident that within the last decade efforts have been made towards creating a global response to climate change and disasters, where a new paradigm for disaster risk reduction focuses on preventing and mitigating risk versus the traditional approach that was focused on readiness for emergency situations and response, so that now human vulnerability and human capacity are at the core of the disaster risk reduction concept (UNDP 2016). However, regarding the treatment of the participation of local and community actors within the three international frameworks for DRR,[4] it is noted that there is a "concerning shift away from valuing local community input and toward promoting technological advances" (Tozier de la Poterie and Baudoin 2015:128).

Floods are hazards that have very clear and obvious links to climatic change. In accordance with the forecasts respecting climate change scenarios, it has been pointed out that floods will threaten between 10 and 25 million people by 2050 or 40 and 140 million by the year 2100 in correlation with the scenario (Nicholls and Lowe 2004). Thus, due to the impact of climate change, planners will face not only natural hazards such as floods, more powerful hurricanes and fires, but also the migration of populations into areas with more favorable climatic conditions. (APA 2011). The climate change predictions refer to global warming with increases of annual temperatures by 1.8 °C. For the Western Balkans region, the annual mean precipitation is expected to decrease by 10–20%. But despite the lack of precipitation, when it does rain it will be more intense (Pollner et al. 2010: 15). Reports for a period of more than a decade indicate that Serbia has been affected by floods once in every two years (UN-ISDR, WB 2008). Spatial plans and regulations (building codes, zoning) that treat these natural hazards, as well as the enforcement of existing building codes and regulations, are of key importance for the mitigation of all hazards, since "mitigation reduces the impact of disasters, saves lives, and reduces economic loss" (Pollner et al. 2010: 35).

Based on reported damages from natural disasters in Europe since 1980, floods and landslides account for 64% of reported damages, while extreme temperatures, including droughts and forest fires account for another 20% (EEA 2018). That natural hazards are becoming more frequent is also seen in a rise in the amount of overall reported damages—from 9 billion euros in the 1980s to over 13 billion euros in the 2000s (Ibid.). In that context, space is becoming "full of risky locations" (Surjan et al. 2016:77) where, inter alia, "dominant institutional characteristics critically affect the steering capacity of organizations/agencies (including their coordination capacity) to address climate-related risks" (Francesch-Huidobro et al. 2017:1). Within spatial and urban planning, the frequent natural hazards throughout the world (floods, hurricanes) have resulted in the development of risk management strategies that represent

---

[4]The Yokohama Strategy and Plan of Action for a Safer World, the Hyogo Framework for Action 2005–2015 and the Sendai Framework for Disaster Risk Reduction 2015–2030 (SFDRR).

"a cluster of concrete and practical strategies and actions on how to handle risks, and in terms of spatial and urban planning, including those risks that are of spatial importance or significant with regard to planning processes" (Birkmann 2018). However, in order to ensure that both local actions and plans have been produced as well as implemented, it is stated that it is essential for "the governance structure of the city and the capacity of the different stakeholders to collaborate on the definition and implementation of adaptation plans" as well to develop more formalized guidance that can steer communities towards holistic integration of climate change into hazards planning (Stults 2017).

Taking into consideration the contemporary international and regional strategic frameworks, as well as recent trends in coping with natural hazards within planning, the goals of this paper are to (a) present the framework for the treatment of natural hazards in Serbia in relation to the climate change issues and (b) provide an overview of measures to mitigate and manage these hazards given within selected planning documents. The paper takes into account that within the last decade, especially after the catastrophic floods in May 2014, significant efforts have been made to innovate the legal and planning framework in this field. This paper also presents the case of the Kolubara lignite basin, the area that was under the influence of floods and that has had, due to its specific purpose and public interest, a special legal and planning treatment regarding natural hazard issues.

Methodology

This research is conducted using qualitative research methods, a document analysis and literature research, analysis of the institutional and regulatory framework and case study analysis. International and national documents, as well as the literature, have been analyzed in order to identify natural hazards issues within planning. The case study analysis involves a comprehensive qualitative assessment of the selected planning documents that are adopted within the period 2000–2014 and the period after 2014. In order to present the issue of natural hazards—the floods that occurred in May 2014—the acquired analytical framework is then used to analyze the case study of the Kolubara lignite basin.

Planning and Natural Hazards in Serbia

In accordance with the climate change scenarios, the outcome for Serbia and the Western Balkans presents an increase in the temperature in future with a great probability of prolonged droughts and open fires and also prolonged and intense precipitation and snow melting, which will result in the appearance of large sites, slopes and floodplains (IPCC 2014; UNDP 2016). In Serbia the average annual air temperature has been increasing within the last decades and, after 1998, the average temperature in Serbia exited out of the normal range and entered into the range of warm weather (UNDP, Agency for Environmental Protection 2007). Changes are also evident within precipitation where a positive trend that is noted until 2040 will decrease to negative values by the end of the century, with the largest deficit during the summer (RS, Ministry of Environmental Protection 2017). In relation to the available data referring to natural hazards, floods and torrents have been occurring in Serbia. However, during the last decade, due to the changes in climatic conditions—an increase of water and air temperature, as well as increased precipitation—2800 natural hazards were recorded, 800 more compared to the decade of 1980–1990 when 2000 were recorded (Official Gazette of the Republic of Serbia 2011).

In the Republic of Serbia, the legal and planning framework within the subject of environmental protection, together with the management of natural disasters/hazards[5] is continuously harmonized in line with international and other relevant frameworks. Special progress has been made since 2012 when the RS received the status of candidate for membership in the EU. So far, a total of 16 chapters have been opened out of 35 (two are temporarily closed), which between them Chap. 27 regulate the environment. Within the Report for 2018 "Communication from the Commission to the European Parliament, the Council, the European Economic and Social Committee and the Committee of the Regions" it was emphasized that next year Serbia will focus its activities towards "implementation of the Paris Climate Agreement, among other things by developing a comprehensive climate change strategy that will be aligned with the EU framework for climate and energy policy for the period up to 2030, and well integrated into all relevant sectors"(EC 2018:87).

The main strategic framework that promotes the climate change issue in Serbia is as follows: (1) the National Strategy for Sustainable Development and the Action Plan for the period 2009–2017 (2008) emphasize the necessity of sectoral actions that would lead to the reduction of GHG emissions and point out the need to develop an adjustment plan for the economic sector; (2) the National Program for Environmental Protection for the period 2010–2019 (2010) defines the activities needed to mitigate the effects of climate change; (3) the National Strategy for Inclusion of the Republic of Serbia in the Clean Development Mechanism (CDM) (2010) for the sectors of agriculture, forestry and waste management identifies the ways and opportunities offered by the flexible mechanism of the Kyoto Protocol in these sectors; (4) the Strategy of Biodiversity for the period 2011–2018 (2011) promotes the issue of climate change within biodiversity and increases the capacity of institutions; (5) the National Strategy for Sustainable Use of Natural Resources and Properties (2012), among other efforts, promotes adaptive management within the plans of protected areas; (6) Energy Strategy of the Republic of Serbia for the

[5]The role of disaster risk management implies a systematic process of the application of appropriate administrative directives, as well as the necessary organizational and operational capacities and capacity to implement strategies and policies in order to reduce the negative impacts of disasters and increase the possibilities of their forecasting (UNISDR: Terminology 2018b). On the other hand, *Disaster risk reduction* (DRR) focuses on "preventing new and reducing existing disaster risk and managing residual risk, all of which contribute to strengthening resilience and therefore to the achievement of sustainable development (Ibid.)".

period until 2025 with projections until 2030 (2015) supports the use of renewable resources. Regarding the issue of climate change, as it is estimated within the Report "Communication from the Commission to the European Parliament, the Council, the European Economic and Social Committee and the Committee of the Regions" RS has achieved a certain level of commitment while the implementation is at an early stage. Also, it is stressed that it is necessary to increase the accuracy of the obtained data. Serbia ratified the Paris Climate Agreement 2017, and in October submitted its Second National Communication of the Republic of Serbia to the UN Framework Convention on Climate Change (UNFCCC). However, the Second National Communication to UNFCCC regarding the issue of climate change states that "the current level of integration of climate change into the sectoral and overall development strategy, the level of knowledge, the institutional and individual capacities, the available technology and financial resources at the national level, together with the involvement of local governments are not sufficient, despite numerous activities and efforts, for an effective and rapid response to climate change issues" (RS, Ministry of Environmental Protection 2017:109).

Serbia is an active member of the EU Civil Protection Mechanism. As estimated, Serbia is moderately prepared in terms of civil protection (EEA 2018). The main framework for the implementation of the Hyogo Framework in Serbia is the *National Strategy for Protection and Rescue in Emergency Situations* (2011), which directs the development of the national system, respecting the EU's frameworks in terms of establishing appropriate institutional, organizational and personal conditions. Together with the *Law on Emergency Situations* (2009) an obligation is established to define protection conditions for the prevention of natural disasters within planning documents (heavy rain, droughts, floods, atmospheric discharges, torrents, storms, hail or landslides, avalanches and snow layers, extreme air temperatures and ice accumulation on the watercourse). The Law on Planning and Construction (2009) sets up the conditions for planning documents (spatial and urban plans) and for presenting areas of vulnerability and activities on environmental protection, as well as protection of natural and cultural heritage. The Law on Planning and Construction also establishes the obligation to arrange the space of interest for national defense and the protection against natural disasters. Further, the legal framework that considers natural hazards also includes the Law on Meteorological and Hydrological Activities (2010), the Law on Fire Protection (2009), the Law on Local Government Act (2007) and the Law on Insurance (2014). As was estimated, "these documents provide solid foundations for integrating disaster risk reduction (DRR) principles into the policies and practices of development at all levels—from the local to the national" where "local governments are seen as the main contributors to DRR" in the creation of development plans for DRR (Crncevic and Lovren-Orlovic 2018:111). In this regard, it was pointed out that progress has been made in the implementation of instruments for the mechanism for civil protection, such as post-disaster risk assessment and flood risk mapping (EEA 2018). Regarding the strategic framework, an action plan for disaster risk management was adopted, although the relevant law has not yet been adopted. Further tasks within the Report primarily stress the need to further strengthen its capacity to prevent disasters, reduce disaster risks and manage disaster risk, including flood risk management (Ibid.).

Treatment of Natural Hazards Within the Planning Documents in Serbia

In Serbia, the Law on Planning and Construction (2009) (LPC) regulates the field of spatial and urban planning and regulates the adoption of planning documents, the documents for the implementation of spatial plans, as well as the urban technical documents. Spatial planning documents include the Spatial Plan of the Republic of Serbia, the regional spatial plans, the spatial plan for the local government unit and the special purpose area spatial plan. The urban plans include the master (general urban) plan, the general regulatory plan and the plan of detailed regulation.

In accordance with the provisions of the LPC, the planning, regulation and use of space is based, inter alia, on the principle of protection from natural disasters and elimination of the causes of climate change.

The issues regarding natural hazards are contained within the planning documents in Serbia. In the spatial planning documents, special conditions and measures for protection of people's lives and health from natural disasters are given within the rules for regulation and construction.

The Spatial Plan of the Republic of Serbia 2010–2020 (SPRS) adopted in October 2010 ambitiously defined proposals and directives for all spatial development issues and measures and instruments for their implementing. The new performance framework of spatial planning output in Serbia proposed the set of 106 national indicators that are outcome-oriented, including three indicators for outcome/target 'Prevention of natural disasters' (indicators: Housing settlements in areas with potential risk from flooding, Settlements in earthquake risk zones, Housing settlements in areas with potential risk from landslides). This should serve as the guide to all regional and local, general and sectoral plans, strategies and programs in their preparation, since the SPRS is on the top of vertical coordination and integration pyramid for all plans and strategies related to spatial development (Dzelebdzic and Bazik 2011).

The master plan also contains the general conditions and measures of protection from natural disasters and conditions of interest in defense of the country. In spite of this, some of the limitations of this paper and the research conducted should be stressed. First, a lack of research studies regarding the treatment of natural hazards within planning documents in Serbia is noted.

Additionally, the first Report on the SPRS Implementation and Spatial Development Status 2011[6] reflected the lack of relevant data and insufficient state of information support for the preparation of planning documents, as well as the need for new model of future institutional cooperation in data collection, storage, processing and distribution (Bazik and Dzelebdzic 2012). Similar results were also presented in the recent Report on the SPRS Implementation and Spatial Development Status 2014 for three indicators related to the objective "Prevention of natural disasters". Further, the number of case studies for the analyzed period (after 2014) is limited to the area in which natural hazards occurred. The necessity of an extensive period for producing and implementing planning documents is a further limitation.

In order to illustrate the presence of the issue of natural hazards in the planning documents in Serbia over the last two decades, an analyses of selected case studies/planning documents has been performed. The criteria applied for the selection of planning documents indicates that they represent spatial and urban planning and are adopted over two periods. The first one is from the year 2000–2014 and the second from 2014, the period after major floods in Serbia, until the present. The analysis was conducted in relation to the representation of the issue of natural hazards/disasters.

As can be seen in Table 1, the presence of the issue of natural hazards within planning documents is evident, in both the spatial and urban planning documents, as well as in both of the analyzed periods. In the General Urban Plan (GUP) for Vrnjacka Spa (2005), within part 7, is the sectoral attachment "Protection from natural disasters and conditions of interest for the defense of the country." This sector contains: (a) conditions for the arrangement of space from the aspect of protection from natural hazards/disasters and of interest for defense such as: putting under a special regime

Table 1 Overview of the presence of natural hazards/disasters

Planning document	Representation of issues	Coverage
GUP Vrnjacka Spa (2005)	Sectoral attachment (conditions, rules and protection measures)	Natural hazards, conditions of interest for the defense of the country
SPASP Stara Planina (2008)	Sectoral attachment (measures and procedures)	Defense of the country and protection of natural hazards
PGR of Bor (2008)	Sectoral attachment (measures: spatial, construction and technical and operational-organizational)	Defense and protection
SPASP of Kolubara lignite basin (2017)	Sectoral attachment	Defense and protection

[6]In December 2011 the Serbian government adopted the Decree on establishment of the Program of Implementation of the SPRS for the period from 2011 to 2015 that elaborates three key areas of implementation monitoring: (i) strategic priorities/projects of the SPRS; (ii) indicators of spatial development as the starting point for monitoring and evaluation of spatial changes; and (iii) model of information system for monitoring and evaluation of the realization of the SPRS.

of use the zone that is protected as a source and a drinking water reserve, and that the non-constructed area between Vrnjacka Spa and other settlements should not be less than 1000 m as well as the conditions for building shelters; (b) the rules of construction include rules in terms of the distance between residential buildings, the rules for access to facilities, the resistance of shelters; and (c) measures that include protection of the population for protection and rescue, protection against war destruction, as well as protection and rescue from fires and explosions, medical first aid and radiological, chemical and biological (RHB) protection.

In the Special Purpose Area Spatial Plan (SPASP) of Stara Planina (2008), it was especially emphasized that the area of the Spatial Plan is exposed to the dangers of atmospheric accidents, escarpments and landslides, fire, earthquakes and other hazards. Also emphasized is the presence of climatic occurrences of storms, during winters with heavy winds, and over the summer with electric discharges, storms and torrents, especially in the peripheral parts of the basins and the foothills of the mountain. Special risks for this area are phenomena related to the protection of hydrogeological resources, i.e. protection of the quality of lake, river and spring waters. Of the remaining natural hazards, the most important are forest, plant and animal pests.

The SPASP of Stara Planina includes objectives in relation to the defense of the country and protection from natural hazards together with a proposal of a strategic framework for the use and arrangement of a space of interest for defense and for the use and arrangement of areas of interest for protection against natural hazards/disasters.

As pointed out in the Plan, "protection from accidents and natural hazards/disasters and the provision of conditions and measures for protection of the area are realized by: defining measures and procedures for possible accidents by a special program for management of water accumulation "Zavoj", by defining measures of protection against natural hazards/disasters and technical disasters, defining measures of protection from winter hazards/disasters (ice, avalanches, snow breaks, windmills and others) and from summer disasters (such as storms and torrents followed by landslides) that will be achieved by: constructing and arranging the planned supra and infrastructure; afforestation and grassing; water regulation; protection against ice and avalanches over winter; anti-erosion regulation of the basins to minimize the danger of torrential floods; planning of settlements, roads, watercourses and other spatial elements." Forest fires are marked as one of the most dangerous natural hazards/disasters that require the creation of a special protection plan. Further, the Plan considers the measures for protection from earthquakes, measures for protection from diseases of forests, harmful insects and rodents, control of wolf populations, foxes and other pests through determined forest and hunting grounds; defining measures of institutional organizations as a prerequisite for the adequate protection against natural hazards/disasters and providing conditions and measures for the protection of the area downstream from the water accumulation "Zavoj".

In PGR Bor (2018), it was pointed out that "a plan approach is based on a strategy of acceptable risk of present forms of risks, that is, in regard to the criterion that the increased costs of design, construction and financial investments that could prevent damages or demolitions and loss of human lives must not exceed the cost of eliminating the consequence of possible accidents." The Plan identifies the obligation to adopt a decision to update operational plans in order to comply with the new regulations and first of all to develop an operational annual plan for the protection from torrential floods, establish a fire protection plan and similar measures. The Plan establishes the following:

- Development of projects for possible consequences and measures for protection from natural hazards/disasters (accidents and disasters) for larger existing and planned industrial and cultural-historical facilities and complexes and facilities of large infrastructure and incorporate them into appropriate protection and rescue plans and individual risk assessment acts at the workplace and in the working environment (Law on Safety and Health at Work 2005);
- the functioning of an integrated system of protection, which implies the unification of all services dealing with the protection of citizens in emergency situations and the implementation of the action plan for the establishment of a unique service for emergency situations;
- development of a unified system of information and monitoring of the occurrence and protection from natural hazards/disasters, elaboration and implementation of a unique methodology for recording, collecting and storing documents on natural hazards/disasters, as well as assessing damage from natural and technical disasters in order to create a cadastral map of natural hazards/disasters within the territory of the municipality of Bor and the Bor administrative district as a whole.

The Plan also provides measures for the mitigation of risks and possible consequences including: (1) spatial measures (such as arranging adequate spaces and accession roads for access and movement of fire vehicles and for construction of a hydrant network), (2) construction and technical measures (to determine the state and the need for the reconstruction of buildings to be in compliance with fire protection) and (3) operational-organizational measures: (raising operational capability and equipping the firefighting unit with a vehicle with auto-mechanic ladders and hydraulic articulated platform, as well as training of residents).

As is evident in all the analysed documents, the issue of natural hazards/disaters is clearly present. Within the SPASP for Stara Planina the issue is especially present due to the plan coverage, which includes the area of the Nature park and therefore a greater possibility of the occurrence of a variety of natural hazards. Taking into account the period analysed, there is no significant difference within the plan content regarding natural hazards/disaters. However, due to the changes within the approach within DRR it can be noted that plans in the period after the year 2014 place emphases on the prevention and active participation of local governments and residents.

The Case of the Kolubara Lignite Basin

The Kolubara lignite basin (KLB) spreads over nearly 600 km^2, and coal from this basin is very important for the production of electrical energy in Serbia. Due to weather-induced natural hazard floods in mid-May 2014, the mining and energy facilities in the KLB suffered enormous damage, which resulted in the interruption of production. Due to heavy rain, the defensive embankments were demolished and the Kolubara,[7] Peštan and Vraničina rivers flooded. On occasion, the open pit mines "Veliki Crljeni" and "Tamnava zapadno polje" (Fig. 1), which in recent years have been providers of coal production throughout the KLB, have also been flooded.

The Kolubara River reached a floodplain of 80–200 m wide through the formerly regulated riverbed of the Vraničina river, and it entered the open pit mine "Tamnava western field". The outflow of the water into the open pit mine "Tamnava western field" was also due to the "Kladnica"[8] dam. The open pit mine "Veliki Crljeni" was flooded with water from the hinterland.

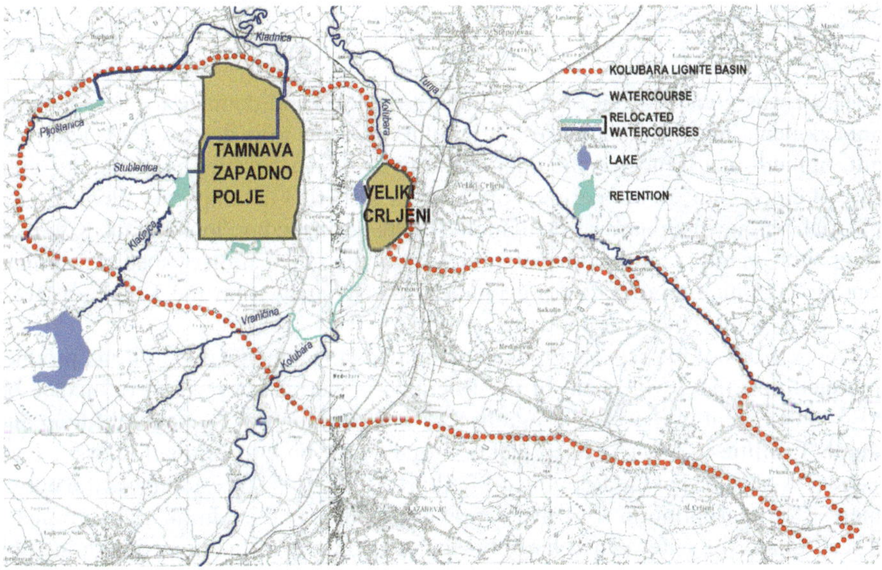

Fig. 1 Position of open pit mines "Veliki Crljeni" and the "Tamnava-western field" in the KLB (source: developed by the authors)

[7] According to *Preliminary assessment of floods*, determined by the competent ministry in accordance with Law on Water (2010) ordinances, the Kolubara River represents a significant flood area throughout its course.

[8] These retentions are part of the river regulation system and the control of its flood waters and their function is to completely or partially stop large waters and reduce flood waves. From the retention, collected water is controlled or pumped out into adjacent river catchments.

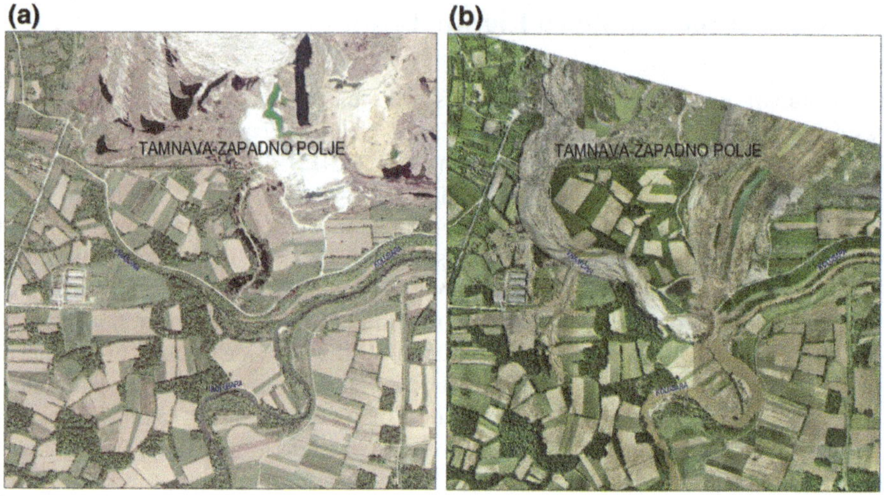

Fig. 2 The open pit mine Tamnava-western field and surroundings. A comparison of the orthophoto map from 2011 (**a**) and from 2014 (**b**)

There was great erosion of the Kolubara riverbed and damage to the defensive embankments, and the regulated estuaries of its affluents were destroyed. A defensive embankment that protected the open pit mines was destroyed. Under the declared state of emergency, in the period from May 17–20 2014, a temporary plug of local material, a stone of unknown granulation, was made on the floodplain of the Kolubara River, with an interventional installation without the satisfaction of special technical conditions, which temporarily prevented further contact from the Kolubara River into the open pit mine "Tamnava western field" (Fig. 2).

On the basis of the reports on the situation and the damage due to flood production facilities, the government of the Republic of Serbia adopted an ordinance on the establishment of the State Program for the Restoration of Damaged Mining and Supply Facilities (2014), in accordance with the Law on the elimination of the consequences of floods in the Republic of Serbia (2014). This law establishes the public interest for expropriation (of privately owned land), i.e. the administrative transfer of real estate (for state-owned land) for the construction of facilities determined by the state program of reconstruction, as well as temporary occupancy of the property for the execution of works covered by the renewal program. At the same time, the right of the user (i.e. a public company) has been determined to invest the works on the reconstruction of facilities owned by the Republic of Serbia, without special government authority.

Before the large floods, these areas were mostly used for agricultural production within individual properties and according to the needs of their owners. Through the development of planning documents for the subject area within the Spatial plan of the special purpose area for the exploitation of KLB (2017), on the covered areas acquired as public property, a new regimen of land use was established, and in connection with this, a new manner of water land use or land use for the construction of planned water facilities.[9] The current cadaster, identifying land parcels, has also been changed and agricultural land has been converted into water land, in accordance with the provisions of the Spatial Plan.

After the floods occurred, there was an increase in the characteristic flows within the entire KLB (which is a significant area of the Kolubara River basin), so that, in practical terms of frequency, 1000-year floods (before the May floods) became a 100-year flood. It can be concluded that for KLB it is necessary to actively apply protective measures in view of the assessment of all the risks involved. The degree of protection of this and the wider area should take acceptable risk into account, that is, the acceptable relationship between the increased investments in achieving a certain degree of protection and the estimated expected damages.

Conclusions and Recommendations for Planning Policy

Given the documented reports and recent evidence of weather trends, climate change induced disasters in Serbia are likely to rise. With climate change contributing to an increase in disaster risk, disaster risk management becomes a vital and urgent component of any climate change adaptation program (Pollner et al. 2010:2). So, in order to implement climate change adaptation policies and investments, Serbia needs to focus on reducing its vulnerability and planning for measures to mitigate natural disaster risks.

Land use planning, as one of the mechanisms that have an impact on the reduction of pressure on land and other resources (Zivanovic Miljkovic et al. 2012, Zivanovic Miljkovic and Popovic 2014), as well as building regulations and enforcement of urban plans and building codes are typically enacted to address mitigation measures. In addition, land use legislation and flood and landslide vulnerability mapping need urgent attention, as Serbia is highly prone to floods and landslides (UN-ISDR, WB 2008).

Continuous monitoring and innovation of the planning and legal basis are a prerequisite for establishing the conditions for an adequate response in case of natural hazard. The increase and more frequent occurrence of extreme temperatures, droughts and wildfires has conditioned the need to establish a system that, in addition to acting as a preventive, has the ability to respond quickly. As pointed out in this paper, the issue of natural hazards/disasters has been treated in the planning documents in Serbia, in accordance with the current legal basis, together with the protection of the country where, in line with the plan coverage, adequate measures and conditions are given. As has been noted, hazard becomes disaster when people and

[9]According to the provisions of the Law on Water (2010), waters are a good of public interest, and water management activities in the regulation of waters and watercourses and protection against harmful effects of water (flood protection) are considered to be activities of public interest, so it is possible to declare a public interest (expropriation) for water facilities.

their livelihoods are endangered. Therefore, in accordance with the current global framework that aims to act preventively, planning frameworks should include the reduction of exposure to hazards and the vulnerability of people and communities. Planning frameworks should also include the improvement of preparedness and an early warning system for potential risks where certainly "effective monitoring will provide not only an effective response and support to process development but also a permanent measure needed to protect vulnerable and risk zones from development and move people away from marginal areas (Crncevic and Lovren-Orlovic 2018).

The efforts that Serbia is making are in line with international and EU frameworks and expressed through activities on the development of its strategic framework. Thus, in the drafting process, they are, inter alia, a national inter-sectoral strategy for climate change, which will be in line with the EU framework for climate and energy policy for the period up to 2030. This, together with the adoption of a relevant law for disaster risk management, will provide conditions for establishing a planning basis that will provide an adequate response in both the prevention of and the response to disasters. Therefore, future prospects in the treatment of natural hazards within spatial and urban planning in Serbia can be identified in the conducting of measures within disaster risk management—with a frequent focus on reducing climate change vulnerability, together with developing guidance on an early warning system for potential risks, along with including all stakeholders and authorities.

Acknowledgements This chapter has been prepared within the scientific projects No. III 47014, TR36035 and TR 36036, financed by the Republic of Serbia Ministry of Education, Science and Technological Development in the period 2011–2019.

References

17 GOALS (2018). http://17goals.org/paris-agreement-sdgs/. Accessed 10 Nov 2018

American Planning Association (APA) (2011) Policy guide on planning and climate change. www.planning.org/policy/guides/pdf/climatechange.pdf. Accessed 15 Oct 2018

Bazik D, Dzelebdzic O (2012) Web-based support of spatial planning in Serbia. Spatium 28:67–73

Birkmann J (2018) Urban and Spatial planning in natural hazard risk management, Oxford research encyclopedia, natural hazard science. oxfordre.com/ naturalhazardscience, Oxford University Press, USA. http://oxfordre.com/naturalhazardscience/view/10.1093/acrefore/407.001.0001/acrefore-9780199389407-e-8?print = pdf. Accessed 15 Jan 2019

Crncevic T, Lovren-Orlovic V (2018) Displacement and climate change: improving planning policy and increasing community resilience. Int J Clim Chang Strat Manag 10(1):ISSN 1756–8692

Dzelebdzic O, Bazik D (2011) National indicators for evaluating the outcome of reinventing spatial planning in Serbia. Spatium 24:27–36

European Commission (EC) (2018) Commission staff working document Serbia 2018 report accompanying the document communication from the commission to the European parliament. The council, the European economic and social committee and the committee of the regions 2018 communication on EU enlargement policy. https://ec.europa.eu/neighbourhood-enlargement/sites/near/files/20180417-serbia-report.pdf. Accessed 15 Oct 2018

European Environment Agency (EEA) (2018) https://www.eea.europa.eu/data-and-maps/indicators/direct-losses-from-weather-disasters-3/assessment-1. Accessed 15 Jan 2019

Francesch-Huidobro M, Dabrowski M, Tai Y, Chan F, Stead D (2017) Governance challenges of flood-prone delta cities: Integrating flood risk management and climate change in spatial planning. Progress in Planning 114:1–27

Intergovernmental Panel on Climate Change (IPCC) (2014) Climate change 2014: synthesis report, in climate change 2014 synthesis report summary for policymakers. www.ipcc.ch/pdf/assessment-report/ar5/syr/AR5_SYR_FINAL_SPM.pdf. Accessed 15 Oct 2018

International Federation of Red Cross and Red Crescent Societies (IFRC). https://www.ifrc.org/en/what-we-do/disaster-management/about-disasters/definition-of-hazard/. Accessed 15 Oct 2018

International Strategy for Disaster Reduction ISDR (2018) https://www.unisdr.org/who-we-are/international-strategy-for-disaster-reduction. Accessed 10 Dec 2018

Johannesburg Plan of Implementation of the World Summit on Sustainable Development (2002) http://www.un-documents.net/jburgpln.htm. Accessed 25 Oct 2018

Nicholls RJ, Lowe J (2004) Benefits of migration of climate change for coastal area. Glob Environ Change 14(3):220–244

Pollner J, Kryspin-Watson J, Nieuwejaar S (2010) Disaster risk management and climate change adaptation in Europe and central Asia. Global facility for disaster reduction and recovery. The World Bank, Washington

Republic of Serbia (RS), Ministry of Environmental Protection (2017) Drugi nacionalni izveštaj Republike Srbije prema Okvirnoj konvenciji UN o promeni klime (Second National Communication of the Republic of Serbia to the UN Framework Convention on Climate Change). http://www.klimatskepromene.rs/wp-content/uploads/2017/12/Drugi-izvestaj-o-promeni-klime-SNC_Srbija.pdf, Accessed 28 Oct 2018

Stults M (2017) Integrating climate change into hazard mitigation planning: opportunities and examples in practice. Clim Risk Manag 17:21–34

Surjan A, Rahman A, Ara Parvin G, Shaw R (2016) Urban regulation and enforcement: a challenge in urban disasters and resilience in Asia. In: Shaw R, Rahman A, Surjan A, Ara Parvin G (eds), Elsevier. https://ac.els-cdn.com/B9780128021699000057/3-s2.0-B9780128021699000057-main.pdf?_tid=62fe2136-3cfd-43b0-a98e-7c9bdd71b74c&acdnat=_ec550601296006c67b32ae715e1dbbef. Accessed 15 Jan 2019

Tozier de la Poterie A, Baudoin MA (2015) From Yokohama to Sendai: approaches to participation in international disaster risk reduction frameworks. Int J Disaster Risk Sci 6:128–129

UN (United Nations)-ISDR, WB (World Bank) (2008) South eastern Europe disaster risk mitigation and adaptation initiative: risk assessment in South Eastern Europe—a desk study review. United Nations, Geneva, Switzerland

UNDP, Agency for Environmental Protection (2007) Indikatori životne sredine u Republici Srbiji, Kratak vodič (Environmental Indicators in Republic of Serbia, Short guide), Beograd

UNFCCC 2018. https://unfccc.int/news/wmo-report-the-escalating-impacts-of-climate-related-natural-disasters. Accessed 10 Nov 2018

UNISDR (2005) Hyogo framework for action 2005–2015, https://www.unisdr.org/we/inform/publications/1037. Accessed 25 Sep 2018

UNISDR (2015). Sendai framework for disaster risk reduction 2015–2030. Available at: www.unisdr.org/files/43291_sendaiframeworkfordrren.pdf. Accessed 10 Nov 2018

UNISDR (2018a). https://www.unisdr.org/we/inform/publications/8241. Accessed 10 Nov 2018

UNISDR (2018b) The united nations office for disaster risk reduction: terminology. www.unisdr.org/we/inform/terminology. Accessed 15 Sep 2018

United Nations (UN), Paris Agreement (2018). http://unfccc.int/paris_agreement/items/9444.php. Accessed 25 Sep 2018

United Nations Development Programme (UNDP) (2016) Human development report 2016: risk proofing the western Balkans: empowering people to prevent disasters, UNDP, Sarajevo. Available at: http://hdr.undp.org/sites/default/files/risk_proofing_the_western_balkans.pdf. Accessed 10 Sep 2018

Zivanovic Miljkovic J, Crncevic T, Maric I (2012) Land use planning for sustainable development of peri-urban zones. Spatium 28:15–22

Zivanovic Miljkovic J, Popovic V (2014) Land use regulation and property rights regime over land in Serbia. Spatium 32:22–27

National Legislation and Planning Documentation

Energy Strategy of the Republic of Serbia for the period until 2025 with projections until 2030 (2015) Official Gazette of the Republic of Serbia, No.101/2015

General Urban plan of Vrnjacka Spa, Official Gazette of the Kraljevo Municipality, No.12/2005

Law on Emergency Situations (2009) Official Gazette of the Republic of Serbia, No.111/2009, 92/2011, 93/2012

Law on Fire Protection (2015) Official Gazette of the Republic of Serbia, No.111/2009, 20/2015

Law on Insurance (2014) Official Gazette of the Republic of Serbia, No.139/2014

Law on Local Government (2007) Official Gazette of the Republic of Serbia, No.129/2007,83/2014

Law on Meteorogical and Hydrological Activities (2010) Official Gazette of the Republic of Serbia, No.88/2010

Law on Planning and Construction (2009) Official Gazette of the Republic of Serbia, no. 72/2009, 81/2009—correction, 64/10- constitutional court decision, 24/11,121/12, 42/13- constitutional court decision, 50/13- constitutional court decision, 98/13- constitutional court decision, 132/14, 145/2014, 83/2018, 31/2019 and 37/2019 - other Law

Law on Safety and Health at Work (2017) Official Gazette of the Republic of Serbia, no.101/2005, 91/2015 and 113/2017 - other Law

Law on Water, Official gazette of the Republic of Serbia (2010) no.30/2010, 93/2012, 101/2016, 95/2018 and 95/2018 - other Law

Law on the Elimination of the Consequences of Floods in the Republic of Serbia (2014) Official Gazette of the Republic of Serbia, no. 75/2014, 64/2015 and 68/2015

National Strategy for Disaster Risk Reduction and Protection and Rescue in Emergency Situations (2011) Official Gazette of the Republic of Serbia, No. 86/2011

National Strategy for Sustainable Development and Action Plan for the period 2009–2017 (2008) Official Gazette of the Republic of Serbia, No. 57/2008

National Strategy for Inclusion of the Republic of Serbia in the Clean Developmnent Mechanism (2010) Ministry of Environmental and Spatial Planning of Serbia

National Program for Environmental Protection for the period 2010-2019 (2011) Official Gazette of the Republic of Serbia, No.13/2011

National Strategy for Sustainable Use of Natural Resources and Goods (2012) Official Gazette of the Republic of Serbia, No.33/2012

Ordinance on the establishment of the state program for the restoration of damaged mining and supply facilities (2015) Official Gazette of the Republic of Serbia, No. 112/2014 and 71/2015

Plan of general regulation for Bor (2018) Official Gazette of Bor municipality, No.1/2018

Program implementacije Prostornog plana Republike Srbije od 2010. do 2020. godine za period od 2011.do 2015. godine (2011) Ministarstvo životne sredine i prostornog planiranja Republike Srbije, Beograd

Spatial Plan of the Special Purpose for the Area of the Nature park Stara Planina (2008) Official Gazette of the Republic of Serbia, No. 115/2008

Spatial plan of the Republic of Serbia 2010–2020 (2010), Official Gazette of the Republic of Serbia, No. 88/2010

Spatial Plan of the Special Purpose area for the exploitation of Kolubara lignite basin (2017) Official Gazette of Republic of Serbia, No. 107/2017

Strategy of Biodiversity for the period 2011–2018 (2011) Official Gazette of the Republic of Serbia, No.13/2011

Occurrence and Characteristics of Flash Floods in Bavaria (Germany)

Maria Kaiser, Marco Borga and Markus Disse

Abstract Due to the global warming, an increase in the number and intensity of flash flood events triggered by heavy rain is expected. Flash flood risk management requires the identification of regions where the occurrence of flash flood events can be expected to become more frequent. In this study, we shed light on the occurrence and characteristics of flash floods in the state of Bavaria, Germany. A discharge time series of 342 gauging stations (average measurement period 1960–2017) was investigated with regard to the past flash flood events. The catchments of the selected gauging sites range from 1 to 500 km^2 and are not affected by flood retention structures. Only half of the chosen gauging stations recorded flood events, which showed a short response time and a significant flood magnitude that were typical for flash floods. This led us to the conclusion that some catchments are more prone to flash floods than others. It was found that the flash flood occurrences are not evenly distributed over Bavaria, but rather form spatial clusters in the alpine region and the Bavarian Forest. The statistical evaluation of 5,732 flash flood hydrographs for different catchment sizes further deepened our understanding of the characteristics of this natural hazard in Bavaria.

Introduction

Flash floods pose a serious threat to humans. Triggered by heavy rain events, flash floods can generate rapidly, typically within a few hours. As the triggering heavy

M. Kaiser (✉) · M. Disse
Chair of Hydrology and River Basin Management, Technical University of Munich, Arcisstraße 21, 80333 Munich, Germany
e-mail: maria.kaiser@tum.de

M. Disse
e-mail: markus.disse@tum.de

M. Borga
Department of Land, Environment, Agriculture and Forestry, University of Padova, AGRIPOLIS, via dell'Università 16, Legnaro (PD) 35020 Padua, Italy
e-mail: marco.borga@unipd.it

W. Leal Filho et al. (eds.), *Climate Change, Hazards and Adaptation Options*,
Climate Change Management, https://doi.org/10.1007/978-3-030-37425-9_16

rain events are usually small-scale and highly variable, the prediction and warning of flash floods is difficult. Due to the excessive hydrodynamic forces of the fast-flowing water, flash floods can cause a substantial amount of damage. Furthermore, the rapidity of flash flood events aggravates the timely protection efforts. In addition, structural protection measures have only a minor effect, since flash floods can also occur far from the watercourse. For these reasons, flash floods are among the most dangerous natural hazards. Although flash floods are rather localized phenomena, the average annual loss from the many flash flood events is comparable to that of a 100-year flood on major rivers (Munich Re 2005).

With climate change and the projected global warming, a general increase in rainfall amounts is expected. According to the IPCC (2014), some regions of the world will have to deal with more intense and frequent heavy precipitation events, which may trigger severe flash flooding. At a regional scale, the risk of flooding may increase as a consequence of increasing the extreme precipitation and discharges in some basins (IPCC 2014). Climate models predict a significant increase in heavy rainfall events for Europe (Kovats et al. 2014). For Northern Europe, the climate projections show increased extreme precipitation for all seasons while for Continental Europe, for all seasons except summer (Kovats et al. 2014). This is in accordance with the findings of a climate study for southern Germany. According to KLIWA (2005), until 2050, the precipitation in Bavaria will increase significantly in winter while smaller changes are expected in the summer months. Due to the impacts of climate change, an increase in the severity and frequency of the flash flood hazard is expected (Kleinen and Petschel-Held 2007; Beniston et al. 2011).

As precise forecasts of flash floods are almost impossible to produce, the understanding of the flash flood generation is all the more important. In terms of precaution, it is crucial to identify endangered catchments. Over the past decade, our knowledge about flash floods has improved. However, the involved processes and the catchment characteristics promoting the flash flood generation are still not fully understood. Due to the small spatial scales, the flash floods are controlled by the local conditions and the climatic region. Gaume et al. (2009) found, for instance, that Mediterranean flash floods tend to last longer and are more intense than flash floods in Continental Europe. Thus, to fully understand the controlling processes in our study region, the past flash floods of this particular region need investigation. To date, several studies have been carried out on past flash flood events in the Euro-Mediterranean region (e.g., Gaume et al. 2009; Boudevillain et al. 2011; Braud et al. 2014; Amponsah et al. 2018). Yet, the Continental flash floods have rarely been included in those studies.

In May and June 2016, a series of flash flood events occurred in the state of Bavaria, Germany that was new in terms of the number and severity. Although flash floods have occurred in Bavaria before, only the flash flood events of 2016 caught the attention of politics and society. Until that time, this natural hazard was not being urgently investigated in Bavaria. To improve understanding of flash floods and to thus contribute to precautionary efforts, the Bavarian State Ministry of the Environment and Consumer Protection funded the HiOS project. The objective of the HiOS project is the development of a method for the evaluation and classification of the endangerment of the Bavarian municipalities by surface runoff and flash floods.

As part of the HiOS project, we aimed at improving our understanding of the natural hazard flash flood, with focus on the state of Bavaria, Germany. Our goal was to find out where, when, and how often flash floods occur and how fast and intense they are.

This paper investigates the occurrence and characteristics of Bavarian flash floods based on a discharge analysis of 342 stream gauges. It is organized as follows. Section "The flash flood events in May and June 2016 in Bavaria" summarizes the occurrences of flash floods in Bavaria in 2016. Section "Flash flood identification based on discharge measurements" explains the methodology of the flash flood identification from discharge measurements. Section "Occurrence and characteristics of flash floods in Bavaria" describes the occurrence and characteristics of Bavarian flash floods. The findings are discussed in Section "Discussion", and concluding remarks are given in Section "Conclusion".

The Flash Flood Events in May and June 2016 in Bavaria

In May and June 2016, the federal state of Bavaria (Germany) was hit by a series of flash flood events. The following summary of the flash flood events is based on the report by the LfU (2017). Over several weeks, local heavy precipitation events occurred in alternation with continuous rain in different Bavarian regions. The period from May 27 to June 26, 2016 was particularly rainy. In these four weeks, the rainfall summed up to more than 300 mm in some southern regions of Bavaria. Due to an exceptional weather situation, small-scale stationary storm cells formed. As the soils had already been saturated by rain in the previous days, large amounts of the heavy rain turned directly into surface runoff. Within no time, flash floods occurred, overflowing the riverbanks. Although the flash floods were spatially very limited, they caused enormous damage to property. Seven people were surprised by the rapidly rising water levels and died. In total, 31 districts across Bavaria reported damages due to the heavy precipitation events.

As a particularly catastrophic flash flood event, the event at Simbach am Inn on June 1, 2016 went down in Bavarian history. Simbach am Inn is a town in Lower Bavaria with approximately 10,000 inhabitants. In this town, the Simbach flows into the Inn River, which is the natural border to neighboring Austria. The Simbach is a torrent with a 33 km^2 large catchment. When the heavy precipitation hit the Simbach catchment, the soils were already saturated by intensive rains from the previous days. Within six hours, it rained 111 mm on the entire catchment. The convective cell stayed above the Simbach catchment for a long time, slowly moving in the flow direction. The combination of saturated soils and heavy rain led to a very rapid concentration of runoff with an extremely high peak discharge in Simbach am Inn. In addition, a sizable water mass ponded at a road embankment outside of town. As the drainage culvert could not pass the water mass through the road embankment rapidly enough, a water reservoir of approximately 120,000 m^3 formed. During the flash flood event, the road embankment was overtopped, eroded, and finally failed.

In addition, a lumber yard area below the road embankment was washed out, leading to blockages and severe damages in the town of Simbach am Inn. The impounding dams along the Inn led to an additional aggravation of the flood situation, as the dams prevented the water masses from flowing into the Inn. The event peak discharge was estimated to be 180–190 $m^3\ s^{-1}$ prior to the dam failure. With the dam failure, the peak discharge reached between 280 and 300 $m^3\ s^{-1}$. To put this discharge in perspective, the Simbach's channel can accommodate 60 $m^3\ s^{-1}$ without flooding. A flood volume of approximately 3.3 million cubic meters was reconstructed for the Simbach catchment, qualifying it as an extraordinary flood event. The maximum measured water level was 5.06 m.

Many people had to be evacuated or rescued from extreme danger. For seven people, the help came too late, and they perished in the floods. Due to the risk of collapse, many people could not return into their houses after the flash flood. The drinking water supply and the sewage disposal system were disturbed. Furthermore, important transportation routes were damaged, causing severe traffic problems.

It was only the large number of flash flood events in May and June 2016 in Bavaria, and especially the Simbach event, that has increased the awareness of natural hazards in Bavarian society and politics. Also, other federal states of Germany were affected by flash floods. The towns of Braunsbach and Schwäbisch-Gmünd in the State of Baden-Württemberg gained sad notoriety, as they were also affected by extreme flash floods. Recently, the flash flood of Braunsbach has been intensively studied regarding the meteorological, hydrological, and geomorphological aspects (Laudan et al. 2017; Bronstert et al. 2018; Lucía et al. 2018; Ozturk et al. 2018).

Flash Flood Identification Based on Discharge Measurements

The discharge measurement of flash floods is difficult. As flash floods are rapid and severe floods, the damage or failure of the measuring instruments during the event is a common occurrence. Furthermore, the validity of the water level-discharge relationship may be questionable in cases of extreme bank overflow. Due to the rapidity of flash flood events, an hourly measurement may not adequately represent the event, especially in very small catchments. An underestimation of the peak discharge is thus probable. In addition, the true shape of the flash flood hydrograph may be missed, as gauging stations are normally installed in bigger catchments. The discharge signal of the event may be weak or absent at the downstream gage due to translation and retention of the flood wave. It is often the case, however, that the catchment affected by the flash flood is ungauged.

Despite all these possible difficulties, a discharge measurement of a flash flood event is still the most valuable information we may have. The event hydrograph provides information about the speed, magnitude, and course of the flash flood. In combination with the precipitation data, detailed information can be obtained not only into the generation of the event, but also into the catchment behavior. In contrast to event information from newspaper articles or eyewitness reports, these discharge measurements are quantitative and unbiased.

In this study, we investigated the characteristics of flash floods in Bavaria (Germany) based on hourly discharge measurements. For this purpose, we separated the flood events from the hydrographs of suitable stream gauges. From this flood dataset, flash floods based on discharge thresholds were selected and analyzed. Our investigation of the flash floods was focused on the description of their occurrence and characteristics. The present study was conducted between August and October 2018.

Selection of Suitable Gauging Stations

The state of Bavaria operates a dense discharge measurement network. More than 700 gauging stations cover an area of 70,550 km^2. Within the scope of the HiOS project, the Bavarian Environment Agency provided the measurement time series of 716 Bavarian gauging stations. To ensure the general validity of the evaluation, the catchments that are unaffected by impoundments and prone to flash flooding were exclusively investigated. Consequently, the gauging stations meeting one or more of the following criteria were disregarded.

- Gauging stations with a catchment area greater than 500 km^2
- Gauging stations of lakes or channels
- Catchments with upstream lakes, which make up more than 10% of the catchment area
- Catchments influenced by water- or energy-supply dams and flood control reservoirs

GKD (2018) and Skublics et al. (2016) provided information about the Bavarian lake gauging stations and flood retention reservoirs. Following the above mentioned criteria, approximately half of the available gauging stations were considered suitable for the investigation (Fig. 1). In total, 342 gauging stations were selected and subdivided into five catchment classes for further analyses.

$\leq$25 km^2	42 Gauging stations
>25–50 km^2	51 Gauging stations
>50–100 km^2	90 Gauging stations
>100–250 km^2	99 Gauging stations
>250–500 km^2	60 Gauging stations

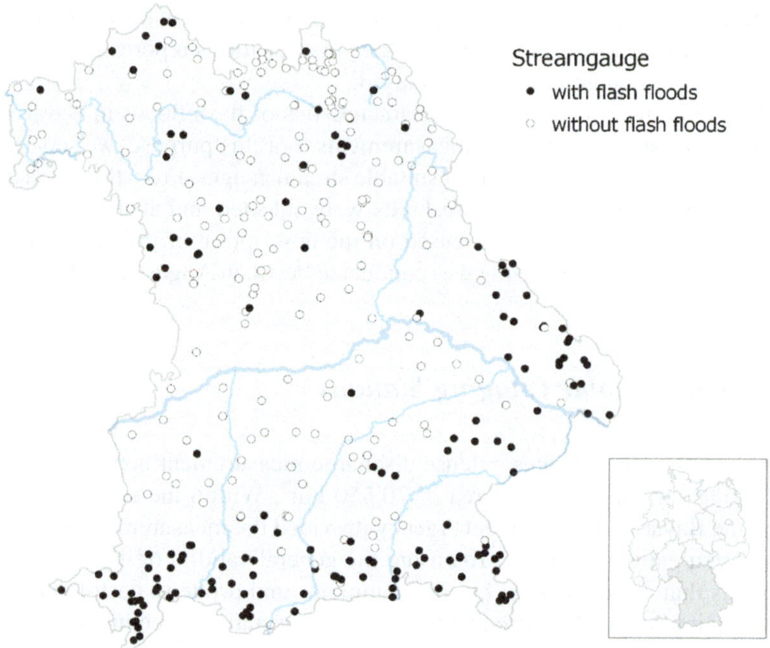

Fig. 1 342 stream gauges in Bavaria (Germany) were considered suitable for the discharge analysis regarding flash floods. Approximately half of the investigated gauging stations measured flash floods corresponding to our thresholds (unit peak discharge $\geq$0.3 m^3 s^{-1} km^{-2}, time-to-peak $\leq$20 h, period of occurrence, April–October)

The majority of the gauging stations are located in medium to large catchments (50–250 km^2). The fewest stream gauges are available for the rather small catchments (up to 50 km^2), which make up only half of the other catchment classes.

Flood Identification from Discharge Time Series

We investigated the hourly discharge data of 342 Bavarian stream gauges regarding flash floods. For each of the 342 gauges, we evaluated the total measurement period for which hourly values were available, up to and including June 2017. On average, the hourly discharge measurements started in 1960, so we were able to analyze approximately 60 years per gauging station. We identified the flood events in the hydrograph following the procedure reported by Merz et al. (2006). First, the hydrograph was slightly smoothed to minimize the influence of erroneous individual values. Then, for the flood identification, the hydrograph was separated into direct runoff and base flow by applying the filter equation described by Chapman and Maxwell (1996). Subsequently, the flood events were determined according to the method of Merz et al. (2006), which involves filtering the hydrograph data and determining the beginning and end of the event by the changes in peak discharge. For further analyses, we derived descriptive parameters, such as the time-to-peak (the time difference between the beginning and peak of the flood event) or the specific peak discharge (the ratio of the peak discharge to the upstream catchment area) for each flood event.

Determination of Flash Flood Events

Flash floods tend to occur on small spatial and temporal scales. The sudden response of a catchment to a short, but high-intensity rain event is a characteristic feature of a flash flood. However, the flash flood regime varies depending on the climatic region. Gaume et al. (2009) stated that Mediterranean flash floods are generally more intense than those in the Continental region of Europe. According to Gaume et al. (2009), the Continental events tend to have a smaller spatial extent and a shorter duration compared to the Mediterranean flash floods. As the threshold for the European HYDRATE flash flood database, Gaume et al. (2009) applied a maximum catchment size of 500 km^2, whereas Marchi et al. (2010) consider basins up to 1,000 km^2 as being potentially affected in their study of the European flash floods. In the EuroMedeFF database, flash floods originating from the catchments up to 3,000 km^2 were included (Amponsah et al. 2018). Regarding the rapidity of flash floods, Braud et al. (2014) assumed a time-to-peak of a few hours or less for catchments up to 100 km^2 and a time-to-peak less than 24 h for catchments of approximately 1,000 km^2. For flash floods in the US, Georgakakos and Hudlow (1984) defined a time-to-peak of 6 h for catchments up to 400 km^2. Other authors founded their definition on the storm duration triggering the flash flood instead of the time-to-peak. Gaume et al. (2009), for instance, set the maximum storm duration for European flash floods to 24 h, and Amponsah et al. (2018) used 48 h.

From all identified flood events, those with certain flash flood characteristics, namely, short response time and significant flood magnitude, were selected. By selecting the gauging stations, the events had already been limited to a catchment size of 500 km^2 or less. The analysis of past Bavarian flash flood events has shown that the upper limit of 500 km^2 is reasonable, and this is also confirmed by the results of this investigation. The determination of the flash flood events was based solely on the discharge measurements. For simplicity, we applied the same thresholds all over Bavaria. The character of a flash flood varies strongly depending on the precipitation event, the catchment properties, and the prevailing catchment conditions. Thus, to capture the possible variability, the applied thresholds should not be too strict. Still, a clear distinction between river flood or flash flood is difficult, as the transition is seamless. Therefore, the determination of a flash flood event remains a matter of definition. Accordingly, an event was defined as a flash flood based on the following criteria:

- Time-to-peak less than or equal to 20 h
- Specific peak discharge greater than or equal to 0.3 $m^3\ s^{-1}\ km^{-2}$
- Occurrence period from April to October

Braud et al. (2014) and Amponsah et al. (2018) applied a unit peak discharge of 0.5 $m^3\ s^{-1}\ km^{-2}$ as the lowest value for defining a flash flood event in Europe. However, we lowered the unit peak discharge to 0.3 $m^3\ s^{-1}\ km^{-2}$, since we noticed that the higher value excluded many events that were considered relevant.

Occurrence and Characteristics of Flash Floods in Bavaria

Flash Flood Occurrence and Frequency

We investigated the hydrographs of 342 stream gauges for flash floods. The analysis showed that approximately half of the gauges measured flash floods in the past (Fig. 1). In total, we discovered 5,732 flash flood events for 157 gauging stations. The stream gauge selection covered Bavaria quite homogeneously. However, the stream gauges having measured flash floods are not distributed equally over Bavaria, but form spatial clusters. One cluster is found in the alpine range, where fast-reacting floods are generated easily due to steep slopes and high precipitation amounts. Also, in the pre-alpine region, south of the Danube, some gauging stations measured flash floods. Another flash flood hot spot is in the low mountain range of eastern Bavaria while northern Bavaria has some affected catchments, as well. Central Bavaria, however, seems to be less affected by flash floods. Still, the gauges are also more sparsely distributed in this region, especially around the Danube River.

The number of stream gauges per studied catchment class was not the same in all catchment classes. We investigated the most stream gauges for catchment areas between 50 and 250 km^2. For the catchments smaller than 25 km^2, the fewest gauging stations were available. Still, we identified the most flash flood events for the smallest catchment class. On average, there were 1.3 flash flood events per year and gauging station in catchments smaller than 25 km^2 (Fig. 2). For catchments larger than 250 km^2, however, a flash flood was measured on average every 5 years. With increasing catchment size, the number of flash floods per year and station decreased significantly. It was noticed that large catchments in Bavaria experience flash floods, but they are clearly less often affected. In this regard, the chosen upper boundary of 500 km^2 for the investigation is considered appropriate. This catchment threshold may also apply to other Inland Continental regions of Europe.

The catchments in which flash flood events have been measured tend to cluster in different natural regions of Bavaria (Fig. 3). Most frequently affected are the catchments in the alpine and pre-alpine regions. The Alps are characterized by small and steep catchments that are fast-reacting to heavy rain, as they encourage rapid runoff concentration. In addition, the amounts of precipitation in the Alps are high due to the orographic effects of the mountains. Therefore, the majority of the alpine catchments

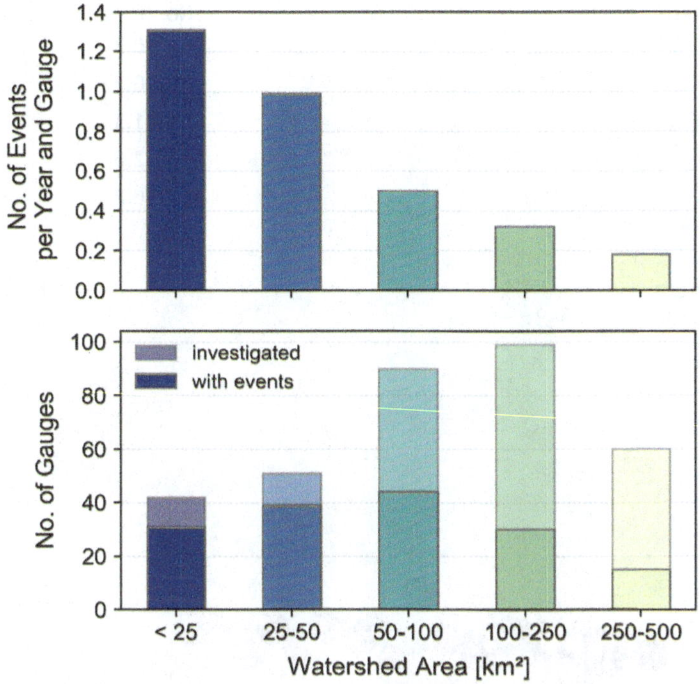

Fig. 2 The number of identified flash floods per year and station decreases with increasing catchment area. On average, there were 1.3 flash flood events per year and gauging station in catchments smaller than 25 km^2. For catchments larger than 250 km^2, however, a flash flood was measured on average every 5 years

experience several flash floods per year. In the Bavarian Forest (Eastern Mittelgebirge), there is also a cluster of affected catchments. Yet, the flash floods occur here significantly less frequently than in the Alps. In the South-Western Mittelgebirge, the affected gauging stations measured a flash flood every three years on average. In general, flash floods occurred more often in regions with high elevation differences, as for example, in the Alps or the low mountain ranges of the Mittelgebirge. However, flash floods were also measured in rather flat areas, such as in northern Bavaria along the Main River.

Classification into the European Context

The catchment area influences the peak discharge. To make the peak discharges of different catchments comparable, the unit peak discharge was used. The unit peak discharge is calculated as the ratio of the peak discharge to the upstream catchment area. The relationship of catchment size to unit peak discharge, usually plotted in

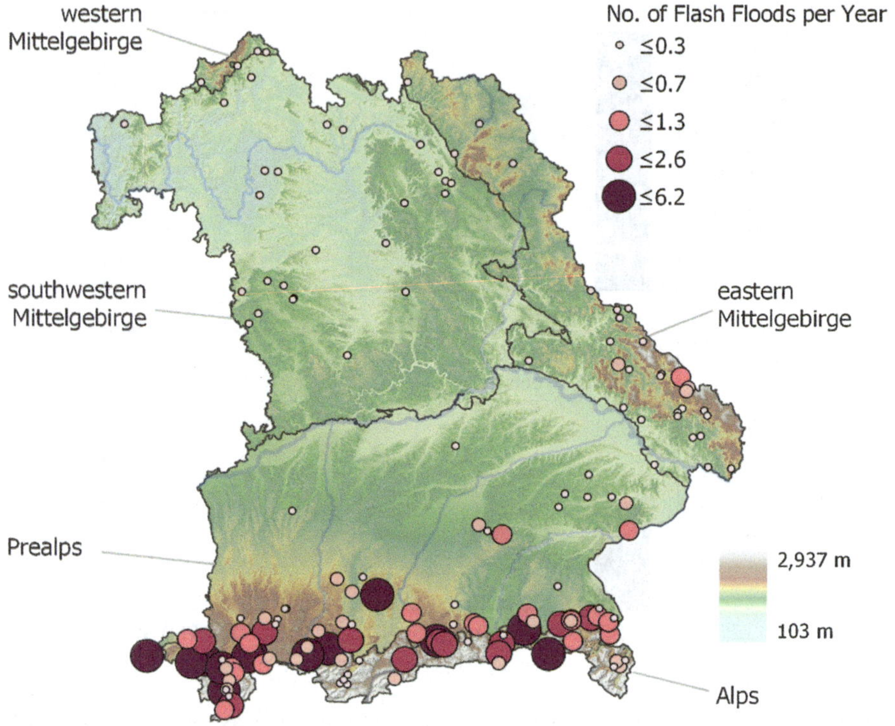

Fig. 3 Number of identified flash flood events per gauging station together with the elevation and the five natural regions of Bavaria. The stream gauges with flash floods reveal spatial clusters of affected regions. By far the most events were found in the alpine region. Flash floods also occurred in the South-Western and Eastern Mittelgebirge. Here the Bavarian Forest is especially affected

a log-log diagram, is often used to investigate the upper boundary of flash floods (Gaume et al. 2009; Borga et al. 2011; Tarolli et al. 2012; Amponsah et al. 2018). Amponsah et al. (2018) empirically derived the upper bounds for flash floods in different climatic regions of Europe based on 680 peak discharge data (Fig. 4). Compared to the upper bounds of the Mediterranean flash floods, the Bavarian events reach significantly lower unit peak discharges. This difference in intensity is due to the different spatial and temporal scales of the triggering rain events (Borga et al. 2011). Our Bavarian dataset approaches the upper boundary of the Inland Continental flash floods reported by Amponsah et al. (2018) and seems to confirm the envelope curve. In the following, we list the maximum unit peak discharge measured per catchment class:

$\leq$25 km^2	4.2 $m^3\ s^{-1}\ km^{-2}$
>25–50 km^2	5.8 $m^3\ s^{-1}\ km^{-2}$

(continued)

(continued)

>50–100 km^2	2.4 m^3 s^{-1} km^{-2}
>100–250 km^2	1.9 m^3 s^{-1} km^{-2}
>250–500 km^2	1.5 m^3 s^{-1} km^{-2}

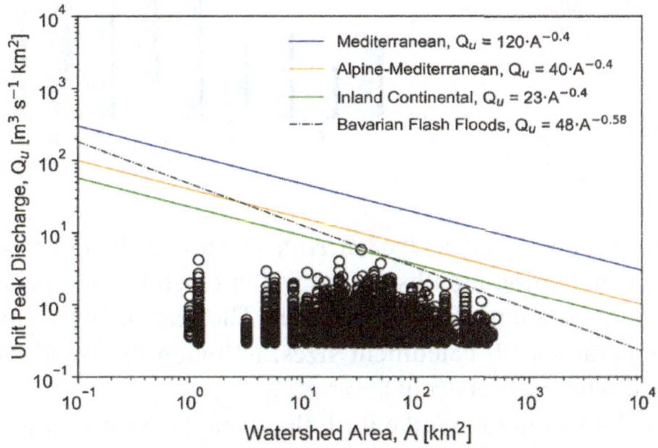

Fig. 4 Unit peak discharge versus watershed area for the studied flash floods based on hourly discharge measurements. Flash floods in Bavaria (Germany) reach significantly lower unit peak discharges at maximum than Mediterranean flash floods. Amponsah et al. (2018) empirically derived the envelope curves for Mediterranean, Alpine-Mediterranean, and Inland Continental flash floods based on the European EuroMedeFF database. The envelope curve for the Bavarian flash floods was derived from the present study results

Our flash flood dataset shows the expected decline in unit peak discharge with increasing catchment size. The highest measured unit peak discharge is the 5.8 m^3 s^{-1} km^{-2} of the Simbach flash flood in 2016. We fitted an envelope curve to the maximum values of our Bavarian sample. However, our envelope curve shows a more pronounced decrease of unit peak discharge with catchment size than the Inland Continental curve proposed by Amponsah et al. (2018).

Seasonal Distribution of Flash Floods

We investigated the seasonality of the Bavarian flash floods based on the derived dataset. Figure 5 shows the number of flood events per month with a short rising time ($\leq$20 h) and a specific peak discharge greater than or equal to 0.3 m^3 s^{-1} km^{-2}. The Bavarian dataset reveals the influence of the season on the flash flood occurrence. It was observed that the flash floods occur mainly from May to September, with most of the flash floods taking place in July. During the winter months, fast floods occur too, but mainly in the catchments up to 100 km^2. However, these floods are probably

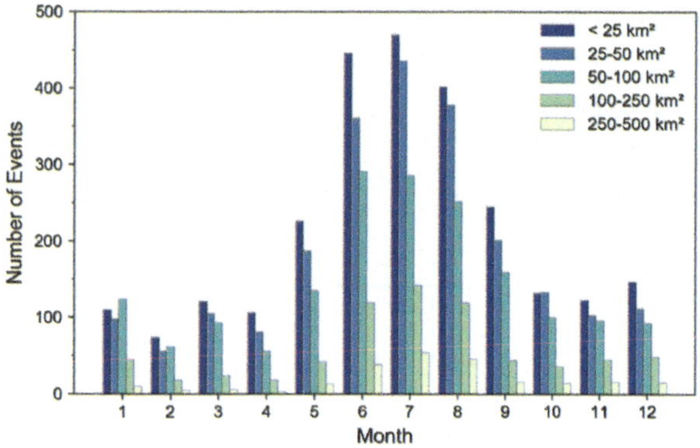

Fig. 5 Number of flood occurrences per month with a short rising time ($\leq$20 h) and a specific peak discharge $\geq$0.3 m^3 s^{-1} km^{-2}. Most of the flash floods occurred between May and September, with April and October as transition months to winter flood events

caused by rain on snow, rain on frozen soil, or snowmelt and thus they are not flash floods. In the transition months, i.e., April and October, the prevailing weather conditions determine if a flash flood will occur. The seasonality signal for Bavarian flash floods is similar for all catchment sizes, although the small catchments also show pronounced winter flood events.

Gaume et al. (2009) found that the flash floods in the Mediterranean and Alpine-Mediterranean region occur mostly in autumn while those in the Inland Continental region commonly occur in summer. Gaume et al. (2009) attributed the shift in flash flood seasonality across Europe to different meteorological flood-triggering events. As the Bavarian flash floods occur mainly in summer, the Bavarian seasonality fits well to that of the other central European countries (Austria, Slovakia, and Romania) investigated by Gaume et al. (2009).

Figure 6 shows the month with the most flash flood events per stream gauge. We note that the seasonality of flash floods varies spatially over Bavaria. The alpine region, for instance, is dominated by flash flood events occurring in June and July, whereas northern Bavaria experiences the most events from May to June. The Bavarian Forest in eastern Bavaria, in contrast, shows no pronounced seasonality. Here, even catchments in proximity can have a very different seasonality. To summarize, northern Bavaria is mainly affected by flash floods in late spring, whereas southern Bavaria experiences more flash floods in midsummer. The difference in seasonality across Bavaria reveals that the flash floods in the different natural regions are not triggered by the same types of meteorological events.

Flash Flood Magnitude and Speed

The magnitude of the flash flood events in Bavaria varies. Figure 7 illustrates the maximum unit peak discharge measured per stream gauge. It was observed that the flash floods with the highest flood magnitude occurred in the alpine and pre-alpine regions. Most Bavarian flash floods, however, do not exceed 1 m^3 s^{-1} km^{-2}. In total,

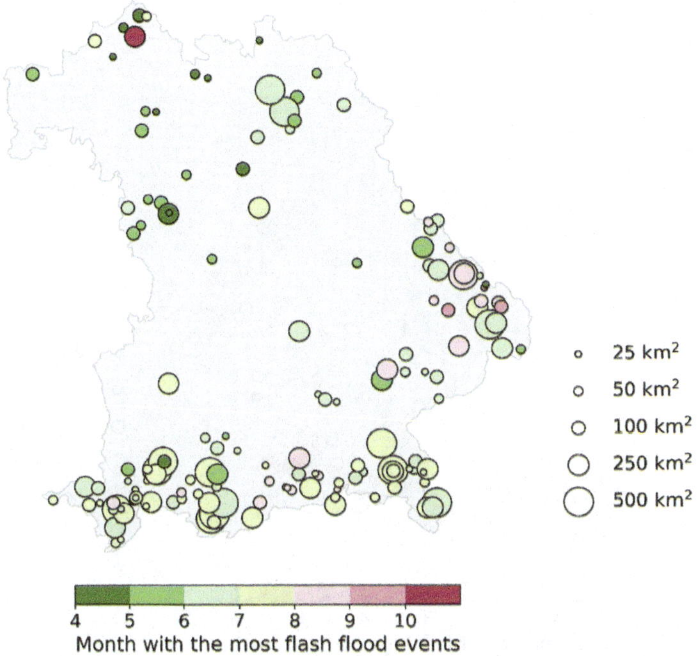

Fig. 6 Month with the most identified flash flood events per stream gauge. The alpine region is predominated by summer events, whereas northern Bavaria shows more events in spring to early summer. The eastern part of Bavaria, the Bavarian Forest, however, is heterogeneous

only 6% of all identified flash flood events had a specific peak discharge greater than 1 $m^3\ s^{-1}\ km^{-2}$. Flash flood events ranging between 1 and 3 $m^3\ s^{-1}\ km^{-2}$ mostly took place in the Alps, rarely in the Bavarian Forest (eastern Bavaria). To date, the discharge of 5.8 $m^3\ s^{-1}\ km^{-2}$ during the Simbach event is the highest ever measured specific peak discharge of a Bavarian flash flood. In fact, only four other stations have recorded events with a unit peak discharge greater than 3 $m^3\ s^{-1}\ km^{-2}$. The four Continental flash floods investigated by Amponsah et al. (2018) ranged from 0.8 to 11 $m^3\ s^{-1}\ km^{-2}$. Compared to these Continental peak unit discharges, the Bavarian flash floods show approximately similar ranges. The Mediterranean flash floods, however, have a much higher flood magnitude. The maximum unit peak discharge of Mediterranean flash floods were 40 and 77 $m^3\ s^{-1}\ km^{-2}$ as reported by Gaume et al. (2009) and Amponsah et al. (2018), respectively.

Rapid occurrence is a characteristic feature of a flash flood. Thus, a flash flood is characterized by a short time-to-peak. The time-to-peak was defined as the time between the onset and the peak of the flash flood event. As expected, the average time-to-peak increased with increasing catchment size (Fig. 8). The time-to-peak in the smallest catchments (<25 km^2) ranged between 4 and 14 h, with a median of 8 h. The catchments greater than 250 km^2 have an average time-to-peak of 7 to 17 h and a median of 13 h. While the small catchments hardly differ regarding their

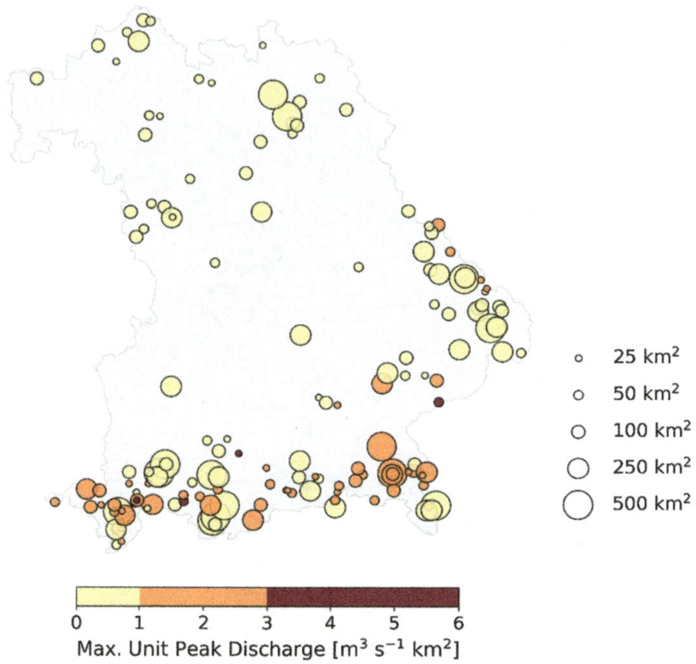

Fig. 7 Maximum unit peak discharge measured for each investigated stream gauge. The most extreme flash floods occurred in the alpine region, as well as occasionally in the Bavarian Forest. However, most of the maximum events do not exceed 1 $m^3\ s^{-1}\ km^{-2}$

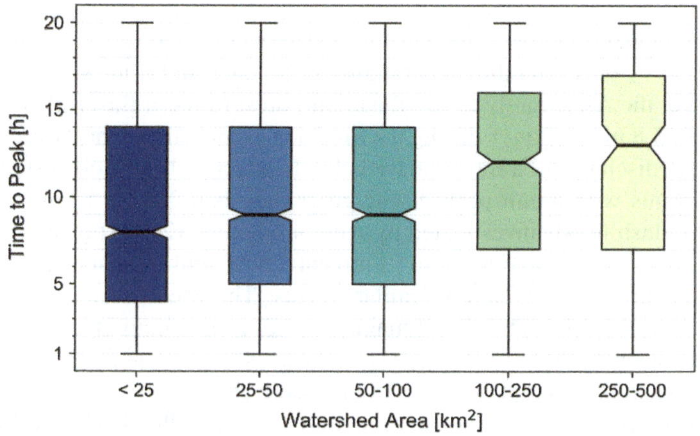

Fig. 8 Boxplots showing the distribution of the time-to-peak for the various catchment classes. The tendency is the bigger catchments to have a longer time-to-peak. However, the biggest catchments also experienced events with a one hour rising time. For the small catchments, the time-to-peak ranged mainly between 4 and 14 h. The boxplots indicate the minimum, maximum, 25th-percentile, 75th-percentile, and median of time-to-peak for the investigated catchment classes. In addition, the notch of a boxplot gives the 95% confidence interval of the median

characteristic time-to-peak, a clear shift is visible for the catchments bigger than 100 km^2. Fast flash flood events with a one hour rise time were found for every catchment size. The investigation shows, however, that a flash flood with a one hour rise time is also possible in big catchments up to 500 km^2. Depending on where the triggering rain event hits the catchment and what conditions prevail, a very fast flash flood is also possible in huge catchments.

Discussion

The discharge analysis of 342 gauging stations gave valuable insights into the occurrence and characteristics of Bavarian flash floods. The analysis revealed that approximately half of the investigated stream gauges measured flash floods. Furthermore, it is striking that the affected catchments were clustered in the Alps, the low mountain ranges (Mittelgebirge), and the Bavarian Forest. The clustering suggests that some basins are more prone to flash flooding than others. Less surprising, on the other hand, is that the small catchments are the most often affected. Contrary to our expectation, it was observed that catchments of approximately 500 km^2 can also generate a flash flood within one hour. Therefore, the flash flood hazard in large catchments should not be underestimated, although they experience flash floods less often than small catchments. In a European comparison, the Bavarian flash floods are significantly less intense (regarding the unit peak discharge) than Mediterranean flash floods. Furthermore, Bavaria shows a pronounced summer flash flood season comparable with other Continental countries like Austria, Romania, and Slovakia. The most extreme events occur in the alpine region and more sporadically, in the Bavarian Forest.

There are also some limitations that must be considered when evaluating the findings of this study. The greatest potential impact on our results is likely due to the determination of flash flood events using thresholds. With the application of uniform thresholds to Bavaria, we disregarded the diversity provided by the natural spatial variability. For example, we applied the same thresholds to large, flat catchments as we did to small alpine catchments. Nonetheless, defining the area-specific thresholds requires knowledge of the flash flood behavior in these regions, which was not given so far. Thus, we defined looser thresholds that allowed us to capture the flash flood variability in Bavaria. On the other hand, these generous thresholds may have led to an overrepresentation of alpine flash flood events, because they may exceed the limits more easily.

Further limitations related to the measurement of the flash floods are also considered. It is worth underlying that not all flash floods are measured whether due to the measurement errors, station failures, or missing stream gauges. Moreover, the station density and location influences which flash floods will be recorded. It was observed that the gauging network is denser in the alpine region and the lower mountain ranges. There, in the mountainous regions, the stream gauges are mostly installed in small catchments (<50 km^2). To increase the validity of the study, the affected catchments were excluded from the analysis. In southern Bavaria, for instance, several stream gauges had to be disregarded due to the numerous lakes and the hydroelectric power plants affecting the hydrographs. Retention basins, especially south of the Danube and in the eastern low mountain range (Bavarian Forest), led to the exclusion of further stations. Thus, due to the network density and the station selection, some Bavarian regions may be underrepresented in our analysis.

To overcome the limitations of the flash flood thresholds, it seems reasonable to apply flash flood thresholds tailored to the natural region and catchment in question. To give an example, the minimum unit peak discharge for alpine catchments could be increased, and the maximum time-to-peak for catchments up to 100 km^2 could be lowered. With both catchment size and natural conditions considered in the threshold choice, more precise results are expected, as the noise in the data sample is reduced.

Conclusion

In this study, the flash floods in Bavaria (Germany) based on discharge measurements were investigated. By applying thresholds, the flood events with certain flash flood characteristics, namely, short rising time and high unit peak discharge, were selected. For the investigation, the catchment size was limited to 500 km^2, the period of occurrence ranged from April to October. In total, 5,732 flash flood events measured by 157 stream gauges exploring their occurrence and characteristics were studied. The key findings are listed below.

- There are flash flood hot spots in Bavaria. Flash floods mainly occur spatially clustered in the alpine region and the low mountain ranges (Mittelgebirge), where high elevation differences prevail. Most flash floods occur in the Alps, with the frequently affected catchments having on average more than two flash floods per year.
- Flash floods occur in all of the investigated catchment sizes, confirming a basin flash flood threshold of 500 km^2 for Continental Europe. However, small catchments (<25 km^2) are the most often affected in Bavaria.
- Bavaria shows pronounced flash flood seasonality, as the flash floods tend to occur from May to September, with most of the events taking place in July. However, the seasonality effect varies regionally. While northern Bavaria tends to have flash floods in late spring to early summer, southern Bavaria experiences more of the events in midsummer.
- Bavarian flash flood events rarely reach unit peak discharges larger than 1 m^3 s^{-1} km^{-2}, except for the alpine catchments. Therefore, Bavarian flash floods show a pronounced Inland Continental character and are much less intense than Mediterranean flash floods.
- In both small and very large catchments, the flash floods can occur within a very short rising time of one hour. In general, the flash floods in the catchments up to 100 km^2 have a median rising time of 8 h and the larger catchments a median of 13 h.

Further knowledge could be gained by connecting the flash flood discharges with gray literature, such as newspaper articles, witness reports, or videos that provide information on the event intensity. Furthermore, the analyses could be expanded to better understand the factors controlling the occurrence and magnitude of a flash flood. In this context, a study of the characteristics of the particularly affected catchments appears promising. The enhancement of the hydrographs with the event precipitation data should deliver further insights into the different catchment behaviors.

Overall, the exploration of this natural hazard will continue to gain importance, as the processes impacting the risk of flash flood are aggravated. Due to the urbanization, changes in land use and hydrographic networks take place, which increases the erosion and surface runoff. At the same time, it can be assumed that climate change will intensify heavy precipitation events and thus aggravate the flash flood hazard. The main goal should be to identify the triggering factors of flash flooding to improve the identification of endangered catchments, ensuring the protection of the local communities. To this end, the present study should serve as a basis for further flash flood investigations in Bavaria.

Acknowledgements The research presented in this paper has been carried out as part of the *HiOS* project (*Hinweiskarte Oberflächenabfluss und Sturzflut)* funded by the Bavarian State Ministry of the Environment and Consumer Protection (StMUV) and supervised by the Bavarian Environment Agency (LfU). Furthermore, this work was supported by Deutsche Forschungsgemeinschaft (DFG) through the TUM International Graduate School of Science and Engineering (IGSSE). The authors would like to thank Davide Zoccatelli from the Hebrew University of Jerusalem for providing the MATLAB code for the flood separation.

References

Amponsah W, Ayral PA, Boudevillain B, Bouvier C, Braud I, Brunet P et al (2018) Integrated high-resolution dataset of high-intensity European and Mediterranean flash floods. Earth Syst Sci Data 10(4):1783–1794. https://doi.org/10.5194/essd-10-1783-2018

Beniston M, Stoffel M, Hill M (2011) Impacts of climatic change on water and natural hazards in the Alps: can current water governance cope with future challenges? Examples from the European "ACQWA" project. Environ Sci Policy 14(7):734–743. https://doi.org/10.1016/j.envsci.2010.12.009

Borga M, Anagnostou EN, Blöschl G, Creutin JD (2011) Flash flood forecasting, warning and risk management. HYDRATE project Environ Sci Policy 14(7):834–844. https://doi.org/10.1016/j.envsci.2011.05.017

Boudevillain B, Delrieu G, Galabertier B, Bonnifait L, Bouilloud L, Kirstetter PE, Mosini ML (2011) The Cévennes-Vivarais Mediterranean hydrometeorological observatory database. Water Resour Res 47(7):98. https://doi.org/10.1029/2010WR010353

Braud I, Ayral PA, Bouvie C, Branger F, Delrieu G, Le Coz J et al (2014) Multi-scale hydrometeorological observation and modelling for flash flood understanding. Hydrol Earth Syst Sci 18(9):3733–3761. https://doi.org/10.5194/hess-18-3733-2014

Bronstert A, Agarwal A, Boessenkool B, Crisologo I, Fischer M, Heistermann M et al (2018) Forensic hydro-meteorological analysis of an extreme flash flood: The 29/05/2016 event in Braunsbach. SW Germany. Sci Total Environ. 630:977–991. https://doi.org/10.1016/j.scitotenv.2018.02.241
Chapman TG, Maxwell AI (1996) Baseflow separation—Comparison of numerical methods with tracer experiments. In: Hydrology and Water Resources Symposium 1996: Water and the Environment 96(5):539–545
Gaume E, Bain V, Bernardara P, Newinger O, Barbuc M, Bateman A et al (2009) A compilation of data on European flash floods. J Hydrol 367(1–2):70–78. https://doi.org/10.1016/j.jhydrol.2008.12.028
Georgakakos KP, Hudlow MD (1984) Design of national real-time warning systems with capability for site-specific flash flood forecasts. Bull Am Meteorol Soc 67:1233–1239
Gewässerkundlicher Dienst (GKD) (2018) Gewässerkundlicher Dienst Bayern. Seen Wasserstand. Online available at https://www.gkd.bayern.de/de/seen/wasserstand//tabellen. Last checked on 19th Sept 2018
IPCC (Intergovernmental Panel on Climate Change) (2014) Climate change 2014: synthesis report. In: Core Writing Team, Pachauri RK, Meyer LA (eds) Contribution of working groups I, II and III to the fifth assessment report of the intergovernmental panel on climate change. IPCC, Geneva, Switzerland, pp 151
Kleinen T, Petschel-Held G (2007) Integrated assessment of changes in flooding probabilities due to climate change. Clim Change 81(3):283–312. https://doi.org/10.1007/s10584-006-9159-6
KLIWA (Klimaveränderung und Wasserwirtschaft) (2005) Der Klimawandel in Bayern für den Zeitraum 2021–2050. Climate change in Bavaria for the period 2021–2050. Kurzbericht
Kovats RS, Valentini R, Bouwer LM, Georgopoulou E, Jacob D, Martin E, Rounsevell M, Soussana JF (2014) Europe. In: Barros VR, Field CB, Dokken DJ, Mastrandrea MD, Mach KJ, Bilir TE, Chatterjee M, Ebi KL, Estrada YO, Genova RC, Girma B, Kissel ES, Levy AN, MacCracken S, Mastrandrea PR, White LL (eds) Climate change 2014: impacts, adaptation, and vulnerability. Part B: regional aspects. contribution of working group II to the fifth assessment report of the intergovernmental panel on climate change. Cambridge University Press, Cambridge, United Kingdom and New York, NY, USA, pp 1267–1326
LfU (Bayerisches Landesamt für Umwelt) (2017) Sturzfluten- und Hochwasserereignisse Mai/Juni 2016–Wasserwirtschaftlicher Bericht. Flash Flood and Flood Events May/June 2016–Water management report. Augsburg
Laudan J, Rözer V, Sieg T, Vogel K, Thieken A (2017) Damage assessment in Braunsbach 2016. Data collection and analysis for an improved understanding of damaging processes during flash floods. Nat Hazard Earth Sys 17(12):2163–2179. https://doi.org/10.5194/nhess-17-2163-2017
Lucía A, Schwientek M, Eberle J, Zarfl C (2018) Planform changes and large wood dynamics in two torrents during a severe flash flood in Braunsbach, Germany 2016. Sci Total Environ 640–641:315–326. https://doi.org/10.1016/j.scitotenv.2018.05.186
Marchi L, Borga M, Preciso E, Gaume E (2010) Characterisation of selected extreme flash floods in Europe and implications for flood risk management. J Hydrol 394(1–2):118–133. https://doi.org/10.1016/j.jhydrol.2010.07.017
Merz R, Blöschl G, Parajka J (2006) Raum-zeitliche Variabilität von Ereignisabflussbeiwerten in Österreich (spatial-temporal variability of event runoff coefficients in Austria). HyWa. 50(1):2–11
Munich Re (2005) Schadenspiegel. Risk factor of water. Spec Featur Issue 48:8–13
Ozturk U, Wendi D, Crisologo I, Riemer A, Agarwal A, Vogel K, López-Tarazón JA, Korup O (2018) Rare flash floods and debris flows in Southern Germany. Sci Total Environ 626:941–952. https://doi.org/10.1016/j.scitotenv.2018.01.172
Skublics D, Blöschl G, Rutschmann P (2016) Effect of river training on flood retention of the Bavarian Danube. J Hydrol Hydromech 64(4):349–356. https://doi.org/10.1515/johh-2016-0035
Tarolli P, Borga M, Morin E, Delrieu G (2012) Analysis of flash flood regimes in the North-Western and South-Eastern Mediterranean regions. Nat Hazard Earth Sys 12(5):1255–1265. https://doi.org/10.5194/nhess-12-1255-2012

Ozone Layer Holes, Regional Climate Change and Possible Ways for Their Forecasting

Aliaksandr Krasouski, Siarhei Zenchanka, Veronika Zhuchkevich, Tsimafei Schlender and Henry Sidsaph

Abstract A regional climate change, as a long-term repeated weather regime typical for a specific area, has become one of the main themes in Climate Change (CC) research. Forecasting the possible climate changes requires knowledge of the variations of natural and anthropogenic climate-forming factors, understanding of the mechanisms that determine the response of the Earth's climate system to these impacts. One of the factors influencing the regional climate is the total ozone content in the atmosphere. The aim of this article is to establish a correlation links between the total ozone content and the short-term regional climatic events. Presented analysis is based on data from orbital observations of the movement of ozone anomalies, data from different databases and researches of the National Ozone Monitoring Research and Education Center.

A. Krasouski · T. Schlender
Geography Faculty, Belarusian State University, 4, Nezavisimosti Avenue, 220030 Minsk, Belarus
e-mail: krasovsky@bsu.by

T. Schlender
e-mail: timajaya@mail.ru

S. Zenchanka (✉)
Minsk Branch of PRUE, 127-2, Mayakovskogo Street, 220028 Minsk, Belarus
e-mail: zench@tut.by

V. Zhuchkevich
National Ozone Monitoring Research and Education Center, Belarusian State University, 7, Kurchatova Street, 220045 Minsk, Belarus
e-mail: ronya@tut.by

H. Sidsaph
Business Research Institute, University of Chester, Riverside Campus, Castle Drive Chester, Chester CH11SL, UK
e-mail: h.sidsaph@chester.ac.uk

W. Leal Filho et al. (eds.), *Climate Change, Hazards and Adaptation Options*,
Climate Change Management, https://doi.org/10.1007/978-3-030-37425-9_17

Introduction

Forecasting the possible climate changes requires knowledge of the variations of natural and anthropogenic climate-forming factors, understanding of the mechanisms that determine the response of the Earth's climate system to these impacts.

For a long time, the discussion of the problem of climate change and its prediction was connected with the lower atmosphere—the troposphere. However, the recent decades the satellite Earth observing has shown that the layers of the upper atmosphere (stratosphere, mesosphere) make a significant contribution in the general circulation of the atmosphere, the changes the thermodynamic parameters of which impact on the nature of the circulation of the troposphere (weather conditions). Satellite images help track the movement of air masses and accordingly predict the weather at various areas of the Earth (Copernicus 2015). Reports of International Panel on Climate Change (IPCC 2007, 2014) and World Meteorological Organization (WMO 2011, 2014) noted that globally the troposphere has warmed and the lower stratosphere has cooled since the mid-20th century. There is medium confidence in the rate of change and its vertical structure in the Northern Hemisphere of extratropical troposphere. This implies the need for further studies of the global circulation of the atmosphere and the factors that influence this circulation.

Changes in ozone are often either not included, or are included in a very simplified manner, in current climate models. This is due to the complexity and the sheer computing power it takes to calculate these changes, an essential deficiency in some studies (Nowack et al. 2015). But recent publications, for example, WMO (2014), EEA (2015), stressed the role of total ozone (TO) as a climate-forming factor.

Methodology

A statistical analysis of the relationships of tropospheric parameters and tropospheric circulation processes with the distribution of the total ozone (TO) field, analyses of long-term changes in the ozonosphere state at global and regional scales, modelling the TO distribution have been conducted using the following set of data:

- OMI/OMPS (Ozone Monitoring Instrument/Ozone Mapping Profiler Suite) instruments from satellite Aura for measurements TO values are represented by average daily TO for the periods of 1997–2017 and by maps of the TO (https://www.esrl.noaa.gov/gmd/grad/neubrew/SatO3DataTimeSeries.jsp). From this data, the time series of TO values have been obtained from several mid-latitude ozone stations. Satellite data was used because the data of ground measurements of the direct sun radiation are discrete and therefore are not suitable for statistical analysis. A comparative analysis of the ground-based and satellite data showed that, despite some differences in the values of TO, their magnitudes and temporal variations (day-to-day changes and trends) are in a good agreement. The long-term averages from 1997 to 2017, obtained for the given regions, are considered as a climatic norm;

- Meteorological data are presented using the NCEP/NOAA Reanalysis data (https://www.esrl.noaa.gov) (from 1979 to 1992) and data of the Hydrometeorological Center of the Republic of Belarus (from 1997 to 2017). The climatic norm for the surface temperature and TO in Minsk has been taken for the period of 1997–2017 from site http://www.pogoda.by. To define the correlation between TO and the surface temperature the values of TO (OMI data) for Minsk and the observed surface temperature for the period of 1997–2017 obtained by the Hydrometeorological Center were used. Based on the daily TO and surface temperature (T_{1000}) data, the average monthly and annual values have been computed. For the correlation links, the deviations of the average daily values of TO and T_{1000} from their climatic norm were considered.

To present and confirm the hypothesis of the ozone mechanism, the method of joint analysis of the data of the tropopause height, TO and the Geopotential of 300 hPa (polar front) were used. The following data were considered:

- Maps of synoptic weather were taken from UKMET (Numerical weather prediction systems) (http://meteocentre.com/numerical-weather-prediction/forecast-systems.php?lang=en&map=eur&mod=ukmet) and DWD (German Weather Service) (http://www2.wetter3.de). Data of temperature and wind speed for the polar region were taken from Goddard Space Flight Center (https://ozonewatch.gsfc.nasa.gov);
- Maps of the distribution of the TO field and the field's variation from multi-year averages have been provided by the World Ozone Monitoring and Mapping Center (Canada) (http://exp-studies.tor.ec.gc.ca). For case studies, the data of TO were taken from AIRS (https://airs.jpl.nasa.gov).

Data of tropopause height, Geopotential 300 hPa, were taken from AIRS (https://airs.jpl.nasa.gov).

The Weather Research and Forecasting Model (WRF) (Skamarock, et al. 2008) was used for calculating the synoptic situation over the territory of Belarus in the period July 10–13, 2016. Next parameters were applied:

- Calculation grid with a step of 15 and 3 km;
- 35 vertical levels, up to 50 hPa;
- Map projection of Lambert (https://gisgeography.com/conic-projection-lambert-albers-polyconic/);
- Microphysics parameterization by the WRF Single moment 6 class Scheme (Hong and Lim 2006);
- Cloud parameterization in accordance with Grell 3D Ensemble Scheme. (Grell and Dévényi 2002);
- Data of ERA-Interim reanalysis.

Ozone Layer and Ozone Hole

Ozone (O_3) was discovered in 1785. In 1913 the "ozone layer", which is 12–50 km above the earth's surface, was discovered. The properties of ozone layer were explored by G.M.B. Dobson starting from 1925. These investigations established the main features of the seasonal variation of ozone, the maximum in the spring and the minimum in the autumn, and also demonstrated the close correlation between ozone amount and the meteorological conditions in the upper troposphere and lower stratosphere (https://www2.physics.ox.ac.uk/research/atmospheric-oceanic-and-planetary-physics/history/biography-dobson). The unit of ozone layer thickness received the name "Dobson Units" (DU). A typical thickness of the ozone layer is 300 DU which corresponds 3 mm. The value of 220 Dobson Units has been chosen for ozone hole determination because total ozone values of less than 220 DU were not found in the historic observations over Antarctica prior to 1979 (https://ozonewatch.gsfc.nasa.gov/statistics/annual_data.html).

The ozone layer protects the earth from harmful ultraviolet (UV-C) radiation of the sun (from 100 to 280 nm). Accordingly, UV-A (315–400 nm) and UV-B (280–315 nm) radiations affect all the elements of the biome including plants, pathogens. Total ozone at any location on the globe is defined as the sum of all the ozone in the atmosphere directly above that location. Most ozone resides in the stratospheric ozone layer, and a small percentage (about 10%) is distributed throughout the troposphere (Hegglin et al. 2015).

The first decreases in the total ozone in Antarctic were observed in the early 1980s over research stations located on the Antarctic continent. The observations showed unusually low total ozone during the late winter/early spring months of September, October, and November. Total ozone was lower in these months compared with previous observations made as early as 1957. Satellite measurements confirmed the spring ozone depletion and further showed that in each late winter/early spring season starting in the early 1980s, the depletion extended over a large region centered near the South Pole. The term "ozone hole" has emerged as a description of very low total ozone levels, which for many weeks were above the Antarctic continent. Currently, the formation and severity of the Antarctic ozone hole are documented each year by a combination of satellite, ground-based, and balloon observations of ozone. (Hegglin et al. 2015).

Significant depletion of ozone has been observed in the Arctic stratosphere in recent decades. Arctic ozone hole also occurs in the late winter/early spring period, but over a somewhat shorter period than in the Antarctic. Besides that, small ozone holes appear periodically in other high latitude locations (Lindsey 2016).

To overcome the ozone layer depletion the Vienna Convention (1985) and the Montreal Protocol (1987) have been adopted. Fulfilling these international treaties helped to decrease releasing of ozone-depleting substances into the atmosphere. Ball et al. (2018) pointed out that total column measurements of ozone between the Earth's surface and the top of the atmosphere indicate that the ozone layer has stopped declining across the globe, but no clear increase has been observed at latitudes

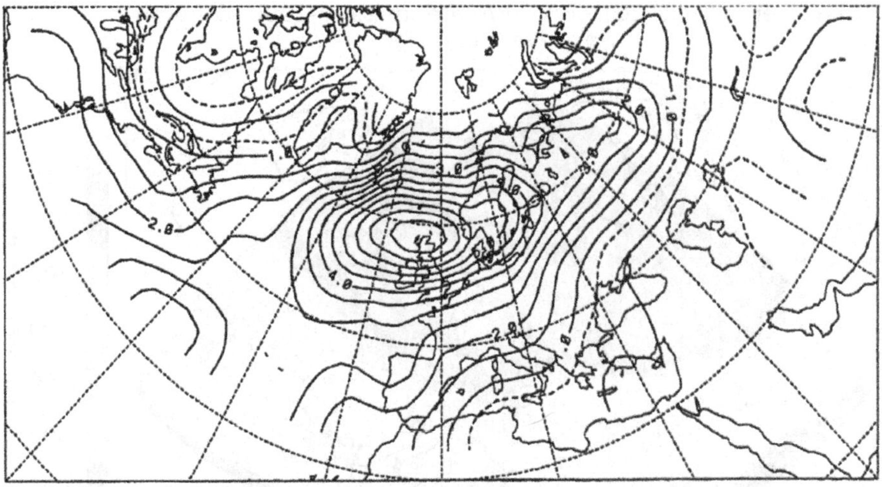

Fig. 1 The average geographical distribution of the frequency of occurrence of ozone “mini-holes” in the period November–March 1979–1993 James (1998)

between 60° S and 60° N outside the Polar Regions (60°–90°). At the same time, there is no significant change in total column ozone between 1998 and 2016, which includes both tropospheric ozone and the stratospheric ozone layer.

A “mini-hole” is a synoptic-scale region of depleted total column ozone. It has been established that ozone “mini-holes” appear in the North Atlantic sector much more often than in the Pacific sector. The most significant number of days with reduced TO values is observed in the European sector of the Atlantic, over the waters of the Northern, Baltic and Norwegian Seas, over Europe north of the 45th parallel (James 1998).

Figure 1 Shows the average geographical distribution of the frequency of occurrence of ozone “mini-holes” in North Atlantic in the period November–March 1979–1993

The contours show the number of days normalised to the surface area with a reduced TO content, as a percentage of the duration of the time period under consideration.

The distributions of negative ozone anomalies in different time periods with various norms and re-analysis were obtained using the method of objective automatic identification of ozone anomalies (Barodka 2016). The result of re-analysis is shown on Fig. 2.

Analysis of the TO fields showed that for the entire analyzed period (1979–1992, 1997–2017) no anomalies over the territory of the Republic of Belarus were formed, but they entered this territory from Western Europe and from the North of the European part of Russia, and anomalies from the North Atlantic mainly observed. It should be noted that the central parts of the anomalies rarely pass over the territory of Belarus. More often, these areas touch Belarus by edge, but the decreasing of the

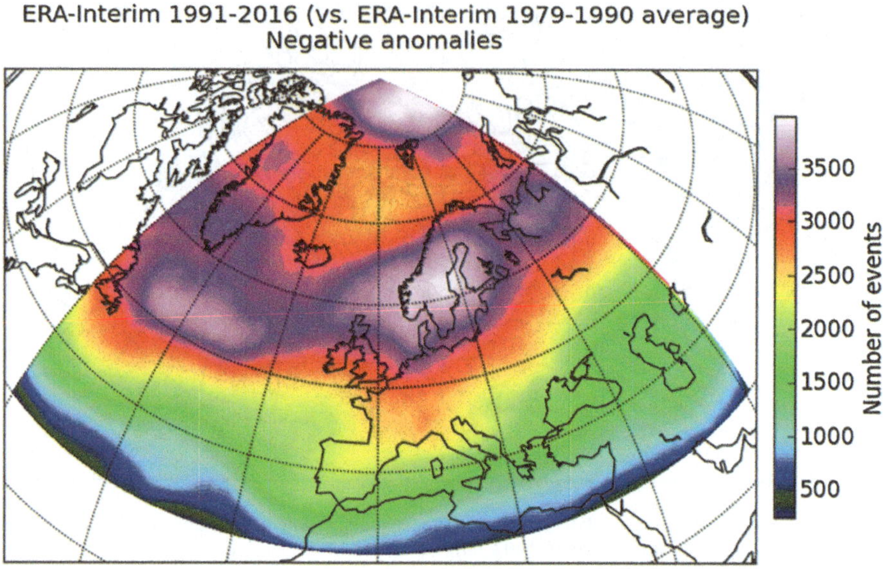

Fig. 2 Spatial distribution of negative ozone anomalies according to ERA-Interim reanalysis data for 1991–2016 (TO norm is averaging of 1979–1990 from the ERA-Interim with a 30-day smoothing)

TO is also significant. The deviations from the perennial means, while the ozone holes are above Belarus averages 20–25% and in some periods they exceed 40%.

ERA-Interim reanalysis data for 1991–2016 reflect a more complete and up-to-date picture of the spatial distribution of negative ozone anomalies compared with data for the period 1979–1993 obtained by James (1998), where only the case of negative anomalies for the winter season was used. The centre of the appearance of the maximum number of negative anomalies has shifted over the past 25 years to the East, in particular, to Scandinavia. The region of an intensive appearance of negative ozone anomalies extends from the southern part of Greenland to the northern part of the Ural Mountains.

Reanalysis data (MACC-Reanalysis, https://confluence.ecmwf.int/pages/viewpage.action?pageId=81010554) for 2003–2012 are presented at Fig. 3.

These data show that the negative ozone anomalies most frequently appear in the autumn-spring period with a maximum in January. The interannual appearance of ozone mini-hole changes from 20 to 45 per year.

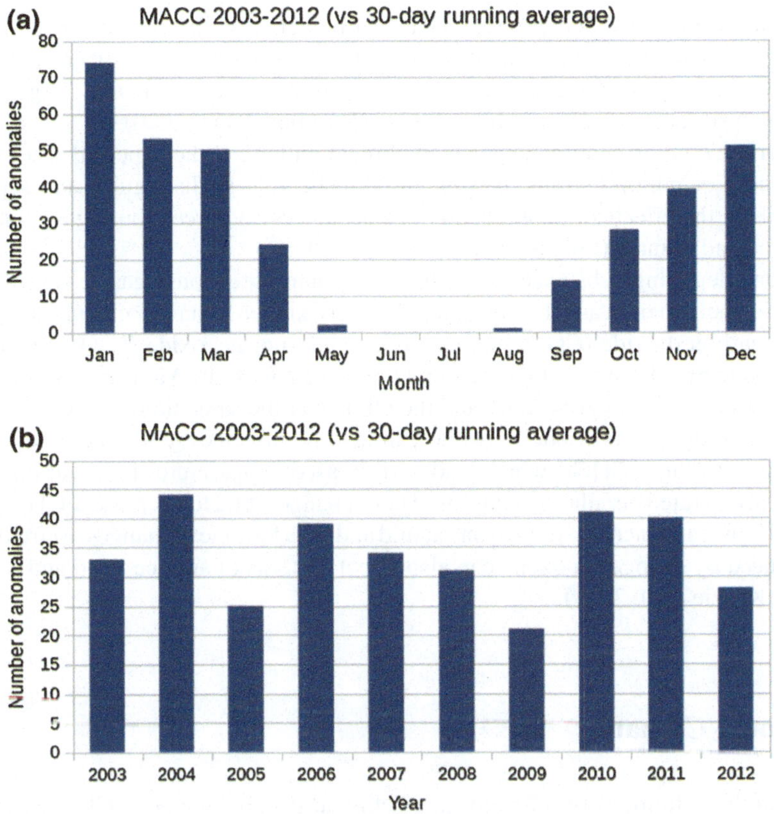

Fig. 3 Season (**a**) and interannual (**b**) distributions of negative ozone anomalies above Europe

Ozone and Its Influence on Earth's Ecosystem

WMO (2014) pointed out that ozone occurs at three levels: the stratospheric ozone, the tropospheric ozone and surface ozone. Most of the ozone was found in the stratosphere between 19 and 30 km up in the Earth's atmosphere forming so-called "ozone layer". The stratospheric ozone arises under the influence of solar radiation, the tropospheric ozone varies with the daylight variations, and ozone near-ground is a pollutant, and its production is enhanced due to air pollutants (Aggarwal 2013).

In accordance with spatial distribution, the ozone layer is an essential factor in biological, environmental and health global phenomena, because without the ozone layer, much ultraviolet (UV) radiation from the Sun will reach the Earth's surface, causing untold damage to most living species (Sivasakthivel and Reddy 2011).

Some investigations pointed mutual influence of total ozone and climate change, for example:

- Stratospheric ozone depletion and GHG warming may both be producing increased meridional temperature gradients in the extratropical lower stratosphere and upper troposphere, and thus acting synergistically to produce surprisingly large trends in both surface and stratospheric climate (Hartmann et al. 2000).
- Over the coming century, changes in climate will affect the expected recovery of ozone. For policy reasons, it is essential to be able to distinguish and separate attribute the effects of ozone-depleting substances and greenhouse gases on both ozone and climate (Shepherd and Theodore 2008).
- Ozone-depleting substances are also very potent greenhouse gases, contributing to the phenomenon as other substances widely known to have a greenhouse effect like carbon dioxide (CO_2), methane (CH_4) and nitrous oxide (N_2O). Therefore, by reducing emissions of ozone-depleting substances, the Montreal Protocol has protected both the ozone layer and the climate at the same time (EEA 2015);
- The amount of stratospheric ozone can also be affected by the increases in the concentration of GHGs, which lead to decreased temperatures in the stratosphere and accelerated circulation patterns. These changes tend to decrease total ozone in the tropics and increase total ozone at mid and high latitudes. Changes in circulation induced by changes in ozone can also affect patterns of surface wind and rainfall (McKenzie et al. 2011).

Regional Climate

Look at the definitions of 'climate" and "climate system", given at Glossary (IPCC 2013):

"Climate in a narrow sense is usually defined as the average weather, or more rigorously, as the statistical description in terms of theme and variability of relevant quantities over a period ranging from months to thousands or millions of years. The classical period for averaging these variables is 30 years, as defined by the World Meteorological Organization. The relevant quantities are most often surface variables such as temperature, precipitation and wind. The climate in a wider sense is the state, including a statistical description, of the climate system".

"The climate system is the highly complex system consisting of five major components: the atmosphere, the hydrosphere, the cryosphere, the lithosphere and the biosphere, and the interactions between them. The climate system evolves in time under the influence of its own internal dynamics and because of external forcing such as volcanic eruptions, solar variations and anthropogenic forcing such as the changing composition of the atmosphere and land use change".

The main factor influencing on the Earth's climate is solar radiance. To explain Earth' climate response to solar variability three mechanisms were suggested (Report 2012):

- The mechanisms connecting variations in total solar irradiance directly to the troposphere;

- The mechanisms that influence indirectly upper parts of the atmosphere, such as variations in solar ultraviolet radiation and possibly solar energetic particles;
- The mechanisms that link variations in galactic cosmic rays to climate change.

Different mechanisms of solar activity influence on the Earth's atmosphere were considered by Loginov and Mikutski (2017). These mechanisms include:

- The mechanisms associated with the variability of solar ultraviolet radiation and its effect on stratospheric ozone and the subsequent transmission of radiation from the stratosphere to the troposphere;
- The mechanisms associated with the influence of galactic and solar cosmic rays on the cloud cover and the global electric circuit;
- The mechanisms associated with the influence of low-energy particles on the upper atmosphere and the effect of planetary waves;
- The mechanisms associated with radiation-chemical processes;
- The mechanisms based on changes in the solar constant and infrared radiation, which increases during geomagnetic storms.

According to Gray et al. (2010) "the most mature Sun-climate mechanism at this time involves the direct effect of the observed variation in solar UV radiation affecting stratospheric ozone, leading to associated temperature variations". The same conclusion was made by Forster et al. (2011) and Arblaster et al. (2014). These authors pointed that "stratospheric ozone changes, together with changes in other stratospheric constituents, directly affect the radiative budget of the stratosphere, causing changes in stratospheric temperature and circulation, which in turn influence tropospheric climate. Changes in stratospheric ozone influence climate both by direct radiative effects and by affecting stratospheric and tropospheric circulation".

It is clear that more and more researchers start to include ozone and ozone layer changing in climate changing. At the same time, it is necessary to clarify the mechanisms of ozone-climate change interaction further.

Climate Change and Total Ozone

Total ozone at any location on the globe is defined as the sum of all the ozone in the atmosphere directly above that location. Most ozone resides in the stratospheric ozone layer, and a small percentage (about 10%) is distributed throughout the troposphere (Hegglin et al. 2015).

Ozone in the upper stratosphere is determined by a combination of photochemical and chemical processes, in the lower stratosphere—by dynamic processes appearing in the stratosphere and upper troposphere. Khrgian (1973) paid attention to the fact that "… the ozone-induced heating is very significant and largely determines the temperature field of the stratosphere … Perhaps, ozone is the main factor in the circulation of the atmosphere".

Discussing the results of TO measurement in 2012–2016, Josefsson et al. (2016) concluded that "the ozone-related processes in the stratosphere are relatively slow and the natural spatial and temporal variations in ozone concentration are large and therefore it cannot be expected to see an immediate rapid response in the stratospheric ozone. Thus, a significant recovery may not be stated within the immediate future of a few years. Instead, a substantial part of a century might be needed, with high quality ground based ozone observations".

One of the possible mechanisms for transferring changes in the upper atmosphere (stratosphere, mesosphere) to the troposphere and surface layer of the atmosphere is an ozone mechanism (Krasouski et al. 2016; Krasouski and Zenchanka 2018).

This mechanism and its influence on regional climate include:

- The combination of stratospheric processes determining the spatial distribution of TO in the stratosphere, which, in turn, forms the height of the tropopause over the cells of global circulation;
- Variation in a position of the tropopause changing a speed of the circulation processes in the troposphere thereby affecting a position of the stationary fronts;
- Stationary fronts, which are the boundaries between the global circulation cells, determine the trajectory of the drift of baric formations in the troposphere.

A relief of the tropopause directly reflects the general picture of the atmospheric circulation and the distribution of the fields of TO: more TO—the tropopause is lower, less TO—the tropopause is higher. As a result, the level of the tropopause is determined by the competition of both underlying and overlying atmospheric processes.

Thus, as was shown by Law et al. (2005), the effect of ozone on the weather and climate in a given region is not due to some direct effect on atmospheric fluxes, but primarily by changing the position of the stationary fronts and, respectively, the characteristics of the global circulation cells themselves (for example, a shift in the tropopause level affects the entire circulation cell).

The degree of influence of tropospheric and stratospheric mechanisms depends on the region and season. In the tropical region, thermodynamic processes (convection) prevail throughout the year in the unstable troposphere, and in the polar latitudes, the stratospheric ozone mechanism prevails. In the middle latitude area in which Belarus is located, during the summer period the tropospheric processes have a significant impact, and in the course of the transition in the winter period, the ozone mechanism impacts (Krasouski et al. 2016).

Regional Hazards and Their Prediction

Statistical Assessment of Total Ozone and Surface Temperature Correlation

A joint analysis of the daily and monthly averaged TO and T_{1000} (temperatures on the isobaric surface of 1000 hPa) for the periods of 1979–1992 and 1997–2017 shown that in the summer the changes in TO over the considered regions occur synchronously with the temperature change at the Earth's surface (see Fig. 4). In

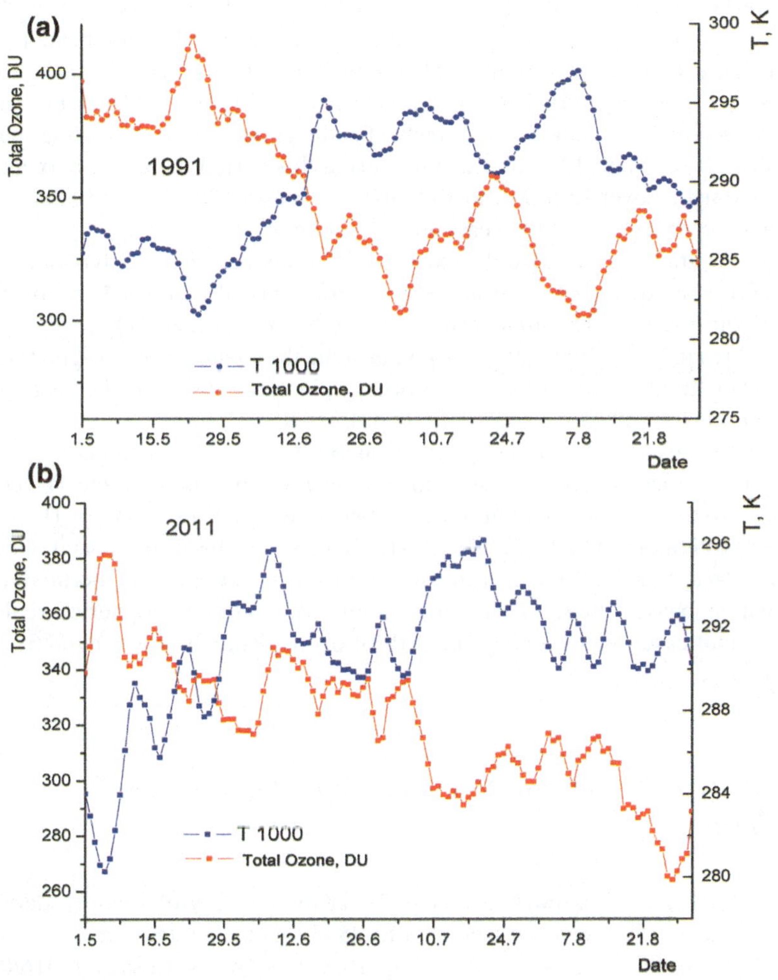

Fig. 4 Daily values of TO and surface temperature smoothed by a 5-day moving average (OMI and hydrometeorological centre data)

winter this link is broken while again being restored in the spring.

For the entire examined period, statistically significant anticorrelation has been obtained between the daily temperatures and TO. Founded relationships between the TO and the surface temperature are statistically significant and more or less stable.

To define the correlation between TO and the surface temperature the values of TO (OMI data) for Minsk and the observed surface temperature for the period of 1997–2016 obtained by the Hydrometeorological Center were used. Based on the daily TO and T_{1000} data, the average monthly and annual values have been computed. For the correlation relationship, the deviations of the average monthly values of TO and T_{1000} from their climatic norm were considered. The results are presented in Table 1.

The average annual correlation coefficient is −0.32. In some years it reaches −0.45. The largest seasonal long-term coefficients are observed in the spring (−0.47) and in summer (−0.64). In individual seasons the minimum coefficient can reach (−0.78), or, on the contrary, maximum positive values of (+0.18). The winter period only has positive correlation coefficients. The breakdown in links during the cold season can be explained by the fact that during this period there is a maximum of ozone anomalies over Europe and Belarus. The reasons for the low values of the correlation coefficient in some years are not yet known to us.

Thus, the studies confirmed the existence of a link between the dynamics of the TO field (or the dynamics of stratospheric ozone) and the tropospheric processes in the Belarus region. The high correlation coefficient between TO and the surface temperatures indicates that in the warm season the tropopause height is greatly influenced by the tropospheric convective processes; in the cold season, the height of the tropopause is formed by the stratospheric processes.

The hypothesis of the effect of stratospheric ozone (ozone mechanism) on the tropospheric processes has been applied to a specific synoptic situation. The considered situation reflects a violation of the statistical relationship between TO and the surface temperature, which reveals (most often), as has been mentioned above, in summer and spring. At the same time, in summer TO undergoes the thermodynamic deformation, but still, there are situations when it affects the tropospheric circulation. Almost in all cases, this is due to the drift of ozone anomalies over the territory of Belarus.

Case study 1: The Ozone Mechanisms on the Territory of Belarus

On July 10–13, 2016, a hurricane passed over the territory of Belarus, damaging 510 settlements. The wind gust was up to 25–30 m/s. This hurricane was one of the strongest in the country's history (http://naviny.by/rubrics/disaster/2016/07/14/ic_articles_124_192104).

Table 1 Correlation coefficients for variability of TO and surface temperature for Minsk

Years	1997	1998	1999	2000	2001	2002	2003	2004	2005	2006	2007
Annual coefficient	−0.39	−0.33	−0.39	−0.30	−0.40	−0.27	−0.28	−0.37	−0.24	−0.33	−0.21
Spring (March–May)	**−0.56**	−0.30	**−0.60**	−0.40	**−0.62**	**−0.55**	−0.37	**−0.50**	**−0.53**	−0.30	−0.47
Summer (June–August)	**−0.56**	**−0.77**	**−0.70**	**−0.78**	**−0.68**	**−0.74**	**−0.73**	**−0.60**	**−0.52**	−0.38	**−0.56**
Autumn (September–November)	**−0.52**	−0.10	−0.47	−0.09	−0.25	−0.16	−0.46	−0.27	−0.09	−0.42	−0.38
Winter (December–January)	**−0.53**	0.15	−0.16	−0.27	−0.36	−0.29	0.13	−0.31	0.18	−0.11	
Years	2008	2009	2010	2011	2012	2013	2014	2015	2016	2017	Means
Annual coefficient	−0.26	−0.24	−0.33	−0.23	0.33	−0.45	−0.33	−0.28	−0.44	−0.32	−0.32
Spring (March–May)	−0.34	−0.40	**−0.56**	**−0.71**	**−0.55**	**−0.66**	−0.39	−0.49	−0.45	**−0.62**	−0.47
Summer (June–August)	**−0.65**	**−0.71**	**−0.60**	**−0.59**	**−0.73**	**−0.67**	**−0.61**	**−0.70**	**−0.71**	−0.49	**−0.64**
Autumn (September–November)	−0.33	−0.46	−0.41	−0.28	−0.31	−0.48	−0.35	−0.28	−0.47	−0.26	−0.33
Winter (December–January)	−0.37	0.14	0.15	−0.18	−0.13	−0.02	−0.11	−0.02	**−0.54**	−0.14	−0.14

Bold data correspond high values of correlation coefficient. The high correlation coefficient between TO and the surface temperature indicates that in the warm season the tropopause height is greatly influenced by the tropospheric convective processes

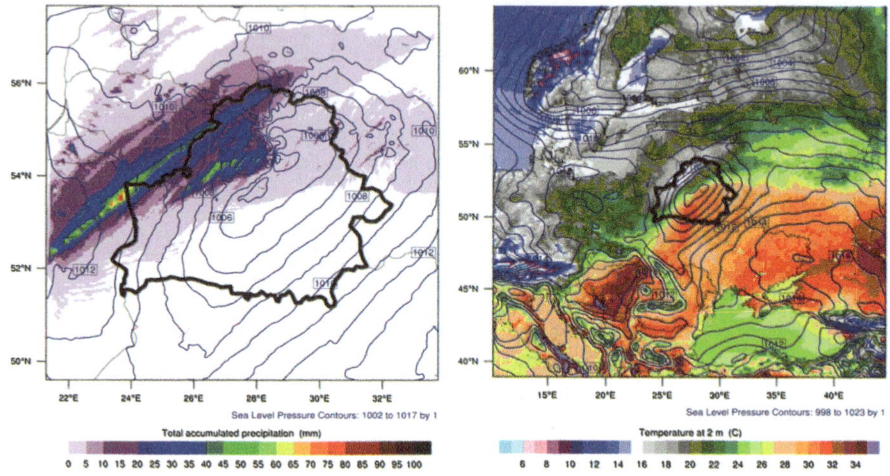

Fig. 5 WRF maps of mini-cyclone passing through the territory of Belarus on July 13. On the left: a map of temperature at 2 m (C) and sea level pressure (hPa) at 12:00 a.m., on the right: a map of precipitation (mm) and pressure (hPa) at 15:00 p.m. (data of ERA-interim reanalysis)

At that time the territory of Belarus was in the intermediate zone between the deep cyclone, whose centre was located above the Kola Peninsula, and the anticyclone, which was located above the Black Sea. Thus, the synoptic situation of July 10–13, 2016 led to a collision of cold and warm masses followed with instability in the atmosphere and resulted in the formation of a mini-cyclone over the territory of the Carpathian Mountains.

The digitised maps of the WRF (Skamarock et al. 2008) were visualized for July 13, 2016, for the territory of Belarus with different resolution steps—3 and 15 km and a time interval of 1 h. It permitted to receive for a more detailed and accurate analysis of the synoptic situation. Figure 5 demonstrates the mini-cyclone passing through the territory of Belarus. Minimum pressure was 1005 hPa in the centre. The main areas, where much rain fell, were the Grodno, Minsk and Vitebsk regions. On July 10, Belarus was in an area of a low height of the tropopause, and on July 13 it was in a high tropopause area.

Analysis of the climatic situation in the Republic of Belarus in the period from July 10 to July 13, 2016, showed the presence of large gradients in the field of total ozone.

There was a large gradient in the field of total ozone over the territory of Belarus in the period from July 10 to July 13, 2016. In the North, South and South-East (see Fig. 6) the TO maximum was observed. This maximum lowered the height of the tropopause and formed a specific narrow spatial corridor from the South-West to the North-East, along which the mini cyclone passed.

The TO maximum generated a decrease in the height of the tropopause in the North and South of the territory of Belarus. As a result, the tropospheric air masses, namely the mini cyclone, which was formed literally during half a day in the Carpathian

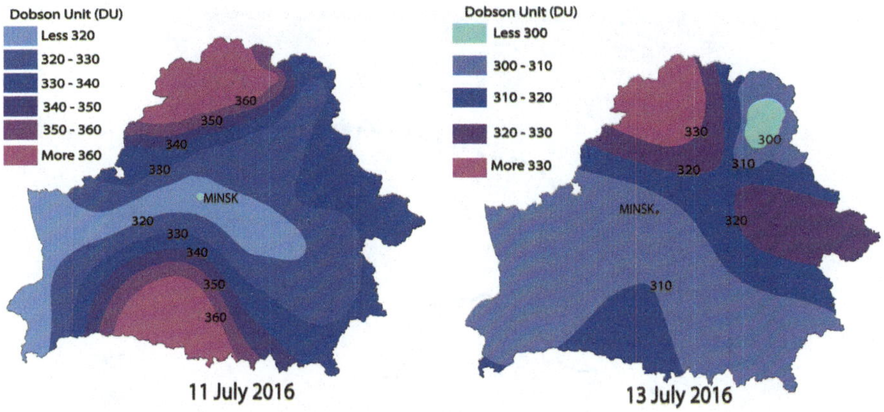

Fig. 6 Distribution of TO (in DU) over the territory of Belarus for July 11–13, 2016 (from OMI/OMPS data) calculated using the program ArcGIS

region, passed through Belarus from the South-West to the North-East. The maximum damage was observed in the settlements located along this line.

Moist-saturated air mass with strong winds, bounded in North and South by a low tropopause height and high TO, had to move along a narrow corridor with a high tropopause height and low TO. This situation resembles the well-known problem of solving the Bernoulli equation when the flow velocity increases at the point of the narrowing of the channel (Fig. 7).

This figure illustrates well the main provisions of the hypothesis of the ozone mechanism influence on the weather processes and regional climate and adds some novelty to our understanding of the development of atmospheric processes: a change in the height of the tropopause over a specific air mass can change the intensity and general character of the tropospheric circulation. The analysis and numerical calculation showed that even in summer, when the TO plays a secondary role in the circulation of the atmosphere, nevertheless, under ozone anomalies (positive or negative), the winter ozone mechanism affecting the tropospheric circulation is involved.

Thus, summer sharp drops in TO can lead to substantial changes in the level of the tropopause, appearance of powerful jet streams and, as a consequence, deterioration of weather conditions in the lower troposphere.

Case study 2: The Ozone Mechanism and Sudden Stratospheric Warming in February 2018

There are sharp rises in temperature and a change of wind in the opposite direction in the polar latitudes in the stratosphere every winter. Such a phenomenon is called

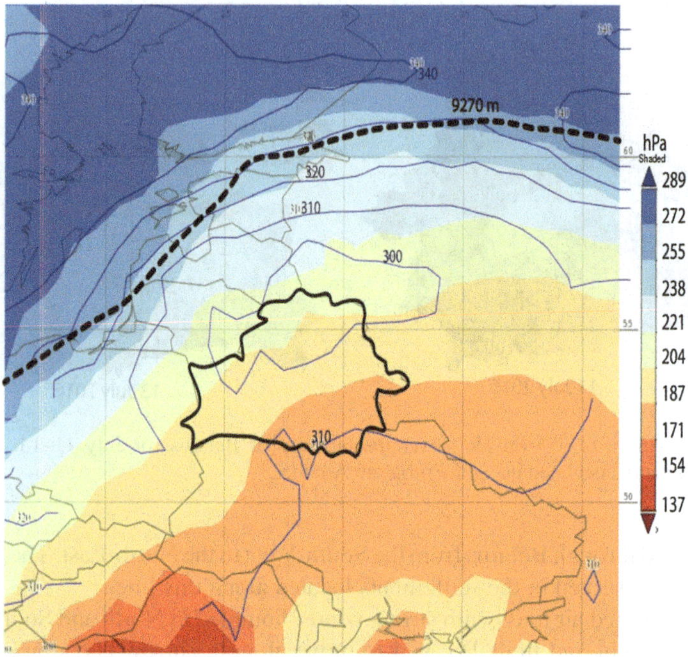

Fig. 7 The height of tropopause, TO and polar front on 13 July 2016 from AIRS (tropopause colour shaded (in Pa), TO—blue contours (in DU), polar front—the black dotted line of geopotential (in metres) on level 300 hPa) (https://airs.jpl.nasa.gov)

sudden stratospheric warming (SSW). According to the definition of WMO (Butler et al. 2015) SSW is characterised by a sharp increase in temperature (40–60 K) for several days in the polar stratosphere (60° latitude and above; 10 hPa and below) with slowing down or changing the direction of the zonal wind flow.

In February and March 2018 (see Fig. 8), the strongest in history sudden stratospheric warming occurred over Eurasia. According to Goddard Space Flight Center (https://ozonewatch.gsfc.nasa.gov), the temperature of the polar stratosphere has risen sharply by 40 K over several days, and the wind has changed direction from West to the East. Sunspots were not significant during these two months (http://omniweb.gsfc.nasa.gov/ow.html).

February was the coldest month of 2017–2018 calendar winter. The peak of the freezing weather occurred this year on the last days of February. The temperature these days has dropped below normal by 7 °C in the Central District of the Russian Federation. The night of February 23 in Moscow and the Moscow region was the coldest in the season—the temperature dropped to −30 °C (https://meteoinfo.ru/novosti/14769-fevral-vnosit-kholodnyj-vklad-v-itogi-zimy). Most of Europe in late February—early March was subjected to heavy snowfall and frost. As a result of the bad weather, more than 60 people and many victims occurred (https://www.theguardian.com/uk-news/gallery/2018/feb/27/siberian-blast-blankets-europe-snow-inpictures).

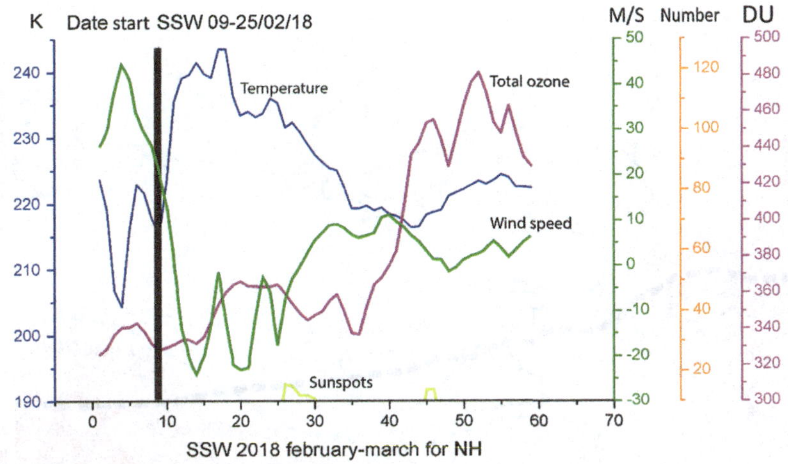

Fig. 8 SSW graph, February–March 2018. The beginning of the SSW is marked with a vertical solid black line on the graph

From the ozone mechanism, such an anomalous cold is explained by the fact that the anomalously high TO for this region and season began to lower the tropopause height to 470 hPa. Usually, in polar conditions, arctic air has a low tropopause height, which begins about 7–8 km. In this case, due to the increasing the TO in the stratosphere, the decreasing in the tropopause has reached the level of 5–6 km (see Fig. 9). This allowed the cold air to move quickly and spread deep into the mainland. Also, a characteristic feature of the terrain of this area contributed to a specific trajectory of the arctic air. The Scandinavian and Ural mountains limited the cold air both in space and in height, creating a kind of pocket where the arctic air accumulated.

The distribution of the TO over the territory of Belarus on February 20, 2018, and for the North Hemisphere is presented on Fig. 10.

Within the western and eastern borders of the country, a distance is about 500 km, the gap in the TO was about 50–60 units. Such high values of TO, about 400 DU and above, is peculiar to Arctic air masses. Usually, the TO accumulates in the Far East and, over time, is transferred further along the latitude to the East, since the stratosphere has a stable state of the atmosphere. However, at the end of February—beginning of March 2018, stratospheric ozone changed its direction of movement to the West and subsequently began to shift meridionally to the south, creating abnormal weather conditions in the lower troposphere.

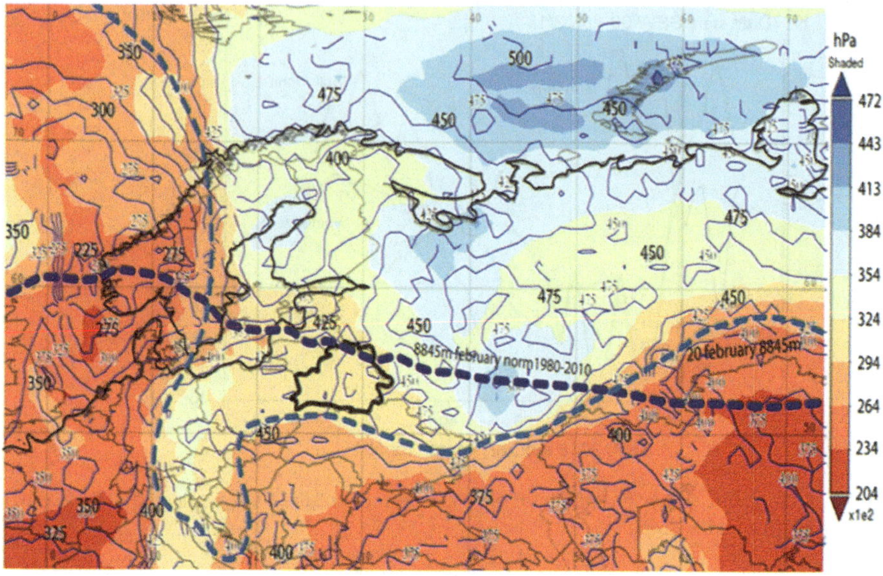

Fig. 9 The height of tropopause, polar front and TO on 20 February 2018 over Europe from data AIRS (tropopause (in hPa) colour shaded, TO—the blue contours (in DU), polar front—blue dotted line of geopotential (in metres) on 20/02/2018 on level 300 hPa and line polar front for February (dark blue) 1980–2010 (from AIRS) (https://airs.jpl.nasa.gov)

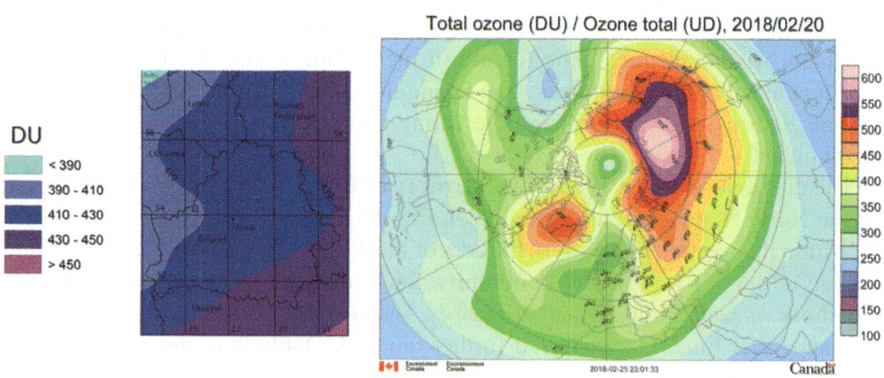

Fig. 10 TO distribution 20 February 2018 over the territory of the republic of Belarus (left) and for Northern hemisphere (right), (OMI/TOMS data)

Conclusion

As a result of statistical processing of multi-year monitoring data, the strong anti-correlation link between fluctuations in surface temperature and TO deviations has been revealed for Minsk covering the period from 1997 to 2017. The average multi-year anticorrelation coefficient is −0.32. The average annual correlation coefficient reached −0.45 from 1997 to 2017. The largest seasonal multi-year magnitudes of the anticorrelation coefficient are observed in the spring and the summer. Those are −0.47 and −0.64, respectively. In some years, the correlation coefficient reaches both −0.77 and +0.18.

The results of the study have confirmed the validity of the lead-in of the ozone mechanism concept in the description of atmospheric processes. The essence of the concept is that the field of total ozone is formed by the competition of two mechanisms, one of which, that may be called "radiative", is responsible for the formation of the ozone layer in the upper stratosphere and lowers the height of the tropopause. The other, the traditional one, presumably called "thermodynamic", acts in the troposphere creating conditions for raising the height of the tropopause resulted in the reduction of the total ozone amount. The mechanism itself does not act directly, but instead by changing the position of global circulation cells (with their characteristic elements—stationary fronts, jet streams and the level of the height of the tropopause).

In the process of the statistical analysis, the breaking of the correlation between TO and the surface temperature during the spring-summer ozone anomalies has been discovered. To clarify the reasons for these deviations, the specific case of the extreme weather phenomenon in the summer has been treated. The case of the hurricane occurred on July 10–13, 2016 in Belarus, illustrated the partial effect of the stratosphere on the troposphere by demonstrating the influence of the jet streams and large gradients in the field of TO over the height of the tropopause. This example improves understanding of the development of atmospheric processes. In particular, the change in the height of the tropopause (or TO) can alter the tropospheric circulation not only in winter but also in summer, when ozone anomalies are observed.

The occurrence of natural hazards from sudden stratospheric warming can also be explained by the ozone mechanism of climate change. Thus, it has been defined statistically, experimentally and by the modelling procedure that even in summer TO can influence the troposphere and contribute to the emergence of dangerous meteorological phenomena such as strong wind gusts. Observation of the ozone layer can predict the appearance of different climate phenomena.

References

Aggarwal A et al (2013) Depletion of the ozone layer and its consequences: a review. Am J Plant Sci 4 (1990–1997). Available at: http://file.scirp.org/pdf/AJPS_2013100915534704.pdf/. Accessed 15th Nov 2017

Arblaster JM, Gillett NP, Calvo N, Forster PM, Polvani LM, Son S-W, Waugh DW, Young PJ (2014) Stratospheric ozone changes and climate. Chapter 4 in Scientific Assessment of Ozone Depletion: 2014, Global Ozone Research and Monitoring Project—Report No. 55, World Meteorological Organization, Geneva, Switzerland

Ball WT, Alsing J, Mortlock DJ, Staehelin J, Joanna D. Haigh JD, Thomas Peter T, Tummon F, Rene Stübi R, Stenke A, Anderson J, Bourassa A, Davis SM, Degenstein D, Frith S, Froidevaux L, Roth C, Sofieva V, Wang R, Wild J, Yu P, Jerald R. Ziemke JR, and Rozanov EV (2018) Evidence for a continuous decline in lower stratospheric ozone offsetting ozone layer recovery. Atmos Chem Phys 18:1379–1394

Barodka S (2016) Climatology, structure and formation mechanisms of local ozone anomalies in Europe. In: Barodka S, Krasouski A, Lapo P, Svetashev A, Shlender T, Yakautsava Y, Zhuchkevich V, Bruchkouski I (eds) WCRP/SPARC workshop: "challenges for climate science—synergies between SPARC and the WCRP grand challenges", Berlin, Germany, 31 Oct—01 Nov 2016

Butler AH et al (2015) Defining sudden stratospheric warming. Bull Am Meteorol Soc 96:1913–1928. https://doi.org/10.1175/bams-d-13-00173.1

Copernicus (2015) Europe's eyes on earth. Directorate-general for communication publications. 1049 Brussels, Belgium, p 25

EEA (2015) Protecting the ozone layer while also preventing climate change. Available at: https://www.eea.europa.eu/downloads/af8c374db73d407a8c708f91a8abc238/1484653105/protecting-the-ozone-layer-while.pdf. Accessible 15 Nov 2018

Forster PM, Thompson DWJ, Baldwin MP, Chipperfeld M, Dameris JD, Haigh DJ, Karoly PJ, Kushner WJ, Randel KH, Rosenlof DJ, Seidel S, Solomon G, Beig P, Braesicke N, Butchart NP, Gillett KM, Grise DR, Marsh C, McLandress TN, Rao SW, Son GL, Stenchikov, Yoden S (2011) Stratospheric changes and climate. Chapter 4 in Scientifc Assessment of Ozone Depletion: 2010, Global Ozone Research and Monitoring Project–Report No. 52, p 516. World Meteorological Organization, Geneva, Switzerland

Gray LJ, Beer J, Geller M, Haigh JD, Lockwood M, Matthes K, Cubasch U, Fleitmann D, Harrison G, Hood L, Luterbacher J, Meehl GA, Shindell D, van Geel B, White W (2010) Solar influences on climate. Rev Geophys 48, RG4001. https://doi.org/10.1029/2009rg000282

Grell GA, Dévényi D (2002) A generalized approach to parameterizing convection combining ensemble and data assimilation techniques. Geophys Res Lett AGU J 29(14):38(1)–38(4)

Hartmann DL, Wallace JM, Limpasuvan W, Thompson DWJ, and Holton JR (2000) Can ozone depletion and global warming interact to produce rapid climate change? PNAS 97(4):1412–1417, 15th Feb 2000

Hegglin MI (Lead Author), Fahey DW, McFarland M, Montzka SA, Nash ER (2015) Twenty questions and answers about the ozone layer: 2014 update, scientific assessment of ozone depletion 2014. World Meteorological Organization, Geneva, Switzerland, p 84

Hong S-Y, Lim J-OJ (2006) The WRF single-moment 6-class microphysics scheme (WSM6). J Korean Meteorol Soc 42(2):129–151

IPCC (2007) Climate change 2007: synthesis report. In: Core Writing Team, Pachauri RK, Reisinger A (eds) Contribution of working groups I, II and III to the fourth assessment report of the intergovernmental panel on climate change. IPCC, Geneva, Switzerland, p 104

IPCC (2013) Annex III: Glossary (Planton S (ed)). In: Stocker TF, Qin D, Plattner GK, Tignor M, Allen SK, Boschung J, Nauels A, Xia Y, Bex V, Midgley PM (eds) Climate change 2013: the physical science basis. Contribution of working group I to the fifth assessment report of the intergovernmental panel on climate change. Cambridge University Press, Cambridge, United Kingdom and New York, NY, USA

IPCC (2014) Climate change 2014: synthesis report. In: Core Writing Team, Pachauri RK, Meyer LA (eds) Contribution of working groups I, II and III to the fifth assessment report of the intergovernmental panel on climate change. IPCC, Geneva, Switzerland, p 151

James PM (1998) A climatology of ozone mini-hole over the Northern hemisphere. Int J Climatol 18:1287–1303

Josefsson W, Löfvenius MO, Löfvenius P (2016) Measurements of total ozone 2012–2015. SMHI Meteorology, p 161

Khrgian A (1973) Atmosphere ozone physics. Hydrometeoizdat, Leningrad (in Russian)

Krasouski AN, Turyshev LN, Svetashev AG, Borodko SK, Zhuchkevich VV (2016) Ozone mechanism of the regional climate and weather control. Sci Innov 9:17–20 (in Russian)

Krasouski A, Zenchanka S (2018) Montreal and Kyoto: needs in inter-protocol communications. In: Leal Filho W et al (ed) Handbook of climate change communication. Theory of climate change communication. series climate change management, vol 1. Springer, Berlin

Law K, Pan L, Wernli H, Fischer H et al (2005) Processes governing the chemical composition of the extratropical UTLS, report from the joint SPARC-IGAC Workshop, 18–20 May 2005. Mainz, Germany

Lindsey R (2016) 4 ways the ozone hole is linked to climate, and 1 way it isn't. Available at: https://www.climate.gov/news-features/understanding-climate/4-ways-ozone-hole-linked-climate-and-1-way-it-isn%E2%80%99t. Accessible 1 Dec 2018

Loginov V, Mikutski V (2017) Climate change: trends, circles, pauses. Minsk. Belarusian science, pp 79 (in Russian)

McKenzie RL, Aucamp PJ, Bais AF, Björn LO, Ilyas M, Madronich S (2011) Ozone depletion and climate change: impacts on UV radiation. Photochem Photobiol Sci 10(2):182–198. https://doi.org/10.1039/c0pp90034f Epub 2011 Jan 20

Montreal Protocol (1987) European commission, Luxembourg: office for official publications of the European communities. Accessed 25th Mar 2017. Available at: https://treaties.un.org/doc/Publication/UNTS/Volume%201522/volume-1522-I-26369-English.pdf. Accessible 1 Dec 2018

Nowack P et al (2015) A large ozone-circulation feedback and its implications for global warming assessments. Nat Clim Chang 5:41–45

Report (2012) The effects of solar variability on earth's climate: a workshop report. The National Academies Press, Washington, D.C

Skamarock WC, Klemp JB, Dudhia J, Gill DO, Barker DM, Duda MG, Huang X, Wang W, Powers JG (2008) A description of the advanced research WRF version 3 (Electronic resource). NCAR Technical Note. NCAR/TN–475 + STR

Shepherd, Theodore G (2008) Dynamics, stratospheric ozone, and climate change. Atmos-Ocean 46(1):117–138. https://doi.org/10.3137/ao.460106, https://doi.org/10.3137/ao.460106

Sivasakthivel T, Siva Kumar Reddy KK (2011) Ozone layer depletion and its effects: a review. Int J Environ Sci Dev 2(1):30–37

Vienna Convention (1985) The vienna convention for the protection of the ozone layer. Available at: http://www.unep.org/ozone/. Accessible 10 Dec 2018

WMO (2011) World meteorological organization, scientific assessment of ozone depletion: 2010, global ozone research and monitoring project–report no. 52, p 516. Geneva, Switzerland

WMO (2014) Scientific assessment of ozone depletion: 2014, world meteorological organization, global ozone research and monitoring project—report no. 55, p 416. Geneva, Switzerland

The Protective Role of Forests to Reduce Rockfall Risks and Impacts in the Alps Under a Climate Change Perspective

Emanuele Lingua, Francesco Bettella, Mario Pividori, Raffaella Marzano, Matteo Garbarino, Marco Piras, Milan Kobal and Frédéric Berger

Abstract Among the functions provided by forests, protection has always been considered one of the preeminent in mountain areas. In order to fulfil, maximize, and sustain this function, specific forest structures should be obtained and maintained through properly designed forest management. A specific management goal should be defined with a well-defined forest target against each natural hazard, based on the protection potentially provided by the forest stands, in either an active (e.g. against avalanches) or passive way (e.g. against rockfall). Climate change is forecast to affect both disturbance regimes and forest ecosystems, leading to new challenging issues concerning protection forest management. This paper describes how a forest

E. Lingua (✉) · F. Bettella · M. Pividori
TESAF, University of Padova, Viale dell'Università, 16, 35020 Legnaro, Italy
e-mail: emanuele.lingua@unipd.it

F. Bettella
e-mail: francesco.bettella@unipd.it

M. Pividori
e-mail: mario.pividori@unipd.it

R. Marzano · M. Garbarino
DISAFA, University of Torino, L.go Braccini, 2, 10095 Grugliasco, Italy
e-mail: raffaella.marzano@unito.it

M. Garbarino
e-mail: matteo.garbarino@unito.it

M. Piras
DIATI, Politecnico di Torino, C.so Duca degli Abruzzi, 24, 10129 Turin, Italy
e-mail: marco.piras@polito.it

M. Kobal
Department of Forestry and Renewable Forest Resources, University of Ljubljana, Večna pot 83, 1000 Ljubljana, Slovenia
e-mail: milan.kobal@bf.uni-lj.si

F. Berger
Mountain Ecosystems Research Unit, IRSTEA—National Research Institute of Science and Technology for Environment and Agriculture, rue de la Papeterie, 2, BP 76, 38402 Saint-Martin-d'Hères Cedex, France
e-mail: frederic.berger@irstea.fr

W. Leal Filho et al. (eds.), *Climate Change, Hazards and Adaptation Options*, Climate Change Management, https://doi.org/10.1007/978-3-030-37425-9_18

stand exerts its protective role against rockfalls and the target profile to be reached for sustaining this function. Potential consequences of climate change on forest ecosystems that management will have to face in the near future are also addressed. New perspectives are provided taking into account the knowledge coming from recent research studies and specifically the results obtained in the RockTheAlps project (ASP462), dealing with the assessment of protection forests against rockfall in the Alps.

Protection Forests

In mountainous areas, forest stands have always played an important role in maintaining valleys safe for living in and transiting. This importance is increasing, since in the last decades settlements have also been built in remote mountain areas and there is the need to access them throughout the year, mostly for tourism (Wehrli and Dorren 2013; Lingua et al. 2017).

The protective role is considered among one of the most important functions provided by a forest stand. All forests, regardless of their location, can accomplish this, for instance protecting the soil from surface erosion and taking part directly and indirectly in the hydrological cycle. However, despite this general protective role, we do not define all forests as protection forests. A protection forest is instead characterized and defined due to the existence of a specific natural hazard against which it offers protection in an active or passive way. It might also fulfil other functions, but its protective role is preeminent. Protection forests are particularly effective against some types of abiotic disturbances, like snow avalanches, rockfalls, debris flows, shallow landslides, surface erosion and floods. Since these are mostly gravity-driven hazards, it is clear that the protective role provided by forests can became fundamental in mountain areas, where steep slopes increase the risk of occurrence (Lingua et al. 2017). By definition, active protection is exerted when the forest helps to prevent the occurrence of a natural hazard. Passive protection occurs instead in those situations in which the presence of the forest contributes to mitigating the effects produced by the disturbance. The distinction between these two kinds of protective roles mostly depends on the characteristics of the considered process. For instance, forests play an important protective role against snow avalanches in the starting zone, preventing their release. For rocks falling down a slope, forests can instead reduce the speed and number of rocks reaching the bottom of the slope.

We can further classify stands based on the presence of a specific object to be protected into direct and indirect protection forests (Brang et al. 2006). A direct protection forest is defined as such when it grows in close proximity to an endangered asset to which it offers protection against natural hazards. This direct protection is generally offered to people, buildings and any other infrastructure that might be exposed to a specific hazard in a mountain area. The direct protective function is usually provided over an area, which is limited in size and located below, and close to the protection forest. A typical example of a direct protection forest is represented by a stand located above a group of houses or a road, which are threatened by snow avalanches.

The indirect protective function is instead exerted by forests, independently of their exact location, simply by their presence at a broader scale (e.g. the landscape level). This could be the case for stands in mountain catchments where they can potentially reduce soil erosion and flooding at the closing section.

Almost any forest can offer some indirect protection, for instance through its effect in intercepting precipitation or affecting the local climate. Only some forests affected by gravitational hazards have a direct protective function.

The distinction between direct and indirect protection forests is particularly important in the context of forest planning, specifically in the definition of intervention priorities and management targets (Berger and Rey 2004).

A stand with a direct protective function should be permanently effective (Lingua et al. 2017). This can be achieved if the stand has high resistance to natural hazards and high persistency. The only way to maintain a stand in this efficiency window is with active forest management.

All forests are anyway subject to stand dynamics, which may modify or limit their protection effectiveness. Most of these dynamics are driven by the occurrence of natural disturbances, which often coincide with the hazards they are meant to prevent or mitigate.

Protection forests might be affected by different kinds of disturbances, including wildfires, storms, snow break, bark beetle outbreaks, which can act at both small and large scale. This last category of events can deeply alter the capacity of the stand to maintain its protective function. The degree and temporal extent in the impairment of the protective function produced after the disturbance is strongly related to the severity and spatial extent of the disturbance itself, as well as the recovery process undertaken naturally by the disturbed system.

At this stage, the quality and quantity of biological legacies (e.g. logs, root plates) can affect both the recovery process, for instance favouring the establishment of natural regeneration, and the residual protection function offered by the disturbed stand (Lingua et al. 2017).

A forest can offer a certain degree of protection, being more or less effective in its protective function, based on two main aspects: (1) the type and characteristics of the natural hazards involved; (2) the main features of the stand, together with its conditions when a damaging event occurs (Brang et al. 2008).

Concerning the disturbance regime, intensity and frequency of occurrence are the two most influential attributes affecting the capacity of a forest stand to provide an effective protective function.

In this context, proper management of protection forests should be oriented towards the maintenance of forest structures with a high degree of resistance and resilience to disturbances, as well as capable of providing effective protection to people and structures at risk. To reach this objective may sometimes require making compromises, since the characteristics needed to increase resistance to a certain disturbance might not necessarily be those that maximise the protective function of the same stand against some specific hazard. For instance, to decrease the probability of a crown fire spreading in the forest canopy, we could decide to reduce the stand density, but this management decision might consequently diminish the protection effectiveness of that same forest against rockfall.

Structure and composition of protection forests should thus be designed based on the type of disturbance that might affect the stand, the disturbance regime, the required protection function and the possibility of guaranteeing stand renewal. Obtaining the stand structures required to reach these management goals is a long-term process requiring decades or even centuries.

Managing Protection Forests Against Rockfall Hazard

In natural forest stand dynamics, some stages will not provide a protective effect, since trees are in turn too small, too sparse, or too few (Motta and Haudemand 2000). To be effective, a protection forest should have defined characteristics, especially concerning tree density and average stem size. In order to guarantee the protective function, a certain structure (i.e., combination of tree density and average size) has therefore to be achieved and maintained over time. As previously stated, appropriate silvicultural management can guide a forest towards its highest protection effectiveness and it should thus be designed and pursued to reach this aim.

Current management of protection forests adopts the term target profile to describe the characteristics of the stand that we want to obtain through silvicultural interventions, according to the natural hazards involved and local site conditions (Brang et al. 2006). A target profile thus describes the state of the forest that is expected to have an effective protective action against natural hazards and that can be permanently maintained with minimum effort.

Depending on the hazard type, different attributes of the forest have to be taken into account in terms of their protective effect (Lingua et al. 2017). Considering snow avalanches, for instance, the forest performs an active protective role in the starting zone impeding the release of an avalanche. In this situation, the canopy cover is the key parameter to consider since it plays the important role of snow interception (Frehner et al. 2007). When dealing with a rockfall hazard, forests mainly act in a passive way, limiting its impact by reducing the runout length. In this case, tree density and size are the main stand parameters of interest.

A rockfall process is the movement of falling rocks and their interaction with the environment (Frehner et al. 2007). On a slope characterized by rockfalls, three distinct areas (that often overlap) can be found: the zone of origin, the transit zone and the runout and deposition zone.

In the zone of origin, where rocks are released (generally corresponding to steep slopes exceeding 30°), the forest does not play a relevant protective role, other than general water regimentation and root holding functions. In some cases trees can actually act as triggering factors. Roots can penetrate into cracks, increasing frost wedging effects. Moreover, roots can produce acid exudates that, together with litter (in the case of conifers), can corrode and weather rocks, and they can release rocks they were previously holding when trees sway or are uprooted by strong winds. Forest management in the zone of origin should aim to avoid the presence of unstable trees potentially prone to uprooting, and perhaps even tall trees, since they are more subjected to wind load (Table 1). If suitable for the site and species, coppicing can be applied, to maintain stands at lower height.

When rocks start moving down the slope, they enter the transit zone. In this zone, rocks can slide, roll and bounce (if the incline exceeds 35°). The potential contribution of the forest in mitigating the impact of rockfall (considering blocks up to a size of 5 m^3) can be significant. Trees act as energy-dissipating elements, making falling rocks lose part of their kinetic energy with each impact on stems, lying deadwood logs and root plates. In addition to dissipating energy, the collision against a tree can cause a falling rock to deviate its trajectory or even to stop (Fig. 1).

The diameter of the stems in relation to the size of the rocks strongly determines the energy reduction efficiency. The larger the trees, the more effective they can be in dissipating energy or stopping rocks. However, big trees usually belong to older age classes, being more prone to falling as a consequence of senescence dynamics, with subsequent short-lived protection effects.

A high density of trees is supposedly desirable since it increases the probability for a falling rock to hit a stem. However, from an ecological point of view, a stand with a high density of large trees is not sustainable. The self-thinning rule explains why in the presence of a large number of stems, their average size will be smaller compared to a stand with lower tree density where trees can grow bigger.

The desired target profile against rockfall can be obtained by silvicultural management, focusing on prominent stand parameters such as basal area and mean diameter at breast height (DBH; corresponding to the diameter at 1.30 m) or stem density.

It has been proven that, to be effective against rockfall, a protection forest should have a minimal length of 250 m along the slope. Moreover, attention should be given not to create gaps between stems larger than 40 m along the maximum slope. In this short distance, rocks can in fact regain high speed, depleting all the protective role of the upslope forest.

Rocks start decreasing their speed when the slope diminishes to less than 30°; they then stop rapidly if the slope reaches less than 25°. In this last area, known as the runout and deposition zone, the forest can play an important role in reducing the length of the rock path, contributing to slowing down the rocks in the same manner as in the transit zone. Here, however, the forest stand can be even more effective, since the rock energy (i.e. its speed) is already reducing because of the diminishing slope steepness. Consequently, in the runout and deposition zone even small trees can stop big rocks.

To increase the protection function against rockfall, silvicultural management can also act on species composition, where possible, favouring the presence of broadleaves (hardwood species), since they are usually more resistant than conifers.

The importance of the protective function of a forest stand against rockfall should be considered looking at the residual hazard, considering the risk reduction (Dorren and Berger 2007). The ability to dissipate the rocks kinetic energy and thus their speed can effectively reduce the intensity of the hazard, leading, if the element

Table 1 Silvicultural guidelines for rockfall mitigation by forest stands. Modified from Berger et al. (2017), Frehner et al. (2007)

	Recommendations	Thresholds
Zone of origin (release zone)	Remove unstable trees (leverage effect due to the wind) at the top of cliffs or outcrops and in the release area	Coefficient of stability value (height/diameter at breast height = H/DBH) Conifers: H/DBH ≤ 65 Broadleaves: H/DBH ≤ 80
	Maintain a high basal area compatible with stand sustainability at the foot of the release area	
	Whenever possible, limit boulder's distance to the beginning of the stand	
	Promote broadleaved trees that are more resistant than conifers with equivalent diameter Maintain more than 30% of broadleaved trees among the largest trees. Depending on site conditions, a certain amount of conifers is needed for stand stability	
	Limit the size of gaps (same thresholds as for transit and runout zones)	
	Harvest trees leaving stumps with a min. height of 1.30 m or, if rockfall can occur, completely remove stump to ground-level (or screed in order to avoid a trampoline effect)	
	Fell/cut trees at an oblique angle to the slope leaving felled trees on the ground in a position from which they cannot be easily moved	
Transit and run out zones	If possible, increase the planimetric length of the forested slope	Recommended horizontal length of forested slope >200 m (ideal >250 m)
	Limit the size of gaps	Length of gap along the steepest slope: High forest <40 m Coppice <20 m, In all cases, recommended value (H = average height of trees): length ≤1.3H with a wooded strip below the gap >2H (recommended >4H)
	Promote broadleaved trees, which are more resistant than conifers with equivalent diameter Maintain more than 30% of broadleaved trees among the largest trees. Depending on the site conditions, a certain amount of conifers is needed for increasing the stand stability	
	Maintain an appropriate basal area for the efficient trees	In the transit zone: the basal area of trees with a DBH ≥15 cm is required to be ≥25 m^2/ha

(continued)

Table 1 (continued)

	Recommendations	Thresholds
		In the run out zone, the basal area of trees with a DBH ≥15 cm is required to be ≥20 m^2/ha
	Maintain an appropriate stem density for the efficient trees Maintain a high density in a band of 25 m on either side of a corridor	In all cases the stem density for trees with a diameter of ≥20 cm is required to be ≥350 stems/ha
	Remove unstable trees along corridors	Value of the coefficient of stability (height/diameter at breast height = H/DBH) Conifers: H/DBH ≤ 65 Broadleaves: H/DBH ≤ 80

Fig. 1 Rock stopped by trees in a protection forest (Auronzo di Cadore, BL, Italy). The stems show evidence of recent wounds produced by bouncing rocks

to be protected requires no marginal risk, to fewer and less impacting permanent infrastructures being built. We should remember that a forest stand provides several ecosystem services besides the protective function (i.e. habitat provision, recreation, aesthetic value), while a rockfall net is only a passive protection.

Managing stands to perform a protective function requires accurate knowledge of their spatial location. The availability of accurate maps with a good resolution is thus highly desirable. When rock sources are known, the area affected by the rockfall processes can be defined by modelling the path of the rocks. Rockfall trajectories modelling is frequently implemented via simulation models with different spatial dimensions (2D, 2.5D, or 3D), normally using topographic layers that are created in most GIS programs (Volkwein et al. 2011; Pradhan and Fanos 2017). The reliability of the results mostly depends on the input data; the accuracy of Digital Elevation Models (DEM) is therefore crucial for a correct rockfall assessment (Žabota et al. 2019). The availability of new remote sensing tools, particularly LiDAR (Light Detection and Ranging), and platforms (i.e. UAV—Unmanned Aerial Vehicle), provides enhanced information that renders the simulation closer to reality, at least concerning slope profile. Forest parameters can be also extracted from remote sensing data with good accuracy (Eysn et al. 2015), but the goodness of the results is still strongly affected by the forest stand complexity (e.g. structural diversity, species composition).

In the Alps, several studies provided detailed maps of protection forests in single municipalities or valleys (see for example Motta and Haudemand 2000), and some maps are also available at the regional or country level. However, the coverage is not complete and the methodologies and spatial scales adopted are not harmonized. The project RockTheAlps (ASP462), carried out in the framework of the Interreg Alpine Space (2014–2020), aimed at filling this gap, identifying and applying innovative methodologies to detect and map rockfall risk and protection forests, providing decision makers and policy makers with harmonized information for the whole Alpine Space. At the broader scale, a model (ROCK-EU) based on more than 10,000 real cases collected from the Alps has been implemented to map the potential runout zone (Fig. 2). The forests located in this area are thus considered protection forests. In order to define protection forests with direct protective function, the presence of endangered assets has been taken into account and added to the model. The effectiveness of the identified protection forests will be assessed by means of TORRID (Toolbox for assessing the protective effect of forests against rockfall and expressing the protective role in a Risk Reduction InDex), a new toolbox developed for the entire Alpine Space. The rockfall risk reduction provided by forest stands will be defined and the gap between optimal characteristics and current situation will be identified, guiding forest managers in the crucial phase of prioritization according to an adaptive management approach. The preliminary map of the protective function obtained after the first version of the model shows that around 20% of forests in the Alpine Space are potentially providing this important ecosystem service. The analyses have been conducted on a 25 m DEM, since this is the resolution available for the entire Alpine Space, but the methodologies applied are flexible and can easily be re-run as soon as a better resolution becomes available for the whole area. Furthermore the results of the simulation at 25 m spatial resolution have proven to be comparable

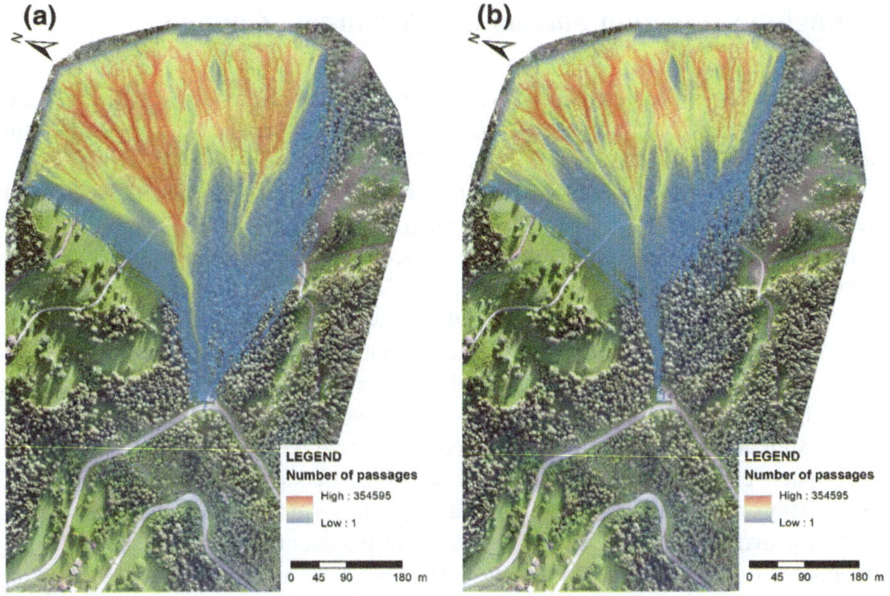

Fig. 2 Simulation of rockfall in the Colcuc study case (Colle Santa Lucia, BL, Italy). The number of passages (size 1.2 m^3; 1000 replication from each source cell) is shown for the scenario without forest (**a**) and with forest (**b**). The reduction in runout length provided by the protection forest is evident

with the results obtained at a finer scale and can therefore be used effectively at the regional level, where only a general overview of the potential risk is needed, and location accuracy is not the main purpose (Žabota et al. 2019).

Several case studies distributed over the Alps have been used to test the large-scale model locally, as well as to evaluate the economic value of the protective function (Bianchi et al. 2018). Some of the case studies were also selected among forest stands recently affected by high severity disturbances, which consequently altered their ability to sustain the provision of ecosystem services and particularly their potential protective function. In the western Alps, during autumn 2017, almost 10,000 ha of forests were affected by large wildfires, and some of these stands were protection forests. In the post-disturbance silvicultural intervention plan, priorities were assigned to the protection forests characterized by high burn severity. The recent storm Vaia, in late October 2018, which affected more than 40,000 ha of forests in the central and eastern Italian Alps, resulted in more than 8.6 M cubic meters of windthrown trees. Several of these forest stands had provided, among other ecosystem services, a relevant protective function.

Managing Protection Forests Under Climate Change

Climate change is predicted to produce important consequences, both direct and indirect, on forest ecosystems and on the disturbance regimes possibly affecting them (Seidl et al. 2017; Seidl and Rammer 2017; Thom et al. 2017).

Forest ecosystems in mountain regions will be greatly affected by climate change. Tree species compositions, forest cover and growth rate are among the parameters of great importance for the fulfilment of the protective function that could be altered in the mid- and long-term perspective (Albrich et al. 2018). Not only rising mean temperatures and changes in precipitation patterns will shape future forests by selecting species more adapted to the new conditions, but also the alteration of natural disturbance regimes will have a crucial role. Disturbance regimes are forecast to be altered by climate change in several parts of the world (Dale et al. 2001; Seidl et al. 2017). Both abiotic and biotic disturbance agents will be affected and their cumulative impacts can result in unprecedented negative effects on forests (Temperli et al. 2013). Appropriate disturbance regimes are in equilibrium with species composition and forest structure that are thus able to resist or persist after disturbances since they are adapted to their occurrence with a specific intensity and return interval. Altering the disturbance regime can instead lead to changes in species arrangement, and the creation of degraded structures.

Extreme events have the potential to affect forests much more than gradual changes in temperature (Lindner et al. 2014). Indeed natural disturbances are discrete events that could produce sudden alterations to forests, while changes in climatic conditions can increase stress resulting in a stand decline that can take several years before causing tree death.

From the perspective of forest management, focusing specifically on protection forests, there are two main topics to tackle: the non-permanence issue and species fitness (Dyderski et al. 2018).

When a high severity disturbance affects a protection forest, forest cover will be partially or totally removed, with consequences on the level of protection offered by the disturbed forest, based on the residual structure and composition. In the case of wildfires or insect outbreaks standing dead trees will dominate the landscape immediately after the event; they will persist standing for a variable amount of time, depending mostly on tree species and size (Marzano et al. 2012). In the case of a windthrow there will instead be mostly logs and uprooted trees covering the ground. The rise in temperatures will result in increasing fire risk, particularly in the driest valleys of the Alps. Rather than the gradual increase in temperature, extreme and prolonged droughts may increase the probability of large stand-replacing fires (Zumbrunnen et al. 2009; Eelkin et al. 2013). Biotic agents can also be highly favoured by the presence of trees stressed by water limitation, and even a moderate increase in temperature or drought length has been found to potentially raise the risk of forest pathogens outbreaks (Bentz et al. 2010). Natural disturbances are often interdependent, and the occurrence of one type can promote the occurrence of another one and increase its intensity. Windthrows or severe droughts are generally followed by insect outbreaks, and standing dead trees killed by pests and diseases can potentially increase fire risk.

Forest regeneration will be the most susceptible stage. Mature trees show a sort of biological inertia; they already have a well-developed and larger root system compared to seedlings, and are therefore able to counteract increasing stresses up to a certain threshold.

Concerning rockfall, one of the most probable consequences of climate change will be an increase in occurrence (Berger et al. 2017). Although rockfall is an almost unpredictable phenomenon, triggering factors are generally related to weather conditions, such as freeze-thaw processes (frost wedging) and precipitations. A high variation of temperature over a short period, and long duration and/or high intensity of precipitations generally precede rockfall events. Indeed, some studies already indicated an increase in rockfall occurrences for years with weather anomalies (Berger et al. 2017). Climate change will not only directly affect rocks release, but also and probably to a greater extent the forests that should mitigate this hazard.

Forest management should take into account climate change effect on forest ecosystems, identifying and mitigating potential threats to resistance and resilience traits of the stands.

Thinning can be a correct silvicultural treatment in order to enhance resistance to increasing water stress. Reducing inter-tree competition and providing more growing space for roots in search of water produced encouraging results also in post-drought recovery (Hlásny et al. 2014). However, concerning rockfall mitigation, reducing tree density is not always a good option, since we need to maintain the high probability of impacts between rocks and stems to dissipate the energy, deviate the path, and eventually stop the rolling rocks. In this case, it could be better to focus on manipulating species composition towards the creation of a stand including a mixture more adapted to the new conditions.

Given the expected alteration of disturbance regimes within mountain areas, particular attention should be paid to the assessment of the most suitable post-disturbance management decisions. After a disturbance affecting a protection forest, management practices to be adopted should be carefully considered. In the Alps the most common post-disturbance management practice is salvage logging (i.e. the felling and removal of affected trees), followed or not by plantation. This practice has proven to act as an additional disturbance on the already disturbed environment, with several negative effects on the ecosystem processes and services (Leverkus et al. 2018; Marzano et al. 2013). Furthermore, removing these deadwood elements from forests that had a protective function can further deplete the residual protection they can still offer after the disturbance. Lying deadwood, especially if formed of large logs, greatly increases the surface roughness of the forest floor, acting as additional obstacles for the rocks (Fuhr et al. 2015). Nevertheless, its role is still generally underestimated, and not considered reliable. After a stand replacing disturbance, if the area is not salvaged, lying logs can exert a protective function for a very long time, despite decaying processes. The long-lasting and beneficial effect produced by the presence of large amounts of lying deadwood resulting in greater ground roughness has already be proven after past high-severity disturbance events. In forests affected by the storm Vivian in 1990, for instance, very few follow-up rockfall events have occurred (Wohlgemuth et al. 2017).

Moreover, deadwood elements can result in enhanced microsite conditions facilitating regeneration establishment and survival, particularly in sites affected by severe disturbances, where the disturbance might have exacerbated already harsh conditions (Marzano et al. 2013; Leverkus et al. 2018). Preferential recruitment is in fact particularly evident in climatically stressed sites. Within areas characterized by high insolation and low precipitation, fallen trees and uprooted stumps create microsites where more shade and moisture are available for seedlings, stabilizing microclimatic conditions (Beghin et al. 2010). The sheltering effects of deadwood elements directly protecting tree regeneration against high radiation, high temperature and high transpiration rates were found to be determinant in arid environments (Callaway 2007; Marzano et al. 2013). These nurse objects can also act as traps for wind-dispersed seeds. In areas with low winter temperatures, the beneficial effect of deadwood material could also result in maintaining higher soil temperatures during the night, positively affecting winter seedling survival, as found by Castro et al. (2011). At the same time, similar results can be obtained in cold environments where deadwood protects seedlings from snow gliding and favours snow melting, locally increasing the length of the growing season.

Leaving standing and lying deadwood after a disturbance can thus produce a double positive effect in the management of protection forests, providing both enhanced microsites for regeneration, and eco-engineering structures. Even if deadwood undergoes natural decaying processes, reducing its protective effects over time, during its permanence standing or lying on the ground it may still perform an important function for the time lapse required by the natural or planted regeneration to establish (Wohlgemuth et al. 2017).

With specific reference to rockfall activities, it could be even more appropriate to discuss global change rather than just climate change. Including land-use and land cover changes will in fact provide a better overview on the foreseen scenarios for this natural hazard (Lopez-Saez et al. 2016) for the Alpine area. Since temperature is rising and land use is shifting from crops or pastures to forests due to the abandonment of marginal lands, forest cover in mountain areas will increase, theoretically providing increased protection against rockfall propagations (Berger et al. 2017). Forests will potentially cover more land at the upper elevations, tree growth will increase, broadleaves will gain a higher share. The increase in lengths of forested slopes, basal area, and broadleaves percentage are all consequences leading to a generic increase in the protective effect against rockfall. In two French Alpine departments (Haute-Savoie and Isère), this expansion of protection forests has been forecast to reach around 20% of the current area (Berger et al. 2017). However, rockfall is a site-specific phenomenon, involving local lithological, geomorphological, climate and forest issues, so we cannot excessively generalize the possible effects of global change.

Conclusions

The management of protection forests under climate change is a challenging task. Unfortunately, there is no "one size fits all" management approach, but it is advisable to adopt a site and case specific tailor-made solution. Forests in the Alps will probably grow better and increase their range in the next decades, but uncertainty related to natural disturbances calls for a more careful assessment of their potential protective role. Since site conditions are changing, and therefore so are species performances, we should also change the forest management approach. Adaptive management, considering both climate and land-use changes, should be promoted. The inclusion of natural hazards and protection function assessment in forest management plans should be supported by guidelines providing factual information derived from scientific knowledge acquired in the recent research projects. Specific attention should be paid to finding evidence from research studies focusing on monitoring protection forests affected by severe natural disturbances in order to define proper post-disturbance management considering their preeminent protection function.

Acknowledgements This work was funded by the European Commission (project ASP 462 RockTheAlps) within the European Territorial Cooperation program "Alpine Space".

References

Albrich K, Rammer W, Thom D, Seidl R (2018) Trade-offs between temporal stability and level of forest ecosystem services provisioning under climate change. Ecol Appl 28:1884–1896

Beghin R, Lingua E, Garbarino M, Lonati M, Bovio G, Motta R, Marzano R (2010) *Pinus sylvestris* forest regeneration under different post-fire restoration practices in the northwestern Italian Alps. Ecol Eng 36:1365–1372

Bentz BJ, Régnière J, Fettig CJ, Hansen EM, Hayes JL, Hicke JA, Kelsey RG, Negron JF, Seybold SJ (2010) Climate change and bark beetles of the western United States and Canada: direct and indirect effects. Bioscience 60:602–613

Berger F, Dorren L, Kleemayr K, Maier B, Planinsek S, Bigot C, Bourrier F, Jancke O, Toe D, Cerbu G (2017). Eco-engineering and protection forests against rockfalls and snow avalanches. Management strategies to adapt alpine space forests to climate change risk, pp 191–210

Berger F, Rey F (2004) Mountain protection forests against natural hazards and risks: new French developments by integrating forests in risk zoning. Nat Hazards 33:395–404

Bianchi E, Accastello C, Trappmann D, Blanc S, Brun F (2018) The economic evaluation of forest protection service against rockfall: a review of experiences and approaches. Ecol Econ 154:409–418

Brang P, Schönenberger W, Frehner M, Schwitter R, Thormann JJ, Wasser B (2006) Management of protection forests in the European Alps: an overview. For Snow Landsc Res 80:23–24

Brang P, Schönenberger W, Ott E, Gardiner B (2008) Forests as protection from natural hazards. In: Evans J (ed) The forest handbook, vol II. Blackwell Science, London, pp 53–81

Callaway RM (2007) Positive interactions and interdependence in plant communities. Springer, Dordrecht, The Netherlands

Castro J, Allen CD, Molina-Morales M, Marañón-Jiménez S, Sánchez-Miranda Á, Zamora R (2011) Salvage logging versus the use of burnt wood as a nurse object to promote post-fire tree seedling establishment. Restor Ecol 19:537–544

Dale VH, Joyce LA, McNulty S, Neilson RP, Ayres MP, Flannigan MD, Hanson PJ, Irland LC, Lugo AE, Peterson CJ, Simberloff D, Swanson FJ, Stocks BJ, Wotton BM (2001) Climate change and forest disturbances. Bioscience 51:723–734

Dyderski MK, Paź S, Frelich LE, Jagodziński AM (2018) How much does climate change threaten European forest tree species distributions? Glob Change Biol 24:1150–1163

Dorren LKA, Berger F (2007) Principles of the tool Rockfor.net for quantifying the rockfall hazard below a protection forest. Schweizerische Zeitschrift für Forstwesen 6:157–165

Eelkin C, Gutierrez AG, Leuzinger S, Manusch C, Temperli C, Rasche L, Bugmann H (2013) A 2 °C warmer world is not safe for ecosystem services in the European Alps. Glob Change Biol 19:1827–1840

Eysn L, Hollaus M, Lindberg E, Berger F, Monnet JM, Dalponte M, Kobal M, Pellegrini M, Lingua E, Mongus D, Pfeifer N (2015) A benchmark of LiDAR-based single tree detection methods using heterogeneous forest data from the alpine space. Forests 6:1721–1747

Frehner M, Wasser B, Schwitter R (2007) Sustainability and success monitoring in protection forests. Guidelines for Silvicultural Interventions in Forests with Protective Functions, FOEN, Bern

Fuhr M, Bourrier F, Cordonnier T (2015) Protection against rockfall along a maturity gradient in mountain forests. For Ecol Manage 354:224–231

Hlásny T, Mátyás C, Seidl R, Kulla L, Merganičová K, Trombik J, Dobor L, Barcza Z, Konôpka B (2014) Climate change increases the drought risk in central European forests: what are the options for adaptation? Lesn Časopis—For J 60:5–18

Lindner M, Fitzgerald JB, Zimmermann NE, Reyer C, Delzon S, van der Maaten E, Schelhaas MJ, Lasch P, Eggers J, van der Maaten-Theunissen M, Suckow F, Psomas A, Poulter B, Hanewinkel M (2014) Climate change and European forests: what do we know, what are the uncertainties, and what are the implications for forest management? J Environ Manage 146:69–83

Lingua E, Marzano R, Berger F (2017) Protection forests and their management: an overview. EFI Tech Rep 101:115–122

Leverkus A, Rey Benayas J, Castro J, Boucher D, Brewer S, Collins B, Donato D, Fraver S, Kishchuk B, Lee E-J, Lindenmayer DB, Lingua E, Macdonald E, Marzano R, Rhoades CC, Royo A, Thorn S, Wagenbrenner JW, Waldron K, Wohlgemuth T, Gustafsson L (2018) Salvage logging effects on regulating and supporting ecosystem services—a systematic map. Can J For Res 48:983–1000

Lopez-Saez J, Corona C, Eckert N, Stoffel M, Bourrier F, Berger F (2016) Impacts of land-use and land-cover changes on rockfall propagation: insights from the Grenoble conurbation. Sci Total Environ 547:345–355

Marzano R, Lingua E, Garbarino M (2012) Post-fire effects and short-term regeneration dynamics following high-severity crown fires in a Mediterranean forest. iForest 5:93–100

Marzano R, Garbarino M, Marcolin E, Pividori M, Lingua E (2013) Deadwood anisotropic facilitation on seedling establishment after a stand-replacing wildfire in Aosta valley (NW Italy). Ecol Eng 51:117–122

Motta R, Haudemand JC (2000) Protective forests and silvicultural stability. An example of planning in the Aosta valley. Mt Res Dev 20:180–187

Pradhan B, Fanos AM (2017). Rockfall hazard assessment: an overview. in laser scanning applications in landslide assessment. Springer, pp 299–322

Seidl R, Rammer W (2017) Climate change amplifies the interactions between wind and bark beetle disturbances in forest landscapes. Landsc Ecol 32:1485–1498

Seidl R, Thom D, Kautz M, Martin-Benito D, Peltoniemi M, Vacchiano G, Wild J, Ascoli D, Petr M, Honkaniemi J, Lexer MJ, Trotsiuk V, Mairota P, Svoboda M, Fabrika M, Nagel TA, Reyer CPO (2017) Forest disturbances under climate change. Nat Clim Chang 7:395–402

Temperli C, Bugmann H, Elkin C (2013) Cross-scale interactions among bark beetles, climate change, and wind disturbances: a landscape modeling approach. Ecol Monogr 83:383–402

Thom D, Rammer W, Seidl R (2017) The impact of future forest dynamics on climate: interactive effects of changing vegetation and disturbance regimes. Ecol Monogr 87:665–684
Volkwein A, Schellenberg K, Labiouse V, Agliardi F, Berger F, Bourrier F, Dorren LKA, Gerber W, Jaboyedoff M (2011) Rockfall characterisation and structural protection—a review. Nat Hazards Earth Syst Sci 11:2617–2651
Wehrli A, Dorren L (2013) Protection forests: a key factor in integrated risk management in the Alps. In: Renaud FG, Sudmeier-Rieux K, Estrella M (eds) The role of ecosystems in disaster risk reduction. UNU Press, Tokyo, pp 321–342
Wohlgemuth T, Schwitter R, Bebi P, Sutter F, Brang P (2017) Post-windthrow management in protection forests of the Swiss Alps. Eur J Forest Res 136:1029–1040
Žabota B, Repe B, Kobal (2019). Influence of digital elevation model resolution on rockfall modelling. Geomorphology 328: 183–195
Zumbrunnen T, Bugmann H, Conedera M, Burgi M (2009) Linking forest fire regimes and climate-a historical analysis in a dry inner alpine valley. Ecosystems 12:73–86

Handling the Impacts of Climate Hazards via Adaptation Methods and Options

Towards a Prioritized Climate Change Management Strategy: A Revisit to Mitigation and Adaptation Policies

Cosmos Nike Nwedu

Abstract Scientific evidence shows that climate change is real and its risky impacts are localized in different parts of the world. A critical analysis of past and current climatic trends from an existentialist lens is equally axiomatic to that effect. Countries all over the world particularly developing countries face manifest extreme weather and other climate-related impacts on asymmetric scales. Therefore, it is now patent that climate change reality is no longer a contested global issue. The problem however remains how to effectively manage it. This concern over the years resulted in two major climate policies: mitigation and adaptation. Mitigation has popularly received more focus and emphasis, as all countries are required to put their best efforts in combating climate change. This is true even as adaptation was originally perceived to be an inconsequential concept likely to preposterously impede global mitigation efforts. Nonetheless, there is high uncertainty with regards to detectable and measurable result from global efforts in reducing greenhouse gas emissions. More so, differences in climate variability and its impacts question how mitigation policy in a global context can achieve inclusiveness, bearing in mind that climate impacts are locally felt. This study re-examines the relevance of mitigation and adaptation policies in a need-based context and scale of experience of some countries or regions with regards to climate change variability and its impacts. Specifically, it argues that disparities in climate impacts trigger a growing inevitable need demanding a more prioritized reactive adaptation option over mitigation policy for certain countries or regions that are ever extremely vulnerable to climate change. The study adopts a qualitative analysis approach by building its argument on existing literature.

Introduction

Until recently, climate change was a divisive issue that took the center stage of global geopolitics, diplomacy and policy-making. Several reasons account for this problem.

C. N. Nwedu (✉)
Department of Criminology and Security Studies, Federal University Ndufu-Alike Ikwo (FUNAI), P.M.B. 1010, Ikwo, Ebonyi, Nigeria
e-mail: cosmosnike.nwedu@gmail.com

W. Leal Filho et al. (eds.), *Climate Change, Hazards and Adaptation Options*, Climate Change Management, https://doi.org/10.1007/978-3-030-37425-9_19

First, there was a lack of global understandability in and unanimity on the science of climate change. This particular problem contributed to incredulity about climate change existence. Second, early global climate debates were traditionally founded upon justice, equity and fairness with regards to in what way and the extent countries were relatively responsible for greenhouse gas emissions triggering climate change. Furthermore, framing an international climate policy suitable for combating climate change raised various scientific, political, economic, social and ethical concerns that indeed further polarized global climate debates. These issues explain how and the extent climate change management and policymaking shaped and affected global climate governance. Despite these foundational concerns, climate change has increasingly continued to be part of the mainstream policy agenda of states, both at national and local government levels including cities and regions (Moser and Dilling 2011). There is good news for climate change regime today. The misconception and disbelief the general public initially held about climate change reality and anthropogenic connection has starkly vacillated. This progress is made because countries all over the world have now accepted that climate change is existent as one of the most current complex environmental issues and unavoidable threat facing humanity (Anderegg et al. 2010; Barbir et al. 2009; Root et al. 2003; Kamau and Mwaura 2013; IPCC 2014a, b). The recently concluded Paris Agreement,[1] has not only significantly quelled global climate debates, but also exemplified how to craft a globally feasible climate change policy with a proven unanimity and acceptability of countries.

However, apparent problems still characterize global climate change governance. The first is disparities in globally distributed climate impacts leading to a very difficult solution to achieving the coveted average global temperature below 2 °C. The crux of the problem is that, while climate change is a global phenomenon (Jasanoff 2010), its risk impacts are unevenly dispersed and directly felt at local and regional levels (Ayers 2010; Kazys et al. 2013). This scenario convolutes global climate risks management. Again, the future of climate change is building with high uncertainty and indeterminate hope for humanity. The Intergovernmental Panel on Climate Change (IPCC),[2] has warned that the severity of climate change impacts is likely to be on increase and developing countries with least capacity to manage the impacts would perhaps be struck most (Schneider et al. 2007). Now, finding real-world, feasible and economical answers to the problems of climate change remains a global priority (Leal Filho 2009), a concern that has resulted in mitigation and adaptation as two corresponding policies for managing climate impacts (Adhhikari 2017; Davoudi 2009). But it appears no actual, practical and economic solution has been proffered,

[1]The Paris Agreement, adopted on 12 December 2015 and entered into force on 4 November 2016 remains hailed as one of the most ambitious international agreements the world has ever agreed upon towards global climate change mitigation. For the first time in the history of global climate change governance, all countries unanimously agreed under this Agreement that climate change is a global problem that must be fought with concerted global efforts in the most possible approach.

[2]The IPCC was a brainchild of the United Nations Environment Programme (UNEP) and the World Meteorological Organization (WMO) being an internationally scientific body tasked with evaluating the state and risk- impacts of global climate change. See www.ipcc.ch.

given a noticeable growing failure of mitigation policy on the part of greenhouse gas emission reduction.

This study generally problematizes global adaptation policy and contributes to its debates by employing a pragmatic and humanistic approach. The rationale for the study is to re-examine the relative relevance of mitigation and adaptation policies in a need-based context, and its need has arisen given that efforts to managing global climate change focus more on the reduction of greenhouse gas emissions with little attention given to adaptation. The study is significant in considering which of the climate change management policies should take preference based on the current realities of disparate climate change impacts on humankind in different countries. Such consideration is inevitable because in order to effectively manage climate change, we must first appreciate what it means to diverse communities in different segments of the world (Leal Filho 2009). It is of this reason that the study analytically argues for a prioritized adaptation policy over mitigation for some countries, and or regions in the context of their specific needs and experiences of climate variability and impacts.

This study is structured into the following sections. The first section begins with a brief introduction on the past and present status of global climate policy debates. This is followed with a discussion on mitigation and adaptation as two climate change management policies, and how each policy evolved to this present time. The next section logically advocates climate change governance that gives more prioritized emphasis on adaptation over mitigation, in the context of differing global climate change impacts and development needs of specific countries, and or regions. This is justified by a clear analysis of patent reasons and advantages inherent in adaptation priority. Conclusively, the study highlights its status in global adaptation debates and offers insights for an effective adaptation policy.

Climate Change Management Policies: A Dilemma Between Two Choices

Mitigation Policy

Mitigation is typically one of climate change management policies. Article 2 of the United Nations Framework Convention on Climate Change (UNFCCC)[3] emphatically recognized mitigation as the very primary intervention by which limitation or stabilization of greenhouse gas concentrations in the atmosphere can be achieved. To

[3]The UNFCCC was the first international climate change framework adopted in New York, USA, on 9 May 1992 and entered into force on 21 March 1994. Though it laid groundwork for climate change governance, the framework has been massively criticized and ridiculed from all angles as the worst international attempt to address climate change for reasons not limited to its focus on bottom-up approach and lack of enforcement mechanism.

further international cooperation on climate change, recurrent emphasis on greenhouse gas emission limitation also manifested at article 2 of the Kyoto Protocol[4] to the UNFCCC, and much of the mitigation responsibility was then credited to Annex 1 countries[5] in accordance with their reduction commitments under article 3 of the Protocol. The institution of the UNFCCC by its conference of the parties (COP) periodically deliberates on the mandate of the Convention to combat climate change, and where necessary, modifies some of the provisions in order to stress contemporary realities of climate change and effectively strengthen global climate efforts.

The IPCC defines mitigation as 'a human intervention to reduce the sources or enhance the sinks of greenhouse gases (IPCC 2014a, b). Mitigation involves economic, social, political, and policy dimensions of efforts that aim to moderate, and where possible, prevent climate change causes, and any processes and activities that activate such causes. The interaction between man and the natural and built environments is considerably at the core of mitigation efforts primarily because such interaction constitutes a key source of emission of greenhouse gases. Mitigation also involves a more technically natural process, and as the case may be, human activity and mechanism that specifically removes greenhouse gases. For example, forest or trees are known to be potential sinks or absorbers of carbon dioxide. Forest generally plays a significant role in climate change mitigation (Ochieng 2017), and its recognition and importance has prominently featured in international climate change treaties though deforestation and forest degradation might perhaps exacerbate carbon dioxide emissions. From all indications, mitigation has so far underscored the causes of climate change and the cutback of such causes (Adhhikari 2017). This consequently determined earlier international efforts, which was thought to be most possibly effective in reducing greenhouse gas emissions and achieving average temperature (Nyong et al. 2007). Because reduction of greenhouse gases was perceived as a major solution to climate change (Ayers 2010), predominant tendency of global climate governance has over time focused on mitigation policy as the most likely inclusive policy to engage every country towards combating climate change. This completely shaped the worldview on climate change management, and occasioned an indisposition that set a historic precedent on how global geopolitics and diplomacy of climate change simultaneously played a role in combating climate change. There are good instances to explain why and how mitigation policy, in particular, dominated international climate change geopolitics in the past decades. First, the

[4]The Kyoto Protocol was adopted at COP3 in Kyoto, Japan, on 11 December 1997, as a follow-up international climate treaty in line with the mandate of the UNFCCC and as provided under article 27 of the Convention. Unlike its mother Convention, the Protocol was widely hailed for its top-down approach that provided countries with a guided blueprint of actions to follow rather than allowing them define their own methodologies in combating climate change.

[5]Annex 1 countries under the UNFCCC simply refer to developed countries or those countries that have achieved robust and significant economic and industrial development or growth. There has been a frequent argument that this category of countries is highly responsible for greenhouse gas emissions triggering global climate change, and for that reason, must take a lead in combating climate change.

UNFCCC fundamentally placed much emphasis on mitigation with little prominence given to adaptation despite the Convention recognizing developing countries' specific needs and special circumstances. Second, it could be argued that protracted past debates over climate change responsibility pretentiously gave credence to a high level of global mitigation support for all countries. Furthermore, it was believed that the absence of mitigation among countries would mean an appealing opportunity for which most countries will prefer due to its inherent potential benefits (Kjellen 2006).

Despite much-admired potentiality of mitigation in combating climate change, a growing uncertainty continues to mount over it. In the first place, in-spite of every country's involvement in mitigation efforts, a critical assessment of progress from such international collective efforts indicates a nearly undetectable and measurable global result. Even the IPCC has maintained that future climate change still poses much uncertainty particularly with regards to costs of mitigation in-spite of scientific progress. The IPCC therefore emphasized that the appropriateness of mitigation efforts has to be measured on the basis of expected reduction in average damages, the value of reduced risks, and uncertainties accompanying such efforts (Halsnaes et al. 2007). Costs of mitigation, in particular, are vague so much that particular countries would likely bear mitigation costs alone whilst some may evade them (Halsnaes et al. 2007). This will ultimately occasion unjust asymmetric distribution of climate change costs and benefits. Again, the fact that climate change vulnerability is locally felt (Ayers 2010), with its disparate impacts measured and experienced on different scales by different countries presents a substantial reason to question the legitimacy of any globally contextualized mitigation efforts. Global mitigation policy essentially invokes an almost indescribable context, as it fails to delineate how we decide what climate risks are acceptable to specific countries, and or regions before justifying such risks. This lack of understanding especially with regards to climate change economics can undermine an effective international response to climate change (Stern 2007), as well as threaten development objectives in a national context. Roberts (2010) has argued that the focus of mitigation on carbon makes the policy abstract and most at times ill understood. Therefore, global mitigation efforts do not in practice better the lives of the really vulnerable and affected populations, and or segments of the world.

Adaptation Policy: From a Dilemma to Choice

Adaptation is now a growing second option for climate change management as much emphasis is being accorded to it through international policy (Hannah et al. 2013). The IPCC defines adaptation as an adjustment in natural or human systems in reaction to real or probable climatic stimuli or their consequences, which regulates damage or uses favorable opportunities (IPCC 2007). According to Lim et al. (2004), it is 'a process by which individuals, communities and countries seek' to manage climate change impacts. In a broader sense, this involves any practice, strategy or result in a system such as ecosystem, sector, community, region, group, or household that enables the system to survive, manage, or optimally adapt to the changing environments, pressures, risks, and or opportunities connected to climate change (Smit and Wandel 2006). While adaptation process appears to have gained long existence, incorporating future climate risk into policymaking is indeed new (Lim et al. 2004). This argument nonetheless arguably informs how adaptation originally suffered a global stereotype as an inappropriate policy. The initial recourse to adaptation was met with bias and disagreeable feeling so much that it became a much-disputed and feared concept in the global climate policymaking debates. This problem makes for an assumption to argue that adaption was not seen as a necessary end to climate change. The general assumption was that adaptation lacked a global application but instead focused on specific vulnerable local groups or places, and as such, would impede global mitigation efforts devoid of 'global good' (Ayers and Forsyth 2009). Such notion was motivated by a widely held view that adaptation represented an eye-catching opportunity which some countries might consider cost-effective as it comes with instant huge benefits (Kjellen 2006). Early research shows further that adaptation largely received insufficient understanding as to what it constituted, which was in part due to deficient responsiveness accorded to it by the scientific community (Kates 1997). It has been argued further that climate change was misconceived under the UNFCCC, and thus created adaptation bias that equally elicited climate change politicization (Pielke 2005).

There were manifest facts or reactions that indeed validated the initial global indisposition towards adaptation policy, both from political and academic lens. In the United States of America (USA) for example, various climate change activists interpreted adaptation support as an effort by which the Republicans would undercut climate change action (Ayers 2010). Gore (1992) once maintained that it is a gullible sort of lethargy to believe we can adapt to anything, an indication of brash confidence in our capability to react timeously. Like other writers, Schumacher (2016) argued that adaptation be given less preference as it promotes 'private good' and not 'public good', and for that reason must be seen as an enormous lost to worldwide benefit. Furthermore, not only that adaptation has marginally undergone a historic oversight as a policy option in managing climate change (Ayers 2010), it was also underrated as a 'poor cousin' of mitigation in the climate policy field (Pielke et al. 2007), despite their coexistent reference by the UNFCCC (Klein et al. 2005). As climate change mitigation has been primarily concentrated on global welfare, considering adapting to existing impacts of climate was by no means any country's priority in spite of the already experienced climate impacts in our natural systems (Parmesan and Yohe 2003; Root et al. 2003). This suggests that the consideration of individual and state welfare on a case-by-case basis did not significantly shape early climate change geopolitics, but was rather a more collective and centralized affair of the world. Thus national interest in and survival approach to climate change were not taken seriously as an adaptation policy concern until lately. Whether a country can prioritize promotion of or investment in 'global good' for mitigation efforts over its survival prospect contingent on unavoidable actions for effective national climate adaptation is an issue arguable. But without much enquiry, a country is unlikely to promote global affairs at the detriment of its sovereign existence and citizens' welfare. Parties to the UNFCCC especially developing country parties, however, someway demonstrated their concern for national interests in early climate change debates. The global geopolitics of climate change that significantly played out from the onset to the 2015 Paris Climate Change Conference clarifies this point. For example, beyond geopolitics, economic interest was a major factor that influenced international climate policy negotiations particularly as the global south believed they would be ripped off their economic potency for sustainable national economic development and prosperity should international climate change treaty be based on a top-down approach. Even developed countries also had similar sentiment. The sensitivity laid a context and motivation to the so-called current bottom-up climate change policy rooted in intended nationally determined contributions (INDICS) of various countries.

Though interestingly, contemporary viewpoints on adaption policy are gradually changing. From an initial exclusion to inclusion in climate change agenda and policymaking process, adaptation is currently attracting a great deal of attention from researchers, policy-makers, and various organizations. It is now viewed as an inevitable option and also a priority for climate change management. Therefore, adaptation can be seen to have gradually gained importance under the UNFCCC, but not as much as mitigation (Burton et al. 2002). With an intense call to mainstream the policy, adaptation is now a highly considered effective option in managing climate change impacts and reducing vulnerability in particular, and is being applied alongside mitigation as a complement (Nyong et al. 2007). There are researchers calling for prioritization of adaptation on the basis of growing need in one country or another, and at different governance levels in such countries (e.g. Champalle et al. 2015; Henderson-Sellers 2012). A shift of attention from a mere adaptation approach to a more pragmatic 'incremental' and 'transformative' adaptation that delivers hope and reveals 'a sense of urgency' is also taking the heart of current adaptation policy research (Pelling 2011; Schlosberg et al. 2017; UK Climate Impacts Programme n.d.). It is believed that adapting to climate change impacts is as significant as mitigating its driving causes (Frohlich and Knieling 2013). Demonstrated attempts by the international community certainly attest to the mounting significance of adaptation. The Cancum Adaptation Framework (CAF)[6] in which state parties to the UNFCCC stressed that adaptation should be prioritized as much in the same way as mitigation is classically remarkable. Prior to the CAF, adaptation fund had already been launched in 2007, and officially established in 2001 at the Seventh Conference of the Parties (COP-7) to the UNFCCC in Marrakech, Morocco. There are different adaptation funds today, which include Least Developed Countries Fund (LDCF) and Special Climate Change Fund (SCCF) recognized by Decision 7/CP.7 under the UNFCCC and Adaptation Fund (AF) under the Kyoto Protocol. The Funds have different innovative design features and mechanisms of operationalization and were established

[6]The CAF was adopted by parties to the UNFCCC as part of the 2010 Cancum Agreements reached at the Climate Change Conference in Cancum, Mexico, 2010.

primarily by the recognition of the high vulnerability of several developing countries to climate change and the resultant need to adapt (Barnett and Dessai 2002; Huq 2002).

There are major reasons for rising local and international preeminence of adaptation. On the one hand, the ever-increasing uncertainty over climate change coupled with its disparate impacts means that adapting to the changing environment is inevitable. On the other hand, and as noted by the IPCC, adaptation is a necessity in order to effectively deal with the already existing impacts of climate change resulting from past emission of greenhouse gases (Mimura et al. 2014). There is a lack of adequate improvement or poor results being recorded from global mitigation efforts and evidence of further escalation of climate change impacts (Ayers 2010; Lonsdale et al. 2015). These facts are inspiring an increased global interest in mainstreaming adaptation strategies especially as prevention of climate change in near decades is unlikely regardless of any rigorousness or littleness of global mitigation efforts (Dessai and Hulme 2001; Pittock and Jones 2000; Wigley 1998). Since climate change risks cannot be completely eliminated (IPCC 2012), adaptation became an inevitable priority in order to address vulnerability and exposure to its occurrence, and to undertake capacity development and resilience to potential adverse impacts of climate change (Schwerdtle et al. 2018). However, adaptation strategies and adoption motivations differ at all levels as most countries approach adaptation in different perspectives. For instance, the need to increase the level of local resilience to increasing climate uncertainty and achieve diverse development and service delivery goals motivated early adaptation strategies in South Africa (Carmin et al. 2009). Such an increasing uncertainty is likely to garner an increasing political support for adaptation (Roberts 2010). Adaptation is thus a needful necessity (Parry et al. 1998), which requires preference over mitigation in order to first and better cope with the existing and prospective impacts of climate change.

Contextualizing Adaptation Priority

There are cogent facts for mainstreaming adaptation priority. These facts provide a logical approach to contextualizing adaption itself, which is crucial for understanding climate change dimensions and adaptation strategies. The first approach is by a perspective on the 'cause-effect' of climate change. As far as everyone is aware, climate change is triggered by certain greenhouse gas emissions. There is stark evidence of existing climate impacts from past emissions of greenhouse gases. These impacts are already threatening humanity and must be cushioned for the living of mankind. There is also likely future impact from increasing emissions. Both empirical and theoretical evidence support the cause-effect analysis of climate change. But the evidence of climate impacts is not just enough until we understand where (for which countries, regions or people) such evidence will be most useful by the appropriate climate change policy. Furthermore, different scientific reports and research, however, regularly acknowledge that developing countries have been hit most, and yet, stand the riskier chance of bearing the brunt of global climate impacts mainly due to lack of adequate technical and financial capacity to mange the impacts (Bolanos et al. 2016). Moreover, more evidence of increasing climate change risks perhaps remain remotely unreported in most developing countries, in most cases, due to lack of knowledge among vulnerable local residents and political support. Reliance on mitigation efforts by such developing countries means their livelihood and development goals are likely to be impacted negatively.

Consequently, there have been stern calls for urgent adaptation strategies as well as international support for developing countries. This call is crucial especially as climate change draws upon interdisciplinary concerns: from sustainable economic development to environmental sustainability. Contemporary research and global climate change governance continue to increasingly recognize and situate developing countries' concerns into context-based development needs. This recognition is pointedly a move towards adaptation priority in order to achieve such needs. So, it is time to consider what works and what not for many vulnerable developing countries and or regions by giving due regards to their different contexts and need-based environments, in the context of climate change management. To that degree, adaptation priority integrating local understanding of climate change risks with global assessment of risks (Ayers 2010) remains a core policy for climate change management.

There are regional or continental comparative analysis of climate change variability and impacts that present typical good case studies in situating adaptation priority. Strong scientific evidence, particularly of the IPCC scientific body makes a good case for any case study for prioritizing adaption. For example, with strong scientific evidence, 75–250 million Africans will relatively face water stress and access to food challenge (IPCC 2007). These concerns will generate high risk of vulnerability to climate change for Africans if serious social and political actions are not taken immediately. There is also a projection that production from agriculture will be severely compromised as a result of climate variability in most African countries and regions (IPCC 2007). Likewise, other research also points to observed similar risks of climate change, and can further be attested to by personal experience. Connolly-Boutin and Smith (2016) have argued that climate induced water-related problem such as drought, or increased demand leading to resource scarcity of water poses serious environmental and socio-economic stressor with severe physical trauma especially for the most vulnerable African communities. This will continue to exacerbate an unimaginable rate of food scarcity as well as poor livelihoods with long run shocks (Akrofi et al. 2012; Swift 1989). Further projection indicates that densely populated sea-lying coastal areas in Africa would gravely be affected towards the end of 21st century with accompanying wide degrading effects on mangroves, costal reefs, fisheries and tourism, and change in seasonal farming length and yield potential mostly in the semi-arid and arid areas will be experienced alongside malnutrition and food insecurity from decline in local food provisions (IPCC 2007). As pointed in IPCC report, novel findings show that Africa remains one of the very susceptible continents to climate change as a result of manifold stressors and little coping capacity

(IPCC 2007). Concern has been expressed to that effect especially with regards to future climate change, which draws on the need to enhance adaptation priority in the continent.

For other continents of the world, climate change story continues to be analogous, yet different in terms of variability and impacts experience. According to IPCC (2007) report, the next two to three decades would mark an ever increasing flooding and landslides of rocks from damaged slopes in Asia, due to the melting of glacier particularly in the Himalayas. This likely event will invariably affect water resources in the continent (IPCC 2007). There are other projected multi-impacts of climate change in Asia. They include decline in agricultural and forestry production in major parts of southern and eastern Australia to east of New Zealand due to likely increase in inferno and water deficiency, freshwater decrease in Central, South, East and South-East Asia notably in large river basins (IPCC 2007). These events will affect a billion number of people by 2050s (IPCC 2007). Projection also revealed great risks of flooding from seas as well as flooding from rivers for some mega-deltas and densely-populated mega-delta districts within coastal areas in South, East and South-East Asia, 30% decrease in crop yields in Central and South Asia by half of 21st century, rise in endemic morbidity and mortality rate from floods and droughts-linked diarrhea in East, South and South-East Asia due to hydrological changes connected with global warming, intensification of water security problems by 2030 in southern and eastern Australia, New Zealand, Northland, and some parts of eastern region due to increased evaporation and decreased precipitation (IPCC 2007). Considerable loss of biodiversity by 2020 in some ecological rich sites involving Great Barrier Reef and Queensland Wet Tropics, Kakadu wetlands, south-west Australia, sub-Antarctic islands, and the alpine areas of each country, and exacerbation of risks from sea level rise and severity, frequent storms and coastline flooding by 2050 in areas like Cairns and South-East Queensland of Australia and Northland to Bay of Plenty in New Zealand will all be experienced (IPCC 2007). The same report further revealed a threatened prospect of sustainable development in Asian countries as a result of multidimensional stresses involving natural resources and the environment, which is concomitant with fast urbanization, industrialization and economic growth (IPCC 2007).

The experience of small island regions cannot be easily forgotten in the history of climate change governance. Research shows that pacific island countries remain considerably receptive to climate health risks of highest priority (McIver et al. 2016). Many of the islands such as Fiji, Solomon Islands, Vanuatu, and Papua New Guinea are already making relocation plans as a result of rising sea level (Schwerdtle et al. 2018). Generally speaking, small islands are characteristically most vulnerable regions facing diverse climatic conditions that among other things include high sea level rise, extreme incidents, deteriorated coastal events such as coral peroxiding and loss of coastlines, which will invariably affect or threaten fisheries, tourist attraction, crucial 'infrastructure, settlements and amenities supporting the island community livelihoods (IPCC 2007). The Caribbean and Pacific regions will, in particular, experience reduced water resources to the extent of insufficiently meeting demand during the periods of low rainfall by mid-century (IPCC 2007). Leal Filho

(2017) has argued for the need to set up adaptation priority efforts for small islands. Leal Filho's argument is indeed a needful call as research shows that small islands have little preparedness for any increased climate vulnerability.

The IPCC (2007) report further evinces the degree of disparate climate impacts for European and American continents. Like other parts of the world, Europe will for the first time in history record lengthier cultivating seasons, changes in varieties of species and extraordinary magnitude of heat-wave related health impacts of climate change, which will also replicate future impacts in the region (IPCC 2007). Whereas projection shows that nearly every European regional economic sector will be adversely impacted by climate change, increased manifest disparities in their natural resources and assets will, however, still be observed and some parts of Europe such as Southern Europe, Central and Eastern Europe and Northern Europe will also experience extreme heats, droughts, decrease in summer precipitation, and forest productivity with greater incidence of increased peatland fires, and regular winter flood coupled with compromised ecosystems, respectively (IPCC 2007). Nonetheless, a study shows evidence of moderate impacts of climate change throughout Europe (Aaheim et al. 2012). This finding supports the observation that most developing countries are ever extremely vulnerable to climate change than their developed country counterparts.

The Americas at large relatively faces both current and future risks of climate change. Latin America, in particular, is projected to face multifarious climate impacts. The tropical forest by grassland in east of Amazonia will be replaced by mid-century as a result of temperature intensification and loss of soil water while arid-land vegetation will likely replace semi-arid vegetation (IPCC 2007). Similarly, there will be a major loss of flora and fauna, increased danger of flooding in littoral zones (IPCC 2007). These events will be triggered by a rise of sea level, damage of Mesoamerican coral reefs and change in the locale of south-east Pacific fish stocks due to also a rise in sea surface temperature, loss of water availability for consumption, farming and energy generation resulting from vicissitudes in precipitation arrangements and fading of glaciers as well as salinization and desertification of farming land for drier areas likely to affect crops and livestock efficiency and activate consequential food insecurity (IPCC 2007). The Northern American counterpart will as a result of heating of western mountains face winter flooding, less snowpack and heightened number too, intensity and length of heat waves in the course of century with likely negative health effects for cities presently facing heat waves (IPCC 2007). Forests in the region will be considerably impacted due to forest diseases, pests and expansive durable wildfire (IPCC 2007). Climate change impacts straddling development and environmental pollution are likely potential stressors for North American coastal communities and habitats, and increasing population alongside infrastructural need within the coastal areas will exacerbate climate change vulnerability for the people (IPCC 2007).

From all indications, whatever climate impacts a country or region faces depends on many elements of climate change, but primarily the topography of such country and region, which in similar sense, defines vulnerability dimension and degree. This equally determines short and long term impacts of climate change to a great extent. In both cases, attempting to focus on mitigation of greenhouse gas emissions is nothing, but a clear exercise in futility due to the trans-boundary nature of climate change effects. So, adaptation remains an immediate non-negotiable policy for countries experiencing major impacts of climate change, which must be systematic. Systematic adaptation is one that recognizes and addresses both current and prospective climate impacts based on determinant elements of climate change and in the context of environments and needs. On the one hand, it is defined by preparedness and provision of an important opportunity for effective adaptation monitoring, evaluation and transition on the other hand.

Benefits of Adaptation Priority

The role and significance of adaptation are increasingly being recognized in academic research and international and national policy debates (Smit et al. 2000). There is also an increasing attention on adaptation priority due to its benefits and increasing urgent need. Ambrette (2017) argued that prioritizing adaptation signifies today's responsible approach towards current and future reduction of climate impacts. Constructively, climate change impacts can be modified by numerous adaptation strategies (Smit 1994; Tol et al. 1998), and adaptation priority is crucial for effective economic and inclusive climate change management. Proper understanding of adaptation techniques and factors leading to climate change are essential in designing climate change policies. This implies that adaptation strategies are mitigation policy enablers. From understanding local vulnerability and impacts-based occurrence rate, impacts-based awareness, development needs to suitable adaptive tendencies; adaptation priority provides a point of departure for climate policy that focuses on effective management of existing climate risks based on informed best available evidence. Adaptation, in this sense, addresses immediate concerns of vulnerable local people, and expresses how much of climate risks are understandably acknowledged. Therefore, the desirability of mitigation policy ought to be a true reflection and measure of local impacts reduction of climate change that must be of value and less uncertain.

Adaptation priority is critical for an analysis of future climate change scenarios on a case-by-case basis, and to gain a profound understanding of adaptive policy process for potential climate variability and impacts through its geographical spread. This scenario provides basis for suitable adaptive strategies and supports from national and international bodies. Smithers and Smit (1997) have argued that a better understanding of distinct and group adaptation provides foresights to estimate potential adaption and also helps to address contemporary issues of sustainable development. Thus, in order to ensure appropriate mitigation measures, vulnerability risk and its dimension must first be evaluated. Furthermore, a defense can be established against exposure to future risk impacts of climate change if exposure to current variability impacts is addressed (Ribot et al. 1996).

A prioritized adaptation regime provides an open, neutral forum for policy dialogue that strengthens cooperative efforts from all stakeholders. Coupled with this, adaptation typically invokes humanitarian and development notion (Morton et al. 2008). Such notion of humanitarian and development invocation could be conceptualized to involve technical, political and economic development assistance for developing countries most vulnerable to climate change impacts. Adaptation is considerably fundamental in modeling and integrating current and future causes and impacts into existing policy frameworks, and to building transparent and robust institutions. There is an underlying argument that a failed adaption policy activates climate change environmental migration (Warner et al. 2008). The failure can occur despite that environmental migration is an intervention risks management per se, and a sort of adaptation to climate change (Schwerdtle et al. 2018). Therefore, it is arguable that a failed adaptation policy arises from a lack of priority, a case in which adaptation can be considered an afterthought. The policy of adaptation is a powerful mechanism to address climate injustice occasioned by unequal construction of climate change in general and climate impacts in particular (Agarwal and Narain 1991; Kartha 2011; O'Brien 2012). Overall, adaptation priority is paramount to a practical and effective climate change management that secures our ecosystems, human security and livelihood from the trial and error of mitigation policy regime.

Conclusion

The deprioritization of adaptation policy at the inception of climate change management regime is not doubtful. This is why adaptation was allegedly taken to be defective and ineffective in fostering the global common good, unlike its mitigation partner. The problem primarily stems from the groundwork of global climate governance under the UNFCCC. The UNFCCC placed more priority on mitigation despite that both adaptation and mitigation policies featured concurrently under the Framework. What is rather clear is that adaptation was recognized in principle than in practice. However, diffusing adaptation priority into a local context in managing climate change especially for developing countries is now increasingly demanding urgency. To that degree, social, political and economic supports from stakeholders are crucial for adaptation to take a competitive place in global climate governance. But any climate change governance seeking adaptation strategies or measures to improve people's adaptive capacity and livelihood need be locally evidence-based in order to provide policymakers with best available evidence for effective and resulted-oriented climate change management. Accordingly, it is remarkable that increasing local and global concern for an increased adaptation is gaining momentum. Thus, the era of adaptation as an afterthought is gradually waning. Nevertheless, the significance of mitigation as a management partner of adaptation is not at all dispelled, aware that each policy can be a substitute or complement of the other depending on certain facts of a particular situation of climate variability and impacts exposure. This awareness overall determines which of the policies can serve the most practical purpose and provide desired maximum outcome.

This study does not claim to examine different adaptation strategies and their effectiveness, or any kind of best adaptation strategies for different countries or regions. Thus, it is limited in terms of explanation as to what adaptation strategies are suitable for particular countries or regions. The major strength of the study is that it argues for adaptation priority of a wider global initiative, however, based on local context. One further strength lies in its comparative approach to understanding what climate change per se, as well as adaptation mean to different countries or regions of the world in terms of variability, vulnerability, degree of impacts-based experience, adaptive need and capacity and the overall understanding of these factors. The study has future potentials for climate change management in general. Therefore, it provides a framework for further research on climate change adaptation and mitigation management policies. Governments, intergovernmental organizations, non-governmental organizations, development organizations, policymakers, researchers, academics and students alike will all find this study invaluable as a source material.

References

Aaheim A, Amundsen H, Dokken T, Wei T (2012) Impacts and adaptation to climate change in European economies. Glob Environ Change 22(4):959–968

Adhhikari AP (2017) Communicating climate change: issues ahead and action needed in himalayan regions. J Polit Govance 7(2):80–96

Agarwal A, Narain S (1991) Global warming in an unequal world: a case of environmental colonialism. New Delhi: centre for science and environment. https://www.eldis.org/document/A60021

Akrofi S, Price LL, Struik PC (2012) HIV and severity of seasonal household food-related coping behaviours in rural ghana. Ecol Food Nutr 51(2):148–175

Ambrette B (2017) Prioritizing local climate adaptation through regional collaboration on Maryland's Eastern shore. A white paper prepared for the Eastern shore climate adaptation partnership by the eastern shore land conservancy. https://www.eslc.org/wp-content-uploads/2015/02/ESCAP-White-Paper_FINAL.pdf

Anderegg WRL, Prall JW, Harold J, Schneider SH (2010) Expert credibility in climate change. Proc Natl Acad Sci 107:12107–12109

Ayers J (2010) Understanding the adaptation paradox: can global climate change adaptation policy be locally inclusive? A Ph.D. thesis submitted to the London school of economics and political science. http://etheses.lse.ac.uk/393

Ayers J, Forsyth T (2009) Community-based adaptation to climate change: strengthening resilience through development. Environment 51(4):22–31

Barbir J, Leal Filho W, Spiric J (2009) An appraisal of measures to cope with climate change in the baltic sea. Manag Environ Qual Int J 20(1):82–91

Barnett J, Dessai S (2002) Articles 4.8 and 4.9 of the UNFCCC: adverse effects and the impacts of response measures. Clim Pol 2(2/3): 231–239

Bolanos TG, Costa MM, Nehren U (2016) Development of a prioritization tool for climate change adaptation measures in the forestry sector: a nicaraguan case study, report 28. Climate Service Center, Germany, Hamburg

Burton I, Huq S, Lim B, Pilifosova O, Schipper EL (2002) From impacts assessments to adaptation priorities: the shaping of adaptation policy. Clim Policy 2(2):145–159
Carmin J, Roberts D, Anguelovski I (2009) Planning climate resilient cities: early lessons from early adapters. Paper Prepared for the World Bank 5th Urban Research Symposium: Cities and Climate Change: Responding to Urgent Agenda, Marseille, France
Champalle C, Ford JD, Sherman M (2015) Prioritizing climate change adaptations in canadian arctic communities. Sustainability 7:9268–9292
Connolly-Boutin L, Smith B (2016) Climate change, food security, and livelihoods in sub-saharan Africa. Reg Environ Change 16(2):385–399
Davoudi S (2009) Framing the role of spatial planning in climate change. School of Architecture, Planning and Landscape, Newcastle University, United Kingdom, p 34
Dessai S, Hulme M (2001) Climatic implications of revised IPCC emission scenarios: the Kyoto protocol and quantification of uncertainties. Integr Assess 2(3):159:170
Frohlich J, Knieling J (2013) Conceptualizing climate change governance. In: Knieling J, Leal Filho W (eds) Climate change governance. Springer, Berlin, pp. 9–26
Gore A (1992) Earth in the balance: forging a new common purpose. Earthscan Ltd, London
Halsnaes K, Shukla P, Ahuja D, Akumu G, Beale R, Edmonds J, Gollier C, Grubler A, Ha Duong M, Markandya A, McFarland M, Nikitina E, Sugiyama T, Villavicencio A, Zou J (2007) Framing issues. In: B Metz, OR Davison, PR Bosch, R Dave, A Meyer (eds) Climate change 2007: mitigation. contribution of working group III to the fourth assessment report of the intergovernmental panel on climate change. Cambridge University Press, Cambridge, United Kingdom and New York, NY, USA
Hannah L, Ikegami M, Hole DG, Seo C, Butchart SHM, Peterson AT, Roehrdanz RP (2013) Global climate change adaptation priorities for biodiversity and food security. PLoS ONE 8(8):e72590
Henderson-Sellers A (2012) adaptation investment tool for local governments to prioritise actions on climate change. National climate change adaptation research facility (NCCARF) showcase: climate change adaptation decisions, new research and tools. Sydney, Australia. https://www.nccarf.edu.au/publications-adaptation-investment-tool-local-governments-prioritise-actions-climate-change
Huq S (2002) The Bonn-Marrakech agreements on funding. Clim Pol 2(2/3):243–246
IPCC (2007) Summary for policymakers. In: Parry ML, Canziani OF, Palutikof JP, van der Linden PJ, Hanson EC (eds) Climate change 2007: impacts, adaptation and vulnerability. Contribution of working group II to the fourth assessment report of the intergovernmental panel on climate change. Cambridge University Press, Cambridge, pp 7–22
IPCC (2012) Managing the risks of extreme events and disasters to advance climate change adaptation. In: Field CB, Barros B, Stocker TF, Qin D, Dokken DJ, Ebi KL, Mastrandrea MD, Mach KJ, Plattner GK, Allen SK, Tignor M, Midgley PM (eds) A special report of working group I and II of the intergovernmental panel on climate change. Cambridge University Press, Cambridge
IPCC (2014a) Summary for policymakers. In: Edenhofer O, Pichs-Madruga R, Sokona Y, Farahani E, Kadner S, Seyboth K, Adler A, Baum I, Brunner S, Eickemeier P, Kriemann B, Savolainen J, Schlomer S, von Stechow C, Zwickel T, Minx JC (eds) Climate change 2014: mitigation of climate change. contribution of working group III to the fifth assessment report of the intergovernmental panel on climate change. Cambridge University Press, Cambridge, United Kingdom and New York, NY, USA
IPCC (2014b) Climate change 2014: synthesis report. In: Pachauri RK, Meyer LA (eds) Contribution of working groups I, II and III to the fifth assessment report of the intergovernmental panel on climate change. IPCC, Geneva, Switzerland, p 151
Jasanoff S (2010) A new climate for society. Theory Cult Soc 27(2–3):233–253
Kamau JW, Mwaura F (2013) Climate change adaptation and EIA studies in Kenya. Int J Clim Chang Strat Manag 5(2):152–165
Kartha S (2011) Discourses of the global south. In: Dryzek JS, Norgaard RB, Schlosberg D (eds) Oxford handbook of climate change and society. Oxford University Press, Oxford

Kates RW (1997) Climate change 1995—impacts, adaptations, and mitigation. Environment 39(9):29–33

Kazys J, Leal Filho W, Stonevicius E, Rimkus E, Valiuskevicius G (2013) Climate change impact on small coastal river basins: from problem identification to adaptation in Klaipeda city. Climate Dev 5(2):1–11

Klein RJT, Schipper ELF, Dessai S (2005) Integrating mitigation and adaptation into climate and development policy: three research questions. Environ Sci Policy 8:579–588

Kjellen B (2006) Forward. In: Adger WN, Paavola J, Huq S, Mace J (eds) Fairness in adaptation to climate change. MIT Press, Cambridge, MA

Leal Filho W (2009) Communicating climate change: challenges ahead and action need. Int J Clim Chang Strat Manag 1(1):6–18

Leal Filho W (2017) Climate change adaptation in the pacific: setting-up priorities in the health sector. In: Leal Filho W (ed) Climate change adaptation in pacific countries. Springer, Cham

Lim B, Spanger-Siegfried E, Burton I, Malone LE, Huq S (2004) Adaptation policy frameworks for climate change: developing strategies, policies and measures. Cambridge University Press, Cambridge

Lonsdale K, Pringle P, Turner B (2015) Transformational adaptation: what it is, why it matters and what is needed. UK Climate Impacts Programme (UKCIP), University of Oxford, Oxford, UK

Mclver L, Kim R, Woodward A, Hales S, Spickett J, Katscherian D, Hashizume M, Honda Y, Kim H, Iddings S, Naicker J, Bambrick H, McMichael AJ, Ebi KL (2016) Health impacts of climate change in pacific island countries: a regional assessment of vulnerabilities and adaptation priorities. Environ Health Prospect 124(11):1707–1714

Mimura N, Pulwarty RS, Duc DM, Elshinnawy I, Redsteer MH, Huang HQ, Nkem JN, Sanchez RA (2014) Adaptation planning and implementation. In: Field CV (ed), Climate change 2014: impacts, adaptation, and vulnerability. Part A: global and sectoral aspects. Contribution of working group II to the fifth assessment report of the intergovernmental panel on climate change. Cambridge University Press, Cambridge, United Kingdom and New York, NY, USA, pp 869–898

Moser S, Dilling L (2011) Communicating climate change: closing the science-action gap. In: Dryzek JS, Norgaard RB, Schlosberg D (eds) The oxford handbook of climate change and society. Oxford University Press, New York, NY, pp 161–174

Morton A, Boncour P, Laczko F (2008) Human security policy challenges. In: Couldrey M, Herson M (eds) Climate change and displacement. Forced Migr Rev 31:5–7

Nyong A, Adesina F, Osman Elasha B (2007) The value of indigenous knowledge in climate change mitigation and adaptation strategies in the African Sahel. Mitig Adapt Strat Glob Change 12(15):787–797

O'Brien K (2012) Global environmental change II: from adaptation to deliberate transformation. Prog Hum Geogr 36(5):667–676

Ochieng RM (2017) The role of forests in climate change mitigation: a discursive institutional analysis of REDD + MRV. Wageningen University, Wageningen, Netherlands. https://www.cifor.org/library/5623/

Parmesan C, Yohe G (2003) A globally coherent fingerprint of climate change impacts across natural systems. Nature 421(6918):37–42

Parry M, Arnell N, Hulme M, Nicholls R, Livermore M (1998) Adapting to the inevitable. Nature 395(6704):741

Pelling M (2011) Adaptation to climate change: from resilience to transformation. Routledge, New York

Pielke RA (2005) Misdefining "climate change": consequences for science and action. Environ Sci Policy 8(6):548–561

Pielke R, Prins G, Rayner S, Sarewitz D (2007) Lifting the taboo on adaptation. Nature 445(8)

Pittock AB, Jones RN (2000) Adaptation to what and why? Environ Monit Assess 61(1):9–35

Ribot JC, Najam A, Watson G (1996) Climate variation, vulnerability and development in the semi-arid tropics. In: Ribot JC, Magalhaes AR, Panagides SS (eds) Climate variability, climate change and social vulnerability in the semi-arid tropics. Cambridge University Press, Cambridge, pp 13–54

Roberts D (2010) Prioritizing climate change adaptation and local level resilience in Durban, South Africa. Environ Urban 22(2):397–413

Root TL, Price JT, Hall KR, Schneider SH, Rosenzweig C, Ponds JA (2003) Fingerprints of global warming on wild animals and plants. Nature 421(6918):57–60

Schumacher I (2016) Climate policy must favour mitigation over adaptation. Ecole polytechnique. https://hal.archives-ouvertes.fr/hal-01279017/document

Schwerdtle P, Bowen K, McMichael C (2018) The health impacts of climate-related migration. BMC Med 16:1

Schlosberg D, Collins LB, Niemeyer S (2017) Adaptation policy and community discourse: risk, vulnerability, and just transformation. Environ Polit 3:413–437

Schneider SH, Semenov S, Patwardhan A, Burton I, Magadza CHD, Oppenheimer M, Pittock AB, Rahman A, Smith JB, Suarez A, Yamin F (2007) Assessing key vulnerabilities and the risk from climate change. In: Parry ML, Canziani OF, Palutikof JP, van der Linden PJ, Hanson CE (eds) Climate change 2007: impacts, adaptation and vulnerability. contribution of working group II to the fourth assessment report of the intergovernmental panel on climate change. Cambridge University Press, Cambridge, pp 779–810

Smithers J, Smit B (1997) Human adaptation to climatic variability and change. Glob Environ Change 7(2):129–146

Smit B (ed). (1994) Adaptation to climatic variability and change. Report of the task force on climate adaptation. Department of Geography, Guelph University, ON, Canada. https://inis.iaea.org/search/citationdownload.aspx

Smit B, Wandel J (2006) Adaptation, adaptive capacity and vulnerability. Glob Environ Change 16:282–292

Smit B, Burton I, Klein RJT, Wandel J (2000) An Anatomy of adaptation to climate change and variability. Clim Change 45(1):223–251

Stern N (2007) The economics of climate change: the stern review. Cambridge University Press, Cambridge

Swift J (1989) Why are rural people vulnerable to famine? IDS Bull 20(2):41–49

Tol RSJ, Frankhauser S, Smith JB (1998) The scope for adaptation to climate change: what can we learn from the impact literature? Glob Environ Change 8(2):109–123

UK Climate Impacts Programme (n.d.) Transforming how we think about adaptation. https://www.ukcip.org.uk/transforming-how-we-think-about-adaptation/

Warner K, Dun OV, Stal M (2008) Field observations and empirical research. In: Couldrey M, Herson M (eds) Climate change and displacement. Forced Migr Rev 31:13–15

Wigley TML (1998) The Kyoto protocol: Co_2, CH_4 and climate implications. Geophys Res Lett 25(13):2285–2288

Urban Heat Island Effect, Extreme Temperatures and Climate Change: A Case Study of Hong Kong SAR

Charles Galdies and Hok Sin Lau

Abstract The Urban Heat Island (UHI) effect is analyzed using LANDSAT8 satellite data acquired on two episodes of heatwaves over Hong Kong and processed using the split-window algorithm to retrieve the Land Surface Temperature (LST) over this area. The in situ ambient air temperatures measured by a number of local weather stations of the Hong Kong Observatory were used to validate the acquired LST. Regional temperature changes for the Hong Kong area for the 21st century generated using the climate scenario generator tool MAGICC/SCENGEN and constrained to SRES A2AIM project a rise in temperatures of between +0.9 and + 5.4 °C. The results show the existence of severe UHI effects between urban and suburban localities during two severe heatwave events. Geospatial analysis of this local UHI problem quantifies how urban parks can minimize the UHI effect and a number of adaptation measures related to urban spatial planning are being recommended in view of a changing climate.

Introduction

Rapid urbanization and the establishment of more megacities have triggered a number of environmental issues. One of the most well-known phenomena is the Urban Heat Island (UHI) effect in which urban regions experience warmer temperatures than their rural surroundings (Roth 2013). The UHI is an artifact of the complex built environment having a high density of human activities but lacks cooling vegetation, resulting thus in energy imbalances between urban and rural environments.

Studies about the relationship between land use cover and the respective local climate (Kubota et al. 2017) has shown that in most cases, UHIs tend to form over built-up areas such as industrial or commercial areas; on the contrary, 'cool islands' have been observed in vegetated areas such as parks and forests. Ongsomwang et al.

C. Galdies (✉) · H. S. Lau
Institute of Earth Systems, University of Malta, Msida MSD 2080, Malta
e-mail: charles.galdies@um.edu.mt

H. S. Lau
e-mail: judy.lauhs@gmail.com

W. Leal Filho et al. (eds.), *Climate Change, Hazards and Adaptation Options*, Climate Change Management, https://doi.org/10.1007/978-3-030-37425-9_20

(2018) used remote sensing and geographic information system to investigate the patterns of the surface temperature profile for different land use types in Thailand. Their study showed how the highest daytime surface temperatures were observed in the central business district area, while the lowest land surface temperatures were observed in agricultural areas surrounding Bangkok. To date numerous studies on UHIs have been conducted in various parts of the world (Santamouris et al. 2015) and issues related to UHIs in countries like China are now getting a lot of scientific attention in view of its rapid social development (Yao et al. 2018).

A changing climate with rising global temperatures bring into focus the sustainability of tropical and subtropical cities, such as Hong Kong with a population of 7 million that has less than 25% of its total area of 1104 km^2 available for building. This condition lends itself to heavy urbanization (Loue and Sajatovic 2012) and to maximize the urban land use, Hong Kong is surrounded by high-rise buildings of 30–80 stories. Nichol and Hang (2012) reported that the air temperature can reach to over 33 °C in a typical summer as a consequence of the urban characteristics, thus making Hong Kong a suitable laboratory for the study of the UHI and its effects. What is equally pressing is that there is very little information yet regarding the future urban climate over this region.

The objective of this study are therefore (1) to study the latest thermal signatures associated with the urban and non-urban areas in Hong Kong, and (2) to qualify the likely impact of UHI phenomena up till 2100. This chapter aims to complement and improve over past methodologies that looked at the spatial and temporal resolution of variations in urban temperatures over Hong Kong (such as the studies conducted by Nichol and Hang 2012; Nichol et al. 2014) by identifying landscape features that are giving rise to harmful microclimates. This is done by fusing high resolution digital thermal, cadastre and demographic data in order to answer the following research questions: (1) how do the urban heat islands over Hong Kong vary spatially during the chosen events? (2) what urban landscape features are driving any of the observed thermal variation? and (3) what are the likely health implications at various districts in Hong Kong? In doing so, we will be looking at some basic strategies for mitigating UHI effects and adapting to a changing climate.

Regional Climate Projections

The exact impact of climate change on the UHI remains uncertain due to many unknown variables (Chapman et al. 2017), but if a greater occurrence of heatwaves occur over megacities such as Hong Kong, the likelihood is that the impact of the UHIs may increase. A generic analysis of regional climate projections can therefore shed more light on this local problem.

The availability of climate models and their products makes possible the study of climate change over time (Santer and Wigley 2010). One climate scenario generator tool that is widely used is MAGICC/SCENGEN 5.3 (www.ucar.edu) which makes use of the full suite of General Circulation Models (GCMs) and several management scenarios as applied to the 4th Assessment Report (AR4) of the IPCC with the ability to produce multi-model ensemble averaged forecasts of temperature, precipitation and mean sea level pressure, for any given year up to 2100.

MAGICC continues to be one of the primary tools made available by the IPCC since 1990 to produce projections of future global-mean temperature and sea level rise (Meehl et al. 2007, Meinshausen et al. 2011a, b). SCENGEN works on the output of MAGICC, combined with AR4's CMIP3 archived GCM database (www.pcmdi.llnl.gov/), and a pattern scaling algorithm (Santer et al. 1990), to produce thematic maps of future changes in temperature, precipitation and atmospheric pressure. These two models work in conjunction with one another and allow the exploration of global and regional patterns of climate change in the form of gridded information for baseline climate (20-year average focused on 1990), climate change from 1990, and future climate (baseline plus change). According to Fordham et al. (2012) some limitations can arise from the choice of climate model, which then tends to influence the output of specific climate projections. Luckily, this can be minimized by looking at the statistical output provided by SCENGEN that is based on an ensemble of climate projections.

For this study, MAGICC's SRES scenario was selected by taking into account one of the likely expected scenario for the region under study. The results of the existing statistical data based on demographics and GHG emissions data for the Hong Kong region is policy scenario WRE650 (equivalent to a stabilized of CO_2 concentrations at 650 ppm along the WRE pathway, with a gradual switch away from carbon-producing fuels) and two likely reference scenarios: B1AIM (Global cooperation and economic, social, and environmental sustainability; AIM modeling) and A2AIM (equivalent to uneven economic growth and regional consolidation; where economic growth and technological advancements are slower and fragmented as regions develop separately and within the input of others). Both reference scenarios were evaluated for two periods centered on 2050 and 2100. The climate sensitivity of 3 °C and a mid-carbon cycle model (equivalent to 1.1 GtC annually) were selected for these two periods, with carbon-cycle climate feedbacks.

Additional SCENGEN model settings included a linear annual scale centered around the periods 2050 and 2100. All 20 SCENGEN GCMs were used individually and as an ensemble to derive seasonal modeling representations of the pattern of GHG-induced warming over the region.

Mapping the Urban Heat Island Effect

In order to give a spatially continuous view of the UHI, this study uses a combination of in situ data from an official meteorological station network and satellite thermal data (Yao et al. 2018). With the ongoing progress made in the field of earth observation and free data access, satellite-based remote sensing has becoming a mainstream approach to monitor patterns of UHIs since the 1970s (Rao 1972). Given its long operational lifetime starting in 1972, the LANDSAT program for example has been widely used for the detection and mapping of UHIs worldwide (Qiao et al. 2014; Fang 2015; Kikon et al. 2016; Singh et al. 2017; Estoque et al. 2017) in a continuous fashion (Mendelsohn et al. 2007).

The Study Area

Located in the continent of Asia, Hong Kong SAR covers 1054 km^2 of land, with a population of 7.392 million (World Bank 2019). According to WMO (2019) the 1961–1990 climate normals for the maximum air temperature at Hong Kong Observatory ranges between 18.6 °C (January) to 31.5 °C (July) while the minimum temperature ranges between 13.6 °C (January) and 26.6 °C (July). Analysis of the annual mean temperature data showed that there was an average rise of 0.12 °C per decade from 1885 to 2017 (https://www.weather.gov.hk/climate_change/obs_hk_temp_e.htm). The rate of increase in average temperature became faster in the latter half of the 20th century, reaching 0.18 °C per decade during 1988 and 2017.

Resulting from this decadal increase in ambient temperature, Nichol and Hang (2012) observed higher air temperature in outdoor spaces during night time which tend to cause further reliability on air conditioners in indoor spaces. Mongkok area, a commercial/residential area in Kowloon (having a population of 2.2 million and land area of 47 km^2) for example, is the most affected area that shows an annual average air temperature of more than 20.5 °C when compared to that of the nearby district of Cheung Sha Wan, which ranges between 17.9 and 18.5 °C (Wong and Lau 2013).

Two Recent Heat Wave Episodes Over Hong Kong

August 2013 Under the dominance of a ridge of high pressure, the weather over Hong Kong between August 4th and 6th was sunny and very hot. Further reinforcement of the ridge led to very hot conditions between the 8th and 12th, during which the highest air temperature (T_a) at the Hong Kong Observatory rose to a maximum of 34.9 °C on August 12th, which was the highest of the month (Hong Kong Observatory 2019). The sharpest gradient in T_a of 2.5 °C was observed on August 8th at 2000UTC between Lau Fau Shan (WMO Station Code: 45035; located at 22° 28′ N; 113° 59′ E) and Ta Kwu Ling station (WMO Station Code: 45032; located at 22° 32′ N; 114° 09′ E).

May 2018 With long hours of sunshine, a prolonged heat wave was formed in the latter half of the month with daytime T_a at the Hong Kong Observatory reaching ≥33.0 °C for 15 consecutive days between 17th and 31st May. On May 19th at 1810UTC local time, a T_a of 34.7 °C was recorded at the Hong Kong Observatory in Tsim Sha Tsui. Readings hit as high as 35.9 °C in Peng Chau, west of Hong Kong

Island, whilst Ta Kwu Ling in the northern New Territories registered 35.8 °C. The highest T_a was registered on May 30 (35.4 °C), which was also the second highest on record for the month May. The "Very Hot Weather" warning remained in force for more than 320 hrs, a record since the introduction of this warning in 2000UTC. The monthly mean T_a for May 2018 was 28.3 °C, which constitutes the highest record for May (Hong Kong Observatory 2019). The temperature variations during the 2018 heatwave episode were stronger than in 2013, with Ta Kwu Ling station registering a T_a of 33.4 °C with a low 4 knots wind blowing from the SW, compared to a lower T_a of 30.5 °C registered at Lau Fau Shan (WMO Station Code: 45035; located at 22° 28′ N; 113° 59′ E) at 0300UTC, at which a refreshing 6 knots wind speed was blowing from the West. The Lau Fau Shan station is close to the coast and is exposed to sea breezes mainly from the West. This might explain the lower temperature observed at the time.

The diurnal variations observed offered further insights into the dynamic nature of these heatwave events. The observed late morning temperatures are indeed considered to represent considerable stress to the local community.

Mapping Temperature Gradients of Hong Kong During the Heatwave Events

The thermal environment of Hong Kong is examined at high spatial resolution using satellite images provided by LANDSAT's Operational Land Imager (OLI) sensor with 9 bands having a spatial of 30 m and the Thermal Infrared Sensor (TIRS) with 2 bands having a resolution of 100 m (Finkl and Makowski 2014) corresponding to daytime (higher solar radiation) when the surface UHI (Yao et al. 2018) can be best observed. We consider these two heatwave events as ideal case studies for this chapter. Available LANDSAT8 satellite overpasses closest to maximum air temperature on August 9th, 2013 (LC81220442013221LGN00) and May 19th 2018 (LC81220442018139LGN00 and LC81220452018139LGN00), were downloaded from USGS EarthExplorer.

The technique used to estimate the Land Surface Temperature (LST), is the Split-Window Algorithm (Becker and Li 1990). The in situ ambient air temperatures measured by official meteorological stations of the Hong Kong Observatory at 2 m above ground on these specific dates and time were obtained to compare with the remotely-sensed LST.

Derivation of the Land Surface Temperature

Satellite remote sensing of the LST includes the correction of the remote measurement for the emission and the absorption of the atmospheric water vapour, as well as for the effect of the surface emissivity (Sobrino et al. 1993). Based on radiative transfer theory, for a cloud-free atmosphere under thermodynamic equilibrium, the channel radiance $B_i(T_i)$ on Top Of the Atmosphere (TOA) can be approximated as follows:

$$B_i(T_i) = \varepsilon_i B_i(T_s)\tau_i + (1 - \varepsilon_i)R_{atm\_i\_down}\tau_i + R_{atm\_i\_up} \tag{1}$$

where,

τ_i	Effective transmittance of the atmosphere in channel i.
ε_i	Channel effective surface emissivity.
B_i	Plank function.
$B_i(T_s)$	Radiance if the surface is a black body with a surface temperature T_s (K).
$R_{atm\_i\_up}$ and $R_{atm\_i\_down}$	Upward and downward atmospheric thermal radiance.

The advantage of the split-window algorithm is that it removes the atmospheric effect through differential atmospheric absorption in the two adjacent LANDSAT8 thermal channels centered at about 11 and 12 μm (Zhang et al. 2006) and the combination of the brightness temperatures is used for LST estimation. This algorithm has been widely used in LST retrieval from several sensors and is considered as one of the most effective and successful estimation method for retrieving LST (Wang et al. 2018).

Information about the emissivity is obtained by calculating the Normalized Difference Vegetation Index (NDVI) for which bands 2–5 were needed; and for the TIR bands 10 and 11 were used to estimate the spectral radiance.

From LANDSAT8 metadata, Eq. 1 above can be used by transforming it as follows:

$$\begin{aligned} LST = {} & TB_{10} + C_1(TB_{10} - TB_{11}) + C_2(TB_{10} - TB_{11})^2 + C_0 \\ & + (C_3 + C_4W)(1 - \varepsilon) + (C_5 + C_6W)\Delta\varepsilon \end{aligned} \tag{2}$$

where,

LST	Land Surface Temperature (K).
C_0–C_6	Coefficient values of SW (Skokovic et al. 2014).
TB_{10} and TB_{11}	Brightness Temperature (K) of Bands 10 and 11.
ε	Mean Land Surface Emissivity of the TIR Bands.
W	Atmospheric water vapour content (g/cm^2).
$\Delta\varepsilon$	Difference in Land Surface Emissivity.

Calculating the Brightness Temperature (TB)

TB is a parameter that indicates both the rate of energy that radiated at a certain wavelength and the surface brightness (or the intensity) of the source (Bhatnagar and Livingston 2005). The calibration process to convert the DN values (Q_{cal}) from each TIR bands to TOA spectral radiance (L_λ) is done as follows:

$$L_\lambda = M_L Q_{cal} + A_L \tag{3}$$

where,

L_λ TOA spectral radiance.
M_L Band specific multiplicative rescaling factor.
Q_{cal} Quantized and calibrated standard product pixel values.
A_L Band-specific additive rescaling factor.

Once calculated, L_λ must be then converted to the At-Satellite Brightness Temperature as follows (Rajeshwari and Mani 2014):

$$T = K_2/(\ln[(K_1/L_\lambda) + 1]) \tag{4}$$

where,

T At-Satellite TB.
K_1 and K_2 Band-specific thermal conversion constants.
L_λ TOA spectral radiance.

Calculating the Land Surface Emissivity (ε)

The surface emissivity refers to the ratio of the radiation emitted by an object at a surface temperature to the blackbody radiation from the object that is at the same surface temperature (Geiger et al. 2012). The NDVI threshold method was used to estimate the normalised **ε** in order to separate the temperature from the emissivity since typically, the LST measurements stand for the radiometric temperature of the sunlit over non-vegetated surface areas (Rajeshwari and Mani 2014):

$$LSE = \varepsilon_s(1 - FVC) + \varepsilon_v FVC \tag{5}$$

where,

LSE Land Surface Emissivity (ε).
ε_s Values of the *soil* emissivity of band 10/band 11 (Skokovic et al. 2014).
ε_v Values of the *vegetation* emissivity of band 10/band 11 (Skokovic et al. 2014).
FVC Fractional Vegetation Cover.

The FVC describes the nature and the amount of the vegetation cover as well as the proportions of the vegetation on the ground surface and the visibility of ground to a sensor (Weng et al. 2004).

Calculating the NDVI

The NDVI is required by the SW algorithm as an indicator of vegetation abundance. It refers to the band ratio of the radiation properties of chlorophyll (which takes up visible light for photosynthesis) and the cell structure of leaves (which reflects the near-infrared radiation) (Markert et al. 2003). Data collected from OLI, specifically the red (band 4) and near-infrared (band 5) images were used to produce the NDVI images. According to Tucker (1979), the equation of NDVI is as follows:

$$NDVI = (B_5 - B_4)/(B_5 + B_4) \quad (6)$$

where,

B4 and B5 Quantized and calibrated standard product pixel values of OLI band 4 and 5 respectively.

For the calculation of the FVC (Eq. 7), the values of the NDVI, the NDVI of bare soil ($NDVI_s$) and the NDVI of vegetation ($NDVI_v$) are needed as follows (Rajeshwari and Mani 2014):

$$FVC = (NDVI - NDVI_s)/(NDVI_v - NDVI_s) \quad (7)$$

where,

$NDVI_s$ NDVI of bare soil.
$NDVI_v$ NDVI of vegetation.

Applying the correct values of the $NDVI_s$ and $NDVI_v$ is a critical task since those values are generally seasonal and regional specific. Sobrino and Raissouni (2000) proposed the values of $NDVI_s$ equivalent to ≤ 0.2 and $NDVI_v$ equivalent to ≥ 0.5 (for global conditions applicable over any region in the Earth; Walawender et al. 2014); while Gutman and Ignatov (1998) proposed the values of $NDVI_s = 0.04 \pm 0.03$ and $NDVI_v = 0.52 \pm -0.03$ (which refers to the minimum and maximum values of the desert and the ever green areas).

The LSE for Band 10 and Band 11 from the TIR were generated and the mean and difference land surface emissivity was found from the following two equations (Rajeshwari and Mani 2014):

$$\varepsilon = (\varepsilon_{10} - \varepsilon_{11})/2 \quad (8)$$

$$\Delta\varepsilon = \varepsilon_{10} - \varepsilon_{11} \tag{9}$$

where,

ε	Mean land surface emissivity.
ε_{10} and ε_{11}	Land surface emissivity of Bands 10 and 11 respectively.
$\Delta\varepsilon$	Difference in land surface emissivity.

Deriving the Atmospheric Water Vapour

This was derived using the closest global atmospheric model reanalysis data valid for 0300UT from the Global Forecast Model (GFS; 0.5° × 0.5°) archived at the National Climate Data Center (http://nomads.ncdc.noaa.gov/data.php#hires_weather_datasets) (Lubchenco 2011). The modeled water vapour content in the atmosphere over Hong Kong area (22.2670° N, 114.1880° E) during satellite overpass was 47.79 kg m^{-2} (for August 9, 2013) and 44.70 kg m^{-2} (for May 19, 2018).

Finally, using Eq. 1, the LST (K) was obtained in raster format for the entire LANDSAT scenes processed. A cloud mask was applied to the final LST raster maps according to Corredor (2018). Cloud cover was much more extensive in the imagery acquired on May 19, 2018.

Spatial Analysis Using GIS

Geographic Information Systems tools were used to include additional contexts to the derived LST, such as streets and road network, topographic and demographic information of the area of interest to provide a comprehensive analysis of the drivers and impacts of the UHI on urban and sub-urban areas in the region. Spatial data produced by DATA.GOV.HK were downloaded from Esri China (HK) Ltd. These consisted of: (1) Population distribution by household sizes in 2016 within the 18 districts of Hong Kong, and (2) Population Distribution by Age Group 2016. They are both subset of the census data 2016 made available by the Census and Statistics Department under the Government of Hong Kong Special Administrative Region.

Relating the Derived LST with Corresponding T_a Measurements

Table 1 shows the weather conditions at time of satellite overpass and the corresponding estimated LSTs of the pixels around the individual station location. The WMO-coded weather stations are located in urban (45039; 45032), sub urban (45035; 45044)

Table 1 Weather observations at 0300UTC and estimated LST on August 9, 2013 (a) and May 19, 2018 (b)

Station (WMO Code)	Wind speed (knots)		Wind direction		Air temperature—T_a (°C)		Estimated land surface temperature (°C)	
Event	(a)	(b)	(a)	(b)	(a)	(b)	(a)	(b)
Hong Kong International Airport (45007)	4	8	WNW	W	31.2	32.3	29.5	30.5
Ta Kwu Ling (45032)	2	4	W	SW	32.8	33.4	Cloud cover present	26.3
Lau Fau Shan (45035)	8	6	SW	W	31.7	30.5	24.1	26.5
Sha Tin (45039)	4	6	ESE	SSW	32.7	32.7	Cloud cover present	Cloud cover present
Cheung Chau (45044)	4	6	W	S	29.8	30.8	25.2	25.1

Source Hong Kong Observatory

and industrial (45007) locations. Unfortunately due to cloud cover restrictions, some of the stations could not be used for comparison between T_a and LST.

Meteorologically speaking, the highest intensity UHIs are expected to occur under low wind speed and low cloud cover conditions (Arnfield 2003; Hoffmann et al. 2012), since low wind speeds would maintain the heat within the city while the low cloud cover (especially over rural areas) would trap more radiation (Ackerman 1985; Bonan 2008) and therefore decreases the temperature gradients between the countryside and the city (Oke et al. 1979).

The temperature gradient between thermal hotspots and suburban environments can be seen from Table 1. This is the case between Hong Kong International Airport and its surroundings (32.3 °C) versus Cheung Chau (30.8 °C) during the May 2018 heatwave episode. As detected by this study, an impressive difference of 5.4 °C can be also seen in the LST between these two locations. The wind speed of 8 knots measured over the International Airport is defined as a gentle breeze according to the Beaufort Scale and therefore it might have led to a decrease in the LST as a result of this surface advection. On the other hand, a light breeze was measured at Cheung Chau, which might have maintained an overall high ambient temperature, thus decreasing the observed UHI gradient at the time of observation. A similar temperature gradient appears to have taken place during the August 2013 heatwave, with an estimated difference in LST of 4.3 °C between these two same locations.

Heat Stress Mapping

There are two factors that commonly affect the occurrence and intensity of UHIs, these being: (1) the above-mentioned meteorological conditions (Chapman et al. 2017), and (2) the urban design (such as density of built up areas, aspect ratio, sky view factor, etc.) and construction materials (Shashua-Bar et al. 2004). What follows is a discussion that takes into account the above two factors by taking into account the various land use cover and their thermal signature encountered in Hong Kong at the time of satellite overpasses.

Urban Green Spaces

Analysis at street level provides spatially explicit descriptions of the distribution of urban heat throughout the city at the time analysis. During late morning, a heat distribution pattern was already observed with evident temperature gradients between city centres and their outskirts.

Urban green areas, such as parks and sideway greenery show a tendency towards cooler temperatures. Areas with lower canopy cover, like those along around freeways and arterial roadways consistently appear hotter, with small and localized 'pockets' of heat within close proximity. These findings are consistent with thermodynamics literature and one of the main reasons why urban parks are the coolest in the city is because of the occurrence of evapotranspiration (Baldinelli and Bonafoni 2015), thus acting like a temperature regulator. Therefore, strategic citing of trees and related green infrastructure can reduce the UHI and cool the air by between 2 and 8 °C, (Doick and Hutchins 2013). Shading is also another direct effect of parks with high canopy and dense tree configuration which can aid in reducing thermal discomfort. The observed mapped results therefore originate from such site-specific and background effects of land cover.

Urban planning measures to minimize the intensity of the heat distribution over them are therefore crucial. The Kowloon peninsula for example, consists of commercial and residential units, busy traffic and roads linkages, and the presence of green spaces can mitigate by the heat generated at the core of the urban city. Results show that the LST over the urban Kowloon Park and neighbouring green infrastructure was lower when compared to other surrounding built-up areas (Fig. 1).

The cooling effect of the Kowloon City green space is seen to extend out into the surrounding area towards the Tsimshatsui Kowloon Park area to the northeast. Givoni (1998) demonstrated how the cooling effect felt at the 0.5 ha Benjamin Park in Haifa, Israel extended up to a 150 m buffer. Similarly, a cooling band of 1–2 °C extended some 20 m around a small (0.24 ha) green space in Kumamoto City, Japan (Saito 1990). Using mathematical models, Honjo and Takakura (1990) calculated that a 100 m wide greenspace cools to a distance of 300 m and a 400 m-wide greenspace cools to a distance of 400 m. Therefore, even for smaller size parks and urban green

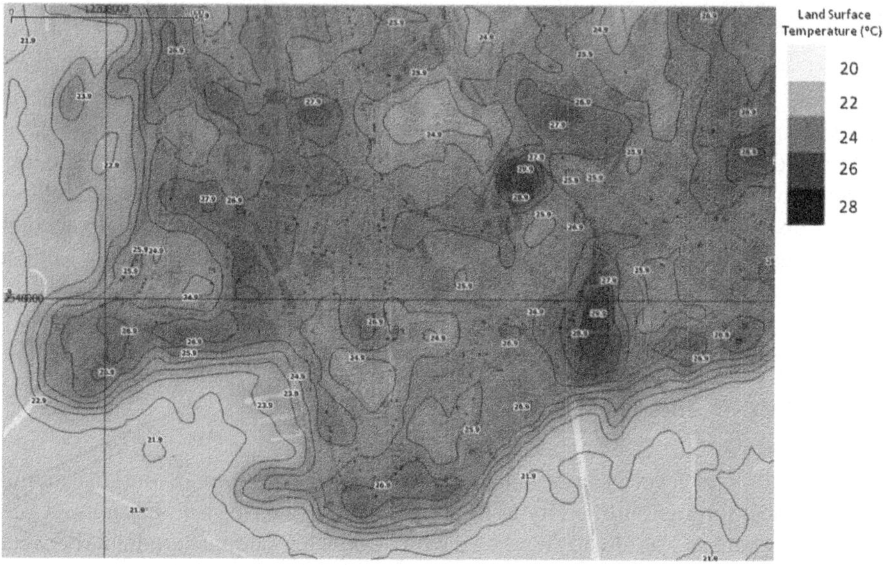

Fig. 1 Example of the LST distribution (in °C) over part of Kowloon City as detected on August 9, 2013 at 0300UTC

landscaping such as along Chatham Road South, the cooling influence can be seen to be measurable.

The susceptible nature of the Kowloon peninsula towards a higher temperature regime stems from the fact that it is topographically sheltered both from the North and the South which according to Yan (2007) very mild wind flows prevalently from the south-southwest during the summer period over this part of Hong Kong. The high topographic nature flanking the peninsula on the north and south of it shelters it from any degree of cooling that can be brought by this southerly wind surface advection.

Landfills and Expressways

UHI behavior have been observed in landfill environments in the past (Plocoste et al. 2014) such as in Guadaloupe, significant enough to generate an urban breeze during the night and pollute nearby towns and villages. In the present study, a very strong thermal signature was detected over the West New Territories Landfill (WENT) located in Nim Wan, Tuen Mun. This landfill has an area of 11 ha and receives municipal, construction and special waste (epd.gov.hk). It seems to be strategically positioned there since it is generally exposed to mild winds from the NE (winter and autumn), Easterly (spring), and Southeasterly (summer) winds (Yan 2007). Strong thermal gradients are observed over this land cover type from the 2018 thermal imagery, most probably sustained by the bordering vegetation. It is good to note the ability of LANDSAT imagery to detect narrow temperature gradients thus allowing for temperature analysis at the street level for the entire study area without any resampling of the. Nowhere is this more important than in the discernible, smooth gradients of temperature surroundings heat-reducing landscapes (such as major parks, natural areas, vegetated terrain, etc.) that can be described only with high spatial resolution data since coarser pixels would obscure these subtleties, such when ASTER satellite data is used.

The Hong Kong International Airport is located in a reclaimed land with a very high proportion of concrete area, as a result contributing its own prominent UHI. The associated terminals and reclaimed land cover to make up for route connectivity (such as the Sham Shui Kok Drive) also show a strong thermal signature.

Residential Areas

When the locations of UHIs were spatially overlaid with Hong Kong's demographic information, a disturbing picture emerges. When considering, for example, the number of households having 6 or more members aggregated by districts, one immediately notices the high numbers of such households that are located in areas with detectable UHIs such as Kowloon City (N = 117,957), Yau Tsim Mong (N = 12,715), Kwun Tong (N = 23,289) and Kwai Tsing (N = 19,246). Inhabitants living in crowded conditions become more susceptible to prolonged heat stress.

Another interesting 2016 demographic statistic is the spatial distribution of age categories. Figure 2 shows the distribution of residents aged 70 and more living in hotspot districts. It is well known that Hong Kong experienced a massive increase in its elderly population in recent decades with an average life expectancy of 80 years. Hui (2000) explains how some 14% of the older persons live on their own or with unrelated persons. This condition makes them even more susceptible to environmental stressors such as heatwaves and the impact of urban quarters as far as UHIs is concerned.

Regional Climate Projections and Implications on UHIs for 2050 and 2100 in Hong Kong

The results of the climate modelling scenarios for Hong Kong for the seasonal mean temperature change by 2050 and 2100 indicate that all 20 GCMs (except one model: INMCM-30 WRE_650 A2AIM) projected increasingly positive temperatures for Hong Kong. Results show increased temperature projections under WRE_650 B1AIM (equivalent to AR5's RCP 4.5) for 2050 that range between +0.8 and +2.4 °C and between +1.5 and +4.3 °C for 2100. A slightly higher but narrower

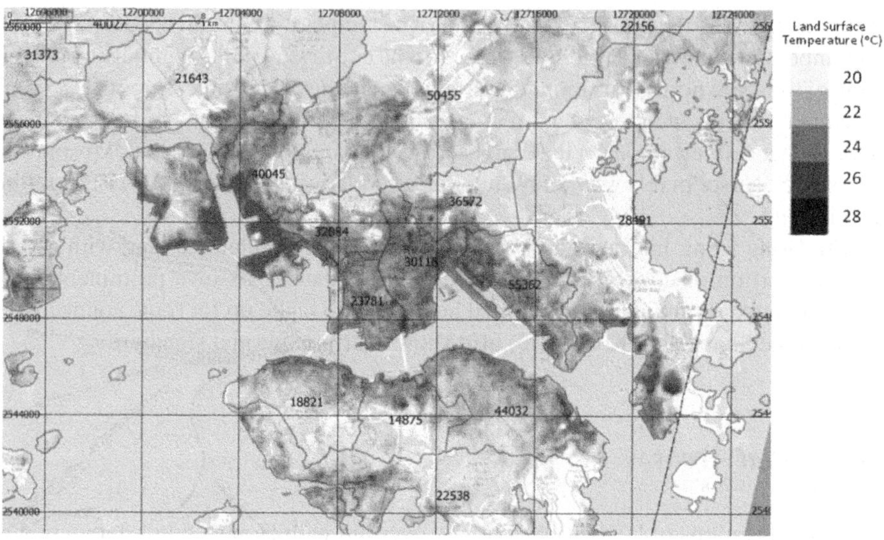

Fig. 2 Distribution of inhabitants aged 70 years and more against a backdrop of the UHI as detected on August 9, 2013 at 0300UTC. Source of Census data: DATA.GOV.HK

range (+1.4 to +3.2 °C) was obtained by Lee et al. (2014) using different conditions as defined by IPCC AR5 CMIP5 for RCP4.5 for Hong Kong.

The largest inter-model fluctuation was seen for Hong Kong's winter season for both 2050 and 2100 projections, indicating some degree of modeling uncertainty for this season. On the other hand, the modeling ensemble seems to be more confident in the projected temperature range for both the summer and autumn season. The most important result here is that all models are in agreement with regards to an overall increase in the ambient air temperature by 2100, especially for winter, by around + 2 °C from the baseline).

For the worst case climate scenario, i.e. WRE_650 A2AIM (equivalent to AR5's RCP 8.5), the temperature projections for 2050 range between −0.2 and +1.75 °C and +0.9 and +5.4 °C for 2100. A similar range was obtained by Lee et al. (2014) using different conditions as defined by IPCC AR5 CMIP5 for RCP 8.5 for Hong Kong, while Nichol et al. (2014) obtained a range of +3.0 to +6.0 °C by 2100.

This study also points to a somewhat different spring climate regime in the future during which higher temperatures are expected to be reached when compared to the least-worst case scenario (B1AIM). The projections for 2100 under the A2AIM SRES have less of a spread than for the other two seasons, indicating a stronger agreement between the models towards higher mean temperatures of around +2.5 °C (for JJA) and of +2.9 °C (for SON).

It should be pointed out that the absence of downscaling processing makes the impact of urbanization on the above-mentioned temperature projections quite challenging. Research has shown that the UHI does not remain the same under climate change due to multi-parameter changes, and so the current UHI cannot be simply added to climate change projections (McCarthy et al. 2009). Here, both urban and rural areas may respond differently to climate change due to differences in weather variables such as cloud cover, wind speed, evapotranspiration and radiative fluxes (Hoffmann and Schluenzen 2013). Where evapotranspiration is concerned, this is a key component that is able to modulate the UHI. This is because an increase in soil dryness due to a warmer climate would lead to a reduction in the rural evapotranspiration, meaning a decrease in the UHI due to lower variation between rural and urban temperatures (Oleson 2012). The relationship can also depend on changes in meteorological conditions. Thus if climate change decreases winds and cloud cover, then the UHI can be expected to increase and vice versa. Examining climate change without considering urban land use patterns could result in under-estimating future increases in urban temperatures, both mean and extreme values. Unlike microscale climate models, the use of MAGICC/SCENGEN models does not attempt to simulate complex climate or weather systems interactions for the creation of long-term forecasting products of the climate. Often, similar (but scarce) studies point to areas where further simulations are needed, as climate model performance can often vary at different locations, scales and computational physics (Taha 1999; Loikith et al. 2015).

Hong Kong's 2030 strategic planning blueprint (www.pland.gov.hk) for example, aims at a change in land use/cover that is directed to meet the future needs of commerce and industry, such as new innovation and technology centres. It also aims at regenerating old and developing new urban centres to meet housing and community needs, conserving the natural landscape and enhancing the townscape. Hence it is quite difficult to see how the expected future climate and the planned future urbanization of Hong Kong will lead to either the improvement or deterioration of livable conditions as far as urban heat stress is concerned.

As Hong Kong is the most densely built city in the world, many additional factors may contribute to the sharp increase in temperature, and is therefore important to continue understanding how urban temperatures will change in order to inform urban climate adaptation strategies for Hong Kong. If the estimations made by Nichol et al. (2014) of an increased UHI (Chapman et al. 2017) and a temperature increase of +3.7 to +6.8 °C from the current baseline (when using a constant urbanization rate over the last 100 years of 0.08 °C per decade) are correct, then the occurrence of extreme weather events such as heat waves will become ever more serious, and urgent and effective adaptation measures need to be put in place in the near future.

Limitations

The accuracy of the radiative transfer equation-based method depends mostly on how well the atmospheric profiles are able to represent the actual atmosphere over the area of interest. This study follows similar ones that make use of modelled data to derive atmospheric profile (Coll et al. 2012) whenever real-time radiosonde data is lacking. As far as the LST algorithm is concerned, Vera et al. (2010) observed that in terms of absolute values, the algorithm used by this study shows the lowest errors when compared to other versions, and therefore it is expected that the presently-derived LST maps are to be considered as sufficiently accurate for the purpose of this study. However, one should highlight the fact that the LST maps are based on the native resolution of the TIRS channels of LANDSAT8 and ideally more in situ readings are required in order to identify more accurate the LST gradients observed especially during heatwaves.

Future Prospects

What follows from this study is the possibility to identify urban landscape characteristics that are both amenable to modification, and of direct relevance to public policy, urban planning, and public health authorities in favour of reducing extreme heat and related fatalities.

It is being recommended that authorities make urgent use of the proposed "smart tools, for smart growth"—an initiative aimed at improving the city's social, economic and environmental resilience to a warmer climate. Such initiative must be strongly data-driven, aimed at a number of actions such as a "greening strategy" and "cool material strategy". While the former would be to preserve and expand the tree canopy to providing cooling on hot summer days to the benefit of people with health conditions or disadvantaged, the latter could promote the extensive use of paving and roofing materials that reflect sunlight.

Such priorities can be mapped and analyzed, district by district, to identify areas that would benefit most from a combination of these interventions. In addition to canopy coverage, the analysis can incorporate income levels, public health information, and temperatures in areas with the highest pedestrian traffic, principally focusing on those districts with the highest elderly and child populations. Street safety can also be included to identify places where "green streets" could improve pedestrian and bicycle safety while also improving water quality and cooling urban heat islands.

Conclusion

Our results help explain the greatest variations in temperatures seen across the many districts of Hong Kong during two prolonged heat wave events. While other researchers have observed similar results (such as Wang et al. 2018), this chapter goes a further by providing UHI snapshots at greater detail and relating them to topographic, meteorological, climatic and demographic data in order to understand better the implications of urban heat at this location.

Essential to this understanding is the ability to track the spatial distribution temperature across the entire metropolitan region. Even though only two time periods were used for this assessment, it is considered to be sufficient to describe fairly conservative sources of heat in view of the urban design, topography and land cover type. These include heat sources such as depots/yards, heavy industrial zones, ports, transit corridors, and the airport have consistently higher temperatures. Equally important are the heat sinks, such as urban parks and suburban vegetation. For example, green spaces and wider green infrastructure should occupy an area of at least 0.5 ha in order to achieve cooling at significant distances beyond the site boundaries.

Regional temperature changes over Hong Kong area for the 21st century are discussed in this chapter. Climate models constrained to SRES A2AIM project a rise in temperatures between +0.9 and +5.4 °C for 2100 showing a somewhat different warmer spring climate regime.

In order to follow on the UHI effect in Hong Kong, more long-term monitoring and research are needed and the results utilized for future urban development such as land reclamation, different land use planning and construction materials, transport and road planning, urban greening, and urban design. Moreover, there is also the need for policy enforcement in favour of the comprehensive plans and design guidelines.

Apart from urban design, the results of this study can be used to inform public health programmes. The high spatial resolution provided by this study provides an unparallel description of potential heat exposure within the various districts of this city. Locations in which the heat is 'trapped' and does not dissipate are important to identify especially for vulnerable city dwellers. By coupling an understanding of exposure data with the demographic information specific to susceptible populations (i.e. older adults) and coping capacities (such as number of people residing in the same household) this study can feed into any heat-health mitigation strategy at municipal scale. This is because spatially precise exposure information increases the likelihood of a successful overall health outcome for urban population (Voelkel et al. 2016).

Environmental change will likely create impacts that go beyond our ability to respond effectively. This study offers a timely and effective approach to address them before they occur. Although restricted to just one city, this methodology can be made transferable to other metropolitan regions, and similar applications are currently underway by the authors.

Acknowledgements The authors are grateful to Climate Research Unit, University of East Anglia, Norwich, UK and the National Communications Support Program, UNDP/GEF, New York, USA for providing MAGICC/SCENGEN 5.3 code.

References

Ackerman B (1985) Temporal march of the Chicago heat Island. J Clim Appl Meteorol 24(6):547–554

Arnfield AJ (2003) Two decades of urban climate research: a review of turbulence, exchanges of energy and water, and the urban heat island. Int J Climatol 23(1):1–26

Baldinelli G, Bonafoni S (2015) Analysis of albedo influence on surface urban heat island by spaceborne detection and airborne thermography. In: Murino V, Puppo E, Sona D, Cristani M,

Sansone C (eds) New trends in image analysis and processing—ICIAP 2015 workshops. Springer International Publishing, Genoa, Italy, pp 95–102
Becker F, Li Z-L (1990) Toward a local split window method over land surface. Int J Remote Sens 3:369–393
Bhatnagar A, Livingston WC (2005) Fundamentals of solar astronomy, vol 6. World Scientific
Bonan G (2008) Ecological climatology: concepts and applications, 2nd edn. Cambridge University Press, New York
Chapman S, Watson JEM, Salazar A, McAlpine MTCA (2017) The impact of urbanization and climate change on urban temperatures: a systematic review. Landscape Ecol. https://doi.org/10.1007/s10980-017-0561-4
Coll C, Caselles V, Valor E, Niclòs R (2012) Comparison between different sources of atmospheric profiles for land surface temperature retrieval from single channel thermal infrared data. Remote Sens Environ 117:199–210
Corredor X (2018) Cloud Masking Qgis plugin (Version x.x), SMByC-IDEAM and FAO. Available: https://smbyc.bitbucket.io/qgisplugins/cloudmasking. Last accessed Apr 2019
Doick K, Hutchings T (2013) Air temperature regulation by urban trees and green infrastructure. Research Note. Forestry Comission, UK. https://www.forestry.gov.uk/forestresearch. Last accessed 8 Dec 2019
Estoque RC, Murayama Y, Myint SW (2017) Effects of landscape composition and pattern on land surface temperature: an urban heat island study in the megacities of Southeast Asia. Sci Total Environ 577:349–359
Fang G (2015) Prediction and analysis of urban heat island effect in Dangshan by remote sensing. Int J Smart Sens Intell Syst 8(4):2195–2211
Finkl CW, Makowski C (2014) Remote sensing and modeling: advances in coastal and marine resources, vol 9. Coastal Research Library. Springer
Fordham DA, Wigley TML, Watts MJ, Brook BW (2012) Strengthening forecasts of climate change impacts with multi-model ensemble averaged projections using MAGICC/SCENGEN 5.3. Ecography 35:4–8
Geiger R, Aron RH, Todhunter P (2012) The climate near the ground. Springer Science and Business Media. ISBN: 3322865827, 9783322865823
Givoni B (1998) Impact of green areas on site and urban climates (chap 9). In: Givoni B (ed) Climate considerations in building and urban design. Wiley, New York, pp 303–330
Gutman G, Ignatov A (1998) The derivation of the green vegetation fraction from NOAA/AVHRR data for use in numerical weather prediction models. Int J Remote Sens 19:1533–1543
Hoffmann P, Krueger O, Schluenzen KH (2012) A statistical model for the urban heat island and its application to a climate change scenario. Int J Climatol 32(8):1238–1248
Hoffmann P, Schluenzen KH (2013) Weather pattern classification to represent the urban heat island in present and future climate. J Appl Meteorol Climatol 52(12):2699–2714
Hong Kong Observatory (2019). https://www.hko.gov.hk/ Last accessed Apr 2019
Hui YG (2000) The case in Hong Kong. Ageing Int 2000:47
Honjo T, Takakura T (1990) Simulation of thermal effects of urban green areas on their surrounding areas. Energy Build 15:443–446
Kikon N, Singh P, Singh SK, Vyas A (2016) Assessment of urban heat islands (UHI) of Noida city, India using multi-temporal satellite data. Sustain Cities Soc 22:19–28
Kubota T, Lee HS, Trihamdani AR, Phuong TTT, Tanaka T, Matsuo K (2017) Impacts of land use changes from the Hanoi Master Plan 2030 on urban heat islands: part 1. Cooling effects of proposed green strategies. Sustain Cities Soc 32:295–317
Lee T, Tong H, Chan H (2014) Climate projections for Hong Kong based on IPCC AR5. Hong Kong Observatory. WMO. http://www.wmo.int/pages/prog/wcp/ccl/ccl16/teco/documents/ppt/TECO-Presentations-PDF/02-07-2014/Session9/1200-1220/1-PostPresentation_ClimProj_tclee_HKO.pdf. Last accessed Apr 2019

Loikith PC, Waliser DE, Lee H, Neelin JD, Lintner BR, McGinnis SA, Mearns LO, Kim J (2015) Evaluation of large-scale meteorological patterns associated with temperature extremes in the NARCCAP regional climate model simulations. Clim Dyn 45:3257–3274

Loue S, Sajatovic M (eds) (2012) Encyclopedia of immigrant health. Springer, Berlin

Lubchenco J (2011) National oceanic and atmospheric administration united states department of commerce: What is NOAA? http://www.legislative.noaa.gov/policybriefs/What%20is%20NOAA%202011%20-%20FINAL.pdf. Last accessed Apr 2019

Markert BA, Breure AM, Zechmeister HG (2003) Bioindicators and biomonitors, vol 6. Gulf Professional Publishing

McCarthy M, Best M, Betts R, Hendry M (2009) Climate change, cities, and the urban heat island. In: AMS 21st conference on climate variability and change, Phoenix

Meehl GA, Stocker TF, Collins WD, Friedlingstein P, Gaye AT, Gregory JM, Kitoh A, Knutti R, Murphy JM, Noda A, Raper SCB, Watterson IG, Weaver AJ, Zhao ZC (2007) Global climate projections. In: Solomon S, Qin D, Manning M, Chen Z, Marquis M, Averyt KB, Tignor M, Miller HL (eds) Climate change 2007: the physical science basis. Contribution of working group I to the fourth assessment report of the IPCC. Cambridge University Press, Cambridge, United Kingdom and New York, NY, USA

Meinshausen M, Raper SCB, Wigley TML (2011a) Emulating coupled atmosphere-ocean and carbon cycle models with a simpler model, MAGICC6—part 1: model description and calibration. Atmos Chem Phys 11:1417–1456. https://doi.org/10.5194/acp-11-1417-2011

Meinshausen M, Wigley TML, Raper SCB (2011b) Emulating atmosphere-ocean and carbon cycle models with a simpler model, MAGICC6—part 2: applications. Atmos Chem Phys 11:1457–1471. https://doi.org/10.5194/acp-11-1457-2011

Mendelsohn R, Kurukulasuriya P, Basist A, Kogan F, Williams C (2007) Climate analysis with satellite versus weather station data. Clim Change 81:71–83

Nichol J, Hang TP, Ng E (2014) Temperature projection in a tropical city using remote sensing and dynamic modeling. Clim Dyn 42:2921–2929. https://doi.org/10.1007/s00382-013-1748-2

Nichol JE, Hang Pui HT (2012) Temporal characteristics of thermal satellite images for urban heat stress and heat island mapping. ISPRS J Photogrammetry Remote Sens 74:153–162

Oleson K (2012) Contrasts between urban and rural climate in CCSM4 CMIP5 climate change scenarios. J Clim 25(5):1390–1412

Oke TR (1979) Review of urban climatology 1973–1976. World Meteorological Organization, Geneva

Ongsomwang S, Dasananda S, Prasomsup W (2018) Spatio-temporal urban heat island phenomena assessment using LANDSAT imagery: a case study of Bangkok metropolitan and its vicinity, Thailand. Environ Nat Resour J 16(2):299–344

Qiao Z, Tian T, Zhang L, Xu X (2014) Influences of urban expansion on urban heat island in Beijing during 1989–2010. Adv Meteorol 11

Plocoste T, Jacoby-Koaly S, Molinié J, Petit RH (2014) Urban climate, vol 10, Part 4, pp 745–757. https://doi.org/10.1016/j.uclim.2014.03.007. Last accessed Apr 2019

Rajeshwari A, Mani ND (2014) Estimation of land surface temperature of Dindigul district using LANDSAT8 data. Int J Res Eng Technol 3(05):122–126

Rao PK (1972) Remote sensing of urban heat islands from an environmental satellite. Bull Am Meteorol Soc 53:647–648

Roth M (2013) Urban heat islands. In: Fernando HJS (ed) Handbook of environmental fluid dynamics volume 2: systems, pollution, modeling, and measurements. Taylor and Francis Group, New York, pp 143–159

Saito I (1990) Study of the effect of green areas on the thermal environment in an urban area. Energy Build 15:493–498

Santamouris M, Cartalis C, Synnefa A, Kolokotsa D (2015) On the impact of urban heat island and global warming on the power demand and electricity consumption of buildings—a review. Energy Build 98:119–124. https://doi.org/10.1016/j.enbuild.2014.09.052

Santer BD, Wigley TML (2010) Progress in detection and attribution research. In: Schneider SH et al (eds) Climate change science and policy. Island Press, pp 28–43

Santer BD, Wigley TML, Schlesinger ME, Mitchell JFB (1990) Developing climate scenarios from equilibrium GCM results. Max-Planck-Inst. für Meteorologie, Report no. 47

Shashua-Bar L, Tzamir Y, Hoffman ME (2004) Thermal effects of building geometry and spacing on the urban canopy layer microclimate in a hot-humid climate in summer. Int J Climatol 24(13):1729–1742

Skokovic D, Sobrino JA, Jimenez-Muñoz JC, Soria G, Julien Y, Mattar C, Jordi C (2014) Calibration and validation of land surface temperature for LANDSAT8 TIRS sensor. Land product validation and evolution, ESA/ESRIN Frascati (Italy), pp 6–9

Singh P, Kikon N, Verma P (2017) Impact of land use change and urbanization on urban heat islands in Lucknow city, central India: a remote sensing based estimate. Sustain Cities Soc 32:100–114

Sobrino JA, Raissouni N (2000) Toward remote sensing methods for land cover dynamic monitoring: application to Morocco. Int J Remote Sens 21(2):353–366

Sobrino JA, Caselles V, Coll C (1993) Theoretical split-window algorithms for determining the actual surface temperature. Il Nuovo Cimento C 16(3):219–236

Taha H (1999) Modifying a mesoscale meteorological model to better incorporate urban heat storage: a bulk-parameterization approach. J Appl Meteorol 38:466–473

Tucker CJ (1979) Red and photographic infrared linear combinations for monitoring vegetation. Remote Sens Environ 8:127–150

Vera L, Parra CJ, Morales L, Mattar C, Jorquera-Fontena E (2010) Comparative analysis of split-window algorithms for estimating soil temperature. RC Suelo Nutr Veg 10(1):35–39

Voelkel J, Shandas V, Haggerty B (2016) Developing high-resolution descriptions of urban heat islands: a public health imperative. Prev Chronic Dis 13:160099. http://dx.doi.org/10.5888/pcd13.160099

Walawender JP, Szymanowski M, Hajto MJ, Bokwa A (2014) Land surface temperature patterns in the urban agglomeration of Krakow (Poland) derived from landsat-7/ETM+ data. Pure Appl Geophys 171(6):913–940

Wang M, Heb G, Zhang Z, Wang G, Wang Z, Yin R, Cuib S, Wub Z, Cao X (2018) A radiance-based split-window algorithm for land surface temperature retrieval: Theory and application to MODIS data International. Int J Appl Earth Observations Geoinf 76:204–217

Weng Q, Lu D, Schubring J (2004) Estimation of land surface temperature–vegetation abundance relationship for urban heat island studies. Remote Sens Environ 89(4):467–483

WMO (2019) https://cpdb.wmo.int/hong-kong-china. Last accessed 8 Dec 2019

Wong JKW, Lau LS-K (2013) From the 'urban heat island' to the 'green island'—a preliminary investigation into the potential of retrofitting green roofs in Mongkok district of Hong Kong. Habitat Int 39:25–35

World Bank (2019) https://data.worldbank.org/indicator/SP.POP.TOTL?locations=HK. Last accessed 8 Dec 2019

Yan YY (2007) Surface wind characteristics and variability in Hong Kong. Royal Meterological Soc 62(11):312–316

Yao L, Wanga L, Huang X, Zhanga W, Lid J, Niua Z (2018) Interannual variations in surface urban heat island intensity and associated drivers in China. J Environ Manage 222:86–94

Zhang J, Wang Y, Li Y (2006) A C++ program for retrieving land surface temperature from the data of LANDSAT TM/ETM + band6. Comput Geosci 32(10):1796–1805

Internal Displacement Due to Disasters in Latin America and the Caribbean

Roberto Ariel Abeldaño Zuñiga and Javiera Fanta Garrido

Abstract Environmental disasters have been identified as a significant cause of human mobility. Particularly in developing regions, climate change is responsible for rising the frequency and intensity of weather-related disasters in the last decades, thus increasing the number of people who migrate within their countries. This chapter examines the magnitude and duration of internal displacements due to disasters in 18 countries of Latin America and the Caribbean between 2013 and 2015, and analyzes the quality of the Global Internal Displacement Database. Overall, 505 events of disaster were identified, which led to the internal migration of 4,217,737 people. The mean of displaced persons per event was 8351 (SD = 69,755) and the mean duration of the displacement was 11.9 (SD = 40.5) days. The primary reason for internal displacement in the countries examined was hydro-meteorological disaster related to climate change (51%). Results conclude that the Global Internal Displacement Database accurately identifies the starting date of internal displacements, but presents limitations to measure the duration of the displacements during the reference period.

Introduction

In the last two decades, a consistent body of evidence has been developed documenting the adverse effects of climate change on sustainable development, including human, environmental and economic dimensions (Caballeros Otero and Zapata Martí 1994; Swart et al. 2003; Adger et al. 2003; Daoud et al. 2016). The increasing frequency and intensity of weather-related disasters and extreme events such as storms, floods, droughts and cold and heat waves, together with the expansion of the geographic distribution of these events, are one of the most immediate and visible effects of climate change.

R. A. Abeldaño Zuñiga (✉)
División de Estudios de Postgrado, Universidad de La Sierra Sur, 70800 Oaxaca, Mexico
e-mail: ariabeldanho@gmail.com

J. Fanta Garrido
Instituto de Investigaciones Gino Germani (UBA-CONICET), 1114 Buenos Aires, Argentina
e-mail: javiera.fanta@gmail.com

W. Leal Filho et al. (eds.), *Climate Change, Hazards and Adaptation Options*,
Climate Change Management, https://doi.org/10.1007/978-3-030-37425-9_21

Meteorological, hydrological and climatological disasters can lead to loss of life, reduced water availability, destruction of infrastructure, loss of biodiversity, and reduced agricultural production, threatening food security and increasing the risk of insect and water-borne diseases. Additionally, internal and external human migration arises as one of the major indirect consequences of climate change. Other non-climate factors, such as government policy, population growth and resilience to disasters, can also drive human migration and displacement (Kumari Rigaud et al. 2018).

Kälin (2010) suggests that nearly 36 million people around the world were displaced within their own countries in 2008, of which 20 million were displaced by disasters related to climate change. According to World Bank estimates (Kumari Rigaud et al. 2018), by 2050 climate change could turn nearly 140 million people into internal migrants in South Asia, Latin America and the Caribbean (LAC) and Sub-Saharan Africa. In LAC, the number of internal migrants could reach over 17 million people, representing 2.6% of the region's population. It is expected that internal climate migrants in this region will move from areas with increased sea level, reduced water availability and, by extension, lower agriculture production, and from areas stricken by storms (Kumari Rigaud et al. 2018).

Human migration and displacements caused by disasters have several implications in terms of human rights, health and sanitary effects and social costs. These aspects have been addressed by global and regional agendas, and prolific evidence on this topic is emerging from scientific research. Aware of the rising incidence and intensity of disasters in the LAC region, and the prospects for internally displaced populations, this chapter examines the volume and duration of internal displacements due to disasters in 18 LAC countries between 2013 and 2015 by analyzing and comparing census data with information from the Global Internal Displacement Database developed by the Internal Displacement Monitoring Centre (IDMC).[1]

There is consensus that the least developed regions, including LAC, are most seriously affected by disasters, as many of them are highly dependent on natural resources and climate-sensitive sectors, such as agriculture; either through participating directly in agriculture production or by providing services to this sector. Also, many countries of these regions are located in parts of the globe where temperature and precipitation changes are more likely to occur, and it is usually harder for them to recover from climate stresses (Bruckner 2012). Such is the case with several LAC countries, whose geographic features make the region prone to a higher incidence of hydrological and meteorological disasters. Tropical storms, cyclones and hurricanes are more likely to occur in countries of the Caribbean region and the Mexican coasts, Central American countries and the Amazonia region are mainly affected by floods and storms, mudslides and landslides are observed in pre-mountain zones, and desert areas in the Andean regions of Peru, Bolivia, Argentina and Chile are mostly affected by droughts (Inzulza-Contardo and Díaz Parra 2016; Caruso 2017; Abeldaño Zúñiga and González Villoria 2018). Although geophysical events are not expected to be impacted by climate change (Bruckner 2012), several LAC countries have also been

[1]The IDMC is part of the Norwegian Refugee Council (NRC). See: http://www.internal-displacement.org/.

affected by these types of disasters in recent years, such as earthquakes and volcanic eruptions.

Given the growing environmental vulnerability of the LAC countries, the regions governments have sought to mitigate the effects caused by disasters, including migration, through two tracks based on the 2030 Agenda for Sustainable Development (Rodríguez Aldabe 2018) as follows:

- Reducing social inequalities in areas like gender, social class and race, education and health-care access.
- Promoting Green Development through policies and activities based on sustainable use, distribution and consumption of environmental resources.

However, despite the concerns expressed by regional governments and the measures to mitigate the effects of disasters, the region is still a long way off from producing accurate data to assess and monitor the effects of disasters, particularly their impact on migration.

Considering the increasing trends on the incidence of disasters in the LAC region and the current volume of internally displaced population, this chapter aims to describe the magnitude and duration of internal displacements due to disasters in 18 countries of Latin America and the Caribbean between 2013 and 2015, and to analyze the quality of the Global Internal Displacement Database.

Forced Displacement Caused by Disasters

Due to the difficulties involved in isolating environmental factors from other factors that trigger migration, there is no consensus on the definition of migration due to environmental degradation or climate change; nor is there clarity about whether this type of movement is forced or voluntary, since it is often a matter of long-term, cumulative and cyclical processes (Goodwin-Gill and McAdam 2017). However, when dealing with disasters, the forced character of the displacement is easily identifiable, since the time lapse between the sudden event and human movement is, in general, brief.

Adverse environmental conditions, including meteorological, hydrological, climatological and geophysical disasters, have been associated with forced displacement in different LAC countries. Profuse evidence has been found in Mexico and Ecuador, where social and demographic factors have been associated with the lack of social networks and problematic environmental contexts that, combined, affect human displacement (Abeldaño Zúñiga and González Villoria 2018). Besides deaths and economic costs, the effects of disasters include damages to community infrastructure and housing. In some cases the scale of these damages makes forced displacement one of the most common and immediate consequences of disasters for the regional population (Kolmannskog and Trebbi 2010).

Forced displacement also limits post-disaster recovery and resettlement. Evidence shows that individuals and families who are forced to leave their homes due to a disaster frequently lose their jobs and, thus, their main source of income. In addition, these people are more likely to experience health problems during their displacement process, even after relocation (Hori and Schafer 2010).

In the Americas, Hurricane Katrina seriously affected the states of Louisiana, Mississippi, and Alabama in 2006, resulting in one of the highest rates of internal displacements due to disasters in recent years (Hori and Schafer 2010). Significant events of this nature have also occurred in Haiti (Lu et al. 2012), Colombia (Shultz et al. 2014; Castellanos Ospina and Peláez 2015), Chile and in certain communities of Argentina (Abeldaño 2015).

In 2015, the United Nations adopted the Sendai Framework for Disaster Risk Reduction, aimed at building resilience and reducing the risk of disasters, including the risk of human displacement (UN Office for Disaster Risk Reduction 2015). In December 2015, the adoption of the Paris Agreement by the United Nations Framework Convention on Climate Change convened a working group aimed at developing integrated recommendations and approaches to avoid, minimize and address displacement caused by adverse impacts of climate change (United Nations Framework Convention on Climate Change 2015) and both instruments have been adopted by most LAC countries.

Complementarily, the World Humanitarian Summit 2016 defined specific commitments and actions towards supporting communities affected by forced displacements due to disasters. More recently, the 24th United Nations Conference on Climate Change, held in Katowice, Poland (2018), established a series of recommendations aimed at implementing the points contained in the Paris Agreement.

However, the multiple discussions and frameworks derived from these efforts do not guarantee consistent and articulated actions towards preventing and lessening the effects of disasters and the subsequent population displacements (Guadagno 2016).

The Duration of Forced Displacements Due to Disaster Situations

Regarding the duration of population displacement caused by disaster situations, Kälin (2010) has proposed at least five possible scenarios:

Scenario 1: This is characterized by sudden onset hazards, such as floods, storms, hurricanes, typhoons, landslides or avalanches that can cause sudden large forced displacements. This kind of movement is not expected to be long term, and depending on the management of the hazard by government agencies (local and national), most people could return to their homes.

Scenario 2: This is a scenario of slow environmental degradation caused by long-term effects of rising sea levels, droughts and soil desertification, with the resultant degradation of drinking water sources. It is linked to threats to the availability of drinking water and irrigation in certain regions, a threat of considerable economic impact on regional economies. This long-term scenario may lead in some cases to

a "voluntary" migration in adaptation to environmental change. However, in cases where the areas cannot be recovered, depending on the level of desertification, it would result in permanent forced displacement.
Scenario 3: This scenario corresponds to the small island states that have a special case of slow onset of hazards. Due to rising sea levels and low altitude, these areas could become uninhabitable without desertification, and since displacement would be urgent, the population would need to move to other countries. El Hinnawi stated more than two decades ago, that this scenario would result in "climatic refugees" (El Hinnawi 1985). It should be noted that this concept of "environmental refugees or climatic refugees" has been heavily criticized, because the terms "refugees", "migrants" and "displaced" involve different causes and legal status. Currently, academics recommend reserving the term "refugee" for cases of active persecution (Kolmannskog and Trebbi 2010).
Scenario 4: This is the case where there is active intervention by local authorities to evacuate homes located in areas considered at risk (riverbanks, hillsides and mountains, etc.). In these cases, unlike those mentioned in scenario 1, there is no possibility of return and this can trigger legal concerns due to permanent displacement.
Scenario 5: This is a scenario of greater complexity, since it relates to forced displacements where there are situations of violence caused by conflicts related to the availability of natural resources (availability of water for drinking and irrigation). In these cases of displacement, return would depend on the solution of the conflict and the availability of natural resources.

Measuring Internal Displacements Due to Disasters

The Internal Displacement Monitoring Centre (IDMC), based in Geneva, together with the Norwegian Refugee Council, monitor internal displacements in the world due to violence (conflicts) and disasters since 1998 (Internal Displacement Monitoring Centre 2017a). The IDMC methodology consists of producing estimates of internal displacements related to disasters based on the monitoring and case reports of disaster-induced displacement "on an event-by-event basis". For each event, IDMC "collect information from different sources, including: national and sub-national government authorities, UN agencies and other international organizations, related global databases, civil society organizations, news media outlets, and other entities" (Internal Displacement Monitoring Centre 2017b).

This chapter focuses on internal displacement due to disasters, using the data source developed by the IDMC for the countries comprised in the Latin American and Caribbean Economic System (SELA) (2017) during the period 2013–2015.

In accord with the operational criteria determined by the Centre for Research on the Epidemiology of Disasters (2016), a disaster situation is defined as the event "which overwhelms local capacity, necessitating a request to national or international level for external assistance [definition considered in EM-DAT]; an unforeseen and often

Table 1 Hazards classification based on their origin

Hazard groups		Hazard subgroups
Climate related	Meteorological	Storms
		Extreme temperatures
	Hydrological	Flood
		Landslide
		Wave action
	Climatological	Drought
		Wildfire
Non-climate-related	Geophysical	Earthquakes
		Volcanic activity
		Mass movement

Source Centre for Research on the Epidemiology of Disasters—CRED, School of Public Health, Université Catholique de Louvain

sudden event that causes great damage, destruction and human suffering. Though often caused by nature, disasters can have human origins" (Centre for Research on the Epidemiology of Disasters 2016). Given that definition of the term "disaster", a hazard is considered as a "threatening event or probability of occurrence of a potentially damaging phenomenon within a given time period and area" (Centre for Research on the Epidemiology of Disasters 2016).

As seen in Table 1, climate-related disasters (Nagy et al. 2018) were classified including meteorological, hydrological, and climatological hazards; on the other hand, there are geophysical hazards.

Overall, 505 situations of internal displacement due to disasters were identified in 17 LAC countries (out of 33 countries considered in ECLAC) during 2013, 2014 and 2015.

Data Sources on Internal Displacement: Comments and Remarks

The Internal Displacement Database from the IDMC considers two units to measure the volume of internal displacements: households and individuals. In order to offer accurate estimates, and to set valid comparisons between countries and over time, the database was harmonized by using the average number of persons per household. As shown in Table 2, this information was derived from the following official sources (considering the "the latest available data source" criterion): National Housing and Population Censuses, Demographic and Health Surveys (DHS) and Demographic Yearbooks from the United Nations, according to the recommendations of the IDMC (Internal Displacement Monitoring Centre 2017b).

Table 2 Data sources for estimating average household size in countries of Latin America and The Caribbean

Country	Data source for estimation of average household size	Latest available data (year)	Average household size
Argentina	National Housing and Population Census	2010	3.3
Bolivia	Demographic Yearbooks from the United Nations	2012	3.5
Brazil	Demographic Yearbooks from the United Nations	2010	3.3
Chile	National Housing and Population Census	2002	3.6
Colombia	Demographic and Health Surveys (DHS)	2015	3.5
Dominican Rep.	Demographic and Health Surveys (DHS)	2013	3.5
Ecuador	Demographic Yearbooks from the United Nations	2010	3.8
El Salvador	Integrated Public Use of Microdata Series (Minnesota Population Center)	2007	4.1
Guatemala	Demographic and Health Surveys (DHS)	2014	4.8
Honduras	National Housing and Population Census	2013	3.9
Mexico	Inter-Census Population Survey	2015	3.7
Nicaragua	National Housing and Population Census	2005	5.2
Panama	National Housing and Population Census	2010	3.7
Paraguay	Integrated Public Use of Microdata Series (Minnesota Population Center)	2002	4.6
Peru	Demographic and Health Surveys (DHS)	2012	3.8
Uruguay	Demographic Yearbooks from the United Nations	2010	2.8
Venezuela	National Housing and Population Census	2011	3.3

Source Elaborated by the authors

In order to assess the quality of data referred to the magnitude of internally displaced people due to disasters, the following aspects were analyzed after standardizing the cited measurement units:

a. The accuracy of the beginning date of each displacement reported in the database
b. The accuracy of the recorded end date of these displacements
c. The global accuracy, comprising both dates (beginning and end date of the displacements)
d. The measurement unit reported for counting each internal displacement included in the database.

Analyses run after the database standardization showed that data registered by the Internal Displacement Database accurately identifies the beginning date (onset) of displacements due to disasters in LAC countries in 96% of the records during the period studied. However, the accuracy of the records decreases to 48.1%, when analyzing the reported end dates of internal displacements. Globally, when combining both dates, it was found that only 48.1% of the events were correctly reported (Table 3).

Regarding the unit of measurement, 77.2% of the recorded displacement events referred to households, which led to estimating average household size in order to measure "individuals" in all of the countries analyzed.

Estimates of Internal Displacement in Latin America and the Caribbean

Of the 505 disaster situations that forced internal displacement in Latin American and Caribbean countries, 216 (42.8%) events were identified in 2013, 153 (30.3%) events in 2014, and 136 (26.9%) events were recorded in 2015.

Overall, a total of 4,217,737 displaced persons were registered. Each disaster situation caused, on average, an internal displacement of 8351 (SD 69,755) people. The average duration of the displacement was only calculated in the 191 events where the accuracy of the start and end date of the reported internal displacement permitted. In those cases, the average duration was 11.9 (SD 40.5) days for each disaster situation.

As can be seen in Table 4, Colombia was the country with the highest number of reported disaster events that caused internal displacement, with a total of 357, which represented 70.7% of the total of events registered. This number of events has caused the displacement of 125,047 people in Colombia, with an average duration of 2.99 (SD 8.88) days. Most of these displacements are due to hydro-meteorological disasters, as can be seen in more detail later in the text. This may be due to Colombia's geographical position, where there is a high risk of hydro-meteorological incidents.

Despite having the highest number of reported disasters in the period under study, Colombia was not the country with the greatest number of displaced persons.

Table 3 Accuracy of dates and measurements of internal displacements due to disasters reported in the Internal Displacement Database, 2013–2015

Country		Accuracy of the start date		Accuracy of the end date		Global accuracy	Unit of measurement			Total
		Yes	No	Yes	No		Household	Persons	Other	
Argentina	N	13	5	13	5	12	1	17	0	18
	%	72.2	27.8	72.2	27.8	70.6	5.6	94.4	0.0	100
Bolivia	N	8	1	6.0	3.0	6	5	3	1	9
	%	88.9	11.1	66.7	33.3	66.7	55.6	33.3	11.1	100
Brazil	N	11	4	8	7	8	2	13	0	15
	%	73.3	26.7	53.3	46.7	61.5	13.3	86.7	0.0	100
Chile	N	13	1	10	4	9	4	10	0	14
	%	92.9	7.1	71.4	28.6	64.3	28.6	71.4	0.0	100
Colombia	N	357	0	149	208	149	353	4	0	357
	%	100.0	0.0	41.7	58.3	41.7	98.9	1.1	0.0	100
Costa Rica	N	3	0	1	2	1	0	3	0	3
	%	100.0	0.0	33.3	66.7	33.3	0.0	100.0	0.0	100
Dominican Rep.	N	8	0	7	1	7	2	6	0	8
	%	100.0	0.0	87.5	12.5	87.5	25.0	75.0	0.0	100
Ecuador	N	5	1	3	3	3	2	4	0	6
	%	83.3	16.7	50.0	50.0	50.0	33.3	66.7	0.0	100
El Salvador	N	4	0	4	0	4	1	3	0	4
	%	100.0	0.0	100.0	0.0	100.0	25.0	75.0	0.0	100
Guatemala	N	8	1	6	3	6	3	5	1	9
	%	88.9	11.1	66.7	33.3	66.7	33.3	55.6	11.1	100

(continued)

Table 3 (continued)

Country		Accuracy of the start date		Accuracy of the end date		Global accuracy	Unit of measurement			Total
		Yes	No	Yes	No		Household	Persons	Other	
Honduras	N	3	1	1	2	1	1	3	0	4
	%	75.0	25.0	25.0	75.0	33.3	25.0	75.0	0.0	100
Mexico	N	10	0	6	4	6	1	9	0	10
	%	100.0	0.0	60.0	40.0	60.0	10.0	90.0	0.0	100
Nicaragua	N	8	0	4	4	4	2	5	1	8
	%	100.0	0.0	50.0	50.0	50.0	25.0	62.5	12.5	100
Panama	N	10	1	6	5	6	2	7	2	11
	%	90.9	9.1	54.5	45.5	60.0	18.2	63.6	18.2	100
Paraguay	N	3	2	3	2	3	2	3	0	5
	%	60.0	40.0	60.0	40.0	60.0	40.0	60.0	0.0	100
Peru	N	14	3	11	6	10	8	8	1	17
	%	82.4	17.6	64.7	35.3	62.5	47.1	47.1	5.9	100
Uruguay	N	5	0	3	2	3	0	5	0	5
	%	100.0	0.0	60.0	40.0	60.0	0.0	100.0	0.0	100
Venezuela	N	2	0	2	0	2	1	1	0	2
	%	100.0	0.0	100.0	0.0	100.0	50.0	50.0	0.0	100
Total	N	485	20	243	262	240	390	109	6	505
	%	96.0	4.0	48.1	51.9	48.1	77.2	21.6	1.2	100

Source Elaborated by the authors based on data of the Internal Displacement Monitoring Centre

Table 4 Frequency and quantitative measures of selected variables by country

Country	Number of events	Percentage of events	Total displaced persons	Displaced time (days)	
			Sum	Mean	Standard deviation
Argentina	18	3.6	75,393	11.56	17.54
Bolivia	9	1.8	685,262	3.50	4.90
Brazil	15	3.0	322,126	13.36	22.76
Chile	14	2.8	2,038,973	9.90	13.48
Colombia	357	70.7	125,047	2.99	8.88
Costa Rica	3	0.6	1122	5.50	6.36
Dominican Rep.	8	1.6	54,844	3.57	2.70
Ecuador	6	1.2	11,644	46.00	51.55
El Salvador	4	0.8	4918	1.25	0.50
Guatemala	9	1.8	15,951	3.57	3.91
Honduras	4	0.8	7960	12.00	–
Mexico	10	2.0	267,472	4.86	5.40
Nicaragua	8	1.6	38,952	5.40	6.02
Panama	11	2.2	2166	5.57	7.16
Paraguay	5	1.0	432,005	21.40	22.96
Peru	17	3.4	58,497	77.58	138.75
Uruguay	5	1.0	29,501	22.00	14.73
Venezuela	2	0.4	45,904	32.00	19.80
Total	505	100.0	4,217,737	11.9	40.5

Global Internal Displacement Database 2013–2015
Source Elaborated by the authors based on data of the Internal Displacement Monitoring Centre

Chile was the country with the largest number of displaced persons, with a total of 2,038,973, representing 48.3% of the total number of persons in the period studied. The greatest percentage of these displacements is due to disasters caused by geophysical risks, which have a very high incidence in this country crossed by the Andes Mountains.

The volume of internal displacements registered in Bolivia should be noted. In a total of 9 disaster situations, Bolivia had 685,262 displaced persons, followed by Paraguay with 432,005, Brazil with 322,126 and Mexico with 267,472. In sum, the internal displacements reported by these five countries (Chile, Bolivia, Paraguay, Brazil and Mexico) represent 89% of the total displacements of the entire Latin American and Caribbean region.

Peru appears as the country where the average duration in days of displacement was greatest in the period from 2013–2015. There were 17 disaster events that led to the displacement of 58,497 people, with an average of 77.58 (SD 138.75) days. Peru is a country where, due to its geophysical characteristics, geophysical and hydro-meteorological events combine, with a greater proportion of hydro-meteorological events. A similar situation occurs in the neighboring country of Ecuador, which has similar characteristics due to its location, and with a higher incidence of hydro-meteorological rather than geophysical phenomena. Ecuador was the country with the second longest duration of internal displacement, with an average of 46.00 (SD 51.55) days.

As can be seen in Table 5, Bolivia, Brazil, the Dominican Republic, Honduras, Panama, Paraguay, Uruguay and Venezuela do not have records of the incidence of disasters caused by geophysical risks in the period analyzed. For this reason the largest number of internal displacements in these countries are due to hydrological events and the second largest are due to meteorological events.

In some countries that did have an incidence of phenomena related to geophysical events (Argentina, Costa Rica, Ecuador, El Salvador, Mexico, Nicaragua, Peru), the greater proportion of internal displacement was related to hydro-meteorological events and in second place, geophysical events.

In Chile, Colombia and Guatemala, the incidence of disasters related to geophysical events led to greater displacements than events related to other types of threats.

Chile had a total of 1,984,882 displacements related to geophysical threats, Bolivia 683,261 persons displaced by hydrological hazards, Paraguay 407,165 displacements due to hydrological threats, while Brazil had a total of 321,126 persons displaced by hydrological events and Mexico had 224,000 displacements due to meteorological threats in the three analyzed years.

Table 6 shows the pattern of impact of climate related disasters versus disasters of geophysical origin, without taking into account the country. Among events related to geophysical threats, earthquakes and tsunamis were responsible for 47.29% of internal displacements. In total, the events of origin not related to climate change have accounted for 49.05 of total displacements. Events caused by threats directly related to climate change were responsible for 50.05% of internal displacements, and the flooding that caused 38.53% of the displacements is notable.

All disasters (independent of the type of hazard involved) can produce population displacement to a greater or lesser extent. However, disasters related to climate change generated the highest volume of displacement for a longer mean duration in the Latin America and The Caribbean region during 2013–2015, which shows a differential impact.

In particular, the LAC scenario is far from encouraging, due to the impact of climate change that often affects the region. In accord with international trends, it is expected that the intensity and frequency of hazards related to climate change in the region will increase in the coming years (Lopez-Calva and Ortiz-Juarez 2009). Kälin (2010) states that global records on disasters increased from 200 to 400 per year between 1990 and 2010, with most related to climate change.

Table 5 Total displaced persons, mean displacement time by country, and hazard origin

Country	Hazard groups	Total displaced persons	Displaced time (days)	
		Sum	Mean	Standard deviation
Argentina	Geophysical	758	2.00	1.41
	Hydrological	67,907	14.08	19.64
	Meteorological	6728	6.00	7.07
Bolivia	Climatological	2000	1.00	–
	Hydrological	683,261	3.86	5.18
Brazil	Hydrological	321,126	14.60	23.59
	Meteorological	1000	1.00	–
Chile	Geophysical	1,984,882	6.40	10.99
	Climatological	20,738	6.75	6.95
	Hydrological	33,352	40.00	–
Colombia	Geophysical	52,664	1.50	0.71
	Climatological	596	–	–
	Hydrological	51,692	3.14	9.29
	Meteorological	20,093	1.40	0.55
Costa Rica	Geophysical	20	–	–
	Hydrological	1102	5.50	6.36
Dominican Rep.	Hydrological	21,635	2.33	1.15
	Meteorological	33,209	4.50	3.32
Ecuador	Geophysical	668	36.67	38.81
	Climatological	2109	1.00	–
	Hydrological	8867	119.00	–
El Salvador	Geophysical	2322	1.00	0.00
	Hydrological	2596	1.50	0.71
Guatemala	Geophysical	10,324	1.00	0.00
	Hydrological	5627	4.60	4.28
Honduras	Hydrological	7960	12.00	–
Mexico	Geophysical	70	–	–
	Hydrological	43,402	6.50	6.86
	Meteorological	224,000	2.67	2.08
Nicaragua	Geophysical	2335	1.00	0.00
	Hydrological	36,618	12.00	0.00
Panama	Hydrological	535	6.80	8.32
	Meteorological	1631	2.50	2.12
Paraguay	Hydrological	407,165	26.50	23.01
	Meteorological	24,840	1.00	–

(continued)

Table 5 (continued)

Country	Hazard groups	Total displaced persons	Displaced time (days)	
		Sum	Mean	Standard deviation
Peru	Geophysical	1723	1.00	–
	Climatological	16,218	11.00	–
	Hydrological	32,544	98.50	165.21
	Meteorological	8012	65.50	81.32
Uruguay	Hydrological	28,501	26.50	17.68
	Meteorological	1000	13.00	–
Venezuela	Hydrological	45,904	32.00	19.80

Global Internal Displacement Database 2013–2015
Source Elaborated by the authors based on data of the Internal Displacement Monitoring Centre

Table 6 Displaced persons and mean displaced time by type of hazard

Hazard subgroups	Displaced persons	Displaced time (days)		Percentage
	Sum	Mean	SD	
Earthquakes and tsunamis	1,994,456	4	9.49	47.29
Volcanic eruptions	61,311	12.89	25.72	1.45
Mass movement	12,942	1.28	1.08	0.31
Not climate change related	2,068,709	6.1	12.1	49.05
Coldwaves	1209	–	–	0.03
Extreme temperatures	18,218	6	7.07	0.43
Flashfloods	20,039	–	–	0.48
Floods	1,625,161	14.69	21.45	38.53
Hail	53	-	–	0.001
Landslides	17,090	75.9	152.93	0.41
Mudflows	29,739	-	–	0.71
River floods	92,561	27.2	25.82	2.19
Storms	195,662	5	4.02	4.64
Storm surges	2264	2	0	0.05
Cyclones	400	–	–	0.01
Hurricanes	97,350	1	0	2.31
Tornados	25,840	1	0	0.61
Wildfires	23,443	5.6	6.54	0.56
Climate change related	2,149,029	15.4	24.2	50.05
Total	4,217,738	13.0	21.2	100.0

Global Internal Displacement Database 2013–2015
Source Elaborated by the authors based on data of the Internal Displacement Monitoring Centre

Discussion

Returning to what was proposed by Kälin (2010), LAC countries are characterized by the incidence of sudden onset hazards that can cause large, sudden forced displacements.

At the global level, available evidence suggests that since the late 1990s, the frequency and severity of disasters related to hydro and meteorological hazards has increased, and that in coming years, this trend is likely to continue (Quarantelli 1999; Coleman 2006; International Federation of Red Cross and Red Crescent Societies 2016). Regardless of its root causes (whether by climatological, geophysical, or technological hazards) and place of occurrence, there is a quantitative and qualitative (in terms of intensity) increase in Latin America and Caribbean countries (Quarantelli 1999; Abeldaño Zúñiga and González Villoria 2018).

It is widely recognized that the increasing number of hazards related to climate change is closely linked to the process of industrialization in the region (Quarantelli 1999; Coleman 2006; Kniivilä 2007; Programa de las Naciones Unidas para el Desarrollo (PNUD) 2013; López Levi and Toscana Aparicio 2016). In this respect, the United Nations has documented this process in some LAC countries from decade of the 1940s until the mid-1980s, when developmental policy was mainly limited to the import substitution industrialization model (Kniivilä 2007).

The policy of growing industrial development leads, necessarily, to other processes related with disasters; urbanization and environmental degradation (Aparicio 2011; Programa de las Naciones Unidas para el Desarrollo (PNUD) 2013; Viand and Briones 2015).

Both phenomena create synergies that deepen the vulnerability of populations and their level of exposure to climate change related hazards. For this reason, some scholars (Gray and Bilsborrow 2013; Nawrotzki et al. 2015; Guadagno 2016) predict that the subsequent population displacement generated by climate change related hazards will accelerate the pace of disaster occurrence, especially in certain Latin-American and Caribbean countries.

The complexity of this matter involves several implications, including the regulatory frameworks referring to the assistance and protection of displaced populations. In 1998, the United Nations High Commissioner for Refugees adopted the Guiding Principles on Internal Displacement in order to provide protection and guarantees to persons displaced by various causes due to violence, political conflicts or disasters. Over the last two decades, a wide range of normative devices have been adopted to address matters of health and humanitarian and human rights guarantees involved in population displacements. Most recently specific instruments focused on disasters, such as the Sendai Framework for Disaster Risk Reduction, the Paris Agreement on climate change, the 2030 Agenda for Sustainable Development and the New Urban Agenda, among others, have been developed.

From 2005 to the present, significant progress has been made in the enhancement of international attention to disasters. Initially, the Hyogo Framework for

Action 2005–2015 ('Building the Resilience of Nations and Communities to Disasters') included a brief reference on displacement due to disaster events in paragraph 19 and stated that population displacement provoked by disasters and conflicts might increase the level of vulnerability and exposure to danger (United Nations International Strategy for Disaster Risk Reduction 2005).

The following debates around the Sendai Framework and the Paris Agreement extensively covered topics related to displacement, including relocations, migration and specific conditions of vulnerability of displaced populations, as well as the transfer of remittances and the resilience factors of displaced persons that shape the multiple effects of disasters (Kälin 2010; Guadagno 2016).

These progressive discussions highlighted the importance and complexity of the different dimensions related to displacements caused by disasters. However, as stated in the introduction, the current regional scenario raises the question of whether these efforts are sufficiently comprehensive and consistent, and if their objectives are attainable in the context of Latin America and the Caribbean, considering that there were more than four million internally displaced persons in the region between 2013 and 2015, who have faced numerous adversities as a consequence. The displaced population due to disasters in LAC countries in this three-year-period exceeds the total population of Panama, Puerto Rico or Uruguay.

A second implication of disasters and their relationship with other structural processes, refers to the technical solutions available to tackle the consequences of displacement. Effective and timely responses are crucial to recover from the effects of disasters (Quarantelli 1999; Alexander 2006; Kilmer and Gil-Rivas 2010). The needs and obstacles faced by displaced populations due to disasters consist of a wide range of factors, including mental and physical health (primarily gastrointestinal and breathing problems), psychological and social stability, nutritional problems, difficulties in accessing services and supplies, economic deprivation, the need for legal protection and the right to housing, among others. As stated by Kolmannskog and Trebbi (2010), all these needs should be integrally considered from a human rights perspective.

A third implication refers to the sustainable development of countries in LAC countries. It is vital to have an in depth discussion of the guidelines of developmental models, how regional governments are implementing them, and the consequences they have on economic activities and subsequent environmental effects, considering that these issues determine the conditions in which future generations will live. Daoud et al. (2016) argues that disasters affect the development of nations in two ways: they can lead to an increase in poverty and secondly, poverty may generate conditions of vulnerability that deepen the effects of disasters.

Fourthly, disasters imply social and economic costs for individuals and families that underlie the situations of displacement. These costs derive from the economic loss provoked by damages to infrastructure and properties, loss of employment, school closures and other non-tangible assets affected by disasters (Cavallo et al. 2010).

A fifth implication is the creation and systematization of continuous records by national agencies, in order to monitor the magnitude of forced displacements. The lack of accurate estimates delays and inhibits effective responses to the needs of displaced populations. Colombia appears to be the country with the highest rate of displacement. This reflects the existence of a national statistical system committed and responsive to quantifying this phenomenon (Castellanos Ospina and Peláez 2015).

Regarding data sources, this chapter analyzed the accuracy of the Internal Displacement Database to measure the number of internal displaced persons and to determine the beginning and end dates of recorded disasters. Despite the fact that no satisfactory indicators were found, this is the only data source available to provide the necessary information to monitor displacements due to disasters at the global level. In this respect, there is agreement on recognizing that the accurate estimation of the volume of internally displaced people is difficult to quantify. Thus, it is hard to assess the magnitude of the damage caused by these events and to fill in the information gaps that still prevail on this matter.

There are some limitations in this study, so the results should be considered with some caution. The first limitation is the time period analyzed through the databases obtained from the IDMC. This could mask potential overestimations (due to an extreme event that motivated a large internal displacement), or conversely, a reverse situation could lead to an underestimation of some of the calculations.

The second limitation has to do with the multiple units of displacement measurement reported by the IDMC database (households and individuals), which is why indirect displacement estimation techniques were used that could introduce a potential bias in the mathematical calculations. These estimates were made according to the methodological recommendations of the IDMC, using international demographic reference standards to estimate the average size of households.

The lack of precision in reporting the dates of beginning, ending or both, should also be assumed as a limitation of this work, since this reduces the accuracy of the database when calculating the average duration of internal displacements due to disasters.

Finally, it should be noted that Colombia does not necessarily represent a country with the largest number of people displaced by disasters, but they have a records system that results in a more accurate monitoring. While the other countries in the region would not necessarily have a lower incidence of disasters and a smaller volume of displacement, they may not have precise mechanisms for monitoring and recording the situation.

Conclusions

According to our analysis, the disasters related to climate change caused the greatest volume of internal displacement in LAC countries between 2013 and 2015, with a total of 4,217,737 displaced persons. The outlook for the region is worrisome due to an increasing trend in climate change related disasters and a potential acceleration in internal migration, with multiple implications for the countries involved.

Vulnerability to climate change effects, including the increasing incidence of disasters, depends not only on the character and magnitude of these effects, but also on the level of exposure to climate risks and the ability of the population to cope with the risks. It is well documented that poorer communities are disproportionately affected by disasters, with less opportunities for local adaption. These conditions shape the level of climate associated migration and affect the nature of movement. Although this work did not address the directionality of internal migration, evidence suggests that internal climate immigrants in LAC countries generally move from areas of increasing climate risks, including small towns and areas affected by hydro-meteorological hazards, to locations with better environmental conditions, particularly urban areas able to provide employment opportunities. Based on the volume of internal migrants in the region, it is expected that large and medium-sized cities will have to adapt their urban infrastructure, social support systems and labor market to increasing migratory flows. Likewise, it should be considered that internal migration has historically been associated with the emergence of social marginality, which threatens the objective of reducing the social inequalities that persist in the region. This aspect hinders mitigating the effects of climate change.

Regarding the data source used in this chapter, its accuracy was analyzed, and although satisfactory indicators were not obtained, it is the only available source of data at a global scale to monitor this problem. Therefore, there is a consensus that the precise estimation of the numbers of the displaced is difficult, so the quantification of the magnitude of the damage has some gaps that must be addressed.

Even with the technical difficulties involved in estimating the numbers of the displaced population, Colombia seems to have a genuine concern about this phenomenon. This strategy should be imitated by the other countries of the region, since the magnitude of this phenomenon has had an important magnitude in the years analyzed, and we believe that it could be greater if the accuracy of the data sources is improved. In that sense, we ask ourselves if the national statistical institutes could strengthen this monitoring at the country level.

Future Prospects

The IDMC admits that in spite of their best efforts, there continues to be a series of technical and methodological gaps that increase the uncertainty of estimates and this, without a doubt, poses a great challenge for policy and program development. These challenges have to do with planning based on outdated data, limited geographic coverage, difficulty distinguishing between new and previous (or secondary) displacements, and challenges to obtain disaggregated data at sub-national and geospatially referenced levels.

In addition to being prepared to provide a humanitarian response according to internal displacement needs, national programs of accurate monitoring must be carried out. The key could be in the role played by national statistical institutes in the producing reliable official statistics, with the support of international agencies and

United Nations technical agencies. This would require a significant investment at the national level in order to have technical capabilities at the local level to make these estimates.

It is also necessary to revise and reorient the national policies of adaptation to climate change and disaster risk reduction at the national and sub-national levels, in order to align with the goals set out in the Sustainable Development Agenda for the year 2030 At this point, it is important that Latin American and Caribbean countries change the discourse and political strategies from the "answers" approach, to a management and risk reduction approach to displacement. In this regard, international experts have proposed measuring the "risk of displacement" and early warning capabilities at the national and local levels.

Acknowledgements The project was funded by the Programa para el Desarrollo Profesional Docente, Tipo Superior, Secretaría de Educación Pública, (PRODEP-SEP (MEXICO) [511-6/17-7463]).

References

Abeldaño R (2015) Percepción del desastre y respuestas psicosociales en la comunidad de Tartagal (Argentina): Aproximaciones cualitativas a partir del alud de 2009. Rev Chil Salud Pública 19:243–250

Abeldaño Zúñiga RA, González Villoria RAM (2018) Desastres en México de 1900 a 2016: patrones de ocurrencia, población afectada y daños económicos. Rev Panam Salud Publica 42:1–8. https://doi.org/10.26633/RPSP.2018.55

Adger N, Huq S, Brown K, Conway D, Hulme M (2003) Adaptation to climate change in the developing world. Prog Dev Stud 3:179–195. https://doi.org/10.1191/1464993403ps060oa

Alexander D (2006) Globalization of disaster: trends, problems and dilemmas. J Int Aff 59:1–23

Aparicio AT (2011) Civil protection, population, vulnerability and risk in Santiago. Investig Geográficas, Boletín del Inst Geogr UNAM 35–47

Bruckner M (2012) Climate change vulnerability and the identification of least developed countries. United Nations, Washington DC

Caballeros Otero R, Zapata Martí R (1994) The impacts of natural disasters on developing economies: Implications for the international development and disaster community. In: World Bank (ed) Economic aspects of natural disaster. Reduction for sustainable development. World conference on natural disaster reduction. National Academy of Science, Yokohama, Japan, pp 11–15

Caruso GD (2017) The legacy of natural disasters: the intergenerational impact of 100 years of disasters in Latin America. J Dev Econ 127:209–233. https://doi.org/10.1016/j.jdeveco.2017.03.007

Castellanos Ospina ÓA, Peláez E (2015) Una aproximación a la medición del impacto poblacional del desplazamiento interno por causa de la ola invernal en Colombia, 2010–2011. Rev ib 4:64–84

Cavallo E, Powell A, Becerra O (2010) Estimating the direct economic damages of the earthquake in Haiti. Econ J 120:298–312. https://doi.org/10.1111/j.1468-0297.2010.02378.x

Centre for Research on the Epidemiology of Disasters (2016) The international disaster database (EM-DAT). School of Public Health, Université Catholique de Louvain. http://www.emdat.be. Accessed 21 Nov 2018

Coleman L (2006) Frequency of man-made disasters in the 20th century. J Contingencies Cris Manag 14:3–11. https://doi.org/10.1111/j.1468-5973.2006.00476.x

Daoud A, Halleröd B, Guha-Sapir D (2016) What is the association between absolute child poverty, poor governance, and natural disasters? A global comparison of some of the realities of climate change. PLoS ONE 11:1–20. https://doi.org/10.1371/journal.pone.0153296

El Hinnawi E (1985) Environmental refugees. United Nations Environment Programme, Nairobi

Goodwin-Gill GS, McAdam J (2017) UNHCR and climate change, disasters and displacement. The United Nations Refugee Agency (UNHCR), Geneva, Switzerland

Gray C, Bilsborrow R (2013) Environmental influences on human migration in rural ecuador. Demography 50:1217–1241. https://doi.org/10.1007/s13524-012-0192-y

Guadagno L (2016) Human mobility in the sendai framework for disaster risk reduction. Int J Disaster Risk Sci 7:30–40. https://doi.org/10.1007/s13753-016-0077-6

Hori M, Schafer M (2010) Social costs of displacement in Louisiana after Hurricanes Katrina and Rita. Popul Environ 31:64–86. https://doi.org/10.1007/s11111-009-0094-0

Internal Displacement Monitoring Centre (2017a) Global Internal Displacement Database. In: Global Internal Displacement Database. http://www.internal-displacement.org/database/. Accessed 10 Mar 2018

Internal Displacement Monitoring Centre (2017b) Methodological annex. In: Internal Displacement Monitoring Centre (ed) Global report on internal displacement 2017, GRID 2017. Internal Displacement Monitoring Centre, Geneva, Switzerland, pp 94–112

International Federation of Red Cross and Red Crescent Societies (2016) World disasters report 2016—resilience: saving lives today, investing for tomorrow. In: World disasters report 2016—Resil. http://www.ifrc.org/en/news-and-media/press-releases/general/world-disasters-report-2016—resilience-saving-lives-today-investing-for-tomorrow/. Accessed 8 Aug 2018

Inzulza-Contardo J, Díaz Parra I (2016) Desastres naturales, destrucción creativa y gentrificación: estudio de casos comparados en Sevilla (España), Ciudad de México (México) y Talca (Chile). Rev Geogr Norte Gd 109–128. https://doi.org/10.4067/s0718-34022016000200008

Kälin W (2010) Conceptualising climate-induced displacement. In: McAdam J (ed) Climate change and displacement: multidisciplinary perspectives, 1st edn. Hart Publishing, Oxford, United Kingdom, pp 81–104

Kilmer R, Gil-Rivas V (2010) Responding to the needs of children and families after a disaster: linkages between unmet needs and caregiver functioning. Am J Orthopsychiatry 80:135–142. https://doi.org/10.1111/j.1939-0025.2010.01016.x.Responding

Kniivilä M (2007) Industrial development and economic growth: implications for poverty reduction and income inequality. In: Industrial development for the 21st century: sustainable development perspectives. United Nations, Department of Social and Economic Affairs, New York, pp 295–332

Kolmannskog V, Trebbi L (2010) Climate change, natural disasters and displacement: a multi-track approach to filling the protection gaps. Int Rev Red Cross 92:713–730. https://doi.org/10.1017/S1816383110000500

Kumari Rigaud K, De Sherbinin A, Jones B, Bergmann J, Clement V, Ober K, Schewe J, Adamo S, McCusker B, Heuser S, Midgley A (2018) Groundswell: preparing for internal climate migration, 1st edn. World Bank Publications, Washington DC

Latin American and Caribbean Economic System (2017) Member States. In: Latin American and Caribbean economic system. http://www.sela.org/celac/. Accessed 20 Feb 2018

López Levi L, Toscana Aparicio A (2016) Vulnerabilidad en Tlatelolco a tres décadas de los sismos de 1985. Política y Cult 125–152

Lopez-Calva LF, Ortiz-Juarez E (2009) Evidence and policy lessons on the links between disaster risk and poverty in Latin America: methodology and summary of country studies. New York

Lu X, Bengtsson L, Holme P (2012) Predictability of population displacement after the 2010 Haiti earthquake. Proc Natl Acad Sci 109:11576–11581. https://doi.org/10.1073/pnas.1203882109

Nagy GJ, Leal Filho L, Azeiteiro UM, Heimfarth J, Verocai JE, Li C (2018) An assessment of the relationships between extreme weather events, vulnerability, and the impacts on human wellbeing in Latin America. Int J Environ Res Public Health 15:1802. https://doi.org/10.3390/ijerph15091802

Nawrotzki RJ, Hunter LM, Runfola DM, Riosmena F (2015) Climate change as a migration driver from rural and urban Mexico. Environ Res Lett 10:114023. https://doi.org/10.1088/1748-9326/10/11/114023

Programa de las Naciones Unidas para el Desarrollo (PNUD) (2013) Informe sobre Desarrollo Humano 2013. El ascenso del Sur: Progreso humano en un mundo diverso. Programa de las Naciones Unidas para el Desarrollo (PNUD), New York

Quarantelli E (1999) Implications for programmes and policies from future disaster trends. Risk Manag 1:9–19. https://doi.org/10.1057/palgrave.rm.8240011

Rodríguez Aldabe Y (2018) Potenciar la resiliencia de las ciudades y sus territorios de pertenencia en el marco de los acuerdos sobre cambio climático y de la Nueva Agenda Urbana. Comisión Económica para América Latina y el Caribe, Santiago, Chile

Shultz J, Gómez Ceballos ÁM, Espinel Z, Rios Oliveros S, Fonseca MF, Hernandez Florez LJ (2014) Internal displacement in Colombia. Disaster Heal 2:13–24. https://doi.org/10.4161/dish.27885

Swart R, Robinson J, Cohen S (2003) Climate change and sustainable development: expanding the options. Clim Policy 3:19–40. https://doi.org/10.1016/j.clipol.2003.10.010

UN Office for Disaster Risk Reduction (2015) Sendai framework for disaster risk reduction 2015–2030, 1st edn. Oficina de las Naciones Unidas para la Reducción del Riesgo de Desastres, Geneve, Switzerland

United Nations Framework Convention on Climate Change (2015) Paris Agreement of the United Nations framework convention on climate change. United Nations Framework Convention on Climate Change, Paris, France

United Nations International Strategy for Disaster Risk Reduction (2005) Hyogo framework for action 2005–2015, COP 21. United Nations International Strategy for Disaster Risk Reduction, Hyogo, Japan

Viand J, Briones F (2015) Riesgos al Sur. Diversidad de riesgos de desastres en Argentina, 1st edn. La Red de Estudios Sociales en Prevención de Desastres en América Latina, Buenos Aires

Climate Change and Southern Hemisphere Tropical Cyclones International Initiative: Twenty Years of Successful Regional Cooperation

Yuriy Kuleshov

Abstract Tropical cyclones (TCs) are the most dangerous weather phenomena to regularly affect countries in the Southern Hemisphere (SH). Historical records demonstrate significant inter-annual variability in TC frequency and spatial distribution of cyclone tracks; consequently, the year-to-year impact of TCs on countries varies. The El Niño-Southern Oscillation is one of the key global climate drivers which affects TC occurrences in the SH; however, there are other important large-scale environmental factors which contribute to inter-annual and intra-annual TC variability. Additionally, climate is changing on a global scale and it is important to understand how a warmer climate may affect TC activity. Since 1999 several academic, research and operational agencies from countries of the SH combined their efforts in improving our understanding of changes in regional TC activity due to natural variability and climate change, and established the "Climate Change and Southern Hemisphere Tropical Cyclones" International Initiative. The Initiative's overarching strategy is to provide science-based information for enhancing TC early warning systems and assist with decision making in climate change adaptation and disaster risk reduction. As a result of these multi-national efforts, the first consolidated regional database of historical TC data for the SH (satellite-era, i.e. 1970 to present) was created; influence of global climate drivers on TC variability was investigated; a comprehensive TC climatology for the SH and skillful TC seasonal prediction models were developed. This publication highlights major achievements of this International Initiative over the past two decades. It also includes a brief description of the Tropical Cyclone Data Portal helping readers to learn quickly how to use this web-based information tool for examining historical TC activity over the regions of interests.

Y. Kuleshov (✉)
Bureau of Meteorology, Docklands, VIC, Australia
e-mail: yuriy.kuleshov@bom.gov.au; yuriy.kuleshov@rmit.edu.au

School of Science, Royal Melbourne Institute of Technology (RMIT) University, Melbourne, VIC, Australia

School of Mathematics and Statistics, The University of Melbourne, Parkville, VIC, Australia

W. Leal Filho et al. (eds.), *Climate Change, Hazards and Adaptation Options*, Climate Change Management, https://doi.org/10.1007/978-3-030-37425-9_22

Introduction

Tropical cyclones (TCs) are the most dangerous weather phenomena to regularly affect countries in the Southern Hemisphere (SH). Multiple hazards associated with TCs (destructive winds, storm surges and torrential rain) can make dramatic impacts on population and property (Kuleshov et al. 2014); e.g. Australia's most notorious cyclone Tracy (Australian Government 1977) which affected Darwin on 25 December 1974 caused 71 fatalities, destruction of more than 70% of the city's buildings and over A$800 million in damage (1974 AUD; estimated as more than A$6 billion 2014 AUD). Similarly, numerous high impact TCs affected island nations in the South Pacific Ocean (SPO) and the South Indian Ocean (SIO), e.g. TC Namu (1986) killed at least 150 people in Solomon Islands and caused about $125 million (1986 USD) in damage and economic losses; TC Gafilo (2004), the most intense cyclone recorded in the South-West Indian Ocean, killed at least 363 people and caused about $250 million (2004 USD) damages in Madagascar (https://emdat.be).

Historical records demonstrate significant inter-annual variability in TC frequency and spatial distribution of cyclone tracks; consequently, the year-to-year impact of TCs on countries varies. To reduce the risk of multiple hazards associated with TCs, it is vital to fully understand influence of natural and anthropogenic forcings on TC spatial and temporal variability. The El Niño-Southern Oscillation (ENSO) is one of the key global climate drivers which significantly affects TC occurrences in the SH (Kuleshov et al. 2008, 2009a); however, there are other important large-scale environmental factors which contribute to inter-annual and intra-annual TC variability. Climate change is another important factor—modelling indicates that it is likely that the global mean TC maximum wind speed will increase by 2100 (IPCC 2014); however, response to climate change in regional TC activity is still not well understood (Knutson et al. 2010).

Since 1999 several academic, research and operational agencies from countries in the SH combined their efforts in improving our understanding of changes in regional TC activity due to natural variability and climate change and established the "Climate Change and Southern Hemisphere Tropical Cyclones" International Initiative (Kuleshov 2013). An overarching strategy for the Initiative is to improve our knowledge about TCs in the SH and enhance early warning systems (EWS) to assist countries in Asia-Pacific, particularly small island developing states (SIDS) and least developed countries (LDCs), with decision making in climate change adaptation (CCA) and disaster risk reduction (DRR). Three key objectives—the development of regional consolidated database, including data portal for dissemination of information; comprehensive TC climatology; and seasonal prediction of TC activity for the regions—contribute to the Initiative's overarching strategy.

The World Meteorological Organization (WMO) established the Regional Specialized Meteorological Centres (RSMCs), and WMO Regional Associations (RAs) I and V recognized Tropical Cyclone Warning Centres (TCWCs) in the South-West Indian Ocean and the South Pacific Ocean with designated areas of responsibility to issue advisories on TCs for their respective regions or countries as well as to archive historical data. In the SH, RSMC La Réunion, France is responsible for the area south to the equator, 30° E–90° E (the South-West Indian Ocean, or SWIO); TCWCs in Perth, Darwin and Brisbane, Australia; Jakarta, Indonesia; and Port Moresby, Papua New Guinea—for the area between 90° E and 160° E (the Australian Region, or AR); RSMC Nadi, Fiji and TCWC Wellington, New Zealand—for the area between 160° E to 120° W (the South Pacific Ocean, or SPO).

Such subdivision into three major regions—SWIO, AR and SPO, with eight designated RSMCs and TCWCs—provides comprehensive coverage for TC advisories for the SH. However, in the past such subdivision sometimes caused discrepancies in information at the regions' boundaries resulting in different positions, intensities and even names for the same TC. At the first stage of implementing the Initiative, meteorological agencies of Australia, Fiji, France, Indonesia, New Zealand and Papua New Guinea worked together on harmonizing historical TC records. This work was focused on retrospective analyses of TC data ensuring consistency of historical records among the agencies and adjusting the records using mutually agreed methodologies if discrepancies were found. As a result of these multi-national efforts, the first consolidated regional database of historical TC data for the SH (satellite-era, i.e. 1970 to present) was created.

Further steps of the Initiative were undertaken through the Pacific Climate Change Science Program (PCCSP), Pacific Adaptation Strategy Assistance Program (PASAP) and Pacific-Australia Climate Change Science and Adaptation Planning Program (PACCSAP)—large scale programs of the Australian Government to assist SIDS and LDCs in Asia-Pacific with CCA and DRR. Under PCCSP, TC-ENSO relationship was examined in detail. For the first time, a comprehensive TC climatology for the SH was developed based on historical data from the created satellite-era TC data archive and relationship between TC spatial variability and ENSO was analysed. Under PASAP and PACCSAP, skilful TC seasonal prediction models were developed and TC data portal was designed to disseminate TC information. Over the years, this International Initiative expanded to the Western North Pacific Ocean; organisations from Japan, China, Korea and USA joined various research and operations components of the Initiative on improving our knowledge about TC historical data, climatology and seasonal prediction in that region.

Significant scientific results achieved in the development of the SH TC archive, climatology and prediction for the regions through this International Initiative have been successfully implemented into operations of National Meteorological and Hydrological Services (NMHSs) of Member countries in WMO RAs I and V. Outcomes of the Initiative directly contribute to enhancing Climate Risk and Early Warning Systems (CREWS) assisting communities at risk, especially in LDCs and SIDS with DRR and CCA. This motivated us to prepare this publication which highlights major achievements of the "Climate Change and Southern Hemisphere Tropical Cyclones" International Initiative over the past two decades, to share our experience with international TC community and encourage further strengthening of such fruitful international cooperation regionally and globally.

In this publication, detailed description of the SH TC historical data archive is provided; impact of ENSO on spatial distribution of TC activity in the SH is discussed; and the developed statistical TC seasonal prediction models are presented. The chapter also includes a brief description of the Tropical Cyclone Data Portal helping readers to learn quickly how to use this state-of-the-art web-based information tool to examine historical TC activity over the regions of interests.

Southern Hemisphere Tropical Cyclone Data Archive

Long-term historical records are essential for analysis of TC activity including deriving TC trends, identifying areas of high TC occurrences and using the data as inputs for statistical models to predict future TC activity. Verification of the historical cyclone records and creation of an accurate and homogeneous consolidated regional database covering the SIO and the SPO was the first objective of the Initiative.

Historical records of TC occurrences in the SPO go back to the 16th century (d'Aubert and Nunn 2008; de Scally et al. 2006). In the AR, list of records produced by the Bureau of Meteorology starts from the beginning of the 20th century (Lourensz 1981). However, early records are scarce and typically confined to areas in the vicinity of islands and coastlines with very limited data over the oceans. Similarly, in the SIO, examining historical TC records from Mauritius Meteorological Service available at http://metservice.intnet.mu/publications/list-of-historical-cyclones.php one can find that only six cyclones are listed between 1892 and 1958.

The quality of the historical database for TC occurrences in the SIO and SPO has been significantly improved since the 1970s due to availability of information obtained from geostationary and polar-orbiting meteorological satellites (Holland 1984; Broomhall et al. 2010). The first version of the consolidated regional database (the SH TC data archive) prepared at the National Climate Centre of the Australian Bureau of Meteorology, in collaboration with meteorological agencies from the region, was released in 2003 (Kuleshov et al. 2008) and then was regularly updated (Kuleshov et al. 2010; Dowdy and Kuleshov 2012).

Data quality control and verification are essential when dealing with complex data sets such as the consolidated SH TC data archive derived from several sources. First step was to collect TC best track data from RSMCs and TCWCs in the SH and produce a complete list of historical TC records from 1969 to 1970 cyclone seasons (satellite-era records) including a cyclone name, position and intensity. Then the collected records were analysed to find discrepancies between the data sources from various agencies. Quality control procedures allowed removing errors and introducing corrections to TC track data if required. Consultations between RSMCs and TCWCs were carried out to make consensus decisions about joining tracks of TCs which crossed boundaries of areas of responsibility, and adjusting TC position and intensity accordingly to ensure homogeneity of the records. The finalised data were converted to the format of the SH TC archive which includes the following parameters: TC name, date, position (latitude and longitude), intensity in terms of minimum central pressure and strength in terms of maximum sustained wind averaged over a ten-minute time span, in accordance with WMO standards.

Southern Hemisphere Tropical Cyclone Climatology

Availability of TC best track data from the consolidated regional database allowed for the first time to develop a comprehensive TC climatology and to analyse relationship between TC spatial variability and the ENSO over the entire SH. First, a list of historical El Niño and La Niña events has been developed based on analysis of a number of ENSO indices including the Southern Oscillation Index (SOI), area-averaged sea surface temperature (SST) anomalies in the Niño 3, Niño 3.4 and Niño 4 regions, and two multivariate ENSO indices—the MEI (Wolter and Timlin 1998) and the 5VAR (an index that combines the mean sea level pressure from Tahiti and Darwin as well as the Niño 3, Niño 3.4, and Niño 4 indices; Kuleshov et al. 2008, 2009a). This allowed to stratify TC historical data accordingly and produce the first maps of TC density over the SH for all years as well as for El Niño, La Niña and neutral years (Kuleshov and de Hoedt 2003). The maps of TC occurrences in the SH is a valuable source of information about cyclones' spatial variability over the regions; they are periodically updated and the most recent maps are available at the Australian Bureau of Meteorology's website http://www.bom.gov.au/climate/maps/averages/tropical-cyclones/.

It was found that TC activity in the SH is higher during La Niña years (29 TCs on average) compared to El Niño years (25 TCs on average). Regionally, TC occurrence in the SIO is higher than in the SPO, especially during La Niña years—on average, 18 TCs form in the SIO compared to 11 TCs in the SPO (Kuleshov et al. 2008).

Examining trends in TCs in the SH and two subregions, the SIO and the SPO, it was found that for the 1981–1982 to 2016–2017 TC seasons there was a decline in the total numbers of TCs (defined as those tropical systems attaining a lifetime minimum central pressure (LMCP) of 995 hPa or lower; Fig. 1) as well as in numbers of severe TCs (i.e. tropical systems attaining LMCP of 970 hPa or lower; Fig. 2) in the SIO and the SPO. These findings support conclusions of our earlier study about overall decrease in TC occurrences in the SH based on shorter time period examined (26 years of data from 1981–1982 to 2006–2007 seasons; Kuleshov et al. 2010).

Significant spatial variability in TC occurrences was observed in both basins depending on the phase of the ENSO. In the SPO, during El Niño years tropical cyclogenesis shifts eastwards while in La Niña years it shifts westwards (Kuleshov et al. 2008; Dowdy et al. 2012); as a result, frequency of cyclones which affect east coast of Australia in La Niña years almost doubles compared to it in El Niño years (Callaghan 2003). This confirmed earlier reports that during El Niño years TC tracks are displaced away from the Australian coast towards the eastern Pacific (Gray 1988; Basher and Zheng 1995). In the SIO, during La Niña years increase in TC activity in the eastern part (regional maximum of TC genesis in Western Australia) and its

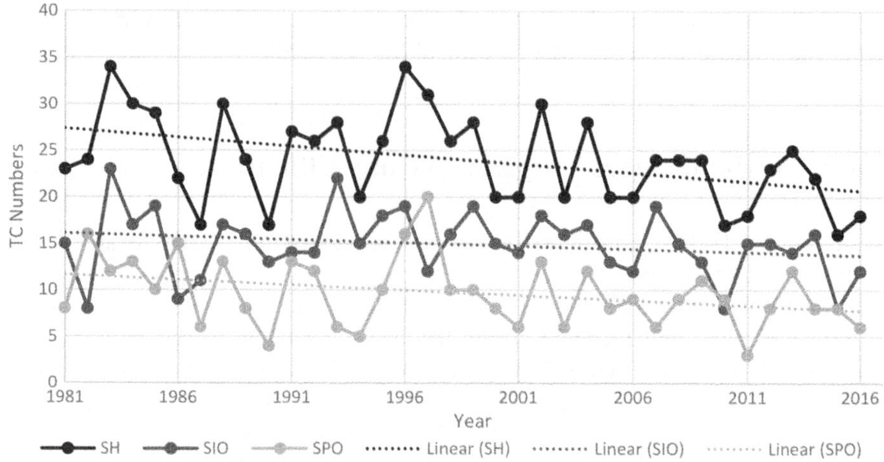

Fig. 1 Annual numbers of TCs with LMCP of 995 hPa or lower for the SH, SIO and SPO in the 1981–1982 to 2016–2017 seasons, with linear trends

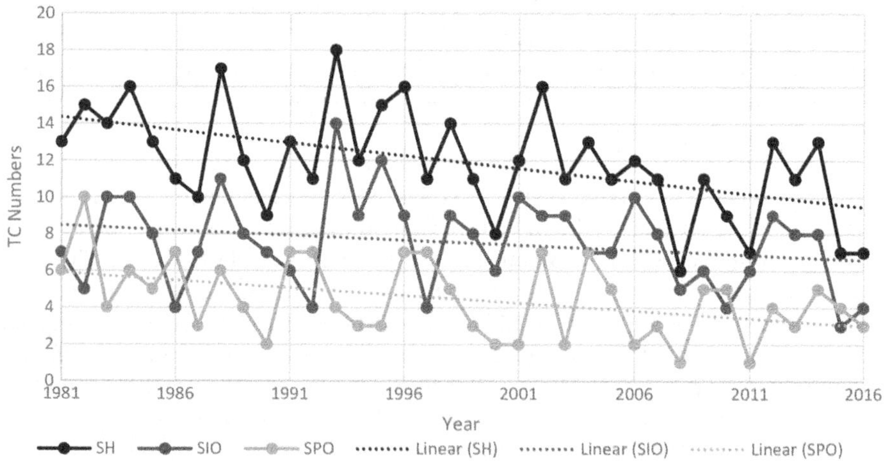

Fig. 2 Annual numbers of TCs with LMCP of 970 hPa or lower for the SH, SIO and SPO in the 1981–1982 to 2016–2017 seasons, with linear trends

reduction in the western part of the basin are observed (Ho et al. 2006; Kuleshov et al. 2008).

Investigating influence of the ENSO on the observed changes in TC genesis and development, it was found that the following large-scale environmental indicators are important contributors to interannual variability of TCs in the SH: sea surface temperature, relative humidity in mid-troposphere, vertical wind shear and lower tropospheric vorticity (Kuleshov et al. 2009b). Examining those indicators, it was demonstrated that the east-west shift in TC genesis in the SIO and SPO as well as displacement of TC activity further away from the equator in La Niña years compared to El Niño years were primarily attributed to changes in relative humidity and vorticity across both basins. Warmer SSTs were an additional contributor to the observed increase in TC genesis in Western Australia (maximum around 120° E) in La Niña years. In the SPO, changes in SSTs and vertical wind shear additionally favoured increased frequency of tropical cyclogenesis in the western part of the basin in La Niña years (Kuleshov et al. 2009b).

Another important driver of TC inter-annual variability in the SH is the Indian Ocean Dipole (IOD) (Saji et al. 1999)—a coupled ocean-atmosphere phenomenon which is similar to the ENSO. During the IOD positive phase, westward shift in positive SST anomalies is observed coupled with shift in atmospheric convection which favours TC genesis. During the IOD negative phase, opposite shift occurs: positive SST anomalies in the eastern part of the SIO and the AR together with associated enhanced atmospheric convection favour TC genesis over this area.

In addition to the ENSO and the IOD which are the main global climate drivers of TC inter-annual variability, the Madden-Julian Oscillation (MJO) is a major contributor to modulation of TC activity in the SH at the intra-annual time-scale (Ho et al. 2006; Leroy and Wheeler 2008; Camp et al. 2018). At shorter time-scales (up to 72 h), potential vorticity at 600 hPa, relative vorticity at 925 hPa and vector vertical wind shear (200–700 hPa) are the key environmental variables to favour TC genesis (Wijnands et al. 2016a).

These findings were central to understanding the environment favourable for TC genesis and development in the regions of the SH; the findings were applied to the development of a number of statistical models to predict seasonal TC activity (Kuleshov et al. 2008, 2009a, 2012b; Wijnands et al. 2014, 2015).

Tropical Cyclone Seasonal Prediction

One of the key objectives of the Initiative was to enhance EWS for TCs, reduce risk from TC hazards and strengthen capacity in countries of Asia-Pacific (Kuleshov et al. 2012a, c). Seasonal outlooks of TC activity are an effective means to assist government authorities and communities with CCA and DRR planning and decision making. Based on the developed SH TC climatology, it was found that the ENSO has significant impact on spatial variability of TCs. As such, TC-ENSO relationship has been used for developing a statistical model to predict TC activity in the coming season (Kuleshov et al. 2009a). TC seasonal outlooks are issued operationally by the Australian Bureau of Meteorology for November to the end of April (SH TC season) for the AR since 2009 and for the SPO since 2010 (Kuleshov et al. 2009a, 2012b).

The statistical TC seasonal forecast model used by the Bureau applies linear discriminant analysis (LDA) to identify the historical relationship between observed numbers of TCs and indices (predictors) describing the state of ENSO. The ENSO predictors employed in the Bureau's statistical prediction model are the SOI and equatorial SST anomalies in the Niño 3.4 region (N34). The LDA technique is applied in the Bureau model to identify the relationship between the July–August–September (JAS) mean of an ENSO predictor (either SOI or N34) and the observed number of TCs in a TC season in a region/sub-region over a training period. The resulting statistical relationship is then applied to a predictor value for a JAS period subsequent to the training period to yield predictions for the probability that the number of TCs in the forthcoming season is greater than the median over the training period and for the number of TCs in that season (Kuleshov et al. 2009a, 2012b).

The use of support vector regression (SVR) models, exploring new explanatory variables and the non-linear relationships between them, the use of model averaging, and lastly the integration of forecast intervals based on a bias-corrected and accelerated non-parametric bootstrap have been investigated, aiming to improve skills of operational TC seasonal forecasts issued by the Bureau for the AR and the SPO, as well as sub-regions therein (Wijnands et al. 2014, 2015). As mentioned above, the ENSO and the IOD are two key drivers of TC intra-seasonal variability; indices which describe the state of the ENSO and the IDO were utilised for developing statistical predictive models. Analysis of the generated models shows that the Dipole Mode Index, the 5VAR index and the SOI were the most frequently selected as explanatory variables for TC seasonal forecasting in all regions. Hindcasting analyses showed that the SVR model outperforms several benchmark methods. For both the AR and the SPO, nRMSE statistics for the hindcasts analyses for 2003/2004 to 2013/2014 indicates that SVR has better skill than LDA: in the AR, nRMSE was 0.93 and 1.25 for SVR and LDA, respectively; in the SPO, nRMSE was 0.83 and 0.94 for SVR and LDA, respectively (Wijnands et al. 2015). Overall, the new SVR methodology is an improvement over the current LDA models and has the potential to increase accuracy of TC seasonal forecasts in the AR and the SPO.

For decades, statistical models were used operationally by NMHSs around the world for producing TC seasonal outlooks. However, in a rapidly changing environment, the skill of statistical climate models which are solely based on past records is expected to deteriorate. Hence dynamical climate models, which are based on laws of physics rather than past relationships, are preferable for forecasting TC activity on time scales from weeks to months. Utilizing outputs of dynamical climate models for TC prediction on sub-seasonal to seasonal time scales is one of the current research priorities for the "Climate Change and Southern Hemisphere Tropical Cyclones" International Initiative.

Tropical Cyclone Wind Field Model

One of the multi-hazards associated with TCs is destructive winds. Accurate modelling of cyclone wind field can help with DRR providing vital information about spatial distribution of damaging winds and in this way enhancing EWS, improving preparedness to cyclone impacts and informed decision making.

As part of the "Climate Change and Southern Hemisphere Tropical Cyclones" International Initiative, research was undertaken to develop a new mathematical model to accurately describe distribution of near-surface TC wind speeds incorporating multiple satellite remote sensing and surface-based wind observations (Wijnands et al. 2016b). Using near-surface wind observations obtained from Australian meteorological observation stations, performance of the developed model was evaluated and it was demonstrated that TC wind speeds over waters can be accurately described by the developed model.

Tropical Cyclone Data Portal

A state-of-the-art Tropical Cyclone Data Portal is a specialised web-based information tool for examining historical TC activity over the regions of interests. The Portal allows to disseminate historical TC data, visualise and analyse TC tracks in the SH and the Western North Pacific Ocean (WNPO). The Portal was developed using OpenLayers platform (Kuleshov et al. 2013) and its functionality includes dynamic map navigation, presenting detailed information for a selected region and the display of changes in TC intensity over the lifetime of a cyclone.

The SH Portal displays TC best track data from the SH TC archive which currently consists of the data from the 1969/1970 TC season to present. Similar web-based information tool has been developed for the WNPO to display TC best track data prepared by RSMC Tokyo from 1970 to present. In this section, a few examples are presented to illustrate the Portals' functionality for a variety of application, e.g. for TC spatial and temporal selection over an area of interest. Detailed description of the Portal's functionality is presented in Appendix: Southern Hemisphere Tropical Cyclones Data Portal.

TC Activity Over a Selected Area

Examining historical TC activity over a selected area during specific time frame is often required by various users. In this example, TCs occurrences in the vicinity of Darwin, Australia are examined. Specifying an area (e.g. within 50 km radius from Darwin) and TC seasons of interest (e.g. TC seasons from 1969/1970 to 2017/2018) users can display tracks of TCs which affected the selected area during that time. An analysis demonstrates that 12 TCs passed through the selected circular area with a radius of 50 km centred on Darwin during the selected seasons (Fig. 3).

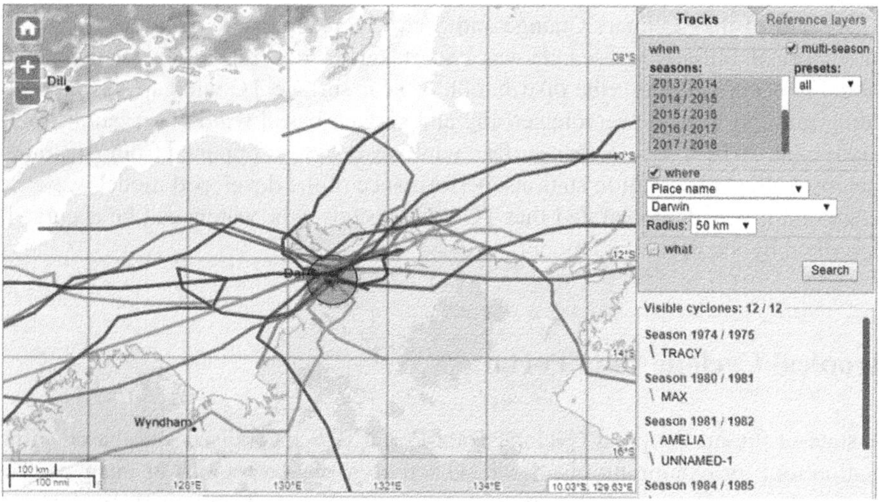

Fig. 3 Tracks of 12 TCs which passed through an area with a radius of 50 km centred on Darwin in the 1969/1970 to 2017/2018 TC seasons

TC-ENSO Relationship

As discussed above, the ENSO is one of the most important global climate drivers which has significant impact on TC activity including changes in cyclones' spatial and temporal distribution over different regions in the SH and the WNPO. It could be of interest for various users to examine relationship between ENSO and TC variability over specified areas.

Example 1

In this example, changes in TC genesis with respect to position of the South Pacific Convergence Zone (SPCZ) are examined. The SPCZ is a part of the Intertropical Convergence Zone and it extends from the Western Pacific Warm Pool at the maritime continent south-eastwards towards French Polynesia. The SPCZ is manifested as a band of low-level convergence, cloudiness and precipitation. Based on four indicators of the SPCZ axis (maximum rainfall, maximum of low level convergence, maximum of the 500 hPa vertical motion and minimum in outgoing longwave radiation, see Folland et al. 2002 for detail), climatological (average over 30 years) position of the SPCZ is presented as a line with coordinates 5° S 150° E–30° S 120° W.

Analysing the ENSO impact on TC activity in the South Pacific, it has been shown that the area favourable for cyclogenesis shifts eastwards in El Niño years in comparison to La Niña years (e.g., Basher and Zheng 1995; Kuleshov et al. 2008). To illustrate these findings, tracks of TCs which formed in an area between the equator and the SPCZ in El Niño and La Niña years are presented in Figs. 4 and 5, respectively (examined seasons are from 1969/1970 to 2017/2018). Significant difference in cyclogenesis in the selected area was observed: 97 TCs were formed during the ENSO warm phase and 20 TCs—during the ENSO cold phase.

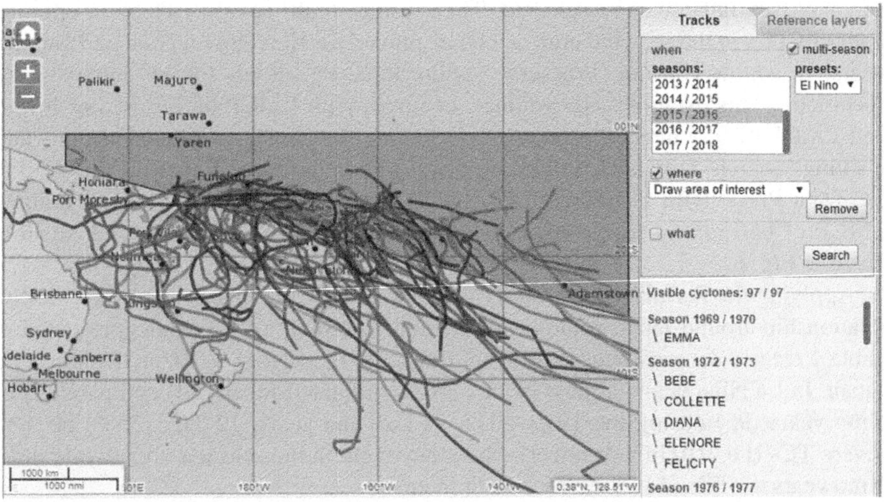

Fig. 4 Tracks of 97 TCs which formed in an area between the equator and the SPCZ in El Niño years recorded in the 1969/1970 to 2017/2018 TC

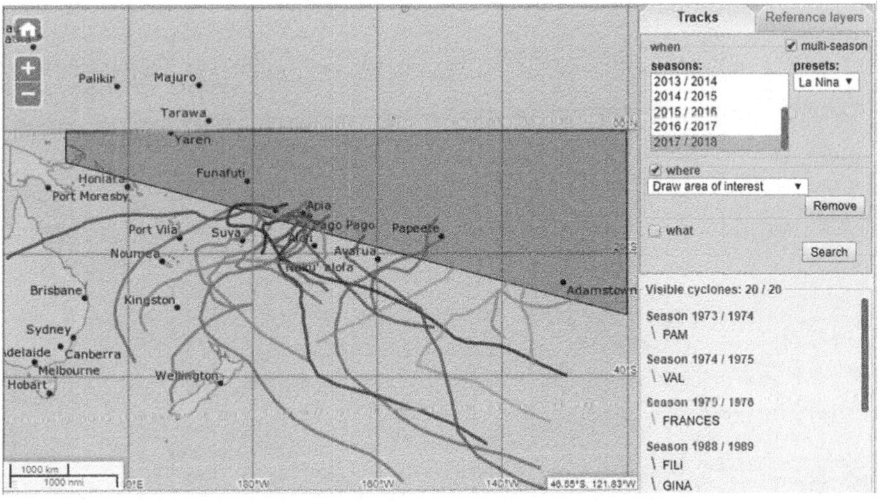

Fig. 5 Tracks of 20 TCs which formed in an area between the equator and the SPCZ in La Niña years recorded in the 1969/1970 to 2017/2018 TC seasons

Example 2

In this example, impact of the ENSO on TC activity in one of the sub-regions in the WNPO (specifically, an area which encompasses Japan) is examined. Firstly, using 'Draw area of interest' option of the Portal's functionality, the area which encompasses Japan can be selected. Then, using 'multi-season' selection options, tracks of TCs in the selected area can be displayed for El Niño, La Niña and Neutral years. In addition, using 'Pressure' option, three thresholds for TC intensity are specified in this example: all cyclones, cyclones with LMCP of 970 hPa or lower, and LMCP of 950 hPa or lower. After selecting the above mentioned options and clicking 'Search' button TC tracks are displayed on the map. A view of the WNPO TC Data Portal with displayed tracks of TCs in El Niño seasons which attained LMCP of 950 hPa or lower when they were in the selected area around Japan is given in Fig. 6.

Applying the Portal's functionality as described above, an analysis of TC-ENSO relationship around Japan could be performed. Summary of the results presented in Table 1 clearly demonstrates significant impact of the ENSO on TC activity around Japan. In La Niña years, almost twice less TCs affected Japan (53) compared to El Niño years (95) during the 1970–2017. In La Niña years, 19 TCs (35%) became severe TCs (i.e. 970 hPa or lower) when they were in the selected area, while in El Niño years 42 TCs (44%) attained such intensity.

TC-ENSO relationship in the WNPO was thoroughly investigated; the results indicate that the timing with which ENSO impacts TC inter-annual variability (frequency and intensity) closely follow the evolution of the ENSO cycle

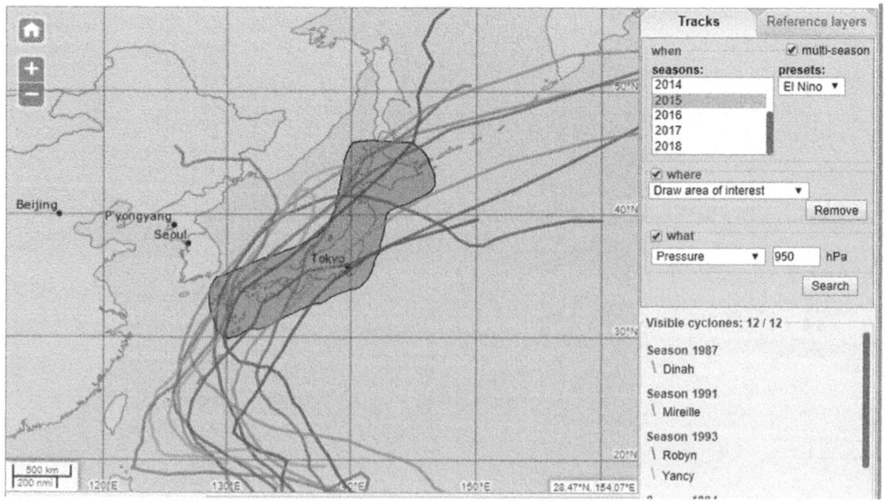

Fig. 6 Tracks of 12 TCs with LMCP of 950 hPa or lower which passed through the selected area around Japan in El Niño years recorded in the 1970–2018

Table 1 Number of tropical cyclones which attained specified intensity (all cyclones, cyclones with LMCP of 970 hPa or lower, and LMCP of 950 hPa or lower) in the selected area around Japan in El Niño, La Niña and Neutral years in the 1970–2017

Cyclone intensity in the selected area	El Niño	La Niña	Neutral
All cyclones	95	53	93
970 hPa	42	19	35
950 hPa	12	6	11

(Goh and Chan 2010; Li and Zhou 2012). Based on TC-ENSO relationship (and other environmental conditions favourable for TC genesis and development), seasonal forecasting of TC activity over the whole Western North Pacific region as well as over its sub-regions such as South China Sea, Japan and Korea has been developed (Chan et al. 2001; Goh and Chan 2010).

Example 3

TCs present significant threat to life, property and infrastructure when they make a landfall and it is of interest for various users to examine TC variability in coastal areas, to make informed decisions ahead of a coming TC season. In this example, impact of the ENSO on TCs activity around west coast of the Australian continent between longitudes of 5° S and 30° S (area shaded in grey in Figs. 7 and 8) is presented. The area is drawn using 'Draw area of interest' functionality of the Portal. There is a

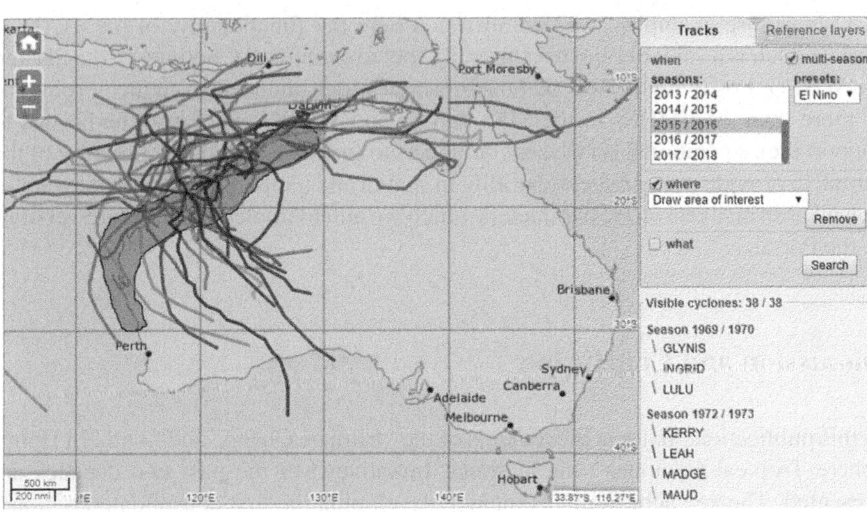

Fig. 7 Tracks of 38 TCs which passed through the selected area around west coast of the Australian continent (area shaded in grey) during El Niño years

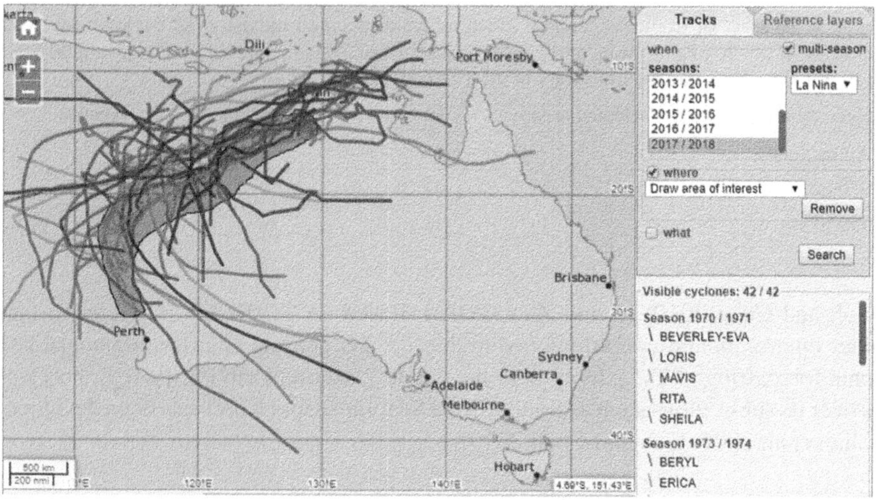

Fig. 8 Tracks of 42 TCs which passed through the selected area around west coast of the Australian continent (area shaded in grey) during La Niña years

slight decrease in TC activity in the selected area during warm phase of the ENSO but it is statistically not significant: 38 TCs passed through the area during El Niño years (Fig. 7) compared to 41 TCs during La Niña years (Fig. 8). This information is important for communities at risk: typically, TC activity in the Western Australia region is higher in La Niña years compared to El Niño years; however, this analysis demonstrates that the communities should stay vigilant and prepared for potential TC impact in each season.

Using those examples, we demonstrated here the functionality of the TC Data Portal which is a valuable and user-friendly tool to examine TC activity in the SH and the WNPO. Positive feedback and useful suggestions obtained from meteorologists of more than 20 NMHSs in the SPO, SWIO and WNPO which use the portals to support their operational service encourage us to further develop functionality of the Portal, to provide even greater flexibility in spatial and temporal selection of cyclones as well as in analysis of TC parameters which we aim to implement in future versions of the Portal.

Discussion and Conclusion

In this publication, major achievements of the "Climate Change and Southern Hemisphere Tropical Cyclones" International Initiative over the past two decades are presented. The key achievements include developing the first consolidated regional database—the Southern Hemisphere Tropical Cyclone historical data archive, and the first tropical cyclone climatology for the Southern Hemisphere. Scientific knowledge developed through the Initiative directly contribute to applications in services e.g. statistical TC seasonal prediction models are used operationally by meteorological agencies in the regions. The chapter also includes a brief description of the Tropical Cyclone Data Portal helping readers to learn quickly how to use this state-of-the-art web-based information tool to examine historical TC activity over the regions of interests. The Initiative is growing over the years, starting with cooperation of countries in the SH and then expanding to countries in the WNPO. Summarising lessons learned from two decades of our productive cooperation and outlining future prospects we would like to highlight the following.

Prior to establishing the Initiative, there was no consolidated SH archive of historical TC tracks. TC best track data were kept in archives of RSMCs and TCWCs in Australia, Fiji, France and New Zealand with responsibilities for issuing TC advisories in the designated regions. In addition, some historical TC data were kept in archives of other meteorological services in the SH but the data were fragmented and confined to vicinity of islands and coastal areas. TCs are large-scale meteorological phenomena typically travelling long distances and often crossing regional boarders. However, archives of NMHSs keep only records for cyclones which occur in their designated areas of responsibility; as such, for numerous TCs these archives had records which only partially described entire tracks of those cyclones which occurred in more than one region. Creation of the archive which consolidated TC data from all RSMCs and TCWCs of the SH for satellite era was beneficial for improving quality and completeness of historical cyclone records in all meteorological services of the region.

Availability of high-quality consolidated historical TC best track data for the SH allowed to advance TC research. Utilizing these records, the first tropical cyclone climatology for the SH was developed. It was a significant achievement in improving our understanding of impact of global climate drivers on TC inter-annual variability clearly demonstrating influence of the El Niño-Southern Oscillation on spatial and temporal variability in TC activity in the SH. It was demonstrated that in El Niño years TC activity shifts away from Australia towards the Eastern SPO; on the other hand, in La Niña years TC activity increases in the Australian region, with TC landfall in eastern coast of the country almost doubling compared to that in El Niño years.

In addition to the objectives of creating the consolidated TC archive for the SH, identifying the key global climate drivers of inter-annual TC variability and producing TC climatology, the objective of developing skilful seasonal prediction for the regions was also the top priority for the Initiative. As a result, the developed statistical model-based methodology for predicting regional TC activity is now implemented operationally for producing TC seasonal outlooks for the AR and the SPO region, providing early warnings about expected TC activity in a coming season and assisting governments and communities at risk with improving preparedness to potential TC impacts. Statistical model-based seasonal TC outlooks proved to be a useful component of the EWS for countries in the regions.

However, it is expected that in the near future skill of statistical models used for climate prediction will deteriorate because an assumption about stationarity of climate is no longer valid. Being aware of the limitations of using the statistical models for TC prediction on sub-seasonal to seasonal time scales, in our future activities we are planning to explore an alternative approach and evaluate skill of dynamical climate models for predicting TC activity in the AR, SPO and SWIO on time scales from weeks to months. Recently, we demonstrated potential of dynamical climate model ACCESS-S (The Australian Community Climate and Earth-System Simulator—Seasonal) for skilful multi-week TC prediction in the AR and the SPO (Camp et al. 2018). These first results are encouraging; prospects for utilizing outputs of ACCESS-S for TC prediction on sub-seasonal (from 1 to 4 weeks) and seasonal (from 3 to 6 months) time scales in operational services of the Australian Bureau of Meteorology will be further explored as part of future activities of the Initiative.

Outcomes of collaborative efforts of organisations from countries in Asia-Pacific undertaken through the "Climate Change and Southern Hemisphere Tropical Cyclones" International Initiative significantly contributed to improvement of our knowledge about TCs as well as enhancement of TC EWS for countries in the SH. The developed regional consolidated TC database and specialised web-based portal for dissemination of information; comprehensive TC climatology; and methodology for seasonal prediction of TC activity for the regions were particularly beneficial for SIDS and LDCs in Asia-Pacific which have limited resources and capacity to deal with enhancing EWS alone greatly assisting the countries with decision making in CCA and DRR. Over the years, translation of scientific achievements of the Initiative to operational services benefited 24 Member countries in WMO RAs I and V strengthening capacity of NMHSs, enhancing EWS and resulting in reduction of disaster risk related to multi-hazards associated with TCs. Prospects of future geographical expansion of the Initiative and strengthening cooperation between developed and less developed countries in other WMO regions will be explored.

The "Climate Change and Southern Hemisphere Tropical Cyclones" International Initiative meaningfully contributes to outcomes of other international activities on enhancing EWS, e.g. Climate Risk and Early Warning Systems (CREWS) initiative which was launched at the 2015 United Nations Climate Change Conference (COP 21) held in Paris in 2015. CREWS initiative is implemented by the World Meteorological Organization, the United Nations Office for Disaster Risk Reduction, the World Bank and the Global Facility for Disaster Reduction and Recovery and focused on improving multi-hazard EWS. These systems proved their value for reducing loss of life and economic hardship caused by meteorological hazards such as tropical cyclones, floods, severe storms, forest fires and heatwaves. Contribution of the "Climate Change and Southern Hemisphere Tropical Cyclones" International Initiative in enhancing CREWS in SIDS and LDCs in the Pacific is well recognized. Future prospects for enhancing CREWS in SIDS and LDCs in other oceanic basins—the North Atlantic Ocean and the North Indian Oceans—will be explored in our future activities.

To conclude, over the past two decades the "Climate Change and Southern Hemisphere Tropical Cyclones" International Initiative, through dedicated efforts of cooperating agencies in Asia-Pacific, contributed significantly to enhancing EWS in the region providing governments and communities at risk with valuable information and assisting with decision making in dealing with multi-hazards associated with tropical cyclones, and adaptation to climate variability and change.

Acknowledgements Support was provided by the Australian Greenhouse Office, the Australian Agency for International Development, the Department of Climate Change and Energy Efficiency, the Department of Foreign Affairs and Trade, the Laboratory of the Atmosphere and Cyclones at the University of La Reunion and Meteo-France and the World Meteorological Organization (WMO). Dr. Taoyong Peng, Chief, Tropical Cyclone Programme, WMO and Dr. Gustavo J. Nagy, Universidad de la República de Uruguay provided useful comments on improving the manuscript.

Afterword

In December 2012, I was working in Fiji with colleagues from RSMC Nadi on revision of TC historical data. Since 1999, when the "Climate Change and Southern Hemisphere Tropical Cyclones" International Initiative was established meteorologists from Fiji were among active supporters of and contributors to this Initiative. High quality of observational data is vital for deriving plausible concussions about past TC activity and we dedicated substantial efforts preparing the SH TC data archive over the years of the Initiative.

December is the second month of a TC season in the SH and we already expected that at least one cyclone will be forming in the South Pacific in that month. When I came to RSMC Nadi's operational division on 10 December 2012 first that I heard from my colleagues was about tropical depression which according to numerical weather prediction models will develop to a stage of tropical cyclone soon. 'Tropical cyclone bench' was activated and for the next ten days we closely monitored formation and development of this system which became the first cyclone in the 2012–13 South Pacific cyclone season.

The system was named as a tropical cyclone Evan on 11 December 2012 and over duration of its life cycle affected Western Samoa, American Samoa, Fiji, Tonga, Wallis and Futuna causing severe destructions in the island countries. We were looking at the historical data and it was fear that Evan'*s* impact may be as bad as impact of 1993 tropical cyclone Kina—one of the deadliest cyclones to affect Fiji in recent history with 23 lives lost. It was expected that Evan will arrive into the Fiji waters around 17 December 2012 as a category 4 or possibly category 5, severe tropical cyclone.

In preparation for the cyclone's impact on Fiji, colleagues from RSMC Nadi regularly issued Media Releases sending a clear message—"All communities in Fiji should be prepared now and act responsibly, to avoid unnecessary loss of lives and property". Fijian authorities took fast and effective actions evacuating residents from areas which could be potentially affected as well as tourists from resorts located in small islands. It was reported that about 8000 people found refuge from Evan in emergency shelters. Such actions were highly efficient and as a result there were no fatalities associated with cyclone Evan in Fiji.

On Monday, 17 December Evan started to affect Viti Levu, the largest island in Fiji. The cyclone was moving to the west of the island with an eye wall (area of a cyclone with most damaging winds) passing over the coastal areas of Viti Levu. It was the most exciting 24 h for a scientist who spent 30 years observing cyclones with satellite instruments, flying around cyclones in the Pacific Ocean on board of Ilyushin-18 aircraft-laboratory and examining historical cyclone data from all regions of the Southern Hemisphere. It was 'my day'—to closely observe increasing strength of winds and intensity of rain when Evan was approaching Nadi, to feel hurricane for real, to witness its power.

By mid-day wind became stronger and firstly branches of trees started to move, and then whole trees came in motion; wind speed was increasing with time and small branches, and then larger branches were broken. Listening to sound of roaring winds, I was watching how leaves striped from palms and broken branches of flame and umbrella trees were flying around; from time to time I heard that broken branches and flying debris were hitting the roof and walls. Sometime these sounds were so loud that I was thinking—is this building strong enough to withstand fury of the cyclone?

Winds and rain further increased and by the evening I observed very widespread damage to vegetation with some trees uprooted as well as roofs damaged and windows broken. By about 9 pm winds reached the maximum strength of about 185 km/h and Evan's damaging impact on Viti Levu continued through the night.

In the morning of 18 December Evan moved further south-west leaving devastated Viti Levu behind. The cyclone caused severe damage to vegetation, structural damage to buildings and infrastructure and recovery operations begun immediately. Once again, measures undertaken by the authorities were efficient, main roads were cleared from fallen trees and power supply was restored in the next few days.

Yes, living through a cyclone is a life time experience. After researching cyclones for decades, analysing changes in their geographic distribution, evaluating trends in cyclone occurrences and estimating their power, I experienced power of a cyclone and its ferocity together with thousands of Fijians who live through cyclones year after year. This experience vividly demonstrated to me vital importance of assistance which the 'Climate Change and Southern Hemisphere Tropical Cyclones' International Initiative provides to people living in small island countries.

Appendix: Southern Hemisphere Tropical Cyclones Data Portal

Introduction

As part of the "Climate Change and Southern Hemisphere Tropical Cyclones" International Initiative, the first consolidated regional database of historical tropical cyclone (TC) data for the Southern Hemisphere (SH) (satellite-era, i.e. 1970 to present) was created. Extending the Initiative to the Western North Pacific Ocean (WNPO), TC best track data from the Regional Specialised Meteorological Centre (RSMC) Tokyo, Japan were also added to the archive of historical data. To display TC best tracks and analyse TC spatial and temporal variability in the SH and the WNPO, the Tropical Cyclone Data Portal has been created. A detailed description of the Portal is presented here, helping readers to learn quickly how to use this web-based information tool for examining historical TC activity over the regions of interests.

The "Climate Change and Southern Hemisphere Tropical Cyclones" International Initiative has been established in 1999 aiming to improving our understanding of changes in regional tropical cyclone (TC) activity due to natural variability and climate change. During twenty years of productive cooperation among 13 organizations from 8 counties in the Asia-Pacific Region, the Initiative produced significant results which include creation of the first consolidated regional database of historical TC data for the Southern Hemisphere (SH) for satellite-era (i.e. 1970 to present). The Initiative was growing over the years extending to the Western North Pacific Ocean (WNPO). As a result, TC best track data from the Regional Specialised Meteorological Centre (RSMC) Tokyo, Japan were also added to the archive of historical data. To display TC best tracks and analyse TC spatial and temporal variability in the SH and the WNPO, the Tropical Cyclone Data Portal has been created. Here, a detailed description of the Portal is presented, helping readers to learn quickly how to use this web-based information tool for examining historical TC activity over the regions of interests.

Tropical Cyclone Data Portal

A state-of-the-art Tropical Cyclone Data Portal is a specialised web-based information tool for examining historical TC activity over the regions of interests. The Portal allows to disseminate historical TC data, visualise and analyse TC tracks in the SH and the WNPO. The Portal was developed using OpenLayers platform (Kuleshov et al. 2013) and its functionality includes dynamic map navigation, presenting detailed information for a selected region and the display of changes in TC intensity over the lifetime of a cyclone.

The SH Portal displays TC best track data from the SH TC archive which currently consists of the data from the 1969/1970 TC season to present. Similar web-based information tool has been developed for the WNPO to display TC best track data prepared by RSMC Tokyo from 1970 to present. While the same functionality is implemented in both portals there are some features specific to each portal due to differences in seasonal TC activity in the Southern and Northern Hemispheres which will be outlined here.

Detailed description the Portal's functionality is presented below helping readers to learn quickly how to use this specialised web-based information tool to examine historical TC activity. Note that the web-browsers Mozilla Firefox and Google

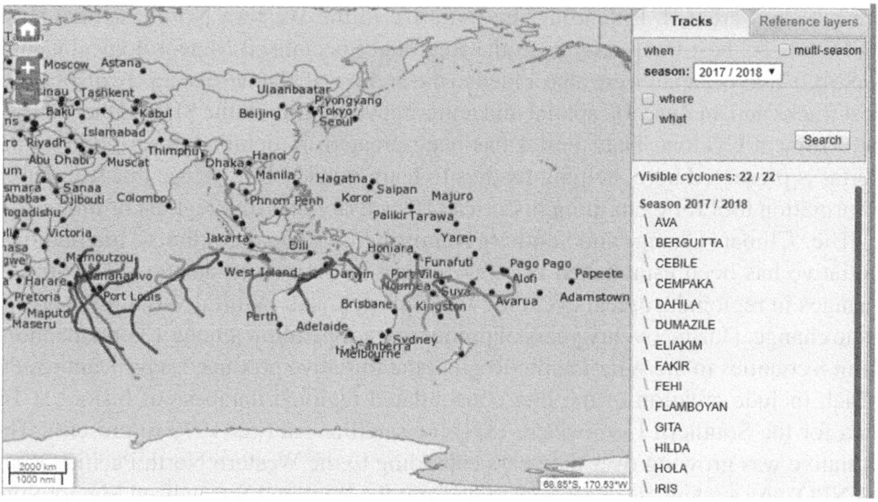

Fig. 9 Initial view of the tropical cyclone data portals for the Southern Hemisphere

Chrome are preferable for using the portals, but others, such as Internet Explorer, are supported.

Navigating a Map

Zooming in and out: Click '+' or '−' buttons of the navigation tool at the top left corner of the map (Fig. 9) to zoom in or to zoom out, respectively. Alternatively, users can zoom using the mouse scroll wheel or by holding the SHIFT key whilst selecting an area by dragging the mouse over the map. Users can zoom to eight different map scales. The scale is shown in kilometres (km) and nautical miles (nmi) on the lower left side of the map. Clicking on 'home' button will zoom out bringing the map back to initial view.

Panning the map area: Users can navigate by clicking (with left mouse button) and dragging the map with the mouse.

Selecting a Time Period

Click on the 'Tracks' tab on the top right of the screen for selecting a time period under 'when' option. Select a single season by clicking on a black arrow (drop down menu will appear) and then selecting a season from the menu (e.g. 1971/1972, Fig. 10). In the SH TC data portal, each cyclone season is listed using two years (e.g. 1971/1972, Fig. 10) because cyclones in the SH occur in the austral summer and a

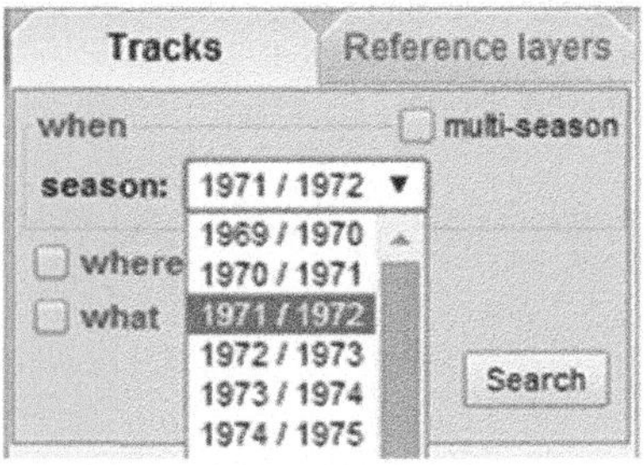

Fig. 10 Selecting a time period for a single season

cyclone season in the SH typically lasts from November to April. In the WNPO TC Data Portal each cyclone season is listed using one year (e.g. 1971). In this region TCs occur during all months, with a minimum in February and March and a peak in September.

Select a range of seasons checking 'multi-season' box by clicking with left mouse button, selecting the first season from drop down menu, then holding SHIFT key and selecting the last season (e.g. seasons from 2008/2009 to 2010/2011, Fig. 11). This option allows users to select a range of consecutive seasons. If using Internet Explorer, try to avoid selecting a large range of seasons as this may significantly slow down the responsiveness of the application.

There is an additional option for selecting a time period which provides users with various sets of pre-selected seasons. Clicking on a black arrow under 'pre-sets' brings the drop down menu with the following options: 'all', 'latest', 'El Niño', 'La Niña' and 'Neutral' (Fig. 12). From the list users can select 'all' to display all tracks of TCs available from the archives, 'latest'—to display TC tracks for the latest available data, and 'El Niño', 'La Niña' and 'Neutral'—to display TC tracks during three different phases of ENSO. For example, clicking on 'El Niño' option (Fig. 12) will automatically select those TC seasons when the ENSO was in its warm phase.

After selection is completed, to display TC tracks on the map for the selected time period (a single season, a range of consecutive seasons or a pre-selected set of seasons), click on 'Search' button.

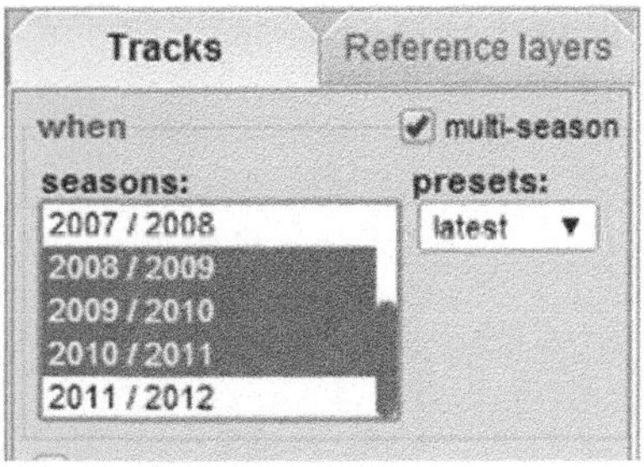

Fig. 11 Selecting a time period for a range of seasons

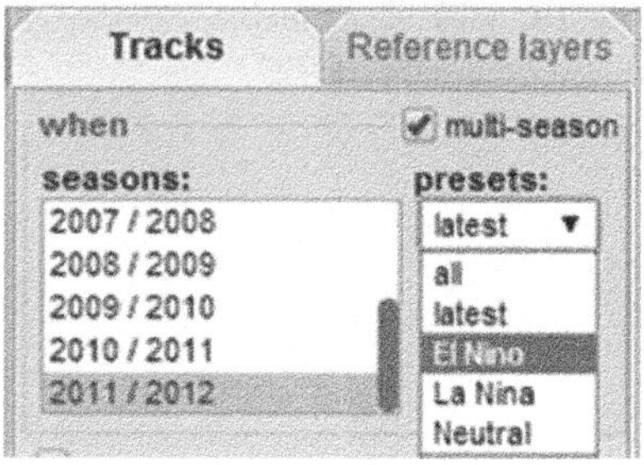

Fig. 12 Selecting a time period for a pre-selected range of seasons

Selecting a Specific Location

Click on the 'Tracks' tab and check 'where' box. From drop down menu, four options to select a specific location are available: 'Place name', 'Coordinates', 'Exclusive Economic Zone' and 'Draw area of interest' (Fig. 13). Selecting one of the options allows users to examine TC activity with relation to a specific location (option 1—place name from a list of selected locations and option 2—any location specified by its geographic coordinates) or over a specific area (option 3—country's Exclusive Economic Zone and option 4—area of interest of any shape which can be drawn by users). Details are given below.

For option 1 ('Place name'), select a place name from the drop down menu (e.g. Darwin, Australia; Fig. 14). Eighty eight places from 13 countries in the SH and 10 places from 5 countries in the WNPO are listed. TC activity can be examined over a circular area centred on a selected place with a specified radius of 50, 100, 200, 300, 400 or 500 km. After selecting the place and the radius, click on 'Search' button to display TC tracks on the map for the selected time period. In Fig. 14, an example of track of TC Tracy which passed through a circular area with a 50 km radius centred on Darwin, Australia during 1974/1975 season is presented.

If a place of interest can't be selected from the drop down menu, option 2 ('Coordinates') can be used. Coordinates of a place can be entered manually in 'latitude' and 'longitude' boxes. Alternatively, click on the map to obtain coordinates. Note a

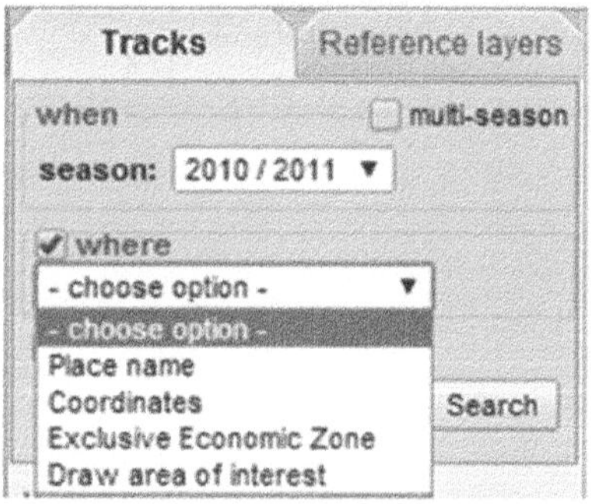

Fig. 13 Selecting a specific location

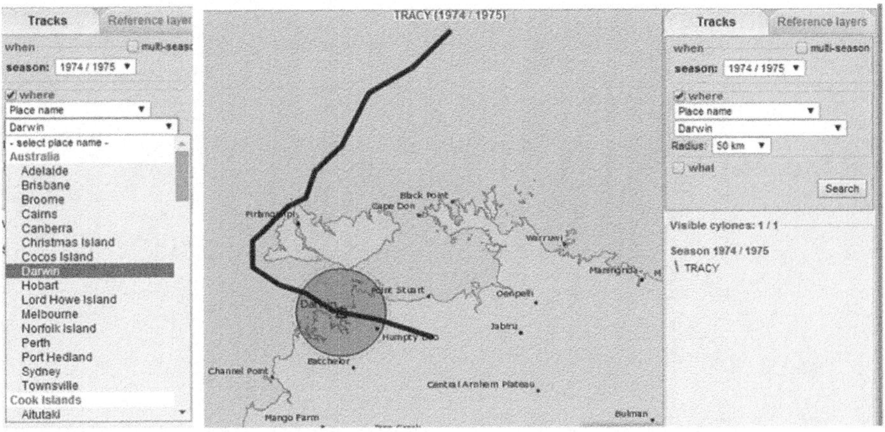

Fig. 14 Selecting a specific location using 'Place name' option; track of TC Tracy is displayed

minus sign which should be entered to specify a latitude in the SH; for the Northern Hemisphere, a latitude is specified with a plus sign. To specify a longitude, use a plus sing for the Eastern Hemisphere and a minus sign for the Western Hemisphere (see an example of TC tracks which affected Samoa, 13° S–172° E, in 2012/2013 TC season, Fig. 15).

An option to examine TC activity over a country's Exclusive Economic Zone is incorporated in the portal (Fig. 16). Select the country from drop down menu (e.g.

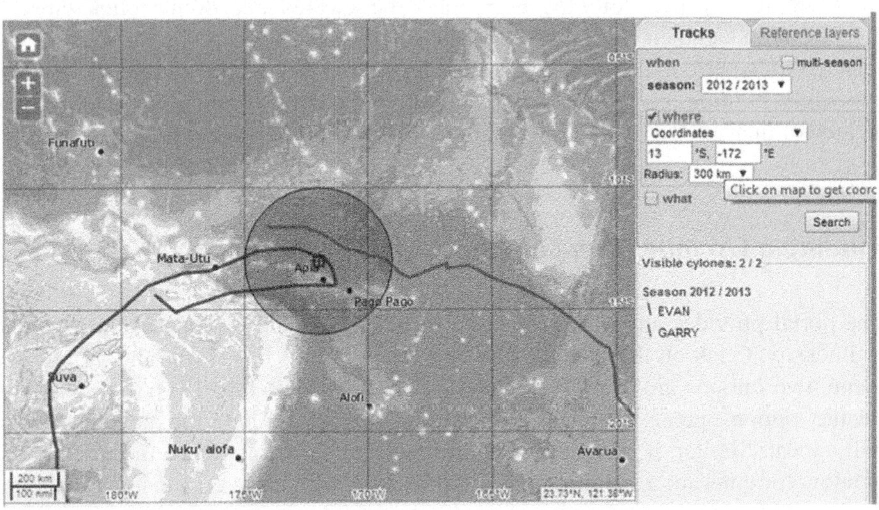

Fig. 15 Selecting a specific location using 'Coordinates' option; tracks of TCs Evan and Garry in the vicinity of Samoa (2012/2013 TC season) are displayed

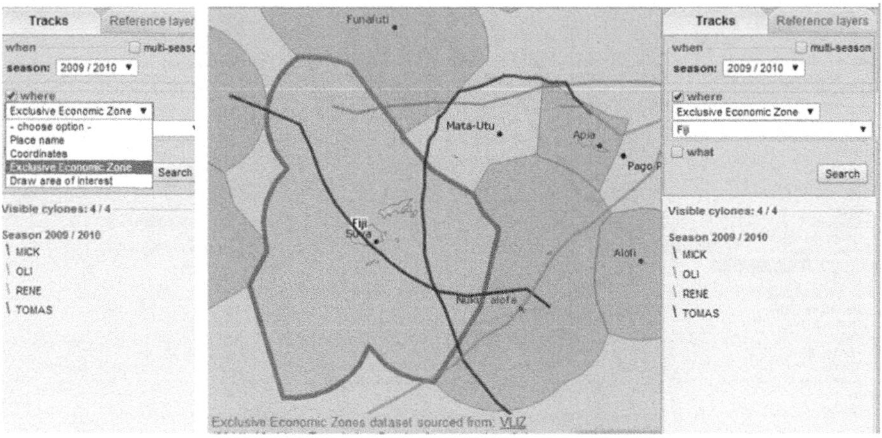

Fig. 16 Selecting a specific location using 'Exclusive Economic Zone' option; tracks of TCs Mick, Oli, Rene and Tomas which passed through Exclusive Economic Zone of Fiji during 2009/2010 season are displayed

Fiji) and click on 'Search' button to display TC tracks on the map for the selected time period (e.g. 2009/2010 season, Fig. 16).

Another useful option for examining TC activity over a specified area is implemented under 'Draw area of interest' portal's functionality. Click on a map (with left mouse button) to select points which encompass an area of interest (e.g. a rectangular area of the Australian region specified as 0–40° S, 90–160° E highlighted in grey colour in Fig. 17). To specify the last point of the selected area, double click with left mouse button and then click 'Search' button to display TC tracks on the map for the selected time period. Users also can draw an area of any shape holding SHIFT key and whilst holding left mouse button and moving a cursor around the map to draw an area of interest. To finish drawing, both keys should be released.

Finding a Cyclone Name or Intensity

The portal provides an option for users to find a specific cyclone using its name or intensity. Click on the 'Tracks' tab and check 'what' box. From the drop down menu, two options are available: 'Pressure' and 'Name' (Fig. 18). After selecting 'Name' option, screen will be modified and a window for entering a cyclone name will appear. Similar, if 'Pressure' option is selected, screen will be modified and a window for entering a specific pressure threshold will appear. To display cyclones with an entered name (or those which attained a specified intensity in terms of minimum central pressure) during a selected season or a number of seasons, click on 'Search' button.

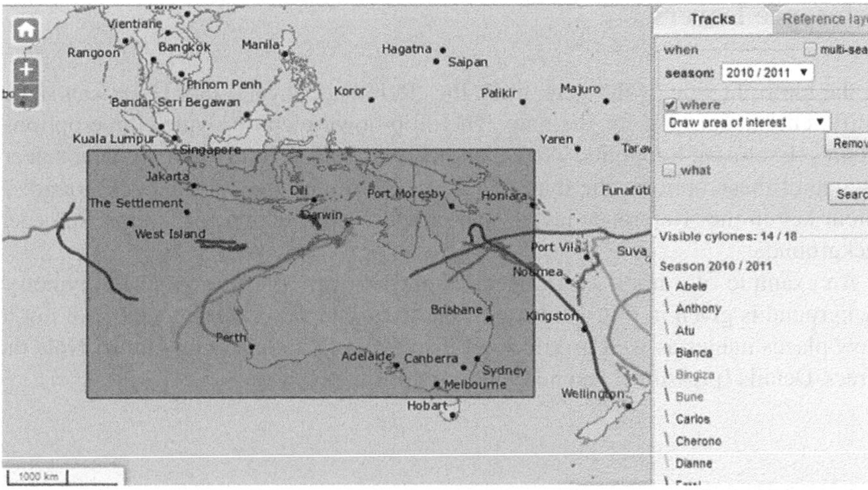

Fig. 17 Selecting a specified area using 'Draw area of interest' option. Tracks of TCs which passed through the selected area (the Australian region) during 2010/2011 season are displayed

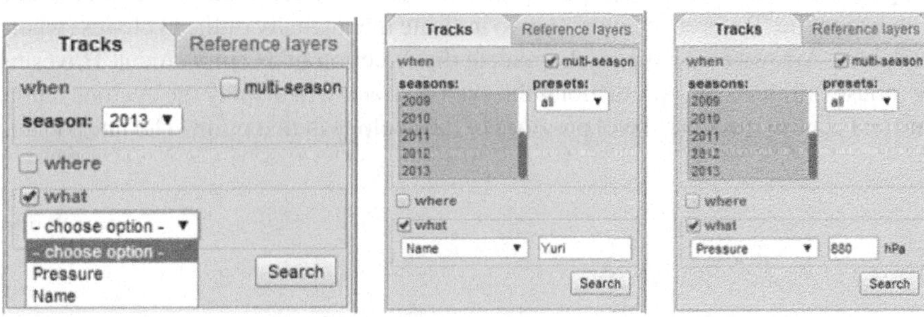

Fig. 18 Selecting a cyclone name or intensity under 'what' option; entering cyclone name if 'Name' is selected; entering pressure threshold if 'Pressure' option is selected

Visible Cyclones

On the bottom right of screen under 'Visible cyclones', users will see two numbers separated with a '/' symbol. These numbers indicate the part of the cyclone tracks visible on screen in respect to the number of cyclones that met the search criteria. Cyclones are grouped by TC seasons which are listed chronologically. By clicking on the name of a cyclone, a pop-up box will appear on the map containing information about the chosen cyclone, including the cyclone name, season and start and end dates and times.

Reference Layers

At the top right side of the map, click the 'Reference layers' tab. Users can select a different background for the map. The drop-down menu provides three options: 'Plain', 'Exclusive Economic Zones' and 'Elevation and Bathymetry'. After selecting one of these options, the map will be redrawn with the chosen background. A colour key in the 'Reference layers' tab provides information for interpreting each background.

An example of a track of TC Heta displayed over 'Elevation and Bathymetry' background is given in Fig. 19. In addition, users can also choose whether or not to show places names as well as grid lines displaying latitude and longitude. Note the 'Track Details (pressure)' legend on the right side of screen.

Cyclone Details

In the pop-up box, select 'show details'. Cyclone tracking points will now appear along the selected cyclone track as shown in Fig. 19. These points are colour coded as shown in the 'Reference layers' tab to indicate the intensity of the cyclones (where available). Arrows along the track indicate the direction of TC movement. Hovering the mouse pointer over a point along the cyclone track will display the cyclone name, time and central mean sea level pressure of the cyclone at that point. The latitude and

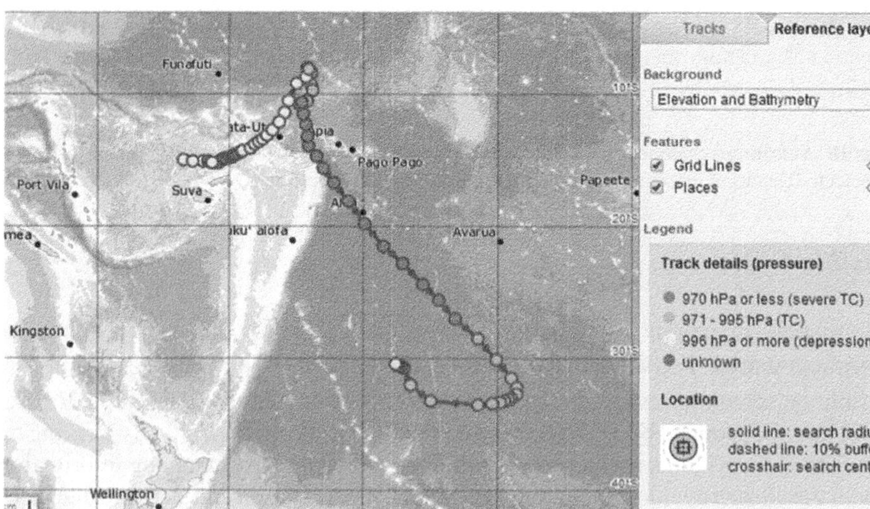

Fig. 19 An example of a track of TC Heta displayed over 'Elevation and Bathymetry' background. Note the track details (pressure) legend on the right side of screen

longitude of the mouse position is also shown beneath the lower left edge of the map underneath the map scale. Clicking on a point on the track will bring up the pop-up information box.

Downloading Information

Users can download detailed information about individual cyclones from the pop-up box described above. The information can be downloaded as a CSV (Comma Separated Value, will open in Microsoft Excel). A CSV file contains best track data across the lifetime of the cyclone. Name, time, position and estimated intensity (in terms of central mean sea level pressure and 10 min averages sustained wind) are indicated. If there is no intensity estimate at particular point, 'no estimates' will be displayed in the data. An example of a 'CSV' file is shown in Table 2.

Table 2 An example of a 'CSV' file download of best track data for TC Yuri

Year	Seq.	Name	Date time	Lat.	Lon.	Pressure	Wind	Grade
1991	28	Yuri	1991-11-22 18:00:00+00	4.6	165.9	1002	0	TD
1991	28	Yuri	1991-11-23 00:00:00+00	4.7	165.8	1000	0	TD
1991	28	Yuri	1991-11-23 06:00:00+00	4.8	165.8	996	35	TS
1991	28	Yuri	1991-11-23 12:00:00+00	5	165.7	992	40	TS
1991	28	Yuri	1991-11-23 18:00:00+00	5.3	165.5	990	45	TS
1991	28	Yuri	1991-11-24 00:00:00+00	5.5	165.3	980	55	STS
1991	28	Yuri	1991-11-24 06:00:00+00	6.1	164.2	975	60	STS
1991	28	Yuri	1991-11-24 12:00:00+00	6.4	162.7	970	65	TY
1991	28	Yuri	1991-11-24 18:00:00+00	6.8	161.3	965	70	TY
1991	28	Yuri	1991-11-26 06:00:00+00	9.4	151.5	915	105	TY
1991	28	Yuri	1991-11-26 12:00:00+00	9.7	149.7	910	110	TY
1991	28	Yuri	1991-11-26 18:00:00+00	10	148.3	900	115	TY

(continued)

Table 2 (continued)

Year	Seq.	Name	Date time	Lat.	Lon.	Pressure	Wind	Grade
1991	28	Yuri	1991-11-27 00:00:00+00	10.7	146.8	895	120	TY
1991	28	Yuri	1991-11-27 06:00:00+00	11.6	145.5	895	120	TY

References

Australian Government (1977) Report on cyclone tracy December 1974. Australian Government Publishing Service, Canberra, Australia, p 82

Basher RE, Zheng X (1995) Tropical cyclones in the Southwest Pacific: spatial patterns and relationships to Southern Oscillation and sea surface temperature. J Clim 8:1249–1260

Broomhall M, Grant I, Majewski L, Willmott M, Jones D, Kuleshov Y (2010) Improving the Australian tropical cyclone database: extension of GMS satellite image archive. In: Charabi Y (ed) Indian Ocean tropical cyclones and climate change. Springer, NY, pp 199–206

Callaghan J (2003) The tropical cyclone risk in Cairns, pp 129–153

Camp J, Wheeler MC, Hendon HH, Gregory PA, Marshall AG, Tory KJ, Watkins AB, MacLachlan C, Kuleshov Y (2018) Skilful multiweek tropical cyclone prediction in ACCESS-S1 and the role of the MJO. Q J Royal Meteorol Soc 144:1337–1351

Chan JCL, Shi JE, Liu KS (2001) Improvements in the seasonal forecasting of tropical cyclone activity over the western North Pacific. Weather Forecast 16:491–498

d'Aubert A, Nunn PD (2008) Furious winds and parched islands: tropical cyclones (hurricanes) 1558–1970 and droughts 1722–1987 in the Pacific, Pacific climate change science program, 293 pp

de Scally FA, Wood GV, Maguire LK, Fournier-Beck MA, Silcocks D (2006) A history of tropical cyclones and their impacts in the Cook Islands, Cook Islands meteorological service, 377 pp

Dowdy A, Kuleshov Y (2012) An analysis of tropical cyclone occurrence in the Southern Hemisphere derived from a new satellite-era dataset. Int J Remote Sens 33(23):7382–7397

Dowdy A, Qi L, Jones D, Ramsay H, Fawcett R, Kuleshov Y (2012) Tropical cyclone climatology of the South Pacific Ocean and its relationship to El Niño-Southern Oscillation. J Clim 25:6108–6122

Folland CK, Renwick JA, Salinger MJ, Mullan AB (2002) Relative influences of the Interdecadal Pacific oscillation and ENSO in the South Pacific convergence zone. Geophys Res Lett 29(13):21-1–21-4

Goh AZC, Chan JCL (2010) An improved statistical scheme for the prediction of tropical cyclones making landfall in South China. Weather Forecast 25:587–593

Gray WM (1988) Environmental influences on tropical cyclones. Aust Meteorol Mag 36(3):127–139

Ho CH, Kim JH, Jeong JH, Kim HS, Chen D (2006) Variation of tropical cyclone activity in the South Indian Ocean: El Niño-Southern oscillation and Madden-Julian oscillation effects. J Geophys Res 111(1–9):D22101

Holland GJ (1984) On the climatology and structure of tropical cyclones in the Australian/southwest Pacific region: I. Data and tropical storms. Aust Meteorol Mag 32:1–15

IPCC (2014) Summary for policymakers. In: Field CB, Barros VR, Dokken DJ, Mach KJ, Mastrandrea MD, Bilir TE, Chatterjee M, Ebi KL, Estrada YO, Genova RC, Girma B, Kissel ES, Levy AN, MacCracken S, Mastrandrea PR, White LL (eds) Climate change 2014: impacts, adaptation, and vulnerability. Part A: global and sectoral aspects. Contribution of working group II to the fifth assessment report of the intergovernmental panel on climate change. Cambridge University Press, Cambridge, United Kingdom and New York, NY, USA, pp 1–32

Knutson TR, McBride JL, Chan JCL, Emanuel K, Holland G, Landsea C, Held I, Kossin JP, Srivastava AK, Sugi M (2010) Tropical cyclones and climate change. Nat Geosci 3:157–163

Kuleshov Y (2013) Climate change and Southern Hemisphere Tropical Cyclones international initiative—progress since the first international conference on Indian Ocean tropical cyclones and climate change. In: Mohanty et al (eds) Monitoring and prediction of tropical cyclones over Indian Ocean and climate change. Capital Publishing Co. and Springer Publications, pp 18–32

Kuleshov Y, de Hoedt G (2003) Tropical cyclone activity in the Southern Hemisphere. Bull Aust Meteorol Oceanogr Soc 16:135–137

Kuleshov Y, Qi L, Fawcett R, Jones D (2008) On tropical cyclone activity in the southern Hemisphere: trends and the ENSO connection. Geophys Res Lett 35:L14S08

Kuleshov Y, Qi L, Fawcett R, Jones D (2009a) Improving preparedness to natural hazards: tropical cyclone prediction for the Southern Hemisphere. In: Gan J (ed) Advances in geosciences, vol 12. Ocean science. World Scientific Publishing, Singapore, pp 27–143

Kuleshov Y, Chane Ming F, Qi L, Chouaibou I, Hoareau C, Roux F (2009b) Tropical cyclone genesis in the Southern Hemisphere and its relationship with the ENSO. Ann Geophys 27:2523–2538

Kuleshov Y, Fawcett R, Qi L, Trewin B, Jones D, McBride J, Ramsay H (2010) Trends in tropical cyclones in the South Indian Ocean and the South Pacific Ocean. J Geophys Res 115:D01101

Kuleshov Y, Jones D, Hendon H, Charles A, Cottrill A, Lim EP, Langford S, de Wit R, Shelton K, Spillman CM, Amjadali A, Pahalad J, Kaniaha S, McClymont D (2012a) Pacific adaptation strategy assistance program: strengthening the capacity for seasonal prediction services in Pacific countries. Bull Aust Meteorol Oceanogr Soc 25(1):7–12

Kuleshov Y, Wang Y, Apajee J, Fawcett R, Jones D (2012b) Prospects for improving the operational seasonal prediction of tropical cyclone activity in the Southern hemisphere. Atmos Clim Sci 2(3):298–306

Kuleshov Y, Spillman C, Wang Y, Charles A, de Wit R, Shelton K, Jones D, Hendon H, Ganter C, Watkins A, Apajee J, Griesser A (2012c) Seasonal prediction of climate extremes for the Pacific: tropical cyclones and extreme ocean temperatures. J Mar Sci Technol 20(6):675–683

Kuleshov Y, de Wit R, Atalifo T, Prakash B, Waqaicelua A, Kunitsugu M, Caroff P, Chane Ming F (2013) Developing an enhanced tropical cyclone data portal for the Southern hemisphere and the Western Pacific Ocean. European Geosciences Union General Assembly, 7–12 Apr 2013, Vienna, Austria

Kuleshov Y, McGree S, Jones D, Charles A, Cottrill A, Prakash B, Atalifo T, Nihmei S, Seuseu FLSK (2014) Extreme weather and climate events and their impacts on island countries in the Western Pacific: cyclones, floods and droughts. Atmos Clim Sci 4:803–818

Leroy A, Wheeler MC (2008) Statistical prediction of weekly tropical cyclone activity in the Southern hemisphere. Mon Weather Rev 136(10):3637–3654

Li CY, Zhou W (2012) Changes in western Pacific tropical cyclones associated with the El Nino-Southern oscillation cycle. J Clim 25:5864–5878

Lourensz RS (1981) Tropical cyclones in the Australian region July 1909 to June 1980. Bureau of Meteorology, Melbourne, Australia, p 94

Saji NH, Goswami BN, Vinayachandran PN, Yamagata T (1999) A dipole mode in the tropical Indian Ocean. Nature 401:360–363

Wijnands JS, Shelton K, Kuleshov Y (2014) Improving the operational methodology of tropical cyclone prediction in the Australian and the South Pacific Regions. Adv Meteorol 2014(Article ID 838746)

Wijnands JS, Qian G, Shelton K, Fawcett RJB, Chan JCL, Kuleshov Y (2015) Seasonal forecasting of tropical cyclone activity in the Australian and the South Pacific Ocean regions. Math Clim Weather Forecast 1:21–42

Wijnands JS, Qian G, Kuleshov Y (2016a) Variable selection for tropical cyclogenesis predictive modelling. Mon Weather Rev 144(12):4605–4619

Wijnands JS, Qian G, Kuleshov Y (2016b) Spline based modelling of near-surface wind speeds in tropical cyclones. Appl Math Model 40(2016):8685–8707

Wolter K, Timlin MS (1998) Measuring the strength of ENSO—how does 1997/98 rank? Weather 53:315–324

Solutions Stories: An Innovative Strategy for Managing Negative Physical and Mental Health Impacts from Extreme Weather Events

Emily Coren and Debra L. Safer

Abstract Health impacts from extreme weather events have massive, deleterious effects on our communities. Plots from mankind's long history of storytelling cannot help us prepare for the unprecedented environmental shifts we are experiencing. Dystopian plots focusing on a catastrophic future caused by climate change often immobilize rather than spur action. Despite over a half-century of evidence that science and health communication strategies, including entertainment-education, can effectively change human behavior, such strategies are rarely applied by our contemporary entertainment or media industries. Tools from social psychology, including social modeling and building self and collective efficacy, can help us to create a new model for current, culturally-relevant stories that can help communities prepare for extreme events. These "solution stories" are hopeful, funny stories that effectively teach climate change adaptation, mitigation, and resilience strategies. Solution stories are set in the present day and frame climate solutions to feel "easy, fun and popular." Up-to-date research on adaptation, mitigation, and resilience strategies can be iteratively utilized in these stories to better equip communities to manage climate change disasters and lessen harmful physical and mental health impacts. An example of a potential climate communication solution story, "Rhythm and Glue," is provided. In summary, solution stories are an innovative management strategy for substantially improving preparedness and, hence, health outcomes among communities facing extreme weather events caused by climate change.

Introduction

Climate change is causing extreme weather which is damaging to human health. Rising temperatures, flooding, drought, and wildfires can lead to physical consequences

---

E. Coren (✉)
Science Communicator, Santa Cruz, USA
e-mail: emilycoren@gmail.com

D. L. Safer
Stanford University, Stanford, USA
e-mail: dlsafer@stanford.edu

W. Leal Filho et al. (eds.), *Climate Change, Hazards and Adaptation Options*, Climate Change Management, https://doi.org/10.1007/978-3-030-37425-9_23

such as higher rates of heat-related illnesses (e.g., heat stroke), communicable diseases (e.g., West Nile virus), and respiratory symptoms related to poor air quality (e.g., asthma). Extreme weather threatens physical well-being as well as psychological health and resiliency. At higher temperatures, for instance, individuals are more likely to behave aggressively towards others and towards themselves (e.g., suicidal thoughts). Extreme weather also leads to greater risks that individuals will experience psychiatric disorders such as depression, anxiety, and substance abuse.

While there have been a few regional entertainment-education programs that have addressed sustainability concerns (Reinermann et al. 2014), given the scope and magnitude of climate change a larger, more systematic approach to applying established health communication practices to climate change is needed. In 2015, Emily Coren and Bill Ryerson, President of Population Media Center, began discussing options to create such a program. The show, Rhythm and Glue is the current version of that work, and this paper shares what we have learned through this process to assist other projects in scaling this work efficiently while continuing to facilitate the requisite media productions. Debra joined our team supporting entertainment-education and climate work by organizing a summit at Stanford in 2018 on Scripted Entertainment's Ability to Move the Needle on Climate Change.

The goal of this chapter is to provide a useful tool-set for constructing efficient, compelling disaster preparedness communication to decrease harmful climate change-related health impacts. After first identifying some of the major impacts of climate change on physical and mental health and reviewing effective actions that individuals and communities can take, we will discuss the potential for storytelling to serve as an effective climate communication strategy to improve disaster preparedness. The importance of framing these narratives as "solution stories" will be explained along with the more general principles of entertainment education, including the use of research and transmedia platforms. We include a specific example of a potential climate communication solution story, "Rhythm and Glue," to effectively teach climate change adaptation, mitigation, and resilience strategies. In the interest of preventative public health promotion, we include climate mitigation within our disaster management strategies. Our rationale is that the degree of future climate change will be largely determined by choices our current society makes about emissions (Jackson et al. 2015; Melillo et al. 2014). Stronger mitigation efforts can reduce the rate of climate change.

Health and Climate

Climate change has immediate negative consequences on human health (Luber and Lemery 2015; Lemery and Auerbach 2017; USGCRP 2016). Health impacts include: illness resulting from extreme temperatures such as heat stroke, respiratory diseases from poor air quality and wildfires, extreme events such as flooding and drought can cause water related infections (*Vibrio* spp.) and food-related infections (*Norovirus*, *Salmonella* spp.), increasing vector-borne infections (Lyme disease, West Nile virus), and mental health impacts. "While often assessed individually, exposure to multiple climate change threats can occur simultaneously, resulting in compounding or cascading health impacts." (USGCRP 2016). Promoting public health and designing our communities with healthy environments and resilient infrastructure will prepare us for our changing climate (Luber and Lemery 2015). "Nearly all avenues of public health promotion can act to reduce the risk of future climate-related disasters," (Keim 2015) (Fig. 1).

Mental Health Impacts from Extreme Weather

The impacts of climate change on mental health are important to recognize. Though less well-known than the impacts on physical health, overlooking these mental health impacts can lead to a serious underestimation of both the short and longer term ill-effects of extreme weather (Clayton et al. 2017). Comprehensive reviews regarding the consequences of climate change on mental health (Hayes and Poland 2018; USGCRP 2016) make clear, for example, the associations that have been found between high temperatures and greater risks for violent behavior, aggression, and suicide. Additionally, extreme weather events such as hurricanes, wildfires, and

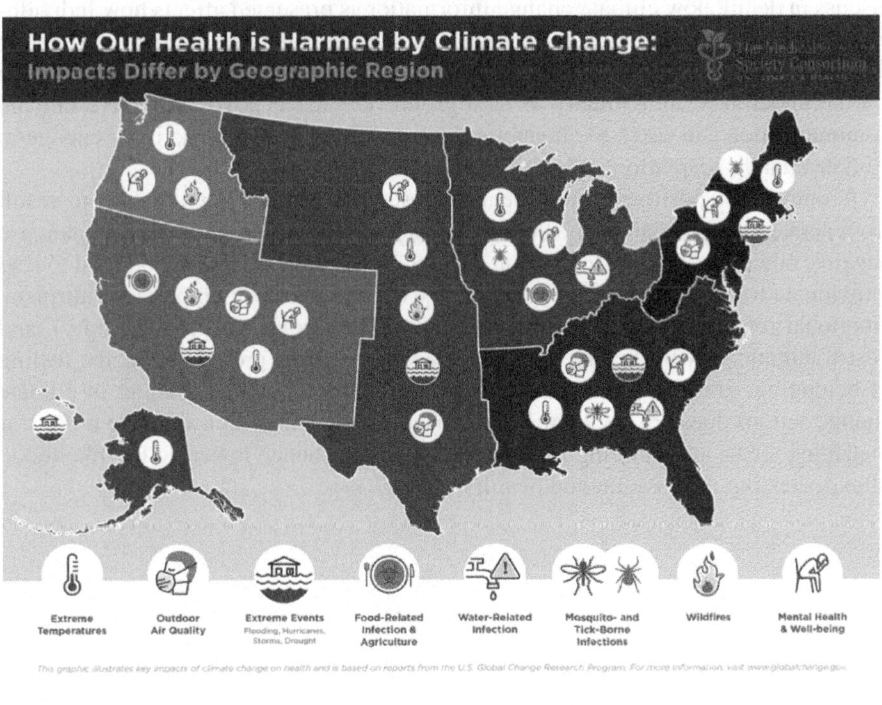

Fig. 1 How our health is harmed by climate change (Sarfaty et al. 2017)

drought have been shown to increase the likelihood of depression, anxiety, post-traumatic stress disorder (PTSD), substance abuse, and suicidal thoughts. Mental health effects include the loss of a sense of identity and place along with feelings of helplessness. Extreme weather also strains relationships, with increased rates of domestic violence reported. Potential outcomes include reduced cognitive functioning and depression resulting from infection with vector-borne diseases such as Lyme disease and West Nile Virus. Compounding these direct impacts of climate change are the related economic (e.g., job loss, displacement, homelessness) and social costs (e.g., isolation, reduced opportunities for exercise and other stress-relieving activities) that can adversely influence mental health. Further, while the mental health implications of climate change can affect everyone, these impacts tend to be greatest among individuals who are the most vulnerable and marginalized, such as those with pre-existing mental health problems, children, seniors, women (especially if pregnant), and/or the socioeconomically disadvantaged (Hayes and Poland 2018; USGCRP 2016).

An often overlooked impact of climate change on psychological well-being involves the chronic stress that results from ongoing exposure to headlines and media images regarding the catastrophic threats posed by extreme weather. Such chronic awareness is associated with generalized anxiety, a heightened perception of risk, sadness, guilt, pessimism, helplessness, and the loss of belief in either or both individual and collective efficacy (USGCRP 2016). Importantly, as this chapter will discuss in detail, how climate change information is presented affects how individuals respond. Coverage that is discouraging or scientifically inaccurate evokes a very different response than coverage that focuses on adaptive and preventive actions that can be undertaken individually or as a group (USGCRP 2016). Effective climate communication can serve to channel awareness into a trigger to actively engage in climate change adaptation and mitigation actions.

A number of positive mental health consequences have been shown to result from active engagement. Just as our changing climate has the potential to increase negative outcomes such as depression and aggression, it also has the potential to lead individuals to work with one another and exhibit empathy, compassion, altruism, emotional resilience (i.e., the ability to adapt and recover despite adversity), and post-traumatic growth (i.e., meeting trauma with personal growth, hope, the feeling of belonging, gratitude, and transformation). While active engagement in climate change actions does not necessarily decrease stress, it can offer a valuable buffer for managing stress and building emotional resilience—hence lowering the likelihood of experiencing negative mental health impacts.

Mental Health Resources to Promote Resiliency and Limit Negative Impacts of Extreme Weather

There are different types of resources to which individuals can turn to help promote their resilience and limit the negative consequences of extreme weather on their mental health. Examples of formal resources include obtaining referrals for psychotherapy/counseling, psychiatric medications, and/or inpatient psychiatric hospitalization. Given the positive psychological effects of active engagement in climate change actions, additional helpful resources include those that offer the opportunity to connect with others in civic action. One such resource is the Transition Town movement, a grass root movement in which citizens join together to help themselves and their communities respond to the changing climate (Transition Town Movement 2018). Such movements offer psychosocial support and help build individual and community-based resilience to help individuals and communities process their fear, despair and other responses to climate change (Christensen 2019; Higgins 2012; Macy and Johnstone 2012; Work that Reconnects Network 2018). A switch from images of pollution, environmental damage and protests to images to including social care check-ins and active emotional and physical community improvements (home repair, habitat restoration, civic participation) will provide constructive social modeling. The skills to model in solutions stories include active engagement in climate change adaptation and mitigation actions, as these not only directly address the threats caused by climate change but also have been linked to positive mental health benefits.

Solutions Stories

What is needed today are climate change stories that model successful conclusions through our own actions. These are solutions stories that give us agency, courage, and hope. Too many climate change stories emphasize inevitability, disaster, and loss (Schneider-Mayerson 2018). Such stories are about the problems created by the impact of climate change, so called ‘impact stories’. In contrast, ‘Solutions Stories’ offer a much-needed alternative. The stories are set in the present time and place and model resourceful people collaborating to solve sustainability issues. The narratives focus on how to succeed at desired tasks such as civic engagement, renewable energy, and habitat restoration. These are hopeful stories for to transitioning our culture toward collaboration, social resilience, and sustainable habits.

Solutions stories feature characters who possess a strong sense of personal and collective efficacy. Efficacy is our ability to succeed at difficult tasks. The daily realities of climate solutions are full of impediments and conflicts. The solutions are not easy. Social psychology teaches that iterative goal setting, or making short and long term goals and periodically reevaluating your success, increases our sense of efficacy. Belief in personal efficacy is not a Pollyannaish mind-set such as “I believe, therefore, I achieve." The most functional belief system for undertaking difficult tasks combines realism about tough odds with optimism that one can beat those odds through self-development and perseverant effort. A resilient sense of efficacy provides the needed staying power" (Bandura 2016). A fictional representation of people facing and overcoming challenges is itself a climate solution.

Solutions stories are hopeful, courageous and funny, demonstrating strategies for how the issues are solvable. Solution stories are specific to a new era in which the problem we face is unprecedented—a rapidly shifting climate. They use humor to help people adjust to new conditions and empathize with each other in problem solving tricky situations. Anxiety reduces participation in climate change mitigation activities. Comedy, especially positive humor, can successfully "soften" scary topics. Making the scope of the problem smaller and the solutions more do-able will provoke interest and engagement in the tasks needed to mitigate climate change.

Stories providing examples of climate solutions being easy, fun, and popular would be ideal (Maibach 2019). We need to switch the message of dystopian films like The Day After Tomorrow (Emmerich 2004), that "Climate change is huge, scary and inevitable," to the message that "We can solve this, and here's how," like the collective advocacy Earth music video by Lil Dicky (Tierney and Heller 2019). Best practice bounces these stories between formats and audiences to reinforce concepts, making them "simple clear messages, repeated often, by a variety of trusted sources," (Maibach 2019). Stories that model how easy and inspiring it is to mitigate climate change might include fictional examples in which characters vote for candidates that support clean energy. Ideally, entertainment programs connect viewers directly to actual social programs that increase voter participation. As Andrew Law, the executive story editor at The Good Place observes, "What this show is about, at its core, is making each other better as a group" (Hanson 2019). Modeling collective efficacy becomes the new narrative structure. These uplifting and funny stories have the commercialization potential to reach broad audiences. Solutions stories begin today, with what's available right now and guide us to a sustainable world (Fig. 2).

Comedy Framing

Fear and humor play integral roles in audience engagement with climate change. Representations of climate change need to be grounded in present realities, neither separated by time and space nor exaggerated (Sakellari 2015). Fear does not motivate long-term behavioral changes towards pro-environmental behaviors (Sakellari 2015). Due to the possible catastrophic outcomes of climate change, mental health impacts, like anxiety (Melillo et al. 2014) paired with fearful framing of the issues, reduce enthusiasm for media content about climate change as well as reducing participation in climate change mitigation activities (Doherty and Clayton 2011; Clayton et al. 2013), and anxiety impairs climate change mitigation behaviors (Doherty and Clayton 2011; Morley Rolls 2013). Comedy, especially positive humor, can be successful in "softening" scary topics (Samson and Gross 2012). Making the scope of

Solutions Stories Framework:

1. **Focus on solutions**, not problems.
2. **Set in present day.**
3. Do the characters model **personal and collective efficacy**? This can be taught through iterative goal setting and periodic reevaluation.
4. Is the tone of the story **hopeful and funny**? Comedy might not be applicable in all news articles.
5. It has to be **specific and do-able**. (simple, clear messages) Is the scope of issue is solvable? What is the do-able action?
6. Message **repeated often, by a variety of trusted sources** - How does your story reinforce the concept with the actions of other collaborators? Who do you interlink with? Who is making these recommendations in your story? This is even better if these messages are delivered by trusted advisors such as friends, family and medical community. Are the do-able actions portrayed as "**easy, fun, and popular**"?
7. **Empathy** - we need the stories to be inclusive and not create "hero/villain/victim stories." What are the community needs that your story is focusing on? How are they shared priorities by many groups?
8. **Images need to match messaging.** Frequently, because of the production constraints such as funding deficits or short turn-around cycles, this is not prioritized, however it really needs to be for effective, efficient, clear communication.

Fig. 2 Solutions stories framework

the problem smaller and the solutions more do-able is likely to provoke interest and engagement in daunting tasks, such as climate change mitigation behaviors (Doherty and Clayton 2011).

One of the founding tenants of entertainment-education is that the narrative content needs to be culturally representative of the target audience; for Americans now, this means comedy (Ryan 2016). Comedy helps make climate change approachable, reducing fear and anxiety (Bore and Reid 2014). Television is a primary source for American audiences for climate change information and is influential in shaping national perceptions of the related issues (Boykoff 2008; Boykoff and Goodman 2015). Television comedy, in particular, is a useful tool for broadening audiences for climate change topics (Brewer and McKnight 2015). While there are a few contemporary examples of comedy in science communication (Pinto et al. 2015; Riesch 2015), there is room for further experimentation, especially using comedy within narrative fiction (not stand-up comedy), for science engagement. For all of these reasons, this chapter's proposed entertainment-education about climate change is framed as a comedy.

Creating stories which model climate change solutions in a context that is funny and engaging, which model specifically how to do mitigation behaviors, increases perceptions of self-efficacy regarding climate change mitigation behaviors, and increase the perception that we can reduce our carbon dioxide emissions and reduce the overall harm caused by climate change. Intrinsic reinforcement, such as perceived choice, competence, and community will be used as rewards, rather than external rewards, such as money (Clayton et al. 2013). "Some behaviors are avoided because people believe, rightly or wrongly, that they do not know how to perform them. In these cases, behavioral skill training is more important than encouragement". People may need to hear how to reduce their carbon emissions, for example, rather than just being exhorted to do it.

Entertainment-Education

While solutions stories are conceptually applicable to many types of content, we would like to propose that entertainment-education, an established health communication mechanism, is an ideal vehicle for conveying these messages. Entertainment-education is a fully integrated communication campaign that uses a narrative story (frequently broadcast via television or radio) as one piece of a larger communication strategy to build audience engagement around important topics of public concern. Entertainment-education narratives are designed to deliver accurate information from reliable sources and model desired change through parasocial interaction, character identification, self-efficacy, and the mediation of interpersonal communication (Singhal and Rogers 1999; Singhal et al. 2004; Singhal 2013; Singhal et al. 2013). Entertainment-education involves production methods for long-running multi-episode narratives or serial dramas and are grounded in theories of narrative persuasion and behavior change with the explicit aim of promoting pro-social messages (Singhal et al. 2004). These projects were commercially successful and led to behavior changes such as higher enrollment in adult literacy classes, adoption of family planning methods, and gender equality (Nairman 1993; Singhal and Rogers 1999; Singhal 2013; Singhal et al. 2013). Entertainment-education begins with formative research that includes demographic and other pertinent information about the target audience and the desired pro-social behaviors the show will model. This research is used by writers to accurately create characters represent the target populations. Positive characters are modeling the desired behavior, negative are modeling what you don't want to happen and transitional characters are learning the new desired behavior. Fictional narratives are created for engaging cognitively, emotionally, and socially, helping to effect attitudinal and behavioral change among individuals and communities (Bandura 2004; Green et al. 2002, 2012; Kincaid 2002). Fictional narratives use implicit persuasion approaches that reduce counter-arguing and psychological resistance (Eveland 2001; Moyer-Gusé 2008). Entertainment-education has proliferated, over the past thirty years, to thousands of projects globally becoming an influential health promotion strategy (Green et al. 2012; Kincaid 2002; Moyer-Gusé 2008; Singhal et al. 2013).

Sustainability themed entertainment-education projects have already been implemented successfully with a broad range of international audiences (Heong et al. 2008; Reinermann et al. 2014; Singhal et al. 2000). Entertainment-education is particularly effective at producing cultural shifts in attitude. It is therefore a well suited mechanism to shape and support personal dialog and engagement with the topic of climate change. Narrative fiction is ideal for engaging with audiences over long periods of time because it maintains audience interest and reduces messaging fatigue. Entertainment-education as a strategy is likely to be effective as a tool for creating healthy long-term behavior patterns for creating cultural resilience. For example, there are many barriers to voting in a US election in support of climate mitigation initiatives and greater public health infrastructure, including: gerrymandering, voter suppression, apathy ("My vote doesn't matter"), and insufficient information. It can be confusing for voters to sort-out which candidates and measures support their values. For instance, the Environmental Voter Project (a non-profit aiming to improve voter turnout of environmentalists) encourages people interested in sustainability issues to vote, but will not recommend what or who to vote for due to its efforts to remain non-partisan (Environmental Voter Project 2018). Many organizations do make endorsements yet there remains no "unified" and "easy" resource for climate election support. A "unified" website that provides information on how to vote, with a climate ticket prepared for each district and multiple storylines modeling the desired behavior could increase civic participation in public health. The television production company, Shondaland, launched a wonderful example of multi-platform civic engagement in the US 2018 election by implementing these principles. For example, an episode of Grey's Anatomy prior to the election modeled one character (Joe Wilson) mailing in a co-worker's ballot (Alex Karev) for him then putting an "I voted" sticker on his scrubs (Rhimes 2005). Transmedia coverage on the campaigns' twitter page attached specific information, such as, "If you're in line before the polls close #stayinline." And, the cast and crew of all the Shondaland shows shared their "I Voted" selfies on election day.

Efficacy

To improve efficacy, we are proposing a narrative structure shift from hero/villain/victim cycles to iterative goal setting. Hero/villain/victim cycles are culturally pervasive in media (Karpman 1968). While often entertaining, they can perpetuate real life conflict (Bandura 1978). There is a notable shortage of cultural models demonstrating the range of climate engagement or problem solving behaviors. Therefore, alternative story structures are important to consider when the goal is modeling how social groups can unite in the face of a common cause. One way to slightly modify the Sabido methodology would be to avoid creating archetypal "bad" characters and place greater emphasis on the positive and transitional characters. This would enable feedback regarding behavioral consequences (e.g., smoking exacerbates asthma) but prevent the trivialization and victimization of an entire group of viewers who may identify with the "bad" character. For example, in Rhythm and Glue, the character Roger starts out as a "negative character" who exhibits a variety of climate denial perspectives (Heald 2017). Instead of portraying him in an exaggerated or overly negative way, the first season aims to establish empathy for him.

Overall, he spends more of his time on the series as a transitional character who eventually becomes an ally for transitioning the community to renewable energy, despite having initial motivations that differ from the story protagonists. Efficacy motivates climate change adaptation behaviors (Bandura 2016; Valkengoed and Steg 2019).

Efficacy is achieved through setting proximal and distal goals that are periodically re-evaluated. A sense of efficacy is important at both the individual level (e.g., deciding to assist in a disaster by offering first aid), as well as at the larger, collective level (e.g., voting in favor of public health measures). Individual level actions such as walking, biking, and public transit are sometimes dismissed as "not having a big enough impact" (Lazarovic 2018). However, making and carrying out individual level decisions both supports a community's overall outcomes and provides positive social modeling of the desired behavior. Ideally, one would not attempt to convince one's peers to carry out the behavior, but one would simply model it. For example, riding a bike serves both to eliminate fossil fuel emissions from travel directly and helps demonstrate to other people more likely to view riding a bike in one's neighborhood as something normal and fun. One is even likely to receive unsolicited questions about bike equipment and routes from fellow community members. Behaviors that help pool resources and build infrastructure, such as voting for supporting transitions to clean, renewable energy and healthcare resources, benefit everyone (Fig. 3).

Empathy

Empathy as a skill reduces conflict and helps individuals navigate the complex decision making processes of attempting to manage limited resources. There are many tactics for teaching empathy, but one especially useful strategy involves a simplification of Nonviolent Communication (Rosenberg 2015). Empathy communication is divided into four parts: Observations, Feelings, Needs, and Requests. Observations include what one sees or hears, as if one were viewing a video. Requests, which must be specific and do-able, can be requests for either action or connection. An example of an action request could be, "Please turn down the temperature on your thermostat,"

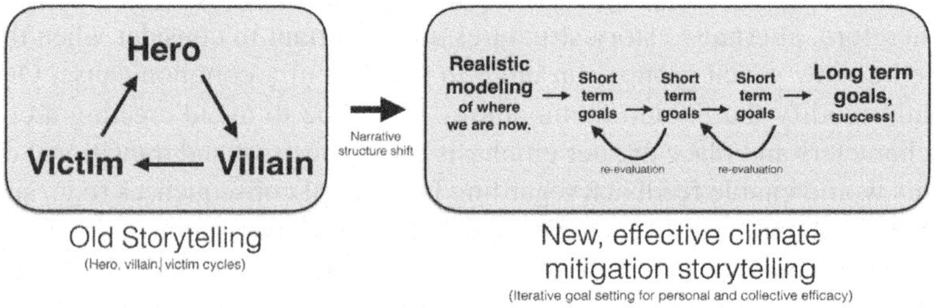

Fig. 3 Solutions stories switch from a Hero, Villain, Victim drama cycle, which narrative cycle perpetuates conflict to iterative goal setting with narratives focused on personal and collective efficacy at overcoming realistic resource conflicts

rather than, "Please reduce your family's energy consumption." The action request should include specific instructions for how the task can be accomplished, including guided mastery through social modeling. The action request should be followed by a connection request. Connection requests provide an opportunity for feedback, such as: "Are you getting stuck? Do you need help knowing how to turn down your water heater thermostat?" Some individuals will require new skills to experience success, as well as access to troubleshooting resources.

Viewing empathy as a skillset that can be taught through a formulaic structure (i.e., making observations, expressing feelings, stating needs, and making requests) has important advantages. In disasters, for example, when people are experiencing increased stress, having a reservoir of tools—such as empathy—or managing conflicts improves the efficiency and effectiveness of responses. Practicing these skills regularly establishes cultural patterns that make these responses "easier" under duress. Modeling the practice of conflict resolution skills is a tool for building cultural resilience (Fig. 4).

Transmedia

Transmedia storytelling uses narrative elements creatively coordinated across different communication platforms to build a story world and enrich the audience's entertainment experience (Jenkins 2007). Transmedia storytelling holds unique advantages for reaching and engaging audiences from otherwise fragmented media markets to participate in the converging popular culture (Jenkins 2006, 2007). Each communication platform carries the story independently but the interwoven tapestry

Feelings:

content	connected	enthusiastic	scared
satisfied	confident	uncomfortable	anxious
relieved	hopeful	impatient	horrified
happy	encouraged	sad	confused
delighted	safe	lonely	torn
proud	eager	frustrated	hesitant
sympathetic	curious	angry	overwhelmed

Needs:

air	belonging	creativity	exercise
food	trust	dignity	nutrition
water	empathy	hope	rest
shelter	competence	purpose	time
touch	ability	celebration	hygiene
sleep	independence	order	movement

Fig. 4 A sample from the nonviolent communication: feelings and needs inventory (Rosenberg 2015)

is designed to make for a richer whole, deepening the audience members' relationship with the characters, the plotlines, and the issues (Davidson 2010).

Transmedia extensions provide specific, detailed information in an entertaining format, making the information interesting, relevant, easy to connect with, and tie directly into peoples own lives. Transmedia tie-ins provide details for "how-to" do the various tasks in a way that relates them directly to characters in more specificity than is appropriate within the narrative directly. Brazilian telenovelas establish precedents for audience participation and script modification for creative societal discussion through television media. These programs include social marketing inserted into narrative programming, similar to conventional product placement, but with pro-social messages instead of commercial products, and definitive endings that permit narrative closure. "Brazilian telenovelas are created in a cyclical, continuous manner with ongoing external, diverse inputs from organizations, the government, and the audience. Brazilian writers, producers, and directors write and shoot episodes for a telenovela only a few weeks before they are broadcast. This process allows for last-minute changes, reflects new social trends, and includes current political events. The storyline and social merchandizing consistently evolve due to community input primarily through audience ratings and focus group interviews" (Pastina et al. 2004). Brazilian telenovelas provide an example of responsive, open-genre media and demonstrate an ideal communication mechanism for facilitating a real societal dialog connected to civic engagement towards social and personal improvement. These stories can be interwoven as a network, reducing the redundancy and cost of producing and directing people to the "how to" information, by connecting it through a unified hub. Ideally, these solutions stories are iterative, evolving real-time cycles supporting our evolving climate and communicating with audiences both about current community needs and providing updated research-based support (Fig. 5).

American Climate Change Beliefs

While these solutions story tactics broadly apply to many target audiences and behaviors, we provide an example focusing on mainstream American audiences. While climate change and global warming are distinct, yet related issues, tracking public option and perceptions of them helps build our understanding of possible target audiences and story framing of these issues (Roser-Renouf et al. 2015). Such as, although a majority of Americans already think that global warming is happening; very few discuss it with friends and family (Leiserowitz et al. 2015). Also, despite the general awareness about climate change, important misunderstandings persist. For example, Americans often perceive climate change as having limited personal relevance (Brechin 2003; Lorenzoni and Pidgeon 2006; Maibach et al. 2015; Reynolds et al. 2010). Climate change is an emotionally charged subject, and peoples' beliefs and experiences strongly shape perceptions of climate change (Leiserowitz et al. 2015; Moser 2010).

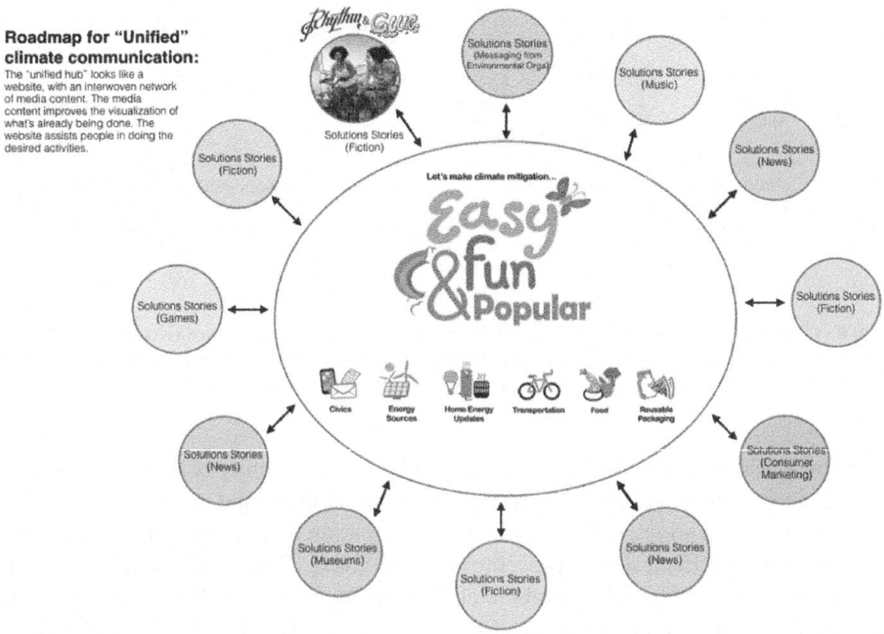

Fig. 5 Transmedia roadmap for unified climate communication

The most effective way to interact with public audiences is through an "engagement model" (Stilgoe and Wilsdon 2009) that conveys scientific information while also addressing issues of ethics and public trust. In short, despite the polarized political discourse on climate change, most Americans now seem to accept the basic tenets that climate change is real and are even willing to support some form of action to address it. However, they struggle to grasp the dimension and timeliness of the threat and have difficulties seeing that they have a personal role in addressing it. For many Americans, the question has shifted from "does it happen" to "what can or should I/we do to address it?" It is, therefore, our job as communicators to assist in building efficacy at the component skills (Fig. 6).

Rhythm and Glue as a Solutions Story Example for American Climate Communication

To provide an example of what a narrative approach to climate communication could look like, we've proposed Rhythm and Glue. The story emphasizes personal and collective efficacy and aims to improve public receptivity via layered examples regarding how climate change is already affecting the health of Americans while modeling how effective actions can be easy and fun.

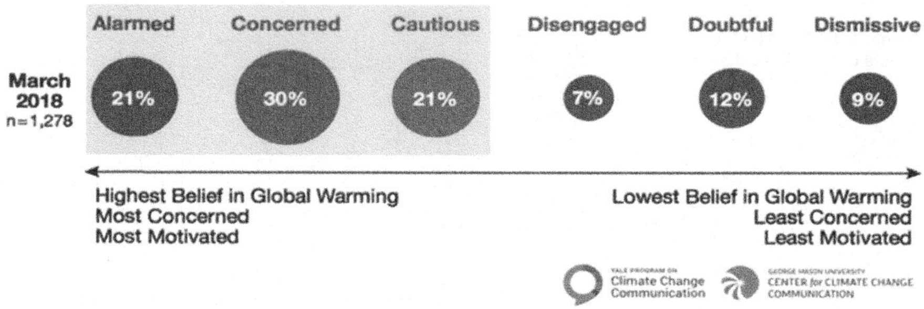

Fig. 6 Global Warming's Six Americas, description of audience perspectives (Roser-Renouf et al. 2015; Yale 2016). Our target audiences are the Alarmed, Concerned and Cautious with the goal of increasing efficacy of climate mitigation behaviors in those audience segments

In the serial comedy-drama, Rhythm and Glue, a community collaborates to build a more sustainable city. Comedy, romantic entanglements, and career advancement anchor the plot, while real-time audience participation drives engagement with sustainability topics. Each season presents an overarching climate change issue. Season One focuses on transportation and the current health effects of climate change, with subsequent seasons addressing; energy, water, food, and waste. The show highlights the day-to-day realities of climate change facing residents of the United States—and role models realistic solutions and behaviors that citizens can take to mitigate those effects for themselves and their communities. Successful entertainment-education programs, public heath communication techniques are applied to climate change mitigation to improve the rate of cultural adoption.

Real-time audience participation shapes the story's conflicts and resolutions engaging a national conversation about how we can collaborate to mitigate climate change. The show provides a model for transitioning sustainable urban transportation internationally and creates a national conversation with a variety of stakeholder perspectives. As a comprehensive media platform it connects viewers with a network of organizations supporting the actionable behaviors that the audience can take locally to mitigate climate change to increase the capacity of existing sustainability programs.

Story Engine

We propose a fixed narrative structure with placeholders for real-time audience participation to facilitate a national conversation through the microcosm of the show. The pilot sets up the basic story and establishes the characters.

All further content has two sections:

- Fixed story checkpoints, revolving around the characters romantic entanglements and career advancement.

- Flexible storylines that incorporate audience participation through transmedia elements. These areas give the audience an opportunity to share their successes and frustrations with climate change mitigation activities. These sections relate to social and engineering challenges experienced by show viewers. Through partnership feedback, we can efficiently identify existing bottlenecks to social-environmental change.

Interspersing these two elements, we build an architecture that allows for narrative character development, while still supporting a real-time national dialog.

Teaching efficacy, the show uses iterative goal setting as a story structure. The characters start with small individual level decisions (learning active and public transit) and as they gradually gain skills, the level of social complexity at the behaviors increases (such as collaborating to implement municipal scale planning). While the characters experience temporary setbacks which lead story conflicts, the trend lines in these characters development gradually trend upward as they gain personal and collective efficacy, building complex, competent characters modeling increasing community resilience.

Transmedia extensions for a program like this include "how-to" videos of characters performing each of the desired tasks (learning to use active and public transit, civic participation, community building activities) while connecting audiences to other programs and detailed information about climate change. For example, a weathercaster character links to climate and weather information sources such as the Climate Central, Climate Matters program (Climate Matters 2018) (Fig. 7).

Solutions Stories as Applied to Disaster Preparedness

Solutions to reduce the harm of disasters include: skills for emergency medical treatment, preventative public health measures, resource management skills, and social/emotional resilience training. Medically, we can provide social modeling of first aid training and triage techniques to share how to care for each other in a disaster. For mental health support, psychotherapy/counseling and peer-to-peer empathy can be modeled through media to provide concrete examples of healthy social strategies. Preventative public health measures include: clean air policies (such as investment in renewable energy), municipal water management, active and public transportation (USGCRP 2016), eating healthy local foods, regular vaccinations and hygiene practices like hand washing and access to toilets. Many of these have co-benefits which improve short term, long term community and personal health. For example, active transportation provides regular exercise which improves individual health as well as reduces the overall air pollution in the community. Community disease prevention measures, like access to toilets, hand washing, and vaccines are grouping points in Healthy Cities plans which require continual reevaluation and continuous public buy-in investment (Rydin et al. 2012). Infrastructure such as water, power

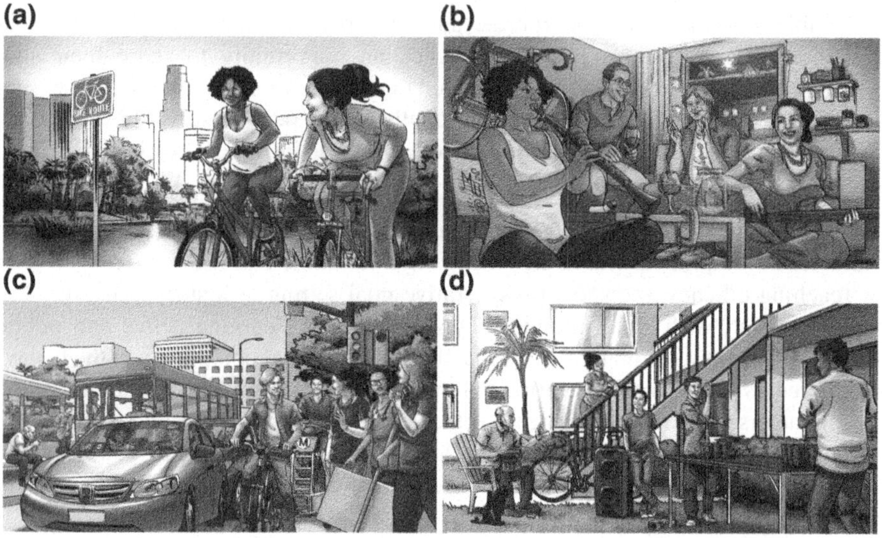

Fig. 7 Rhythm and Glue concept illustrations. **a** Sarah and Jessica exploring Los Angeles together on bicycles. This subplot provides personal and collective efficacy at one of the new skills as well as guided mastery for a complex skill. **b** Jose and Daniel join Sarah and Jessica at their apartment to drink and play music in the evenings. The community music plot thread teaches community building skills. **c** Season one highlights transitions to active and shared transit, including: bicycling, walking, bike share, rideshare, public transit and updates to electrified transit (electric bikes and cars). **d** Andy, Josh, and Matt playing beer pong in their apartment courtyard while Roger scolds them for the noise and clutter. The informal community games evolve within the story into community building opportunities to improve local level discussions and problem solving for climate engagement. These concept images were drawn by Liberum Donum (http://liberumdonum.com/)

and supply chains often breaks in disasters. Therefore teaching staff how to maintain medical treatment in disasters, functioning with limited resources and without technology provides resilience (Luber and Lemery 2015). Social modeling should include resource management techniques, such as water treatment procedures and how to prioritize limited resources through clear communication and collaboration creating cultural precedence for these skills. Social/emotional skills such as empathy, check-ins add embedded emotional resilience techniques. Many of the individual solutions will vary by region, so providing many specific interrelated examples is important. Solutions stories can be crafted to convey all of these individually, and also some tools such as entertainment-education methods can creatively interweave them to teach many of these in parallel. Arvind Singhal proposes using Positive Deviance to identify the successes of outliers who have already found solutions to complex problems by overcoming great odds and subsequently modeling that behavior for the rest of the at risk population (Singhal and Dura 2017). Communication of solutions stories can be used to amplify the successful techniques of these Positive Deviance examples, shifting them into culturally normative practices.

There is ongoing research in social resilience strategies, which seems to unify under "hope" (Cousin-Frankel 2018). Social resilience appears to be a complex, multifaceted issue where we need to be simultaneously addressing individual, family, and community levels. A strong, "sense of agency, or control over one's fate," helps children transcend difficult circumstances (Underwood 2018). Camaraderie and a perceived ability to adapt to stressors were predictive factors in recovery (Servick 2018). While research on resilience strategies is still ongoing and there's much more to learn as we refine these techniques, continuing to study and then iteratively feed the recommendations from these studies into media products will support the cultural integration of disaster preparedness solutions.

An example of applying a solution story to teach disaster preparedness involves a proposed subplot in Rhythm and Glue, in which several of the characters work in a medical office. One of these characters, a medical assistant, decides to help the office go "green." To achieve this, the character faces and overcomes bureaucratic challenges while also forming new friendships and gradually building self-confidence. Over the course of episodes spanning several seasons, the medical assistant builds a career that includes taking on a public health advocacy position. With the help of new and old friends, the character helps create a community-wide public health program. Transmedia extensions include a variety of first-aid training, community organization techniques, and tools for advocacy of public health and disaster prevention planning. We give this sub-plot example to show how real-life programs, such as My Green Doctor, which assists healthcare professionals in implementing environmental practices (My Green Doctor 2018), can be incorporated. Specific emergency medical training would be incorporated throughout, with a massive fire as the primary conflict in Season Six, in which the characters implement the training and, together, model the procedures for managing the disaster.

Limitations

It is important to acknowledge the limitations of this work and the constraints of this paper. The productions described are expensive and time consuming to produce (compared to typical science communication budgets). Therefore, scalability is dependent on the creation of active partnerships with existing media production companies. While entertainment-education as a tool-set has proven effective for health communication, the potential for serial dramas to impact widescale climate mitigation actions have not yet been tested at the scale and coordination we are proposing.

In addition, while climate solutions can be framed within a comedy context, the consequences of climate change are serious. Climate solutions such as using renewable energy sources, engaging in active transportation, eating local food, improving hygiene practices, and strategies to increase social support should be framed as "easy, fun and popular" (Maibach 2019) and the overall tone of the artwork be kept light, yet it is important to interweave subplots addressing the effects of climate change (e.g., the current health impacts of extreme weather). Finding this balance artistically is nuanced, and may be considered a potential constraint of this work. Iterative content evaluation can provide constructive feedback to measure and improve audience interaction with media products and advise media production. Stories illustrating role models capable of completing difficult sustainability tasks must demonstrate not only their courage but also their vulnerability (Brown 2019). Such examples of persistence in the face of challenges create a cultural reservoir of knowledge that viewers can model.

Conclusions

Despite the global evidence showing that entertainment-education stories are effective at changing behavior, they have not yet been created in the U.S. for climate change. This chapter proposes that Solutions Stories are an innovative vehicle for teaching collaboration and resilience strategies with the potential to reduce the harm from extreme weather disasters caused by climate change while also better preparing communities to respond to them when they do occur. There are nearly an infinite number of possible Solutions Stories capable of pertaining to a myriad of target audiences while teaching a variety of resilience strategies.

The main lessons are that

- We can use established entertainment-education procedures to create engaging content to reduce the harm from climate change. There are enormous, undeveloped areas here for expanding content creation.
- Switching from hero/villain/victim cycles to collective advocacy/iterative goal setting story structures will improve our cultural capacity for solving complex social and environmental problems efficiently (Fig. 3).
- Comedy framing and setting stories in the present day with a public health focus that climate solutions will improve adoption and cultural relevance of climate stories.

Organizing efforts for collaborating to create a system of content creation that supports climate change mitigation will improve the efficiency of content production. Future prospects hopefully include the rapid development of the show Rhythm and Glue, expanding into both a myriad of related content developed for other target audiences and regions as well as the application of these principles into established shows to create a unified team of media increasing the rate of sustainable cultural practice adoption. The checklist (Fig. 2) is a tool for creating stories that frame climate communication set in the present day, with characters modeling efficacy at problem solving climate solutions through engaging in do-able behaviors. Each story can teach specific interventions that become interwoven building an effective communications infrastructure. Employing this set of storytelling techniques will assist cultural adaptation patterns and reduce the massive, deleterious effects on our communities caused by the negative physical and mental health impacts caused by climate change. We can do it, let's get started.

References

Bandura A (1978) Social learning theory of aggression. J Commun 28(3):12–29

Bandura A (2004) Social cognitive theory for personal and social change by enabling media. In: Singhal A, Cody MJ, Rogers EM, Sabido M (eds) Entertainment-education and social change: history, research, and practice. Erlbaum, Mahwah, NJ, pp 75–96

Bandura A (2016) Moral disengagement: how people do harm and live with themselves. Worth Publishers, New York, NY

Bore IK, Reid G (2014) Laughing in the face of climate change? Satire as a device for engaging audiences in public debate. Sci Commun 36(4):454–478

Boykoff MT (2008) Lost in translation? United States television news coverage of anthropogenic climate change 1995–2004. Clim Change 86:1–11

Boykoff MT, Goodman MK (2015) Science (and policy) friction: how mass media shape US American climate discourses. In: Sommer B (ed) Cultural dynamics of climate change and the environment in North America. Brill, Boston, pp 189–205

Brechin SR (2003) Comparative public opinion and knowledge on global climatic change and the Kyoto Protocol: the U.S. versus the world? Int J Sociol Soc Policy 23:106e34

Brewer PR, McKnight J (2015) Climate as comedy: the effects of satirical television news on climate change perceptions. Sci Commun 37(5):635–657

Brown B (2019) Brené Brown: the call to courage. Netflix, United States. https://www.imdb.com/title/tt10050766/?ref_=ttfc_fc_tt. Accessed 2 May 2019

Christensen J (2019) Climate anxiety is real but there's something you can do about it. CNN. https://www.cnn.com/2019/05/07/health/climate-anxiety-eprise/index.html. Accessed 8 May 2019

Clayton S, Litchfield C, Geller ES (2013) Psychological science, conservation, and environmental sustainability. Front Ecol Environ 11:377–382. https://doi.org/10.1890/120351

Clayton S et al (2017) Mental health and our changing climate: impacts, implications, and guidance. American Psychological Association and ecoAmerica, Washington, D.C.

Climate Matters (2018) Climate central: researching and reporting the science and impacts of climate change. http://medialibrary.climatecentral.org/. Accessed 30 Dec 2018

Cousin-Frankel J (2018) The roots of resilience. Science 359:970–971

Davidson D (2010) Cross-media communications: an introduction to the art of creating integrated media experiences. ETC Press, Pittsburg, PA

Doherty TJ, Clayton S (2011) The psychological impacts of global climate change. Am Psychol 66(4):265–276

Emmerich R et al (2004) The day after tomorrow. 20th Century Fox, United States. https://www.imdb.com/title/tt0319262/?ref_=fn_al_tt_1. Accessed 30 Dec 2018

Environmental Voter Project (2018) https://www.environmentalvoter.org/. Accessed 30 Dec 2018

Eveland WP Jr (2001) The cognitive mediation model of learning from the news: evidence from non-election, off-year election, and presidential election contexts. Commun Res 28:571–601

Green MC, Strange JJ, Brock TC (2002) Narrative impact: social and cognitive foundations. Psychology Press, New York

Green MC, Chatham C, Sestir MA (2012) Emotion and transportation into fact and fiction. Sci Study Lit. 2(1):37–59. https://www.jbe-platform.com/content/journals/10.1075/ssol.2.1.03gre

Hanson P (2019) Afterlife lessons. Sitcom ethics: Michael Schur on the good place. Retrieved from https://www.mydigitalpublication.com/publication/?i=550563#{%22issue_id%22:550563,%22page%22:26}

Hayes K, Poland B (2018) Addressing mental health in a changing climate: incorporating mental health indicators into climate change and health vulnerability and adaptation assessments. Int J Environ Res Public Health 15(9):pii: E1806. https://doi.org/10.3390/ijerph15091806
Heald S (2017) Climate silence, moral disengagement, and self-efficacy: how Albert Bandura's theories inform our climate-change predicament. Environ Scie Policy Sustain Dev 59:4–15. https://doi.org/10.1080/00139157.2017.1374792
Heong KL, Escalada MM, Huan NH, Ky Ba VH, Quynh PV, Theit LV, Chien HV (2008) Entertainment-education and rice pest management: a radio soap opera in Vietnam. Crop Prot 27:1392–1397
Higgins L (2012) Community music: in theory and in practice. Oxford University Press, New York
Jackson RB et al (2015) Two or three degrees: CO2 emissions and global temperature impacts. The summer issue of the bridge (National Academy of Engineering) on Energy. Environ Clim Change 45(2):16–21
Jenkins H (2006) Convergence culture: where old and new media collide. New York University Press, New York, NY
Jenkins H (2007) Transmedia storytelling 101. http://henryjenkins.org/2007/03/transmedia_storytelling_101.html. Accessed 19 Mar 2016
Karpman S (1968) Fairy tales and script drama analysis. Trans Anal Bull 7:39–43
Keim M (2015) Extreme weather events: the role of public health in disaster risk reduction as a means for climate change adaptation. In: Luber G, Emery J (eds) Global climate change and human health, pp 35–76
Kincaid DL (2002) Drama, emotion, and cultural convergence. Commun Theor 12:136–152
Lazarovic S (2018) Small works: little actions everywhere. Yes! Magazine. https://www.yesmagazine.org/issues/mental-health/5-ways-small-actions-have-huge-power-20180910. Accessed 30 Dec 2018
Leiserowitz AA, Maibach E, Roser-Renouf C, Feinberg G, Rosenthal S (2015) Climate change in the American mind. Yale University and George Mason University, Yale Program on Climate Change Communication, New Haven, CT
Lemery J, Auerbach P (2017) Enviromedics: the impact of climate change on human health. Rowman and Littlefield, Lanham, MD
Lorenzoni I, Pidgeon NF (2006) Public views on climate change: European and USA perspectives. Clim Change 77(1e2):73e95
Luber G, Lemery J (2015) Global climate change and human health: from science to practice. Jossey-Bass, San Francisco, CA
Macy J, Johnstone C (2012) Active hope: how to face the mess we're in without going crazy. New World Library, CA
Maibach E (2019) Increasing public awareness and facilitating behavior change: two guiding heuristics. In: Hannah L, Lovejoy T (eds) Climate change and biodiversity, 2nd edn. Yale University Press, New Haven, CT
Maibach EW et al (2015) Do Americans understand that global warming is harmful to human health? Evidence from a national survey. Ann Glob Health 81(3):396–409
Melillo JM, Richmond TC, Yohe GW (2014) Climate change impacts in the United States: the third national climate assessment. U.S. global change research program, p 841. https://doi.org/10.7930/j0z31wj2
Morley Rolls J (2013) Why we delay: how human cognition impairs our response to climate change. Int J Clim Change Impacts Responses 4(3):173–186
Moser S (2010) Communicating climate change: history, challenges, process and future directions. Wiley Interdiscip Rev Clim Change 1(1):31–53
Moyer-Gusé E (2008) Toward a theory of entertainment persuasion: explaining the persuasive effects of entertainment-education messages. Commun Theor 18:407–442
My Green Doctor (2018). https://www.mygreendoctor.org/. Accessed 30 Dec 2018
Nariman H (1993) Soap operas for social change. Praegar, Westport, CT

Pastina A et al (2004) Chapter 14: social merchandizing in Brazilian telenovelas. Entertainment-education and social change: history, research, and practice. Routledge, New York
Pinto B, Marcal D, Vaz S (2015) Communicating through humour: a project of stand-up comedy about science. Public Underst Sci 24(7):776–793
Reinermann J-L, Lubjuhn S, Bouman M, Singhal A (2014) Entertainment-education: storytelling for the greater, greener good. Int J Sustain Dev 17(2):176–191
Reynolds TW, Bostrom A, Read D, Morgan MG (2010) Now what do people know about global climate change? Survey studies of educated laypeople. Risk Anal 30(10):1520e38
Rhimes S (2005) Grey's anatomy. ABC, United States. https://www.imdb.com/title/tt0413573/?ref_=nv_sr_1. Accessed 30 Dec 2018
Riesch H (2015) Why did the proton cross the road? Humor and science communication. Public Underst Sci 24(7):768–775
Rosenberg M (2015) Nonviolent communication: a language of life. PuddleDancer Press, Encinitas, CA
Roser-Renouf C, Stenhouse N, Rolfe-Redding J, Maibach E, Leiserowitz A (2015) Engaging diverse audiences with climate change: message strategies for global warming's six Americas. In: Cox R, Anders H (eds) Handbook of environment and communication
Ryan M (2016) All hail comedy's takeover of TV. Online, Mar 15. Variety. https://variety.com/2016/tv/features/comedy-takes-over-tv-1201725403/. Accessed 30 Dec 2018
Rydin Y et al (2012) Shaping cities for health: complexity and the planning of urban environments in the 21st century. Lancet 379(9831):2079–2108
Sakellari M (2015) Cinematic climate change, a promising perspective on climate change communication. Public Underst Sci 24:827–841
Samson AC, Gross JJ (2012) Humour as emotion regulation: the differential consequences of negative versus positive humour. Cogn Emot 26(2):375–384
Sarfaty M et al (2017) Medical alert! Climate change is harming our health. Medical society consortium for climate and health march 2017, p 1. www.medsocietiesforclimatehealth.org/reports/medical-alert. Accessed 29 Jan 2019
Schneider-Mayerson M (2018) The influence of climate fiction: an empirical survey of readers. Environ Humanit 10(2):473–500
Servick K (2018) After the deluge: twelve years after Hurricane Katrina, social scientists seek lessons from its survivors. Science 359(6379):972–975
Singhal A (2013) Fairy tales to digital games: the rising tide of entertainment-education. Crit Arts 27:1–8
Singhal A, Dura L (2017) Positive deviance: a non-normative approach to health and risk messaging. In: Oxford research encyclopedia of communication: communication theory, health and risk communication. https://doi.org/10.1093/acrefore/9780190228613.013.248
Singhal A, Rogers EM (1999) Entertainment-education: a communication strategy for social change. Routledge, New York
Singhal A, Pant S, Rogers EM (2000) Environmental activism through an entertainment-education radio soap opera in India. In: Oepen M, Hamacher W (eds) Communicating the environment: environmental communication for sustainable development. Peter Lang, Frankfurt, pp 176–183
Singhal A, Cody MJ, Rogers EM, Sabido M (2004) Entertainment-education and social change: history, research, and practice. Routledge, New York
Singhal A, Wang H, Roger EM (2013) The entertainment-education communication strategy in communication campaigns. In: Rice RE, Atkins C (eds) Public communication campaigns, 4th edn. Sage, Beverley Hills, CA, pp 321–334
Stilgoe J, Wilsdon J (2009) The new politics of public engagement with science. In: Holliman R, Whitelegg E, Scanlon E, Smidt S, Thomas J (eds) Investigating science communication in the information age: implications for public engagement and popular media. Oxford University Press, Oxford, pp 18–34
Tierney N, Heller F (2019) Earth: Lil Dicky music video. https://welovetheearth.org/video/. Accessed 5 May 2019

Transition Town Movement (2018). http://transitionus.org/transition-town-movement. Accessed 30 Dec 2018

Underwood E (2018) Lessons in Resilience: in war zones and refugee camps, researchers are putting resilience interventions to the test. Science 359(6379):976–979

USGCRP (U.S. Global Change Research Program) (2016) The impacts of climate change on human health in the United States: A scientific assessment. In: Crimmins AJ, Balbus JL, Gamble CB, Beard JE, Bell D, Dodgen RJ, Eisen N, Fann MD, Hawkins SC, Herring L, Jantarasami DM, Mills S, Saha MC, Sarofim J, Trtanj J, Ziska L (eds) U.S. global change research program, Washington, DC, 312 pp. http://dx.doi.org/10.7930/J0R49NQX

Valkengoed A, Steg L (2019) Meta-analyses of factors motivating climate change adaptation behaviour. Nat Clim Change 9:158–163

Work that Reconnects Network (2018). https://workthatreconnects.org. Accessed 30 Dec 2018

Yale Program on Climate Change Communication (2016). https://climatecommunication.yale.edu/about/projects/global-warmings-six-americas/. Accessed 9 May 2019

Transforming Gendered Lives and Livelihoods in Post-disaster Settings in the Bangladesh Sundarbans Forest

Sajal Roy

Abstract This chapter surveys literature on gendered livelihoods, disasters and community resilience, and contributes to the application of intersectionality in research on the post-cyclone influences of Cyclone Aila on gendered livelihoods in the Sundarbans forest communities of south-west Bangladesh and Aila's intersectional impacts on the forest communities' resilience. Overall, the literature has established Aila's severe impacts on the gendered livelihood-seeking behaviours of the forest-dependents in the Sundarbans. This review chapter examines the resilience-building of the rural poor communities of the Sundarbans forest areas, and patterns of livelihood support services given to the Aila survivors by the four national NGOs (Shushilan, BRAC, SAMS and LEADRS). Thereafter, the paper investigates the intersectional dimensions of gender, religion and marital status influencing the gendered relations of the Aila survivors. Finally, the study demonstrates the utility of intersectionality as a theoretical tool for delving into livelihood transformation and gendered relations of two contrasting Sundarbans forest communities.

Introduction

Bangladesh is frequently acknowledged as one of the more disaster-prone countries in the world (Mallick et al. 2017; Asaduzzaman 2016). The south-western coastal areas of this country cover the districts of Satkhira, Bangerhat, Khulna and Barguna, which adjoin the Sundarbans mangrove forest and the Bay of Bengal (Roy 2013). Annually, millions of people living in coastal areas experience weather events, which include flooding, drought, cyclones, tidal waves, river erosion and salinity intrusion (Mallick and Vogt 2011). Amongst these extreme weather events, tropical cyclones hit coastal areas of the country every three years on average (Mallick et al. 2017;

S. Roy (✉)
Institute for Culture and Society, Western Sydney University, Building EM, Parramatta Campus, Locked Bag 1797, Penrith, NSW 2751, Australia
e-mail: s.roy@westernsydney.edu.au

Department of Gender and Development Studies, Begum Rokeya University, Rangpur, Bangladesh

W. Leal Filho et al. (eds.), *Climate Change, Hazards and Adaptation Options*, Climate Change Management, https://doi.org/10.1007/978-3-030-37425-9_24

Ali 2003). Between 2007 and 2019, the coastal inhabitants have experienced varied intensity and scales (frequency, length, both short-term and long-term socio-economic impacts and vulnerabilities) of tropical cyclones, such as 'Sidr' (2007), 'Aila' (2009), 'Mohasen' (2013), 'Royanu' (2016), 'Mora' (2017) 'Foni' (2019) and most recently 'Bulbul' (2019). According to Auerbach et al. (2015) and Ahmed et al. (2016), tropical cyclones trigger severe casualties and potentially damage people's livelihood.

Livelihood activities, particularly, at the Sundarbans forest region at the Shyamnagar sub-district of Satkhira, predominantly include shrimp aquaculture (Gher), crab-catching, rice cultivation and the gathering of forest resources (Abdullah et al. 2016a, b; Asaduzzaman 2016). According to Abdullah et al. (2016a, b), during Cyclone Aila, the local infrastructure (houses, road networks, institutions) and natural resources at the coast were acutely damaged. Due to the devastation of Aila, the poor lost their permanent houses, and were obliged to be temporarily accommodated at the limited number of cyclone centres or to migrate to the district town, Satkhira (Roy 2013). This severely affected local people's livelihoods, since they were dependent on Sundarbans forest resources. Thus, lacking a source of livelihood both during and after Aila, coastal inhabitants were compelled to seek an alternative livelihood from aid sources and social networks (Mallick et al. 2017). Globally, poor women and men, children, physically handicapped people and religious minority groups are the most affected by a disaster. Post-disaster, the earning capacity of communities dependent on natural resources is broadly shaped by their caste, ethnicity, marital status, religion, gender and class-based relations (Adeola and Picou 2017). This situation applies to the Sundarbans-forest-dependent communities of the south-western coastal zones of Bangladesh.

In the post-Aila context, there is a key gap in the literature in the Shamnagar sub-district region that addresses the measures taken for livelihood rehabilitation for disaster survivors (Kamal 2013). In this sub-district, there is a lack of statistical data showing the livelihood status of males and females. Because of these knowledge gaps, the existing literature published both before and after Aila has not adequately shown how livelihood assets (e.g., household members, knowledge and skills, livestock, access to land, forest and water, income and credit) are protected for the pursuit of the poor people's rural livelihoods. To address these knowledge gaps, this paper primarily documents how livelihoods of those dependent on the Sundarbans forest are gendered, and explores the associations between gendered livelihoods and a weather event, such as Aila. It also documents the intersectional impacts of Cyclone Aila on community resilience. In doing so, it particularly focuses on two contrasting forest-dependent coastal communities: Shora (Muslim) and Munda (Adivasi). The current study contributes to the literature by offering fresh perspectives in the areas of gender, forest and post-disaster, predominantly in the south-western coastal regions of Bangladesh.

Intersectionality as a Theoretical Framing

Race, class, gender, caste and religion intersect with socio-economic vulnerabilities instigated by climate change impacts and weather events (Osborne 2015). Osborne (2015) further emphasises how race, class, gender, ethnicity and marital status cut across the social position, lived experience and vulnerabilities of disaster-affected people. Intersectionality addresses 'the interaction between gender, race, other categories of differences in individual lives, social practices, institutional arrangements and cultural ideologies and outcomes of these interactions in terms of power' (Davis 2008, p. 68). Addressing four intersectional dimensions of gender, marital status, religion and caste of the Shora and Munda people documents the ways in which diverse individuals and groups of Shora and Munda forest-dependents experienced the damage to their traditional livelihoods and the transformation of gendered relations post-Aila. Examples of the diverse individuals and groups of adult Aila survivors are 'married men and women', 'divorced women', 'tiger widows' and 'Adivasi minorities' (Roy 2017). However, these groups are not discrete; some members may belong to more than one group.

Jonson (2011) points out that poor women in the Global South are more vulnerable than men in relation to environmental calamities. Impacts of an extreme weather event place women in the trap of socio-economic vulnerabilities. This, in turn, eventually downgrades women's 'power relations' with their husbands and family members in the household and community. 'Power relations' in the coastal regions of Bangladesh are reproduced through social practices in daily life (Roy 2013). The social practices consist of household decision-making, access to resources and markets, gendered labour force participation and performing a managerial role in the community (Roy 2018).

Law (1991) views 'power relations' as 'power to', 'power over', 'power storage', and 'power discretion'. The circulation of power impacts individuals and groups of people in many ways, such as injustices in material conditions and normative expressions within a social structure. Both the Shora and Munda people regard 'power' as the capacity to deal with sociocultural difficulties in searching for livelihoods and managing the dynamics of displacement posed by a weather event (Perucca 2011). Such difficulties both constitute and are constituted by gendered relations. Both the raw challenges of existence and the navigation of these 'power relations' have been crucial for Aila survivors. 'Power' affects all human 'relations' (e.g., gender relations, financial relations, political relations), from institutional practices (e.g., household, family, community) to individual actions (de los Reyes and Mulinari 2005; Lykke 2009 quoted in Kaijser and Kronsell 2013). Power and wealth allow rural people to mitigate the risks posed by an extreme weather event (IFRC 2016).

The social categorisation of individuals and marginal groups like the Munda in rural coastal regions of Bangladesh are deeply rooted in gender, religious identity, caste, class, marital status, and ethnicity, as well as economic status (Roy 2017; Perucca 2011). These social categorisations lead to the construction of multiple hierarchies, among them patriarchal hierarchies which rank men above women in most spheres of life within the Shora and Munda communities. Women are traditionally assigned different social roles and responsibilities (Nasreen 2012). Additionally, a lot of husbands escape from the villages during times of crisis, while wives are left behind taking care of their children and the elderly (Arntzen 2015). In a related context, in rural India it is mostly men who are the heads of the family, breadwinners and decision-makers in the public realm, whereas women are responsible for reproductive tasks and performing domestic activities (Nielsen and Waldrop 2014).

In the view of Carastathis (2013, p. 945), the intersectionality lens claims identity categories as 'actual coalitions' or 'potential coalitions'. This research study points out the nature of identity, what identities are, how identities are constructed and reconstructed from time to time, and the possibilities of collaborating despite their differences. Conceptualising identity categories as 'actual coalitions' focuses simultaneously on intragroup (women vs. women/men vs. men) and intergroup (men vs. women) differences (Carastathis 2013). Perceiving 'actual coalitions' as intragroup and intergroup (women vs. women, men vs. men, and women vs. men), Shora and Munda community members subconsciously adopt decision-making stances which relate to their access to the forest, its resources, and control of the income gained from the harvested forest resources (Roy 2017).

The greatest burden of an extreme weather event is experienced by the most vulnerable and poor of the global south countries (Hankivsky 2014). Aila survivors who have been living in vulnerable situations lead their everyday lives in hierarchised, casteist settings and are sharply divided in the sociocultural context of Bangladesh. Kapadia (2014, p.235) claims that 'this is as true of gender relations as it is of caste, class, and religion-based relations'.

Hoffman (2016) suggests including aspects of 'social structure', 'religion', 'art and literature', 'gender', and 'law' in the research related to disaster studies. Caste, religion, age, marital and educational status cut across women's access to and control of environmental resources, rights, and responsibilities in the coastal regions of Bangladesh (Sultana 2010). The varied impacts of Aila have generated gendered power differentials amongst the Shora and Munda community people (marginalised women and men) since they have experienced and coped with the hazards and risks of Aila differently. It draws on what Sultana (2010, p. 44) terms 'the relational nature of gender construction and the power relations that are constructed through norms, practices, and beliefs as it is important in the ways that gendered identities are understood and experienced'.

Examining gender relations requires the dominant sociocultural and political structures of a specific geographical location to be considered (Kaijser and Kronsell 2013). Cho et al. (2013) highlight two aspects of intersectionality: structural intersectionality and political intersectionality. Structural intersectionality examines 'power' through the analysis of multilayered power (self-capacity in getting access to the forest, obtaining ownership of cash income gained from forest resources) and the interplay between categories of identities (socio-environmental identity) versus structures of socio-economic inequalities.

'Power relations' intersect with gender, marital status and class relations of the 'disadvantaged' and 'vulnerable populations' in the post-Aila Sundarbans context. In their seminal study on 'front-line recovery workers' after hurricane Katrina, Weber and Messias (2012, p. 1834) observe: 'Intersectional attention to power and gender, race, and class relations highlights the perceptions of the disadvantaged and vulnerable populations as well as the ways in which disadvantage is tied to the privilege and control of more powerful groups'. Likewise, in the Bangladesh post-Aila social context, the Sundarbans Shora and Munda people have experienced both 'macro-level' and 'micro-level' social inequalities in pursuing traditional livelihoods, and this happens due to different factors, including key dimensions of marital status, caste, gender and religion (SAMS 2016). There is a key gap in existing literature regarding the consequences of Aila and how these pre-existing social forces have caused livelihood-related disadvantages in both communities (Roy 2018). Thus, this study explores how the disadvantaged and marginalised men and women of both communities suffer the effects of Aila more acutely than those who have access to greater economic and social status and resources.

Political intersectionality addresses some socio-economic and cultural forces, which collectively structure 'differences' between individuals and groups of marginalised people (Cho et al. 2013). Similarly, Sultana (2014) suggests how disadvantaged rural people understand, respond to and cope with variability and uncertainties posed by a weather event. Post-Aila, seeking financial support and obtaining temporary shelter from local institutions (e.g., local government, NGOs) and the local elite (businessmen, NGO owners, local politicians, hatchery owners and forest officers) are common concerns. The marginalised Shora and Munda people face sociopolitical struggles for livelihood and survival. Cho et al. (2013, p. 800) note that 'Political intersectionality provides an applied dimension to the insights of structural intersectionality by offering a framework for contesting power and thereby linking theory to existent and emergent social and political struggles'.

The application of both political and structural intersectional ties is a significant theoretical lens for developing a broader understanding of the intersectional dimensions of gender, religion, marital status and caste in livelihood-seeking behaviours. Along with this, such an approach indicates the ways of reconstructing the gendered relations amongst Aila survivors. In future it will develop in far more depth as the coverage of literature expands and the extent of empirical data increases.

The Influence of Aila on Gendered Livelihoods

The Sundarbans is the largest mangrove forest in the world. The United Nations Educational Scientific and Cultural Organization (UNESCO) inscribed the Bangladesh Sundarbans as a 'World Heritage' site in 1997 (Chowdhury et al. 2008; Ortolano et al. 2016). This forest adjoins numerous rivers, including Meghna, Brahmaputra, and Kholpatuaya on the Bay of Bengal. Around 3.5 million forest-dependent inhabitants in the coastal areas obtain their livelihoods from the Sundarbans Mangrove Forest (Roy et al. 2013). This mangrove forest is undergoing many ecological hazards: climate-change-related extreme events, excessive rainfall in summer, increasing temperature, intense flooding, cyclones, tidal surges, and sea level rise (Ortolano et al. 2016). These hazards are gradually increasing and are causing mangrove ecosystem and livelihood damage (Roy 2018). As a result, the 'regenerative function and land productivity' of the Sundarbans are endangered by the increase of 'salinity intrusion' and 'cyclonic inundation' (Nandy et al. 2013). In addition, most farmers are considered to be particularly vulnerable, resulting in loss of livelihoods, financial assets, agricultural yield and the worsening of debt problems (Chandra et al. 2017). The Sundarbans, as a reserved forest ecosystem, directly delivers a large number of services to neighbouring communities.

Accordingly, research (Karim 1994; Kathiresan and Rajendran 2005; Barbier 2007; Walters et al. 2008 quoted in Uddin et al. 2013, p. 88) indicates that: 'Mangroves ecosystems, lying along the coastlines in the tropics and subtropics, provides [sic] a number of ecosystem services such as provisioning services (e.g. typical forest products, fisheries), cultural services (e.g. tourism, worship, educational research), regulatory services (e.g. protection from cyclones and storm surges, flood regulation, climate regulation) and supporting services (e.g. nursery ground of fish, nutrient cycling, habitat of biodiversity)'. The aggregated services provided by the Sundarbans are crucial, not only for the rural economy within the coastal regions, but also for generating revenue for the national economy of Bangladesh (Uddin et al. 2013). In addition, the forest continually provides resources to support the livelihoods of forest-dependent communities throughout their life-cycle.

The mangrove ecosystem of the Sundarbans is an important part of Bangladesh's broader ecological future. The ecological future of coastal areas in South Asia is uncertain, due to the increasing impacts of floods, cyclones and droughts since the late 1990s. For instance, an increase of one metre in the sea level will lead to 15 million landless people being environmentally displaced (Dasgupta et al. 2011). The World Bank (2016) identifies three climate changed-related major challenges: 'cyclonic inundation', 'river salinity' and 'soil salinity'. Such challenges, emerging from extreme weather events, will severely impact on many poor people's livelihood capacity, crop productivity, and cash income opportunities in the coastal regions.

In particular, extreme weather events, like cyclones as a direct result of climate change, visibly destroy lives and rural livelihood patterns. Four cyclones—Aila, Mahasen, Royanu and Mora—occurred between 2009 and 2018. These destroyed mangrove resources and caused massive loss of life, depleting large tracts of the Sundarbans forest resources. Notably, the long-standing impacts of Aila led to the deterioration of the lives of the forest-dwellers, in particular impacting upon their resilience and livelihood capacities (Roy 2013).

As a direct result of climate change, cyclones have significant and varying impacts on regions, societies, and individuals. Djoudi and Brockhaus (2016) report a number of climate change impacts that they term 'multiple stresses'. These include extensive poverty, failures of governance, inadequate access to livelihoods and degradation of natural resources (Djoudi and Brockhaus 2016). Collectively, these stresses develop overlapping 'socio-ecological vulnerabilities', according to the Intergovernmental Panel on Climate Change (IPCC 2014). The increasing impacts of cyclones and floods lead rural people of the south-west coastal regions to experience a high level of gender inequalities and power differentials, in which women are more disadvantaged (Chandra et al. 2017). These are related to sources of living from natural resources, including forests and rivers. The connection between gender lines and climatic hazards in South Asia is particularly characterised by patriarchal norms and socio-economic inequities (Sultana 2014). Elmhirst (2011) suggests that 'gender' is involved in a large number of ecological aspects: environmental struggles around knowledge and power, degradation, conservation, resilience, access to and control of environmental resources.

Amongst the south-western coastal areas, some villages (Shora, Dumuria, Gabura, and Paddapukur) of the Shamnagar, the sub-district of Satkhira adjoining the Bay of Bengal, have a 'subsistence peasant economy' (Guhathakurata 2008). Guhathakurata's study of pre-Aila communities shows that poor women of these areas occupy socially and economically disadvantaged positions due to the prejudices and inequality effected by 'gender, ethnicity and work'. While her work provides a significant baseline from which to understand pre-Aila communities, it does not look at the impacts of Aila itself. Theoretically, Guhathakurata's study does not examine the intersection of different dimensions of inequality. This clearly indicates that there is a key gap in the literature showing how inequality is compounded, leading to socio-economic and material disadvantage among women and men in the Sundabarns region.

The coastal poor undergo further socio-economic vulnerabilities throughout the year. This is due to impending natural disasters, such as floods, cyclones, seasonal droughts, river erosion in the monsoon season, and extensive cold in the winter (WARPO 2006). Having no land of their own, the poor earn livelihoods working in shrimp industries, harvesting rice in paddy fields, rearing livestock, and serving as day labourers in road and dam construction (Guhathakurata 2008). Dev (1998) further notes that commercial shrimp industries developed as a great source of employment, becoming actively visible in the late 1980s because shrimp were in demand in international food markets. People also get involved in *gher* farming, which entails joint cultivation of freshwater fishes, prawn, dike crops and rice cultivation in fields with high, peripheral dikes made of boards (Ahmed et al. 2008).

Livelihood rehabilitation after a disaster event encompasses relief distribution and the recovery of former livelihoods. For disaster survivors, this livelihood rehabilitation has to establish a 'sustainable and desirable state of affairs' (Longley 2006). Pomeroy et al. (2006), quoted in Kamal (2013, p. 39), suggest: 'Rehabilitation of coastal livelihoods after a natural disaster should be seen as an opportunity to strengthen and revitalise coastal communities. The focus of rehabilitation efforts should be on building the economic basis of livelihoods rather than on physical reconstruction and giving the coastal people the skills and resources for self-recovery'.

Livelihood diversifications seem to be central to the construction of sustainable livelihoods (Hussein and Nelson 1999). The livelihood services of Sundarbans, which include traditional fishing (prawn, shrimp and crabs) in the forest surrounded by small canals, and gathering timbers and non-timber forest resources (honey), were fundamentally affected by Aila (Abdullah 2014). Abdullah's research does not, however, show how Aila affected men's and women's capacity to earn a daily living from the Sundarbans. The livelihood impacts of Aila reflect a broader process by which gender is connected to environmental circumstances: 'gendered livelihoods encompass the material realities and ideological processes that shape and are shaped by economic strategies in diverse geographical locations' (Oberhauser et al. 2004, p. 205). For example, immediately after Aila, Shora women who were cyclone survivors reduced their forest-going activities of collecting wood and catching shrimps and prawns from the Sundarbans waterbody (Roy 2013). The Shora Aila survivors (forest-dependent Muslim men and women) were then involved in gathering relief and cash support, given by national and international development agencies, including BRAC,[1] Shushlian,[2] LEDARS,[3] SAMS[4] and Oxfam (Roy 2017).

The *Sendai Framework for Disaster Risk Reduction 2015–2030* was evolved at the Third UN World Conference On Disaster Risk Reduction (WCDRR) in Sendai, Japan.[5] This recent framework suggests that each UN member state should initiate practical action plans for reducing future disaster risks. The effective implementation of such plans may protect persons, communities and their livelihoods, health, cultural heritage, socio-economic assets and ecosystems. They may also strengthen the livelihood resilience capacity of communities living in disaster-prone areas.

For a disaster-affected community, the role NGOs play as powerful agents is vital. Disaster-affected communities get two types of support services from them: short-term (relief support immediately after a disaster event) and long-term (livelihood assistance for many years). These support services, according to Allen (2006), include providing money, disaster management skills and knowledge, all of which assist households to increase their monetary resilience by undertaking income-generating activities (e.g., tailoring). Islam and Walkerden (2015) note that there are three kinds of NGOs—locally created NGOs, National NGOs and International NGOs—working actively in post-disaster recovery to build livelihood resilience in rural areas of Bangladesh. Islam and Walkerden (2015) further note that local NGOs distribute food, water, medicine, household utensils and shelter-building materials for constructing new houses as 'immediate relief'. They also find that NGOs provide microcredit, cropping seeds, livestock, fishing boats and nets, and long-term 'livelihood assistance' to disaster survivors. In short, it seems evident that BRAC, Shushilan, LEADRS, World Vision and SAMS provided both 'immediate relief' and 'livelihood assistance' to the Aila survivors.

The effects of Aila possibly reduced the Shora and Munda forest-dependents part-time paid work at hatcheries, where they were involved in shrimp packaging and maintenance activities (Roy 2017). In the pre-Aila context, there is no ethnographic study documenting specific links between livelihood and gendered relations amongst Munda forest dependents living around the Sundarbans forest regions. There is no study in South Asia focusing on the comparative aspects of the sexual division of labour and gender relations of two contrasting forest-dependent communities (Shora and Munda) before the Aila event (Roy 2017).

In Sundarbans, adult 'men and women' of the Shora Muslim community possess a conservative and adverse societal outlook towards women's increasing participation in the labour market of the coastal regions of Bangladesh (Roy 2013). An ethnographic study after Aila explores how this outlook marginalises women, along with their forest use experiences and financial contributions to the household (Roy 2013). Consequently, this further tends to ignore, overlook and downplay their crucial place and role in meeting and dealing with severe livelihood shocks caused by an environmental event like Aila. Development programmes and research that are mainly biased against females (Aguilar et al. 2011) compound such marginalisation processes.

Gender relations in forest-dependent environments involve a complex set of material and ideological aspects. This includes household and gender 'division of labour and resources' between men and women (Agarwal 1997). Gender relations also include 'ideas and representations of women's and men's different abilities, attitudes, desires, personality traits and general behavioural patterns' (Agarwal 1997, p. 1). Gender roles (economic, social and political roles) in forest-dependent societies significantly contribute to the loss of livelihood activities, along with the material and social resources of women after a disaster event (Bradshaw 2013). For example, after Aila, women in Shora and Munda who worked to reconstruct their houses found their domestic responsibilities conflicting with the time and effort required to earn livelihoods from Sundarbans (Roy 2017). After a disaster socio-economic status determines people's livelihood choices (Bradshaw 2013). After Cyclone Aila there might be shifts in the gender-specific roles and responsibilities (economic role as earner, social role as care-giver and community role as manager) amongst the disaster survivors.

Earning livelihoods after a natural disaster like Aila depends on how people in different places live and seek a minimal means to provide their daily needs. Livelihood is defined as 'the means of gaining a living' (Chambers 1995) or 'a combination of resources used and the activities undertaken in order to live'.[6] Livelihood of the Sundarbans forest communities can be considered 'a complex web of activities and interactions that emphasises the diversity of ways people make a way of living' (Scoones 2009, p. 172).

Livelihoods and social changes are further, and increasingly impacted by processes of extreme weather events in the coastal regions of Bangladesh (Kabir 2015). Pham et al. (2016) point out that the impacts of climate-change-related weather events increase the workload of women far more than that of men. There are knowledge gaps with regard to how Aila has triggered social changes: transformation of local labour, increased dependency on NGOs for livelihoods and enhancement of people's mobility for employment (Roy 2017).

In terms of earning a livelihood, there is traditionally a gendered division of labour amongst the forest dependents of Shora and Munda (SAMS 2016). This is because these two communities have poor literacy rates, negative perceptions of women doing paid work, limited transport options (which reduces their physical mobility) and, notably, large-scale dependency on religious faith. The Shora Muslim area inhabitants (the majority) allow adult married women hardly any access to the forest, whilst men are key drivers of earning livelihoods, due to their access to the dense section of the Sundarbans (Roy 2013).

Munda is an almost unknown and under-researched indigenous community living in Satkhira (Perucca 2011). This community, with a Hindu religious background, is a minority in the coastal areas of Satkhira (SAMS 2016). The Munda community members have experienced structural vulnerabilities (employment disparity, the wage gap, landlessness, insecurity inside the forest and community) and uneven power relations for decades, as they are positioned in a marginalised social class. In spite of dwelling in male-headed households, the Hindu Munda women have liberty within the family, as their husbands and in-laws mutually share household chores (SAMS 2016). In contrast, Shora women barely dare to do paid work outside their households due to the difficulties faced by their marginalised social class and insecurity about self-protection (Roy 2013).

Intersectional Impacts of Cyclones on Community Resilience

The poor rural communities of the south-western coastal areas have endured around 159 typhoons, along with 48 cyclonic storms and 68 tropical depressions (Sultana and Mallick 2015). These areas of the country experience 'one damaging cyclone each three years' (Dasgupta et al. 2011; Paul and Dutt 2010). During the cyclonic hit of Aila, on 9 May 2009, the coastal village communities of the Sundarbans forest regions were worst hit, with 2.3 million people losing their houses, 237 fatalities and a considerable amount of livestock resources washed away (Paul and Dutt 2010). Moreover, farming lands with crops, cultivable lands, embankments and fishing ponds were severely damaged, impacting upon the normal pattern of the livelihoods (Sultana and Mallick 2015). Despite Sultana and Mallick (2015) investigating both the immediate (moving into shelter homes) and long-term coping strategies (migration) of Aila-affected people, their study does not address gendered livelihood-related complex crises.

As a large-scale extreme weather event, Aila produced a series of complex crises (Roy 2013). These included both small and significant cultural, economic, environmental and social stocks and tensions. Accordingly, Smyth and Sweetman (2015, p. 405) note that 'Complex crises put a cumulative, and enormous, strain on individual women and men, their households and communities, taking a toll on mental and psychological health, lives and traditional livelihoods'.

Immediately after Aila, the Shora and Munda forest people were displaced haphazardly (SAMS 2016). Afterwards, they were shifted to a cyclone centre for temporary shelter. This displacement and substitute livelihood provision by NGOs *alters* their gender specific roles and responsibilities (Salehin 2014). In the end, most of the displaced community individuals returned to their own villages and traditional forest-based livelihood pursuits, but the old order has never been fully restored. Their traditional gender relations might have changed, as indeed traditions are always changing (Otto and Pedersen 2005). There have been distinct moments of change in the livelihood-seeking behaviours of Shora and Munda communities' members from

Sundarbans, both before and after Aila (Roy 2017). This largely unexamined consequence of the cyclonic event, related to the shift in livelihood-seeking behaviours and gendered relations amongst Aila survivors from 2009 to 2017, has not been documented.

During the emergency precipitated by Cyclone Aila, NGOs (BRAC, Shusilan, LEARDS, SAMS) distributed emergency relief supplies, such as dry foods and bottled water, to the Aila-affected people at cyclone centres. Aila survivors depended on these emergency provisions to support themselves and their families because the financial assistance given by NGOs was inadequate (Roy 2013). Apart from the damage to traditional forest-based livelihoods caused by Aila, the forest-dependents were homeless, shocked and increasingly vulnerable (Roy 2013). The isolation of the adult men and women of the Shora and Munda community might have negatively affected their traditional socio-environmental identities[7] in terms of earning a livelihood and cash income from the Sundarbans. Arntzen (2015) notes that people who have no livestock and whose earnings are low are the most vulnerable to the impacts of a climatic hazard. Interestingly, households with a large amount of livestock have greater access to livelihood resources and this makes it much easier for them to adapt to the impacts of a hazardous event (Arntzen 2015).

In the post-Aila landscape, no previous research shows the extent of the psychological and economic stress under which Aila survivors were placed due to the uncertainty of livelihood support, tensions over personal security, and physical health hazards triggered by diarrhoea and cholera. These diverse factors affect not just the livelihoods of Aila survivors, but also their gendered relations and the household division of labour.

The level of vulnerability arising from Aila's social impact on Shora and Munda communities varies from high to low, depending on political and social factors that have placed individuals and groups of people from these contrasting communities in different positions (Roy 2018). More data needs to be collected in order to obtain a better understanding of the largely unexamined livelihood-related social impacts and gendered vulnerabilities imposed by Aila on both the Shora and Munda communities (Roy 2017).

Canon (2002) and Blaikie et al. (1994) suggest four components of social vulnerability: (a) 'initial condition of a person' (e.g., how well fed disaster survivors are during and after a disaster event, what their physical and mental health condition and movements are, and their abilities for building self-confidence); (b) 'resilience of livelihood' (e.g., how rapidly and easily disaster survivors can recommence activities for earning money to support their basics such as food and shelter); (c) 'opportunities for self-protection'(e.g., protected accommodation constructed above flood levels); and (d) 'access to social protection and social capital' (e.g., form of preparedness to adjust to a disaster event, warning system and ability to adopt disaster-recovery procedures). The level of vulnerability experienced by disaster survivors is, of course, different for wealthy and poor victims, and 'differs also according to gender, age, class and other aspects of identity' (Smyth and Sweetman 2015, p. 406). Accordingly, investigating the comparative aspects of social vulnerabilities experienced by the Shora and Munda Aila survivors might tease out the reconfiguration of gender relations in terms of the forest-based livelihood practices from an intersectional perspective that sees gender intertwined with age, class, ethnicity and religious identity.

Livelihood strategies of the Sundarbans forest society are determined by material realities and the position of women and men, in the light of before, during and after a disaster event (Nasreen 2012). Nasreen (2012) notes that women in south-western coastal zones are generally more vulnerable than men to extreme weather events due to their social status and cultural norms—especially their poor access to and control over natural resources and lack of participation in decision-making processes. Her studies (Nasreen 1995, 2008a, b, 2012) further demonstrate that destitute and vulnerable women in Bangladesh bear numerous responsibilities at home. Their general responsibilities include food preparation, provision of cooking fuel, health care, and caring for children and their education. To combat post-disaster effects on women, BRAC (2016) reports that providing 'enterprise development and life-skill training' (self-management, handling business), livelihood-related 'asset transfer' (livestock, poultry, agriculture farming, horticulture nursery), and 'soft loans with low interest and tailor-made essential health care service' (health education, basic healthcare, pregnancy-related care, immunisation, and vitamin A tablets for children under five) to ultra-poor women builds up their self-determination, confidence, social networking and asset management skills.

Aila hit the poorest communities hardest and, within them, individuals and groups of women already experiencing marginalisation and socio-economic inequalities were the most vulnerable (Roy 2013). After Aila, little attention has been paid to whether there might be changes in the long-term resilience-building of individuals in Shora and Munda communities. This resilience-building covers 'boosting well-being, realizing rights and investing in resources, activities and social relationships which reduce risk' (Smyth and Sweetman 2015, p. 406). Particularly among women, due to their more significant economic roles in post-disaster communities, changes may also include the enhanced participation of Aila survivors in the local and urban labour force and NGO dependency for micro-credit to build homes and generate new income opportunities (Roy 2017). Aila, therefore, can be understood to have contradictory effects: immiserating the most poor even further, yet also disrupting previous norms regarding the gender relations of Shora and Munda forest-dependents.

More specifically, it is evident that, for survivors of a natural disaster, 'gender shapes vulnerabilities and resilience' (Drolet et al. 2015 p. 437). After Aila, the resilience and livelihoods capacity of Shora and Munda community members might have declined. The adults of these communities either migrated to urban areas (district towns) to seek alternative employment or stayed in their own community by extending their dependency on Sundarbans for their daily livelihood. The existing adult men and women of the Shora and Munda communities have possibly been coping differently with—and mitigating—the socio-economic vulnerabilities (e.g., limited cash income and fragile communication networks which diminish the capacity to recover) posed by Aila. There is a potential knowledge gap concerning the intersectional links between gender and the household division of labour, and the social vulnerabilities posed by a natural calamity in the post-disaster context of South Asia (Enarson and Pease 2016).

Adult women in Asian rural forest communities derive half of their household income from forest resources, while men derive only a third (World Bank 2009). Moreover, males dominate policy design, implementation, and management by excluding women from forestry governance (Aguilar et al. 2011). Case studies from Canada, Thailand, Nepal, Kenya and Sweden indicate that forest-connected decision-making typically restricts women's participation (Lidestav and Reed 2010). Reed and Christie (2008) point out that a household focus and income gained from forest resources is a relatively new but significant site for 'gender and environmental scholarship'. Their research has four key aims: (i) 'to explicitly focus on differential life changes of women and men by identifying their relative efficacy and vulnerability' to an extreme weather event as part of environmental changes, (ii) 'to grapple with new constructs' of gender and use them to 'analyse policy formation', (iii) 'to draw insights from developing countries', and (iv) 'to address feminist epistemologies more directly', by addressing the gap in the forest, gender relations and disaster discourse (Reed and Christie 2008).

Concluding Notes

The greater south-western coastal region of Satkhira, Khulna, Bagherhat and Barguna has become both socio-economically and ecologically vulnerable. Here, the inhabitants have faced severe cyclones, floods, drought and salinisation since 1991. The varied consequences of such weather events have led to livelihood marginalisation of the Sundarbans forest-dependent communities. The surveyed literature identifies the missing links in understanding the influence of Cyclone Aila on gendered livelihoods and the intersectional impacts of cyclones in community resilience. This paper advocates further development of political and structural intersectionalities in future research studies on the gendered livelihoods of marginalised social groups in the Sundarbans forest regions. These livelihood marginalisation processes of Shora and Munda forest-dependents are both gendered and culturally discrete as they intersect with their age, 'socio-environmental identity', religion, marital status, caste, ethnicity and educational status. Experiencing such marginalisation processes over decades of survival on the coast, both the Shora and Munda communities in the Sundarbans region have done their best to demonstrate resilience in coping with climatic disasters like Aila but continue to face an uphill struggle.

Notes

1. BRAC is an international non-governmental organisation (NGO) which mainly works to reduce rural poverty in Bangladesh and nine other countries. Available: http://www.brac.net/.

2. The NGO Shushlian has been working in the south-west coastal region of Bangladesh since 1992. Available: http://shushilan.org/.
3. LEDARS, which stands for Local Environment Development and Agricultural Research Society, is a local NGO based in Shamnagar usb-district. Available: http://ledars.org/.
4. SAMS stands for Sundarban Adibasi Munda Sangstha, working with Munda community's health, education and livelihoods. Available: http://www.samsbd.org/.
5. https://www.unisdr.org/we/coordinate/sendai-framework.
6. http://www.glopp.ch/B7/en/multimedia/B7_1_pdf2.pdf.
7. I use the concept 'socio-environmental identities' to refer to the social constructions of the Cyclone-Aila survivors' relationship with Sundarbans through their labour and livelihoods, in a specific sociopolitical context of the post-cyclonic landscape of Aila.

Acknowledgements This paper is an outcome of my ongoing doctoral thesis at the Institute for Culture and Society, Western Sydney University. This study was funded by the International Postgraduate Research Scholarship of the Western Sydney University. I am grateful to Dr. Liam Magee, my primary supervisor of my doctoral thesis, for reading and offering extremely helpful comments in the initial draft of this paper. Many thanks to Dr. Penelope Rossitier, my co-supervsior, for her comments on earlier draft of this chapter.

References

Abdullah ANM (2014) Livelihood strategies of people surrounding the Sundarbans mangrove forest. Ph.D. thesis, Charles Darwin University. http://espace.cdu.edu.au/view/cdu:42400 (last accessed on 24 Feb 2016)

Abdullah ANM, Zander KK, Myers B, Stacey N, Garnett ST (2016a) A short- term decrease in household income inequality in the Sundarbans, Bangladesh, following Cyclone Aila. Nat Hazards (83):1103–1123

Abdullah ANM, Stacey N, Garnett ST, Myers B (2016b) Economic dependence on mangrove forest resources for livelihoods in the Sundarbans, Bangladesh. For Pol Econ (64):15–24

Adeola OF, Picou SA (2017) Hurricane Katrina-linked environmental injustice: race, class, and place differentials in attitudes. Disasters 41(2):228–257

Agarwal B (1997) "Bargaining" and gender relations: within and beyond the household. Feminist Econ 3(1):1–51

Aguilar L, Quesada-Aguilar A, Shaw DMP (eds) (2011) Forests and gender. IUCN, Switzerland

Ahmed N, Allison EH, Muir JF (2008) Using the sustainable livelihoods framework to identify constrains and opportunities to the development of the freshwater prawn farming in Southwest Bangladesh. J World Aquaculture Soc 39(7):598–611

Ahmed B, Kelman I, Fehr HK, Saha M (2016) Community resilience to cyclone disasters in coastal Bangladesh. Sustainability (8):805

Ali A (2003) Impacts of climate change on tropical cyclones and storm surges in Bangladesh. In: Quadir DA, Prasad K, Hussain MA (eds) Proceedings of the SAARC seminar on climate variability in the South Asian region and its impact. SAARC Meteorological Research Centre (SMRC), Dhaka, 10–12 Dec 2002, pp 130–136

Allen KM (2006) Community-based disaster preparedness and climate adaptation: local capacity building in Philippines. Disaster 30(1):81–101
Arntzen B (2015) Attention to gender and climate change: transformation of gender roles and adaptive capacity of rural women in two villages in Meatu District Tanzania. Master thesis, Department of International Environment and Development Studies, Norwegian University of Life Sciences
Asaduzzaman M (2016) Livelihood vulnerability of women in the context of climate change impacts: insights from coastal Bangladesh. Ph.D. thesis, The University of Newcastle, Australia
Auerbach LW, Goodbred SL Jr, Mondal DR, Wilson CA, Ahmed KR, Roy K, Steckler MS, Small C, Gilligan JM, Ackerly BA (2015) Flood risk of natural and embanked landscapes on the Ganges–Brahmaputra tidal delta plain. Nat Clim Change (5):153–157
Barbier EB (2007) Valuing ecosystem services as productive inputs. J Econ Policy 22(49):177–229
Blakie P, Cannon T, Davis I, Wisner B (1994) At risk: natural hazards, people's vulnerability, and disasters. Routledge, London
BRAC (2016) Programme components. http://brac.net/targeting-ultra-poor/item/749-programme-components (last accessed on 13 July 2016)
Bradshaw S (2013) Gender, development and disasters. Edward Elgar, UK
Canon T (2002) Gender and climate hazards in Bangladesh. Gend Dev 10(2):45–50
Carastathis A (2013) Identity categories as potential coaltions. J Women Cult Soc 38(4):941–965
Chambers R (1995) Poverty and livelihoods: whose reality counts?. ID discussion paper. 347. IDS, Brighton
Chandra A, McNamara KE, Dargusch P, Caspe AM, Dalabajan D (2017) Gendered vulnerabilities of smallholder farmers to climate change in conflict-prone areas: a case study from Mindanao, Philippines. J Rural Stud 50:45–59
Cho S, Crenshaw KW, McCall L (2013) Toward a field of intersectionality studies: theory, applications and praxis. J Women Cult Soc 8(4):785–810
Chowdhury AN, Mondal R, Brahma A, Biswas MK (2008) Eco-psychiatry and environmental conservation: study from Sundarban Delta, India. Environ Health 2:61–76
Dasgupta S, Huq M, Masud MS, Khan ZH, Mukherjee N, Pandey K et al (2011) Climate proofing infrastructure in Bangladesh: the incremental cost of limiting future flood damage. J Environ Dev 20(2):167–190
Davis K (2008) Intersectionality as buzzword: a sociology of science perspective on what makes a feminist theory successful. Feminist Theory 9(1):67–85
de los Reyes P, Mulinari D (2005) Intersektionalitet. Liber, Malmö
Dev AK (1998) Fake blue revolution: environmental and socio-economic impacts of shrimp culture in the coastal areas of Bangladesh. Ocean Coast Manag 41:63–88
Djoudi H, Brockhaus M (2016) Unveiling the complexity of gender and adaptation: the "feminization" of forests as a response to drought-induced men's migration in Mali'. In: Carol J, Colfer P, Basnett BS, Elias M (eds) Gender and forest. Taylor and Francis, USA
Drolet J, Dominelli L, Alston M, Mathbor G, Wu H (2015) Women rebuilding lives post-disaster: innovative community practices for building resilience and promoting sustainable development. Gend Dev 23(3):433–448
Elmhirst R (2011) Introducing new feminist political ecologies. Geoforum 42:120–132
Enarson E, Pease B (eds) (2016) Men, masculinities and disaster. Routledge, London
Guhathakurata M (2008) Globalization, class and gender relations: the shrimp industry in southwestern Bangladesh. Development 51:212–219
Hankivsky O (2014) Intersectionality 101. The Institute for Intersectionality Research and Policy, Canada
Hoffman SM (2016) The question of culture continuity and change after disaster. Ann Anthropol Pract 40(1):39–51
Hussein K, Nelson J (1999) Sustainable livelihoods and diversification. IDS Worldng Paper 69

IFRC (2016) World Disasters Report 2016 Resilience: saving lives today, investing for tomorrow. International Federation of Red Cross and Red Crescent Societies (IFRC), CH-1211 Geneva 19, Switzerland

IPCC (2014) Climate change 2014: synthesis report. Fifth assessment report of the intergovernmental panel on climate change. IPCC, Switzerland

Islam R, Walkerden G (2015) How do links between households and NGOs promote disaster resilience and recovery?: a case study of linking social networks on the Bangladeshi coast. Nat Hazards 78:1707–1727

Jonson SA (2011) Virtue and vulnerability: discourse on women, gender and climate change. Glob Environ Change 21:744–751

Kabir HM (2015) Gendered poverty and social protection for women in Bangladesh. J Soc Sci 8(10):1–12

Kaijser A, Kronsell A (2013) Climate change through the lens of intersectionality. Environ Polit 23(3):417–433

Kamal MA (2013) Livelihood coping and recovery from disaster: the case of coastal Bangladesh. Curr Res J Soc Sci 5(1):34–44

Kapadia K (2014) Caste and class in gendered religion: Dalit women in Chennai's Slums. In: Nielsen KB (ed) Women, gender and everyday social transformation in India. Anthem Press, London and New York, pp 235–251

Karim A (1994) The physical environment. In: Hussain Z, Acharya G (eds) Mangroves of the Sundarbans, II. IUCN, Bangladesh

Kathiresan K, Rajendran N (2005) Coastal mangrove forests mitigated tsunami. J Estuar, Coast Shelf Sci 65(3):601–606

Law J (1991) Strategies of power: power, discretion and strategy. In: Law J (ed) A sociology of monsters. Routledge, London, pp 165–191

Lidestav G, Reed M (2010) Preface: gender and forestry. Scand J For Res 25(9):1–5

Longley C (2006) Agricultural rehabilitation: mapping the linkage between humanitarian relief, social protection and development. HPG Report, Overseas Development Institute, London

Lykke N (2009) Genusforskning – en guide till feministisk teori, metodologi och skrift. Liber, Stockholm

Mallick B, Vogt J (2011) Social supremacy and its role in local level disaster mitigation planning in Bangladesh. Disaster Prev Manag 20:453–556

Mallick B, Ahmed B, Vogt J (2017) Living with the risks of cyclone disasters in the South-Western coastal region of Bangladesh. Environments 4(13):1–17

Nandy P, Ahammad R, Alam M, Islam A (2013) Coastal ecosystem based adaptation: experience from Bangladesh. In Shaw R, Mallick F, Islam A (eds) Climate change adaptation actions in Bangladesh. Springer, Japan, pp 277–303

Nasreen M (1995) Coping with floods: the experience of rural women in Bangladesh. Unpublished Ph.D. dissertation, Messey University, New Zealand

Nasreen M (2008a) Impact of climate change on food security in Bangladesh: gender and disaster perspectives. In: International symposium on climate change and food security in South Asia, Dhaka 25–30 Aug 2008. http://www.wamis.org/agm/meetings/rsama08/S203-Nasreen-Bangladesh.pdf (last accessed on 29 June 2016)

Nasreen M (2008b) Impact of climate change on agriculture and food security. Action Aid Bangladesh, Bangladesh

Nasreen M (2012) Women and girls: vulnerable or resilient?. Institute of Disaster Management and Vulnerability Studies, University of Dhaka, Bangladesh

Nielsen KB, Waldrop A (eds) (2014) Women, gender and everyday social transformation in India. Anthem Press, London and New York

Oberhauser AK, Mandel JK, Hapke HM (2004) Gendered livelihoods in diverse global contexts: an introduction. Gend Place Cult J Feminist Geogr 11(2):205–208

Ortolano L, Trina SE, Ferdausi SA (2016) Strategy for adapting to climate change and conserving biodiversity in the Bangladesh Sundarbans. Clim Dev 1–12

Osborne N (2015) Intersectionality and kyriarchy: a framework for approaching power and social justice in planning and climate change adaptation. Plann Theory 14(2):130–151

Otto T, Pedersen P (2005) Disentangling traditions: culture, agency and power. In: Otto T, Pedersen P (eds) Tradition and agency: tracing cultural continuity and invention. Aarhus University Press, Denmark

Paul BK, Dutt S (2010) Hazard warnings and responses to evaluation orders: the case of Bangladesh's cyclone Sidr. Geogr Rev 100(3):355–366

Perucca C (2011) Social water management among Munda people in the Sundarban. University of Liberal Arts Bangladesh

Pham P, Doneys P, Donna LD (2016) Changing livelihoods, gender roles and gender hierarchies: the impact of climate, regulatory and socio-economic changes on women and men in a Co Tu community in Vietnam. Women's Stud Int Forum 54:48–56

Pomeroy RS, Ratner BD, Hall SJ, Pomiljinda J, Vivekanandan V (2006) Coping with disaster: rehabilitating coastal livelihoods and communities. Mar Policy 30(6):786–793

Reed MG, Christie S (2008) Environmental geography: we're not quite home- reviewing the gender gap. Prog Hum Geogr 33(2):246–255

Roy S (2013) Connectedness with Mangrove forest "Sundarbans": perceptions and interactions of the forested community 'Shora' in Bangladesh. M.Phil. thesis, University of Bergen, Norway

Roy S (2017) Long-term effects of extreme weather events on rural livelihoods and gender relations: a case study of the cyclone Aila's impacts on the Shora (Muslim) and Munda (Hindu) Sundarbans forest communities of Bangladesh. Unpublished doctoral research project proposal, Institute for Culture and Society, Western Sydney University

Roy S (2018) Livelihood resilience of the indigenous Munda community in the Bangladesh Sundarbans forest. In: Filho WL, Ayal D (eds) Handbook of climate change resilience. Springer Nature, Germany, pp 1–22

Roy AK, Alam K, Gow J (2013) Community perceptions of state forest ownership and management: a case study of the Sundarbans mangrove forest in Bangladesh. J Environ Manage 117:141–149

Salein MM (2014) Piety, gender relations and muslim women's empowerment: the case of islam NGOs in Bangladesh. In: Clarke M, Tittensor D (eds) Islam and development: exploring the invisible aid economy. Taylor and Francis, pp 173–196

Scoones I (2009) Livelihoods perspectives and rural development. J Peasant Stud 36(1):171–196

Smyth I, Sweetman C (2015) Introduction: gender and resilience. Gend Dev 23(3):405–414

Sultana F (2010) Living in hazardous waterscapes: gendered vulnerabilities and experiences of floods and disasters. Environ Hazards 9(1):43–53

Sultana F (2014) The right to water. In: Debra Rowe (Ed), Achieving sustainability: visions, principles, and practices. Macmillan References USA, Detroit, pp 668–670

Sultana Z, Mallick B (2015) Adaptation strategies after cyclone in Southwest coastal Bangladesh-pro poor policy choices. Am J Rural Dev 3(2):24–33

Sundarbans Adibasi Munda Sangsta (SAMS) (2016) Themes and programmes. http://www.samsbd.org/index.php?page=themes (last assessed on 13 May 2016)

Uddin MS, Steveninck ER, Stuip M, Shah MAR (2013) Economic valuation of provisioning and cultural services of a protected mangrove ecosystem: A case study on Sundarbans reserve forest, Bangladesh. Ecosyst Serv 5:88–93

Walters BB, Rönnbäck P, Kovacs JM, Croma B, Hussain SA, Badola R et al (2008) Ethnobiology, socio-economics and management of mangrove forests: a review. Aquat Bot 89(2):220–236

Water Resources Planning Organisation (WARPO) (2006) The state of coast 2006. Dhaka, Bangladesh

Weber L, Messais DKH (2012) Mississippi front-line recovery work after Hurricane Katrina: an analysis of the intersections of gender, race, and class in advocacy, power relations, and health. Soc Sci Med 74(11):1833–1841

World Bank (2016) Climate change poses urgent threat to poor of coastal Bangladesh. http://www.worldbank.org/en/news/feature/2016/06/24/climate-change-poses-urgent-threat-to-poor-of-coastal-bangladesh (last viewed on 27 June 2016)

Strengthening Africa's Adaptive Capacity to Climate Change: African Union Law and Implications of China's Belt and Road Policy

Michael Addaney

Abstract People in Africa are particularly vulnerable to the adverse effects of climate change due to factors such as agro-based economies, poor infrastructure, low technology and industrialization. This phenomenon has weakened Africa's adaptive capacity to climate change. In an interconnected global economy, effective and inclusive development demands effective partnerships. On the other hand, the Belt and Road Initiative (BRI) adopted by the Chinese Government propagates infrastructural investment in participating countries and logistic connectivity. This chapter discusses the strategic goals and scope of the BRI in relation to Africa's priority adaptation sectors such as agricultural modernization, industrialization, energy and infrastructural development. It argues that the BRI has huge potential in promoting agricultural modernization, infrastructural development and industrialisation in Africa which will strengthen the continent's adaptive capacity. In conclusion, it suggests that the African Union and African governments must develop and/or revise their multi-sectoral climate change adaptation policy frameworks to tap into the opportunities that the BRI offers.

Introduction

The changes in the global climatic system may increase the occurrence of extreme weather and environmental events such as floods, storm surges, droughts, and rise in sea level, which could possibly exacerbate hunger and malnutrition, injuries, illnesses, water stress, forced displacements and deaths across the human population (Conference of the Parties of the United Nations Framework Convention on Climate Change 2007; Jegede 2016a, b). The consequence of climate change on the African continent is dire. For instance, the Fifth Assessment Report of the IPCC in 2014 estimated that, of all the continents, Africa will be the severely hit by the adverse impact

M. Addaney (✉)
Research Institute of Environmental Law, School of Law of Wuhan University, Wuhan, China
e-mail: addaneym@whu.edu.cn

Research Unit on Law, Justice and Sustainability of the Faculty of Law, North-West University, Potchefstroom, South Africa

W. Leal Filho et al. (eds.), *Climate Change, Hazards and Adaptation Options*,
Climate Change Management, https://doi.org/10.1007/978-3-030-37425-9_25

of climate change due to the continent's weak capacity for adaptation (Intergovernmental Panel on Climate Change [IPCC] 2014). Africa's major economic sectors are vulnerable to current climate sensitivity, with huge economic impact, and this vulnerability is exacerbated by existing developmental challenges such as endemic poverty, poor infrastructure, complex governance and institutional dimensions, limited access to capital, poor energy and low technology, ecosystem degradation, and complex disasters and conflicts (IPCC 2014). These in turn have contributed to continent's weak adaptive capacity, increasing vulnerability to projected climate change.

The impacts of climate change will therefore mount pressures on Africa's limited resources and threaten the attainment of the sustainable development goals and aspirations. As an outcome of socio-economic disruptions (Lin 2018), responding to climate change in Africa requires fundamental economic, legal and social transformation in a manner that involves multi-level governance and various actors at multiple scales. In 2013, the Chinese Government adopted the Belt and Road Initiative (BRI) which propagates logistic connectivity through infrastructural development in participating countries, promotion of regional trade, investment as well as social and cultural cooperation (Kuo and Kommenda 2018). The BRI involves China funding infrastructure investments worth billions of dollars and linking countries in Asia, Europe and Africa (Xinhua 2015). The plan covers two physical routes: the Silk Road Economic Belt which stretches from Asia to Europe; and the Maritime Silk Road that starts from China through the Indian Ocean littoral to East Africa and then Europe (Xinhua 2015). The strategic character of this economic scheme seeks to build a community of shared prosperity, in which participating countries can share mutual benefits and coexist peacefully along the trade routes (Kitaba 2017). This initiative may prove opportune to African countries in relation to diversifying their economies, building strategic and much needed infrastructure to drive industrialization in order to strengthen their adaptive capacities.

This chapter therefore discusses the intersection of the BRI and African Union's (AU) climate change adaptation policy and how they can harmoniously strengthen Africa's adaptive capacity. It explores the gaps in AU adaptation law and policy and how they can be remedied by the benefits of the BRI especially in the areas of energy and infrastructural development, agricultural modernization and industrialization. Following this introduction, Section "Climate Change Vulnerability and Adaptation in Africa: Legal and Policy Implications" conceptualizes the key terms associated with Africa's vulnerability to climate change and priority areas for strengthening its adaptive capacity. It further draws out the legal and policy implications of the climate crisis in Africa. Section "Climate Change Adaptation in Africa: AU Law and Policy Responses" discusses the inherent weaknesses and implementation gaps of the AU adaptation policy. Section "The BRI as a Solution to the Implementation Gaps in AU's Adaptation Policy in Africa" explores the prospects and challenges of the BRI as a practical policy solution to the implementation gaps in AU's adaptation law and policy in Africa. Section "Conclusion" summarizes and concludes the discussion.

Climate Change Vulnerability and Adaptation in Africa: Legal and Policy Implications

Although climate change is a global challenge, it is geographically and locally different in terms of impact, vulnerability and adaptation (Addaney and Cobbinah 2019). Therefore, understanding the different adaptation challenges and priorities across different geographical locales and sectors in Africa is fundamental for examining AU policy responses in the context of its cooperation with China. Specifically, the section draws from evidence in key sectors such as agriculture, infrastructure and energy to exposit that there are already signs of challenging barriers to adaptation on the African continent that require multi-level legal and policy action.

Conceptualization of Adaptation, Vulnerability and Adaptive Capacity in Relation to Climate Action in Africa

The conceptualization of adaptation is often constrained by societal and organizational perceptions and misperceptions, institutional structures, legal and policy principles and processes resulting in varied definitions of climate change adaptation (Adger et al. 2006). Turner (2010) describes adaptation in terms of the capacity of systems to self-organize, learn and acclimatize to abrupt or gradual changes, alterations, and disturbances in their environments. Cannon and Müller-Mahn (2010) explain adaptation as 'the capacity of a society or a community to coordinate decision-making, to act collectively, and to give collective action some stability by means of an institutional framework'. The IPCC (2007) defines it as 'adjustment in natural or human systems in response to actual or expected climatic stimuli or their effects, which moderates harm or exploits beneficial opportunities'. Deviating from this definition, Moser and Ekstrom (2010) argue that adaptation should go beyond climate change and may be instigated or carried out in a system of non-climatic windows of opportunity. Adaptation can also be viewed as a process of steering society away from such thresholds or at least in curtailing potential losses and damages arising from extreme weather events (Nalau and Leal Filho 2018). These varied definitions strengthen the legal conception of adaptation by offering diverse perspectives and strategies to deal with climate change (O'Neil 2010). The multiple conceptualizations provide pathways for mainstreaming adaptation policy into climate action at different levels and across scales. According to Karlsson et al. (2007), multiple conceptions help policymakers to avoid policy skewness and the marginalization of certain adaptation concepts that can make climate action comprehensive and effective.

Climate change impacts in Africa are largely apparent in the increase in and intensity of extreme weather events such as floods, drought, high temperatures and violent storms and lightning. This include the recently observed low water levels in most lakes across Africa such as Lake Volta, Lake Chad, Lake Malawiand Lake Kariba, which threatens hydro-electricity power (HEP) generation in the affected countries.

The IPCC (2014) indicated that lower temperature increases may aggravate existing vulnerabilities concerning food supply, infrastructure, health, water resources, and coastal systems. Intermediate estimates point to significant losses in biodiversity, decreasing global agricultural productivity and glacial melt (Otto-Zimmermann 2013). The socio-economic repercussions are already telling through frequent electricity cuts to industry, institutions, households and farmers (Otto-Zimmermann 2013). This cumulatively affects agricultural productivity and economic performance of most African countries as a result of the reliance of their economies on agriculture. On the whole, the legal and policy responses to adaptation require good governance, robust institutional capability and financial capacity at the regional, national and local levels (Paavola and Adger 2006).

Other words associated with adaptation such as vulnerability and resilience are also not defined in the international climate agreements and Conference of Parties decisions. The concept of vulnerability has been widely defined in different contexts. Liverman (1990) equates vulnerability to concepts such as 'resilience, marginality, susceptibility, adaptability, fragility, and risk'. In the field of climate change, vulnerability is conceptualized as 'the degree to which geophysical, biological and socio-economic systems are susceptible to and unable to cope with the adverse impacts of climate change' (IPCC 2007). Tyler and Moench (2012) argue that rather than focusing on disconnected measures to adapt to specific projected climate risks and vulnerabilities, it may be more effective and valuable for countries to focus their laws and policies on building resilience. The IPCC (2007) conceptualizes resilience as 'the ability of a social or ecological system to absorb disturbances while retaining the same basic structure and ways of functioning, the capacity of self-organization, and the capability to adapt to stress and change'. Although the origins of 'resilience' denote strength and resistance, its recent applications in socio-ecological systems, disaster management and urban sustainability is understood to entail flexibility, learning and change (Antrobus 2011). Klein et al. (2003) argue that the definitions of resilience incorporating notions of learning and adaptive capacity embody normative legal and social objectives that are applicable to developing regions. The legal aspect of a resilience approach thus promotes innovation and change to support recovery from stresses and shocks that may or may not be predictable. The linkages between the policy implications and legal aspects of vulnerability, adaptive capacity, and adaptation are often circular rather than linear in nature. Consequently, the ability of regions, countries and communities to control the variables that determine vulnerability might be translated into their capacity to adapt (Kelly and Adger 2000). The AU and member states may therefore strengthen their adaptive capacity by adopting and/or revising and implementing suitable legislation and policies relating to resilience and adaptation.

The Impact of Climate Change in Africa's Priority Sectors: Agriculture, Energy and Other Critical Infrastructure

Located in a region that is vulnerable to the adverse effects of climate change, the countries in sub-Saharan Africa are already faced with the drastic effects of climate change (IPCC 2014). The Food and Agriculture Organization (2008) contends that due to the fundamental connection between climate change and food production, agriculture becomes one of the most climate sensitive sectors in Africa. The World Bank (2003) has warned that temperature extremes in Africa are likely to affect the production of maize, rice, wheat, and other important crops. This development will negatively affect food production and aggravate existing food insecurity on the continent. Other studies have also confirmed that some African countries' highly productive agro-ecological regions are shrinking in size at the expense of drier regions (Mugandani et al. 2012). This projected shrinkage may affect the agricultural sector gravely as most farmers in Africa depend on rain-fed agriculture. This may eventually affects food security and livelihood options and weaken the adaptive capacity of countries and communities. Most economies in Africa are largely agro-based with bulk of the peasantry populated in rural areas and reliant on climate-sensitive livelihoods such as livestock rearing, arable farming and other ecosystem services. Climate change is therefore a major threat and boosting adaptive capacities in this sector cannot be ruled out (Chanza 2018).

Adaptation laws, policies and strategies in the agricultural sectors of most African countries include irrigation farming, use of drought tolerant crops such as small grains, conservation agriculture, and an assortment of indigenous based adaptation, among others (Dube and Phiri 2013). Nonetheless, Patt and Gwata (2002) argues that using these measures and practices has a tendency of being influenced by access to environmental information, scale of climatic event, institutional support and societal or individual behavior. In a study that examined the prospect of transitioning from rain-fed agriculture to irrigation farming in selected countries in Sub-Saharan Africa (including Ghana, Nigeria, Kenya and Zimbabwe), Amjath-Babuet al. (2016) observed complex interface between factors such as the availability of groundwater and energy, cost of labor, post-harvest technology and access to market, infrastructure, population density and political variables. Conversely, Ogwu (2019) and Matenga (2019) observe that local knowledge in most African countries is woefully inadequate in responding to the scale and intensity of potential climate change risks and impact. For instance, the 2015/16 El-NINO-induced drought affected more than 30% of the population of Zimbabwe leading to food shortage; forcing the government to declare it as a national disaster (Chigogo 2016). Arguably, this course of adaptation in the agricultural sector is not sustainable and demonstrates a clear policy and practice gap that needs to be addressed.

In addition, climate change is projected to have huge impact (directly and indirectly) on almost all infrastructural sectors (Goyol and Pathirage 2017). Climate change impacts on various critical infrastructure including road and transport (Nemry and Demirel 2012; Neumann et al. 2015), power and energy (Panteli and Mancarella 2015; Van Vliet et al. 2016), water resources (Olmstead 2014), and agriculture and industry (Ghile et al. 2014). In particular, the energy and infrastructure sector has been identified as one of the most vulnerable sectors to the adverse effects of climate changes in Africa (Field et al. 2012). Extreme climatic events pose adaptation challenges to an already strained public infrastructure in most African countries such as roads, bridges and dams (Kabo-Bah and Diji 2018). Low-lying areas in selected coastal African countries such as Ghana, Nigeria, Kenya and Somalia are also vulnerable to climate change-induced flooding (Amoako 2016). These events pose serious risks to the countries and communities through infrastructure damages, injuries and exposure to diseases, loss of assets, homes and lives (Amoako and Boamah 2014). This situation tends to be compounded by the combined effects of inadequate infrastructure funding and limited budgetary support (Okem 2019). For instance, it has recently been reported that, a group of 10 African nations including Morocco, Ethiopia, Ivory Coast, Senegal, Egypt, Ghana, Tunisia, Benin, Guinea and Rwanda, will fall $1 trillion short of the infrastructure financing required to meet the UN sustainable development goals by 2030 (Business and Financial Times 2018). This underscores the infrastructural development challenges facing Africa in terms of developing roads, railways, airports, sea ports, electricity, water and physical telecommunications infrastructure. It also highlights the opportunities for strategic investors willing to take the thrust.

The energy sector also features prominently in climate change adaptation discourses in Africa. This is mainly due to its sensitivity to climate change and the ramification of the effects of energy crisis on the industrial sector. Already, energy supply challenges are cited as a hindrance for propelling sustainable development on the continent (African Union Commission 2014; Addaney 2018). For instance, it has been revealed that nearly 620 million people in Africa are afflicted with energy poverty, limiting economic and social opportunities (Sy and Copley 2017). Without access to secure and reliable sources of energy, businesses, households, and public institutions cannot operate effectively, weakening adaptive capacity and resources. According to Sy and Copley (2017), about 70% of the continent's energy supply is generated from hydroelectric power sources. Further, most African countries' power sources are currently performing at below generation designed capacity owing to lack of adequate financing for operations, high transmission losses, obsolete and poor maintenance of equipment (African Progress Panel 2015). During peak periods, there are reported power generation shortfall of between 600 and 1600 megawatts (MW) in countries including Ghana, Kenya, Nigeria, Uganda and Ethiopia (African Progress Panel 2015). The African Progress Panel (2015) reports that financing Africa's energy infrastructure needs is estimated around $8 billion in 2013. This comprised of 50% domestic financing while external funding such as private participation in infrastructure, official development finance, and Chinese investments constituted for the rest. In 2013 alone, the estimated cost of contending with Africa's energy infrastructure needs was approximately $63 billion with a financing gap of approximately $55 billion (African Progress Panel 2015). The continent's proclivity to transition to clean energy sources such as HEP, solar energy and biofuel is mainly inhibited by inadequate funding (African Progress Panel 2015). Although concepts such as green energy and energy efficient infrastructure are also being promoted, adoption of such policy tends to be limited by financial and technical constraints (Chanza 2018).

As discussed above, the impacts of climate change in Africa have already prompted adaptation responses that touch on many aspects of law and policy decision-making. Adapting to the adverse impacts of climate change will continue to raise legal and policy issues. The regional law and policy responses governing climate change adaptation in Africa must therefore be carefully devised considering the continent's unique vulnerabilities and complex realities such as poverty and poor infrastructure. In anticipation of the inevitable shift from adaptation planning and policy to adaptation action and the critical role of law and policy in this shift, the next section discusses the role of law in facilitating adaptation and building resilience in Africa.

Climate Change Adaptation in Africa: AU Law and Policy Responses

Notwithstanding the increasing attention on climate adaptation at the global level, the variety of climate impacts and regional conditions point to a major role for regional actors. Within the AU, climate change is framed as a socio-technical problem which requires practical legal and policy solutions. However, due to the uniqueness of Africa's social, legal and political aspects to this multifaceted issue, there is the need for unique African solution. Chanza (2018) argue that supranational bodies provide an important fiscal and regulative context for regional governance. The AU as a regional actor therefore faces the challenge of developing and coordinating the implementation of adaptation options to strengthen the adaptive capacity of its member states so that expected or unexpected impacts of climate change can be sustainably addressed. This section therefore addresses how the AU can leverage its climate change adaptation policy to harness opportunities offered by the Chinese BRI to boost the adaptive capacity of member states. The focus on BRI (through, FOCAC) is a deliberate effort to guide the design of multi-level governance arrangements that will contribute to developing and implementing adaptation options to strengthen the adaptive capacity of AU member states.

Africa's Climate Change Adaptation Policy: Priority Sectors and Implementation Challenges

The United Nations has adopted conventions and treaties to govern climate change adaptation actions at the international level as expressly set out in the United Nations Framework Convention on Climate Change (UNFCCC 1992). As a multilateral forum, the UNFCCC is important in pushing regional and national adaptation agendas forward. The UNFCCC (1992) provides that 'the developed country parties included in Annex II shall assist the developing country parties that are particularly vulnerable to the adverse effects of climate change in meeting costs of adaptation to those adverse effects'. Furthermore, article 4(3) read with articles 4(1)(e) and 4(4) obliges that 'developed states shall provide new and additional financial resources to meet the agreed full costs of adaptation by developing states, especially African states, and other states that are particularly vulnerable to climate change'(UNFCCC 1992). The recently adopted Paris Agreement (2015) also appeals to Parties to strengthen regional cooperation on adaptation where appropriate and, where necessary, establishes regional centers and networks, in particular in developing countries. It further establishes 'the global goal on adaptation of enhancing adaptive capacity, strengthening resilience and reducing vulnerability to climate change, with a view to contributing to sustainable development and ensuring an adequate adaptation response …'. Parties are also encouraged to strengthen cooperation on enhancing action on adaptation, taking into account the Cancun Adaptation Framework. The Cancun Adaptation Framework (2010) agrees that '… enhanced action and international cooperation on adaptation is urgently required to enable and support the implementation of adaptation actions aimed at reducing vulnerability and building resilience in developing country Parties, taking into account the urgent and immediate needs of those developing countries that are particularly vulnerable'. These provisions particularly emphasize the importance of international cooperation and multi-level governance in the successful implementation of adaptation policy (De las Heras 2015). Scholtz (2010) however notes that industrialized states have in general proved to be hesitant in honoring their commitments under the international climate change regime. It is therefore essential for African states to utilize other multilateral and bilateral fora to secure the much needed investment to fund adaptation priorities in order to strengthen their adaptive capacities.

In Africa, the policy and scientific interest on adaptation has gained prominence, fundamentally due to the continent's unique vulnerability to the adverse effects of climate change (Addaney and Cobbinah 2019). Keenan (2012) posits that adapting to climate change is a continually evolving legal and policy process, requiring new approaches and tools. McDonald (2011) opines that 'each climate change impact requires appraisal of adaptation options involving consideration of legal, political, economic and technological factors, and a re-think of policy and institutional arrangements across a wide range of sectors. Underscoring the importance of a supranational policy framework in governing adaptation, Keskitalo (2010) argues that 'the successful implementation of adaptation policies in any region or country demands coherent coordination between the different levels of political responsibility and integration of appropriate policy measures in the sectors affected'. An effective policy regime can therefore provide stability in times of rapid change and uncertainty, a principled base from which to produce the needed restructuring, and a mechanism by which to design and implement adaptation actions (McDonald 2011). Conversely, the advantages of developing an AU-wide climate change policy were not clearly understood in the early 2000s (Scholtz 2010). More than 18 years later, the AU has still proven incapable of developing a coordinated policy action for its member states.

In the AU, the African Ministerial Conference on the Environment (AMCEN) has contributed significantly to the development of African response to climate change. This includes Africa's adoption of a common position on climate change and an inclusive agenda on African climate change programmes. In May 2009, the Nairobi Declaration on the African Process for Combating Climate Change was adopted to serve as a unified demonstration of the continent's determination to play pivotal role in addressing the challenge of climate change. The Declaration underscores the priorities for Africa including adaptation, capacity building, financing and technology development and transfer (AMCEN 2009). It consequently urges the international community to consider these priorities in delivering its support for the continent (AMCEN 2009). Despite these developments, the 2014 Draft African Climate Change Strategy (Draft Strategy) remains the basic policy that distils the international adaptation commitments and governs how African countries adapt to the adverse effects of climate change. As a draft policy, it still contains vital guidelines on adaptation in Africa.

In July 2009, at the African Union Summit, the Assembly requested the Commission to work with partners including Member States, Regional Economic Communities, and UN agencies to develop an African Strategy on Climate Change. The AU Draft Strategy (2014) seeks to guide the continent to achieve 'climate-smart' socio-economic development through providing the AU, Member States and other actors with a reliable source of strategic guidance to enable them to effectively deal with the challenges associated with climate change. Africa's priorities in adaptation are therefore meant to assuage poverty and to integrate climate change adaptation measures into regional and national law and sustainable development policies, programmes, plans and strategies. Another priority is building socio-economic resilience through the diversification of African economies to reduce dependence on climate-sensitive sectors. In realizing these, for example, Goal 17 focuses on agriculture development and seeks to enhance capacity for developing and implementing effective water harvesting, irrigation, and sustainable water management for agriculture and other uses. It also underscores the links between water, energy and food security. In Goal 9, the Draft Strategy encourages the AU to provide leadership for effective climate change policy, institutional, management and legal frameworks at continental, regional and national levels. To realize this, Action two provides for the AU to coordinate provision for synergies of bilateral and multilateral agreements related to climate change. It also acknowledges that nearly all aspects of the strategy can only be realized through partnerships and close collaboration between various institutions and stakeholders at the international, regional and national levels. Despite the significance of the Adaptation Strategy as a more comprehensive and more ambitious policy framework, the most important challenge facing the AU is undoubtedly funding (Jegede 2016a, b; Hallegate et al. 2016; Lawlor et al. 2015). Scholtz (2010) argues that the meager capacity of the AU and member states to respond to the adverse effects of climate change makes it indispensable for them to access external assistance for adaptation.

The realization of global and national climate change goals depends greatly on the availability of funds (Diko 2019). Pickering et al. (2017) note that climate finance enables national and sub-national agencies to effectively mainstream and implement climate actions. The United Nations Environmental Program (UNEP) (2012) asserts that climate finance takes up most of the international financing agreements on environment. Presently; there are general funds like the Global Climate Finance and Adaptation Funds (Persson and Remling 2014). The adaptation funds seek to assist developing countries to effectively deal with the impact of climate change. These funds take the form of both bilateral and multilateral agreements (UNEP 2012). There has been a dedicated financing arrangement for resilience such as the US$1.3 billion Pilot Program for Climate Resilience (PPCR). For African countries, financing adaptation is a challenge as it poses another layer to the already woefully inadequate development finance. Conversely, accessing these funds has been a challenge for most African countries (Diko 2019). This situation limits their capacity to respond to the adverse effects of climate change as Bak et al. (2017) argue that the success of climate action depends on the quantity and type of climate finance. Another challenging aspect is ensuring that the limited available climate funds can effectively match up with the scale and magnitude of the impacts of climate change (Miller 2008). Although developed countries have committed to increase climate finance to some US$100 billion per year by 2020, the fulfilment of this pledge has been discouraging. Another blow is the pulling out of the Paris Agreement by the US. Although, it is yet to take effect, it is undeniably going to affect the flow of climate financing in Africa as the US is one of the largest contributor of official development assistance to Africa.

The BRI as a Solution to the Implementation Gaps in AU's Adaptation Policy in Africa

The Chinese government announced the BRI as a development scheme to advance cooperation and connectivity between China and the rest of the world (Jinchen 2016). Kuo and Kommenda (2018) compare the multibillion dollar initiative to the Marshall Plan. Through this initiative, China plans to play an increasing role in international affairs (Chaturvedy 2017). Since its announcement in 2013, the BRI has become a general catchword to describe almost all aspects of Chinese engagement abroad. Connecting countries in Asia, Europe and Africa, the BRI covers 71 nations, accounting for 50% of the global population and about 35% of global GDP (Chaturvedy 2017) (Fig. 1).

Most of these countries are emerging and developing ones (Lu 2018). The Chinese government argues that the main objective of the initiative is to support the participating countries to improve their transportation, energy infrastructure, logistics and trade (McBride 2015). The Chinese President, Xi Jinping, emphasized during the 10th BRICS Summit held in South Africa in July 2018 that the BRI would 'create new opportunities for social and economic development for participating countries' (Jinping 2018). African countries including Djibouti, Egypt, Kenya, Madagascar, South Africa and Mozambique have officially pledged support (Jinping 2018). The

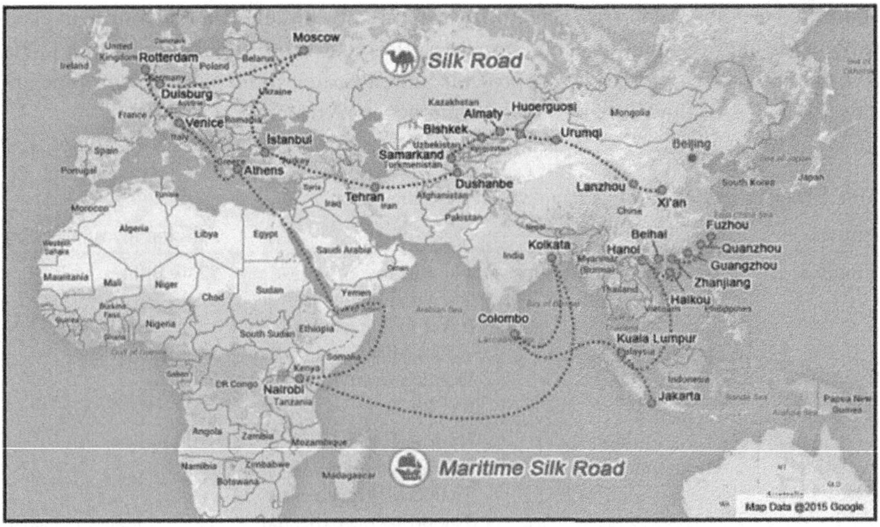

Fig. 1 The Map of the belt and road initiative. *Source* Xinhua News, http://en.xinfinance.com/html/OBAOR

HSBC (2018) estimates the BRI to cost more than US$1tn (£760bn). The Asian Infrastructure Investment Bank (AIIB) headquartered in Beijing and its auxiliary Silk Road Fund is dedicated to lending for projects that are associated with the BRI (Kuo and Kommenda 2018). In 2014, the Silk Road Fund was incorporated as a company in Beijing with an initial fund of US$10 billion involving a prospective total funding of US$40 billion (Chen 2016). Although there are conflicting estimates of how much money has been invested; it has been argued that China has invested more than $210bn with majority of the investment in Asia (Chen 2016). However, the Chinese authorities have been cautioned to ensure that the plan produces mutual benefits for the participating countries (Biswas and Tortajada 2018). On the face of it, the BRI seems ready to revolutionize a number of economic, social and strategic landscapes that African countries can tap into to strengthen their ailing infrastructure, industrial and energy sectors in order to boost their resilience to climate change. As discussed in the previous sections, both the AU climate change policy and literature review identify agricultural modernization and industrialization, energy generation and infrastructure development as the priority areas for building Africa's adaptive capacity and resilience. With the most significant challenge facing the AU being adaptation funding, the BRI is well-timed as it offers the much needed financial capital for the implementation of priority adaptation actions.

Although the African leg of the BRI is still work in progress, the Chinese government asserts that it will hold continuing discussions with the participating African countries to make consensus decisions based on the economic, social and political feasibility of individual projects (Biswas and Tortajada 2018). It is imperative to highlight that the BRI is not starting from the scratch in Africa. China is already committed to numerous large-scale infrastructure investments on the continent especially in the energy generation and transportation sectors as indicated in Table 1. The table catalogs a selection of larger projects and indicates some of the geographic and sectoral depths of the Chinese's investment across African.

Table 1 Selected African infrastructure built by China

Nations	Projects	Description	Amount invested (US$)
Ethiopia	Addis Ababa Port of Doraleh in Djibouti	The railway crosses 752 km and cuts the travel time between Addis Ababa and Port of Doraleh from three days to 12 h	$2.49 billion
	Grand Ethiopian Renaissance Dam	Hydroelectric power of an average output of 39,000 MW per year. The project is expected to be completed by 2025	$100 billion
Chad	Chad-Sudan Railway	A 1344-km being constructed in three phases and will also connect the two countries with Cameroon	$5.6 billion
DR Congo	Infrastructure for Mines (Barter deal)	The deal was to develop the mine fields in Mashamba and Dima basins and Kolwezi	$7.16 billion
Kenya	Standard Gauge Railway	A 609-k railway linking Mombasa's port to the capital Nairobi (completed in 2017)	$3.8 billion
Mozambique	Mphanda Nkuwa Dam and Hydroelectric Station Project	Generate 1500 megawatts of power to national electricity grid and includes construction of Moamba major Dam to supply drinking water to residents of Maputo	$3.1 billion
Nigeria	Coastal Railway	1402-km railway to connect Lagos with Calabar (to east), passing through 10 states and connecting cities with oil-rich Niger Delta states	$12 billion

(continued)

Table 1 (continued)

Nations	Projects	Description	Amount invested (US$)
	Dangote Cement PLC Expansion	Dangote Cement PLC expansion in Nigeria, and into Cameroon, Ethiopia, Ivory Coast, Kenya, Mali, Senegal and Zambia. A boost to cement production of 25 million tones and taking production to over 70 mn tone per year	$4.3 billion
South Africa	Modderfontein New City	A housing and entertainment precinct being built in outer Johannesburg, South Africa's largest city	$7 billion
Sudan	Port Sudan Khartoum Railway	The project was completed in 2012 and connects Port Sudan to the nation's capital, Khartoum in a 762-km of railway	$1.38 billion
Tanzania	Bagamoyo Port	20 million (annual) container ports, which would be the largest in East Africa	$7 billion

Source Mail and Guardian (2015)

As argued by Bak and others, the success of climate action is reliant on the quantity and nature of climate finance, it can therefore be argued that the Chinese investment in Africa directly and indirectly play critical role in the realization of the AU's climate change adaptation policy. As indicated in the table, the Chinese government is already providing considerable investment support in enhancing connectivity and infrastructural development across the African continent particularly in the countries involved in the initiative. For example, the Export-Import Bank of China funded the 759-km long Addis Ababa-Djibouti Railway, the first transboundary and longest electrified railway line in Africa (Mail and Guardian 2015). The railway project commissioned in 2016 links landlocked Ethiopia to the maritime trade routes of the Gulf of Aden and the Red Sea (Xinhua 2018). China is also credited for the construction of the Madaraka Express connecting the Kenyan capital city of Nairobi to the port of Mombasa, a distance of 489 km (Lu 2018). This railway is reportedly being extended to Naivasha in Kenya's northwest with plans of extending it even further to interconnect Kenya with Burundi, Ethiopia, Rwanda, South Sudan, Tanzania and Uganda (Lu 2018). This project has already reduced the transportation cost and time between the two most important cities in Kenya (Biswas and Tortajada 2018); consequently, improving trade and boosting economic growth.

Furthermore, the importance of the Forum on China-Africa Cooperation (FOCAC) held in Johannesburg in South Africa from 4th to 5th December 2015 for the first time originates from the context and outcome (Viswanathan 2015). The two-day Summit culminated into the Johannesburg Declaration and Action Plan with a massive US$60 billion pledge from the Chinese government (FOCAC 2015). It has been reported that Chinese companies signed side deals worth US$930 million with 50 South African companies in sectors such as energy, fruits, medicine, textiles and wines (FOCAC 2015). The Chinese Ministry of Commerce clarified that $35 billion of the $60 billion official assistance was to be dispersed through preferential loans and export credit lines, $15 billion to serve as capital for China-Africa Development Fund and another $5 billion as loans to develop African Small and Medium scale enterprises (FOCAC 2015). The entire package included ten cooperative programmes and projects in critical sectors such as energy and infrastructure development, agricultural modernization and industrialization, financial services and investment facilitation (FOCAC 2015). In addition, the Chinese government also pledged to import commodities amounting to 10 trillion from Africa alone in the next five years (Ministry of Commerce of China 2017). Although this seemed to be over ambitious, just a fraction of this amount could be real boosts for the struggling African economies to enable them industrialize, modernize the agricultural sector and to invest in critical infrastructure to strengthen their adaptive capacities as revealed in the adaptation policy. Meanwhile, it has been reported that Chinese investment in Africa especially in the energy and transportation sectors increased after the 2015 FOCAC Summit in South Africa where the Chinese government pledged $60 billion in investment to the continent (Sow 2018) (Fig. 2).

With about 600 million people lacking access to electricity in Sub-Saharan Africa, the continent's need for infrastructure investment is overwhelming with the energy sector alone requiring some US$450 billion through 2030 (Mungai 2015). Mungai (2015) estimates that Africa needs to invest US$38 billion annually on infrastructure to maintain its current level of development. With their abundant natural resources and infrastructure development opportunities, African countries can harness the opportunities offered by the BRI to address their infrastructure gap (Wekesa 2015). This is a necessary step for promoting agricultural modernization and industrialization in Africa in order to strengthen the continent's adaptive capacity and resilience.

In the area of agricultural development and modernization, during the 2015 Johannesburg Summit of the FOCAC, the FAO was identified as key partner for the cooperation in agriculture and food security (FOCAC 2015). The $80 million FAO-China South-South Cooperation (SSC) Programme was mentioned in the Action Plan as an entry point for the implementation of the agricultural cooperation between China and Africa (FOCAC 2015). Through this multilateral and bilateral cooperation, China pledged to share its development solutions, knowledge, experiences and agricultural technologies with African countries. Since then, the FAO-China SSC Programme has provided support for agricultural and rural development in 28 African countries by fielding over 1000 Chinese experts (FAO in China 2017). In 2017, the Chinese

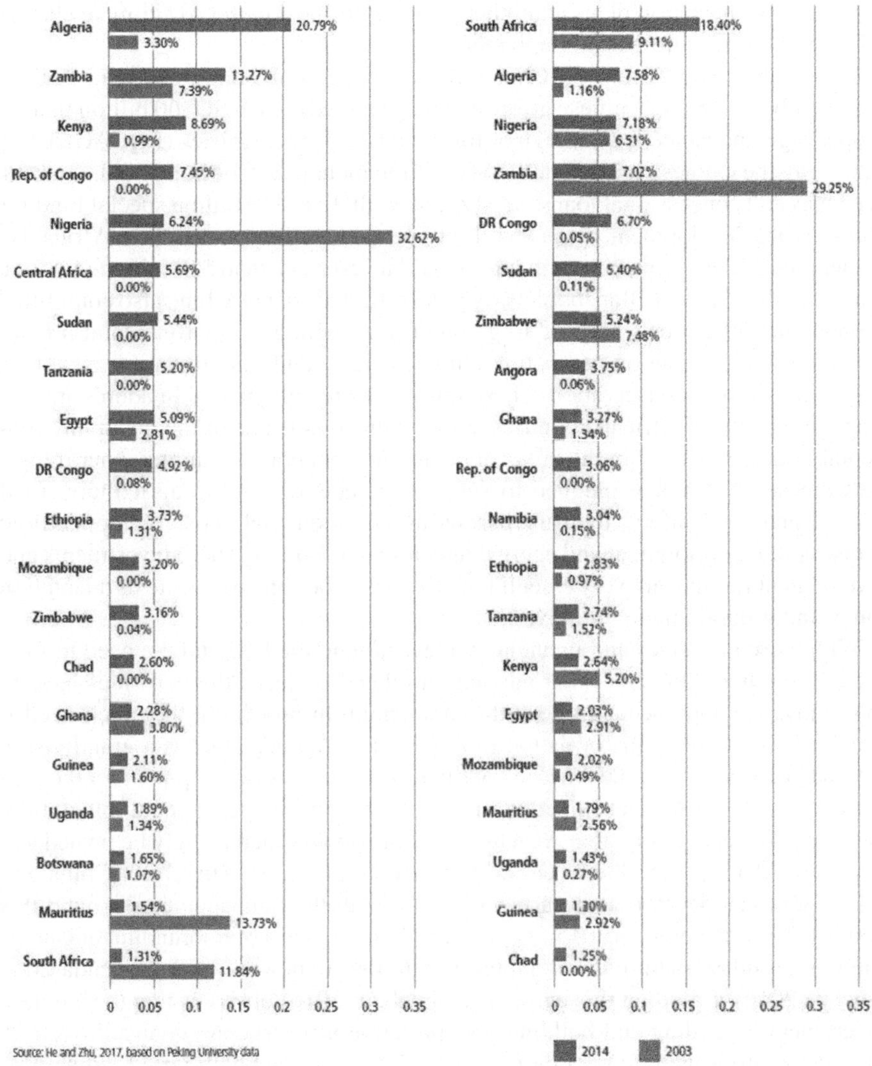

Fig. 2 Top African destinations of Chinese FDI flows (**a**) and stocks (**b**). *Source* UN-Habitat and IHS-Erasmus University Rotterdam (2018). The State of African Cities 2018: The Geography of African Investment

government convened the 4th Conference on China-Africa Agricultural Cooperation to implement the consensus during the Johannesburg Summit of the FOCAC (FAO in China 2017). Nine agreements on agricultural cooperation were signed between research institutes and agro-enterprises from China and Africa during the Conference. All these programmatic initiatives associated with the BRI are critical in filling the funding gap in the AU climate change policy. Without these initiatives from the

Chinese BRI, the action plans and other projects outlined in the AU climate change policy will be very difficult to implement.

During the latest Forum on China-Africa Cooperation summit held in Beijing in September 2018, the Chinese government again pledged a USD$60 billion financing package and cancelled some debt for poorer African countries (FOCAC 2018). According the Chinese president, the $60 billion include $15 billion of aid, interest-free loans and concessional loans, $20 billion credit line, $10 billion special fund for China-Africa development, and a $5 billion special fund for imports from Africa. He further pledged that Chinese companies would invest more than $10 billion in Africa in the next three years (Blanchard-Wrigglesworth et al. 2018). China also committed to undertake 50 projects relating to green development and environmental protection in Africa focusing on combating climate change and desertification as well as promoting wildlife conservation. These initiatives are very pivotal in addressing the adaptation needs of African countries as desertification is one of the major environmental issues in Africa. In the areas of economic cooperation, among other things, the Chinese government pledged to support projects relating to agriculture, food security and food safety; industry partnering and industrial capacity cooperation; infrastructure development and energy and natural resources. These investments and development funding are very essential if Africa is to be able to execute its adaptation policy and programmatic initiatives.

With these initiatives and financial pledges around the BRI and the need to fund Africa's Agenda 2063 and other adaptation related projects, the two sides have to find innovative approaches to keep the momentum in producing the much needed social and economic transformation in Africa. The Agenda 2063 is a grand vision adopted in 2013 to guide the African continent's sustainable development for the next 50 years (AUC 2014). Developed around pan-Africanism and regional integration, the agenda seeks to ensure that African development will increasingly be owned and driven by Africans (DeGhetto et al. 2016; Addaney 2018). Therefore, China and other traditional development partners have to plan their engagements around this parameter. For example, in 2015, the AU and China signed a Memorandum of Understanding on infrastructure development within the framework of the Agenda 2063 (Jinchen 2016). Based on this agreement, Jinchen (2016) observes that the Chinese government is funding and building transportation infrastructure connecting all 55 African countries including modern airports, highways, and high-speed railways.

The pivotal role of industrialization in building a resilient Africa has been underscored in the AU's climate change policy due to its potential to promote inclusive prosperity, create new jobs and better incomes across the African continent (Tafirenyika 2016). Lin and Goldstein (2017) observe that since the first Industrial Revolution, labor-intensive light manufacturing such as textiles, garments, shoes, and associated tools and machinery has played a key role in facilitating economic growth. On the other hand, Africa is currently less industrialized than it was 40 years ago and this has weakened it socio-economic reliance and adaptive capacity (Lin and Goldstein 2017). For instance, the contribution of Africa's manufacturing sector to its gross domestic product declined from 12% in 1980 to 11% in 2013 and has since remain stagnant (Economic Commission of Africa 2016). Lin and Goldstein

(2017) observe that, in 2015, the apparel exported by all of Sub-Saharan Africa was less than that of El Salvador. Africa therefore urgently needs an industrial revolution to generate jobs for its fast-growing youthful population in order to build socio-economic resilience. In a 2016 report published by the Economic Commission of Africa (2016), 'Economic Report on Africa 2016: Greening Africa's Industrialization', it unanimously concluded that the only practicable option for Africa's sustainable development is industrialization. The continent however has to define and design its own pathway based on its realities and learning from its history. It has to also learn from the experiences of other regions particularly Asia in order to overtake traditional, carbon-intensive methods of economic growth and champion a low-carbon development path as outlined in the AU climate change policy (World Resource Institute 2016). Sun (2015) argues that addressing these challenges requires the assistance of external actors through effective partnerships—a reality that China, understands better than most developed countries. The BRI can thus open new opportunities for African countries as China has been a major source of private investment in Africa (Jinping 2018).

Despite the development opportunities presented by the BRI to enable the AU and its member states to implement their climate change policy, the debate on its impact is usually framed around the winners and losers (Lin and Ngai 2017). Conversely, Wheeler (2018) argues that the BRI should be seen as a 'win-win initiative instead of as a threat to replace the US's hegemony in global politics or a form of voracious economics through which China exports its infrastructure development model that heavily relies on state-owned enterprises'. It should also be noted that the BRI incorporates key objectives that ensure that infrastructure projects are socio-economically sustainable and deliver prosperity across the participating countries through integrated development plans. This was clearly articulated during the 2017 Beijing BRI summit (AFP/EC 2017). At BRI Summit, President Xi Jinping underscored that the Confucian principles of 'win-win thinking' is underpinning the initiative. He further noted that 'China is willing to share its development experience with the rest of the world, but will not intervene in other nations' internal affairs, export its social system and development model, nor force others to accept them' (AFP/EC 2017). Critics may also point to the fundamental economic and power imbalance between China and African countries to brand the cooperation as neo-colonial. Wu and Alden (2018) however argue that the AU and African governments exercise more agency than they are given credit for. This includes frequently playing China and traditional Western development partners off against each another. The AU has also played a key role in ensuring that Africa takes a common position at international negotiations and evolving multilateral environmental agreements (Scholtz 2010). Indeed, since the dawn of the 21st Century, Africa has enhanced its organizational effectiveness at various international fora acting in unison during international conferences; building helpful strategic economic partnerships with the powerful emerging states of China, India, Russia and Brazil etc. For example, a substantive show of African solidarity on environmental affairs was confirmed at the 2012 Earth Summit 2012 in Rio de Janeiro. This demonstration of togetherness would need to be brought into play as the AU and African countries ready themselves to tap the opportunities offered by the BRI to boost their adaptive capacities against the climate crisis.

Conclusion

This chapter explored how the AU can leverage the opportunities offered by the Chinese BRI to effectively implement its climate change adaptation policy. It revealed that agricultural modernization, infrastructure development and energy generation are the priority adaptation sectors. However, realizing regional and national climate change goals depends greatly on the availability of funds which are woefully insufficient in Africa. Although developed countries through the UNFCCC regime have committed to increase climate finance to some US$100 billion per year by 2020, fulfilling this pledge has been disappointing. The prevailing situation limits the capacity of African countries to respond to the adverse effects of climate change despite having a policy in place. The chapter therefore argues that addressing the adaptation gaps require multi-level legal and policy action. More broadly, the chapter established that the AU's legal and policy responses to adaptation require good governance, robust institutional capability and financial capacity at the regional and national levels.

The chapter further argues that the BRI does indeed offer many prospects for the AU and African countries to modernize agriculture, develop their infrastructure and energy resources as well as to build their industrial capacity to transform their economies. It further underscored that adaptation to climate change in the face of dwindling international support requires forging and renewing strategic partnerships to invest responsibly and sustainably in priority sectors to deliver practical adaptation results. To realize this, the AU and African governments need to ensure policy coherence as well as create conducive legal environment to attract BRI investments in those priority adaptation sectors. This will ensure that AU takes leadership in collaborating with development partners, RECs and Member States in all its climate change adaptation priority areas including agriculture and industrialization, energy and infrastructure development, multi- and bilateral agreements, and finance etc. The chapter emphasizes that although there are benefits to be shared among African countries participating in the BRI, the inherent risks such the relocation of pollution-oriented industries to Africa and unfair infrastructure loan deals should be avoided so that they do not hinder the progress of Africa's infrastructure development and agricultural modernization.

References

Addaney M (2018) The African union's agenda 2063: education and its realization. In: Onuora-Oguno A, Egbewole W, TE Kleven TE (eds) Education law, strategic policy and sustainable development in Africa. Springer, Cham

Addaney M, Cobbinah PB (2019) Climate change, urban planning and sustainable development in Africa: the difference worth appreciating. In: Cobbina PB, Addaney M (eds) Geography of climate change adaptation in urban Africa. Springer, London

Addaney M, Boshoff E, Oyetola B (2017) The climate change and human rights nexus in Africa. Amsterdam Law Forum 9:5–28

Adger WN, Pavoola J, Huq S, Mace MJ (2006) Toward justice in adaptation to climate change. In: Fairness in adaptation to climate change. MIT Press, Cambridge, MA

AFP/EC (2017) China hosts summit on new Silk Road. Channel News Asia, 14 May. https://www.channelnewsasia.com/news/asia/china-hosts-summit-on-new-silk-road-8844922

Africa Progress Panel (2015) Africa progress report 2015: power, people, planet. Africa Progress Panel. http://app-cdn.acwupload.co.uk/wpcontent/uploads/2015/06/APP_REPORT_2015_FINAL_low1.pdf

African Ministerial Conference on the Environment (2009) 15th ordinary session of the Executive Council 24–30 June, Sirte, Libya EX CL/Dec 502(XV) Decision on the Report of the African Ministerial Conference on the Environment (AMCEN) (2009) Special Session on Climate Change Doc EX CL/519(XV)

African Ministerial Conference on the Environment (2014) Draft African union strategy on climate change. AMCEN-15-REF-11. http://www.un.org/en/africa/osaa/pdf/au/cap_draft_auclimatestrategy_2015.pdf

African Ministers Council on Water (2012) Water security and climate resilient development

African Union Commission (2014) Agenda 2063: the Africa we want, draft document 6. http://www.nepad.org/system/files/Agenda%202063%20%20English.pdf

Amjath-Babu TS, Krupnik TJ, Kaechele H, Aravindakshan S, Sietz D (2016) Transitioning to groundwater irrigated intensified agriculture in Sub-Saharan Africa: an indicator based assessment. Agric Water Manag 168:125–135

Amoako C (2016) Brutal presence or convenient absence: the role of the state in the politics of flooding in informal Accra, Ghana. Geoforum 77:5–16

Amoako C, Boamah EF (2014) The three-dimensional causes of flooding in Accra, Ghana. Int J Urban Sustain Dev 7:109–129

Antrobus D (2011) Smart green cities: from modernization to resilience? Urban Res Pract 4:207–214

Bak C, Bhattacharya A, Edenhofer O, Knopf B (2017) Towards a Comprehensive approach to climate policy, sustainable infrastructure and finance. economics: the open-access. Open-Assess E-J 11:1–14

Biswas AK, Tortajada C (2018) How new silk road will cement China as major trading partner for Africa. The Conversation, 20 Aug. https://theconversation.com/how-new-silk-road-will-cement-china-as-major-trading-partner-for-africa-100533

Blanchard-Wrigglesworth E, Webster MA, Farrell SL, Bitz CM (2018) Reconstruction of Snow on Arctic Sea Ice. J Geophys Res: Oceans 123(5): 3588–3602

Blue G (2016) Framing climate change for public deliberation: what role for interpretive social sciences and humanities? J Environ Planning Policy Manage 18:67–84

Business and Financial Times—Ghana (2018) 10 African nations face US$1 trillion infrastructure funding gap. 5 July. https://thebftonline.com/2018/headlines/10-african-nations-face-us1trillion-infrastructure-funding-gap/

Cancun Agreements (2010) Report of the conference of the parties on its sixteenth session, held in Cancun from 29 Nov to 10 Dec, Decision 1/CP.16, FCCC/CP/2010/7/Add.1, FCCC/CP/2010/7/Add.1

Cannon T, Müller-Mahn D (2010) Vulnerability, resilience and development discourses in the context of climate change. Nat Hazards 55:621–635

Chanza N (2018) Limits to climate change adaptation in Zimbabwe: insights, experiences and lessons. In: Filho WL, Nalau J (eds) Limits to climate change adaptation. Springer, Cham

Chaturvedy RR (2017) The 21st century maritime Silk Road. Observer research foundation, 10 Feb. http://www.orfonline.org/research/the-21st-century-maritime-silk-road/

Chen H (2016) China's 'one belt, one road' initiative and its implications for Sino-African investment relations. Transnational Corporations Rev 8:178–182
Chigogo S (2016) 12,000 Boreholes run dry, Minister. Herald (Harare), 01 Feb
Conference of the Parties of United Nations Framework Convention on Climate Change (2007) Report of the conference of the parties on its thirteenth session, held in Bali from 3 to 15 Dec 2007 Addendum—part 2: action taken by the conference of the parties at its Thirteenth Session. UN Doc FCCC/CP/2007/6/Add.1
De las Heras BP (2015) Adaptation to climate change in the European union: between supranational action and state reluctance. IUCNAEL E-J 6:1–21
DeGhetto K, Gray JR, Kiggundu MN (2016) The African union's agenda 2063: aspirations, challenges, and opportunities for management research. Afr J Manag 2:93–116
Diko SK (2019) Missed opportunities? Financing climate change adaptation in urban Ghana and Uganda. In: Cobbinah PB, Addaney M (eds) Geography of climate change adaptation in Urban Africa. Springer, Cham
Dube T, Phiri K (2013) Rural livelihoods under stress: the impact of climate change on livelihoods in South Western Zimbabwe. Am Int J Contemp Res 3:11–25
Economic Commission of Africa (2016) Economic report on Africa 2016: greening Africa's industrialization
FAO (2008) Climate change adaptation and mitigation in the food and agricultural sector. ftp://fao.org/docrep/fao/meeting/013/ai782e.pdf
Field CF, Barros V, Stocker T (2012) Managing the risks of extreme events and disasters to advance climate change adaptation. IPCC/Cambridge University Press, Cambridge, UK/New York
Fields B (2009) From green dots to greenways: planning in the age of climate change in post-Katrina New Orleans. J Urban Des 14:325–344
Food and Agriculture Organization in China (2017) China-Africa agricultural cooperation heading for a sustainable future' FAO in China, 11 Dec. http://www.fao.org/china/news/detail-events/en/c/1073804/
Forum on China-Africa Cooperation (2018) Forum on China-Africa cooperation Beijing action plan (2019–2021), 5 Sept. https://www.fmprc.gov.cn/mfa_eng/zxxx_662805/t1593683.shtml
Forum on China–Africa Cooperation (FOCAC) (2015) The forum on China–Africa cooperation Johannesburg action plan (2016–2018), 10th Dec 2015. https://www.fmprc.gov.cn/mfa_eng/zxxx_662805/t1323159.shtml
Ghile YB, Taner MÜ, Brown C, Grijsen JG, Talbi A (2014) Bottom-up climate risk assessment of infrastructure investment in the Niger River basin. Clim Change 122:97–110
Goyol S, Pathirage C (2017) Impacts of climate change on agrarian infrastructure and cascading effects on human and economic sustainability in Nigeria. In: Conference/workshop item. http://usir.salford.ac.uk/43561/
Hallegatte S, Bangalore M, Bonzanigo L, Kane T, Narloch U, Rozenberg J, Treguer D, Vogt-Schilb A (2016) Shock waves: managing the impacts of climate change on poverty. World Bank, Washington D.C.
HSBC (2018) A new future for the world economy, 14 June. https://www.business.hsbc.com/belt-and-road/a-new-future-for-the-world-economy
Intergovernmental Panel on Climate Change (IPCC) (2014) Climate change 2014: synthesis report. https://www.ipcc.ch/pdf/assessment-report/ar5/syr/AR5_SYR_FINAL_SPM.pdf
IPCC (2007) Climate Change 2007: Working Group II: impacts, adaptation and vulnerability, glossary. https://www.ipcc.ch/pdf/assessment-report/ar4/wg2/ar4_wg2_full_report.pdf
Jegede AO (2015) Climate change: safeguarding indigenous peoples through 'land sensitive' adaptation policy in Africa. In: Leal Filho W (ed) Handbook of climate change adaptation. Springer, Cham
Jegede AO (2016a) The climate change regulatory framework and indigenous peoples' lands in Africa: human rights implications. Pretoria University Law Press, Pretoria
Jegede AO (2016b) Rights away from home: climate-induced displacement of indigenous peoples and the extraterritorial application of the Kampala convention. Afr Hum Rights Law J 16:58–82

Jinchen TJ (2016) One belt and one road: connecting China and the world. In: McKinsey voices on infrastructure, July 2016. https://www.mckinsey.com/industries/capital-projects-and-infrastructure/our-insights/one-belt-and-one-road-connecting-china-and-the-world
Jinping X (2018) Speech at the BRICS business forum in Johannesburg, South Africa, Wednesday, 25 July. http://www.brics2018.org.za/speech-11
Kabo-Bah A, Diji C (2018) Sustainable hydropower in West Africa: planning, operation, and challenges. Academic Press
Karlsson S, Srebotnjak T, Gonzales P (2007) Understanding the North-South knowledge divide and its implications for policy: a quantitative analysis of the generation of scientific knowledge in the environmental sciences. Environ Sci Policy 10:668–684
Keenan R (2012) Adapting to climate change: more questions than answers!' The conversation, 28 June. https://theconversation.com/adapting-to-climate-change-more-questions-than-answers-7898
Kelly PM, Adger WN (2000) Theory and practice in assessing vulnerability to climate change and facilitating adaptation. Clim Change 47:325–352
Keskitalo HC (ed) (2010) Developing adaptation policy and practice in Europe: multi-level governance of climate change. Springer, Cham
Kitaba M (2017) Why African Nations welcome China. The diplomat, 16 Feb. www.thediplomat.com/2017/02/why-african-nations-welcome-china/
Klein RJT, Nicholls RJ, Thomalia F (2003) Resilience to natural hazards: how useful is this concept? Glob Environ Change Part B: Environ Hazards 5:35–45
Kuo L, Kommenda N (2018) What is China's belt and road Initiative? The guardian, 30 July. https://www.theguardian.com/cities/ng-interactive/2018/jul/30/what-china-belt-road-initiative-silk-road-explainer
Lawlor K, Handa S, Seidenfeld D (2015) Cash transfers and climate resilient development: evidence from Zambia's child grant programme. UNICEF Office of Research, Florence
Lin J (2018) Governing climate change: global cities and transnational lawmaking. Cambridge University Press, Cambridge
Lin JY, Goldstein A (2017) Africa's role in the global manufacturing hub of the future. World Econ Forum. https://www.weforum.org/agenda/2017/02/africa-global-manufacturing-hub
Lin DY, Ngai JL (2017) One belt, one road gaining traction but unanswered questions leave funding uncertain. Channel News Asia, 22 Apr. https://www.channelnewsasia.com/news/asia/commentary-one-belt-one-road-gaining-traction-but-unanswered-8743648
Liverman DM (1990) Vulnerability to global environmental change. In: Kasperson RE and others (eds) Understanding global environmental change: the contributions of risk analysis and management. Earth Transformed Program
Lu J (2018) How one belt one road benefits developing countries. The Borgen Project Blog, 11 Apr. https://borgenproject.org/one-belt-one-road-benefits/
Luers AL (2005) The surface of vulnerability: an analytical framework for examining environmental change. Glob Environ Change: Hum Policy Dimensions 15:214–223
Mail and Guardian (2015) 'What Crisis? 16 of China's biggest projects in Africa—it is all billion dollar territory in Mail and Guardian, 18 Sept. http://www.mgafrica.com/article/2015-09-18-multi-billion-dollars-deals-chinas-27-biggest-active-projects-in-africa
Matenga L (2019) Placing climate change in wetland conservation and urban agriculture contestations in harare Zimbabwe. In: Cobbinah PB, Addaney M (eds) Geography of climate change adaptation in urban Africa. Springer, Cham
Mayer B (2018) The international law on climate change. Cambridge University Press, Cambridge
McBride J (2015) Building the new silk road. Council on Foreign Relations, 22 May. https://www.cfr.org/backgrounder/building-new-silk-road
McDonald J (2011) The role of law in adapting to climate change. WIREs Clim Change 2:283–295
McNicoll L et al (2017) Estimating publicly-mobilised private finance for climate action: a South African case study. OECD Publishing, Paris

Miller AS (2008) Financing the integration of climate change mitigation into development. Clim Policy 8:152–169
Ministry of Foreign Affairs of PR China (2018) Forum on China-Africa cooperation Beijing Action Plan, 5th Sept. https://www.fmprc.gov.cn/mfa_eng/zxxx_662805/t1593683.shtml
MOFCOM (2017) 2016 business review XXVI: China–Africa trade and economic cooperation makes steady progress, 10 Feb. http://english.mofcom.gov.cn/article/newsrelease/significantnews/201702/20170202515699.shtml
Moser SC, Ekstrom JA (2010) A framework to diagnose barriers to climate change adaptation. www.pnas.org/cgi/doi/10.1073/pnas.1007887107
Mugandani R, Wuta M, Makarau A, Chipindu B (2012) Re-classification of agro-ecological Regions of Zimbabwe in conformity with climate variability and change. Afr Crop Sci J 20:361–369
Mungai C (2015) Could drones be a solution to Africa's infrastructure problems? World Economic Forum. https://www.weforum.org/agenda/2015/09/could-drones-be-a-solution-to-africas-infrastructure-problems/
Nalau J, Leal Filho W (2018) Introduction: limits to adaptation. In: Leal Filho W, Nalau J (eds) Limits to climate change adaptation. Springer, Cham
Nemry M, Demirel H (2012) Impacts of climate change on transport: a focus on road and rail transport infrastructures. European Commission Joint Research Centre
Neumann JE, Price J, Chinowsky P, Wright L, Ludwig L, Streeter R, Jones R, Smith JB, Perkins W, Jantarasami L, Martinich J (2015) Climate change risks to US infrastructure: impacts on roads, bridges, coastal development, and urban drainage. Clim Change 131:97–109
O'Neill SJ (2010) Disciplines, geography and gender in the framing of climate change. Bull Am Meteor Soc 91:997–1002
Ogwu M (2019) Towards sustainable development in Africa: the challenge of urbanization and climate change adaptation. In: Cobbinah PB, Addaney M (eds) Geography of climate change adaptation in urban Africa. Springer, Cham
Okem EA (2019) Connecting climate change to poverty reduction: poverty reduction co-benefits of climate change-related projects in eThekwini municipality, South Africa. In: Cobbinah PB, Addaney M (eds) Geography of climate change adaptation in urban Africa. Springer, Cham
Olmstead SM (2014) Climate change adaptation and water resource management: a review of the literature. Energy Econ 46:500–509
Otto-Zimmermann K (2013) Building the global adaptation community. In: Otto-Zimmermann K (ed) Resilient cities: cities and adaptation to climate change proceedings of the global forum 2010. Springer, Cham
Paavola J, Adger WN (2006) Fair adaptation to climate change. Ecol Econ 56:594–608
Panteli M, Mancarella P (2015) Influence of extreme weather and climate change on the resilience of power systems: impacts and possible mitigation strategies. Electr Power Syst Res 127:259–270
Paris Agreement (adopted 12 December 2015, entered into force 4 November 2016) 55 ILM 740
Parry ML, Canziani OF, Palutikof JP, van der Linder P, Hnason CE (eds) (2007) Climate change 2007: impacts, adaptation and vulnerability. Contribution of working group II to the fourth assessment report of the intergovernmental panel on climate Change. IPCC/Cambridge University Press, Cambridge
Patt A, Gwata C (2002) Effective seasonal climate forecast applications: examining constraints for subsistence farmers in Zimbabwe. Glob Environ Change 12:185–195
Persson A, Remling E (2014) Equity and efficiency in adaptation finance: initial experiences of the adaptation fund. Clim Policy 14:488–506
Pickering J, Betzold C, Skovgaard J (2017) Managing fragmentation and complexity in the emerging system of international climate finance. Int Environ Agreements Polit Law Econ 17:1–16
Scholtz W (2010) The promotion of regional environmental security and Africa's common position on climate change. Afr Hum Rights Law J 1:1–25
Smit B, Wandel J (2006) Adaptation, adaptive capacity and vulnerability. Glob Environ Change 16:282–292

Sow M (2018) Figures of the week: Chinese investment in Africa. Brookings Africa in focus, 6 Sept. https://www.brookings.edu/blog/africa-in-focus/2018/09/06/figures-of-the-week-chinese-investment-in-africa/

Sun Y (2015) Inserting Africa into China's One belt, one road strategy: a new opportunity for jobs and infrastructure. Brookings Institute, 2 Mar. www.brooking.edu/blog/africa-in-focus/posts/2015/03/02-africa-chinajobs-infrastructure-sun

Sy A, Copley A (2017) Closing the financing gap for African energy infrastructure: trends, challenges, and opportunities. Africa growth initiative at Brookings Policy Brief. https://www.brookings.edu/wp-content/uploads/2017/04/global_20170417_africa-energy-infrastructure.pdf

Tafirenyika M (2016) Why has Africa failed to industrialize? Experts call for bold and creative policies. Africa Renewal, Aug–Nov 2016. https://www.un.org/africarenewal/magazine/august-2016/why-has-africa-failed-industrialize

Turner MG (2010) Disturbance and landscape dynamics in a changing world. Ecology 91:2833–2849

Tyler S, Moench M (2012) A framework for urban climate resilience. Clim Dev 4:311–326

UNEP (2012) Global environment outlook 5 (GEO 5): environment for the future we want. UNEP

UN-Habitat and IHS-Erasmus University Rotterdam (2018) The State of African cities 2018: the geography of African investment. In: Wall RS, Maseland J, Rochell K, Spaliviero M (eds) United nations human settlements programme (UN-Habitat)

United Nations Framework Convention on Climate Change (2007) 1771 UNTS 107

United Nations Framework Convention on Climate Change (UNFCCC) (adopted 4 June 1992, entered into force 21 March 1994) 1771 UNTS 107

Van Vliet MT et al (2016) Power-generation system vulnerability and adaptation to changes in climate and water resources. Nat Clim Change 6:375–380

Viswanathan HHS (2015) FOCAC 2015: consolidating China-Africa relations. Observer research foundation, 16 Dec. https://www.orfonline.org/research/focac-2015-consolidating-china-africa-relations/

Wekesa B (2015) Forum: China's Silk Road economic belt: African perspectives and implications. Afr East-Asian Aff 1–2:144–161

Wheeler A (2018) Belt road initiative a vision of a new global economic order, win-win collaboration. Channel News Asia, 28 June 2018. https://www.channelnewsasia.com/news/commentary/belt-road-initiative-china-sri-lanka-port-10474816

World Bank (2003) Turn down the Heat: Why a 4 °C warmer world must be avoided, climate extremes, regional impacts, and the case for resilience. The World Bank, Washington D.C.

World Resource Institute (2016) COP 22—greening industrialization in Africa. 15 Nov. https://www.wri.org/events/2016/11/cop-22-greening-industrialization-africa

Wu YS, Alden C (2018) Ties between African countries and China are complex: understanding this matters. The Conversation, 15 Oct. https://theconversation.com/ties-between-african-countries-and-china-are-complex-understanding-this-matters-104700

Xinhua (2015) China unveils action plan on Belt and road initiative. China Daily, 3 Mar. www.chinadaily.com.cn/business/2015-03/28/content_19938124.htm

Xinhua (2018) Chinese-built Ethiopia-Djibouti railway begins commercial operations, 1 Jan. http://www.xinhuanet.com/english/2018-01/01/c_136865306.htm

Legal Recognition of Women's Role in Combating Desertification in Africa: The Case for Uganda

Charlotte Kabaseke

Abstract Environmental degradation is fast becoming a world wide problem as a result of desertification, especially in the face of climate change. The African continent is specifically susceptible to desertification because of its low adaptive capacity to climate change. Uganda is no exception. Women have been noted to be more affected in the face of desertification because of their socially constructed roles which require them to rely on natural resources. Because of their experience dealing with natural resources, women have acquired traditional knowledge on how to preserve these resources in the face of environmental changes. The United Nations Convention to Combat Desertification (UNCCD) recognises women's role in combating desertification. Against this background, this paper seeks to examine the effect of desertification on women and the role women have played in combating the same. The paper also analyses the extent to which the law at international, regional levels recognises women's role in combating desertification and to what extent these laws have been domesticated and implemented in Uganda.

Introduction

Women have been identified to be a group very vulnerable to land degradation because of their heavy reliance of natural resources like land, forests and water bodies. This has been attributed to their socially constructed role of caretaking their households (Lau et al. 2013) which requires them to provide food, water and firewood among other necessities (Abebe 2014). Specifically, women's role in agriculture as the world's leading food producers has been under scored (Mor 2018; Skinner 2011). Women in Sub-saharan Africa carry out 70–80% of subsistence farming. When resources are depleted therefore, women have to move longer distances further away from their homes, in search of water, food and firewood (Aditya 2016). This not only eats into the time they would have spend on other valuable projects

C. Kabaseke (✉)
International Environmental Law, School of Law, Research Institute of Environmental Law, Wuhan University, 299 Bayi Road, Wuhan 430072, People's Republic of China
e-mail: charlottekabaseke@gmail.com

W. Leal Filho et al. (eds.), *Climate Change, Hazards and Adaptation Options*, Climate Change Management, https://doi.org/10.1007/978-3-030-37425-9_26

like education, but it also exposes them to other dangers like sexual harassment. These factors among others have been noted to make women more zealous towards conservation of natural resources (Aditya 2016). Through their continued reliance on natural resources, women have acquired knowledge and experience taking care of these resources especially in the face of climate change effects like drought and desertification (Mor 2018). Despite their experience, contribution and knowledge, however, women have continued to be excluded from natural resource conservation and management as well as decision making processes that concerns natural resources (Mor 2018). This has been attributed to discriminatory cultural practices, which still view men as sole decision makers and also deny women legal rights over property, specifically land (BRAC and Landesa 2017). As a result, women are unable to access credit which would enable them to invest in some anti-desertification projects like soil conservation, drought resistant seeds, afforestation, as well as modern agricultural technology (BRAC and Landesa 2017). As result, in the face extreme weather changes like drought, soil gets degraded and therefore continued production of poor crop yields, if any (BRAC and Landesa 2017). Women's wide exclusion from natural resource conservation management has contributed to the slow arrival at a lasting solution to environmental risks and disasters. This position underscores the significance of a strong legal framework for the recognition of women's role in combating environmental degradation (Broeckhoven and Cliquet 2015). Scholars have emphasised the need for this recognition at the International, regional as well as national levels. They deem it crucial for purposes of environmental conservation and hence curbing environmental degradation (Prior et al. 2013).

During the United Nations Conference on Environment and Development (UNCED) in 1992, the idea to enact the United Nations Convention to Combat Desertification in those Countries Experiencing Serious Drought and/or Desertification, Particularly in Africa (UNCCD) (1994) was birthed. The UNCCD recognises the critical role played by women in environmental conservation and calls upon states parties to involve women in taking decisions that concern combating land degradation and drought (Preamble, article 5 (d), article 10 (f)). Despite these provisions however, women in arid areas have been noted to continually be left out of decision making processes as a result of discriminatory cultural practices. The issue raised by this state of affairs and which forms the crux of this chapter is whether the provisions of the UNCCD are adequate enough in ensuring the achievement of women's contribution and involvement in combating desertification and how these provisions have been implemented in Uganda. The chapter is divided into four parts. The first part is the introduction; the second part discusses women's vulnerabilities in the face of desertification and their contribution in combating the same. This part gives value to the need for effective legal provisions at all levels to buttress the need for women's contribution in finding solutions to desertification. Part three discusses the adequacy of the legal provisions in the UNCCD in ensuring women's participation in combating desertification and the fourth part discusses the implementation of the UNCCD in Uganda. A conclusion is drawn and recommendations are made on to adequately ensure recognition of women's role in curbing degradation. Uganda is used as a case study because it is rapidly facing land degradation, with women being the most affected, yet they are not widely being included in finding solutions to desertification. Further, it would be unfeasible to conducted studies of all nations that are currently facing desertification. The conclusions drawn from this study however will be applicable to other countries in a similar situation like Uganda. In achieving the objectives of this chapter, the doctrinal method of research will be employed. Laws at international, regional and national levels as well as credible reports and other secondary literature on women and desertification were analysed.

Combating Desertification: The Role and Contribution of Women

Degradation of resources has been noted to be a world wide concern especially in the rural areas. Africa has been noted to be the worst affected by the degradation because two thirds of the continent is made of deserts or dry lands (Angelsen 1997). The world's poor have also been noted to be the worst affected especially because they live in the most vulnerable ecological areas. Women have been noted to make up 70% of the world's poor. This has been attributed to among other reasons, discriminatory practices which deny access and ownership of productive resources like land. This makes women unable to mitigate disaster (Levitt 2015).

Women are key players in the agricultural sector. The International Labour Organisation notes that, 'agriculture remains the most important source of employment for women in low-income and lower middle-income countries' (ILO 2016). As already pointed out, Women in Sub-saharan Africa carry out 70–80% of subsistence farming and 52% of agricultural land has been said to be affected by drought (Mor 2018). Through agricultural practices like soil conservation, inter cropping and crop rotation, soil fertility and growing drought resistant crops among other practices, women have been noted to contribute to mitigation and adaptation to desertification (Climate Change and Agroecological Approaches (LEISA) 2017; Kelkar 2009). A study conducted in Botswana, Western Zambia, Trinidad and Peru confirmed that women's experiences in agricultural production are similar. In Botswana for instance, women do not quickly access oxen for ploughing like the men. Women can access oxen more easily through male relatives. It was observed that the few women that attended the meeting were more knowledgeable about agricultural practices than the men and that they were more committed to attending the meetings (Jiggins 1994).

There is growing global water crisis which has been attributed to among other reasons, climate change, over population and urbanisation (UN 2005). This among other effects, has led to poor crop yields and water borne diseases because populations have to resort to contaminated water sources (UN 2005). Measures have been put in place at international level to mitigate the water crisis. Involvement of women in the development and implementation of these measures has been noted to be of critical importance (UN 2005). Lack of adequate water facilities has been noted to be a critical problem in developing countries. Mobile water is scarce and the available water sources are contaminated due to poor sanitary conditions. These water sources have been a source of diseases like cholera and typhoid. Women, by virtue of their domestic roles come face to face with these water bodies and are hence more likely to be contaminated with the diseases (Melchoir 1989). As a result, the World Health Organisation (WHO) noted women's role in water management. Women have been noted to be experts on how and where to get water in terms of a crisis. Women also know how to preserve and conserve water through traditional methods.

Africa was reported to be one of the regions that lost the largest forest cover between 2000 and 2010 (UN 2012). Women have played a significant role in forest conservation. Indigenous women have specifically been noted to be care takers of forests and minerals like oil and gas from exploitation. In the face of scarcity, women are the most affected because they are the dominant users of wood fuel (Dankelmann and Davison 1988). Further, women have been noted to spend about 9 h per day searching for firewood and Africa today continues to face wood shortage (Levitt 2015). Deforestation has also been noted to affect food security. As women spend more time looking for firewood, they spend less time on agriculture and cooking and as a result, they opt for food that takes less time to cook, which food is usually not very nutritious. In addition, firewood emits biomass fuels which not only contribute to pollution but also affect the health of women once they are inhaled (WHO 1984). Women's groups have been significant in advancing forest conservation and tree planting. A case in point is the 1977 green belt movement of Kenya, a women's social movement that aimed to plant tees across Kenya (Hyma and Nyamwange 1993). It has however been argued that involving women in tree planting may not yield much especially if they continue to have limited access to ownership of resources like land (Levitt 2015). Levitt further notes that development policy has still failed to curtail the structural issues which promote the discrimination against women in terms of resource ownership and control.

Women's domination in environmental management arguably places them in a better position to participate in planning and implementing policies. Unfortunately, women continue to constitute the minority in political and legislative structures. The purpose of women empowerment hence gets defeated especially in line with environmental conservation. Women hence need to be fully integrated in environmental conservation projects, programmes and plans needs deliberate effort Scholars like Kirk and Okazawa-Rey (1998) however do not agree that women are more caring towards the environment than the men are. They argue that by conserving the environment, women are performing their natural duties as daughter, mothers and sisters. That whereas this role has largely been bestowed upon them by nature, it is nothing special. This section set out to discuss the impact of desertification on women, their role and contribution in combating the same, as well as the challenges they face in their effort to combat desertification. Merely documenting women's contribution is not enough. States need to put sufficient programmes and policies that are all inclusive and do away with patriarchal practices that hinder women from contributing to important decisions in the fight against desertification. Legal recognition of the women's contribution will go a long way in buttressing the role of women as well as involving them in seeking solutions to the challenging that affects them the most.

The next section therefore seeks to analyse the extent to which women's role and contribution is recognised under international law, specifically under the UNCCD.

Legal Recognition of Women's Role in Combating Desertification: International Law and Practice

International Environmental Law lacks a codified set of rules or principles to govern the field. It is largely governed through collective effort with different instruments, each responding to different environmental risks. It has been argued that as a result, the global environmental concerns of different persons are not addressed in an equitable manner (Levitt 2015). Nonetheless, certain concepts exist under other international bodies of law and instruments that supplement the operation of international environmental law. A case in point is International Human Rights Law.

International Human Rights Law

Although the UNCCD is the main international agreement on combating desertification and makes reference to the role of women in combating desertification, it does not operate in a vacuum. International human rights law guarantees women's right to gender equality and non discrimination (Universal Declaration on Human Rights (UDHR), Article 2; International Covenant on Civil and Political Rights (ICCPR), article 3; International Covenant on Economic, Social and Cultural Rights (ICESCR), article 3; Convention on the Elimination of All Forms of Discrimination against Women (CEDAW), article 2, Article 5; CEDAW Committee 2010; African Charter on Human and Peoples' Rights (ACHPR), Articles 2 and 3; Protocol to the African Charter on Human and Peoples' Rights (Maputo Protocol), Article 2). The CEDAW Committee has recommended the need for states parties to recognise women's right to land and natural resources. This recommendation is also provided for in the Beijing Declaration and Platform for Action (1995) (Para. 35) as well as the 2030 Agenda for Sustainable development. SDGs 1, 2 and 5 on poverty eradication, food security and improved nutrition and gender equality respectively have been advanced as key for eradicating land degradation through promotion of women's right to land and natural resources (Mor 2018). Having noted women's role and the challenges they face in agriculture as well as in dealing with natural resources in the previous section, it is trite to point out that guaranteeing gender equality in natural resource ownership and management is key for curbing desertification.

The Convention on Bio Diversity (CBD) and the United Nations Framework Convention on Climate Change (UNFCCC)

The CBD, the UNFCCC and the UNCCD together make up the legal framework that governs the changing natural environment. They are interconnected and a joint group (Joint Liaison Group) was established in 2001 in order to coordinate the three conventions and coordinate programmes (UNCCD n.d.b). The link between the three RIO conventions is crucial because biodiversity loss leads to desertification especially in a continuously changing climate. The CBD and the UNFCCC do not have direct reference to gender in their main texts. Entry points have however been created for having gender incorporated into the framework. These entry points include the different entry points under the Conventions as well as different programmes. The CBD only mentions the vital role of women in biological diversity conservation in its preamble. The CBD has however had various Gender Action Plans (GAPs) under it, the current one being the GPA 2015–2020 which ensures gender mainstreaming within the frameworks of the CBD Secretariat (UNEP/CBD/COP/DEC/XII/7). Further the CBD guidelines on the implementation of gender mainstreaming in National Biodiversity Strategies and Plans of Action and the NBSAPS ensures gender considerations in planning, budgeting and development of legal frameworks concerning NBSAPS (CBD 2012).

In 2009, the UNFCCC established the Women and Gender Constituency (WGC) whose main purpose is to ensure that the rights and voices of women are embedded and into UNFCCC (WGC 2017). UNFCCC COP 20 of 2014 adopted the Lima work programme on gender, whose main aim is to advance gender responsive policies and negotiations. The COP 21 further adopted the 2015 Paris agreement which is the first climate convention to acknowledge gender equality as critical to seeking climate change solution. Although the provision of gender equality is preambular, it is a ray of hope for gender and women empowerment discussions to find their way into climate change law. Decision 18 of COP 23 of 2017 has been noted for having introduced gender as agenda item. As a result, gender equality is discussed at the subsequent annual COPs. The dedicated gender item has been said to place the UNFCCC above the two sister Rio Conventions. Further, the COP 23 adopted the Gender Action Plan to ensure effective implementation of the gender mandates under UNFCCC.

United Nations Convention to Combat Desertification (UNCCD)

The UNCCD was birthed as result of many countries globally but specifically in Africa bearing the brunt of drought, desertification and land degradation (UNCCD n.d.a). These accounted for the environmental, social and economic difficulties of the affected countries. As a result, the United Nations Conference on Desertification (UNCOD) (1977) adopted a Gender Plan of Action to Combat Desertification (PACD) in what has been referred to as the first attempt ever to combat desertification (UNCCD n.d.a). With limited success, the problem of desertification continued to be a challenge hence discussions on the issue continued to the United Nations Conference on Environment and Development (UNCED) (1992) held in Rio de Janeiro. The conference came with a new approach of ensuring sustainable development at the community level. The conference called upon the United Nations General Assembly (UNGA) to establish a Intergovernmental Negotiating Committee (INCD) whose mandate was to come up with a convention to combat desertification. The UNGA by resolution adopted this decision (resolution 47/188). The UNCCD was adopted on 15th June, 1994 after five rigorous sessions and it came into force in 1996. The major objective of the UNCCD is to combat desertification and land degradation. It is further aimed at mitigating drought effects in affected countries especially in Africa (UNCCD article 2).

In its preamble, the UNCCD recognises the role played by women in the in regions affected by desertification and drought as well the need for ensuring women's full participation in every area in the quest for solutions to desertification (Preamble, UNCCD 2017). The UNCCD has been recognized as the only RIO Convention that has provisions on gender and women's role in combating desertification within the body of the Convention text itself. Articles 5 (d), 10 (f) and 19 (1) (a) encourages states parties to ensure the participation of resource users who include women, among others in the fight against desertification (UNCCD 2017; Poulsen 2003; decisions 8/COP9, 9/COP10, 9/COP11, 36/COP11, and 3/COP12). The UNCCD has also been lauded for the gender equality land mark achievement during the COP 13 held in Ordos, Inner Mongolia, China in 2017. During this COP, various declarations and decisions came into force and most notably the Ordos Declaration (2017) which provided for the need to address gender inequalities which hinder the implementation of the UNCCD. Similarly, the Namibia declaration (2013 paragraph 6) and the Ankara Initiative (COP 2016 paragraph E 1 (b)) emphasise the need for gender mainstreaming in the implementation of the UNCCD. COP 13, through its decisions recognises the role of gender equality and the implementation the UNCCD together with its 2018–2030 strategic framework as well as the 2030 Agenda for sustainable development (COP 13 Decision 30). Further, the states parties at the COP came up with a Gender Action Plan (GAP) and the COP recognised that gender mainstreaming is key for the implementation of the action. The GAP was accompanied by a what has been referred to as a land mark decision on "gender equality and women's empowerment for the enhanced and effective implementation of the Convention" (COP 13 decision 30). The GPA aims to have gender mainstreaming in the implementation on the 2018–2030 strategic framework as brought into force by COP 7. The GPA further puts in place strategies on how to include women in the fight against desertification. These include sensitization (para. 6), women participation in programme design and implementation as well as decision making processes. Gender mainstreaming as a mechanism to curb gender inequality is recommended. This is key for achieving Land Degradation Neutrality (para. 8).

The GPA further emphasises women's economic empowerment through access to improved technology (para. 17). The GPA has been lauded for its gender perspective in seeking solutions to desertification and land degradation generally through the implementation on of the GPA. It has however been criticised for heavily leaning towards ensuring women's access to resources as the major solution for implementing the UNCCD. The GPA does not pay attention to the gaps in the implementation structure and institutions especially as regards non efficient representation of women (Mor 2018). It has been recommended that a women specific body or group be established to oversee the effective implementation of the UNCCD as well as overseeing some of the programmes under UNCCD, such as reviews of the LDN as well as the National Action Plan (NAP) implementation (Mor 2018). Having noted that women are barred from access and ownership of resources like land due to discriminatory practices (para. 17 (c)), the GPA should contain recommendations on how to curb such barriers to women for example through legal and policy recommendations.

The UNCCD is linked to SDG 15 on combating land degradation. This includes desertification as mentioned in target 15.3 and its indicator in 15.3.1, 'proportion of land degraded over total land area.' Although the indicators for SDG 15 lack sex-disaggregated data, they can be linked with the indicators for SDG 5 on gender equality. These indicators provide governments with guidance on how to tally data on women's land rights both in law and in practice and how those land rights can be secured and how critical this is in curbing land degradation (Mor 2018). Further, the Food and Agricultural Organisation (FAO) (2013) recommends gender equality as key in governing land tenure, fisheries and forests. It implores states parties to, 'ensure that women and girls have equal tenure rights and access to land, fisheries and forests independent of their civil and marital status' (2013: 3). Implementation of the Policy Advocacy on Drought (Decision 29/COP.13) and Policy Advocacy Framework to Combat Sand and Dust Storms (Decision 31/COP.13) need to be informed by the GAP. GAP will offer critical areas for gender mainstreaming in the implementation of these policies. Further more, there is need to put in place effective measures like staff, resources and structures capable of ensuring proper implementation of the policies. A human rights based approach against gender inequality and non discrimination should be considered in light with the principles of participation, accountability and transparency.

Further, COP 13 in decision 7 adopted the UNCCD strategic framework (2018–2030) which enables states parties and the secretariat to realise gender responsive implementation. States parties are further mandated to ensure full and equal participation of both men and women in the planning, decision making as well as the implementation of the UNCCD. States parties are also required to empower women, girls and youths in the affected areas. The implementation of the strategic framework will be supported by the GAP. The framework specifically seeks to empower local populations affected by desertification specifically women and youth to participate in decision making (objective 2:3). The framework lays down expected impacts and these include, 'Land productivity and related ecosystems services are maintained or enhanced' (1.1), 'Measures for sustainable land management' (1.4), 'food security and adequate access to water' (2.1), 'improved livelihoods' (2.2), 'Synergies with other multilateral environmental agreements and processes are enhanced' (3.2). Having noted that women deal with natural resources in their every day lives, it can be argued that if the already identified barriers that hinder women's ownership and adequate access of land and natural resources are broken, the listed expected impacts will be easily met or enhanced. It is interesting to note that the strategic plan does not mention land tenure or land rights any where in the text of the document which are crucial to the implementation of the 'land use and degradation, and desertification, land degradation and drought (DLDD)' polices and programming. This is contrary to the implementation efforts under the COP 13, including but not limited to the Ordos Declaration (2017 preamble) which point to land tenure and specifically the land rights of women as crucial to implementation of the UNCCD.

The UNCCD in decision 8 COP 13 requires states parties to strengthen the capacity for gender mainstreaming, through partnerships. This can be done through enhancing the role of women in combating desertification. In a bid to achieve this, the UNCCD secretariat has developed strategies like enhancing staffing, developing a 'tool box' as well allocating a budget. The challenge however has been that all this has to be achieved within the available resources which resources have already been noted not be sufficient. In addition, UNCCD secretariat has widely published briefs and reports on the need for the empowerment of women in drylands. The UNCCD recognition of women's efforts to fight against desertification is crucial for women especially African women and more specifically having seen their crucial role in natural resource management. Despite its robust provisions on the role and participation of women and girls in combating desertification, the UNCCD lacks a strong approaches to gender inequality. Overall, the UNCCD provisions on women's participation in finding solutions to desertification are progressive, they are rendered redundant by the discriminatory practices against women. In addition, international law can only have effect if it implemented at national level. The next section therefore analyses the regional provisions on women and desertification and the extent to which these provisions together with the UNCCD provisions have been implemented in Uganda.

Draft African Union Strategy on Climate Change (AUSCC) and the East African Community Climate Policy (EACCP)

Africa is largely arid and semi arid and is experiencing extreme weather changes among them drought, especially in the face of climate change (African Union Climate Change Strategy (AUCCS) 2014). It has also been noted that Africa is the most vulnerable continent to climate change impacts (IPCC 2001). Africa has put in place some measures to curb climate change and its effects, among them, African Ministerial Conference on the Environment (AMCEN) (1985) whose main purpose among others is to provide advocacy for environmental protection in Africa. In addition, the African Union (AU) has in place the Climate Change and Desertification Unit

(CCDU) whose main purpose is to curb climate change through putting in place measures to curb its diverse impacts effects on the African continent and its people as well (AUCCS 2014). Although still under development, Africa has in place the Draft African Union Strategy on Climate Change (2014). Its main vision and objective is 'to provide the AU as a whole, the RECs, member States and other stakeholders with a reliable source of strategic guidance to enable them effectively address climate change challenges' and 'to enable the continent achieve "climate-smart" socio-economic development' respectively (AUCCS 2014: 3, 16). Most importantly, ASCC incorporates gender and recognises that women are more affected by climate change than men, especially because they are poor and that because women stand in the face of climate change and therefore they have a lot of knowledge on how to curb climate change. Women should there fore be involved in finding solutions to climate change. The AU has committed itself to reduce the climate change impacts on gender through education, reviewing existing strategies, incorporating gender mainstreaming into climate change framework and promoting gender sensitive adaptation measures (Mburia 2015). How this will achieved is however not spelled out and there is no mention of how this will be funded in other words there is not budget proposal of how this will done. Though promising, the strategy is still in draft form and even if it were to be passed, it will still not have binding effect since it is not law.

The global impacts of climate change are not strange to the East African region. East Africa just like the wider African region largely depends on rain fed agriculture and hence is highly susceptible to climate change impacts (EACCCP 2010). In 2009, the East African heads of state gave a directive for the process to formulate a climate change policy to be initiated. This was during the 11th Summit of the Heads of State held in Arusha, Tanzania on 20th November 2009. The policy objective is to guide EAC member states as they put in place strategies to deal with climate change impacts, through adaptation and mitigation in order to enhance the adaptive capacity of the region and reduce vulnerability. The policy notes that climate change destroys natural resource and specifically it worsens land degradation. It is further noted that droughts have become rampant in the East African region. States parties are called upon to therefore to promote actions that reduce land degradation (EACCCP 2010). The policy however only recognises women's role in water conservation and calls upon states parties to involve women in the management of water sources (EACCCP 2010). This provision alone is not sufficient especially in light women's role as has been noted in the preceding sections.

Legal Recognition of Women's Role and Contribution in Combating Desertification at National Level: The Case for Uganda

AU has noted Uganda to be one of the African countries that faced the highest rain fall deficit in 2010/2011. As a result, Uganda has lost billions of dollars equivalent to 7.5% of its Gross Domestic Product (GDP) in 2010 (AU Strategy 2014). Like else where in the world, women and girls in Uganda have been noted to bear the brunt of drought. During drought, girls are given away in marriage, in exchange for food, commonly termed as famine marriage (Government of Uganda 2007a, b). This is appalling especially seeing as women dominate in agricultural production and labour force (76%) in Uganda, hence contributing to desertification control (Bowen 2015). For an international instrument to be applicable in Uganda, it has to first be ratified and domesticated. This is because Uganda is a dualist state (Constitution of the Republic of Uganda (1995), Article 123 (2); Ratification of treaties Act, Section 2 and 4). Objective XX111 of the Constitution further mandates the government of Uganda to put in place measures to handle disaster and natural calamities. Article 245 provides that "Parliament shall, by law, provide for measures intended: (a) to protect and preserve the environment from abuse, pollution and degradation …" Gender balance and fair representation of marginalised groups on Constitutional and other bodies, as well as recognition of the role women play in society is a principle that is recognised under the Constitution (Principles VI and XV, National Objectives and Directive Principles of State Policy). This could be interpreted to mean that the role of women in the fight against desertification needs to be recognised and that women need to be represented on bodies taking decisions on desertification. The Constitution further recognises gender equality in articles 21 and 33 and mandates the government to put in place measures to empower women so that they can realise their potential.

Having noted that women in Uganda are involved in agriculture and contribute 76% of the agricultural work force, it paradoxical that rural women face limited land ownership and only access it for their agricultural production through a male relative (Bowen 2015). Women in Uganda have been noted to own only 7% of the land, and hold 8% of the leasehold land although 97% of the women have access of the land (Hori et al. 2011; UNCCD n.d.a). Women are therefore afraid of investing in desertification control practices like soil conservation even when they dominate the agriculture sector (UNCCD n.d.a). This is because they are afraid of losing the land to the owner after they have made investments. Although women are widely involved in agriculture, they do not take decisions concerning land or their produce (CBD Secretariat 2012). Women are also kept out of decision making processes concerning land and their contribution in production is rarely acknowledged. This has affected productivity and food security (Lanyero 2012). Uganda's dependence on rain fed agriculture has been said to put it at higher risk of desertification. This because agricultural production has been noticed to be continuously declining. In accordance with the constitutional provision therefore, women in Uganda need to be empowered through having their right to land ownership strengthened and realised. This way, they will be able to realise their full potential in agricultural production and continue to contribute to combating desertification through modern agricultural practice like soil conservation.

The National Environment Act (2019) is the major environmental conservation law in Uganda. It recognises the need to prohibit and to stop soil and environmental degradation practices (Section 3 (g), Section 56 (1), (2) (e), Section 107 (2) (e)).

It further recognises that rangelands are at risk of desertification (Section 66 (2) (C) (c)). Although the Act recognises the need to encourage people's participation in environmental management (Section 5 (2) (a)), it does not specifically recognise women's contribution in seeking solutions to desertification. It only vaguely notes the need for 'gender responsive use of the environment and natural resources' (Article 5 (2) (b)). This is so, despite the fact that it recognises that Uganda has obligations under international that need to be honoured (Section 9 (2) (r) and (s)). By implication, the obligations under the UNCCD in respect of recognising and realising women's contribution in combating desertification need to be recognised and implemented accordingly. The Act only recognises the need to have at least a third of women as members of the National Environment Management Authority Board (Section 12 (4)). Similarly, the Forestry and Tree planting Act, 2003 does not make mention of women's role in tree planting and forest management. Uganda has been said to be facing a 'deforestation crisis' (CARE International 2008: 21). Uganda's forest cover constituted 40% of the country, only 50 years ago. Currently, Uganda's forest cover is only 20% and loses 2.2% of forest cover per year, making it one of the world's fastest forest cover losses. This has affected food security and women have to move longer distances searching for firewood. Women in Uganda have been noted to be instrumental in forest conservation (CARE International 2008). The Forestry Policy (2009) recognises that women use forests more sustainably than the men. It recognises that women only access for medicine, firewood and agro forestry. Men on the other hand deplete the forests for timber for commercial purposes.

Women's role in water and sanitation management in Uganda has been documented (National Environmental Management Authority (NEMA) 2014). Women have been noted to appear on water and sanitation committees and the improvement in water and sanitation management (NEMA 2014). Due to their role in agricultural production and the household roles, women in Uganda have been noted to participate in the maintenance of fresh water sources. It is thus surprising that the Water Act, Cap 152 is silent on the role of women in ensuring water conservation. Despite women's contribution as highlighted above, women in dryland areas in Uganda continue to be left out of water governance initiatives (Government of Uganda 2004).

Although the recognition of women's contribution in fighting desertification is not sufficiently captured in a binding law, a wealth of policies and programmes which attempt to recognise the need to involve women in environment management are in place. These include the National Climate Change Policy which recognises the role of women in addressing climate change. The National Environment Management Policy (Uganda) (1994), Section 2.3 (iv)) recognises the need for the recognition of women's role in natural resource management and decision making. Similarly, the National Water Policy (1999, Section 5.4.3 (1)) recognises the need to integrate women into improved water supply strategies as well as training programmes. The National Action Programme (NAP) and the National Agriculture Research System are in place in Uganda to facilitate the UNCCD and its programmes. The Government of Uganda (GoU) acknowledges the low presence of women in all its plans and programmes for example the Integrated Drylands Development Program (IDDP). This is so, despite the fact that the government has a standing policy which requires women to form 30% of the committee compositions (GoU 2004). This has been attributed to discriminatory cultural practices that view women as non decision makers (GoU 2016).

In order to mitigate the rapid land degradation, Uganda adopted the LDN strategy. The strategy also serves as proof of commitment by Uganda to achieve 'Sustainable management of the environment and natural resource base' (GoU 2018a, b). Uganda has put in place mechanisms to implement the LDN. Among other targets, Uganda envisions a Land Degradation Neutral Uganda by 2030, a 21% and a 12% increase in forest and wetland cover respectively (GoU 2018a, b). Uganda is recognised for having been among the first countries to 'come up with development plans in line with SDGs at the national and local government levels in 2015/16–2019/20. 76% of the SDGs are reflected in the plans' (GoU 2018a, b). SDGs 1, 2, 8 and 15 on no poverty, zero hunger, decent work and economic and life on land have been noted to be in line with SDG target 15.3 on Land Degradation Neutrality. The SDGs are also connected to the CBD, UNFCCC, UNCCD (GoU 2018a, b). In addition to the already outlined legal and policy framework, Uganda has in place regulatory framework for operationalising the LDN. It includes the Uganda Vision (2040); National Development Plan (2015–2020); Uganda Strategic Investment Framework for Sustainable Land Management (2010–2020); National Agriculture Policy (2010); National Land Policy (2009) and National Land Use Policy (2007) among others. The strategies employed by Uganda to achieve LDN have been noted to be very vital for ensuring sustainable development; preservation, protection, management. These different frameworks focus on the restoration of degraded natural resources among them rangelands; wetlands, and ecosystems (GoU 2018a, b). Although the LDN approach has been hailed to be inclusive of as well to protect the land tenure rights of vulnerable groups, this can only be achieved when the existing laws and policies on environmental protection are amended in order for them to be more inclusive of women participation (GoU 2018a, b).

Again, in its vision 2040, Uganda has acknowledged that its natural resources are being rapidly degraded and the environmental laws and policies in place are not being enforced (para. 246–248). To this end, Uganda aspires to put in place more strict measure to have the laws and policies enforced (para. 250). Uganda also aspires to put in place better institutions to ensure enforcement of the law (para. 258). Uganda further acknowledges and recognises the role of women in natural resources conservation in Uganda. Uganda further aspires to involve women in programme and policy formulation for environmental and natural resources protection (para. 258). With this recognition, questions are raised as to why Uganda has not acted upon the inclusion of women in natural resources protection yet. Uganda's vision 2040 aspires a Uganda where both men and women are equal partners in development (para. 265). This implies that the government acknowledges disparities in the inclusion of women in Uganda's development agenda. Further, the government of Uganda acknowledges the limited access and ownership of productive resources like land, by women and aspires to erase all forms of gender based discrimination (para. 263, 265, 267; GoU 2004, 2018a, b). Overall, the legal framework in Uganda lacks adequate recognition of women's roles in environmental management. Neutral legal provisions miss the opportunity to appreciate the unique and special contribution of women in natural resource conservation. Additionally, most of the provisions on women are integrated into policy provisions, which have no binding effect. Women still lack the opportunity to share their experience as well as their knowledge. There is still urgent need for women to be included in project planning and policy formulation.

Conclusion

This chapter set out to examine the adequacy of the legal provisions on women's contribution and involvement in combating desertification and how these provisions have been implemented in Uganda. Specifically, the chapter analysed the UNCCD as well as the legal framework in Uganda. The chapter confirms that women have indeed made significant contribution towards combating desertification globally and in Uganda specifically. The chapter further establishes that whereas the UNCCD recognises women's contribution in combating desertification and the necessity to have them participate in environmental decisions concerning desertification control, the provisions have not been adequately enforced. The substantive environmental laws of Uganda do not recognise women's role in combating desertification. The scanty provisions which recognise women's contribution largely fall under policy framework which is not legally binding hence making enforcement even harder. There is therefore urgent need to incorporate the contribution of women in desertification control within the law, especially since evidence of their contribution exists, as set out in this paper. The recognition of women's contribution is vital, especially in the face of fast land degradation which in fact affects women disproportionally compared to other members of society. It is however important to note that mere recognition is not enough, women have to be empowered to make significant contribution especially through ensuring their ability to adequately access and own productive resources like land. This will enable them to invest in desertification control mechanisms like drought resistant crops and soil conservation, given their significant role in agriculture. This will go a long way in contributing to communities' safety from the severe effects of land degradation.

Despite these findings, the chapter was not without limitations and constraints. There was scarcity of empirical data on which to assess women's contribution and low participation in desertification control specific to Uganda. Further, there was scarcity of existing academic research on the area of legal recognition of women's contribution to the fight against desertification. Therefore it is important for countries, Uganda specifically, to invest into in-depth and evidence based research on the actual contribution of women in desertification control and the need for their participation in decision making concerning the same. This will enable the availability of gender disaggregated data on the subject of desertification control. Facilitating and ensuring the improved understanding of women's knowledge and contribution will also provide a foundation for the formulation of legislation that adequately recognise this knowledge and contribution in a bid to ensure combating of desertification. In order to achieve this, efforts need to be geared towards consultation of these women at all levels in order for their knowledge and contribution to be well captured. Involvement of women will enhance programme development and implementation. Finally, Uganda needs to put in place finance mechanisms and strategies in order to finance women's efforts to fight desertification as well as address women's historical constraints. There is therefore need to recognise women's contribution and enhancement of their participation in the fight against desertification in Uganda.

References

Abebe MA (2014) Climate change, gender inequality and migration in East Africa. Wash J Environ Law Policy 4(1):114

Aditya SK (2016) Role of women in environmental conservation. Int J Polit Sci Dev 4(4):140–145

African Charter on Human and Peoples Rights (Adopted 27 June 1981, OAU Doc. CAB/LEG/67/3 rev. 5, 21 I.L.M. 58 (1982), entered into force 21 Oct 1986)

African Union (2014) Draft African Union Strategy on Climate Change. https://www.un.org/en/africa/osaa/pdf/au/cap_draft_auclimatestrategy_2015.pdf. Accessed 15 Apr 2019

Angelsen A (1997) The poverty—environment thesis: was Brundtland wrong? Forum Dev Stud 24(1):135–154

Beijing Declaration and Platform for Action, UN The Fourth World Conference on Women (1995) http://www.un.org/womenwatch/daw/beijing/pdf/BDPfA%20E.pdf. Accessed 25 Jan 2019

Bowen DH (2015) Leveling the field for women farmers in Uganda. World Bank. http://blogs.worldbank.org/voices/leveling-field-women-farmers-uganda. Accessed 8 Feb 2019

BRAC and Landesa (2017) Issue brief: land tenure as a critical consideration for climate change-related displacement in slow-onset disaster zones (Hilton Prize Coalition). http://prizecoalition.charity.org/wp-content/uploads/2017/05/HPC_CollaborativeIssueBrief_Landesa_BRAC_053117.pdf. Accessed 24 Jan 2019

Broeckhoven N, Cliquet A (2015) Gender and ecological restoration: time to connect the dots. J Soc Ecol Restor 23(6):729–736

CARE International (2008) Reclaiming rights and resources: women, poverty and environment. https://www.care-international.org/files/files/publications/Natural-Resources-Environment-Reclaiming-Rights-Women-Poverty-Environment-2008.pdf. Accessed 8 Feb 2019

CBD Secretariat (2012) Guidelines for mainstreaming gender into national biodiversity strategies and action plans. CBD technical series no. 49. https://www.cbd.int/doc/publications/cbd-ts-49-en.pdf. Accessed 8 Feb 2019

CEDAW Committee, CEDAW, General Recommendation No. 34 on the Rights of Rural Women (2016) UN Doc. CEDAW/C/GC/34, para. 5. https://tbinternet.ohchr.org/Treaties/CEDAW/Shared%20Documents/1_Global/INT_CEDAW_GEC_7933_E.pdf. Accessed 10 Feb 2019

CEDAW Committee, General Recommendation No. 28, para. 31, 47th Sess. (2010) https://www2.ohchr.org/english/bodies/cedaw/docs/CEDAW-C-2010-47-GC1.pdf. Accessed 10 Apr 2019

Climate Change and Agroecological Approaches (LEISA) (2017) Low external input sustainable agriculture (India), issue no. 19.2, June 2017. http://leisaindia.org/magazines/english/climatechange-and-agroecological-approaches-june-2017-issue-19-2/. Accessed 31 Jan 2019

Conference of the Parties (2017) Ordos Declaration, Decision 27/COP.13 (15 September), see in particular para. 12. http://www2.unccd.int/sites/default/fles/sessions/documents/2017-11/cop21add1_eng.pdf. Accessed 10 Feb 2019

Constitution of the Republic of Uganda (1995) Adopted by the Constituent Assembly on 22nd September, 1995. https://www.parliament.go.ug/documents/1240/constitution. Accessed 31 Jan 2019

Convention on Biodiversity United Nations Treaty Series, vol 1760, no 30619
Convention on Biological Diversity (2012) NBSAPs and national reports. https://www.cbd.int/gender/nationalreports/159. Accessed 1 Feb 2019
Convention on the Elimination of all forms of Discrimination Against Women (CEDAW) adopted by the UN General Assembly on December 18th, 1979
Dankelmann I, Davison J (1988) Women and environment in the third world: alliance for future. Earthscan Publications Ltd, London
Decision adopted by the Conference of the Parties to the Convention on Biological Diversity. Mainstreaming gender considerations 2015–2020 gender plan of action
East African Community (2010) East African Community climate change policy. http://www.rema.gov.rw/rema_doc/Climate%20change/EAC%20Climate%20Change%20Policy_April%202011.pdf. Accessed 15 Apr 2019
FAO (2013) Governing land for women and men: a technical guide to support the achievement of responsible gender-equitable governance of land tenure. FAO, Rome
Government of Uganda (2004) Third national report to the Conference of the Parties on the implementation of the United Nations Convention to Combat Desertification in Uganda (GoU). http://www.unccd-prais.com/Uploads/GetReportPdf/1de39836-8fc5-483d-b974-a0fa014a4a88. Accessed 31 Jan 2019
Government of Uganda (2007a) National adaptation programmes of action. https://unfccc.int/resource/docs/napa/uga01.pdf. Accessed 27 Apr 2018
Government of Uganda (2007b) Uganda vision 2040. https://www.gou.go.ug/content/uganda-vision-2040. Accessed 15 Apr 2019
Government of Uganda (2016) Environment and natural resources sub-sector gender mainstreaming strategy (2016–2021). https://www.mwe.go.ug/sites/default/files/library/ENR%20GENDER%20STRATEGY%20%202016%20-2021.pdf. Accessed 31 Jan 2019
Government of Uganda (2018a) The LDN target setting programme Uganda country report. https://knowledge.unccd.int/sites/default/files/ldn_targets/Uganda%20LDN%20TSP%20Country%20Report.pdf. Accessed 10 Feb 2019
Government of Uganda (2018b) Water and environment sector performance report 2018 (Ministry of Water and Environment). https://www.mwe.go.ug/library/sector-performance-report-2018. Accessed 8 Feb 2019
Hori Y, Stuhlberger C, Simonett O (2011) Desertification: a visual synthesis. UNCCD and Zoe Environment Network
Hyma B, Nyamwange P (1993) Women's role and participation in farm and community tree-growing activities in Kiambu District, Kenya. In: Momsen JH, Kinnaird V (eds) Different places, different voices: gender and development in Africa, Asia and Latin America. Routledge, London and New York, pp 30–45
ICCD/COP(11)/23, 1 Nov 2013. Available at https://www.unccd.int/sites/default/files/sessions/documents/ICCD_COP11_23/23eng.pdf. Accessed 26 Jan 2019
ICCD/COP(12)/20, 21 Jan 2016. https://www.unccd.int/sites/default/files/sessions/documents/2017-09/ICCD_COP%2812%29_20-1520824E.pdf. Accessed 26 Jan 2019
ICCD/COP(13)/8-ICCD/CRIC(16)/2, 21 June 2017. https://www.unccd.int/sites/default/files/sessions/documents/2017-07/ICCD_COP%2813%29_8-1710706E.pdf. Accessed 27 Jan 2019
International Covenant on Civil and Political Rights. Adopted and opened for signature, ratification and accession by General Assembly resolution 2200A (XXI) of 16 December 1966 entry into force 23 March 1976
International Covenant on Economic, Social and Cultural Rights. Adopted and opened for signature, ratification and accession by General Assembly resolution 2200A (XXI) of 16 December 1966 entry into force 3 January 1976
International Labour Organization (2016) Women at work: trends 2016. Geneva. http://www.ilo.org/wcmsp5/groups/public/---dgreports/---dcomm/---publ/documents/publication/wcms_457317.pdf. Accessed 24 Jan 2018

IPCC (2001) Climate change 2001: impact, adaptation, and vulnerability. Summary for policymakers. A report of working group II of the Intergovernmental Panel on Climate Change. Cambridge University Press
Jiggins J (1994) Changing the boundaries: women-centered perspectives on population and the environment. Island Press, Washington, DC
Kelkar G (2009) Adivasi women: engaging with climate change. UNIFEM. http://www.refworld.org/docid/4ab37acb2.html. Accessed 31 Jan 2019
Kirk G, Okazawa-Rey M (eds) (1998) Women's lives: multicultural perspectives. Mayfield Publishing Co., Mountain View, CA
Lanyero F (2012) Flirting with desertification. Daily Monitor Newspaper Online. https://www.monitor.co.ug/artsculture/Reviews/Flirting-with-desertification/691232-1449294-ydxkslz/index.html. Accessed 8 Feb 2019
Lau WKM, Wu HT, Kim KM (2013) A canonical response of precipitation characteristics to global warming from CMIP five models. Geophys Res Lett 40(12) 3163–3169
Levitt JI (2015) Black women and international law: deliberate interactions, movements and actions. Cambridge University Press
Mburia R (2015) Africa climate change policy: an adaptation and development challenge in a dangerous world. Climate Emergency Institute. http://admin.indiaenvironmentportal.org.in/files/file/AFRICA%20CLIMATE%20CHANGE%20POLICY-CEI.pdf. Accessed 15 Apr 2019
Melchoir S (1989) Woman and sanitation or counting tomatoes as well as pumps. PROWWESS/UNDP technical series. UNDP, New York
Module 9 (B series), mainstreaming gender considerations in the development and implementation of biodiversity strategies and action plans. https://www.cbd.int/doc/training/nbsap/b9-traingender-en.pdf. Accessed 15 Apr 2019
Mor T (2018) Towards a gender-responsive implementation of the United Nations Convention to Combat Desertification. 2018 research paper. United Nations Entity for Gender Equality and the Empowerment of Women (UN Women). http://www.unwomen.org/media/headquarters/attachments/sections/library/publications/2018/towards-a-gender-responsive-implementation-of-un-convention-to-combat-desertification-en.pdf?la=en&vs=3803. Accessed 24 Jan 2019
National Environmental Management Authority (NEMA) (2014) State of the environment report for Uganda 2014. NEMA, Kampala
National Water Policy (Uganda) (1999) https://www.nutrition.opm.go.ug/wp-content/uploads/2017/11/National-Water-Policy-1999.pdf. Accessed 15 Apr 2019
Poulsen L (2003) COP 6 and the promotion of women. IUCN. https://portals.iucn.org/union/sites/union/fles/doc/cop6_and_the_promotion_of_women.pdf. Accessed 20 Apr 2019
Prior T, Duyck S, Heinämäki L, Koivurova T, Stępień A (2013) Addressing climate vulnerability: promoting the participatory rights of indigenous peoples and women through Finnish Foreign Policy. Juridica Lapponica 38
Protocol to the African Charter on Human and People's Rights of Women in Africa (2003) Adopted by the African Union on July 11, 2003 and entered into force on November 25, 2005
Secretariat of the Convention on Biodiversity. Guidelines for mainstreaming gender into national biodiversity strategies and action plans. CBD technical series no. 49. https://www.cbd.int/programmes/cross-cutting/gender/cbd_ts49_gender_guidelines.pdf. Accessed 10 Feb 2018
Skinner E (2011) Gender and climate change overview report. BRIDGE cutting edge pack on gender and climate change. Institute of Development Studies, Brighton. https://gsdrc.org/document-library/gender-and-climate-change-overview-report/. Accessed 13 Feb 2019
Stockholm Declaration on the Human Environment, U.N. Doc. A/Conf. 48/14/Rev. 1 (1973)
The National Environment Management Policy (Uganda) (1994). Available at http://enr-cso.org/wp-content/uploads/2014/12/Draft-Final-National-Environment-Management-Policy-Dec-2014.pdf. Accessed 15 Apr 2019
The National Environment Management Act no. 5 of 2019 (Uganda)
The Ratification of Treaties Act, Chapter 204 Laws of Uganda 2000

Transforming our world: the 2030 agenda for sustainable development. A/RES/70/1. Available at: https://sustainabledevelopment.un.org. Accessed 25 Jan 2019
UN (2005) Women and water (UN). http://www.un.org/womenwatch/daw/public/Feb05.pdf. Accessed 8 Feb 2019
UNCCD (n.d.a) Gender and desertification. UNCCD thematic fact sheet series no. 4. http://catalogue.unccd.int/23_loose_leaf_Desertification_gender.pdf. Accessed 31 Jan 2019
UNCCD (n.d.b) UNCCD, CBD and UNFCCC Joint Liaison Group. https://www.unccd.int/convention/about-convention/unccd-cbd-and-unfccc-joint-liaison-group. Accessed 7 Feb 2019
UNCCD (2017) UNCCD convention (2017). http://catalogue.unccd.int/936_UNCCD_Convention_ENG.pdf. Accessed 14 Apr 2019
UNEP/CBD/COP/DEC/XII/7, 17 Oct 2014. https://www.cbd.int/doc/decisions/cop-12/cop-12-dec-07-en.pdf. Accessed 1 Feb 2019
United Nation General Assembly (UNGA) (1993) G.A. Res. 188, U.N. GAOR, 47th Sess., Agenda Item 79, U.N. Doc. A/RES/47/188
United Nations (2012) Millennium development goals report, 2012. UN, New York. Available at http://mdgs.un.org/unsd/mdg/Resources/Static/Products/Progress2012/English2012.pdf. Accessed 6 Feb 2019
United Nations Convention to Combat Desertification in those countries experiencing serious drought and or desertification particularly in Africa. http://catalogue.unccd.int/936_UNCCD_Convention_ENG.pdf. Accessed 15 Apr 2019
United Nations Framework Convention on Climate Change (UNFCCC) (1992) https://unfccc.int/resource/docs/convkp/conveng.pdf. Accessed 15 Apr 2019
WGC (2017) Welcome letter to new members, 16 July (on file with Landesa); Women and Gender Constituency. About us. http://womengenderclimate.org/about-us/. Accessed 31 Jan 2019
WHO (1984) Biomass, fuel combustion and health. World Health Organization, Geneva

Women in a Climate Changing World. The Need of a Policy Solution for Cross-Border Displacement

Beatriz López-Fanjul Díez del Corral

Abstract An international framework that addresses displacement across borders due to sudden events does not exist, and a policy embracing a gender perspective on the topic is lacking. This absence of specific regulations on the matter leaves women in a vulnerable situation. The purpose of this work is to advance knowledge in order to identify international policy conventions and agreements that consider climate change disasters, gender and migration, to provide support for women and girls of developing countries when a sudden disaster strikes their communities, a situation where cross border displacement is the only option. This report strives to clarify the importance of having an equity perspective to create equality between genders in a migration process. This is not equivalent to favor one gender upon the other, but simply stresses the fact that the "starting line" for women and girls is not the same as that for men. It is important to consider this positive discrimination of women as a first step to create gender equality by building empowerment through resilience and coping capacity.

Introduction

Currently, climate change is no longer a hypothesis, as indicated by the IPCC confirmation of the last three decades being the warmest since 1850 (IPCC 2014). As a result, tidal storms and extreme weather events actually have a wider range and produce more severe effects (Ware and Doig 2014). The debate is focused on the implement of effective measures to fight the detrimental impact.

Climate-related disasters have been classified by the UNFCCC as slow-onset and sudden-onset events (wind storms, floods, wild fires, tropical cyclones, landslides, avalanches, earthquakes and volcanic eruptions), which have been continuously increasing between 1950 and 2012 (FAO 2016). When the economic possibilities of a group of people are reduced, they may be forced to live in regions that are greatly affected by natural hazards (Wisner et al. 2003).

B. López-Fanjul Díez del Corral (✉)
Independent Researcher, Guisando 34, Madrid 28035, Spain
e-mail: lopezfanjul@gmail.com

W. Leal Filho et al. (eds.), *Climate Change, Hazards and Adaptation Options*,
Climate Change Management, https://doi.org/10.1007/978-3-030-37425-9_27

A 2006 IPCC study considered the Gross Domestic Product (GDP) and the per capita impact in agriculture, a climate-sensitive economic sector, revealing that developing countries will be those that suffer most from damage due to climate change (Mendelsohn et al. 2006). Inadequate diversification of economic resources, large geographical location of exposure to climate change, and absence of safety nets, differ significantly between developed and developing countries. The IPCC highlights that least developed countries (tropical and subtropical areas) are highly vulnerable to the impact of climate change in comparison with developed countries (IPCC 2014). However, the degree of disaster impact on a given community and on each individual will vary depending on inequalities concerning the levels of wealth and education, disability, and health status, as well as gender, age, class, and other social and cultural characteristics (IPCC 2012). The strong impact of climate change on developing countries further increases the vulnerability of their citizens, and limits their opportunities to cope with the adverse effects of change. Sudden environmental changes and the corresponding degradation affect security, resulting in human displacement (IOM 2007). People migrate as a way of adapting to climate impacts after a disaster event, and migrants need to find a way to integrate in their new environment and build resilience. Environmental degradation, combined with lack of resources and infrastructure, limits men and women's capacity to cope and migrate.

In 2015, the Internal Displacement Monitoring Center estimated that 19.3 million people were displaced because of disasters (IDMC 2015) and the estimation for 2050 varies between 25 million and 1 billion environmental migrants according to International Organization of Migration (IOM 2015a). These statistics show how human mobility in the climate change context is a fact[1] that can be voluntary or forced.[2] It is necessary to consider that, in the context of sudden events, universal criteria to define a cross border migration as voluntary or forced is not in place (The Nansen Initiative 2015a). Between 2008 and 2014, it was estimated that 26.4 million people were displaced by disasters, a figure increasing in the last year (2015) adding 19.2 million people in 100 different countries. In other words, around 62,000 people are displaced by weather-related disasters every day (IDMC 2015).[3]

---

[1] Human mobility: population mobility [in the context of environmental change] is probably best viewed as being arranged along a continuum ranging from totally voluntary migration [...] to totally forced migration (IOM 2015a).

[2] Forced Migration: the movements of refugees and internally displaced people (those displaced by conflicts) as well as people displaced by natural or environmental disasters, chemical or nuclear disasters, famine, or development projects (IOM 2014a).

[3] The displacement of people refers to the forced movement of people from their locality or environment and occupational activities. It is a form of social change caused by a number of factors, the most common being armed conflict. Natural disasters, famine, development and economic changes may also be a cause of displacement (UNESCO 2016).

Disaster risk[4] depends on three factors: disaster events, vulnerability[5] and exposure.[6] The terms exposure and vulnerability are dynamic and directly depend on eight different factors: economic, social, geographic, demographic, cultural, institutional, governance, and environmental (IPCC 2012).

In 2015, the Organization for Economic Co-operation and Development (OECD), stated that 48% of international migrants are women (OECD 2013). The literature has highlighted different reasons to justify why women are more vulnerable than men when they are forced to migrate: "women tend to be poorer, less educated, have a lower health status and have limited direct access to or ownership of natural resources" (Masika 2002), but these handicaps do not prevent women to migrate. The biological differences after birth between females and males are located in just one chromosome but, during growth, the distinction between genders is increased by socio-cultural stresses. When these push-pull factors become risky, girls and women have less access to their rights due to their gender role and position in society (Women's Refugee Commission 2016). There is a growing disaster vulnerability literature showing that inequalities in sensitivity to risk, access to resources, exposure, capability and opportunities systematically put women at disadvantage (Neumayer and Plümper 2007). Masika (2002) detected that women face the challenge of not being protected in countries of transit and destination due to inexistent and/or inadequate policy frameworks addressing the status of climate migrants. In addition, migrants who decide to leave their home of origin because of disasters and/or climate change impacts, face the need to adapt to their new environment and cope with new challenges such as language barriers.

This article aims to answer the following question:

Are the vulnerabilities and necessities of girls and women, internationally displaced by sudden disasters, taken into account in the international framework conventions and agreements?

To do that, women's vulnerability factors when sudden disaster strikes their communities and cross border displacement is the only option, have been considered during three stages: (1) pre-migration, (2) border crossing, and (3) post-migration. In addition, the pertinent international policy conventions and agreements providing support for females of developing countries in those extreme situations are analyzed in detail, emphasizing the need to create equality between genders in a migration process.

---

[4]Disaster Risk: the likelihood over a specified time period of severe alterations in the normal functioning of a community or a society due to hazardous physical events interacting with vulnerable social conditions, leading to widespread adverse human, material, economic, or environmental effects that require immediate emergency response to satisfy critical human needs and that may require external support for recovery (IPCC 2012).

[5]Vulnerability: The propensity or predisposition to be adversely affected (IPCC 2012).

[6]Exposure: The presence of people; livelihoods; environmental services and resources; infrastructure; or economic, social, or cultural assets in places that could be adversely affected (IPCC 2012).

Methodology

The methodology observed in this paper has been to identify and analyze those international conventions, agreements and UN agencies specifically covering three areas: (1) climate-related disasters, (2) resulting migration movements, and (3) the consideration of a gender perspective in these situations. To do that, a desktop literature search was performed to gather an array of different publications from various web pages. This includes international organization policy reports, academic journals, books and papers, as well as other relevant literature.

During the research, six face to face interviews with relevant organizations were conducted: (1) the International Organization of Migration (IOM), (2) the United Nations Refugee Agency (UNHCR), (3) CEAR (Comisión Catalana de Ayuda al Refugiado), (4) Barcelona City Council (Barcelona City of Refugees), (5) the Institute of Studies on Conflicts and Humanitarian Action (IEACH), and (6) the Autonomous University of Barcelona (Investigation group Antígona). Those interviews were utilized to support the analyses of the different conventions, agreements and UN agencies examined.

A Women's Perspective

In this section, women's position when facing cross border migration induced by sudden climate-related disasters will be analyzed from a gender perspective.

Neumayer and Plümper (2007) reflected on the biological and physiological social constructs, norms and roles imposed by society and gender discrimination based on scarcity of resources, and confirmed the existence of a gender gap in mortality rates. The term woman comprises different stages of the life cycle, including childhood, adolescence and adulthood. As mentioned before, women are the key actors to have into consideration for protection, but 'girls'—children and adolescents—are in some occasions underlooked when a sudden-onset disaster strikes.

The starting line for women in these situations is not the same as that for men, and the added risk of being young increases the possibilities of being victims of harmful practices such as rape, abuse, sexual slavery, early forced marriage, and abduction. In addition, after a disaster strikes family structures are broken, leading to households headed by orphans lacking adult protection, who are likely victims for mafias and human trade (Women's Refugee Commission 2014). An orphan girl with siblings is forced to assume roles, take responsibilities and make decisions without adult capacities and skills (Women's Refugee Commission 2014). In 2014, the Women's refugee commission concluded that it is necessary to link girls and disaster relief resources to increase their capacity to recovery (Women's Refugee Commission 2014). This report also highlights that 48% of girls were forced in their first sexual experience and nearly 50% of sexual assaults were committed to girls under 16 years.

Migration Process: Women's Vulnerability Factors

During the process of cross border migration, women need to go through three stages. The first is that of pre-migration, the second occurs when the border is crossed, and the third when the country of destination is reached.

Figure 1 shows the process followed by a woman once she decides to migrate. In the aftermath, community members or individuals make an evaluation of the post-disaster scenario, the "real" status of living conditions, and future possibility to bounce back to normal. When there are not enough natural, economic and material resources in a country, women are forced to migrate, either inside or across the country.

During the migration process, women face many vulnerabilities. The first are physical and biological, intrinsic to the female condition. In addition, there are cultural and social vulnerabilities imposed by their role in the community and their religion,

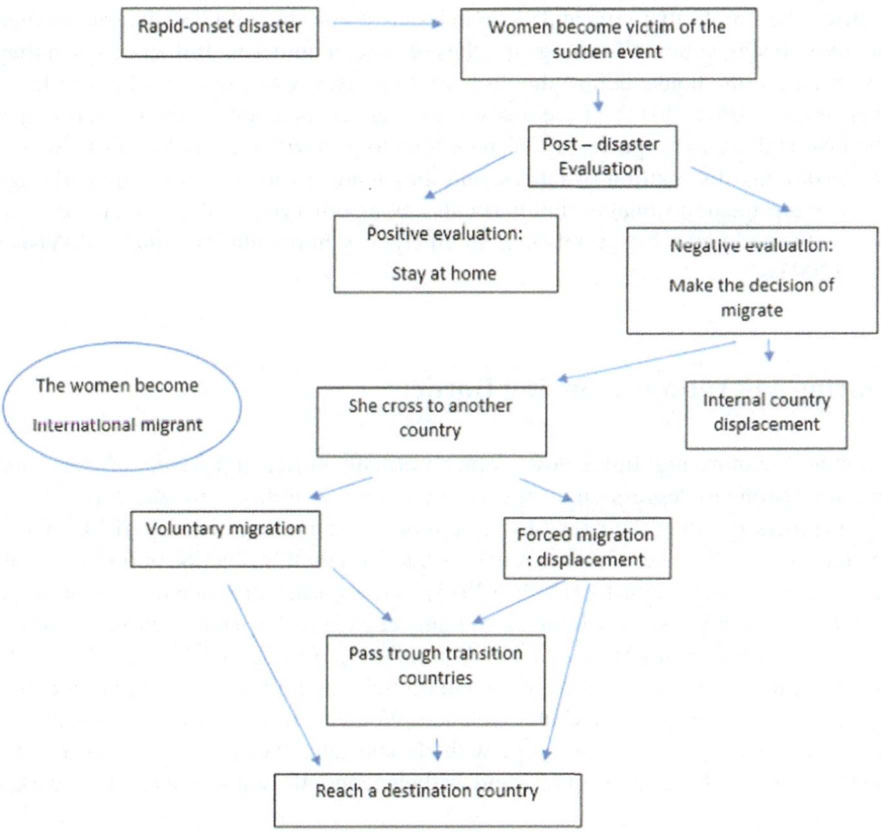

Fig. 1 Decision making process of women to migrate across border. Graph assembled by the author

leading to acceptance of traditional tasks. Finally, there are gender discrimination vulnerabilities, usually linked with those of the second group.

Before Crossing the Border: Pre-migration

The pre-migration stage (Table 1) starts just after the sudden hazards strike a country, and until women make the decision to migrate.

Physical and biological vulnerability plays a key role when a disaster strikes, since women tend to have less strength, endurance and ability (Masika 2002). In addition, women need more hygiene than men to combat infections and diseases, particularly when women are pregnant or lactating. Women are used to work directly with natural resources, carrying water and working in agriculture; the loss of land and resources adds stress and increases the level of sensitivity that can, in turn, lead to depression (Chindarkar 2012). Women's dress code, a reflection of cultural and religious norms, restricts their mobility to run and swim in a self-rescue situation. In addition, women are used to stay at home for long stretches of time, a hindering influence when they try to escape the house before the disaster. Moreover, women are told not to leave their homes (IPCC 2014). If we analyze the gender inequality, men normally own the land and are more powerful. Women tend to perform householder tasks within the traditional distribution of labors, and they tend to leave school at an early age (early marriage,traditional initiation rituals among others). Finally, women have less access to food and relief supplies in an emergency humanitarian situation (Wisner et al. 2003).

During Migration: Crossing Border

Existing literature highlights how women normally suffer high levels of stress and are more prone to depression, as their primary responsibility is to take care of family members ignoring their own losses (cultural assumption of role) (IPCC 2014). Women are likely to be economically dependent on their husbands, or on male family members. According to UNDP (2015), cross-border displacement is often an unsafe way of migration for females, as many engage in desperate measures, such as trading sex for crossing the border, and/or enduring treacherous routes. As a result, many women have been victims of human trafficking, mutilation or amputation, and associated health impacts of sexual violence (Women's Refugee Commission 2016). Women's capacity to adapt and cope with climate-related disasters depends directly on their control of natural resources and on their economic independence (Chindarkar 2012) (Table 2).

Table 1 Women vulnerability when the disaster strikes before the migration

Pre-migration	Physical and biological	– Strength, less self-rescue (women are less strong than men) – Physical size (on the average, women can run and climb less) – Endurance (on the average, women run less and at lower speed) – Physiological (stress, high levels of sensitivity, depression). Women can suffer physical affliction after a hazard (amputation, mutilation …)
		– Need of hygiene, increasing in the reproductive stage (pregnancy/lactation) – Women are more susceptible to disease – Lack of nutrition (on the average, women have less access to resources) – Family charge (on the average, women take care of the children). A mother lost will increase children mortality and the number of orphans – On the average, women are more dependent on natural resources – On the average, women have a higher dependence on the community and family
	Social norms and role	– Cultural role dress code (restrict mobility) and long hair (which can sting during a tsunami or flood) – Women are prevented to leave their homes – Marital status – Family factors (size, age/sex composition) – Unequal distribution of power inside the family – Traditional distribution of labour and household tasks – Place to work (on the average, women spend long times at home). Isolation and lack of external information (radio, TV or internet) – Economic dependence (absence of income and properties-land)
	Gender inequality	– Patterns of gender discrimination because of social and cultural norms – Domestic and sexual violence – On the average, men own the land. Power relation – When a disaster strikes, relief efforts are usually managed, controlled and distributed by men as heads of the family, resulting in discrimination access to relief supplies – On the average, women are overcharged of tasks (transportation of water and wood, family charge, householder tasks …) – After a disaster strikes, it is complicated to sustain law and order and often the assault to women increase – Lack of privacy and resources during female hygienic care (if they are in camps)

Table assembled by the author

Table 2 Women vulnerability during the process of migration

During migration	Physical and biological	– Need of hygiene, increasing in the reproductive stage (pregnancy/lactation) – Women are more susceptible to disease – Victims of human traffic or sexual/physical violence – Lack of nutrition (less access to resources)
	Social norms and role	– The status of the family (both parents are still alive/single parent) – Family charge – Absence of education (lack of skills) – Place to work (Average women spend longer times at home). Isolation and lack of external information (radio, TV or internet) – Women are prevented to leave their homes – Behaviour restrictions of dependence to their husbands and brothers
	Gender inequality	– Economic dependence – Sex gender-based violence

Table assembled by the author

After Crossing the Border: Post-migration

When the decision to migrate is made, it is necessary to consider the relevant factors from a global perspective, including income, health, education and natural resources (Table 3): "These factors influence not only whether a woman moves but they also influence the countries to which women migrate" (Brookings et al. 2015). When a woman reaches a new country she can be involved in several compelling situations such as pregnancy, in charge of family (with/without husband), orphan, victim of human trafficking, dealing with a mutilation or amputation, relevant health problems, or being survivor of sexual/physical violence (rape), are common states of affairs (Women's Refugee Commission 2016).

Cultural and gender inequality vulnerabilities, as the absence of education for young girls, leads them to being less trained than men, this affecting their vocational skills and their knowledge of other languages (Chindarkar 2012). As a result, they have limited access to job positions and worthy salaries. Lack of resources can increase the risks of exploitation such as prostitution, sexual trafficking, early marriage, poverty, and unhealthy or economic dependent situations (Jolly and Reeves 2011). Females need to cope with their personal situations (family charge, lack of education, physical and psychological burdens) while also needing to adapt to a new environment and deal with new challenges.

If migrant women cross borders, they may end in mix migration camps (as they cannot be direct to a refugee camp, as they are not asylum seekers) with no access to employment opportunities, therefore women lose the opportunity to be economically

Table 3 Women vulnerability when they reach the destination country

Post-migration	Physical and biological	– Physiological (stress, trauma, high levels of sensitivity, depression, loss of self-confidence, feeling of loneliness, acceptation of the possibility of no return, loss of people, loss of home) – Possible physically affected (sexual abuse, rape, amputation, mutilation) – Deterioration in nutritional status
	Social norms and role	– Family charge – Lack of opportunity to find a job (lack of education, family charge, partner power relation, language) – Adaptation to women role in the country of destination – Adaptation to a new religion, culture, policies and laws – Learn another language and skills
	Gender inequality	– Limit access to a job position – Women can resort to sexual work, trafficking, smuggling or forced marriage because of lack of resources (language, network, skills and incomes) – Inequality of salaries – Lack of education to access a better job – The family charge and the power relationship with husbands or family males will exclude women from integration programs

Table assembled by the author

independent. It is also necessary to consider the limited participation in decision-making processes that affects their social independence. The householder and cultural roles burden women with domestic tasks, exposing them to sexual and gender-based violence attacks, including domestic violence (Brookings et al. 2015).

International Policy Frameworks for Women Facing Cross Border Displacement by Sudden Events

The main challenge in climate induced cross border migration policies is to include a gender perspective to fill the existing gap between the creation of procedures to face the problem and to implement them. To address this challenge, particularly for climate-induced migration, two problem-solving approaches need to be considered. First, to accept the absence of an international policy framework addressing climate induced displacement by sudden events. Second, to consider the equity perspective of gender in the policy process of international migration. In the following, the international conventions and agreements that consider climate change, migration and gender perspective are analyzed, together with the labor of the agencies inside the United Nations that deal with aspects of the problem and the new state-led consultative process.

International Conventions and Agreements

There is a lack of an international framework embracing women as displaced persons because of sudden disasters, but there are several international conventions and agreements that concern the main topics of this article (gender, climate-related disasters and migration), as shown in Tables 4 and 5.

The Global Compact on Migration signed on July 2018 is the stronger international agreement embracing gender (dissociated in women, men, girls and boys), forced migration and climate-related disasters. The agreement identifies 23 objectives to combat migration. It is important to highlight objective 2, including the deterioration

Table 4 Conventions and agreements that include: climate-related disasters and/or gender and/or migration

Year	Convention/agreement	Gender approach	Climate change approach	Migration approach
1967	Declaration on the elimination of discrimination against women	X		
1995	Beijing Declaration and Platform for Actions	X		
2001	Kyoto protocol		X	
2007	The global gender and climate alliance	X		X
	Bali action plan	X		
2010	Cancun agreement	X		X
2013	The Warsaw International Mechanisms	X		X
2014	Lima call for action	X	X	
2015	Sendai Framework for Disaster Risk Reduction	X	X	X
2018	The Global Compact for Migration	X	X	X

Table assembled by the author

Table 5 Summaries all the international conventions and declarations created by the United Nations considering gender in a climate change or migration perspective

Conventions and declarations	Articles/objectives
Rio declaration	20
Beijing Declaration and Platform for Actions	22, 124 (g),147 (h), 158
Kyoto protocol Doha amendment	23
Cancun agrement	12, 14 (f)
Lima call for action	Lima work programme on gender
Sendai framework 2015–2030	4, 19 (d), 32, 36 (a) and (i)
Paris Agreement	7 (5), 11 (2)
Global Compact for Migration	2, 3, 5, 7, 11, 15, 18, 20

Table assembled by the author

of the environment as a drive to leave a country and the importance of integrate displacement in the preparedness strategies (IOM 2018).

The Sendai Framework for Disaster Risk Reduction (substitute for the Hyogo Framework for Action) and the Paris Agreement, both signed in 2015, consider the three main topics, but only the first was ratified by 195 member states. The Sendai Framework highlights gender perspectives as a key factor for the empowerment of women, promoting gender equality through effective response, recovery, rehabilitation and reconstruction. It also highlights the need to integrate gender into policies and practices. On the other hand, the Paris Agreement focuses on the adaptation strategy on gender responsive, participatory and transparent approaches for new policies. It also considers the necessity of taking actions to address climate change and to respect, promote and consider the obligations of migrants, in a human rights perspective.

The Conference of Parties (COP) is the supreme decision-making body inside the UNFCCC (United Nations Framework Convention on Climate Change). The last COP 24 was hold in Katowice, Poland. During the COP 24 presented the Rulebook, a guidance to implement the bottom-up lends of the Paris Agreement.

The Paris Agreement is the most recent convention to fight the adverse actions of climate. This convention was signed at COP 21, where a gender approach has been considered, but international displacement of women due to sudden events has not been addressed. In the seventh COP, parties highlighted the necessity to increase women's participation in climate-related concerns (UNDP 2015). In 1995, the Beijing Declaration and Platform for Action recognized "the need for women's participation in the decision-making process to combat, mitigate, and adapt to climate change" (UNDP 2015). In 2014, the COP 20 included a specific work program to reach gender equality at the policy-making level. The Paris Agreement COP 21 (ratified by 23 parties of 197) recognizes the need to acknowledge the rights of migrants (not specifically women) in the context of climate change action, and sets out plans to establish a task force on climate change displacement. The agreement reached a historical moment by including climate-related loss and damage in the decision. There are two main debates concerning loss and damage. The first gives more importance to loss and damage in the UN climate debate. The second concerns the legal issue between developed and developing countries on liability and compensation. From the point of view of developing countries, loss and damage needs to be part of a legal binding in a deal, subject to universal laws. On the other hand, developed countries are not comfortable in inscribing loss and damage under an international law, preferring an agreement focused on ensuring international compensation to the climate victims of the future. The reference to gender and women's empowerment appear in the text but is not integrated into all areas of the agreement (The Advisory Group on Climate Change and Human Mobility 2015). The first document that relate gender, migration and climate change is the "Gender Action Plan" (UNFCCC 2017).

Over the years, women have been taken into consideration at the conference of parties (Table 6a). After analyzing the gender perspective throughout COP, four of them should be highlighted: COP 7, COP 18, COP 20, and COP 21. These consider the need to improve women's participation, while also promoting balance. In the COP 20, the discourse over gender gained momentum, and a work program focusing on women and their needs was carried out. The COP 20 implemented the decision to represent women in all aspects of the Convention. Participation and balance were still considered important, but gender equality and gender-responsive decisions for mitigation and adaptation in climate policies were added. The last COP 21, the Paris Agreement, focuses its attention on the need for gender responses during adaptation actions through environmental and socio economic policies. In addition, the COP 21 pushes all levels of action (national, sub-national and local) to build capacity in gender responsive in adaptation actions.

COP's 22 and 23 were key to establish the Gender Action Plan (GAP). As mentioned before, previous COP's pointed out gender as a priority approach, but GAP analyzed the importance of women's role in climate action for the first time. The plan considers five objectives to improve the effectivity to effective inclusion of gender approach in climate change negations.

Table 6a Analysis of the Conferences of Parties (COP) from a gender approach

Conference of Parties	Location	Year	Gender approach
7	Marrakech	2001	Decision 36
18	Doha	2012	Decision 23
20	Lima	2014	Lima work programme on gender
21	Paris	2015	Articles 7–5 Article 11
22	Marrakech	2016	Gender Action Plan
23	Bonn	2017	
24	Katowice	2018	The Paris rulebook

Table assembled by the author

Table 6b Analysis of the Conferences of Parties (COP) from a migration approach

Conference of Parties	Location	Year	Migration approach
16	Cancun	2010	Paragraph 14 f
18	Doha	2012	Decision 3 paragraph 7 vi
19	Warsaw	2013	Warsaw International Mechanisms (WIM)
20	Lima	2014	Article 8, 2, 4
21	Paris	2015	Article 8
22	Marrakech	2016	– Marrakech Partnership for Global Climate Action – Adoption of the Warsaw International Mechanism (WIM) on loss and damage
24	Katowice	2018	The rulebook of the Paris Agreement

Table assembled by the author

Analyzing the Conferences of Parties from a migration approach (Table 6b), climate-related migration appears for the first time in the COP 14. The COP 16 highlights the need to enhance adaptation action to understand, coordinate and cooperate in climate change induced displacement, migration and planned relocation at all levels. In Doha, the COP 18 recognized how the impact of climate change is affecting migration patterns, displacement and human mobility. The Warsaw International Mechanisms were established in the COP 19 in the context of loss and damage because of climate change. The Paris Agreement addressed the need to understand the loss and damage associated with the adverse effects of climate change. Loss and damage has been considered in the agenda since COP 19 trough Warsaw International Mechanisms, reviewed at COP 22. At COP 23 the Marrakech Partnership for Global Climate Action was adopted to support the implementation of the Paris Agreement. Gender, climate change and migration started to be considered together at the Paris Agreement.

United Nations Agencies

There is a negative 'feedback loop' between defining an international displacement person within the context of climate change and the need to identify existing International Conventions. Literature uses different names instead of a global term that adequately addresses those particular groups of people. The lack of an international framework for the displacement of women means that it is also important to analyze the labor of the executive agencies that are responsible for implementing the programs dealing with the international displacement by sudden climate-related disasters.

Inside the UN, there are four different agencies (UNHCR, UNDP, UN Women, and IOM) carrying out work within the scope of this paper. Currently, the IOM (International Organization of Migration) started to be part of the UN in 2016 (UN

News 2016) being the most prominent body in pushing the agenda forward, and its terms and definitions are as follows:

> Environmentally displaced person: "Persons who are displaced within their country of habitual residence or who have crossed an international border and for whom environmental degradation, deterioration or destruction is a major cause of their displacement, although not necessarily the sole one. This term is used as a less controversial alternative to environmental refugee or climate refugee [in the case of those displaced across an international border] that have no legal basis or raison d'être in international law, to refer to a category of environmental migrants whose movement is of a clearly forced nature" (IOM 2014b).

The IOM[7] has changed their research interests, adapting to social priorities and needs. In the last decade, streams of research on climate change and migration have been intensified after the coordination of several camps in the aftermath of a climate-related disasters event. The IOM focuses on research, establishing projects and coordinating seminars to raise awareness about climate change and migration. Notwithstanding, the IOM does not have a mandate to protect displaced people, but operates within the UNFCCC (United Nations Framework Convention on Climate Change) to address the impacts of climate change and migration (Hall 2016). As a result, the IOM became the first institution that defined environmental migrants, outlining a series of principles for the UNFCCC member states to address environmental migration in their national policies. One example of this collaboration was the side–event at the Conference of Parties (COP) 15 in Copenhagen addressing "Climate Adaptation Continuum, Migration and Displacement: Copenhagen and Beyond", where different organizations, including the UNFCCC and the IOM, worked together (IOM 2015b).

Since 1989, the IOM collaborated with the UNHCR (The Office of the United Nations High Commissioner for Refugees) with voluntaries, economics migrants and individuals in need of international assistance (grey zone) such as stranded migrants, IDPs (Internally displaced persons) and those displaced by disasters or potentially by climate. IOM member states did not try to overlap with other humanitarian agencies, and did not consider emergency work as a priority. Over the following years, the agency focused its activity on research regarding the environment and migration (Hall 2016). On the other hand, the IOM continues to conduct research on the topic of environmental migration and, in 2016, published "The Atlas of Environmental migration" (Ionesco et al. 2016). Until December 2016, the organization ran a project called "Migration, Environmental and Climate Change: Evidence for Policy (MECLEP)", to fill gaps in the formulation of new policies through analyzing environmental migration data in six different countries. Furthermore, as part of the IOM, the Migrants in Countries in Crisis Initiative (MICIC) launched its 2016 guidelines to protect migrants in countries experiencing conflict or climate-related disasters, to provide the stakeholders.

[7]The IOM has been an international organization independent of United Nations until 25th of July of 2016, when it became a part of the United Nations (UN News Service Section 2016). Prior to this, the IOM was a committee that had been involved with the environmental migration since 1989 (Hall 2016). The IOM was created as a provisional agency focused on European migration (Hall 2016).

The United Nations Convention related to the Status of Refugees of 1951 defined in article 1 the term refugee, as well as enumerating in section A2 the 'reasons of persecution'[8] to seek asylum in other countries (UNHCR 1951). The UNHCR has a specific policy for women refugees and five commitments for them, with particular attention to the international approach. In the last years, international dialogues have push the term "Climate refugee" or "environmental refugee", as a possible name to define people who are internationally displaced by climate-related disasters (UNHCR 2011a). However, climate-related disasters or environmental catastrophes are not considered a reason for protection under this convention. On the other hand, the UNHCR conducts research and produces on the topic, producing reports such as "Guidance on Protecting people form disaster and environmental change through planned relocation" (Brookings et al. 2015).

The UNDP focuses on sustainable human development and, in 2007, climate change became a priority. This resulted in the issue of the 2008 report "Human development, the climate change strategy". In September 2015, the new agenda of sustainable development goals was published, including goal 5—gender equality and goal 13—climate action. Like others United Nations agencies, the UNDP created a guidebook ("Gender, climate change and community-based adaptation") for designing and implementing gender sensitive policies based in community adaptation programs and projects (UNDP 2010). The 2014 UNDP report "Migration, Resettlement and Climate Change in Viet Nam" (UN 2014), put forward visible results in migration as a way of adaptation within a bottom-up approach that can be extrapolated into the creation of the international convention.

In the last decade, UN Women considered a gender approach for labor migration to promote efficient policies in developing countries. Since the signature of the Global Compact for Migration, UN Women has participated in different high-level events to ensure that the agreement achieves equality and empowers females (UN Women 2018).

The Nansen Initiative: Advance the New Policies

This Initiative was launched in 2012 by the governments of Norway and Switzerland as a response to a protection gap for cross-border displacement in the context of disasters and climate change. The process of recommendation relays on consultations between all stakeholders involved in building consensus between states. The Initiative enhanced these efforts by providing a bottom-up approach with a large network of

[8]As a result of events occurring before 1 January 1951 and owing to well-founded fear of being persecuted for reasons of race, religion, nationality, membership of a particular social group or political opinion, is outside the country of his nationality and is unable or, owing to such fear, is unwilling to avail himself of the protection of that country; or who, not having a nationality and being outside the country of his former habitual residence as a result of such events, is unable or, owing to such fear, is unwilling to return to it (OHCHR 2015).

members, which enables a platform for gender and social equity declarations based on knowledge and experiences (Schrepfer and Kälin 2012).

In 2015, the Protection Agenda (Protection of Cross-Border Displaced Persons in the Context of Disasters and Climate Change) described how to protect a cross-border disaster-displaced person, how to reduce risk in the original country, and pointed out the priority areas for future actions. The first chapter of volume I of the Agenda analyzes how to identify these kinds of migrants, the need for preparedness through disaster risk management, and how humanitarian protection laws need to be based on immigration and human rights laws. Moreover, it identifies the rights of migrants and their responsibilities during the stay in the residence country, while also analyzing the no-return situation to the country of origin. In a gender perspective, volume I also addresses how women are subjected to specific risks when migrating, such as trafficking or exploitation, often being the ones most seriously affected by disasters. The agenda emphasizes the need to engage women for a successful relocation, involving them in decision-making.

The second chapter questions disaster risk management in the country of origin, to find a way to reduce vulnerability and build resilience. Finally, the third chapter identifies the priority areas for future actions and the need for collecting data on the use of humanitarian protection measures to strengthen disaster risk management plans in countries affected by climate change (The Nansen Initiative 2015b).

The volume II of the Agenda analyses regional human movement dynamics due to climate change through projections and examples, highlighting that South Asia is one of the most affected areas by displacement risk (The Nansen Initiative 2015c). The Platform on Disaster Displacement was established in May 2016 to follow up the Nansen Initiative.

Discussion of Results

The three main points of this article are gender, migration (displacement), and climate-related sudden disasters events. However, even though there are many organizations that can deal with these problems, there is no consensus about which should be the one taking responsibility. The lengthy working process in the UN and its way to establish conventions also creates momentum, as well as delays progress. However, various efforts from national initiatives, such as the Nansen Initiative, have been discussing this policy gap in scientific articles (Ransan-Cooper et al. 2015).

The international conventions focus their effort on finding better practices and creating policies on adaptation and mitigation in order to tackle climate change impacts. According to the present international dialogue in the context of climate change, three sets of obligations have been stated, namely: mitigation, i.e. the task of curbing the degree of greenhouse gas (GHG) emissions; adaption, i.e. the challenge of how best adapt to the impacts of climate change; and protection, i.e. the exigency to secure the rights and address the (humanitarian) needs of people affected by negative effects of climate change (Schrepfer and Kälin 2012).

Although the term environmental displacement has been accepted in the migration terminology for the UN and stakeholders involved in the climate change topic, it has been considered a direct consequence of a set of factors acting after a strong climate exposure increasing vulnerabilities after a disaster. The 1951 Refugee Convention, through the definition of the term "refugee" (Article I), gives a legal status to these migrants (UNHCR 1951). The Global Compact for Migration has drawn a line between two groups: migrants and refugees (IOM 2018). This distinction may be the foundation for future policies on migration.

In Section "Women perspective", women were identified as being vulnerable when directly affected by disasters and climate change impacts, as well as by other processes that leading to displacement. Some vulnerabilities directly depend on the physiognomy of women, negatively affecting their security and increasing their exposure to dangerous situations, such as rape or human traffic. Young girls and adolescents are the most vulnerable after a sudden disaster. On the other hand, some vulnerability factors can be further imposed by the society in the countries of origin and destination. Furthermore, the role of women in developing countries is closely linked with cultural and religious traditions. One example of this is the impossibility to have a paid job given the time consuming informal labor conducted at the household level, which is unpaid and unrecognized. However, it is deemed necessary to move from vulnerabilities, understood as socio-cultural roles, to advantages conferred on women to cope with adverse situations and build resilience. An example of this occurred during the Tsunamis in South and South-East Asia in 2004, where women climbed trees, and those who could swim rescued people. In addition, women can cope better with food shortages because of their lower nutritional requirements and higher body fat, unless they are pregnant (Chindarkar 2012).

Often, women were identified as a minor group or as victims in environmental migrations. This marginalized position does not allow women to be seen as a key victim in the adaptation to climate change, without emphasizing how women (after a disaster) become the head of the household if the family disintegrates, or how they learn useful abilities if they have a chance to be educated. Women need to be seen as a main member of the family, independently of a cultural role that puts them in charge, made responsible of agricultural tasks and of the household labors in general.

The analysis of the different international frameworks shows how the absence of international policies leaves women in a vulnerable position. This makes it difficult for them to cope in a post disaster scenario, and reduces their possibilities to create resilience and empowerment.

There is growing awareness about gender approaches in the creation of policies. However, this consideration only operates at a superficial level. The international conventions intend to involve women in participatory processes and decision making, but in fact in some cultures and countries women are not equal to men (officially or not). The conventions also stress the necessity to have gender responsive policies, but in practice it is complicated to implement these at a lower level. The only international legal instrument that quotes several vulnerabilities that women face through the migration process, is Recommendation No. 26 on women migrant workers against sex- and gender-discrimination (The Committee on the Elimination of Discrimination against Women 2008). These recommendation intends was created to fill the gaps of the International Convention on the protection of the Rights of all migrants workers and members of their families, that protects individual human but not specifically women.

The work of UN agencies was conducted to evaluate if there is a mandate organism to conduct research on the topic, and how different actions deal with the absence of an international framework. The existing gap between the lack of universal recognition and definitions for international migrants by hazards and the corresponding want of international policies was identified. The four agencies of the UN currently carrying out research on the topic of women's displacement have produced several guidelines that help in the implementation of the general international conventions. Two examples of this is the "Guidance on protecting people from disasters and environmental change through planned relocation" (Brookings et al. 2015) and the Guidelines for prevention and response "Sexual and Gender-Based Violence against Refugees, Returnees and Internally Displaced Persons" (UNHCR 2003) that in the future can be extrapolated to cross border displacement. Furthermore, the Global Compact for Migration opens a new path for the creation of policies that include the three pillars proposed in this paper.

The need for a bilateral approach (top down and bottom up) is urgent for the implementation of new policies. Good practices also need to be considered relevant for the creation of an international framework. Some countries like Brazil or Kiribati have dealt with climate change displacement since long time ago.

After the 7.3 magnitude earthquake hit Haiti in 2010 (300,000 people injured, 220,000 killed and 1.5 million people homeless), Brazil faced a Haitian forced migration looking for new opportunities. The National Committee for Refugees (CONARE) based on the Cartagena Declaration on Refugees, decided that the Haitians displaced could not be granted refugee status because the convention does not consider hazards and climate factors as a cause for asylum. Simultaneously, the CONARE and The National Immigration Council in Brazil (CNIg) in collaboration with UNHCR, gave to the Haitians citizens a special permit (humanitarian visa) to stay in Brazil, as a complementary protection until the situation in their country was improved (UNHCR 2011b). In June 2011, UNHCR and the OACNUDH made an official call to the Brazilian's government to suspend the involuntary repatriation of Haitian citizens because of the situation of the country. Since 2012, Brazil adopted a special humanitarian migration policy that continues until now, giving special visas to Haitians. In November 2015, the Brazilian Ministry of Justice enlarged the Haitian rights, giving to 43,871 migrants the opportunity to request for a permanent residence in the country (Governo Federal do Brasil 2015).

The fifth assessment report of the IPCC had a specific regional chapter analyzing the impact of climate change in small islands and how they may adapt and manage the pertinent risks (IPCC 2014). The Republic of Kiribati is an example of good practice. This country carried out a three-phase adaptation program (Preparation, Pilot Implementation and Expansion) from 2003 to 2016. Currently, the president of the country had bought land in the Fiji Island to support the relocation of the citizen of Kiribati when forced to abandon the place (Caramel 2014).

Conclusion: The Need of an Interdisciplinary Policy Pathway

The main aim of this article is to assess if current international conventions and agreements regarding the increasing environmental displacement of people, continuously experienced in recent times, include a gender equity approach in the development and implementation of their corresponding policies. This consideration is deemed essential to supplement, protect and enforce basic gender rights. Although dissociation between gender has been included in international policies such as the Global Compact on Migration, and most existing conventions recognize the necessity of establishing equality between genders, none of them fully implement policies or actions addressing this issue through enactment. Recommendation 26th of the Committee on the Elimination of Discrimination against Women recognizes that labor migration can be a threat for women's human rights and a risk for their security, and should be taken into consideration for the creation of gender sensitive policies related with environmental migration. Nevertheless, any international agreement or convention must consider females as persons displaced by sudden events, which should be protected from all possible vulnerabilities experienced during the migration process. Thus, it is fundamental to handle the displacement of females from an interdisciplinary perspective. In this respect, it is indispensable to create a multilateral approach (bottom-up, top-down) for the adaptation of existing frameworks or the formation of new ones, which should be developed at the international, regional, sub-regional and national levels.

Women, due to the fact of being women, need to be protected by specific international policies to ensure equality and equity between genders. However, women should not be identified as a special group which should be privileged in comparison to males, but as one to which affirmative actions should be assigned in the pertinent policies in order to reach a balance between genders. In an international framework that embraces women in a displacement situation, it is required to move from socio-cultural vulnerabilities to proceed to give advantages to women in order to empower them to cope with adverse conditions and to build resilience.

References

Brookings, UNHCR, George Town University (2015) Guidance on protecting people from disasters and environmental change through planned relocation. https://www.brookings.edu/research/guidance-on-protecting-people-from-disasters-and-environmental-change-through-planned-relocation/

Caramel L (2014) Besieged by the rising tides of climate change, Kiribati buys land in Fiji. The Guardian. https://www.theguardian.com/environment/2014/jul/01/kiribati-climate-change-fiji-vanua-levu

Chindarkar N (2012) Gender and climate change-induced migration: proposing a framework for analysis. Environ Res Lett 7(2). https://doi.org/10.1088/1748-9326/7/2/025601

FAO (2016) FAO's emergency activities: technical handbook series. http://www.fao.org/docrep/003/X6868E/x6868e00.htm#d

Governo Federal do Brasil (2015) Governo brasileiro garante direitos para imigrantes haitianos. Ministério da Justiça e Cidadania. http://www.justica.gov.br/noticias/governo-brasileiro-garante-direitos-para-imigrantes-haitianos
Hall N (2016) Displacement, development, and climate change: international organizations moving beyond their mandates. https://doi.org/10.4324/9781315639758
IDMC (2015) Global estimates 2015. http://www.internal-displacement.org/assets/library/Media/201507-globalEstimates2015/20150713-global-estimates-2015-en-v1.pdf
IOM (2007) Discussion note: migration and the environment. Ninety-fourth session. https://www.iom.int/jahia/webdav/shared/shared/mainsite/about_iom/en/council/94/M_INF_288.pdf
IOM (2014a) Glossary migration, environment and climate change: evidence for policy (MECLEP). http://publications.iom.int/system/files/pdf/meclep_glossary_en.pdf?language=en
IOM (2014b) IOM oultlook on migration, environmnent and climate change. https://publications.iom.int/system/files/pdf/mecc_outlook.pdf
IOM (2015a) Migration and climate change. https://www.iom.int/migration-and-climate-change-0
IOM (2015b) IOM at UNFCCC conference 15, Copenhagen. https://www.iom.int/jahia/webdav/shared/shared/mainsite/activities/env_degradation/migration_displacement_flyer.pdf
IOM (2018) Global compact for migration. https://refugeesmigrants.un.org/sites/default/files/180711_final_draft_0.pdf
Ionesco D, Mokhnacheva D, Gemenne F (2016) Atlas des migrations environnementales. https://environmentalmigration.iom.int/sites/default/files/01-AtlasMigration-ta%CC%80p-18MARS-WEB.pdf
IPCC (2012) Managing the risks of extreme events and disasters to advance climate change adaptation. Special report of the intergovernmental panel on climate change. https://www.ipcc.ch/report/managing-the-risks-of-extreme-events-and-disasters-to-advance-climate-change-adaptation/
IPCC (2014) Summary for policymakers in climate change 2013—the physical science basis: working group I contribution to the fifth assessment report of the intergovernmental panel on climate change. Cambridge University Press. https://doi.org/10.1017/cbo9781107415324.004
Jolly S, Reeves H (2011) Gender and migration. Bridge: development-gender. http://www.bridge.ids.ac.uk/sites/bridge.ids.ac.uk/files/reports/CEP-Mig-OR.pdf
Masika R (2002) Gender, development, and climate change. https://oxfamilibrary.openrepository.com/bitstream/handle/10546/121149/bk-gender-development-climate-change-010102-en.pdf;jsessionid=A4B7684A68798E3F98EE003B9DB62E18?sequence=1
Mendelsohn R, Dinar A, Williams L (2006) The distributional impact of climate change on rich and poor countries. Environ Dev Econ 11(2):159–178. https://doi.org/10.1017/S1355770X05002755
Neumayer E, Plümper T (2007) The gendered nature of natural disasters: the impact of catastrophic events on the gender gap in life expectancy, 1981–2002. Ann Assoc Am Geogr 551–566. https://doi.org/10.1111/j.1467-8306.2007.00563.x
OECD (2013) World migration in figures. http://www.oecd.org/els/mig/World-Migration-in-Figures.pdf
OHCHR (2015) Understanding human rights and climate change. https://www.ohchr.org/Documents/Issues/ClimateChange/COP21.pdf
Ransan-Cooper H, Farbotko C, McNamara KE, Thornton F, Chevalier E (2015) Being(s) framed: the means and ends of framing environmental migrants. Glob Environ Change 35:106–115. https://doi.org/10.1016/j.gloenvcha.2015.07.013
Schrepfer N, Kälin W (2012) Protecting people crossing borders in the context of climate change normative gaps and possible approaches. https://www.unhcr.org/4f33f1729.pdf
The Advisory Group on Climate Change and Human Mobility (2015) Human mobility in the context of climate change UNFCCC—Paris COP-21. https://www.unhcr.org/protection/environment/565b21bd9/human-mobility-context-climate-change-unfccc-paris-cop-21-recommendations.html
The Committee on the Elimination of Discrimination against Women (2008) General recommendation No. 26 on women migrant workers. https://www2.ohchr.org/english/bodies/cedaw/docs/GR_26_on_women_migrant_workers_en.pdf

The Nansen Initiative (2015a) The Nansen Initiative global consultation. https://www.nanseninitiative.org/global-consultations/
The Nansen Initiative (2015b) Agenda for the protection of cross-border displaced persons in the context of disasters and climate change, vol 1. https://nanseninitiative.org/wp-content/uploads/2015/02/PROTECTION-AGENDA-VOLUME-1.pdf
The Nansen Initiative (2015c) Agenda for the protection of cross-border displaced persons in the context of disasters and climate change, vol 2. https://www.nanseninitiative.org/wp-content/uploads/2015/02/PROTECTION-AGENDA-VOLUME-2.pdf
UN (2014) Migration, resettlement and climate change in Viet Nam. https://www.undp.org/content/dam/vietnam/docs/Publications/Migration%20&%20Climate%20change%20-%20Eng.pdf
UN News (2016) UN News—General Assembly decision makes International Organization for Migration part of UN system. http://www.un.org/apps/news/story.asp?NewsID=54545#.V5nn0fmLTIU
UN Women (2018) Promoting gender-responsive migration governance through the global compact for migration. http://www.unwomen.org/en/news/events/2018/09/event-promoting-gender-responsive-migration-governance-through-the-global-compact-for-migration
UNDP (2010) Gender, climate change and community-based adaptation. https://www.undp.org/content/dam/aplaws/publication/en/publications/environment-energy/www-ee-library/climate-change/gender-climate-change-and-community-based-adaptation-guidebook/Gender%20Climate%20Change%20and%20Community%20Based%20Adaptation%20(2).pdf
UNDP (2015) Human development report 2015, work for human development. http://hdr.undp.org/sites/default/files/2015_human_development_report_0.pdf
UNESCO (2016) Displaced person/displacement. http://www.unesco.org/new/en/social-and-human-sciences/themes/international-migration/glossary/displaced-person-displacement/
UNFCCC (2017) Gender action plan. https://unfccc.int/sites/default/files/cp23_auv_gender.pdf
UNHCR (1951) Convention and protocol relating to the status of refugees. https://cms.emergency.unhcr.org/documents/11982/55726/Convention+relating+to+the+Status+of+Refugees+%28signed+28+July+1951%2C+entered+into+force+22+April+1954%29+189+UNTS+150+and+Protocol+relating+to+the+Status+of+Refugees+%28signed+31+January+1967%2C+entered+into+force+4+October+1967%29+606+UNTS+267/0bf3248a-cfa8-4a60-864d-65cdfece1d47
UNHCR (2003) Sexual and gender-based violence against refugees, returned and internally displaced person-guidelines for prevention and response. https://cms.emergency.unhcr.org/documents/11982/51689/UNHCR%2C+Sexual+and+GenderBased+Violence+against+Refugees%2C+Returnees+and+Internally+Displaced+Persons.+Guidelines+for+Prevention+and+Response%2C+2003/9591f62d-4f6a-4faf-8c03-df0d6bacf871
UNHCR (2011a) Summary of deliberations on climate change and displacement. https://www.unhcr.org/4da2b5e19.pdf
UNHCR (2011b) Haitianos reciben residencia permanente en Brasil. http://www.acnur.org/noticias/noticia/haitianos-reciben-residencia-permanente-en-brasil/
Ware J, Doig A (2014) Taken by storm: responding to the impacts of climate change. https://www.christianaid.org.uk/sites/default/files/2017-08/taken-by-storm-responding-impacts-climate-change-march-2014.pdf
Wisner B, Blaikie P, Cannon T, Davis I (2003) At risk: natural hazards, people's vulnerability and disasters. https://www.preventionweb.net/files/670_72351.pdf
Women's Refugee Commission (2014) I'm here: adolescent girls in emergencies approaches and tools for improved response. https://www.womensrefugeecommission.org/images/zdocs/I-m-Here-report-FINAL.pdf
Women's Refugee Commission (2016) Falling through the cracks: refugee women and girls in Germany and Sweden. https://www.womensrefugeecommission.org/gbv/resources/1308-protection-germany-sweden

Re-naturing Cities: Impact of Microclimate, Human Thermal Comfort and Recreational Participation

Ruzana Sanusi and Sheena Bidin

Abstract With rapid urbanisation, changes to the urban environment and climate are inevitable. This exacerbates the risks for meteorological hazard such as heatwaves where this is one of the common issues faced by major cities worldwide. Heat related issues pose an impact not only to the environment but also lead to heat related human health problems, economic loss to businesses, damages to infrastructures and in extreme cases, cause deaths. Poor urban designs that are characterised by the abundance of grey infrastructures increase the impact on urban climate with the potential of intensifying heat waves and urban heat island (UHI) effects. Additionally, these grey designs also limit recreational participation. Re-naturing cities using green infrastructure can provide nature-based solutions that can be useful in mitigating the abovementioned meteorological hazards. Therefore, understanding the impact of re-naturing cities using green infrastructures (GI) on microclimate mitigations, human thermal comfort improvement and recreational participation is beneficial in improving the urban environment and ensuring the wellbeing of urban citizens. This paper provides justifications on implementing nature based solutions using green infrastructures in cities, thus improving future urban planning that potentially increases cities' resilience to environmental hazards.

Introduction

Globally, urban areas are expanding due to the expansion of populations living in cities. Reportedly in 2018, 55% of the world's populations are living in urban areas and expected to continue (UN DESA 2018). Within the next 20 years, the world's urban population is projected to reach 68% (UN DESA 2018). Many countries will face challenges in achieving sustainability and resilience of the urban areas with this

R. Sanusi (✉) · S. Bidin
Faculty of Forestry, Universiti Putra Malaysia, 43400 Serdang, Selangor, Malaysia
e-mail: ruzanasanusi@upm.edu.my

R. Sanusi
Institute of Tropical Forestry and Forest Products, Universiti Putra Malaysia, 43400 Serdang, Selangor, Malaysia

W. Leal Filho et al. (eds.), *Climate Change, Hazards and Adaptation Options*, Climate Change Management, https://doi.org/10.1007/978-3-030-37425-9_28

expansion. In particular, countries such as India, China and Nigeria are expected to have highly concentrated urban population with the projected population growth in these three countries are expected to rise 35% by 2050 (UN DESA 2018). This imposes greater responsibilities in minimising the implications of urban expansion and finding strategies in ensuring urban sustainability and resiliency, especially in fast-growing cities.

This expansion increases pressures on cities as many of the natural resources have been converted into the built environment to support the demands for residential, commercial and industrial areas. Given the increase of urban population, large fraction of the population will be exposed to climate driven hazards as the changes in the cities imposes a greater proportion of sealed surfaces in these areas. These sealed surfaces built with artificial materials such as concrete, asphalt and other impervious materials have been widely reported to be one of the factors that lead to Urban Heat Island (UHI) effect. UHI is a common phenomenon in urban area as the air and surface temperatures, especially during night time in cities that are characterised with extensively built infrastructures is greatly higher than the rural areas with minimal grey infrastructures (Block et al. 2012). The UHI is often formed as a result from the combination of factors such as less vegetation that reduces the evaporative cooling, anthropogenic heat emissions introduce by human activities and changes of the surface roughness, albedo and urban canyon geometry due to the altered urban environment. Moreover, UHI exists in almost all cities regardless of the city size (Debbage and Shepherd 2015) and climate (i.e. warm or cold) (Wang et al. 2016).

Due to the escalation of climate change and urbanisation, UHI introduces more challenges to urban citizens in the future (Rizwan et al. 2008). The combination with urbanisation, poor urban design and UHI as well as current climate change progression and future prediction, it is more likely this will lead to more intense and prolonged climate hazards such as extreme temperatures, changes in precipitation patterns and drought (IPCC 2014). This potential climate hazards, can influence human health and wellbeing of urban citizens in terms of thermal perception as well as participation in outdoor recreational activities as the climate hazards may introduce human heat related health problems and in extreme cases can eventually cause mortality (Chen et al. 2015; Joe et al. 2016; López-Bueno et al. 2019). Given the impact of the urbanisation on urban climate and human health wellbeing, it is therefore imperative to have a better understanding of the complex interaction of the climate, thermal stress and the recreational participation of urban citizens as well as addressing the mitigation strategies to ensure urban sustainability and resiliency. In Malaysia particularly, where cities such as the Klang Valley has seen rapid urbanisation and effects of climate changes, discussions and research that can contribute to improving the liveability of urban residents through natural mitigations are needed.

Urbanisation: Microclimate, Human Thermal Comfort, Recreational Participation

Urban Microclimate

Urbanisation has changed the urban fabric that resulted from the land use land cover (LULC) changes that alter the urban environment. Land cover is characterised by the physical elements of the earth surface such as vegetation, water, soil and ground surfaces while land use usually represents the use of land carried by humans where these two terms of LULC can sometimes be used interchangeably although they have different meanings (Rawat and Kumar 2015). The shift caused by LULC changes driven by human activities for urbanisation contributed to the deterioration the urban landscapes. These LULC changes have changed the biophysical characteristics of the urban areas such as the ground surfaces, albedo and soil moisture due to the activities and the materials used (i.e. asphalt and concrete) for buildings, road and paved surfaces (de Noblet-Ducoudré et al. 2012; Ishola et al. 2016). This alteration of urban landscapes affects the CO_2 and greenhouse gas emissions to the atmosphere, climate, biodiversity and water quality (Baur et al. 2015; Le Quéré et al. 2015; Mohan and Kandya 2015; Wilson 2015; Miranda et al. 2017).

Specifically, the LULC changes driven by urbanisation affect the localised hydrometeorological processes that consequently changes the urban microclimate (Pathirana et al. 2014). Microclimate is referred as the climate of a specific area at a micro level close to the ground however; the microclimate condition is greatly affected by the local climate and environmental conditions of the area (Camuffo 2013; Aussenac 2000). The urban microclimate is characterised to have greater air temperature, wind speed and lower relative humidity compared to the rural areas, especially during night time and as previously discussed, these characteristics of urban area are referred as the UHI effect. Multiple factors are causing the UHI effects where Grimmond (2007) stated that in the urban environment, the physical features influence the UHI as:

(a) Surface area is increased due to high rise buildings that contribute to the increase in vertical surfaces. Moreover, with the changes of urban canyon geometry, it changes the Sky View Factor (SVF, defined as fraction of sky seen at a specific point at ground level) due to the changes of height to width ratio in which, the high rise buildings restricted the solar radiation penetration to ground due to the restriction of SVF (Carrasco-Hernandez et al. 2015). SVF is also known to influence the solar and terrestrial radiation interception, thus has an impact on the cooling or heating of an area. With large vertical surfaces and reduction of SVF, the shortwave radiation absorption is increased and reduces the longwave radiation loss, total turbulent of heat transport and wind speed.

(b) The artificial surface that dominates the urban area is one of the main reasons that alter the urban microclimate that lead to urban warming. The materials of these artificial surfaces change the thermal characteristics of the area as they have greater heat capacities and conductivities where these materials also act as a heat storage "platform" during the daytime.

(c) It is important to note that the artificial materials store heat at day time and release the heat to the surrounding at night (Qi et al. 2019). For instance, Jim (2015) reported that the heating of artificial materials such as roof surface starts to be heated up as early during sunrise and peaked right after solar noon time and ends at sunset time where the stored heat takes time to be released to the atmosphere thus extending the air warming effect at night time which aggravates the UHI effect. In contrary to the rural area that has more vegetation and less artificial materials, cools faster at night time compared to the urban area (Yow 2007).

(d) Loss of vegetation increases the impervious surfaces due to the extended use of artificial materials compared to natural elements has changed the moisture characteristics in urban areas as it has large areas with impervious surfaces. It also increases stormwater runoff, reduces relative humidity and evapotranspirative cooling. For instance, Chen et al. (2017) noted that the urbanisation introduced more impervious surfaces that contribute to a total of average annual runoff of 3.3 billion cubic meters from 2001 to 2011 thus highlights its impact on the reduction of groundwater recharge, water quality and even flooding. This further related to the decrease in evapotranspiration leading to a more prominent increase in air and surface temperature as well as reduction in relative humidity (Hass et al. 2016).

In the urban environment that is characterised by the built environment, dense population and lack of vegetation, urban development has adversely affected the environment by altering the urban microclimate. Consequently, many studies have focused on the influence of urban development and vegetation loss on UHI and microclimate. Urbanisation attributes to the significant increase of temperature as observed in many cities and climate change exacerbates the differences of extreme hot night conditions between rural and urban areas (McCarthy et al. 2010). For instance, in Nairobi, Kenya, it was observed that over 47 year period from 1961 to 2007, there were significant minimum temperature increase of 0.43 °C and the maximum temperature increase of 0.11 °C per decade (Ndolo et al. 2018). While in Putrajaya, Malaysia, this city was observed to the temperature increase at 1.66 °C per decade, with a prominent UHI intensity of approximately 2.1 °C (Morris et al. 2016). On the other hand, a study by Radhi et al. (2013) showed that city in Bahrain was also affected by urbanisation when this city experienced mean air temperatures increase within 2–5 °C where newly developed artificial islands had 2–3 °C increase while the hot and arid deserted urban lands with 3–5 °C increase. The authors further noticed that the UHI intensity was varied according to the land cover type and land surface, where it was largely affected by the on-going development, vegetation loss and sea reclamation.

Furthermore, the influence of urbanisation on microclimate and UHI is significantly dependant on the urban profile of each city. A study by Qaid et al. (2016) that looked at intra-city air temperature within a planned tropical city showed that maximum UHI appeared to happen in the afternoon that reached up to 3 °C in low-rise residential buildings, however, the high-rise residential buildings and the Boulevard were 4 °C lower than low-rise buildings. This indicates that different building heights determine the microclimate condition of an area, thus influence the UHI effect. In the urban context, the ratio of building height to the street width or usually referred to as Height to Width (H:W) ratio determines the characteristics of the street canyon where this H:W ratio has an influence in the solar radiation interceptions of an area thus, will have an influence on the microclimate conditions of the street canyon. The H:W ratio is usually termed as the "aspect ratio" where the canyon with the aspect ratio value of 1 is indicated as uniform street canyon while aspect ratio of 0.5 and 2 are referred as shallow and deep street canyon, respectively (Shishegar 2013).

On the other hand, as previously explained, the SVF also determines the microclimate where urban canyon with lower SVF typically has cooler condition in the morning but exhibits warmer condition during night time due to lower and restricted longwave radiative cooling from the street surfaces to the atmosphere (Shahrestani et al. 2015). It is also important to note that the magnitude of UHI and microclimate conditions significantly vary according to the location within a city, time of day, shape of a building, street orientation, building block configuration and albedo of materials used (Mirzaei 2015). For instance, the street design can influence the cooling of a street canyon due to the wind flow, especially at night, simultaneously mitigates the UHI and microclimate conditions (Shishegar 2013). This shows that how we design our city can influence the environmental conditions, thus in future planning, this aspect should be considered as key factors in designing the city to minimise the adverse effect from urbanisation.

Heatwaves, Mortality and Human Thermal Comfort

Other than the effect on global energy consumption, UHI and microclimate, the street design also influences the human thermal comfort (Shishegar 2013). Rapid urbanisation decoupled with global warming, the urban climate is ultimately changing and predicted to be more severe as the heat load in cities increases parallel with urban development. According to Mishra et al. (2015), the thermal load increase as a result of the urban built environment in addition to the materials used in the development that resulted in adverse thermal conditions that lead to poor thermal comfort. The thermal load issues also affected the newly developed areas where this area experiences a significant impact on thermal stress (Radhi et al. 2013).

The intensification of UHI added by global warming lead to occurrences of extreme climate events such as heatwaves. According to Yang et al. (2019), the most appropriate definition of a heatwave is an event with daily maximum temperature at $\geq$ 92.5th percentile that occurred for a duration of 3 days or more. The heatwave events have implication on human thermal comfort, human health and wellbeing, where this impact is depending on heatwaves intensity and its duration (Yang et al.

2019). Moreover, with the increase and continuation of climate change, the heatwaves are predicted to be more intense and frequent in the future, particularly in the area at higher latitudes and large cities that are not well adapted to extreme temperature events (Luber and McGeehin 2008). From the observation of 217 cities in a study by Mishra et al. (2015), it was recorded that the amount of heatwave event was significantly increased but showed a decline in the frequency of cold waves in the year of 1973 until 2012. This study further elaborates that 48 and 63% of the studied cities showed increase in the frequencies of extremely hot days and hot nights, respectively.

This is a worrying situation as more intense and frequent heatwaves will have more detrimental impact, especially to the urban citizens as the heat wave event leads to heat related human health such as heat stress and heat stroke and in extreme cases, mostly ended with mortality (Hopp et al. 2018). Extreme weather related mortality has been a concern for the public health issues as heatwave event is considered as the deadliest health problem in the United States compared to hurricanes, flood and tornados (Luber and McGeehin 2008; Klinenberg 2015). Many worldwide cases of heatwaves had been reported and these events not only pose health threats but also recorded high number of mortality. From a study that looked at the extreme heat event in 43 cities in United States from 1987 to 2005, mortality was reported to increase by 3.74% during heatwaves compared to non-heatwave days with 2.49 and 0.38% of mortality risk increase with 1 °F of temperature increase and 1 day increase in heat wave duration, respectively (Anderson and Bell 2010).

A remarkable event of heatwave in Chicago, United States in year 1995, showed excess death of more than 700 people where among these deaths, it was determined that people with existing health problems, socially isolated and no access to air conditioning were at the greatest risk for heat-related death occurrences (Semenza et al. 1996). Another example in France, during the heat wave event in 2003, nearly 15,000 excess deaths due to the heatwave related health problems where it was observed that the mean maximum temperature exceeded 11–12 °C of the typical local climate condition for nine consecutive days (Lagadec 2004; Fouillet et al. 2006). It was further reported that the excess death was significantly related to age where women at comparable age of 45 years had 15% greater mortality than men and the overall mortality from this heatwave event was related to heatstroke, hyperthermia, dehydration, cardiovascular diseases, ill-defined morbid disorders, respiratory diseases and nervous system diseases (Fouillet et al. 2006). In addition, a study by Guo et al. (2018) predicted that cities with no adaptation will have heatwave related excess mortality increase in near equator regions, mostly tropical and subtropical countries compared to European countries and the United States. Moreover, this study also expected that the future heatwave related excess mortality will increase as the population and greenhouse gas emissions increase.

From all these occurrences of heat waves events indicate the importance of human thermal comfort, especially in cities during summer time. The thermal stress can be the trigger for heat related human health problems and in severe cases, can lead to mortality. As heat related issues are a major concern for people living in harsh urban condition, many studies have focused on this problem to address the importance of human thermal comfort. Moreover, since the environmental condition of the outdoor environment is hard to be controlled compared to the indoor environment (provided that artificial cooling by fan or air conditioner is installed indoor to improve human thermal comfort), many studies are currently focused on the outdoor human thermal comfort.

Human thermal comfort is dependent on two factors, which are the environmental conditions and personal factor. The environmental factors include environmental parameters such as air temperature, relative humidity, solar radiation, wind speed and mean radiant temperature, while personal factors are clothing and metabolic rate. Since the personal factor can be controlled by individual, the environmental factor is the focus as it is hard to be controlled in the outdoor environment. As discussed previously, the urban areas are known to have harsh microclimatic conditions. Therefore, human thermal comfort is largely depending on the microclimatic condition of an area. For instance, a study in Central Europe revealed that the thermal index of Physiologically Equivalent Temperature (PET) was strongly affected by the solar radiant interception, thus influencing the radiation heat of mean radiant temperature (Mayer et al. 2008). The thermal stress limits also dependent on the local climate of an area for instance, the normal thermal range of PET in Brazil and Germany are ranged from 16 to 30 °C and 18 to 28 °C, respectively, thermal stress of hot category starts from 32 to 35 °C, respectively and thermal stress of very hot category starts from 36 to 38 °C, respectively (da Silveira Hirashima et al. 2018).

The urban designs affect the microclimate, therefore this can also influence the human thermal comfort. The albedo of the materials used and side of the street are important factors that influence the human thermal comfort, where in albedo of 0.2, human thermal comfort of PET was 26% lower in North facing sidewalk than the South facing sidewalk with the reduction of 42% in mean radiant temperature (Lee and Mayer 2018). Materials with lower albedo such as asphalt and concrete usually have greater surface temperature compared to other materials with higher albedo such as soil and grass and thus influence the radiation heat that most likely to worsen the thermal comfort conditions during day time (Yilmaz et al. 2007). Moreover, the SVF, aspect ratio and street orientation determine the human thermal comfort (Ali-Toudert and Mayer 2006; Rodríguez-Algeciras et al. 2018). According to Ali-Toudert and Mayer (2007a), the larger the SVF, the greater human thermal stress while the East-West street shows greater heat stress suggesting that deviation from this street orientation can improve the human thermal condition. A study by Rodríguez-Algeciras et al. (2018) also supported this as they found that the East-West street had the highest thermal discomfort with PET values of 36 °C where it saves up to 4 h of heat stress when deviated from this street orientation. Qaid et al. (2018) also further suggest that the most suitable daytime micrometeorological and thermal comfort conditions can be achieved when the visible sky is positioned in a North East-South West or North-South street direction and to the South West of a North West-South East street direction.

Moreover, the solar radiant interception of a standing body under cloudless summer weather can absorb heat of 74% for long wave irradiance and 26% for short wave irradiance thus suggests that by providing shading to human or any urban surfaces can ameliorate the effect of human thermal stress (Ali-Toudert and Mayer 2007b). This shows how considerations of the urban and human factors can help to mitigate extreme heat stress especially in relation to planning that is suited for pedestrian use and may influence the frequency of usage of the urban spaces (Ali-Toudert and Mayer 2006).

Recreational Participation in Cities

In addition to the increase of urban dwellers population, the transformation and trends of the digital technology hinder people from partaking in outdoor recreational activities. People are keen to spend time indoors, increasing the global pandemic of physical inactivity (Brownson et al. 2005; Lee et al. 2012). However, despite physical inactivity becoming a global concern, research shows that urbanites are more interested to partake in recreational activities as compared to those living in rural settings (Chen et al. 2017). But do urban people actually participate in recreational activities?

Previous research done using Australian samples in the city of Perth shows that 59.2% of the population participated in physical activities to keep them healthy, while only about 9% was completely sedentary (Giles-Corti and Donovan 2002). Zhang et al. (2015) study found that majority of the young urban population in Beijing, China participated in at least one or two activities, with walking being the most popular form of activity. Only 1% of the sample indicated that they do not participate in any physical activities (Zhang et al. 2015). Similarly, research by Fernandez et al. (2015) in recreational participation among Latinas shows that the populations majorly partake in recreational activities, although there were differences in partaking in passive and active activities. In Asian samples, it was also found that majority of the urban population visits parks at least 1–2 times a week, indicating the prevalence of park usage and partaking in recreational activities (Sreetheran 2017).

These findings show that despite the increased concerns for physical inactivity, there are valid reasons to suggest that the urban population is keen on participating in recreational activities, possibly due to the increased realisation of the vast benefits of engaging in physical outdoor activities. Hence, the need for recreation becomes increasingly important is vital to ensure psychological and physiological health. However, there are some challenges in recreational participation.

Without a doubt, the design of urban environment is an important aspect that influences people's participation in recreational activities. Research has found that people are more likely to be physically active with the provision of green spaces (Ellaway et al. 2005; Tzoulas et al. 2007; van Herzele and de Vries 2012). However, urban designs are not necessarily the only important aspects to be considered. One of the main issues in recreational participation in some countries is the risk of exposure to air pollutions (Woodcock et al. 2014). These are particularly a serious issue in cities that reportedly have high threats of pollutions such as Beijing. Zhang and Yang (2014) discuss that recreational participations are tied to the accessibility to green spaces in urban areas and the quality of vegetation, all of which are tied to the planning and management of urban spaces. Indeed, climate plays an important factor in recreational participation.

As discussed, people are more likely to engage in recreational activities with the availability of green spaces. This is due to the fact that green spaces, apart from being more aesthetically pleasing to the eyes, gives individuals physiological comfort as green spaces provide human thermal comfort. However, many urban spaces lacked the necessary planning and design of green spaces that maximise the benefits of recreational participation. In many cases, the needs of the residents for recreational participation are not being met due to poor planning and design of recreational areas that do not provide green spaces. The question remains, what and how is the best way to mitigate the problem of the design of green spaces in urban areas to maximise the benefits?

Mitigating the Problem: An Introduction to the Solutions

In recent years, the term Urban Green Infrastructure has been widely discussed. Green Infrastructure (GI) designs represent a way to mitigate problems associated with urbanisation such as air pollution, urban heat island and flooding (Connop et al. 2016). Proper use and design of GI benefits urban citizens in so many, consequently improving their psychological and physiological wellbeing. The term GI is widely associated with terms such as "re-naturing cities", nature-based solutions and smart cities, all of which used in the planning and design of the modern urban cities (Harrison and Donnelly 2011; Anthopoulos and Vakali 2012; Connop et al. 2016; Lafortezza et al. 2018). The concept of GI focuses on working with natural solutions that address environmental and social challenges in cities. The concept of GI is tied closely with the concept of nature-based solutions for cities; defined as "living solutions inspired by, continuously supported by and using nature, which are designed to address various societal challenges in a resource-efficient and adaptable manner and to provide simultaneously economic, social and environmental benefits" (Bauduceau et al. 2015).

Fundamentally, the concept of GI is an integration of many components to make up for a design of urban spaces that has more "green solutions", as opposed to the design of cities that are dominated by "grey elements" (i.e., buildings, glasses, steels). Historically, the concept of GI has long been implemented in various countries. In Japan, the provision of green walkways or accessible green areas is an important aspect to be included in the planning of temples and shrines (Sorensen 2002).

The main objective of GI is to improve cities in becoming more liveable in the long-term and enhances cities resiliency to the environmental challenges in urban spaces. Consequently, these approaches directly benefit the environment and the people living within the space. Most importantly, in ensuring cities resiliency to environmental challenges, it is very important to look at how this concept of green cities is designed.

Connop et al. (2016) illustrate a framework that consists of five main components of GI, which are: landscapes, water bodies, parks and gardens, pocket parks and green structures. Various cities in different parts of the world implemented GI using a combination of these components. Suffice to say that the planning and management for urban cities incorporated GI. In fact, many cities main objectives are trending towards the concept of "greening the cities" to improve liveability. For example, in Ludwigsburg, Germany, planners and managers constructed a free-standing green-wall structure with living plants, as an effort to maximise greenery in small spaces (Connop et al. 2016). These living walls were constructed with considerations of species of plants that were adaptable. Naturally, the concept of living walls is also applied on ground, where selected vegetation is included in urban landscaping, parks and gardens, pocket parks as well as in roofs. These inclusions of living walls and enhancement of greeneries in urban landscaping have proved to significantly cool urban areas, which act as a mitigation to urban heat island (Zhou and Shepherd 2010).

A useful illustration of the application of the concept of GI and using nature-based solutions as a way to urban cooling and human thermal comfort can be referred to Coutts et al.'s (2013). They illustrate the importance of incorporating different components of GI such as green roofs and walls, increased canopy coverage through vegetation planting, and stormwater harvesting and bio-retention; all of which result in a cooler environment that improves human thermal comfort.

Re-naturing: What Is the Impact on Microclimate, Human Thermal Comfort and Recreational Participation?

One of the best efforts in minimising the influence of rapid urbanization and climate related issues is to emphasise on the concept of re-naturing the cities through the implementation of GI and other nature based solutions (i.e. blue infrastructures) in urban landscapes. The GI and nature based solutions such as street trees, parks and water resources provide ecosystem services benefits that are important for microclimate regulation, carbon storage, stormwater runoff regulation, recreation, aesthetics and biodiversity (Wang and Banzhaf 2018; Aida et al. 2016; Dobbs et al. 2011). In specific, areas in cities that have high components of GI such as areas with high tree densities (i.e. parks) can improve the climate condition provided by tree cooling benefits through the reduction of air temperature and increase in relative humidity (Hutyra et al. 2011; Georgi and Zafiriadis 2006; Hamada and Ohta 2010). GI can reduce heat load in cities through the process of evapotranspiration where the atmosphere is cooled through the latent cooling by the water vapour in addition to provides shading of the ground and buildings especially from trees. Streets with greater coverage of canopies have lower air temperature and solar radiation interception compared to the streets with lower canopy cover percentage (Sanusi et al. 2016). According to Brown and Gillespie (1995), functions of leaves are to absorb 80% of incoming radiation, reflecting 10% and the remaining is transmitted below the canopy hence reduce the convective heating and heat storage of the ground surfaces. Thus this will influence the urban microclimate by lowering the surface and air temperatures and consequently decreasing the UHI intensity.

Increase in thermal load and UHI in cities raises health concerns of the urban citizens, thus urging the needs to improve the human thermal comfort. This human thermal comfort can be improved through re-naturing the urban landscapes that function as climate regulators, thus mitigating urban heat related health issues. Some studies have associated the functions of GI with its human thermal comfort benefits (Sanusi et al. 2016, 2017; Morakinyo et al. 2018). For instance, it has been proved that the urban shade trees provide positive effects with the reduction of mean radiant temperature and PET in areas with trees compared to the areas without trees (de Abreu-Harbich et al. 2015).

The microclimatic and human thermal benefits of GI are also dependent on the street design. For instance, Sanusi et al. (2016) reported that different street orientation could influence the microclimate, thus human thermal comfort where the high-percentage canopy cover in East-West streets had greater reductions in air temperature with 2.1 °C compared to the North-South streets with 0.9 °C. Moreover, it was further reported that side of the streets and time of the day greatly influence the microclimatic and human thermal benefits as the East sidewalks in North-South street had greater air temperature, mean radiant temperature, and solar radiation interception in early morning while during midday, they were greater in West sidewalks.

On the other hand, other studies also reported that the species of the trees installed in the cities also play an important determinant factor for its microclimatic and human thermal benefits. Species of *Ulmus procera*, *Platanus × acerifolia* had greater air temperature, relative humidity, solar radiation, mean radiant temperature and wind speed benefits compared to *Eucalyptus scoparia* due to different canopy characteristics (Sanusi et al. 2017). (Example of tree species with different canopy characteristics is presented in Fig. 1.) A study by de Abreu-Harbich et al. (2015) further emphasised that the shading by trees that filters the solar radiation interception is greatly influenced by the canopy form and density where trees with plagiotropic trunk and large tree canopies have higher cooling capacities in terms of thermal comfort. Takács et al. (2016) also reported that *Sophora japonica*, had the greatest air temperature reductions with 0.8 °C followed by in *Celtis occidentalis* with 0.7 °C reduction due to their greater canopy spread. In detail, Shahidan et al. (2010) pointed out that *Mesua ferrea* L. intercepted greater solar radiation in comparison to *Hura crepitans* L. due to its branching, twigs and leaf cover densities.

Comfort is an integral aspect for people to participate in recreation, particularly in cities with high heat temperatures and exposures to air and noise pollutions. Research shows that people are more likely to rate higher comfort levels when they participate in recreational activities in green urban spaces (Klemm et al. 2015). Previous research in the medical field found that patients recuperate faster when they have views of greenery, as opposed to bricks and walls (Ulrich 1984). The positive effects and benefits of increase in green spaces are profound as the green spaces improve social connections between residents (Klemm et al. 2015). As such, the design to create a

Fig. 1 Different species may show different canopy characteristics. These are the images for *Platanus* × *acerifolia* (top), *Ulmus procera* (middle) and *Eucalyptus* sp. (bottom) trees

comfortable environment for recreational areas plays a pivotal role in maintaining the wellbeing of the people.

To this end, urban areas must be designed with the recreational needs of the residents. Past research shows that walking is one of the most common physical activity that is carried out by all walks of life. Hence, the planning and design of walkways need to be prioritised based on the needs of the local climate. For example, in cities with high temperature and humidity such as Hong Kong and Kuala Lumpur, urban design which cools the environment plays a pivotal role. Therefore, planning and design for green walkways and green corridors must be focused on tree arrangements and selection that reduces the temperature. Additionally, the arrangements of shadows when planning are also equally important (Ng and Cheng 2012). Moreover, when planning and design using nature based solution, it is necessary to acknowledge the importance of recreation participation and its integration with the microclimate mitigation and human thermal stress conditions. The incorporation of these three elements in the re-naturing effort is still lacking, and therefore, necessary to be considered as one of the strategies in the urban planning and design especially, to improve the liveability of cities.

Most importantly, the impact on the implementation of re-naturalisation in cities must be carefully studied in ensuring the resiliency and sustainability of the nature based approach. For instance, Connop et al. (2016) tested the conceptual framework of the functional role of GI in the re-naturing strategy in some cities. They selected three sites as case studies using appropriately designed GI to show its multifunctional ecosystem service benefits. These three selected study cases were the: (1) Barking Riverside brownfield landscaping situated in London Borough of Barking and Dagenham which was selected to illustrate the implementation of GI in the masterplanning; (2) Green Living Room in Ludwigsburg, Germany which was selected to apply adaptation measures from architectural to re-naturalisation strategy using nature based solution of GI via innovative living walls structure; (3) Derbyshire St Pocket Park, Bethnal Green, London which was selected to apply urban GI design principles into sustainable drainage systems design. From these study cases, it was highlighted that the contribution of GI from the innovative re-naturing effort can optimise its multifunctional benefits in cities in terms of environmental, biodiversity and communities in which, more extensive studies can be strategically incorporated with the planning policy of urban areas. Moreover, further investigations on the impact of re-naturing efforts are needed to ensure more strategic GI approach delivery to improve the liveability of cities in the future.

Conclusion

Urbanisation influences the microclimate and consequently causes negative impacts on urban residents in terms of environment, health, economic, and social aspects especially to heat related health problems to human. Nature based solution of GI such as trees and other vegetation can be used as one of the strategies to provide ecosystem services in urban area to mitigate the climate impact especially extreme temperature events on urban citizens. However, regardless of the effort in re-naturing the cities through GI and other nature-based solutions, there is still lack of incorporation between microclimate mitigation, human thermal improvement all together with recreational participation in urban planning process. This is also a key issue that needs to be addressed so that the effort in re-naturing the cities will be fully beneficial for microclimate mitigation, improvement of human thermal comfort and recreational participation.

This paper has discussed the variables that relate to issues regarding re-naturing and strategies involved in planning for the urban environment. However, to some extent, other factors may affect the execution and planning for these re-naturing strategies that are not covered in this paper. For example, in some urban environments, adequate budgeting to introduce these mitigation strategies, which may be expensive, plays a major setback, particularly in countries that are still economically developing. Additionally, factors such as the natural landscapes, extreme temperatures, differentiation of seasons that may have an entirely different approach to re-naturing the urban environment is not discussed specifically in this paper.

Future planning by incorporating nature based solutions need to be the forefront of every discussion held to plan and design urban current and future urban spaces. With nearly 68% of the world population living in urban environment by 2050, there is an urgent need to divert our attention to provide solutions that mitigate long term impact of urbanisation, while focusing current needs of creating a stable and safe environment for those living in urban areas. To this end, it is hoped that the discussed issues and mitigation strategies provided in this paper justify the importance of re-naturing urban environment for future urban landscape planning. Without doubt, these strategies lead to the optimisation of the urban ecosystem services in ensuring urban sustainability and resiliency.

References

Aida N, Sasidhran S, Kamarudin N, Aziz N, Puan CL, Azhar B (2016) Woody trees, green space and park size improve avian biodiversity in urban landscapes of Peninsular Malaysia. Ecol Ind 69:176–183

Ali-Toudert F, Mayer H (2006) Numerical study on the effects of aspect ratio and orientation of an urban street canyon on outdoor thermal comfort in hot and dry climate. Build Environ 41(2):94–108

Ali-Toudert F, Mayer H (2007a) Effects of asymmetry, galleries, overhanging facades and vegetation on thermal comfort in urban street canyons. Sol Energy 81(6):742–754

Ali-Toudert F, Mayer H (2007b) Thermal comfort in an east–west oriented street canyon in Freiburg (Germany) under hot summer conditions. Theoret Appl Climatol 87(1–4):223–237

Anderson GB, Bell ML (2010) Heat waves in the United States: mortality risk during heat waves and effect modification by heat wave characteristics in 43 US communities. Environ Health Perspect 119(2):210–218

Anthopoulos LG, Vakali A (2012) Urban planning and smart cities: interrelations and reciprocities. In: The future internet assembly. Springer, Berlin, Heidelberg, pp 178–189

Aussenac G (2000) Interactions between forest stands and microclimate: ecophysiological aspects and consequences for silviculture. Ann For Sci 57(3):287–301

Bauduceau N, Berry P, Cecchi C, Elmqvist T, Fernandez M, Hartig T et al (2015) Towards an EU research and innovation policy agenda for nature-based solutions & re-naturing cities: final report of the Horizon 2020 expert group on 'nature-based solutions and re-naturing cities'. Publications Office of the European Union, Bruxelles, 76 pp

Baur AH, Förster M, Kleinschmit B (2015) The spatial dimension of urban greenhouse gas emissions: analyzing the influence of spatial structures and LULC patterns in European cities. Landsc Ecol 30(7):1195–1205

Block AH, Livesley SJ, Williams NS (2012) Responding to the urban heat island: a review of the potential of green infrastructure. Victorian Centre for Climate Change Adaptation Research, Melbourne, Australia

Brown RD, Gillespie TJ (1995) Microclimatic landscape design: creating thermal comfort and energy efficiency. Wiley, New York

Brownson RC, Boehmer TK, Luke DA (2005) Declining rates of physical activity in the United States: what are the contributors? Annu Rev Public Health 26:421–443

Camuffo D (2013) Microclimate for cultural heritage: conservation, restoration, and maintenance of indoor and outdoor monuments. Elsevier, 526 pp

Carrasco-Hernandez R, Smedley AR, Webb AR (2015) Using urban canyon geometries obtained from Google Street View for atmospheric studies: potential applications in the calculation of street level total shortwave irradiances. Energy Build 86:340–348

Chen K, Bi J, Chen J, Chen X, Huang L, Zhou L (2015) Influence of heat wave definitions to the added effect of heat waves on daily mortality in Nanjing, China. Sci Total Environ 506:18–25

Chen J, Theller L, Gitau MW, Engel BA, Harbor JM (2017) Urbanization impacts on surface runoff of the contiguous United States. J Environ Manage 100(187):470–481

Connop S, Vandergert P, Eisenberg B, Collier MJ, Nash C, Clough J, Newport D (2016) Renaturing cities using a regionally-focused biodiversity-led multifunctional benefits approach to urban green infrastructure. Environ Sci Policy 62:99–111

Coutts AM, Tapper NJ, Beringer J, Loughnan M, Demuzere M (2013) Watering our cities: the capacity for water sensitive urban design to support urban cooling and improve human thermal comfort in the Australian context. Prog Phys Geogr 37(1):2–28

da Silveira Hirashima SQ, Katzschner A, Ferreira DG, de Assis ES, Katzschner L (2018) Thermal comfort comparison and evaluation in different climates. Urban Clim 23:219–230

de Abreu-Harbich LV, Labaki LC, Matzarakis A (2015) Effect of tree planting design and tree species on human thermal comfort in the tropics. Landsc Urban Plan 138:99–109

de Noblet-Ducoudré N, Boisier JP, Pitman A, Bonan GB, Brovkin V, Cruz F et al (2012) Determining robust impacts of land-use-induced land cover changes on surface climate over North America and Eurasia: results from the first set of LUCID experiments. J Clim 25(9):3261–3281

Debbage N, Shepherd JM (2015) The urban heat island effect and city contiguity. Comput Environ Urban Syst 54:181–194

Dobbs C, Escobedo FJ, Zipperer WC (2011) A framework for developing urban forest ecosystem services and goods indicators. Landsc Urban Plan 99(3–4):196–206

Ellaway A, Macintyre S, Bonnefoy X (2005) Graffiti, greenery, and obesity in adults: secondary analysis of European cross sectional survey. BMJ 331(7517):611–612

Fernandez M, Shinew KJ, Stodolska M (2015) Effects of acculturation and access on recreation participation among Latinos. Leis Sci 37(3):210–231

Fouillet A, Rey G, Laurent F, Pavillon G, Bellec S, Guihenneuc-Jouyaux C et al (2006) Excess mortality related to the August 2003 heat wave in France. Int Arch Occup Environ Health 80(1):16–24

Georgi NJ, Zafiriadis K (2006) The impact of park trees on microclimate in urban areas. Urban Ecosyst 9(3):195–209

Giles-Corti B, Donovan RJ (2002) Socioeconomic status differences in recreational physical activity levels and real and perceived access to a supportive physical environment. Prev Med 35(6):601–611

Grimmond S (2007) Urbanization and global environmental change: local effects of urban warming. Geogr J 173(1):83–88

Guo Y, Gasparrini A, Li S, Sera F, Vicedo-Cabrera AM, Coelho MDSZS et al (2018) Quantifying excess deaths related to heatwaves under climate change scenarios: a multicountry time series modelling study. PLoS Med 15(7):e1002629

Hamada S, Ohta T (2010) Seasonal variations in the cooling effect of urban green areas on surrounding urban areas. Urban For Urban Green 9(1):15–24

Harrison C, Donnelly IA (2011) A theory of smart cities. In: Proceedings of the 55th annual meeting of the ISSS-2011, Hull, UK, Sept 2011, vol 55, no 1
Hass AL, Ellis KN, Mason LR, Hathaway JM, Howe DA (2016) Heat and humidity in the city: neighborhood heat index variability in a mid-sized city in the southeastern United States. Int J Environ Res Public Health 13(1):117
Hopp S, Dominici F, Bobb JF (2018) Medical diagnoses of heat wave-related hospital admissions in older adults. Prev Med 110:81–85
Hutyra L, Yoon B, Alberti M (2011) Terrestrial carbon stocks across a gradient of urbanization: a study of the Seattle, WA region. Glob Change Biol 17(2):783–797
IPCC (2014) Climate change 2014: synthesis report. Contribution of working groups I, II and III to the fifth assessment report of the intergovernmental panel on climate change (eds: Core Writing Team, Pachauri RK, Meyer LA). IPCC, Geneva, Switzerland, 151 pp
Ishola KA, Okogbue EC, Adeyeri OE (2016) Dynamics of surface urban biophysical compositions and its impact on land surface thermal field. Model Earth Syst Environ 2(4):208
Jim CY (2015) Assessing climate-adaptation effect of extensive tropical green roofs in cities. Landsc Urban Plan 138:54–70
Joe L, Hoshiko S, Dobraca D, Jackson R, Smorodinsky S, Smith D, Harnly M (2016) Mortality during a large-scale heat wave by place, demographic group, internal and external causes of death, and building climate zone. Int J Environ Res Public Health 13(3):299
Klemm W, Heusinkveld BG, Lenzholzer S, Jacobs MH, van Hove B (2015) Psychological and physical impact of urban green spaces on outdoor thermal comfort during summertime in The Netherlands. Build Environ 83:120–128
Klinenberg E (2015) Heat wave: a social autopsy of disaster in Chicago. University of Chicago Press, Chicago
Lafortezza R, Chen J, van den Bosch CK, Randrup TB (2018) Nature-based solutions for resilient landscapes and cities. Environ Res 165:431–441
Lagadec P (2004) Understanding the French 2003 heat wave experience: beyond the heat, a multi-layered challenge. J Contingencies Crisis Manag 12(4):160–169
Le Quéré C, Moriarty R, Andrew RM, Canadell JG, Sitch S, Korsbakken JI et al (2015) Global carbon budget 2015. Earth Syst Sci Data 7(2):349–396
Lee H, Mayer H (2018) Thermal comfort of pedestrians in an urban street canyon is affected by increasing albedo of building walls. Int J Biometeorol 62(7):1199–1209
Lee IM, Shiroma EJ, Lobelo F, Puska P, Blair SN, Katzmarzyk PT (2012) Effect of physical inactivity on major non-communicable diseases worldwide: an analysis of burden of disease and life expectancy. Lancet 380(9838):219–229
López-Bueno JA, Díaz J, Linares C (2019) Differences in the impact of heat waves according to urban and peri-urban factors in Madrid. Int J Biometeorol 63(3):371–380
Luber G, McGeehin M (2008) Climate change and extreme heat events. Am J Prev Med 35(5):429–435
Mayer H, Holst J, Dostal P, Imbery F, Schindler D (2008) Human thermal comfort in summer within an urban street canyon in Central Europe. Meteorol Z 17(3):241–250
McCarthy MP, Best MJ, Betts RA (2010) Climate change in cities due to global warming and urban effects. Geophys Res Lett 37(9):L09705
Miranda A, Altamirano A, Cayuela L, Lara A, González M (2017) Native forest loss in the Chilean biodiversity hotspot: revealing the evidence. Reg Environ Change 17(1):285–297
Mirzaei PA (2015) Recent challenges in modeling of urban heat island. Sustain Cities Soc 19:200–206
Mishra V, Ganguly AR, Nijssen B, Lettenmaier DP (2015) Changes in observed climate extremes in global urban areas. Environ Res Lett 10:024005
Mohan M, Kandya A (2015) Impact of urbanization and land-use/land-cover change on diurnal temperature range: a case study of tropical urban airshed of India using remote sensing data. Sci Total Environ 506:453–465

Morakinyo TE, Lau KK, Ren C, Ng E (2018) Performance of Hong Kong's common trees species for outdoor temperature regulation, thermal comfort and energy saving. Build Environ 137:157–170
Morris KI, Chan A, Salleh SA, Ooi MCG, Oozeer MY, Abakr YA (2016) Numerical study on the urbanisation of Putrajaya and its interaction with the local climate, over a decade. Urban Clim 16:1–24
Ndolo JI, Muthama NJ, Oludhe C, Ng'ang'a JK, Odingo RS (2018) Influence of urbanisation on minimum and maximum temperature characteristics over Nairobi City. J Clim 1(2):73–80
Ng E, Cheng V (2012) Urban human thermal comfort in hot and humid Hong Kong. Energy Build 55:51–65
Pathirana A, Denekew HB, Veerbeek W, Zevenbergen C, Banda AT (2014) Impact of urban growth-driven landuse change on microclimate and extreme precipitation—a sensitivity study. Atmos Res 138:59–72
Qaid A, Lamit HB, Ossen DR, Shahminan RNR (2016) Urban heat island and thermal comfort conditions at micro-climate scale in a tropical planned city. Energy Build 133:577–595
Qaid A, Lamit HB, Ossen DR, Rasidi MH (2018) Effect of the position of the visible sky in determining the sky view factor on micrometeorological and human thermal comfort conditions in urban street canyons. Theoret Appl Climatol 131(3–4):1083–1100
Qi JD, He BJ, Wang M, Zhu J, Fu WC (2019) Do grey infrastructures always elevate urban temperature? No, utilizing grey infrastructures to mitigate urban heat island effects. Sustain Cities Soc 101392
Radhi H, Fikry F, Sharples S (2013) Impacts of urbanisation on the thermal behaviour of new built up environments: a scoping study of the urban heat island in Bahrain. Landsc Urban Plan 113:47–61
Rawat JS, Kumar M (2015) Monitoring land use/cover change using remote sensing and GIS techniques: a case study of Hawalbagh block, district Almora, Uttarakhand, India. Egypt J Remote Sens Space Sci 18(1):77–84
Rizwan AM, Dennis LY, Chunho LIU (2008) A review on the generation, determination and mitigation of urban heat island. J Environ Sci 20(1):120–128
Rodríguez-Algeciras J, Tablada A, Matzarakis A (2018) Effect of asymmetrical street canyons on pedestrian thermal comfort in warm-humid climate of Cuba. Theoret Appl Climatol 133(3–4):663–679
Sanusi R, Johnstone D, May P, Livesley SJ (2016) Street orientation and side of the street greatly influence the microclimatic benefits street trees can provide in summer. J Environ Qual 45(1):167–174
Sanusi R, Johnstone D, May P, Livesley SJ (2017) Microclimate benefits that different street tree species provide to sidewalk pedestrians relate to differences in Plant Area Index. Landsc Urban Plan 157:502–511
Semenza JC, Rubin CH, Falter KH, Selanikio JD, Flanders WD, Howe HL, Wilhelm JL (1996) Heat-related deaths during the July 1995 heat wave in Chicago. N Engl J Med 335(2):84–90
Shahidan MF, Shariff MK, Jones P, Salleh E, Abdullah AM (2010) A comparison of *Mesua ferrea* L. and *Hura crepitans* L. for shade creation and radiation modification in improving thermal comfort. Landsc Urban Plan 97(3):168–181
Shahrestani M, Yao R, Luo Z, Turkbeyler E, Davies H (2015) A field study of urban microclimates in London. Renew Energy 73:3–9
Shishegar N (2013) Street design and urban microclimate: analyzing the effects of street geometry and orientation on airflow and solar access in urban canyons. J Clean Energy Technol 1(1):52–56
Sorensen A (2002) The making of urban Japan: cities and planning from Edo to the twenty-first century. Routledge, London
Sreetheran M (2017) Exploring the urban park use, preference and behaviours among the residents of Kuala Lumpur, Malaysia. Urban For Urban Green 25:85–93
Takács Á, Kiss M, Hof A, Tanács E, Gulyás Á, Kántor N (2016) Microclimate modification by urban shade trees—an integrated approach to aid ecosystem service based decision-making. Procedia Environ Sci 32:97–109

Tzoulas K, Korpela K, Venn S, Yli-Pelkonen V, Kaźmierczak A, Niemela J, James P (2007) Promoting ecosystem and human health in urban areas using green infrastructure: a literature review. Landsc Urban Plan 81(3):167–178
Ulrich RS (1984) View from a window may influence recovery from surgery. Science 224:420–421
UN DESA (2018) Revision of world urbanization prospects. Multimedia Library—United Nations Department of Economic and Social Affairs. Retrieved 7 Feb 2019, from https://www.un.org/development/desa/publications/2018-revision-of-world-urbanization-prospects.html
van Herzele A, de Vries S (2012) Linking green space to health: a comparative study of two urban neighbourhoods in Ghent, Belgium. Popul Environ 34(2):171–193
Wang J, Banzhaf E (2018) Towards a better understanding of green infrastructure: a critical review. Ecol Ind 85:758–772
Wang Y, Berardi U, Akbari H (2016) Comparing the effects of urban heat island mitigation strategies for Toronto, Canada. Energy Build 114:2–19
Wilson CO (2015) Land use/land cover water quality nexus: quantifying anthropogenic influences on surface water quality. Environ Monit Assess 187(7):424
Woodcock J, Tainio M, Cheshire J, O'Brien O, Goodman A (2014) Health effects of the London bicycle sharing system: health impact modelling study. BMJ 348:g425
Yang J, Yin P, Sun J, Wang B, Zhou M, Li M et al (2019) Heatwave and mortality in 31 major Chinese cities: definition, vulnerability and implications. Sci Total Environ 649:695–702
Yilmaz S, Toy S, Yilmaz H (2007) Human thermal comfort over three different land surfaces during summer in the city of Erzurum, Turkey. Atmósfera 20(3):289–297
Yow DM (2007) Urban heat islands: observations, impacts, and adaptation. Geogr Compass 1(6):1227–1251
Zhang W, Yang J (2014). Development of outdoor recreation in Beijing, China between 1990 and 2010. Cities 37:57–65
Zhang W, Yang J, Ma L, Huang C (2015) Factors affecting the use of urban green spaces for physical activities: views of young urban residents in Beijing. Urban For Urban Green 14(4):851–857
Zhou Y, Shepherd J (2010) Atlanta's urban heat island under extreme heat conditions and potential mitigation strategies. Nat Hazards 52(3):639–668

Natural Hazards and Climate Change: Lessons and Experiences from Kerala Flood Disaster

Neha Goel Tripathi and Nidhin Davis

Abstract During August 2018, Kerala experienced unprecedented rainfall, which resulted in the worst ever flooding in the state's history since 1924. Economic losses from the 2018 floods in the Indian state of Kerala reduced the GDP growth by 1%. Apart from the loss of human and animal lives, the severe flooding also destroyed buildings, roads, and infrastructural systems like water and sanitation, highlighting the magnitude of the natural disaster. With the growing need to have a better response mechanism to the challenge of climate change, it is pertinent for policy makers and planners to understand the interlinkage and complexity between development and environment. It is especially important to understand the human and natural factors that exacerbated and contributed to this disaster in Kerala for future preparedness. In this paper, the natural and anthropogenic factors behind the Kerala floods based on existing scientific information and research were established. Furthermore, Kerala ranks first among Indian states on the Human Development Index (HDI). It has made significant advances in terms of education and healthcare as compared to rest of India. Aspects that contribute to the uniqueness of Kerala is the successful decentralised governance and strong citizen participation. With this context, the paper reviews the post disaster response of the government, in terms of governance and intervention. Emerging insights from post disaster impacts can help in understanding the gaps and assist in formulating guidelines for future adaptation and risk reduction policies. Finally, the way forward has been suggested as an integrated approach for mainstreaming climate resilience (Climate resilience refers to the capacity of a socio-ecological system to adapt, reorganize, and evolve to be better prepared for future disasters and climate change impacts.) for the State of Kerala.

N. G. Tripathi (✉)
Environmental Planning, School of Planning and Architecture, New Delhi, India
e-mail: nehagtripathi@gmail.com

N. Davis
State Emergency Operations Centre, Kerala State Disaster Management Authority, Thiruvananthapuram, India
e-mail: davisknidhin@gmail.com

W. Leal Filho et al. (eds.), *Climate Change, Hazards and Adaptation Options*, Climate Change Management, https://doi.org/10.1007/978-3-030-37425-9_29

Introduction

Climate change is both a global and a local challenge, having direct and indirect impact on both natural and man-made systems. Within the man-made systems, cities and urbanised regions are more vulnerable as they have a higher concentration of human life. The association between climate change and natural disasters is a complex challenge. At the same time, it cannot be denied that all over the world climatic data trends indicate that extreme weather events are taking place more frequently than before. Statistical data trends show that the increase in disasters due to natural hazards over the past 20 years has, for the most part, been caused by meteorological and hydrological events (Thomas and López 2015).

India's Vulnerability to Climate Change

India as a country is prone to extreme climatic events, such as heavy rainfall leading to floods, or monsoon failure leading to droughts and water scarcity. There are many instances of years with a flood (strong monsoon) or drought (weak monsoon) during which India as a whole receives excess or deficient seasonal rainfall respectively (Krishnamurthy and Kinter 2003). India lies within the tropical monsoon zone and has a high spatial variation in rainfall and its frequencies. Consequently, the spatial distribution of rainfall pattern is complex.

A study by Guhathakurta et al. (2011) considered daily rainfall data for the period 1901–2005 for more than 6000 rain gauge stations all over India. This study concluded that there were noticeable changes in the extreme rainfall events in the past century. Moreover, the trend analysis of the frequency of typical rainy days and heavy rainfall days showed significant decreasing trends over central and many parts of north India; it also displayed increasing trends over peninsular India.

The fact that the number of extreme rainfall events has gone up by threefold has also been supported by Indian Institute of Tropical Meteorology, Pune. The findings of their study indicate that there have been 285 reported flooding events in India over 1950–2017, affecting about 850 million people, leaving 19 million homeless and killing about 71,000 people (Choudhary 2018).

In the year 2018 itself, according to latest government official figures, 1428 people died due to extreme weather events with heavy rains and subsequent flooding, dust storms, and thunderstorms.

Ecological Setting of Kerala

Kerala is a coastal state in Southern India on the South West tip of the Indian Peninsula that borders the Arabian Sea (Fig. 1). Kerala is described as 'God's own country' in

Fig. 1 Location of Kerala state and districts with in Kerala state. *Source* Kerala State Disaster Management Authority 2019

India, with magnificent natural landscapes and fertile valleys. It is one of the most visited places in India.

Kerala is a state with a unique geography. Located between the Western Ghats Mountains and the Arabian Sea, the 570 km long state is about 120 km at its broadest. Lengthwise, Kerala is divided into three distinct ecosystems—the highlands, the midlands, and the coastal plains. The state's population (in 2011) of 33.3 million accounts for 3.01% of India's population. The population density of the state is 859 persons per km, one of the highest in the country (Government of Kerala 2014).

The data on land use pattern of Kerala for the year 2009–10 reveals that out of a total geographical area of 3.8 million, the net sown area is about 56% and forest occupies around 28% (Government of Kerala 2007).

Kerala stands first in the UNDP's ranking of states in the Human Development Index for Indian States (Government of Kerala 2005). Some of the human development indicators for Kerala are at par with those of developed countries. It has the highest literacy rate among all states and was declared the first fully literate state in India about a decade back. Life expectancy, both male and female, is very high, close to that of the developed world. Other parameters such as fertility rate, infant and child mortality are among the best in the country.

Kerala has a vibrant economy. The service sector (including tourism, public administration, banking and finance, transportation, and communications) and the agricultural and fishing industries contribute largely to the economy of Kerala (Government of Kerala 2017).

High levels of social indicators comparable to the developed countries have attracted a lot of attention to the developmental process in Kerala. It is referred to as the 'Kerala model of development' (Government of Kerala 2005).

The State of Kerala is popularly known as the "Gateway of summer monsoon" over India. The onset of rain in Kerala is the most anxiously awaited weather singularity in the Indian subcontinent, as it heralds the rainy season and marks the end of the hot summer. Kerala is almost the entry point of South West monsoon rainfall in the Indian sub-continent. It is also one of the wettest places in the world, where annual rainfall is 3000 mm (Government of Kerala 2007).

Kerala consists of 44 rivers of small and medium category, interconnecting backwaters and canals, several lakes and ponds of diverse capacities, streams, springs, wells, extensive wetlands, and paddy fields. Most of these rivers have their source in the Western Ghats and drain into the Arabian Sea. One of the striking features of the state is the continuous chain of lagoons or backwaters existing along the coastal region (Government of Kerala 2007).

Trends, Climate Variability and Change in the Kerala's Monsoon

The analysis done by the Indian Meteorological Department (IMD) of historic rainfall data of Kerala indicates cyclic trends in annual rainfall, with a declining trend of South West Monsoon rainfall during the last 60 years. According to IMD, during the last 43 years, the mean maximum temperature has risen by about 0.8, the minimum by 0.2 and the average by 0.6 °C over Kerala (27.3–27.9), indicating a clear upward trend in surface air temperature (Government of Kerala 2014). The year 1918 recorded a minimum rainfall of 115 cm and the year 1924 received a maximum rainfall of 311.5 cm. In 2016, both the southwest and the northeast monsoon failed in Kerala, and resultantly, there was an unprecedented drought in the summer of 2017.

The first report on "Impact of climate change in four regions of the country" submitted to the Government of India by the Indian Network for Climate Change Assessment (INCCA)[1] has stated that reduced rainfall, increased atmospheric temperature, and flooding due to rising sea levels are the climate change scenarios for the Western Ghats and Kerala in the next 20 years (Sharma and Chauhan 2011; Government of Kerala 2014).

A detailed study was done by Joesh and Menon about the monsoonal trend of Kerala for a period of 100 years (Government of Kerala 2014). The rainfall of Kerala shows the unique phenomena of monsoon behaviour and has shown a decreasing trend in the last 100 years. Further research by Anie (2017) has also confirmed the decreasing trend in southwest monsoon, based on long term data analysis (112 years). The trend analysis of the normal, early and delayed onset years revealed a decreasing trend in rainfall, which agrees with the nature of climate change.

Extreme Event and Climate Change in Kerala

Contrary to the above long term rainfall trends, Kerala experienced unusually high rainfall from 1st June 2018 to 19th August 2018. Kerala recorded more than 37% excess rainfall in just two-and-a-half months. As per IMD data, Kerala received 2346.6 mm of rainfall from 1st June 2018 to 19th August 2018, in contrast to an expected 1649.5 mm of rainfall. Further, the rainfall over Kerala during June, July, and 1st to 19th of August was 15, 18 and 164% above normal respectively (Government of India 2018). This resulted in severe flooding in 13 out of 14 districts in the State (Fig. 2).

The graph computed by the Indian Meteorological Department at Thiruvananthapuram Centre on August 20, Fig. 3 shows that the rainfall departure from normal for the southwest monsoon season starting June 1 was by 41.09%. The recent flooding

[1] INCCA has been conceptualized as network based scientific program designed to: assess the drivers and implication of climate change through scientific research in 2009, by Government of India.

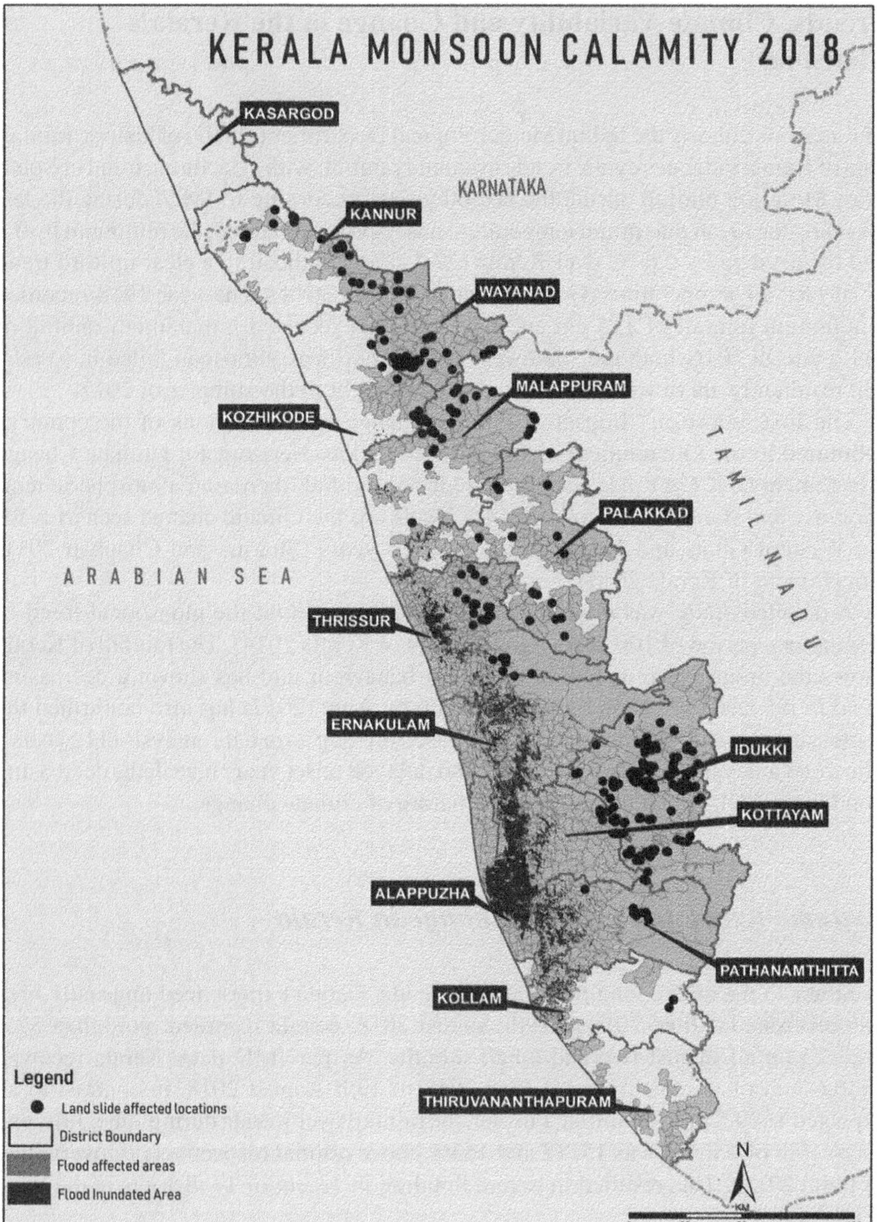

Fig. 2 Locations of landslides and flood affected areas post 2018 disaster. *Source* Kerala State Disaster Management Authority 2019

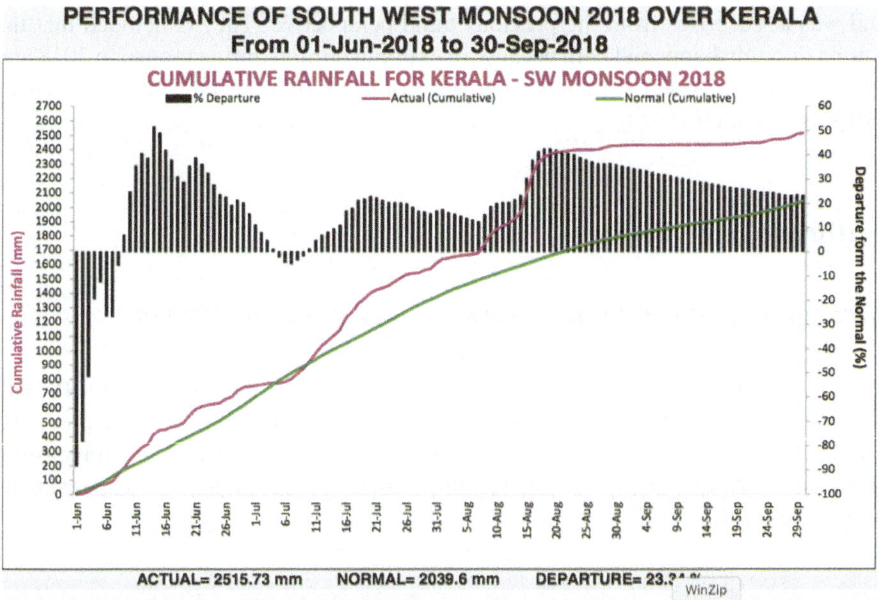

Fig. 3 Performance of South West monsoon 2018 over Kerala from 1 June 2018 to 30 September 2018. *Source* Graph by IMD Thiruvananthapuram 2018

was due to a low-pressure zone above the Southern peninsula, causing prolonged rainfall in the week commencing 13 August 2018 (Warrier 2018).

Furthermore, the rainfall distribution across districts was uneven. Idukki, which is a mountainous district, got an excess of 93.38%. This district received rainfall of 3521.43 mm, against a normal of 1821 mm for the 11-week period of the southwest monsoon, which is almost double. Seven districts had excess that was more than the state average. Palakkad had an excess of 73.45%, Kollam 54.27%, Kottayam 51.26%, Malappuram 50.81%, Ernakulam 48.43%, Pathanamthitta 47.01% and Thiruvananthapuram 45.33%. On the other hand, the northernmost district Kasargod had a deficit of 11.74%. These numbers indicate a significant spatial variation of rainfall in the state of Kerala. They also highlight the complex challenge being faced and the need for more scientific information in order to plan for such areas. The main conclusion based on various studies is that the Kerala flood was a result of short term extreme weather and climate event.

This hydro-climatological perspective on Kerala floods was confirmed by Mishra and Shah (2018) through a comprehensive analysis for this event. Previously, they had evaluated the long term rainfall, temperature and extreme events data in Kerala between 1951 and 2017, together with model simulations. The research then also analysed the changes in the 1 and 2-day annual maximum precipitation for the previous years and compared it to the 2018 rainfall. The finding, according to the study, was that the event was driven by a short-term natural variability of climate and was a 'surprise' from the previous trend perspective. They concluded that the Kerala flooding happened both because of extreme rainfall during monsoon 2018 and large amounts of water stored in catchments of reservoirs, which had to be released (Mishra and Shah 2018).

Anthropogenic Factors

Inadequate Information and Awareness About the Flood

Even though IMD issued district level rainfall forecasts from 8th August 2018 with Yellow, Orange and Red alerts in several districts, a regional analysis of predicted rainfall against actually realised rainfall revealed the gaps in existing forecasting systems. Owing to this, there was not much warning for local authorities and residents to prepare for the flood.

Changes in the Landform

Kerala is very unique in settlement pattern also. In most of its parts, it is a continuous spread of habitation without much open land or many fields separating habitations. The spatial distribution of shows that urban areas are concentrated in coastal and mid-land regions. Absence of risk informed urban planning, non-compliance of design standards, and non-incorporation of resilient features to critical infrastructure in urban areas have also aggravated the flood risk. Improper land use pattern and practices in the past decade in the state also compounded the multi hazard vulnerability of the state, specifically urban floods and landslides in hilly areas.

As per present urbanisation pattern of Kerala, fertile agricultural lands in the mid-land region are being converted for non-agricultural purposes. Land under forests has reduced from 44.2% in 1905 to 28% currently (Government of Kerala 2017). The high intensity rainfall resulted in heavy discharges in all the rivers. The continuous and heavy precipitation during the monsoon months that occurs in the steep and undulating terrain would find its way into the main rivers through innumerable streams and water courses. The changes in the geographical and topographical features due to man-made interventions blocked the normal waterways affecting the drainage patterns (Paniyil 2018). Further, with forests being cleared, less rainfall is being intercepted and more water is rapidly running into already overflowing streams and rivers. This has been also being flagged as a serious concern by the previous technical committee, the Gadgil Committee.[2]

[2]Gadgil commission—An expert panel formally known as Western Ghats Ecology Expert Panel (WGEEP) constituted to suggest strategy for conserving Western Ghats, an ecological rich resource in India.

Management of Water Release from Dams

While an extreme rainfall event (ERE) was the primary reason for this flood, there was criticism at various levels that the authorities failed to manage the flood effectively through reservoir operations. The rainfall that occurred from May to August 2018 in Kerala contributed to above-normal reservoir storage. Resultantly, reservoirs did not have the capacity to accommodate the additional flow generated by extreme rainfall and the water levels in several reservoirs were almost near their Full Reservoir Level (FRL). Because of this, water had to be released from several dams (Government of India 2018).

The nature of the event's extremeness gets underlined when analysis indicates that 35 out of the 46 dams within the state were opened for the first time in history. To add to the already critical situation, 5 overflow gates of the Idukki Dam were opened at the same time, for the first time in 26 years (Mishra et al. 2018).

In this context, however, the modelling and simulation studies conducted in one of the most impacted River Basin studied the role of water's released from the reservoirs. The research stated the current reservoir operation policy is based on water conservation so as to maintain the reservoir levels close to FRL throughout the monsoon season to ensure maximum power generation rather than flood management. The analysis suggested that reservoir operations could not have helped in avoiding the flood situation as only 16–21% peak attenuation was possible by emptying the reservoir in advance (Sudheer et al. 2019).

This emphasizes the need for better information and actions outlined in the case an event of this nature happened through understanding and projecting climate change scenarios. Sudheer et al. (2019) suggests revisiting the rule curves of the reservoirs by considering the dams as multi-purpose and multi-reservoir water resources systems, and developing integrated reservoir operation policies so as to maintain the balance between flood control and other system objectives, such as hydropower generation and irrigation.

Impact of Kerala Floods

Socio-physical Impacts

Uninterrupted rains from 8th to 18th of August 2018 resulted in wide spread destruction in all the major sectors of the state. In a span of 30 days, 445 human lives were lost and over 1.5 million people were moved to relief camps. Kerala has suffered huge economic losses on account of the floods. According to a conservative estimate, close to 2.6% of Kerala's gross state domestic product (GSDP) got washed away by the floods instantly. The devastating floods and landslides caused extensive damage to houses, roads, railways, bridges, power supplies, communications networks, and other infrastructure; washed away crops and livestock, and affected the lives and livelihoods of millions of people in the state. The five worst-affected districts of the state's 14 districts i.e. Idukki, Ernakulam, Kollam, Kottayam, and Pathanamthitta have an estimated population of 11.09 million which accounts for nearly 30% of the state's total population (Fig. 2).

Cochin International Airport which is one of the busiest International airports of the country got flooded and suspended its operations from 15th to 29th of August 2018. The Post Disaster Need Assessment[3] estimates the total damages to be around INR 10,557 crore and total losses to be around INR 16,163 crore amounting to a total disaster effect of around INR 26,720 crore (USD 3.8 billion) (Government of Kerala 2018).

Health Impacts

Kerala has the highest life expectancy (almost 75 years), and the highest sex ratio (1084 women per 1000 men), across all Indian states. Kerala has made significant advances in all the three components of health transition, namely demographic, epidemiological, and healthcare. Although there was no epidemic outbreak following the floods, health impact was substantial as most health facilities were fully or partially destroyed. Field assessment by Sphere India,[4] 2018 indicated that patients of non-communicable diseases like Diabetes and Hypertension lost their medications and medical records. In Wayanad and Palakkad district, some children were found prone to malnutrition. Sleep disturbance and mental health problem in people were noted in multiple districts (Government of Kerala 2018).

Educational Sector Impacts

Education has played an important part in Kerala's tremendous transition from a rigidly caste-divided society into one of India's most egalitarian states. In January 2016, it became the first Indian state to achieve 100% primary education. However, the disaster brought about the loss of learning materials of students and teachers, disruption of the learning cycle, loss of teaching days and damage to school infrastructure. Schools were used as relief camps and transition/temporary shelters, which

---

[3]The PDNA for the floods and landslides was made possible due to the collaborative efforts of the Government of Kerala, the Kerala State Disaster Management Authority, the United Nations agencies, the European Commission, Swiss Agency for Development and Cooperation (SDC), and European Union Civil Protection and Humanitarian Aid (ECHO).

[4]Sphere India coordinated a Multi-Sectoral Joint Detailed Needs Assessment (JDNA) in the 10 worst affected districts namely Alappuzha, Ernakulam, Idukki, Kottayam, Kozhikode, Malappuram, Palakkad, Pathanamthitta, Thrissur and Wayanad in 11 sectors each led by one of the Sphere India sector expert organizations.

disrupted routine academic schedules. There were concerns regarding loss of mark sheets and trauma to students caused by the event (Government of Kerala 2018).

Agricultural Sector and Livelihoods Impacts

Nearly 52% of Kerala's population lives in rural areas, and 17.15% of the population depends on the agricultural sector (including crops, livestock and fisheries) for its livelihood. Locations like Kuttanad, have been declared part of the global heritage in agriculture.

All three subsectors (crops, livestock, and aquaculture/fisheries) have suffered losses and damages in the flooding and landslides of 2018. Crops were most heavily affected, followed by livestock and fishery/aquaculture. Subsistence agriculture is an important source of income for rural communities (Government of Kerala 2018).

Among the worst affected were workers in the informal sector, who constitute more than 90% of Kerala's workforce. Thousands of casual workers and daily wage earners, such as agriculture labourers, workers in the coir, handloom, and construction sector and in the plantations, have experienced wage loss for 45 days or more. Hence, there is a critical need for sustained action on livelihood restoration and mainstreaming resilient livelihoods (Government of Kerala 2018).

Damage to Human Lives and Property

Massive damage to the roads, houses, and other infrastructure occurred in the northern districts of the state because of the landslides due to heavy rainfall. The devastating floods and landslides affected 5.4 million people, displacing 1.5 million to relief camps during the floods, as their homes were inundated with flood water.

Many buildings have been damaged either fully or partially, potentially affecting 0.75 million people. Housing damage during the flooding and landslides was caused by the scouring of foundations, settlement of soil, and inundation for several days. Houses in low-lying areas with low plinth heights were damaged much more than houses with high plinths and disaster-resistant features such as plinth and lintel bands. Resultantly, people had to live in temporary shelters or other temporary accommodation until their homes are safe enough for them to return.

Damage to Infrastructure

Access to piped water was disrupted for 20% of the state's population. Shallow wells were damaged and water was contaminated in six worst affected districts directly affecting people. The Kerala Water Authority and Irrigation departments run a large network of water supply for drinking as well as for irrigation purposes. Both of these sectors had a broad network of pipelines and canal systems which were passing through the urban and rural areas in Kerala. Irrigation canals, public taps, pipelines, pump houses, check dams, bunds, irrigation pumps and other irrigation machinery and structures were damaged due to floods, and landslides. Most of the engineering structures washed away by huge landslides and inundated by flooding. Huge losses in machinery, equipment, structural and non-structural assets have been estimated by concerned authorities.

Most of the houses had access to sanitation before the disaster; post-disaster, the houses, along with bathrooms, were washed away by the landslides or floods.

Additionally, the Public Works Department (PWD) has suffered unprecedented losses, as evidenced by the damage to physical infrastructure, especially roads and bridges. Some roads and bridges have even been completely washed away due to floods, and a total of 9538.45 km of roads have been damaged in Kerala. As many as 510 bridges have been damaged due to the calamity (Government of Kerala 2018).

Impact on Women

In Kerala, women are dependent on natural resources based livelihood activities, e.g. agriculture, fisheries, forest related activities or livestock rearing. Post disaster, with many crops and other natural resources, severely affected, there was considerable insecurity and stress w.r.t to food security and livelihood opportunities.

Though climate change is gender neutral, it affects men and women differently on a number of levels, including economically, socially, psychologically, and in terms of exposure to risk and risk perception. During the field survey by Sphere India (2018), 39% of women expressed privacy and dignity issues due to lack of private space for latrine and personal hygiene.

Kerala State Action Plan Climate Change, 2014

In response to the climate change impact, India released the National Action Plan for Climate Change (NAPCC) in October 2008. Additionally, the State Action Plans for Climate Change (SAPCC) are formulated to implement the national plan at the state level and find local-global solutions for the state-specific challenges. Kerala SAPCC was formulated by the Department of Environment and Climate Change with the technical support of UNDP India as Government of Kerala's strategy for action in the state in response to climate change in 2014 (Government of Kerala 2014).

In its current form, the SAPCC has undertaken vulnerability assessment for climate sensitive sectors like agriculture, fishery, forestry, etc. SAPCC recognises the role of local government in addressing climate change impacts in the state. It even states that it crucial to implementing for adaptation and mitigation strategies at grass root level i.e. every Gram Sabha, District Panchayat, Municipality, and Corporation should have area-specific climate resilient plans. But this needs to be taken up further and supplemented with detailed action plans.

The SAPCC admits the policy framework to align human development and climate change response efforts through adaptation is largely lacking in the state. It acknowledges the local self-governments structure in the state has not yet started any concrete steps for dealing with climate change. The strategies to address climate change challenges are limited to identifying the broad strategies, identify the department involved and the financial budget provision for the next five years.

Kerala State Disaster Management Plan, 2016

With the motto "Towards a Safer State", the Kerala State Disaster Management Authority has approved the State Disaster Management Plan 2016 as mandated by the Disaster Management Act, 2005. The Kerala State Disaster Management Plan has identified 39 hazards, categorised as naturally triggered hazards (natural hazards), and anthropogenically triggered hazards (anthropogenic hazards) (Government of Kerala 2016).

Nevertheless, floods are the most common of all natural hazards in Kerala. Nearly 14.5% of the state's land area is prone to floods, and the proportion is as high as 50% for certain districts. It outlines disaster preparedness and mitigation initiatives and responsibilities of stakeholders along with disaster response and relief.

In reference to climate change vulnerability, the report notes there is observed decreasing trend based on rainfall data for the last 100 years however extreme events expected to increase in frequency. Further, it is assumed that the hazard footprint may not increase beyond the worst case scenarios mapped and hence separate hazard footprint assessment in light of climate change scenarios were not conducted (Government of Kerala 2014).

State Government's Response to the Disaster

The state administration, in response to the disaster, decentralized the resources and duties down to the district and local body level. All precautionary measures were taken to meet the floods in accordance with the Kerala State Disaster Management Plan, 2016. According to the Disaster Management Plan, the levels of disasters have already been categorised and disseminated as L0, L1, L2 and L3, based on the ability of various authorities to deal with them.

When the flood situation started assuming the dimensions of an L3 disaster[5] (a catastrophe that overwhelms the state government) precautionary measures of disaster risk reduction as outlined in the standard operating protocol of the Kerala state Emergency operations of the Kerala State Disaster Management Authority (KSDMA) were taken. This included all measures supported by the allocation of responsibilities, budgetary provisions, and the guidelines, covering both structural and non-structural measures for risk reduction in the state. Several central government agencies like the Natural Disaster Management Authority, Indian Meteorological Department, the Army, the Navy, Central Water Commission, and the Home Ministry and Defence Ministry came together to execute flood evacuation, rescue, and relief operations in Kerala.

The National Disaster Response Force flew in 58 rescue teams in one of the largest rescue missions in India that were conducted in the decade. In addition, the Army, the Airforce and the Navy stepped into the rescue mission with their own teams. The assistance of the National Disaster Response force was sought for additional help for search and rescue operations, and for ensuring relief supplies reached affected areas. For the rescue operation, the forces airlifted stranded people and built temporary bridges for evacuation. The Southern Naval Command deployed 92 teams with Gemini boats and seamen, accompanied by expert divers from the Navy for the search and rescue operation (Sen 2018).

Local Government

The decentralisation process, which started in 1996 in Kerala, has empowered local governments with funds, functions, and functionaries. But when it comes to strategic climate resilience planning, the role of local governments remains understated. This can be attributed to a lack of experts and knowledge tools at the local level, affecting their decision-making capabilities.

Local Community

The community came together to organize rescue and relief work, led by different types of peoples' organizations, ranging from trade unions and student-youth organizations to self-help groups, charitable organizations, sports clubs, and other collective bodies.

This was one of the largest civilian rescue operations the state had ever undertaken, with 4500 fishermen in 700 boats involved in rescue operations. The fishermen had

---

[5] L3 disaster situations arise from large where districts and the state may not have the capacity to respond adequately and require assistance from central government for reinstating the state and district machinery.

the benefit of local knowledge of habitats and understanding of the rivers to their advantage. Also, they used twin engine, locally designed fishing boats that can move in shallow waters (less than 2 feet deep) and restrictive spaces. Working with the local authorities, volunteer groups and the police, they were able to rescue over 65,000 of people stranded in their homes (Sen 2018).

Voluntary Partnerships

Citizen volunteers also played a critical role in helping manage the rescue operations. People's support in the various forms of monetary donations and technological support helped. Volunteers set up informational technology networks to register calls for help and guide rescue workers, supplementing the official call centre and helplines. An online portal KeralaRescue.in was started by citizens that not only collected donations, but helped enlist volunteers, recorded requests for help, and contributed other needed services.

Discussion

Several studies have already projected that the impact of climate change will increase the intensity of hydro-meteorological hazards such as floods and cyclones. There is a general agreement among experts that disaster risk is bound to increase in the coming decades on account of climate variability, climate change, and environmental degradation (Government of India n.d). With the 2018 extreme weather event, this became a reality in Kerala. Various reports and studies indicated that the deluge in Kerala was an outcome of a set of interacting atmospheric, oceanic, and human factors (Mishra et al. 2018).

There are a number of fundamental issues that need to be addressed while dealing with climate change challenges, especially its complexity. Contrary to the monsoon trend studies, which established that rainfall pattern in recent years in Kerala as decreasing rainfall, there was a high intensity of rainfall in a short duration. Consequently, rainfall in August 2018 caused floods in multiple locations. The anthropogenic drivers, like modified land use and land cover, reservoir storage and operations, and the encroachment of flood plains also indirectly or directly aggravated this disaster.

Kerala mainly took actions according to the disaster management protocol. While disaster management was in place, there were no detailed climate change action plans, especially at the local level outlining the actions to be undertaken in case of such a disaster. There were knowledge gaps about climate change impacts and actions in the Kerala State Climate Action Plan. In its current form, the Kerala SAPCC lacks detailed climate vulnerability analyses, state-specific climate research, and evidence-building, including time series data mechanisms. Sector-specific priorities and programs need re-alignment through the adoption of risk-informed planning for improved response.

Currently, the assessments of flood risk and response is a static, steady-state system. What is needed, however, is a dynamic response. It makes more sense to take a proactive approach (rather than a reactive one) by making scientific information-based choices from the very beginning. This implies the disaster plan, policies, climate management strategies need to be systematically reviewed, modified, and integrated. The disaster highlighted the need to integrate Disaster risk reduction (DRR)[6] and Climate Change Adaptation[7] (CCA) in planning and policy processes.

Way Forward

Based on the analysis, a framework has been suggested for improving resilience specifically for Kerala State. The framework suggested can serve as a guide for future adaptation, risk reduction policies and increased climate resilience for the State of Kerala. In fact, the lessons and experiences post disaster can serve as a guide for future adaptation, risk reduction policies and increased climate resilience for the State of Kerala (Fig. 4).

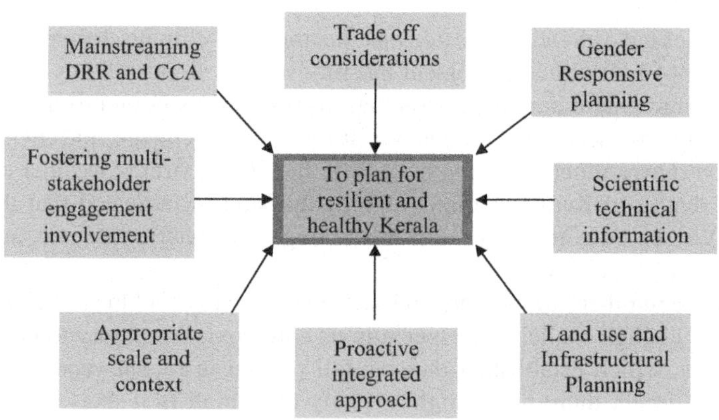

Fig. 4 Approach for strengthening Kerala's climate change resilience. *Source* Authors

[6]Disaster risk reduction is defined by the International Strategy for Disaster Reduction (ISDR) as, 'Action taken to reduce the risk of disasters and the adverse impacts of natural hazards, through systematic efforts to analyse and manage the causes of disasters, including through avoidance of hazards, reduced social and economic vulnerability to hazards, and improved preparedness for adverse events'.

[7]Whereas the term adaptive capacity originates in the Third Assessment Report of the IPCC Working Group 2, Chap. 18 and describes the ability of a society to successfully respond to changes in its environment, in this case, hydrological risks.

Comprehending Disaster Risk and Climate Change Impacts and Taking the Proactive Integrated Approach

Disaster risk reduction (DRR) and Climate Change Adaptation (CCA) are two ways of reducing the risk posed by natural hazards and climate change. DRR emphasizes prevention, mitigation, preparedness, and recovery from geological hazards and focuses more on the humanitarian sector. CCA, on the other hand, mainly aims at reducing vulnerability due to climate variability through adaptation to climate change. However, both DRR and CCA include vulnerability and risk assessments and both aim to reduce vulnerabilities and improve capacities so that the impacts of disasters are reduced. Both are relevant from the national level to the local level and need policies, guidelines, and laws to affect needed change. Both are concerned with reducing vulnerability, monitoring hazards, and raising societal capacities to reduce and manage risks. The overlap between DRR and CCA is in managing disaster risk related to climate variability and climate extremes and preparing for risks related to climate change. There is an increasing recognition that these two approaches share a common focus: reducing the vulnerability of communities and contributing to sustainable development. Consequently, the need for systematic integration, or functional linking, of DRR and CCA in activities and mainstreaming efforts on the ground is an immediate and pressing concern (UNISDR, UNDP 2012; ASEAN and JICA 2017).

Appropriate Measure at Appropriate Spatial Scale and Context

One of the major challenges faced frequently for the integration of DRR and CCA is the spatial scale. The DRR focuses primarily on local vulnerabilities and hazards, whereas CCA starts from a global to local scale (Birkmann and Teichman 2009). Taking into account the spatial scale in the case of Kerala—there exists a Climate Change Plan at the State level, which has done vulnerabilities and risk assessment at the State level.

For instance, the Kerala study shows that there was substantial rainfall variation between the 14 districts in Kerala State. In this context, the strategies will have to respond to the challenges that are context specific and scale specific. There is a need for study of spatial variability of extreme rainfall events helps to identify zones of high and low values of extreme rainfall events. A detailed regionalized study at the right scale is useful for planners and other users. Detailed Climate Action Plans needs to be prepared for various spatial scales.

Improving Scientific Technical Information and Knowledge Based Decision Making

It is clear that first and foremost, the knowledge and information base on climate change impacts needs to be improved in the case of Kerala. Lack of flood forecasting stations led the disaster to assume a larger magnitude. The first step is to have a robust forecasting and prediction system. Weather forecast and satellite data needs to be used in synergy with ground data from reservoirs and dams to issue flood forecasts and warnings. Past experiences also indicate that appropriate and timely risk communication is critical for effective adaptation and disaster risk management.

The next task is to enhance the capacity for scientific assessment and awareness among government institutions and all stakeholders through renewed understanding, dialogue, and partnerships among key institutions regarding climate risks and disaster risks (Mall and Attri Kumar 2011). Currently, the understanding of these issues is limited to individually or separately assessing the disaster and climate risks. Being a climate vulnerable state, Kerala needs to understand how to deal with multi-hazards and climate change impacts in addition to natural hazards.

Keeping in Mind the Trade-off During Decision Making

The Stockholm Environment Institute's Initiative on Transforming Development and Disaster Risk, launched in 2015, aims at focusing on the relationship between development and disaster risk. The Initiative's fundamental precept is the recognition that decision-making at all levels—from the individual up to the national government—involves trade-offs. The approach advocates that from the very start, there should be recognition of the multiple policy-related goals involved. Impact assessments of potential alternatives should include direct and indirect social, economic, and environmental aspects (Stockholm Environment Institute 2018).

The complexity with respect to climate impacts need to be acknowledged, understood then responded to. In the case of Kerala, the long term climate change projections rainfall trend analysis indicates that the rainfall is decreasing over the years. Rainfall is just one risk associated with trend analysis. There are more challenges with respect to temperature and sea levels rising affecting various sectors such as agriculture, water, energy, transport. Hence, the proper response will include the prioritization of certain objectives in decision-making while allocating resources, keeping in mind the trade-offs. Another important issue to consider is the alignment of risk prioritization at various levels of decision-making, for example, between local authorities and citizens. Hence, an integrated approach is important in minimising the use and costs of limited technical, administrative, and financial resource while considering the trade-off during decision making.

Fostering Multi Stakeholder Involvement Especially the Local Community to Promote Community Resilience

To facilitate local action based on local knowledge and awareness, decision-making at the local level needs to be promoted. A mechanism that provides a platform for stakeholder participation during the stages of design, appraisal, implementation, and review is important mainstreaming resilience (Government of India n.d).

Fisherman, non-governmental organisations, volunteers and many others played a key role post disaster and the importance of their role cannot be negated. But this has to be done systematically through community resilience[8] based planning. The local communities have to be involved in conducting vulnerability and risk assessments and preparing a management plan. Integrating multi-level stakeholder engagement into the early stages of planning and development processes lays a foundation for inclusive and efficient engagement.

Land Use and Infrastructural Planning

The pattern of urbanisation and high density of population of the State are huge challenges for the creation of infrastructure. Moreover, the infrastructural facilities available to the population, in general, do not vary much between rural and urban, especially in the case of access to educational and health care facilities.

It is suggested that delineation of climatic risk zones based on historical weather data, land capability, and employing GIS techniques should be undertaken. It is recommended that the land-use on steep slopes, low-lying areas where water can converge and riparian zones and ecologically sensitive regions needs to be strictly regulated. With this analysis complete, there is a need to ensure local governments have accurate knowledge of the practical issues during implementation for disasters.

Gender Responsive Planning

Kerala is a state where the status of women is better compared to the rest of the country on indicators like literacy, Infant Mortality Rate (IMR), Infant Mortality Rate (MMR) and other social and health indicators. There are very strong groups of women as self-help groups and neighbourhood groups. Moreover, the Gender Equality and Women Empowerment Policy was approved by the Government of Kerala in 2015.

---

[8]Community resilience includes resistance (the ability of a community to absorb perturbation), recovery (the speed and ability to recover from the stressors) and creativity (the ability of a social system to maintain a constant process of creating and recreating as the community not only responds to adversity, and efficient functioning).

During the field assessment, the challenges faced by women post disaster were noted. The main recommendations would be to promote greater engagement of women, their specific concerns and commitment to gender equality in the context of disasters and climate. Women can play a decisive role in the design and implementation of the SAPCC (Climate Resilient Kerala 2017). The SAPCC must recognize the vulnerabilities of children and women and ensure their inclusion in design and implementation of all future mitigation, adaptation and disaster risk reduction programmes. Focussed intervention and livelihood protection for women should also be encouraged.

Conclusion

The negative impacts of the disaster in 2018 have severely affected the gains made by Kerala State in terms of development. The experiences from the disaster have contributed to understanding the gap between disaster risk and climate adaptation approaches. Once the risks are clearly understood, it can be followed by an integrated well-defined programme implementation and management mechanism. The key message here is that climate risk management and its integration in projects, plans, and policies are to be perceived as continuous and ongoing dynamic processes. Mainstreaming climate change resilience will help reduce vulnerability to climate impacts and variability and offers a path forward, both to strategically respond to the climate change threat and to protect the communities in this region. With further improved scientific understanding, community-based and participatory governance and proactive approach Kerala can manage both climate change risks as well as a natural risk even better to ensure long-term sustainability and minimising losses.

Acknowledgements The authors would like to acknowledge the support of the entire team of Kerala State Disaster Management Authority led by Dr. Sekhar L. Kuriakkose (Member Secretary, KSDMA and Head Scientist, Kerala State Emergency Operations Centre), Mrs. Parvathy S. (Hazard and Risk Analyst, Kerala State Emergency Operations Centre, KSDMA), Mr. Pradeep G. S., Mr. Sathyakumar C. J., Mr. Fahad Marzook, Ms. Anupama (Hazard Analysts, Kerala State Emergency Operations Centre). The authors would like to acknowledge the Department of Information and Public Relations, Government of Kerala.

References

Anie S (2017) Onset of South West monsoon over Kerala with reference to climate change. Int J Eng Dev Res 5(4)

ASEAN and JICA (2017) One against disaster and climate risks: a repository of good practices for strengthening DRR and CCA integration in ASEAN. ASEAN and JICA, Bangkok, Thailand

Birkmann J, Teichman K (2009) Integrating disaster risk reduction and climate change adaptation: key challenges—scales, knowledge, and norms. Sustain Sci 5:171–184

Choudhary S (2018) Rise in extreme weather events in India raises concerns over climate change impact. Livemint. Retrieved from https://www.livemint.com/Politics/0lEa4TGyXdIadjyvArOxzL/Rise-in-extreme-weather-events-in-India-raises-concerns-over.html

Climate Resilient Kerala (2017) Stakeholder recommendations for Kerala SAPCC. [Ebook]. Thiruvananthapuram, Kerala. Retrieved from http://www.cansouthasia.net/wp-content/uploads/2018/02/CRK-REPORT.pdf

Government of India (n.d) A study on mainstreaming DRR and CCA in development planning. New Delhi. Retrieved from https://www.ndmindia.nic.in/images/pdf/Mainstreaming%20Report_Final_sept7.pdf

Government of India, Central Water Commission (2018) Study report Kerala floods of August 2018

Government of Kerala (2005) Human development report. Thiruvananthapuram

Government of Kerala (2007) State of environment report Kerala. Thiruvananthapuram

Government of Kerala (2014) Kerala state action plan on climate change. Thiruvananthapuram

Government of Kerala (2016) Kerala state disaster management plan. Thiruvananthapuram

Government of Kerala (2017) Economic review 2016. Thiruvananthapuram

Government of Kerala (2018) Post disaster needs assessment floods and landslides, Aug 2018. Thiruvananthapuram

Guhathakurta P, Sreejith O, Menon P (2011) Impact of climate change on extreme rainfall events and flood risk in India. J Earth Syst Sci 120(3):359–373. https://doi.org/10.1007/s12040-011-0082-5

Krishnamurthy V, Kinter J (2003) The Indian monsoon and its relation to global climate variability. In: Rodo X, Comin F (eds) Global climate: current research and uncertainties in the climate system. Springer, Berlin, Heidelberg

Mall R, Attri Kumar S (2011) Extreme weather events and climate change policy in India. J South Asian Disas Stud 4

Mishra V, Shah H (2018) Hydro climatological perspective of the Kerala flood of 2018. J Geol Soc India 92(5):645–650. https://doi.org/10.1007/s12594-018-1079-3

Mishra V, Aaadhar S, Shah H, Kumar R, Pattanaik D, Tiwari A (2018) The Kerala flood of 2018: combined impact of extreme rainfall and reservoir storage. Hydrol Earth Syst Sci Discuss 1–13. https://doi.org/10.5194/hess-2018-480

Paniyil M (2018) Extreme rainfall and human interventions devastate Kerala. Retrieved from https://indiaclimatedialogue.net/2018/08/22/extreme-rainfall-and-human-interventions-devastate-kerala/

Sen S (2018) Improved disaster management saves Kerala, despite lack of preparedness. ORF. Retrieved from https://www.orfonline.org/expert-speak/43901-improved-disaster-management-saves-kerala-despite-lack-of-preparedness/

Sharma S, Chauhan R (2011) Climate change research initiative: Indian network for climate change assessment. Curr Sci 101(3):308–311. Retrieved from http://www.jstor.org/stable/24078509

Sphere India (2018) Kerala floods joint detailed needs assessment report. New Delhi. Retrieved from https://reliefweb.int/sites/reliefweb.int/files/resources/jdna-kerala-report_1st-draft.pdf

Stockholm Environment Institute (2018) How do we prioritize when making decisions about development and disaster risk?

Sudheer K, Bhallamudi S, Narasimhan B, Thomas J, Bindhu V, Vema V, Kurian C (2019) Role of dams on the floods of August 2018 in Periyar River Basin, Kerala. Curr Sci 116(5):780–794. https://doi.org/10.18520/cs/v116/i5/780-794

Thomas V, López R (2015) Global increase in climate related disaster. [Ebook]. Asian Development Bank. Retrieved from https://www.adb.org/sites/default/files/publication/176899/ewp-466.pdf

UNISDR, UNDP (2012) Disaster risk reduction and climate change adaptation in the Pacific: an institutional and policy analysis. UNISDR, UNDP, Suva, Fiji, p 76

Warrier GS (2018) Kerala floods: time for introspection and amends as water recedes. Retrieved from https://india.mongabay.com/2018/08/24/commentary-kerala-floods-time-for-introspection-and-amends-as-water-recedes/

The Impact of Extreme Floods on Rural Communities: Evidence from Pakistan

Ali Jamshed, Joern Birkmann, Joanna M. McMillan, Irfan Ahmad Rana and Hannes Lauer

Abstract This chapter studies impacts of flood events on the livelihoods of rural communities in Pakistan. It explores the adverse effects of extreme flooding, as well as how communities learn and adapt their livelihoods in the wake of such events. Three flood-affected districts in the province of Punjab were selected as case study areas. A household survey was conducted and analyzed by using descriptive statistics. The results suggest that various livelihood assets were adversely impacted by flooding. The most significant impact was the loss of assets in farming communities. The impact of floods has led to several changes in the communities—notably, households changed how they make a living and their relationship with the nearest major city. Other changes include changes in construction techniques and a range of behavioural changes. Differences among communities were found, particularly in how they responded to flood events. It was observed that rural communities near large city more rapidly adopt changes that make them better adapted to future flooding as they have access to more avenues for livelihood diversification and more services, as well as better and more innovative markets. Observing and understanding these changes that the surveyed rural households made in response to flooding, is an important contribution to understanding how communities can adapt to extreme events and how this adaptation can be supported.

A. Jamshed (✉) · J. Birkmann · J. M. McMillan · H. Lauer
Institute of Spatial and Regional Planning (IREUS), University of Stuttgart, Pfaffenwaldring 7 (3rd Floor), 70569 Stuttgart, Germany
e-mail: ali.jamshed@ireus.uni-stuttgart.de

J. Birkmann
e-mail: joern.birkmann@ireus.uni-stuttgart.de

J. M. McMillan
e-mail: joanna.mcmillan@ireus.uni-stuttgart.de

H. Lauer
e-mail: hannes.lauer@ireus.uni-stuttgart.de

I. A. Rana
Department of Urban and Regional Planning, National University of Science and Technology Islamabad (NUST), Main Campus, H-12 Sector, Islamabad, Pakistan
e-mail: iarana@nit.nust.edu.pk

W. Leal Filho et al. (eds.), *Climate Change, Hazards and Adaptation Options*, Climate Change Management, https://doi.org/10.1007/978-3-030-37425-9_30

Introduction

Climate extreme events, especially in the form of hydro-metrological disasters, have wreaked havoc around the world and in Pakistan. Fluvial flooding is the most common and damaging of all natural hazards (IPCC 2012; Kinoshita et al. 2018) and are in many areas of the world becoming more frequent and intense due to climate change (Kundzewicz et al. 2013; Jongman et al. 2018). Floods have caused widespread loss of life as well as economic losses and physical damages. Between 1980 and 2016, total direct damages worldwide due to flooding exceeded 1.6 trillion USD, and at least 225,000 people lost their lives (EM-DAT 2018). Winsemius et al. (2016) indicate that damages from fluvial flooding could be 20 times greater by the end of this century. Furthermore, Bilak et al. (2016) contend that flood events were the primary reason for internal displacement between 2008 and 2016. These increasing losses due to flooding are not only due to an increase in the frequency and intensity of the flood hazard, but also the exposure and vulnerability of the population and their assets to flooding. Exposure to flooding has significantly increased in many areas (Jongman et al. 2012), which is considered as one of the main drivers of flood risk (Muis et al. 2015). Asian countries will be severely impacted by this increase in flood risk (Winsemius et al. 2016; Willner et al. 2018), most probably due to both climate change and socio-economic development (O'Neill et al. 2014; Winsemius et al. 2016; Alfieri et al. 2017). The impact of such extreme flood events can bring several formal and informal changes in the society which can influence their vulnerability and risk profile (Birkmann et al. 2008). However, the extent of impact and following changes depends on socio-economic status and physical development of communities and societies. Limited attention has been paid to study the changes following extreme flood events.

The impact of flooding can be particularly destructive in developing countries like Pakistan, due to low levels of flood management, both in urban and rural areas (Doocy et al. 2013a; Willner et al. 2018). Pakistan, a developing country in South Asia, is faced with high flood risk due to diverse climatic and hydrological conditions. According to Pakistan's National Disaster Management Authority (NDMA), 67 out of 156 districts are at high flood risk (see Fig. 1). Frequency and intensity of floods have increased and climate change is considered as one of the main reasons (GoP 2012). According to the Climate Risk Index (CRI), Pakistan is ranked as the 5th most affected by climate change in 2014. The country's rank was 6th in 2013, 3rd in 2012 and 2011, 1st in 2010 (Kreft et al. 2016). Willner et al. (2018) contend that Pakistan is already highly affected and will observe almost a doubling in high-end flood risk at subnational level.

Pakistan can be divided into three major hydrological landscapes: Indus Basin (mainly fluvial flooding), Kharan Basin (flash flooding), and Makran Coastal drainage area (coastal flooding) (Tariq and van de Giesen 2012). Flooding in the Indus basin has caused major economic and human losses in the past. Indus basin is very fertile for agriculture and plays a huge part in the country's economy. As a result, it is highly populated, and population pressures have led to development in

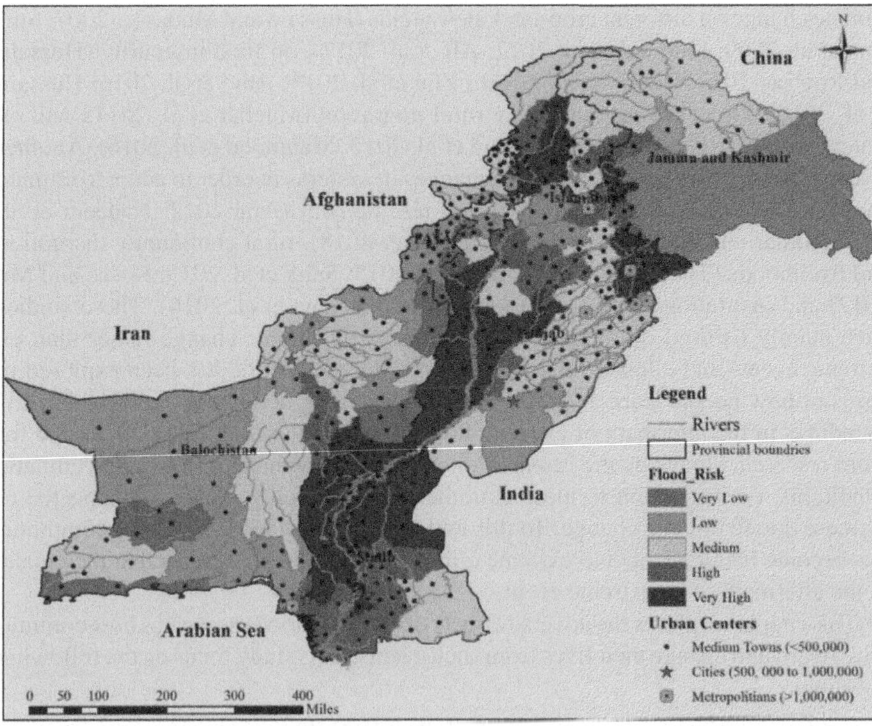

Fig. 1 Flood risk map of Pakistan with the spatial distribution of urban centres. *Source* Own construct based on NDMA (2012)

floodplains, with no regard to the resulting increase in flood risk due to higher exposure. The government has spent a vast amount on structural engineering works to counter flood risk but these measures could not prevent large-scale disastrous floods in recent years (Tariq and van de Giesen 2012). These disastrous floods affected both rural and urban areas. The few urban areas which were affected quickly recovered from the flood events, whereas recovery of rural areas was slow due to lower socioeconomic, physical and institutional capacities (Kirsch et al. 2012). This supports the assumption that the impacts of extreme events are spatially and socially uneven and depend not only on the physical and natural characteristics of human settlements, but also on socio-economic characteristics (Jamshed et al. 2017). Thus, it has become imperative to understand social, economic and physical impacts of extreme events, in order to invoke appropriate actions among communities and institutions to reduce vulnerability.

There is a significant body of research focusing on the local impact of climate change on agriculture and the water resources sector in rural settings. Studies have been conducted on farmers' perceptions regarding climate change and their subsequent risk attitudes and adaptation strategies (Abid et al. 2015; Ullah et al. 2015; Arshad et al. 2016; Saqib et al. 2016). Other research focused on the impact of climate change on different crops and their yields (Hussain and Mudasser 2007; Sultana et al. 2009; Siddiqui et al. 2012; Ali et al. 2017), on food insecurity (Hussain and Routray 2012; Doocy et al. 2013b; Zhu et al. 2013; Abid et al. 2016; Hussain et al. 2016; Ali and Erenstein 2017), rural migration (Mueller et al. 2014), and on water resources (Qureshi 2011; Ahmad et al. 2012; Mahmood et al. 2016). Another research priority has been focused on changes in systems in order to adapt to climate change. These include rehabilitation and resettlement (Arai 2012; Nadeem et al. 2014; Akbar and Aldrich 2018; Jamshed et al. 2018), rural community disruption and livelihoods (Thomas 2014; Arshad et al. 2015; Salik et al. 2015; Kosec and Mo 2017) and institutional and legal frameworks (Cheema et al. 2016). These studies have mainly focused on the slow-onset impacts of climate change rather than on extreme events and climate-related disasters. Moreover, little has been explored in terms of how people learn and change their behaviour, actions as well as their surroundings in the aftermath of extreme flood events in Pakistan. There is a need for more research, alongside the research on adaptation to changes in average climatic conditions, on adaptation to more extreme climatic events, which are expected to increase due to climate change. To this end, this research looks into how communities become better adapted to extreme climate events due to changes that they make in the aftermath of an extreme event.

This chapter explores the adverse effects of extreme floods as well as how communities learn and change their lives from such events. This study focus on the following questions:

(a) What are the impacts of floods on rural communities in Pakistan?
(b) How do communities change their livelihoods to cope and adapt to floods?

Based on these questions, the paper aims to provide insights into how the impact of the extreme event can bring changes in the communities' way of life. It focuses particularly on how communities adapt their livelihoods after flood disasters. The chapter is based on both an extensive literature review to provide the conceptual understanding of the study and to explain the flood hazard profile of the case study areas, as well as empirical work done by the lead author in three different flood affected communities in Punjab province of Pakistan.

Theoretical and Conceptual Understanding

Extreme climate events (e.g. droughts) and extreme weather events (e.g. excessive rainfall) are often referred to as climate extremes or climate-related hazards. They are defined as "the occurrence of a value of a weather or climate variable above (or below) a threshold value near the upper (or lower) ends of the range of observed values of a variable" (IPCC 2012). This understanding implies that if an extreme event, such as heavy rainfall during the monsoon season, oversteps a certain threshold value, it might have hazardous consequences—for example in the form of a flood hazard when the riverbeds cannot hold back the water. Nevertheless, the thresholds are dynamic across time and space and are highly location specific. Essentially, not every weather or climate extreme results in a disaster.

What makes an extreme event a disaster? The core concept of the IPCC's Special Report on Managing Risks of Extreme Events and Disasters to Advance Climate Change Adaptation (SREX) indicates that the potential of a certain climate extreme event to be disastrous, depends on the magnitude of the event itself and of the interplay with exposure and the vulnerability of human and natural systems (IPCC 2012). In disaster risk literature, disaster is defined as an event causing "serious disruption of the functioning of a community or a society involving widespread human, material, economic or environmental losses and impacts, which exceeds the ability of the affected community or society to cope using its own resources" (UNISDR 2009).

Extreme events and their impacts that cause disastrous damages and losses can lead to changes in the societies that they affect. In their framing of extreme disaster events, impacts, and associated changes, Birkmann et al. (2008) differentiate between impacts of and changes caused by extreme events. Impacts are the result of a disaster event and can be direct (in the form of losses and damages) and indirect (influencing socio-ecological systems like markets, health, income etc.). Whereas change is regarded as a response to minimize the current and future impacts. The resilience of a society to disaster events is determined both by their ability to cope with the immediate impacts of the event, as well as their ability to change to become better adapted to new conditions and to future such events (Birkmann et al. 2009; Birkmann 2011). In this context changes refers to the concept of capacity and capacity leads to reduce vulnerability (Jamshed et al. 2019). The conceptual framework of this chapter is based on the framework of Birkmann et al. (2008), modified to make it specific to the context of flooding in Pakistan (see Fig. 2).

The aftermath of a disaster is seen as a window of opportunity in which social, economic and environmental conditions can be modified. Berkes et al. (2003) and Folke (2006) identify change after a disaster as a key factor in the resilience of coupled socioecological systems. These changes might be formal or informal and have negative, positive, or neutral effects. Moreover, they might be proactive, intending to build a different—a better—future, or they might be rather responsive, initiated out of pure necessity. This chapter focuses on the informal changes undertaken by the flood-affected households in rural areas.

Methodology

The first step of this research was the case study selection, which was based on a literature review of past flood disasters in Pakistan and judgements from experts. Three flood affected areas of Punjab province in Pakistan were selected. In these case study areas, quantitative empirical research was conducted to capture information about the flood impacts and changes in response to those flood events. Descriptive statistics were used to analyse the data and draw conclusions. In addition to these

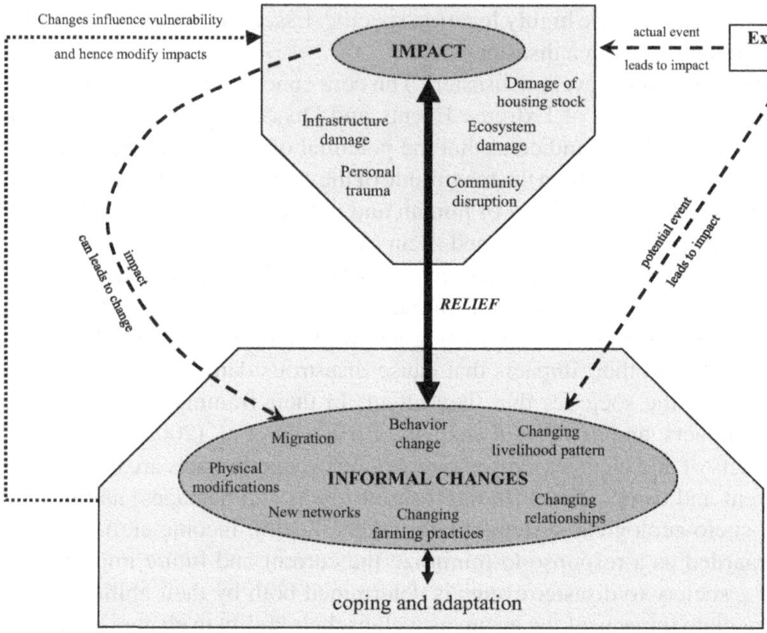

Fig. 2 The conceptual framework of the study. A feedback loop leading from changes to impact emphasises that societal changes in response to the impact extreme events impacts influence the vulnerability of the system. *Source* Modified from Birkmann et al. (2008)

quantitative methods, key informants interview were conducted to understand the impacts and changes that occurred after flooding.

Case Study Areas

Flood hazards, predominantly fluvial floods, are common throughout Pakistan, causing widespread socio-economic damage and loss of life. The province of Punjab is particularly affected by flooding and was selected as the focus of this case study. It is the most populous province of the country, accommodating about 53% of the population. According to the recent census, the population of Punjab was 110 million, with approximately 36% living in urban areas (PBS 2017; Rana 2017). The literal meaning of Punjab is "land of five rivers" consisting of Indus, Jhelum, Chenab, Ravi, and Sutlej. Punjab contributes to 57% of the total cultivatable area in the country, 53% share in agriculture gross domestic product and 74% of total cereal production in Pakistan (Hanif et al. 2010; Abid et al. 2015). This makes Punjab an important entity in the local and national economy since it provides raw materials to urban-based industries, as well as provides food and services to its inhabitants.

Historically, Punjab has experienced numerous fluvial floods from different rivers in rural areas. However, the higher frequency of fluvial floods between 2010 and 2014 is seen as an impact of the changing climate, particularly the increased and more intense monsoon rainfall (Hussain and Mumtaz 2014). These flood events have displaced millions of people, destroyed thousands of houses, and damaged the agriculture base, social services, transport networks, small businesses and an extensive irrigation system. According to an estimate, the flood of 2010 heavily impacted the livelihood of people by damaging standing crops on 7.5 million acres of land, and destroying more than 40,000 small shops and 41 industrial units (NDMA 2010). In 2014, the flood caused the livelihood loss of 73% farmers and 50% daily wage labourers. Almost 750 schools were damaged and hampered the schooling of children. It has also been reported that more than 50% of access roads were critically damaged (NDMA 2014). Table 1 shows the characteristics of these flood events. These flood impacts have serious repercussions not only for the livelihood of rural households but also for urban settlements whose economy is closely linked with urban areas.

Three districts within Punjab Province were selected as case study areas—namely Districts Bhakkar, Muzaffargarh and Multan. Within these three districts, one sub-district was selected, in each case the sub-district contains the major city. Two union councils[1] within each of these three sub-districts were selected for the empirical research (see Fig. 3). A brief description of these case studies is given in following paragraphs.

The western side of District Bhakkar, where the small major city Darya Khan lies, is characterised by fertile and irrigated plains along the Indus River. During the summers, high water discharge in the river Indus causes flooding in the district and influences the local economy. The district is predominantly characterized by agricultural function with few agro-industries. The city is characterized by services like agriculture inputs, food items, building material, a food storage house, a public hospital, education centres, and a large sugar mill.

District Muzaffargarh is bounded by two major rivers. Indus flows in the west and Chenab in east making the district a *doab*, a land between two rivers. A large area of the district consists of sand dunes and barren land (DoIP 2012). The two major rivers caused yearly large-scale flooding from 2010 to 2014. The district is specialized in agro-industrial function and several large, medium and small-scale agriculture-based industries operate in the district (DoIP 2012). Muzaffargarh is a medium sized city which offers services like large public hospital, vegetable and fruit markets, agriculture input services, construction materials, rescue services, and various large and small scale industries providing non-farm employment to rural population.

District Multan is a heartland of southern Punjab. The river Sutlej flows in the south and river Chenab in the west. In the monsoon, flood occurs in River Chenab cause widespread socio-economic and physical losses in the district. The district has mixed functions and provides a leadership role in agriculture, industry, education,

[1]Union council is the lowest administrative unit in Pakistan.

Table 1 Characteristic of flood events between 2010 and 2014 in Punjab

Indicators	2010	2011	2012	2013	2014
Time	End of July	Beginning of August	Beginning of August	Middle of August	Beginning of September
Causes	Extreme rainfall in monsoon, snowmelt, dike breach	Extreme rainfall in monsoon	Extreme rainfall in monsoon	Extreme rainfall in monsoon	Extreme rainfall, snowmelt, transboundary water discharge, dike breach
River bodies	Indus, Chenab, Kabul	Indus and Sutlej	Indus and Chenab	Chenab and small tributaries	Chenab, Jhelum, Ravi
Exposure					
Districts affected	12	7	7	13	15
Settlements affected	3000	335	1512	2946	3500
Population affected (thousand)	7000	26	887	634	3000
Area submerged (thousand acre)	3500	137	1491	746	3000
Loss and damages					
Deaths	112	4	60	109	284
House damaged (thousand)	400	1	26	20	100
Loss of crops (thousand acre)	7500	126	474	555	2400

Source Nadeem et al. (2014), NDMA (2014), PDMA (2014), Jamshed (2015)

health and commerce and various other sectors in southern Punjab (Ahsan and Noreen 2009). Thus, the district offers various services and livelihood opportunities for the people. Multan is a large city which provides multiple services to the rural population.

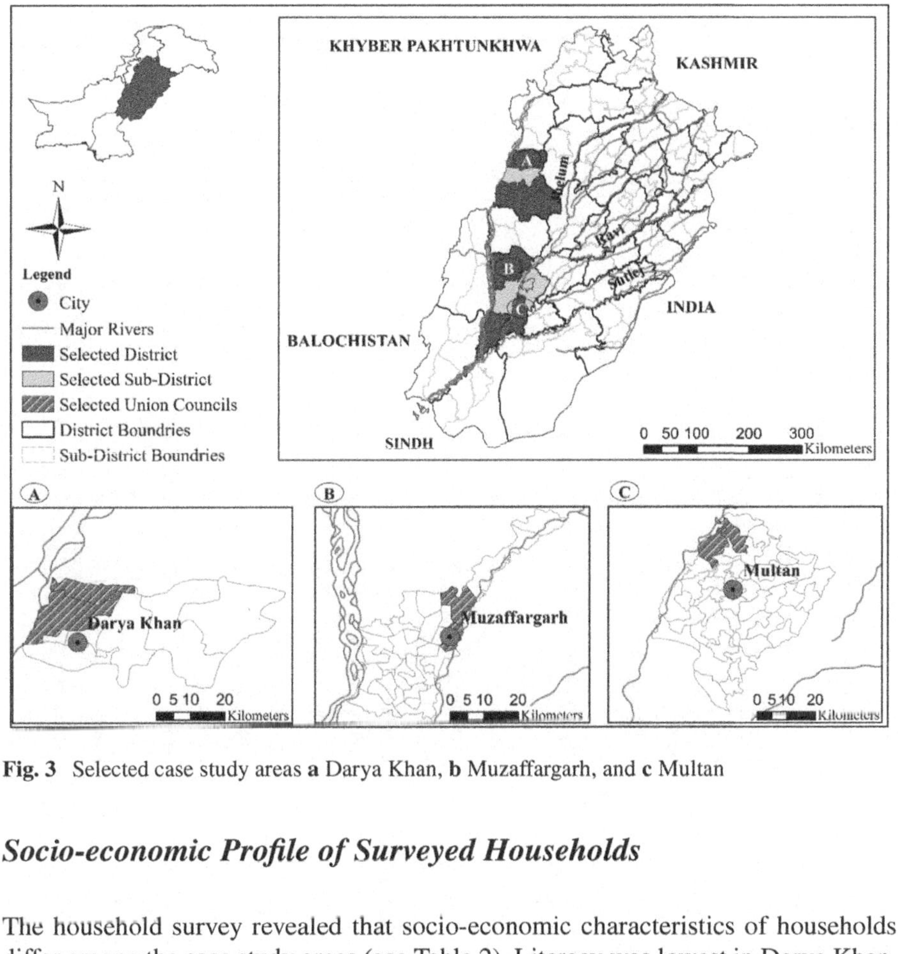

Fig. 3 Selected case study areas **a** Darya Khan, **b** Muzaffargarh, and **c** Multan

Socio-economic Profile of Surveyed Households

The household survey revealed that socio-economic characteristics of households differ among the case study areas (see Table 2). Literacy was lowest in Darya Khan, where only 33% of respondents were literate followed by Muzaffargarh. The highest number of literate people live in Multan (60%). Similarly, households in Darya Khan and Multan were also characterized by the large family size compared to Muzaffargarh. The predominant primary source of income was farming in all study areas, where more than 70% of rural households were involved in crop or livestock farming. In Muzaffargarh, remittance was also seen as a source of income for 15% of the surveyed population. Income levels among rural communities were varying. Majority of rural households in Darya Khan and Muzaffargarh were in lower income bracket i.e., less than PKR 20,000 whereas almost 50% of households in Multan had more than PKR 20,000. Mean family income was significantly higher in Multan compared to other case study areas. Regarding housing conditions, the majority of residents lived in semi-pecca or katcha houses in Darya Khan and Muzaffargarh. The situation was better in Multan where almost 42% of people live in pecca houses. There was a large number of households who own agriculture assets like farmland,

Table 2 Socio-economic characteristics of case study areas

Socio-economic characteristics	Classifications	Darya Khan (N = 114)		Muzaffargarh (N = 115)		Multan (N = 96)		Overall (N = 325)	
		Freq.	%	Freq.	%	Freq.	%	Freq.	%
Literacy	Illiterate	77	67.5	66	57.4	38	39.6	181	55.7
	Literate	37	32.5	49	42.6	58	60.4	144	44.3
Family size	>10	32	28.1	28	24.3	27	28.1	87	26.8
	5–10	76	66.7	75	65.2	63	65.6	214	65.8
	<5	6	5.3	12	10.4	6	6.3	24	7.4
Primary source of income	Government	4	3.5	4	3.5	6	6.2	14	4.3
	Remittance	6	5.3	17	14.8	5	5.2	28	8.6
	Commerce/trade	10	8.8	2	1.7	1	1.04	13	4
	Farming	88	77.2	84	73.0	74	77.1	246	75.7
	Wage labourers	6	5.3	8	6.9	10	10.4	24	7.4
Family income (in PKR) (1 US dollar is Pakistani Rupees (PKR) 139 in March 2019)	<10,000	17	14.9	24	20.9	13	13.5	54	16.6
	10,000–19,999	52	45.6	50	43.5	35	36.5	137	42.2
	20,000–29,999	24	21.1	20	17.4	26	27.1	70	21.5
	30,000–39,999	9	7.9	6	5.2	9	9.4	24	7.4
	>39,999	12	10.5	15	13.0	13	13.5	40	12.3
Type of house	Katcha	22	19.3	20	17.4	13	13.5	55	16.9
	Semi pecca	60	52.6	68	59.1	43	44.8	171	52.6
	Pecca	32	28.1	27	23.5	40	41.7	99	30.5
Ownership of farmland	Landless	13	11.4	2	1.7	5	5.2	20	6.2
	1–5 acres	52	45.6	74	64.3	42	43.8	168	51.7

(continued)

Table 2 (continued)

Socio-economic characteristics	Classifications	Darya Khan (N = 114)		Muzaffargarh (N = 115)		Multan (N = 96)		Overall (N = 325)	
		Freq.	%	Freq.	%	Freq.	%	Freq.	%
	6–10 acres	27	23.7	26	22.6	41	42.7	94	28.9
	>10 acres	22	19.3	13	11.3	8	8.3	43	13.2

however, size of land varies among the case studies. In Muzaffargarh, around 65% are small landholders. Overall, the differences in the mentioned characteristics were due to the opportunities and services provided by their cities. In general, Muzaffargarh was more deprived in terms of socio-economic characteristics as compared to Darya Khan and Multan. This finding was consistent with Jamal (2012) where Muzaffargarh ranked 34, Darya Khan 27 and Multan 22 out of 36 districts of Punjab (Jamal 2012).

Methods

Sampling Process

A probabilistic sampling technique was used to calculate the sample size. Cochran (1977) sampling formula (see Eq. 1) was used to define the sample size, which was calculated as 384. In the next step, the correction formula was applied which give the sample size 380 households, as the number of households was less than 50,000. The sample size was then proportionately distributed among the three case study area. As a result, required number of samples came to be 136, 135 and 109 in Darya Khan, Muzaffargarh and Multan respectively. However, extra samples were collected from each community. After the completion of survey, incomplete questionnaires were eliminated from the sample which gave the final sample of 325. Final samples were 114, 115 and 96 for Darya Khan, Muzaffargarh and Multan respectively.

$$\mathrm{SS} = \frac{\mathrm{Z}^2(\mathrm{p})(1-\mathrm{p})}{\mathrm{e}^2} \tag{1}$$

SS = Sample Size

Z = confidence interval (taken as ±1.96 at 95% for this study)

p = percentage picking choice (expressed as a decimal and is taken 0.5 for the sample size needed to get the maximum value of sample size needed)

e = error value (0.05 = ±5% error is used for calculating the sample)

$$\mathrm{SS} = \frac{(1.96)^2(0.5)(1-0.5)}{(0.05)^2}$$

SS = 384.

Data Collection and Analysis

A structured questionnaire was prepared keeping in view the objectives and questions of the research. Broadly, the questionnaire was divided into area profile, socio-economic and demographic characteristics of households, hazard profile and flood impacts, the influence of flood events on livelihood, goods and services, and flood response measures. Considering the cultural and language barrier, local surveyors were hired and trained to conduct face-to-face interviews from flood affected rural households. Descriptive statistics were used to analyse the data. Frequency analysis and cross-tabulation was carried out. In addition, means values were compared for the questions. The results were presented according to each case study area as well as with respect to the whole sample size.

Results and Discussion

Impacts of Rural Flooding

Impacts of Rural Flooding 2007–2017

Flooding has a varied impact in the case study areas (see Fig. 4). Households have experienced more than one flood event in the period between 2007 and 2017, inclusive. The rural communities of Multan experienced the most incidents of floods. Here more than 80% of people experienced more than four flood events in the ten years studied. In Muzaffargarh, 57% of households experienced more than four floods whereas in Darya Khan 76% experienced only one flood. Flood entered inside the houses once in Darya Khan and more than twice in Muzaffargarh and Multan. Flood events in the last ten years also heavily affected livelihoods. These livelihoods include

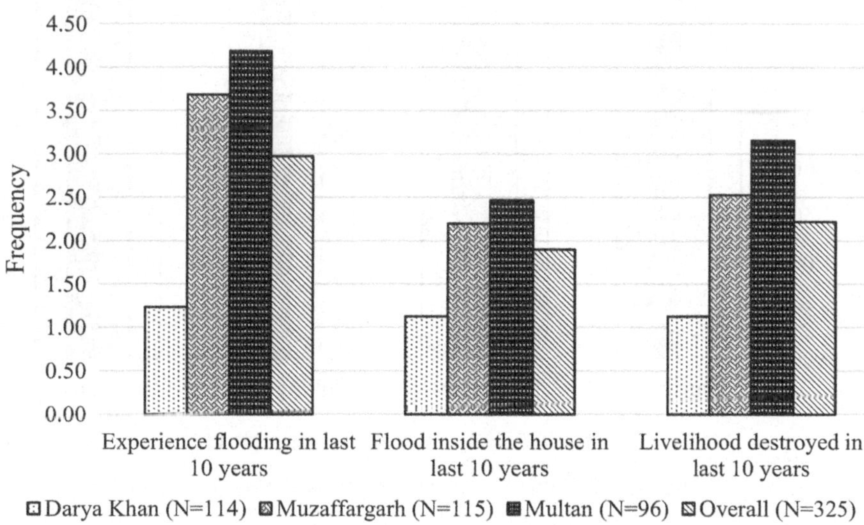

Fig. 4 Average number of floods experience between 2007 and 2017, the number of times flood reached inside the house between 2007 and 2017, and number of times livelihood destroyed by flooding between 2007 and 2017

damage/loss of crops, tools, livestock, shop etc. On average, livelihood was destroyed once in the last ten years in Darya Khan, 2.5 times in Muzaffargarh and 3.2 times in Multan. Results indicate that all the households and their livelihoods were exposed to flood hazard.

Impacts of Last Worst Flood Event

A flood is particularly challenging as it affects services of rural communities on which their livelihood depends. Households reported that services and facilities like transport, electricity, drinking water and schooling were severely disrupted for almost one month after the flood receded. In Muzaffargarh and Multan, the situation was found much worse. The services disruption exceeded more than a month in both study sites (see Fig. 5). Households reported that they were depending on relief organizations for food and drinking water. They had difficulty as the access for relief organizations was affected due to the damage to transport infrastructure. Unavailability of electricity resulted in communication problems with emergency services.

The assets of households were adversely impacted in all three case studies, by the most recent extreme flood event they had experienced. In Darya Khan, all the households reported that the 2010 flood event was the worst, whereas, in Muzaffargarh and Multan, some households reported the 2014 as the worst flood event. All households had experienced severe damages to their physical and livelihood

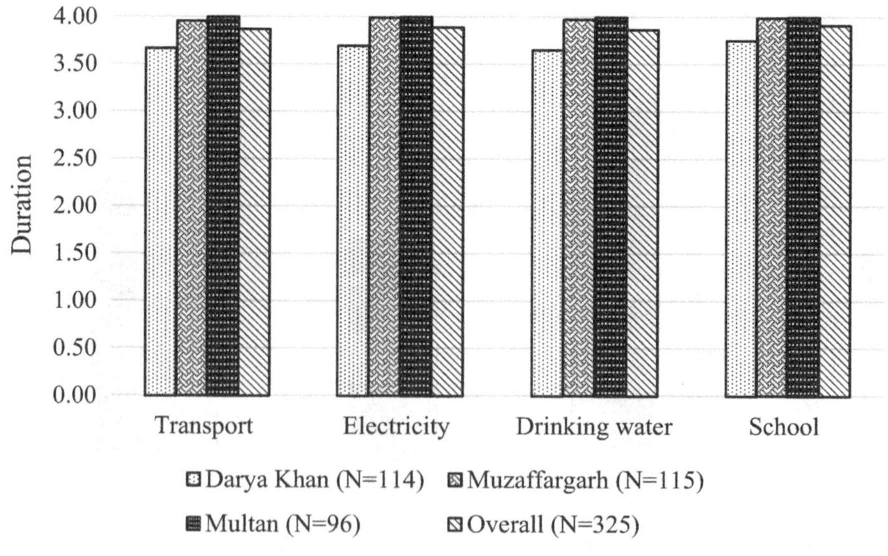

Fig. 5 Average of the duration of disruption of various infrastructure services and facilities in last worst flood event (0 = did not disrupt; 1 = less than a week; 2 = two to three weeks; 3 = almost a month; 4 = disrupted more than 1 month after the flood receded)

assets (see Table 3). Damages to the housing structure were highest in Multan and Darya Khan where 72 and 70% of houses, respectively, were completely damaged. In Muzaffargarh, all the respondents experience damages to the houses. Damage and loss of household belongings were highest in Darya Khan. In the other two cases, the majority experienced partial losses of household belongings and 31% of respondents in Muzaffargarh and Multan faced no damages. Loss of harvested crops/stored grains for sale or household consumption and livestock were limited in all case study areas and the majority experienced no losses especially in Muzaffargarh and Multan. Loss of standing crops was high in all study areas in the last worst flood events. All the respondents experienced damage to standing crops. Since the majority of households were engaged in farming, this loss of standing crops had a significant impact on the overall livelihood. Overall, it was noticed that flood impacts were highest in Darya Khan, as compared to Muzaffargarh and Multan. Most of the damages were associated with immovable assets like a house and standing crops. The highest losses being observed in Darya Khan, which is probably because this area has less experience of flooding than the other two areas, meaning people were less aware of issues related to flood events. In general, flood critically damaged various assets of rural communities, especially their major source of income i.e., cropland.

Response to the Flood

Immediate Response

In the event of extreme flooding, the immediate response came from the government in the form of flood warnings. A large proportion of households reported that they received a flood warning an hour or two before the flood struck their settlements. Majority of people were evacuated to safe locations, with only a few left trapped in the flood (see Table 4). Almost 76% of households in Darya Khan, 71% in Muzaffargarh and 79% in Multan were displaced for more than a month. This indicates the severity of an extreme flood event where flood water remained in the area for more than a month and in some places for more than two months.

Affected and displaced households took refuge upon flood dykes, in relief camps and with relatives and friends. In Darya Khan and Multan, most of the people went to relatives and friends in nearby villages and cities. In Muzaffargarh, flood-affected households stayed on flood dykes and relief camps in the nearby city (see Fig. 6). Overall, most of the people moved to flood dykes and government relief camps in a quick response to flood warnings. The results also suggest that many people, especially in Muzaffargarh and Multan, chose to stay on flood dykes closer to their homes. Results also highlight the strength of social relationships of flood-affected households with their relatives and friends who helped them for the duration of the flood event.

Both governmental and non-governmental relief organizations responded during and after the flood to provide relief to the flood victims. These organizations first

Table 3 Loss and damage profile of household assets in the last worst flood event in case study areas

Damages to/loss of	The extent of damage/loss	Darya Khan		Muzaffargarh		Multan		Overall	
		Freq.	%	Freq.	%	Freq.	%	Freq.	%
Housing structure	Complete	80	70.2	67	58.3	69	71.9	216	66.5
	Partial	32	28.1	48	41.7	26	27.1	106	32.6
	No damage	2	1.8	0	0.0	1	1.0	3	0.9
	Mean	1.32		1.42		1.29		1.34	
	Std. dev	0.503		0.495		0.479		0.495	
Household belongings	Complete	38	33.3	7	6.1	5	5.2	50	15.4
	Partial	58	50.9	72	62.6	61	63.5	191	58.8
	No damage	18	15.8	36	31.3	30	31.3	84	25.8
	Mean	1.82		2.25		2.26		2.10	
	Std. dev	0.682		0.560		0.548		0.635	
Harvested crops/stored grains	Complete	43	37.7	21	18.3	11	11.5	75	23.1
	Partial	50	43.9	47	40.9	44	45.8	141	43.4
	No damage	21	18.4	47	40.9	41	42.7	109	33.5
	Mean	1.81		2.23		2.31		2.10	
	Std. dev	0.727		0.738		0.670		0.746	
Livestock	Complete	7	6.1	1	0.9	0	0.0	8	2.5
	Partial	43	37.7	27	23.5	16	16.7	86	26.5
	No damage	64	56.1	87	75.7	80	83.3	231	71.1
	Mean	2.50		2.75		2.83		2.69	
	Std. dev	0.613		0.456		0.375		0.515	

(continued)

Table 3 (continued)

Damages to/loss of	The extent of damage/loss	Darya Khan		Muzaffargarh		Multan		Overall	
		Freq.	%	Freq.	%	Freq.	%	Freq.	%
Standing crops	Complete	111	97.3	111	96.5	94	97.9	316	97.2
	Partial	3	2.7	4	3.5	1	1.0	8	2.5
	No damage	0	0.0	0	0.0	1	1.0	1	0.3
	Mean	1.02		1.03		1.03		1.03	
	Std. dev	0.132		0.184		0.227		0.182	

Table 4 Duration of community displacement in response to a flood event

Duration of displacement	Darya Khan		Muzaffargarh		Multan		Overall	
	Freq.	%	Freq.	%	Freq.	%	Freq.	%
Trapped in flood	1	0.9	3	2.6	6	6.3	10	3.1
<15 days	3	2.6	4	3.5	0	0.0	7	2.2
15–30 days	23	20.2	26	22.6	14	14.6	63	19.4
31–45 days	12	10.5	17	14.8	20	20.8	49	15.1
46–60 days	35	30.7	23	20.0	40	41.7	98	30.2
>60 days	40	35.1	42	36.5	16	16.7	98	30.2
Total	114	100.0	115	100.0	96	100.0	325	100.0

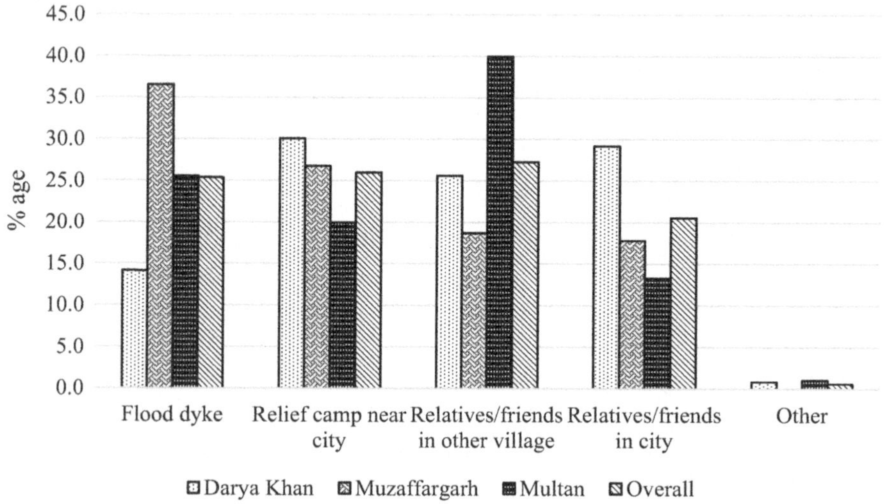

Fig. 6 Locations where households lived during the duration of the flood

provided for basic needs like food and non-food items, then helped in the recovery of losses by providing flood victims with financial aid, building material, as well as seeds and fertilizer. More than 90% of households in Darya Khan, 87% in Muzaffargarh and 80% in Multan received at least one kind of support from relief or donor organizations. Food items for relief and financial aid, building material, and seeds/fertilizers were widely distributed (see Fig. 7). However, the extent of relief and recovery support differs among the case study areas, especially for the provision of building material and seeds/fertilizers for cropping. These differences can be attributed to the level of damages experienced by households and the focus of non-governmental relief and donor organization on particular settlements.

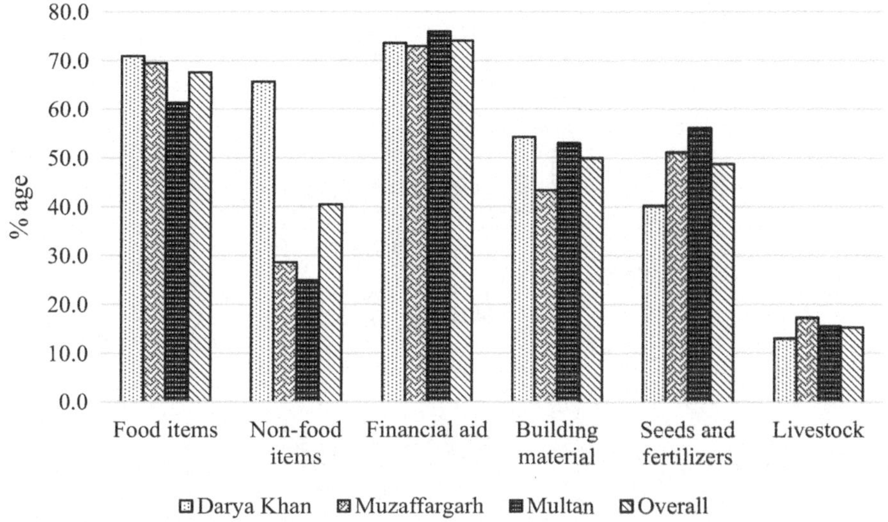

Fig. 7 Percentage of households who got multiple relief and recovery items during and after the flood

Changes After Extreme Flood Events

Physical Changes

Rural households have made physical changes to their houses following the flood events by changing construction material or raising the base of their houses (see Table 5). Regarding the change of the construction material, 50% of households in Darya Khan, 48% in Muzaffargarh and 63% in Multan implemented this adaptation measure. Changing construction material mainly implies a roof with TR Garders, and walls with adobe and semi adobe bricks, which are more resilient to floods as compared to mud houses. Moreover, a large number of rural houses in Muzaffargarh (47%) and Multan (48%) raised their plinth (base) level when reconstructing their homes. The majority raised their plinth level between 2 and 4 m. The respondents implied that previous flood height had motivated them to undertake this structural change. It was also reported that those who have changed construction material or raised the base of their houses have experienced no or partial damages in the following flood events.

Changes in Livelihood Practices

The extreme flood events in the last ten years led to several changes and transformations of the livelihood practices of the rural communities (see Table 6). As the primary sources of income were adversely affected by the flood, the households had to adopt

Table 5 Physical and structural changes in the house by households after the flood events

Physical/structural changes	Response	Darya Khan		Muzaffargarh		Multan		Overall	
		Freq.	%	Freq.	%	Freq.	%	Freq.	%
Change construction material	Yes	57	50.0	55	47.8	60	62.5	172	52.9
	No	57	50.0	60	52.2	36	37.5	153	47.1
	Total	114	100.0	115	100.0	96	100.0	325	100.0
Raised house platform	Yes	28	24.6	54	47.0	46	47.9	128	39.4
	No	86	75.4	61	53.0	50	52.1	197	60.6
	Total	114	100.0	115	100.0	96	100.0	325	100.0

Table 6 Changes in livelihood practices to adapt to future floods

Livelihood practices		Darya Khan		Muzaffargarh		Multan		Overall	
		Freq.	%	Freq.	%	Freq.	%	Freq.	%
Change of primary occupation after the flood event	Yes	9	7.9	27	23.5	17	17.7	53	16.3
	No	105	92.1	88	76.5	79	82.3	272	83.7
Multiple income sources	Before flood	86	75.4	61	53.0	61	63.5	208	64.0
	After flood	106	93.0	107	93.0	85	88.5	298	91.7
Members working in the city	Before flood	28	24.6	26	22.6	28	29.2	82	25.2
	After flood	50	43.9	51	44.3	44	45.8	145	44.6

Frequency and percentage values represent the proportion of the total sample of 325 households

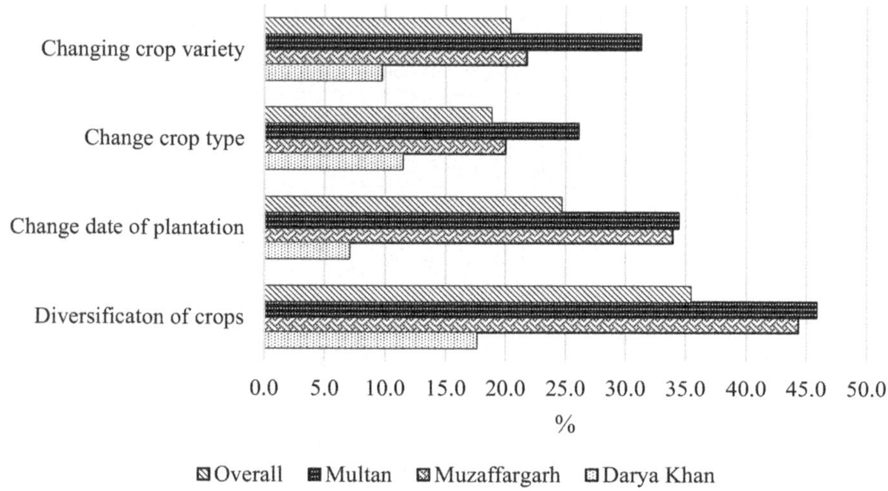

Fig. 8 Changes in cropping patterns in order to adapt to future flood

other sources. In Muzaffargarh and Multan, 24 and 18% of households, respectively, changed their primary source of income. The majority working in farming adopted various other occupations after the flood, such as labourers or in commerce. Table 6 shows that after the flood there was an increase in the number of households having multiple income source. A huge difference could be seen among the households of Muzaffargarh where flood victims were involved in daily wage jobs, such as selling milk, farm labour or commerce. Diversification of income was also associated with the migration pattern of households. The increase in the number of households working in the city after the flood suggested that flooding has triggered migration for income diversification. This migration is probably due to household member leaving his village in order to search for additional income sources because the extreme flood events destroyed incomes so severely that they could not fully recover. It is interesting to note than the difference between the three areas was higher before the flood but there was relatively more migration to cities after the flood in the other two areas than in Multan. Since, Multan city is characterized by mixed functions, large urban centre and offered more opportunities for income diversification, rural communities surrounding Multan commute rather than migrating to city. Therefore, relative change of migration is less in Multan compared to other two case studies.

A major change was seen in cropping patterns of rural households in order to adapt to future flood events (Fig. 8). Almost all the households surveyed were involved in farming as a primary, secondary or tertiary occupation and had diversified their crops since experiencing flooding. Households opted for changing the type of crop as well as changing the variety. Major changes were seen in Muzaffargarh and Multan, where 44 and 46% of households, respectively, had grown multiple crops. In addition to traditional crops, farmers were growing vegetables, which have a shorter maturity period and can be harvested before the flooding season. However, the use of

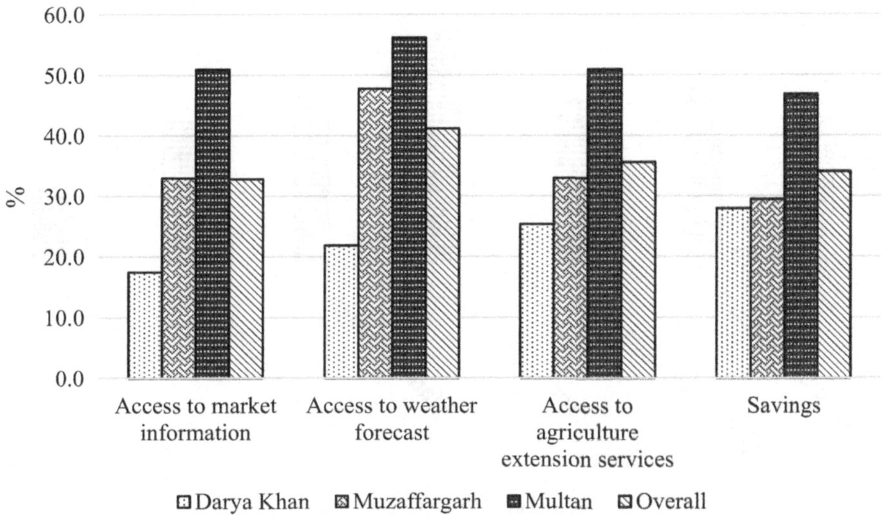

Fig. 9 Behavioural and habitual changes in households after the flood

this adaptation option remained limited especially in Darya Khan where only 11% diversified their crops. A significant number of farmers in Muzaffargarh and Multan shifted the plantation and sowing dates of crops to take into account the potential of future flooding. Households started growing the same crops, for example, sugarcane, wheat, rice, maize, and cotton, but with different varieties which can be sown in late monsoon period and grows more quickly. Overall, Multan has the highest proportion in adopting all cropping pattern adaptations.

Behaviour Changes

In addition to diversification of income sources, households might also change their behaviour and habits to cope with and adapt to floods, thus enhancing their livelihood security. These behavioural changes include more frequent access to market information, weather forecast, agriculture extension services, and financial savings. The results show that the aforementioned changes were highest in Multan followed by Muzaffargarh and Darya Khan (see Fig. 9). More than 50% of rural households in Multan frequently accessed information and agriculture extension services and opted to save money in response to floods. The proportion of use of these strategies ranged between 30 and 50% in Muzaffargarh. Moreover, flood-affected households become more informed and prepared for future flood events. Highest mean scores can be found for Multan (8.2), followed by Muzaffargarh (7.8) and the lowest was found in Darya Khan (5.9). Less frequent use of these strategies and changes in behaviour was seen in Darya Khan as compared to other two case studies. This might be explained by the lower number of floods experienced as well as by the small size of the nearest

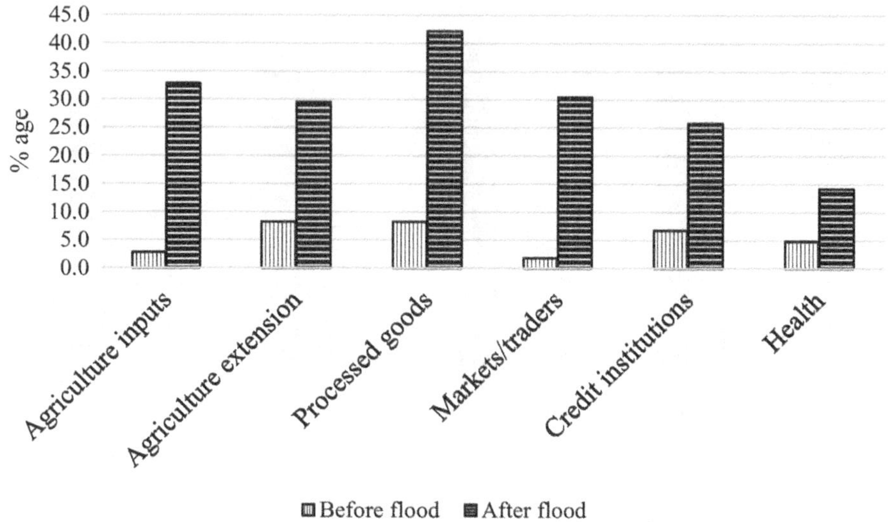

Fig. 10 Households availing services from village markets/centres before and after the flood

city, which was less specialized and provided limited opportunities for information and agriculture extension services. This can also be seen as one of the main reason behind the limited change of cropping pattern in Darya Khan as explained in the previous section.

Changing Relationship with the City

Extreme flood events have also modified the relation of rural areas with their cities. Rural areas depend on cities for various facilities, goods and services like agriculture inputs, extension services, a market to sell and buy goods, credit services and health facilities. However, due to the disruption caused by flooding, there has been a shift towards more local services in some areas, especially in Muzaffargarh and Multan. Before flooding, there was a limited number of rural households making use of services like purchasing agriculture inputs and processed goods, or selling their harvest to rural traders and getting credit services from small village markets (see Fig. 10). The majority of households used to go to the nearest cities for availing such services. The floods caused destruction of road infrastructure and this, combined with expensive transport services, has resulted in the increased use of small rural markets for such services. After the flood events, small local shops opened for providing goods of daily need, agriculture inputs, trade, and informal credit services. In addition, more frequent agriculture extension programs by governmental and non-governmental organizations, as well as the improvement and construction of basic health centres, have changed the flow of services from urban to rural areas. Nevertheless, the majority still depends on cities, especially for health, credit and agriculture extension and input services. However, the extreme flood events can influence the flow of services and facilities between urban and rural areas and thus can change the relationship of rural households with the cities on which they depend on for their livelihood development.

Conclusions

The impact of extreme events and disasters stimulate formal and informal changes that can influence capacities and vulnerabilities of individuals, communities, regions, or countries. These formal and informal changes are considered as 'window of opportunity' for better management and reduction of future disaster risk. This study focuses on informal changes on the municipal level in the aftermath of extreme flood events in Pakistan. Observed changes in response to extreme floods took place both internally (i.e. within rural communities) and externally (i.e. in relationship with the city). In the immediate fallout of flooding, impacts have instigated pressures on local administration for provision of relief and reconstruction activities. In the aftermath of flooding, more long-term changes were undertaken by rural households. The major types of changes reported by surveyed households were: Physical changes to their houses, particularly by shifted from traditional building construction practices to more flood resilient development; Significant changes in their livelihood practices to make them more diverse, such as by seeking work in the city or by diversifying cropping patterns; An increased usage of information and support services due to being more aware of the risk of flooding; And an increase in the provision of more localised services. All these changes refers to capacities and can reduce vulnerability.

The magnitude of these changes differs between the case study areas. The flood experiences have instigated more changes in the rural communities of Multan than in the others. The larger size of the city of Multan has made it easier for its rural inhabitants to frequently access information and attain better and diverse quality of services. This has helped these people to better adapt to future flood hazards through changing livelihood strategies. That the rural households of Multan made the most adaptations and Multan is also a large city with a higher centrality function, indicates that the size and range of services provided by the city to its surrounding rural areas is one factor that determines how much rural households change and adapt in the aftermath of flood events. Interestingly, the relationship between the rural areas and their city changes after a flood event. Though the flow of services and facilities from urban to rural areas is hampered after the flood event due to damaged roads, the flow of people for income diversification and information gets more intense. Overall, the findings of the paper indicate that people learn and change their lives following the extreme disaster events. Communities undertake structural, behaviour and livelihood changes in the long run. Such extreme events can also influence the relationship between rural areas and their cities.

Extreme flood events provide a prospect for learning and to initiate change in communities. These lessons and changes are imperative for rural households to become better adapted to future flood hazards, and hence reduce their vulnerability. This research did not explicitly quantify if the observed changes made by rural households after flooding affected their vulnerability—this should be the focus of further research. Furthermore, more in-depth research should be carried out to examine why more significant adaptation was observed in rural areas around larger cities, whether this is due to the size of the city or due to socio-economic or other factors. The topic permits further investigation on how changing dependence on cities can influence rural-urban linkages, and hence affect the local economic structure of rural areas as well as small and medium cities. The study does not intend to provide generalized suggestions for other regions and countries, as people in each region make informal changes depending upon their social, cultural, economic and geographic conditions and hazard severity. However, this research has provided an entry point to identify various informal changes, which can be further investigated using inferential tests to find more concrete evidences. Despite limitations in the study, results can guide policymakers and disaster managers to frame their policies and appropriate actions to reduce local vulnerability. The findings can be useful in looking at the most significant changes and providing support accordingly.

Acknowledgements This research work is a part of the Ph.D. process at the Institute of Spatial and Regional Planning (IREUS), University of Stuttgart, Germany. The authors would like to thank Higher Education Commission (HEC), Pakistan (SAP-50020940) and German Academic Exchange Service (DAAD) (PIN = 91549672) for providing funding to the first author to carry out research work.

References

Abid M, Scheffran J, Schneider UA, Ashfaq M (2015) Farmers' perceptions of and adaptation strategies to climate change and their determinants: the case of Punjab province, Pakistan. Earth Syst Dyn 6(1):225–243

Abid M, Schneider UA, Scheffran J (2016) Adaptation to climate change and its impacts on food productivity and crop income. Perspectives of farmers in rural Pakistan. J Rural Stud 47:254–266

Ahmad Z, Hafeez M, Ahmad I (2012) Hydrology of mountainous areas in the upper Indus Basin, Northern Pakistan with the perspective of climate change. Environ Monit Assess 184(9):5255–5274

Ahsan R, Noreen I (2009) District profile and sector assessments—Multan District, Islamabad

Akbar MS, Aldrich DP (2018) Social capital's role in recovery. Evidence from communities affected by the 2010 Pakistan floods. Disasters 42(3):475–497

Alfieri L, Bisselink B, Dottori F, Naumann G, de Roo A, Salamon P, Wyser K, Feyen L (2017) Global projections of river flood risk in a warmer world. Earth's Future 5(2):171–182

Ali A, Erenstein O (2017) Assessing farmer use of climate change adaptation practices and impacts on food security and poverty in Pakistan. Clim Risk Manag 16:183–194

Ali S, Liu Y, Ishaq M, Shah T, Abdullah, Ilyas A, Din IU (2017) Climate change and its impact on the yield of major food crops: Evidence from Pakistan. Foods 6(6)

Arai T (2012) Rebuilding Pakistan in the aftermath of the floods. Disaster relief as conflict prevention. J Peacebuild Dev 7(1):51–65

Arshad M, Amjath-Babu TS, Kächele H, Müller K (2015) What drives the willingness to pay for crop insurance against extreme weather events (flood and drought) in Pakistan? A hypothetical market approach. Clim Dev 8(3):234–244

Arshad M, Kächele H, Krupnik TJ, Amjath-Babu TS, Aravindakshan S, Abbas A, Mehmood Y, Müller K (2016) Climate variability, farmland value, and farmers' perceptions of climate change. Implications for adaptation in rural Pakistan. Int J Sustain Dev World Ecol 24(6):532–544

Berkes F, Colding J, Folke C (eds) (2003) Navigating social-ecological systems. Building resilience for complexity and change. Cambridge University Press, Cambridge

Bilak A, Cardona-Fox G, Ginnetti J, Rushing EJ, Scherer I, Swain M, Walicki N, Yonetani M (2016) Global report on internal displacement. Geneva

Birkmann J (2011) Regulation and coupling of society and nature in the context of natural hazards. In: Brauch HG (ed) Coping with global environmental change, disasters and security, vol 5. Springer Berlin Heidelberg, Berlin, pp 1103–1127

Birkmann J, Buckle P, Jaeger J, Pelling M, Setiadi N, Garschagen M, Fernando N, Kropp J (2008) Extreme events and disasters. A window of opportunity for change? Analysis of organizational, institutional and political changes, formal and informal responses after mega-disasters. Nat Hazards 55(3):637–655

Birkmann J, Teichman K, Aldunce P, Bach C, Binh NT, Garschagen M, Kanwar S, Setiadi N, Le Thach N, Oliver-Smith A (2009) Addressing the challenge. Recommendations and quality criteria for linking disaster risk reduction and adaptation to climate change. In: Birkmann J, Telzlaff G, Zentel K-O (eds). DKKV, Bonn

Cheema AR, Mehmood A, Imran M (2016) Learning from the past. Disaster Prev Manag 25(4):449–463

Cochran WG (1977) Sampling techniques, 3rd edn. Wiley series in probability and mathematical statistics. Wiley, New York, London

Directorate of Industries Punjab DoIP (2012) District pre-investment study—2012. Bhakkar. http://www.doi.punjab.gov.pk/system/files/Bhakkar_6.pdf. Accessed 21 Mar 2018

Doocy S, Daniels A, Murray S, Kirsch T (2013a) The human impact of floods. A historical review of events 1980–2009 and systematic literature review. PLoS Curr 5

Doocy S, Leidman E, Aung T, Kirsch T (2013b) Household economic and food security after the 2010 Pakistan floods. Food Nutr Bull 34(1):95–103

EM-DAT (2018) Disaster damages. https://www.emdat.be/emdat_db/. Accessed 27 Nov 2018

Folke C (2006) Resilience. The emergence of a perspective for social–ecological systems analyses. Glob Environ Change 16(3):253–267

Government of Pakistan GoP (2012) National disaster risk reduction policy-2012. Government of Pakistan, Islamabad

Hanif U, Syed SH, Ahmad R, Malik KA (2010) Economic impact of climate change on the agricultural sector of Punjab. Pak Dev Rev 49(4):771–798

Hussain SS, Mudasser M (2007) Prospects for wheat production under changing climate in mountain areas of Pakistan—an econometric analysis. Agric Syst 94(2):494–501

Hussain M, Mumtaz S (2014) Climate change and managing water crisis. Pakistan's perspective. Rev Environ Health 29(1–2):71–77

Hussain A, Routray JK (2012) Status and factors of food security in Pakistan. Int J Dev Issues 11(2):164–185

Hussain A, Rasul G, Mahapatra B, Tuladhar S (2016) Household food security in the face of climate change in the Hindu-Kush Himalayan region. Food Secur 8(5):921–937

Intergovernmental Panel on Climate Change IPCC (2012) Summary for policy makers. In: Field CB, Barros V, Stocker TF, Dahe Q, Dokken DJ, Ebi KL, Mastrandrea MD, Mach KJ, Plattner GK, Allen SK, Tignor M, Midgley PM (eds) Managing the risks of extreme events and disasters to advance climate change adaptation. Special report of the Intergovernmental Panel on Climate Change. Cambridge University Press, Cambridge, pp 3–24

Jamal H (2012) Districts' indices of multiple deprivations for Pakistan, 2011. Research report. Sustainable Policy Development Centre, Islamabad

Jamshed A (2015) Assessing vulnerability and capacity of flood affected communities in Punjab, Pakistan. Case study: district Jhang and Muzaffargarh. Master, University of Stuttgart, Stuttgart

Jamshed A, Rana IA, Birkmann J, Nadeem O (2017) Changes in vulnerability and response capacities of rural communities after extreme events. Case of major floods of 2010 and 2014 in Pakistan. J Extrem Events 4(3):1750013

Jamshed A, Rana IA, Khan MA, Agarwal N, Ali A, Ostwal M (2018) Community participation framework for post-disaster resettlement and its practical application in Pakistan. Disaster Prev Manag 27(5):604–622

Jamshed A, Rana IA, Mirza UM, Birkmann J (2019) Assessing relationship between vulnerability and capacity. An empirical study on rural flooding in Pakistan. Int J Disaster Risk Reduct 36:101109

Jongman B, Ward PJ, Aerts JCJH (2012) Global exposure to river and coastal flooding. Long term trends and changes. Glob Environ Change 22(4):823–835

Jongman B, Winsemius HC, Fraser SA, Muis S, Ward PJ (2018) Assessment and adaptation to climate change related flood risks. Nat Hazard Sci 1–29

Kinoshita Y, Tanoue M, Watanabe S, Hirabayashi Y (2018) Quantifying the effect of autonomous adaptation to global river flood projections. Application to future flood risk assessments. Environ Res Lett 13(1):14006

Kirsch TD, Wadhwani C, Sauer L, Doocy S, Catlett C (2012) Impact of the 2010 Pakistan floods on rural and urban populations at six months. PLoS Curr 4:e4fdfb212d2432

Kosec K, Mo CH (2017) Aspirations and the role of social protection. Evidence from a natural disaster in rural Pakistan. World Dev 97:49–66

Kreft S, Eckstein D, Melchior I (2016) Global climate risk index 2017. Who suffers most from extreme weather events? Weather-related loss events in 2015 and 1996 to 2015. Germanwatch Nord-Süd Initiative e.V, Bonn

Kundzewicz ZW, Kanae S, Seneviratne SI, Handmer J, Nicholls N, Peduzzi P, Mechler R, Bouwer LM, Arnell N, Mach K, Muir-Wood R, Brakenridge GR, Kron W, Benito G, Honda Y, Takahashi K, Sherstyukov B (2013) Flood risk and climate change. Global and regional perspectives. Hydrol Sci J 59(1):1–28

Mahmood R, Jia S, Babel M (2016) Potential impacts of climate change on water resources in the Kunhar River Basin, Pakistan. Water 8(1):23

Mueller V, Gray C, Kosec K (2014) Heat stress increases long-term human migration in rural Pakistan. Nat Clim Change 4:182–185

Muis S, Güneralp B, Jongman B, Aerts JCJH, Ward PJ (2015) Flood risk and adaptation strategies under climate change and urban expansion. A probabilistic analysis using global data. Sci Total Environ 538:445–457

Nadeem O, Jamshed A, Hameed R, Anjum GA, Khan MA (2014) Post-flood rehabilitation of affected communities by NGOs in Punjab, Pakistan. Learning lessons for future. J Fac Eng Technol 21(1):1–20

National Disaster Management Authority NDMA (2010) Flood 2010. http://www.ndma.gov.pk/new/aboutus/flood_2010.pdf. Accessed 26 May 2015

National Disaster Management Authority NDMA (2012) National disaster management plan-2012. National Disaster Management Authority, Islamabad

National Disaster Management Authority NDMA (2014) Multi-sector initial rapid assessment (MIRA) PDMA/NDMA/HCT Punjab floods. National Disaster Management Authority. www.ndma.gov.pk/new/Documents/mira_2014.pdf. Accessed 26 Jan 2015

O'Neill BC, Kriegler E, Riahi K, Ebi KL, Hallegatte S, Carter TR, Mathur R, van Vuuren DP (2014) A new scenario framework for climate change research. The concept of shared socioeconomic pathways. Clim Change 122(3):387–400

Pakistan Bureau of Statistics PBS (2017) Province wise provisional results of census. http://www.pbs.gov.pk/sites/default/files/PAKISTAN%20TEHSIL%20WISE%20FOR%20WEB%20CENSUS_2017.pdf. Accessed 13 Feb 2018

Provincial Disaster Management Authority PDMA (2014) Punjab disaster response plan. http://pdma.gop.pk/wp-content/uploads/2014/08/PDMA-Contingency-Plan-2014.pdf. Accessed 26 Jan 2015

Qureshi AS (2011) Water management in the Indus Basin in Pakistan. Challenges and opportunities. Mt Res Dev 31(3):252–260

Rana S (2017) 6th census findings: 207 million and counting. The Express Tribune, Islamabad. https://tribune.com.pk/story/1490674/57-increase-pakistans-population-19-years-shows-new-census/. Accessed 19 Mar 2018

Salik KM, Jahangir S, Zahdi WZ, Hasson S (2015) Climate change vulnerability and adaptation options for the coastal communities of Pakistan. Ocean Coast Manag 112:61–73. https://doi.org/10.1016/j.ocecoaman.2015.05.006

Saqib SE, Ahmad MM, Panezai S, Rana IA (2016) An empirical assessment of farmers' risk attitudes in flood-prone areas of Pakistan. Int J Disaster Risk Reduct 18:107–114

Siddiqui R, Samad G, Nasir M, Jalil HH (2012) The impact of climate change on major agricultural crops: evidence from Punjab, Pakistan. Pak Dev Rev 51(4):261–274

Sultana H, Ali N, Iqbal MM, Khan AM (2009) Vulnerability and adaptability of wheat production in different climatic zones of Pakistan under climate change scenarios. Clim Change 94(1–2):123–142

Tariq MAUR, van de Giesen N (2012) Floods and flood management in Pakistan. Phys Chem Earth Parts A/B/C 47–48:11–20

Thomas A (2014) Rising waters, broken lives. Experience from Pakistan and Colombia floods suggests new approaches are needed. In: Martin S, Weerasinghe S, Taylor A (eds) Humanitarian crises and migration: causes, consequences and responses. Routledge, London, pp 53–76

Ullah R, Shivakoti GP, Ali G (2015) Factors effecting farmers' risk attitude and risk perceptions. the case of Khyber Pakhtunkhwa, Pakistan. Int J Disaster Risk Reduct 13:151–157

United Nations International Strategy for Disaster Reduction UNISDR (2009) UNISDR terminology on disaster risk reduction. Geneva

Willner SN, Levermann A, Zhao F, Frieler K (2018) Adaptation required to preserve future high-end river flood risk at present levels. Sci Adv 4(1):eaao1914

Winsemius HC, Aerts JCJH, van Beek LPH, Bierkens MFP, Bouwman A, Jongman B, Kwadijk JCJ, Ligtvoet W, Lucas PL, van Vuuren DP, Ward PJ (2016) Global drivers of future river flood risk. Nat Clim Change 6(4):381–385

Zhu T, Ringler C, Iqbal MM, Sulser TB, Goheer MA (2013) Climate change impacts and adaptation options for water and food in Pakistan. Scenario analysis using an integrated global water and food projections model. Water Int 38(5):651–669

Crop-Livestock Inter-linkages and Climate Change Implications for Ethiopia's Agriculture: A Ricardian Approach

Zenebe Gebreegziabher, Alemu Mekonnen, Rahel Deribe Bekele, Samuel Abera Zewdie and Meseret Molla Kassahun

Abstract There have been few attempts to look into the economic impacts of climate change in the context of Ethiopia. Although mixed crop-livestock farming is a dominant farming style, most of the studies on climate change, at least in the context of Ethiopia, have emphasized only crop agriculture and disregarded the role of livestock. In this research, we analyze climate change and agricultural productivity in Ethiopia in its broader sense, inclusive of livestock production. We employ a Ricardian approach, estimating three modified versions of the Ricardian model. Results show that warmer temperature is beneficial to livestock agriculture, while it is harmful to the Ethiopian economy from the crop agriculture point of view. Moreover, increasing/decreasing rainfall associated with climate change is damaging to both agricultural activities.

Introduction

Climate change is expected to have serious impacts on agriculture in Africa in general and sub-Saharan Africa in particular (Challinor et al. 2007). Ethiopia is an agrarian country. Agriculture contributes about 40% of Ethiopian GDP (Gross Domestic Product), about 80% of employment, more than 80% of commodity export earnings and 70% of raw material supply for agro-based industries. Except for the lowlands

Z. Gebreegziabher (✉)
Department of Economics, University of Mekelle, Mekelle, Ethiopia
e-mail: zenebe.gebreegziabher@mu.edu.et

A. Mekonnen
Department of Economics, Addis Ababa University, Addis Ababa, Ethiopia

R. D. Bekele
Center for Development Research, University of Bonn, Bonn, Germany

S. A. Zewdie
Environment and Climate Research Center, Policy Studies Institute, Addis Ababa, Ethiopia

M. M. Kassahun
Private Consultant, Addis Ababa, Ethiopia

W. Leal Filho et al. (eds.), *Climate Change, Hazards and Adaptation Options*, Climate Change Management, https://doi.org/10.1007/978-3-030-37425-9_31

and pastoralist areas, mixed crop-livestock farming is the dominant farm type in the country. Eighty-one percent of the peasant farmers—particularly those concentrated in the Ethiopian highlands—practice mixed farming. Crop production contributed about 65% of agricultural GDP during 2007/08; the rest comes mainly from the livestock sub-sector (CSA 2008). About 95% of the total agricultural output comes from about 11.7 million individual smallholder peasant farmers (MoARD 2010a, b). Cereals—mainly *teff*, maize, sorghum, wheat, and barley—are the most important crops in the country in terms of land area. These crops accounted for about 80% of the total crop area in 2010/11 (CSA 2011). The fact that agriculture is largely traditional and rain fed, making it dependent on weather conditions (Diao and Pratt 2007), makes the issue of climate change particularly important for Ethiopia. The dimensions of climate change that need to be considered are also diverse. For example, how would external perturbations such as climate change affect crop and livestock production? How would this, in turn, affect Ethiopia's agricultural productivity and food security situation? Thus, in this study, using a Ricardian approach, we analyze the impact of climate change on agriculture production, i.e., crop net revenue; livestock net revenue; and whole farm net revenue (taking agriculture as a whole inclusive of livestock). We find that warmer temperature is beneficial to livestock agriculture, but harmful to crops. In addition, changes in rainfall associated with climate change are damaging to both agricultural activities.

Ethiopia is a country of more than 1.1 million square kilometers, located in the Horn of Africa. With more than 100 million inhabitants, Ethiopia is the second-most populous nation in Africa after Nigeria. The country has a sustained record of strong economic growth, which, during the last decade, contributed significantly to the sustainable development agenda: GDP has nearly tripled since 1992 with a corresponding reduction in head count poverty from 56% in 1992 to 29.5% in 2011 (MoFED 2012). Ethiopia has witnessed double digit growth (i.e., 11.2% growth in real GDP) (MoFED 2010). This growth performance effectively surpasses the 7% annual rate required for attaining the MDG of halving poverty by 2015.

Climate change has recently attracted the attention of various stakeholders in Ethiopia. The first conference organized by a newly established national Climate Change Forum, which was held in January 2009, was opened by the late Prime Minister Meles Zenawi. Besides his leading role in the international climate negotiations, he was co-chair of the High-Level Advisory Group on Climate Change (UNFCCC). A civil society network on climate change has also been established recently. In addition, Addis Ababa is part of the C40, a group of 40 large cities committed to tackling climate change. A national adaptation program of action (NAPA) and a nationally appropriate mitigation action (NAMA) have been submitted to UNFCCC. Impacts of current climate variability identified in the NAPA document include food insecurity due to drought and floods; outbreak of diseases such as malaria, water borne diseases and respiratory diseases; and land degradation due to heavy rainfall.

Since February 2011, Ethiopia has been designed a Climate Resilient Green Economy (CRGE) strategy, under the leadership of the Prime Minister's Office, the Environmental Protection Authority, and the Ethiopian Development Research Institute, in order to build a green economy (FDRE 2011). Under seven sectoral teams, more than fifty experts from twenty leading government institutions have been involved. The strategy identifies green economy opportunities and levers that could help Ethiopia reach its ambitious growth targets while keeping greenhouse gas emissions low. Preparation of the climate resilient part of the study is also currently underway. However, only a few studies have looked at vulnerability and impacts of climate change in the context of Ethiopia in a rigorous fashion. These include Deressa (2007), Deressa et al. (2008a, b), Deressa and Hassan (2009), and Yesuf et al. (2008). Even though climate change is expected to have an impact on both crops and livestock production, most of the studies concentrate only on crop agriculture, disregarding the role of livestock.

Thus, the purpose of this study is to analyze the impact of climate change on agriculture inclusive of livestock production. Specifically, the objectives of this study are threefold: (i) assess the impact of climate change on crop farming; (ii) analyze the impact of climate change on livestock farming; and (iii) analyze the combined (overall) impact of climate change on agriculture, that is, for crop and livestock farming combined. In doing so, by broadening and extending our understanding of climate change and agricultural productivity in Ethiopia and enhancing informed policy/decision making at various levels, this research will contribute to poverty reduction and environmental sustainability.

The paper is organized as follows. The next section presents a review of climate change, agriculture and the Ricardian model. The third section presents the theory and model, while the fourth section presents the empirical model and data. The fifth section presents results and discussion. The paper ends with conclusions and policy implications.

Climate Change, Agriculture and Ricardian Model: Literature Review

The Ricardian approach became popular in assessing the impact of climate change on agriculture following Mendelsohn et al. (1994) seminar work. Since then, there have been quite a few studies on climate change and agriculture that employed the Ricardian model or technique. The model is amenable to analyzing cross-sections of farms under different climatic conditions and examines the relationship between the value of land (Sanghi et al. 1998; Mendelsohn et al. 1994) or net revenue (Kumar and Parikh 1998; Ouedraogo et al. 2006; Molua and Lambi 2006; Kabubo-Mariara and Karanja 2006; Eid et al. 2006; Benhin 2006; Sene et al. 2006; Jain 2006; Mano and Nhemachena 2006; Deressa 2007) and climatic factors, soils and socio-economic variables. It has been applied to value the contribution of environmental factors to farm income by regressing farm performance, with land values or net revenue taken as dependent variables, on a set of independent variables, including environmental factors, traditional inputs (land and labor) and support systems (infrastructure). Besides measuring the contribution of each factor, the Ricardian approach is also used to detect the effects of long-term climate change on farm values (Mendelsohn et al. 1994; Mendelsohn and Dinar 1999). The Ricardian approach was named after David Ricardo (1772–1823) owing to his original observation that land rents would reflect the net productivity of farmland (Mendelsohn and Dinar 2003).

The Ricardian model was initially applied in the context of developed countries in general and US agriculture in particular (Mendelsohn et al. 1994; Adams et al. 1998; Polsky and Esterling 2001; Polsky 2004). It has then recently been applied in specific developing countries contexts. Some of these recent studies include Sanghi (1998), Kumar and Kavi (2009), Deressa and Hassan (2009), Benhin (2006), Gbetibouo and Hassan (2005), Ouedraogo et al. (2006), Sene et al. (2006), Seo et al. (2005), Kurukulasuriya and Mendelsohn (2008), Kabubo-Mariara and Karanja (2006), Molua and Lambi (2006), Eid et al. (2006), Jain (2006), Mano and Nhemachena (2006) and Zhai et al. (2009). Not surprisingly, most of these studies have shown that agriculture in developing economies is very vulnerable to climate change. Most of these studies also reveal that the magnitude and direction of the impact may differ from region to region.

Deressa (2007), Deressa and Hassan (2009) and Kassahun (2009) are among the few studies that have investigated impacts of climate change on agriculture in the context of Ethiopia using a Ricardian approach. Deressa and Hassan (2009) find that the climate variables have a significant impact on net crop revenue per hectare of farmers under Ethiopian conditions. They also find that, whereas marginally increasing seasonal precipitation during spring would significantly increase net crop revenue per hectare, marginally increasing seasonal temperature during summer and winter would significantly reduce net crop revenue per hectare. Moreover, their analysis of impact of predicted climate scenarios from three models (i.e., CGM2, HaDCM3 and PCM) for the years 2050 and 2100 show that there would be a reduction in net crop revenue per hectare by the years 2050 and 2100 and that the reduction in net revenue per hectare by the year 2100 would be higher than the reduction by the year 2050, suggesting that the damage posed by climate change will increase with time unless this negative impact is countered through adaptation efforts. Their results also indicate that the net revenue impact of climate change is not uniformly distributed across the different agro-ecological zones of Ethiopia. However, these studies looked at the impacts of climate change on crop agriculture only.

The Ricardian approach has also been applied to a broader African context. Some of these attempts include Seo and Mendelsohn (2006, 2008a, b), Kurukulasuriya and Mendelsohn (2008) and Seo et al. (2009). These studies suggest that hot and dry climate scenarios would reduce net crop revenues in Africa. Most of these studies also reveal that the magnitude and direction of the impact may differ from region to region. For example, Kurukulasuriya and Mendelsohn (2006) find that climate change affects African countries with a dry climate, such as Niger and Burkina Faso, more than those with a relatively cooler climate. Moreover, in relation to livestock, earlier African studies suggest that small landholders would turn to livestock in response to, warming and that farmers with livestock would do much better than if they depended on crops alone (Seo and Mendelsohn 2008a, b). These studies also imply that omitting livestock in the analyses would overestimate climate damages.

The following issues stand out from the foregoing review. Firstly, climate change is a real threat to countries like Ethiopia. It also turns out that both the relatively least-developed, semi-arid, and arid regions and the relatively well endowed regions are vulnerable to climate change, although the level of vulnerability varies. Thus, there is a real need for adaptation policies that are tailored to the anticipated impacts for that specific country; quantifying the impact of climate change on agriculture can therefore provide an important guide to policy. Secondly, whereas the mixed crop-livestock farming is the dominant farming style, rigorous studies of impacts of climate change on agriculture, at least in the context of Ethiopia, have looked at crop agriculture only, disregarding the role of livestock. Intuitively, however, the effect of climate change would be expected to be different with inclusion of livestock production of farmers, thereby affecting farm incomes. Hence, disregarding livestock production may misrepresent the impact of climate change. Moreover, and perhaps more importantly, there appear to be significant inter-linkages (and trade-offs) between the crop and livestock subsystems within the farm household system (Fafchamps et al. 1998; Scoones and Wolmer 2000; Kazianga and Udry 2004; Erenstein et al. 2007), particularly at times of stress, which is another dimension of interest in climate change study. Therefore, it would be important to look at climate change and agricultural productivity in a broad sense, inclusive of livestock production.

Theory and Model

To analyze the impact of climate change and climate variation on Ethiopian agriculture, this study uses a Ricardian analysis following Mendelsohn et al. (1994). Crops and livestock are likely to react to climate change differently. This study attempts to analyze the impact of climate change on crop farming and livestock farming, as well as on agriculture (crop and livestock farming combined). In order to capture the return from crop production, we specify land value, V_L, as:

$$V_L = \int_0^\infty [(P_i Q_i * (K_i, S) - C_i(Q*, \Omega, S))/L_i]e^{-rt}dt \qquad (1)$$

where P_i is the price of output of good i, in our case, crop i; Q_i is the quantity of output of good i, in our case, crop i; $K_i = (k_{i1}, k_{i2}, \ldots, k_{ij}, \ldots k_{iJ})$ is a vector of all purchased inputs j used to produce Q_i; k_{ij} is the purchased input $j(1, 2, \ldots, J)$ in the production of good (crop) i. $S = (S_1, S_2, \ldots, S_m, \ldots, S_M)$ is a vector of site specific exogenous environmental factors such as climate (e.g., temperature and precipitation) and soil in M sites. C_i is the cost of production of good (crop) i and Ω (ω_1, ω_2, …, ω_n) is a vector of all other factor prices except land. L_i is the land employed for the production of crop i (in hectares), r is interest rate and t is time period. Note also that the $-rt$ term represents discounting. The farmer is assumed to choose K to maximize crop net revenues given farm characteristics and market prices.

Now let's turn to the framework for capturing returns from livestock farming (production). Assuming the farmer maximizes net revenue by choosing which livestock to purchase and which inputs to apply:

$$Max(\pi) = P_{qj} Q_j(L_G, S, F, K, H, \Psi, Z) - P_F F - P_W H - P_K K \quad (2)$$

where π is net revenue from livestock (animal) j, P_{qj} is market price of animal j, Q_j a production function for animal j, L_G is grazing land, F is feed, H is a vector of labor inputs, K is a vector of capital such as barns and milking equipment, S is a vector of climate variables, Ψ is available water, Z is a vector of soil characteristics of grazing land, P_F a vector of prices for each type of feed, P_W is a vector of prices for each type of labor, and P_K the rental price of capital.

Suppose that the farmer chooses rearing the species (animal) j and the number of animals that maximize profits. Then, the profit maximizing condition to the representative farmer's problem can be specified as:

$$\pi* = \pi(P_q, S, F, \Psi, Z, P_F, P_W, P_K) \quad (3)$$

Note that Eq. (3) is the Ricardian equivalent to livestock production; it explains how profits change across all the exogenous variables the representative farmer is facing (Seo and Mendelsohn 2006).

Empirical Model and Data

Empirical Model

In this section, we outline the empirical model we employed in our analysis. As noted, the Ricardian model is based on a set of explanatory variables such as climate, soils and socio-economic variables that affect farm value or revenue. The model uses actual observations of farm performance (Mendelsohn et al. 1994). Note also that the standard Ricardian model relies on a quadratic formulation of climate variables. Therefore, we specify the empirical model as:

$$V = \beta_0 + \beta_1 S + \beta_2 S^2 + \beta_3 Z + \beta_4 G + u \quad (4)$$

where V stands for land value/net revenue, S is vector of state of the environment/climate variables, S^2 is square of vector of climate variables Z is the set of soil variables, G is the set of socio-economic variables, u is an error term, and S and S^2 capture linear and quadratic terms for temperature and precipitation, β's are parameters to be estimated with β_0 standing for the constant term and the rest are coefficients. The introduction of quadratic terms for temperature and precipitation reflects the non-linear shape of the response function between crop net revenue and climate. From past studies, one expects that farm revenues will have a U-shaped or hill-shaped relationship with temperature. When the quadratic term is positive, the net revenue function is U-shaped, and when the quadratic term is negative, the function is hill shaped. The idea is that for each crop, there is a known temperature range in which that crop grows best across the seasons, although the optimal temperature varies from crop to crop (Mendelsohn et al. 1994).

Given Eq. (4), one can derive the marginal impact of climate variables (f_i) on crop revenue evaluated at mean, which can be specified as follows:

$$\begin{aligned} E\left[\frac{dV}{dF_i}\right] &= E[\beta_{1i} + 2\beta_{2i} * S_i], \\ &= \beta_{1i} + 2 * \beta_{2i} * E(S_i), \end{aligned} \quad (5)$$

where E is the expectations operator β's are parameters to be estimated as in above. The change in economic welfare, ΔW, resulting from a change in the state of the environment from A to B, which causes environmental inputs to change from S_A to S_B, can be measured as in Eq. (5) (Kurukulasuriya and Mendelsohn 2006). Here one can analyze the impact of exogenous changes in environmental variables on net economic welfare (ΔW). The net economic welfare is the change in welfare induced or caused by a change in the state of the environment from a given state, i.e., from A to B, which causes environmental inputs to change from S_A to S_B. The change in annual welfare from this change in the state of the environment is, therefore, given by:

$$\begin{aligned} \Delta W &= W(S_B) - W(S_A) \\ &= \int_0^{Q_B} [(P_i Q_i(K_i, S_B) - C_i(Q_i, \Omega, S_B))/L_i] e^{\sigma t} dQ \\ &\quad - \int_0^{Q_A} [(P_i Q_i(K_i, S_A) - C_i(Q_i, \Omega, S_A))/L_i] e^{\sigma t} dQ \end{aligned} \quad (6)$$

If market prices do not change as a result of the change in S, then the above equation, that is, Eq. (6), reduces to:

$$\begin{aligned} \Delta W &= W(S_B) - W(S_A) \\ &= \left[P Q_B(K_i, S_B) - \sum_{i=1}^{n} C_i(Q_i, \Omega, S_B) \right] \\ &\quad - \left[P Q_A(K_i, S_A) - \sum_{i=1}^{n} C_i(Q_i, \Omega, S_A) \right] \end{aligned} \quad (7)$$

Substituting $P_L L_i = P_i Q_i * - C_i(Q_i*, \Omega, S)$ into (7), we have that:

$$\Delta W = W(S_B) - W(S_A) = \sum_{i=1}^{n} (P_{LB} L_{Bi} - P_{LA} L_{Ai}) \tag{8}$$

where P_{LA} and L_A are at S_A and P_{LB} and L_B are at S_B.

The present value of welfare change is thus:

$$\int_0^\infty \Delta W e^{\sigma t} = \sum_{i=1}^{\infty} (V_{LB} L_{Bi} - V_{LA} L_{Ai}) \tag{9}$$

The Ricardian approach takes either (8) or (9), depending on whether data are available on annual net revenues or capitalized net revenues (land values, V_L). However, data on land prices were not available for the selected samples, as these are often difficult to find in developing countries. Therefore, model (8) was employed for this study to measure the impact of climate change on crop, livestock and agriculture for the whole of Ethiopia.

The empirical estimation of the Ricardian model for Ethiopia follows the work of Seo and Mendelsohn (2006) and Kabubo-Mariara (2008) and takes into account changes in temperature and precipitation patterns in the study area resulting from climate change.

Letting $S = T, R$, with T and R respectively standing for temperature and precipitation, and noting that the standard Ricardian model relies on a quadratic formulation of climate variables, the final empirical model is then specified as:

$$\pi = \beta_0 + \beta_1 T + \beta_2 T^2 + \beta_3 R + \beta_4 R^2 + \beta_5 Z + \beta_6 G + \varepsilon \tag{10}$$

where T and T-squared capture levels for temperature, while R and R-squared capture levels for precipitation. Z and G respectively stand for a vector of soil characteristics and socio-economic variables, as above.

From Eq. (10), the expected marginal impact of, for example, changes in temperature and rainfall on livestock net revenue can be derived in the following form:

$$E\left(\frac{\partial \pi}{\partial T}\right) = \beta_1 + 2\beta_2 E(T) \tag{11}$$

$$E\left(\frac{\partial \pi}{\partial R}\right) = \beta_3 + 2\beta_4 E(R)$$

The change in welfare, ΔW, resulting from changes in the state of the environment, such as climate change, from some initial state S_A to S_B can be measured as follows (Kurukulasuriya and Mendelsohn 2006):

$$\Delta W = NR(S_B) - NR(S_A), \tag{12}$$

where *NR* stands for net revenue. Note that if the change in welfare ΔW in Eq. (12) is positive, implying increases in livestock net revenue, it will be beneficial. Otherwise, the change will be harmful; i.e., if it is negative, it implies decreases in livestock net revenue.

Accordingly, climate change impact using the modified Ricardian model for the Nile basin of Ethiopia is specified in terms of net revenue obtained separately from crop and livestock, as well as for agriculture as a whole. Note that, in our case, net revenue, such as that for crops, is defined as gross crop revenue less the total associated cost of production calculated for each agricultural household. Then, this net revenue is used as a dependent variable and specified as a function of the following set of regressors: (i) climate variables (temperature and precipitation); (ii) soil types/characteristics, and (iii) socio-economic variables, such as household characteristics (and characteristics of the farm firm), institution support and infrastructure. We run three modified Ricardian models that consider crop net revenue, livestock net revenue and (whole) farm net revenue, inclusive of livestock, as dependent variables, which are regressed separately. STATA statistical software and its econometric package were used to estimate the Ricardian model.

Outliers, heteroscedasticity, multicollinearity and endogeneity of explanatory variables are major econometric problems often faced with cross-sectional data. In light of the fact that these econometric issues will likely affect the robustness of the regression results, the following remedies were undertaken: leverage, Cook's D, DFITS and DFBETA for unusual and influential data; White's general heteroscedasticity test for heteroscedasticity; a variance inflation factor for continuous variables; and chi-square tests for independence of dummy variables for multicollinearity.

The Data

The dataset we used for this study comes from a cross-sectional survey of 1000 farm households in the Nile Basin of Ethiopia. The survey was carried out during the 2004/05 production year by Environmental Economics Policy Forum for Ethiopia (EEPFE) at the Ethiopian Development Research Institute (EDRI) in collaboration with the International Food Policy Research Institute (IFPRI), with the objective of analyzing the potential impact of climate variability and climate change on household vulnerability and farm production. The household survey covered five regional states of Ethiopia and 20 districts. The sample districts were purposely selected to include different attributes of the basin, such as agro-ecological zones in the basin, the degree of irrigation activity (percent of cultivated land under irrigation), average annual rainfall, rainfall variability (coefficient of variation for annual rainfall), and vulnerability. One peasant association was selected from each district, making a total of 20, by a purposive sampling method designed to include households that irrigate their farms. Once the peasant associations were chosen, 50 farmers were selected at random from each peasant association, making the total sample size 1000 farmers.

The Nile basin of Ethiopia is suitable for the cultivation of a large variety of crops. A total of 48 annual crops were grown in the basin. The main crops of the basin are *teff*, maize, wheat, barley, and beans, which cover 65% of the plots during the study period (Yesuf et al. 2008). The data covered farm production, climate, and soils, as well as socio-economic information. As explained earlier, this study employs three main sets of the modified Ricardian model for crop net revenue; livestock net revenue, and whole farm net revenue (i.e., taking agriculture as a whole, inclusive of livestock). The respective net revenues are taken as dependent variables. Respective net revenue variables are calculated (generated) from the dataset. In this paper, net revenue (gross margin) is calculated as total revenue less costs/expenses involving cash outlay. Moreover, the Ricardian model for the case of crop agriculture analyzed land value or net revenue per hectare. However, it is very difficult to measure the amount of land that farmers use for livestock, because they tend to rely on open, communal and public land. Therefore, the dependent variable in our model is livestock net revenue per farm. It is also worth mentioning that the data we used have limited information about costs of raising livestock. Thus, the definition of livestock total cost includes the cost of feed, tools, machinery and medicines. But we didn't include the household's own labor or hired labor as part of the total cost. In the case of crop, total cost is calculated by adding cost for fertilizer, seed, compost, manure and all other chemical inputs. Net revenues are a function of three sets of regressors: (i) climate variables (temperature and precipitation); (ii) soil types/characteristics, and (iii) socio-economic variables, such as household characteristics, institution support and infrastructure. The independent variables considered in the estimated models also included the linear and quadratic temperature and precipitation terms for the four seasons: summer (the average for June, July and August), winter (the average for December, January and February), spring (the average for March, April and May) and fall (the average for September, October and November). The effects of the seasonal climate variables vary across the three models. Table 1 presents descriptive statistics of the variables used in the analysis. As can be seen from the table, for the year 2004/05, the mean total net farm revenue of the study area was around 3521 Birr; the larger percent, 83%, from crop agriculture (2870 Birr) and a much smaller percent, 17%, from livestock (615 Birr). Most of the study area was covered with red soil (61%). A very small portion of the Nile Basin had access to electricity during the study period (19%). About nine-tenths of the households in the Nile Basin own livestock. Nearly 50% of them have farm extension and credit access. Households in the study area, on average, have 23 years of farming experience.

Table 1 Descriptive statistics of variables used in the three regressions (crop, livestock and total agriculture)

Variable	Mean	Std. dev.	Min	Max
Crop net revenue	2870.31	3574.76	31.98	34,867.09
Livestock net revenue	651.10	2252.29	(9281.00)	18,967.00
Net farm revenue	3521.42	4438.08	(7695.81)	34,937.09
Summer temperature	18.25	2.1649	12.86	26.83
Winter temperature	18.94	3.2203	11.84	26.44
Spring temperature	20.70	2.6685	16.01	25.85
Fall temperature	18.16	2.4439	13.11	24.18
Summer precipitation (mm)	265.12	72.4330	3.90	368.18
Winter precipitation (mm)	9.66	11.4851	0.07	105.21
Spring precipitation (mm)	89.54	48.6278	17.65	196.36
Fall precipitation (mm)	100.86	49.3686	1.42	190.38
Clay soil (0/1)	0.19	0.3885	0	1
Sandy soil (0/1)	0.17	0.3767	0	1
Dark soil (0/1)	0.59	0.4926	0	1
Red soil (0/1)	0.61	0.4873	0	1
Electricity (0/1)	0.189	0.39170	0	1
Irrigation (0/1)	0.17	0.3732	0	1
Crop land area (ha)	2.06	1.2585	0.05	11.07
Distance to output market (km)	5.70	3.8682	0.05	45
Distance to input market (km)	5.66	4.1970	0	50
Livestock ownership (0/1)	0.92	0.2775	0	1
Crops extension (0/1)	0.54	0.4986	0	1
Livestock extension (0/1)	0.48	0.4997	0	1
Farm extension (0/1)	0.48	0.4998	0	1
Credit availability (0/1)	0.50	0.5002	0	1
HH heads years of education	1.71	2.7979	0	14
Farming experience (years)	23.42	12.9218	1	68

Results and Discussion

This study employs three main sets of the modified Ricardian model for crop net revenue; livestock net revenue; and whole farm net revenue (i.e., taking agriculture as a whole inclusive of livestock), with the respective net revenues taken as dependent variables. In this paper, the Ricardian approach for the case of crop agriculture analyzed land value or net revenue per hectare. Net revenues are a function of three sets of regressors: (i) climate variables (temperature and precipitation);

(ii) soil types/characteristics, and (iii) socio-economic variables such as household characteristics, institution support and infrastructure.

The empirical question is to find out how and to what extent would the three sets of explanatory variables, i.e., the soil, climate, and socio-economic variables, help explain variability in net revenue. Table 2 depicts the regression estimates of crop net revenue model. The coefficient of the soil variables (clay, dark and red) are positive and significant in explaining the variability in crop net revenues across farmers, indicating the importance of soil types for determining the value of land. Among the socio-economic variables, access to irrigation and household size significantly enhanced crop net revenue. Surprisingly, access to extension programs and access to credit have a negative and significant correlation to net revenue from crop agriculture; perhaps this is because the programs might not be demand-driven. Unexpectedly, distance from input market also has a positive impact on crop net revenue. Size of land for crop cultivation is also positively associated with crop net revenue, suggesting that larger farms are more productive on a per hectare basis than smaller ones.

The results for the climate variables also show that both linear and squared terms are significant in all seasons (except linear winter precipitation and squared fall precipitation), implying that climate has a nonlinear effect on crop net revenues. There is also a connection between the interaction terms and crop net revenue (except the winter interaction term). Concerning the linear terms, results suggest that an increase in seasonal temperatures reduces crop net revenue per hectare, specifically for the summer, winter and spring seasons, whereas increases in fall temperature increases net revenue. An increase in precipitation, particularly for summer and spring, also has positive effects on crop net revenues. In addition, an increase in precipitation in the fall leads to higher crop net revenue. However, the effect of quadratic seasonal climate variables on crop net revenue is not obviously determined by looking at the coefficients, as both the linear and the squared terms play a role (Kurukulasuriya and Mendelsohn 2006). Therefore, the climate coefficients will have to be interpreted based on the marginal effects of climate variables (temperature and precipitation).

The Ricardian model results for livestock net revenue show that the coefficients of the soil variables (red and clay) have a significant and positive effect (Table 3). Among the socio-economic variables, access to formal extension (for livestock) and years of education of head of household turned out significant and positive, supporting the fact that increased access to extension services and education are associated with improved farming information, which is important for livestock productivity and revenue. As expected, livestock ownership is significant and positively associated with net revenue from livestock. Distance to the nearest input market is negative and very significant, perhaps because farmers incur more transaction costs in terms of money and time as the market place becomes further from their farm gates/plots. On the contrary, distance from output market is positively related to net revenue, perhaps because, as distance from output market increases, the product price might become very high and thus add revenue for the seller. Similar to the results for crop net revenues, access to credit is negatively and significantly related to livestock net revenue. Irrigation access also turned out to be significant and positive. Irrigation increases the number of crop harvests per year, which also means more crop residues

Table 2 Ricardian regression estimates of crop net revenue model

Variables	Coefficients	t-statistics
Summer temperature	−18,165.05***	−3.85
Summer temperature sq	431.74***	3.78
Winter temperature	−4977.89***	−2.92
Winter temperature sq	104.75*	2.18
Spring temperature	−15,875.88***	−3.07
Spring temperature sq	433.88***	3.62
Fall temperature	25,354.25***	4.76
Fall temperature sq	−763.56***	−4.93
Summer precipitation	−416.82***	−4.21
Summer precipitation sq	0.18***	3.01
Winter precipitation	184.65	0.63
Winter precipitation sq	−4.37***	−4.50
Spring precipitation	−397.78***	−4.26
Spring precipitation sq	0.22***	1.74
Fall precipitation	515.07**	5.25
Fall precipitation sq	−0.15	−1.37
Summer temp × precip	21.49***	4.89
Winter temp × precip	6.75	0.52
Spring temp × precip	16.19***	3.87
Fall temp × precip	−30.15***	−5.53
Soil—clay	353.48*	2.02
Soil—sandy	273.16	1.45
Soil—dark	368.58*	2.27
Soil—red	319.75*	2.07
Livestock ownership	−319.05	−1.17
Crop land area	451.08***	5.60
Distance from output market	−29.21	−1.23
Distance from input market	44.70*	1.83
Extension program (crop)	−317.53*	−2.02
Credit	−353.00***	−2.74
HH head years of education	10.84	0.39
Family size (log)	630.72***	3.68
Farming experience (years)	−5.74	−1.17
Irrigation	520.88**	2.47
Constant	178,954.60***	4.80
F(30, 890)	19.19	
Prob > F	0.000	

(continued)

Table 2 (continued)

Variables	Coefficients	t-statistics
R-sq	0.531	
Root MSE	1911	
n	925	

*$p < 0.05$; **$p < 0.01$; ***$p < 0.001$

available as feed for livestock. Irrigation also allows grasses to grow on plot boundaries and ridges. Farm size turns out significant and positive, suggesting that large farms are more productive than small farms. Not surprisingly, access to irrigation and years of experience in farming increase livestock net revenue. Regarding the climate coefficients, both linear and squared terms are significant in most of the seasons, implying that climate has a non-linear effect on net revenue. The interpretations of the signs and magnitudes of impacts are further explained in the marginal analysis below.

In the case of Ricardian estimates of total farm net revenue (i.e., net revenue from agriculture as a whole inclusive of livestock), among the soil variables considered in the model, clay turned out to be significant and positive (Table 4). Similar to those of the crop and livestock revenue regression results reported above, access to credit is negatively and significantly correlated with agricultural revenue. The result also suggests that access to irrigation increases net revenue from agriculture, because using irrigation water increase agricultural production, productivity, and the number of times farmers produce per year. Family size is significantly and positively associated to agricultural revenue, indicating that crop as well as livestock agriculture is labor demanding. Surprisingly, distance from input market turned out positive and significant. In line with prior expectations, farming experience increases net revenue from agriculture. The positive and significant effect of farm size suggests the economics of scale that large farms have as compared to small ones. Surprisingly, the result depicts that livestock ownership leads to less agriculture net revenue.

Considering linear, squared and interaction terms, the climate coefficients reveal that agricultural net revenue is generally sensitive to climate variables. However, as explained earlier, the effects of the climate variables on agriculture net revenue cannot be determined by looking at the coefficients, as both the linear and the squared terms play a role. Hence, the climate coefficients need to be interpreted based on the marginal effects of climate variables (temperature and precipitations).

The Fisher-Snedecor test is used to validate the total significance of the models and the Student t test for the individual significance of each coefficient or parameter estimate. The Fisher-Snedecor test shows that all three regressions are significant at the 1% level. The coefficient of determination (R^2) is 53, 24, and 53% for the crop, livestock and total agriculture models, respectively.

Table 3 Ricardian regression estimates of net livestock revenue model

Variables	Coefficients	t-statistics
Summer temperature	−10,078.2***	−4.66
Summer temperature sq	238.88***	4.57
Winter temperature	381.24	0.35
Winter temperature sq	1.03	0.03
Spring temperature	−5776.53*	−1.87
Spring temperature sq	137.48**	1.99
Fall temperature	10,560.03***	4.06
Fall temperature sq	−294.64***	−3.84
Summer precipitation	−140.83***	−3.06
Summer precipitation sq	−0.003	−0.09
Winter precipitation	−992.70***	−3.71
Winter precipitation sq	2.23***	3.93
Spring precipitation	44.17322	1.14
Spring precipitation sq	0.130071	1.51
Fall precipitation	118.6232**	2.34
Fall precipitation sq	−0.04298	−0.58
Soil—clay	373.8652***	3.02
Soil—sandy	72.39039	0.62
Soil—dark	−53.1719	−0.55
Soil—red	174.4503*	1.77
Livestock ownership	251.146*	1.65
Crop land area	66.83165*	1.74
Distance from output market	89.21277***	5.16
Distance from input market	−63.9982***	−3.52
Extension program (livestock)	348.5848***	3.61
Credit	−143.589*	−1.78
HH head years of education	37.98163**	2.32
Family size (log)	96.93869	0.92
Farming experience (years)	6.818176**	1.99
Irrigation	300.4892**	2.5
Constant	6389.63***	3.04
F(30, 921)	8.54	
Prob > F	0.000	

(continued)

Table 3 (continued)

Variables	Coefficients	t-statistics
R-sq	0.235	
Root MSE	1207.5	
n	952	

* $p < 0.05$; ** $p < 0.01$; *** $p < 0.001$

Marginal Effect of Climate Variables

The estimated marginal effects of temperature and precipitation on crop net revenue, livestock net revenue, and total farm net revenue are presented in Table 5. The table shows the net annual and seasonal marginal effect of increases in temperature by 1 °C and increases in precipitation by 1 mm per month.

The marginal impact analysis indicated that a 1 °C increase in annual temperature will lead to a change in net revenue of −1577.72 Birr from crop agriculture, 282.09 Birr from livestock production, and −694.15 Birr from total agriculture inclusive of livestock. All these changes are statistically significant at the 5 and 10% level, respectively.

The marginal impacts of precipitation on net revenues indicate that an annual increase of 1 mm/month of precipitation will have significant positive effects on crop net revenues and net farm revenue, but negative impact on livestock net revenue. For the Nile basin of Ethiopia, an annual net gain of 183.19 Birr and 88.75 Birr is expected from crop agriculture and from total agriculture, respectively. A net loss of 163.78 Birr is expected from livestock production. The seasonal impacts of temperature and precipitation are also indicated in Table 5. The results are consistent with Seo and Mendelsohn (2006): an increase in temperature is related to lower net revenue from crops because farmers shift from crops to livestock, natural ecosystems shift from forests to grasslands, and diseases become less prevalent. Moreover, from an adaptation point of view, livestock becomes an option, as long as there is enough precipitation to support grassland. On the other hand, livestock net revenue decreases as precipitation increases. Farmers shift from livestock to crops, grasslands shift to forests, and the prevalence of animals' diseases increases.

Regarding seasonal differentiation, marginally increasing temperature during summer and spring seasons increases crop net revenue per hectare by 2496.01 Birr and 2042.11 Birr, respectively. Marginally increasing temperature during the winter and fall seasons reduces crop net revenue per hectare by 617.71 Birr and 5498.13 Birr, respectively. During spring season, a slight increase in temperature with the same level of precipitation enhances germination; it is well-known that spring is the planting season in Ethiopia. Increasing precipitation during the summer and winter seasons increases crop net revenue per hectare by 68 Birr and 193 Birr, respectively. Marginally increasing precipitation during spring and fall reduces net revenue per hectare very slightly, by 24 Birr and 54 Birr, respectively. The increase in net revenue per hectare with increasing summer precipitation indicates that the existing current

Table 4 Ricardian regression estimates of whole farm net revenue model

Variables	Coefficients	t-statistics
Summer temperature	−21,725.88***	−4.9
Summer temperature sq	499.52***	4.67
Winter temperature	−4691.91**	−2.48
Winter temperature sq	114.01**	2.12
Spring temperature	−23,244.23***	−4.14
Spring temperature sq	605.83***	4.66
Fall temperature	30,648.82***	5.96
Fall temperature sq	−912.59***	−6.12
Summer precipitation	−498.13***	−5.38
Summer precipitation sq	0.14**	2.46
Winter precipitation	−389.66***	−1.02
Winter precipitation sq	−1.35***	−1.4
Spring precipitation	−379.14**	−3.77
Spring precipitation sq	0.32**	2.05
Fall precipitation	632.24***	6.18
Fall precipitation sq	−0.25*	−2
Summer temp × precip	26.27***	6.44
Winter temp × precip	26.05	1.55
Spring temp × precip	15.9***	3.54
Fall temp × precip	−35.22***	−6.05
Soil—clay	598.77**	2.71
Soil—sandy	259.89	1.22
Soil—dark	231.81	1.21
Soil—red	207.83	1.11
Livestock ownership	−750.28**	−2.69
Crop land area	640.99***	7.21
Distance from output market	−84.48	−1.57
Distance from input market	124.66*	2.15
Extension program	−229.55	−1.28
Credit	−516.47***	−3.4
HH head years of education	59.32*	1.82
Family size (log)	655.78***	3.5
Farming experience (years)	2.19	0.36
Irrigation	654.51**	2.61
Constant	243,993.70***	6.05
F(34, 845)	23.39	
Prob > F	0.000	

(continued)

Table 4 (continued)

Variables	Coefficients	t-statistics
R-sq	0.5274	
Root MSE	2200.8	
n	880	

*$p < 0.05$; **$p < 0.01$; ***$p < 0.001$

Table 5 Marginal effects of climate variables based on coefficients from Tables 2, 3 and 4

Climate variables	Crop	Livestock	Agriculture (total)
Temperature	−1577.72***	**282.09****	**−694.15*****
Summer	*2496.01****	537.16**	*3166.97*
Winter	*−617.71***	733.49**	*−273.37****
Spring	*2042.11***	−249.85**	*2455.61***
Fall	*−5498.13****	−738.71	*−6043.36****
Precipitation	183.19**	**−163.78**	**88.75***
Summer	*67.84****	−11.71*	*55.2****
Winter	*192.72***	−183.72**	*83.2*
Spring	*−23.59*	29.55*	*7.4*
Fall	*−53.78****	2.1	*−57.05****

Note *Significant at 10% level, **Significant at 5% level, ***Significant at 1% level

level of precipitation is not enough for planting. The reduction in net revenue per hectare with an increase in precipitation during the fall is due to crops' reduced water requirements during the harvesting season.

Concerning livestock, a marginal increase in temperature increases income by 537 Birr and 733 Birr during summer and winter season, respectively. However, warmer temperature during spring and fall reduces livestock net revenue by 249 Birr and 738 Birr, respectively. A marginal increase in precipitation reduces livestock net revenue per farm both during summer and winter.

The Impacts of Forecasted Uniform Climate Scenarios

We also analyze the impact of forecasted uniform climate scenarios. The uniform climate scenarios used are based on the projections made by IPCC (2001). According to these projections, the global average surface temperature will increase by 1.4–5.8 °C during the period from 1990 to 2100. In the Sahel, of which Ethiopia is part, the trend during the past decades shows a reduction in precipitation. For this reason, it is imperative to simulate this reduction in precipitation and increase in temperature in our analysis. The simulations for this study are based on scenarios prepared for

Table 6 Impacts of forecasted uniform climate scenarios (based on coefficients from Tables 2, 3 and 4)

Climate change scenarios	Change in net revenue (in ETB)		
	Crop	Livestock	Agriculture
+2.5 °C temperature	−1812.43	−274.02	−1035.46
+5 °C temperature	−3039.67	368.52	363.8
−7% in precipitation	−1609.01	−1112.45	−645.08
−14% in precipitation	−2676.71	−1833.42	−1141.63

Ethiopia (Deressa 2006) from a study carried out under a Global Environmental Facility (GEF)-World Bank project (Kurukulasuriya and Mendelsohn 2006). On the basis of this information, this study examined the effect of climatic change for the following scenarios: an increase in temperature by 2.5 and 5 °C and a reduction in the average rainfall by 7 and 14%.

Note that the impacts of uniform climate change scenarios are analyzed using coefficients in Tables 2, 3 and 4, and assume uniform climate changes. The results are presented in Table 6. As is obvious from the tables, the results indicate that an increase in temperature of 2.5 °C will reduce agricultural net revenue by 1035 Birr, livestock net revenue by 274 Birr and crop net revenue by 1812 Birr on a per hectare basis. Similarly, a loss of 3040 Birr from crop agriculture, and a gain of 364 Birr from total agriculture and 368 Birr from livestock agriculture, will be expected with a 5 °C increase in temperature in the Nile basin of Ethiopia. A 7% decrease in precipitation will reduce net revenue by 1609 Birr, 1112 Birr and 645 Birr for crop, livestock and total agriculture, respectively. A 14% reduction in precipitation will reduce net revenue by 2677 Birr from crop, 1833 Birr from livestock and 1142 Birr from total agriculture. All the above explained outcomes are consistent with the marginal analyses reports in Table 5.

Specific Climate Change Scenarios—Atmospheric-Oceanic Global Circulation Models (AOGCMs)

A set of specific climate change scenarios predicted by AOGCMs (Atmospheric-Oceanic Global Circulation Models) is also used to see the likely impact of climate change for about five and ten decades in the future. In particular, we depend on three scenarios which are consistent for Ethiopia. Specifically, we used the scenarios from the following models: CGCM2, HadCM3, and PCM. The predicted values for the scenario analysis are taken from Strzepek and McCluskey (2007) and Deressa and Hassan (2009).

Table 7 summarizes impacts on net revenue of the specific climate scenarios from the different three AOGCM models for the years 2050 and 2100. These three different scenarios reflect a realistic range of climate outcomes. Because of warming, an

Table 7 Impacts of specific scenarios on net revenue (based on coefficients from Tables 2, 3 and 4)

Model		Crop (%)		Livestock (%)		Farm (%)	
		2050	2100	2050	2100	2050	2100
		A2 scenarios					
PCM	Temperature change (0 °C)	−58	−159	26	110	−75	−169
HadCM3		−98	−145	149	188	−148	−136
CGCM2		−84	−170	102	159	−121	−172
		B2 scenarios					
PCM		−58	−106	26	53	−75	−117
HadCM3		−98	−208	149	168	−148	−217
CGCM2		−73	−144	68	93	−101	−154
		A2 scenarios					
CGCM2	Precipitation change (%)	−37	73	−157	−78	−64	36
PCM		−24	−32	−206	−112	−50	−41
HadCM3		−27	−30	−190	−100	−56	−47
		B2 scenarios					
CGCM2		−32	47	−157	−78	−47	−32
PCM		−27	−25	−206	−112	−33	−13
HadCM3		−24	−14	−190	−100	−39	−19

increasing loss in net revenue from crops will occur in all of the A2 and B2 scenarios of the three models. However, when we come to net revenue from livestock, all the models predict large gains from increasing temperature. In particular, the HadCM predicts the largest gains of all three models. Livestock net revenue is predicted to increase 149% by 2050 and 188% by 2100.

In the case of reduction in precipitation, all the models predict the largest loss for both crop and livestock net revenue. For crops, the net revenue will decrease at an increasing rate (except in case of PCM and HadCM3 in the B2 scenarios). However, for livestock, the three models predict an initial loss in 2050, but this loss eventually declines by 2100. The overall effect of the warming is a decline in net revenue from agriculture as a whole. However, the reduction of precipitation will lead to decreasing agricultural net revenue at a decreasing rate. It is especially worth mentioning the case of the impact of temperature change in different scenarios for agriculture as a whole. Most of the figures (percentage changes) of the losses are three digit numbers. From these results, we can clearly conclude that, unless appropriate adaptation options are adopted, climate change will have a negative impact on the Ethiopian economy.

The effect of climate change on net revenue for different agro-ecological zones using A1B scenarios is presented in Table 8. As can be seen from the table, there will be a decline in crop net revenue throughout the years in all agro-ecological zones, due to temperature increase. It will decrease at an increasing rate. For example, in the case of the moisture-reliable humid lowlands agro-ecological zone, the net revenue

Table 8 Effects of climate change on net revenue for different agro-ecological zones using A1B scenarios

Agro-ecological zones	Crop (%)			Livestock (%)			Mixed (%)		
	2030	2050	2030	2030	2050	2080	2030	2050	2080
	Temperature effect								
Moisture sufficient highlands-cereals based	−9	−13	−31	124	−47	−48	29	−46	−86
Moisture sufficient highlands-enset based	−19	−32	−58	169	−48	−15	77	−83	−102
Pastoralist (arid lowland plains)	−40	−56	−89	145	8	93	56	−59	−97
Humid low lands moisture reliable	−97	−134	−145	614	279	253	43	−47	−85
	Precipitation effect								
Moisture sufficient highlands-cereals based	26	47	62	26	−79	−153	20	33	15
Moisture sufficient highlands-enset based	24	−17	13	24	−154	−215	−10	−50	−45
Pastoralist (arid lowland plains)	−8	−26	−35	−8	−240	−206	−61	−64	−65
Humid low lands moisture reliable	−11	−27	−17	−11	−399	−362	−120	−47	−64

from crop agriculture will decrease by 97% in 2030, by 134% in 2050 and then by 145% in 2080. However, the result is completely different when it comes to the effect of temperature increment on livestock net revenue. Overall, net revenue will increase up to 2030 and then will decline for both the cereals-based and enset-based moisture sufficient zone, but will increase at a decreasing rate for the pastoralists and moisture-reliable humid lowlands. Similar to the result for livestock, the effect of climate change on net revenue of agriculture will be positive only up to 2030, but will decline at an increasing rate afterward.

Likewise, the effect of precipitation decline on crop net revenue for different agro-ecological zones is mixed. For instance, whereas crop net revenue will increase at an increasing rate for the cereal-based moisture sufficient highlands, it will decrease throughout the years in the case of pastoralist and moisture-reliable humid lowlands. Unlike the result for the temperature effects on different agro-ecological zones, a decline in precipitation results in a huge and continuous loss of livestock net revenue, with most of the figures being three digit numbers. The effect of precipitation on agriculture net revenue of the cereals-based moisture-sufficient highlands will be positive. However, decreased precipitation will lead to a decline in net revenue at an increasing rate in both the moisture sufficient *enset*-based highlands and the arid lowland plains.

Conclusions and Policy Implications

This study is an attempt to assess the economic impact of climate change on crop production, livestock production and agriculture as a whole, inclusive of livestock, in the context of Ethiopia using the Ricardian model. Annual net revenue was regressed on climate, socio-economic and soil variables. The regression results were then applied to possible future climate scenarios.

The effects of soil characteristics and socio-economic variables varied across the regressions. The effects of the seasonal climate variables also varied across the three models. Results also suggest that climate has a nonlinear effect on net revenues from crop, livestock and agriculture as a whole.

The results of marginal impact, uniform climate scenarios and specific climate change scenarios from AOGCMs show that the magnitude and direction of climate change impacts crop production, livestock production and agriculture as a whole.

Annual marginal impact analysis of increasing temperature and precipitation also indicate that a unit increase in temperature would reduce crop net revenue per hectare, unlike the effect on livestock revenue. On the other hand, marginal increase in precipitation would enhance crop net revenue. On the contrary, a marginal warming temperature leads to increased revenue from livestock. Though rainfall generally increases crop and pasture productivity, the result of this study reveals that precipitation reduces net income from livestock. The explanation of this result may lie in three possible reasons. First, farmers shift to crops as rainfall increases. Second, grassland shifts to forests as rain increases. This reduces the quality of natural grazing for most animals. Third, increases in precipitation increase the incidence of certain animal diseases.

Uniform climate scenarios are also used to analyze the likely impacts of climate change on net revenues from crops, livestock and total agriculture. The results of these scenarios indicate that increasing temperature, as well as reducing precipitation, reduces crop revenue very substantially in magnitude. A warming of temperature by 2.5 °C and a decline of rainfall by 7 and 14% result in a remarkable decline in net revenue from livestock. The 5 °C increase in temperature decreases crop net revenue. By contrast, it increases livestock and agricultural net revenue, perhaps because crops have very limited optimal temperature range as compared to livestock, and a 5 °C increase is beyond the tolerable limit for crop growth.

Forecasts from three different climate models (PCM, HadCM3 and CGCM2) are also considered in this study to examine the impacts of climate change on net revenue from crop, livestock and total agriculture. The results depict that, in all scenarios, warming temperature is beneficial for net revenue from livestock over the years 2050 and 2100. However, for the rest of the scenarios, future climate change would damage the agriculture sector of the Ethiopian economy. Results from different agro-ecological zones using A1B scenarios also support the previous results. Due to temperature increase, there will be a decline in crop net revenue throughout the years. When we come to the effect of temperature increment on livestock net revenue, the result is completely different. On the other hand, the effect of precipitation decline on crop net revenue for different agro-ecological zone is mixed. Unlike the effect of temperature on different agro-ecological zones, the effect of a decline of precipitation on net revenue of livestock results in a huge and continuous loss, with most of the figures three digit numbers.

Therefore, the following recommendations and policy implications are drawn in relation to crop, livestock and agriculture as a whole in Ethiopia. Concerning crop agriculture, recommendations include: enhancing R & D and introduction of new crops/varieties that are more appropriate to hot and dry conditions and that will give farmers a hand in adapting to harsh climatic conditions; encouraging profitable micro-irrigation systems that will lessen the effects of climate change; providing meteorological information, which will also likely help farmers to adapt; and finding ways to make credit work better for farmers. Regarding livestock agriculture, important issues of policy consideration are: encouraging production and use of local breeds which are adapted to local climatic stress and feed sources; improving local genetics via cross breeding with heat and disease tolerant breeds; improved animal health; and improved water and soil management.

Acknowledgements The authors gratefully acknowledge financial support for this work from the Swedish International Development Cooperation Agency (Sida) through the Environment for Development Initiative (EfD), Department of Economics, University of Gothenburg.

References

Adams R, McCarl BA, Segerson K, Rosenzweig C, Bryant KJ, Dixon BL, Conner R, Evenson RE, Ojima D (1998) The economic effects of climate change on US agriculture. In: Mendelsohn R, Neumann J (eds) The economics of climate change. Cambridge University Press, Cambridge

Benhin JKA (2006) Climate change and South African agriculture: impacts and adaptation options. CEEPA discussion paper no. 21. Centre for Environmental Economics and Policy in Africa, University of Pretoria

Challinor A, Wheeler T, Garforth C, Craufurd P, Kassam A (2007) Assessing the vulnerability of food crop systems in Africa to climate change. Clim Change 83:381–399

CSA (2008) Statistical abstract 2007. CSA, Addis Ababa, Ethiopia

CSA (Central Statistical Agency) (2011) Report on area and production of major crops: private peasant holdings, meher season. Agricultural sample survey 2010/11, vol 1. Addis Ababa, Ethiopia

Deressa TT (2007) Measuring the economic impact of climate change on Ethiopian agriculture: Ricardian approach. World Bank policy research working paper no. 4342. World Bank, Washington, DC

Deressa TT, Hassan RM (2009) Economic impact of climate change on crop production in Ethiopia: evidence from cross-section measures. J Afr Econ 18(4):529–554

Deressa TT, Hassan RM, Ringler C (2008a) Measuring Ethiopian farmers' vulnerability to climate change across regional states. IFPRI discussion paper 00806. IFPRI, Washington, DC

Deressa TT, Hassan RM, Alemu T, Yesuf M, Ringler C (2008b) Analyzing the determinants of farmers' choice of adaptation methods and perceptions of climate change in the Nile basin of Ethiopia. IFPRI discussion paper 00798. IFPRI, Washington, DC

Diao X, Pratt AN (2007) Growth options and poverty reduction in Ethiopia: an economy-wide model analysis. Food Policy 32(2):205–228

Eid H, El-Marsafawy S, Ouda S (2006) Assessing the economic impacts of climate change on agriculture in Egypt: a Ricardian approach. CEEPA discussion paper no. 16. Centre for Environmental Economics and Policy in Africa, University of Pretoria

Erenstein O, Thorpe W, Singh J, Varma A (2007) Crop-livestock interactions and livelihoods in the Trans-Gangetic Plains, India. ILRI (International Livestock Research Institute) research report 10. ILRI

Fafchamps M, Udry C, Czukas K (1998) Drought and saving in West Africa: are livestock a buffer stock? J Dev Econ 55:273–305

FDRE (Federal Democratic Republic of Ethiopia) (2011) Ethiopia's climate-resilient green economy: green economy strategy. Addis Ababa, Ethiopia

Gbetibouo G, Hassan RM (2005) Economic impact of climate change on major South African field crops: a Ricardian approach. Glob Planet Change 47:143–152

IPCC (Intergovernmental Panel on Climate Change) (2001) Climate change 2001: impacts, adaptation and vulnerability. Contribution of working group II to the third assessment report of the IPCC. Cambridge University Press

Jain S (2006) An empirical economic assessment of impacts of climate change on agriculture in Zambia. CEEPA discussion paper no. 27. Centre for Environmental Economics and Policy in Africa, University of Pretoria

Kabubo-Mariara J (2008) Climate change adaptation and livestock activity choices in Kenya. An econometric analysis. Nat Resour Forum 32:132–142

Kabubo-Mariara J, Karanja F (2006) The economic impact of climate change on Kenyan crop agriculture: a Ricardian approach. CEEPA discussion paper no. 12. Centre for Environmental Economics and Policy in Africa, University of Pretoria

Kassahun MM (2009) Climate change and crop agriculture in the Nile basin of Ethiopia: measuring impacts and adaptation options. MSc thesis, Addis Ababa University

Kazianga H, Udry C (2004) Consumption smoothing and livestock in rural Burkina Faso. Working paper 898. Economic Growth Center

Kumar K, Parikh J (1998) Climate change impacts on Indian agriculture: the Ricardian approach. In: Dinar A, Mendelsohn R, Everson R, Parikh J, Sanghi A, Kumar K, McKinsey J, Lonergan S (eds) Measuring the impact of climate change on Indian agriculture. World Bank technical paper 402. Washington, DC

Kumar K, Kavi S (2009) Climate sensitivity of Indian agriculture: do spatial effects matter? SANDEE working paper no. 45-09. SANDEE, Kathmandu

Kurukulasuriya P, Mendelsohn R (2006) Crop selection: adapting to climate change in Africa. CEEPA Discussion Paper No. 26. Pretoria, Centre for Environmental Economics and Policy in Africa, University of Pretoria

Kurukulasuriya P, Mendelsohn R (2008) A Ricardian analysis of the impact of climate change on African cropland. CEEPA discussion paper no. 8. Centre for Environmental Economics and Policy in Africa, University of Pretoria

Mano R, Nhemachena C (2006) Assessment of the economic impacts of climate change on agriculture in Zimbabwe: a Ricardian approach. CEEPA discussion paper no. 11. Centre for Environmental Economics and Policy in Africa, University of Pretoria

Mendelsohn R, Dinar A (1999) Climate change, agriculture, and developing countries: does adaptation matter? World Bank Res Obs 14:277–293

Mendelsohn R, Dinar A (2003) Climate, water, and agriculture. Land Econ 79(3):328–341

Mendelsohn R, Nordhaus W, Shaw D (1994) The impact of global warming on agriculture: a Ricardian analysis. Am Econ Rev 84:753–771

MoARD (Ministry of Agriculture and Rural Development) (2010a) Ethiopia's agriculture sector policy and investment framework (PIF) 2010–2020. Addis Ababa, Ethiopia

MoARD (2010b) The five years (2010/11–2014/15) growth and transformation plan. MoFED, Addis Ababa, Ethiopia

MoFED (Ministry of Finance and Economic Development) (2010) Growth and Transformation Plan (GTP) 2010/11-2014/15, Addis Ababa: MoFED, September 2010

MoFED (Ministry of Finance and Economic Development) (2012) Ethiopia's progress towards eradicating poverty. an interim report on poverty analysis study. 2010/11

Molua E, Lambi C (2006) The economic impact of climate change on agriculture in Cameroon. CEEPA discussion paper no. 17. Centre for Environmental Economics and Policy in Africa, University of Pretoria

Ouedraogo M, Somé L, Dembele Y (2006) Economic impact assessment of climate change on agriculture in Burkina Faso: a Ricardian approach. CEEPA discussion paper no. 24. Centre for Environmental Economics and Policy in Africa, University of Pretoria

Polsky C (2004) Putting space and time in Ricardian climate change impact studies: agriculture in the US Great Plains, 1969–1992. Ann Assoc Am Geogr 94(3):549–564

Polsky C, Esterling WE (2001) Adaptation to climate variability and change in the US Great Plains: a multi-scale analysis of Ricardian climate sensitivities. Agric Ecosyst Environ 85(3):133–144

Sanghi A (1998) Global warming and climate sensitivity: Brazilian and Indian agriculture. Ph.D. dissertation, Department of Economics, University of Chicago

Sanghi A, Mendelsohn R, Dinar A (1998) The climate sensitivity of Indian agriculture. In: Dinar A, Mendelsohn R, Everson R, Parikh J, Sanghi A, Kumar K, McKinsey J, Lonergan S (eds) Measuring the impact of climate change on Indian agriculture. World Bank technical paper no. 402. Washington, DC

Scoones I, Wolmer W (eds) (2000) Pathways of change: crop, livestock and livelihoods in Africa lessons from Ethiopia, Mali and Zimbabwe.IDS (Institute of Development Studies), University of Sussex

Sene IM, Diop M, Dieng A (2006) Impacts of climate change on the revenues and adaptation of farmers in Senegal. CEEPA discussion paper no. 20. Centre for Environmental Economics and Policy in Africa, University of Pretoria

Seo SN, Mendelsohn R (2006) The impact of climate change on livestock management in Africa: a structural Ricardian analysis. CEEPA discussion paper no. 9. Center for Environmental Economics and Policy in Africa, University of Pretoria

Seo SN, Mendelsohn R (2008a) Measuring impacts and adaptations to climate change: A structural Ricardian Model of African livestock management. Agric Econ 38:1–15
Seo SN, Mendelsohn R (2008b) Climate change impacts and adaptations on animal husbandry in Africa. Afr J Agric Res Econ 2:65–82
Seo SN, Mendelsohn R, Munasinghe M (2005) Climate change impacts on agriculture in Sri Lanka. Environ Dev Econ 10:581–596
Seo SN, Maddison D, Dinar A (2006) Will African agriculture survive climate change? World Bank Econ Rev 20:367–388
Seo SN, Mendelsohn R, Dinar A, Hassan RM, Kurukulasuriya P (2009) A Ricardian analysis of the distribution of climate change impacts on agriculture across agro-ecological zones in Africa. Environ Resour Econ 43:313–332
Strzepek K, McCluskey A (2007) The impacts of climate change on regional water resources and agriculture in Africa. World Bank policy research working paper no. 4290. The World Bank, Washington, DC
Yesuf M, Di Falco S, Ringler C, Köhlin G (2008) Impact of climate change and adaptation to climate change on food production in low-income countries: household survey data evidence from the Nile basin of Ethiopia. IFPRI discussion paper 00828. IFPRI, Washington, DC
Zhai F, Lin T, Byambdorj E (2009) A General Equilibrium analysis of the impact of climate change on agriculture in the People's Republic of China. Asian Dev Rev 26(1):206–225

Post-cyclone Aila and Mobility Rights of the Shora Muslim Women of the Bangladesh Sundarbans Forest

Sajal Roy

Abstract The Shora Muslim community in the south-west coastal region of Bangladesh has been earning its livelihood in the Sundarbans forest for centuries. For this community, the traditional livelihood activities in the forest include harvesting wild honey, cutting down trees for timber, gathering firewood, and netting fishes and crabs in the rivers and channels contiguous to the forest. There are key knowledge gaps in the literature regarding the ways in which patriarchy, funnelled through misinterpretation of Islam regarding the veil (purdah), and negative labelling of women (such as in reductive terms like 'wives', 'divorcees', 'tiger widows'), significantly restricts their freedom to earn a livelihood from the denser part of Sundarbans forest. Since the 1980s, the earning of traditional livelihoods in the forest has been affected by the consequences of extreme weather events, such as floods, salinity intrusion in the forest land, tropical cyclones, heatwaves and famines. In particular, Cyclone Aila (2009) caused severe humanitarian displacement of the Shora men and women, and damaged traditional livelihoods in the Sundarbans. As a result of Aila's long-term socio-economic and pre-existing social forces, the mobility of the Shora women in the earning of livelihoods in the Sundarbans and within the community has been affected. Considering these aspects and using auto-ethnography as a method, this paper documents the mobility constraints, which have been an obstacle to Shora women's full enjoyment of their human rights. These are related to gendered identity and social justice in accessing the forest since Cyclone Aila in order to earn a living. The findings suggest that post-Aila there has been a significant increase in women's mobility in the earning of livelihoods. Women are now not only aware of their mobility rights, but are also more informed about the traditional injustices that have impaired their livelihoods.

S. Roy (✉)
Institute for Culture and Society, Western Sydney University, Building EM, Parramatta Campus, Locked Bag 1797, Penrith, NSW 2751, Australia
e-mail: s.roy@westernsydney.edu.au

Department of Gender and Development Studies, Begum Rokeya University, Rangpur, Rangpur, Bangladesh

W. Leal Filho et al. (eds.), *Climate Change, Hazards and Adaptation Options*, Climate Change Management, https://doi.org/10.1007/978-3-030-37425-9_32

Introduction: The Case Study of the Shora Forest Community

Disaster as an extreme weather event impacts on human beings (Perry 2018). A disaster is regarded as an interface between an extreme physical event and a vulnerable human population (Wisner et al. 2004). During a disaster, normal day-to-day activities become severely disrupted (Cisin and Clark 1962). Although disaster affects every gender, the studies of Enarson and Pease (2016) and Roy (2013) claim that women are the most vulnerable and disproportionately affected, particularly during a calamity and post-disaster recovery (Mehta 2007). Due to their poor socio-economic status, rural women in the south-west coastal region of Bangladesh occupy a marginalised social class. The assigned roles—productive role, reproductive role, and involvement in community politics and community management, which rural women play in their families and communities often result in this disproportionate disaster effect between men and women (Roy 2017). The aftermath of the disaster is even more atrocious as slow recovery happens from loss and damage. This recovery, along with the existing pre-disaster gendered roles, has been contributing to the increased socio-economic vulnerabilities of women in the south-west coastal regions in Bangladesh (Roy 2018). While understanding the impact of disasters, the havoc wrought by the consequences which affect communities are always taken into consideration for designing and formulating development programmes and policy interventions. Conversely, disasters can also broaden horizons and offer a window of opportunity, increasing local adaptive capacities and resulting in a gradual systemic improvement (McSweeney and Coomes 2011; Roy 2019).

Shora is an island village whose inhabitants were already dwelling on the banks of the river Kholpetuaya (adjacent to the Sundarbans forest) before British colonists arrived in the Indian subcontinent. This village, in the south-west coastal region of the *upazila* (sub-district) Shyamnagar, is within the administration of the district of Satkhira, Bangladesh. The inhabitants in Shora are predominantly Muslim. The Muslims in Shora hold a conservative social outlook and lack education and adequate life-skills, such as the adaptive behaviour and positive mindset which would enable them to deal effectively with the demands and challenges of life (Roy 2013). For centuries the livelihoods of Shora's inhabitants have been earned from the Sundarbans forest and its adjacent rivers (such as Kholpetuaya) and canals (Roy 2019). Traditional income-generating activities in the Sundarbans forest include harvesting wild honey and gathering wood, bamboo and timber from the forest, as well as netting fish and crabs from the small forest rivers (Islam 2011, Roy 2018). However, since the 1980s, the village and forest have been severely impacted by climate-change-related extreme weather events, such as tidal surges, tropical cyclones, flooding and salinity intrusion. The villagers in Shora epitomise a marginalised social class with a low standard of living, fragile transportation facilities and communication systems, coupled with poor sanitation, drainage and road infrastructures (Mitlin and Satterthwaite 2013). The inhabitants have been living inside mud houses without electricity, and there have been acute socio-economic inequalities between males and females with regard to their participation in the local labour force and the allocation of their income (Islam 2011; Roy 2019).

Every society has typically determined the functional attributes of males and females according to their gender. "Gender" is embedded in every dimension of society from the individual to the interactional and institutional levels. It shapes and reinforces our daily patterns of physical movement. Women have traditionally been regarded the weaker sex, and their mobility has been determined accordingly. Women in rural societies in the Global South are less physically mobile and have a more limited spatial range than men (Li et al. 2004). Particularly in Bangladesh, the daily mobility of females is lower than that of males due to social practices, beliefs and stigma, and males' negative perceptions of females' mobility rights (Adeel 2017). In addition, the socially constructed pre-established ideas of mobility also shape gender relations between men and women in rural societies (Hanson 2010). However, for a disaster-affected Muslim community like Shora, mobility and gender are even more inseparable social constructs (Hanson 2010). Complex practices and social beliefs about lives and sociocultural relations and positioning between males and females shape the gendered roles and responsibilities in Shora (Roy 2013). The treatment that each gender receives varies according to those practices, and mobility differs considerably as a result of unequal gender treatment. This discrimination ultimately determines the physical movement of women in this region and largely decides their social status. In this study, *mobility* is used to denote the physical movement of females from their households to the Sundarbans forest, the local market (bazar), the *upazila* and the district town of Satkhira. Hanson (2010) asserted that mobility can also indicate personal travel, the movement of goods or the long-distance migration of a person, but for an individual it includes performing daily activities, such as paid and unpaid work, leisure activities, socialising and shopping. However, the existing situation in the Shora community reflects the discrepancy between male-female mobility in livelihood earning (Roy 2013). Their limited access to the market has also determined how female mobility is constrained by male chauvinism and the existing religious faith in Shora (Gilbert et al. 2008).

According to Hanson (2010), the 'power' holder within a particular society determines the mobility levels of the other members of the community. Traditionally, males, as the dominant group, hold 'power' in Bangladesh, creating a patriarchal-dominant society (Chowdhury 2009). This dominance is much harsher in rural areas, where the power-holder controls female mobility. In the Shora community, which is not only patriarchal but also characterised by an extremely conservative Muslim caste, men have limited the mobility of the womenfolk to move freely from their household to the community, community to market and market to onward locations such as deeper parts of the forest. The men of the house have been determining the roles of females for centuries. These roles include earning livelihoods through home-based activities, cooking food, rearing children and providing care services to elderly family members (Nasreen 2012). According to a study undertaken by Roy (2013), married and unmarried adult females are discouraged—even prohibited—from going beyond their homesteads to become part of the local labour force

(Roy 2013). Within the hierarchical power structure, the supremacy of males as husbands and breadwinners has traditionally prevented females enjoying the same access to economic resources as the men within their community. However, a change has occurred since Cyclone Aila (2009) hit the area. Aila has forced women to reject their caged existence and step out of their houses in order to earn a living and avoid starvation. The mobility of a particular group of women has increased the participation of Shora women in livelihood earning.

According to Roy (2013), women in the Shora community are divided into four different categories: married women, divorced women, unmarried young women and tiger widows (women whose husbands have been killed by predators while gathering resources from the forest). The tendency in Shora is for both males and females to marry young. Polygamy is not unusual, although many men divorce their previous wives when they remarry. This means there is a high number of divorced women. There are also a lot of tiger widows. Tiger widows are considered to be "ominous" by the whole community, which blames them for the death of their husbands and labels them 'husband eaters'. In order to earn a living for themselves and their other family members, divorced women and tiger widows are obliged to work outside the home, sometimes within the Sundarbans forest. Due to the prejudice against women going out to work, people always look down on them. As a result, their decisions within the family do not matter. They are not even allowed to keep the money they earn, and they stay at the bottom of the social hierarchy. Being a woman who is both highly vulnerable and held responsible for her husband's death makes tiger widows doubly marginalised.

This section has highlighted the enhanced mobility of Shora females since Cyclone Aila (2009). For a long time, unequal constraints prevented Shora women enjoying basic human rights, such as mobility, education, empowerment and freedom of speech. However, Cyclone Aila has opened a door of opportunity, forcing women who were previously trapped in the household to move forward in their struggle for a better life. Their need to eradicate hunger and poverty has forced them to go beyond the traditional social stigmas of their community and contributed to their improved physical mobility. These women are also now aware of the other social injustices they have endured, such as unequal participation in family decisions (for example, the number of children to have), being kept out of discussions about financial matters, and physical violence inflicted by their husbands. Even though they cannot say "No" to these injustices yet, at least they now know that they are victims of these phenomena. Previously, they regarded these unjust acts as inescapable social customs and took it for granted that their lot in life would never improve.

Disaster, Gender and Mobility from Literature

This section looks at the interlinkage between disaster, gender and mobility. Gender as a sociocultural construct addresses the ways in which material and economic aspects of men's and women's everyday lives are shaped by society (Nasreen 2012).

The gendered roles between females and males are based on their differences, including age, ethnicity, physical ability, marginalisation, sexual identity and class-based relations. Every day, the mobility of an individual is reflected while performing gendered roles and responsibilities for the household and community (Hanson and Pratt 1991). Society as a macro-institution is constructed based on the unequal power relations between them (Haraway 1991) and the perceived differences between their performance of domestic and community responsibilities (Scott 1986). In Shora, socio-economic activities are clearly for men, whose only responsibility is to earn a living, while all the domestic chores, from cooking to maintaining the household and rearing children, are performed by women. Even during extreme events, such as natural disasters, women are responsible for taking elderly members of the family to evacuation centres.

Disaster has various impacts on a community and, in most cases, the negative aspects are highlighted in a broader way in research studies. Gender and mobility have a reciprocal influence on each other, and always play a vital role in determining socio-cultural status. In Shora's case, we have discovered how a cyclone or natural disaster can alter and shape the mobility of the people belonging to the community, changing the livelihood earning system and the socio-economic status of people. After Cyclone Aila the people of Shora faced several negative impacts, such as psychological consequences due to the traumatic experience, health issues, lack of job opportunities, reduced income, damage to households and valuable assets, increased salinity and loss of land (Kabir et al. 2016). All these vulnerabilities brought changes to their social and cultural practices and their ways of earning a living (Kabir et al. 2016).

For Shora women, male decision-makers determine gender roles, household responsibilities and movement from house to market, to the forest and the upazila. In most cases, religious customs and beliefs are also used as a tool to impose this unnecessary restriction on female mobility. The mobility of women in other spheres of life apart from Sundarbans (for livelihood earning) is also affected due to the veil (purdah) practice of their conservative Muslim society. Often women's mastering of quality is seen as a challenge to patriarchal dominance (Willard 1895). Probably, in order not to hamper Shora's male chauvinistic society, the (mis)interpretation of religious values was established in such a way that women were kept confined within the household. However, this practice of confining women inside the home has changed during the post-Aila phase.

In a male-dominated rural society, the role of male members during disaster and post-disaster phases is always highlighted and appreciated. On the other hand, because of traditional gender roles, women are presented as victims of disasters instead of agents of resilience (Gaillard et al. 2017). The vulnerability of women during a disaster is well-acknowledged, but during the post-disaster phase, the roles and responsibilities of women can change dramatically (Ariyabandu and Wickramasinghe 2003). Though they are often defined as the worst victims of disasters, they can become the most powerful agents in the reconstruction of the affected area during the post-disaster phase. For example, in the aftermath of Hurricane Mitch that devastated Honduras and Nicaragua in 1998, women organised the disaster recovery effort. They made temporary shelters and latrines and worked effectively to restore livelihoods (Bradshaw 2013). There are similarities in the case of the Shora community, where the women took part in rebuilding the riverbank affected by Cyclone Aila (Roy 2013). They even worked harder than the men did at rebuilding their shattered homes. These women are the same women who have been kept inside the household for a long time, but during the catastrophic situation, they worked outdoors alongside men.

In a report published on 7 August 2018 by the UN Office for Disaster Risk Reduction, Cindy Patricia Quijada Robles emphasised the enhanced mobility of women during the post-disaster phase. This report highlighted cases from both underdeveloped and developing countries where women's mobility from the house to the marketplace and in crowds is usually restrained. However, the report showed how these women participate equally and effectively with male members of the community to restore destroyed areas, demonstrating that even though disasters have an extreme adverse effect on everything, they can bring opportunities for empowerment as well. This is why disaster can be designated a 'window-opener' for many restrictive communities. This driving force occurred in Shora, prompting both men and women to go out and look for any available job opportunities to earn money and rebuild their lives. Such a tendency opened the door to empowerment opportunities for women, and enhanced their mobility.

The livelihood pattern of local dwellers also tends to change after disasters. This has certainly been the case after cyclones Sidr (2007) and Aila (2009). Devastation occurred from different directions because of these extreme hydrometeorological events (Roy 2013). Dams were damaged, and saline water infiltrated agricultural lands. The main occupation of these populations was agriculture-based, but, due to increased salinisation, they could not grow crops and were forced to explore other ways of earning a living (Kabir et al. 2016). During Aila, the Shora community faced the same consequences and lost their age-old agricultural occupations. However, later on, they used saline water for *gher* (salt-water fishing pond) cultivation. Right after Aila, both males and females started working to rebuild the riverbank. The most deprived female groups, the tiger widows and divorced women, were left alone and remained solely responsible for earning a livelihood for their families. So they also started going to Sundarbans. Due to the lack of employment, many began migrating to other districts as well (Kabir et al. 2016). Overall, the Aila-affected people started to explore different occupations as a second job for survival. People with a better socio-economic status coped with the situation more easily than those of lower socio-economic status (Roy 2013). The case study of the Munda community similarly showed how they adopted new approaches to earning a living in order to improve their resilience and regain their previous normal lives (Roy 2018). They started depending on Sundarbans for earning a living and created other working opportunities in which female members of the community also began to participate (Roy 2018). This change in the social system occurred during the post-Aila phase. The case study showed how many women started earning a livelihood on their own through these new working opportunities.

Rural poor communities are often dependent on natural resources for their survival. For a resource-dependent community like Shora, mobility is vital to provide food and earn a living. Shora males have been exploring the denser part of the Sundarbans forest to harvest honey and catch fish (Roy 2019). In contrast, females, due to physical insecurity, net fish in the shallow part of the rivers and canals contiguous to the Sundarbans. They work at *gher* and net fishing to maintain a sound family life. Hanson (2010) equates post-Aila survival in Shora with the consumption of forest resources. Before Aila, women were absent from economic participation while men explored the dense forest resources and consumption rates tended to be lower than in the post-disaster phase. Post-Aila, women also tend to go to Sundarbans to extract resources for their families and to sell.

Methods and Materials

Auto-ethnography, an approach which seeks to describe and systematically analyse (graphy) personal experience (auto) in order to understand cultural experience (ethno) (Ellis et al. 2011), was utilised as a research method, along with two focus groups (one with Shora females, and another with mixed group participants) and twenty-four life story interviews. In addition, direct observation was conducted in the Sundarbans Forest and households of the Shora people. A gatekeeper was recruited before the fieldwork to introduce the researcher to the Shora community. This allowed the researcher to establish a rapport and build trust with the villagers. Twenty-four participants (twelve male and twelve female) aged between twenty-five and forty-five were recruited through a purposive sampling technique which ensured that all four categories of women (tiger widows, divorcees, married women and young women) and men with diverse professions were represented. This age group was selected to allow the participants to share their memories of Cyclone Aila, along with its associated consequences for livelihoods and the local infrastructures.

The researcher conducted three months of fieldwork in Shora (from 20 November 2018 to 19 February 2019). Before beginning the focus groups and interviews, the researcher clearly explained the study aims to participants and recorded their consent. Each focus group lasted for 90 min, whereas a life story interview took around 50 min. The researcher conducted the life story interviews and moderated the focus groups. An audio recorder was used to tape conversations in focus groups and life story interviews. Participants in life story interviews shared livelihood-earning stories and described the impacts of Aila on livelihood activities in the Sundarbans. The transcribed focus group and interview data (Bengali texts, direct quotes) were coded. Afterwards, thematic analysis (Braun and Clarke 2006) was utilised to develop key themes and associated subthemes of the study.

Detailed Discussion of the Findings

An Overview of Shora Women's Lives: Before and After Cyclone Aila

This section details sociocultural aspects of the lives and mobility of Shora women. The society of Shora is Muslim and male-dominated, with early marriage, polygamy and divorce—problematic customs which have existed since the independence of Bangladesh in 1971. As men are the head of their families and community, women cannot always speak about the many problems they face within their family. The majority of inhabitants have been highly dependent on the Sundarbans forest for earning their livelihoods. There is a discrepancy in terms of male and female participation in earning a living from the forest. As Shora is a patriarchal community, women have mostly been confined inside the home. Even when some women were obliged to emerge from their houses and work to survive, they were kept away from Sundarbans forest for both justified (security-related) and non-justified reasons. Due to the interpretation of the Muslim religion by the patriarchal society, the mobility of men and women has varied for a very long time.

Overall, the Shora community is of low socio-economic status and not very solvent. According to participants in life story interviews, their monthly income varies from TK. 3000 to TK. 8000.[1] In most cases, the families are extensive, and maintaining a large family on such a small income is a tough job. Due to their financial difficulties, most of the family members struggle to get three square meals a day. They mostly eat vegetables grown in nearby fields and small fish caught in the shallow part of the river and canals adjacent to the Sundarbans. Beef and venison are cooked once or—barely—twice a year. Their basic meals reveal much about their poverty and poor socio-economic situation. Participants reported robbery inside the Sundarbans forest, and frequent attacks by tigers have been hindering their efforts to earn a livelihood since the 1980s. Consequently, women are never encouraged to enter the denser part of the forest. Focus group participants noted that when tiger widows and helpless divorcees attempt to go inside of forest, they frequently become the victim of robbers and other unwanted incidents. As a result, unless someone is exceptionally compelled, they would not go into the Sundarbans forest to earn a living. These security issues hinder the mobility of women inside the forest and curtail their opportunities to earn a living from its resources. The number of tiger widows and divorcees in Shora is also extremely high for the same reasons, with many men losing their lives to tigers or robbers while working inside Sundarbans. According to LEADRS, a local NGO working with tiger widows, there are around 325 tiger widows in Shora.

---

[1]The main unit of currency in Bangladesh—the Taka—is worth about 0.017 Australian dollars at the time of writing, making the monthly income range of Shora inhabitants roughly equivalent to 50–135 Australian dollars.

This Muslim community in Shora follows religion strictly, and its extreme moral values can be regarded as the reason for the confinement of women inside four walls, in accordance with the Islamic veil custom. This patriarchal society frequently uses religion as a tool to suppress the female community, and therefore, even though it can be difficult to support the family when there is only one breadwinner, men have not allowed their female family members to go out to work. This confinement of women and the mobility barrier have prevented women participating equally with men in the business of earning a living. However, after Aila, a change in female participation has occurred, as they are now sewing and getting involved in raising poultry.

Tiger widows were one of the most despised sectors of society in Shora. They were termed 'witches' and 'husband eaters', and held responsible for their husband's death. Because of many other superstitions in the community, married women whose husbands went to Sundarbans forest also faced much illogical torture from their community. As wives were held responsible for their husband's safety, they were forced to eat only *pantavaat* (water rice) with green chilli. They were also prohibited from touching broomsticks or using soap when their husbands were going to Sundarbans. Although the grounds for these practices have still not been identified, married women would not dare to break these rules because of the fear of losing their husbands. However, tiger widows and divorcees were the two categories of women who had livelihood mobility, even before Cyclone Aila. Being the family member responsible for earning money, they had to emerge from their confined life and participate in paid work.

According to participants, during Aila, as the people could not remain in their houses anymore, they had to go to shelters as evacuees. Even if the female members felt uncomfortable staying under the same roof as unknown men, they had no other alternative. Probably, this is when the myth that women are meant to stay inside the house no matter what started to get blurred. For survival, the women had to stay at the shelter, ignoring their discomfort. Leaving their home to go to the shelter is an example of female mobility. They additionally acquired political mobility when they did not wait for their husband's permission to go to the shelter during Aila, and when they started working outside to fight poverty after losing the head of the family. Their access outside the household demonstrates the restructuring of their mobility rights.

The experience of Aila (2009) was a nightmare, as they lost their loved ones, family members, cattle, houses, household goods, and cash. According to one male participant, after Aila they had to live under the open sky for weeks, without any food or shelter. A female participant described how her family lost their most valuable assets during Aila. Some of the other participants said that women and young girls were the worst victims. However, Cyclone Aila also operated as a forced mobiliser of women. During Aila, they had to take refuge in a shelter, but there was no adequate supply of food or drinking water. The shelters even lacked sanitation, which is why women became the worst sufferers, as they did not have access to a washroom during the emergency. So, not only did they starve and have to stay in a congested place or often under the open sky, they also had to face difficulties in meeting sensitive sanitation needs. One female participant from the Shora community, who works for an insurance company, said: "*During Aila, we faced extreme sanitation problems and skin disease problems because of the unhygienic environment.*" From their shared experience, it is apparent that Aila, like previous natural calamities, has been a disastrous low-point in the life of the community. However, the way Aila's enormous force shattered the whole village also brought an opportunity for positive change in the case of female mobility and earning potential.

During the post-Aila phase, women had to go out in public to collect relief supplies from voluntary organisations and obtain money or other help from different NGOs. Many families lost their male members and, as a result, the female members had to go outside to rebuild their homes and earn money to provide food for the rest of their family. Thus, the stream of female participation in activities outside the household began, and gradually reached a level where they are now getting access to political and even virtual mobility. [Virtual mobility means access to the internet and digital technology (Gilbert et al. 2008).] Post-Aila, there has been an increase in the use of mobile phones, which Shora women use to enhance their access to livelihood opportunities. They are now not only exploring physical mobility but are also more aware of being forward-thinking. The Shora data also reveals that women have started attending NGO meetings as they become more confident about going out and doing their work by themselves. Expressing their point of view and participating in meetings alongside men demonstrates their determination to exercise freedom of speech, which was absent during the pre-Aila time span.

In the case of the Shora Muslim female, gender and mobility are associated with their livelihood activities. This is not only because mobility has shaped their livelihoods but also because Shora women were previously economically dependent on male family members as a result of patriarchal obstacles to equality which prevented them holding a strong position within their families and society. After Aila, the mobility of the Shora women was restructured because of their increased participation in work outside the home. Even though, due to disaster destruction, they were obliged to emerge from their homes; it was their movement to the local domain to earn a livelihood that consolidated the process of breaking down traditional barriers to female mobility. Previously, to conform to traditional gender roles, they were confined inside the home. Now, as a consequence of Aila's impact, that same-gender role of taking care of their family has made them mobile in the outer space.

Comparative Pictures of Women's Mobility Before and After Aila

This section compares Shora women's mobility, considering the before-and-after effects of Cyclone Aila. Shora men, who believed that women were not physically capable of working outside, viewed Pre-Aila, the sight of women working outside negatively. One male participant noted, "*Women working outside in the forest cannot do any actual work. They just go because they're forced to.*" This attitude indicates that not only religion but also physical differences have been used to justify the unequal mobility rights of men and women. Nevertheless, post-Aila, female mobility has been enhanced, both physically and in terms of their income.

On being asked to compare female mobility before and after Aila, another male participant said:

> Our villagers used to be extremely religious in the past. Our women were only housewives and never stepped outside their home. They used to maintain their veil fully. Even if they had to starve or their husbands could not earn much, they would not sacrifice their religious values. However, after Aila all this tradition is shattered. During Aila both males and females stayed together under the open sky or the same shelter, and thus the religious practice of the veil began to be ignored. And after Aila extreme poverty affected most of the families, and gradually female members started coming out of their households to earn a living.

Another participant noted:

> Nowadays, women do not wear the Burqa and do not cover their chest correctly. Other men look at these sights of women, which is very disgusting. Previously, one person did not know the face of another man's wife, but now they know what they look like. Everything is messed up now. Since Aila, women cannot give birth normally and go for surgery. The hassle of an operation in a remote area like Shora is enormous. After Aila female empowerment has increased in order to support their families. However, alongside that, they have got economic independence and now they are aware of what is happening outside and, as a result, they do not pay much heed to what others say. They have extreme moral deterioration, and they forget to value everything nowadays.

The above quotes show that there has undoubtedly been a change in female mobility and empowerment. It is a fact that the male community of Shora does not regard this change in female mobility positively. However, if we ignore the self-interested complaints of men who are aggrieved at the loss of their unquestioned dominance, we can say that Aila has opened a window of opportunity for Shora female mobility. Even though, during the early phase of post-Aila, female participation was not voluntary but forced by fate, the result of this force is positive. The women of Shora nowadays are well aware of their minimum rights, and they are no longer blinded by social restrictions.

Participants working in NGOs opined about women's participation in the local labour force in the following way:

> Before Aila the overall community was very conservative and females lacked interest in participating outside. For example, they hardly ever attended NGO meetings or participated in their discussions. This massive change in female involvement in the wider community took place during the aftermath of Aila as they gained confidence after emerging from their confined lifestyles.

The above quotes demonstrate that, due to Aila, several changes in female mobility have taken place. Even though the men do still not see this positively and other conservative community members, working women are not paying much heed to their negative comments. They still have many challenges to face before gaining equal mobility, but their situation is much better. The continuous involvement of women might bring about more positive changes in the upcoming days. This incident can be compared to the tsunami-impacted rural areas of Tamil Nadu. Women who lost their husbands (and sole earners in their households) ultimately shouldered the responsibilities of maintaining the house, raising small children and taking care of elderly parent-in-laws. Over time, this changed the status of women as they started earning, and this ultimately brought them respect within their communities. From this case study, it can be seen that women are not always passive victims of disaster calamities. Rather, if they can be provided with any responsibilities which society has previously not let them handle, they can perform these responsibilities better than men can. Men rarely perform the duties of a caregiver as women do, but, in addition to this duty, post-disaster women get involved in other duties, and can thus establish their own social identity (Roy 2017).

In a restrictive Muslim community such as Shora, all the prevailing injustices, restrictions and social constraints against women are validated by religious customs. Shora female community members have always been subject to double standards in terms of gender roles and mobility rights. Religious values have always been used to justify such double standards. Even women in a poverty-stricken family could not engage in outside paid work. On the other hand, tiger widows and divorced women—the marginalised groups occupying the lower positions of the social hierarchy—had already been working outside to maintain their families before Aila. However, they did not officially have the mobility right to work outside. During the post-Aila phase, women from other spheres of life also started moving outside their households. They did so because otherwise they would die of hunger, but that force of poverty made them step beyond their confined lifestyle. So in this respect, Cyclone Aila actually worked as a 'window opener' in the lives of the Shora female community.

Analyses of Hindrances to Shora Women's Mobility Before and After Aila

Pre-Aila, women in Shora were subjected to restrictions in their households and their day-to-day gender roles, which were determined by their male partners. According to Nasreen (2012), women in a rural society in Bangladesh are generally regarded as home-makers. Muslim men have always discouraged their women from going beyond their household domain to engage in economic activities, such as paid work. They are mainly recognised only for their reproductive role (giving birth). The social forces (gender role, socio-economic status) which have created gender differences are based on age-old patriarchal traditions and values. These values still prevail in most of Bangladesh. As a result, gender roles are strictly taught and maintained by both genders from a very early age. Because it defies the cultural and moral values of female chastity, shame and domestic confinement, their participation and presence in public places like fields, roads and the market place, is viewed negatively by the community and discouraged. These moral values of the patriarchal society ultimately make these women vulnerable to a range of abuses (Bleie 2005).

For years, the Shora Muslim female has faced several types of obstacle with regard to participating equally with men in Sundarbans forest. It was clear from the research data that female members of the community also have a stereotypical idea of gender roles. Their idea of gender has been made and shaped by the dominant male society. They accept the mocking tagline given by this society, whereby females are termed a *powerless ship* and men a *strong horse*. That is why they do not even question the baseless imbalanced wage system, where men get twice the amount of money for the same amount of labour. They also see female sexual identity as something very degrading, and said to the interviewer that "*Women would not have been women if they had everything equal to men.*"

Another participant from Shora described the villagers'—and especially male members'—outlook on women stepping outside their household domain. According to her:

> The Shora male community, especially the elderly male members, dislike women coming out of their house and going to market to shop or earn a living. The women who work outside are often termed 'bad women', and everyone looks down on them. They believe that when women go to the marketplace, they do not care about religious values like the veiling custom and that they disturb the environment as well. Even though everyone knows these are superstitions, men still say this. This is why, unless someone is absolutely obliged to, no one comes outside. Women who work outside are not safe from harassment, and they cannot even protest against this, as they are very helpless.

The above quote indicates that male members of the community use religion, and often physical harassment, as a tool to suppress women. Working outside is a stigma in the eyes of the Shora community. Even merely going to market to shop is stigmatised and discouraged. Such social labelling poses a challenge to women's mobility in Shora. Nevertheless, the situation can be regarded as favourable, because even if these women are looked down upon for going out to perform errands or work, they are no longer staying inside the household after Aila. They are well aware of such negative categorisation, but are nevertheless participating in the workforce. So, compared to the previous situation, this change shows that, post-Aila, women have managed to obtain some mobility. With the change in practice, mentality might also gradually change. So while today the 'rebellious' women are facing taunts and other social challenges, at some point people's perception of female's mobility and progress in livelihood-earning can be changed.

Female research participants also talked about their barriers towards earning a living, blaming their social duties of homemaker, caregiver, child-rearing, and the taking care of their husbands. They noted:

> Working women need to perform the triple duty of caregiver, homemaker and wage earner. Unlike men, they are obliged to finish their household chores first, then attend to their outside work. Also, because they need to prioritise their social responsibilities, everyone takes it for granted that their labour in the workplace is less than that of men. Moreover, the injustice does not end here. People call the women working outside 'goons'. It is believed that they do not value their social responsibilities or husband, but roam around outside the whole day without any justified reason.

Men of the Shora community have been using religious beliefs as a tool to confine women inside the house, but in many cases, they have been found not to abide by the religious rules themselves. One participant from the male-focused group from Shora said, "I do not let my wife work outside, and I hope I will never have to allow that. Muslim women should stay inside their house and take care of their family. It is not their duty to go outside and, being Muslim, I cannot permit it." He used Islam to justify his decision, but the next moment he was asked about performing his daily prayers, and, interestingly, he said that he does not perform them. Whereas in Islam it is mandatory to perform prayers five times a day, this man disregards that, but he does not hesitate to drag the veil issue up to an intense level. So, such an act exemplifies the community's patriarchal politics and double standard of religious beliefs. Consequently, the community's interpretation of religious writings ultimately affects the mobility of the women in a negative way.

The Shora community has been conservative for a long time, especially in the matter of female mobility and earning a living. However, after Cyclone Aila there was a noticeable change in the social system, especially in the socio-economic activity that took place when the female members of the community were forced to emerge from their houses and work outside to rebuild their homes or embankment. They also had to leave their homes to overcome the new challenge named 'poverty'. Other case studies and reports confirm that poverty strikes people affected by climatic disasters very hard. In a report by Islamic Relief UK, the author proved that the deterioration of economic conditions caused by Aila affected people badly, a scenario exemplified by the post-Aila financial condition of the Shora community. Because of Aila, most of families became homeless; they lost resources, many people died, and in some families, the head family member (male) died. After losing the primary breadwinner of the family, women began participating in local paid work. This participation provided them with the opportunity to enjoy greater mobility, as they had to move from their confined household to the outside community in order to work. Cyclone Aila, in this case, worked as a serendipitous window-opener in the lives of the poor female community of Shora.

Post-Aila Involvement of NGOs in Enhancing Women's Mobility

After the destruction wrought by Aila, NGOs such as World Vision, Shushilon, Islamic Relief, Ahsania Mission, and LEADRS came forward to help the entire community recover from the impact of the cyclone. Their efforts contributed to the greater mobilisation of women. Some of the NGOs had a specific focus—for example, LEADRS mainly worked with the tiger widow community and young girls. During that time, they gave supplies, clothes and plastic bags to build a new shelter. To collect this aid, women had to go to the relief centres alone. This movement outside their households provides evidence of the increased mobility of women. Local NGOs in the post-Aila phase also contributed to the recovery of the community by increasing the Shora women's political mobility. For instance, some of the NGOs provided rice and other dry foods monthly. To collect the food, the women had to take a specific card provided by the NGOs. Only female members of the community, as the owners of these cards, could collect the food provided. As a result, they needed to go to the centre by themselves. This physical movement from the house to an NGO office is an example of political mobility, and this mobility has come because of these organisations.

These NGOs also started offering work to both men and women, giving women the chance to earn TK. 100–150 per day. Due to belonging to a poverty-stricken family, women started working outside the home to earn their living. The NGOs also offered the female community work such as doll-making and sewing, providing women with the opportunity to earn money at home. They even gave empowerment training to some tiger widows, which enhanced their mobility. NGOs additionally offered micro-credit loans to the affected women to enable them to start their own business and become financially self-dependent. Women were invited to meetings to help them use this money effectively. Those who took the loan attended these meetings in the NGO office. Previously, female participation in such meetings beyond the boundaries of their home was not regarded as worthy, but after Aila, as they were desperate to earn money and be solvent, they started attending these meetings with great enthusiasm. This change is undoubtedly an example of the NGOs' involvement in enhancing female mobility with regard to earning a living, from the immediate post-Aila phase to the present time.

The workers of NGOs gave other information after Aila took place. Female-headed families increased to 90%, and this data is the most significant evidence for their mobility. They not only gained the mobility to earn a livelihood but also reached another level of political mobility. Even now, in forming groups for meetings, it is mandatory to include women. The NGO sector pointed out many factors, which worked towards the mobility of the women in the Shora community. In a talk about female mobility and increased participation in NGO programmes, a programme officer of Islamic Relief said that several issues played a role in the increasing mobility of women. He said that forming a separate meeting for men and women forced the women to come and attend as a group in order to obtain the necessary information, whereas previously they obtained information from the male head of household who attended. He mentioned that the World Disaster awareness programme created awareness among women. Lastly, he said that women's empowerment brought mobility to their lives.

Even though this extensive spread of NGO activity took place primarily to ensure female empowerment, hidden barriers still lie within the system. One participant, who is also a tiger widow, said that she had never received any tiger widow allowance from the Government. She received an alternative payment from an NGO. Another tiger widow said that even though the Government and NGOs had offered tiger widows an allowance, they could not get the money because of the illegal interference of the chairman and members of their community. Those powerful unethical groups used to take the money forcefully and enjoy it instead of giving to its rightful owners.

Eventually, the tiger widows protested against such an unacceptable act, and now they receive 300 taka and 30 kg rice a month. Even though it is not enough for their families to survive on, for them this allowance is an important source of income.

The Shora women reported that though a few NGOs only provide economic empowerment loans to women, their husbands use the loan money instead. A boatman revealed this information when talking about his post-Aila struggle. He said:

> My boat was crushed after Cyclone Aila, and I had no money to repair it. Then I sent my wife to an NGO in order to take a loan, because those NGOs do not provide a loan to men. So my wife brought me the money, and I used that loan to repair my boat.

To some extent, he cannot be blamed for such an act. He and his family would have died of starvation after Aila if he had not been able to stabilise his means of earning. Here, our concern is about the Shora men's tendency to assume ownership of the money earned or gained by their wives. In this respect, the goal of NGOs to empower women remains unfulfilled. The question of mobility also occurs. Just going to her centre and collecting the money does not adequately ensure that a woman's mobility right is secured. Instead, we could say that these women may manage to get better mobility, but only if they use the money to do something on their own. So, the challenges regarding mobility remain to some extent, even though the situation is much better than before.

Concluding Notes

The analyses above provide evidence that the women of Shora lacked mobility rights for a long time. Even though the situation of dis-mobility has not yet changed sufficiently, women are getting better access to their mobility rights since Aila. Shora women still lack education facilities and many more services and resources, but the post-Aila female mobility situation is undoubtedly better than before Aila. The outlook of males to towards females has still not entirely changed, but if the female groups who now go out to work can definitively create an excellent example of their mobility right to earn a living, other women might also be encouraged to follow suit.

However, there are still a few things that remain questionable. To what extent have the Shora women managed to obtain full mobility? If simply going outside is the primary concern, then, undoubtedly, they have gained it. However, if we look deeper, we can see that the presence of patriarchal politics still deprives them of their full rights. The social customs and beliefs of the male community oil this unacceptable activity. Proprietorship of wages, loans and allowances is not in the hands of these women. Mostly, the male partner takes their money and uses it for his own needs. The scenario of the tiger widows is even worse. As they are socially stigmatised, they lack many rights over their own earned money, and either their sons or their sons-in-law often take it away from them; in the absence of both, their other relatives assume they have a right to the money. So how the money is used does not depend on the rightful owner's decision, but on the male members of the family.

Even if we overlook these issues, many other problems persist. Nevertheless, over time, and especially after Aila, there has been a huge change in women's mobility and their opportunities to earn a living. At least now women are going to Sundarbans forest and doing other types of work. They are attending awareness programmes held by different organisations and becoming more knowledgeable about their many rights. Now they go shopping in the market and bargain with the shopkeepers, and this act of bargaining also shows their awareness of their right to freedom of speech. A few female participants from Shora also mentioned that now they participate equally in family decision-making with their husbands, and in some cases, though their husbands do not fully listen to them, they at least make their points in front of everyone. Airing their points of view and making decisions show their political mobility. Now they are also aware of the fact that superstitions and strict religious values were invented to suppress them. They are also aware of the fact that religion—or, more accurately, the misinterpretation of religious beliefs—operates as a barrier to their mobility and efforts to earn a living. Thus, the female groups are now more concerned about their other rights and the social injustices which impose them. Through his studies of tornadoes in Texas, Moore (1958) realised that disasters make people adopt new behaviour patterns as a defining feature. Similarly, in Bangladesh, Aila brought many miseries to the lives of Shora's inhabitants, but, at the same time, it has created a window of opportunity for the women of the community to improve their mobility and consequently their ability to earn a living. Over time, this mobility will ensure a better social position for the women of Shora.

References

Adeel M (2017) Gender inequality in mobility and mode choice in Pakistan. Transportation 44(6):1519–1534

Ariyabandu MM, Wickramasinghe M (2003) Gender dimensions in disaster management. ITDG South Asia Publication, Colombo

Bleie T (2005) Tribal peoples, nationalism, and the human rights challenge. The adivasis of Bangladesh. The University Press Limited, Dhaka

Bradshaw S (2013) Gender, development and disasters. Edward Elgar, UK

Braun V, Clarke V (2006) Using thematic analysis in psychology. Qual Res Psychol 3(2):77–101

Chowdhury F (2009) Theorising patriarchy: the Bangladesh context. Asian J Soc Sci 37:599–622

Cisin IH, Clark WB (1962) The methodological challenge of disaster research. In: Baker G, Chapman D (eds) Man and society in disaster. Basic Books, New York, NY, USA, pp 23–54

Ellis C, Adams TE, Bochner AP (2011) Autoethnography: an overview. Hist Soc Res 36(4):273–290

Enarson E, Pease B (eds) (2016) Men, masculinities and disaster. Routledge, London

Gaillard J, Sanz K, Balgos BC, Dalisay SNM, Gorman-Murray A, Smith F, Toelupe V (2017) Beyond men and women: a critical perspective on gender and disaster. Disasters 41(3):429–447

Gilbert M, Masucci M, Homko C, Bove A (2008) Theorizing the digital divide: information and communication technology use frameworks among poor women using a telemedicine system. Geoforum 39:912–925

Hanson S (2010) Gender and mobility: new approaches for informing sustainability. Gend Place Cult 17(1):5–23. https://doi.org/10.1080/09663690903498225

Hanson S, Pratt G (1991) Job search and the occupational segregation of women. Ann Assoc Am Geogr 81:229–253

Haraway D (1991) Situated knowledge, gender for a Marxist dictionary. In: Simians, cyborgs and women: the reinvention of nature, Chap 7. Routledge, New York

Islam MS (2011) Biodiversity and livelihoods: a case study in Sundarbans Reserve Forest, World Heritage and Ramsar site (Bangladesh). A Master thesis submitted of the requirements for the degree of Master of Science (M.Sc.) in Management of Protected Areas at the University of Klagenfurt, Austria

Kabir R, Khan H, Caldwell K, Ball E (2016) Climate change impact: the experience of the coastal areas of Bangladesh affected by cyclones Sidr and Aila. J Environ Public Health 2016. https://doi.org/10.1155/2016/9654753

Li H, Guensler RL, Ogle J (2004) Comparing women's and men's morning commute trip chaining in Atlanta, Georgia, by using instrumented vehicle activity data. In: Conference on research on women's issues in transportation, Chicago, Illinois, United States. Transportation Research Board, pp 14–20

McSweeney K, Coomes OT (2011) Climate-related disaster opens a window of opportunity for rural poor in northeastern Honduras. Proc Natl Acad Sci USA 108:5203–5208. https://doi.org/10.1073/pnas.1014123108

Mehta M (2007) Gender matters: lessons for disaster risk reduction in South Asia. The International Centre for Integrated Mountain Development (ICIMOD), Kathmandu, Nepal

Mitlin D, Satterthwaite D (2013) Urban poverty in the global south: scale and nature. Routledge, New York

Moore HE (1958) Tornadoes over Texas. University of Texas Press, Austin, TX, USA

Nasreen M (2012) Women and girls vulnerable or resilient?. Institute of Disaster Management and Vulnerability Studies, University of Dhaka, Bangladesh

Perry RW (2018) Defining disaster: an evolving concept. In: Rodríguez H et al (eds) Handbook of disaster research. Handbooks of sociology and social research. Springer International Publishing AG. https://doi.org/10.1007/978-3-319-63254-4_1

Roy S (2013) Connectedness with Mangrove forest "Sundarbans": perceptions and behaviors of the forested community 'Shora' in Bangladesh. The University of Bergen

Roy S (2017) Women and disasters in South Asia: survival, security and development. Gend Place Cult. https://doi.org/10.1080/0966369X.2017.1396679

Roy S (2018) Livelihood resilience of the indigenous Munda community in the Bangladesh Sundarbans forest. In: Leal Filho W (ed) Handbook of climate change resilience. Springer Nature Switzerland AG. https://doi.org/10.1007/978-3-319-71025-9_10-1

Roy S (2019) Climate change impacts on gender relations in Bangladesh: socio-environmental struggle of the Shora forest community in the Sundarbans Mangrove forest. Springer Nature, Singapore

Scott J (1986) Gender: a useful category of historical analysis. Am Hist Rev 91:1053–1075

Willard F (1895) A wheel within a wheel: how I learned to ride the bicycle, with some reflections by the way. Applewood Books, Bedford, MA (originally published by Fleming H. Revell Co)

Wisner B, Blaikie P, Cannon T, Davis I (2004) At risk: natural hazards, people's vulnerability and disasters. Routledge, New York, NY, USA

Sanctuary in the City? Climate Change and Internally Displaced Persons in Harare, Zimbabwe

Vincent Itai Tanyanyiwa

Abstract Environmental migration is a multi-causal phenomenon occurring both regionally and internationally, yet one in which environmental drivers play a significant and increasingly determinative role. There has always been a fundamental interdependency between migration and the environment, but the reality of climate change adds new complexity to this nexus. Both gradual environmental change and slow- or sudden onset natural disasters influence population migration patterns. Natural disasters may be atmospheric or hydrological in nature, such as tropical storms or floods, with secondary impacts such as landslides. Some of the latter may be exacerbated by climate change. This study sought to explain how climate change has affected the welfare of the internally displaced populations in Mukuvisi Woodlands, Harare, as well as assessed the measures that have been put in place by both local and central government in integrating internally displaced populations. The study used an exploratory design focusing on key consultations and deliberations with different stakeholders. Purposive sampling was used to identify respondents who are known experts in the field. Respondent driven sampling was used initially to identify respondents who in turn provided referrals to a set number of other potentially informative respondents. The approach was used to select 20 respondents. Internal displacement is a significant consequence of global climate change processes of both a rapid and slow-onset nature, but there is a need for better understanding and research of these processes as well as the impacts and scale of displacement related to climate change. The relationship between environmental and climate change and migration is often complicated by the multifaceted associations with other factors, such as population growth, poverty, governance, human security and conflict. The complexities of the migration-environment nexus call for a comprehensive approach in policy, practice and research.

V. I. Tanyanyiwa (✉)
Department of Geography and Environmental Studies, Faculty of Science, Zimbabwe Open University, P.O. Box MP 1119, Mt Pleasant Harare, Zimbabwe
e-mail: tanyanyiwavi@yahoo.com

W. Leal Filho et al. (eds.), *Climate Change, Hazards and Adaptation Options*, Climate Change Management, https://doi.org/10.1007/978-3-030-37425-9_33

Background and Problem Statement

People flee their own lives for different reasons. Amongst a host of the current global problems is that of internally displaced persons/people (IDPs). One of the objectives of the 2030 Agenda for Sustainable Development is to reduce and or eradicate internal displacement. Arguably, IDPs are some of the most insecure, marginal, vulnerable and poor people who often lose educational opportunities and are unemployed. On average, 26 million people are internally displaced by disasters annually (IOM 2017). Internal displacement is set to continue globally due the rising risks of environmental degradation, climate change, sudden-impact or slow-onset disasters and lack of political will to reduce the drivers of displacement. Thus, the present landscape presents challenges to resolve internal displacement due to limitations in current approaches. An all-inclusive, contextually flexible and multidisciplinary approach is important to solve the IDPs crisis.

Climate change remains one of the biggest threats facing mankind today. Public concern was raised in the 1980s due to scientific evidence linking greenhouse gas (GHG) emissions with human activities as well as the risk of global climate change. Today, several studies and extracts from the Intergovernmental Panel on Climate Change Fifth Assessment Report (IPCC 2014) demonstrate that climate change is unequivocal. The report indicates that the impacts of climate change are projected to impede economic growth, erode food security, prolong existing poverty traps and create new ones which may further increase the levels and dimensions of poverty through displacing people. The most vulnerable are the urban populations in low-and-middle income nations which are likely to be disproportionately affected by the direct and indirect impacts of climate change (Dodman and Satterthwaite 2008).

Extreme weather events such as drought, flood and tropical storms such as the recent Cyclone Idai threaten development gains across a variety of sectors and intensify existing natural hazard burdens for at-risk populations. Climate change presents risks to lives and livelihoods at the individual level and to the economy and infrastructure at regional and national levels (Hellmuth et al. 2007). In Zimbabwe, historical climate analysis, recent observations and model projections are all showing increased rainfall variability, increases in extreme events, thus exacerbating the existing challenges that are already being faced by Zimbabwe (Chagutah 2010; Mudimu 2008). These impacts will have a widespread effect on social and economic development. Climate sensitive sectors such as agriculture, energy, environment, health, housing, livestock, and water are affected the most. In Zimbabwe, climate-induced displacements are glaring. Resettlement and relocation programmes for IDPs are difficult to implement due to limited financial resources and statistics. While the accepted line of argument has been that most of Zimbabwe's IDPs are political and economic refugees, some are a result of climate change. Generally, crops are failing, clean water is becoming scarce and livestock is dying. The hope is that better conditions are found in urban areas especially Harare. From 2008, about half a million people moved into Harare as a result of climate change. Harare's landscape is similar to the source region.

Aim and Objectives of the Study

Study Aim

The aim of the study is to create a general awareness of the problem of IDPs as prompted by natural hazards such as climate change. The government does not take the IDPs issue seriously because it also causes internal displacement. The research was guided by the research objectives and questions below:

Specific Objectives

(a) Describe how climate change has contributed to the phenomenon of IDPs in Harare?
(b) Explain how climate change affected the IDPs in Harare.
(c) Assess the measures that have been put in place by both local and central government in integrating IDPs in Harare.

Research Questions

(a) To what extent has climate change affected the welfare of the IDPs in Harare?
(b) How have IDPs lives been affected by climate change?
(c) How effective are the measures put in place in integrating IDPs in Harare?

Internal Displacement and Forced Migration: An Overview

There is consensus in literature that Sub-Saharan African cities are among the most vulnerable in the world due to widespread poverty, limited coping capacity and a highly variable climate (Carlson and Shumba 2011; Madzwamise 2010 and Abebe 2010). Couldrey and Herson (2008) propose that 26 million people are displaced within their countries as a result of armed conflict and many more are displaced due to natural and man-made disasters. With the advent of climate change these numbers are certainly likely to increase. Climate change leads to the excessive use of wetlands, reduction of water supply services, sewage and solid waste management. Environmental migrants are persons or groups of persons who, for reasons of sudden or progressive changes in the environment that adversely affect their lives or living conditions, are obliged to leave their habitual homes, or choose to do so, either temporarily or permanently, and who move either within their territory or abroad (IOM 2017). This working definition recognizes that

- environmental migrants are not only those displaced by extreme environmental events, but are also those whose migration is triggered by deteriorating environmental conditions;
- environmentally-induced movement can take place within as well as across international borders;
- Population movements triggered by environmental factors can be forced or a matter of choice.

Related to environmental migrants are what are known as IDPs. Many words can be used to describe IDPs including dislocated populations, mobile and vulnerable populations, evacuees or migrants (IOM 2017). Africa is estimated to have at least 12 million IDPs and of this number half is found in urban areas and in protracted displacement (Zetter 2012; Asplet 2013). Both government and international organisations have not given much attention to the problem of IDPS. The term IDPs is not fully embraced by the UN Convention. Rather, the term IDP is used in the context where people fled their homes for reasons similar to those of refugees, but remain within their own national territory and are subject to the laws of that state (Secretary General's Decision No. 2011/20 on Durable Solutions).

An IDP is *someone who is forced to flee his or her home but who remains within his or her country's borders, mistakenly referred to as refugees.* A refugee is "*a person who is outside his/her country of nationality or habitual residence; has a well-founded fear of persecution because of his/her race, religion, nationality, membership in a particular social group or political opinion; and is unable or unwilling to avail himself/herself to the protection of that country, or to return there, for fear of persecution*" (The 1951 UN Refugee Convention). Generally, people are familiar with the term "refugee". An IDP is different from a refugee. The term IDP remains an illusion to many. IDP refers to a "refugee" who could not flee to a neighbouring country but remained within ones' country (Zetter 2012). IDPs in Zimbabwe are refugees on their own land. The UN Guiding Principles on Internal Displacement defines IDPs as *persons or groups of people who have been forced or are obliged to flee or leave their house or places of habitual residence; in particular, as a result or in order to avoid the effects of armed conflicts, situations of generalized violence, violations of human rights or natural or human-made disasters and who have not crossed an internationally recognized state border.*

The implication is that the state, its agents thereof, or some other organised groups such as a political party has carried out the act of violence or persecution. The guidelines on IDPS were specifically formulated at the request of the then Secretary General of the UN, Boutros Boutros-Ghali since there is no international treaty which applies specifically to IDPs (Secretary General's Decision No. 2011/20 on Durable Solutions). For IDPS, it is vital to note that the movement is involuntary and the person remains within the country as opposed to movement by refugees which can be forced or is voluntary and involves crossing national boundaries (The 1951 UN Refugee Convention). Environmental factors have long had an impact on global migration flows, as people have historically left places with harsh or deteriorating conditions (Global IDP Project 2004). The scale of such flows, both internal and cross-border, is expected to rise as a result of accelerated climate change, with unprecedented impacts on lives and livelihoods. Such migration can have positive and negative effects on both the local coping capacity and the environment in areas from which these migrants originate, as well as in their temporary or permanent destinations (Ferris 2016). Migration, climate change and the environment are interrelated. Just as environmental degradation and disasters can cause migration, movement of people can also entail significant effects on surrounding ecosystems. This complex nexus needs to be addressed in a holistic manner, taking into account other possible mediating factors including, inter alia, human security, human and economic development, livelihood strategies and conflict (Zetter 2012).

More than half of the world's population currently lives in urban areas, with most of them situated in low-and middle-income nations (Chagutah 2010). One third of this urban population (1 billion people) lives in precarious informal settlements and slums, which exacerbates their vulnerability to humanitarian crises (IDMC 2014, 2015). By 2030, it is estimated that urban populations will surpass 5 billion and that 80% of the world's urban population will live in towns and cities in the developing world (IPCC 2014). Over the next twenty years, more intense climatic hazards are expected.

Many other factors influence individual vulnerability, such as HIV/AIDS, access to public services, environmental degradation, inadequate housing, conflict and insecurity. Particularly at risk are those communities located in areas prone to floods, cyclones and drought. Suffering repeated climatic shocks depletes their resources and makes them reliant on external assistance. The most vulnerable are people with insufficient assets or resources, who are less prepared or equipped to cope with major climate disruption (Spiegel 2014). Internal displacement is a security problem and its solution lies in establishing effective property repatriation mechanisms for the displaced as well as creating economic opportunities (Chitekwe-Biti 2009). Much emphasis is put on refuges yet the major problems faced by IDPs are similar if not worse off than those faced by refugees especially in the Zimbabwean context (Secretary General's Decision No. 2011/20 on Durable Solutions).

Conceptual Framework—Framework for Durable Solutions on IDPs

The protection of refugee's convention came into existence in 1951 through recognition of refugees. For IDPS, the protocol came very late through the non-binding framework: The 1998 Guiding Principles (30 Guiding Principles) on IDPs. The principles are based on actions to stop displacement, the right to protection from arbitrary displacement as well as support and security of IDPs (Cohen 2004; Marion and

Maurice 2008).The IDPs crisis was brought to the fore through the following instruments: (a) Principles on housing and property restitution for refugees and displaced persons (the "Pinheiro Principles"), focus was on right to return; (b) African Union Convention for the Protection and Assistance of IDPs in Africa (the African Union Convention for the Protection and Assistance of IDPs in Africa (2009) "Kampala Convention") which was promulgated in 2012. It was the first global enforcement convention containing State obligations and rights of IDPs; (c) International Conference on the Great Lakes Region Protocol on the Protection and Assistance to IDPs and Protocol on the Property Rights of Returning Populations (2006), the first instruments to carry legal requests for the domestication of the Guiding Principles on IDPs in national legislation; and d) Secretary General's Decision No. 2011/20 on Durable Solutions. This was a follow up to the UN Secretary General's 2009 report on peacebuilding. Structures and procedures for developing and applying strategies on durable solutions to the IDPs crisis were proposed. It was the later that was used as a conceptual framework for this study.

The Framework for Durable Solutions and the Guiding Principles on IDPs offers integration, relocation, and return as durable solutions (Kalin 2008). The UN Secretary General representative on the Human Rights of IDPs states that:

> *as citizens of their country, IDPs remain entitled to all guarantees of international human rights and international humanitarian law subscribed to by the State or applicable as customary international law. They do not lose, as a consequence of their displacement, the rights of the population at large. At the same time, IDPs have needs and vulnerabilities distinct from the non-displaced population, which must be addressed by specific protection and assistance measures.*

Rights that are supposed to be enjoyed by IDPs are not legally binding, though they are based on international humanitarian law (Mapiko and Chinyoka 2013). These rights are an important international framework for the protection of IDPs as reflected in national laws and policies (Kalin 2008). The Guiding Principles on Internal Displacement state that: *"IDPs shall enjoy, in full equality, the same rights and freedoms under international and domestic law as do other persons in their country. They shall not be discriminated against in the enjoyment of any rights and freedoms on the grounds that they are internally displaced"* (UNOCHA 2004). This is the basis of the rights and freedoms of IDPs. The state is bound to allow IDPs to return voluntarily to their place of origin or to relocate to another part of the country (UNOCHA 2004 ; IDMC 2007, 2008, 2009). The rights would only be enjoyed where security, compensation or reparation and participation are guaranteed. Long lasting solutions only function within a given environmental context in which IDPs live, especially the security aspect. The right to return, resettlement and reintegration are the three main solutions to challenges faced by IDPs (Secretary General's Decision No. 2011/20 on Durable Solutions). The three durable solutions are described below:

Return as a permanent solution is possible based on:

(1) whether the IDPs would be willing to return if they have houses and land in the place of origin,
(2) if they would be willing to return if they have only land but no houses, and

(3) if they were willing to return, but have no houses or no land in their place of origin.

Integration as a solution prospers when:

(1) the IDPs are unwilling to return as they have both a home and land in the present location;
(2) the IDPs are unwilling to return as they have land in the present location but no house, and;
(3) the IDPs are unwilling to return and prefer the present location though they have neither land nor house.

Relocation as a solution would have to consider that IDPs may not be willing to return to their original places or integrate in the present location (IASC Framework on Durable Solutions for IDPs: Brooking institution-University of Bern Project on Internal Displacement 2010). These people may not have houses or land in either their origin or their present location.

Methodology

Study Design

This study adopted a qualitative exploratory research design. Focus was on key consultations and deliberations with different stakeholders but more specifically the Mukuvisi Woodlands IDPs. The design was more analytical than statistical in nature. Babbie and Mouton (2001) assert that *"exploratory research is conducted to provide a better understanding of a situation. It isn't designed to come up with final answers or decisions. Through exploratory research, researchers hope to produce hypotheses about what is going on in a situation"*. Based on the definition above, this study was premised on generating evidence on the feasibility which will inform how the challenge of IDPs is exacerbated by climate change.

Sampling

Purposive sampling was used for this study to suit the subjective nature of IDPs in Zimbabwe. Purposive sampling occurs when a researcher "intentionally" selects individuals and sites to understand a phenomenon (Creswell 2012:206). Its emphasis is on *"the researcher's prior knowledge of potentially helpful respondents"* (Babbie and Mouton 2001). Two variants of purposive sampling were used and these are:

1. Expert Case sampling is where the research team sampled respondents who are known experts in the field of study. This approach facilitated identification two experts from the City of Harare (CoH) the Environmental Management Agency (EMA), as well as from Ministry of Local Government, Public Works and National Housing. Expert case sampling therefore, helped in the selection of respondents that were likely to give the most valuable information.
2. Respondent driven sampling is where initially identified respondents are requested to provide referrals to a set number of other potentially informative respondents. The approach was used to select 15 respondents who are illegal settlers in Mukuvisi Woodlands.

Study Area: The Mukuvisi Woodlands

The Mukuvisi Woodlands (latitude 17°50’35S and longitude 31°5’42E) are Zimbabwe’s national treasure opportunely located on the Highveld offering an initial introduction to Zimbabwe’s game life (Fig. 1). They are stunningly beautiful. The 2.74 km^2 (263 hectares) Mukuvisi Woodlands are located 5 km south east of Harare’s CBD (Muboko et al. 2014). They are situated in the Harare suburb of Hillside along Hillside Road. Mukuvisi Woodlands are found between the Mukuvisi and Chiraura Rivers (Bulton 1995). Rainfall averages between 650 and 850 mm per annum. Mean annual temperatures are 19 °C in winter and 27 °C in summer. The

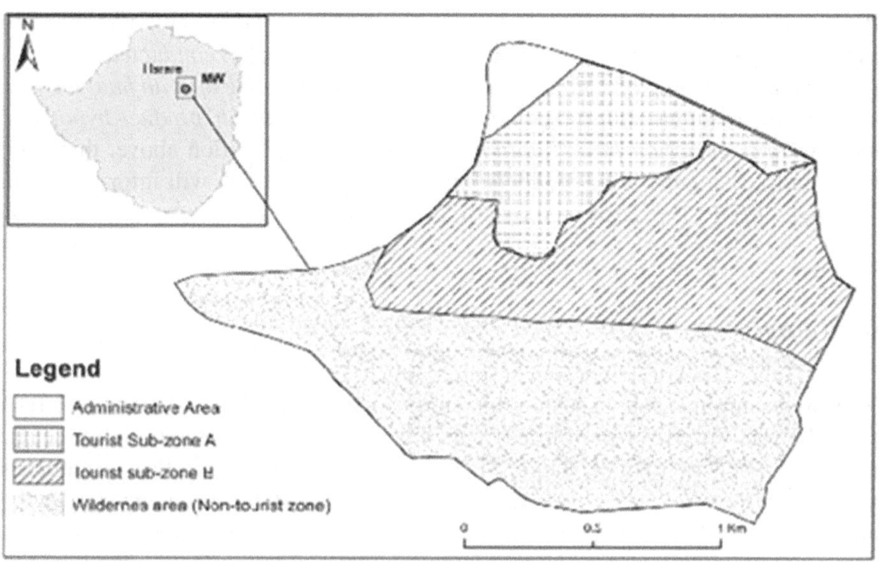

Fig. 1 Map of Mukuvisi woodlands. *Source* (Tanyanyiwa 2018)

woodlands provide one of the few remaining large green spaces that are open to the public where people enjoy nature's beauty within the city limits. Mukuvisi Woodlands environs include low to medium density suburbs and light industrial areas. The woodlands are Miombo and comprise of over 140 indigenous trees species. The main species include *Brachystegia spiciformis*—Igonde (Ndebele), Musasa (Shona); *Parinari curatellifolia*—Umkhuna (Ndebele), Muhacha (Shona) and *Julbernardia globiflora*—Munhondo (Shona), Umtshonkwe (Ndebele) and over 300 bird species. The Woodlands became a green belt protected area in 1910 against suburban development till around 1979. In 1980, the Mukuvisi Woodlands Association was formed. The association was given a lease by Harare Municipality which will expire in 2090. The aim of the association is to conserve and utilise the woodlands; to provide and develop facilities for sustainable development education along the principles and practice of nature conservation, environmental studies and research, and to encourage the public to appreciate the natural environment. These endeavors are under threat from the squatter settlements. Mukuvisi Woodlands are now a sanctuary for ex-convicts, homeless, pariahs as well as those at the bottom rung of the food chain especially environmental refugees. This is due to the vast tracts of land that are available although illegal settlements are outlawed by the government.

Data Collection and Analysis

Data were collected through semi structured interviews with the respondents, City of Harare (CoH) and Environment Management Agency (EMA) (as described in 5b) as well as the IDPs within the woodlands. Doing qualitative research means putting oneself in another person's shoes (Creswell 2007). Thus, this involves seeing the world from another person's perspective. Data analysis entailed being ethical by truly reporting what participants said.

Results and Discussion

Origins of Mukuvisi Woodlands IDPs

After Operation Murambatsvina many people became homeless. Operation Murambatsvina/"Operation Restore Order" literally meaning "getting rid of the filth" was a large scale government initiative to compulsorily remove slum areas within the country. UN estimated that at least 300,000–one million people were displaced after the government destroyed homes and informal business (Tibaijuka 2005). About 200 of these bear the brunt of having to live in Mukuvisi Woodlands. Mukuvisi Woodlands became a sanctuary for the homeless, pariahs, ex-convicts and those at the bottom rung of the food chain in governments view, a claim refuted by all respondents. The

Woodlands have turned into a fully functional squatter camp despite government laws against illegal settlements. The Mukuvisi community is made up of a harmonious and heterogeneous group of people whose age ranges from one month to 80 years and from both sexes. The community's population has been increasing over the years due to natural increase and more IDPs are attracted by the peace that comes with staying within the Woodlands (less than 5 km from Harare's CBD) and along the Mukuvisi River. There is no governance body hence the perceived "peace of Mukuvisi". Police only come to flush out criminals who use this place as a hideout. NGOs occasionally visit to give humanitarian aid.

The perennial Mukuvisi River which is the lifeblood of the IDPs originates in the wetlands near Cleveland Dam near Msasa on the north-eastern limits of Harare. The river meanders through to Manyame River via Graniteside, Msasa industrial sites, Eastlea, Sunningdale, Mbare, Highfield and Waterfalls and to Glen Norah on the southern part of the city. Mukuvisi River is a major source of water for Lake Chivero, Harare's main water source. It is heavily polluted by industrial and sewage effluent. The river banks are a perennial verdant to the homeless but it is now dying due to rainfall variability. The consequences are dire for Harare. Harare generally lacks potable water; yet it is ironic that Mukuvisi is a "life giver" to families.

Perceived Advantages of Living Within the Mukuvisi

IDPs in Mukuvisi are not willing to return to their places of origin. All of then argued that they been living in this area for an average of 15 years and they have built their social capital and therefore established relations with their fellow folks. In addition, there life is relatively inexpensive because they don't pay rentals since they in homes not regulated by CoH. For water provision, they have drilled deep wells for drinking water and Mukuvisi River water for their laundry. In recent years' water along Mukuvisi has been drying up due rainfall intermittent. Residents have been allowed to fetch water at the Mukuvisi Woodlands Association and nearby garages. Two middle aged man in their late 40s argued that their houses had collapsed; they no longer have houses in the communal areas where they came from. One of the two said:

"*I have also been away for 20 years and chances are that my relatives will not accept me. I cannot go back to my rural area Mutoko. I heard that my house has collapsed and that my elder brother has taken over my field. I cannot go and die as I can no longer farm productively due to rainfall variability*". For this reason, he preferred to stay at Mukuvisi. They also have menial jobs in and around Harare and especially at Magaba and Siyaso Complexes where informality thrives in almost every aspect of manufacturing, retailing and services. Because of the acute shortage of potable water in Harare, people also bring in their cars along the Mukuvisi to have them washed. This is proving to be a thriving source of source of livelihood. Dead wood is normally used as energy although as population expands green wood is increasingly being used. Generally, the Mukuvisi residents are not willing to go anywhere. They prefer some integration and rehabilitation rather than being relocated. Their proximity to Harare's proximity to Harare's CBD means they can walk into town. Their argument is that they cannot imagine starting a new life elsewhere.

Migration as an Adaptation Strategy to Internal Displacement

In the words of the then UN Secretary-General Ban Ki-moon, "displacement remains arguably the most significant humanitarian challenge that humanity faces." It is estimated that up to 50 million people are displaced in any given year by natural disasters including climate change and variability (IDMC 2015). Least developed countries and countries with particularly susceptible geographies i.e. economically and socially marginalized groups within the affected communities—the poor, the elderly, women and children are most vulnerable. In Mukuvisi, migration to the woodlands has been an adaptation strategy to climate and environmental change because water, wood and general information spread by word of mouth of the existence of open land that is very close to the CBD. Migration as a coping strategy is not open to everyone as it depends on resources, information and other social and personal factors therefore, it is not necessarily the most vulnerable and the most severely affected by environmental and climatic factors who migrate. In this case the IDPs are a mixed bag, about 80% were a result of migrating from rural areas where rainfall is becoming increasingly scarce whilst the remainder were driven into the area by Operation Murambatsvina Reliable estimates of climate change induced migration are unavailable in Zimbabwe and let alone Harare. Gradual and sudden environmental changes are already resulting in substantial population movements within greater Harare. The number of storms, droughts and floods has increased threefold over the last 30 years with devastating effects on vulnerable communities, particularly in the developing world (IDMC 2015). In 2018, 30 million persons have been displaced by extreme weather events (UNHCR 2018). UNHCR raises awareness on climate change as a driver of displacement and the protection of IDPs in disaster situations through four main areas i.e. (1) Legal advice to enhance the rights of IDPs affected by disasters and climate change (2) policy advocacy coherence to safeguard IDPs and to mainstream across appropriate areas (3) research to promote operational and policy work and (4) field based activities to guarantee sustainable responses to internal displacement (UNHCR, 2018).

Migration often seems to be misperceived as a failure to adapt to a changing environment. It is an essential component of socio-environmental interactions that need to be managed. Migration is a complex phenomenon as it can be a coping mechanism and survival strategy for those who move. At the same time, migration, and mass migration in particular, can also have significant environmental repercussions for areas of origin, areas of destination, and the migratory routes in between and contribute to further environmental degradation. In Mukuvisi, IDPs have contributed to deforestation as wood is the main source of energy used by 76% of the population. In the same vein, laundry is done along the river and together with discharges from industry, water is polluted. Globally, up to 50 million people are displaced in any given year by natural disasters including climate change (OCHA 2018). IOM helps to reduce the vulnerability of populations exposed to environmental risk factors; assists populations on the move as a result of environmental causes; and builds the capacities of governments and other actors to face the challenge of environmental migration. The Zimbabwean government does not recognise IDPs hence little or no funding comes from IOM (Mapiko and Chinyoka 2013). Zimbabwe does not have domestic laws or policies on IDPs; IDPs are therefore unaware of their rights. There are numerous obstacles to the realization of their rights. The United Nations High Commission for Refugees (UNHCR) (2011) proposes an urgent need for studies in relation to climate-related displacement with further empirical research in vulnerability to displacement. In Zimbabwe, more than 1 million are IDPs due to climate change, political violence, slum clearance, civil unrest, human rights violations, hunger, land reform and a socio-economic and political collapse of the country.

The joint effects of hyperinflation, unemployment levels above 90%, food shortages, fuel shortages, power cuts, water cuts, and the breakdown of education and health services make the problem of IDPS more prevalent. During an assessment mission in June 2004 prior to the onset of Operation Murambatsvina, Refugees International found out that IDPs were abandoned. This is due to the Zimbabwean government's obstruction of assistance efforts by international agencies and local non-governmental organizations (UNHCR 2011). In Zimbabwe, there is no official government statistics relating to IDPs. The government has never acknowledged both the reality of displacement, and that its policies such as land reform cause internal displacement. No humanitarian agency has been able to conduct a comprehensive survey to determine the number of IDPs. IDPs are a sensitive issue in Zimbabwe; they are often referred to as mobile and vulnerable populations. Due to the absence of a law or policy, IDPs are more vulnerable than their refugee counterparts because of this lack of international and local protection and assistance. In Zimbabwe, when people are displaced inside their own country, there is no urgent effort by humanitarian or relief groups to provide shelter or source and distribute food for the affected persons. IDPs are left in the hands of the same authorities that have been responsible for their displacement (Human Rights Monitor 2002).

The refusal by the government to allow foreign assistance to internally displaced persons exposes such people to famine and disease. As remarked by UNHCR (2011), Operation Murambatsvina worsened the problem of internal displacement in Zimbabwe to a critical level, having caused hundreds of thousands of people to join the ranks of the country's "abandoned" IDPs. The crisis of IDPs in Zimbabwe has been given little international and local attention and thus most of the victims have gone unprotected and unassisted in Mukuvisi. Local non-governmental organizations such as the Amani Trust and Zimbabwe Agricultural Welfare Trust (ZAWT) have, with limited, funding attempted to assist Zimbabwe's displaced persons (Zvigadza 2008). Support is minimal because of the legal connotations and lack of a supporting and enabling policy framework to assist IDPs.

Early Warning Systems (EWS)

While early warning systems can minimize damage, loss of life and displacement, early warnings alone are insufficient. EWS need to be followed by timely action. In Harare, Mukuvisi Woodlands people do not have access to weather information. For those who have access, more than 80% fail to interpret it. The information is too scientific and therefore not palatable. In the same vein similar studies in famine and resulting displacements in Somalia in July and August 2011 show that while a disaster was forecast as early as November 2010 by the Famine Early Warning Systems Network, the famine was not declared until July 2011, and even then, the response by donors to funding appeals by the United Nations was insufficient (IOM 2017). Technical mechanisms such as early warning systems need to be supported by effective action based on commitment, political will and response mechanisms. The political will is not there in Zimbabwe. Laws and policies should include measures to maintain and repair family unity especially during evacuation, replace personal documentation through rapid and simplified procedures, protect housing, land and property rights through registration and keeping land titles safe, resolve property disputes after a disaster, ensure that there is no discrimination based on gender, age or ethnicity during disaster relief interventions (Chitekwe-Biti 2009).

Impacts of Environmental and Climate Change on Human Mobility

Climate change as a phenomenon does not directly displace people or cause them to move instead it produces environmental effects and exacerbates current vulnerabilities which makes their survival difficult where people are (Matarira et al. 1995). Extreme weather events, notably flood, drought and tropical storms are also expected to increase in frequency and intensity across the African continent (IPCC 2007). These changes, in turn, will likely result in further population movements. According to the UN International Strategy for Disaster Reduction (UNISDR) storms, floods and droughts have increased threefold over the past 30 years. Recent reports by the IPCC in 2007 project that sub-Saharan Africa will experience increased water stress, decreased yields from rain-fed agriculture, increased food insecurity and malnutrition with increase in arid and semi-arid lands.

Climate change is not a distant future threat but it is the main driver behind rising humanitarian needs and we are seeing its impact. The number of people affected and the damages inflicted by extreme weather has been unprecedented. People who are particularly vulnerable include those whose lives are already threatened by abject poverty, HIV/AIDS, environmental degradation, inadequate housing and insecurity. Those living in disaster hot spots such as flood plains or cyclone tracks, are exposed to repeated climatic shocks that compound their vulnerability. With the threat of extreme weather in the future, the demand for disaster response can only rise, as will the costs.

Floods, violent storms and droughts don't have to become disasters. In 2005, more than 700,000 people were displaced during operation Murambatsvina (Operation Clean Up) [Zimbabwe Human Rights NGO Forum 2005]. In addition, people have always been internally displaced e.g. through mining ventures of diamond fields in Marange, dam construction and to pave way for urban development. Displacement of people in urban areas presents both problematic and political challenges. It was only in 2009 that UNHCR accepted and recognized urban refugees hence raising the flag high for their recognition.

Harare has been seriously affected by flooding due to coagulation of the drainage system. Waste collection has been erratic; in the CBD waste is supposed to be collected daily and in the residential areas at least once a week, the situation is so dire that in some residential areas waste can go uncollected for weeks. Flash floods have wreaked havoc for IDPs along the Mukuvisi. Residents have resorted to digging shallow wells along the drying wetlands which contain contaminated water as a survival alternative. Areas such as refills and wetlands previously condemned before housing are now being put into the same use. One respondent complained; "*We are getting sick here. There is no water for us to drink.*" Rising temperatures and increasing urban population in Harare have led to water shortages and this has been exacerbated by drought in recent years e.g. in 2002, 2003, 2005, 2007, 2008, 2009, 2010 and 2018 (Chagutah 2010).

The environmentally displaced form a group that is in danger of being left without protection as their plight is tangled up with regular migration, voluntary environmental migration, and climate change migration. The distinctions that need to be drawn are the actual causal links between the environmental factors and migration and the extent to which migration is forced. This is a particularly complex task because of the myriad of factors that play a role in forced and indeed voluntary migration in the world today. The task is important, however, as a large proportion of environmentally internally displaced persons become urban IDPs.

Obstruction of Humanitarian Aid to IDPs by the Zimbabwean Government

Humanitarian agencies face a myriad of problems when in operating in urban areas, they have less control, and more authority is vested in the urban authority i.e. the CoH. Livelihoods and protection are important for intervention. One of the main problems for intervention in Harare is the selection of beneficiaries from a pool of equally or even poorer host populations. There is need to involve the IDPs in identifying those who need help the most. It is the duty of the government to provide humanitarian assistance to IDPs (Naidu and Benhura 2015). Nevertheless, the government has abdicated on this responsibility and this has led to a displacement crisis because IDPs are largely excluded from government-controlled aid e.g. food aid. They are excluded because they are presumed to be a supporter of the opposition political parties as CoH is run by the opposition political party, the MDC. From a government perspective, there is no special focus on IDPs because all citizens face common challenges such as breakdown of education and health services; food and fuel shortages; hyperinflation; power cuts; unemployment and water cuts. The UN Guiding Principles on Internal Displacement are clear on the role of humanitarian organisations. The UN posits that:

International humanitarian organizations and other appropriate actors have the right to offer their services in support of the IDPs. Such an offer shall not be regarded as an unfriendly act or an interference in a State's internal affairs and shall be considered in good faith. Consent thereto shall not be arbitrarily withheld, particularly when authorities concerned are unable or unwilling to provide the required humanitarian assistance. All authorities concerned shall grant and facilitate the free passage of humanitarian assistance and grant persons engaged in the provision of such assistance rapid and unimpeded access to the internally displaced.

Zimbabwe has violated this UN guiding principle with impunity. Humanitarian organisations are not allowed access to IDPs. Where access is granted, it can just be revoked on arbitrary grounds. In 2008, all NGO operations were suspended. The welfare of the people was jeopardised. This was worsened by government's insistence that there are no IDPs in Zimbabwe. Humanitarian agencies cannot conduct a comprehensive survey of IDPs in Zimbabwe if there is no permission and cooperation from Zimbabwean authorities. As a result, the profiling of Zimbabwe's IDP population remains an illusion. The level of vulnerability of IDPs' cannot be ascertained making it difficult for humanitarian agencies who have to plan for interventions with the barest information.

Political Challenges in Helping IDPs in Harare

The challenges faced by IDPs should be addressed by politicians through economic participation. A new narrative that challenges vulnerability and victimhood as the main cause of IDPS should be challenged. Issues such as fortitude, ingenuity, human rights, development action, resilience, political participation, opportunity, inclusion and participation help in recognizing IDPs in areas where they take refuge. This can be done through elevating their social and economic rights, a herculean task for the government of Zimbabwe. The IOM argues that there should be detailed information on IDPS to inform policy so as to influence intervention mechanisms (IOM, 2017). Currently, the international community and CoH in particular do not seem to have the capacity to address challenges presented by IDPs. Kadane (2011:49) contends that internal displacement is "a symptom of state dysfunction". In Zimbabwe, this is near truth as politicians' rescue IDPs when campaigning, but after the election cycle they are a totally forgotten people. One Mukuvisi resident said:

We are harassed by the police left, right and centre, sometimes they beat us or ask us to bribe them. They ask us to go back to our rural homes and we say we can't go back because our source of livelihood is in Harare. In the rural areas there is no land and some of us do not even have a rural home. The government should therefore help us rather to use us for elections.

The condition in which IDPs at Mukuvisi are in is not life threatening. The provision of decent accommodation is a human rights issue hence there is need on the part of both local and central government to act. This can only have done if the government acknowledges the IDPs crisis. It is still contentious on whether IDPS call for international intervention. In the case of Zimbabwe, the current socio-economic challenges inhibit the government to channel aid from development partners to IDPs. The government has abdicated its responsibility of ensuring that that IDPs have access to and use of decent, permanent shelter to maintain a family unit and privacy. In the absence of a permanent home, the government should at least should provide temporary but durable shelter depending on family size. This is in order to enable them to comply with Principle 18 of the UN Guiding Principles on IDPs which provides that the state has an obligation to ensure at the minimum and regardless of the circumstances and without discrimination, that competent authorities shall provide IDPs with essential food and potable water, basic shelter and housing, appropriate clothing and essential medical services and sanitation. The provision of food and educational facilities (enshrined in section 75 of Zimbabwe's Constitution) is important. At the moment, the Department of Civil Protection is under-resourced and cannot implement such initiatives. The government can do better if it upholds and strengthens its relations with charitable agencies to ease the plight of Mukuvisi Woodlands IDPs as guided by Article 9 (3) of the Kampala Convention which posits that States should be helped by global, humanitarian and civil society organisations in cases of emergencies. In addition, article 9 (2) (c) of the Kampala Convention, posit that the elderly, the disabled, people living with HIV/AIDS, expecting mothers, mothers with young children should be cared for by the state in terms of access to food, health facilities and information. The Parliament of Zimbabwe, through section 243 (1) (i) of the Constitution, should enact an Act of Parliament to cater for IDPs in terms of who they are and the associated support to conform to regional and international standards.

Climate-related displacement responses need to be guided by fundamental principles of humanity, human dignity, human rights and international cooperation (Bagshaw and Paul 2004). They need furthermore to be guided by consent, empowerment, participation and partnership and to reflect age, gender and diversity aspects. IDPs living in Mukuvisi Woodlands are vulnerable to sexual and gender-based violence or other threats to their personal safety as reported by all women in this study. The factors that make durable solutions for IDPs in Harare more likely include:

- Ensuring a transition from a humanitarian assistance phase to early recovery and reconstruction, thus allowing IDPs to return to their places of origin and rebuild their lives after the disaster. Focus should be on basic humanitarian needs;
- Guaranteeing that humanitarian assistance and development are mutually reinforcing from the beginning of the emergency;
- Participation by IDPs in policy formulation that include return and reintegration; Re-establishing local economies and livelihoods, encouraging self-reliance and

promoting IDP participation in all activities, from humanitarian assistance delivery to engagement with development actors;

- Implementing planned relocations, which include strategies related to land, housing and livelihoods. Taking into account the needs of receiving communities, including mechanisms which address possible integration challenges and tensions.
- Strong national leadership and political commitment to reduce the IDPs crisis.
- Availability of technical support, funds for preparedness, national resources, international donor investment and related preparedness;
- Active involvement of communities, civil society and agencies in the implementation of disaster preparedness through construction of local infrastructure such as roads and schools;
- Strong international and national collaborative preparedness;
- Availability of adequate funds during the response and restitution phases i.e. through preparatory measures to regain assets and properties lost.

Conclusion and Recommendations/Policy Implications

Conclusion

In the past few years the problems of IDPs has been has been multifaceted, widespread and usurious in Harare and Zimbabwe in general. The physical and economic displacement has been due to climate change, political violence, land reform. This has led to a critical humanitarian situation. There has not been much effort to solve this problem owing to Zimbabwe's government reluctance to admit that there is a crisis. Notwithstanding the obvious prevalence of the IDPs phenomenon in Zimbabwe, they have not benefited from international policies and norms that are meant to protect and assist them because of the government lackadaisical attitude. This paper has argued that there are the three durable solutions to address IDPs challenges. These solutions should be backed by structures and conditions that alleviate their plight. The participants preferred two permanent solutions i.e. the right to return and the right to integrate. Only two people opted to relocate. Any preferred solution should always insist on the participation of IDPs in the process. Mobilization of support from the international community has always been belated due to Zimbabwe's egregious obstruction of assistance efforts by international agencies and NGOs. Zimbabwe does not have a meaningful register of IDPs in order to assess their needs. Climate change adaptation funds should be used to support adaptation measures related to climate change-induced internal displacement and other forms of human mobility. In this respect, Harare is not providing enough sanctuary to the internally displaced persons. Financing bodies such as IOM should support national governments, and engage with communities, and insist on a comprehensive approach which encompasses prevention to durable solutions. Financing bodies and research institutions should also continue to support, and engage in research and policy which addresses human mobility and climate change especially in Harare.

Recommendations/Policy Implications

Both central and local government should systematically collect data on the number of IDPS in Harare and Zimbabwe. Data on IDPs shows only numbers immediately after a crisis though enumeration should be periodic. There is need for data to trace subsequent developments i.e. returns and durable solutions offered. More research is needed on the scope and scale of IDPs in Zimbabwe. Disaster risk reduction and disaster preparedness measures should seek to prevent displacement to a greater extent. National disaster management systems, laws and policies should draw on human rights approaches and encourage local community involvement. This is in congruence to the guiding principles on IDPs by the UN. Urgency is needed to provide shelter, food, water and sanitation and medical services with priority being given to the chronically ill, elderly, children, women and then men in emergency humanitarian situations. Zimbabwe's national climate change policy should be fully operationalised in order to improve environmental sustainability and sustainable livelihoods with the ultimate goal of achieving resilience. The policy was crafted with a view to overcome Zimbabwe's critical situations such as climate illiteracy, deforestation, flooding matters, gender and climate inequalities, homelessness, inorganic agricultural practices, poverty levels and water scarcity.

In line with the United Nations Guiding Principles on IDPs, Zimbabwe should provide protection and assistance to IDPs in the form of food, medical services, shelter, water and sanitation. Vulnerable people, especially women, children, elderly and chronically ill persons should be given priority on a need basis. Government should allow national and international humanitarian agencies unconditional access to IDPs. IDPs should be encouraged to voluntarily return to their original places of habitual residence thereby ensuring their full reintegration. The African Union (AU) should compel Zimbabwe to permit full and unhindered access to IDPs by national and international humanitarian agencies. In the same vein, the African Commission on Human and Peoples' Rights (ACHPR) should condemn the obstruction of international humanitarian assistance for displaced persons by the Zimbabwean government. The commission should also encourage Zimbabwe to take concrete measures to provide assistance and protection for IDPs, and allow unrestricted international assistance. IDPs should not be viewed as a political punch bag to be used by political parties during campaigns but rather as a humanitarian crisis. The United Nations (UN) and its allied agencies operating in Zimbabwe should do a countrywide needs assessment of IDPs. This assessment could include numbers, conditions and locations of IDPs, their registration, shelter, food, health services, water, sanitation and other forms of assistance (Sen 2005). The UN and its officials should impress upon Zimbabwe to take full responsibility in assisting and protecting IDPS and the unacceptability of obstructing efforts from the international community. The donor governments should ensure that the programmes they fund are not used by the Zimbabwean government to infringe upon the rights of IDPs e.g. by manipulating food assistance to impel relocation to the rural areas.

References

Abebe A (2010) The African union convention on internally displaced persons: its codification background, scope and enforcement challenges. Refug Surv Q 29(3):28

African Union Convention for the Protection and Assistance of Internally Displaced Persons in Africa (Kampala Convention) [2009] available at: https://www.refworld.org/docid/4ae572d82.html [accessed 27/03/2019]

Asplet M (2013) Internal displacement: responsibility and action. UNHCR. Retrieved from http://www.ipu.org/PDF/publications/Displacement-e.pdf

Babbie E, Mouton J (2001) The practice of social research. Oxford University Press, Oxford

Bagshaw S, Paul D (2004) Protect or neglect: toward a more effective united nations approach to the protection of internally displaced persons. (Washington D.C. and New York: the brookings–SAIS project on internal displacement and the UN office for the coordination of humanitarian affairs, 2004), [accessed online at www.brookings.edu, on 15 Jan 2015]

Bulton R (1995) The makabusi historical background. Mukuvisi woodlands association, Harare, Quick Print Publishers

Carlson A, Shumba E (2011) Status of and response to climate change in Southern Africa: case studies in Malawi, Zambia and Zimbabwe. WWF-World Wide Fund for Nature

Chagutah T (2010) Climate change vulnerability and preparedness in Southern Africa, Zimbabwe country report. Henrich Boell Co., Cape Town

Chitekwe-Biti B (2009) Struggles for urban land by the Zimbabwe homeless people's federation. Environ Urban 21(2):347–366

Cohen R (2004) The guiding principles on internal displacement: an innovation in international standard setting. Glob Gov 10(4):459–480

Couldrey M, Herson M (2008) Climate change and displacement. Special issue of forced migration review (31)

Creswell JW (2007) Qualitative inquiry and research design: choosing among five traditions, vol 17. Sage Publications, California

Creswell JW (2012) Educational research, 4th edn. Pearson Education Inc, pp 1–67

Dodman V, Satterthwaite D (2008) Institutional capacity, climate change adaptation and the urban poor. Inst Dev Stud Bull 39(4):67–74

Ferris E (2016) Ten years after humanitarian reform: how have IDPs fared? Brookings-LSE: project on internal displacement, (2014), https://www.brookings.edu/wp-content/uploads/2016/06/Introduction-to-final-report-IDP-Study-FINAL.pdf. Accessed 15 May 2018

Global IDP Project (2004) Internal displacement: global overview of trends and developments. (Geneva: Norwegian Refugee Council 2005), Accessed online at www.idpproject.org, on 17 Jan 2019

Hellmuth ME, Moorhead A, Thomson MC, Williams J (2007) Climate risk management in Africa: learning from practice, international research institute for climate and society (IRI). Columbia University, New York, USA

IASC Framework on Durable Solutions for internally displace persons Brookings Institution-University of Bern Project on international displacement (2010), April pp 11–14

IDMC (2014) The National responsibility to protect internally displaced people: the Kampala Convention. Dec 2014

IDMC (2015) Global overview. NRC May 2015

Internal Displacement Monitoring Center (IDMC) (2009) The rights of IDPs to return home and property restitution. Training on the protection of IDPs, internal displacement monitoring centre and Norwegian Refugee Council

Internal Displacement Monitoring Centre (2007) Addressing urban displacement—a project description. IDMC, Norwegian Refugee Council, Geneva

Internal Displacement Monitoring Centre (2008) Internal displacement: global overview of trends and developments in 2007. IDMC, Geneva

Kalin W (2008) The guiding principles on internal displacement – annotations, transnational legal policy, no. 32. The American society of international law and the brookings institution project

Madzwamise M (2010) Climate governance in Africa, Adaptation strategies and institutions henrich boell Co., Cape Town

Mapiko E, Chinyoka K (2013) The plight of internally displaced children: a case of Zimbabwe. Greener JS 3(9) 168 Vol. 30. No. 1

Marion C, Maurice H (2008) Ten years of guiding principles on internal displacement http://www.fmreview.org/climatechange.htm. Accessed 17 Feb 2019

Matarira CH, Makadho JH, Mwamuka FC (1995) Zimbabwe: climate change impacts on maize production and adaptive measures for the agricultural sector. Interim report on climate change country studies Harare, Zimbabwe

Muboko N, Chigumira T, Mashapa C, Gandiwa E, Chibememe G, Muposhi VK (2014). Impacts of wood poaching on vegetation structure and composition in Mukuvisi woodland, Zimbabwe. J Environ Prot 5:156–163

Mudimu G (2008) Zimbabwe food security issues paper for forum for food security in Southern Africa, department of agricultural economics, Harare University of Zimbabwe

Naidu M, Benhura A (2015) Internal displacement and forced migration within Zimbabwe: an overview. J Soc Dev Afr 30(1):147–169

Sen A (2005) Human rights and capabilities 6(2) Journal of Human Development 151. Displaced and forgotten: IDPs in Zimbabwe report from global IDP project Published on 16 Jul 2003 available on www.idpproject.org Accessed 18 Feb 2019

Spiegel SJ (2014) Legacies of a nationwide crackdown in Zimbabwe: operation Chikorokoza Chapera in gold mining communities. J Mod Afr Stud 52(04):541–570. https://doi.org/10.1017/S0022278X14000469

Tanyanyiwa VI (2018) Weather forecasting using local traditional knowledge (LTK) in the midst of climate change in Domboshawa, Zimbabwe. In: Leal Filho W, Manolas E, Azul A, Azeiteiro U, McGhie H (eds) Handbook of climate change communication. Climate Change Management, vol 2, pp 1-20. Springer, Cham. https://doi.org/10.1007/978-3-319-70066-3_11. Online ISBN 978-3-319-70066-3

Tibaijuka A (2005) Report of the fact-finding mission to Zimbabwe to assess the scope and impact of operation Murambatsvina by the UN special envoy on human settlements issues in Zimbabwe. http://www.un.org/News/dh/infocus/zimbabwe/zimbabwe_rpt.pdf

Zetter R (2012) Lands of no return, population exchange and forced displacement in the 20th century. Athens, Anemon Productions

Zvigadza S (2008) Climate change vulnerability mapping: the case of the city of Harare, Zimbabwe. Paper presented at Cities and Climate Change Workshop, 5 December 2008, Poznan, Poland

Climate Change, Extreme Events and Human Mobility in Latin America: Exploring the Links Through National Laws and Policies

Fernanda de Salles Cavedon-Capdeville, Erika Pires Ramos, Andrea Cristina Godoy Zamur, Diogo Andreola Serraglio, Ignacio Odriozola, Luiza de Moura Pallone, Fernanda Dalla Libera Damacena, Lilian Yamamoto and Giulia Manccini Pinheiro

Abstract The article focuses on the impact of climate change and extreme events on human mobility in Latin America, verifying to what extent and by what means climate national law and policy strategies have incorporated (or not) the human mobility dimension, contributing to the reduction, prevention and better management of the risks and impacts of climate extreme events. The relevance of this aim is justified by the ability to illustrate whether national climate strategies integrate human mobility and are aligned with international guidelines and recommendations on this issue, thereby also leading to the identification of existing gaps and inconsistencies. Relations between quality and availability of reliable data on climate induced human mobility and the level of awareness of the phenomenon and its integration into climate strategies are considered. Latin America was chosen as the research subject because it is one of the most vulnerable regions to the effects of climate change and exposure to extreme events that have impacted human mobility, and the previous experience and research carried out by RESAMA in the region.

Introduction

Latin America presents a broad mosaic of interconnected human and natural systems in a fragile equilibrium: the Earth's greater biodiversity and high concentration of traditional populations.[1] However, the region is under pressure as a result of extractive industry, rapid and unplanned urban growth, large development projects, connected

[1] In 2014 there were 826 indigenous peoples in the region, corresponding to 45 millions of persons (2010 census data) (CEPAL 2014).

F. de Salles Cavedon-Capdeville (✉) · E. P. Ramos · A. C. G. Zamur · D. A. Serraglio · I. Odriozola · L. de Moura Pallone · F. D. L. Damacena · L. Yamamoto · G. M. Pinheiro
South American Network for Environmental Migrations (RESAMA), Florianópolis, Brazil
e-mail: cavedon.capdeville@gmail.com

W. Leal Filho et al. (eds.), *Climate Change, Hazards and Adaptation Options*, Climate Change Management, https://doi.org/10.1007/978-3-030-37425-9_34

with socio-economic vulnerabilities.[2] Even if progress has been made in reducing poverty, the population in a situation of economic vulnerability has increased in Latin America (UNDP 2016), which remains the most unequal region on the planet (CEPAL 2015, p. 18). Poverty and disadvantage are expected to increase as global warming increases, and countries in the tropics and Southern Hemisphere subtropics are projected to experience the largest impact on economic growth (IPCC 2018). Latin America is one of the most vulnerable regions to climate change.[3] Poverty and economic inequalities results in high vulnerability and increasing risk to climate change and low adaptive capacities (IPCC 2014). The combination of socio-environmental vulnerabilities and extreme events result in disasters that exacerbate preexisted vulnerabilities, leading to a vicious circle. These factors affect patterns of human mobility, [4] to move represents a measure to adapt, to run off, to respond to environmental changes.

From 2015, progresses were made in recognizing the links between human mobility, climate change and disasters, through the adoption of new international frameworks. A new international governance is set up around current central themes, representing a historic opportunity to recognize and reinforce protection and assistance to people exposed to climate risks (UNGA 2016).

The South American Network for Environmental Migrations (RESAMA), established in 2010, has been addressing this topic in the region through research, capacity building, policy and advocacy. The network's engagement can be seen through: (i) support of international processes and participation as a member of the Platform on Disaster Displacement (PDD) Advisory Committee; (ii) engagement with the regional agenda, such as the South American Conference on Migration (SACM); (iii)

---

[2]Considering the Human Development Index 2018 (UNDP), no Latin-american country is among the 10, 20 or 30 top countries in human development, the best ranked is Chile (38th) that, with Argentina (45th) and Uruguay (55th) are the only countries in the region to achieve a very high HDI. The countries with the lowest rates are Honduras (130), Guyana (127), Guatemala (125) Nicaragua (124) and Haiti (163) http://hdr.undp.org/en/content/human-development-index-hdi.

[3]Central America is the second region most exposed to climate risk, behind Africa, and South America is in the 6th position according to Climate Change Vulnerability Index 2017, https://reliefweb.int/sites/reliefweb.int/files/resources/verisk%20index.pdf.

[4]Different terms related to human mobility are cited in the article, such as "forced displacement", "forced migration", "environmental migration", since it was chosen to maintain the terminology originally adopted in international, regional and national documents, legal texts and other documents cited.

attendance of international and regional events[5] and publications[6]; and (iv) projects dealing with data compilation and production, as well as on strategies for monitoring the regional and national policies alignment with international recommendations. It is to be noted the launch of *Latin American Observatory on Human Mobility, Climate Change and Disasters* developed in partnership with the University for Peace (UPEACE).

The article focuses on national climate strategies in Latin America, verifying to what extent and by what means they have incorporated (or not) the human mobility dimension. Although the object of the research is related to disaster risk reduction and migration frameworks, it was chosen to address it only through climate policies and laws.[7] It is organized in four sections: (i) the impacts of extreme events and disasters on human mobility in Latin America and data challenges; (ii) the recent international recommendations on human mobility in the context of climate change; (iii) regional initiatives on human mobility in the context of disasters and extreme climate events; (iv) the national climate strategies, focused on National Adaptation Plans, Nationally Determined Contributions, climate policies and laws adopted after the Paris Agreement. In conclusion, it highlights the remaining gaps and challenges, as well as recommendations.

This study is not exhaustive, but a knowledge step for further research, it complements past RESAMA's research and, thus, is also to be boosted through complementary studies.

---

[5] 'UNFCCC Task-Force on Displacement Stakeholder Meeting—Recommendations for integrates approaches to avert, minimize and address displacement related to the adverse effects of climate change' (2018); 'Data Knowledge Working Group Expert Meeting—"Improving data on human mobility, environment and climate change: Priorities for action"—PDD/IOM-GMDAC/IDMC' (2018); 'Expert Panel and Intersessional Panel discussion on human rights, climate change, migrants and persons displaced across international borders'—OHCHR/HRC (2017); VI Plataforma Regional para la Reducción de Riesgo de Desastres en las Américas—UNISDR (2018); 'Reunión Regional Latinoamericana y Caribeña de Expertas y Expertos en Migración Internacional preparatoria del Pacto Mundial para una Migración Segura, Ordenada y Regular'—CEPAL (2017).

[6] For example, Ramos et al. (2018) Informe de Posición RESAMA n. 01: La migración ambiental en el Pacto Mundial para una Migración Segura, Ordenada y Regular—Desafíos y aportes para América Latina y el Caribe, https://resamadotnet.files.wordpress.com/2018/03/informe-de-posiciocc81n-resama-n-1-pacto-mundial.pdf; Ramos and Cavedon-Capdeville (2017), Yamamoto et al. (2018) Human mobility in the context of climate change and disasters: a South American approach, International Journal of Climate Change Strategies and Management, https://doi.org/10.1108/IJCCSM-03-2017-0069; RESAMA (2016) Migración, medio ambiente y cambio climático: agenda 2030, buenas prácticas y desafíos para la región suramericana", http://csm-osumi.org/sites/default/files/documento_de_referencia_-_migracion_medio_ambiente_y_cambio_climatico.pdf; IOM (2017) Brazilian Case Study for the IOM Research Project "Migraciones, Ambiente y Cambio Climático: estudios de caso en América del Sur". https://robuenosaires.iom.int/sites/default/files/publicaciones/Migraciones%2C%20ambiente%20y%20cambio%20clim%C3%A1tico.PDF.

[7] A complete mapping of initiatives related to human mobility in the context of climate change in Latin America through a migratory and DRR approach has been carried out by RESAMA in two studies mentioned in the previous note: (i) Regional responses to climate change and migration in Latin America and (ii) Human mobility in the context of climate change and disasters: a South American approach.

Impacts of Extreme Events and Disasters on Human Mobility in Latin America: Current Efforts on Data and Evidence

Out of 335 disasters registered globally in 2017, 93 took place in the Americas (CRED 2018). If the number of disasters in the region can not be overlooked, their impact yields even more reason for alarm. Extreme events have been taking a progressively increasing toll in the continent. In 2016, 17.3% of people affected by disasters lived in the region, a far larger figure than the average of 5.1% found in the period between 2000 and 2015 (CRED 2017).

In the period 1998–2017, Latin America was the continent with the highest level of economic losses due to climate-related disasters, representing 53% of the global figure (CRED and UNISDR 2018). In the Americas, 4.5 million people were displaced by disasters in 2017, representing 23.8% of the global total. The three major hurricanes of that year—Harvey, Irma and Maria—displaced over 3 million people. Floods were a main cause of displacement throughout the continent. Among the most affected countries by floods in 2017 figure Argentina, Bolivia, Guatemala, Nicaragua, Paraguay, Peru and Uruguay. As an example, the flooding in Peru was the worst in 20 years, displacing almost 300,000 people (IDMC 2018a). Colombia, Chile, Haiti and Brazil joined the list of the 20 countries with the highest number of people displaced by disasters from 2008 to 2014 (IDMC 2015). In 2018, 1.7 million people were displaced by disasters in the Americas. Even if the number of disaster displacement in 2018 is lower than in 2017, climate-related events continue to be the main cause of internal displacement in the Americas (IDMC 2019b). Climate change impacts in the world's three most densely populated regions (Sub-Saharan Africa, South Asia and Latin America) can cause the internal displacement of more than 143 million people by 2050 due slow-onset impacts of climate change. In Latin America, estimates reach the number of 17 million new internally displaced persons (World Bank 2018).

Disaster displacement "is a reality and among the biggest humanitarian challenges facing States and the international community in the 21st century" (The Nansen Initiative 2015). One of the reasons is the difficulty in obtaining accurate, up-to-date data, as well as monitoring it overtime. The Advisory Committee of the Platform on Disaster Displacement (PDD) has highlighted the lack of data and evidence regarding displacement dynamics, particularly in the context of slow-onset events associated with climate change (PDD 2016, p. 4). There are four main ways in which slow-onset events originate displacement: (i) as they lead to a scarcity of vital resources; (ii) as they may turn into a disaster prompted by a rapid-onset event; (iii) as their underlying existence weakens the populations' capacities to cope with other slow- or rapid-onset events; and (iv) as they may become catalysts and multipliers to other crisis (IDMC 2018b).

In 2018, 2.5 million people were affected by drought in Central America, including migration hotspots Guatemala, Honduras, El Salvador and Nicaragua (UN News

2019). The data concerning people affected by this type of event is difficult to obtain and presents gaps and weaknesses, as recognized by the Centre for Research on the Epidemiology of Disasters (CRED): "The direct impact of climate change on human populations will increasingly be felt through catastrophic phenomena, such as drought and extreme temperatures. The human repercussions of these events, as experienced by the EM-DAT team, are typically poorly reported, especially from low income countries. This is partly due to methodological difficulties in registering deaths and the severe consequences caused by droughts and extreme temperatures" (CRED 2019). These are usually connected to a lack of consistency in definitions, in distinguishing displacement from other migratory patterns and in associating displacement to droughts when other stressors are present (IDMC 2018a).

Migration connected to the environment commonly takes place within the borders of the same country. In these cases the data challenge is even more complex, as data availability depends on national policies and methodologies. In an effort to overcome challenges concerning information on internal displacement, the Internal Displacement Monitoring Centre (IDMC) has proposed a methodology that aims to facilitate progress monitoring on internal displacement over time at the national, regional and global levels (IDMC 2019a).

In Brazil, one of the main obstacles to obtain data on disaster displacement is conceptual. Official databases and legal framework on disaster do not adopt the concept of "displaced" person. From the categories accounting for "human damage", only two are closest in meaning to the situation of displacement: "dislodged" and "unhoused/unsheltered" persons,[8] which do not cover the full extent and complexity of disaster displacement. This conceptual gap has direct consequences in the quality of the data collected, as well as in the possibility of monitoring the evolution of disaster displacement.

The data challenge is conceptual, institutional and operational. The first step to advance in normative and policy responses to human mobility in the context of climate change is to generate reliable evidence. This requires that governments act on local, regional and international levels, as well as in cooperation with international organizations, to adopt comprehensive and harmonized data collection systems. Data collectors, analysts and researchers must work in coordination to provide a more harmonized understanding of concepts and definitions to provide better, up-to-date, disaggregated data on the topic (PDD 2016, p. 14).

[8]The full official database is publicly available in the Integrated System of Information on Disasters (S2ID), https://s2id.mi.gov.br.

Recent Recommendations on Human Mobility in the Context of Climate Change in the International Agendas

The UNFCCC Task Force on Displacement

Human mobility due to the effects of climate change was first expressly addressed in the international climate regime in 2010, on the 16th Conference of the Parties (COP16) to the UNFCCC. During the COP18, the topic was raised once again on discussions related to the adoption of a loss and damage mechanism to tackle the negative impacts of climate change (Decision 3/CP.18, UNFCCC 2012). The inclusion of the theme in the Climate Agenda was enshrined with the adoption of the Paris Agreement (COP21). Decision 1/CP.21 (section 49) requests the Executive Committee of the Warsaw International Mechanism on Loss and Damage (WIM) to establish a Task Force on Displacement to develop recommendations to prevent and reduce forced migration processes caused by the adverse effects of climate change (UNFCCC 2015).

The results of the Task Force on Displacement (TFD) were made available at the 24th Conference of the Parties (COP24, 2018), concluding that the inclusion of human mobility and concerns related to climate change in the international scenario have increased since 2015. This trend is particularly significant in agendas dealing with human mobility, climate change and environmental issues. 53% out of 66 countries analyzed mentioned climate change and/or other environmental factors as drivers of human displacement in their migratory legal framework. Likewise, 81% out of 37 countries that have presented their National Adaptation Plans or similar strategies refer to human mobility. The document reinforces the need to strengthen institutional frameworks associated with human mobility and/or climate change at the national level (UNFCCC/WIM-TFD 2018, p. 23).

To the national level, TFD recommends the adoption of specialized legislation ensuring the coordination among actors dealing with human mobility and climate change; engage in partnerships with affected communities and other stakeholders; integrate human mobility challenges and opportunities into the process to formulate and implement National Adaptation Plans, and communicate efforts undertaken through Nationally Determined Contributions. The TFD also invites governments to "enhance research, data collection, risk analysis, and sharing of information to better map, understand and manage human mobility related to the adverse impacts of climate change". A special focus is given for advancing the state of knowledge of displacement in the context of slow onset events. Moreover, the TFD underlines the need of technical guidance to support country mainstreaming displacement in their NAPs and NDCs (UNFCCC/WIM-TFD 2018, pp. 11, 14 and 31).

Considering the regional tools, the TFD recognizes legal and policy-related guidance tools that acknowledge human mobility issues, but do not necessarily provides practical advice and support for specific situations of displacement The recommendations concentrate on developing and enhancing regional strategies for migration and planned relocation, assistance, protection and durable solutions in cases of cross border displacement (UNFCCC/WIM-TFD 2018, pp. 29–30). Lastly, the UNFCCC should develop a process to increase capacity building for regional and national actors to better implement regional tools and guidance.

Global Compact for Safe, Orderly and Regular Migration (GCM)

The *New York Declaration for Refugees and Migrants* (2016) points out the relevance of the causality among migration, environment and climate change, stressing that environmental factors may lead to displacement. Commitments to be pursued in order to formulate a Global Compact for Safe, Orderly and Regular Migration were established.

Regarding the consultation period,[9] a thematic session entitled *Modalities for the intergovernmental negotiations of the GCM* was held with the aim of addressing drivers of migration, including adverse effects of climate change and natural disasters (UNGA 2017a). The compilation of the recommendations was presented in the report *Making migration work all*, recognizing the role played by climate change in migration and the indispensability of enhancing resilience to climate risks. This would ensure such process as an option, and not as a last resort when faced with the exhaustion of other alternatives. The adoption of preventive measures through the expansion of legal instruments that deal with the theme would be desirable (UNGA 2017b). In terms of mass migration processes, it proposes: (i) a human rights based approach and the strengthening of humanitarian assistance; (ii) instruments that provide a legal status to vulnerable migrants after emergency humanitarian response; (iii) temporary and/or long-term alternatives for the receiving and stay of migrants that, despite not qualifying as refugees, remain unable to return to their places of origin. The *Zero Draft* (February, 2018), besides highlighting the guidelines from the previous report, points out the need of reliable data and information related to the connections between migration and climate change. This document was used as background for the discussions that culminated in the Global Compact for Safe, Orderly and Regular Migration (GCM), adopted in December 2018.

Objective 2 of the GCM deals with the drivers of migration, including natural disasters and the adverse effects of climate change. It focuses on mapping and developing strategies to minimize the impact of environmental factors on the decision to move (GCM 2018). Disasters are a factor to consider when developing national and regional practices to enhance availability and flexibility of pathways for regular migration (Objective 5). States commit in Objective 23 to "increase international and regional cooperation to accelerate the implementation of the 2030 Agenda for

[9]RESAMA was invited by ECLAC and IOM to attend in August 2017 the UN regional consultation for GCM through the Latin American and Caribbean Regional Meeting of Experts on International Migration preparatory to the GCM.

Sustainable Development in geographic areas from where irregular migration systematically originates due to consistent impacts of (…) climate change and disasters (…)" (GCM 2018). The *Final Draft* was endorsed by 176 Member-States and has a specific title dealing with migration associated with natural disasters, environmental degradation and the adverse effects of climate change.

While the TFD recommendations detailed on how to avert, minimize, and address the effects of climate change on human mobility, they are also more focused on the environmental factor as a driver of sudden-onset displacement. The GCM has a broader approach to what are the environmental factors, including slow-onset disasters, and it encompasses all phases of the movement, calling for understanding, prevention, regional harmonization, coherence, and policies of appropriate duration. It has a special focus on national, sub-regional and regional strategies for cooperation. Both documents can be considered innovative and complementary. The GCM Preamble mentions the UNFCCC and the TFD invites governments to review climate and human mobility national strategies in connection to the GCM. Both spheres emphasizes the need to develop an appropriate methodology to obtain better data and information, mainly regarding regional migration trends. The next steps will be the alignment of the regional and national levels to these documents by concrete climate and migratory strategies.

Recent Regional Initiatives to Address Human Mobility in the Context of Climate Change and Disasters

There are still few actors who incorporate human mobility into their agendas and strategies to cope with climate change and disasters.[10] A binding regional framework for recognising and protecting people displaced for environmental reasons is nonexistent. Nevertheless, referral to the topic has been gradually increasing in some regional forums through non-binding instruments. Only the most recent initiatives dealing specifically with human mobility in the context of disasters and/or climate change will be considered in this study.

The Nansen Initiative and its follow-up, the Platform on Disaster Displacement (PDD), plays an important role in developing regional guidelines supporting the Regional Consultation Processes (RCPs).[11] In the context of the Regional Conference on Migration (RCM *or Puebla Process*), Central American countries, Mexico,

[10] In South America, this subject was analysed in a previous study of RESAMA (Yamamoto et al. 2018) and will be updated in an upcoming study on DRR strategies in Latin America.

[11] "Regional Consultative Processes on Migration (RCPs) are state-led, ongoing, regional information-sharing and policy dialogues dedicated to discussing specific migration issue(s) in a cooperative manner among States from an agreed (usually geographical) region, and may either be officially associated with formal regional institutions, or be informal and non-binding. RCPs are Inter-State Consultation Mechanisms on Migration (ISCMs) at the regional level." https://www.iom.int/regional-consultative-processes-migration.

Colombia, Haiti and the Dominican Republic, recognized the challenges of displacement in the context of disasters and climate change in the 2013 Regional Consultation.[12] In 2015, most of the RCM members endorsed the Nansen Agenda for the protection of displaced people across borders in the context of disasters and climate change.

In 2016 the Guide "Protection for persons moving across borders in the context of disasters" (CRM 2017) was approved, with the objective to share information about the use of humanitarian protection measures that States can apply temporarily to respond to the needs of immigrants affected by disasters. It approaches three categories of people: (1) immigrants who get into a foreign country seeking temporary protection and assistance during or after a disaster that occurred or might occur in their country of origin; (2) immigrants which are already in another country, when their country of origin is affected by a disaster and they seek to stay temporarily to avoid the consequences and (3) migrants which live or are in transit in a country that suffered a disaster, and seek protection and assistance because they were affected by it. Based in common regional and subregional agreements, the Guide reminds to the States that they could be bounded to guarantee free movement to certain categories of people and that it may also be applicable in disaster situations, in order to protect those who cross an international border seeking protection. It calls the countries' migratory authorities to avoid discretionality when exercising entry permissions, encouraging the application of common-used and favorable categories labeled under "humanitarian reasons", and remembers that in circumstances of disasters, the principle of *non refoulement* should be respected. The member countries should use objective measures, such as the declaration of disaster status, to justify the activation of humanitarian protection measures for immigrants affected.

The South American Conference on Migration—SACM (CSM)[13] has integrated the issue in the regional agenda since 2010 (X Conference) until 2018 (XVIII Conference), where the SACM approved the "Regional Guidelines on Protection and Assistance to people displaced across borders and migrants in countries affected by disasters of natural origin".[14] This document presents guidelines to ensure that responses to cross-border displacement in the context of disasters are more effective and consistent with national regulations. It tackles two scenarios: (1) protection measures for people looking at safety and assistance when crossing an international

[12]Desastres y desplazamiento transfronterizo en América Central: Necesidades emergentes, nuevas respuestas, Conclusiones: Consulta Regional de la Iniciativa Nansen, San José, Costa Rica, del 2 al 4 de diciembre de 2013, www.nanseninitiative.org/central-america-consultations-intergovernmental/.

[13]The SACM/CSM is a Regional Consultation Processes that provides a platform for consultation in the following fields: development, diasporas, rights of migrants, integration, information exchange, migration statistics among others. For more information, see: http://csm-osumi.org/.

[14]For the XVI SACM/CSM Meeting in 2016, RESAMA prepared for the CSM Technical Secretariat a document of reference "Migration, environment and climate change: agenda 2030, good practices and challenges for the South American region". In its contributions as member of PDD Advisory Committee RESAMA reinforced the need of a regional guide for South America. RESAMA took part in the regional meetings with CSM Members to discuss the document before its final approval last year during the XVIII CSM Meeting.

border and (2) protection measures for migrants in transit or residents in the hosting country, when a disaster of natural origin occurs. It provides a set of recommendations to reduce and avoid the risk of displacement in the country of origin (CSM 2018).

The SACM encourages States to consider cross-border cooperation and to share resources to enhance resilience and reduce disaster risk impacts including displacement risk, to establish mechanisms and fast procedures for the admission of displaced people arriving from a disaster impacted country and to identify people facing a real risk to their life or safety, or extreme difficulties in their country of origin, with the purpose of internally activating the mechanisms to adequately respond to the protection needs of such persons. Recommendations are supported by different principles, in which solidarity and regional coordination are highlighted (CSM 2018).

In the framework of the 30-year anniversary of the 1984 Cartagena Declaration on Refugees (Cartagena +30)[15], in which the "Brazil Declaration and Plan of Action" was adopted, due to the challenges raised by climate change and disasters and its impacts on the transboundary movement of people, Latin American and Caribbean countries required a study from UNHCR on the subject (ACNUR 2014). The UNHCR, supported by the Platform on Disaster Displacement, developed the study "Transboundary displacement, climate change and disasters: Latin America and the Caribbean", focused on legal and regulatory measures to the transboundary movement of people in the context of disasters associated with natural hazards and climate change (Cantor 2018, not yet published).

The report concludes that Latin American and Caribbean countries have one of the world's most developed response capacities to tackle disaster displacement. At the regional level, it calls to understand displacement collectively, through the collection and exchange of government-produced data or to exempt visa requirements in the context of disasters. It also suggests regional orientation to avoid discretionality when applying immigration laws, and regional or bilateral approaches to prevent and address disaster displacement, through cross-border preparedness and contingency mechanisms that integrate disaster risk management, immigration, refugee issues and climate change. At the national level, it encourages to gather accurate and standardized official data on (i) the magnitude of the movements involved and (ii) its dynamics, including, in the context of disasters and climate change, the relationship between the internal and international movement, while it promotes flexible visa programs and exceptional categories, including those for "humanitarian reasons" in order to provide responses to disaster displacement and foreigners affected by disasters in their own country.

[15]RESAMA is member of the Regional Working Group for the Brazil Plan of Action (GAR-PAB), an initiative to articulate civil society to the implementation and follow-up of the 2014 Brazil Plan of Action, "a Framework for Cooperation and Regional Solidarity to Strengthen the International Protection of Refugees, Displaced and Stateless Persons in Latin America and the Caribbean". *Brazil Plan of Action is* "a Common Roadmap to Strengthen Protection and Promote Sustainable Solutions for Refugees, Displaced and Stateless Persons in Latin America and the Caribbean within a Framework of Cooperation and Solidarity".

Human Mobility in National Climate Political and Normative Strategies

National Adaptation Plans and Nationally Determined Contributions as well as climate policies and norms adopted after Paris Agreement,[16] were selected to analyze how climate political and normative strategies are coping with human mobility. This temporal delimitation was chosen since this research aims to consider only policies and norms which are aligned with the main process and international documents, in special, the Paris Agreement. NAPs, established within the Cancun Adaptation Framework (UNFCCC 2010), and NDCs, set forth in the Paris Agreement[17] were chosen as climate instruments that can integrate human mobility. All NAPs and NDCs submitted by Latin American countries to the UNFCCC were reviewed and those which mentioned the topic of human mobility were analysed.

National Adaptation Plans (NAPs)

NAPs are key instruments to integrate human mobility in national adaptation planning and advance in considering migration as an adaptation strategy, as there is a demand in this sense in the Global Climate Agenda, (Warner et al. 2014; Wilkinson et al. 2016). The research about NAPs is based on the National Adaptation Plan Central of UNFCCC[18]—NAP Central. Only three States from Latin América forwarded their plans to NAP Central: Brazil, Chile and Colombia.[19] Another States, as Argentina, Uruguay and Mexico, started the process of elaboration of their NAPs.

Brazil's NAP launched in May 2016, mention migration as a possible adaptation strategy to the adverse effects of climate change, recognizing that the increase of temperature may increase the risk of extreme events in Brazil, such as droughts and floods, which may cause displacement. It highlights the possibility of new migratory flows in response to the negative consequences of climate change. The Plan considers the need to adopt measures to enhance understanding, coordination and cooperation with regard to climate change induced human mobility (Yamamoto et al. 2018), in the context of sectoral strategy for vulnerable population. It also adopts a strategy

[16]Sectoral climate policies and norms, are not included in this analysis. Only general climate policies and norms were selected to the study.

[17]The NDCs were provided for in Article 4, paragraph 2, of the Paris Agreement. States Parties shall prepare, communicate and maintain successive Nationally Determined Contributions which they intend to achieve.

[18]Forwarded NAPs of developing countries are available at: https://www4.unfccc.int/sites/NAPC/News/Pages/national_adaptation_plans.aspx.

[19]Brazilian NAP Vol. II, http://www.mma.gov.br/images/arquivo/80182/BOOK_PNA_Volume%20II%20v4.pdf; Chilean NAP, https://www4.unfccc.int/sites/NAPC/Documents/Parties/Chile%20NAP%20including%20sectoral%20plans%20Spanish.pdf; Colombian NAP, https://www4.unfccc.int/sites/NAPC/Documents/Parties/Colombia%20NAP%20Spanish.pdf.

for disaster risk reduction reinforcing the coordination between climate and disaster risk reduction policies.

The Chilean NAP, posted on September 2017, on the UNFCCC NAP Central, recognizes that extreme events can increase migration but does not effectively integrate human mobility in the proposed measures or sectoral plans. Its general objective considers that adaptive measures could reduce exposure to extreme events. Among its principles figures the priority to adaptation measures that consider the most vulnerable. It considers that extreme events can affect the availability of water and food and the impacts of these events on the urban environment will be especially felt by vulnerable and low resilient populations. These elements can be related to measures of prevention and reduction of displacement, and the impacts of climate change pointed out can be considered as vectors of these displacement. The NAP foresees coordination with other policies, especially disaster management policies. In the context of institutional strengthening, it provides the elaboration of a legal and institutional framework on adaptation to climate change.

For Colombia, the NAP posted on February, 2018, on the UNFCCC NAP Central, does not bring specific measures related to human mobility, but recognizes forced displacement and migration as possible consequences of extreme events. It mentions the vulnerability of displaced populations, often settled in precarious settlements and risk zones, due to the greater frequency of extreme events. Colombian NAP adopts strategies and measures that contribute to reduce the risk of displacement by seeking to address environmental, socioeconomic and territorial vulnerabilities that can generate greater exposure to extreme events and, potentially, boosting migratory movements. One example is the community-based approach to adaptation, which aims to increase the adaptability of the most vulnerable communities. It establishes coordination between climate adaptation measures and disaster risk management.

Nationally Determined Contributions (NDCs)

Nationally Determined Contributions describe what measures governments will take in order to reduce emissions under the Paris Agreement. To date, 184 Parties have submitted their first NDCs to the UNFCCC and one Partie has submitted its second NDC. A survey conducted by the International Organization for Migration indicates that, by August 2016, human mobility had been referenced in 33 NDCs, most of them in the regions most affected by climate change. Latin America ranks among these regions, with 21% of NDCs mentioning human mobility (IOM 2016). The NDCs submitted or reviewed after the IOM research were analysed. The survey in the UNFCCC database allowed to identify 22 Latin American states that submitted their NDCs.[20] Colombia, Ecuador, Haïti, Mexico, Suriname and Uruguay refer to human mobility in their NDCs.[21]

Colombia addresses human mobility in the context of the pacification process, associated to economic, social and environmental challenges. However, migration is identified more as a problem that increases the pressure on the environment than an adaptation strategy. NDC from Ecuador is much more proactive, adopting the development of policies and strategies to cope with permanent or temporary migration motivated by factors related to climate change as an adaptation measure. Mexico, Suriname, Uruguay and Haiti refer to planned relocation as an adaptation strategy. In the Mexican NDC, relocation is adopted as a strategy to cope with irregular human settlements in zones prone to disasters. In the case of Suriname, the concern is related to the exposure of coastal zone to the impacts of climate change, and the need to invest in adaptation or promote the relocation away from the threat of the rising sea. Venezuela has highlighted a good practice in its INDC related to the response to disasters loss and damages, but the submitted revised NDCs did not retain the references to human mobility. NDC from Uruguay refer to planned relocation, as it adopted in 2010 a National Relocation Plan to resettle socially disadvantaged persons living in flood-prone and/or contaminated areas. Among the main priorities to address the adverse effects of climate change in social policies, figures: (i) to make available georeferenced information of social vulnerabilities associated to adverse climate events, looking at migrants; (ii) to relocate households in flood or contaminated zones. Haiti's NDC addresses human mobility in different aspects, such as the integration of displaced persons and planned relocation as adaptation strategies, as well as the consideration of internal displacement in the context of disasters in urban and sustainable development planning for cities exposed to flooding.

This analysis indicates that human mobility is most considered in NDCs than in NAPs and, especially, considered not only as a negative consequence of adverse effects of climate change, but presenting concrete measures to prevent and manage displacement and considering human mobility as an adaptation measure.

National Climate Policies and Norms After the Paris Agreement

Climate policies and norms adopted by Latin American States from 2015 were identified and selected in a first moment. It was also verified in which States processes for proposing new climate norms are underway. In a second moment, the policies

[20] Argentina, Bolivia, Brazil, Chile, Colombia, Costa Rica, Cuba, Dominican Republic, Ecuador, El Salvador, Guatemala, Guyana, Haití, Honduras, México, Nicaragua, Panamá, Paraguay, Perú, Suriname, Uruguay, Venezuela.

[21] All the NDCs cited in this item can be accessed at the UNFCCC NDC Registry, https://www4.unfccc.int/sites/NDCStaging/Pages/All.aspx.

and norms identified were analysed to verify to what extent they take into account human mobility.

South America

In South America, four States adopted new climate norms after Paris Agreement[22]: (i) Colombia (2018), (ii) Perú (2018), (iii) Paraguay (2017) and (iv) Uruguay (2017). Ecuador did not adopted a specific climate law, but climate change issues have been incorporated into the Environmental Organic Code from 2017.[23]

Two other States have initiated processes to elaborate new climate norms. On July 5, 2018, the Chilean government announced the elaboration of a new climate law, intending to finalize the phases of elaboration and public consultation until July 2019. Chile created, by Decree n. 52 on January 2018,[24] the Climate Change Permanent Presidential Advisory Committee to advise the President in matters related to the identification and formulation of policies, plans, programs and measures related to climate change, and in the elaboration of a national climate public policy. Venezuela develops a draft new climate law since 2016. Only the Peruvian and Uruguayan norms have provisions related to human mobility. The Climate Change National Policy from Uruguay adopts a human rights and intra and intergenerational equity perspective (section 1) and is more focused in relocation strategies for vulnerable communities living in climate risk zones, in the context of the promotion of communities adaptation and resilience capacity against climate change and extreme events (section 8). The Peruvian Climate Change Law goes much further, adopting a concept of environmental migrant and providing measures of attention to these migrants.[25]

---

[22]Argentina has adopted a new climate law at this month of november (2019), nothing relevant related to human mobility was identified.

[23]Colombia, Ley n. 1931 por la cual se establecen directrices para la gestión del cambio climático, 27 July 2018, http://es.presidencia.gov.co/normativa/normativa/LEY%201931%20DEL%2027%20DE%20JULIO%20DE%202018.pdf; Perú, Ley Marco Sobre Cambio Climático n. 30754, 2 April 2018, https://busquedas.elperuano.pe/normaslegales/ley-marco-sobre-cambio-climatico-ley-n-30754-1638161-1/; Paraguay, Ley Nacional de Cambio Climático n. 5875, 15 September 2017, http://redrural.org.py/wp-content/uploads/2017/09/LEY-5875-CC.pdf; Uruguay, Decreto del Poder Ejecutivo n. 310/017—Politica Nacional de Cambio Climático, 27 April 2017, http://www.mvotma.gub.uy/politica-planes-y-proyectos/politica-nacional-de-cambio-climatico; Ecuador, Codigo Organico del Ambiente, 6 April 2017, http://www.ambiente.gob.ec/wp-content/uploads/downloads/2018/01/CODIGO_ORGANICO_AMBIENTE.pdf.

[24]Chile, Decreto n. 52—Crea Comisión Asesora Presidencial Permanente de Cambio Climático, 22 November 2017, https://www.leychile.cl/Navegar?idNorma=1113261&idParte=0. For more information about public consultation on the new climate change law, see: http://portal.mma.gob.cl/proceso-de-consulta-publica-del-anteproyecto-de-ley-marco-de-cambio-climatico/.

[25]In the glossary, "migrant for environmental reasons" are considered as "persons or a group of persons who, because of unavoidable, sudden or progressive environmental changes that negatively affect their lives or their living conditions, are forced to leave their habitual homes, or decide to do so voluntarily. Displacement can be temporary or permanent, inside the country or abroad." (translation from the original text in spanish).

The federal government must issue an action plan to prevent and respond to forced migration caused by the effects of climate change. It is therefore the first climate normative strategy in Latin America to specifically address environmental migration. The norms analysed establish coordination points with strategies for disaster risk reduction, which is also an important step towards an integrated assessment and response to climate extreme events and consequences on human mobility. Colombian post-2015 climate law provides the articulation and complementarity among the adaptation to climate change and disaster risk management. Climate Change National Policy from Uruguay provides support to policies related to climate change and variability, including the National Disaster Risk Management Policy, strengthen the management of disaster risk associated with climate to reduce vulnerability and strengthen articulated response to extreme weather and climate events (sections 3, 10).

The States with a climate law post-2015 have adopted policies and strategies too.[26] Colombia's Climate Change National Policy (2017) and Paraguay's Climate Change National Policy (2011) make no reference to human mobility. Peru adopted a Climate Change National Strategy (2015) pointing out forced migration as a consequence of climate change, and has also an Action Plan for Adaptation and Mitigation to Climate Change (2010). Uruguay has a National Climate Change Response Plan (2010) providing measures to assist evacuees in consequence of extreme climate events. Other States adopted policies and plans after Paris Agreement. Chile adopted a National Action Plan on Climate Change 2017–2022. Argentina is engaged in the process of elaboration of its NAP, and has adopted in 2017 sectoral climate plans (forest, energy and transport).

Guyana and Suriname adopted climate strategies in 2015 which predate the Paris Agreement, mentioning human mobility. Climate Resilience Strategy and Action Plan for Guyana (2015) mention human mobility in the context of climate risks for indigenous people as sea level rise and flooding can result in relocation of coastal and riverine communities, and for housing as community displacement can result from flooding in urban areas. Surinam adopted the Final National Climate Change Policy, Strategy and Action Plan 2014–2021 (2015) considering migration as a behavioural option of climate resilience action. Ecuador adopted climate plan and strategy before Paris Agreement, with no mention to human mobility.

---

[26]Colombia, Politica Nacional de Cambio Climático, 2017, http://www.minambiente.gov.co/images/cambioclimatico/pdf/Poli_CC_A2_B16_C6_WEB_Resumen_de_la_PNCC_dirigido_a_tomadores_de_decision.pdf; Paraguay, Politica Nacional de Cambio Climático, 2011, http://www.py.undp.org/content/dam/paraguay/docs/Politica%20Nacional%20CC.pdf; Perú, Estrategia Nacional ante el Cambio Climático, 2015, http://www.minam.gob.pe/wp-content/uploads/2015/09/ENCC-FINAL-250915-web.pdf and Plan de Acción de Adaptación y Mitigación frente al Cambio Climático, 2010, https://sinia.minam.gob.pe/documentos/plan-accion-adaptacion-mitigacion-frente-cambio-climatico; Uruguay, Plan Nacional de Respuesta al Cambio Climático, 2010, http://www.mvotma.gub.uy/politica-planes-y-proyectos/plan-nacional-de-respuesta-al-cambio-climatico/item/10009500-plan-nacional-de-respuesta-al-cambio-climatico; Chile, Plan de Acción Nacional de Cambio Climático, http://portal.mma.gob.cl/wp-content/uploads/2017/07/plan_nacional_climatico_2017_2.pdf.

Central America and Mexico

Costa Rica, Mexico and Nicaragua adopted new climate norms or revised preexistent norms after Paris Agreement.[27] Costa Rica National Policy for Climate Change Adaptation 2018–2030 mention human mobility in the context of human rights approach to adaptation measures and make reference to climate internal and cross-border displacement. It integrates adaptation to climate change and disaster risk management to reduce the exposition to extreme climate events. In 2018, Mexican Congress approved the reforms to its General Climate Law (2012), with several references to human mobility, specially related to internal displacement. In the formulation of climate national policy, State must respect human rights of migrants (art. 26). Internal displacement induced by climate change must be considered in the elaboration of adaptation policies, programmes, strategies and actions (art. 28, VII). The administrative authorities must prevent and address the possible internal displacement caused by climate events (art. 30, II), and relocate human settlements most vulnerable to the effects of climate change (art. 30, III). Nicaragua has adopted the National Policy of Mitigation and Adaptation to Climate Change and the National System of Response to Climate Change (2019). Human mobility is mentioned as a climate impact and displacement as a possible consequence of extreme climate events. There are references to planned relocation and evacuation in the context of the assistance to persons affected by extreme events. Complementarity of climate policy with disaster risk management is a principle.

Other States have climate norms adopted before Paris Agreement (as Guatemala and Honduras) and/or climate policies, strategies and plans post-2015 (as Guatemala and El Salvador).[28] Guatemala, in its climate law (2013) did not mention human mobility. The Climate Change National Action Plan (2016) has a human rights approach that considers the rights of migrants, and recognizes the impacts of extreme climate events in migration. Relocation and reinforcement of early warning systems to evacuate vulnerable populations are measures do cope with climate change in the

[27]Costa Rica, National Adaptation Policy to Climate Change 2018–2030, http://www.pgrweb.go.cr/DocsDescargar/Normas/No%20DE-41091/Version1/Politica_ADAPTACION_24_abril.pdf and Decreto Ejecutivo n. 41091-MINAE, http://extwprlegs1.fao.org/docs/pdf/cos178000.pdf; Mexico, Ley General de Cambio Climático, 6 June, 2012 (last review published on 13 July, 2018), http://www.diputados.gob.mx/LeyesBiblio/pdf/LGCC_130718.pdf; Nicaragua, Decreto Presidencial n. 07-2019, 1st February 2019, http://legislacion.asamblea.gob.ni/Normaweb.nsf/164aa15ba012e567062568a2005b564b/2aa845f404d355c6062583a0005a2819?OpenDocument.

[28]Guatemala, Ley Marco para regular la reducción de la vulnerabilidad, la adaptación obligatoria ante los efectos del cambio climático y la mitigación de gases de efecto invernadero - Decreto 7-2013, 5 September 2013, http://www.marn.gob.gt/Multimedios/2682.pdf and Plan de Accion Nacional de Cambio Climático, 2016, http://sgccc.org.gt/wp-content/uploads/2016/10/Plan-de-Acción-Nacional-de-Cambio-Climático-ver-oct-2016-aprobado-1.pdf; Honduras, Ley de Cambio Climático—Decreto n. 297-2013, 14 March 2014, https://observatoriop10.cepal.org/sites/default/files/documents/hn_-_ley_del_cambio_climatico_y_otros_decretos.pdf and Estrategia Nacional de Cambio Climático, 2010, http://extwprlegs1.fao.org/docs/pdf/hon148589.pdf; El Salvador, Plan Nacional de Cambio Climático, 2015 (revised in 2017), http://www.marn.gob.sv/download/Plan%20Nacional%20de%20Cambio%20Climático.pdf.

coastal zone. Honduras' climate law (2014) makes no reference to human mobility, but its National Climate Change Strategy and its Plan of Action (2010), recognize that climate change may affect migration patterns and provide measures to prevent, reduce and address temporary or permanent displacement. El Salvador has not adopted a specific climate norm, but its environmental law was reformed in 2012 to provide mechanisms to facilitate climate adaptation measures. In the same year it was approved the new environmental policy with the objective to reduce vulnerability to climate change. The Climate Change National Strategy was adopted in 2013. A National Climate Change Plan adopted in 2015 and revised in 2017, recognizes forced displacement as a result from climate slow-onset events. Panama has adopted a Climate Change National Policy in 2007, and an Action Plan for its implementation. A Management Strategie to the Mitigation and Adaptation to Climate Change was adopted in 2015. These documents do not mention human mobility.

Human mobility in the context of climate change is still not a key component of national climate agendas in Latin America, even if progress was made after the Paris Agreement. To reduce displacement risk, protect and assist those already displaced and help them to achieve durable solutions, countries must be in the driving seat. Therefore, data, regulation and national policies are fundamental. Informative data is the first step to demonstrate the increasing connection between climate change and human mobility, and promote a better consideration of this issue in political and normative strategies.

For a Human Mobility Approach in Climate Political and Normative Strategies: Conclusive Notes

Taking into account the current efforts on data and evidence of the impacts caused by climate change on human mobility in Latin America, although the nexus between climate change and human mobility is becoming clearer, the national climate policies and normative strategies in the region are still partial. It is possible to realize a greater coordination between human mobility and disaster risk management agendas. As displacement is increasingly one of the first consequences of climate change and disasters, policy and normative strategies should place this issue in the center of the national agenda. This input is complex and demands a series of normative and political articulations that consider the importance of the issue. The lack of information compromises communication about the topic in the region. Information about climate policies, normative strategies and initiatives is not organised and compiled, making difficult their identification and analysis.

Albeit the lack of specific mechanisms, the recommendations from the TFD and the GCM are crucial to reach satisfactory policy and normative strategies at the national level. TFD recommendations are more detailed on how to avert, minimize, and address the effects of climate change on displacement, and focused on sudden-onset displacement. GCM has a broader approach, including slow-onset disasters, and it encompasses all phases of the movement. Both documents interact and complement each other, they must be the main references to guide regional and national climate strategies in the integration of human mobility. The documents claim for the consideration of adapting existing national human mobility policies and legal frameworks to reflect the principles agreed on them.

A low coordination between climate change, disaster risk reduction and migration agendas still remains in the region. It is important to highlight the two initiatives in Central and South America that established guidelines for cross-border displacement by disasters as first steps in integrating these agendas and assisting States in the adoption and/or adequacy of policy and norms. Regional and subregional bodies are key players in strengthening the alignment of national strategies with international guidelines either because they act as a bridge between international and national levels and can stimulate States to strengthen their strategies through recommendations and seeking joint commitments from governments on climate change. It seems to persist a lack of articulation between the Missions of the States participating in these bodies and the institutions that deal with the issue nationally. The data gaps and inconsistency make more difficult to States to visualize and understand the phenomenon and the necessary measures to cope with it. Regional bodies can contribute to improve and harmonize data collection.

Among the national political strategies after the Paris Agreement, NAPs and NDCs are not sufficiently explored as climate instruments to consider the challenges and responses to climate-induced human mobility, notably as an adaptation strategy. Three NAPs from Latin America were found in the UNFCCC NAP Central, and they mention human mobility, especially forced migration, as a possible consequence of climate change. Only brazilian NAP considers migration as an adaptation measure. No concrete measures to prevent and respond to climate induced human mobility or to planned and safe migration to adapt to climate change is mentioned. NDCs seemed to be more developed, as from six Latin American States to refer to human mobility in their NDCs, four of them (Haiti, Mexico, Suriname and Uruguay) refer to human mobility as an adaptation strategy. Uruguay makes a special consideration of migrants as a vulnerable group to climate risk.

As to climate national specific norms, six States have adopted a climate norm after 2015, Mexico revised its climate norm from 2012 and Ecuador has incorporated climate issues in its Environmental Organic Code. Five of them mention human mobility (Costa Rica, Mexico, Nicaragua, Peru and Uruguay). Climate specific policies, plans and strategies were adopted after Paris Agreement, two of them mentioning human mobility (Guatemala and El Salvador). Other States have adopted climate policies, plans and strategies before Paris Agreement, three of them mentioning human mobility (Guyana, Honduras, Peru, Suriname and Uruguay). The references to human mobility are still generic, efforts were made to recognize migration and internal displacement as a consequence of climate change, but specific measures to promote migration as an adaptation strategy, to prevent and manage displacement, to recognize, protect and assist climate-induced displaced people are insufficient.

As recommendations to human mobility consideration in climate strategies, we propose:

Data and Knowledge Gaps

- Strengthen the production of data, information and evidence on human mobility in the context of climate change, both for sudden and slow events, to support policies and measures;
- Protocols for data collection, shared databases, inclusion of displacement data in disaster data collection and global and national indicators related to the Sendai Framework for Disaster Risk Reduction.

Alignment with International and Regional Agendas

- Reinforce the alignment of national legal frameworks to international and regional guidelines, identifying gaps and potential tools, through legal indicators [29];
- Develop, review, adapt and/or strengthen climate policy and normative strategies to reinforce/introduce considerations about human mobility in the context of climate change;
- Ensure that human mobility is considered at climate policies and norms as an adaptation and resilience strategy, integrate human mobility in adaptation plans and strategies and report existing practices and advancements in NDCs.

Regional, Sub-regional and National Recommendations

- Develop open and evolutive information platforms with compiled and analysed information about climate policy and normative strategies, actors and initiatives in the region;
- Make effective the normative and political strategies related to climate-induced human mobility, adopting specific regulations when necessary, and guaranteeing the institutional, legal and financial support;
- Develop human rights and community based approaches to climate policies, with the engagement of displaced people, communities at risk and migrants;
- Promote and support regional, sub-regional and national initiatives and agendas on climate change, disaster risk reduction and human mobility, encouraging state and non-state actors towards better responses, governance and cooperation;

[29]RESAMA has been working on indicators for monitoring and evaluating the implementation of international recommendations related to human mobility in the context of climate change and disasters at the national level. A set of legal indicators was developed to assess the level of alignment of national legal frameworks to international recommendations and targets, tested in Brazil.

- Promote capacity-building for regional and national stakeholders and other actors involved in normative and political processes, to qualify their actions, decisions and participation in climate and migration governance.

References

ACNUR (2014) Brazil declaration and plan of action. Adopted in Brasília, Dec 3rd. https://www.acnur.org/cartagena30/pt-br/declaracao-e-plano-de-acao-do-brasil/

Cantor DJ (2018) Cross-border displacement, climate change and disasters: Latin America and the Caribbean. Study prepared for UNHCR and PDD at the request of governments participating in the 2014 Brazil declaration and plan of action. Geneva: UNHCR-PDD. https://www.unhcr.org/protection/environment/5d4a7b737/cross-border-displacement-climate-change-disasters-latin-america-caribbean.html

CEPAL (2014) Los pueblos indígenas en América Latina: avances en el último decenio y retos pendientes para la garantía de sus derechos. Comisión Económica para América Latina y el Caribe, Santiago de Chile, p 128

CEPAL (2015) América Latina y el Caribe: una mirada al futuro desde los Objetivos de Desarrollo del Milenio. Informe regional de monitoreo de los Objetivos de Desarrollo del Milenio (ODM) en América Latina y el Caribe, 2015 (LC/G.2646). Santiago de Chile: Comisión Económica para América Latina y el Caribe, p 90

CRED (2017) Annual disaster statistical review 2016: the numbers and trends. In: Guha-Sapir D, Hoyois Ph, Wallemacq P, Below R (eds) Centre for Research on the Epidemiology of Disasters, Brussels, p 91

CRED (2018) Natural disasters 2017: lower mortality, higher cost. Université Catholique de Louvain (UCL), Brussels, p 8

CRED, UNISDR (2018) Economic losses, poverty & disasters 1998–2017. In: Wallemacq P, House R (eds) Université Catholique de Louvain (UCL), Brussels, Belgium, p 33

CRED (2019) 2018 review of disaster events. Université Catholique de Louvain (UCL), Brussels, p 6

CRM (2017) Protección para personas que se desplazan a través de fronteras en el contexto de desastres: Guía de Prácticas Eficaces para los Países Miembros de la CRM. Documento elaborado por la Iniciativa Nansen para los países miembros de la CRM. San José: Conferencia Regional sobre Migración, p 33. http://www.rcmvs.org/sites/default/files/publicaciones/es_guia_practicas_eficaces_crm_marzo_2017.pdf

CSM (2018) Lineamientos regionales en materia de protección y asistencia a personas desplazadas a través de fronteras y migrantes en países afectados por desastres de origen natural. Adoptado en Sucre en Noviembre 2018. Bolívia: Conferencia Suramericana sobre Migraciones, p 64. https://robuenosaires.iom.int/sites/default/files/publicaciones/CSM_Lineamientos_regionales_personas_desplazadas_por_desastres.pdf

GCM (2018) Final draft: global compact for safe, orderly and regular migration. New York, USA, 11 July 2018. https://refugeesmigrants.un.org/sites/default/files/180711_final_draft_0.pdf

IDMC (2015) Global estimates 2015: people displaced by disasters. Internal Displacement Monitoring Centre, Geneva, p 109

IDMC (2018a) Global report on internal displacement 2018—GRID 2018. Internal Displacement Monitoring Centre, Geneva, p 122

IDMC (2018b) No matter of choice: displacement in a changing climate—research agenda and call for partners. Internal Displacement Monitoring Centre, IDMC Thematic Series. Geneva, p 9

IDMC (2019a) Monitoring progress to reduce internal displacement: building a global internal displacement index. Thematic series beyond the numbers. Internal Displacement Monitoring Centre, Geneva, p 25

IDMC (2019b) Report on internal displacement 2019—GRID 2019. Internal Displacement Monitoring Centre, Geneva, p 146

IPCC (2014) Climate change 2014: impacts, adaptation, and vulnerability. Part B: regional aspects. Contribution of working group II to the fifth assessment report of the intergovernmental panel on climate change. In: Barros VR, Field CB, Dokken DJ, Mastrandrea MD, Mach KJ, Bilir TE, Chatterjee M, Ebi KL, Estrada YO, Genova RC, Girma B, Kissel ES, Levy AN, MacCracken S, Mastrandrea PR, White LL (eds) Cambridge University Press, Cambridge, United Kingdom and New York, NY, USA, p 688

IPCC (2018) Summary for policymakers. In: Masson-Delmotte V, Zhai P, Pörtner H-O, Roberts D, Skea J, Shukla PR, Pirani A, Moufouma-Okia W, Péan C, Pidcock R, Connors S, Matthews JBR, Chen Y, Zhou X, Gomis MI, Lonnoy E, Maycock T, Tignor M, Waterfield T (eds) Global warming of 1.5 °C. An IPCC special report on the impacts of global warming of 1.5 °C above pre-industrial levels and related global greenhouse gas emission pathways, in the context of strengthening the global response to the threat of climate change, sustainable development, and efforts to eradicate poverty, World Meteorological Organization, Geneva, Switzerland, p 32

IOM (2016). Migration in the INDCs/NDCs. Environmental migration portal. International organization for migration. http://environmentalmigration.iom.int/migration-indcsndcs

PDD (2016). Advisory committee 2016 workshop report. Platform on Disaster Displacement, Geneva, p 58

Ramos EP, Cavedon-Capdeville FS (2017) Regional responses to climate change and migration in Latin America. In: Mayer B, Crépeau F (eds) Research handbook on climate change, migration and the law. Edward Elgar Publishing, Cheltenham, pp 262–287

The Nansen Initiative (2015) Agenda for the protection of cross-border displaced persons in the context of disasters and climate change, vol 1. The Nansen Initiative, Geneva, p 56

UN News (2019) Extreme weather hit 60 million people in 2018, no part of the world spared, 24 Jan 2019. https://news.un.org/en/story/2019/01/1031182

UNDP (2016) Informe regional sobre desarrollo humano para América Latina y el Caribe—progreso multidimensional: bienestar más allá del ingresso. United Nations Development Programme, New York, p 359

UNFCCC (2010) Decision 1/CP. 16 the Cancun agreement: outcome of the work of the ad hoc working group on long-term cooperative action under the convention, Cancun

UNFCCC (2012) Report of the conference of the parties on its eighteenth session held in Doha from 26 November–8 December 2012: Addendum—part two: action taken by the conference of the parties at its eighteenth session, Doha. http://unfccc.int/resource/docs/2012/cop18/eng/08a01.pdf

UNFCCC (2015) Report of the conference of the parties on its twenty-first session: held in Paris from 30 November–13 December 2015, Paris. http://unfccc.int/resource/docs/2015/cop21/eng/10a01.pdf

UNFCCC/WIM-TFD (2018) Report of the task force on displacement. Geneva. 17 Sep 2018. https://unfccc.int/sites/default/files/resource/2018_TFD_report_17_Sep.pdf

UNGA (2016) International cooperation on humanitarian assistance in the field of natural disasters, from relief to development—report of the secretary-general. UN Doc. A/71/329, New York

UNGA (2017a) Resolution 71/280—modalities for the intergovernmental negotiations of the global compact for safe, orderly and regular migration. UN Doc. A/RES/71/280, New York

UNGA (2017b) Making migration work for all—report of the secretary-general. UN Doc. A/72/643, New York

Warner K, Kälin W, Martin S, Nassef Y, Lee S, Melde S, Chapuisat H E, Franck M, Afifi T (2014) Integration human mobility issues within national adaptation plans. UNU-EHS publication series—policy brief n. 9, June. United Nations University Institute for Environment and Human Security, Bonn

Wilkinson E, Kirbyshire M, Mayhew L, Batra P, Milan A (2016) Climate-induced migration and displacement: closing the policy gap. Briefing, Overseas Development Institute (ODI), London, p 12

World Bank (2018) Groundswell: preparing for internal climate migration. Kumari RK; de Sherbinin A, Jones B, Bergmann J, Clement V, Ober K, Schewe J, Adamo S, McCusker B, Heuser S, Midgley A (eds) World Bank, Washington DC, p 122. https://openknowledge.worldbank.org/handle/10986/29461

Yamamoto L, Serraglio DA, Cavedon-Capdeville FS (2018) Human mobility in the context of climate change and disasters: a South American approach. Int J Clim Chang Strat Manag 10(1):65–85

Coping with Extreme Weather in Arid Areas, a Case Study of Uzumba Maramba Pfungwe District, Zimbabwe

Juliet Gwenzi, Emmanuel Mashonjowa and Paramu L. Mafongoya

Abstract The agricultural sector in Zimbabwe is dominated by smallholder rainfed production systems that are characterised by low productivity due to their sensitivity to rainfall fluctuations among other factors such as farm infrastructure, low fertility and access to financial capital. Initiatives to reduce vulnerability at the local scale demand an understanding of the exposure, sensitivity and adaptive capacity of the community. In this study we profile the extreme weather events by analysing climatic events observed by smallholder communities through key informant interviews, focus group discussions and household surveys and analysing measured rainfall data. Further we outline the coping mechanisms in the district and the challenges impinging on the resilience of the farmers. The results show that droughts, dry spells and extreme hot days were the major climatic hazards. The hazards combined caused yield losses of more than 90% in the worst-case scenario. Nearly once in 10 years, did the farmers get the best yields compared to once in 5 years prior to 1990s. Farmers responded by adopting water harvesting strategies, conservation agriculture and crop diversity. However, these adaptation strategies need to be supported by good policies, microfinancing, crop diversity and infrastructure development.

Introduction

Developing countries have contributed less to the annual global greenhouse gas emissions, with Africa contributing 3.8% (Amadou 2015), yet they suffer the most from its effect, the real injustice of climate change and climate variability (Ullah et al. 2018). Climate change and increased climate variability presents considerable

J. Gwenzi (✉) · E. Mashonjowa
Physics Department, University of Zimbabwe, Harare, Zimbabwe
e-mail: julievimbai@gmail.com

E. Mashonjowa
e-mail: emashonjowa@gmail.com

P. L. Mafongoya
University of KwaZulu Natal, Durban, South Africa
e-mail: mafongoya@ukzn.ac.za

W. Leal Filho et al. (eds.), *Climate Change, Hazards and Adaptation Options*, Climate Change Management, https://doi.org/10.1007/978-3-030-37425-9_35

risks to communities now and into the future more so for developing countries. Risk of climate extremes has increased resulting in loses in livestock and rainfed agricultural production systems. Rainfed agriculture which is highly sensitive to weather accounts for 70% of food production in Africa (Smale et al. 2013). Much like most of Sub-Saharan Africa, the agricultural sector in Zimbabwe is dominated by smallholder rainfed production systems that are characterised by low productivity due to their sensitivity to extreme weather and climatic events, among other factors, with projected yield loses of 18% for Africa (Zinyengere et al. 2013) and at least 30% for Zimbabwe in particular (Schlenker and Lobell 2010). Seasonally farmers face new and unfamiliar conditions resulting in increasing vulnerability to climatic shocks (Brown et al. 2012). Maize is the most widely cultivated crop and one of the crops whose performance is strongly affected by extreme weather and climate extremes yet having direct effect on the livelihoods of people (Smale et al. 2013, Tesfaye et al. 2014). Extreme weather events are those rare events at a particular place and time of the year and when they occur, they serve as the basis for lessons about climate change (Spence et al. 2011).

Climate change and increased climate variability impacts are experienced in Zimbabwe with varying magnitudes especially in semi-arid areas such as Uzumba Maramba Pfungwe (UMP), a district to the northeast of Zimbabwe (Fig. 1). El Nino induced droughts, frequent and long dry spells, spike in high temperatures and unpredictable rainfall patterns are chief among the extreme weather events experienced in UMP (Kwenda 2014). Climate extremes pose substantial burden for households, businesses and governments and therefore coping to these events can no longer be

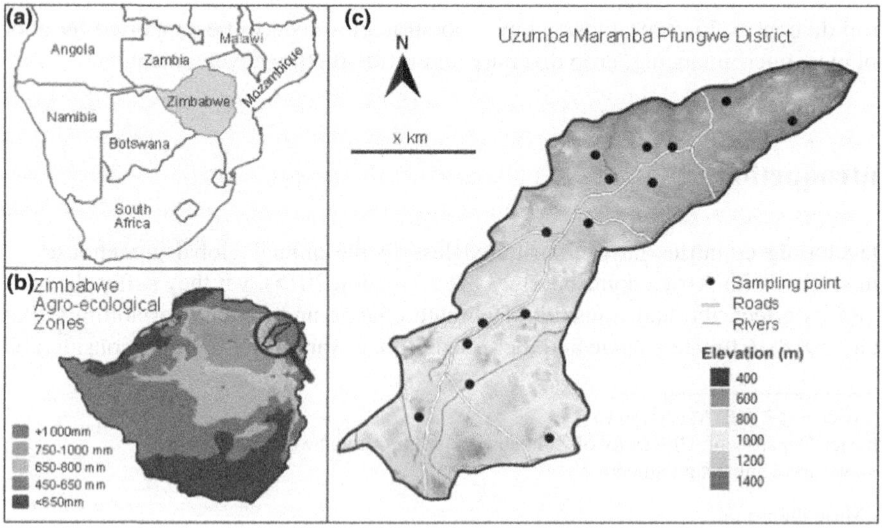

Fig. 1 Map of UMP district in Zimbabwe showing **a** the location of Zimbabwe in Southern Africa region, **b** the Agroecological regions in the country and **c** the variation of rainfall over the sampling points which were used in the project within UMP district

taken as a secondary issue. Although climate change will have significant impacts on food production systems, the risk on individual communities differs substantially. As a result, climatic extremes and their impacts must be understood and interpreted in relation to what smallholder farmers perceive as good or bad (Leonard et al. 2013). While climate science and model projections can only offer a rough guide for localised actions, communities define their responses to climate extremes from their shared experiences and understanding of local environment (Desai et al. 2009). Communities have responded to climate change and climate variability through a cocktail of self-driven options derived from changes in weather patterns, coping mechanisms and capacity to mitigate the impacts.

Community Response to Climate Change and Variability

In order to ensure food security, adapting the agricultural systems to climate change is critical. Identifying community response to climate change and understanding associated socioeconomic impacts at different spatial scales are important steps towards the development and implementation of appropriate adaptation measures (Tesfaye et al. 2014). Given the multiple actors and experiences, community level responses are based on capacity, potential for collective action and actionable climate information (Granderson 2014). The smallholder farmers in UMP have responded to the effects of climate shocks through staggered planting, mulching, use of drought tolerant crop varieties and crop diversification among others, however these methods have not been enough to stabilise crop production. Further options need to be explored and implemented in order to improve agricultural productivity in UMP. One such option is 'response farming', an approach in which farmers are helped to adapt crops to ongoing rainy season, using experiences of the past from interpretations of meteorological rainfall records and current weather and climate forecasts. Response farming approaches were introduced in Indonesia during climate field shops (Stigter 2011), however such approaches have not been tested in most of southern Africa.

This paper contributes to literature in two principal ways: first, it provides empirical evidence on the levels of extreme events experienced in UMP. Secondly, it demonstrates new approaches to coping with climate extremes which were illustrated through use of tailored weather forecasts in choosing suitable seed varieties for the area. Response agriculture approaches are fused into existing agricultural practices and social networks. Further the paper highlights the challenges faced by smallholder farmers which exacerbate their vulnerability to climate extremes underpinned by lack of resources. The paper highlights the importance of accounting for hybridity of climate risks which provides an opportunity for community wide dialogue and development of solutions using participatory approaches. The paper ends by pointing out policy gaps which if addressed would make provisions to support communities' coping mechanisms at local level thereby increasing resilience.

Study Area

The study was carried in Uzumba Maramba Pfungwe district (Fig. 1), a part of Mashonaland East province in Zimbabwe. The choice of the district was its unique spatial extent which covers three agro-ecological regions along its longitudinal extent. Therefore this would allow for comparison of the impacts of extreme events on dry and wet locations which are often exposed to the same strategies. UMP has 3 major divisions as its name suggest. Uzumba (upper UMP) is a high rainfall (>700 mm) area however the sand soils require investment in fertilizers and pesticides. Apart from cereals and legumes grown during the rainy season, horticulture is a characteristic of Uzumba with tomatoes and leaf vegetables being the main horticulture crops grown. Seasonal fruits (mangoes and sugar plums (*Uapaca kirkiana*)) are abundantly available and sold during the months of November to February. The main road to Uzumba is tarred making upper UMP accessible. Maramba is in the central part of UMP exhibiting growth point characteristics and therefore vending is the major economic activity at Mutawatawa business center. Rainfall received in Uzumba is often between 600 and 700 mm. The business center houses the administrative offices for the district and is linked to upper UMP by a tarred road and therefore most vegetables are supplied from Uzumba. Not much in terms of agricultural production occurs in Maramba, except on the peripheral areas. Pfungwe (lower UMP) is a low rainfall area (<600 mm) with loamy soils that require some management in order to support crop production. A gravel road connects Pfungwe to Maramba and therefore making it difficult to access. Cereals and legumes are grown in the area and additionally small livestock production. All the areas have experienced changes in agriculture, however the study focuses on upper and lower UMP where agricultural production is the major livelihood for smallholder farmers.

Methods and Data

The study used a mixture of approaches which combined quantitative and qualitative methods. Qualitative data was obtained through participatory tools such as household interviews (HHI), focus group discussions (FGDs) and key informant interviews. All interviews were conducted based on shared research principles and ethics (Bogner et al. 2009). The questions focused on extracting information on rainfall patterns, extreme events, their frequency and severity, coping strategies and challenges faced by the communities. Field tests were conducted from 2014/15 to 2016/17 season to validate community perceptions of extreme events, to identify suitable seed systems for upper and lower UMP, to identify the best strategies and to identify policy gaps for smallholder farmers. Automatic weather stations were installed at Mashambanhaka in upper UMP and Zunzanyika in lower UMP. Parameters measured from the automatic weather stations included rainfall, minimum and maximum temperature, wind speed and direction and humidity among others. Other simple rain gauges for rainfall measurement only, were supplied to farmers in several locations, represented as sampling points on Fig. 1. These rain gauges were used during farmer training on rainfall measurement and the data provided evidence of extreme events during the study period. The following sections provide details of the methods used in the study.

Participatory Rural Appraisals

The general questionnaire had four sections, one on rainfall patterns and changes, second on hazard occurrence and frequency, third on coping mechanisms and challenges and lastly on crop diversity and seed systems in the area. Two hundred (200) HHI were conducted, two at the beginning of the season and two at the end of the season in 2013/14 season. The rainfall season starts from October of one year to April of the following year. The FGDs comprised of equal numbers of men and woman aged over 25 years whose period of residence in the area exceeded 20 years. The age limit was necessary in order to obtain information from participants who were familiar with the local environment and having experience in agriculture. Households who participated in the survey were randomly chosen and the respondent would be any member of the household who met the criteria. Key informant interviews (Bernard 2011) were held with the district head, head of social services and amenities, livestock production department, head of agriculture in the district and supervisors at ward level were also included. Three village heads and the elderly who were considered to be the knowledge hub for the communities were purposively selected as key informants. FGDs and key informant interviews were repeated during the 2014/15 rainfall season both at the start and at the end of the season to validate the findings from the previous season.

Crop Diversity Trend Analysis

Having obtained information on farmers' perception on the character of extreme events, it was necessary to obtain evidence on the performance of seed systems in UMP versus farmers' response to the changes. One strategy used by farmers was crop diversity. Farmers listed all the crops grown in the area. An understanding of the crop diversity was completed using the four cell analysis/diversity wheel. The participatory four cell analysis was developed by Local Initiatives for Biodiversity, Research, and Development (LI-BIRD) and Bioversity International in Nepal. The four cell analysis helps communities understand on-farm diversity, richness, evenness, unique, endangered and rarity of varieties. The concept started with assessing varietal diversity within one species or one genus that thrive in UMP using two indicators. The varieties were categorized into four groups based on abundance and spread as follows:

(i) few farmers and small area
(ii) many farmers and small area
(iii) few farmers and large area
(iv) many farmers and large area.

Each of the listed crops was placed in one of the cells depending on how it satisfied the conditions. From the analysis it was possible to identify the varieties grown for food security, these were of high value and were grown by many farmers on large areas. Least performing crops were grown by few farmers on small areas for the purposes of having propagating seed. High values crop varieties were tested for suitability in UMP. The different varieties were planted on plots measuring 5 m × 5 m and given same treatments throughout the season until physiological maturity using conservation agriculture techniques. Weeding and fertilizer application were done on the same day on the different plots. Harvests were done from a net plot measuring 3 m × 3 m. Instead of prescribing which varieties to grow, a participatory approach was used to rank the varieties using different traits. Having listed as many traits as possible which included yield, heat tolerance, resistance to pests and diseases, taste, easy of processing among others, the farmers were asked to choose five traits out of the whole list, which they considered the most important and pairwise ranking was carried out. Low ranking was given a score of zero and highest ranking a score of 5. Intermediate rankings were given values between 1 and 5.

Analysis of Characteristics of Frequency and Severity of Extreme Events

In order to understand response farming approaches, the concept of Climate Field School (CFS) (Winarto and Stigter 2013, Winarto and Stigter 2015) was introduced to agricultural extension staff and the farmers. The CFS is a school without walls meant to increase farmers' knowledge on the climate and improve their response through action, make farmers more aware of how to use climate information for managing crops, soils and water resources for the maximum benefit while reducing climate related risks. Agricultural extension officers were trained to become intermediaries who help the farmers interpret weather information and to use the information in response farming. Farmers learnt relationships between weather and plant and the interaction between weather and pests and diseases. Through two CFS established, one in upper UMP (Mashambanhaka) and one in lower UMP (Zunzanyika), farmers participated in rainfall data collection while automatic weather stations in Uzumba and Pfungwe provided additional data on other parameters such as minimum and maximum temperatures. Weekly farm activities were planned based on rainfall measurements and tailored forecasts supplied to the farmers. There being no synoptic weather station in UMP remote sensed data (rainfall and temperature) from NASA POWER Agrometeorology website (http://power.larc.nasa.gov) was used to extend the time period under study. This data is derived in a location specific way by inputting of the GPS coordinates of each location to retrieve data. Analyses of rainfall and extreme event characteristics were achieved using the tool Instat 3.37. Data was analysed using Julian days, taking 1 July as day 1 and 30 June of the following year as day 365/6. The rainfall season in Zimbabwe overlaps between years starting from October of one year to April of the following year.

The famers indicated a change in the rainfall and temperature patterns with more rainfall in the period January to March and extreme heating periods from November. Therefore, data was analysed for start of season (SOS), end of season (EOS), length of growing period (LGP), wet and dry spells, number of rain days and extreme temperatures. The following definitions in (Stern et al. 2006) were used to characterise the events.

Definition of Terms

Start of the Rainy Season (SOS): At least 20 mm received in 1 or 2 consecutive days after 1 October after which there should not be a dry spell of more than 10 days in the next 30 days. (Stern et al. 2006). A dry day was defined as a day on which less than 0.95 mm was received.

End of the Rainy Season: A water balance method was used as defined in Instat+. In this calculation, a default soil of field capacity of 60 mm was used. Given that different crops have different actual evepotranspiration (ETa) values associated with them (Allen et al. 1998), therefore the first day after 1 February that had a water balance of less than or equal to 0 (zero) was defined as the end of season.

Length of Growing Period: The length of growing period (LGP) was defined as the number of calendar days between SOS and EOS.

Rainy day: A day that receives at least 4 mm

Heavy rainfall event was defined as one in which at least 50 mm of rainfall was received in 24 h and a wet spell period was defined as one in which the amount of rainfall exceeded the total rainfall for the month. A threshold of at least 35 °C was defined for extreme temperatures. The results were used to validate the perceptions of the farmers.

Results and Discussion

In total 200 HHI were held of which 61% (n = 122) were male and 39% (n = 78) were female. More men participated in the survey because it is cultural that a visitor in the home speaks to the head of the family if available in the home. Given the large amount of quantitative and qualitative data generated in the analyses, the details will be given by key findings relevant to the scope of this study. This section details the perception of communities on extreme events, crop suitability in UMP, challenges in coping to climate extremes and proposed policy gaps.

Levels of Extreme Events in UMP

Survey results indicated that farmers were familiar with changes in their environment and the effects on agricultural production. Observations included erratic nature in rainfall, change in start of season (SOS), end of season (EOS), length of growing period (LGP), extreme heating periods and severe dry spells. They indicated that rainfall had become more localised and differed from place to place. Rainfall data was analysed to determine changes highlighted by the community (Table 1).

The rainfall over UMP exhibited different characteristics. Famers indicated that the season srated in the last week of November, however during the period 2014–2017 the season started nearly halfway through into December (Table 2), however starting a bit earlier in Uzumba than Pfungwe. The period October to December is characterised by westerly cloud bands bringing rainfall over the country. The waves associated with the cloud bands have low influence over lower UMP. From January, the Inter Tropical Convergence Zone (ITCZ) is well established and influences rainfall over the entire country. The analysis seems to suggest that Lower UMP benefits more from the ITCZ than from upper westerly waves. Variation in seasonal rainfall from 2013/14 season is shown in Fig 2.

Table 1 Extreme event analysis during the 2014/15 Season

	Mashambanhaka	Mutawatawa	Zunzanyika
SOS (date)	13 December	12 December	11 December
End of main rains	10 March	12 March	2 March
Date of last rains	17 April	18 April	19 April
EOS (date)	20 March	14 March	2 March
LGP (days)	100	94	83
Dry spell periods	27 November–12 December 18–29 January 11 March–12 April	19–29 January 11 March–16 April	20–28 January 2 March–17 April
Wet spell periods	30 December–4 January (225.5 mm)	30 December–4 January (215 mm)	30 December–4 January (111 mm)
Number of rainy days	33	32	20
Rainfall > 50 mm (in 24 h)	20 December–55 mm, 30 December 68 mm, 4 January–72 mm, 4 February 52 mm, 15 February–62,5 mm, 28 February–63,5 mm	3 January–75 mm, 15 February–80 mm	1 January–50 mm, 4 January–48 mm, 1 February–(56 mm)
Seasonal total (mm)	755 (788.3–April)	672 (694 mm April)	493 (502.8 mm April)

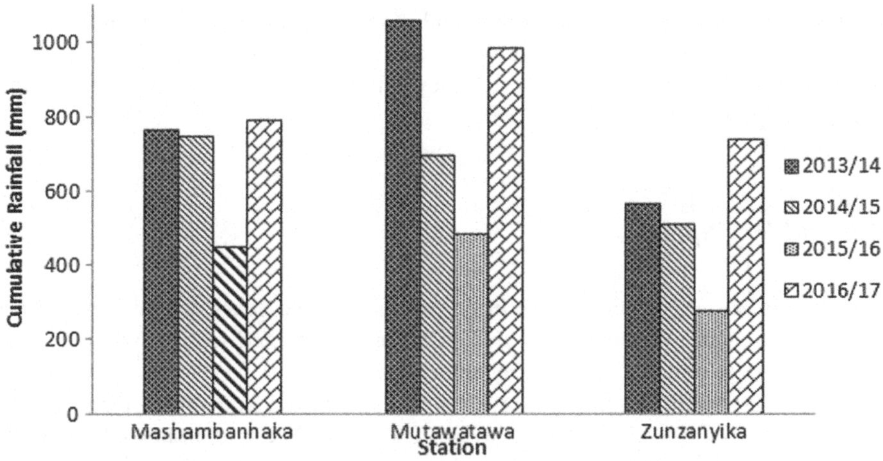

Fig. 2 Cumulative seasonal rainfall over upper and lower UMP from 2013/14 season–2016/17 season

Table 2 Pairwise ranking of crop varieties

Variety	Trait	Yield	Heat tolerance	Resistance to pests and diseases	Maturity	Taste	Total	Rank
SC403		4	4	2	3	3	16	4
PHB3253		5	4	3	3	2	17	3
SC301		5	4	2	5	2	19	1
SC513		5	3	3	4	3	18	2
Pan413		3	2	2	4	2	13	5

Rainfall analyses and observations from the seasons 2013/15–2016/17 revealed the complex nature of extreme events. False start to the season occurred in both 2014/15 and 2015/16 seasons when rains occurred between 22 and 26 November. This agreed with SOS prior to 2012 as suggested by farmers. Then rains were perceived to start by 27 November. Farmers indicated SOS had shifted by up to 3 weeks. In 2013/14 season SOS was between 5 and 12 December. In the season 2014/15 SOS occurred between 9 and 18 December. During the season 2015/16 SOS was 23 November for upper UMP and 17 Dec for central and lower UMP. The 2015/16 season came at the backdrop of an El Nino event. Similar observations were made during the 2016/17 season. The shift in SOS has a strong bearing on crop production as there is a 2 to 3 week delay in planting. The farmers highlighted the complexity to farming following the premature EOS. Prior to 2010 the EOS occurred in early April, however recent years saw a shift backwards by nearly 3 weeks to mid-March. Table 1 shows LGP of less than 110 days for the season 2014/15 season. The El Nino event in 2015/16 season resulted in total crop loss in the district. While medium to short season crop varieties were recommended and adopted, extreme events caused a breakdown in coping strategies developed over time. For example staggered planting and early planting have limits to which they can be used as experienced during the 2015/16 season.

Wet and Dry Spells

The farmers observed the increase in dry spells over the years, and at least 3 such spells occurred every season. An analysis of data obtained from UMP, indicated that long and severe dry spells were punctuated by short wet spells (Table 1). A mid-season dry spell, less than 10 days, normally occurred from end of December to early January as the westerly waves gave way to ITCZ. The results showed dry spells longer than 10 days which affected crops at the critical stage of flowering. Rainfall more than 50 mm (severe event) was experienced more in Uzumba (6 events) than in Maramba (2 events) and Pfungwe (3 events) during 2014/15 season. The number of rainy days reduced to an average of 29 days for the district, supporting the observation that wet days have become shorter, fewer and interspersed by long dry periods. This characteristic was most common during the 2015/16 season following the drought.

Temperature spikes were observed during the same periods causing high evaporative loses. Moisture was not lasting long and therefore negatively affecting crop production. Data from the automatic weather stations showed an average increase of 1, 5–2 °C during extreme heating periods more so in the months of October to January, with the high figures being experienced in lower UMP. Analysis of data from automatic weather stations gave an average of 33.4 °C for Uzumba and 35.1 °C for Pfungwe. However during the period 2014–2017, temperatures reached slightly more than 37 °C. Zimbabwe temperatures have altered by at 0.3–0.6 °C over the last century (ZIMSTAT 2014) while IPCC project a global mean surface temperature change of 0.3–0.7 °C for the period 2016–2035 (IPCC 2014).

During CFS sessions, farmers learnt the relationship between extreme temperatures with pests and diseases. During the dry periods, farmers scouted for pests and diseases. A common pest was aphids on legumes especially cow peas decimating production by at least 15%. Pests on livestock increased and this coupled with poor pastures, resulted in losses. Farmers took collective action of chemically controlling the pests. Finances often became the barrier to effective control of pests and diseases. Selling livestock didn't provide a reprieve to the farmers because the price offers for livestock were too low to allow recovery during the best years. In lean years, farmers' incomes were eroded as farmers spend more on buying food. Given the changes aggressive ways of stabilizing crop production and improving livestock health cannot be ignored.

Understanding Crop Suitability to Improve the Future Production

Farmers grew a diversity of crop varieties which included maize, sorghum, millets and legumes, both short and long season varieties. Some crop varieties had been lost along the years due to a variety of reasons which included unpredictable rainfall patterns as discussed, being long season varieties, lack of propagating material, pests and diseases and the promotion of hybrid seeds. Long season maize varieties included hickory king, long season hybrid seeds and some traditional maize varieties. The farmers argued that while these varieties were high yielding, the length of growing season had reduced significantly with few rainy days. The maize varieties had large grain size which was hard and therefore were not easily attacked by pests such as the large grain borer. The seed was available to a few farmers who grew the crops in their gardens. Traditional long season sorghum and millet varieties were abandoned due to high failure rate given erratic rainfall patterns. Eventually some of them went extinct due to lack of adoption and multiplication by seed companies. Production and processing of small grains was tedious and labour intensive and therefore many families preferred maize. All households (100%) preferred maize to small grains. Additional problems arose from quilea birds which would reduce harvests significantly. Long season varieties of legumes were also abandoned due to poor performance coupled with market prices which were not lucrative. Rice was often grown in swampy portions of the fields. Most of the wetlands had dried with time and therefore rice was now extinct in UMP. The farmers grew mostly hybrid seeds which they believed were tried and tested for different conditions. Average maize consumption in most Southern African countries is 100 kg/capita/year which represent at least 43% of total calories in Zimbabwe (Smale et al. 2011).

Medium to short maize and sorghum were the most common cereals grown on large pieces of land (0.5–1 Ha). The most common and preferred varieties included Seedco (SC513, SC403, SC301), Pannar seed (Pan413), and Pioneer seed (PHB3253). Farmers listed these as the ones they will definitely grow but needed to rank them basing on the most important traits. While famers in Pfungwe grow maize they are shifting to grow more of sorghum because of low rainfall experienced in lower UMP, while those in upper UMP prefer maize. Field experiments followed by pairwise ranking using identified traits produced results in Tables 2 and 3. In the initial year SC301 produced higher yield (3.12 ton/ha), SC513 produced 2.64 ton/ha while PHB3253 produced 2.52 ton/ha. Though SC301 had higher yield farmers finally ranged SC513 higher, indicating that they have other considerations they make apart from high yield. The small grain size coupled with proneness to larger grain borer made it a second best choice. Farmers preferred to grow it as a second crop because of its short growing period. While sorghum was grown in the area, it was not included in the study given the popularity of maize. None of the long season maize varieties was prioritised in the ranking. Using the seasonal and daily forecasts, farmers indicated they would grow equal areas of SC301 and SC513 but would reduce land size for PHB3253.

Table 3 Performance of selected maize varieties during the 2014/15 season

Maize variety	SC513	SC301	PHB3253
Physiological Maturity	105	89	120
Yield/25 m^2	6.6 kg	7.8 kg	6.3 kg
Number of cobs per plant	2	1	1
Grain size	Large	Small	Medium
Weakness	Prone to large grain borer	Attacked by large grain borer more than other varieties	Prone to larger grain borer Attacked by cut worm at germination
Ranking	1	2	3

Coping with Extreme Events

UMP is a challenging environment for agricultural production. Farmers have managed the unpredictable and harsh environment by using concepts of climate smart agriculture. Conservation agriculture (CA) a farming method developed and promoted by the Food and Agricultural Organisation was adopted by most farmers, together with use of short season and drought tolerant varieties, intercropping and intensive agriculture. At least 73% of the farmers used CA on pieces of land up to 0, 1 ha for both maize and sorghum. This method was labour intensive and therefore could not be used on large pieces of land. Farmers made use of weekly weather forecasts and advisories supplied to them through the lead farmers and responsible agricultural extension staff. The forecasts helped them make decisions on planting dates, weeding, fertilizer application and other farm activities. Farmers also used other moisture conservation techniques such as mulching and tied ridges. Given the short season, staggered planting was now a difficult option to pursue. Intercropping as well was now being practised by few farmers (16%) as it created a difficulty to maneuver in the fields. The farmers tested a combination of the techniques as a way of improving their copping strategies. The results are given in Table 4.

Uzumba (upper UMP) has more of sandy soils and require effective management of agronomic practices that promote high yields. Of the three seasons, 2016/17 had

Table 4 Comparison of yield (ton/ha) for different strategies in Uzumba (upper UMP)

Variety	2014/15 (CA)	2015/16 (CA), mulched plot	2015/16 (CA), mulched and water supplement	2016/17 (CA)
SC301 (ton/ha)	2.8	0.72	1.12	5.0
SC401 (ton/ha)	2.1	0.56	0.88	1.6
PHB3253 (ton/ha)	2.5	1.08	1.40	3.2

Table 5 Comparison of yield (ton/ha) for different strategies in Pfungwe (lower UMP)

Variety	2014/15 (CA)	2015/16 (CA), mulched plot	2015/16 (CA), mulched and water suppliment	2016/17 (CA)
SC301 (ton/ha)	3.12	0.05	0.06	12.2
SC513 (ton/ha)	2.60	0.06	0.0.8	13.1
PHB3253 (ton/ha)	2.50	0.10	1.13	12.4

the highest yields (Table 4) as the location received rainfall amount of nearly 800 mm. In the same season Pfungwe (lower UMP) surpassed its average rainfall (600 mm) and the yields were high (Table 5). However during the drought season of 2015/16 the poorest yields were obtained (<1 ton/ha in Uzumba and less than 0.1 ton/ha in Pfungwe). Combining the strategies improved the yields however differentiated according to location. Results also show the importance of rainfall in crop production. Where a cocktail of strategies were combined and water supplemented, yields increased. The results demonstrate that no one solution can account for all the stressors of climate variability. Systematic observations show that the spatial pattern and behaviour of extreme events go beyond the bounds of estimated natural variability.

Farmers indicted that other coping strategies suffer from the ripple effects of climate change and increased climate variability. In Uzumba, horticulture and sell of both indigenous and exotic fruits is common. In drought years and during dry spell periods, the quality of leafy vegetables was significantly reduced. Extreme heating periods caused high evaporative loses. Vegetables harvested during such conditions were often sour. Tomatoes were equally affected being small in size. Fewer fruits pushed prices up. Sales decreased during drought periods. In Pfungwe non-farm strategies such as brick making were adopted. Fresh water resources were limited and human wildlife conflict increased. Sale of livestock and small grains was common in 2015/16 season, however the prices were low. The road infrastructure was in a poor state and only the middle men would reach the farmers in Pfungwe and therefore demanded low prices. Livestock form the backbone of smallholder farmers together with crop production. Any sale should provide a platform for improved livelihoods. The position of farmers in dry areas such as Pfungwe, requires non-farm strategies which are backed by infrastructure development so that smallholder farmers reach lucrative markets.

The farmers have been trying to realign their farming systems to current climatic patterns, however the changes outpace the ability to cope given poor resources and limited scientific knowledge to back their observations. Significant strides can be made in improving agricultural productivity through a combination of strategies. No one size fits all, and therefore there is need for innovation and a continuous development of methods being used in agriculture (Leichenko and O'Brien 2002). A lot of time and energy is invested in agricultural activities. These activities provide food crops and horticultural crops provide income moreso over upper UMP where market gardening is one of the major livelihood systems. Maize and sorghum are commonly grown in UMP, however the demands of the farmers usually don't match prescribed solutions. Whilst SC301had high yield most of the time, the grain size was not appealing and its proneness to the large grain borer made it a lesser variety to SC513. Therefore all solutions require input and participation from the users rather than a top—down approach. The lack of support on non-farm coping strategies makes upper UMP more vulnerable during lean periods. Policy formulation should promote inclusiveness and address the plight of people in semi-arid areas. No previous study exists in detail on extreme events in UMP and how farmers are responding to the climatic changes, therefore development agents together with government have a role to play to improve the livelihoods of such communities through policies that support the provision of financial mechanisms that are pro poor.

Conclusion

The farmers' perceptions and observations of climate extremes in UMP agree with empirical data and other studies on farmers' perceptions within the Southern African region (Gwenzi et al. 2016, Obert and Mafongoya 2018). Extreme events affect smallholder farmer communities differentially in UMP. Crop yields were high in lower UMP when high and evenly distributed rainfall was received, however yields were low in upper UMP, the results being driven by soil types, rainfall received and heat units. No consistent rainfall pattern exists but fluctuations of climatic events are highly variable and of different magnitude across the district. Farming is an important livelihood system for the district, however, the complexity of climate change is bringing ever greater uncertainty and exposure to multiple climate stresses on Zimbabweans living in UMP. Droughts, dry spells, short LGP and pests and diseases are increasing in frequency and therefore eroding the adaptive capacity of farmers. Crops have been lost and only drought tolerant are being grown however minor varieties are still being grown in very small portions. Developing demand driven and bottom up adaptation strategies based on better weather and climate information leads to improved resilience and food security and reduction of vulnerability to climate shocks. This will require support from government to provide proper infrastructure. Strategies for coping with climate extremes cannot be a one size fits all but differ by location. Some areas may require options backed by provision of water sources (lower UMP), while some may require an improvement in soil structure (upper UMP). Small grains have been embraced by communities however marketing models tend to promote maize and therefore there is no incentive in growing small grains. Non-farm livelihood systems need urgent promotion backed by infrastructure development and financial mechanisms that are pro-poor.

Limitations

While we present results for 2 locations in UMP, the results cannot be generalised to every location in the district or province. The findings only suggest that extreme weather events affect farmers differentially and therefore solutions have to be developed at a local scale to me more effective. There is need to identify existing or potential pathways in a participatory manor and limitations to the success of the new options. Given the road infrastructure in UMP, some locations could hardly be reached. Whilst rain gauges were given to farmers in these areas, monitoring processes in replicated experiment plots was not possible and the data collected from the same areas had a lot of gaps to be used in the study. However, insights from crop suitability and diversification processes are a valuable foundation for the design of programs targeting improved crop productivity.

Way Forward

A dense rainfall network provides for the better resolution for monitoring microclimates. Increasing the number of meteorological data points for understanding the characteristics of extreme events and leveraging on assistance to access lucrative markets will increase the resilience of communities in semi-arid areas.

References

Allen RG, Pereira LS, Raes D, Smith M, Ab W (1998) Crop evapotranspiration guidelines for com puting crop water requirements-FAO irrigation and drainage paper 56. Irrigation and Drainage, 300

Amadou S (2015) Africa: financing adaptation and mitigation in the world's most vulnerable region. Africa Growth Initiative [Online]. [Accessed 23/06/2019]

Bernard RH (2011) Research methods in anthropology: qualitative and quantitative approaches, Oxford, AltaMira Press

Bogner A, Littig B, Minz W (2009) Introduction: expert interviews—an introduction to a new methodological debate. Interviewing experts. Palgrave Macmillan, UK

Brown D, Chanakira RR, Chatiza K, Dhliwayo M, Dodman D, Masiiwa M, Muchadenyika D, Mugabe P, Zvigadza S (2012) Climate change impacts, vulnerability and adaptation in Zimbabwe. In: Reid H (ed) Fisher S. International Institute for Environment and Development (IIED), London

Desai S, Hulme M, Lempert R, Pielke R (2009) Climate prediction: a limit to adaptation. In: Adger WN, Lorenzoni I, O'Brien KL (eds) Adapting to climate change: thresholds, values, governance. Cambridge University Press, Cambridge

Granderson AA (2014) Making sense of climate change risks and responses at the community level: a cultural-political lens. Clim Risk Manag 3:55–64

Gwenzi J, Mashonjowa E, Mafongoya PL, Rwasoka DT, Stigter K (2016) The use of indigenous knowledge systems for short and long range rainfall prediction and farmers' perceptions of science-based seasonal forecasts in Zimbabwe. Int J Clim Chang Strat Manag 8:440–462

IPCC (2014) Emergent risks and key vulnerabilities. In: B. FC, Barros VR, Dokken DJ, Mach KJ, Mastrandrea MD, Bilir TE, Chatterjee M, Ebik L, Estrada YO, Genova RC, Girma B, Kissel ES, Levy AN, MacCracken S, Mastrandrea PR, White LL (eds) Climate change 2014: impacts, adaptation and vulnerability. Part A: Global and sectoral aspects. Contribution of working group II to the fifth assessment report of the intergovernmental panel on climate change. Intergovernmental Panel on Climate Change

Kwenda B (2014) An Evaluation of the vulnerability of Uzumba Maramba Pfungwe (UMP) district, Zimbabwe to agrometeorological hazards and extreme events for improved farmer resilience. Master of science degree in agricultural meteorology, University of Zimbabwe

Leichenko RM, O'Brien KL (2002) The dynamics of rural vulnerability to global change: the case of Southern Africa. Mitig Adapt Strat Glob Chang 7:1–18

Leonard S, Parsons M, Olawsky K, Kofod F (2013) The role of culture and traditional knowledge in climate change adaptation: insights from East Kimberley, Australia. Glob Environ Chang 23:623–632

Obert J, Mafongoya PL (2018) Managing vulnerability to drought and enhancing smallholder farmers resilience to climate change risks in Zimbabwe. In: Filho WL (ed) Handbook of climate change resilience. Springer International Publishing AG

Schlenker W, Lobell DB (2010) Robust negative impacts of climate change on African agriculture. Environ Res Lett 5

Smale M, Byerlee D, Jayne T (2013) Maize revolutions in Sub-Saharan Africa. In: Otsuka K, Larson DF (eds) An African green revolution: finding ways to boost productivity on small farms. Springer, Dordrecht

Smale M, Byerlee D, Jayne TS (2011) Maize revolutions in Sub-Saharan Africa. Tegemeo Institute of Agricultural Policy, Nairobi, Kenya

Spence A, Poortinga W, Butler C, Pidgeon NF (2011) Perceptions of climate change and willingness to save energy related to flood experience. Natl Clim Chang 1:46–49

Stern R, Rijks D, Dale I, Knock J (2006) Instat climatic guide. Statistical Services Centre, The University of Reading, UK

Stigter CJ (2011) Rural response to climate change in poor countries: ethics, policies and scientific support systems in their agricultural environment. In: Sauer T, Norman J (eds) Sustaining soil productivity in response to global climate change—science, policy and ethics. Wiley Inc, Wisconsin, Iowa, USA

Tesfaye K, Gbegbelegbe S, Cairns JE, Shiferaw B, Prasanna BM, Sonder K, Boote K, Makumbi D, Robertson R (2014) Maize systems under climate change in sub-Saharan Africa: potential impacts on production and food security. Int J Clim Chang Strat Manag 7:247–271

Ullah W, Nihei T, Nafees M, Zaman R, Ali M (2018) Understanding climate change vulnerability, adaptation and risk perceptions at household level in Khyber Pakhtunkhwa, Pakistan. Int J Clim Chang Strat Manag 10:359–378

Winarto Y, Stigter K (2013) Science field shops to reduce climate vulnerabilities: an inter- and trans-disciplinary educational commitment. Collab Anthropol 4:419–441

Winarto YT, Stigter K (Year) Incremental learning and gradual changes: "science field shops" as an educational approach to coping better with climate change in agriculture… In: Promoting climate change awareness through environmental education, 2015. IGI Global, Hershey, PA, USA

ZIMSTAT (2014) Facts and figures—Zimbabwe. Gorvernment Printers, Harare

Zinyengere N, Crespo O, Hachigonta S (2013) Crop response to climate change in southern Africa: a comprehensive review. Glob Planet Chang 111:118–126

Climate Change Induced Soil Compaction: Evaluating the Adaptation Measures to Enhance Maize Yields in a Tropical Humid Acidic Soil, Nigeria

Chukwudi Nwaogu, Teowdroes Kassahun and Patrick U. S. Eneche

Abstract Compaction of agricultural soils is a great concern for farmers within the tropics as climate change and poor farm practices severely affect the soil and exacerbate food insecurity in the area. The study investigated the responses of soil nutrients and maize yields to soil compaction parameters (bulk density, surface penetration resistance, gravimetric, soil water content) and management (tillage, mulching, fertilizer) in acidic soil, in Nigeria. We hypothesized that climate change and tillage promote soil compaction, whereas mulching increases mineralization of organic carbon, N, P, and K contents which consequently elevates maize yields. The study comprises of seven treatments: mulch (M), fertilizer (F), tillage (T), mulch and fertilizer (MF), tillage and mulch (TM), tillage and fertilizer (TF), and unmanaged (U) under early and late growing seasons. The mineral fertilizer NPK 12:24:12 used were urea, calcium superphosphate, and potassium sulfate respectively. Data were collected and analyzed based on 4 replicates using appropriate methods. Results showed that in the prevailing weather status, tillage had the highest dry bulk density and lowest soil water content, whereas unmanaged plots recorded the highest resistance to root penetration. Joint treatment (such as tillage and fertilizer, tillage and mulching, and fertilizer and mulching) revealed the highest soil nutrients concentrations and maize yields, relative to unmanaged and tillage. The applied treatments significantly affected all the investigated soil nutrients in 2014 growing season. A thorough knowledge of the relationship between climate change, soil compaction and nutrient processes is fundamental in solving the problems of poor yields in Nigeria.

C. Nwaogu (✉) · T. Kassahun
Department of Ecology, Faculty of Environmental Sciences, Czech University of Life Sciences, Kamýcká, 129, CZ 165 00 Prague 6-Suchdol, Czech Republic
e-mail: cnwaogu@gmail.com

T. Kassahun
e-mail: teka@fzp.czu.cz

C. Nwaogu
Institute of Agroecosystems, Crops and Grassland Research Station, Liberec, Czech Republic

P. U. S. Eneche
Department of Geography and Environmental Studies, Faculty of Social Sciences, Kogi State University, Anyigba, Kogi State, Nigeria
e-mail: Enelche.psu@ksu.edu.ng

W. Leal Filho et al. (eds.), *Climate Change, Hazards and Adaptation Options*, Climate Change Management, https://doi.org/10.1007/978-3-030-37425-9_36

This study has demonstrated that improved soil fertility and food could be secured by mitigating the impacts of climate change induced soil compaction through the adoption of proper conservation agricultural practices in the study area.

Introduction

Climate change has been identified as the most serious environmental threat in the fight against starvation, malnutrition, disease and poverty in Africa, through its impact on soil and agricultural productivity (Ziervogel et al. 2006). However, records prove that climate change and its impacts are global, but the most adverse effects are suffered in developing countries, mainly those in Sub-Sahara Africa (SSA) because of poor resources for adaptation (Nwafor 2007; Jagtap 2007). Studies have predicted that crop yield in Africa may fall by 10–20% by 2050 or even up to 50% due to climate change (Jones and Thornton 2003), particularly African agriculture that is typically rain-fed and hence basically dependent on the unpredictability of weather which affects the soil.

Soil is an indispensable resource for agricultural production and sustainable food supply. In a rapid population growing region of Sub-Sahara Africa (SSA) such as Nigeria, soil degradation caused by climate change is a major problem which needs urgent attention in order to increase crop yields (Anikwe et al. 2003; Mamman et al. 2007). High soil nutrients depletion due to extreme weather and activities leading to soil compaction have been reported as a factor responsible for the poor yields of the cereal crops in the region (Anikwe et al. 2003). Several authors in the region have affirmed an increase in the annual soil nutrient loss per hectare due to poor management (Manyong et al. 2001; Sanchez 2002; Anikwe et al. 2003). And these have consequently led to decrease in crop yields (especially maize) from 3 t ha^{-1} to less than 1 t ha^{-1} in some areas (Sanginga et al. 1995).

As one of the main challenges, soil compaction is a soil structure degradation with increasing dry bulk density (BD) due to external or internal loading, causing poor root penetration, soil hydraulic conductivity and shoddy aeration (Chen 1999). Compaction-induced soil deterioration is a globally reported problem to over 65 million hectares of land (Flowers and Lal 1998). Soil compactness has severe effects on both soil properties and crops growth. For instance, Nawaz et al. (2013) reported that soil compaction impedes the germination and growth of many plants by altering the nitrogen and carbon cycles, and consequently decreased soil biodiversity, microbial biomass, enzyme activity and soil fauna. Dry BD, gravimetric soil water content, total porosity and surface resistance to root penetration have been studied as key parameters of soil compaction (Singh et al. 2015) and are significantly related to crop and soil management practices (Anikwe et al. 2003; Nyeki et al. 2017). The occurrence of soil compactness is crucial because both natural and human factors could be responsible for the densification of surface soil. For instance, extreme weather can cause excess soil moisture, leading to flooding, increased erosion or leaching. Similarly, limited precipitation could intensify temperature and elevate moisture evaporation which consequently creates hardpans and crusts on the soil surface. Acute intermittent wetting and drying cycles is a typical characteristic of the tropical climate, thus, the need for proper soil management.

Many studies on the human-induced soil compaction and its effects have focused on mechanized agricultural lands (Zhang et al. 2006; Singh et al. 2015), whereas minimal attention has been given to the non-mechanized agricultural systems (Anikwe et al. 2003). And these later systems are highly prevalent in most SSA countries especially in Nigeria. For example, soil management such as tillage, mulching and use of fertilizer impact on mass and energy movement in soils, could influence soil organic matter (SOM) and compactness. Mulching and fertilizer applications can tangibly mitigate the effects of nutrient leaching, by protecting the top soil and increasing SOM contents and crop yield. On the other hand, tillage might cause soil degradation by exposing the soil to erosion, moisture evaporation and compaction leading to poor root penetration and low yield. Fertilizer and mulching are widely applied globally to increase soil nutrients deficiencies, but in Nigeria like most SSA regions their usage remains virtually below average requirement with less than 7 kg ha^{-1} yr^{-1} on average (Vanlauwe 2005; Smaling 2006). This is because the cost of mineral fertilizers is above the income of many farmers (Sanchez 2002; Ogundijo et al. 2017), and crop residues are limited due to alternative use for various domestic needs (livestock feed, house roofing, craft work, herbal medicines) (Manyong et al. 2001). Degraded soil quality and unavailability of fertilizer coupled with climate change exacerbated the problems of poor crop yield and low food production. Climate change led to excessive soil erosion, leaching, evaporation, and acute soil compaction which were predominant features in the study area. The farmers traditional farming practices could neither enhance agricultural output nor improve the peoples' livelihood. In this context, we conducted this study as to help in addressing the threaten challenges of food insecurity and instable economy in this region which has more than 200 million people (Chikoye et al. 2000; Sanchez 2002). Conservation agriculture including mulching (Dube et al. 2012) was identified as one of the best management options for soil fertility enrichment, and this was adopted in this research. In this study, we aimed at appraising the responses of soil nutrients and maize yields to soil compaction, tillage, mulching, and fertilization. It is hypothesized that climate change and tillage promoted soil compaction, whereas mulching increased mineralization, Corg, N, P, K contents which consequently elevated maize yields. Within this context, we attempted to address the following research questions: (i) What are the effects of tillage, mulching and inorganic fertilization on the soil nutrients and maize yields? (ii) Does soil compaction and maize yields differ among the treatments? (iii) Does seasonality in climate have significant impact on soil properties and maize yields?

Materials and Methods

Study Site

The study was conducted in southern Nigeria in one of Obibi community farmlands located between lower Imo River watershed and Ogeochie River. Before the start of the experiment mixed soil samples (0–15 cm depth) were randomly collected and analyzed. The result showed that the site has the following soil properties: Organic carbon (Corg) 9.1 kg ha^{-1}, total nitrogen (Ntot) 1.2 kg ha^{-1}, P 82.4 mg ha^{-1}, K 73.5 mg ha^{-1}, Ca 96.1 mg ha^{-1}, Mg 65.8 mg ha^{-1}, pH 5.8, sandy 6.1%, silt 17.3%, and clay 21.5%. The experiment was performed under rain fed conditions and the mean monthly rainfall (cm) and temperature (°C) of the study site varied (Fig. 1) with 2012 representing dry year, 2014 normal rainfall year, and 2016 wet year. The area has no month without rainfall, but the rainfall varied, creating a form of bimodal rainfall distribution pattern. The early season rain starts February and ends in July, and we classified this as the early growing season. On the other hand, the late season rain commences in August and ends December (Chikoye et al. 2000).

This prompted two-cropping seasons per year for maize (Babatunde et al. 2012). The site is dominated with elephant-grasses, though it is surrounded by relics of rain-forest vegetation which have been deforested by several human activities (domestic firewood, timber, slash-burn).

Experimental Design and Treatments

Land preparation for the experiment took place in 2011 when the study site was entirely mowed, marked, and allowed to fallow till 2012 (the first experimental year). The experiments were performed in every even year during the experiment period, whereas the odd years marked the fallow periods following the tradition of the local farmers' in the region. The experiment was established on a completely randomized block design with four replicates (Fig. 2).

The field experiment was located on a gentle sloping (<2.3%) local farmer's field and it run for three even-years (with 6 growing seasons: early and late). There are two main experimental plots, one for early growing season and the other for late growing season. The experiment was 2520 m^2 (72 m × 35 m) in area with 8 m × 5 m per treatment plot, and separated by 2 m column buffer and 3 m wide row buffer which demarcated the replicates. The experimental plots included seven treatments: mulch (M), fertilizer (F), tillage (T), mulch and fertilizer (MF), tillage and mulch (TM), tillage and fertilizer (TF), and unmanaged (U) under early and late growing seasons.

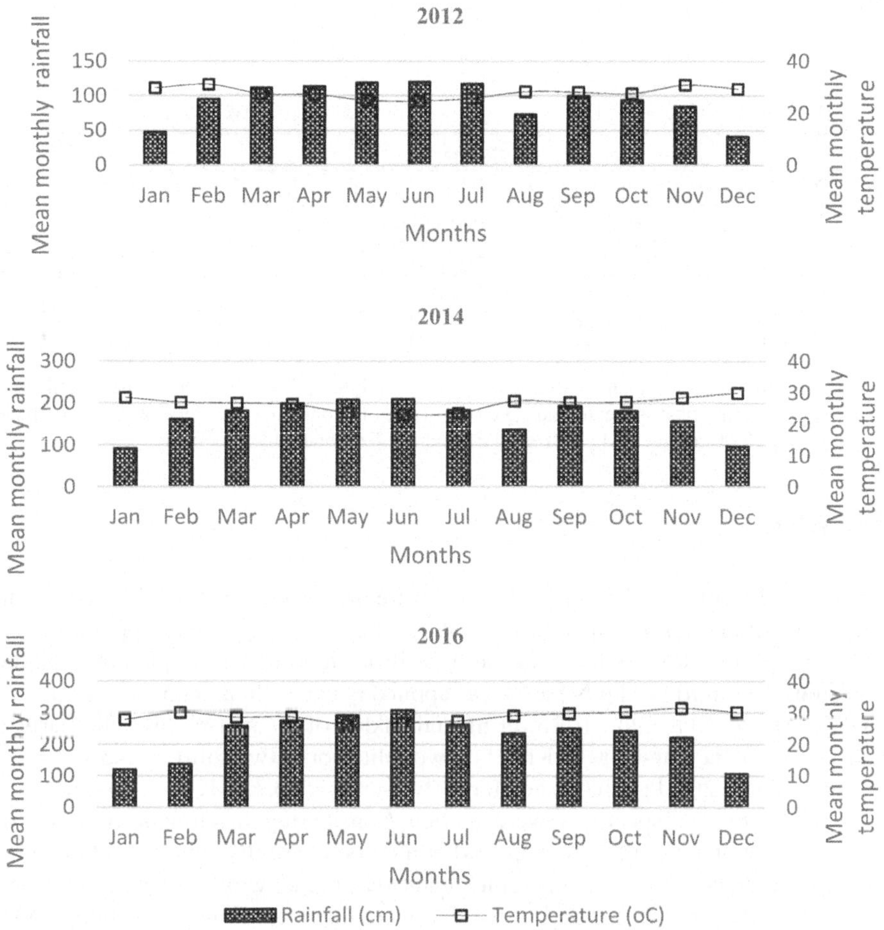

Fig. 1 Mean monthly rainfall and temperature seasonal variations in the study site/period

Mulching

The *P. purpureun* plants were cut using cutlasses (matches) at the height of 1 cm. The cut grass was crushed into smaller pieces using the hand-driven crop cutter machine (CE ISO; 9Z-10.0, Shanghai). Between 2.9 and 3.2 t ha^{-1} dry matter of the plant residues was applied 7 days after sowing the maize seeds. Second phase of mulching (though was not performed at the first growing season) involved the abandonment of about 10% of harvested maize straws and leaves on the soil surface, and these were in addition to the cut elephant-grass used as mulch materials for subsequent growing season. However, these would have served as organic fertilizer source though were not considered in this study as the rate was not substantial. *Mulching (M)* means 'only mulching treatment was applied' and no tillage.

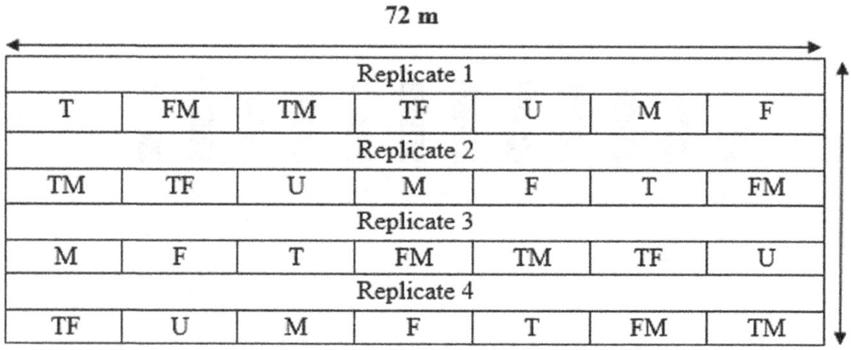

Fig. 2 Experimental design and layout of plots at the study site. Seven treatments and 4 replicates. The treatment abbreviations are: unmanaged (U), tillage (T); mulching (M); fertilization (F); tillage and mulching (TM); tillage and fertilizer (TF), and fertilizer and mulching (FM)

Fertilizer

The mineral fertilizer NPK 12:24:12 used were urea, calcium superphosphate, and potassium sulfate respectively. A total of 120 kg N ha^{-1} urea was applied in two splits: two-third (80 kg N ha^{-1}) as basal fertilizer dressing before planting and the remaining one-third (40 kg N ha^{-1}) was applied between 40 and 45 days after crop emergence (WAE) to support flowering and silking of the maize plant. Similarly, a total of 80 kg P ha^{-1} and 100 kg Kha^{-1} of two splits: for P (two-third = 53.3 kg P ha^{-1} and one-third = 26.7 kg P ha^{-1}) and for K (two-third = 66.7 kg K ha^{-1} and one-third = 33.3 kg K ha^{-1}) respectively were applied. Considering the climatic condition of the site, more quantity of N was applied at the first phase than the second phase of the experiment because: (i) temperature tends to be higher with decrease rainfall and as to avoid wasting the N away in the dry air (ii) greater N enhances more leaves than grains, thus to balance the development proportion between the leaves, stem and grain. All treatments received a uniform NPK fertilizer application rate. The fertilizer application was as recommended by the Ministry of Agriculture, Nigeria and IITA. *Fertilizer (F)* means that only fertilizer treatment was applied, no tillage and no mulch.

Tillage (T)

Conventional tillage method with less than 15% opening of the soil as a form of conservation agriculture was employed where locally hand-held hoes were used to dig and plough planting holes of 20–25 cm in diameter. On the other hand, at the plots where tillage was not applied, a pointed stick (dibble stick) was used to open the soil while 2 seeds were sown by dropping into the hole. *Tillage (T)* represented the application of only tillage treatment.

Fertilizer and Mulching (MF)

This treatment involved the combination of mineral fertilizer application and mulch following the procedures earlier stated. Maize was planted by using a pointed stick in rows. No tillage.

Tillage and Mulching (TM)

The conventional tillage was minimized only at the points where the seeds were to be sown. This was integrated with mulching by following the afore-specified steps.

Tillage and Fertilizer (TF)

After the conventional tilling of the top soil, fertilizer was applied before sowing.

Unmanaged (U)

These plots were never disturbed during the experimental period. Though, the plant biomass were cut and removed to ease planting. Neither tillage, fertilizer, nor mulching was applied before or after sowing the seeds. Maize was planted using a pointed stick (dibble stick) to open the soil in rows and no tillage was applied.

Data Collection and Laboratory Analyses

To analysis the dry BD, four soil core samples were randomly collected from 0 to 15 cm depth in each management plot. The samples were air-dried, visible biomass residues removed, ground in a mortar to pass a 2 mm sieve and used for analyses. Similarly, for the determination of gravimetric soil water content, four auger samples were also collected. The soil core samples were collected using 100 cm^3 cores. They were analyzed independently and mean results used. On the other hand, the auger samples were mixed, and a composite sub-sample was taken for analysis. Soil surface resistance to penetration measurements were also taken from four points in each plot using a pocket penetrometer with 5 cm in diameter 'blunt ended' tip (Model 06:01, Eijkelkamp Agrisearch). The resistance to penetration measurement was made to the depth of 5 cm at angle 90°. Measurements for dry bulk density, total soil porosity, penetration resistance and gravimetric soil water content were made at 45 and 90 days after sowing (DAS). Other soil properties analyzed were Corg, Ntot, P, K, and pH (KCl). Corg was determined by using the Walkely-Black wet oxidation method (Nelson and Sommers 1982). Ntot was measured using micro-Kjeldahl procedure (Bremner and Mulvaney 1982). Available P was determined using Bray II method as stated by Page et al. (1982). Soil pH (KCl) was determined by the glass electrode pH meter (McLean 1982). Available K concentrations were extracted using Mehlich III Solution (Mehlich 1984) regent and determined using ICP-OES. Ca and Mg concentrations were derived using Atomic Absorption Spectrometer. Furthermore, particle-size distribution was determined by the hydrometer method (Gee and Bauder 1986). Dry bulk density was determined by the core method (Blake and Hartge 1986). The soil total porosity was derived from the dry bulk density data as the fraction of total volume not occupied by soil considering a particle density of 2.65 Mg m^{-3}. Available water capacity was estimated using a pressure plate apparatus at 10 kPa (field capacity) and 1500 kPa (PWP-permanent wilting point) (Stolte 1997).

Maize Varieties and Yield Measurements

The varieties of maize were uniform for all plots. The seeds were of very high yield (VHY) and early maturity varieties with mixture of IITA Hybrid EEWH-21 (Ife Maize hyb-5), IITA Hybrid EEWH-26 (Ife Maize hyb-6), and the SAMMAZ 13 which were developed by IITA and CIMMYS funded by Bill and Melinda Gates foundation. The varieties have been extensively tested in Nigeria with assistance of Institute of Agricultural Research and Training (IAR and T) in support of the Drought Tolerant High and Fast Yielding Maize for Africa Project. The maize varieties take between 95-110 days to mature. The sowing was done in a spacing of 0.3 m within rows and 0.9 m between rows, and this was applicable at each treatment plot. At each point, two seeds were planted and 2 weeks after crop emergence (WAE) the seedlings were thinned to one per point. Weeds were manually cleared 3 times from all plots during each season. No artificial water supply was employed; it was rainfed. The kombat (2.5% Carbaryl) was used to treat the maize crop against maize stalk borer (*Busseola fusca*) at the rate of 2.5–3.5 kg ha^{-1}.

Twenty five maize plants were randomly selected, tagged and sampled from 4 m × 3 m (12 m^2) in the four center rows of each treatment plot to prevent the border effects. Maize stand height (cm) was also measured at 45, 90 and 110 DAS, whereas grain and above ground biomass yields were measured during harvest at physiological maturity (115 DAS). Grains and aboveground biomass (including stover, cobs, straw, stems and leaves) were collected. The weight of the samples was determined in the field using a digital scale while subsamples were collected to find grain moisture content.

Grain yield was computed on a per hectare basis at 12.5% moisture content. The grains were shelled manually from the cobs using hands and removed from stover. The biomass collected was oven dried at 75 °C till constant mass was reached to convert fresh biomass yields determined in the field to dry biomass.

Statistical Analyses

To evaluate the effects of treatment on soil nutrients concentrations and maize yields in years, a one-way ANOVA was applied, and followed by a post hoc comparison using Tukey HSD test which was used to measure the significant difference bet the means. In addition, a repeated measure ANOVA was performed to find the interactions and effects of treatments and season in relation to the soil nutrients. We also examined the how the maize yields correlate with the soil nutrients using linear regression analysis. Climate data was expressed using spread sheet. All analyses were conducted using the IBM SPSS Statistics Version 20 (IBM, NY, USA) (www.ibm-spss-statistics.soft32.com) and STATISTICA 13.0 software (Statsoft, Tulsa, OK, USA). All data were expressed as means of four replicates, and were examined for homogeneity of variance and normality using Levene and Shapiro-Wilk tests.

Limitations and Constraints of the Study

The work had some limitations and constraints including difficulties in accessing climatic data covering the whole study period, challenges of ensuring that the experimental sites were well secured from human and animal destructions, and finance needed to execute the entire study from experimental design to site establishment, data collection and analysis. Other constraint involved during the study was overcoming the rough terrain of the road leading to the experimental sites. Besides being too narrow for mobility, the roads were either muddy, bushy and/or sandy, thus making it difficult to access the research sites.

Results

Soil Nutrients

The effects of the different treatments on the concentrations of soil Corg, Ntot, P and K significantly differ in 2014 growing seasons (Tables 1 and 2), whereas no difference was recorded in 2016 growing season.

Table 1 Result of a one-way ANOVA showing mean soil concentrations of Corg, and Ntot

	Corg (kg ha^{-1})			Ntot (kg ha^{-1})		
Treatment	2012	2014	2016	2012	2014	2016
Early growing season						
U	8.8 ± 0.4a	11.9 ± 1.8a	10.2 ± 2.2	1.0 ± 0.2	1.1 ± 0.1a	0.9 ± 0.0
T	9.6 ± 0.7a	10.3 ± 3.3a	9.7 ± 0.7	1.2 ± 0.1	1.3 ± 0.1a	1.0 ± 0.1
M	15.9 ± 1.2b	19.7 ± 4.6c	11.4 ± 4.6	1.8 ± 0.2	2.2 ± 0.3b	1.1 ± 0.1
F	20.0 ± 3.2c	14.1 ± 1.9b	12.9 ± 1.1	1.7 ± 0.3	2.1 ± 0.2b	1.2 ± 0.1
TM	22.1 ± 0.6c	30.6 ± 2.3d	13.8 ± 3.5	1.8 ± 0.1	3.9 ± 0.5c	1.6 ± 0.3
TF	27.5 ± 4.1d	34.3 ± 0.8d	18.1 ± 5.1	1.9 ± 0.4	3.1 ± 0.8c	1.6 ± 0.2
FM	30.6 ± 2.7d	45.9 ± 6.5e	19.2 ± 8.2	2.0 ± 0.1	5.3 ± 0.7d	1.7 ± 0.3
P-value	0.041	0.002	ns	ns	0.017	ns
Late growing season						
U	10.0 ± 1.1	12.2 ± 0.8a	11.8 ± 0.3	1.0 ± 0.1	1.3 ± 0.4a	1.0 ± 0.0
T	12.7 ± 0.5	19.9 ± 2.3b	10.9 ± 2.2	1.2 ± 0.0	1.5 ± 0.1a	1.1 ± 0.0
M	11.9 ± 1.0	13.6 ± 3.7a	11.3 ± 0.9	1.3 ± 0.3	2.0 ± 0.1b	1.3 ± 0.1
F	12.4 ± 2.7	12.5 ± 4.1a	10.1 ± 1.5	1.1 ± 0.0	2.2 ± 0.2b	1.4 ± 0.0
TM	14.8 ± 0.9	21.1 ± 0.6b	13.5 ± 0.6	1.3 ± 0.2	3.0 ± 0.2c	1.1 ± 0.1
TF	15.4 ± 1.4	20.5 ± 1.5b	14.8 ± 3.8	1.2 ± 0.1	3.1 ± 0.2c	1.2 ± 0.1
FM	15.2 ± 2.1	34.6 ± 4.4c	15.3 ± 2.1	1.3 ± 0.0	4.4 ± 0.3d	1.3 ± 0.2
P-value	ns	0.023	ns	ns	0.014	ns

Numbers represent average of four replicates. ± is the standard error of mean. Significant differences ($P < 0.05$) among treatments in accordance with the Tukey's post hoc test is shown by different letters in column (a < b < c < d < e < f). ns denotes not significant. Treatment abbreviations (U, T, M, F, TM, TF, FM) are described in Fig. 2

The highest concentrations of the soil nutrients were found in 2014 under the combined treatment (TM, TF, and FM), while the unmanaged (U), and sole tillage treatment (T) had the lowest nutrient concentrations. Generally, the early growing season revealed the highest nutrient concentration relative to the late season. The concentrations of the soil nutrients in the different treatments were in the order of: FM > TF > TM > F = M > T > U.

The result of the repeated measurement ANOVA showed that Corg concentration was significantly affected by T × M, T × F, and F × M interactions (Table 3). All the investigated soil nutrients were significantly affected by mulch (M), fertilizer-mulch (F × M), and management-season interactions (Tables 3, and 4).

Table 2 Result of a one-way ANOVA showing mean soil concentrations of Phosphorus (P) and K

	P (mg ha^{-1})			K (mg ha^{-1})		
Treatment	2012	2014	2016	2012	2014	2016
Early growing season						
U	30.3 ± 6.1a	84.6 ± 11.4a	98.1 ± 16.9	45.9 ± 11.2a	81.7 ± 6.5a	86.2 ± 12.3
T	41.7 ± 9.4a	65.2 ± 9.2a	100.5 ± 10.1	52.7 ± 9.5a	99.9 ± 7.1a	91.7 ± 22.5
M	95.5 ± 21.7b	217.1 ± 40.0b	204 ± 18.3	125.3 ± 13.6b	236.5 ± 42.2b	110.9 ± 35.6
F	104.1 = 33.2b	231.5 ± 23.6b	109.3 ± 15.5	117.8 ± 23.7b	271.4 ± 31.3c	104.3 ± 17.6
TM	167.9 ± 28.1c	280.7 ± 21.8c	112.6 ± 22.6	151.9 ± 10.3c	297.1 ± 25.7c	112.1 ± 19.9
TF	185.2 ± 46.3 cd	296.6 ± 17.3c	115.5 ± 24.8	126.2 ± 8.8b	302.2 ± 46.4c	102.1 ± 26.1
FM	201.4 ± 35.9d	342.9 ± 29.5d	101.1 ± 37.1	216.5 ± 29.5d	364.9 ± 29.1d	108.6 ± 38.4
P-value	0.032	<0.001	ns	0.006	0.021	ns
Late growing season						
U	43.9 ± 9.2a	73.5 ± 6.8a	67.8 ± 13.5	41.0 ± 5.5a	96.2 ± 14.0a	57.8 ± 7.1
T	77.8 ± 12.6a	90.2 ± 11.1a	83.7 ± 9.2	98.5 ± 7.9a	128.9 ± 17.3a	96.3 ± 9.5
M	146.5 ± 25.3c	184.7 ± 33.9b	91.1 ± 18.6	204.1 ± 26.2b	221.6 ± 19.8b	100.5 ± 12.3
F	153.1 ± 18.8c	170.3 ± 27.0b	111.3 ± 22.6	251.8 ± 20.3c	268.8 ± 26.6c	85.7 ± 10.0
TM	107.1 ± 20.0b	222.6 ± 14.3c	106.2 ± 15.1	300.3 ± 21.1d	281.9 ± 31.4c	99.4 ± 13.5
TF	142.9 ± 23.7c	236.4 ± 31.1c	119.1 ± 20.7	271.5 ± 28.9c	317.0 ± 35.1 cd	103.5 ± 11.9
FM	188.4 ± 19.1d	271.8 ± 28.7d	125.5 ± 19.4	326.2 ± 30.0d	343.7 ± 32.6d	127.1 ± 16.5
P-value	0.035	<0.001	ns	0.041	<0.001	ns

Numbers represent average of four replicates. ± is the standard error of mean. Significant differences ($P < 0.05$) among treatments in accordance with the Tukey's post hoc test is shown by different letters in column (a < b < c < d < e < f). ns denotes not significant. Treatment abbreviations (U, T, M, F, TM, TF, FM) are described in Fig. 2

Table 3 Result of repeated measures ANOVA for management and season

	Corg (g kg^{-1})			Ntot (g kg^{-1})		
Treatment	2012	2014	2016	2012	2014	2016
Factor	*P*-value (f)	*P*-value (f)	*P*-value (f)	*P*-value (f)	*P*-value (f)	*P*-value (f)
T	ns (120.1)	<0.001 (21.7)	ns (70.6)	ns (92.3)	<0.001(14.1)	<0.001 (6.7)
M	<0.001 (75.4)	<0.001 (48.1)	<0.001 (9.9)	0.087 (66.7)	<0.001 (2.5)	0.042 (29.6)
F	ns (149.3)	0.009 (27.4)	ns (83.2)	0.011 (71.1)	ns (94.7)	ns (85.8)
T × M	0.002 (6.9)	<0.001 (2.8)	0.041 (10.6)	<0.001 (9.9)	ns (0.02)	ns (0.1)
T × F	0.018 (0.5)	<0.001 (4.3)	<0.001 (5.7)	ns (1.8)	ns (0.71)	<0.001 (3.5)
F × M	<0.001 (1.8)	0.035 (6.1)	<0.001 (2.2)	0.046 (0.1)	<0.001 (0.9)	0.013 (2.0)
T × M × F	ns (1.2)	<0.001 (4.7)	ns (0.5)	<0.001 (2.6)	ns (4.4)	<0.001 (1.3)
Treatment × season	<0.001 (3.3)	<0.001 (8.5)	0.027 (6.2)	<0.001 (4.4)	<0.001 (3.8)	0.03 (6.7)

P-value is the probability value ($P < 0.05$), f (in brackets) is the value derived from F-statistics. Treatment abbreviations are described in Fig. 1

Table 4 Result of repeated measures ANOVA for treatment and season

	P (mg kg^{-1})			K (mg kg^{-1})		
Treatment	2012	2014	2016	2012	2014	2016
Factor	*P*-value (f)	*P*-value (f)	*P*-value (f)	*P*-value (f)	*P*-value (f)	*P*-value (f)
T	0.011 (143.8)	<0.02 (129.5)	ns (136.5)	ns (9.2)	<0.001(0.6)	0.07 (1.9)
M	<0.001 (72.1)	<0.001(55.3)	<0.001 (18.1)	<0.001(4.7)	<0.001(11.1)	<0.001(7.1)
F	0.023 (161.2)	<0.001 (124)	0.031 (58.9)	<0.001(6.6)	0.001 (5.8)	0.710(9.2)
T × M	<0.001 (4.8)	0.028 (5.1)	<0.001 (0.2)	0.038 (8.4)	<0.001(2.6)	<0.001(2.1)
T × F	ns (7.7)	0.007 (2.6)	ns (0.6)	ns (5.7)	0.010(7.8)	0.216(0.7)
F × M	0.006 (1.4)	<0.001 (0.5)	<0.001(1.1)	0.004 (39.2)	<0.001 (0.1)	0.981(8.4)
T × M × F	ns (0.8)	<0.001 (2.7)	ns (2.4)	<0.001(13.5)	ns (7.2)	0.452(1.5)
Treatment × season	<0.001 (1.7)	0.032 (0.9)	<0.001 (0.8)	0.017(5.8)	<0.001 (5.9)	0.01(0.07)

P-value is the probability value ($p < 0.05$), f (in brackets) is the value derived from F statistics. Treatment abbreviations are described in Fig. 1

Soil Bulk Density (BD), Penetration Resistance and Water Content

At 45 and 90 DAS, the different treatments influenced the soil bulk density in the maize plots in 2012 and 2014 growing seasons, whereas 2016 showed no significant differences (Table 5). The highest soil dry BD was found in the T treatment. At 45 DAS in the 2014 growing season, the soil BD in the maize tillage plots decreased by 42%, and slightly increased by 2% in 2016. On the other hand, at 90 DAS, the soil dry BD in T treatment decreased by 41% in 2014 and later increased by 21% in 2016 growing season.

At 45 and 90 DAS, the surface soil resistance to maize roots penetration was significantly affected by the soil treatments in 2012 and 2014 growing seasons (Table 5). Though no statistical difference was recorded in 2016, yet it had high values of soil resistance to penetration. Lower measures of resistance to penetration were recorded in the tillage and mulch treatment plots (T, TM) relative to no-tillage and unmanaged plots. At 45 and 90 DAS in 2012 and 2014, the penetration resistance values of 1.11 and 1.32 kg cm^{-2} were recorded in tillage treatment, and these were more than 50% lower than the values for the unmanaged.

The gravimetric soil water content was significantly affected by the different treatments at 45 and 90 DAS in 2014 and 2016 seasons, whereas no significant difference in the treatments was found in 2012 (Table 5). Higher gravimetric soil water content was obtained in the no-tillage (U, F, M, FM) treatment plots relative to tillage (T, TM, TF) plots. The gravimetric soil water content rapidly increased in the unmanaged treatment plots by 58 and 93% between 2012 and 2016 at 45 and 90 DAS respectively. On the other hand, the results for the tillage plots showed less than 5% increase in the gravimetric soil water content at 45 and 90 DAS.

Maize Plant Heights

The highest mean height of the plant (264.2 cm) was recorded in 2014 growing season under TF treatments, whereas the lowest was 57.0 cm in 2016 under U treatment (Fig. 3). In general, the combined management treatments had higher crops growth relative to U and singly managed treatments (F, M, T). Therefore, the plant growth height in the different treatments was in the order of: FM > TF > TM > F > M > T > U.

Maize Yields and Soil Nutrients

The treatments significantly affected the maize grain and biomass yields in 2012 and 2014 growing seasons (Fig. 4). The treatments showed no significant difference in the

Table 5 Mean effects of treatment practices on soil dry density (Mg m^{-3}), surface soil resistance to penetration (kg cm^{-2}), and gravimetric soil water content (kg kg^{-1}) in maize plots

	DAS	45			90		
Treatment	Year	2012	2014	2016	2012	2014	2016
Soil bulk density $(Mg\ m^{-3})$							
U		1.27	0.97	1.22	1.36	1.2	1.31
T		1.43	1.25	1.26	1.49	1.29	1.35
M		1.19	1.08	1.21	1.28	1.16	1.3
F		1.23	1.17	1.23	1.32	1.22	1.31
TM		1.38	1.24	1.25	1.41	1.3	1.32
TF		1.4	1.21	1.26	1.54	1.31	1.34
FM		1.26	0.88	1.24	1.4	1.25	1.3
F-ratio		1.02	1.1	0.99	0.21	0.72	0.46
P-value		<0.001	0.031	ns	0.02	0.007	ns
Surface soil resistance to penetration $(kg\ cm^{-2})$							
U		2.38	2.14	1.53	2.44	1.89	1.72
T		1.11	0.86	1.57	1.32	0.75	1.65
M		1.3	1.02	1.55	1.09	1.04	1.57
F		1.45	0.75	1.59	1.36	1.17	1.55
TM		1.73	0.98	1.54	2.58	1.59	1.69
TF		1.81	1.07	1.51	1.95	1.21	1.74
FM		1.27	1.01	1.42	2.61	0.96	1.83
F-ratio		0.81	0.11	0.65	1.89	0.47	1.01
P-value		0.039	0.442	ns	< 0.001	0.019	ns
Gravimetric soil water content $(kg\ kg^{-1})$							
U		19.4	20.2	26.7	20.3	26.1	39.1
T		19.1	16.4	19.9	19.7	19.4	23.1
M		20.5	24.3	31.2	20.8	31.1	40.7
F		20.7	28.9	33.5	20.9	22.2	39.5
TM		19.9	20.1	27.7	19.1	20.6	35.3
TF		18.9	21.6	25.8	18.3	18.8	36.6
FM		20.5	23.9	29.2	19.7	24.1	38.5
F-ratio		3.1	5.4	8.7	1.1	1.6	9.3
P-value		ns	0.016	<0.001	ns	0.034	< 0.001

Days after sowing (DAS) represents days after sowing. ns indicates 'not significant'. *F*-ratio is the value derived from *F*-statistics. *P*-value is the probability value ($P < 0.05$). Treatment abbreviations (U, T, M, F, TM, TF, FM) are described in Fig. 2

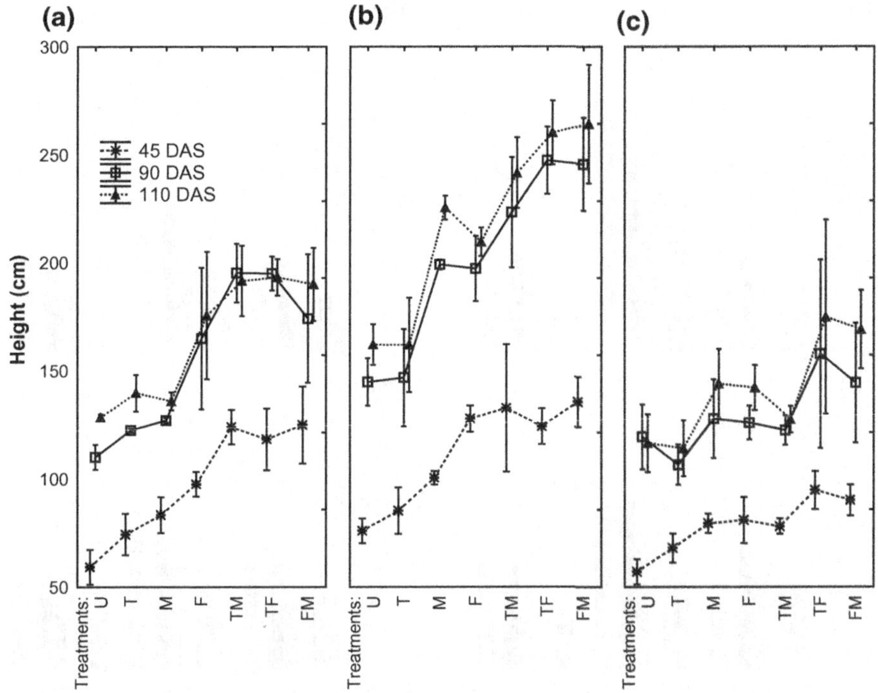

Fig. 3 Mean height of maize plants (n = 25) at 45, 90, and 110 days after sowing (DAS) in the different treatments. **a** 2012 growing season, **b** 2014 growing season, **c** 2016 growing season. Error bars represent standard error of the mean (SEM); treatment abbreviations (T, M, F, TM, TF, FM, U) are described in Fig. 2

yields was found in 2016. The highest yields were dictated in the joint management (TM, TF, FM) plots relative to the unmanaged and single management treatment (T, M, F) plots. In terms of seasonal yields variations, early growing season of 2014 recorded the highest yields. Cumulatively, the grain yields in the different treatments were more than 20% higher in 2014 season compared to the values for 2012 and 2016 seasons (Fig. 3a, b). The maize DM biomass in 2014 was at least 14% higher than the values for 2012 and 2016 growing seasons (Fig. 3c, d). The yields in the different treatment were in the order of: FM > TM > TF > F = M > T = U.

The soil concentrations of Corg, Ntot, and P were significantly positive with maize yields (Fig. 5). The relationship between C: N ratio and maize yields were significantly negative, whereas no significant relation was recorded between yields and K concentration. Furthermore, a significantly negative correlation was obtained between maize yields and soil dry bulk density at 45 and 90 DAS (Table 6). On the other hand, the measures of penetration resistance and soil water content were not significantly correlated with the maize yields.

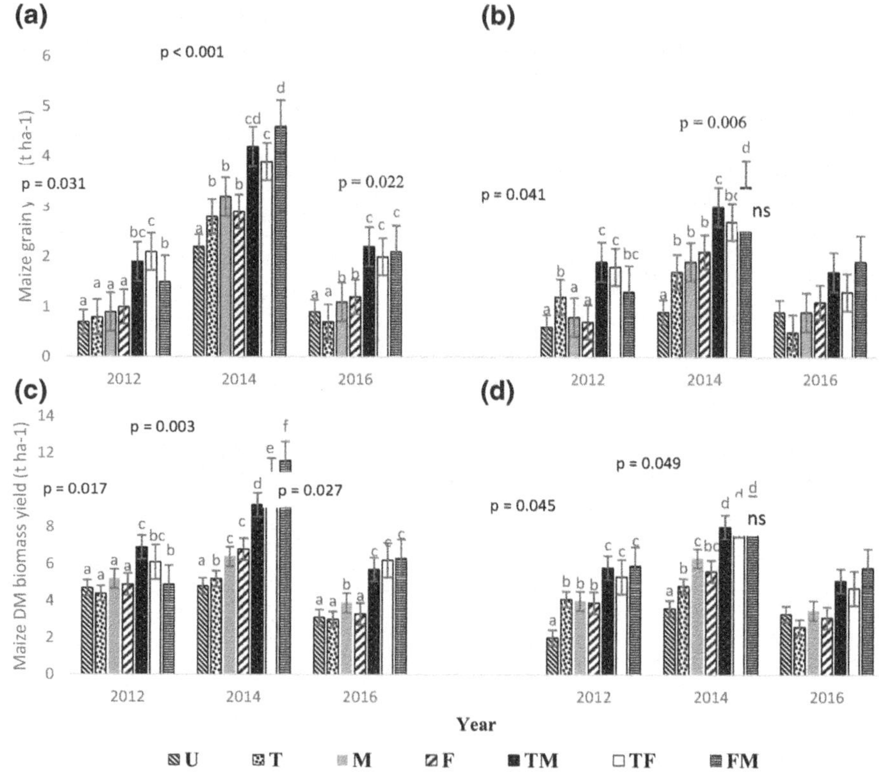

Fig. 4 Mean maize yields, **a** grain yield for early growing season, **b** grain yield for late growing season, **c** biomass yield for early growing season, **d** biomass yield for late growing season in different treatments. p-value represents corresponding probability value. Error bars represent standard error of the mean (SEM); significance differences ($P < 0.05$) between treatments in accordance with the Tukey's post hoc test is shown by lowercase letters (a < b < c < d < e < f). ns denotes not significant. Treatment abbreviations (U, T, M, F, TM, TF, FM) are described in Fig. 2

Discussion

Soil Nutrients

The normal rainfall year (2014 growing season) revealed significant different effects on the concentrations of soil Corg, Ntot, P, and K in the applied treatments. It also recorded the highest concentrations of the soil nutrients relative to either the dry year (2012) or the wet year (2016). This was attributed to the optimal and evenly distributed rainfall which occurred in 2014 contrary to the 2012 dry year, and 2016 wet year (Fig. 1). This result was consistent with other findings in the region where heavy rainfall was reported to have caused decrease in top soil nutrients by severe erosion (Iwara 2014), and high infiltration (Kayombo et al. 1991). Ojo and Olufolaji (2009)

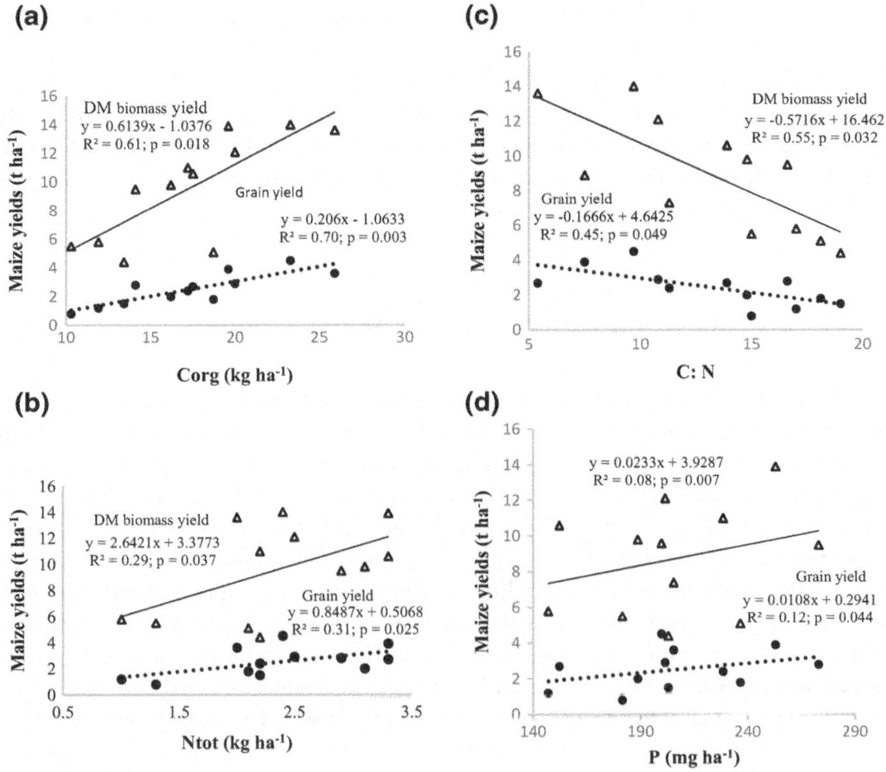

Fig. 5 Relationships among maize yields and soil nutrients, **a** organic carbon (Corg), **b** total nitrogen (Ntot), **c** C: N ratio, and **d** phosphorous (P) in all applied treatments in the study years

Table 6 Result of the relationship between maize yields as dependent variables and soil compatibility indices as independent variables

Soil compatibility indices	Maize grain yield		Maize DM biomass yield	
	Corr. (R^2)	*P*-value	Corr. (R^2)	*P*-value
Dry bulk density (Mg m^{-3}) at 45 DAS	0.63	0.021	0.67	0.019
Dry bulk density (Mg m^{-3}) at 90 DAS	0.75	0.007	0.81	0.034
Penetration resistance (kg cm^{-2}) at 45 DAS	0.04	ns	0.37	ns
Penetration resistance (kg cm^{-2}) at 90 DAS	0.18	ns	0.21	ns
Gravimetric water content (kg kg^{-1}) at 45 DAS	0.52	ns	0.46	ns
Gravimetric water content (kg kg^{-1}) at 90 DAS	0.49	ns	0.33	ns

DAS denotes days after sowing. DM denoted dry matter. ns represents not significant ($P < 0.05$)

in the same region also elucidated that optimum soil moisture promotes nutrients enrichment, while soil moisture-stressed conditions impede nutrients availability.

The integration of the treatments (TM, TF and FM) increased the soil nutrients relative to the U or T treatment. This was because mulching reduces soil sediments loss by erosion and leaching, fertilization adds nutrients, while combination of tillage enhances soil biota. Thus, incorporating the treatments improves the soil nutrients (Wang et al. 2011).

Soil BD, Penetration Resistance and Water Content

At 45 and 90 DAS, the soil BD was significantly affected by the different management treatments in 2014 and 2012 but not in 2016 growing season. Heavy storm which caused intermittent flooding and making it difficult for variations to occur in the managed plots was responsible for no differences in the 2016. The T treatment recorded the highest soil bulk density at 45 and 90 DAS. This was because tillage (if applied alone) removed the soil vegetative cover and exposed the soil to external factors (raindrops, solar heat, trampling) leading to higher compaction and increased BD. This finding was coherent with the report by Anikwe et al. (2003) who concluded that maize tillage plots have higher soil dry BD relative to the 'no-tillage' plots in the South-Eastern Nigeria. Several authors have found elevated soil BD in the tillage system than in the no-tillage (da Silva et al. 1997). The 2014 growing season recorded decreased soil BD when compared with the other seasons possibly because the climatic conditions were favorable for the soil microbes (Abatenh et al. 2018). Thus, cases of soil moisture stresses, soil thawing and votilization never occurred.

At 45 and 90 DAS, the surface soil resistance to maize roots penetration was not significantly influenced by treatment in 2016 season, though the season have high values of soil resistance to penetration. The masking (or coverage) of the field variations by the soil water content of the acidic soil, and the extreme rainstorm impeded the chances of observing any differences in the treatments during the wet year (Nawaz et al. 2013). The T and M treatments showed relatively low surface soil resistance to penetration when compared with the U treatment. This was because tillage and mulching opens the soil-pore spaces for air, decreases evaporation, promotes rapid decomposition which increases soil nutrient (Sainju et al. 2005). This finding is supported by the work of Ehlers et al. (1983) who revealed that in conventionally tilled soil, root growth was constrained at a penetration resistance of 3.6 MPa whereas, the comparable limit in no-tilled soil was about 2–4 MPa higher. Another study in Western Nigeria found that mulching decreased bulk density from 1.17 to 0.98 Mg m^{-3}, and penetration resistance from 1.54 to 1.07 kg cm^{-2} (Lal 2000).

The gravimetric soil water content was significantly affected by the different treatment at 45 and 90 DAS in 2014 and 2016 growing seasons. However, no significant difference was found in the 2012 growing season because 2012 was dry year and soil water content was substantially influenced by limited amount of precipitation (Meilke and Weihelm 1999). Higher gravimetric soil water content was achieved in the no-tillage plots relative to the tillage plots. Tillage promotes proper aeration of soil, and allows the plants roots to penetrate and absorb excess water to be released in photosynthetic and respiration processes. In the central great plains of USA, the work of Lyon et al. (1999) concurred that water storage was higher in the no-tillage treatment relative to the tillage.

Maize Plant Heights

Maize crop was more than one-fifth taller in 2014 growing season and under the combined treatments (TM, TF and FM) relative to the height found in 2012 and 2016 under the U treatment. Variations in rainfall and soil compactness might be responsible for this difference. This result was coherent with the work by Mari et al (2006) who concluded that maize plant growth was better in slightly compacted soil relative to heavily compacted soils.

Treatments, Yields and Soil Nutrients

The different treatments significantly affected maize yields in 2012 and 2014 growing seasons, whereas no difference was observed in the 2016 season. This was attributed to the extremely high and unevenly distributed rainfall in 2016. The storm in the wet year restrained soil nutrient enrichment, caused low mineralization, and poor soil microbial activities which consequently led to poor soil fertility and low yields. Higher yields were found in the integrated management systems (TM, TF, FM) compared to the unmanaged. This result was consistent with the work of Tao et al (2015) who found mulching and subsoil tillage more beneficial for maize production than either only mulching or only tillage. Combined soil management increases nutrients. enhances biological activities and improves soil water retention capacity (Upawasa 1997; Anikwe et al. 2003; Mo et al. 2017; Abatenh et al. 2018). Contrary to our finding, the result from the ferralitic soil in Southern Benin concluded that the combination of fertilizer and mulching never produced higher maize yield (Saidou et al. 2003). Dissimilarity in soil properties might be the reason for these differences since the present study was performed in an acidic soil.

Year 2014 growing season had relatively higher maize yields because it recorded optimal rainfall with higher Corg, Ntot and P concentrations, whereas the dry and wet years (that is 2012 and 2016 respectively) had unfavorable climate with lower soil nutrients which led to poor maize growth and low yields. Besides having low nutrients, extreme weather conditions cause soil compaction which changes the morphology, biological and chemical compositions of plant root systems (Taylor and Brar 1991). C:N ratio was negatively related with maize yields because higher C: N ratio means low mineralization which in turn inhibits the soil microbial activities and nutrients availability. Maize yields were negatively correlated with soil BD in our study because increased soil BD is associated with higher soil compaction, poor water and air permeability, conductivity and diffusivity (Greenland 1977), creating greater root penetration resistance (Zhang et al. 2006), and low organic matter (Verbist et al. 2007). As the soil type in our study, sandy loam had 34% maize yield reduction in dry season (Ijoyah et al. 2012), and yield decreases in seasons with excessive and poorly distributed rain due to waterlogging, root decaying and soil compaction (Nyeki et al. 2017; Masvaya et al. 2017).

Conclusion

Maize yields were significantly correlated with Corg, Ntot, P, and C: N ratio. At 45 and 90 DAS, the different treatments adopted affected the soil BD, resistance to penetration, and gravimetric soil water content in the normal rainfall year, 2014. Tillage treatment had the highest dry BD and the lowest gravimetric soil water content, whereas unmanaged plots recorded the highest resistance to root penetration. Joint treatment (TF, TM, FM) had the highest soil nutrients concentrations and maize yields, relative to U and T treatments. The treatments significantly affected all the investigated soil nutrients in 2014 growing season. The wet year, 2016 recorded the lowest maize yields relative to the normal rainfall year, 2014. The concentrations of the soil nutrients, the plant growth height, and the maize yields in the different treatments were in the order of: FM > TF > TM > F = M > T > U, FM > TF > TM > F > M > T > U, and FM > TM > TF > F = M > T = U respectively. Climate change had substantial impacts on the soil and consequently agricultural production. For instance, the dry year 2012 and the wet year 2016 had the lower maize yields when compared with the normal rainfall year 2014. A thorough knowledge of the relationship between climate change, soil compaction and nutrient processes is fundamental in solving the problems of poor yields in Nigeria. This study has demonstrated that improved soil fertility and food could be secured by mitigating the impacts of climate change induced soil compaction through the adoption of proper conservation agricultural practices in Nigeria. Apart from encouraging conservation agriculture, long-term studies are recommended in order to consolidate the findings of this research, while it is also ideal to extend similar investigations to other major crops in the study area.

Acknowledgements We appreciate the support from the Czech University of Life Sciences (CULS) for its continuous help in terms of regular trip to Nigeria to conduct the research. We also appreciate the effort of the Nigerian-based volunteers who assisted in collecting data on our behalf. The anonymous reviewers who made valuable suggestions are also acknowledged.

Disclosure Statement Authors declare no conflicts of interest.

References

Abatenh E, Gizaw B, Tsegaye Z, Tefera G (2018) Microbial function on climate change—a review. Environ Pollut Clim Chang 2:147

Anikwe MA, Obi ME, Agbim NN (2003) Effect of crop and soil management practices on soil compatibility in maize and groundnut plots in a paleustult in Southeastern Nigeria. Plant Soil 253:457–465

Babatunde FE, Dantata IJ, Olawuyi OJ (2012) Performance of sweet potato and soyabeans as affected by cropping sequence in the Northern Guinea Savanna of Nigeria. J Agron 11:22–26

Blake GR, Hartge KH (1986) Bulk density. In: Klute A (ed) Methods of soil analysis, physical and mineralogical methods. American Society of Agronomy Inc., Madison, WI, pp 363–375

Bremner JS, Mulvaney CS (1982) Nitrogen-total. In: Page AL, Miller RH, Keeney DR (eds) Methods of soil analysis, chemical and microbiological properties. American Society of Agronomy Inc., Madison, WI, pp 595–624

Chen ZN (1999) Selecting indicators for soil quality. Food and fertilizer technological centre. Extension bulletin 473, Taiwan

Chikoye D, Manyong VM, Ekeleme F (2000) Characteristics of speargrass (*Imperata cylindrica*) dominated elds in West Africa: crops, soil properties, farmer perceptions and management strategies. Crop Prot 19:481–487

Da Silva AP, Kay BD, Perfect E (1997) Management versus inherent soil properties effects on bulk density and relative compaction. Soil Tillage Res 44:81–93

Dube E, Chiduza C, Muchaonyerwa P (2012) Conservation agriculture effects on soil organic matter on a Haplic Cambisol after four years of maize–oat and maize–grazing vetch rotations in South Africa. Soil Tillage Res 123:21–28

Ehlers W, Kopke U, Nesse F, Bohm W (1983) Penetration resistance and root growth of oats in tilled and untilled loess soil. Soil Tillage Res 3:261–275

Flowers M, Lal R (1998) Axle load and tillage effects on soil physical properties and soybean grain yield on a mollic ochraqualf in northwest Ohio. Soil Tillage Res 48:21–35

Gee GW, Bauder JW (1986) Particle size analysis. In: Klute A (ed) Methods of soil analysis Part I. Agronomy 9. American Society of Agronomy Inc., Madison, WI, pp 383–411

Greenland DJ (1977) Soil damage by intensive arable cultivation: temporary or permanent? Philos Trans R Soc B Biol Sci 281:193–208

Iwara AI (2014) Evaluation of the variability in runoff and sediment loss in successional fallow vegetation of Southern Nigeria. Soil Water Res 9:77–82

Ijoyah MO, Iorlamen T, Idoko JA (2012) Yield response of intercropped maize (Zea mays L.) and okra (*Abelmoschus esculentus L. Moench*) to seasonal conditions at Makurdi Nigeria. J Nat Sci Res 2:79–85

Jagtap S (2007) Managing vulnerability to extreme weather and climate events: implications for agriculture and food security in Africa. In: Proceedings of the international conference on climate change and economic sustainability held at Nnamdi Azikiwe university, Enugu, Nigeria, 12–14 June 2007

Jones PG, Thornton PK (2003) The potential impacts of climate change in tropical agriculture: the case of maize in Africa and Latin America in 2055. Glob Environ Chang 13:51–59

Kayombo B, Lal R, Mrema GC, Jensen HE (1991) Characterizing compaction effects on soil properties and crop growth in southern Nigeria. Soil Tillage Res 21:325–345

Lal R (2000) Mulching effects on soil physical quality of an Alfisol in western Nigeria. Land Degrad Dev 11:383–392

Lyon DJ, Stroup WW, Brown RE (1999) Crop production and soil water storage in long term winter wheat-fallow tillage experiments. Soil Tillage Res 49:19–27

Mamman E, Ohu JO, Crowther T (2007) Effect of soil compaction and organic matter on the early growth of maize (Zea mays) in a vertisol. Int Agrophysics 21:367–375

Manyong VM, Mankinde KO, Sanginga N, Vanlauwe B, Diels J (2001) Fertilizer use and definition of farmer domains for impact-oriented research in the Northern Guinea Savanna of Nigeria. Nutr Cycl Agroecosystems 59:129–141

Mari GR, Changying J, Jun Z, Bukhari FS (2006) Effect of tillage machinery traffic on soil properties, corn root development and plant growth. Agric Eng Int CIGR Ejournal. Manuscript PM 06 026 vol VIII

Masvaya EN, Nyamangara J, Descheemaeker K, Giller KE (2017) Tillage, mulch and fertilizer impacts on soil nitrogen availability and maize production in semi-arid Zimbabwe. Soil Tillage Res 168:125–132

McLean EO (1982) Soil pH and lime requirement. In: Page AL, Miller RH, Keeney DR (Eds.), *Methods of Soil Analysis*. American Society of Agronomy Inc., pp. 199–234. Madison, WI

Mehlich A (1984) Mehlich III soil test extractant: a modification of Mehlich 2 extractant. Commun Soil Sci Plant Anal 15:1409–1416

Mielke LN, Wilhelm WW (1999) Comparisons of soil physical characteristics in long-term tillage winter wheat—fallow tillage experiments. Soil Tillage Res 49:29–35

Mo F, Wang JY, Zhou H, Luo CL, Zhang XF, Li XY, Li FM, Xiong LB, Kavagi L, Nguluu SN, Xiong YC (2017) Ridge-furrow plastic-mulching with balanced fertilization in rainfed maize (Zea mays L.): an adaptive management in east African Plateau. Agric For Meteorol 236:100–112

Nawaz M, Bourrié G, Trolard F (2013) Soil compaction impact and modelling. A Rev Agron Sustain Dev 33:291–309

Nelson DW, Sommers LE (1982) Total carbon, organic carbon and organic matter. In: Page AL, Miller RH, Keeney DR (eds) Methods of soil analysis. American Society of Agronomy Inc., Madison, WI, pp 539–579

Nwafor M, Adenikinju A, Ogujiuba K (2007) The impact of trade liberalization on poverty in Nigeria: dynamic simulations in a CGE model. RePEc:lvl:mpiacr:2007–16. Retrieved on 26th February 2019 from File URL: https://econpapers.repec.org/paper/lvlmpiacr/2007-16.htm

Nyeki A, Milics G, Kovacs AJ, Nemenyi M (2017) Effects of soil compaction on cereal yield: a review. Cer Res Commun 45:1–22

Ogundijo DS, Adetunji MT, Azeez JO, Arowolo TA (2017) Integrated fertilizer management: influence on soil nitrogen, available phosphorus, potassium, nutrient uptake and maize yield. Commun Soil Sci Plant Anal 48:943–954

Ojo OD, Olufolaji AO (2009) Effect of soil moisture stress on the emergence, establishment and productivity of Amaranthus (*Amaranthus cruentus L.*). In: Wesonga J, Kahane R (eds) 1st all African horticultural congress. *Acta Horticulturae*, 911: 113–121

Page AL, Miller RH, Keeney DR (1982) Methods of soil analysis: chemical and microbiological properties, 2nd edn. Agronomy Series 9. American Soc. of Agronomy, Madison, Wisconsin, USA

Saidou A, Janssen BH, Temminghoff EJM (2003) Effects of soil properties, mulch and NPK fertilizer on maize yields and nutrient budget on ferralitic soils in southern Benin. Agric Ecosyst Environ 100:265–273

Sainju UM, Whitehead WF, Singh BP (2005) Bi-culture legume-cereal cover crops for enhanced biomass yield and carbon and nitrogen. Agron J 97:1403–1412

Sanchez PA (2002) Soil fertility and hunger in Africa. Science 295:2019–2020

Sanginga N, Vanlauwe B, Danso SKA (1995) Management of biological N2 fixation in alley cropping systems: estimation and contribution to N balance. Plant Soil 174:119–141

Singh J, Salaria A, Kaul A (2015) Impacts of soil compaction on soil physical properties and growth: a review. Int J Food Agric Vety Sci 5:23–32

Smaling EMA, Dixon J (2006) Adding a soil fertility dimension to the global farming systems approach, with cases from Africa. Agr Ecosyst Environ 116:34–46

Stolte J (1997) Manual of soil physical measurements. Version 3. Wageningen, D.L.0 Staring Centre, Tech. Doc. 37

Tao Z, Li C, Li J, Ding Z, Xu J, Sun X, Zhou P, Zhao M (2015) Tillage and straw mulching impacts on grain yield and water use efficiency of spring maize in Northern Huang–Huai–Hai Valley. Crop J 3:445–450

Taylor JH, Brar GS (1991) Effect of soil compaction on root development. Soil Tillage Res 19:111–119

Upawansa GK (1997) New Kekulam rice cultivation: a practical and scientific ecological approach. Rebuilding lost soil fertility. Leisa, Ileia Newsletter 13 no. 3 Leusden, Netherlands. pp 20–21

Vanlauwe B (2005) Long-term integrated soil fertility management in South-western Nigeria: crop performance and impact on the soil fertility status. Plant Soil 273:337–354

Verbist K, Cornelis WM, Schiettecatte W, Oltenfreiter G, Van MM, Gabriels D (2007) The influence of a compacted plow sole on saturation excess runoff. Soil Tillage Res 96:292–302

Wang S, Wilkes A, Zhang Z, Chang X, Lang R, Wang Y, Niu H (2011) Management and land use change effects on soil carbon in northern China's grasslands: a synthesis. Agric Ecosyst Environ 142:329–340

Zhang XY, Cruse RM, Sui YY, Jhao Z (2006) Soil compaction induced by small tractor traffic in Northeast China. Soil Sci Soc Am J 70:613–619

Ziervogel G, Nyong A, Osman B, Conde C, Cortes S, Dowing T (2006) Climate variability and change: implications for household food security. Assessments of impacts and adaptations to climate change (AIACC) working paper No. 20, January 2006. The AIACC Project Office, International START Secretariat, Washington DC, USA

Knowledge Dialogues and Climate Change: Integrating Participatory Approaches in the Design of Ecosystem-Based Adaptation Measures in the Peruvian Andes

Florencia Zapata and Erin Gleeson

Abstract Communities throughout the Andes—and indeed much of the world—are increasingly challenged by climate change and other modern pressures such as migration and globalization. Ecosystem-based Adaptation seeks to address human vulnerability to climate change using the natural environment in ways that simultaneously increase human well-being and ecosystem resilience. This paper examines the role of participatory research methods in promoting adaptation measures that are robust, i.e., effective under both current and potential future conditions, as well as embraced by the local people.

Introduction

As an adaptation concept that promotes the use of biodiversity and ecosystem services to reduce the vulnerability of people to the adverse effects of climate change (CBD 2009), Ecosystem-based Adaptation (EbA) is gaining increasing attention by the international community (IPCC 2014; FEBA 2017). EbA measures, however, are very context-dependent: solutions must be tailored to the specific social, cultural, and environmental situation to be effective (Lo 2016). In the Andes, some frequent challenges to developing EbA actions at the territorial scale are:

(i) the high uncertainty regarding climate changes and the lack of resources and data to carry out analyses of vulnerability and climate risk. Such analyses are important for identifying and designing adaptation measures that respond to current conditions and can readily be adapted to future conditions;
(ii) tensions between local management practices and sustainable management objectives; and

F. Zapata (✉)
The Mountain Institute, Andean Program, Calle Gral, Vargas Machuca 408, Lima, Peru
e-mail: florenciaz@mountain.org

E. Gleeson
The Mountain Institute, 1725 I Street NW, Suite 300, Washington, D.C 20006, USA
e-mail: egleeson@mountain.org

W. Leal Filho et al. (eds.), *Climate Change, Hazards and Adaptation Options*,
Climate Change Management, https://doi.org/10.1007/978-3-030-37425-9_37

(iii) ensuring the long-term sustainability of adaptation measures (Dourojeanni et al. 2016; Lo 2016; UNDP et al. 2016).

Can participatory research help overcome these challenges? In data-scarce environments with a high degree of uncertainty, such as in the Peruvian Andes, research that incorporates the knowledge of local people and actions that are shaped by deliberate dialogue between technical specialists and local people may help reduce uncertainty and overcome a lack of data. This paper addresses the role of participatory research by systematizing the results of two participatory research experiences conducted within the framework of robust EbA measures implemented in the communities of Canchayllo and Miraflores, located in the Nor Yauyos-Cochas Landscape Reserve (NYCLR) in the Peruvian Andes.

The cases of Canchayllo and Miraflores demonstrate the value of local knowledge and participation in the development of effective EbA solutions. Participatory research can help fill data gaps regarding vulnerability, as well as promote local participation and generate a sense of ownership for the measures, which are designed based on the dialogue between local knowledge and scientific knowledge. This sense of local ownership is fundamental for the sound management and sustainability of any adaptation measure (IIED 2009). The cases also illustrate the challenges of local participation and reveal the need to deepen and continue the knowledge dialogue beyond design and implementation of the EbA measures.

We begin this paper by presenting the concepts of Ecosystem-based Adaptation, low-regret adaptation, and Participatory Action Research (PAR) and Transdisciplinary Research, and discussing the articulation between these concepts. We then describe the project, the NYCLR, the communities of Canchayllo and Miraflores, and the EbA measures that were implemented in each community. We analyze and describe the participatory research experiences that were carried out in the two communities, including the assessment, co-design, and implementation of measures and the collaborative reflection on the knowledge exchange process. Finally, we present the main results and conclusions.

Conceptual-Methodological Framework

Ecosystem-Based Adaptation: A People-Centered Approach

Ecosystem-based Adaptation (EbA) is defined as "the use of biodiversity and ecosystem services as part of a larger strategy to help people adapt to the adverse effects of climate change" (CBD 2009). EbA refers to the "sustainable management, conservation, and restoration of ecosystems as part of an adaptation strategy that takes into account the multiple social, economic, and cultural co-benefits for local communities" (CBD 2010). EbA aligns human well-being with that of the environment, and therefore recognizes the importance of integrating and maintaining local knowledge and cultural values (Lo 2016). In the context of sustainable development, EbA

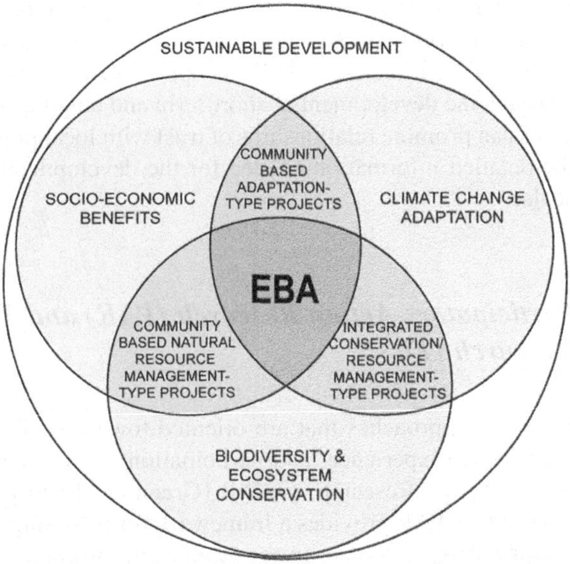

Fig. 1 EbA and its relationship with other adaptation strategies, conservation, and development *Source* FEBA (2017, adapted from Midgley et al. 2012)

results from the synergy of three end goals: conservation of biodiversity and ecosystems, adaptation to climate change, and generation of socio-economic benefits for the population (Fig. 1).

Low-Regret Adaptation

In a context of climate uncertainty, low-regret or robust adaptation measures enable action that helps to reduce climate risks whatever the extent of future climate change (Lo 2016; Rizvi et al. 2014). "Low-regrets EbA interventions minimize trade-offs and optimize benefits across many sectors to achieve multiple goals, including reducing vulnerability to climate change for people and nature" (Lo 2016:26). EbA has the potential to be a low-regret strategy because measures harness the social, environmental, and economic potentials that already exist in the system, rather than attempting to introduce new species or costly technologies. EbA measures may, however, complement infrastructural measures such as dikes, or offer alternatives to conventional agricultural or forestry practices (FEBA 2017).

Such robust measures are especially important in environments like the Andean puna, which is a highland ecoregion dominated by natural grasslands and characterized by heterogenous landscapes, as well as a scarcity of climate observations. EbA measures require analyses of future climate vulnerability and risk (Dourojeanni et al. 2016); however, it is particularly difficult to predict the impacts of climate change at the scale of a peasant community, which is the scale most relevant to human well-being in the highland Andes. An alternative is an analysis based on "expert" information derived from: (1) local knowledge about the occurrence, location, and timing/variability of climate events and patterns; (2) literature published on trends of change at local scales, and (3) the opinions of specialists (Dourojeanni et al. 2016). Through the development of short-term and tangible actions, robust adaptation measures can promote relationships of trust with local populations while also generating the detailed information needed for the development of EbA activities on a larger scale.

Participatory Action Research (PAR) and Transdisciplinary Research (TR)

Research approaches that are oriented toward social change and are based on the knowledge, experience, and participation of local actors are referred to as Participatory Action Research, or PAR (Greenwood and Levin 1998; Reason and Bradbury 2001). PAR provides a framework for involving the local knowledge needed to design robust EbA measures, as well as for promoting local engagement throughout the analysis, design, implementation, and monitoring process.

PAR refers to a set of approaches that have three common pillars:

i. Knowledge-based investigation;
ii. Stakeholder participation and horizontal relationships; and
iii. Action that seeks to improve local conditions (modified from Greenwood and Levin 1998).

PAR is a broad family of ways of doing research for social change with the active participation of people, as opposed to prescribing actions to passive subjects. PAR has materialized in a number of ways around the world, from Popular Education in Denmark and Latin America to Organizational Learning within companies to Participatory Rural Appraisal or Cooperative Research in the Northern Hemisphere, to mention just a few (Fals Borda 2016; Greenwood and Levin 1998). All of these schools of thought are oriented towards action, are participatory, and are based on knowledge and experience (Reason and Bradbury 2001).

Thus, PAR is implemented by a group of people who work to solve a question or a problem that affects their own community. These local researchers, in the process of investigating, can be accompanied by one or more people who facilitate the process and who fulfill the double role of facilitators and external researchers. During PAR, other external researchers may also be invited to intervene, usually specialists or scientists who share their knowledge with local researchers.

PAR can use the methods, techniques, and tools that are necessary and approved by the group of local and external researchers (Greenwood 2016). For example, techniques may include: social maps, open and semi-structured interviews, focus groups, participatory analysis of secondary sources, mental maps, collective constructions of timelines, discussion groups, participatory monitoring and evaluation, collective elaboration of texts and narratives, etc. (Zapata and Rondán 2016). However, it is not the methodology or techniques used that distinguish PAR from other ways of doing research, but the commitment to social change and to empowering vulnerable groups to decide how to manage their own change. More important than the techniques and tools used is the advancement of research via a collective knowledge production process that constantly involves spaces for reflection and action (Fals Borda 2016; Greenwood and Levin 1998).

For its part, Transdisciplinary Research (TR) refers to "a form of research that transcends disciplinary boundaries to address and solve problems related to the real world" (Mittelstraß 1992, cited by Hirsch Hadorn et al. 2008: 19). TR emerged in the 1970s as a proposal for the reorganization of higher education in response to concerns about the continued specialization of science. In the 1990s, the concept gained new currency as a result of the search for solutions to new global concerns, such as sustainable development (Bernstein 2015; Hirsch Hadorn et al. 2008; Lang et al. 2012). Like PAR, TR seeks solutions to complex real-world problems and does so by actively engaging academic, government, private sector, and civil society perspectives (Pohl and Hirsch Hadorn 2007).

Despite their different origins, PAR and TR are closely related (Stokols 2006). PAR emerged as a call to democratize the production of knowledge, whereas TR evolved in response to a call to de-silo science. The resurgence of TR in the 90s, emphasizing solutions to complex problems and the search for the common good, highlights the crucial role of civil society actors and the need for PAR. In the search for EbA measures that are environmentally and socially appropriate and that are effective under both current and future conditions, PAR and TR play key roles in helping design a process that engages both the local community and expertise from a range of disciplines.

The Context and the Measures

In 2011, the United Nations Environment Program (UNEP), the International Union for the Conservation of Nature (IUCN), and the United Nations Development Program (UNDP) started a collaborative initiative to enhance the resilience of mountain ecosystems and their local communities in Peru, Nepal, and Uganda. The "Ecosystem-based Adaptation in Mountain Ecosystems" Program ran until 2016 and was financed by the Federal Ministry of the Environment, Nature Protection, Public Works and Nuclear Safety of the German Government (BMUB). In Peru, the program was commissioned by the Ministry of the Environment (MINAM) and was implemented in the Nor Yauyos-Cochas Landscape Reserve (NYCLR), with support from the National Service of Natural Protected Areas (SERNANP). The Mountain Institute (IM), commissioned by IUCN and in coordination with the RPNYC, was responsible for the development and implementation of robust EbA measures in the rural communities of Canchayllo and Miraflores in NYCLR.

Currently, the program "Scaling up Mountain Ecosystem-based Adaptation" (2017–2020) is being implemented by TMI and IUCN in six countries: Peru, Uganda,

Nepal, Colombia, Kenya, and Bhutan. This new program seeks to generate evidence on the impacts of EbA and to expand the scale at which EbA is applied. It consolidates lessons from the previous project and replicates successful measures in new sites, as well as promotes the integration of the EbA approach into policy instruments at different levels, from the local to the international.

The Nor Yauyos-Cochas Landscape Reserve and the Communities of Canchayllo and Miraflores

The NYCLR was created in 2001 and is located in the central Andes of Peru, in the regions of Lima and Junín. It spans an area of 221, 268 ha and ranges from 2500 to 5860 m above sea-level (a.s.l.) The highest point of the Reserve is the summit of the Pariacaca mountain, which has been regarded as a tutelary deity since pre-Inca times and is venerated to this day. The main objective of the Reserve is the conservation of the upper Cañete River basin and the Pachacayo River basin (Mantaro river tributary), both of which are important for water regulation, hydroelectric energy production, and other ecosystem services (INRENA 2006). Due to its status as a Landscape Reserve, NYCLR is classified as a protected area of direct use, where "the use or extraction of resources is allowed in those areas and places and for those resources defined by the management plan" (INRENA 2006: 21). It was because of this designation and its importance as a supplier of ecosystem services that the Peruvian Ministry of the Environment requested that the project focus on the Reserve.

The mountainous landscape of NYCLR encompasses a complex hydrological system of glaciers, waterfalls, and 485 lagoons. The climate is cold: the average annual temperature is 8 °C, with an average maximum of 19 °C and an average minimum of 0 °C. The region receives 650–750 mm of precipitation per year, and experiences distinct wet and dry seasons. It is also home to great floristic diversity, with 330 plant species cataloged to date (MINAM 2011). High Andean grasslands (including pajonal, puna grass, and bofedales) above 3800 m a.s.l. are the predominant vegetation, covering about 70% of the Reserve's surface.

Today, the Reserve is home to about 15,000 people (INEI 2007). However, humans have molded the landscapes of the Reserve for thousands of years, as evidenced both by the distribution and characteristics of its flora and fauna and by its wealth of intangible and material cultural values (INC 2009, INRENA 2006). Ancient technologies such as platforms, terraces, agricultural and livestock practices, canals, and pre-Hispanic roads are still in use and reflect a deep knowledge of the environment and the development of very specific knowledge and skills. Together with the Agricultural Society of Social Interest (SAIS) "Tupac Amaru", 19 peasant communities make decisions about the management of the territory and its resources.

According to the NYCLR Master Plan (INRENA 2006) and the project assessment (TMI 2014), the Reserve faces multiple pressures. The challenges stem from changes in demographics and production patterns, overgrazing, and changes in rain patterns and extreme weather events. Access to water and healthy pastures is essential to the well-being of local communities, who depend primarily on agropastoral activities for their livelihoods. In the last decades, there has been a general decrease in agricultural production (particularly of native crops such as potatoes) associated with climatic uncertainty, a decline in agricultural prices, a lack of labor due to migration, and an increase in livestock activities. The increase in livestock husbandry is largely due to an economic tradeoff and migration trends, as livestock is more lucrative and less labor intensive (FDA 2013; Podvin et al. 2014; TMI 2014). However, the increase in livestock ranching, coupled with a weakening of the communal organization, generates unsustainable management practices that are causing the degradation of puna grassland ecosystems, which represent about 70% of the Reserve's surface. According to the inventory and evaluation of the natural heritage in the Reserve made by MINAM in 2011, the problem of overgrazing is widespread and urgently requires a change in the system of resource use.

According to the "Study of the Vulnerability and Impact of Climate Change on the Nor Yauyos-Cochas Landscape Reserve" (VIA) (FDA, 2013), there is a high level of uncertainty regarding future trends and climate scenarios in the Reserve. The study reports that temperatures increased between 0.21 and 0.32 °C per decade between 1950 and 2010, and that a further increase of 0.61–1.12 °C between the years 2011 and 2030 is likely. The study also predicts that while overall precipitation volumes will remain fairly stable, distribution patterns will change significantly. Similarly, surface runoff is expected to decrease as the water volume stored in glaciers and snowpack declines. These predictions are in line with climate scenarios prepared by the Geophysical Institute of Peru—IGP (2005), which also predicts more intense frosts and declining precipitation by 2050. The future scenarios proposed for the Reserve suggest changes that could affect pastures and water, vital resources for peasant communities that depend on agricultural activities (Podvin et al. 2014). Local residents have already noted changes in seasonal rainfall patterns, a decrease in rainfall, more intense frosts, greater radiation, and increased disease in animals (TMI 2014) (Fig. 2).

The Peasant Community of Miraflores

The community of Miraflores is located between 3000 and 5400 m a.s.l. in the upper part of the Cañete River basin, whose flow throughout the year is one of the most regular on the Pacific slope. The community was officially recognized in 1925 and has an area of 13,031 ha. Approximately 80 families (about 320 people) live in Miraflores, more than 70% of whom depend mainly on agro-pastoral activities for their livelihoods (FDA 2013).

Until a few decades ago, the population was engaged mainly in the cultivation of native potatoes and corn. However, due to migration, the market, and other drivers of change, many families began to invest more intensively in livestock (TMI 2014 and 2015). According to testimonies collected in 2014 and 2015, young people migrate

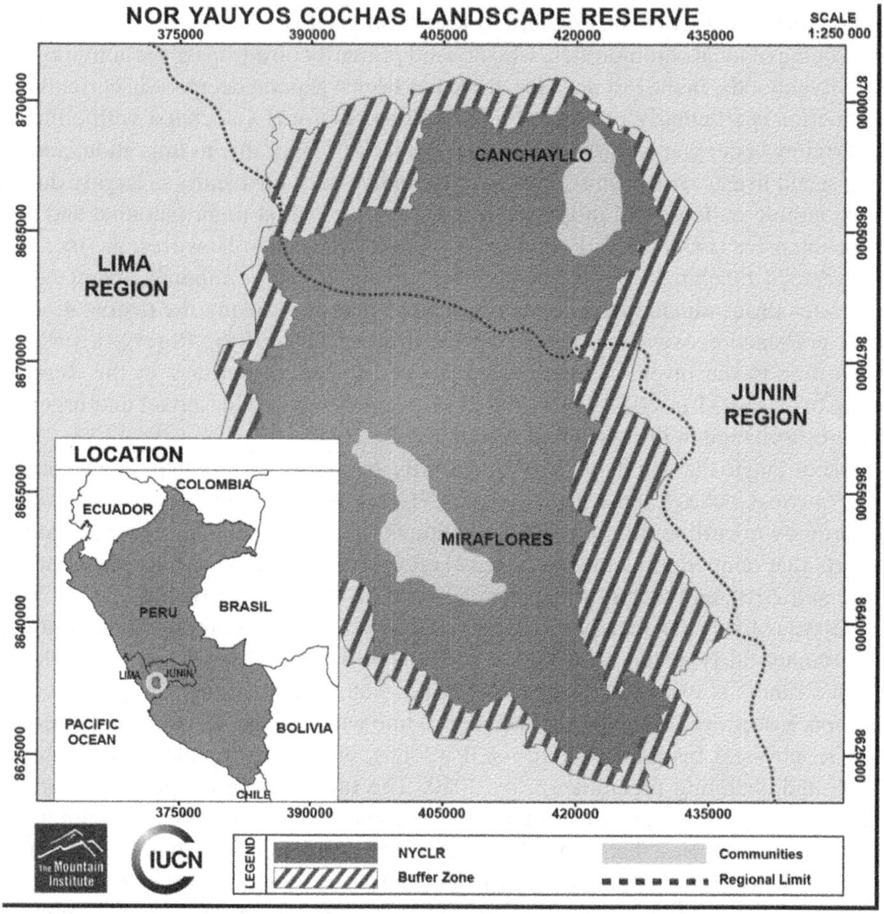

Fig. 2 Map of the communities of Canchayllo and Miraflores in NYCLR

out of the Reserve in search of better educational opportunities and income because agricultural activities are no longer as profitable as before, and because success is increasingly associated with life in cities.

Livestock activities include the production of alpacas and sheep in areas above 4000 m, and cattle in middle and lower areas. Cattle ranching is increasing because it is both more profitable than agriculture and less labor intensive than either agriculture or camelid production (TMI 2014 and 2018). However, the topography is very rugged and only 44.5% of the grassland area is suitable for grazing (Ruiz 2013). According to a participatory assessment that was carried out in 2013, the condition of the community's most important pastures was rated as poor or very poor. In the absence of change, these conditions will result in a reduction of native seed banks, as well as a reduction in native pasture for grazing (Ñaupari et al. 2013).

Fig. 3 Pre-Inca dam of Yanacancha, still in use after *circa* 700 years *Source* Ramírez and Herrera (2013)

Annual rainfall in Miraflores averages between 700 and 800 mm per year (Cárdenas 2013), and the district has three important glaciated mountains (Umán, Ancovilca, Cutuni) and 13 lagoons. Although there is sufficient water available at higher elevations, the lower sectors, where most of the livestock are concentrated, lack water as a result of poor distribution. However, there is strong potential to remedy the situation, as the community not only has sufficient water resources, but also an extensive network of ancient water management systems. An archaeological study by Ramírez and Herrera (2013) indicated that the area has a high cultural and archaeological value, although dams, canals, and terraces are deteriorating due to a lack of maintenance. In addition, the study identified six pre-Inca dams in an area above the community of Miraflores called Yanacancha. Together, these dams are an impressive example of technology for the retention and filtration of water (Fig. 3). However, during the assessment it was found that both the dams and the associated canal were deteriorating due to a lack of maintenance. During the participatory process, the community and the specialists identified the recovery of some of these technologies as a robust EbA measure.

The Peasant Community of Canchayllo

The community of Canchayllo was founded in 1942 and covers an area of 7650 ha with elevations ranging between 3600 and 5700 m a.s.l. The territory, which includes four mountains and 28 lagoons, is dominated by extensive areas of grassland. About

800 people live in Canchayllo, many of whom practice traditional livestock husbandry, although many also supplement their income with other activities (TMI 2018). For example, a significant percentage of the population worked for the local power generation company until 2014, when construction was completed. Since the completion of the hydropower plant, there has been some demand to reinitiate mining activities that occurred prior to the incorporation of the community into the Reserve. However, such activities conflict with the objectives and provisions of the protected area.

Currently, 60 families depend directly on livestock for their incomes and some young people are interested in continuing to raise livestock. However, the community is aware that water availability is decreasing and there is a high degree of concern regarding future availability (Cárdenas 2013). Similarly, the participatory assessment of 2013 and subsequent evaluations reveal that the condition of the pastures in all three zones ranges from "poor" to "regular" following the Parker method (Ñaupari et al. 2013; TMI 2014 and 2018). Thus, improving water and grassland management to facilitate better livestock production is a priority for the community (TMI 2014 and 2018).

The participatory assessment identified three productive zones: (i) the upper zone (upper 4200 m a.s.l), which is where communal livestock graze; (ii) the middle zone (3800–4200 m a.s.l), which includes the sectors of Yanaotuto and Pumapanca and is privately owned pastureland with native grasses; and (iii) the lower zone (below 3800 m a.s.l), which includes the town and its surroundings. The lower zone is used primarily for rain-fed and irrigated agriculture and for grazing livestock on introduced grasses on a rotational calendar (Segura 2013, TMI 2014 and 2018). There is evidence of overgrazing in most of the territory due to inadequate livestock management, as well as of soil degradation due to the frequent burning of pastures, which is a common practice by local people to call for rains and to encourage the growth of tender young grasses but one that degrades the vegetation and the soil. The assessment also revealed a lack of capacity to regulate water and grassland use; there is no clear water management authority, and existing regulations are ignored and unenforced (Segura 2013; TMI 2014).

As in Miraflores, the assessment identified a set of technologies, both pre-Hispanic and modern, relevant for the management of pastures, water, and livestock (Ramírez and Herrera 2013). For example, contemporary water management structures were identified near the Chacara lagoon, including a dam and a system of canals. According to the inhabitants, this infrastructure was installed in 1938 by Cerro de Pasco Cooper Corporation (TMI 2014), which ran cattle in the grasslands until 1970. The main canal, which starts in the Chacara lagoon, carried water to the communal farm. From there, a system of canals, possibly pre-Hispanic, helped to maintain the wetlands and distribute water to the sectors of Yanaotuto and Pumapanca, where the livestock families are settled. After the Agrarian Reform, the system was operated by the community of Canchayllo but eventually fell into decline as the labor needed to maintain the system decreased.

EbA Measures Implemented in Canchayllo and Miraflores

The process of implementing the EbA measures in Miraflores and Canchayllo was divided into several phases:

I. An initial phase of consultation, assessment, and design, which ran from April to November 2013 and included: (a) Consultation with internal partners and potential communities and the identification of possible measures; and (b) Participatory assessment, selection, and design of the EbA measures.
II. The implementation phase of the adaptation measures, which began in December 2013 and ended in October 2015.
III. The systematization and transfer phase, which overlapped in part with phase II and ran from September 2014 to December 2015.
IV. The consolidation phase, which began at the end of 2017 and will continue until the middle of 2020.

Although they are tailored specifically to the local context, the EbA measures in both communities focus on the expansion and conservation of wetlands, on the communal management of native grasslands, and on the restoration of ancestral and contemporary technologies for water management in the puna. All measures help communities adapt to climate change by (1) strengthening institutional and community organization, (2) enhancing local capacities and knowledge, and (3) improving infrastructure (Fig. 4).

Efforts to strengthen local organization and build capacity included the development of grazing and water management plans (CCC et al. 2015b; CCM et al. 2015b) to promote the integrated management of pastures, water, and livestock. These plans were collaboratively developed by the communities themselves, which

Fig. 4 Components of the EbA measures of Miraflores and Canchayllo *Source* TMI (2014)

not only helped improve the condition of the pastures, but also strengthened the organizational framework of the communities. Efforts also included courses, workshops, and training in the evaluation and application of techniques for pasture recovery. These actions were accompanied by communication activities such as participatory videos and Community Theater, which were catalysts for local participation.

At the same time, measures were implemented to improve the water infrastructure in both communities. In Canchayllo, the Chacara lagoon dam was repaired and the canal restored, which in turn allowed the recovery of the network of surface canals. This infrastructure now supplies water to 560 ha of pastures, as well helps to recharge temporary and underground aquifers (Hidroandes 2015). Fences were installed to recover pastures that are monitored by NYCLR with community support. In Miraflores, efforts concentrated on the Curiuna grassland area and the pre-Inca dams of Yanacancha, which are the main source of water for both the village and the livestock grazing in Curiuna. The wetlands, or *bofedales,* surrounding the dams were enclosed to prevent contamination and degradation by livestock, and the Yanacancha—Curiuna—Huaquis water pipeline was repaired. Fences were installed in Curiuna to create a new grazing area and five livestock watering troughs were repaired and/or constructed, actions that resulted in the improved management of 165 ha of pastures.

As we will see in more detail in the next section, the inhabitants of both communities played a leading role in all of the activities. The selection, design, and implementation of the measures were direct outcomes of identifying and valuing local knowledge, voices, and perspectives, and merging those perspectives with the available technical and scientific knowledge. It should also be noted that the community and the municipality also made financial and labor contributions to the implementation of the measures, which both signaled and reinforced their ownership of the process.

Process

Knowledge Dialogue: Participatory Action Research for EbA

The initial phase of consultation, assessment, and design was an 8-month process (April to November 2013) that involved the communities and their authorities, NYCLR staff, project partners, and specialists from multiple disciplines. First, TMI staff worked with Reserve staff and other project partners to identify project sites (the communities of Miraflores and Canchayllo) based on environmental, social, ecological, political, and operational criteria (Podvin et al. 2014). The process of identifying robust EbA measures then included initial consultations with the communities, a participatory assessment of the needs and priorities of local stakeholders, and the co-design of the EbA measures (Rizvi et al. 2015).

Initial consultations with local inhabitants took place in the field and via workshops and focused on understanding local interpretations of vulnerability (climate and otherwise) and building strong working relationships. Participants worked together to propose solutions to identified vulnerabilities that they then prioritized according to five criteria: potential to reduce local vulnerability, co-benefits, overall support from the community, potential for long-term sustainability, and cost-benefit ratio (TMI 2014). Both communities expressed strong interest in improving the availability of water in the grazing areas and improving the management of pastures and livestock.

To navigate the unprecedented uncertainties posed by climate change, a combination of locally specific traditional knowledge and globally relevant modern science is needed. To integrate these two types of knowledge, The Mountain Institute developed an assessment methodology based on the Rapid Rural Appraisal and Participatory Rural Appraisal methods (Chambers 1994 and 1997; Zabala 2000). Called the "Integrated Participatory Rural Appraisal (or DRPI in Spanish)", this methodology facilitated a process of social learning and co-design in which robust adaptation measures result from the dialogue between local actors and external specialists (TMI 2014).

The DRPI focused on the two issues prioritized by the communities during the initial consultation phase, water availability and grassland management. To carry out the DRPI, a transdisciplinary team of external researchers worked together with a team of local researchers and park rangers. The transdisciplinary team was comprised of a hydrologist, an anthropologist, an archaeologist, a grassland specialist, and a specialist in agropastoral systems, whereas the team of local researchers was comprised of men and women from the community recognized for their interest, leadership, curiosity, critical thinking, and knowledge about the environmental and social realities of their community. Both groups worked with the TMI facilitation team who helped promote the horizontal exchange and integration of knowledge, as well as the collective analysis of climate risks and EbA options (TMI 2014).

In the case of both Canchayllo and Miraflores, the two teams worked separately and in concert to collect field data, develop analyses, plan activities, and organize logistics. Throughout the process, the two teams came together at specific intervals, both in the field and in workshops, to exchange knowledge and refine ideas, with the facilitation team working to promote a horizontal exchange. During this time, the external specialists prepared reports about the hydrology (Cárdenas 2013), grassland management (Ñaupari et al. 2013), archaeology (Ramírez and Herrera 2013), productive systems (Ruiz 2013), and social organization and culture (Segura 2013) of the communities.

Finally, all project stakeholders, including the local and external researchers, NYCLR staff, and representatives of the EbA in Mountains Project participated in a collaborative 2-day workshop to analyze EbA alternatives, select the measures to be implemented, complete detailed measure designs, and generate agreements regarding their implementation. The final set of EbA measures were selected for implementation using to a set of criteria agreed upon by all participants: improve climate resilience, robustness, socio-environmental relevance, cultural adequacy, financial and operational feasibility.

Once the measure plans were finalized, local researchers and TMI and NYCLR staff presented the plans to the broader community in both Miraflores and Canchayllo for final discussion and approval. Afterward, TMI integrated the results of the sector-specific reports and the dialogue process into assessment documents for each community, which were then delivered to the families in each community (CCC et al. 2015a; CCM. et al. 2015a).

Although local researchers had been organized specifically to participate in the assessment process, they continued to play a key role in the implementation of the EbA measures, including coordination, field work, preparation and implementation of management plans, communication of results, and analysis of project progress and results (Zapata et al. 2016). Since the start of the new project in mid-2017, local researchers continue to play a central role in driving the measures forward by evaluating the effectiveness of the original measures, involving new authorities and local leaders, and identifying the actions required to further improve the measures. However, as we will see in the discussion section, the outcomes in the two communities have been quite different, despite the use of a common approach.

Reflecting on the Knowledge Dialogue Process

Our analysis for this paper is based on the systematization of the EbA project in Miraflores and Canchayllo. Systematization is a method for knowledge generation developed in Latin America in which a group of people that lived a common experience analyze it in a systematic way: "It is an individual and collective learning process performed and facilitated by the stakeholders of the experience themselves" (Herout and Schmid 2015:65, see also Falkembach and Carillo 2015).

Maintaining the participatory approach that characterized the whole process of consultation, assessment, design, and implementation, the project also sought to gather lessons learned by reflecting with project partners about their experiences. As with the DRPI, the TMI team facilitated the systematization of the project and developed the methodology for its implementation (Zapata et al. 2016) and again, local researchers from Canchayllo and Miraflores and the NYCLR played a central role in both reconstructing the facts of the project and in identifying lessons learned from the experience. The perspectives and opinions of the various actors linked to the experience were gathered via 29 interviews, 16 focus groups, 4 life stories, 2 multi-actor workshops, and 1 institutional systematization workshop. In total, 107 people took part in the systematization process, which yielded deeper insight into the project experience of selecting, designing and implementing the robust EbA measures in Canchayllo and Miraflores (UNDP et al. 2016; Zapata et al. 2016).

In this paper, the results of the 2015 report of the systematization process by Zapata et al. (2016) are complemented by the situational analysis developed at the beginning of the "Scaling up Mountain EbA" program in 2017 (TMI 2018). As with the systemization in 2015, the situational analysis developed between September 2017 and March 2018 consisted of semi-structured interviews, participatory analysis workshops, formal and informal conversations, and field visits.

Discussion

As noted, common challenges associated with developing EbA actions include high climate uncertainty, tensions between local management practices and sustainable management objectives, and ensuring measure longevity and sustainability. Here, we discuss the results of using a participatory research approach in Miraflores and Canchayllo in light of its effectiveness in responding to these challenges, as well as look at some of the key challenges and factors for success.

As we've shown above, the methods, tools, and format of participatory assessments provide a means for integrating local perspectives with specialized scientific knowledge. This finding is in line with those of Dourojeanni et al. (2016), who compared three strategies for assessing climate vulnerability in the NYCLR context (i.e., participatory assessments, Vulnerability and Impact Assessments (VIA), and analysis of secondary data) and found that participatory assessments were most effective at describing the socio-environmental context and vulnerabilities at the community level. Interesting, the study found that all three approaches recommended very similar adaptation measures. However, the Dourojeanni et al. (2016) study found that the participatory approach made it possible to obtain sufficient information (e.g., about livelihoods, ecosystems, land management practices, etc.), even without having the necessary resources or data to carry out a detailed vulnerability and impact study. Thus, participatory approaches can help overcome the challenges of working in places where data is scarce and climatic uncertainty is particularly high.

In addition, the results of the project systematization (Zapata et al. 2016) coincide with both an independent evaluation (Vela 2017) and research carried out by the IIED (Reid et al. 2018) in pointing out the importance of participatory processes to achieve the results of the project. When consulted about the assessment process, the villagers and project partners highlighted the effectiveness of the methodology and the participatory format for assessment and its implications for the development of subsequent activities (Zapata et al. 2016). According to the testimonies and evaluations, the constant dialogue and coordinated work between local experts and external specialists helped ensure that the solutions were suited to the social and environmental context. The EbA measures implemented in Canchayllo and Miraflores are designed to address the high climatic variability *currently* experienced by the population, as well as be effective in a range of *future* climate scenarios. The measures also respond to current social challenges, such as the weakened community organization and reduced labor availability, and are designed to accommodate future trends, such as an increase in migration (Zapata et al. 2016; Reid et al. 2018).

More broadly, the presence of a team specialized in facilitating a participatory process was essential to bridging differences between local knowledge and external expertise, as well as between different disciplines. Thoughtfully facilitated horizontal knowledge exchanges helped to foster a culture of trust and respect among participants that enabled a productive co-design experience (Zapata and Gómez 2015). The participatory format also created space for discussing concerns about unsustainable management practices from the perspective of greater climate vulnerability and helped motivate participants to seek contextualized solutions, thereby fostering the social and organizational capacity to manage their natural resources and the EbA measures in a sustainable way (UNDP et al. 2016; Vela 2017).

However, the experiences in the two communities were not without obstacles. For example, achieving and maintaining good participation levels throughout the course of the project was, as anticipated by Wiesmann et al. (2008: 433), one of the main challenges. Likewise, involving and ensuring the recognition of sectors and community groups that are typically under-represented (e.g., women, young people, older adults) was difficult. Other, more subtle barriers—such as prejudices, power relations, and disciplinary boundaries—were also present. In this respect, it is important to reflect on the attitudes and behavior of external facilitators and researchers (Vernooy 2006) and to ensure that they "do not dominate the process but give control to the population to catalyze and facilitate their participation" (Zabala 2000). Nevertheless, the participatory process was essential for legitimizing the chosen course of action and obtaining the long-term commitment of the community, thereby ensuring the sustainability of the solutions (Reid et al. 2018).

The start of the "Scaling up Mountain EbA" project in 2017 offered the opportunity to revisit the measures in Canchayllo and Miraflores two years after their implementation was concluded. The resulting situational report (TMI 2018) revealed that, in the case of Miraflores, the measure catalyzed further action within the community. The measures implemented in 2013–2015 were designed to improve the management of 165 ha of grasslands; as of 2018, 6000 ha were under new communal management and livestock rotation plans. According to the villagers, the success of the earlier work encouraged leaders and other community members to enforce the statutes and make additional agreements regarding better livestock management. In Canchayllo, however, the measures have not had such a positive ripple effect. There, community members are more strongly divided between those who are dedicated to livestock and those who derive their main income from salaried work. Currently, there is strong pressure to develop the mineral resources in the area, which is directly at odds with the interests of livestock herders and the Reserve itself (TMI 2018).

Conclusions

The systematization of the experiences in Canchayllo and Miraflores confirms the capacity of Participatory Action Research to integrate the knowledge of local inhabitants and the technical and scientific knowledge of various disciplines, and to bring local management practices and sustainable management objectives closer together. PAR helps to identify and understand, from both local and scientific perspectives, the current conditions, trends, possible future scenarios, and drivers of change that, together, are key to determining the vulnerability of communities and ecosystems and thus selecting and designing adequate adaptation measures. PAR also fosters a dialogue of knowledge that helps ensure that measures are designed to be robust under both current conditions (climatic, environmental, sociocultural) and a range of possible future conditions. This is especially relevant for EbA, which focuses on helping people adapt to climate change, both now and in the future, with ecosystem health as a means to adaptation.

Importantly, PAR also lays the foundation for sustainable measures by generating a sense of ownership within the community. Local researchers play a fundamental role in the assessment stage and throughout the process of designing, implementing, monitoring, and maintaining EbA measures. Local partners assume leadership roles and contribute to the costs of labor and materials, which further consolidates the sense of ownership. It is therefore advisable to extend the PAR cycle to the whole intervention process, as well as to institutionalize the role of local researchers to facilitate the continuity of the measures after the life of the project. This, together with agreements regarding measure maintenance and monitoring (through actions defined in local policy instruments such as management plans, community regulations, etc.), provides the basis for the sustainability of the measures.

However, it should be noted that PAR is a terrain of knowledge, meanings, and agendas in dispute. The measures described here were not the result of consensus (understood as the agreement of all people) but of a mediated dialogue and a process of collective decision-making. Agreement is reached as a result of exploring differences in a reflective manner, respecting the contributions of different perspectives and disciplines, and creating solutions together. In this respect, it is important to pay attention to the attitudes and behavior of external facilitators and researchers and to ensure that they do not dominate the assessment and EbA design process. Thus, we recommend that knowledge dialogues be mediated and facilitated by a team trained in participatory approaches and methodologies. The team must foster a reflective, conscious process that is cognizant of the many barriers to the dialogue of knowledge (prejudices, power relations, gender gaps, etc.), and guide participants in actively identifying strategies to overcome these obstacles.

Ultimately, local researchers should be at the center of the investigation and the decisions that are made. It is fundamental that we continually ask ourselves who controls and makes decisions, who undertakes the activities, and who benefits from the results. It takes time for local people to take control of a research process, and is a process that is neither easy nor automatic. Nor does it depend solely on the good intentions or skill of the facilitator. It is a process that involves disarming patterns of relationships and beliefs that are deeply rooted. It requires time, perseverance and, above all, constant reflection, both individually and as a group. The experiences in Canchayllo and Miraflores corroborate both the potential and the challenges of participatory research, and highlight the need to deepen and continue the dialogue of knowledge throughout the entire EbA experience.

References

Bernstein JH (2015) Transdisciplinarity: a review of its origins, development, and current issues. J Res Pract 11(1)

Secretariat of the Convention on Biological Diversity (CBD) (2009) Connecting biodiversity and climate change mitigation and adaptation: report of the second ad hoc technical expert group on biodiversity and climate change. Technical series No. 41. CBD, Montreal

Secretariat of the Convention on Biological Diversity (CBD) (2010) X/33 Biodiversity and climate change, decision adopted by the conference of the parties to the convention on biological diversity at its tenth meeting. Naoya, Japan. CBD, Montreal

Cárdenas A (2013) Reporte del especialista en Hidrología. Diagnóstico Rural Participativo Integrado para el diseño de medidas robustas de adaptación al cambio climático en las comunidades campesinas de Canchayllo y Miraflores (Reserva Paisajística Nor-Yauyos Cochas—Perú). Instituto de Montaña, Lima. Internal report

Comunidad Campesina de Canchayllo, Instituto de Montaña, UICN y Reserva Paisajística Nor Yauyos Cochas (CCC, IM, UICN y RPNYC-SERNANP) (2015a) Diagnóstico participativo para el manejo de pastos y agua de la Comunidad Campesina de Canchayllo. The Mountain Institute, Huancayo

Comunidad Campesina de Canchayllo, Instituto de Montaña, UICN y Reserva Paisajística Nor Yauyos Cochas (CCC, IM, UICN y RPNYC-SERNANP) (2015b) Plan de Manejo de Pastos y Agua de la Comunidad Campesina de Canchayllo. The Mountain Institute, Huancayo

Comunidad Campesina de Miraflores, The Mountain Institute, IUCN and Nor Yauyos Cochas Landscape Reserve (CCM, IM, IUCN and RPNYC-SERNANP) (2015a) Diagnóstico participativo para el manejo de pastos y agua de la Comunidad Campesina de Miraflores. The Mountain Institute, Huancayo

Comunidad Campesina de Miraflores, The Mountain Institute, IUCN and Nor Yauyos Cochas Landscape Reserve (CCM, IM, IUCN and RPNYC-SERNANP) (2015b) Plan de Manejo de Pastos y Agua de la Comunidad Campesina de Miraflores. The Mountain Institute, Huancayo

Chambers R (1994) The origins and practice of participatory rural appraisal. World and development, vol. 22, No. 7, Elsevier Science Ltd, London, pp 953–969

Chambers R (1997) Whose reality counts? Putting the First Last. ITDG, London

Dourojeanni P, Fernandez-Baca E, Giada S, Leslie J, Podvin K, Zapata F (2016) Vulnerability assessments for ecosystem based adaptation: lessons from the Nor Yauyos Cochas landscape reserve in Peru. In: Salzmann N, Huggel C, Nussbaumer S, Ziervogel G (eds) Climate change adaptation strategies—an upstream-downstream perspective. Springer, Cham. https://doi.org/10.1007/978-3-319-40773-9_8

Falkembach E, Carillo A (2015) Systematization of experiences: a practice of participatory research from Latin America. In: Bradbury, H. The sage handbook of action research. SAGE Publications Ltd, 55 City Road, London, pp 76–82. https://doi.org/10.4135/9781473921290

Fals Borda O (2016) Por un conocimiento vivencial. In: Damonte G, García M (eds) La Investigación Acción Participativa: referente inspirador de investigación y docencia sobre el agua en América Latina. Serie Agua y Sociedad 23 No 9. PUCP, Lima

FDA (2013) Evaluación del Impacto y Vulnerabilidad del Cambio Climático de la Reserva Paisajística Nor Yauyos Cochas y áreas de amortiguamiento. UNEP, Lima

Friends of Ecosystem-based Adaptation (FEBA) (2017) Making ecosystem-based adaptation effective: a framework for defining qualification criteria and quality standards (FEBA technical paper developed for UNFCCC-SBSTA 46). IIED, IUCN, GIZ, London, Bonn, Gland

GIZ (2014) The vulnerability sourcebook concept and guidelines for standardised vulnerability assessments. GIZ, Bonn

Greenwood, D (2016) Investigación Acción Pragmática. In: Damonte, G, García M (eds) La Investigación Acción Participativa: referente inspirador de investigación y docencia sobre el agua en América Latina. Serie Agua y Sociedad 23 No 9. PUCP, Lima

Greenwood D, Levin M (1998) An introduction to action research. Sage Publications, California

Herout P, Schmid E (2015) Doing, knowing, learning: systematization of experiences based on the knowledge management of HORIZONT3000. Knowl Manag Dev J 11(1): 64–76. http://journal.km4dev.org/
Hidroandes (2015) Estudio hidrogeológico conceptual del parea de influencia del proyecto Chacara-Jutupuquio (RPNYC, Perú). IUCN, Quito Internal report
Hirsch Hadorn G, Hoffmann-Riem H, Biber-Klemm S, Grossenbacher-Mansuy W, Joye D, Pohl C, Wiesmann U, Zemp E (eds) (2008) Handbook of transdisciplinary research. Springer, Dordrecht
IGP (2005) Vulnerabilidad Actual y Futura ante el Cambio Climático y medidas de adaptación en la Cuenca del Río Mantaro. CONAM, Lima
International Institute for Environment and Development (IIED) (2009) Special issue: community-based adaptation to climate change. Participatory learning and action 60. IIED, London
INC (2009) Apu Pariacaca y el Alto Cañete: Estudio de Paisaje Cultural. INC, Lima
INEI (2007) Censo Nacional: IX de Población y VI de Vivienda. Sistemas de Consulta de Datos de Centros Poblados (CCPP) y Población Dispersa. INEI, Lima
INRENA (2006) Plan Maestro 2006-2011. Reserva Paisajística Nor Yauyos Cochas, INRENA, Lima
Intergovernmental Panel on Climate Change (IPCC) (2014) Synthesis report. Contribution of working groups I, II and III to the fifth assessment report of the intergovernmental panel on climate change. Core Writing Team, Pachauri RK, Meyer LA (eds). IPCC, Geneva
Lang DJ, Wiek A, Bergmann M (2012) Transdisciplinary research in sustainability science: practice, principles, and challenges. Sustainability science, vol 7, Supplement 1. Springer, pp 25–43. https://doi.org/10.1007/s11625-011-0149-x
Lo V (2016) Synthesis report on experiences with ecosystem-based approaches to climate change adaptation and disaster risk reduction. Technical series No. 85. CBD, Montreal
Midgley G, Marais S, Barnett M, Wågsæther K (2012) Biodiversity, climate change, and sustainable development—harnessing synergies and celebrating successes. The World Bank, Cape Town Final technical report
MINAM (2011) Inventario y evaluación del Patrimonio Natural en la Reserva Paisajística Nor Yauyos-Cochas. MINAM, Lima
Ñaupari J, Flores E, Yalli B (2013) Reporte del especialista en agrosto-edafología. Diagnóstico Rural Participativo Integrado para el diseño de medidas robustas de adaptación al cambio climático en las comunidades campesinas de Canchayllo y Miraflores (Reserva Paisajística Nor-Yauyos Cochas—Perú). Laboratorio de Ecología y Utilización de Pastizales de la Universidad Nacional Agraria La Molina and Instituto de Montaña, Lima. Internal report
Podvin K, Cordero D, Gómez A (2014) Climate change adaptation in the Peruvian Andes: implementing no-regret measures in the Nor Yauyos-Cochas landscape reserve. In: Murti R, Buyck C (eds) Safe havens: protected areas for disaster risk reduction and climate change adaptation. IUCN, Gland
Pohl C, Hirsch Hadorn G (2007) Principles for designing transdisciplinary research. Oekom Verlag, Proposed by the Swiss Academies of Arts and Sciences, Munich
Ramírez F, Herrera A (2013) Reporte de arqueología en el Diagnóstico Rural Participativo Integrado para el diseño de medidas robustas de adaptación al cambio climático en las comunidades campesinas de Canchayllo y Miraflores (Reserva Paisajística Nor-Yauyos Cochas—Perú). The Mountain Institute, Lima Internal report
Reason P, Bradbury H (eds) (2001) Handbook of action research. Sage Publications, California
Reid H, Podvin K, Segura E (2018) Ecosystem-based approaches to adaptation: strengthening the evidence and informing policy research results from the mountain EbA project. IIED, Peru, London
Rizvi A, Barrow E, Zapata F, Cordero D, Podvin K, Kutegeka S, Gafabusa R, Khanal R, Adhikari A (2014) Ecosystem based adaptation: building on no regret adaptation measures. In: Technical paper 20th session of the conference of the parties to the UNFCCC and the 10th session of the conference of the parties to the Kyoto protocol, IUCN, Gland, Lima, Peru, 1–12 Dec 2014

Rizvi A, Barrow E, Zapata F, Gómez A, Podvin K, Kutegeka S, Gafabusa R, Adhikari A (2015) Learning from participatory vulnerability assessments—key to identifying EbA options. Technical paper. In: 21st session of the conference of the parties to the UNFCCC and the 11th session of the conference of the parties to the Kyoto protocol, IUCN, Gland, Paris, France, 30 Nov–11 Dec, 2015

Ruiz CJ (2013) Reporte del especialista en sistemas productivos. Diagnóstico rural Participativo Integrado para el diseño de medidas robustas de adaptación al cambio climático en las comunidades campesinas de Canchayllo y Miraflores (Reserva Paisajística Nor-Yauyos Cochas—Perú). The Mountain Institute, Lima. Internal report

Segura JE (2013) Reporte del especialista en organización social y cultura. Diagnóstico Rural Participativo Integrado para el diseño de medidas robustas de adaptación al cambio climático en las comunidades campesinas de Canchayllo y Miraflores (Reserva Paisajística Nor-Yauyos Cochas—Perú). The Mountain Institute, Lima. Internal report

Stokols D (2006) Toward a science of transdisciplinary action research. Am J Community Psychol 38:63–77

The Mountain Institute (TMI) (2014) Diagnóstico rural participativo Integrado para el Diseño de Medidas Robusta de Adaptación basada en Ecosistemas en las comunidades campesinas de Canchayllo y Miraflores (Reserva Paisajística Nor-Yauyos Cochas). Internal report

The Mountain Institute (TMI) (2015) Diagnóstico Participativo para el plan de Manejo de pastos y Agua de la comunidad de Miraflores, Yauyos, Lima. TMI and IUCN, Huancayo

The Mountain Institute (TMI) (2018) Análisis participativo de la Situación Actual de las Medidas de Adaptación basada en Ecosistemas de Montaña, implementadas en la reserva Paisajística Nor Yauyos Cochas. TMI, Huancayo Internal report

United Nations Development Programme (UNDP), United Nations Environment Programme (UNEP), International Union for Conservation of Nature (IUCN), The Mountain Institute (TMI) (2016) El futuro ancestral: la adaptación basada en ecosistemas. UNDP, Lima

Vela C (2017) Terminal evaluation of the UN environment project ecosystem based adaptation for mountain ecosystems. Country paper Peru. Final report. UNEP [http://wedocs.unep.org/bitstream/handle/20.500.11822/22413/00609-1866_2017_te_unvironment_global_spcc_Peru_country_paper_eba_mountain.pdf?sequence=4&isAllowed=y]

Vernooy R (2006) La Calidad de la Participación: Reflexiones Fundamentales Acerca de la Toma de Decisiones, Contexto y Metas". In: Gonsalves el at. (eds) Investigación y Desarrollo Participativo para la Agricultura y el Manejo Sostenible de Recursos Naturales: Libro de Consulta. Volumen 1: Comprendiendo. "Investigación y Desarrollo Participativo. Perspectivas de los Usuarios con la Investigación y el Desarrollo Agrícola". Centro Internacional de la Papa and Centro Internacional de Investigaciones para el Desarrollo, Ottawa and Laguna

Wiesmann U, Biber-Klemm S, Grossenbacher-Mansuy W, Hadorn GH (2008) Enhancing transdisciplinary research: a synthesis in fifteen propositions. In: Hadorn H et al (eds) Handbook of transdisciplinary research. Springer, Dordrecht

Zabala N (2000) Diagnóstico rural participativo. Diccionario de Acción humanitaria y cooperación al desarrollo, Icaria and Hegoa, Bilbao

Zapata F, Gómez A (2015) Adaptación basada en Ecosistemas de Montaña: experiencia y lecciones aprendidas en la restauración de tecnologías ancestrales y contemporáneas para el manejo de los pastos y el agua en la puna. Presentation at SEPIA XVI, Arequipa

Zapata F, Rondán V (2016) La Investigación Acción Participativa: Guía conceptual y metodológica del Instituto de Montaña. The Mountain Institute, Lima

Zapata F, Torres M, Gómez A, Podvin K (2016) Informe de sistematización de la experiencia: implementación de las medidas robustas de Adaptación basada en Ecosistemas en las comunidades campesinas de Canchayllo y Miraflores (Reserva Paisajística Nor Yauyos Cochas)". The Mountain Institute and IUCN, Lima

Effects of Climate Change on the Morphological Stability of the Mediterranean Coasts: Consequences for Tourism

Federica Rizzetto

Abstract Over the last decades, the Mediterranean coasts have experienced intense development, mainly due to the growing role of tourism, maritime trade, and energy industry. This situation has favoured a rapid expansion of urban coastal settlements and infrastructures, thus producing negative effects, such as an irreversible reshaping of the littorals, a transformation of their natural resources, and a decrease of their resilience. At present, climate change is worsening this condition by causing modifications on both weather and hydrodynamic processes. In this respect, the past and present impacts of climate change on the morphological stability of the Mediterranean coasts and their effects on tourism have been recently analysed. Data interpretation has shown that the littorals differently respond to this occurrence in relation to their characteristics and to local weather and hydrodynamic conditions. The results have proved that erosion and flooding are two major threats to the preservation of the coasts; they are endangering the beach-based tourism mainly causing land loss and serious damages to human settlements and activities. As in the Mediterranean coasts population is growing and social and economic interests are increasing, the implementation of flood and coastal erosion risk management strategies is necessary to face future challenges related to climate change and to increase coastal resilience.

Introduction: Present Conditions of the Mediterranean Coasts

Mediterranean coasts are among the most threatened ones in the world. Over the last decades, they have undergone intense development, mainly due to the growing role of tourism and recreational activities, maritime trade, industry, agriculture, and fishing. These conditions have produced a rapid poorly planned urbanization and an overexploitation of coastal resources that have been responsible for a number of negative impacts, such as a remarkable irreversible reshaping of the littorals,

F. Rizzetto (✉)
Istituto Di Scienze Marine-Consiglio Nazionale Delle Ricerche, Arsenale-Tesa 104, Castello 2737/F, Venice 30122, Italy
e-mail: federica.rizzetto@ismar.cnr.it

W. Leal Filho et al. (eds.), *Climate Change, Hazards and Adaptation Options*, Climate Change Management, https://doi.org/10.1007/978-3-030-37425-9_38

environmental degradation of ecosystems and habitats, and loss of coastal natural landscapes.

At present, a large percentage of people living in Mediterranean regions is concentrated near the coasts and population densities are much greater in coastal than in non-coastal areas, especially in the southern Mediterranean (UNEP/MAP/MED POL 2005). However, urbanization does not refer only to the increase of people living in coastal areas, but also to a significant expansion of other activities, first of all, tourism, which is the number one contributor to the Mediterranean's economic production (Randone et al. 2017). Indeed, owing to its attractive landscapes, favourable environmental conditions (e.g. mild climate, availability of good quality beaches, and clear seawater), rich biodiversity, and cultural heritage, the Mediterranean regions are leading tourism destinations (European Environment Agency 2014) and half of tourist arrivals are in coastal areas (Fosse and Le Tellier 2017), accounting for about one-third of all global tourism (Randone et al. 2017). Overall, tourism provides a significant contribution to the gross domestic product, even if its value varies from one Mediterranean country to another (Randone et al. 2017): in fact, alongside countries like Italy, Spain, France, Greece, and Turkey that have been tourist destinations for a long time, there are nations like Croatia, Montenegro, Albania, Syria, Tunisia, and Algeria in which tourism is emerging or growing (Eurostat 2001; Randone et al. 2017). Unfortunately, littoralization has also strongly altered structure and functioning of the coasts, thus increasing their vulnerability and reducing their resilience.

Owing to the economic value of the coasts, actions aimed at stabilizing and protecting the littorals from erosion, flooding, and wave attack have been widely carried out in the last decades. As a result, a large amount of shoreline has been armoured. However, these interventions have produced modifications on local hydro- and morphodynamics, which have not always solved erosion problems, but have displaced them to other coastal stretches. Climate change is expected to worsen this situation owing to the great impacts that it could produce on coastal setting and evolution, mainly due to erosion and flooding.

The actions needed for the correct management and protection of coastal zones have to face the new emerging environmental challenges and require an accurate and updated knowledge of the ongoing processes, particularly of those responsible for modifications of beach morphology (Pranzini and Wetzel 2008).

In this context, the present chapter provides an overview of coastal morphodynamics in Mediterranean regions, focusing both on possible impacts of climate change on the morphological stability of their coasts and the resulting effects on tourism and recreational activities. It explains how climate and weather events have affected the different coastal stretches over the last decades in relation to the characteristics of the littorals and what climate change is expected to cause in the future. This knowledge is necessary for hazard identification and risk assessment and, consequently, to define correct future planning and mitigation strategies, mainly in view of the projected climate change.

Methods of Analysis

The effects of climate change on the morphological stability of the Mediterranean coastal areas have been investigated through the analysis of hundreds of multidisciplinary data coming from different sources (Rizzetto and Vacca 2017a; 2017b). Particular attention has been paid to the Mediterranean coastal regions (including islands) characterized by urbanization, concentration of economic activities, and presence of natural and cultural resources.

Firstly the most recent existing studies on coastal morphology and driving physical processes characterizing the Mediterranean littorals at different scales have been selected on the basis of their content, data quality, and reference period. They were realized by scientists, national and international authorities, and working groups within projects or other initiatives aimed at risk assessment and coastal zone management, mainly in view of climate change. The results of these researches are currently available as articles, reports, and maps (sometimes published as interactive web maps).

Secondly data and information derived from these investigations have been stored and organized into a GIS (Geographic Information System) database mostly considering their category, precision, collection period, and quality. The use of this method has made easier and faster data management, analysis, and interpretation. It has given the possibility (i) to visualize the distribution of the different types of coasts on the basis of their geological and geomorphological settings and sedimentological characteristics, (ii) to identify the present conditions of the littorals and their evolution, hydrodynamic processes, sea-level rise trends, local weather and climate conditions (including extreme events), subsidence or uplift, land use, coastal defence measures, and other human activities or interventions, and (iii) to distinguish coastal areas at risk.

Data collection has been carried out from Mediterranean to NUTSIII scale (Eurostat 2015), allowing the morphodynamic analysis of the coasts at different levels of detail.

On the whole, the problems occurred during this analysis have been the scarcity of local and regional studies in non-European countries, irregular and patchy distribution of available information, and the different reference periods for collecting data, which have resulted in a non-homogeneous update of the present conditions of the littorals and their evolutionary trends. In regions lacking in detailed information, these gaps have been partially filled using general observations gathered from articles or reports on the entire Mediterranean coasts.

The identification of the morphological and sedimentological characteristics of the littorals and their evolution in the last years has been particularly useful to distinguish areas that are expected to be more exposed to erosion (e.g. sandy beaches) and flooding (e.g. deltas and estuaries, and tidal flats) under climate change and consequent sea-level rise. Impacts on population and tourism have been identified considering the conditions of the littorals, their use for recreational activities, and the ongoing and foreseen processes.

Therefore, results coming from the integrated data analysis have allowed defining all the factors threatening touristic activities.

Impacts of Climate Change on the Morphological Stability of the Mediterranean Coasts

Climate change represents a serious threat for the Mediterranean coastal areas because it produces remarkable modifications on local weather and hydrodynamics, thus inducing not only a significant reshaping of beach morphology, but also land loss, soil degradation, and desertification. In particular, the major impacts on the coasts are due to changes in temperature (which is one of the main causes of sea-level rise), modifications in rainfall regime, general increase of extreme events (in particular, more intense and frequent storms, characterized by strongest winds and higher waves), and increase in frequency and amplitude of tides. The effects may be magnified because the processes responsible for their occurrence are often interrelated and act concurrently. However, as littorals differently respond to climate change in relation to their characteristics (e.g. geology, geomorphology), conditions (e.g. littoralization, level of artificialization), position, and orientation, and to different local processes occurring along the various coastal stretches (e.g. local hydrodynamics and vertical land movements), impacts could be different under the same scenarios. Besides these factors, socio-economic characteristics also make coastal zones more or less vulnerable to climate change effects.

Sea-level rise and the increase in frequency and intensity of extreme weather events have serious implications on coastal hazards because they mainly produce erosion, flooding, modifications to coastal morphodynamics and salinization, and pollution of coastal aquifers with serious irreversible damages to coastal environments.

In tectonically active areas, rising sea-level can be also offset or amplified by uplift or subsidence, respectively. It is the case, for example, of Croatia and Greece, where different sea-level rise trends measured along the coast are probably the result of differential local uplift and subsidence (Baric et al. 2008; Kontogianni et al. 2012; Hellenic Republic—Ministry of Environment, Energy and Climate Change 2014). In subsiding low-lying areas, vulnerability to rising sea-level and flooding increases. This condition frequently occurs in deltas, such as those of the Ebro (Spain), Var and Rhone (France), Po and Arno (Italy), Nile (Egypt), and Medjerda (Tunisia) rivers (e.g. Stanley 1997; Besset et al. 2017).

The effects of climate change vary over time and also depend on the resilience of the coastal systems. Impacts are more severe on coastal environments that have lower adaptive capacity and this situation may worsen if human actions are not managed properly. In this regard, with a large and growing population in the coastal zones and a low ability to adapt mainly due to low national wealth, coastal countries of Africa appear to be highly vulnerable to sea-level rise (Brown et al. 2011). Based on the findings of the Intergovernmental Panel on Climate Change (IPCC) and the Arab Forum for Environment and Development (AFED), El Raey (2010) stated that the Arab countries are among the most vulnerable in the world to the potential effects of climate change, particularly to the increased average temperatures, less and more erratic precipitation, and sea-level rise, as these territories already suffer from aridity, drought, and water scarcity. These processes have also negative effects on agriculture and, consequently, on economy of the North Africa regions because croplands in surge zones are expected to increase (Dasgupta et al. 2009). In particular, Dasgupta et al. (2009) ranked Egypt in the top-ten most impacted countries (out of 84 developing coastal countries considered world-wide) under sea-level rise scenarios and intensified storm surges, in the light of the conditions of coastal population, settlements, economic activity, and wetlands existing at the time of the study. In this top-ten list, the authors also included Libya, Tunisia, Morocco, and Algeria. In particular, with respect to the economic impact induced by the expected climate change in the southern Mediterranean regions, Algeria could become one of the most vulnerable African nations under the A1B mid-range scenario (Brown et al. 2011), i.e. the emission scenario based on a balance across all sources (IPCC 2007). In this country, sea-level rise only due to climate change could also produce the greatest residual damage cost of any African nation.

Erosion

Erosion is seriously affecting the entire Mediterranean coasts on both low-lying areas and coastal cliffs and is producing shoreline retreat, resulting in a progressive narrowing of the littorals and landward migration or complete disappearance of beaches and marshlands. In coastal stretches characterized by intense anthropization, the situation is more critical owing to the reduction of the beach width also deriving from the seaward expansion of the urban fronts. Therefore, littoralization has diminished the area for wave energy dissipation.

A tentative evaluation of the evolutionary trends of the entire Mediterranean coast has been carried out using the most recent available data (Table 1). In the countries belonging to the European Union, information derived from the EUROSION Project (Salman et al. 2004) has been considered as the starting point for the present analysis. Up to now, data from more recent studies have made it possible to update EUROSION results along 11% of the European Mediterranean littorals, whereas, along the non-European Mediterranean coasts, analysis of existing information has allowed defining the evolutionary trends of 22% of the littorals (Rizzetto and Vacca 2017b). However, as data are constantly updated, these results undergo continuous improvements. At present, lack of information mainly related to coastal stretches having minor socio-economic importance or located in countries characterized by lower national wealth has prevented the knowledge of the evolutionary trends of 44% of the Mediterranean coast. Analysis of existing data along the remaining 56% has shown that about 30% of the littorals is affected by erosion and progradation occurs

Table 1 Percentages of Mediterranean coast subjected to erosion and progradation or under stable conditions calculated using the presently available data

	Total length (km)	No data (%)	Erosion (%)	Progradation (%)	Stability (%)	Artificial coastline (%)
Mediterranean coast	46,441	44	29	5	52	14
European Mediterranean coast	35,346	35	30	5	55	10
Non-European Mediterranean coast	11,095	78	28	6	28	38

only along 5%. Stability and progradation generally refer to armoured shorelines. Despite these gaps, Table 1 provides significant information on the extent of coastal areas in which climate change could cause remarkable impacts.

Vulnerability to erosion is extremely variable as a result of the different characteristics of the littorals; anyway, sandy beaches are the most exposed.

Erosion can be consequence of both natural and anthropogenic processes. In the last decades, human interventions have had a major role in triggering erosion along the entire Mediterranean coast. Supply from rivers, shelf, adjacent embayments, cliff degradation, and biogenic production are the main sources of sediments delivered to the coasts (Hughes 2016). Anthropogenic reduction of sediment inputs (mainly through river mining, canalizations, damming, and decrease of sediment transport efficiency by lowering water discharges due to reduced water availability), human-induced modifications in circulation patterns, and wave-shore interactions have drastically diminished the amount of material necessary for natural beach preservation. Anthropogenic alteration of coastal hydrodynamics has been generally caused by the construction of coastal defences, harbours, marinas, and other offshore structures that have changed coastal circulation. Consequently, the natural flow of sand along the littorals has been obstructed and the longshore transport has been impeded or intercepted, thus preventing the nourishment of the beaches in the downdrift zones. In other cases, engineering structures have increased the nearshore slope, also favouring an intensification of wave energy on the littorals and, consequently, the propagation of erosion waves along the coast. In the future, this condition could be exacerbated by an amplification of wave-induced impacts due to climate change.

On the other hand, significant modifications in sediment supply could be also caused by climate change as a consequence of modifications in precipitation and temperature regimes that could be responsible for variations in river discharge and vegetation cover. In fact, temperature and humidity influence chemical weathering of continental rocks to produce sediments, whereas rainfall and runoff from the mainland have an effect on the rates of sediments provided to the littorals. Consequently, the Mediterranean coastal regions, already affected by severe erosion problems, could be exposed to more severe risks if their expected climate conditions were unfavourable for sediment production and the amount of sediment supply was lower than the rate of material removed by coastal processes.

As erosion also increases the exposure to storm-induced hazards and to the risk of flooding, it can be indirect cause of adverse impacts on human settlements located in the backshore.

On cliffs, erosion causes instability of coastal slopes, thus generating slumps and rockfalls, which represent serious risks for the road and railway networks, the stability of buildings, and the safety of people, including beach tourists, swimmers, sailors, and other persons travelling by boats near the coast. In turn, even if erosion is mainly due to wave-shore interaction and is controlled by the geological structure and the lithological characteristics of the rocks, it is frequently enhanced by the destabilisation induced by the presence of infrastructures (e.g. roads, railways) or by increased load over the underlying rock (e.g. buildings). Climate change could accelerate the process through an intensification of wave and storm surge actions, more intense precipitations, and/or high values of rainfall accumulation. Beaches located in front of the cliffs could reduce their impacts, but if they were eroded, their function would fail.

Highly erodible lithologies are particularly prone to landslides, as observed, for example, in some Slovenian, Croatian, Montenegrin, and Italian flyschoid coastal stretches. However, flysch erodibility varies depending on the resistance of its forming rocks, as it is mainly characterized by alternations of marls, siltstones, sandstones, and calcareous layers. In general, cliffs with wave notches evolve where lithologies are more resistant; on the contrary, the destructive impact of waves, which causes mass movements, earth flows, and rockfalls, is more pronounced in coasts composed by less resistant rocks, even in sheltered zones where wave energy is low, as locally observed by Juračić et al. (2009) along the Croatian coasts. In this case, the increasing frequency of rockfalls that have endangered roads and beaches has been favoured not only by unfavourable rock characteristics and weathering in combination with heavy precipitations, but also by the rapid development of littoral urbanization and tourism that have led to an improper land use (Arbanas et al. 2012; Pikelj et al. 2013).

Another example of rocky coast destabilization due to the concurrence of climate factors and anthropogenic pressures can be observed on the Andalusian coasts (Spain), where very narrow coastal plains originated in front of the slopes of the Betic Mountains. Here, urban development has progressively extended from the plains onto the adjacent areas characterized by steep slopes and high elevations, which are at risk owing to the presence of pre-existing and dormant landslides (Mateos et al. 2017). Nevertheless, they are highly urbanized because their elevations offer excellent sea views and a gentler slope than the adjacent stable zones. In this case, an essential role for landslide reactivation is played by the concentration of load due to urbanization on the upper parts of the landslides, with consequent damages to the buildings and other anthropogenic structures (Notti et al. 2015; Mateos et al. 2017). Such critical conditions could be worsened by intense precipitations and/or high values of rainfall accumulation.

Overall, erosion is also responsible for a general depletion of environmental resources and for damages to nature reserves, archaeological sites, and historical remains. These occurrences, together with the reduction of the available beach surface, have negative effects on tourism and recreation industry. In this regards, many Mediterranean coastal plains have been affected by human activities for hundreds or thousands of years, which have left a rich cultural heritage, including a number of archaeological evidences distributed along the littorals or submerged in shallow waters. Erosion can not only damage or destroy these sites, but also lead to the exposure of the remains that were previously buried, thus causing additional serious problems to the cultural coastal landscapes.

Therefore, the effects of rising temperatures and increasing frequencies and intensity of extreme events could worsen erosion and, consequently, deteriorate the quality of life for local population in general and make the coasts less attractive for tourism.

Flooding

The vulnerability to coastal flooding depends on width, morphological characteristics, and spatial variability of the coasts (including the presence of dune systems) in relation to the intensity of the ocean's forcing (Wright and Short 1984; Sallenger 2000; Sallenger et al. 2003; Duran et al. 2016).

The areas considered as the most vulnerable to flooding are flat and low-lying coastal plains, deltas, estuaries, and sandy littorals characterized by a gentle sloping beach face.

Flooding mainly produces a variety of negative impacts, such as inundation and displacements of lowlands and wetlands, modifications of coastal morphology, shoreline retreat, disappearance of beaches, lagoon and other coastal environments, inundation and damage of human settlements, infrastructures, industrial facilities, harbours, and marinas (thus increasing the risk of pollution), threat to human health and safety, and habitat loss. These damaging effects can intensify as a result of the combination of sea-level rise, storm surges, high tides, and heavy precipitations.

Inundations may be temporary, if they are caused by extreme events (flash floods), or permanent, if they are consequences of sea-level rise. Although flash floods affect the entire Mediterranean regions, their frequency and impact are not homogeneous; for example, in southern Mediterranean this type of flooding is less frequent, but it has usually catastrophic effects with a very high number of casualties (Llassat et al. 2010). Syria is one of the southern countries particularly exposed to flash floods, which are due to very heavy rains (sometimes exceeding 50 mm/hour) and are responsible for inundations that cause severe damages to humans, livestock, and agriculture, and produce soil degradation (Meslmani 2010; Saleh and Allaert 2014). In general, devastating flash floods are more frequent in areas of the Mediterranean region in which atmospheric circulation favours the formation of short intense cloudbursts (Gaume et al. 2009, 2016). Their frequency, magnitude, and impact seem to decrease from west to east with the increasing distance from the Atlantic Ocean (Gaume et al.

2016), which has more influence on weather and climate conditions of the western Mediterranean countries.

Subsiding deltas, such as those of the Ebro, Var, Po, and Nile rivers, are at higher risk of flooding due to their critical surface elevation. The Ebro river delta, which represents one of the most threatened social-ecological Spanish systems, can be cited as an example of coastal environment that is facing an extremely uncertain future (Fatoric and Chelleri 2012). Deficit of sediment supply, sea-level rise (also favoured by both natural and human-induced subsidence), wave action, and storms are the main responsible for its erosion and are exposing its lower zones to an increasing risk of flooding and shoreline retreat. Formerly susceptible to climate change owing to its topographic characteristics, over the last decades it has become more vulnerable also because urbanization and economy have strongly modified its structure and functioning.

Coastal zones characterized by deepening nearshore waters owing to land subsidence can experience a major intensification of wave activity. Similarly, the impacts of flooding on coastal areas subjected to erosion may be higher if erosion causes the disappearance of beaches and wetlands, i.e. coastal environments that commonly function as natural buffers against inundation. For example, such a condition has occurred in the north-western Adriatic coast, where the combined effects of sea-level rise and subsidence, started in the 1970s and strongly decreased after the halt of groundwater withdrawals (Carbognin et al. 2009), have threatened the industrial areas, the urban zones, and the surrounding vast reclaimed marshland, which have become more prone to being submerged.

Saline Intrusion and Pollution of Coastal Aquifers

Other negative impacts of climate change on coastal areas are saline intrusion and pollution of coastal aquifers, which make water undrinkable or unusable for agricultural purposes.

Saline intrusion occurs when saltwater is pushed through the ground into the aquifer. Rising sea level, storm events, and changes in precipitation regime can favour this process. Local geological and geomorphological structures, such as relict features characterized by high-permeability sediments, may control the propagation of seawater and contaminants landward as they represent preferential pathways of penetration (e.g. Rizzetto et al. 2003).

One of the most common conditions that induce saltwater intrusion is the lowering of the water table. This occurrence could worsen owing to a decrease in rainfall and river runoff and to increased water withdrawals in response to future higher demand, which are two situations strictly related to climate change.

In coastal areas of arid and semi-arid regions (e.g. North Africa) the problem is crucial. In fact, the increase of salt-water intrusion due to rising sea level on already scarce groundwater resources could create a great damage on vital water resources and could worsen after significant overexploitation or during droughts because of rising temperatures (e.g. Alfarrah and Walraevens 2018, and references therein).

Another problem that can affect coastal aquifers and can be made worse by climate change conditions is pollution. In fact, the quality of coastal waters may be threatened by heavy precipitations that increase runoff from industrial, urban and agricultural areas, or by warmer temperatures in inland mountain regions and consequent snow melting. In addition, contaminants may be dispersed into the ground during flood and storm events.

The present analysis has shown that the entire Mediterranean coast is experiencing both seawater intrusion and pollution of coastal aquifers. However, unconfined aquifers located in low-lying coasts, also subjected to erosion and flooding, are the most vulnerable.

Consequences of Climate Change for Coastal Tourism in the Mediterranean Regions

In the Mediterranean regions, climate change is expected to cause a number of serious hazards, irreversible modifications of the littorals, and severe damages to economic activities, habitats, biodiversity, and other resources occurring in coastal areas.

In general, climate and weather exert influence on the choice of tourist destinations and affect the success of this type of business. In particular, climate change can limit recreation opportunities by inducing environmental modifications and can be responsible for possible displacement of tourism flows from the coasts in summer owing to extremely hot temperatures and increased frequency of heat waves, also exacerbated by increasing shortage of water resources.

Economic benefits to tourism and recreation industry are mainly given by sandy littorals. Consequently, erosion and flooding represent two major threats to the development of these activities if they reduce the available surface of beaches and have negative impacts (i.e. destruction, damage or submersion) on urban settlements, infrastructures, marinas, cultural sites, natural protected areas, and other tourism opportunities. When the effects of erosion and flooding threaten social and economic activities, special defence measures are usually taken to induce coastal stabilization or progradation. However, if these interventions are incorrectly planned they are ineffective and not solve the problems, but displace them to other zones.

The availability of wide beaches makes the littorals more attractive and increases their carrying capacity. This condition leads to the expansion of urbanization and housing and to the consequent growth of the value of coastal territory. Therefore, the narrowing or disappearance of beaches can cause severe impacts on economy from the local to the national scale. On the other hand, tourism growth frequently triggers an uncontrolled construction of buildings that amplifies the vulnerability of the coasts to sea-level rise and extreme events. This situation represents a serious problem in the coastal regions that have narrow beaches because of their geological or geomorphological configuration. Such a condition occurs, for example, in Syria, where the high concentration of human settlements and economic activities in the coastal strip is mainly due to the physical characteristics of the country, whose hinterland is unsuitable for intense urbanization and land use as it is mountainous, mostly rocky, and with slopes covered by forests (UNEP 2009).

However, geomorphological modifications of coastal features, beach erosion or submersion, and damages to natural resources due to climate change and consequent sea-level rise could have adverse effects not only on common beach uses, but also on other types of recreation activities, such as sportfishing, bird watching, hiking, biking, and scuba diving that are popular in many Mediterranean coastal regions. In particular, negative impacts on nature-based tourism could be mainly due to environmental changes caused, in the long term, by variations of temperature, wind, rainfall patterns, and frequency and intensity of extreme events, as well as by pollution and saltwater intrusion.

Moreover, sea-level rise and modification of sediment dynamics could lead to changes in sailing conditions, induce problems in the functioning of coastal infrastructure, such as ports and marinas, and affect the safety of harbours and activities related with the sea, thus negatively interfering with tourism and other recreation opportunities based on boating.

Rising sea-level, intensification of storm events, modification of precipitation patterns, increase of water withdrawals due to future higher demand and possible decrease in river runoff could worsen salt water intrusion into the aquifers and reduce the availability of drinking water and freshwater for agricultural uses. This occurrence would also favour desertification and, consequently, could make coastal areas less attractive to tourism.

The contamination of coastal aquifers, together with the inundation of low-lying areas, could cause impacts on food supplies and economy because of damages to the local productivity. For example, in countries like Egypt, in which tourism is one of the main contributors to the local economy (Eurostat 2001), climate change could be responsible for devastating effects. Considering that Egypt already suffers from a number of serious problems (e.g. unplanned development, land subsidence, excessive erosion rates, water logging, salt-water intrusion, soil salinization and ecosystem degradation), new climate change impacts on human settlements and infrastructures, agricultural activities, industry, and environmental and cultural resources would represent serious threats to tourism (Egyptian Environmental Affairs Agency 1999). Indeed, sea-level rise on the low coast of Egypt could definitely lead to losses of beaches and cultivated lands, to an exacerbation of water salinization, and to growing health problems. In addition, the effects of rising temperatures and increasing frequencies and severity of extreme events would negatively impact the archaeological Egyptian heritage too (Egyptian Environmental Affairs Agency 2010) and induce the loss of coastal attractiveness.

Erosion, flooding, and scarcity of water supply could be also crucial for islands and coastal areas characterized by the prevailing presence of pocket beaches (i.e. a type of beach particularly common in the Mediterranean regions) because tourism represents the main economic resource of most of them. Moreover, a potential decline in rainfall and river runoff, responsible for a decrease of water availability, could trigger conflicts of interest between local population and tourism authorities on the use of water reserves, particularly where water shortages already occur (Nicholls and Hoozemans 1996; Perry 2000).

Conclusions

The investigations carried out at different scales during the present study have given the possibility to identify threats to the littorals and to recognize climate and weather impacts and their consequences on tourism and recreation activities. Therefore, the huge amount of data and information collected from different sources have provided an updated knowledge of coastal morphodynamics at Mediterranean scale in relation to the physical and morphological characteristics of the coasts and to natural and anthropogenic processes. It also represents the "starting point" for future analyses aimed at coastal safeguarding and conservation.

Data interpretation has made it possible to detect the major problems that coastal areas are already facing and to understand how they could worsen under climate change conditions. Consequently, the results obtained from this study can give important support to all actions aimed at coastal protection, urban planning, management of coastal tourism, environmental restoration, and habitat protection.

As in the future the increase of people living in Mediterranean coastal areas is expected to continue, the population exposed to climate change impacts will increase too. Similarly, economic activities, such as tourism that represents one of the main uses of the coast and a substantial value at both local and national scales, will seriously suffer from these negative effects. Understanding how coastal systems respond to natural events and anthropogenic pressures and if they are able to counteract and survive the effects of climate change under the same forcing is essential to assess coastal risk, to implement the best measures to mitigate its negative effects, and to develop adaptation strategies to increase coastal resilience. Without a correct land-use planning based on a proper knowledge of coastal morphodynamics, these effects could adversely affect social and economic life. Therefore, knowledge is the foundation necessary to guarantee the conservation of the environment and the safety and protection of coastal population and activities.

The results obtained from the present study have also attested that urgent actions are needed to make sure that Mediterranean economy can continue to benefit of tourism in the future. According to Pranzini and Wetzel (2008), since the coasts stretch across different countries characterized by dissimilar socio-economic reality, transboundary policies should be put into practice to reach a coherent and effective approach to coastal management.

In the last years, the awareness of the value of coastal areas considered as tourist destinations is improved. Therefore, many actions have been undertaken to protect these territories and maintain as much as possible their characteristics despite the uncertainties on the future increase of temperature and sea-level rise that still exist and make the implementation of effective mitigation strategies difficult to define on the long term.

Unfortunately, as observed in non-European countries, needed information on coastal morphodynamics is not always available or is inadequate to predict possible future coastal trends. So, major gaps persist in the identification of the behaviour of some coastal stretches and their vulnerability. As a consequence, further studies are necessary to provide the missing data necessary for the correct planning and management of the entire Mediterranean littorals.

Acknowledgements This study has been performed in the frame of the Interreg Mediterranean Project "CO-EVOLVE—Promoting the co-evolution of human activities and natural systems for the development of sustainable coastal and maritime tourism", co-financed by the European Regional Development Fund.

References

Alfarrah N, Walraevens K (2018) Groundwater overexploitation and seawater intrusion in coastal areas of arid and semi-arid regions. Water 10(2), 143: 1–24

Arbanas Z, Grošić M, Udovič D, Mihalić S (2012) Rockfall hazard analyses and rockfall protection along the Adriatic coast of Croatia. J Civ Eng Arch 6(3), Serial No. 52: 344–355

Baric A, Grbec B, Bogner D (2008) Potential implications of sea-level rise for Croatia. J Coastal Res 24(2):299–305

Besset M, Anthony EJ, Sabatier F (2017) River delta shoreline reworking and erosion in the Mediterranean and Black seas: the potential roles of fluvial sediment starvation and other factors. Elementa—Science Anthropocene 5(54): 1–20

Brown S, Kebede AS, Nicholls RJ (2011) Sea-level rise and impacts in Africa, 2000–2100. University of Southampton, UK

Carbognin L, Teatini P, Tosi L (2009) The impact of relative sea level rise on the Northern Adriatic Sea coast, Italy. WIT Trans Ecol Environ 127:137–148

Dasgupta S, Laplante B, Meinsner C, Wheeler D, Yan J (2009) The impacts of sea level rise on developing countries: a comparative analysis. Clim Change 93(3–4):379–388

Durán R, Guillén J, Ruiz A, Jiménez JA, Sagristà E (2016) Morphological changes, beach inundation and overwash caused by an extreme storm on a low-lying embayed beach bounded by a dune system (NW Mediterranean). Geomorphology 274:129–142

Egyptian Environmental Affairs Agency (1999) The Arab Republic of Egypt: initial national communication on climate change. Prepared for the United Nations framework convention on climate change UNFCCC

Egyptian Environmental Affairs Agency (2010) Egypt. Second national communication under the United Nations framework convention on climate change. Arab Republic of Egypt

El Raey M (2010) Impact of sea level rise on the Arab region. University of Alexandria and Regional Center for Disaster Risk Reduction, Egypt

European Environment Agency (2014) Horizon 2020 Mediterranean report. Toward shared environmental information systems. EEA-UNEP/MAP joint report, Publications Office of the European Union, Luxembourg

Eurostat (2001) Tourism trends in Mediterranean countries. Office for Official Publications of the European Communities, Luxembourg

Eurostat (2015) Regions in the European Union—nomenclature of territorial units for statistics—NUTS 2013/EU-28. Publications office of the European Union, Printed in Italy

Fatoric S, Chelleri L (2012) Vulnerability to the effects of climate change and adaptation: the case of the Spanish Ebro Delta. Ocean Coast Manag 60:1–10

Fosse J, Le Tellier J (2017) Sustainable tourism in the Mediterranean: state of play and strategic directions. Plan Bleu paper 17, Plan Bleu, Valbonne, France

Gaume E, Bain V, Bernardara P, Newinger O, Barbuc M, Bateman A, Blaškovičová L, Blöschl G, Borga M, Dumitrescu A, Daliakopoulos I, Garcia J, Irimescu A, Kohnova S, Koutroulis A, Marchi L, Matreata S, Medina V, Preciso E, Sempere-Torres D, Stancalie G, Szolgay J, Tsanis I, Velasco D, Viglione A (2009) A compilation of data on European flash floods. J Hydrol 367:70–78

Gaume E, Borga M, Llassat MC, Maouche S, Lang M, Diakakis M (2016) Mediterranean extreme floods and flash floods. In: Allenvi (ed) The Mediterranean region under climate change. A scientific update. Coll. Synthèses, IRD Editions, Marseille, France: 133–144

Hellenic Republic—Ministry of Environment, Energy and Climate Change (2014) 6th national communication and 1st Biennial report. Under the United Nations framework convention on climate change

Hughes M (2016) Climate change impacts on beaches and estuary sediments. CoastAdapt Impact Sheet 1, national climate change adaptation research facility, Gold Coast

IPCC (2007) Summary for policymakers. In: Solomon S, Qin D, Manning M, Chen Z, Marquis M, Averyt KB, Tignor M, Miller HL (eds) Climate change 2007: the physical science basis. Contribution of working group I to the fourth assessment report of the intergovernmental panel on climate change. Cambridge University Press, Cambridge, United Kingdom and New York, NY, USA

Juračić M, Benac Č, Pikelj K, Ilić S (2009) Comparison of the vulnerability of limestone (karst) and siliciclastic coasts (example from the Kvarner area, NE Adriatic, Croatia). Geomorphology 107:90–99

Kontogianni A, Tourkolias C, Skourtos M, Papanikolaou M (2012). Linking sea level rise damage and vulnerability assessment: the case of Greece. In: Young SS, Silver SE (ed) International perspectives on global environmental change, Section 19: 375–398

Llasat MC, Llasat-Botija M, Prat MA, Porcù F, Price C, Mugnai A, Lagouvardos K, Kotroni V, Katsanos D, Michaelides S, Yair Y, Savvidou K, Nicolaides K (2010) High-impact floods and flash floods in Mediterranean countries: the FLASH preliminary database. Adv Geosci 23:47–55

Mateos RM, Azañón JM, Roldán FJ, Notti D, Pérez-Peña V, Galve JP, Pérez-García JL, Colomo CM, Gómez-López JM, Montserrat O, Devantèry N, Lamas-Fernández F, Fernández-Chacón F (2017) The combined use of PSInSAR and UAV photogrammetry techniques for the analysis of the kinematics of a coastal landslide affecting an urban area (SE Spain). Landslides 14:743–754

Meslmani Y (2010) Initial national communication of the Syrian Arab Republic. Ministry of state for environment affairs (MSEA), in collaboration with United Nation development program (UNDP) in Syria, and the global environmental facility (GEF), submitted to the United Nations framework convention on climate change (UNFCCC), Damascus, Syria

Nicholls RJ, Hoozemans FMJ (1996) The Mediterranean: vulnerability to coastal implications of climate change. Ocean Coast Manag 31(2–3):105–132

Notti D, Galve JP, Mateos RM, Montserrat O, Lamas-Fernández F, Fernández-Chacón F, Roldán-García FJ, Pérez-Peña JV, Crosetto M, Azañón JM (2015) Human-induced coastal landslide reactivation. Monitoring by PSInSAR techniques and urban damage survey (SE Spain). Landslides 12:1007–1014

Perry AH (2000) Impacts of climate change on tourism in the Mediterranean: adaptive responses. Fondazione Eni Enrico Mattei, Working Paper No. 35.00

Pikelj K, Dragnić V, Malovrazić N (2013) Eastern Adriatic: Slovenia, Croatia and Montenegro. In: Pranzini E, Williams A (ed) Coastal erosion and protection in Europe. Routledge, Taylor & Francis Group, London, UK, pp 324–344

Pranzini E, Wetzel L (2008) Managing Mediterranean beaches: the need for quality and standardised data in beach monitoring at different scales. In: Pranzini E, Wetzel L (eds) Beach erosion monitoring: results from BEACHMED-e/OpTIMAL project. Nuova Grafica Fiorentina, Italy, Firenze, pp 9–11

Randone M, Di Carlo G, Costantini M (2017) Reviewing the economy of the Mediterranean sea: actions for a sustainable future. WWF Mediterranean Marine Initiative, Rome, Italy
Rizzetto F, Tosi L, Carbognin L, Bonardi M, Teatini P (2003) Geomorphological setting and related hydrogeological implications of the coastal plain south of the Venice Lagoon (Italy). In: Servat E, Najem W, Leduc C, Shakeel A (eds) Hydrology of Mediterranean and Semiarid regions, IAHS Publication no. 278: 463–470
Rizzetto F, Vacca C (2017a) Coastal morphodynamics in Mediterranean touristic areas under climate change conditions. Interreg Mediterranean CO-EVOLVE project, activity 3.2 "Threats to co-evolution—Mediterranean scale: climate changes and morphological stability", Deliverable 3.2.1., Unpublished report
Rizzetto F, Vacca C (2017b) Mapping of coastal morphodynamics descriptors in Mediterranean touristic areas. Interreg Mediterranean CO-EVOLVE project, activity 3.2 "Threats to co-evolution—Mediterranean scale: climate changes and morphological stability", Deliverable 3.2.2., Unpublished report
Saleh HA, Allaert G (2014) Disaster management and risk reduction: impacts of sea level rise and other hazards related to tsunamis on Syrian coastal zone. In: Tang DL, Sui GJ (eds) Typhoon impact and crisis management, advances in natural and technological hazards research 40: 481–537
Sallenger AH Jr (2000) Storm impact scale for barrier islands. J Coastal Res 16:890–895
Sallenger AH Jr, Krabill WB, Swift RN, Brock J, List J, Hansen M, Holman RA, Manizade S, Sontag J, Meredith A, Morgan K, Yunkel JK, Frederick EB, Stockdon H (2003) Evaluation of airborne topographic LiDAR for quantifying beach changes. J Coastal Res 19:125–133
Salman A, Lombardo S, Doody P (2004) Living with coastal erosion in Europe: sediment and space for sustainability. EUROSION project report, service contract B4-3301/2001/329175/MAR/B3, "Coastal erosion—Evaluation of the need for action", directorate general environment, European Commission. http://www.eurosion.org
Stanley DJ (1997) Mediterranean deltas: subsidence as a major control of relative sea-level rise. In: Briand F, Maldonado A (eds) Transformations and evolution of the Mediterranean coastline, Bulletin de l'Institut Océanographique, Monaco 18, CIESM Science Series 3: 35–62
UNEP (2009) Syria's coastal zone and its desired integrated management. Proposed vision and policy. UNEP/MAP-METAP SMAP III project: promoting awareness and enabling a policy framework for environment and development integration in the Mediterranean with focus on integrated coastal zone management. Priority actions programme, Regional Activity Centre, Split, Croatia
UNEP/MAP/MED POL (2005) Transboundary diagnostic analysis (TDA) for the Mediterranean sea. UNEP/MAP, Athens
Wright L, Short A (1984) Morphodynamic variability of surf zones and beaches: a synthesis. Mar Geol 56:93–118

Chennai City and Coastal Hazards: Addressing Community-Based Adaptation Through the Lens of Climate Change and Sea-Level Rise (CBACCS)

A. Saleem Khan, M. Sabuj Kumar, R. Sudhir Chella and B. Devdyuti

Abstract Low-lying coastal cities are chronic to the bleakness of the coastal hazards and climate change exacerbates these hazards into multifold. Sea-level rise (SLR), intensified storms and storm surges, etc. will have a profound impact on coastal regions and coastal communities. In particular, densely populated coastal cities are at high risk to the impacts of changing climate and rising sea-levels. However, adaptation is considered as one of the appropriate methods to address these challenges and community-based adaptation (CBA) of urban regions is seldom discussed and warrants urgent attention. Grounding on these premises and based on the United Nations Sustainable Development Goals (SDG 11-Sustainable Cities and Communities and SDG 13-Climate Action), this paper explores CBA through the lens of climate change and SLR of the hazard prone coastal city such as Chennai city (India). In this context, this paper reviews the nature of Chennai city's exposure and vulnerability to coastal hazards and outlines the impacts and challenges of a changing climate and rising sea-levels. This study has made based on a combination of qualitative and quantitative methodology in which primary and secondary sources are tapped for information and perspectives and secondary literatures are collected based on keyword search technique. The findings outline (a) Chennai city is vulnerable to coastal natural hazards and climate change; (b) SLR poses serious threats to Chennai city's coastal infrastructures, coastal natural resources and coastal communities; (c) Adaptation is considered as one of the suitable response strategies to SLR and CBA warrants urgent attention. Importantly, this study introduces "CBACCS (Community Based Adaptation for Chennai city to Climate change and Sea-level rise)" approach and emphasizes the need of CBA for Chennai city to address these challenges. Thus, the information gleaned from this study will provide insights to local level policy planners, decision makers and research penchants to understand and address CBA for coastal cities like Chennai through the lens of climate change and SLR.

A. Saleem Khan (✉) · M. Sabuj Kumar · R. Sudhir Chella · B. Devdyuti
Department of Humanities and Social Sciences and Indo-German Centre for Sustainability, Indian Institute of Technology Madras (IIT-M), Chennai, Tamil Nadu 600036, India
e-mail: asaleemkhan.cc@gmail.com

W. Leal Filho et al. (eds.), *Climate Change, Hazards and Adaptation Options*,
Climate Change Management, https://doi.org/10.1007/978-3-030-37425-9_39

Introduction

Coasts are highly dynamic and geo-morphologically complex systems, which respond in various ways to extreme weather events. Coastal floods are regarded as among the most dangerous and harmful of natural disasters (Douben 2006; Balica et al. 2012). Climate change is one of the foremost emerging challenges, the impacts of which are increasingly manifesting themselves through highly erratic instances of weather deviations and induced extreme events. In addition to all anthropological stresses that coastal cities are facing now the implications of exacerbated pressures from the prospects of climate change and associated sea-level rise (SLR) will be of serious concern in the future (Kumar and Kunte 2012).While both urban and rural areas are vulnerable to climate change, its impacts on cities and towns are of particular concern due to high concentrations of people and infrastructure in these areas (TERI 2014). The lesser developed countries have a significantly higher proportion of their total populations and their urban populations in the low elevation coastal zone (McGranahan et al. 2007), suggesting that the impacts of climate change, including the heavy socio-economic costs created by climate-related hazards and disasters, will probably be greater on coastal regions of developing countries with fewer resources for mitigation and adaptation (Nicholls et al. 2007; Oliver-Smith 2009). Many Asian cities are also particularly at risk. Asia has many of the world's largest cities/metropolitan areas that are in the floodplains of major rivers and cyclone-prone coastal areas (Huq et al. 2007). Accelerated SLR poses a major threat to low-lying coastal cities and coastal communities. Rising sea- levels have many adverse impacts, including inundation of coastal plains, increased beach and coastal erosion, removal of protective sand dunes and vegetation, and intrusion of salt water into fresh water supplies, already concern in many Asian cities (Fuchs 2010). Though, cities are on the climate change frontline in terms of experiencing the impacts of a changing climate, it also stands first in terms of taking measures to adapt to these impacts and build resilience (Archer et al. 2017).

Adaptation to climate change is a process whereby individuals and communities seek to respond to "...actual or expected climatic stimuli or their effects" (Mukheibir and Ziervogel 2007). As many studies demonstrate, cities have begun to proactively address climate risks as 'opportunities to develop responsive climate change adaptation measures, both in the near- and long-term, that protect their citizens and the infrastructure on which they depend' (Funfgeld 2010). Furthermore, the adaptation process must be localized, so that communities can reorganize to address increased environmental uncertainty. Community-based adaptation (CBA) has increasingly gained momentum and it is 'a community-led process, based on communities' priorities, needs, knowledge, and capacities, which should empower people to plan for and cope with the impacts of climate change' (Reid and Huq 2014). CBA draws on participatory approaches and methods developed in disaster risk reduction and community development work as well as sectoral approaches. Innovative participatory methods also emerging to help communities analyze the causes and effects of climate change (SLR), to integrate scientific and community knowledge of climate change (SLR) and to plan adaptation measures (Reid et al. 2009). However, to build resilience among these vulnerable communities, a city can incorporate CBA into city plans, and cities may find it all the more beneficial to pursue a participatory process of prioritization that integrates both top-down and bottom-up, community-based perspectives (The World Bank Group 2011).

Meanwhile, the international community made an unprecedented set of commitments to pursuing a sustainable future. As a result three landmark global agendas were produced: the Paris Agreement; the Sustainable Development Goals (SDGs) within the framework of the 2030 Agenda for Sustainable Development; and the Sendai Framework for Disaster Risk Reduction 2015–2030. The world sets course for a transition to low-carbon, climate-resilient societies and economies, with countries working towards common goals while focusing on national circumstances, challenges and opportunities. Adapting to climate change is a key objective of the three post-2015 agendas integrating adaptation with the SDGs and the Sendai Framework can be very beneficial for building resilience comprehensively across societies (UNFCCC 2017). The Goal 11 of SDG's i.e. "Sustainable Cities and Communities" outlines that many cities around the world are facing acute challenges in managing rapid urbanization from ensuring adequate housing and infrastructure to support growing populations, to confronting the environmental impact of urban sprawl, to reducing vulnerability to disasters. It sets its targets (i) enhance inclusive and sustainable urbanization and capacity for participatory, integrated and sustainable human settlement planning and management in all countries; (ii) to substantially increase the number of cities and human settlements adopting and implementing integrated policies and plans towards inclusion, resource efficiency, mitigation and adaptation to climate change, resilience to disasters, and develop and implement, in line with the Sendai Framework for Disaster Risk Reduction 2015–2030, holistic disaster risk management at all levels; (iii) the proportion of cities with a direct participation structure of civil society in urban planning and management that operate regularly and democratically; and (iv) other targets (UNSDG Report 2018). Likewise, the Goal 13 of SDGs i.e."Climate Actions" states that the world continues to experience rising sea levels, extreme weather conditions and increasing concentrations of greenhouse gases. This calls for urgent and accelerated action by countries as they implement their commitments to the Paris Agreement on Climate Change. It sets its targets (i) to strengthen resilience and adaptive capacity to climate-related hazards and natural disasters in all countries; (ii) to improve education, awareness-raising and human and institutional capacity on climate change mitigation, adaptation, impact reduction and early warning; (iii) the proportion of local governments that adopt and implement local disaster risk reduction strategies in line with national disaster risk reduction strategies; and (iv) other targets (UNSDG Report 2018). Responding to the call of UN SDG's, this study is intended to address the targets of the UN SDG Goal 11 and Goal 13.

In this context, the aim of this study is to address the CBA through the lens of climate change and SLR for Chennai city. Chennai is one of the coastal cities lie along the Bay of Bengal, is a cyclone-prone area and is likely vulnerable to natural hazards such as coastal hazards and impacts of changing climate such as SLR. Occurrence of Tsunami in 2004, frequent floods due to cyclone especially during North East Monsoon lead to lot of structural changes along the shores of the city. Changing climate scenario and resultant SLR in Chennai are also reported to be one of the factors for the coastal vulnerability hazard (Marriappan and Devi 2012). Realizing the need of the hour with respect to Chennai city and coastal hazards, this study has set three objectives viz., (1) to outline the coastal natural hazards of Chennai city; (2) to review SLR and its impacts on Chennai city; (3) to evolve an approach to address CBA for Chennai city to rising sea-levels. For this purpose, this study has adopted a combination of qualitative and quantitative methodology in which primary and secondary sources are tapped for information and perspectives and, secondary literature are collected based on keyword search technique.

Chennai City: A Profile

Chennai is also known as Madras is the capital city of the Indian state of Tamil Nadu. Located on the Coromandel Coast off the Bay of Bengal, it is a major commercial, cultural, economic and educational center in South India. Chennai, is also referred to as the "Gateway to South India," is located on the southeastern coast of India in the north–eastern part of Tamil Nadu on a flat coastal plain known as the Eastern Coastal Plains (DES-GoTN 2017). Geographically, it is located between 12.75°–13.25°N and 80.0°–80.5°E on the southeast coast of India and in the northeast corner of Tamil Nadu (Fig. 1). It is India's fourth largest metropolitan city covering an area of 1177 km^2. The city stretches nearly 25 km along the bay coast from the southern part to the northern part of the city (Boominathan et al. 2008). As per the 1971 census, Chennai's population was 0.3 million, which has increased to more than 8 million as per census 2011 (Aithal and Ramachandra 2016). According to Census 2011 the total population of Chennai city is, 4,646,732 (Census of India 2011). Thus, the Chennai metropolitan area is the fourth most populated agglomeration in India and 31st largest urban area in the world (DES-GoTN 2017). Chennai city is governed by the Corporation of Chennai erstwhile, Corporation of Madras, established in 1688. The area of jurisdiction of the Corporation of Chennai expanded to 426 sq kms from 174 sq kms in 2011. The expanded Chennai Corporation is divided into three regions—North, South and Central (MSME 2016) which covers 200 wards (previously 155 wards and 10 zones) with 15 zones. The Chennai Metropolitan Development Authority (CMDA) is the nodal agency responsible of planning and development of the Chennai Metropolitan Area (CMA), which is spread over an area of 1,189 km^2 (459 sq mi), covering the Chennai district and parts of Thiruvallur and Kanchipuram districts.

The city is intersected by two streams, viz, Coovam river and the Adyar. The Coovam river runs through the heart of the city and enters the sea. The Adyar river runs through the south part of the city and enters the sea. A third river, the Kosasthalaiyar, flows through the northern fringes of the city before draining into the Bay of Bengal, at Ennore. The Buckingham canal runs almost parallel to the Coromandal coast within the limits of 5 km from the coast. It joins up with a series of natural backwaters and

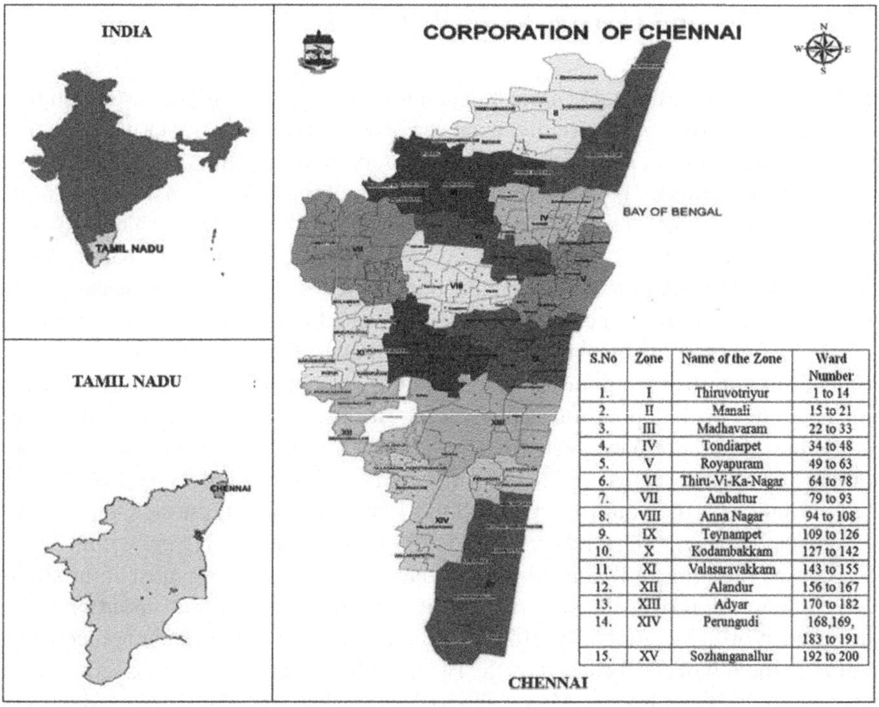

S.No	Zone	Name of the Zone	Ward Number
1.	I	Thiruvotriyur	1 to 14
2.	II	Manali	15 to 21
3.	III	Madhavaram	22 to 33
4.	IV	Tondiarpet	34 to 48
5.	V	Royapuram	49 to 63
6.	VI	Thiru-Vi-Ka-Nagar	64 to 78
7.	VII	Ambattur	79 to 93
8.	VIII	Anna Nagar	94 to 108
9.	IX	Teynampet	109 to 126
10.	X	Kodambakkam	127 to 142
11.	XI	Valasaravakkam	143 to 155
12.	XII	Alandur	156 to 167
13.	XIII	Adyar	170 to 182
14.	XIV	Perungudi	168,169, 183 to 191
15.	XV	Sozhanganallur	192 to 200

Fig. 1 Study area—Chennai city, India. *Source* zone wise map—corporation of Chennai

connects all the coastal districts from Guntur to Cuddalore (MSME 2013). Several lakes of varying size are located on the western fringes of the city. Chennai is classified as being in Seismic Zone III, indicating a moderate risk of damage from earthquakes (DES-GoTN 2017). The average elevation of the city is about 8 m above the mean sea level (Aithal and Ramachandra 2016). However, Chennai's terrain is very flat with most of the area at an average elevation barely 2 m above mean sea level and certain areas are even below the mean sea level, which makes it difficult to manage inundation and de-watering during monsoon and floods (Chennai Corporation 2017). Chennai has a tropical wet and dry climate. The weather is hot and humid for most of the year. The hottest part of the year is late May to early June with maximum temperatures around 35–40 °C. The coolest part of the year is January, with minimum temperatures around 15–22 °C (ENVIS 2015). The analysis of mean annual temperature data from 1951 to 2010 of Chennai city reveals that the mean annual temperature has increased 1.3 °C since the last 60 years (Jeganathan and Andimithu 2012). The average annual rainfall is about 140 cm (55 inches). The city gets most of its seasonal rainfall during the northeast monsoon period from mid–October to mid–December. Prevailing winds in Chennai are usually southwesterly between April and October and northeasterly during the rest of the year (ENVIS 2015).

Chennai City and Coastal Natural Hazards

The topography of Chennai is almost flat and the ground level in the district slightly rises up to 22 ft. (6.7 m) above the mean sea level and favors devastating impact on this low lying coastal city to any coastal disasters including the threat of rising sea levels (TNENVIS 2017). Chennai's coastline, which includes tourist resorts, ports, hotels, fishing villages, and towns, has experienced threats from many disasters such as storms, cyclones, floods, tsunami, and erosion (Kumar and Kunte 2012). The city experiences floods during monsoon, which is the dominant season of the year. During floods, the water level of different water bodies rise and flow to different floodplains submerging them (Manohar and Muthaiah 2016). There were several past instances of catastrophic floods in Chennai (1943, 1976, 1984/85) caused by heavy rain associated with depressions and cyclonic storms, led to flooding in major rivers and failure of and drainage systems (Gupta and Nair 2010). Chennai was severally flooded due to heavy rains (16–20 cm, attributed to a trough of low pressure from the Gulf of Mannar to the South-west bay off the Tamil Nadu coast) during October 30 to November 2 during 2002. On November 5, 2004, a heavy rainfall (6 cm within 24 h or less) caused flooding and water-logging in many areas, inundating most of the slums. A deep depression over the Bay of Bengal brought 42 cm rainfall in around 40 h during the North-east monsoon of 2005 (Gupta and Nair 2010). The unprecedented rainfall in the fall of 2015 caused by the El Nino phenomenon had collapsed Chennai massively with flash floods (Manohar and Muthaiah 2016). This devastating flood that hit Chennai city and other parts of Tamil Nadu during November-December 2015 have claimed more than 400 lives and caused enormous economic damages (Narasimhan et al. 2016).

On the other hand, Chennai is prone to coastal hazards and it was severely affected by the 2004 Indian Ocean Tsunami. Marina beach present in this region witnessed maximum impact due to its flat topography, resulting in an inundation of about 300 m landward with a high flow velocity of the order of 2 m/s. (Usha et al. 2012). In Tamil Nadu 7965 people lost their lives, whereof most of the fatalities (6051) were recorded in Nagapattinam and 206 in Chennai (Sheth et al. 2006; Gopinath et al. 2014) due to 2004 Tsunami. The 2005 North Indian Ocean cyclone season, namely Cyclone Baaz was destructive and deadly to southern India, despite the weaker storms. Chennai recorded 420 mm (17 in) of rainfall. The storm brought several days of heavy rainfall to southern India, forcing 50,000 people to evacuate. The city experienced number of other tropical cyclones such as Cyclone Ogni 2006 (Ogni formed south of Kanyakumari, and followed the coast of Tamil Nadu throughout its lifetime as it moved northeastwards. It passed over Chennai as a deep depression with winds of 55 km/h); Cyclone Nisha in 2008; cyclone Jal 2010 (this storm is the first to have directly impacted Chennai in this list. Maximum winds 110 km/h. Winds during landfall 55 km/h); cyclone Laila in 2010 (It did not hit Chennai directly, but peaked about 150 km east of Chennai. Maximum winds 100 km/h and battered the Coastal areas in Northern parts of Tamil Nadu, Chennai city and its suburbs); cyclone Thane in 2011; cyclone Nilam in 2012 (originated over the Bay of Bengal,

Nilam moved close to Sri Lanka, recurved northwards and hit Mahabalipuram, Tamil Nadu with peak winds of 85 km/h. Significant impact to Chennai); cyclone Nada and cyclone Vardha in 2016. It crossed the eastern coast of India over Chennai on December 12, 2016. Winds were estimated at 100 mph (160 km/h) during landfall. Cyclone Vardha is one of the worst hit cyclones that Chennai city has ever faced and left a trail of destruction (TNENVIS 2017). An outline of the important and extreme coastal events of Tamil Nadu and Chennai city is given in Table 1.

Chennai City and Climate Change Induced Sea-Level Rise

Cyclones and storm surges could have a devastating impact on large coastal urban centers, including the megacities such as Chennai. Besides this, SLR can put a number of regions and cities at risk (Sharma and Tomar 2010). The recorded relative sea-level trend for Chennai City (Chennai Port) based on tide gauge data from Permanent Service for Mean Sea Level (PSMSL) is 0.52 mm/year with a 95% confidence interval of $\pm$0.35 mm/yr based on monthly mean sea level data from 1916 to 2013 (with the data gap from 1921 to 1952) which is equivalent to a change of 0.17 ft. in 100 years (Fig. 2) (NOAA 2018). Nevertheless, the Fifth Assessment Report of Intergovernmental Panel on Climate Change (IPCC) based on climate models has projected future global SLR for 2081–2100 relative to 1986–2005 will likely be in the range of 55–82 cm (IPCC 2013a). IPCC has also projected a regional SLR of about 50 cm in the Bay of Bengal by 2100 from 2006 base level (IPCC 2013b; SoER-TN 2017). However, Ramachandran et al. 2017 has projected local SLR based on IPCC AR5 for Tamil Nadu coast as it ranges from 36.98 to 78.15 cm. The study has also project SLR for Chennai, city with the range of 07.10 cm in 2025 and 36.87 in 2100. It is important to note that the accelerated SLR will have impacts on all coastal regions. The effects of SLR on the coasts are not uniform, but vary considerably from region to region and over a range of temporal scales (Nicholls et al. 2007; Khan 2013).

The SLR is likely to have the following impacts on the coast of Tamil Nadu such as inundation, flooding and storm damage; wetland loss, beach erosion, saltwater intrusion and, rising water tables impeding drainage (SoER-TN 2017). Byravan et al. (2010) has calculated the land area of 1241 acres in Chennai district are within the elevation of up to 1 m (in other words, the above said acres of land area are at risk of inundation to 1 m SLR) and also estimated the Probable Maximum Storm Surge (PMSS) above the tidal level in Chennai as 4.19 m (S100 is 2.9 m, S100 = Current 1 in 100 m surge height) with areas below 5 m are at risk. A number of coastal infrastructures and coastal resources that fall below at 10 m elevation are also exposed to the risk of rising sea levels. For instance, Ennore Port Trust, Ennore Minor Port, Kattupalli Port (Thiruvallur District adjacent to Chennai) and Chennai Port Trust are less than 1 m elevation level; Chennai east coast road cumulative length of 0.4 km up to 1 m; North Chennai thermal power station of 3–5 m elevation, Ennore thermal power station and Basin bridge power station are in the elevation

Table 1 List of some of the important extreme coastal events of Tamil Nadu and Chennai City

S.No	Year	Name of the extreme coastal events (cyclone/storms/flood/tsunami)	Short description
1.	2018	Gaja	Severe cyclonic storm Gaja crossed Tamil Nadu and Puducherry coast between Nagapattinam and Vedaranniyam with an average wind speed of 120 km/h during 10–20 Nov 2018. Though cyclone Gaja lashed Tamil Nadu with moderate to heavy rains, but the state capital of Chennai escaped the fury of the storm and only managed with light showers, with few intense spells over some isolated pockets
2.	2017	Ockhi	Ockhi is one of the strong tropical cyclone formed around 21 November, 2017 in the eastern Andaman Sea. It passed just south of Puthalam, the southern tip of mainland India and caused havoc and destruction in the southernmost districts of Tamil Nadu particularly Kanyakumari. However, Chennai did not experience much the wrath of Ockhi
3.	2016	Vardah	Vardah is one of the intense tropical cyclone of the 2016 North Indian Ocean cyclone season. It crossed the eastern coast of India over Chennai on December 12, 2016 Winds were estimated at 100 mph (160 km/h) during landfall
4.	2016	Nada	Cyclone Nada with wind-speed of 45–54 km per hour lost its strength and slowed down to 35 km per hour before making landfall near Karaikal which is roughly about (350 km away from Chennai) on 2nd Dece 2016
5.	2015	2015 Chennai floods	2015 Chennai floods resulted from heavy rainfall generated by the annual northeast monsoon in November–December 2015. Chennai received 1049 mm (41.3 in) of rainfall in November, the highest recorded since Nov 1918 when 1088 mm (42.8 in) in of rainfall was recorded. The flooding in Chennai city was described as the worst in a century

(continued)

Table 1 (continued)

S.No	Year	Name of the extreme coastal events (cyclone/storms/flood/tsunami)	Short description
6.	2012	Nilam	Originated over Bay of Bengal, Nilam moved close to Sri Lanka, recurved northwards and hit Mahabalipuram, Tamil Nadu with peak winds of 85 km/h. Significant impact to Chennai. 28 Oct-1 Nov 2012
7.	2011	Thane	Very severe cyclonic storm Thane was the strongest tropical cyclone of 2011 within the North Indian Ocean. This storm formed over central Bay of Bengal, and crossed the coast of Tamil Nadu near Cuddalore with winds of 140 km/h. 25–31 Dec 2011
8.	2010	Jal	This storm is the first to have directly impacted Chennai in this list. Jal originated over the South China Sea, crossed the Malay Peninsula, intensified over Bay of Bengal and directly hit Chennai as a deep depression. Maximum winds 110 km/h. Winds during landfall 55 km/h. 18 Nov 2010
9.	2010	Laila	Laila was the first cyclonic storm to affect southeastern India in May since the cyclone. It did not hit Chennai directly, but peaked about 150 km east of Chennai. Maximum winds 100 km/h and battered the Coastal areas in Northern parts of Tamil Nadu, Chennai city and its suburbs. 17–21 May, 2010
10.	2008	Nisha	Cyclonic storm Nisha formed close to north Sri Lanka and moved inland near Karaikal. Maximum winds 85 km/h. The storm brought heavy rainfall which resulted in 189 deaths in Tamil Nadu. 25–29 Nov 2008
11.	2006	Ogni	Ogni formed south of Kanyakumari, and followed the coast of Tamil Nadu throughout its lifetime as it moved northeastwards. It passed over Chennai as a deep depression with winds of 55 km/h. 29–30 Oct 2006

(continued)

Table 1 (continued)

S.No	Year	Name of the extreme coastal events (cyclone/storms/flood/tsunami)	Short description
12.	2005	Fanoos	This storm, like the previous one originated near the Andaman and Nicobar Islands. Moving westwards, it attained peak intensity with winds of 85 km/h and made landfall near Vedaranyam, Tamil Nadu. The storm was known for its excessive rainfall. 6–10 Dec 2005
13.	2005	Baaz	The 2005 North Indian Ocean cyclone season was destructive and deadly to southern India, despite the weak storms. Maximum winds 85 km/h over Bay of Bengal, 55 km/h during landfall over Tamil Nadu. 28 Nov-2 Dec 2015
14.	2004	Tsunami	The 2004 Indian Ocean Tsunami had a significant impact on the Tamil Nadu coast. In Chennai mainly the beach area and fishing facilities were destroyed. Marina beach present in this region witnessed maximum impact due to its flat topography, resulting in an inundation of about 300 m landward with high flow velocity of the order of 2 m/s. 26 Dec, 2004
15.	2002	Severe cyclonic storm BOB 04	The storm started as a depression off the coast of Chennai. It then tracked northeast and crossed the coast over West Bengal. Maximum winds 100 km/h during landfall. 10–12 Nov, 2002

level of 5–10 m (Byravan et al. 2010). Likewise, Anushiya (2015) has quantified the area of inundation of Chennai LULC to the predicted impact of 0.5 m SLR. The areas of 21.75 sq. km of Chennai LULC with wetlands, water bodies, forest areas, settlements, etc. are at high risk of inundation to SLR. The main wetland namely Pallikaranai wetland, plays a vital role in urban ecosystem with rich biodiversity of both flora and fauna and it is identified now as more vulnerable to projected SLR of 0.5 m. It has been recorded that the inundation area is greater in southern and northern zones of Chennai city with nearly 120,000 person per sq.km in these zones and poses serious risks to its inhabitants (Anushiya 2015). The Climate Disaster and Resilience Initiative have conducted research on the administrative zones within Chennai. Results indicate that citizens living in the slum areas in the North and in the older, densely populated zones in the center are often more vulnerable. These zones

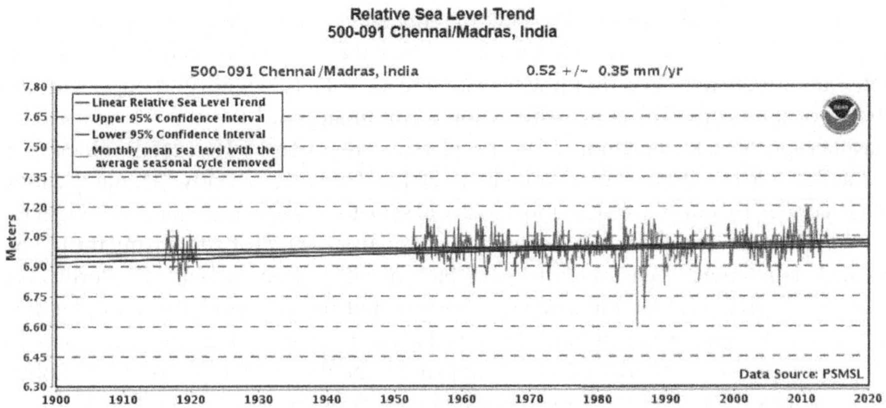

Fig. 2 Relative sea-level trend of Chennai/Madras, India. *Source* NOAA

have a greater potential to be exposed to natural disasters such as storms, floods etc. Slum dwellers in Chennai are particularly vulnerable because slums are located in areas that are environmentally hazardous and lack protective infrastructure. Many of these slums are in low-lying coastal zones and thus, increasingly vulnerable to SLR, extreme weather events, and flooding (USAID 2016).

Chennai City Through the Lens of Coastal Adaptation

The task of adapting cities to the impacts of climate change such as SLR is of great importance because urban areas are hotspots of high risk given their concentrations of population and infrastructure; their key roles for larger economic, political and social processes; and their inherent instabilities and vulnerabilities. Yet, the discourse on urban climate change adaptation has only recently gained momentum in the political and scientific arena (Birkmann et al. 2010). IPCC (2014) has defined adaption as the process of adjustment to actual or expected climate and its effects. In other words, it may be persuasive to state that one of the effective ways to address climate change is creating adaptation mechanisms to boost resilience and the ability to cope with the anticipated impacts (Nicol and Kaur 2008; Khan et al. 2012; Khan et al. 2016). Adaptation depends greatly on the adaptive capacity or adaptability of an affected system, region or community to cope with the impacts and risks of climate change (Smith et al. 2003). However, Cities are often the first responders to climate impacts. Because cities are dynamic systems that face unique climate impacts, their adaptation must be location specific and tailored to local circumstances (The World Bank Group 2011).

Nevertheless, the IPCC has classified coastal adaptation strategies broadly as retreat, accommodation, and protection (Nicholls et al. 2007), these strategies are widely used and applied in both developed and developing countries (Boateng 2010;

Linham and Nicholls 2012). "Protect" to reduce the risk of the event by decreasing its probability of occurrence, "Retreat" to reduce the risk of the event by limiting its potential effects and "Accommodate" to increase society's ability to cope with the effects of the event (Tol et al. 2008). Some combination of these three approaches may be appropriate, depending on the physical realities and societal values of a particular coastal community (Wong et al. 2014). However, worldwide, many of the urban centers that need to adapt most to avoid serious impacts have large deficiencies in all of these preconditions for successful adaptation—and for addressing the development deficiencies that underpin their lack of adaptation capacity (Satterthwaite et al. 2007). In particular, cities in developing countries lack guidelines for planning capacity to address urban climate change adaptation. One fundamental condition for adaptation planning in urban contexts appears to be the availability of a good planning capacity, including the knowledge and participation base needed for informing decision making for the urban future (Johnson and Breil 2012). Importantly, the practitioners and various stakeholders in the urban space not only need an understanding of climate science but also the linkages to climate impacts to city systems, such that this information could be logically applied to decision-making processes and could be inbuilt into the development paradigm (APN 2014). In the near term, the focus in Indian cities, neighborhoods and communities will need to be on adaptation to climate change (Revi 2008). They can also lead on neighborhood pilot projects to test new methods of CBA, specifically for slum and informal settlements and vulnerable populations (Revi 2008). The efforts and attention for CBA in urban coastal cities are often ignored and not given much importance (Khan 2017). Until recently, most adaptation efforts have been top-down, and little attention has been paid to communities' experiences of climate change and their efforts to cope with their changing environments (Reid et al. 2009). Thus, there is a need for adaptation of cities in developing countries (such as Chennai City) with an additional discussion of challenges facing informal settlements, the urban poor, and other vulnerable groups (The World Bank Group 2011).

Chennai City and Coastal Community-Based Adaptation

Adaptation is considered one of the most appropriate response strategies to address the challenges of climate change such as SLR at the local level and community-based adaptation (CBA) begins by identifying the communities most vulnerable to climate change (Ziervogel et al. 2006). CBA focuses attention on empowering and promoting the adaptive capacity of communities. It is an approach that takes context, culture, knowledge, agency, and preferences of communities as strengths (IPCC 2014). CBA draws on participatory approaches and innovative participatory methods to help communities analyze the causes and effects of climate change (Ried et al. 2009). CBA experiences emphasize that it is important to understand a community's unique perception of their adaptive capacities in order to identify useful solutions and that scientific and technical information on anticipated coastal climate impacts need to be translated into a suitable language and format that allows people to be able to participate in adaptation decision-making and planning (IPCC 2013a). To support urban adaptation to climate change, community representatives should be included in decision-making processes to understand and influence local climate actions—in terms of both information dissemination and how to improve and adapt infrastructure (Lenhart 2009). Thus, community capacity to understand climate risk issues, effectively use of available information, develop the necessary institutions and networks, plan and build appropriate climate change adaptation actions and, evaluate and monitor these to learn from experience is an essential prerequisite for effective adaptation (IFRC 2009). Therefore, it is important to provide an avenue to engage citizen participation and citizen action within social networks to understand, learn from and develop local adaptation strategies (Lenhart 2009). In particular, it is important to engage in community mobilization and awareness raising through designing activities that are tailored to local practices and establish strong relationships with the communities to enable sustainable actions to involve the key stakeholders in adaptation action and enhance capacity building (Khan 2013).

For a city like Chennai that is chronic to the bleakness of the coastal hazards and climate change impacts such as SLR, there is an urgent need to address these challenges particularly through the lens of adaptation. The SLR is a significant challenge that requires a coordinated response from all members of the community, particularly in the urban context. While SLR can be a sensitive topic for public engagement, research has shown that approaches with an emphasis on learning, collaboration and openness can help the City to achieve wide public support for action (Barisky 2015; Khan 2017). But, cities in developing countries like this are struck between the challenges of climate change at one end and ongoing developmental challenges at the other. Not all cities are well prepared to face these challenges; it requires knowledge, ideas, plans, strategies, guidelines etc., in different perspectives and from various sources. In this context, this study introduces "CBACCS (Community Based Adaptation for Chennai city to Climate change and Sea-level rise)" approach. The approach integrates theoretical and empirical knowledge of the factors contributing to CBA to SLR with processes for translating those concepts into practice. The approach also includes characteristics of urban systems, the agents (people and organizations) that depend on and manage those systems, institutions that link systems and agents, and patterns of exposure to climate change (Tyler and Moench 2012). As an approach, it lays out a perspective of CBA and its different dimensions to address SLR. It acts as a starting point and simple guidelines for policy planners, decision makers, researchers, and practitioners to understand CBA and SLR for low-lying coastal cities such as Chennai city in this case (Fig. 3).

CBACCS address six major dimensions of the CBA and SLR with reference to Chennai city. It includes

(1) *About the city*

To outline the nature of the city, population, and geography, past and present climate, sea-level trend, future climate change and SLR projections, its predicted impact, identifying vulnerable coastal resources, infrastructure, communities, etc.

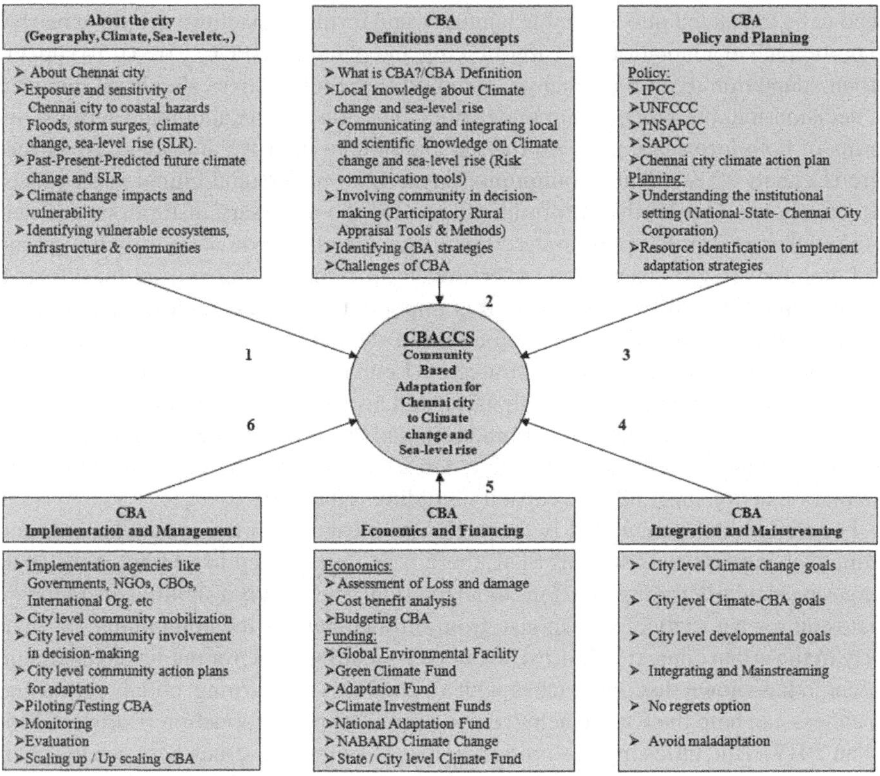

Fig. 3 Schematic representation of CBACCS (community based adaptation for Chennai city to climate change and sea-level rise) approach

This will give an idea about the nature of Chennai city and its exposure, vulnerability to coastal hazards and changing climate.

(2) *CBA definitions and concepts*

CBA begins by identifying the communities most vulnerable to climate change. CBA experiences emphasize that it is important to understand a community's unique perception of their adaptive capacities in order to identify useful solutions and that scientific and technical information on anticipated coastal climate impacts need to be translated into a suitable language and format that allows people to be able to participate in adaptation planning.

(3) *CBA Policies and planning*

United Nations Framework Convention on Climate Change, National Action Plan on Climate Change (NAPCC), State Action Plan on Climate Change (SAPCC), and City Action Plan on Climate Change etc. gives guidelines to plan suitable CBA strategies for Chennai city at the local level. Local governments have become major policy players in the area of climate change policy to set and implement a wide range of policy decisions (OECD 2008). At the local scale, city governments represent a core institutional unit that is increasingly recognized as having a critical role to play in climate adaptation (Naess et al. 2005; Smith et al. 2009; DCCEE 2010, Measham et al. 2011).

(4) *CBA Integration and Mainstreaming*

Mainstreaming is integrating climate concerns (SLR) and adaptation responses into relevant policies, plans, program and projects at the national, sub-national and local scales. It is to advocate strongly for climate change (SLR) adaptation and for bundles of adaptation measures to address priority issues within the scope of development goals (USAID 2009; Khan et al. 2012). For an example, 2005 Indian Ocean Tsunami led to a series of Integrated Coastal Zone Management Plans (ICZMPs), along with a review of the principles and norms to manage Coastal Regulation Zones (CRZs), which identified SLR and other climate risks as critical issues to factor into management plans and policies (Sharma and Tomar 2010).

(5) *CBA Economics and Financing*

Economics plays a major role in adaptation finance. There are a number of methods to assess economic loss and non-economic loss due to climate change and one such assessment is loss and damage. It also plays a key role in budgeting CBA for implementation. Additionally, there are a number of financing mechanisms such as funds from the Global Environmental Facility (GEF). For an example, through the GEF's Small Grants Programme (SGP), the GEF has invested $450 million and leveraged similar levels of co financing supporting over 14,500 community based projects in over 125 countries. In addition, there are other funding mechanisms such as the Special Climate Change Fund, the Adaptation Fund, Green Climate Fund, Climate Investment Funds etc. at the international level. However, at the national level National Adaptation Fund from Government of India, Climate Change Fund from NABARD (National Bank for Agricultural and Rural Development), State and City level fund for climate change can be tapped for the purpose of the CBA to climate change and SLR.

(6) CBA Implementation and Management

Implementing and managing CBA actions are daunting task and it can be achieved only by the collective partnership of different players from the Government bodies/Departments, International Developmental Organizations, Non-Governmental organizations (NGOs); Community-Based Organization (CBOs) etc. In addition, a number of monitoring and evaluation mechanisms are available; however, it needs to be customized to the local level. Importantly, successful implementation, monitoring and evaluation of CBA opens window of opportunities to scale CBA in a bigger and wider scale. Thus, based on the above six dimensions CBACCS approach torched some insights about addressing CBA and SLR at the city level.

Climate change exacerbates coastal hazards into multifold and Chennai city has no exception to these risks. The most severely affected sectors are expected to be the residential and recreational areas, ports and fishery facilities and the natural ecosystem. These are to be protected through strict enforcement of the Coastal Regulation Zone (CRZ) Act, and any further coastal developmental activities and protection work along the Chennai coast should be based upon an Integrated Coastal Zone Management (ICZM) approach for long-term sustainable development (Kumar and Kunte 2012). The most important ongoing climate change-related risk mitigation programme is the National Cyclone Risk Mitigation programme, which is being implemented across India's coastal states. A similar initiative needs to be developed for urban areas that combine natural hazard risk mitigation and climate change adaptation (Revi 2008). As cities build more flood-management infrastructure to adapt to the effects of a changing climate, they must go beyond short-term flood protection and consider the long-term effects on the community, its environment, economy, and relationship with the water (Stefan 2018). Chennai city can evolve urban CBA based on the above CBACCS approach together with the support of urban CBA capacity building and SLR risk communication tools such COREDAR (Communicating Risk of climate change and Engaging stakeholDers in framing community-based Adaptation stRategies) (Khan et al. 2015; USIEF 2015; UNAI 2016; State of the planet 2016) and BASIEC (Building capacities for Adaptation to Sea-level rise through Information, Education and Communication for coastal communities) and others. The city can evolve a number of CBA strategies such as building capacities for the communities through information, communication, awareness and education; diversified livelihood, insurances, relocation and or migration to elevated region, better housing, etc. However, more detailed analysis and assessment are required to evolve suitable CBA that are placed-based, location specific and community centric within the city level to changing climate and rising sea-levels.

Conclusion

Accelerated SLR, increased frequency and intensity of coastal and inland storms, and more riverine flooding because of increased precipitation would exacerbate the already considerable vulnerability of coastal communities to natural hazards (IPCC 2001). Importantly, the rising sea levels lead to increased inundation of low-lying areas and put global cities and urban, coastal communities at greater risk (Khan 2017; Khan et al. 2018). Cities are home to over half of the world's people and are at the forefront of the climate change issue. In particular, SLR poses a particularly ominous threat because 10% of the world's population (634 million people) lives in low-lying coastal regions within 10 m elevation of sea level (McGranahan et al. 2007). However, climate change will have unavoidable impacts on urban systems and populations, especially in Asia, where many large cities are exposed (Tyler and Moench 2012). Furthermore, it will place unique burdens on the urban poor, residents of informal settlements, and other vulnerable groups, such as women, children, the elderly and disabled, and minority populations (The World Bank 2011). Thus, coastal cities face a specific set of challenges that require a unique set of adaptation strategies due to their concentration of people and critical infrastructure in low-lying coastal zones, inability to easily shift locales, overlapping regulatory jurisdictions, and especially the variety and complexity of the infrastructure and the population's dependence on it (Rosenzweig et al. 2011). CBA is one of the broad adaptation strategies and it recognizes that environmental knowledge, vulnerability and resilience to climate impacts are embedded in societies and cultures. This means the focus is on empowering communities to take action based on their own decision-making processes (BCAS 2011). In recent years, urban CBA has been receiving attention widely. CBA, a bottom-up and citizen-based approach of governance, emphasizes how to take actions that suit local contexts through the local inhabitants (Chen and Tung 2018). A growing number of development NGOs and international organisations have started to engage with efforts to assist towns and cities to adapt to climate change, with increasing interest in the potential of CBA to build adaptive capacity among low-income urban residents. However, there has been relatively little discussion about the ways in which the particular nature of urban societies and economies might shape the implementation of CBA, or how to scale-up lessons learned within and across cities (Katarina et al. 2014). This is particularly important to densely populated low-lying coastal cities in Asia, and among Asian cities, Chennai is one of the vulnerable cities to coastal hazards and climate change impacts. In this purview, this study addressed the CBA of Chennai city through the lens of climate change and SLR. The findings of this study revealed that

(a) *Vulnerability of Chennai city to coastal natural hazards and climate change*

The low-lying topography of Chennai city is vulnerable to coastal natural hazards such as storm surges, cyclones, floods etc. Changing climate add more burden to the city and places the city at high risk.

(b) *Climate change induced SLR poses serious threats to Chennai city*

Scientific evidences based on climate models project a future rise of SLR with the range of 07.10 cm in 2025 and 36.87 in 2100. This predicted SLR poses serious threats to the Chennai city's coastal infrastructures (e.g. Ennore port), coastal natural resources (e.g. Pallikaranai wetland) and coastal communities (e.g. Fishing communities).

(c) *Community-based adaptation of Chennai city to SLR*

Adaptation is considered as one of the suitable response strategies of Chennai city to SLR and CBA warrants urgent attention. CBACCS approach proposed in this study emphasizes the need of CBA for Chennai city to address climate change and SLR. It plays an important to understand the different dimensions of CBA and paves the way to evolve suitable place-based, location specific, CBA adaptation strategies of Chennai city to climate change and SLR. Thus, the information gleaned from this study will provide insights to local level policy planners, decision makers and research penchants to understand and address CBA for coastal cities like Chennai through the lens of climate change and SLR.

Acknowledgements The authors are thankful to the Institute Postdoctoral Fellowship 2018-2019 (Ref,No.F.ARU./IPDF/R3/2018), Department of Humanities and Social Sciences, Indian Institute of Technology Madras (IITM) for their support and encouragements. The authors are indeed grateful to Indo-German Centre for Sustainability (IGCS) of IITM and Department of Science and Technology (DST), Government of India, India for their support (Ref.No: DST/CCP/CoE/141/2018(C) by Department of Science & Technology under SPLICE—Climate Change Programme).

References

Aithal BH, Ramachandra TV (2016) Visualization of urban growth pattern in Chennai using geoinformatics and spatial metrics. J Indian Soc Remot 44(4):617–633

Anushiya J (2015) Impact and vulnerability of emerging climate change scenario on natural resources in Chennai metropolis India. Ph.D. Thesis, Anna University, Chennai, Tamil Nadu, India

APN (2014) Comprehensive capacity building program on urban climate change resilience in India. Asia-Pacific network for global change research (APN). Online http://www.apn-gcr.org/resources/items/show/1966 Accessed 31 Jan 2019

Archer D, Monteith W, Scott H, Gawler S (2017) Developing city resilience strategies: lessons from the ICLEI–ACCCRN process. Working paper series 41, Asian cities climate resilience. IIED, London, UK

Balica SF, Wright NG, van der Meulen F (2012) A flood vulnerability index for coastal cities and its use in assessing climate change impacts. Nat Hazards 64:73–105

Barisky T (2015) A public engagement toolkit for sea-level rise. Prepared for the Sustainability Group, City of Vancouver. Vancouver, British Columbia, Canada, p 72

BCAS (2011) Proceedings of the 5th international conference community based adaptation (CBA), 24–31 March 2011, Dhaka, Bangladesh

Birkmann J, Garschagen M, Kraas F, Quang N (2010) Adaptive urban governance: new challenges for the second generation of urban adaptation strategies to climate change. Sustain Sci 5(2):185–206

Boateng I (2010) Spatial planning in coastal regions: facing the impact of climate change. FIG Publication No. 55, International federation of surveyors (FIG), Copenhagen, Denmark, p 60

Boominathan A, Dodagoudar GR, Suganthi A, Uma Maheswari R (2008) Seismic hazard assessment of Chennai city considering local site effects. J Earth Syst Sci 117(S2):853–863

Byravan S, Rajan SC, Rangarajan R (2010) Sea level rise: impacts on major infrastructure, ecosystems and land along the Tamil Nadu coast. Institute of Financial Management and Research, Chennai, India. Online http://www.ifmrlead.org/wp-content/uploads/2015/OWC/Sea_Level_Rise_Full-Report.pdf Accessed 31 Jan 2019

Census of India (2011) District census handbook: Chennai. Directorate of census corporation, Tamil Nadu, Ministry of Home affairs, Rajaji Bhavan, Besant Nagar, Chennai, Tamil Nadu, India. Online http://www.censusindia.gov.in/2011census/dchb/DCHB_A/33/3302_PART_A_DCHB_CHENNAI.pdf Accessed 31 Jan 2019

Chen W, Tung. C (2018) Urban community-based adaptation in Taipei. Geophys Res 20: EGU2018-3611-1

Chennai Corporation (2017) Greater corporation of Chennai (GCC) Disaster management plan 2017. Corporation of Chennai, Chennai, Tamil Nadu, India. Online http://www.chennaicorporation.gov.in/NorthMonSoon2017/cdmcb.pdf Accessed 31 Jan 2019

DCCEE (2010) Climate change adaptation actions for local government. Department of Climate Change and Energy Efficiency, Canberra

DES-GoTN (2017) District statistical handbook: Chennai district. Department of economics and statistics, Government of Tamil Nadu (GoTN), Chennai, Tamil Nadu, India

Douben N (2006) Characteristics of river floods and flooding: a global overview, 1985–2003. Irrig Drain 55:S9–S21

ENVIS (2015) Chennai district. Online http://tnenvis.nic.in/files/CHENNAI.pdf Accessed 31 Jan 2019

Fuchs RJ (2010) Cities at risk: Asia's coastal cities in the age of climate change. Asia Pacific issues, East-West Centre. Online https://scholarspace.manoa.hawaii.edu/bitstream/10125/17646/api096.pdf Accessed 31 Jan 2019

Funfgeld H (2010) Institutional challenges to climate risk management in cities. Curr Opin Env Sust 2:156–160

Gopinath G, Løvholt F, Kaiser G, Harbitz CB, Raju KS, Ramalingam M, Singh B (2014) Impact of the 2004 Indian Ocean tsunami along the Tamil Nadu coastline: field survey review and numerical simulations. Nat Hazards 72(2):743

Gupta AK, Nair SS (2010) Flood risk and context of land-uses: Chennai city case. J Geogr Reg Plan 12:365–372

Huq S, Kovats S, Rreid H, Satterthwaite D (2007) Editorial: reducing risks to cities from disasters and climate change. Environ Urban 19(1):3–15

IFRC (2009) Climate change adaptation strategies for local impact key messages for UNFCCC negotiators. Prepared by the technical paper for the inter agency steering committee, task force on climate change. International federation of red cross and red crescent societies, p 11 Online http://unfccc.int/resource/docs/2009/smsn/igo/054.pdf Accessed 31 Jan 2019

IPCC (2001) Summary for policymakers: climate change 2001: impacts, adaptation, and vulnerability. A report of working group II of the intergovernmental panel on climate change (IPCC), Gland, Switzerland

IPCC (2013a) Chapter 5: Coastal systems and low-lying areas. In: Poh Poh W, Iñigo JL (eds) Impacts, adaptation and vulnerability. Contribution of working group II to the fifth assessment report of the intergovernmental panel on climate change. Cambridge University Press, Cambridge, UK, pp 1–28

IPCC (2013b) Climate Change 2013: The Physical Science Basis. In: Stocker TF, Qin D, Plattner GK, Tignor M, Allen SK, Boschung J, Nauels A, Xia Y, Bex V, Midgley PM (eds) Contribution of working group I to the fifth assessment report of the intergovernmental panel on climate change. Cambridge University Press, Cambridge, United Kingdom and New York, NY, USA, p 1535

IPCC (2014) Glossary. In: Climate change 2014: impacts, adaptation and vulnerability. Working group II to the fifth assessment report of the intergovernmental panel on climate change. Cambridge University Press, NY

Jeganathan A, Andimuthu R (2012) Temperature trends of Chennai City. India Theor Appl Climatol 111(3–4):417–425

Johnson K, Breil M (2012) Conceptualizing urban adaptation to climate change. Findings from an applied adaptation assessment framework. Research papers issue RP0131, climate impacts and policies, an economic assessment division, Fondazione Eni Enrico Mattei, Centre Euro Mediterraneo. Online https://papers.ssrn.com/sol3/papers.cfm?abstract_id=2077476 Accessed 31 Jan 2019

Katarina S, Anna B, Ashvin D, David D (2014) Community participation in urban adaptation to climate change, vol 12, Routledge in association with GSE Research, pp 214–225

Khan AS (2013) Climate change induced sea level rise and its impacts on the Tamil Nadu coast, India: framing ecosystem and community based adaptation strategies. Ph.D. Thesis, Anna University, Chennai, Tamil Nadu, India

Khan AS (2017) COREDAR for cities: developing a capacity building tool for sea-level rise risk communication and urban community-based adaptation. In: Leal Filho W, Keenan JM (eds) Climate change adaptation in North America: fostering resilience and the regional capacity to adapt. Climate change management, Springer, Berlin, pp 137–157

Khan AS, Ramachandran A, Selvam V, Punitha S (2012) Predicted impact of sea level rise at Vellar-Coleroon estuarine region of Tamil Nadu coast, India: mainstreaming adaptation as coastal zone management option. Ocean Coast Manag 69:327–339

Khan AS, Chen RS, De Sherbinin A, Ramachandran A, Palanivelu K (2015) COREDAR framework-communicating risk of sea-level rise and engaging stakeholders in framing community-based adaptation strategies. American geophysical union (AGU), fall meeting, 16 Dec 2015, San Francisco, California, USA. Online https://agu.confex.com/agu/fm15/webprogram/Paper78074.html

Khan AS, Ramachandran A, Palanivelu K, Selvam V (2016) Climate change induced sea level rise projections for the Pichavaram mangrove region of the Tamil Nadu coast, India: a away forward for framing time-based adaptation strategies. Indian J Geo-Mar Sci 45(2):296–303

Khan AS, MacManus K, Jane M, Madajewicz M, Ramasubramanian L (2018) Building resilience of urban ecosystems and communities to sea-level rise: Jamaica Bay, New York City. In: Leal Filho W (ed) Handbook of climate change resilience—towards integrated responses to the causes and impacts of climate change. Climate change management, Springer, Berlin, pp 1–24

Kumar AA, Kunte PD (2012) Coastal vulnerability assessment for Chennai, east coast of India using geospatial techniques. Nat Hazards 64:853–872

Lenhart J (2009) Empowering social networks to localize climate strategies in impoverished urban communities: examining the Basic Urban Services Strategy. In: Fifth urban research symposium, Embu, Kenya

Linham MM, Nicholls RJ (2012) Adaptation technologies for coastal erosion and flooding: a review. In: Proceedings of the institute of civil engineers maritime engineering, vol 165, pp 95–111

Manohar L, Muthaiah KT (2016) Towards resilience in Chennai. In: Hein C (ed) International planning history society proceedings, 17th IPHS conference, history-urbanism-resilience, TUDelft 17–21 July 2016, TUDelft Open 3:251

Marriappan VE, Devi RS (2012) Chennai coast vulnerability assessment using optical satellite data and GIS techniques. Int J Rem Sense and GIS 1(3):175–182

McGranahan G, Balk D, Anderson B (2007) The rising tide: assessing the risks of climate change an human settlements in low elevation coastal zones. Environ Urban 19(17):17–37

Measham GT, Preston BL, Smith TF, Brooke C, Gorddard R, Withycombe G, Morrison C (2011) Adapting to climate change through local municipal planning: barriers and challenges. Mitig Adapt Strat Gl 16:889–909

MSME (2013) Brief industrial profile of Chennai district 2012–2013. Micro, small and medium enterprises (MSME) development institute, government of India, ministry of MSME, Guindy, Chennai, India. online http://dcmsme.gov.in/dips/ips%20chennai_revised.pdf Accessed 31 Jan 2019

MSME (2016) Brief industrial profile of Chennai district 2015–2016. Micro, small and medium enterprises (MSME) development institute, government of India, ministry of MSME, Guindy, Chennai, India. Online http://dcmsme.gov.in/dips/2016-17/DIP.CHENNAI.2015.16.pdf Accessed 31 Jan 2019

Mukheibir P, Ziervogel G (2007) Developing a municipal adaptation plan (MAP) for climate change: the city of Cape Town. Environ Urban 19(1):143–158

Naess LO, Bang G, Eriksen S, Vevatne J (2005) Institutional adaptation to climate change: flood responses at the municipal level in Norway: adaptation to climate change: perspectives across scales. Glob Environ Chang 15(2):125–138

Narasimhan B, Bhallamudi SM, Mondal A, Ghosh S, PMujumdar P (2016). Chennai Floods 2015: a rapid assessment. Interdisciplinary centre for water research Indian institute of science, Bangalore, India. Online http://www.icwar.iisc.ernet.in/wp-content/uploads/2016/06/Chennai-Floods-Rapid-Assessment-Report.pdf Accessed 31 Jan 2019

Nicholls R, Wong P, Burkett V, Codignotto J, Hay J, McLean R, Ragoonaden S, Woodroffe C (2007) Coastal systems and low-lying areas. In: Parry P, Canziani O, Palutikof J, Van der Linden P, Hanson C (eds) Climate change 2007: impacts adaptation and vulnerability, contribution of working group II to the fourth assessment report of the intergovernmental panel on climate change. Cambridge University Press, Cambridge, UK, pp 315–356

Nicol A, Kaur N (2008) Climate change: getting adaptation right. Opinion. Overseas Development Institute, London, United Kingdom

NOAA (2018) Relative sea level trend, 500-091 Chennai/Madras, India. Products-tides and currents, national oceanic and atmospheric administration (NOAA), USA
OECD (2008) Competitive cities and climate change. In: The organization for economic co-operation and development (OECD) conference proceedings, Milan, Italy, 9–10 Oct 2008. Online https://www1.oecd.org/gov/regional-policy/50594939.pdf#page=172 Accessed 31 Jan 2019
Oliver-Smith A (2009) Sea level rise and the vulnerability of coastal peoples responding to the local challenges of global climate change in the 21st century. UNU Institute for environment and human security (UNU-EHS), UN Campus, Bonn, Germany
Ramachandran A, Khan AS, Prasnnahvenkatesh R, Palanivelu K, Jayanthi N (2017) Projection of climate change induced sea level rise for the coasts of Tamil Nadu and Puducherry, India using SimCLIM: a first step towards planning adaptation policies. J Coast Conserv 21:731–742
Reid H, Huq S (2014) Mainstreaming community-based adaptation into national and local planning. Clim Dev 6:291–292
Reid H, Alam M, Berger R, Cannon T, Huq S, Milligan A (2009) Community-based adaptation to climate change: an overview. Theme section: community-based adaptation to climate change, participatory learning and action 60, International institute for environment and development, Russell Press, Nottingham, UK p 210
Revi A (2008) Climate change risk: an adaptation and mitigation agenda for Indian cities. Environ Urban 20(1):207–229
Rosenzweig C, Solecki WD, Blake R, Bowman M, Faris C, Gornitz V, Zimmerman R (2011) Developing coastal adaptation to climate change in the New York city infrastructure-shed: process, approach, tools, and strategies. Clim Chang 106(1):93–127
Satterthwaite D, Huq S, Pelling M, Reid H, Lankao PR (2007) Adapting to climate change in urban areas: the possibilities and constraints in low- and middle-income nations. International institute for environment and development, discussion paper, 124. Online http://pubs.iied.org/10549IIED.html Accessed 31 Jan 2019
Sharma D, Tomar S (2010) Mainstreaming climate change adaptation in Indian cities. Environ Urban 22(2):451–465
Sheth A, Sanyal S, Jaiswal A, Gandhi P (2006) Effects of the December 2004 Indian Ocean tsunami on the Indian mainland. Earthq Spectra 22(S3):435–473
Smith JB, Klein RJT, Huq S (2003) Climate change, adaptive capacity and development. p 356
Smith TF, Preston B, Brooke C, Gorddard R, Abbs D, McInnes K, Withycombe G. Morrison C, Beveridge B, Measham TG, (2009) Managing coastal vulnerability: new solutions for local government. In: Moksness E, Dahl E, Støttrup JG (eds) Integrated coastal zone management, Wiley-Blackwell, West Sussex
SoER-TN (2017) Chapter 12: Coastal resources. State of environment report-Tamil Nadu. ENVIS Centre, Department of environment, government of Tamil Nadu, Chennai, India
State of the planet (2016) Helping communities respond to climate change. Earth Institute, Columbia University, New York, USA. Online https://blogs.ei.columbia.edu/2016/02/23/new-communications-tool-to-help-communities-respond-to-climate-change Accessed 31 Jan 2019
Stefan A (2018) Adapting cities to sea-level rise: green and grey strategies. Island Press. Online https://islandpress.org/books/adapting-cities-sea-level-rise Accessed 31 Jan 2019
TERI (2014) Climate proofing Indian cities: a policy perspective. Policy brief. The Energy and Resource Institute (TERI), New Delhi, India
The World Bank Group (2011) Guide to climate change adaptation in cities. Online http://siteresources.worldbank.org/INTURBANDEVELOPMENT/Resources/336387-1318995974398/GuideClimChangeAdaptCities.pdf Accessed 31 Jan 2019
TNENVIS (2017) Chennai City. Tamil Nadu environment information system. Online http://tnenvis.nic.in/files/CHENNAI.pdf Accessed 31 Jan 2019
Tol RSJ, Klein RJT, Nicholls RJ (2008) Towards successful adaptation to sea-level rise along Europe's coasts. J Coast Res 24(2):432–442
Tyler S, Moench M (2012) A framework for urban climate resilience. Clim Dev 4(4):311–326

UNAI (2016) New tool helps communities better understand climate change. United nations academic impact, United Nations Headquarters, New York, USA. Online https://academicimpact.un.org/content/new-tool-helps-communities-better-understand-climate-change Accessed 31 Jan 2019

UNFCCC (2017) Opportunities and options for integrating climate change adaptation with the sustainable development goals and the sendai framework for disaster risk reduction 2015–2030. United Nations Framework Convention on Climate Change, Bonn, Germany. Online https://unfccc.int/resource/docs/2017/tp/03.pdf Accessed 31 Jan 2019

UNSDG Report (2018) Sustainable development goal 11: make cities and human settlements inclusive, safe, resilient and sustainable. United Nations Sustainable Development Goals (UNSDGs) Knowledge Platform. Online https://sustainabledevelopment.un.org/sdg11 Accessed 31 Jan 2019

USAID (2009) Adapting to coastal climate change: a guidebook for development planners. U.S. Agency for International Development, Washington, DC, USA, U.S, p 162

USAID (2016) ANNEX I. CASE STUDIES case study. Chennai, India. Online https://www.mercycorps.org/sites/default/files/Annexes_1-2-3-4_Curriculum_2016_FINAL.pdf Accessed 31 Jan 2019

Usha T, Murthy MVR, Reddy NT, Mishra P (2012) Tsunami vulnerability assessment in urban areas using numerical model and GIS. Nat Hazard 60(1):135–147

USIEF (2015) COREDAR Framework-Communicating risk of sea-level rise and engaging stakeholders in framing community-based adaptation strategies. New-Delhi, India

Wong PP, Losada IJ, Gattuso JP, Hinkel J, Khattabi A, McInnes KL, Saito Y, Sallenger A (2014) Coastal systems and low-lying areas. In: Field CB, Barros VR, Dokken DJ, Mach KJ, Mastrandrea MD, Bilir TE, Chatterjee M, Ebi KL, Estrada YO, Genova RC, Girma B, Kissel ES, Levy AN, MacCracken S, Mastrandrea PR, White LL (eds) Climate change 2014: impacts, adaptation, and vulnerability. Part A: global and sectoral aspects. Contribution of working group II to the fifth assessment report of the intergovernmental panel on climate change, Cambridge University Press, Cambridge, United Kingdom and New York, NY, USA. pp 361–409

Ziervogel G, Bharwani S, Downing ET (2006) Adapting to climate variability: pumpkins, people and policy. Nat Resour Forum 30(4):294–305

Selection of Five Rice Varieties (*Oryza sativa*) Under Salinity Stress in Climate Field Schools

Neni Rostini, M. Khais Prayoga, Tualar Simarmata, Mieke Rochimi Setiawati, Silke Stoeber and Kustiwa Adinata

Abstract Rising sea levels as one of the major climate change lead to salinity stress of rice farming on coastline area. The study to enable farmers to select the saline tolerant rice varieties through climate field schools. Weekly meetings took place between farmers and university researchers. The university provided training to help farmers to identify their soil health. The experiment as field farmer school was carried out using randomized block design consisted five treatments and provided with replications. Farmers and researchers jointly tested performance indicators of salt-tolerant (Inpari 34, Inpari 41 and Inpara Pelalawan) and flood-tolerant varieties (Inpara 02 and Mendawak). The level of salinity raised from 1.49 to 7.36 dS m^{-1} (high) during the season. The response to salinity stress showed significant differences of agronomic performance indicators, such as panicle length, number of grain per panicle, weight per plant, weight per plot, and yield per hectare. Contrary to all expectations, the flood-tolerant variety performed better, outperforming the salt-tolerant varieties in many agronomic traits, despite the high level of salinity. Based on a participatory evaluation, farmers decided for Mendawak as best performing variety under salinity stress. The results serve as entry point for upscaling climate-resilient rice farming in this region.

N. Rostini (✉) · M. Khais Prayoga · T. Simarmata · M. R. Setiawati
Faculty of Agriculture, Universitas Padjadjaran, Jl. Raya Bandung Sumedang KM.21, Jatinangor, Sumedang Regency, Bandung, West Java 45363, Indonesia
e-mail: neni.rostini@unpad.ac.id

S. Stoeber
Centre for Rural Development (SLE) Humboldt-Universität zu Berlin, Robert-Koch-Platz 4, 10115 Berlin, Germany

K. Adinata
Indonesian Farmers Community Network (JAMTANI), Jl. Raya Babakan, Ciamis Regency, Pangandaran, West Java 46396, Indonesia

W. Leal Filho et al. (eds.), *Climate Change, Hazards and Adaptation Options*, Climate Change Management, https://doi.org/10.1007/978-3-030-37425-9_40

Introduction

Climate change belongs to the most relevant global risks for the majority of 26 nations that represented the world in a recent global risk study (Poushter and Huang 2019). Indonesia is one of the countries that are highly exposed to the impacts of climate change (Hoegh-Guldberg et al. 2018). Regional-scale phenomena ocean atmospheric phenomena like El-Nino Southern Oscillation (ENSO) and Indian Ocean Dipole (IOD) have great impact on the regional climate and include tropical cyclones, sea and land breezes, monsoons. As their interactions are affected by climate change, weather anomalies caused by these phenomena occur more often, leading to more frequent temperatures extremes and precipitation shifts with more erratic rainfall patterns (Evans et al. 2012). Increased trends of rising sea levels are threatening Indonesia, which it is the world's largest archipelago with over 81,000 km of coastline (Leal Filho 2018). The long-term average of Indonesia's sea level rise is 0.8 mm/year. However, it has increased to 1.6 mm/year since 1960 and then to 7 mm/year since 1993 (GoI 2013a).

Due to its dependency on rainfall and temperature, the agricultural sector, particularly the rain-fed agriculture, is highly sensitive to climate change (Juanda 2015). Moreover, as agriculture plays a vital role in Indonesian's economy with more than 30% of the population employed in this sector (ILO 2018) or 37 million smallholder farmers, Indonesia has given high consideration in its climate change policies to agriculture. Law No. 19/2013 on the protection and empowerment of farmers addresses this issue and grants farmers an array of improvements, among which a financial compensation in the case of natural disasters features most prominently. The central ordering power for Indonesia's climate change efforts is the National Climate Change Adaptation Plan (RAN API), in which food security is one overarching element in order to strengthen economic resilience (GoI 2013a). RAN API seeks to decrease the loss of agricultural and fisheries food production due to climate change. The means to achieve this are an adjustment of farming practices to climate change and development and implementation of climate adaptive technology. Coastal areas should become climate resilient through capacity building measures. A supporting system aims to increase research and development of technologies related to climate change adaptation.

Climate change has a high negative impact on the southern coastal region of Java, where this study takes place (Cilacap Regency). While coastal area land has been always influenced by the characteristics of the sea such as tides, sea breeze and seepage of salt water (Dahur et al. 1996), this influence has been greatly intensified by climate change. Rawaapu village is largely affected by the sea level rise, increase of waves heights and salt water intrusion. Most of the rice fields in the village of Rawaapu are affected by tidal flooding. Seawater floods have a negative impact on rice production as they cause higher levels of salinity in rice fields. High salinity levels may result in nutritional disorders and osmotic stress (Asch et al. 2000; Zhu 2001). Detrimental effects can be so severe, that the growth and development of disturbed plants can even lead to death (Maas and Nieman 1978).

Even though the Paris Agreement's central aim is to limit global warming to below 1.5°, it is so evident, that some of the effects of global warming are already irreversible and therefore farmers are required to adapt to climate change. One widespread adaptation strategy are stress-tolerant, i.e. flood-tolerant or saline-tolerant, varieties (Thamrin et al. 2013). It is important to find out which variety shows the best performance in the climate change affected region in order to further increase productivity under changing agricultural conditions.

The selection of superior varieties for dissemination in a region can be done by involving local farmers. This is generally achieved through participatory action research. Participatory action research is a process in which farmers are fully involved in planning, implementing, monitoring and evaluating research programmes (Bhargav and Meena 2014). One approach of participatory action research is the use of farmer field schools (FFS) that were already popularized in Indonesia in the 90s (FAO 2016). According to Ernawati et al. (2015) FFS is a motor for improving farmer competencies along the concept of integrated crop management, by using the principles of adult education and under consideration of traditional knowledge and local wisdom. The farmer field school program aims to build awareness of farmers, to motivate, to enable and to empower them to develop their farms evolutionary and self-sustaining. In a multi-stakeholder effort, the Indonesian government has further developed the FFS concept into climate field schools, considering the main entry points for climate mitigation and adaptation (GoI 2013b). Demonstration plots are an additional feature used under this concept. The purpose of this paper is to demonstrate the assessment process of stress-tolerant varieties with farmers by conducting climate field schools and to analyse the major results gained from this process.

Description of the Study Areas and Methodology

Study Area

The experimental farm as Climate field school was established in Rawaapu village, Patimuan subdistrict, Cilacap district, Central Java (7° 37′ 59.0″ S 108° 45′ 51.5″ E, Altitude: 4 m) on a site with typical characteristics of smallholder rice production in the region. In order to ensure visibility and outreach the demonstration plot is located near the main road. The site has been under rice cultivation for the past 30 years. During that period, rice has been cultivated twice a year by use of chemical fertilizer. The climate belongs to the category B, which means according to the classification by Schmidt-Ferguson a subhumid ago-climatic zone. The soils are broadly classified as silty clay. The plot has been rented by JAMTANI (Jaringan Masyarakat Tani Indonesia), the Indonesian Farmers Community Network. JAMTANI also acts as main partner in this research with its members acting as farmer researchers.

Research Methods

Experimental Design and Treatments

The randomized block design experiment consisted five treatments and provided with five replications has been conducted from 30th May to 20th September 2018 in Rawaapu Village. Farmers and researchers jointly tested performance indicators of stress-tolerant varieties. The saline-tolerant varieties comprise Inpari 34, Inpari 41 and Inpara Pelalawan and the flood-tolerant varieties are Inpara 02 and Mendawak. The treatments used 5 kg of seed per hectare. As each replication had a size of 200 m^2 (20 m $\times$ 10 m) with a 0.5 m buffer in between, the size of the experimental plot was 1000 m^2. The plot followed the intermittent irrigation management combined with the principles of organic farming. The land was prepared by mechanical ploughing. Prior planting, 10 tons/ha goat manure was applied. The seedlings 21 days old is planted with plant spacing about 30 $\times$ 30 cm. The seedling is planted by slipping in sideways rather than plunging it into the soil vertically makes the shape of the transplanted seedling more like an L than like a J. With an L shape, it is easier for the tip of the root to resume its growth downward into the soil (Simarmata et al. 2011). In addition, organic liquid fertilizers of azolla were sprayed to plant at 21 and 42 day after planting using a knapsack sprayer.

Data Sources and Collection Methods

This study follows participatory farmer field research, involving farmers in the design, implementation, data recording and assessment of the results. The research aimed to test the stress-tolerant varieties is based on farmers' preferences. Farmers recorded all of the observations. Farmer field labs were established by following the principles of the climate field schools. 16 Farmer researchers (of which 6 were women) were trained to perform field research tasks according to a standard research protocols. The major observations were conducted in the generative growth phase (96 days after planting (dap) until the post-harvest period (Soemartono and Hardjono 1980). The observed characters were the number of productive tillers, panicle length, number of grain per panicle, percentage of filled grain per panicle, weight per plant, weight per plot and yield per hectare. The observations were made every morning at 8 am weekly. Observations started at 12 dap (vegetative phase) and ended 82 dap (generative final phase). In addition, the environmental components such as temperature, humidity, soil pH, salinity and inundation height were recorded regularly.

Data Analysis

Observation data on each character is recorded by the farmer, then collected by the group leader for further analysis. Differences between the varieties were tested via an ANOVA post hoc analysis using the Tukey's honest significance test or Tukey's HSD. The data was computed using the PKBT stat software.

Results and Discussion

Environmental Conditions

Environmental conditions are very influential on the growth and development of rice plants (Satake and Yoshida 1978). Consequently, development of climate-resilient superior varieties in Indonesia is higly important to mitigate the climate change impact and to increase the rice productivity (Baihaki and Wicaksana 2005). Therefore, the observations were made on the state of the environment, namely temperature, humidity, soil pH, salinity and inundation height.

According to Satake and Yoshida (1978) for the growth and development of rice the optimum temperature ranges between 25 and 33 °C, while the optimum humidity ranges from 60 to 80%. The results of the season show that the temperature was predominantly in the range of the optimal temperature. The same applies to the humidity range. The average temperature ranged from 27.3 to 33.85 °C, while the average humidity ranged from 56 to 85% (Figs. 1 and 2). Satake and Yoshida (1978)

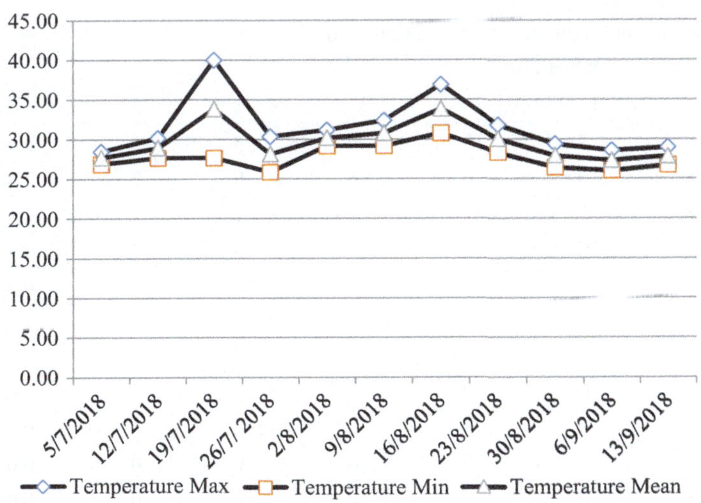

Fig. 1 Temperature during observation

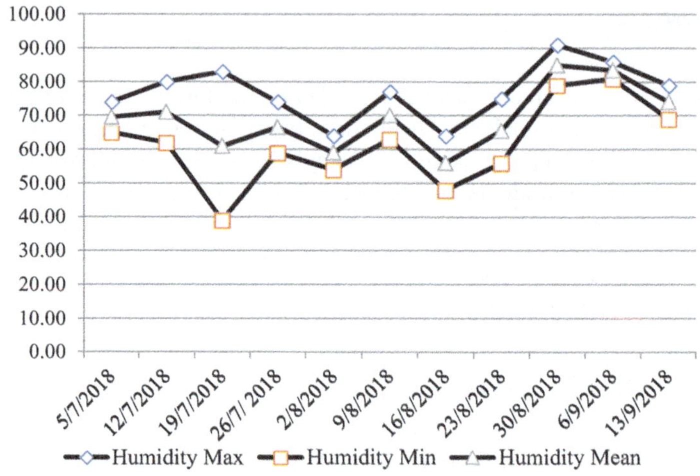

Fig. 2 Humidity during observation

assessed these values for temperature and average humidity as suitable for the growth of rice plants.

Demonstration plots in Rawaapu Village are tidal rice fields. Rising sea levels will have a direct impact on plot demonstration land. During the experiment, the average inundation height of rice fields caused by sea tides ranged from 0 to 10.8 cm (Table 1). The water that comes from the sea brings along salt particles so that it increases salinity levels in the field. The observations show that the average level of soil salinity ranges from 1.49 to 7.36 ds/m (Table 2).

Table 1 Inundation height during observation (cm)

Date	Repetitions					
	1	2	3	4	5	Average
05/07/2018	7	10	6	5	5	6.6
12/07/2018	10	6	6	8	7	7.4
19/07/2018	5	1	5	2	1	2.8
26/07/2018	10	5	4	5	3	5.4
02/08/2018	12	15	10	10	7	10.8
09/08/2018	7	0	2	0	0	1.8
16/08/2018	5	0	5	5	0	3.0
23/08/2018	0	0	0	0	0	0.0
30/08/2018	0	0	0	0	2	0.4
06/09/2018	0	0	0	0	0	0.0
13/09/2018	0	0	0	0	0	0.0

Table 2 Soil salinity level during observation (ds/m)

Date	Repetitions					
	1	2	3	4	5	Average
05/7/2018	3.77	4.28	3.28	3.30	2.70	3.47
12/7/2018	2.28	1.92	1.91	2.65	3.20	2.39
19/7/2018	2.74	3.30	3.30	2.90	2.90	3.03
26/7/2018	2.23	2.90	2.16	3.01	3.93	2.85
02/8/2018	1.44	1.29	1.49	1.49	1.72	1.49
09/8/2018	2.77	2.53	2.50	2.74	3.92	2.89
16/8/2018	3.30	2.83	3.27	3.62	4.36	3.48
23/8/2018	4.85	4.70	5.11	5.33	5.23	5.04
30/8/2018	4.65	6.99	6.99	9.61	8.54	7.36
06/9/2018	5.77	6.57	7.16	7.17	3.45	6.02
13/9/2018	3.45	3.67	5.51	3.27	4.18	4.02

The highest level of salinity was observed on 30th August 2018. At that time the lowest salinity level was found in replication 1 plots of 4.65 ds/m and the highest level of salinity was found in replication plot 4 which reached 9.61 ds/m. If we connect with the high inundation, on 2nd August 2018 there will be an increase in inundation caused by rising sea levels. At that time the salinity level was only 1.49 ds/m, but in subsequent observations the salinity level tended to increase. This reinforces the suspicion that the inundation that enters the demonstration plot carries salt particles, because when the inundation begins to recede because the particles evaporate the salt still remains in the soil and causes the level of salinity to be high.

Smith (1975) classified the level of salinity into five namely very low (<1 ds/m), low (1–2 ds/m), medium (2–3 ds/m), high (3–4 ds/m), and very high (>4 ds/m). The results showed that the salinity level in the demonstration plot ranged from low to very high. The highest salinity stress occurs from 23rd August to 13rd September 2018. During the weeks of salinity the salinity level is very high.

Other environmental variables that influence the growth and development of rice plants are soil pH. According to Ilagan et al. (2014) the optimum pH for rice plants is 5.5–7.0. The results of the average soil pH observed during the observation ranged from 3.4 to 5.1 (Table 3). According to the USDA (1998) soil with a pH of 3.5–4.4 is classified as extremely acid, and 4.5–5.0 classified as very strong acid. The soil pH at the demonstration plot is almost exclusively acidic and very acidic throughout the growing period, which indicates a major environmental stress factor for the rice plants.

Acidity is one of the main obstacles in tidal rice fields. High acidity affects the increasing solubility of Al, Fe and Mn. The acidity of land in tidal rice fields is caused by the oxidation of pyrite compounds which are aided by oxidizing iron and sulfur bacteria. The balance of chemical reactions in the soil and the availability of

Table 3 Soil pH level during observation

Date	Repetitions					
	1	2	3	4	5	Average
05/7/2018	3.8	3.6	3.9	4.4	4.5	4.0
12/7/2018	3.8	4.2	4.2	4.2	4.9	4.3
19/7/2018	3.3	3.5	3.5	4.9	3.7	3.8
26/7/2018	5.3	4.0	3.0	4.0	3.0	3.9
02/8/2018	5.2	5.4	4.2	4.4	4.2	4.7
09/8/2018	4.7	3.5	3.6	3.0	4.2	3.8
16/8/2018	4.2	3.2	3.8	3.1	3.4	3.5
23/8/2018	3.3	4.8	3.1	3.3	3.2	3.5
30/8/2018	3.3	3.7	3.7	3.7	2.7	3.4
06/9/2018	4.6	4.2	3.7	2.8	5.6	4.2
13/9/2018	5.6	5.0	4.8	5.4	4.8	5.1

nutrients in the soil, especially phosphate, is affected by high soil acidity (Mariana et al. 2007).

Problems that occur during the research in the field are discussed between farmers and university and JAMTANI. Guidance on further steps to be taken is given by the university researchers. Every week after having observed the plants in the fields, the conditions in the field were discussed. Farmers presented their findings including the problems that occured. The university researchers provided recommendations to solve the problems. The recommendations were always based on or at least considered the local wisdom and the solutions provided by the farmers.

Before starting the climate field school, the farmer researchers had been provided with simple soil analysis training. The training aims to increase farmers' knowledge about the characteristics of fertile soil and teach farmers to be able to operate measuring instruments such as pH meters (to measure pH) and EC meters (to measure salinity). The results of the training made farmers understand about the conditions of the land in the demonstration plot. Strained soil conditions are very suitable for selecting stress resistant rice varieties. One of the varieties is expected to be able to provide the best appearance in a gripped condition.

Appearance of Five Rice Varieties

The selection of varieties was primarily measured on the basis of productivity and yield criteria. This is in accordance with the results of Prayoga et al. (2018) who had already noticed in a previous season that yield and yield components determined the selection of the best preforming varieties.

Table 4 Analysis of variance for each characters

Characters	Varieties	C.V. (%)
Number of productive tillers	ns	22.29
Panicle length	**	3.55
Number of grain per panicle	**	14.2
Percentage of filled grain per panicle	ns	3.60
Weight per plant	**	19.98
Weight per plot	**	12.77
Yield per hectare	**	12.77

*Significant in $P < 0.05$; **significant in $P < 0.01$; *ns* not significant

The yield very much depends on the yield component of which number of productive tillers, panicle length, total grain, percentage of filled grain, weight per plant, weight per plot and yield per hectare was tested.

The rice varieties used are rice varieties that have been released by the Indonesian Ministry of Agriculture. The five varieties that were selected for testing were based on the recommendations of the university and preferences of the farmers involved. Farmers had the power to make their own choices on which varieties to be tested.

The results of the Analysis of Variance (ANOVA) indicate that the coefficient of variation (CV) of each character ranges from 3.55 to 22.29% (Table 4). CV values indicate the level of accuracy of the treatments. CV value is also a good index of research conditions. The high and low values of CV depend on the type of research, plants, and the properties measured (Gomez and Gomez 1995). According to Gomez and Gomez (1995), the CV value of less than 25% is good. A CV value of less than 25% indicates a trial error for the observed characters is relatively small. The greater the CV value, the greater the uncertainty of a study. Thus, in this study trial errors are minor and the results can be regarded as rather robust.

The character of the yield and yield components of the five rice varieties tested showed significant differences in panicle length, number of grain per panicle, weight per plant, weight per plot and yield per hectare. The character of number of productive tillers and percentage of filled grain per panicle are not significantly different. On the number of productive tillers characters, on average each variety ranges from 17.40 (Pelalawan) to 23.71 (Mendawak).

Similar to the number of productive tillers, the percentage of filled grain per panicle character also does not show a significant difference between varieties. The average percentage of filled grain per panicle of each variety was 87.83%. According to Horrie et al. (2006) percentage of filled grain is more influenced by genetic factors than nongenetics. Norman et al. (1984) states that a percentage of filled grain is influenced by the ability of plants to maximize photosynthesis. A percentage of filled grain above 85% shows that rice plants are capable of maximizing photosynthesis.

Another character that affects rice production is panicle length. The analysis results show that the Inpari 34 variety has the longest panicle length (24.20 cm)

Table 5 Average of each observation characters

Varieties	Number of productive tillers	Panicle length (cm)	Number of grain per panicle	Percentage of filled grain per panicle (%)	Weight per plant (g)	Weight per plot (kg)	Yield per hectare (ton)
Inpari 34	17.86[a]	24.20[a]	124.54[a]	91.07[a]	39.47[abc]	15.92[ab]	3.98[b]
Inpari 41	23.26[a]	21.34[b]	118.46[a]	86.93[a]	46.42[ab]	17.09[a]	4.28[a]
Pelelawan	17.40[a]	20.66[bc]	119.96[a]	85.88[a]	30.27[c]	12.32[bc]	2.95[c]
Inpara 02	21.71[a]	19.83[c]	82.42[b]	89.26[a]	33.41[bc]	11.78[c]	3.08[bc]
Mendawak	23.71[a]	20.40[bc]	111.62[ab]	86.02[a]	50.89[a]	17.28[a]	4.32[a]
Average	20.79	21.29	118.65	87.83	40.09	14.88	3.72

Note Letters after numbers indicate the presence or absence of differences between each variety. Number followed by the same letter is not significant ($P < 0.05$)

compared to other varieties. The panicle length determines the productivity level of a variety. Long panicles will give more grain (Yoshida 1981). This is in accordance with the results of the study which showed that Inpari 34 variety had the highest number of grains per panicle (124.54) but which is not significantly higher than the varieties of Inpari 41 (118.46), Pelalawan (119.96) and Mendawak (111.62). The number of grains per panicle does not guarantee that the varieties have high yields. With regard to the character weight per plant, the Mendawak variety has the highest weight per plant (50.89 g) compared to the other varieties even though the number of grains per panicle is with only 111.62 less than the number of grains per panicle of Inpari 34 (124.54).

Table 5 also shows that the more number of productive tillers there are, the higher the weight per plant will be. In accordance with this, the harvest weight of the plot also increases. Mendawak variety has the highest weight per plot (17.28 kg), although statistically, the results of the follow-up tests show no significant difference with Inpari 41 (17.09), as well as the yield per hectare. The yield per hectare is the calculated result of the weight per plot. The results suggest that Mendawak and Inpari 41 varieties are able to adapt well to the rice farming environment in Rawaapu Village.

The results of this study serve as reference for farmers in Rawaapu Village in choosing the rice varieties to be planted in the next season. The participatory action research jointly conducted by farmer researchers and the university researchers accelerates the process of technology transfer, for the following reasons. Field testing of varieties in demonstration plots allow farmers to see and adjust the results directly in their farm. Through participatory research the capacity of farmers can be improved. Observations and recording of results carried out jointly with farmers allow to increase the independence of farmers. Farmers, who have carried out participatory research, usually serve as example for other farmers who were not involved in the activity. This research reveals, that farmer researchers are seen as risk takers and innovators and increasingly serve as point of orientation for farming decisions to be taken by the community.

Conclusion

The paddy fields in Rawaapu Village belong to tidal rice fields which are strongly influenced by sea level rise. Climate change induced stresses prevalent in the rice growing area are high levels of soil salinity and acidity. In general, temperature and air humidity are rather suitable in supporting the growth of rice plants. The results showed that Mendawak and Inpari 41 varieties gave the best appearance compared to other varieties. The participatory evaluation revealed, that farmers decided to rate the flood-resistant Mendawak as the best performing variety under salinity stress. The results and the process serve as future entry points for climate resilient rice farming in this region. The participatory action research approach is subject to a paradigm in which the farmer himself becomes the main actor in knowledge generation compared to his role as mere receiver of technologies in the transfer of technology approach. The climate field school and climate field lab process are instruments of a co-production of knowledge approach, which can be regarded as innovative system within the global framework of the agricultural knowledge and innovation system approach.

Climate change results in environmental changes for rice growth thus affecting yield. Climate field school needs to be carried out continuously to select other varieties that are well adapted to the climate conditions of the Rawaapu village, so that farmers have many choices of the best varieties for their rice cultivation. More research with farmers in the centre of attention could further strengthen the adaptive capacity of farmers and contribute to the growth of independent, confident and smart farmers able to adapt to the adverse effects of climate change.

Acknowledgements This study is part of the Climate-resilient Investigation and Innovation Project (CRAIIP) funded by the German Non-Governmental Organisation Bread for the World (second phase: 2018), which is implemented by JAMTANI. The authors are equally grateful to the farmers from the Taruna Tani Rawa Jaya Farmer Group, Rawaapu Village, Patimuan Subdistrict, Cilacap District, for their active participation in design, implementation and discussion of results and for the close collaboration throughout the experimental period.

References

Asch F, Dingkuln M, Dörffling K, Miezan K (2000) Leaf K/Na ratio predicts salinity induced yield loss in irrigated rice. Euphytica 113:109–118

Baihaki A, Wicaksana N (2005) Genotypic interaction x environment, adaptability, and results stability, in the development of superior varieties in Indonesia. Zuriat 16(1):25–38

Bhargav DK, Meena HP (2014) Participatory plant breeding: farmers as breeders. Popular Kheti 2(1):7–14

Dahur R, Rais J, Ginting SP, Sitepu MJ (1996) Integrated management of coastal and oceanic resources. PT. Saptodadi, Jakarta

Ernawati L, Djafar, Sudirman (2015) Integrated crop management field school program as an effort to increase farmer capacity. J Tesis PMIS-UNTAN 1(1):1–21

Evans J, McGregor J, McGuffie K (2012) Future regional climates. In: Henderson-Sellers A, McGuffie K (eds) The future of the world's climate, 2nd ed, Chapter 9. Elsevier

FAO (2016) Farmer field school guidance document—planning for quality programmes, Rome

GoI (2013a) National action plan for climate change adaptation (RAN-API) synthesis report. Ministry of National Development Planning/National Development Planning Agency (Bappenas), Ministry of Environment (KLH), National Council on Climate Change (DNPI), Meteorological, Climatological, and Geophysical Agency (BMKG), Jakarta

GoI (2013b) Climate field school foundation module. Government of Indonesia, Ministry of Agriculture, Directory of food crops. Jakarta

Gomez KA, Gomez AA (1995) Statistical procedures for agricultural research. Universitas Indonesia, Jakarta

Horrie T, Homma K, Yoshida H (2006) Physiological and morphological traits associated with high yield potential in rice. Abstracts. In: 26th international rice research conference second international rice congress 2006, pp 12–13

Hoegh-Guldberg O, Jacob D, Taylor M, Bindi M, Brown S, Camilloni I, Diedhiou A, Djalante R, Ebi KL, Engelbrecht F, Guiot J, Hijioka Y, Mehrotra S, Payne A, Seneviratne SI, Thomas A, Warren R, Zhou G (2018) Impacts of 1.5 °C global warming on natural and human systems. In: Masson-Delmotte V, Zhai P, Pörtner H-O, Roberts D, Skea J, Shukla PR, Pirani A, Moufouma-Okia W, Péan C, Pidcock R, Connors S, Matthews JBR, Chen Y, Zhou X, Gomis MI, Lonnoy E, Maycock T, Tignor M, Waterfield T (eds) Global warming of 1.5 °C. An IPCC special report on the impacts of global warming of 1.5 °C above pre-industrial levels and related global greenhouse gas emission pathways, in the context of strengthening the global response to the threat of climate change, sustainable development, and efforts to eradicate poverty. In Press

Ilagan LA, Tablizo RP, Barba Jr RB, Marquez NA (2014) Soil fertility evaluation for rice production in catanduanes province, Philippines. Int J Sci Technol Res 3:81–87

ILO (2018) International labour organisation: employment by sector. ILO modelled estimates, Genf, May

Juanda BR (2015) Environment for production and sustainable food security. Agrosamudra Res J 2(2):61–70

Leal Filho W (2018) Impacts of climate change in coastal areas: lessons learned and experiences. In: Leal Filho W (ed) Climate change impacts and adaptation strategies for coastal communities. Springer International Publishing, Cham, pp 471–478

Maas EV, Nieman RH (1978) Physiology of plant tolerance to salinity. In Jung GA (ed) Crop tolerance to suboptimal land conditions, pp 277–299 (ASA Spec)

Mariana ZT, Razie F, Septiana M (2007) The activity of oxidizing iron and sulfur acidophil bacteria on tidal land in South Kalimantan. J Agritek 15(4):888–895

Norman MJT, Pearson CJ, Searle PGE (1984) The ecology of tropical food crops. Cambridge University Press

Poushter J, Huang C (2019) Climate change still seen as the top global threat, but cyberattacks a rising concern, Pew Research Center

Prayoga MK, Rostini N, Setiawati MR, Simarmata T, Stoeber S, Adinata K (2018) Preferences of farmers to superior rice (*Oryza sativa*) for rice fields in Pangandaran and Cilacap regions. J Kultivasi 17(1), 523–530

Simarmata T, Joya B, Turmuktini T (2011) Water saving and organic fertilizers based technology to remediate the health of paddy soils and to increase rice productivity in Indonesia. In: Proceeding of conference on international research on food security, natural resource management and rural development http://www.tropentag.de/2011/abstracts/full/694.pdf

Smit CM (1975) Salty soils and saline seep definitions, identification. Mont St U Coop Ext Ser Bull 1166

Soemartono B, Hardjono R (1980) Rice planting. C.V. Yasaguna, Jakarta
Thamrin M, Ruchjaningsih, Nappu MB (2013) Climate change and anticipation of technology in the management of dry land corn plants. In: Proceedings of the cereals national seminar, pp 353–370
USDA (1998) Soil quality indicators: pH. USDA natural resources conservation service. U.S. Department of Agriculture. Washington, D.C
Yoshida S (1981) Fundamental of rice science. International Rice Research Institute, Los Baños, Laguna, Philippines
Zhu JK (2001) Plant salt tolerance. Trends Plant Sci 6:66–71

SMART Agriculture and Rural Farmers Adaptation Measures to Climate Change in Southeast Nigeria: Implications for Sustainable Food Security

P. C. Obasi and C. Chikezie

Abstract The study evaluated SMART agriculture and rural farmers' vulnerability to climate change, adaptation measures and determinants of choice of adaptation strategies to climate change in Southeast, Nigeria. Data used for the study covered the period 1984–2014 and involved 1240 participatory rural farmers. Data collected were analysed using bivariate and multivariate analyses and multinomial logit model. The findings showed unsteady distribution of climate variables with peak points across the period under review. Rainfall volume, rain days, and relative humidity showed decreasing trends, while temperature and sunshine duration showed increasing trends. Late commencement of planting, use of fertilizers and pesticides, tree planting, use of soil conservation techniques and use of resistant crop varieties were identified to be the most effective adaptation measures to climate change adopted by farmers in the area. The result further shows that age of farmer, land tenureship, extension contact, gender, farm size, value of assets, farming experience and access to credit were the major determinants of choice of adaptation measures. The average per capita food energy consumption estimated for the farmers (2100 kcal < 2500 kcal) show that the farmers were food insecure. It was further observed that the sustainability values estimated for the farmers over the time period showed increasing trend.

Introduction

Climate-SMART Agriculture (CSA) as an approach that helps to guide the actions of farmers in transforming and reorienting agricultural systems, effectively support development and guarantee food security in a changing climate. It is an approach for developing agricultural strategies to secure sustainable food security under climate change. Climate-SMART Agriculture is derived from the acronym SMART, with S standing for specific, M for measureable, A for achievable, R for reliable and T

P. C. Obasi (✉) · C. Chikezie
Department of Agricultural Economics, Federal University of Technology Owerri, P. M. B. 1526, Owerri, Imo State, Nigeria
e-mail: polycarp.obasi@futo.edu.ng

W. Leal Filho et al. (eds.), *Climate Change, Hazards and Adaptation Options*, Climate Change Management, https://doi.org/10.1007/978-3-030-37425-9_41

for timely (McCarthy et al. 2012). Agricultural production in Southeast, Nigeria is largely rain-fed and strongly influenced by annual variations in rainfall and other climatic factors. While drought is a recurring problem in the northern part of the country, floods cause serious damage to livelihoods and agriculture in the South (Oyewole et al. 2014). Climate change as a global phenomenon has become a threat to humanity. In the last ten years, evaluation of Nigeria's policy response to climate change shows a non-realization of the full potentials of environmental policies such as mitigation and adaptation to climate change. In order for the effect of climate change and its shock to be cushioned as well as enhance food production and food security, farmers' knowledge on the forms of climate change is very essential especially when comparison is made with the trend of climatic variables in the past years. Therefore, knowing the vulnerability of farmers' to climate change and their adaptive capacity will help to reduce exposure to climate impacts. Changes in climate have been known to profoundly affect natural resources, and thus increase human vulnerability to harshness, hunger and infection (Loon et al. 2007).

Nigerian farmers, who constitute bulk of the poor face tragic crops failure, reduced agricultural productivity and increased hunger. It is projected that crop yield in Africa may fall by 10–20% by 2050 or even up to 50% due to climate change (Jones and Thornton 2009). Similarly, Food and Agriculture Organization (FAO 2011) estimated 20.5% loss of cereals, 54% loss of root and tubers, 31% loss of oil seeds and pulses and 66% loss of fruits in developing world as a result of adverse weather conditions and production practices. Given the above statistics, climate change has indeed become a global issue. It is very likely that climate change can aggravate losses in food production and slow the pace of progress along sustainable development pathways, either directly through increased exposure to adverse impact or indirectly through erosion of the capacity to adapt. Attempts have been made by researchers to assess the causes of such down turn in crop yield and their results show that climate change is a major contributing factor (Ayinde 2010; Ayinde et al. 2011). In spite of these, there are limited studies on effects of climate change variables on crop output in Southeast, Nigeria (Aondoaka 2012). Report by Molua and Lambi (2007) attested to the fact that there has been a gap in knowledge on the effects of climate change on crop production in most agro-ecological zones of Nigeria. Odekunle et al. (2007) demonstrated effect of climate change on maize. Related studies on selected food crops were carried out by Agboola and Oyeleye (2007), Oluwasegu and Olaniran (2010). However, there is the dearth of literature on SMART agriculture and rural farmers' adaptation measures to climate change in southeast Nigeria; hence this study covers the gap in knowledge.

Objectives of the Study

The broad objective of this study was to evaluate climate SMART Agriculture in relation to rural farmers' vulnerability to climate change, adaptation measures and determinants of choice of adaptation strategies to climate change in Southeast, Nigeria. The specific objectives of the study were to:

i. determine the trends in climate variables in Southeast, Nigeria for the period 1984–2014
ii. evaluate the determinants of vulnerability to climate change among farmers in the area,
iii. ascertain the adaptation measures to climate change practiced by farmers in the area,
iv. estimate the sustainability of agricultural production system as well as the food security status of the farmers in the area.

Materials and Method

The Study Area

The study was conducted in Southeast, Nigeria. The area which comprises of Abia, Imo, Enugu, Anambra and Ebonyi States is located between Latitudes 5°N and 6°N, and longitudes 6°E and 8°E with a land mass of about 28,987 km^2 (NBS 2010). The Southeast states fall within the rainforest ecological zone of Nigeria. It has a belt of tall trees dominated by climbing plants. The zone experiences two seasons, the rainy and dry seasons. The rainy season usually begins in April and ends in October or November while the dry season usually starts from November and ends in March. The mean annual rainfall in the area is between 2000 and 3000 mm while the mean annual temperature is 28 °C (BNRCC 2011a). The people of the area are mainly traders, farmers, civil servants and artisans. The major crops grown in the area comprise of food crops such as yam, cassava, cocoyam and maize, while the cash crops are cocoa and oil palm. Land use types in the area are shifting cultivation and mixed cropping. The predominant soil of the area is deep well drained sandy loam soil. Data used for the study were obtained from the National Root Crops Research Institute (NRCRI) Umudike, Umuahia and covered the period 1984–2014.

Sample Selection and Data Collection

The multi-stage sampling technique was used in selecting the sample. In the first stage, three out of the five states that make up the area (Abia, Ebonyi and Anambra) were purposively selected based on differences in topography, vegetative covers and effects of climate change. Abia State has three agricultural zones and seventeen Local Government Areas (LGAs), Ebonyi State has three Agricultural zones and thirteen Local Government Areas while Anambra State has four agricultural zones and twenty-one Local Government Areas. Based on the differences in the number of LGAs in each agricultural zone of the selected States, the proportionate sampling technique was used to select 6 LGAs from Abia State, 5 LGAs from Ebonyi State and 8 LGAs from Anambra State. The selection was based on 40% of the total number of LGAs in each agricultural zone. Furthermore, Lists of communities in each selected LGA were obtained from the community development officer at the local government headquarters. The selected LGAs in Abia, Ebonyi and Ananbra have 57, 41 and 67 communities respectively. At the third stage, 24, 16 and 27 communities were proportionately selected from Abia, Ebonyi and Anambra states respectively. At the fourth stage, 64 villages from Abia, 57 villages from Ebonyi and 78 villages from Anambra state were selected. At the final stage, the proportionate sampling technique was used to select 110 farmers from Abia State, 79 farmers from Ebonyi State and 132 farmers from Anambra State. This gave a sample size of 321 farmers.

Data used for the study were collected from both primary and secondary sources. Secondary data used include information on average climate variables, crops yield and optimal climatic conditions. Data on weather indices were collected from the Nigerian Metrological station (NIMET) of the National Root Crops Research Institute Umudike Umuahia. Average values of temperature, rainfall volume, rain day, and relative humidity as well as sunshine hours covering the period 1984–2014 were used for this study. Primary data collected include socio-economic characteristics of the respondents such as age, sex, level of education, household size, years of farming experience, extension contact and adaptation measures to climate change by farmers. Data collection lasted from September, 2014 to February, 2015.

Data Analyses

Analysis of Trend of Climate Variables and Farmers' Perception of Climate Change

To evaluate the changes in major climate variables such as temperature, sunshine hours, rainfall volume, rainfall days and relative humidity, the trend analysis was carried out using bivariate regression model and line graphs. Following Hageback et al. (2005), the bivariate regression model estimated is of the form:

$$Y_i = f(t) \tag{1}$$

where

Y_i climate change related hazards;
t time (years).

Analysis of Levels and Determinants of Vulnerability to Climate Change

To determine farmers' vulnerability to climate change, suitable indicators based on literature (Nelson et al. 2010), were chosen and normalized. Normalization was done because of the different units of the indicators. For indicators that are related positively to vulnerability and climate change, the standardization was achieved following Wannasai et al. (2013) by employing the formula:

$$Y_{ij} = \frac{x_{ij} - MinX_{ij}}{MaxX_{ij}\_MinX_{ij}} \quad (2)$$

For indicators negatively related to vulnerability and climate change, the formula is:

$$Y_{ij} = \frac{MaxX_{ij}\_X_{ij}}{MaxX_{ij} - MinX_{ij}} \quad (3)$$

where

X_{ij} the *i*th vulnerability indicator in the *j*th metric set.

The next step was to assign weight to the normalized indicators. Following Gbetibouo and Ringler (2009), Principal Component Analysis (PCA) was used for this purpose. In this study, the number of times drought, flood and windstorm occurred in 5 years represented the exposure metrics. Sensitivity of households to climate induced stresses was represented by its associated impacts such as number of times a member of household was sick, number of crops damaged and percentage of non-farm income. The adaptive metrics include type of roofing sheet, source of information, source of water, dependency ratio, fertilizer use, farm income, assets value, social or age group membership, number of livestock owned and number of economic trees owned. Eigenvalues of the correlation matrix were estimated and these represent a partitioning of the total variation accounted for by each principal component. The proportion (share of variance explained by each eigenvalue) and cumulative proportion (the variance of both current and all preceding principal components) were calculated. After the identification of factors to be retained, the variables were transformed into indices for each responding farmers using STATA 13.0 software, with values ranging from negative to positive. The values more negative show greater vulnerability and those more positive indicate less vulnerability (Opiyo et al. 2014).

Analysis of the Determinants of Choice of Adaptation Measures Among Farmers

To analyse the determinants of choice of adaptation measures to climate change among rural farmers, the multinomial logistic model (MLM) was estimated following Gujarati and Sangeetha (2007). The multinomial logistic model used in this study to estimate the odds in favour of adapting to an option is stated as follows:

$$\mathrm{Ln}\left(\frac{P_i}{1-P_i}\right) = \beta_1 + \beta_2 \mathrm{X_i} + \mathrm{e_i} \tag{4}$$

where; $\mathrm{P_j}$ = Probability to adapt an option

Y = Adaptation options defined by j = 1, 2, …, 7

1 = Crop diversification (mixed cropping)

2 = Changing planting dates

3 = Soil conservation (mulching)

4 = Use of resistant variety

5 = Use of land augmenting input (fertilizer)

6 = Irrigation/watering

7 = Diversifying from farming to non-farming activity

$\mathrm{X_i}$ = independent variables (i = 1, 2, …, 15)

$\mathrm{X_1}$ = Farm size (hectares)

$\mathrm{X_2}$ = Savings (naira)

$\mathrm{X_3}$ = Access to credit (1 = access, 0 = otherwise)

$\mathrm{X_4}$ = Extension contact (1 = access, 0 = otherwise)

$\mathrm{X_5}$ = Farming experience (years)

$\mathrm{X_6}$ = Sex (1 = male, 0 = female)

$\mathrm{X_7}$ = Value of assets (naira)

$\mathrm{X_8}$ = Primary occupation (1 = farming, 0 = otherwise)

$\mathrm{X_9}$ = Age (years)

$\mathrm{X_{10}}$ = Social group membership (1 for membership, 0 if otherwise)

$\mathrm{X_{11}}$ = Land fragmentation (number of lands fragmented)

$\mathrm{X_{12}}$ = Land tenure (1 = owned, 0 = otherwise)

$\mathrm{X_{13}}$ = If household farm in Anambra State (1 = yes, 0 = no)

$\mathrm{X_{14}}$ = If household farm in Abia State (1 = yes, 0 = no)

$\mathrm{X_{15}}$ = If household farm in Ebonyi State (1 = yes, 0 = no)

e = error term.

$\mathrm{B_j}$ = β_0, β_1, β_2, β_3, β_4, β_5, β_6, β_7, β_8, β_9, β_{10}, β_{11}, β_{12}, β_{13}, β_{14} and β_{15} are the parameter estimates of the independent variables. In this study, the estimation of the multinomial logistic model was undertaken by normalizing one adaptation option, which is normally referred to as the "base option". For this analysis, changing planting dates was chosen as the base option. The base category is expressed by the following formula:

$$P_{i0} = \frac{1}{1 + e^{\beta j X}} \tag{5}$$

where P_{i0} is the probability of being in base category β_j was estimated using STATA 13

The natural logarithm of the odd ratio gave the estimated equation as follows:

$$P_{(y)} = \frac{\beta j x}{p_{io}} \tag{6}$$

This denotes the probability of each group relative to the probability of base category and based on the review of literature on adaptation to climate change studies, a range of household characteristics and other factors that describe local conditions were hypothesized to influence farmers' adaptation choice in the study area. The a priori expectation is stated mathematically as follows: X_1, X_2, X_3, X_4, X_5, X_6, X_7, X_8, X_{10}, X_{12}, X_{13}, X_{14} and $X_{15} > 0$; X_9, $X_{11} < 0$. Following Deressa et al. (2008a), Chi Square from the Likelihood Ratio Statistics was used to test the overall significance of the multinomial model.

Measurement of Sustainability of the Agricultural System in Southeast Nigeria

Following Ehui and Spencer (1990) and Lyman and Herdt (1989), the TFP model was used an index of productivity while the ordinary least square regression model was used to measure the influence of climate change on sustainability. Following Onyeagocha et al (2018), the Total Factor Productivity (TFP) model estimated is given as:

$$\text{TFP} = \frac{V_{T_o}}{V_{T_n}} = \frac{\sum_{j=1}^{m}\sum_{i=1}^{n} P_{qij} Q_{ij}}{\sum_{j=1}^{m}\sum_{t=1}^{n} P_{xtj} X_{tj}} \tag{7}$$

where

- V_{T_o} Value of Total Output in Naira/annum
- V_{T_n} Value of Total Input in Naira/annum.
- P_q Price of output in Naira
- Q Quantity of Output
- i Type of output (i ranges from 1 – nth output type)
- j Farmers (j ranges from 1 – m^{th} farmer)
- P_x Price of input
- X Quantity of input
- t Type of input (t ranges from 1 – nth input type).

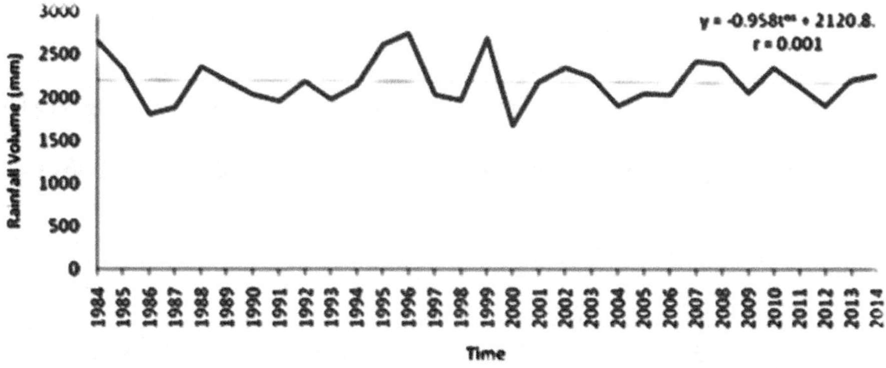

Fig. 1 Mean annual rainfall variation over the time period 1984–2014. *Source* National Root Crops Research Institute, Umudike, 2014

Results and Discussion

Trends in Climate Change in Nigeria

Mean Annual Rainfall Variation Over the Time Period 1984–2014

The distribution of rainfall volume in southeast Nigeria for the period 1984–2014 is shown in Fig. 1.

Figure 1, rainfall variation showed a very unstable pattern with high volatility over the time period. There were unsteady rise and fall in rainfall volume between 1994 and 2000 with a peak value of 2850 mm in 1996 and the lowest value of 1700 mm in year 2000. This is consistent with the findings of Ammani et al. (2012) who found that rainfall distribution patterns were as a result of the climate change experienced in the country. The period between 2001 and 2008 witnessed fair distribution of rainfall with moderate fluctuation. The line graph showed a decreasing trend with a correlation coefficient of 0.001 and a negative trend coefficient of −0.958. Though not significant, the decreasing trend suggests that low levels of soil water associated with such trend are in most cases detrimental to crop growth and survival. In general, declining rainfall at some points brought about delayed cultivation of food crops such as maize, cassava and yam.

Trend of Mean Annual Rainy Days Over the Time Period 1984–2014

The distribution of number of rain days in southeast Nigeria for the period 1984–2014 is shown in Fig. 2

Figure 2 shows the distribution of rainfall days in southeast Nigeria for the period 1984 to 2014. Number of rain days was steady for 150 days from 1989 to 1998, rose

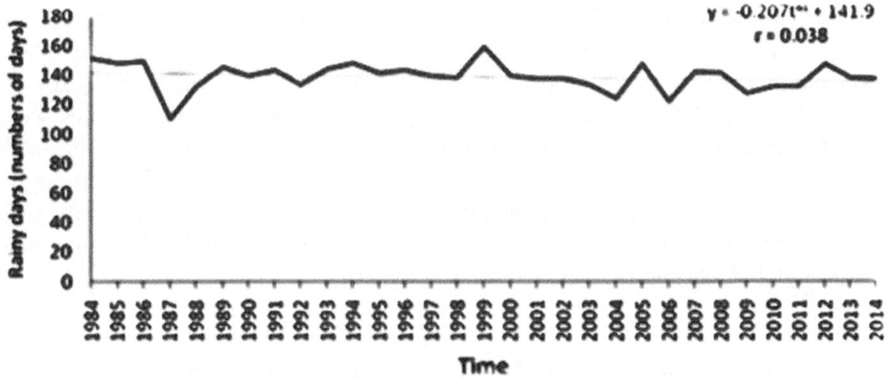

Fig. 2 Trend of mean annual rainy days over the time period 1984–2014. *Source* National Root Crop Research Institute, Umudike, 2014

in the 160th day in 1990 and declined. The highest rain day period was 160 days in 1999 and lowest in 1987 with mean rain day period of 110 days. This shows a negative trend coefficient of −0.207. The coefficient of correlation between rain days and time was 0.038 and not statistically significant. This observation also indicated high volatility in the movement of rain day in the area and posed great difficulty in prediction.

Trend of Mean Annual Temperature Over the Time Period 1984–2014

Figure 3 shows the temperature variation for three decades in the study area.

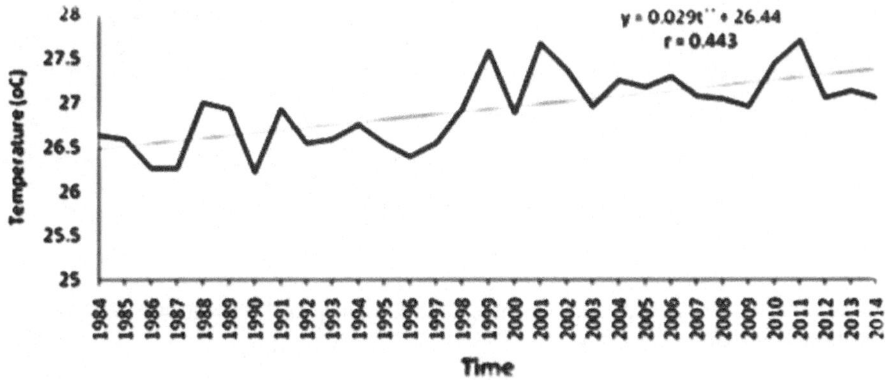

Fig. 3 Trend of mean annual rainy days over the time period 1984–2014. *Source* National Root Crop Research Institute, Umudike, 2014

The figure shows that temperature averaged about 26.7 °C within a period of ten years (1984–1994), rose to 27 °C from 1995 to 2004 and became 27.5 °C between 2005 and 2014. This indicates an increasing trend. The trend coefficient was 0.029 and was statistically significant at 1%. The minimum and maximum values were 26.1 °C and 27.6 °C occurring in 1990 and 2011 respectively. This is in line with the findings of Chidiebere-Mark et al. (2014); Nwaiwu et al. (2014). The correlation coefficient was 0.443 and was statistically significant at 1%. This positive significance of time in temperature indicated a rising trend of temperature with time which thus signifies global warming effects of climate change.

Trend of Mean Annual Relative Humidity Over the Time Period 1984–2014

Figure 4 shows the distribution of variations in relative humidity in the area over the period.

Figure 4 shows the occurrence of high volatility in relative humidity over the period with substantial fluctuations. A maximum relative humidity of 78 mmHg was recorded in 1989 followed by unsteady movements and a record low relative humidity of 65 mmHg in 2010. The graph has a negative trend coefficient of −0.068 and coefficient of correlation value of 0.070. Though this is a weak and insignificant relationship, it is capable of causing crop failure. The result obtained here is consistent with Nwaiwu et al. (2014), Tologbonse et al. (2010).

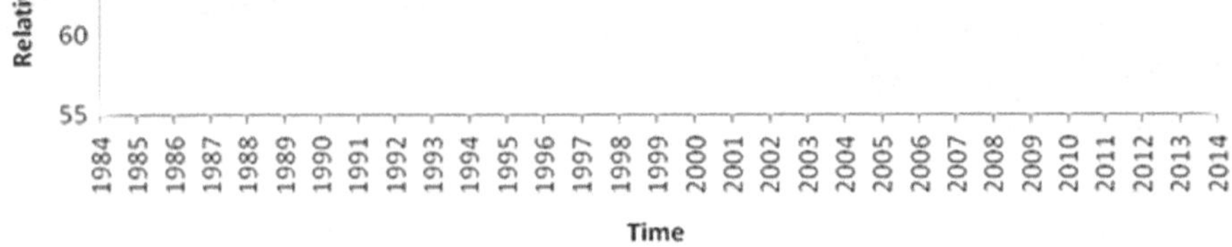

Fig. 4 Trend of mean annual relative humidity over the time period 1984–2014. *Source* National Root Crop Research Institute, Umudike, 2014

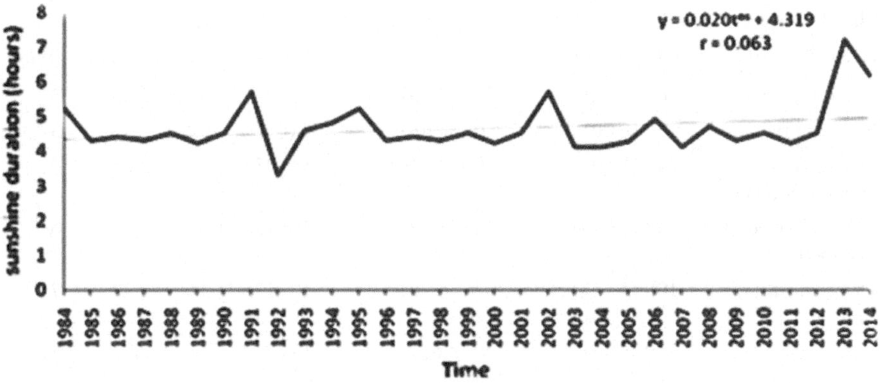

Fig. 5 Trend of mean annual sunshine duration over time period 1984–2014. *Source* National Root Crop Research Institute, Umudike, 2014

Trend of Mean Annual Sunshine Duration Over the Time Period 1984–2014

The trend of sunshine variation over the 30 years period is shown in Fig. 5.

The line graph in Fig. 5 indicates an unsteady change in sunshine duration. The maximum and minimum hours of sunshine were 7.5 h/day in 2013 and approximately 3 h/day in 1992 respectively. The coefficient of correlation between sunshine duration and time was 0.063 thus indicating an increasing trend. Also, a positive trend coefficient value of 0.020 was recorded. Although this is not statistically significant, it reflects an increase in temperature.

Forms of Climate Change Perceived by Farmers in the Study Area

Table 1 shows the different forms of climate change and their degree of impact as perceived by farmers in the study area.

Based on the 3-point likert type scale adopted to quantify the variables, the mid value calculated was 2. Table 1 show that, 86.2% of the respondents perceived the incidence of flooding as the major form of climate change with weighted mean of 2.85. This is in line with the findings of Bhavnani et al. (2008) who observed that droughts and floods account for 80% of the loss of life and peoples means of livelihood in sub-Saharan Africa. The Table also shows that 64.65% of the respondents saw the rise in sea level as a form of climate change with a mean of 2.63. Furthermore, 43.10% of the respondents saw drying of streams/ponds to a little extent as a form of climate change with a mean of 2.07. Also, 63.79%, 23.27% and 12.94% of the respondents perceived longer period of dry spell with a mean of 2.51 to be a form

Table 1 Forms of climate change perceived by farmers in Southeast Nigeria

Events/hazards	Great extent (3)	Little extent (2)	Not likely (1)	Mean	Rank
Flooding	200 (86.2)	30 (12.93)	2 (0.87)	2.85	2
Sea level rise	150 (64.65)	80 (34.48)	2 (0.87)	2.63	5
Drying of streams/ponds	74 (31.89)	100 (43.10)	58 (25.01)	2.07	8
Longer period of dry spell	148 (63.79)	54 (23.27)	30 (12.94)	2.51	6
Delay in start of rain	108 (46.55)	84 (36.20)	40 (17.25)	2.29	7
Heat stress	200 (84.2)	20 (8.62)	12 (7.16)	2.81	3
Greater intense of sunshine	232 (100)	0 (0)	0 (0)	3.00	1
Windstorm	180(77.58)	50(21.55)	2 (0.87)	2.77	4
Landslide	32 (13.80)	50(21.55)	150 (64.65)	1.49	9

Source Field survey data, 2015; Figures in parentheses are percentages

of climate change. Similarly, delay in the start of rain, heat stress and windstorm with weighted means of 2.29, 2.81 and 2.77 respectively were seen by farmers as other forms of climate change. The result obtained here is in line with the findings by Anselm et al. (2011) who noted that erratic rainfall pattern, long period of dry season, thunderstorms, heavy winds, intense heat wave were on the increase.

Household Vulnerability to Climate Change in Southeast, Nigeria

The categorization of households vulnerability to climate change based on normal distribution is presented in Table 2.

Table 2 shows that the majority of households are within the moderately vulnerable category, with 68.97% of the households having indices of −2.07912 to 1.95995. The less vulnerable households constitute 15.08% of the respondents with indices ranging

Table 2 Distribution of Households by levels of vulnerability to climate change

Vulnerability level	Vulnerability indices	Frequency	Percentage (%)
Highly vulnerable	−7.65285 to −2.079115	37	15.95
Moderately vulnerable	−2.07912 to 1.95995	160	68.97
Less vulnerable	1.96000–4.899319	35	15.08
Total		232	100.00

Source Field survey data, 2014; principal component analysis (PCA) result from 1.96000 to 4.899319, while the highly vulnerable households had indices of −7.65285 to −2.079115 and constitute 15.95% of the total households sampled. When a farmer is vulnerable to climate change, it means that his exposure and sensitivity to climate change are more than his ability to cope with vagaries of weather. In addition, high level of flooding and windstorm may entail an increased outmigration of farmers to a more conducive area. However, with the majority being moderately vulnerable may suggest that the farmers need temporary assistance (Opiyo et al. 2014).

Determinants of Farmers' Vulnerability to Climate Change

The vulnerability indices were subjected to further parametric analysis using the ordinary least square method (OLS) using four functional forms. Based on econometric, statistical and economic a priori expectation, the linear functional form was chosen as the lead equation and used for further analysis of the data. The results are presented in Table 3.

The coefficient of multiple determination (R^2) value of 0.4694 as presented in Table 4 shows that 46.94% of the variations in the level of household vulnerability were explained by the factors X_3, X_5, X_6, X_7, X_9, X_{14} and X_{16}.

The Variance Inflation Factor (VIF) was computed because of the likelihood of the existence of multi collinearity and the result produced a VIF value of 1.87. As a rule-of-thumb the VIF should not exceed 5 or 10. Various recommendations for acceptable levels of VIF are common in literature. Perhaps most commonly, a value of 10 has been recommended as the maximum level of VIF (O'Brien 2007; Zuur et al. 2010). VIF is used to describe how much multi-collinearity exists in a regression analysis. Multi-collinearity is a problem because it can increase the variance of the regression coefficients, making them unstable and difficult to interpret (Gujarati and Porter 2009). In the model, there was no redundancy because of relatively high tolerance of the variables as shown in Table 4. The results show that out of the included variables, savings is statistically significant ($p < 0.01$) and negatively related to vulnerability to climate change. As the amount saved increases by 1 unit, vulnerability to climate change decreases by 0.0000178. This implies that, with savings, food crop farmers could invest in alternative businesses, thereby reducing the impact of climate change. Constantly changing climate reduces agricultural productivity and make farmers food insecure. Therefore, increasing farmers' savings especially during bumper harvests would help to give them adequate security against impending negative climate events. Harvey et al. (2014) observed that savings is a determinant of vulnerability to climate change by noting that less than 2% of farmers surveyed had personal savings accounts. Nepal Climate Change Management Centre (NCCMC 2014) also observed that planting fodder trees, saving of grains and money, off-farm activities and credits are not sufficient and sustainable to cope with climatic vagaries. According to NCCMC (2014) strategies are not enough to keep farmers from being vulnerable to climate change.

Table 3 Results of multiple regression analysis on vulnerability to climate change

Variables	Linear	Semi-log	Double-log	Exponential
Sex (X_1)	.0341619 (0.16)	−.0065431 (−0.06)	.0104461 (0.09)	.0468834 (0.20)
Farm size (X_2)	−.1148595 (−0.46)	−.0835563 (−0.62)	−.1963992 (−0.70)	−.2975815 (−0.56)
Saving (X_3)	−.0000178*** (−6.19)	−8.61e−06*** (−7.09)	−.0457943** (−2.13)	−.1035546** (−2.41)
Credit (X_4)	−.2611589 (−1.12)	−.1669389 (−1.47)	−.0876767 (−0.64)	−.0723877 (−0.25)
Extension (X_5)	−.7089335*** (−3.14)	−.3745739*** (−3.24)	−.4700842*** (−3.78)	−.930319*** (−3.76)
Household exp. (X_6)	−.0174134*** (−3.60)	−.0061973** (−2.66)	−.1284661** (−2.12)	−.3022214** (−2.33)
Value of crop (X_7)	−.0176954*** (−3.22)	−.0088478*** (−3.41)	−.0948276* (−1.76)	−.2040219* (−1.85)
Education (X_8)	.0158128 (0.61)	.0059262 (0.46)	.1672498* (1.82)	.3522846* (1.94)
Age (X_9)	−.0192366** (−2.41)	−.0077628* (−1.94)	−.4434184** (−2.42)	−1.074452** (−2.86)
Cooperative mgt. (X_{10})	−.1542046 (−0.70)	−.0683969 (−0.61)	−.1440576 (−1.21)	−.313033 (−1.34)
Household size (X_{11})	.0306452 (1.09)	.0065936 (0.46)	.0699409 (0.48)	.3101798 (1.09)
Land frag. (X_{12})	.3081883 (1.34)	.0861553 (0.76)	.1836503 (1.53)	.5055387** (2.05)
Non-farm income (X_{13})	−.0763276 (−0.26)	−.0658966 (−0.48)	−.0313635 (−0.22)	−.0139753 (−0.04)
Land ownership (X_{14})	−.4484502** (−2.00)	.2014819* (1.74)	.1361641 (1.05)	.2831158 (1.10)
Ebonyi (X_{15})	−.4058287 (−1.20)	−.1404747 (−0.88)	−.3293275** (−2.02)	−.778692** (−2.25)
Anambra (X_{16})	−.5953051** (−2.05)	−.1976701 (−1.33)	−.2962029* (−1.92)	−.7647458** (−2.54)
Constant	2.278511*** (3.33)	1.173199*** (3.31)	2.624454*** (3.19)	5.668215*** (3.39)
R^2	0.4694	0.4408	0.3095	0.3300
R^{-2}	0.4415	0.4113	0.2732	0.2947
F-value	16.16	19.00	7.02	5.90
Standard error	.6843022	.3544531	.8220838	1.669729

Source Field survey data, 2014; values in parenthesis are t- ratios
N/B *** = significant at 1%; ** = significant at 5% * = significant at 10%
Dummy variable for Anambra and Ebonyi States were included as State effect to take care of clustering, Abia state served as base category

Table 4 Result of linear regression analysis on vulnerability to climate

Variables	Coefficients	Standard error	T-value	Tolerance
Sex	.0341619	.2171196	0.16	0.813795
Farm size	−.1148595	.2472338	−0.46	0.832646
Saving	−.0000178***	2.88e−06	−6.19	0.707423
Credit	−.2611589	.2337443	−1.12	0.810123
Extension	−.7089335***	.225674	−3.14	0.802450
Household exp.	−.0174134***	.0048393	−3.60	0.810598
Value of crop	−.0176954***	.0054977	−3.22	0.845352
Education	.0158128	.0259924	0.61	0.739686
Age	−.0192366**	.0079954	−2.41	0.621859
Cooperative m.	−.1542046	.2197787	−0.70	0.830954
Household size	.0306452	.0280241	1.09	0.873631
Land frag.	.3081883	.229734	1.34	0.870454
Non-farm income	−.0763276	.2980897	−0.26	0.839388
Land ownership	−.4484502**	.2242275	2.00	0.801536
Ebonyi	−.4058287	.3381653	−1.20	0.489777
Anambra	−.5953051**	.2905845	−2.05	0.451023
Constant	2.278511***	.6843022	3.33	
R^2 = 0.4694, F(16, 215) = 16.16				

Source Field survey data, 2014
N/B *** = significant at 1%; ** = significant at 5%

Extension contact is negative to vulnerability to climate change and statistically significant at $p < 0.01$. With the vulnerability decreasing by 0.7089335, extension contact increases by one unit. More access to extension services ensures farmers' adoption of improved techniques and enhances their knowledge of imminent climatic impact and adaptation strategies. Also, adequate extension service delivery affords the farmers the opportunity of accessing improved and resistant varieties which may help to enhance their adaptive capacity. Household expenditure is statistically significant at $p < 0.01$ and negatively related to vulnerability to climate change. As household expenditures increases by 1 unit, the vulnerability to climate change decreases by 0.0174134. This suggests that farmers were living beyond their monthly income. However, when farmers spend more on activities that enhances adaptive capacity, they could be spared of the agony of losing so much to climate hazards.

Land ownership status is significant ($p < 0.05$) and negatively correlated to vulnerability to climate change. It also shows that as ownership right to land increases by 1 unit the climate vulnerability decreases by 0.4484502. The more ownership right to land a farmer has, the less vulnerable he becomes to climate change. For instance, adaptation to flooding and erosion require long term investments and this is only possible where there is permanent ownership of land. Again, this is similar to findings of Abayomi (2012).

The Determinants of Farmers' Choice of Adaptation Methods

Table 5 presents the parameter estimates of the multinomial results.

In this analysis, changing planting date was taken as the "base category." The data were tested for the validity of the independence of the irrelevant alternatives (IIA) assumptions by using the Hausman test for IIA (Chmelarova 2007). For the Hausman test, the chi-Square (χ^2) ranged from −9.206 to 12.893 with *p*-value of 1.00. Based on the results obtained, the null hypothesis of independence of the climate change adaptation options was accepted. This further implies that the application of the MNL specification on the determinants of adaptation measures was justified. The likelihood ratio statistic as indicated by $\chi^2 = 162.57$ and statistically significant ($P < 0.00001$). The coefficient of farm size in the equation gave −3.986129 and an exponential value of 0.01857. This suggests that if farm size increases by 1 unit, the odd ratio for changing planting distance would decrease by a factor of 0.0186.

Estimation of Sustainability of the Agricultural System and Food Security Status of Farmers in Southeast Nigeria

Applying Eq. (7) to the data, the maximum, minimum and mean sustainability values estimated for the area are 3.47. 0.48 and 1.35 respectively. The maximum value occurred in 1987 while the minimum value occurred in 2008 with a starting value of 1.5 in 1984. The value declined steadily reaching a value of 1.00 in 2012 and further declined to 0.98 and 0.89 in 2013 and 2014 respectively. Since the sustainability values estimated for the farmers over the time period showed increasing trend, we observe that the production system in the area is sustainable.

With respect to the food security status of the farmers, the average per capita food energy consumption estimated was 2100 kcal. This value was arrived at by calculating the calorific output of food commodities consumed by members of each household. Following Gaman and Sherrington (1977), 1 g of carbohydrate provides 3.75 kcal, 1 g of fat provides 9 kcal, 1 g of protein provides 4 kcal. The value (2100 kcal) estimated for the farmers' is less than the recommended value of 2500 kcal for non-industrialized countries, thus suggesting that the farmers are food insecure.

Table 5 Parameters of the multinomial logistic regression results

Explanatory variables	Watering	Planting dif. varieties	Mulching	Resistant variety	Use of fertilizer	Farm to non-farm
Farm size	−.986129** (0.038)	.1645484 (0.642)	−.5678452 (0.312)	−1.489221 (0.225)	−.2171792 (0.750)	−6.944313* (0.001)
Saving	4.52e−06 (0.550)	−.0000102 (0.201)	−.0000115 (0.106)	1.39e−06 (0.831)	−1.62e−06 (0.834)	1.27e−06 (0.874)
Acc. credit	−.1021086 (0.874)	.5324011 (0.353)	−.0182468 (0.974)	.7308574 (0.298)	−.4258146 (0.490)	.5432507** (0.0466)
Extension contact	−1.6492* (0.009)	−1.026541** (0.042)	−1.326901** (0.016)	−1.484836** (0.012)	−1.478153** (0.014)	−1.331764*** (0.094)
Farm Experience	.0124643 (0.344)	−.0056008 (0.639)	−.0240187 (0.122)	−.0293327** (0.027)	−.0104004 (0.339)	−.0024919 (0.884)
Sex	1.64899* (0.010)	1.292336* (0.010)	1.635399* (0.007)	.8168656 (0.143)	.9717192 (0.116)	2.072341*** (0.051)
Value of assets	−4.42e−07** (0.027)	−7.40e−08 (0.316)	−1.12e−07 (0.318)	−2.58e−07 (0.117)	3.68e−08 (0.632)	−9.03e−08 (0.592)
Pri. occup	−.1059811 (0.878)	−.4907385 (0.330)	−.4810921 (0.402)	−.32886 (0.594)	−.498072 (0.428)	.7984404 (0.377)
Age	−.0396428*** (0.083)	−.0004376 (0.978)	.0119455 (0.504)	−.001232 (0.951)	.0068131 (0.735)	−.0537832*** (0.059)
Social group	.7485912 (0.437)	−.4651746 (0.467)	−.731721 (0.247)	.112532 (0.892)	−.6914659 (0.330)	−.088206 (0.916)
Land fragmentation	.0508944 (0.943)	.1948878 (0.705)	.7788427 (0.156)	−.0306211 (0.962)	.1496099 (0.811)	.8045909 (0.277)
Land tenure	1.875057** (0.033)	1.102988** (0.029)	1.759747* (0.009)	.0911737 (0.862)	1.047357*** (0.061)	1.783638 (0.171)

(continued)

Table 5 (continued)

Explanatory variables	Watering	Planting dif. varieties	Mulching	Resistant variety	Use of fertilizer	Farm to non-farm
Anambra State	1.76023** (0.029)	.1660077 (0.804)	1.013317 (0.167)	.9478004 (0.183)	1.321015 (0.117)	2.493213** (0.044)
Ebonyi State	−1.00018 (0.334)	−1.45115** (0.034)	−.9071518 (0.257)	−.4097275 (0.666)	.1937447 (0.824)	.4247518 (0.771)
Constant	.2281074 (0.892)	.7495721 (0.514)	−.3992499 (0.790)	.9232884 (0.523)	−.1840781 (0.891)	−.8980216 (0.693)

Notes ***, **, * = significant at 1%, 5%, and 10% probability level, respectively
Base category = changing planting dates
Number of observation = 232
Log pseudo likelihood = −364.17651
Wald chi2 (84) = 162.57
Log likelihood = −364.17651
Prob > chi2 = 0.0000
Pseudo R^2 = 0.154
Source Field survey data, 2014; the figures in parenthesis are *P*-values

Conclusions

Based on the results of the study we conclude that the more ownership right to land a farmer has, the less vulnerable he becomes to climate change because adaptation to flooding and erosion require long term investments and this is only possible where there is permanent ownership of land. Also it is established that the production system in the area is sustainable, and that the farmers are food insecure. The farmers are said to be food insecure because 2100 kcal estimated for the farmers' is less than the recommended value of 2500 kcal for non-industrialized countries.

References

Abayomi SO (2012) Vulnerability of peasant cocoa farmers to climate change in South-West Nigeria. Kamla-Raj 2012. J Hum Ecol 40(1):33–41. Retrieved from http://www.krepublishers.com/02-Journals/JHE/JHE-40-0-000-12-Web/JHE-40-1-12

Agboola T, Ojeleye D (2007) Climate change and food crop production in Ibadan, Nigeria. In: African crop science society conference proceeding 8, pp 1423–1433. Retrieved from https://www.cabdirect.org/cabdirect/abstract/20103268531

Ammani AA, Ja'afaru AK, Aliyu J A, Arab AI (2012) Climate change and maize production: empirical evidence from Kaduna State, Nigeria. J Agric Extension 16(1):1–8. Retrieved http://dx.doi.org/10.4314/jae.v16il.1

Anselm AE, Ignatius IM, Josephat CM, Anthony NO, Elizabeth AO, Fidelis E (2011) Indigenous agricultural adaptation to climate change: Study of Imo and Enugu States in South-East Nigeria African technology policy studies. Network working paper, vol 53, pp 6–23. Retrieved from https://searchworks.stanford.edu/view/10322614

Aondoaka SC (2012) Effects of climate change on agricultural productivity in the federal capital territory (fct), Abuja, Nigeria. Ethiop J Environ Stud Manag (EJESM) 5(4). Retrieved from https://doi.org/10.4314/ejesm.v5i4.s16

Ayinde OE (2010) Empirical analysis of agricultural production and climate change: a case study of Nigeria. J Sustain Dev Afr 12(6):275–282. Retrieved from https://www.researchgate.net/publication/263228332_empirical_analysis_of_agricultural_production_and_climate_change_a_case_study_of_Nigeria

Ayinde OE, Muchiea M, Olatunji GB (2011) Effect of climate change on agricultural productivity in Nigeria: a co-integration model approach. J Hum Ecol 35(3):189–194. Retrieved from http://www.krepublishers.com/02-Journals/JHE/JHE-35-0-000-11-Web/JHE-35-3-000-11-Abst-PDF/JHE-35-3-189-11-2147-Ayinde-O-E/JHE-35-3-189

Bhavnani R, Vordzorgbe S, Owor M, Bousquet F (2008) Report on the status of disaster risk reduction in the sub-Saharan Africa region commission of the African Union, United Nations and the World Bank. Retrieved from http://www.unisdr.org/files/2229DRRinSubSaharanAfricaRegion.pdf

Building Nigeria's Response to Climate Change (BNRCC) (2011a) National adaptation strategy and plan of action on climate change for Nigeria (NASPA-CCN); draft for stakeholder review, September 2011. Retrieved from http://csdevnet.org/wp-content/uploads/national-adaptation-strategy-and-plan-of-action.pdf

Building Nigeria's Response to Climate Change (BNRCC) (2011c) Gender dimensions and indigenous knowledge for adaptation to climate change in South-East Nigeria research report women and children development initiative (WACDI)

Chmelarova V (2007) The Hausman Test, and some alternatives, with heteroskedastic data. Louisiana State University and Agricultural & Mechanical College, 2007. Retrieved 1/6/2007 from here (http://etd.lsu.edu/docs/available/etd-01242007-165928/unrestricted/Chmelarova_dis.pdf)
Chidiebere-Mark NM, Eze CC, Nwankwo OO, Onyeneke RU, Ukwandu JC, Ejike RD, Enyia CO (2014) Perception and adaptation of food crop farmers to climate change in Imo State, Nigeria. Paper presented at the 14th annual national conference of the Nigerian Association of Agricultural Economists, Federal University of Technology, Akure, 24th–27th Feb
Deressa T, Hassan RM, Alemu T, Yesuf M, Ringler C (2008a) Analyzing the determinants of farmers' choice of adaptation methods and perceptions of climate change in the Nile basin of Ethiopia. International Food Policy Research Institute (IFPRI) Discussion paper No 00798 environment and production technology division. Washington D.C: IFPRI. Retrieved from https://www.africaportal.org/documents/5406/IFPRI_Discussion_Paper_00798.pdf
Ehui SK, Spencer DSC (1990) Indices for measuring the sustainability and economic viability of farming systems. RCMP research monograph No.3. IITA Ibadan, Nigeria
Food and Agriculture Organization FAO (2011) Global food losses and food wastage: Study conducted for international congress of save food. Rome, Italy. Retrieved from http://www.fao.org/docrep/014/mb060e/mb060e00.pdf
Gaman PM, Sherrington KB (1977) The science of food, pp 179–181
Gbetibouo GA, Ringler C (2009) Mapping South African farming sector vulnerability to climate change and variability a sub-national assessment. International food Policy Research Institute (IFPRI) discussion paper 00885. Retrieved from https://cgspace.cgiar.org/handle/10568/21657
Gujarati DN, Sangeetha N (2007) Basic econometrics, 4th ed. Tata McGraw–Hill, New Delhi
Gujarati DN, Porter DC (2009) Basic econometrics, 5th ed. Mc Graw–Hill, New Delhi
Hausman JA (1978) Specification tests in econometrics. Econometrica 46:1251–1271
Hageback J, Sundberg J, Ostwald M, Chen D, Yun X, Knutsson P (2005) Climate variability and land use change in Danagou, watershed, China—examples of small scale farmers' adaptation. Clim Change 7(2):189–212. https://doi.org/10.1007/s10584-005-5384-7
Harvey CV, Rakotobe Z, Rao NS, Dave R, Razafimahatratra H, Rabarijohn RH, MacKinnon JL (2014) Extreme vulnerability of smallholder farmers to agricultural risks and climate change in Magadascar 369:1639. https://doi.org/10.1098/rstb.2013.0089
Jones PG, Thornton PK (2009) Croppers to livestock keepers: livelihood transition to 2050 in Africa due to climate change. Environ Sci Policy 12(4):427–437. https://doi.org/10.1016/j.envsc.2008.08.006
Loon HV, Meehl GA, Shea DJ (2007) Coupled air-sea response to solar forcing in the Pacific region during Northern winter. J Geophys Res Atmos 1(12):1–8. https://doi.org/10.1029/2006JD007378
Lynam JK, Herdt RW (1989) Sense and Sustainability: sustainability as an objective in international agricultural research. Agric Econs 3:381–398
McCarthy N, Winters P, Linares AM, Essam T (2012) Indicators to assess the effectiveness of climate change projects. Inter-American Development Bank. Impact-evaluation guidelines: technical. Notes, No. IDB-TN-398:1-37
Molua EL, Lambi CM (2007) The economic impact of climate change on agriculture in Cameroon. World Bank development research group (sustainable rural and urban development team). Policy research working paper no. 4364. Retrieved from https://elibrary.worldbank.org/doi/pdf/10.1596/1813-9450-4364
National Bureau of Statistics (NBS) (2010) Annual abstract of statistics Abuja, Nigeria. Retrieved from http://www.nigerianstat.gov.ng/pdfuploads/ANNUAL%20ABSTRACT%20STATISTICS%20VOLUME-1.pd
Nepal Climate Change Management Centre (NCCMC) (2014) Climate change vulnerability, impacts and adaptation of agriculture in a mountain region of Western Nepal. A dissertation submitted in partial fulfilment of the requirement for Master of Sci. Degree in environment and natural resources. Department of Environmental Science, Kathmandu

University, Nepal. Retrieved from http://www.ncckmcnast.org.np/publications/climate-change-vulnerability-impacts-and-adaptation-agriculture-mountain-region-western

Nelson R, Kokic P, Crimp S, Martin P, Meinke H, Howden SM, Nidumolu V (2010) The vulnerability of Australian rural communities to climate variability and change. Part ii—integrating impacts with adaptive capacity. Environ Sci Policy 1(3):18–27. Retrieved from www.sciencedirect.com/science/article/pii/S1462901109001324

Nwaiwu IUO, Orebiyi JS, Ohajianya DO, Ibekwe UC, Onyeagocha SUO, Henri-Ukoha A, Tasie CM (2014) The effects of climate change on agricultural sustainability in Southeast Nigeria–implications for food security. Asian J Agric Extension Econ Sociol 3(1):23–36. Retrieved from http://www.sciencedomain.org

O'Brien RM (2007) A caution regarding rules of thumb for variance inflation factors. Qual Quant 41(5):673–690. https://doi.org/10.1007/s11135-006-9018-6

Odekunle TO, Orinmoogunje IOO, Ayanlade A (2007) Application of GIS to assess rainfall variability impacts on crop yield in Guinean Savanna part of Nigeria. Afr J Biotechnol 6(18):2100–2113. Retrieved from http://www.academicjournals.org/AJB

Oluwasegun OA, Olaniran JM (2010) Academic journals effects of temporal changes in climate variables on crop production in tropical sub-humid South-Western Nigeria. Afr J Environ Sci Technol 4(8):500–505. Retrieved from http://www.academicjournals.org/AJEST

Onyeagocha SUO, Nwaiwu IUO, Obasi PC, Korie CO, Ben-Chendo NG, Ellah GO, Okpeke MY (2018) Encouraging climate SMART agriculture as part solution to the negative effects of climate change on agricultural sustainability in Southeast, Nigeria. Int J Agric Rural Dev (IJARD) 21(2):3600–3610. www.ijard.com

Opiyo FEO, Wasonga OV, Moses MN (2014) Measuring household vulnerability to climate-induced stresses in pastoral rangelands of Kenya: implications for resilience programming. Pastoralism J: Res Policy Pract 4(10). https://doi.org/10.1186/s13570-014-0010-9

Oyewole JA, Thompson AM, Akinpelu JA, Jegede OO (2014) Variation of rainfall and humidity in Nigeria. J Environ Earth Sci 4(2):29–37. Retrieved from http://www.iiste.org/Journals/index.php/JEES/article/download/10586/10950

Tologbonse EB, Auta SJ, Bidoli TD, Jaliya MM, Onu RO, Issa FO (2010) Farmers' perception of the effects of climate change and coping strategies in three agro-ecological zones of Nigeria. J Agric Extension 14(1):125–136. Retrieved from https://www.ajol.info/index.php/jae/article/view/64080

Wannasai N, Sasiprapa W, Suddhiyam P, Koshawatana C, Prasertsak P, Kumsueb B, Singh NP (2013) Vulnerability to climate change: adaptation strategies and layers of resilience–quantifying vulnerability to climate change in Thailand. Int Crops Res Inst Semi-Arid Trop (ICRISAT) (9):1–11. Retrieved from http://oar.icrisat.org/7235/1/ICRISAT_MIP_RP_9_Vulberability_2013.pdf

Zuur AF, Leno EN, Elphick CS (2010) A protocol for data exploration to avoid common statistical problems. Methods Ecol Evol 1(1):3–14. https://doi.org/10.1111/j.2041-210X.2009.00001x

University's Inclusion in Providing Climate Services to Farmers: Is It Possible Without Agricultural Agents and Farmer Facilitators?

Yunita T. Winarto, Sue Walker, Rhino Ariefiansyah, Iqbal H. Lisan, Maudy Y. Bestari and Tiara Audina

Abstract Since the early 1970s, agricultural extension services in Indonesia have been developed based on the "Training and Visit" model to reach targets of increasing crop productivity. Until recently, extension services have been neither modified nor improved to accommodate farmers' needs to respond to the consequences of climate change. Accordingly, farmers have been left to fend for themselves, often helpless to survive the impacts of increasing climate variability. Significant reductions in yields and harvest failures have been common in many parts of Indonesia. Two crucial dimensions have been missing in the existing extension approach, nurturance and trust. This paper examines recent agrometeorological improvements by farmers, through Science Field Shops (SFSs) provided by anthropologists and their students from Universitas Indonesia, in collaboration with agrometeorologists from the University of the Free State, and earlier from C. J. Stigter (*Agromet Vision*). Based on the lessons learned over the past decade, where agricultural extension practitioners provided climate services without "nurture and trust" built into the ongoing collaboration between scientists, students, farmers, and agricultural agents; the educational commitment did not yield effective results. The strategies and processes developed by

Y. T. Winarto (✉) · R. Ariefiansyah · I. H. Lisan · M. Y. Bestari · T. Audina
Department of Anthropology, Faculty of Social and Political Sciences, Universitas Indonesia, Kampus UI, Depok 16424, Indonesia
e-mail: yunita.winarto@gmail.com

R. Ariefiansyah
e-mail: rhino.ariefiansyah@gmail.com

I. H. Lisan
e-mail: labsiqbalaj@gmail.com

M. Y. Bestari
e-mail: maudyyuliabs@gmail.com

T. Audina
e-mail: tiaraaudina91@gmail.com

S. Walker
Department of Soil, Crop & Climate Sciences, Faculty of Natural and Agricultural Sciences, University of the Free State, PO BOX 339, Bloemfontein, South Africa
e-mail: WalkerS@arc.agric.za

W. Leal Filho et al. (eds.), *Climate Change, Hazards and Adaptation Options*, Climate Change Management, https://doi.org/10.1007/978-3-030-37425-9_42

the university's interdisciplinary team show how those two dimensions were incorporated when SFSs were introduced in a new location, Sumedang Regency in West Java Province. SFSs included rainfall observers, scientists and students, as well as high-ranking local agricultural officials and farmer facilitators. They proved to be vital to the furtherance of farmers' lifelong education and livelihoods.

Introduction

After a long drought that began in July 2018, with no significant rainfall through the end of that September, farmers' rice fields in two regencies in West Java Province—Indramayu and Sumedang in Indonesia—were left parched, with deep fissures. In such condition, without any available irrigation, farmers were not able to begin their planting in the rainy season period. As reported elsewhere, farmers are confused of the changes in both the timing of rainy seasons and rain patterns due to the unpredictable weather as the consequences of climate change. Global warming, increasing climate variability, more frequent and more severe weather events have been affecting people's livelihood, human health, agricultural and ecosystem productivity, and global food production (Fischer et al. 2002; Parry et al. 2004; Howden et al. 2007; Schmidhuber and Tubiello 2007; Brown and Funk 2008; Jennings and Margrath 2009; Nelson et al. 2009; Winarto et al. 2018a). Under the increasing risks and vulnerability, only those who have been engaged with agrometeorological learning in Indonesia were able to develop their coping strategies. Facing the threat of increasing El Niño in the late 2018, some farmers decided to change their commodities from rice to other crops, or postpone their land preparations.

Making decisions in anticipation of climate conditions indicates farmers' ability to develop "response farming to consequences of climate change" (cf. Winarto and Stigter 2013, 2016). This capability is not common. Those farmers are members of an association of rainfall observers who joined agrometeorological learning provided in the so-called "Science Field Shops" (SFSs). SFSs were organized by a collaborative inter-disciplinary team from Universitas Indonesia and Agromet Vision (Netherlands, Indonesia and Africa) from 2008 to 2016, and the University of the Free State from 2016 to the present (Stigter and Winarto 2016; Winarto et al. 2018a).[1] Other farmers, who did not participate in such learning, began making nursery beds once they had experienced several days of rain. They were unaware of the "false start of rainy season," when a dry spell follows a brief period of rain (Winarto et al. 2018a). Accordingly, their seedlings failed to grow when the rains were insufficient to wet the nursery beds.

[1]The first agrometeorologist was the late Cornelis (Kees) J. Stigter, the founder of International Society of Agricultural Meteorology, who passed away in May 2016 while leading a training session for farmers in Indramayu Regency. His substitute, Sue Walker, was one of his fellow lecturers at the University of the Free State, in Bloemfontein, Free State, South Africa.

Traditionally, rice farmers in Indonesia used cosmology to determine when to plant. Javanese farmers consulted their traditional agricultural calendars while observing bionatural indicators for particular months of the year (Hidayat 2011; Rovihansyah 2014). Such traditional cosmology and agricultural calendars no longer became the prime reference for farming when the Green Revolution intensification program was introduced. Farmers' decisions were influenced by the state's irrigation schedules as referral points of when to start seasonal planting. Therefore, in the past 50 years, farmers were unaccustomed to making decisions based upon their own choices. They have been "trapped" in the cage of the Green Revolution paradigm with high-levels of inputs (Winarto et al. 2017, 2019). In such situations, the agrometeorological learning and climate services provided in the SFSs became a significant means for them to develop their analytical capacities as the bases of their coping strategies. Once they were able to anticipate possible climate conditions in the near future, while learning from their past and recent experience, they were better equipped to make decisions. It is expected that once they are able to gain better yields, avoid reduced yields or harvest failures from their decisions, the new strategies under the same climate condition would be part of their adaptation in the future.

Such capability was developed by a number of rainfall observer farmers in the Regency of Sumedang, West Java Province, who began their agrometeorological learning in the early days of September 2018. Only by measuring rainfall over two-to-four months, daily observation of their fields' ecosystems, and documenting the data, were they able to understand the need to develop appropriate coping strategies in response to ongoing El Niño conditions. Considering the need to "nurture" the learning process and build up "trust" between scientists, farmers, and government agencies (Winarto et al. 2018b), the burning question is: 'how can these two dimensions be developed at Sumedang, in a relatively short period of time?'

Consider the historical chain of events in other locations. The first introduction of the SFSs in Indramayu Regency in West Java Province was in early 2009. It was another five years before the next one in East Lombok Regency in West Nusatenggara Province. In the latter case, the government agencies were quite supportive and gave a good indication to facilitate the scaling-out of the SFSs to many farmers in their regency. However, as found in Indramayu Regency, the real engagement with the extension agents was minimal. Farmers were for all practical purposes left to their own devices, to advance their agrometeorological knowledge.

Universitas Indonesia (UI) scientists provided climate services to farmers as they had for the last ten years when they faced constraints due to "missing or lack of on-time and appropriate services" by government agencies (Stigter 2016; Winarto et al. 2017). With the same objectives of facilitating farmers improving their decision making capability as found in various other climate services, the SFSs position farmers as active researchers, analysists, and decision-makers, not merely as receivers and users (Buontempo et al. 2014; Stigter and Winarto 2015; Sitgter et al. 2016; Clarkson et al. 2017; Tall et al. 2018; Dayamba et al. 2018; Haines 2019). Yet, by having some limitations as researcher-facilitators on the ground, the role of extension intermediaries is necessary. Stigter, the founder of the SFSs in Indonesia in collaboration with Winarto, the anthropologist from Universitas Indonesia, argued for the engagement of extension intermediaries as the mediators between the two parties (farmers and scientists), particularly with regard to relevant information and passing along farmer vulnerabilities (Stigter and Winarto 2015).

Unfortunately, this did not prove to be the case in practice. As part of the ongoing reflection and intersubjectivity, we learned that relying too heavily on extension intermediary involvement did not yield real progress. Therefore, from 2013 onwards, we decided to provide training for farmer facilitators in Indramayu Regency. During 2016 and 2017, this expanded in Indramayu and East Lombok to include extension intermediaries. Without significant changes in the extension system to enable staff to assist farmers in response to unexpected risks of climate change, again, there was insignificant follow-up with the extension intermediaries and their commitments to farmer education. In response to this failure, we decided to approach the regency agricultural officials as soon as we introduced the SFSs at Sumedang Regency, to the south of Indramayu Regency, West Java Province.

This paper examines the latest outcomes of introducing the SFSs in Sumedang Regency by involving agricultural extension staff and farmer-facilitators from Indramayu Regency. In the arguments presented in this paper, we have highlighted the extent to which "nurture and trust" have been developed in providing climate services in the SFSs, as well as its constraints and its effects on the farmers' education, in particular on the changes in their knowledge and practices of "response farming to the consequences of climate change."

Description of the Study Areas and Methodology

Study Area

The Regency of Sumedang is located between latitude 6° 44′ S and 70° 83′ S, and longitude 107° 21′ E and 108° 21′ E, in the province of West Java, Indonesia. The widest area measures 152,220 ha. The regency consists of 26 districts with a total of 279 villages/smallest administrative units, in the mountainous and hilly areas with the range of 25–1667 m above sea-level (MASL). Gunung Tampomas is the highest mountain in the area (1684 MASL). Diverse forms of topography are found in the regency: (a) 12.2% flatlands; (b) 5.4% undulating zone; (c) 51.7% undulating up to hilly areas; (d) 31.6% hilly up to the mountainous areas; and (e) 11.4% mountainous areas (Badan Pusat Statistik 2018).

Such topography differentiates Sumedang regency from the other two regencies in Indonesia where the SFSs were introduced earlier—Indramayu Regency in West Java Province and East Lombok Regency in West Nusatenggara Province, which mainly consist of lowlands and undulating areas. Accordingly, diverse ecosystems are spread across the region of Sumedang with dry rain-fed fields, and irrigation water resources from springs, streams, and rivers. Those are (1) dry rain-fed rice fields; (2) flatlands of rice with medium height of mud and springs as the water resource; (3) fields with deep layer of mud (*cengkar*) and irrigated water schemes planted with rice, secondary crops or horticultural crops; and (4) the dry rain-fed fields for horticulture. In the past two years, a large reservoir was constructed in the south of Sumedang to irrigate rice fields in the northern areas of West Java province, including Indramayu Regency.

Farmers' Existing Knowledge and Experience

The main premise underlying the need for SFSs, as argued by Stigter and Winarto (2015), is the constraints farmers have on their traditional and empirical knowledge when faced with the uncommon risks and consequences of climate change. Such is the dilemma. On the one hand, farmers are knowledgeable people who have detailed empirical knowledge of their habitat. When they join the SFSs, some have been working as farmers for up to 20 years. That said some have fewer years, even one or two, or, in any event, less than ten. They learned farming from their parents, including traditional cosmology and weather lore, as well as practical experience enriched with individual experience from multifarious sources. The latter may include formal schooling in high school and university, introduced technological packages and recommendations in crop intensification programs based on the Green Revolution paradigm, as well as various training sessions conducted by the regency government (e.g. diverse kinds of Farmer Field Schools and other training programs). On the other hand, their existing knowledge is based heavily on empirical observation and subjective interpretation. Such education can no longer be used to respond to recent unprecedented risks and threats emerging in their habitats (Winarto et al. 2018a).

The University's program of enriching farmers' knowledge of climate change and its impacts on farming through the SFSs was thus perceived by the regency government as a strong opportunity for farmers to develop sustainable farming strategies. Inviting local farming community leaders was a way to engage farmers in the SFSs. Almost all of them joining the SFSs have positions as leaders of local farmer groups or hamlets, or they may be village officials, religious leaders, as well as holding other jobs as teachers, traders, and local businessmen. In those positions, they have been used to building close relationships with local agricultural representatives, extension staff and pest/disease observers.

Prior to joining the SFSs, those farmers had been continuously learning from daily experience in response to natural threats. For example, when farmers from East and South Sumedang Regency faced wetter than normal dry-seasons known as "Twin Rainy Seasons" (e.g. La Niña in 2016), severe pest outbreaks of brown planthoppers (BPH) *Nilaparvata luges stal.*, and viruses, such as grassy-stunt and ragged-stunt viruses with BPH as the vector, infested the rice fields. Consequently, farmers experienced significant yield reductions and harvest failures. However, they tended to perceive these problems as "bad luck" or "fate." Thus, they did not critically question the threats or take the lessons learned to improve their farming strategies in the future.

Cropping Patterns and the Agrometeorological Zones

When considering how to manage the situations of farmers in diverse ecosystems, we learned from the range of farmers' cropping patterns and cultivation strategies. Two main patterns were practiced by farmers, namely multiple cropping of horticultural crops in western Sumedang, and monoculture cropping of either rice or maize in the northern, eastern, and southern parts of Sumedang, and the surrounding city. The main horticulture crops consisted of cabbage and chili planted in mountainous areas, yet in smaller dry rain-fed fields, to optimize production.

As soon as the rainy season arrived, farmers planted cabbage along the edges of their fields and left the middle part for other crops, such as chili, tomato and eggplants, or onions and shallots at a later stage. In fields where irrigated water was available in addition to rainfall, farmers preferred to plant paddies during the rainy season, followed by secondary or horticultural crops in the dry season. In the latter, farmers preferred relay cropping to optimize the use of nutrition, and avoid pest and disease infestation.

In other parts of Sumedang where irrigation water were available throughout the year, the cropping pattern consists of triple planting, as in paddy-paddy-paddy or paddy-paddy-secondary crops (maize or groundnuts). For dry rain-fed fields or half-irrigation schemes, farmers used to plant paddies twice: in the rainy season and in the first dry season. In general, paddy planting in Sumedang Regency depends heavily on rainfall for both dry rain-fed fields and irrigation schemes from rivers and streams. To cope with limited water availability in the rainy and dry seasons, farmers adjusted their choice of rice varieties according to their maturing age of either short season (*hawara,* less than 90 days of age) or long season (*leuir,* up to more than 100 days). Both varieties are planted in either rainy or dry seasons in line with the farmer's individual choice.

By considering such diversity it would not be possible for farmers from distant places to regularly meet once a month in SFS evaluation meetings. Therefore, following the consensus achieved in the first workshop when introducing the agrometeorological learning in the SFSs, the farmers who agreed to join the SFSs were divided into 5 zones. The division into zonation was also based on the need to have an effective dialectical learning through dialogue and discussion among farmers who cultivate the same commodities in the similar landscape and ecosystem. The main aim was to enable farmers to share their experience and to learn from one another of the vulnerabilities found in their fields. See Table 1 of the nature of landscape and commodities in each zone.

Due to the limited number of farmers attending the regular evaluation meetings in two zones, the East and City Zones, farmers of these two zones were integrated into one evaluation meeting covering the two zones.

Table 1 Zones for agrometeorological learning in the SFSs in regency of Sumedang

No.	Zone	Landscape	Commodities
1	North Zone	Flat, undulating, hilly and mountainous areas; dry rain-fed and irrigation from streams/rivers	Rice and others (maize, cassava, groundnuts)
2	West Zone	Undulating, hilly and mountainous areas; dry rain-fed and irrigation from streams/rivers	Rice and others (maize, cassava, sweet potatoes, yams, groundnuts), horticultural crops (tomato, chili, cabbage), annual crops (coffee, melon), and trees
3	East Zone	Flat and undulating areas; dry rain-fed and irrigation from streams/rivers	Rice and maize
4	Zone of City	Flat and undulating areas; dry rain-fed and irrigation from streams/rivers	Rice and maize
5	South Zone	Flat areas; dry rain-fed and irrigation from streams/rivers	Rice and maize

Methodology of the Collaborative Works of Participatory Research and Engagement

The initiative to introduce agrometeorological learning to farmers would not be possible without the collaboration between two scientists from different disciplinary backgrounds: agrometeorology and anthropology. Both had expertise in different domains that could not have worked alone to provide routine climate services to meet the farmers' needs. The agrometeorologist did not have any training in carrying out ethnographic fieldwork, with its principal method of participant observation, nor did he have any training in developing community engagement with subjects. By contrast, the anthropologist did not have substantial knowledge of climate change, nor any background in agrometeorology. Thus, an interdisciplinary approach between the two fields was the foundation of the university's contribution in developing services for farmers (Winarto and Stigter 2011, 2017). Reflecting on the realities found on the ground and their further developments became the basis of the conversation between the agrometeorologist and anthropologist (Winarto and Stigter 2013, 2016). Gradually, each party improved his and her understanding of one another's domain of knowledge, as well as considering the issues in advancing the methods and approach of providing climate services.

Without including the farmers themselves as researchers and active learners, however, the farmers would not have internalized the lessons into their own knowledge and actions. In an earlier paper, Winarto et al. (2017) argued for the efficacy of trans-disciplinary approach in engaging farmers in agrometeorological learning. By actively researching more structured techniques for scientifically measuring rainfall and observing the agroecosystem of their fields, the learning process developed much faster than it had in prior government training at Farmer Field Schools (FFSs). In Integrated Pest Management (IPM) FFSs and Climate Field Schools, farmers were trained to observe special, selected plots and not their own fields. Their learning was thus constrained by not being able to analyze the relationship between ecosystem indicators in their own fields. Nor were they able to learn from personal experience the interactions between their own management strategies, crop growth, rainfall data and its impacts on their fields and plants, and finally the yield they obtained (Winarto 2004; Siregar and Crane 2011).

SFSs encourage farmers to document and consider the relationships between all these elements that forms the basis for future management decisions. A feedback loop in the learning process was thus a valuable tool for enriching farmers' knowledge and practices. In such an inter- and trans-disciplinary collaboration, the dimensions of "trust" and "nurture" played significant roles (Winarto et al. 2018a).

By contrast, farmers in Indramayu Regency, had lost their trust in extension staff who were supposed to assist them in their work and with any problems they may have in their fields. In fact, the extension staff seldom visited them. If an extension meeting was held, the one-way communication of delivering the state's recommendation for a particular technological package, or promoting new chemical fertilizers or pesticides as delivered by the company's sales representatives, were given prominence. Farmers preferred the two-way communication and dialog of the SFS meetings than the one-way communication with extension. Farmers also trusted the scientists more than the extension staff in terms of their educational background and experience (also see Prihandiani 2017).

The "nurturance" dimension had not been part of the training program, either, whether with or without further follow-up activities (Winarto et al. 2018b). Consequently, farmers could easily revert to their previous habits without maintaining their educational commitment (see Siregar and Crane 2011). By their training in the ethnographic method, the anthropologists and students, who must immerse themselves in the subjects' everyday lives, helped the farmers while observing the learning process. The ongoing lessons learned by both parties, farmers and ethnographers, provided a rich opportunity to discover any problems in the farmers' learning and to find solutions together. "Nurturing" the learning process and improving the climate services was made possible through such collaborative work.

Seven climate services form the hallmark of the SFSs and were introduced gradually. See Box 1 for the seven climate services of the SFSs.

Box 1 Seven Climate Services of the Science Field Shops (SFSs)

1. Daily rainfall measurements by farmers in their own fields.
2. Daily agroecological observations.
3. Measurement of yield and understanding differences between fields, seasons, and years.
4. Organization of the SFSs.

5. Development and exchange of seasonal climate scenarios in the form of seasonal rainfall scenarios.
6. Exchanging new knowledge.
7. Establishment of field-experiments to develop best practices and obtain on-farm answers for urgent local questions.

Source Winarto et al. (2018a).

A particular approach involving the regional and local government officials, however, should be seriously considered. The first step was to introduce the aims of replicating the SFSs in the new regency, Sumedang, to their agricultural officials. Inviting them to be involved from the beginning would be a good start to obtain their commitment and support. Involving the existing farmer facilitators from the neighboring Indramayu Regency to share their knowledge and experience was the method used as we expected this to secure the new learners' confidence and trust. The interactions with the Indramayu farmer facilitators gives credibility to the UI team of scientists.

In addition, following farmers' daily activities as they pursued their agrometeorological learning and observing their practices was a vital part of the ethnographic methods used while being immersed in their daily life. Discovering their new and previous knowledge and farming strategies on a comparative basis, also formed a valuable part of the ethnographic methods used in this program. Farmer presentations and discussions at the regular meetings and the participation of both the farmer facilitators and extension staff became part of the observation and documentation records. This ethnographic fieldwork on a collaborative basis with the farmers in Sumedang Regency was carried out from the end of August 2018–March 2019.

Research and Discussion

Engaging Extension Agents and Involving Farmer Facilitators

Upon entering the field site in Sumedang Regency, the first step was to introduce the SFSs to the representatives of the regency agricultural office. Surprisingly, as soon as the officials understood the aims of the program, they decided to invite the heads of district offices, the extension staff, and several farming group leaders from each village. More than 100 people (including some pest/disease officials) attended the initial workshop introducing the SFSs concepts and principles. Within a short time period, as many as 54 farmers from 26 districts registered as rainfall observers. Toward the end of 2018, some farmers withdrew from and some newcomers joined the SFSs, making 44 stations from five zones. The number has continued fluctuating until the present time. Our invitation to the extension staff to also measure rainfall at their own site was declined, as they preferred to participate only in the monthly evaluation meetings. Unfortunately, for various reasons, not all extension agents and pest/disease observers had committed to facilitating farmers both in the fields and in the regular monthly meetings. The lack of commitment from the extension staff was found in both Indramayu and East Lombok Regencies. Some extension staff joined the monthly meetings only when they did not have other responsibilities, and would only be available if they were approached for special cases, such as finding solutions for problems discovered in farmers' fields (see 3.3.).

When it came to designing the program for the Regency of Sumedang, inviting Indramayu farmer facilitators was thought to produce significant results to:

(a) gain the trust of the new rainfall observers who could learn directly from the farmer facilitators' knowledge and experience; and
(b) strengthen the facilitators' confidence and trust in their own abilities to transfer their knowledge and experience to other fellow farmers.

About 16 farmer facilitators from Indramayu Regency were assigned the task of instructing the new Sumedang rainfall observers in how to observe their fields, measure, document, and draw simple rainfall data graphs for presentation and discussion of the rainfall data and emerging vulnerabilities in their fields. For every introduction by each farmer facilitator, they mentioned the excitement of joining the rainfall observers association, which, for the new learners, was something amazing. Only two of the farmer facilitators (12.5%) joined the SFSs in Indramayu Regency less than five years. The rest have been engaged in the SFSs from six to nine years. Among their number were some young farmers under 35 years old. Their persistence as rainfall observers over such a long period of time was something novel and noteworthy, in particular their strong commitment and enthusiasm to noting daily rainfall measurements, especially compared to the other state training projects.

Based on Winarto's observations of various FFSs since the early 1990s, farmers would easily accept new ideas on the basis of "seeing is believing" (Winarto 2004). Without "seeing", farmers were skeptical about the providers' ideas. Since farmers knew that the extension staff lacked their practical experience as farmers, they were reluctant to believe them. Farmers often complained that the extension staff were "just talking" without practice or proof of what they tried to convey. On the other hand, Sumedang farmers could not argue with the farmer facilitators' long-term experience and knowledge in practical rice farming systems.

Farmers' Own Initiatives by Integrating New Knowledge

It had only been four months after introducing the SFSs among Sumedang farmers, when new ideas introduced by the UI team and Indramayu farmer facilitators—under the increasing El Niño conditions toward the end of 2018—were being internalized by Sumedang farmers. Examples of those ideas were:

- the ways of presenting rainfall data in the form of a simple graph;

- the need to pay attention to the number of wet-days and dry-days,
- the patterns of rainfall distribution and intensity;
- "the false start of rainy season" following the first rainfall recieved;
- making dry-nursery beds instead of wet-nursery ones, with a modification in shortening the "grain-soaking" period;
- the significance of selecting rice varieties in terms of their maturing age for a particular climate condition;
- the need to avoid the unnecessary use of chemical pesticides; and
- the relevance of making decisions by anticipating future climate conditions.

By referring to the uncommonly low rainfalls of the early rainy season of 2018–2019, seasonal climate predictions the farmers received, namely the mild El Niño, proved to be accurate. Farmers also observed the low water level of rivers and streams, as well as the very dry and cracked soil in their fields for the long period from June until November 2018. These examples confirm the crucial need and efficacy of introducing the "witnesses" (farmers themselves), to the concepts of developing response farming to address the impacts of climate change.

The reception of new ideas among the horticulturalist farmers, however, was peculiar in comparison to rice farmers who could not start any planting activities, given the dry condition of their fields. In the beginning, the horticulturalists perceived the El Niño as providing positive impacts for their crops by avoiding the unnecessary water abundance which would damage them. They were referring to the severe crop damage they experienced during the wetter than normal dry-season of La Niña of 2016. Growing vegetable crops in dry weather is preferable to wet weather, since one could still use ground-water pumps or purchase water if necessary. Along with the agrometeorological insights, the farmers realized that the long period of dry days under El Niño would harden the soil and constrain the growth of crops. Since horticultural crops also need water, one farmer began to calculate the available water left in the soil under continuous dry days with no rain. As a result, one farmer determined the period of making nursery bed of Kiwano (*Cucumis metuliferus*) in early December 2018 that would enable the crop to grow with enough rainfall at a later stage (in January 2019). It was expected that the rainfall would not be as abundant in January as it would be in February 2019. For another crop, taro/yam, the same farmer decided to wait until the rainfall was sufficient for the plant to grow. Too little water during the vegetative stage or too much water during the reproductive stage would damage the crops.

One rice farmer decided to plant eggplant instead of rice during the long drought at the beginning of rainy season. He was able to incorporate the seasonal scenarios and the recent condition of his field of red-clayish soil without deep mud (*sawah cengkar*). This condition would not be good for growing rice if water were lacking. Prior to the SFSs, he used to begin making nursery beds when the first rains fell, without thinking of a probable dry-spell period to come. Another case is the choice of a short maturing age variety based on the farmer's new understanding of the probable abbreviated rainy scason by having a long drought under the El Niño that delayed the planting schedule. Instead of planting the variety he usually cultivated in the rainy season by also considering the need to start planting soon for the 2019 dry season, he chose *Situ Patenggang*. This variety matures in less than 90 days after transplanting. He also thought of avoiding the risks of pests and disease infestation at a later stage by knowing that *Situ Patenggang* is a resistant variety. In the mid-planting season, March 2019, the farmer presented his findings at the monthly SFS evaluation meeting. He proudly reported that his variety grew well without any pest or disease infestation.

These are examples of farmers' own initiatives after gaining new knowledge and understanding from their daily observations of rainfall and the status of their agroecosystems, together with receiving seasonal climate scenarios, and having dialogs with fellow farmers and scientists during the monthly evaluation meetings. Only in the meetings did the extension staff and pest/disease observers come to join the evaluation and discussion. None of them followed the farmers' learning process via the observations, anticipation and decision-making. The learning progression was thus entirely in the hands of the farmers themselves, by gaining new ideas and practices through the whole SFS interaction procedure.

Farmers' Anticipation and Decision-Making with the Assistance by Local Extension and Government Officials

It was a surprise for the UI team that only in the first evaluation meeting held in one zone, City Zone, after hearing the seasonal scenarios predicting an increasing El Niño during winter 2018–19, did one farmer say that he was thinking of planting maize instead of paddy in that rainy season. How could he have decided so soon, while his fellow farmers were still thinking of when to start planting rice?

The answer was that, in the prior three years, 2015–17, he had changed his cropping pattern from paddy-paddy-paddy to paddy-paddy-maize, while introducing the state's program of planting paddy-maize-soybean (*Padi-Jagung-Kedelai, PAJALE*) in an area up to 10 ha.[2] The government provided the maize seeds free to the farmers. At that time, this farmer had just been nominated as the head of the farming group in his area. Even though he was not able to cultivate as much as 10 ha of maize, the yields he and his fellows achieved were perceived as reaching the government's target. Under the guidance of the extension staff, the planting area target for maize was widened by providing another maize hybrid variety. Based on that experience,

[2]The state's national program for self-sufficiency in food crops and to reduce imports of rice-maize-soybean in terms of production and size of lands is known as Upaya Khusus (Upsus) PAJALE (Padi-Jagung-Kedelai). See Upaya Khusus (UPSUS) Swasembada Pangan 2015–2017, http://biogen.litbang.pertanian.go.id/2015/02/upaya-khusus-upsus-swasembada-pangan-2015-2017/ retrieved on 07 March 2019.

the farmer felt sufficiently confident to plant maize under the ongoing El Niño conditions at the end of 2018. His decision was firmly supported by the low availability of water in his area that was then entirely dependent on future rainfall.

In 2008, there was an accident that damaged the irrigation canals along a steep area at a higher elevation in his village. Owing to its location, the damage was not easily repaired, so water was no longer available to irrigate the fields. For a village known as a rich producer of rice, it soon lost its good reputation. Instead, maize gradually become an alternate crop. By his decision as the head of the hamlet and leader of a farming group, the farmer played a significant role in motivating his fellow farmers to plant maize. His intention was made possible by support from the government program and subsidy for maize seeds.

The data the farmer was collecting from 1st September 2018 showed no rainfall and very dry soil in his fields. This strengthened his logical thought process that water would not be available to plant rice during that rainy season. He confirmed his decision by changing his cropping pattern from paddy-paddy-maize into maize-paddy-paddy. If the rainfall in the third planting season was still not available, then he might plant maize again instead of paddy for that season also.

However, how could he operationalize his intention to plant maize rather than rice in the 2018/19 rainy season? He was thinking of replanting the previous seeds that he had received from the extension staff and informed his fellow farmers, the members of his farming group, as well as the village leader. He soon formed a proposal to plant maize across the areas cultivated by his farming group members. This was accompanied by a request for subsidies for new maize seeds, a machine to plant them, and to decob the maize later on. He submitted his proposal to the head of district agricultural office while also informing the extension staff of his plan.

His proposal was accepted. His farming group was able to receive a new variety of maize seeds along with the machine to peel the dry maize after harvest. In this case, the assistance of local extension staff and district agricultural officials, as well as support from the village leader played a significant role in the success of his individual decision. His strength as a leader in persuading and motivating the members of his farming group was also prominent. Without the officials' assistance and his fellows' agreement, a collective action in implementing his anticipation and decision would not have become a reality. This shows that his fellow farmers trusted his decisions as he had previous experience and now also had the newly acquired knowledge and confidence from the SFS and seasonal scenario outlooks from UI.

Farmers' Struggles to Solve Vulnerabilities with the Assistance of Pest/Disease Observers

The main objective of the SFSs is to assist farmers who discover vulnerabilities in their fields and find solutions through knowledge exchange with their fellow farmers, between farmers and scientists; or between farmer and agricultural staff. Besides lack of water needed for crop growth, pest/disease infestation is also a threat to farmers. With every presentation, the rainfall observers are asked to also describe the results of their observations of pests/diseases conditions and threats to their crops. Discussions during the SFS meetings can thus focus on solving the problems based on farmers' experience or scientists' and pest/disease observers' advice. In the present case, knowledge exchange through establishing networks in the SFSs was indeed beneficial for farmers by assisting them to find appropriate solutions which would help them move away from the unnecessary use of or over dependence on chemical pesticides. Avoiding further unexpected risks to their fields and sustaining their ecosystems were part of the aim of the agrometeorological learning (Winarto et al. 2017).

A horticulturalist farmer used to face threats of pests infesting the roots of his taro crops. Since he could not see the pests, its spread and infestation, he could not prevent the damage to his crop. Quite often, he only received advice from his fellows to control the pests by using chemical insecticides in the form of carbofuran granules. After learning agrometeorology in the SFSs, he realized the toxicity of chemical pesticides that could affect the ecosystems and people's own health. He developed his understanding that not only the producers, but also the customers will benefit by avoiding the negative impacts of poisons. As a religious leader, he was afraid and felt guilty for producing poisonous food crops. He thus tried to find alternate solutions without using chemical pesticides.

He enquired from his fellow farmers and agricultural staff, but the application of chemical pesticides were the invariable response. He sent a message to to Winarto—via What's App messaging app with photos of the pests, and their infestation to the tubers of the crop. Winarto sent those photos to an entomologist at another university (Gadjah Mada University) while also advising the farmer to contact the expert directly. From both people, the farmer received an identification of the pest and the way to control it by using a bio-agent: *metharizium.* The question was how to obtain and/or produce that bio-agent? He soon consulted a pest/disease observer who used to regularly attend SFS evaluation meetings. From her, he obtained the knowledge of how to produce that bio-agent, and other one also—*Pseudomonas Fluorescens* to prevent a disease infesting leaves, and *Bacillus Subtilis* to stimulate the growth of crops. The pest/disease observer lent him the equipment and helped the farmer produce those three bio-agents with the assistance of other staff. From the pest/disease observers' viewpoint, this strategy was a novel one in the farmers' tradition which could yield a sustainable ecosystem in the future.

This case is proof of how an inventive idea and the nurturing of an enquiring mind can use existing networks which have been expanded through the SFSs by involving experts from various disciplinary backgrounds. Here, not only an entomologist and agrometeorologist played a role, but also a soil expert who voluntarily came to share her knowledge and expertise in understanding the important role of fertile and healthy soils for sustaining ecosystems and increasing yields.

If time is available to gather more cases of farmers' innovative coping strategies based on their learning and inventive ideas, the paper could present wider and deeper understanding of farmers' adaptation practices in their response to the consequences of climate change. An intensive collaboration with the agricultural officials and extension staff was also not being able to be carried out in a limited period of time.

Conclusion

Introducing the SFSs through the formal network of regional agricultural officials and involving rainfall observers from elsewhere, who had been trained as farmer-facilitators, proved beneficial for establishing the SFSs among new rainfall observers. Some constraints did exist however, based on experiences from introducing the SFSs in the Indramayu and East Lombok Regencies. The existing system of extension services in agriculture with a wide area to cover and excess work for the extension staff to do to promote the state recommended programs and technologies, limited their time available for regular visits to farmers. The offer for the extension staff to measure rainfalls in someones field was available but was refused by referring to their hectic agenda of other commitments in implementing various other programmes. Assisting farmers in regular meetings was by their choice, or as instructed by their superiors. Unfortunately, very few extension staff were able to allocate their time to join the SFS meetings. However, without practical experience and additional knowledge of climate-related issues and problems, they could not effectively participate in the agrometeorological discussions with farmers. Only pest/disease observers with substantive and practical knowledge of how to control pests and diseases could share their knowledge in those areas.

In such situations, farmers' agrometeorological sophistication gradually evolved through mastering the methods of measuring rainfall, observing the agroecosystem, documenting the data, and presenting the results in monthly evaluation meetings. Although some rainfall observers withdrew from participating, for whatever reasons, the confidence and trust in the newly-introduced methods strengthened the rest of the rainfall observers' motivation to persist in their data collection and expertise. They experience the SFSs meetings as a safe place to share ideas and receive nurturing. Some cases of farmers' decisions to carry out collective action or find sustainable alternatives to chemical agents proved the efficacy of the existing network between farmers and agricultural officials, which enlargely due to the inclusion of experts and scientists. Such enrichment of farmers' knowledge and practices strengthened by their belief and trust on their learning are the fundamental "seeds" for adaptive strategies to occur in the long-run. Despite those advantages acquired by farmers, organizing the collective action in the monthly regular meeting was constrained by lack of communication means owned by farmers and agricultural extension staff, and the other farming and societal responsibilities that had to be carried out by farmers.

The experience of Sumedang rainfall observers prove that the SFSs with its seven climate services provided in two-ways communication is promising to be expanded and replicated to other farming communities elsewhere in Indonesia. However, to expand the knowledge base by involving a larger number of farmers and replicate the SFSs, it is vital to learn from the ongoing activities of how to more effectively engage the agricultural officials. Another alternative is to find strategic solutions to produce capable farmer-facilitators for future generations of rainfall observers. This was proved beneficial by involving senior farmer facilitators from another regency to share their knowledge in the SFSs regular meetings. Still another crucial issue is the extent to which university staff will be able to sustain the institutional and financial support of such actions. Expanding the learning and replicating the SFSs in the other places and farming communities in the future need to seriously address those issues.

Acknowledgements The authors are grateful for the support of Universitas Indonesia through its Directorate for Innovation and Business Incubation (2018–2019), which enabled the SFSs to be implemented in the Regency of Sumedang. Our sincere gratitude also to: the late Prof. C. J. Stigter, who initiated the establishment of the SFSs in Indonesia; the following previous donor agencies (2013–2018): Australia Research Council; Knowledge Sector Initiative (Australian and Indonesian bilateral agreement); and Indonesia Climate Change Trust Fund, of the Ministry for Planning and Development. Without the collaboration of Sumedang Agricultural Regency Office, the farmers, research assistants and students from the Department of Anthropology, Faculty of Social and Political Sciences, Universitas Indonesia, and The South African National Research Foundation via the University of the Free State, Bloemfontein for funding Prof Sue Walker's international travel, the work would not be accomplished. Our gratitude to their agreement to collaborate.

References

Brown ME, Funk CC (2008) Food security under climate change. Science 319(5863):580–581

Buontempo C, Hewitt CD, Doblas-Reyes FJ, Dessai S (2014) Climate service development, delivery and use in Europe at monthly to inter-annual timescale. Clim Risk Manage 6:1–5

Clarkson G, Dorward P, Kagabo D, Nsengiyumva G (2017) Climate services for agriculture in Rwanda: initial findings from PICSA monitoring and evaluation. Climate change, agriculture, and food security info note, 1–4 Oct

Dayamba DS, Ky-Dembele C, Bayala J, Dorward P, Clarkson G, Sonogo D, Mamadou LP, Traore I, Diakite A, Nenkam A, Binam OM, Zougmore R (2018) Assessment of the use of participatory integrated climate services for agriculture (PICSA) approach by farmers to manage climate risk in Mali and Senegal. Clim Serv 12:7–35

Fischer G, Shah M, van Velthuizen H (2002) Climate change and agricultural vulnerability. IIASA Publications Department, Vienna

Haines S (2019) Managing expectations: articulating expertise in climate services for agriculture in Belize. Clim Change. https://doi.org/10.1007/s10584-018-2357-1

Hidayat B (2011) The sky and the agro-bio-climatology of Java: need critical reevalution due to environmental change. In: Nakamura T, Orchiston W, Soma M, Strom R (eds) Proceedings of the seventh international conference on oriental astronomy, national astronomical observation of Japan, Tokyo, Japan

Howden SM, Soussana JF, Tubiello FN, Chhetri N, Dunlop M, Meinke H (2007) Adapting agriculture to climate change. Proc Natl Acad Sci 104:19691–19696

Jennings S, Magrath J (2009) What happened to the seasons? OXFAM research report, October. https://www.oxfam.org.uk/page-not-found?aspxerrorpath=/resources/policy/climate_change/downloads/research_what_happen%20ed_to_seasons.pdf. Retrieved on 25 Sept 2015

Nelson GC, Rosegrant MW, Koo J, Robertson R, Sulser T, Zhu T, Ringler C, Msangi S, Palazzo A, Batka M, Magalhaes M, Valmonte-santos R, Ewing M, Lee D (2009) Climate change: impact on agriculture and costs of adaptation., International Food Policy Research Institute Washington, DC

Parry ML, Rosenzweig C, Iglesias A, Livermore M, Fischer G (2004) Effects of climate change on global food production under SRES emissions and socio-economic scenarios. Glob Environ Change A 14:53–67

Prihandiani AF (2017) Diseminasi pembelajaran agrometeorologi: variasi luaran dari keagensian petani pengukur curah hujan Indramayu. Unpublished undergraduate thesis. Universitas Indonesia, Depok, Indonesia

Rovihansyah UP (2014) Dinamika variasi pengetahuan petani Klub Pengukur Curah Hujan Indramayu dalam mengantisipasi kondisi cuaca pada musim *rendheng* 2013–2014, Unpublished undergraduate thesis. Universitas Indonesia, Depok, Indonesia

Schmidhuber J, Tubiello FN (2007) Global food security under climate change. Proc Natl Acad Sci 104:19703–19708

Siregar RP, Crane TA (2011) Climate information and agricultural practice in adaptation to climate variability: the case of climate field schools in Indramayu, Indonesia. Cult Agric Food Environ 33(2):55–69. https://doi.org/10.1111/j.2153-9561.2011.01050

Stigter CJ (2016) Climate crisis in the livelihood of Indonesian farmers. In: Winarto YT (ed) Krisis Pangan dan "Sesat Pikir": Mengapa Masih Berlanjut? Yayasan Obor Indonesia, Jakarta, Indonesia, pp 33–40

Stigter K, Winarto YT (2015) Agrometeorological learning: exchanges of traditional, more recent empirical knowledge and scientific knowledge between farmers, extension intermediaries and scientists. Manuscript presented in the workshop and exhibition of Perlu Solusi Baru untuk Masalah Baru karena Perubahan Iklim: Belajar Agrometeorologi dalam Warung Ilmiah Lapangan. Center for Anthropological Studies, Faculty of Social and Political Sciences, Universitas Indonesia & Indramayu Rainfall Observers Club & KONSEPSI & Pandan Wangi Rainfall Observers Group, Indramayu and Selong, Indonesia, 5 Nov

Stigter K, Winarto YT (2016) Science field shops with farmer extension intermediaries for climate services in agriculture. Training of trainers (ToT) roving seminar training guidance (English and Bahasa Indonesia), Unpublished manuscript, Universitas Indonesia. Depok, Indonesia & Bruchem, the Netherlands/Bondowoso, Indonesia

Stigter K, Winarto YT, Wicaksono MT (2016) Science field shops: an innovative agricultural extension approach for adaptation to climate change, applied with farmers in Indonesia. In: Filho WL, Adamson K, Dunk RM, Azeiteiro UM, Illingworth S, Alves F (eds) Implementing climate change adaptation in cities and communities: integrating strategies and educational approaches. Springer, Switzerland, pp 391–405

Sumedang BPSK (2018) Kabupaten Sumedang dalam Angka. BPS Kabupaten Sumedang, Bandung, Indonesia

Tall A, Coulibaly JY, Diop M (2018) Do climate services make a difference? A review of evaluation methodologies and practices to assess the value of climate information services for farmers: implication for Africa. Clim Serv 11:1–12

Upaya Khusus (UPSUS) Swasembada Pangan 2015–2017, Badan Penelitian dan Pengembangan Pertanian, Kementerian Pertanian, Jakarta, Indonesia http://biogen.litbang.pertanian.go.id/2015/02/upaya-khusus-upsus-swasembada-pangan-2015-2017/. Retrieved on 07 Mar 2019

Winarto YT (2004) Seeds of knowledge: the beginning of integrated pest management in Java. Monograph #53. Yale Southeast Asian Studies, New Haven, USA

Winarto YT, Stigter K (eds) (2011) Agrometeorological learning: coping better with climate change. LAP Lambert Academic Publishing, Saarbrücken

Winarto YT, Stigter K (2013) Science Field Shops to reduce climate vulnerabilities: an inter- and transdisciplinary educational commitment. Collaborative Anthropol 6:419–441

Winarto YT, Stigter K (2016) Incremental learning and gradual changes: 'science field shops' as an educational approach to coping better with climate change in agriculture. In: Wilson L,

Stevenson C (eds) Promoting climate change awareness through environmental education. IGI Global Publisher, Hershey, pp 60–95

Winarto YT, Stigter K (2017) Anthropology teaming up with agrometeorology: getting university science to assist farmers with climate resilience. Asian Polit Policy 9(3):479–492

Winarto YT, Stigter K (CJ), Ariefiansyah R (2017) Anticipating the future. Inside Indonesia 127, Jan–March

Winarto YT, Walker S, Ariefiansyah R (2017) People, clouds and roots: farmers learning to adapt. Unpublished manuscript presented in the international workshop on agriculture in the anthropocene. Cambridge University, Cambridge, 27–28 Oct

Winarto YT, Walker S, Ariefiansyah R, Prihandiani AF, Taqiuddin M, Nugroho ZC (2018a) Institutionalizing science field shops: developing response farming to climate change. Global alliance for climate smart agriculture (GACSA), Rome, Italy. http://www.fao.org/3/I8454EN/i8454en.pdf

Winarto YT, Ariefiansyah R, Prihandiani AF, Taqiuddin M (2018b) Nurturance and trust in developing agrometeorological learning in a changing climate: the science field shops in Indonesia. Indo Malay World 46(136):360–381. https://doi.org/10.1080/13639811.2017.1502514

Winarto YT, Walker S, Ariefiansyah R (2019) People, clouds and roots: between the unseen, the seen, and the unforeseen. Nat Cult (in press)

The Usage of Circular Economy Strategies to Mitigate the Impacts of Climate Change in Northern Europe

Janis Zvirgzdins, Kaspars Plotka and Ineta Geipele

Abstract Climate change has been addressed from climatic, economic and social aspects in Northern Europe as well as worldwide, and evaluation of measures that might prevent consequences of climate change on global economic processes is required. Therefore, the paper evaluates circular economy as a successful strategy to mitigate climate change. Literature review and desk research were used to estimate climate change impact on the economy of Northern European countries and their social processes with special focus on passenger vehicles. The results confirmed relations between human behaviour and global warming, which is a threat for the entire humanity. In addition, the study reveals that circular economy car sharing strategy applied to the European Union countries (excluding Bulgaria, Cyprus, and Malta) has a potential to cut emissions by 358.6 $MtCO_2$ eq. and save 7.61 billion EUR annually. The authors suggest promoting the transition to shared, instead of owned vehicle ownership model. Therefore, the study shows that circular economy consists of features that may mitigate some negative consequences of climate change and, in addition, demonstrates potential to provide a basis for sustainable development.

Introduction

Background

Climate change is an issue that has gained attention of media, academic literature, industries, governments and households. It clarifies changes in global air and water

J. Zvirgzdins (✉) · K. Plotka · I. Geipele
Institute of the Civil Engineering and Real Estate Economics, Riga Technical University, Kalnciema Str. 6, 210, Riga 1048, Latvia
e-mail: Janis.Zvirgzdins_1@rtu.lv

K. Plotka
e-mail: Kaspars.Plotka@rtu.lv

I. Geipele
e-mail: Ineta.Geipele@rtu.lv

W. Leal Filho et al. (eds.), *Climate Change, Hazards and Adaptation Options*, Climate Change Management, https://doi.org/10.1007/978-3-030-37425-9_43

temperatures caused by greenhouse gas (GHG) emissions. Given the magnitude of the low carbon production problem, the climate policy must look at reducing the demand for new materials. Circular economy can achieve getting more from the materials already existing. This will allow humans to meet the needs of modern economy and reduce consumption of new primary materials (Enkvist and Klevnas 2018).

The term 'climate' is referring to the observations of climatic variables: temperature, rainfall, drought, water availability and others. Society experiences changes in these climatic variables and responds by creating relevant short-term events and economically relevant features of the climate. For example, hot temperatures increase the probability of riots in cities—even when extreme temperatures occur only for few hours. This is important for understanding of climate impact because the frequency of these momentary events may change if the distribution of daily temperatures changes (Burke et al. 2015). The influence of climate change will be noticeable earlier in less developed countries, since their economies are more dependent on sectors sensitive to climate change as, for example, agriculture, forestry, fishery and tourism.

In addition, it is crucial to understand that climate changes every day and the human behaviour regarding these changes can also differ from day to day. A great example of how climate change influences society was experienced in the summer of 2018 in Northern Europe. In the last several years, new heat records, droughts, increased fire risks and damages to agriculture characterise summer months in Latvia. All this raise the following questions: Is the climate change obvious now? Will people start thinking and acting in a more sustainable way?

In addition, global conflicts occur mostly due to dissatisfaction—resource insufficiency, political ambitions, rapid changes in climate etc. As far as the recent global warming is concerned, the likelihood of global conflicts related to climate is increasing. All scientific facts point at the need to slow down the rapid increase of global warming. Otherwise, the prediction made by famous physicist Stephen Hawking (1942–2018) in 2017 that humans would need to leave the Earth in 600 years in order to survive, might turn out to be true (Gohd 2017). The major concern of Stephen Hawking and other scientists is that climate change is already causing rapid sea level rise. It is possible that if this progression is not mitigated efficiently, a significant percentage of land will be under water. This is the main reason why the Kyoto protocol and the Paris Agreement aim to reduce CO_2 emissions. The Fifth Assessment Report of the Intergovernmental Panel on Climate Change quantified the global 'carbon budget', the amount of carbon dioxide that we can emit while still having a likely chance of limiting global temperature rise to 2 °C above the pre-industrial levels.

Closing the gap between the current climate policy and the long-term goal of the Paris Agreement—an economy with net zero emissions—has promoted many debates. Discussions regarding the European climate policy have mainly focused on increasing energy efficiency and deploying low-carbon energy sources. Both are essential, but when it comes to industry, which is a major source of emissions, only partial solutions are offered. Therefore, additional strategies are needed. The concept of the 'circular economy' offers precisely this opportunity through various strategies such as recirculating a larger share of materials, reducing waste in production, light-weighting products and structures, extending the lifetime of products, and deploying new business models based on sharing of cars, buildings etc. (Enkvist and Klevnas 2018). By reducing, and perhaps even avoiding, the tremendous waste of resources using synergistic effects of bioeconomy and circular economy, it is possible not only to achieve a greater level of sustainability in various European industrial sectors, but also to contribute to a substantial reduction in CO_2 emissions, and climate change mitigation (Leal Filho 2018; Zvirgzdins et al. 2018; Geipele et al. 2018). The largest components of the ecological footprint are food, transportation and buildings, based on the case of Metro Vancouver (Moore et al. 2013). In the present article, the authors emphasise a circular economy strategy of car sharing in the automotive industry, which is aimed to intensify use of cars and reduce demand for them.

Description of the Study Area and Methodology

Study Area

Up to this day, the industry has focused on supply-side emissions: reducing steel, cement, chemicals etc. Less attention has been devoted to the demand-side emissions: how circular economy could reduce emissions by making better use and re-use of the existing materials in the economy. The present study aims at showing a reduction in the possible CO_2 emissions using the strategies of circular economy, including more effective material use and circular economy strategy application to automotive and building sectors.

This article discusses the most important material flows and value chains, considers circular economy and its economic efficiency emphasising a car sharing strategy in the European Union. It also discusses obstacles that may hinder the transition to a more circular economy, as well as policies and initiatives that can overcome them.

Research Methods

Study Site Selection and Sampling Methods

Desk research methodology was conducted in this study, including information and data from foreign, European and Latvian sources. Current development of circular economy and the development potential of circular economy in the near and distant future were considered. Within the study data from previous studies, statistics and various opinions were collected. Secondary analysis of research data was carried out—an additional analysis of data from previous studies was used in the context of circular economy as a tool for climate change mitigation.

The desk research was based on the collection of data from existing resources, so it is often considered a low-cost method compared to field research, as the main costs are linked to the author's time. Desk research studies take into account existing economic, social, and environmental conditions; current and projected climate change; climate danger; and climate change in the regions. The present study emphasises the region of Northern Europe in the context of climate change and shows the potential benefits of circular economy contributing to climate change mitigation.

This study was conducted in Riga, Latvia, from August 2018 to March 2019. Research was limited to geographical scale of Europe emphasising Northern Europe in the context of climate change. The research considered a reduction in CO_2 emissions using different circular economy strategies (food waste reduction, regenerative and healthy food chains, passive houses, urban planning, renewable energy, recovery and reuse, lifetime extension, circular design, digital platforms), emphasising a passenger vehicle sharing strategy in the European Union countries, excluding Bulgaria, Cyprus, and Malta.

Data Sources and Collection Methods

This study has investigated reduction strategies of greenhouse gas emissions by using the concept of circular economy. The authors provided the literature review of qualitative and quantitative studies to reflect the climate change mitigation potential of circular economy. Desk research method was applied to data collection, including such data sources as Eurostat, national and global reports from 2008 to 2018.

Data Analysis

The study used qualitative and quantitative data analysis methods. The authors worked out formulas to calculate a reduction in emissions, using circular economy strategies. Calculations of potential reduction of car sharing strategy were carried out.

$$\Delta \mathrm{E}_{CEi} = n_c \times e \times k_{CEi} \tag{1.1}$$

where: ΔE_{CEi}—reduction in emissions using circular economy (CE) strategies; n_c—number of passenger vehicles registered in territorial unit; e—average emissions per passenger vehicle from territorial unit (tonnes of CO_2 eq.); k_{CEi}—circular economy coefficient (0–1).

$$\Delta \mathrm{C}_{CEi} = \Delta \mathrm{E}_{CEi} \times P_E \tag{1.2}$$

where: ΔC_{CEi}—reduction in cost using circular economy (CE) strategies (EUR); ΔE_{CEi}—reduction in emissions using circular economy (CE) strategies; P_E—price of EU emission allowances (EUR/tCO_2).

In calculations, k_{CEi} is 0.231, which is a derived beneficial ratio of shared cars necessary versus owned cars to cover the same distance (Enkvist and Klevnas 2018). Circular economy coefficient (k_{CEi}) can differ based on the exploited circular economy strategy (food waste reduction, recovery and reuse, recycling, circular design, renewable energy etc.). The reduction in emissions in percentage can be calculated as the difference between 1 and k_{CEi} ($1 - k_{CEi}$). However, the detection of the correct value of the circular economy coefficient requires justification from different studies based on the circular economy strategy exploited.

The authors used Pearson Correlation Formula (1.3) to calculate the correlation coefficient and determine the relationship strength between: (1) purchasing power adjusted GDP per capita and average age of passenger vehicles; (2) purchasing power adjusted GDP per capita and purchased passenger vehicles per capita; (3) purchasing power adjusted GDP per capita and average number of passenger vehicles per capita. The authors executed calculations of correlation between the previously mentioned factors using Microsoft Excel.

$$r = \frac{\sum_{i=1}^{n}(x_i - \bar{x})(y_i - \bar{y})}{\sqrt{\sum_{i=1}^{n}(x_i - \bar{x})^2}\sqrt{\sum_{i=1}^{n}(y_i - \bar{y})^2}} \tag{1.3}$$

where: x, y—independent factors; $\bar{x}$, $\bar{y}$—average values of independent factors; n—number of data pairs.

When evaluating the strength of the association using Pearson correlation coefficient, for absolute values of r, 0–0.19 is regarded as very weak, 0.2–0.39 as weak, 0.4–0.59 as moderate, 0.6–0.79 as strong and 0.8–1 as very strong correlation (Alston et al. 2016).

Results and Discussion

Climate Change and Its Influence on Northern Europe

The central aim of the Paris Agreement is to strengthen the global response to the threat of climate change by keeping a global temperature rise in the 21st century below 2 °C above pre-industrial levels and to pursue efforts to limit the temperature increase even further to 1.5 °C (United Nations Framework Convention on Climate Change 2019a, b). Up to 12 March 2019, 185 of 197 Parties have ratified the Paris Agreement (United Nations Framework Convention on Climate Change 2019a, b). However, countries, which are highly dependent on manufacturing, decline to ratify the Paris Agreement or set appropriate targets regarding Paris Agreement. Manufacturing countries use the argument that it is not proven that global warming is related with the human activity. They point out that global warming is a result of natural processes.

In the Earth's history before the Industrial Revolution, the Earth's climate changed due to natural causes that were not related to human activity. Most often, global climate has changed because of variations in sunlight. Variations in the Sun itself have increased and decreased the amount of solar energy reaching the Earth. Volcanic eruptions have generated particles that reflect sunlight, brightening the planet and cooling the climate. Volcanic activity has also increased greenhouse gases over millions of years, contributing to episodes of global warming. These natural causes are still in play nowadays, but their influence is too little to explain the rapid warming observed in recent decades. Volcanoes are the biggest natural CO_2 emission creators. On average, volcanoes emit between 130 and 230 million tonnes of carbon dioxide per year. By burning fossil fuels, people release over 100 times more—about 26 billion tonnes of carbon dioxide into the atmosphere every year. As a result, human activity overshadows any contribution volcanoes may have made to recent global warming (NASA Earth Observatory 2010).

Figure 1 clearly illustrates human activity dominance stated above. Since the start of the Industrial Revolution, human emissions of carbon dioxide from fossil fuels and cement production have risen to more than 35 billion metric tons per year, while volcanoes produce less than 1 billion metric tons annually. Human activities—burning of coal and other fossil fuels, cement production, deforestation and other landscape changes—emitted approximately 40 billion metric tons of CO_2 in 2015. Since the start of the Industrial Revolution, more than 2 trillion metric tons of CO_2 have been added to the atmosphere by human activities, according to the Global Carbon Project (Scott and Lindsey 2016).

Many companies worldwide have to pay for CO_2 emissions from the production process. Increasing payments for CO_2 emissions will increase costs of the company. Consequently, the cost of goods or services will also increase. The total expenditure

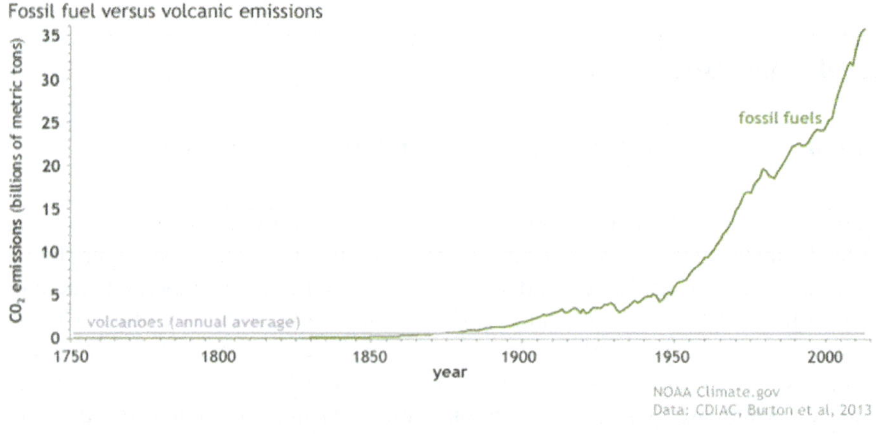

Fig. 1 Comparison of fossil fuels and volcanic emissions (Scott and Lindsey 2016)

for the reduction of pollution varies greatly from country to country, and it is highly dependent on the reduction policies of national greenhouse gas emissions that are influenced by national economic processes.

Increasing global temperatures might lead to glacier and sea ice melting, rising sea levels and increased coastal flooding, heavier precipitation, expansion of deserts and more severe droughts. Higher spring and summer temperatures and earlier spring snow-melt result in forests lead to forests being hotter and drier for longer periods of time, which provide conditions for wildfires to catch fire and spread. While hurricanes are a natural part of our climate system, recent research indicates that their destructive power has grown since the 1970s, particularly in the North Atlantic region. Climate change has significant implications for human health. Rising temperatures will likely lead to increased air pollution, a longer and more intense allergy season, more frequent and dangerous heat waves and heavier rainstorms and flooding. These changes pose serious and costly risks to public health. Another impact is ocean acidification, which can be characterised as a chemical reaction between carbon dioxide and water. This factor and shifting temperature regimes can lead to species extinction, which can harm the biggest treasure of the Earth—biodiversity. Due to the rising sea levels, threats to food security from decreasing crop yields and the abandonment of populated areas are real. The resulting dry conditions will increase the pressure on groundwater supplies as more is pumped to meet demand. Our aging electricity infrastructure is increasingly vulnerable to the growing consequences of global warming, including sea level rise, extreme heat, higher wildfire risk, drought and other issues (Union of Concerned Scientists 2019).

The threats of global warming are obvious—crop losses, price changes. It is also clear that climate change is causing more serious changes in countries whose economies are based on agriculture. Changes in the food market and price shocks may not only lead to a change in the number of people in a particular country, but also destabilize the internal political and economic situation. Problems of one country can spread to neighbouring countries, for example, as the European experience with refugees shows. National borders or the possibility of perishing during migration cannot restrict desperate people. They leave their home countries and seek for a better life elsewhere. This refugee crisis is associated with political motives. A potentially large increase in the number of refugees is possible in the context of climate change in Northern Europe. However, these will not be economic refugees, but rather climate refugees (environmental migrants). The situation can develop critically, and all the inhabitants of the Earth will have to adapt to the new circumstances. In Northern European countries, there will probably be less severe consequences of climate change than in countries of the Southern regions. However, these effects will burden protection in sectors such as agriculture, energy, water supply and others. These global warming consequences could create challenges to national security. The inhabitants of Northern Europe have a high level of tolerance towards immigrants. Therefore, in the future, particular attention should be paid to total protection aspects.

It would be important for Europe to devote attention to design tools of the national climate policy and their implementation because the national solution regarding climate change combating requires a detailed analysis of the national economy and its political system. Tax policy related to green thinking should be developed. This would advance the formation of a co-responsible society.

Circular Economy as a Tool for Climate Change Mitigation

On the planet with limited material resources, circular economy is essential to maintain and improve human lives and well-being. However, fossil energy continues to provide humanity with goods: heat, transport, electricity and power to transform raw materials into the products we use on a daily basis (Deloitte Sustainability 2016).

Every week, the global population increases by 1.5 million people (Worldometers 2017), 3 million people enter the global middle class (Kharas 2017), and great amount of population move from rural to urban areas. This rapid demographic shift has one inevitable consequence: all over the world, there is an increasing demand for basic goods, housing and transportation (Climate-KIC 2018). All of these processes are burdening the Earth and contributing to climate change. Climate change is a growing concern for governments, companies, non-governmental organisations (NGOs) and the civil society in general. From 1880 to 2012, the average global temperature increased by 0.85 °C. Oceans have warmed, the amounts of snow and ice have diminished and the sea level has risen (Deloitte Sustainability 2016).

Circular economy is a growing concept whose goal is to lead to a more sustainable economy, contributing to recycling and reuse of products, and, consequently, to make them 'circular'. In addition to several economically attractive effects, circular economy is aimed to reduce greenhouse gas emissions and, therefore, help to slow down global warming. The goal of circular economy in Europe is to achieve greater sustainability by maximising the use of products, raw materials and waste through recycling and reuse as frequently as possible. Circular economy is becoming a growing topic because natural resources, such as coal and gas, are finite and are still used in unsustainably large quantities based on 'linear consumption'. Linear 'take-make-dispose' model is not only a driving force behind the lack of resources that has developed over the decades of intensive exploitation of natural resources, but it also makes a significant contribution to environmental pollution and loss of biodiversity.

Circular economy is an essential future alternative to a traditional linear model of 'take-make-dispose'. Its key aim is to decouple global economic development from the consumption of finite resources, and to find utility and value in what today is considered 'waste'. Circular business models extend product life, maximise asset utilisation, and create multiple value loops in supply chains. By spreading circular business models that circulate materials and use them more efficiently, it has been estimated that CO_2 emissions of the European Union industry can be cut by 56%. If digitalisation is implemented to maximise the potential of circular economy, cut of 296 million tonnes of annual CO_2 emissions in Europe can be achieved. Digital opportunities are already helping accelerate circular innovations, moving the growing circular economy from a sometimes slow, manual process to a streamlined, efficient business for a sustainable future (Enkvist and Klevnas 2018; Climate-KIC 2018).

Circular economy is an alternative to a traditional linear economy, where resources are stored for as long as possible, using their maximum value while they are in use, and then recovering and regenerating products and materials at the end of their lifetime. Consequently, the use of waste and resources is minimised, and when the product enters the end-of-life phase, it is reused to create benefit. Therefore, circular economy can bring great economic benefits, stimulating innovation, growth and job creation. In addition, circular economy can play an important role in achieving the objectives of the Paris Agreement. It is estimated that the circular economy strategies have a potential to reduce carbon emissions from products by 33% (Janssens 2017).

With the development of the concept, circular economy is increasingly focused on the possibility of greenhouse gas (GHG) mitigation. There are many reasons why circular economy takes a major role in the fight against climate change. First, reducing, reusing and recycling eliminate greenhouse gasses by decreasing emissions from traditional waste management strategies (energy recovery, landfill). Second, circular economy strategies help reduce GHG emissions by cutting the amount of energy needed in industrial production processes to convert primary raw materials into products. Deloitte Sustainability experts have assessed the potential for reducing GHG emissions for four highly contributing sectors: food, construction, vehicles, and electrical and electronic equipment. These four sectors combined account for a third of the total emissions of 3.2 Gt CO_2 eq. (Deloitte Sustainability 2016).

Several studies with different scales have been carried out regarding a decrease in greenhouse gas emissions using circular economy strategies. CE Delft has calculated that if two thirds of all municipal waste alone are recycled, instead of landfilled or incinerated, the annual global greenhouse gas emissions will be reduced by six percent (Bergsma et al. 2016). According to the study of Ellen MacArthur Foundation, 17,000 $MtCO_2$ eq. global emission reduction can be achieved by 2030 using such circular economy strategies as electric, shared, and autonomous vehicles, food waste reduction, regenerative and healthy food chains, passive houses, urban planning, and renewable energy (MacArthur et al. 2015). Another global study shows potential of 7500 $MtCO_2$ eq. global emission reduction in 2030 (Blok et al. 2016; Schroeder et al. 2018). The study considers the following circular economy strategies: recovery and reuse, lifetime extension, sharing and service model, circular design, and digital platforms. In addition, one study in Europe shows that Waste Directives Impact Assessment has a potential to cut greenhouse gas emissions by 62 $MtCO_2$ eq. (European Commission 2014). Material efficiency, which is related to substituting half of the virgin materials used with recycled materials, and doubling the product-life-time of long-lived consumer products could cut carbon emissions by 3–10% (~75 $MtCO_2$ eq.) in Finland, France, the Netherlands, Spain and Sweden by 2030 (Wijkman and Skånberg 2015). Several local scale studies have been carried out. For example, France achieved 2.1 $MtCO_2$ eq. reduction in GHG emissions via household packaging recycling in 2013. It is emphasised that this outcome has been achieved through the involvement of all stakeholders (Dépoues and Bordier 2015).

Food waste is one of the main aspects for reducing greenhouse gas emissions. Approximately a third of the food for human consumption is wasted worldwide (Food and Agriculture Organization of the United Nations 2015). In addition, wasting food is unethical behaviour in the world where almost 1 billion people remain undernourished.

Food is an important source of greenhouse gas emissions (~470 $MtCO_2$ eq. per year). Circular economy strategies include the recirculation of essential nutrients (N, P) from food waste or wastewater. However, the greatest potential is the reduction of food waste (approximately one third of the food produced). If food waste is to be reduced by 50% and 30% of the nutrients come from organic waste or wastewater, the sector's emissions can be reduced by 13% (Deloitte Sustainability 2016).

Figure 2 illustrates food use hierarchy, which is divided in 2 general strategies: (1) waste avoidance; (2) waste management. Waste avoidance includes reduction strategies with the greatest potential to be achieved on the demand side. Shopping lists and plans can be good and simple tools to start with. Additionally, there is a potential for a positive synergy effect as the unpurchased food creates economic savings. However, if there are some food leftovers, people in need should be fed. Therefore, creation of 'leftover banks' should be considered, where the people in need can receive food. However, if food does not qualify for human consumption, then feeding livestock can be the next step down the food value chain. If food cannot be used by livestock due to expiration dates or other obstacles, the waste management strategies are considered. They are mainly aimed at composting and creating biomass, of which 100% renewable bioenergy can be gained. If the previously described strategies are not adapted to food waste, it ends up being disposed at the landfill and degrading the ecosystems, which is opposite of what circular economy is implemented for.

Circular economy in the automotive industry stems from remanufacturing, refurbishment and recycling. These strategies are aimed at improving the efficiency of resources throughout the product and the material life cycle. Other strategies focus on intensifying use (for example, car sharing). Deloitte Sustainability experts note that increasing the recycling of materials could decrease greenhouse gas emissions of the automotive sector by 45%. In order to obtain this circular economy potential, the following strategies should be implemented: (1) vehicle eco-design in order to

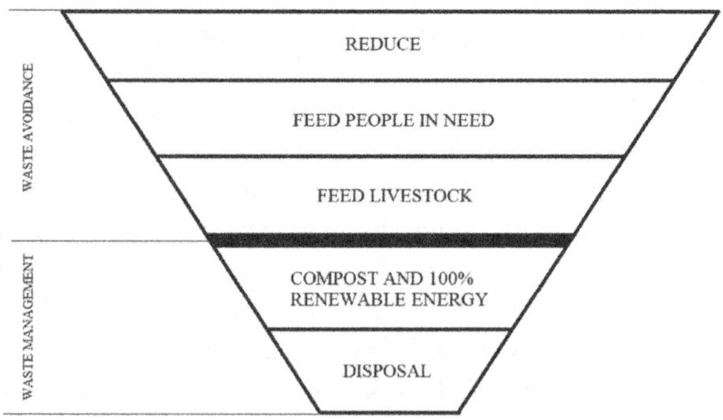

Fig. 2 Food use hierarchy (Deloitte Sustainability 2016)

facilitate the reuse and recycling of parts; (2) implementation of a system to systematically dismantle the reusable parts of a vehicle; (3) to ensure the traceability of parts and materials in order to enable their reuse in closed loops; (4) to find and develop infrastructure of economic outlets for materials retrieved from used cars; (5) investment in research and development (R&D), training of stakeholders, alternatives in transport modes etc. (Deloitte Sustainability 2016).

Approximately 4/5 of the greenhouse gas emissions of vehicles occur during the use phase due to the combustion of fuel (Nemry et al. 2008). Therefore, the industry is keeping its focus on decreasing these emissions.

The authors have carried out analysis of empirical data (ACEA 2018; Purchasing Power Adjusted GDP Per Capita 2019; Passenger Car Sales in Europe 2019) applying the circular economy vehicle sharing strategy to the countries of the European Union (except for Bulgaria, Cyprus, and Malta) to assess the environmental and economic benefits. The data is summarised in Tables 1 and 2. Table 1 reflects dynamics of passenger vehicle count in the European Union from 2012 to 2016. It can be seen (Table 1) that from 2012 to 2016 the number of passenger cars in the majority of the European Union countries increased, which led to an increased air pollution. The greatest increase in passenger vehicle count is observed in Romania (18%), Estonia (14.4%) and Slovakia (14.1%).

Circular economy strategies are used for the efficient use of resources and prolonging the lifetime of any product, which in this case is a passenger vehicle. In order to find out which European Union countries are mainly affecting and increasing the demand for new passenger vehicles, the authors studied the correlation between the purchasing power adjusted GDP per capita and average age of passenger vehicles among the EU member states (except Bulgaria, Cyprus, and Malta) in 2016 (Fig. 3). It can be seen that the average age of passenger vehicles in the European Union is from 8 to 16 years. The assumed pattern is that people from more developed countries, with greater GDP per capita, tend to change their cars more frequently than people from lower income countries. This can be seen in Fig. 3, where an average age of passenger vehicles in developed economies is lower than the average age of passenger vehicles in less developed economies. This has been approved by the Pearson correlation coefficient, which shows a significant negative correlation ($r = -0.78$) between these factors.

The case of vehicle lifetime provides perhaps the clearest example of why this model of ownership and operation limits circularity. Cars are driven approximately 200,000 km before being scrapped, but as each car is driven a relatively short distance, it takes many years to reach that margin. It has been estimated that vehicles are, in average, 14 years old when scrapped. The current ownership model does not allow extending the vehicle lifetime, for example, to 500,000 km before being scrapped. It would take 35 years to reach that margin, and by that time vehicle could be lagging behind technology, safety and design (Enkvist and Klevnas 2018).

It can be stated that the limited lifetime of vehicles today is not based on technical limitations, but on a low rate of utilisation deriving from the current ownership model. Therefore, it should be changed and car sharing should be promoted using different

Table 1 Dynamics of passenger vehicle count in the European Union from 2012 to 2016

Country	Passenger vehicles (2012)	Passenger vehicles (2013)	Passenger vehicles (2014)	Passenger vehicles (2015)	Passenger vehicles (2016)	% change 2016/2012 (%)	Average age of passenger vehicles in years (2016)	Purchasing power adjusted GDP per capita (2016)	Passenger vehicles purchased per capita (2016)
Austria	4,584,202	4,641,308	4,694,921	4,748,048	4,821,557	4.9	8.2	37,600	0.038
Belgium	5,392,909	5,439,295	5,511,080	5,587,415	5,669,764	4.9	8.9	34,500	0.047
Croatia	1,445,000	1,433,563	1,458,149	1,489,338	1,540,260	6.2	14.3	17,900	0.011
Czech Republic	4,698,800	4,787,849	4,893,562	5,115,316	5,368,660	12.5	14.5	25,600	0.024
Denmark	2,225,164	2,265,349	2,320,982	2,391,755	2,477,478	10.2	8.4	36,900	0.039
Estonia	602,133	628,562	652,949	676,592	703,151	14.4	14.6	22,500	0.017
Finland	2,560,190	2,575,951	2,595,867	2,612,922	2,629,432	2.6	11.8	31,800	0.021
France	31,600,000	31,650,000	31,799,000	31,915,493	31,999,953	1.2	8.8	30,500	0.031
Germany	43,431,124	43,851,230	44,403,124	45,071,209	45,803,560	5.2	9.3	36,300	0.041
Greece	5,138,745	5,109,435	5,102,203	5,104,908	5,126,024	−0.2	15.0	19,800	0.007
Hungary	2,978,745	3,035,764	3,101,752	3,192,132	3,308,495	10.0	13.9	19,500	0.010
Ireland	1,882,550	1,910,165	1,943,868	1,985,130	2,026,977	7.1	8.8	51,700	0.030
Italy	37,078,274	36,962,934	37,080,753	37,351,233	37,876,138	2.1	10.8	28,400	0.031
Latvia	618,000	634,214	657,487	677,561	663,091	6.8	16.0	18,800	0.009
Lithuania	1,753,000	1,837,661	1,113,445	1,153,859	1,190,146	−47.3	16.9	22,000	0.007
Luxembourg	344,951	355,358	362,879	372,538	380,860	9.4	6.3	76,200	0.085
Netherlands	8,142,000	8,154,000	8,192,570	8,336,414	8,439,318	3.5	10.4	37,400	0.022

(continued)

Table 1 (continued)

Country	Passenger vehicles (2012)	Passenger vehicles (2013)	Passenger vehicles (2014)	Passenger vehicles (2015)	Passenger vehicles (2016)	% change 2016/2012 (%)	Average age of passenger vehicles in years (2016)	Purchasing power adjusted GDP per capita (2016)	Passenger vehicles purchased per capita (2016)
Poland	18,744,412	19,389,446	20,003,863	20,723,423	21,675,388	13.5	13.6	19,900	0.011
Portugal	4,497,000	4,480,000	4,496,000	4,538,000	4,600,000	2.2	12.5	22,600	0.020
Romania	4,485,148	4,693,651	4,905,630	5,153,182	5,470,578	18.0	16.2	17,400	0.006
Slovakia	1,826,393	1,882,577	1,952,002	2,037,772	2,124,972	14.1	13.5	22,500	0.016
Slovenia	1,080,001	1,085,347	1,096,920	1,116,006	1,143,218	5.5	11.3	24,100	0.031
Spain	22,247,528	22,024,538	22,029,512	22,355,549	22,876,247	2.7	11.9	26,700	0.025
Sweden	4,447,165	4,495,473	4,585,519	4,669,063	4,768,060	6.7	9.9	35,700	0.037
United Kingdom	31,481,823	31,917,885	32,612,782	33,542,448	34,378,386	8.4	7.8	31,200	0.040

[a]Excluding Bulgaria, Cyprus, Malta

Table 2 Potential decrease in emissions (tCO_2 eq.) and potential savings (EUR) from passenger vehicles using a circular economy car sharing strategy

Country	Passenger vehicles per capita	CO_2 emissions from passenger vehicles (tonnes of CO_2 eq.)	Costs of GHG emissions, EUR	CO_2 emissions from passenger vehicles after implementing CE strategies (tonnes of CO_2 eq.)	Costs of GHG emissions after implementing CE strategies, EUR	Potential emission cut (tonnes of CO_2 eq.)	Potential savings using CE strategies, EUR
Austria	0.55	10,848,503	231,073,119	2,506,004	53,377,891	8,342,499	177,695,229
Belgium	0.49	13,323,945	283,800,037	3,077,831	65,557,809	10,246,114	218,242,228
Croatia	0.37	3,773,637	80,378,468	871,710	18,567,426	2,901,927	61,811,042
Czech Republic	0.51	9,932,021	211,552,047	2,294,297	48,868,523	7,637,724	162,683,524
Denmark	0.43	6,131,758	130,606,446	1,416,436	30,170,089	4,715,322	100,436,357
Estonia	0.54	1,353,566	28,830,949	312,674	6,659,949	1,040,892	22,171,000
Finland	0.47	7,296,674	155,419,152	1,685,532	35,901,824	5,611,142	119,517,328
France	0.49	60,159,912	1,281,406,118	13,896,940	296,004,813	46,262,972	985,401,305
Germany	0.56	93,897,298	2,000,012,447	21,690,276	462,002,875	72,207,022	1,538,009,572
Greece	0.46	5,766,777	122,832,350	1,332,125	28,374,273	4,434,652	94,458,077
Hungary	0.34	5,376,304	114,515,283	1,241,926	26,453,030	4,134,378	88,062,253
Ireland	0.42	6,486,326	138,158,752	1,498,341	31,914,672	4,987,985	106,244,081
Italy	0.64	56,814,207	1,210,142,609	13,124,082	279,542,943	43,690,125	930,599,666
Latvia	0.35	1,475,377	31,425,540	340,812	7,259,300	1,134,565	24,166,240
Lithuania	0.42	1,666,204	35,490,154	384,893	8,198,226	1,281,311	27,291,928
Luxembourg	0.64	1,989,994	42,386,862	459,688	9,791,365	1,530,305	32,595,497
Netherlands	0.49	16,456,670	350,527,073	3,801,491	80,971,754	12,655,179	269,555,319

(continued)

Table 2 (continued)

Country	Passenger vehicles per capita	CO_2 emissions from passenger vehicles (tonnes of CO_2 eq.)	Costs of GHG emissions, EUR	CO_2 emissions from passenger vehicles after implementing CE strategies (tonnes of CO_2 eq.)	Costs of GHG emissions after implementing CE strategies, EUR	Potential emission cut (tonnes of CO_2 eq.)	Potential savings using CE strategies, EUR
Poland	0.57	22,759,157	484,770,053	5,257,365	111,981,882	17,501,792	372,788,170
Portugal	0.45	7,590,000	161,667,000	1,753,290	37,345,077	5,836,710	124,321,923
Romania	0.28	6,701,458	142,741,056	1,548,037	32,973,184	5,153,421	109,767,872
Slovakia	0.39	3,028,085	64,498,213	699,488	14,899,087	2,328,597	49,599,126
Slovenia	0.55	3,258,171	69,399,049	752,638	16,031,180	2,505,534	53,367,868
Spain	0.49	45,752,494	974,528,122	10,568,826	225,115,996	35,183,668	749,412,126
Sweden	0.47	10,012,926	213,275,324	2,312,986	49,266,600	7,699,940	164,008,724
United Kingdom	0.51	64,459,474	1,372,986,791	14,890,138	317,159,949	49,569,335	1,055,826,842

[a]Excluding Bulgaria, Cyprus, Malta

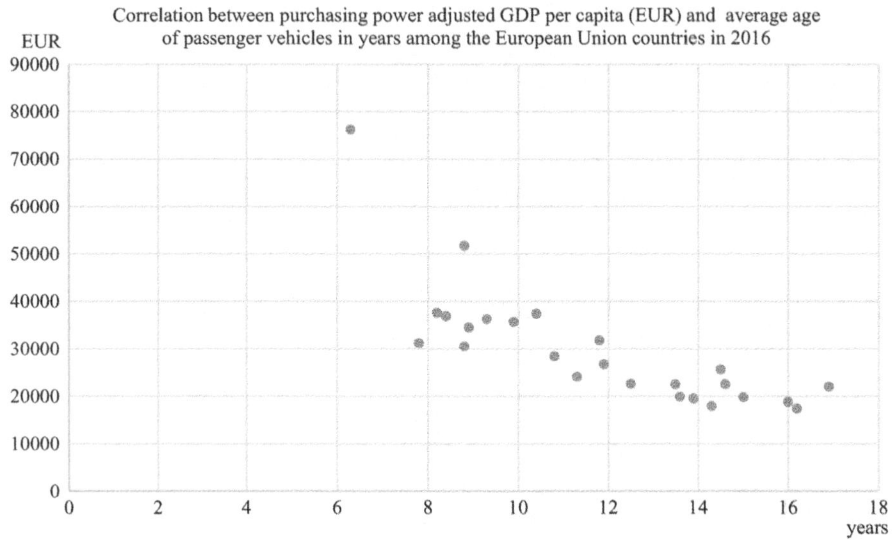

Fig. 3 Correlation between purchasing power adjusted GDP per capita and average age of passenger vehicles among the EU countries in 2016

tools, for example, mobile applications for building a bridge between passengers and drivers with free space from point 'A' to point 'B'.

In order to confirm the assumption that economically developed countries are contributing more to an increase in CO_2 emissions through the purchase of new cars compared to less developed economies, the authors studied correlation between purchasing power adjusted GDP per capita and purchased passenger vehicles per capita among the European Union countries. Figure 4 shows a very strong positive correlation ($r = 0.88$) between these factors, which means that more developed economy members are purchasing newer passenger vehicles than less developed economy members. On average, 0.0262 new vehicles are purchased by each inhabitant of the European Union (1 vehicle per 38.1 inhabitants).

Passenger vehicle sharing strategy relies on a shared vehicle ownership model. It can be stated that the number of vehicles is directly related to the levels of air pollution within the automotive industry. In order to distinguish which countries are contributing more to the climate change in the context of greenhouse gas emissions from passenger vehicles, the authors studied correlation between purchasing power adjusted GDP per capita and average number of passenger vehicles per capita among the European Union countries in 2016. Results (Fig. 5) reflect a moderate correlation ($r = 0.48$). It means that members of developed economies tend to own more passenger vehicles than members of less developed economies. Therefore, the developed economies have greater responsibility of caused air pollution in the context of passenger vehicles than developing economies. The authors assume that countries with a greater number of passenger vehicles per capita will be less flexible in giving up the

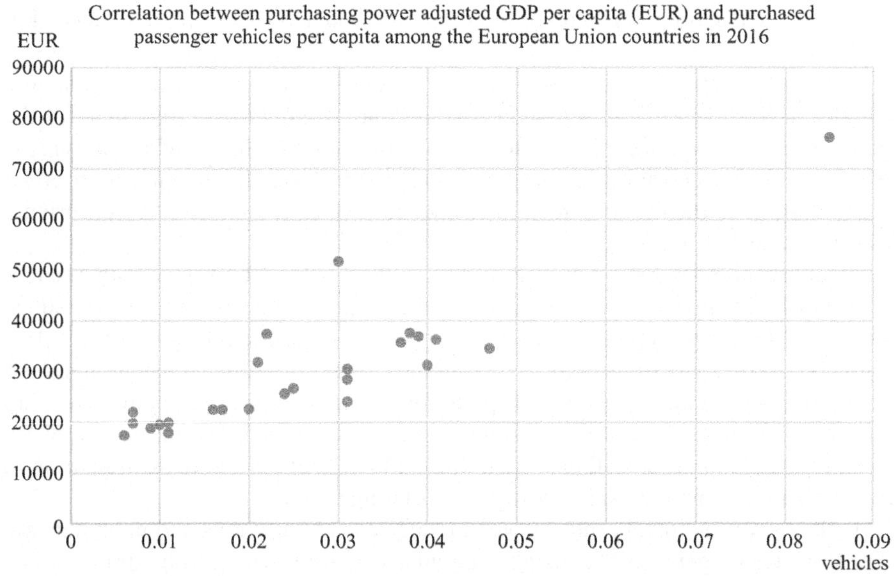

Fig. 4 Correlation between purchasing power adjusted GDP per capita and purchased passenger vehicles per capita among the EU countries in 2016

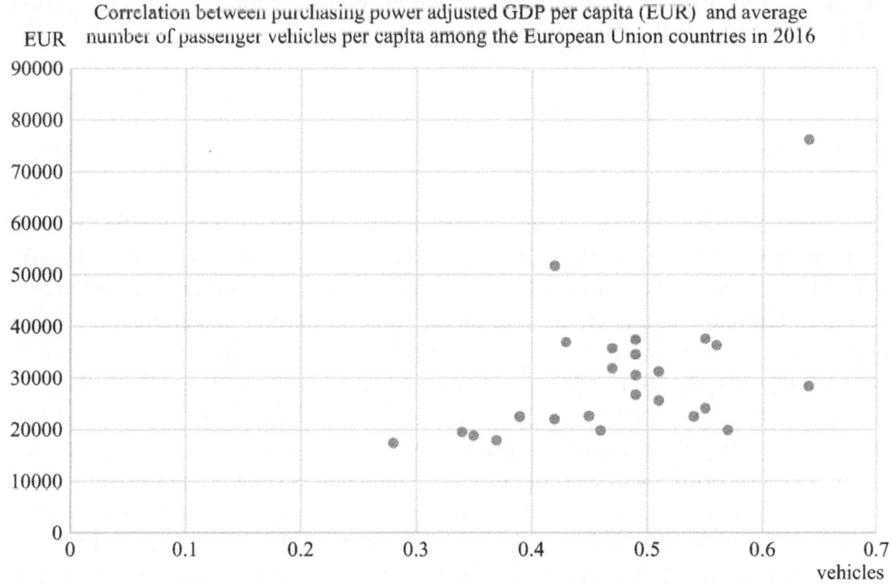

Fig. 5 Correlation between purchasing power adjusted GDP per capita and average number of passenger vehicles per capita among the EU countries in 2016

current car ownership model and promoting transition to a shared vehicle ownership model.

The number of purchased passenger vehicles and the average lifetime of passenger vehicles have shown that developed economies are driving the development of the automotive sector. As one of the factors that characterises higher standard of living is vehicle ownership and ability to use it, developed economies should put in more effort in contributing to the mitigation of consequences, which are followed by ensuring a higher standard of living.

Green economists have raised initiative to establish a Tobin tax, with tax rate 1% of all international deals, and to move those funds to poor countries to reduce growing differentiation between developed and developing countries (Barbier 2011; Patomäki 2001; Zvirgzdins et al. 2018). In the context of climate change, it should be stated that CO_2 emissions and air pollution are cross-border problems related to standard of living. Therefore, the authors suggest using a Tobin tax as a tool for obtaining funds in order to finance activities, which could promote the transition to circular economy practices, including a car sharing strategy.

Table 2 gives an insight into a potential decrease in emissions and potential savings from passenger vehicles using a circular economy car sharing strategy. The authors estimated that applying a car sharing strategy to the European Union countries (excluding Bulgaria, Cyprus, and Malta) could cut emissions by 358.6 $MtCO_2$ eq. saving 7.64 billion EUR annually. Additionally, the number of passenger vehicles per capita varies from 0.28 to 0.64 within the European Union countries (excluding Bulgaria, Cyprus, and Malta).

Conclusion

The issues addressing climate change and measures combating it should be a priority of the policy makers in Northern Europe as well as all around the globe. The governments creating budgets in the future should provide expenditures on adaptation related to climate change. Being one of the most important stakeholders, governments must switch to green public procurement and circular process promotion. In addition, it is crucial to monitor and control long-term goals by ensuring that purchasing institutions contribute to green and circular procurement practices instead of focusing on buying at the lowest price. It would be important to invest in science, research and institutions related to climate change and economic processes.

Circular economy is a wide concept with a variety of strategies that can contribute to climate change mitigation—food waste reduction, regenerative and healthy food chains, passive houses, urban planning, renewable energy, recovery and reuse, lifetime extension, circular design, digital platforms etc. Main aim is to create values, resource efficiency and CO_2 reduction. Desk research has indicated that authors from different studies estimate climate benefits of circular economy strategies, which can result in a decrease of up to 17,000 $MtCO_2$ eq. GHG by 2030. Circular economy car sharing strategy applied to the European Union countries (excluding Bulgaria,

Cyprus, and Malta) has a potential to cut emissions by 358.6 $MtCO_2$ eq. and save 7.64 billion EUR annually. The study has demonstrated that developed economies create greater environmental effects in the context of CO_2 emissions from passenger vehicles than less developed economies. Therefore, a Tobin tax concept can be adapted as a tool for obtaining funds from developed economies and moving these funds to financing activities, which could promote the transition to circular economy practices, including a passenger vehicle sharing strategy. The authors suggest promoting the transition to the shared instead of owned vehicle ownership model because it will also open the door to circular design practices in the automotive sector as the vehicles with greater lifetime than 500,000 km will become demanded. Thus, a great amount of used materials could be saved. As long as the information sharing is a crucial aspect of development of the circular economy concept, the authors emphasise the role of digitalisation in promoting the transition. Implementation of car sharing supportive application could be a good step forward as long as society lives in the touchscreen and application epoch.

The greatest challenge regarding the transition to circular economy is the society's inflexibility towards change, which could be more flexible by educating and building awareness of climate change. The European Union could exploit circular economy strategies, gaining the social, environmental and economic benefits, while showing the rest of the world good examples on how to combat climate change. Further research should aim to extend our study to a more complex analysis of circular economy strategies in the context of climate change mitigation.

References

Alston GL, Battise DM, Neville MW (2016) A 10-year study of the academic progress of students identified as low performers after their first semester of pharmacy school. Am J Pharm Educ 80(7):118

Barbier E (2011) The policy challenges for green economy and sustainable economic development. Nat Resour Forum 35(3):233–245. https://doi.org/10.1111/j.1477-8947.2011.01397.x

Bergsma G, Bijleveld M, Nusselder S (2016) The circular economy as a key instrument for reducing climate change. CE Delft. Retrieved from: https://www.cedelft.eu/en/publications/1803/the-circular-economy-as-a-key-instrument-for-reducing-climate-change

Blok K, Hoogzaad J, Ramkumar S, Ridley A, Srivastav P, Tan I, Terlouw W, de Wit M (2016). Implementing circular economy globally makes paris targets achievable. Circle Economy, Ecofys. Retrieved from: https://www.circle-economy.com/wp-content/uploads/2016/06/ircle-economy-ecofys-2016-implementing-circular-economy-globally-makes-paris-targets-achievable.pdf.pdf

Burke M, Hsiang SM, Miguel E (2015) Climate and conflict. Ann Rev Econ 7(1):577–617. https://doi.org/10.1146/annurev-economics-080614-115430

Climate-KIC (2018) Digitalisation—unlocking the potential of the circular economy. European Institute of Innovation and Technology. Retrieved from: https://www.climate-kic.org/wp-content/uploads/2018/08/ClimateKICWhitepaperFinalDigital_compressed.pdf

Deloitte Sustainability (2016) Circular economy potential for climate change mitigation. Deloitte Conseil: Member of Deloitte Touche Tohmatsu Limited. Retrieved from: https://www2.deloitte.com/content/dam/Deloitte/fi/Documents/risk/Deloitte%20-%20Circular%20economy%20and%20Global%20Warming.pdf

Dépoues V, Bordier C (2015) Mitigating climate change and waste recycling: household packaging case study. CDC climate research. Retrieved from: http://www.cdcclimat.com/IMG/pdf/15-06_climate_report_no50_mitigating_climate_change_and_waste_recycling-2.pdf

Enkvist PA, Klevnas P (2018) The circular economy–a powerful force for climate mitigation: transformative innovation for prosperous and low-carbon industry. Material Economics Sverige AB, Stockholm, Sweden

European Automobile Manufacturers Association (2018) ACEA report vehicles in use Europe 2018. ACEA. Retrieved from: https://www.acea.be/uploads/statistic_documents/ACEA_Report_Vehicles_in_use-Europe_2018.pdf

European Commission (2014) Commission staff working document: executive summary of the impact assessment. Retrieved from: https://eur-lex.europa.eu/legal-content/EN/TXT/PDF/?uri=CELEX:52014SC0208&from=DE

Food and Agriculture Organization of the United Nations (2015) Global initiative on food loss and waste reduction. Fao.org. Retrieved from: http://www.fao.org/3/a-i4068e.pdf

Geipele I, Plotka K, Wirzhbitskis Y, Zvirgzdins J (2018) The synergy in circular economy. In: Third international conference on economic and business management (FEBM 2018), Atlantis Press, pp 65–68. https://doi.org/10.2991/febm-18.2018.15

Gohd C (2017) Stephen Hawking: humans must leave earth within 600 years. Futurism.com. Retrieved from: https://futurism.com/stephen-hawking-humans-must-leave-earth-within-600-years/

Janssens C (2017) What role can circular economy play in delivering the Paris Agreement? Institute for European Environmental Policy. Retrieved from: https://ieep.eu/news/what-role-can-circular-economy-play-in-delivering-the-paris-agreement

Kharas H (2017) The unprecedented expansion of the global middle class. Brookings.edu. Retrieved from: https://www.brookings.edu/research/the-unprecedented-expansion-of-the-global-middle-class-2/

Leal Filho W (2018) Bioeconomy meets the circular economy: the RESYNTEX and FORCE projects. In: Towards a sustainable bioeconomy: principles, challenges and perspectives. Springer, Cham, pp 567–575. https://doi.org/10.1007/978-3-319-73028-8_29

MacArthur E, Zumwinkel K, Stuchtey MR (2015) Growth within: a circular economy vision for a competitive Europe. Ellen MacArthur Foundation. Retrieved from: https://www.ellenmacarthurfoundation.org/assets/downloads/publications/EllenMacArthurFoundation_Growth-Within_July15.pdf

Moore J, Kissinger M, Rees WE (2013) An urban metabolism and ecological footprint assessment of metro Vancouver. J Environ Manage 124:51–61. https://doi.org/10.1016/j.jenvman.2013.03.009

NASA Earth Observatory (2010) Is current warming natural? Earthobservatory.nasa.gov. Retrieved from: https://earthobservatory.nasa.gov/features/GlobalWarming/page4.php

Nemry F, Leduc G, Mongelli I, Uihlein A (2008) Environmental improvement of passenger cars. European Commission. Retrieved from: http://ec.europa.eu/environment/ipp/pdf/jrc_report.pdf

Passenger Car Sales in Europe (2019) Theglobaleconomy.com. Retrieved from: https://www.theglobaleconomy.com/rankings/passenger_cars_sales/Europe/

Patomäki H (2001) Democratising globalisation: the leverage of the tobin tax. Zed Books, London

Purchasing Power Adjusted GDP Per Capita (2019) Eurostat. Retrieved from: https://ec.europa.eu/eurostat/tgm/table.do?tab=table&plugin=1&language=en&pcode=sdg_10_10

Schroeder P, Anggraeni K, Weber U (2018) The relevance of circular economy practices to the sustainable development goals. J Ind Ecol 23(1):77–95. https://doi.org/10.1111/jiec.12732

Scott M, Lindsey R (2016) Which emits more carbon dioxide: volcanoes or human activities? Climate.gov. Retrieved from: https://www.climate.gov/news-features/climate-qa/which-emits-more-carbon-dioxide-volcanoes-or-human-activities

Union of Concerned Scientists (2019) Global warming impacts: the consequences of climate change are already here. Ucsusa.org. Retrieved from https://www.ucsusa.org/our-work/global-warming/science-and-impacts/global-warming-impacts#.W2lga9UzbDd

United Nations Framework Convention on Climate Change (2019) Paris Agreement—status of ratification. Retrieved from: https://unfccc.int/process/the-paris-agreement/status-of-ratification

United Nations Framework Convention on Climate Change (2019) What is the Paris Agreement? Retrieved from: https://unfccc.int/process-and-meetings/the-paris-agreement/what-is-the-paris-agreement

Wijkman A, Skånberg K (2015) The circular economy and benefits for society. Club of Rome, Zurich

Worldometers (2017) Current world population. Retrieved from: http://www.worldometers.info/world-population/

Zvirgzdins J, Plotka K, Geipele S (2018) Eco-economics in cities and rural areas. Baltic J Real Estate Econ Constr Manag 6(1):88–99. https://doi.org/10.2478/bjreecm-2018-0007

Public Energy Preferences from the Perspective of Climate Change Mitigation

Agne Budzyte

Abstract Laws and plans for environmental protection and mitigation of climate change is a tough challenge for governmental organizations at the highest levels. A need to ensure a more sustainable future requires strict actions backed by constant support from the public, and yet there is always a risk that the public will not accept any new strategies, if they do not reflect their personal concerns. This paper presents an investigation into how climate change concerns affect the public's energy preferences. It also reports energy security concerns and the political preferences related to climate change, and how this has to be considered in increasing the support for new political tools in the energy sector. The analysis is based on the data gathered from a representative public survey conducted during October–November 2018 in Lithuania and reveal how concerns about climate change affect the public's preference for energy generated from renewable sources also lessening the preference for fossil fuels and other non-sustainable options.

Introduction

Background

The Global Energy Trends report shows that a +3.7% economy growth led to a +2% increase in energy consumption, and a +2% increase in CO2 emissions in 2017 (Enerdata 2018). Most energy consumed in Europe was produced using non-renewable energy sources, even though an increase in its usage was also noted (Eurostat 2018a, b). The stable growth in the demand for energy has a higher impact on the environment, however, and the European Environment Agency (EEA) notes that "*Rising energy consumption slows EU progress on renewables and energy efficiency targets*" (European Environment Agency 2018). This kind of slow-down in the development of the renewable energy sector puts energy efficiency and de-carbonisation targets at

A. Budzyte (✉)
Civil Society and Sustainability Research Group, Faculty of Social Sciences, Arts and Humanities, Kaunas University of Technology, A. Mickeviciaus str. 37, 44244 Kaunas, Lithuania
e-mail: a.budzyte@ktu.lt

W. Leal Filho et al. (eds.), *Climate Change, Hazards and Adaptation Options*, Climate Change Management, https://doi.org/10.1007/978-3-030-37425-9_44

risk, as the demand for energy is fulfilled by non-renewable energy sources. Climate change mitigation strategies pose new challenges for policy makers, and aims such as keeping the Earth's temperature increase below 2 °C, reducing the air and water pollution caused by human activity, increasing the requirements for environmental protection, and reducing particulate emissions, etc., become difficult to implement.

It is noted that international organizations such as the European Union (EU) have a long standing history of the political initiatives to reduce climate change and its impacts by supporting robust policy-making in the EU member states. The cost-effective manner of reaching the targets of the Kyoto treaty and the Paris Agreement meant that the development of renewable energy sources still has a long way to go. There has been a reduction of greenhouse gases due to EU emissions trading systems (ETS) and the price set on carbon policy works, but further steps need to be taken to make industry and the power sector invest in innovations, and make the low-carbon transition happen. In a period of transition towards more sustainable energy and the implementation of climate change mitigation strategies, there has also been an increase in public awareness on global issues (Vasi 2018). The need to reconsider energy policies against the reality of climate change increases doubts about global reliance on current energy sources (Galloway McLean et al. 2012). Indigenous knowledge, while being aware of such energy and environment related threats occupies an important place in the minds of the public. Lowan-Trudeau (2017) noted that indigenous communities are attempting to position themselves as an influential force so as to engage in the increased development of renewable energy sources. This knowledge, accumulated according to cultural background, economic and political factors, forces communities to demand more environmentally responsible solutions. At this point indigenous knowledge and public preferences in the energy sector could become the most valuable roadmap for decision makers. Extreme weather events, natural disasters and other consequences, that are being faced in current times, generate increased concerns about the need to mitigate the effects of climate change and minimize the impact of energy use on the local environment. Within these circumstances there is grown public demand for more applicable governmental initiatives and individual actions (Tvinnereim et al. 2017).

Social movements, like the one inspired by Greta Thunberg in November 2018, began massive strikes to demand strict action to prevent further global warming and climate change. This movement, named "*Fridays for Future*" or "*Youth for Climate*", shows that the public (especially the younger generations) is ready to fight climate change and possibly provide new kind of politics regarding environmental issues and the future use of energy. This movement crossed the boundaries of a single state and became global, which could lead to greater pressure on policy makers to create climate change mitigation strategies. This could lead to the increased acceptability of renewable energy sources, but actions towards the de-carbonisation of energy and climate change mitigation also increases concerns about the so-called energy trilemma (World Energy Council 2013). This trilemma involves concerns about the reliability, security and affordability of energy, which are equally important for both policy makers and public.

This paper is based on the Public Perceptions of Climate Change: Lithuanian case in a European Comparative Perspective research project, funded by a grant (No. S-MIP-17-126) from the Research Council of Lithuania and contributes to the previously conducted analysis and literature on energy preferences, by presenting an analysis of the public's energy preferences and main expectations for energy production. It analyses the most preferable energy sources (renewables and non-renewables), risk perception and the priorities in solving energy production issues (reliability, security and affordability). These concepts were investigated through the public's perception of climate change. Overall, this research aims to determine how concerns about climate change affect the public's preferences for energy production and the risk perception of energy technologies. This analysis could help policy makers to better understand how climate change concerns connect with the public's preferences for energy security and production, and what decisions will be accepted by the public.

Theoretical Framework

The rapid growth of industry and the whole energy infrastructure in the twentieth century led to the failure of environmental risk management (Janmaimool and Watanabe 2014). Contemporary society is facing the side effects of uncontrollable development in the form of global change, especially in the environmental sphere. More and more people around the world are being affected by climate change and the catastrophic events caused by it. Policy makers and other sectors are thus trying to mitigate the effects by implementing specific tools to reduce the impact (mitigation programmes include various initiatives, from agriculture, forests, manufacturing, waste, cities and transport, to power). Despite the increased threats, however, most of the population still lacks the empathy required to understand the need for these mitigation strategies. It is important to understand what kinds of an impact do concerns about change have on such initiatives in order to find the best practices. This research focuses on the climate change impact on public energy preferences, perceived risks and energy related issues.

The theoretical framework for further analysis is presented below (see Fig. 1). The model combines five main concepts, all of which are correlated. These concepts are going to be tested in further research using collected data and evaluating the validity of such a hypothetical model for the research.

The model accepts climate change concerns as the baseline for the investigation of energy technologies, risk perception and public preferences, by assuming the ability to perceive environmental risks (climate change in this particular case) as dependent on the fact on what kind of impact they have on environmental problems (Anilan 2014). Greater sensitivity about environmental issues should decrease the level of comfort with the risk in the energy sector and increase the expectations of risk

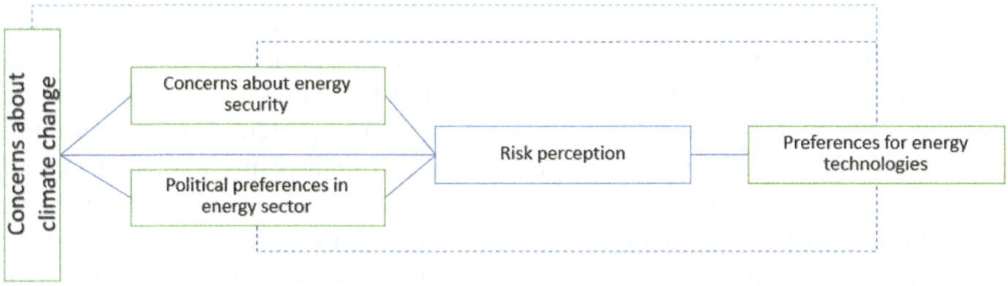

Fig. 1 Hypothetical model of the research

perception, while also increasing the preference for more environmentally friendly energy sources (Paço and Lavrador 2017).

Renn, however, noted that the risk perception of energy technologies is also linked to attitudes towards energy related issues (in this case, the energy trilemma of affordability, vulnerability, accessibility). It implies that cultural preferences, the judgments of peer groups and many individual factors influence the perception and evaluation of technologies within the public (Renn 1994). As the energy sector cannot be dissociated from energy security related issues, the concept of this research was enriched by factoring in the aforementioned issues. Previous analysis of risk perceptions showed different points of view between the experts and public (Sjöberg 1999). A way to explain such a huge variation is that the public conceptualize risk from a broader perspective, and the experts miss the importance of risk for local communities (Chowdhury and Haque 2011). As most decisions are being made by policy makers, especially when talking about the energy sector and mitigation of climate change interaction, the public's support is crucial. The lack of adequate responses to policy can be traced to perceptions of low risk, demonstrating that risk perception is related to policy support (Mayer et al. 2017). Schneider and Ingram (1990, pp. 513–515) identified three different types of political tools: Incentive Tools—"*assume individuals will not be positively motivated to take policy-relevant action unless they are influenced, encouraged, or coerced by manipulation of money, liberty, life, or other tangible payoff*"; Authority Tools—"*assume agents and targets are responsive to the organizational structure of leader-follower relationships and that lower level agents usually will do as they are told*"; and Capacity Tools—"*assume that the target groups will have sufficient incentive or motivation to participate in the activity, or change their behavior, if they are properly informed and have the necessary resources*". Political preferences are thus considered in this research to identify the political tools used and how their support or neglect affects risk perception.

Risk perception also refers to people's subjective judgments about the likelihood of negative effects (Paek and Hove 2017). In this case, energy sectors risks could be caused by intentional human activity (sabotage or terrorism), extreme weather events, technical failures of the system, and other dangers (Holmgren and Thedéen 2010). It could also be assumed that public opinion on various energy technologies partially mediates the effect of perceived usefulness (Kim et al. 2009), multiple cultural and contextual factors, and perceived benefits and risks. This is how general knowledge and opinion about energy sources becomes a crucial part of risk perception and technology preferences. Public opinion, as represented by opinion polls, may also increase politicians' knowledge of citizen preferences (Burstein 2003), and the knowledge of public opinion could increase the accountability of voters by also shaping the future development of the national energy sector.

The benefits of modern technologies always involve new kinds of risk, which can cause severe difficulties for society (Fischhoff et al. 2004). Nevertheless, renewable energy is expensive although the benefits are obvious, and society could evaluate the cost-benefit balance for future generations and choose the most secure option to guarantee the elimination of risks. That means energy use and climate change mitigation strategies interconnect and influence each other in particular ways. The majority of the public has the opportunity to increase concerns at the policy maker level, as modern communication infrastructure allows people to process information faster than ever before, and also provides an opportunity to maintain a connection between public and policy makers (Burstein 2003). This study aims to identify the main factors which shape the general public's position regarding energy preferences and risk perception from a broader perspective—climate change. It is necessary to ensure public support for all mitigation strategies and the investments in new kinds of energy sources, and so a clear evaluation of energy related concerns, political preferences' regarding the energy sector and general public opinion needs to be assessed and analysed as mediators for risk perception and energy source preferences.

Description of the Area Context and Methodology

Context of Study Area

This study was conducted in Lithuania, where the energy sector and contextual factors have interesting shapes. Lithuania has not used nuclear energy as a country since 2009, as the Ignalina nuclear power plant was shut down and the decommissioning process is still in progress (Genys and Krikštolaitis 2017). In 2012 Lithuania held a referendum, and 62% of the public said "No" to the construction of a new nuclear power plant in the country, which stopped the further development of this project.

To more clearly understand the context of Lithuania's energy sector it is necessary to note that gross inland energy consumption in Lithuania is mostly based on solid fuels, petroleum products, renewable energy and electricity imports (Eurostat 2018a, b). Importing electricity is a major issue, which leads to dependency on other countries and puts energy security at risk. Energy security in Lithuania is similar to that in other countries, however, because the main challenges are the same. The Lithuanian Energy Security Annual Review 2013–2014 stated that attention should be paid to the country's ability to ensure a stable supply of energy, provide acceptable energy prices and ensure that national interests are being targeted (Augutis et al. 2015).

In the 2010s, Lithuania invested in the development of alternative energy sources [i.e. Liquid Natural Gas Terminal in Klaipėda (LNG), which provides slightly more than 20% of the required energy in Lithuania's energy market (Tučkus 2015)] to prevent total dependency on energy imports. Investment in the LNG terminal only temporarily ensured stable gas prices in the local market and these began to increase at the beginning of 2019, which could be the consequence of a decrease in gas consumption. The decrease in consumption could be a good sign for climate change mitigation strategies, and so other issues, such as the affordability of energy, may now be debated. The prices of energy is a sensitive topic among the public in Lithuania, even though the price of electricity for household consumers was 0.1097 EUR per kWh, which was almost half the average for the EU-28—0.2049 EUR per kWh (Eurostat 2018a, b).

As the project for a local nuclear power plant was terminated, and the liquid gas terminal cannot sufficiently supply all the required energy, the government of Lithuania issued a new National Energy Independence Strategy (the Strategy) on June, 2018. The Strategy aims to ensure the further development of the energy sector, increasing the efficiency of energy consumption, and providing innovative technologies, and it aims to pursue environmentally friendly energy in future development. To do so "*it is planned that by 2030, electricity generated in Lithuania will make up 70 percent of total final consumption, and that all electricity will be produced in the country by 2050*" (Ministry of Energy of the Republic of Lithuania 2018). As Lithuania lack the local resources to produce energy, most of the energy should be produced by renewable energy sources, which also helps to meet EU and global climate targets. However, the local infrastructure for renewable energy grids are still under development, which means that the targets to ensure that all electricity and heat in Lithuania is produced from renewable and other clean sources by 2050 is ambitious. Such targets for the future development strategy will be difficult to achieve without the approval of both the public and private sectors (Bosnjakovic and Haber 2015).

To increase the chances of public approval, the public needs to be concerned about environmental issues in order to approve national policies such as the Strategy. European Social Survey 8th Wave (2018) revealed that only 40% of all respondents in Lithuania believed that the world's climate is changing. The Special Eurobarometer on Climate Change 459 (2017) also showed that people in Baltic countries (including Lithuania) do not perceive the seriousness of the issue of climate change, and rank such issues as less important than international terrorism, poverty or armed conflict. Concern about this issue is surprisingly low, especially as a comprehensive study conducted by GreenMatch (2019) showed that Lithuania is the country most affected by climate change in the whole of Europe. A lack of concern or recognition of the problem could make it very difficult to implement the Strategy. This study was begun after the official presentation of the Strategy, and so it can be assessed whether the attitudes of the public and the government, regarding development of the energy sector in a more environmentally friendly manner, collide, and whether concerns about climate change plays an important role.

Methods

Data Sampling Methods

Data for the analysis of public energy preferences was gathered during October and November 2018 in Lithuania, using face to face interviews and a semi-structured questionnaire. The main respondents in this quantitative survey were members of Lithuanian households aged 18 and over. The data sampling excluded homeless people, prisoners, hospitalized people and people who officially declared that they were going to leave Lithuania. Selection was based on data from the national Centre of Registers, and stratified proportional selection was used. This data sampling method ensures that the optimum error of results does not exceed more than $\pm 3.1\%$, and makes the survey representative.

Data Sources and Methodology

Quantitative data was gathered from 1029 Lithuanian residents (18 years of age and older) who were interviewed face to face during the study. The whole survey combined questions on attitudes to and beliefs about climate change, with questions on energy related issues and public preferences. Most of the questions used Likert scale items that measured the respondent's opinions on various climate change and energy risk perceptions and related concepts. The questionnaire was adapted (in collaboration with Cardiff University) from The European Perceptions of Climate Change Project (EPCC) survey conducted in four European countries in 2016 (Steentjes et al. 2017), by including more specific research questions for Lithuania. The data collection was conducted within the scope of the *Public Perceptions of Climate Change: Lithuanian Case from a European Comparative Perspective* research project, funded by the Research Council of Lithuania. The variables in this analysis were determined using the following questions (EPCC questions marked 'Q' with original numbering in questionnaire) (see Table 1).

These questions revealed the key trends in the public's preferences and their main concerns, but a new dependent variable was derived to more clearly analyse the trends in the way that concerns about climate change affect risk perception. As there was a huge diversity of energy sources, the overall risk perception index was used for a further overview of its effect on energy preferences. The Cronbach's alpha (0.744) of 10 items (containing all energy technologies used in this survey), with internal consistency in the range from $0.7 \leq \alpha < 0.8$, shows that the risk perception of energy technologies is acceptable for use in further research.

The same method was applied to determine energy related concerns; creating a new variable which was internally consistent. The Cronbach's alpha (0.904) of 5 items (containing all energy production related concerns: environment and climate

Table 1 Questions and variables used for investigation of hypothetical model

Climate change concerns	Q3**. How worried are you, if at all, about climate change?	Not at all worried; Not very worried; Fairly worried; Very worried; Extremely worried; Don't know
Concerns on energy security	Q29**. To what extent do you agree or disagree with the following statements? Energy for Lithuania should be produced in a way that… • Means the environment and climate are being protected • Provides a stable supply of energy • Does not impact negatively on economic development • Means energy is affordable • Means we do not depend on energy imports from other countries	1. Strongly disagree … 5. Strongly agree
Political preferences in energy sector	Q19**. Various policies might be used to reduce climate change or deal with its effects. To what extent do you support or oppose the following policies in Lithuania? • Increasing taxes on any use of fossil fuels (such as coal, oil, diesel, petrol, gas) • Including nuclear power in the energy mix • Using public money to subsidize renewable energy such as wind and solar power • Increasing the price of electricity to reduce our consumption • Using public money to subsidize insulation of homes • A law banning the sale of household appliances that are not energy efficient • Spending public money now to prepare the country for the impacts of climate change (e.g. building flood defences)	1. Strongly oppose … 5. Strongly support

(continued)

Table 1 (continued)

Risk perception	What level of threat, if any, do these energy sources pose? Biomass (wood, plant or animal waste); Coal; Natural gas; Hydropower; Nuclear energy; Oil; Solar energy; Wind energy; Shale gas; Waste incineration	1. Absolutely no threat … 5. Poses a very serious threat
Preferences for energy technologies	Which 3 of the following energy sources do you believe must be used the most to produce energy in Lithuania?	Biomass (wood, plant or animal waste); Coal; Natural gas; Hydropower; Nuclear energy; Oil; Solar energy; Wind energy; Shale gas; Waste incineration*
	Q28**. What is your general opinion about the following methods of energy generation for Lithuania? Please indicate how positive or negative your opinion is?*	1. Very negative … 5. Very positive

*Waste incineration used as an additional technology for Lithuanian sample
** Original EPCC questionnaire numbering

protection, stability, affordability, independency, impact on economy), being in the range from $0.9 \leq \alpha$, allows a newly created variable to be used for further research.

Additional variables (preferences for renewable energy; preferences for fossil fuel; preferences for other options) and (capacity political tools; incentive political tools) were created after the implementation of factor analysis.

Data Analysis

The study used quantitative data analysis methods, including descriptive and inferential practices. The analysis aimed to evaluate the most preferable energy sources, the highest perceived risk from energy sources, and the impact of climate change concerns. Factor and correlation analysis was used to identify the general trends in energy preferences, risk perception, energy production issues in Lithuania and climate change. The results of the data analysis provided insights and conclusions which could help government institutions to prevent public rejection of mitigation tools in the energy sector. The following research questions were empirically analysed:

1. How are climate change concerns related to energy security concerns, political preferences in the energy sector, risk perception and public energy preferences?
2. What kind of political preferences in the energy sector most strongly shape the preference for energy sources?

Results and Discussion

Energy Preferences in Lithuania

The aim of the study was to identify the most preferable energy sources for producing electricity in Lithuania. The national government has presented the National Energy Independence Strategy (2018), which aims not only to ensure energy security in Lithuania, but to reduce greenhouse gases and other particulate emissions, by providing more sustainable energy production methods. This kind of strategy for future development creates debate about the main public concerns about energy, and the reliability, affordability and security of energy, due to decreasing dependency rate from other countries. Huge investments are needed to provide a stable energy supply, but on the other hand, most of the investments are likely to be divided between the users of energy, so a price increase could be expected.

The real difficulty in avoiding the rejection of most climate change mitigation strategies for energy involves not a lack of political initiative, or a lack of awareness, but the general concerns in society regarding local environmental damage and energy prices (local rather than global issues). It was noted, that for a new strategy to be approved and successful, it is important to find out what the public finds acceptable, and the most popular solutions (Ansolabehere and Konisky 2014). Ansolabehere and Konisky (2014) also noted that it is important to adapt public policy according to the costs of what the public will accept today, and what will still be wanted tomorrow, so the public's preferences and concerns should be clearly appraised. Objections due to an individual attitude-behaviour 'gap' means that people who are generally in favour of new kinds of power production methods (i.e. wind or sun), change their minds when they are confronted with a real proposal (Wolsink 2012).

The respondents were asked to name the three most preferred energy sources for producing energy in Lithuania. The results of the survey about energy preferences for different energy sources are summarized in Table 2. The data shows differences in the choice.

It is notable that the first choice of energy source was solar power (60.52%) which has been ranked as the least effective source of renewable energy (Veazey 2017) because toxic compounds are produced in the manufacturing process and solar panels only retain 200% of the original input. This kind of energy is cleaner than any other non-renewable source, but the least efficient renewable source. Such preferences mean that pro-environmental attitudes are predominant in the consciousness of Lithuanian's, but their lack of knowledge about energy efficiency should be considered as well.

The second most preferable option selected by the public—was wind power (59.00%). The technology for wind turbines demands only steel, copper and concrete, and although the processing of these materials has a high production energy cost, it later provides more than 1000% of its original energy input (Veazey 2017). Ranking such an efficient technology second does not change the fact that renewable energy sources are most preferred. These results also connect to previous Eurobarometer

Table 2 Selection of three most favourable energy sources[a]

Energy source	% of respondents selected this source
Sun/solar power	60.52
Wind power	59.00
Natural gas	32.87
Hydroelectric power	32.74
Biomass	30.52
Waste incineration	9.25
Oil	7.32
Nuclear power	6.44
Hydraulic fracturing	6.27
Coal	4.99

[a]Respondents were allowed to choose up to three possible answers

research, where it was found that Europeans were widely supportive of expanding the use of both solar and wind energy (2006), and the ESS (2018) survey where 35% of Lithuanians stated that a very large amount of energy should be generated from wind, and 32% of the respondents specified that a large amount should be produced from wind; this was also the preferred energy source in all the countries surveyed by ESS.

The most interesting part was the selection of natural gas, which was quite frequent in the selection of energy sources (32.87%). This can be "*mainly attributed to the ease of use, established usage habits, and the perception that natural gas is a "clean energy source*"" (Ediger et al. 2018). There was also a difference between natural gas selection in different areas: respondents living in cities gave this kind of energy as one of the main three more frequently than those living in small towns. The reason behind such selection may be based on the different infrastructure between large cities and smaller regions, and the differences in use among households. Respondents living in cities are more likely to be using natural gas in their households, rather than people living in rural areas, so the better-known technology may have an advantage over the others as regards increased perceived usefulness and ease of use.

Hydro energy was selected by 32.74% of respondents, who said that most energy should be generated from this source. Another option, which was frequently mentioned, was biomass sources, however this source had a lower percentage (30.52%) than the other most preferred energy sources. Such a selection also shows the progress towards a modern mix of energy options in Lithuania, after years of public rejection. The debates on biomass as an environmentally friendly option for energy production are real, but more and more scientists and more of the public agrees that biomass helps reduce the amount of GHG, which leads most to believe that it is a renewable energy source which cannot be depleted as long as there are plants on this planet. The location of power plants using biomass in local neighbourhoods was not widely approved by the local public.

The survey results also revealed that only 5.9% of all respondents in Lithuania are extremely worried, and (~25%) very worried about the climate change: the most common answers were fairly worried (48.7%), not very worried (~13%), and not worried at all (~6%). The lack of concern about climate change may be the reason that the least efficient renewable energy source was most popular, and natural gas was ranked in the top three.

Main Energy Related Concerns and Risk Perception of Energy Technologies

After assessing the main trends in energy preferences among the public of Lithuania it is necessary to overview the main concerns. The question asked to understand the main energy security concerns was about the way in which energy should be produced (see Table 3).

The main concern of respondents regarding energy production is its affordability. Most (over 50%) strongly agreed with the statement that energy for Lithuania should

Table 3 Public concerns on energy security in the Lithuanian population according to EPCC Q29

	Strongly disagree	Tend to disagree	Neither disagree nor agree	Tend to agree	Strongly agree
…means energy is affordable	0.9	1.7	7.1	37.6	52.6
…provides a stable supply of energy	0.4	3	6.6	43.3	46.6
…means the environment and climate are being protected	0.1	2.1	8.7	47.4	41.7
…does not impact negatively on economic development	1.1	4	11.6	42.9	40.3
…means we do not depend on energy import from other countries	0.5	5.4	16.2	41.3	36.7

be produced in a way that means energy is affordable. Over 35% of all respondents tend to agree with this statement, making affordability the most important concern for the public in Lithuania. Most households can thus be called "fuel poor", because the cost of all of the energy services requires more than ten percent of their disposable finances (Heindl and Schüssler 2015). A rapid increase of expenditure on energy could thus lead to increased fuel poverty.

The second highest concern was about the stable supply of energy. The International Risk Governance Council (IRGC) notes that it is impossible to see electrical power as just infrastructure, because it also meets the criteria and definition of being critical infrastructure. This is based on the idea that large-scale systems, which are built by humans and supply continual services that are essential to the functioning of society, are the same as critical infrastructures, which makes them the subject and source of numerous threats (Matthewman and Byrd 2014). A stable energy supply is thus one of the main expectations, not only of public policy makers, but of the general public as well.

The third main concerns among the public were protection of the environment and the impact of climate change. These results are a bit different than the results of the Standard Eurobarometer 90 published on 2018 results (European Commision 2018), which showed that various issues concerning the environment, climate and energy sectors were only mentioned by 14% of EU citizens, and were not the highest concerns among the European public. This difference may be due to increased media responses to climate change actions, regarding what the policy makers and the general public need to do to reduce the impact caused by polluting the environment, and to mitigate climate change.

Exploratory factor analysis was used to develop an understanding of the individual subjective classification of political energy preferences. It helped to identify two different attitudes towards political tools. Multiple political tools associated with the environment and climate change are presented in Table 4. The results of the factor analysis indicate that there are some trends in grouping political preferences according to the impact on the public.

The factor analysis revealed two different groupings in political preferences in the energy sector. Factor 1: "*Capacity political tools*" involves support for increased public money spending to upgrade the infrastructure for renewable energy, the better insulation of homes, which should increase energy efficiency, the ban of inefficient household appliances and preparation for further impacts. Capacity tools aim to affect the adoption of innovative programs—in this case related to mitigation on climate change impacts through inefficient energy use. Factor 2: "*Incentive political tools*" removes the opportunity for the public to avoid mitigation tools, by increasing taxes, reducing consumption and including nuclear energy. The public would all be forced to adopt new behaviour models for choosing eco-friendlier alternatives and a more efficient use of energy, by also reducing their impact to the environment.

Having identified the general public's concerns about energy related issues and the types of political preferences, the mean scores of risks on energy sources are presented below in Table 5. It is notable that the main dangers are associated with nuclear power, which despite its low probability, may cause severe damage. It is

Table 4 Classification of political preferences in the energy sector in the Lithuanian population according to EPCC Q19[a]

Variables	Loadings		Communalities
	Factor 1	Factor 2	
Increasing taxes on any use of fossil fuels (such as coal, oil, diesel, petrol, gas)		0.769	0.611
Including nuclear power in the energy mix		0.552	0.338
Using public money to subsidize renewable energy such as wind and solar power	0.781		0.615
Increasing the price of electricity to reduce our consumption		0.772	0.601
Using public money to subsidize insulation of homes	0.753		0.572
A law banning the sale of household appliances that are not energy efficient	0.691		0.550
Spending public money now to prepare the country for the impacts of climate change (e.g. building flood defences)	0.721		0.522
Explained variance	31.90%	22.51%	
Total variance	54.41%		

[a]KMO score is 0.708, Bartlett's test: sig. = 0.000, thus factor analysis is applicable to the given data; Rotation converged in 3 iterations

Table 5 Mean averages of risk perception from energy technologies

Energy technologies	Mean
Nuclear power	4.11
Oil	3.67
Coal	3.61
Waste incineration	3.45
Hydraulic fracturing, otherwise known as fracking	3.26
Natural gas	2.90
Biomass, that is wood, plant and animal waste	2.67
Hydroelectric power	2.50
Sun/solar power	1.95
Wind power	1.93

Lithuanian sample data. N = 1029; Possible answers were: 1. Absolutely no threat ⟨…⟩ 5. Poses a very serious threat

important to note that the mean averages of the top three most favoured energy sources differ. The perceived risks of natural gas are higher compared to renewable energy sources; however, natural gas is less likely to be associated as a potentially threatful option in the energy mix. The derivative variable was created by using these perceptions of risk from energy technologies.

Factor analysis was once again implemented to determine whether it is possible to group energy source preferences among the public. The results presented in Table 6 show that extracting three different groups of energy sources is clearly possible: Fact 1-*renewable energy*; Fact 2-*fossil fuels* and Fact 3-*other options* which cannot be attached to any other groups.

After identifying all possible groupings among the variables, the correlations were produced to identify significant bonds between variables (see Fig. 2). It is notable, that the concerns about climate change, had a significant correlation with energy concerns (0.162) and support for the implementation of both capacity (0.310) and incentive (0.193) political tools. This means that the higher the level of concern about climate change, the higher the support for clear political action. The level of concern about climate change also significantly affected the preferences for renewable energy sources (0.226), which shows that increased concerns about climate change, leads to the public preferring more environmentally friendly solutions by also decreasing the preference for fossil fuels (−0.119). These results suggest that the there is a clear connection between the levels of concern about climate change and the preferred energy sources, however, it is important to note that the concerns on climate change do not affect the risk perception.

Public concern about energy security also has a positive effect on support for capacity political tools (0.403) and preferences for renewable energy sources (0.437), which shows that the energy trilemma and environment protection could be combined for the more productive development of renewable-energy sources by also reducing dependence on other countries and the impact on the environment. More interesting

Table 6 Classification of energy source preferences in Lithuanian population by EPCC Q28[a]

Variables	Loadings			Communalities
	Factor 1	Factor 2	Factor 3	
Biomass, that is wood, plant and animal waste	0.413			0.411
Coal		0.796		0.640
Natural gas		0.621		0.518
Hydroelectric power	0.546			0.500
Nuclear power			0.665	0.542
Oil		0.667		0.612
Sun/solar power	0.927			0.863
Wind power	0.926			0.863
Hydraulic fracturing		0.619		0.386
Waste incineration			0.755	0.586
Explained variance (%)	22.88	20.53	15.79	
Total variance (%)	59.20			

[a]KMO score is 0.688, Bartlett's test: sig. = 0.000, thus factor analysis is applicable to given data; Rotation converged in 4 iterations

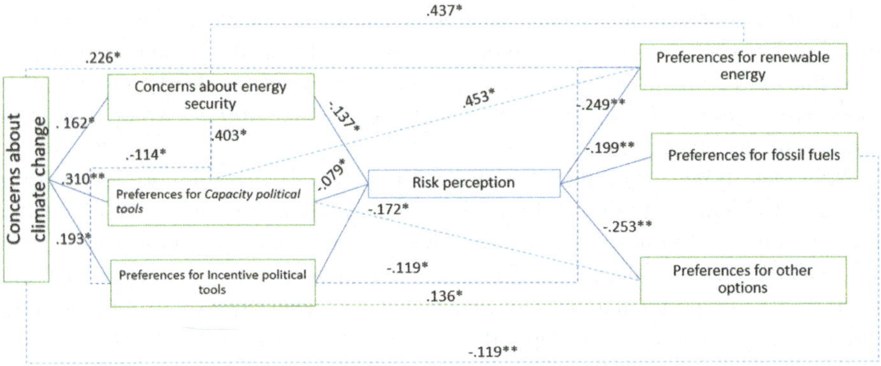

Fig. 2 Correlations of variables (only significant correlations are marked). *Significance level of $\alpha = 0.05$. **Significance level of $\alpha = 0.01$

trends have also been revealed, however: it was noted that members of the population who are more likely to prefer incentive political tools (increased taxes and prices etc.) are less likely to prefer renewable energy sources, while the respondents who preferred capacity tools (subsidize insulation of homes, ban of energy inefficient appliances etc.) were more likely to do so. Such results show that climate change mitigation plans in the energy sector must involve a variety of capacity tools for the public, in order to increase their acceptance of, and preference for, renewable energy sources. As such preferences are subjective and based on personal beliefs, the need to increase personal benefits could also shape preferences and support.

The risk perception part of the hypothetical model of the research contains some more explanatory findings. It was noted that higher support for capacity political tools significantly reduces the risk perception of energy technologies. Support for investments of public money may possibly increase trust in governmental institutions as well as increase the support for future developments of climate change mitigation strategies in the energy sector, which will lead to a decrease in the publics risk perception. On the other hand, it was also noted, that lower risk perception creates some side effects in the preferences of energy sources. In this case, low levels of risk perceptions among the public increase the acceptance of all kinds of energy sources, including the ones perceived as the most threatful (i.e. nuclear energy and coal). So, the responsible authorities need to ensure that the public would be well informed about the possible negative outcomes of each technology, so that it would be possible to reach the required support for future development of renewable energy sources.

Before beginning any climate change mitigation strategies, policy makers must evaluate the need to provide all the information for the public in order to increase knowledge about specific concerns and provide a useful mechanism with which to decrease the possible rejection of new strategies. As energy security related concerns are mostly based on energy affordability issues, capacity tools, which minimize individual impact, must be used prior to tax increases and other forced actions. Bearing in mind that nuclear power is the riskiest option in the energy mix, this needs to be left aside until all the other possible alternatives are being used, because the highly perceived risk increases the threat of public rejection of such development plans. The new energy strategy in Lithuania, which aims to provide more renewable energy sources in the energy mix, so far meets the public's expectations, but mechanisms to maintain this support should be used to ensure that these solutions will still be wanted in the future.

It is important to note that this study contains some context-specific results that apply to Lithuania only, and cannot be generalized to other countries until further comparison with EPCC survey data can be implemented. Cultural and contextual biases regarding climate change and energy preferences among the public in Lithuania could also be influenced by a lack of public awareness about global issues. The public may also be focused on the geopolitical situation and local challenges, which may also be a limitation of this research. The risk perception of energy sources, and its unexpected effects on people's preferences and concerns, require more clear evaluation in future research to find more clearer model of trends, as the current results suggest that factors shaping risk perception are unpredictable.

Conclusions

Consensus between the public and policy makers in the time of transition towards more sustainable energy sources is a process combining multiple concerns. The need to provide stable and affordable energy supply requires well balanced governmental initiatives and the clear evaluation of local community concerns. This research aimed to assess how climate change concerns connect with the public's preferences on energy security and production, and what decisions could be willingly accepted by the public. It was noted that political preferences are affected by climate change concerns and have a serious impact on energy preferences. Different kinds of political actions could create different responses among the public. The capacity political tools that help maximize personal benefits increase the preference for renewable energy sources, while incentive tools reduce the preference for renewables and increase preferences for such alternatives as nuclear power or waste incineration.

The general level of climate change concerns are shaping public energy preferences, but the mediation effect via political preferences and concerns on energy security increases the selection of more sustainable energy solutions, so policy makers must clearly evaluate the need to increase general knowledge on energy and climate issues among the public. Providing additional information on the advantages and disadvantages of all possible solutions can create a clearer understanding for local communities, and prevent the rejection of new strategies, which aim to mitigate the side effects of human activity.

References

Anilan B (2014) A study of the environmental risk perceptions and environmental awareness levels of high school students. Asia-Pac Forum Sci Learn Teach 15(2):1–22

Ansolabehere S, Konisky D (2014) Cheap and clean: how Americans think about energy in the age of global warming. MIT Press, London

Augutis J, Krikštolaitis R, Leonavičius V, Pečiulytė S, Genys D, Česnakas G, Martišauskas L, Juozaitis J (2015) Lithuanian energy security. Annual review 2013–2014. Energy Security Research Centre, Kaunas

Bosnjakovic B, Haber MI (2015) Climate changes and adaptation policies in the Baltic and the Adriatic regions. Univ Tour Manag Skopje J Econ 6(1):21–39

Burstein P (2003) The impact of public opinion on public policy: a review and an agenda. Polit Res Q 56(1):29–40. https://doi.org/10.2307/3219881

Chowdhury PD, Haque E (2011) Risk perception and knowledge gap between experts and the public: issues of flood hazards management in Canada. J Env Res Dev 5(4):1017–1022

Ediger V, Kirkil G, Çelebi E, Ucal M, Kentmen-Çin C (2018) Turkish public preferences for energy. Energy Policy 120:492–502. https://doi.org/10.1016/j.enpol.2018.05.043

Enerdata (2018) Global energy trends. Grenoble: enerdata. Retrieved 25 Jan 2019, from https://www.enerdata.net/system/files/publications/global-energy-trends-2018-edition-enerdata.pdf

ESS Round 8, European Social Survey (2018) ESS-8 2016 documentation report. Edition 2.1. European Social Survey Data Archive, Norwegian Centre for Research Data, Bergen. https://doi.org/10.21338/nsd-ess8-2016

European Commission (2017) The special Eurobarometer on climate change 459. Office for Official Publications of the European Communities, Luxembourg. https://doi.org/10.2834/92702

European Commission (2018) Standard Eurobarometer 90—Autumn 2018 "public opinion in the European Union, first results". Office for Official Publications of the European Communities, Luxembourg. https://doi.org/10.2775/104

European Environment Agency (2018) Rising energy consumption slows EU progress on renewables and energy efficiency targets. Retrieved 26 Jan 2019, from https://www.eea.europa.eu/highlights/rising-energy-consumption-slows-eu

Eurostat (2018a) Electricity prices for household consumers (taxes included), first half 2018 (EUR per kWh). Retrieved 1 Feb 2019, from https://ec.europa.eu/eurostat/statistics-explained/index.php/Electricity_price_statistics

Eurostat (2018b) Gross inland energy consumption by fuel, 2016. Retrieved 1 Feb 2019, from https://ec.europa.eu/eurostat/statistics-explained/index.php?title=File:Gross_inland_energy_consumption_by_fuel,_2016_(%25).png

Fischhoff B, Slovic P, Lichtenstein P, Read S, Combs B (2004) How safe is safe enough? A psychometric study of attitudes toward technological risk and benefits. In Slovic P (ed) The perception of risk. Risk, society and policy series. Earthscan, London, Sterling

Galloway McLean K, Ramos Castillo A, Barret B (2012) Energy innovation and traditional knowledge. Traditional knowledge bulletin: topical issues series

Genys D, Krikštolaitis R (2017) Switching political nuclear energy preferences, changing public attitude. Balt J Law Polit 10(2):225–243. https://doi.org/10.1515/bjlp-2017-0018

GreenMatch (2019) Mapped: impact of climate change on European countries. Retrieved 13 Apr 2019, from https://www.greenmatch.co.uk/blog/2019/04/climate-change-europe

Heindl P, Schüssler R (2015) Dynamic properties of energy affordability measures. German Socio-Economic Panel Study (SOEP), Berlin

Holmgren ÅJ, Thedéen T (2010) Risk analysis. In Holmgren Å, Grimvall G (eds) Risks in technological systems. Springer, London, pp 199–224. https://doi.org/10.1007/978-1-84882-641-0

Janmaimool P, Watanabe T (2014) Evaluating determinants of environmental risk perception for risk management in contaminated sites. Int J Env Res Publ Health 11(6):6291–6313. https://doi.org/10.3390/ijerph110606291

Kim JY, Chun U, Song J (2009) Investigating the role of attitude in technology acceptance from an attitude strength perspective. Int J Inf Manage 29(1):67–77. https://doi.org/10.1016/j.ijinfomgt.2008.01.011

Lowan-Trudeau G (2017) Indigenous environmental education: the case of renewable energy projects. Educ Stud 53(6):601–613. https://doi.org/10.1080/00131946.2017.1369084

Matthewman S, Byrd H (2014) Blackouts: a sociology of electrical power failure, vols 2014-01. Social space. http://hdl.handle.net/2292/22764

Mayer A, O'Connor TS, Chiricos T, Gertz M (2017) Environmental risk exposure, risk perception political ideology and support for climate policy. Soc Focus 50(4):309–328. https://doi.org/10.1080/00380237.2017.1312855

Ministry of Energy of the Republic of Lithuania (2018) Seimas approves progressive and innovative lithuanian energy strategy. Retrieved 28 Jan 2019, from https://enmin.lrv.lt/en/news/seimas-approves-progressive-and-innovative-lithuanian-energy-strategy

Ministry of energy of the republic of Lithuania (2018) National energy independence strategy. Ministry of Energy of the Republic of Lithuania, Vilnius. Retrieved 2 Mar 2019, from https://enmin.lrv.lt/uploads/enmin/documents/files/Nacionaline%20energetines%20nepriklausomybes%20strategija_2018_LT.pdf

Paço A, Lavrador T (2017) Environmental knowledge and attitudes and behaviours towards energy consumption. J Env Manag, 384–392. https://doi.org/10.1016/j.jenvman.2017.03.100

Paek HJ, Hove T (2017) Risk perceptions and risk characteristics. In: Oxford research encyclopedia of communication. Oxford University Press, USA. https://doi.org/10.1093/acrefore/9780190228613.013.283

Renn O (1994) Public acceptance of energy technologies. In: European strategy for energy research and technological development. Seminar proceedings, Fondazione Giorgio Cini, Venezia, November 18–20, 1993. Luxemburg: Office for official publications of the European Community. ISBN 92-825-9199-9 pp 77–93

Schneider A, Ingram H (1990) Behavioral assumptions of policy tools. J Polit 52(2):510–529. https://doi.org/10.2307/2131904

Sjöberg L (1999) Risk perception: experts and the public. Eur Psychol, 1–12. https://doi.org/10.1027//1016-9040.3.1.1

Steentjes K, Pidgeon N, Poortinga W, Corner A, Arnold A, Böhm G, Mays C, Poumadère M, Ruddat M, Scheer D, Sonnberger M, Tvinnereim E (2017) European perceptions of climate change: topline findings of a survey conducted in four European countries in 2016. Cardiff University, Cardiff

Tučkus D (2015) LNG impact on lithuanian natural gas market. LITGAS, Klaipėda

Tvinnereim E, Fløttum K, Gjerstadb Ø, Johannessonc MP, Nordø AD (2017) Citizens' preferences for tackling climate change. Quantitative and qualitative analyses of their freely formulated solutions. Glob Env Change, 34–41. https://doi.org/10.1016/j.gloenvcha.2017.06.005

Vasi IB (2018) Social movements and energy. In: Davidson D, Gross M (eds) The oxford handbook of energy and society. Oxford University Press, New York, pp 405–422

Veazey S (2017) What is the most efficient renewable energy source? Retrieved 10 Feb 2019, from Phoenix energy: http://www.phoenixenergygroup.com/blog/what-is-the-most-efficient-renewable-energy-source

Wolsink M (2012) Wind power basic challenge concerning social acceptance. In Meyers R (ed) Encyclopedia of sustainability science and technology, vol 17. Springer, New York, pp 12218–12254. https://doi.org/10.1007/978-1-4419-0851-3

World Energy Council (2013) World energy trilemma: time to get real—the case for sustainable. World Energy Council, London. Retrieved 1 Mar 2019, from http://www.worldenergy.org/wpcontent/uploads/2013/09/2013-Time-to-get-real-the-case-for-sustainable-energy-investment.pdf

Addressing Climate Change Through Education and Research in Maritime Energy Management: The Case of World Maritime University

Momoko Kitada, Alessandro Schönborn, Aykut I. Ölçer, Fabio Ballini and Dimitrios Dalaklis

Abstract Climate actions interrelating with the "Paris Agreement" have created a heavy pressure on the maritime industry to mitigate negative environmental impacts of shipping activities. Industry needs are rapidly shifting towards green and ecological operation of ships and ports; therefore, new types of education and research are in demand, forming a domain under the term "Maritime Energy Management" (MEM). This demand was first responded to by the World Maritime University (WMU) of the International Maritime Organization (IMO), a Specialized Agency of the United Nations (UN), in the form of establishing a series of postgraduate programmes focusing on MEM related issues. The analysis at hand presents a case study of WMU to showcase how universities can play a pivotal role in creating an interdisciplinary knowledge platform with specific focus on maritime energy. Various indicators to measure the impact are presented such as the evaluation of students' data from their enrollment to career progressions, conference-seminar organisation and participation, research and publications, and stakeholder engagement. Tangible outcomes achieved in a relatively short time by WMU will demonstrate the leading role of universities in responding to the pressing issue of climate change. Among other issues, highlighting the opportunities for universities to build capacity for sustainable development through education and research is clear stands out.

M. Kitada (✉) · A. Schönborn · A. I. Ölçer · F. Ballini · D. Dalaklis
World Maritime University, Fiskehamnsgatan 1, 21118 Malmö, Sweden
e-mail: mk@wmu.se

A. Schönborn
e-mail: as@wmu.se

A. I. Ölçer
e-mail: aio@wmu.se

F. Ballini
e-mail: fb@wmu.se

D. Dalaklis
e-mail: dd@wmu.se

W. Leal Filho et al. (eds.), *Climate Change, Hazards and Adaptation Options*,
Climate Change Management, https://doi.org/10.1007/978-3-030-37425-9_45

Introduction

International shipping plays an essential role in the facilitation of world trade because it is the most cost-effective and energy-efficient mode of mass cargo transport. In the contemporary business environment, sea-going vessels are rightly considered the most important means of transport; vessels engaged in maritime transport activities hold the so-called "comparative advantage" over all other modalities. It is not a coincidence that associated statistics indicate that about 90% of the total volume of global trade is borne (exclusively—or, at least partially) by sea (UNCTAD 2018).

The maritime industry has for a long time been economy-driven in all its activities. This nature is typically illustrated by the ways in which the maritime industry is affected by the global economic conditions inflated by oil price (Pain et al. 2008). Ironically, climate change, in the context of shipping, is not merely viewed as an environmental threat but also as a new business opportunity for Arctic shipping and maritime operations, where the melting of Arctic sea ice as a result of global warming and climate change is providing shorter sea routes, reduced shipping costs, and potentially longer navigation seasons (Mukherjee and Liu 2018). Such a macro-economic viewpoint brought by globalization, and later climate change, has appeared to be incompatible with the concept of "sustainability" in recent years. The development of the International Code for Ships Operating in Polar Waters (Polar Code), which entered into force on 1 January 2017, was one of the international responses from the maritime industry to adapt to climate change in terms of environmental and safety requirements (Brigham 2013).

Environmental concerns, especially, climate change, have been acknowledged as part of the global agenda. The United Nations (UN) 2030 Agenda for Sustainable Development was adopted at the UN Sustainable Development Summit on 25 September 2015 with 17 Goals, including Goal 13, "Take urgent action to combat climate change and its impacts". On 12 December of the same year, Parties to the UN Framework Convention on Climate Change (UNFCCC) adopted a landmark agreement, the so called "Paris Agreement", to combat climate change and to accelerate and intensify the actions and investments needed for a sustainable low carbon future. The IMO aligned itself with the Paris Agreement by adopting the: 'Initial IMO Strategy on Reduction of GHG Emissions from Ships' in resolution MEPC.304(72) on 13 April 2018. The aim of this resolution is to reduce greenhouse gas (GHG) emissions from international shipping by at least 50% by 2050, compared with the 2008 emission level. The maritime industry must contribute its fair share of GHG reduction in those extremely important activities that are already underway and aim to mitigate a risk that can threaten even humanity's own existence.

This significant policy change has made it crucial for the maritime industry to explore different avenues of improving its environmental impact and even consider the use of alternative fuel sources (Ölçer 2018). This trend is not restricted solely to the maritime domain; other means of transport, such as air, road and rail, are also under continuous scrutiny for their levels of gas emissions (Dimoula et al. 2016; FAA Office of Environment and Energy 2015; Merkisz et al. 2016). Many high-level initiatives and regulations are pushing towards a framework of restricting these dangerous emissions and a tighter control on pollution in relation to internal combustion engines. In any case, global consensus strengthened by UN Sustainable Development Goal (SDG) 13 (climate action) and the Paris Agreement has created pressure on the maritime industry to mitigate the negative environmental impacts of shipping activities (Ölçer et al. 2018). Industry needs are rapidly shifting towards green and ecological operation of ships and ports; therefore, new types of education and research are in demand, forming a domain under the term "Maritime Energy Management" (MEM). This demand was first responded to by the World Maritime University (WMU) of the International Maritime Organization (IMO), a Specialized Agency of the UN, having established a series of postgraduate programmes focusing on MEM related issues (Ölçer et al. 2017).

The paper presents a case study of WMU to showcase how universities can play a pivotal role in creating an interdisciplinary knowledge platform with specific focus on maritime energy. Various indicators to measure the impact are presented, such as the evaluation of student data from their enrollment to career progressions, conference-seminar organisation and participation, research and publications, and stakeholder engagement. Tangible outcomes achieved in a relatively a short time by WMU will demonstrate the leading role of universities in global levels of responses to climate change. The conclusions section highlights the opportunities for universities to build capacity for sustainable development through education and research.

Mainstreaming Sustainable Development in Higher Education

Quality education is considered to be the key to breaking free from the cycle of poverty, to reducing inequalities and promoting gender equality, and to empowering people everywhere to live healthier and more sustainable lives. Indeed, education is considered an investment in development both at individual and country levels. For example, OECD countries spend 5.2% of their GDP on average on educational institutions from primary to tertiary, including expenditure on both public and private schools (OECD 2016).

While an earlier stage of education is explicitly stated in UNSDG 4 (quality education) as crucial to poverty eradication and inequality reduction, higher education will have a different role to play within the context of the UNSDGs. Indeed, higher education, including universities, is situated closer to industries and decision makers than earlier levels of education, focusing on children or younger pupils.

To distinguish the role of universities from that of an earlier stage of education, it is noticeable that universities often produce high-level research. Universities have evolved through the transition from labour-intensive economy to "knowledge economy" in the modern information society via innovation. Though today's information society allows individuals to publish knowledge, universities are still privileged to produce knowledge for the most important decision making.

The adoption of the UNSDGs was a turning point for higher education, including universities. Universities traditionally consisted of a number of schools which specialize in their own disciplines, such as medicine, mathematics, bio science, laws, social sciences, literature, and so on. The 17 Goals of the UNSDGs reflect complex global challenges to be interlinked with each other, which encourages all the stakeholders and various experts to come together and find solutions. This attitude recalls the notion of interdisciplinary approaches within and beyond the university. Indeed, the interdisciplinary nature of universities also allows talented young people to be exposed to various perspectives and values in different scholarships. However, it is not always a simple matter to keep a university research-oriented and interdisciplinary because of its tradition, organisational culture, and management and leadership (Watson 2001).

The case of WMU was not an exception. WMU was established in 1982 by IMO, a Specialized Agency of the UN, to build capacity in the maritime domain, with a particular focus on developing countries. As of 2018, the number of WMU graduates reached 4919 from 168 countries around the world. Until recently, WMU's M.Sc. programme in Maritime Affairs consisted of six specialisations: maritime safety and environmental administration (MSEA); maritime education and training (MET); maritime law and policy (MLP); port management (PM); shipping management and logistics (SML); and ocean sustainability, governance, and management (OSGM). Each specialisation stood alone and was distinguished from the others by its own unique characteristics. However, WMU tradition and organisational culture have changed since a new specialisation, Maritime Energy Management (MEM) was launched in 2017.

MEM was designed and proposed even before the adoption of the UNSDGs by foreseeing future needs of the maritime industry. Responding to the rise in environmental concerns, especially climate change, in the maritime industry, WMU created the MEM specialisation to play an active role in educating future maritime leaders. Since its establishment in 2017, the MEM specialisation has entered its third year of intake.

In this paper, a case study on WMU is used in order to elaborate the roles of universities in terms of their contribution to the UNSDGs, in particular, Goals 7 (energy efficiency) and 13 (climate action). The theoretical framework used in this paper is the 5P approach to sustainable development shown in Fig. 1. In this approach, 5P stand for the following:

1. Planet: Protect our planet's natural resources and climate for future generations.
2. People: End poverty and hunger in all forms and ensure dignity and equality.
3. Prosperity: Ensure prosperous and fulfilling lives in harmony with nature.
4. Peace: Foster peaceful, just and inclusive societies.
5. Partnership: Implement the agenda through a solid global partnership.

In this approach, the five P's are considered to be interdependent. Any particular intervention by a university in one of the five P areas of sustainable development will

Fig. 1 The 5P approach to sustainable development. *Source* UNSSC n.d.

inevitably affect other areas. This indicates that integrated and balanced action on all five P's is important for effective capacity building. A case study on WMU will therefore be presented, focusing on these five P's, in the following sections.

Planet: The Development of "Maritime Energy Management"

The initiative to develop the MEM specialisation arose from the direct and urgent need to provide education for the reduction of greenhouse gas emissions, and for the development of sustainable forms of energy in the maritime field. The university responded to a global need for capacity building, which is crucial for achieving the UNSDGs, in particular Goal 7 (Affordable and clean energy) and Goal 13 (Climate action). Providing this education is a powerful means of supporting the development of affordable and clean energy, as well as supporting climate action, by disseminating skills amongst future leaders in the maritime field. Education may thus be considered to have a positive effect on climate change mitigation.

WMU has as its core the important dimension of providing a global environment to the students, in which the exchange of ideas and perspectives is accelerated. The difference in student backgrounds adds to the learning experience and knowledge to which the students are exposed, though it is transmitted from student to student, rather than being provided by the teachers.

All WMU study programmes are designed to respond to the current needs of the maritime industry through a balance of academic study and practical experience. Emphasizing the practical application of knowledge, WMU programmes provide a unique insight into the range of maritime endeavors.

Figure 2 illustrates schematic representation of the Post Graduate Pathway in Maritime Energy Management at WMU. Maritime Energy Management (MEM) is an interdisciplinary subject where breadth of knowledge, such as technology and innovation, human factors, and economics, meets depth knowledge to respond to the mitigation needs of GHG resulting from maritime operations (Ölçer et al. 2017). It is not an M.Sc. programme only; it is a strategic direction for WMU including both teaching and research elements. The teaching pillar includes a spectrum ranging from Professional Development Courses (PDCs) to Post Graduate Diploma by Distance Learning, to a full time M.Sc. programme in the area of MEM. The M.Sc. programme begins with foundation studies (Term 1) to cover the basics of maritime and ocean studies with all the specialisation students. The programme proceeds to specialisation

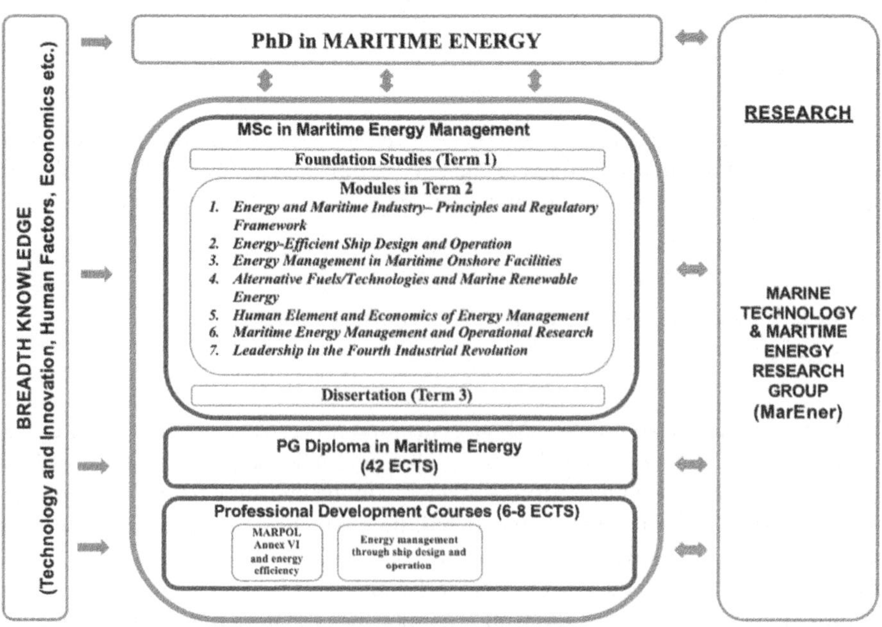

Fig. 2 The postgraduate pathway in the maritime energy stream. *Source* Authors

studies (Term 2) where students are divided into their chosen specialisation course and study in depth. Finally, students write a dissertation (Term 3).

The research pillar comprises of life-cycle management of internationally and nationally funded research projects to M.Sc. and Ph.D. dissertations to the dissemination of the research results in highly reputable academic publications. In addition to UNSDGs 7 and 13, the MEM academic and research focuses are relevant to several UNSDGs, such as Goal 4 (quality education) (Ölçer et al. 2017; Swe et al. 2018); Goal 5 (gender equality) and Goal 12 (responsible consumption and production) (Kitada et al. 2018); and Goal 14 (Ölçer et al. 2018). These research agendas are developed by a WMU research group, called the Marine Technology & Maritime Energy Research Group (MarEner).

Clearly, universities and their wider knowledge communities have a responsibility to future generations and to the planet to minimize the negative impacts of climate change. Energy management should be viewed as a key enabler of energy efficient and environmentally friendly shipping while ensuring profitability within a tightening regulatory framework.

People: Future Maritime Leaders

WMU has been providing high level postgraduate education in maritime affairs since 1983. Today, the University's mandate has extended to ocean affairs and its postgraduate diploma programmes can be accessed in developing and developed countries across the globe. As a world leader in competence-based maritime and oceans education and training, WMU has a demonstrated capacity to provide support under Goal 4 (Ensure inclusive and equitable quality education and promote lifelong learning opportunities for all), in particular to the following targets:

- 4.3 By 2030, ensure equal access for all women and men to affordable and quality technical, vocational and tertiary education, including university.
- 4.4 By 2030, substantially increase the number of youth and adults who have relevant skills, including technical and vocational skills, for employment, decent jobs and entrepreneurship.
- 4.7 By 2030, ensure that all learners acquire the knowledge and skills needed to promote sustainable development, including, among others, through education for sustainable development and sustainable lifestyles, human rights, gender equality, promotion of a culture of peace and non-violence, global citizenship and appreciation of cultural diversity and of culture's contribution to sustainable development.

For target 4.3, WMU increased the number of female students from less than 5% in the late 1990s to 37% in the academic year of 2017. The ratio of female students in the MEM specialisation has also increased from 12.5% in 2017 (when the specialisation was launched) to 25.0% in 2019. WMU will continue to encourage women to study in all the maritime and ocean specialisations, including MEM.

The evaluation of target 4.4 will usually take several years to see the impact of education on students' career development including employment, decent jobs, and entrepreneurship. Though it is not sufficient with only two graduated groups in 2017 ($n = 16$) and 2018 ($n = 14$), since it is a new specialisation course, it is possible to observe a relatively immediate impact on career development after graduation (Table 1). Some graduates received promotions at work and others found new jobs or Ph.D. positions relevant to their study in MEM at WMU.

Towards the achievement of target 4.7, WMU has been integrating the UNSDGs in its curriculum. In the case of the MEM specialisation, a wide spectrum of subjects are taught from regulations and policies to ship design and operation, from port and shipyard to alternative fuels/technology and renewable energy, and further human element and economics (shown in Fig. 2). The interdisciplinary nature of the MEM curriculum has been strengthened by a project-based subject focusing on UNSDGs; the subject was previously called as 'Applied Energy Research' but now updated as 'Leadership in the Fourth Industrial Revolution'.

Prosperity: Redefining the Roles of Universities

Responding to the global need for a shift in education towards issues of sustainability is an opportunity for universities to redefine their roles as facilitators of global action towards achieving the UNSDGs. Shifting the focus towards sustainable development is an opportunity for universities to strengthen their position in society, by becoming an essential part of a more sustainable society. The term, 'global citizenship' well articulates the essence of redefining the roles of universities beyond their academic territory. The participation of universities in building peaceful and resilient societies through partnerships is foremost in improving sustainability for a prosperous society in the long term.

A clear example in the maritime sector is a paradigm shift from the conventional maritime domains (e.g., safety, security, and environmental protection) to the maritime-ocean nexus. Oceans imply a much wider context than simply maritime. Recent maritime activities have been extended to the exploitation of ocean resources, including oil and gas, ocean energy, tourism, and Arctic development. From the prosperity viewpoint, oceans are extremely important for the existence of human beings and flora and fauna. Without oceans, the whole ecosystem will collapse.

In 2015 during the inauguration of the WMU new campus, Dr. Yohei Sasakawa, Chairman of the Nippon Foundation announced the foundation's contribution to the establishment of the WMU-Sasakawa Global Ocean Institute (GOI). The GOI was officially inaugurated in May 2018 as an independent focal point for ocean related dialogue and capacity-building (Fig. 3). In support of the long-term sustainability of our Oceans, GOI will serve as a centre of excellence to advance ocean research, and capacity-building on global ocean policies that are cross-sectoral, interdisciplinary and trans-regional in scope.

Table 1 Summary of MEM graduates' career development after graduating from WMU M.Sc. in 2017–2018

Country	Employer	Current position[a]
M.Sc. in 2017		
China	Dalian Maritime University	Lecturer
Egypt	Arab Academy for Science & Technology & Maritime Transport	Maritime Lecturer
Egypt	Arab Academy for Science & Technology & Maritime Transport	Third Marine Lecturer
Gambia	Gambia Ports Authority	Electrical Manager
Indonesia	PT Terminal Teluk Lamong	PMO Engineering Staff
Iran	World Maritime University	Ph.D. student
Iran	World Maritime University	Researcher
Japan	Ministry of Land, Infrastructure, Transport and Tourism	Deputy Director
Myanmar	Department of Maritime Administration	Staff Officer
Nigeria	Nigerian Maritime Administration And Safety Agency Apapa—Lagos	Chief Shipping Development Officer (Wet Cargo & Gas)
Nigeria	Nigerian Maritime Administration And Safety Agency (NIMASA)	Senior Information Technology Officer
Panama	Bureau Veritas Inspectorate	Operations Manager Assistant
Philippines	Philippine Merchant Marine Academy	Dean (College of Marine Engineering)
Uganda	Ministry of Works & Transport	Transport Officer, Water & Rail Transport Regulation
M.Sc. in 2016		
Bangladesh	Eaglestar	Manager, Energy Efficiency
China	Dalian Maritime University	Lecturer
Egypt	Arab Academy for Science & Technology & Maritime Transport	Maritime Lecturer
Egypt	Arab Academy for Science & Technology & Maritime Transport	Head of the International Maritime Cooperation Unit
Gambia	Gambia Maritime Administration	Marine Surveyor
Ghana	Ghana Maritime Authority	Assistant Administrative Officer
Indonesia	Biro Klasifikasi Indonesia	Surveyor Research & Development
Iran	University of Genoa, Italy	Ph.D. student
Kenya	Kenya Ports Authority	Engineer
Kenya	Kenya Ports Authority	Radio Inspector
Korea	Korean Registry of Shipping	Senior Surveyor

(continued)

Table 1 (continued)

Country	Employer	Current position[a]
Malta	Transport Malta—Merchant Shipping Directorate	Flag & Port State Control Inspector
Nigeria	Nigerian Ports Authority	Pollution Control Officer I
Nigeria	Federal Ministry of Transportation	Mechanical Engineer I
Philippines	Maritime Industry Authority (MARINA)	Shipping Operations Specialist I
Vietnam	Arctic University of Norway (UiT)	Ph.D. student

[a]Current position indicates as of 31 March 2019

Fig. 3 The inauguration of the WMU-Sasakawa Global Ocean Institute (GOI) by Dr. Cleopatra Doumbia-Henry, President, WMU, Dr. Yohei Sasakawa, Chairman of The Nippon Foundation, Mr. Kitack Lim, Secretary-General of IMO, and Ms. Isabella Lövin, Deputy Prime Minister, Sweden. *Source* WMU

This significant milestone clarifies the role of WMU in clearly supporting the UNSDGs, in particular Goal 14 (life below water) and Goal 13 (climate actions) through education and research. The MEM specialisation is also motivated by Goal 13, but to promote Goal 7 (clean energy), which also matches the scope of a maritime-ocean nexus.

Peace: Towards Inclusive Society Through Education and Research

Since its establishment, WMU has focused on capacity building in postgraduate maritime education and research worldwide, in particular in developing countries. This core philosophy contributes to building peace within and beyond the maritime community.

As IMO's centre of excellence for postgraduate maritime education, WMU has been striving to meet the needs of the highest quality of maritime education and research for years. Traditional diploma courses require more time and money to invest in one's education than distant learning solutions. With experience and know-how, WMU offers a range of distance learning programmes, for example on maritime safety and security, executive maritime management, international maritime law, marine insurance, and maritime energy.

In the last four years, a post-graduate diploma in Maritime Energy has been delivered by distance learning in collaboration with Lloyd's Maritime Academy, London. This programme is now further updated by WMU's in-house sources and will be launched in September 2019. This new e-learning programme (PG Diploma in Maritime Energy) includes five modules, which represent a small scale of the MEM M.Sc.: Module 1: Maritime Energy and Sustainable Development; Module 2: Ships and Energy Efficiency; Module 3: Alternatives and Future Technologies; Module 4: Energy Conservation in Ports and Shipyards; and Module 5: Best Practices and Life-Cycle Perspective.

This new academic programme is expected to reach out to a wider audience of potential learners around the world and include everyone who is interested in studying the subject. The promotional video was created with gender-sensitive contents and approaches to attract women as well as men to study. The impacts that universities can make on societies will be huge. Maritime Energy Management is a methodology to approach climate change as a universal concern through the engagement of all people towards an inclusive society.

Partnership: Leading for Actions

Partnerships are a form of agreement of collaboration amongst partners, and are forged with the purpose of strengthening the partners. Partnerships allow the individual members to address problems that lie beyond their individual struggles, and are thus essential for tackling global problems such as climate change, and may be considered a catalyst for forging motions of joint action.

In this regard, to launch the MEM specialisation, WMU hosted the International Conference on Maritime Energy Management (MARENER 2017) on 24–25 January 2017. More than 300 participants from 83 countries engaged in active discussions over the challenges and opportunities surrounding MEM. Keynote speakers included

Mr. Kitack Lim, IMO Secretary-General; Ms. Anna Johansson, Sweden's Minister for Infrastructure; industry representatives from ABS, Turkish Classification Society, Stena AB, NAPA group, INTERTANKO, Copenhagen Malmö Port, and the German Society of Maritime Technology; US Coast Guard; and the University of Strathclyde. Though it was an academic conference, the topic attracted the participation of a range of industry stakeholders. It proved a form of partnership opportunities among academia, industry, and policy makers. During the MARENER 2017 conference, the Maritime Energy Management Research Strategy was launched. The conference outcome was published in the form of the WMU book series of Maritime Affairs, 'Trends and Challenges in Maritime Energy Management' (Ölçer et al. 2018).

It is worth noting that WMU's MEM specialisation programme offers not only campus-based education but also field studies in which students visit various professional sites, relating to their specialized topics. The field studies so far include selected countries such as Italy, Finland, United Kingdom, Sweden, Denmark, and Turkey where students can visit ports, shipyards, ice-breakers, engine manufacturers, software companies, shipping companies, maritime administrations, coast guards, vessel traffic centres, maritime universities, and UN organisations (e.g., IMO and Food and Agriculture Organization (FAO)). Students also travel by sea and experience first-hand the energy efficient operation of a ship. Generous support by various stakeholders enables WMU to embed these on-site experiences into the curriculum as field studies, which complement students' theoretical knowledge with more practical and applicable learning (Fig. 4). Furthermore, field studies help students to see

Fig. 4 Field study to visit LNG-fueled ice breakers in Helsinki, June 2018. *Source* WMU

various perspectives of stakeholders and their relationships in the big picture, emphasising the importance of partnerships in MEM. The host countries and organisations to the WMU field studies also benefit the students' visits as well; students from various countries will remember the services and know-how that they have learned during the visits, and in the future when the graduates are posted in decision-making positions of their countries, there may be possible investment and business cooperation with those FS supporters. This kind of long-term relationship between WMU and the industry stakeholders is seen as a leverage for capacity building in the field of maritime affairs, including MEM.

WMU plays an active role in various global professional networks. For example, IMO and EU have established five regional Maritime Technology Corporation Centres (MTCC) in Africa, Asia, Caribbean, Latin America, and the Pacific with the purpose of furthering efforts to combat climate change. They are in the Global MTCC Network (GMN), where WMU MEM faculty provides technical assistance for capacity building through lectures, seminars, workshops and conferences in close cooperation with IMO. At the academic level, WMU is part of the International Association of Maritime Universities (IAMU), a global maritime university network that facilitates academic and research cooperation on maritime education and training around the world. IAMU is, thus, an ideal platform for cooperation on global issues such as affordable and clean energy and climate action (Kitada et al. 2017). Recently, under the leadership of its Director of Maritime Research, WMU has become part of the research consortium to prepare a proposal for the Fourth IMO GHG Study to support the IMO Strategy on reduction of GHG emissions from ships (MEPC 2018). Being actively involved in knowledge creation in such an influential report is important for WMU as per its mission statement, *to be the world centre of excellence in postgraduate maritime and oceans education, professional training and research, while building global capacity and promoting sustainable development.*

Conclusion

Climate change action has been recognised as a common global agenda from the twentieth century and has provided a dynamic discourse across different sectors, including universities. While a university's role is mainly academic, contemporary needs of education and research have inevitably been reflected into the re-definition of existing curriculum and the development of new academic subjects. One of the subjects that climate change shone a light on was "energy". In the maritime context where ships and ports can contribute to GHG reduction, the Maritime Energy Management (MEM) specialisation was born as a response from a university—WMU.

This paper highlighted various actions and outcomes in which the WMU has contributed to UNSDGs, in particular, Goal 7 (clean energy) and Goal 13 (climate change). Of particular importance is the fact that the WMU has, in a timely manner, integrated the sustainable development agenda into its strategic vision and as a result invested in a series of initiatives to streamline necessary actions. Under the theoretical framework of the 5P approach to sustainable development, all aspects—Planet, People, Prosperity, Peace, and Partnership—are examined as a tool to analyze the role of universities within the case of WMU. A balanced approach across 5P's exhibits a healthy enabling environment wherein a university can make robust progress with various partners to achieve multidisciplinary goals and targets.

MEM is still a very young concept in academia. Its initial focus was on climate change "mitigation" rather than "adaptation" because the economic activities from ships and ports are producing GHG and MEM has been developed to reduce emissions (and possibly achieve zero emissions) by adopting regulations and policies, clean energies and technologies. A very limited account of the "adaptation" side of MEM will be explored along with global climate change discourses. This will be an area for universities, like WMU, to lead a discussion and develop a new scientific linkage between climate change adaptation and MEM.

GHG reduction from ships and ports aims to mitigate the impact of climate change from the maritime sector. It requires policy makers, industries, NGOs, and educators and researchers who play their own roles and share knowledge to build a network of enablers. The fact that WMU graduates represent a wide geographical range also helps to create knowledge from diversity. Research tends to rely on those who possess advanced research capacities, often in developed countries. However, through WMU's capacity building in developing countries, knowledge creation is also looking into the unpublished side of the world. As such, universities can support a fair, accurate, and inclusive overview of "the world" in which we all live together.

References

Brigham L (2013) IMO polar code for ships operating in polar waters. Produced by arctic climate change economy and society (ACCESS), policy brief no. 4. https://www.aeco.no/wp-content/uploads/2013/06/2015-04-ACCESS-PolicyBrief-Polar-Code.pdf. Accessed 18 May 2019

Dimoula V, Kehagia F, Tsakalidis A (2016) A holistic approach for estimating carbon emissions of road and rail transport systems. Aerosol Air Qual Res 16:61–68

FAA Office of Environment and Energy (2015) Aviation emissions, impacts & mitigation: a primer. Federal Aviation Administration Office of Environment and Energy, Washington, DC

Kitada M, Dalaklis D, Drewniak M, Ölçer AI, Ballini F (2017) Exploring the frontiers of maritime energy management research. International Association of Maritime Universities (IAMU) general assembly. Varna, 11–14 Oct 2017

Kitada M, Rado K, Toua OO, Nervale T (2018) The role of maritime transport from the perspective of energy and gender: the case of the Pacific Islands. In: Ölçer AI, Kitada M, Dalaklis D, Ballini F (eds) Trends and challenges in maritime energy management. Springer, Heidelberg. https://doi.org/10.1007/978-3-319-74576-3_26

Merkisz J, Lijewski P, Pielecha J (2016) PEMS-based investigations into exhaust emissions from non-road and rail vehicles. Combust Eng 166(3):46–53. https://doi.org/10.19206/CE-2016-339

MEPC (2018) Reduction of GHG emissions from ships: preparation for the fourth IMO GHG study, MEPC 73/7. IMO, London

Mukherjee PK, Liu H (2018) Legal regime of marine insurance in arctic shipping: safety and environmental implications. In: Hildebrand LH, Brigham LW, Johansson TM (eds) Sustainable shipping in a changing arctic. Springer, Cham

OECD (2016) Education at a glance 2016: OECD indicators. OECD, Paris, https://doi.org/10.1787/eag-2016-en

Ölçer AI (2018) Introduction to maritime energy management. In: Ölçer AI, Kitada M, Dalaklis D, Ballini F (eds) Trends and challenges in maritime energy management. Springer, Heidelberg. https://doi.org/10.1007/978-3-319-74576-3_1

Ölçer AI, Ballini F, Kitada M, Dimitrios D (2017) Development of a holistic maritime energy management programme at the postgraduate level: the case of WMU. In: Proceedings of INTED2017 conference. Valencia, 6–8 Mar 2017, https://doi.org/10.21125/inted.2017.0474

Ölçer AI, Kitada M, Dalaklis D, Ballini F (eds) (2018) Trends and challenges in maritime energy management. Springer, Heidelberg. https://doi.org/10.1007/978-3-319-74576-3. ISBN 978-3-319-74575-6

Pain N, Koske I, Sollie M (2008) Globalisation and OECD consumer price inflation. OECD economic studies no. 44, 2008/1. ISSN 0255-0822

Swe TH, Kitada M, Ölçer AI (2018) The need for education and training in maritime energy management in Myanmar. In: Ölçer AI, Kitada M, Dalaklis D, Ballini F (eds) Trends and challenges in maritime energy management. Springer, Heidelberg, https://doi.org/10.1007/978-3-319-74576-3_25

Watson N (2001) Promising practices: what does it really take to make a difference? Educ Can 40(4):4–6

UNCTAD (2018) Review of maritime transport 2018. In: United nations conference on trade and development. Paris. https://unctad.org/en/PublicationsLibrary/rmt2018_en.pdf. Accessed 31 Mar 2019

UNSSC (n.d.) The 2030 Agenda for sustainable development. United Nations Staff Systems College, Bonn

Is Green Manure (*Azolla pinnata* and *Sesbania rostrata*) a Climate-Resilient Strategy for Rice Farming?

M. Khais Prayoga, Neni Rostini, Tualar Simarmata, Mieke Rochimi Setiawati, Silke Stoeber and Kustiwa Adinata

Abstract Rice farming is one of the largest sources of greenhouse gas emissions in Indonesia. The climate-resilient agriculture investigation and innovation project (CRAIIP), Universitas Padjadjaran and the members of the Indonesian Farmers Community Network (JAMTANI) promoted solutions for climate-resilient rice farming through experiments in so-called climate field labs (CFL). This study contributes to the climate-resilience research by testing the nutrient uptake efficiency of three rice varieties using green manure. The farmers were involved in designing the research, selecting the rice varieties, collecting and analyzing the obtained data from climate farmer field schools. The CFL was set up as a factorial randomized block design. The first factor was green manure (10 tons/ha of cattle compost, 10 tons/ha of *Azolla pinnata* or 2 tons/ha of *Sesbania rostrata* combined with 10 tons of cattle compost) and the second factor was rice variety (the local Bangir, the widely cultivated Ciherang and the saline-tolerant Inpari 34). The CFL results revealed that the local variety Bangir with Sesbania enriched cattle manure tends to be the most efficient strategy for uptaking the nutrients. Climate-resilient policies to improve the resilience of smallholder farmers with minimal trade-offs for the environment should include the propagation of suitable varieties and promising green manure application.

M. K. Prayoga (✉) · N. Rostini · T. Simarmata · M. R. Setiawati
Faculty of Agriculture, Universitas Padjadjaran, Jl. Raya Bandung Sumedang KM.21, Jatinangor, Sumedang Regency, Bandung, West Java 45363, Indonesia
e-mail: mkhaisprayoga@yahoo.com

S. Stoeber
Centre for Rural Development (SLE), Humboldt-Universität zu Berlin, Robert-Koch-Platz 4, 10115 Berlin, Germany

K. Adinata
Indonesian Farmers Community Network (JAMTANI), Jl. Raya Babakan, Ciamis Regency, Pangandaran, West Java 46396, Indonesia

W. Leal Filho et al. (eds.), *Climate Change, Hazards and Adaptation Options*, Climate Change Management, https://doi.org/10.1007/978-3-030-37425-9_46

Introduction

Rice is one of the most important agricultural commodities in Indonesia and it is cultivated on an estimated area of 10.90 million ha. Most of the rice is produced in Java as many as 29.69 million tons or around 52.51% of the total production in Indonesia which is 56.54 million tons. West Java Province belongs to the major rice producers in Indonesia (BPS 2017). Rice cultivation is also one of the biggest sources of greenhouse gas emissions by mainly contributing to CH4 (methane) emissions (IPCC 2015).

Methane emissions occur in ecosystems with dominant anaerobic conditions, which is in rice cutivation mainly due to flooding. Rice cultivation is responsible for approximately 11% of global anthropogenic CH4 emissions. They are estimated to reach 20–120 million tons per year or around 12.5% of the global methane emissions of 470–650 million tons per year and are the highest share among the main food crops. The level of methane emissions from rice cultivation in Indonesia reaches 1.7 million tons per year (Carlson et al. 2017; IPCC 2013; Liquist et al. 2012). Therefore, efforts to reduce methane emissions from paddy fields in Indonesia would have a high impact on the reduction of global greenhouse gas emissions.

Intermittent flooding system of rice fields has been widely promoted for mitigating methane emissions (Minamikawa et al. 2014) however, it main trade-off is the risk of releasing more nitrous oxide, particularly when chemical fertilizers are applied excessively, a common practice among rice smallholder farmers in Java, Indonesia. Most paddy farmers in West Java still rely on inorganic fertilizers to increase grain yields, and often they use large quantities of inorganic fertilizers regardless of environmental aspects (Wahyuni et al. 2017).

Inorganic fertilizers can bring undesirable changes to the chemical structure of the soil, and endanger the survival of organisms both in the soil (Ogbodo 2013). The long-term application of inorganic fertilizers has a real effect on soil biochemical properties such as causing microbial population shifts, changes in soil organic carbon (SOC), nitrogen (N) content, pH, and humidity. This changes the water holding capacity of the soil. This impacts the water availability in the soil and exacerbates the drying out of soils after long dry seasons that occur more frequently due to climate change (Prashar and Shah 2016). Therefore, a climate resilient strategy which also mitigates the greenhouse gases is the reduction of using inorganic fertilizers by farmers.

The climate resilient agriculture investigation and innovation project (CRAIIP) aims at technology transfer from the university to farmers by including the knowledge of farmers in a co-creation of knowledge process. It is a joint project by the members of the Indonesian Farmers Community Network (JAMTANI), the Universitas Padjadjaran and Humboldt-Universitat zu Berlin. CRAIIP provides solutions for climate-resilient rice farming through experiments conducted in so-called climate field labs or climate farmer field schools. The farmers act as active subjects of the research (farmer oriented experiment) and follow the principles of participatory research. This approach enables to accelerating the process of technology development between academia and farmers (Falsafi et al. 2015).

Local adaptive capacities are taken into account when desgning technology development to increase climate resilinece. This includes the adjustment of materials, approaches, and processes in accordinace with local conditions (Filho and Mannke 2014). Accordingly, rice farmers are advised to investigate climate-resilient cultivation techniques, such as the use of green manure and the use of suitable rice varieties. Green manure is expected to substitute inorganic fertilizers that are widely used by the farmers. With the combination of green manure and suitable rice varieties it is expected to increase the rice grain yield in a sustainable way.

The green manures tested were *Azolla pinnata* and *Sesbania rostrata*. Through Azolla green fertilizer the soil fertility can be improved as it provides a high amount of N. Azolla is symbiotic with the cyanobacteria Anabaena azollae that binds free nitrogen from the air. Under good growing conditions, the Azolla-Anabaena symbiosis can fix up to 100–170 kg N/ha per year (Setiawati et al. 2017). Sesbania is an ideal plant as green manure because it is fast-growing, decomposes easily, retends soil moisture, produces organic matter and adds a considerable amout of nitrogen to the soil. Sesbania produces 5.2 tons/ha of dry matter which is equivalent to 135 kg N/ha (Ehsan et al. 2014).

The varieties were chosen in a participatory process between researchers and farmers. In this study the local variety Bangir, the commonly used Ciherang, and the saline-tolerant variety Inpari 34 have been tested.

This study aims at promoting climate resilient agriculture by testing suitable varieties and potential green manure applications. Sustainable ecological intensification has been operationalised in this study by the indicators nutrient uptake and rice productivity (yield). The most promising green manure-variety mix in terms of highest nutrient uptake points to a good trade-off between economic benefits and nitrous oxide emisions. The most favorable variety-green manure combination is then disseminated in the regions through the climate farmer field school approach.

Description of the Study Areas and Methodology

Study Area

The experimental farm as a representative of smallholder rice cultivation (nearby the main road in order to ensure generalizability, visibility, and outreach) has been established in Ciganjeng village, Padaherang subdistrict, Pangandaran district, West Java (Fig. 1). The site has been under rice cultivation for the past 20 years. During that period, rice has been cultivated twice a year by the use of chemical fertilizer. The location has a type B climate according to Schmidt-Ferguson classification. The soils are broadly classified as silty clay.

Fig. 1 Location of experimental farm

Research Methods

Experimental Design and Treatments

The field trial has been designed as a randomized block design with a factorial pattern with three replications. Observations have been done from October 2017 until February 2018 in Ciganjeng village. The green manure was placed as first factor and contained three treatments (P_1 = 10 tons/ha of cattle compost; P_2 = 10 tons/ha of cattle compost + 10 tons/ha of Azolla pinnata; and P_3 = 10 tons/ha of cattle compost + 2 tons/ha of Sesbania rostrata). Varieties were chosen as second factor. They were selected in an informed dialogue between researchers and farmers. At first, the university recommended several varieties and provided information on these varieties to the participating farmers. Then, farmers chose three varieties to be used in the study. The choice was based on farmers' preferences that have been collected by a questionnaire. Three varieties have been tested V_1 = Bangir (local variety); V_2 = Ciherang (popular variety); and V_3 = Inpari 34 (saline-tolerant variety). The size of the experimental farm was 1000 m^2, the size of a plot per treatment was 16 m^2 (8 m × 2 m) with a 0.5 m buffer zone in between.

The plot management used an intermittent irrigation system and followed the principles of organic farming. The production system focused on improving soil biology through microorganisms and organic matter, organic fertilization, appropriate water management through irrigation and crop management (Simarmata et al. 2015). The twin seedling method was used. Two eight-days old rice seedlings were planted at a distance of 5 cm per hill and the distance between the hills was 30 × 30 cm. The young seedlings were not pushed vertically into the soil but carefully inserted sideways. This makes the shape of the transplanted seedling more like an L than like a J. This L shape makes it easier for the roots to find their way down into the soil (Simarmata et al. 2011). The seedlings were planted in the muddy puddle land. Liquid organic fertilizer (one L dissolved into 100 L water) was sprayed 21 days after planting (DAP) and 42 DAP using a knapsack sprayer.

Sources and Data Collection

This study followed the principles of participatory farmer research and involved ten farmers of the farmer organization JAMTANI. A farmer-oriented research design involved farmers not only in the design, but also in data collection and analysis in weekly climate field schools and exchanging the research results at the end of each season. Farmers recorded various plant parameters in their weekly observations. In this climate field lab, which follows the principles of farmer field schools, the farmers were trained to perform their observation and analysis tasks as research farmers. To ensure the comparability of the data, standard research protocols and monitoring sheets were developed, and discussed with research farmers. The observed parameters were the leaf chlorophyll content, the 100-seed-weight, and the grain yield. The leaf chlorophyll content was observed at 82 DAP, while the 100-seed-weight and the grain yield was observed during harvest. In addition, temperature, humidity, and soil pH were measured at 8 am at weekly intervals. These measurements started in the vegetative phase (12 DAP) to the final generative phase (82 DAP). The soil chemical properties (organic carbon and nitrogen) were analysed before planting and after the rice harvest.

Data Analysis

The nutrient uptake efficiency (NUE) is calculated using a formula as follows:

$$E_e = \frac{Ydf - Yef}{F} \times 100\%$$

explanation:

E_e Nutrient uptake efficiency (%);
Ydf Yields of plants that have received additional green manure (ton/ha);
Yef Yields of plants that have not received additional green manure (ton/ha);
F Additional amount of green manure applied (ton/ha).

Differences between the varieties were tested via an ANOVA post hoc analysis using the Tukey's honest significance test or Tukey's HSD. The data were computed using the PKBT stat software.

Results and Discussion

Environmental Conditions

Environmental factors largely influence the growth and development of plants, as the appearance of the plant is the result of interactions between genetics and environment (Fehr 1987; Poehlman 1995; Xiao et al. 2019). The maximum yield potential of rice varieties is obtained, when rice is planted in an environment that is in accordance with its optimal genetic needs. Temperature, humidity and soil pH are important factors in this regard. The results of the observations made by farmers during the climate field lab are presented in the Tables 1, 2, and 3. On 3 January, 2018 the highest daytime temperature (44 °C) and the lowest humidity (23%) was recorded. The mean temperature during the growing season ranged from 30.3 to 37.7 °C and the humidity from 41–62%. The optimum temperature for the growth and development of rice is 25–33 °C, while the optimum humidity ranges from 60 to 80% (Rathnayake et al. 2016). In this study, the mean temperature were ideal enough to support the growth of the rice plants in nine weeks and too hot during two weeks. The humidity in the growing season was suboptimal except during one week, as most of the time the humidity was less than 60%.

The soil pH for all treatments ranged from 5.8 to 6.7, which means slightly acid to slightly neutral. This is a suitable acidity level for rice plants (Ilagan et al. 2014).

Table 1 The temperature during the observation

Date	Temperature (°C)		
	Maximum	Minimum	Mean
29/11/2017	33.4	30.4	31.9
06/12/2017	33.0	30.0	31.5
13/12/2017	34.0	29.6	31.8
20/12/2017	33.0	30.0	31.5
27/12/2017	34.8	33.6	34.2
03/01/2018	44.0	31.4	37.7
10/01/2018	34.0	29.6	31.8
17/01/2018	35.0	28.2	31.6
24/01/2018	32.7	29.4	31.1
31/01/2018	33.4	27.1	30.3
07/02/2018	34.4	30.4	32.4

Table 2 Humidity during the observation

Date	Humidity (%)		
	Maximum	Minimum	Mean
29/11/2017	37	68	53
06/12/2017	46	71	59
13/12/2017	45	70	58
20/12/2017	42	60	51
27/12/2017	47	55	51
03/01/2018	23	59	41
10/01/2018	46	65	56
17/01/2018	36	68	52
24/01/2018	44	69	57
31/01/2018	57	67	62
07/02/2018	37	54	46

Table 3 Soil pH during the observation

Date	Treatments		
	P1	P2	P3
29/11/2017	6.6	6.2	6.2
06/12/2017	6.7	6.7	6.0
13/12/2017	6.4	6.3	6.1
20/12/2017	6.4	6.4	6.3
27/12/2017	6.4	6.4	6.4
03/01/2018	6.0	6.3	6.2
10/01/2018	6.4	6.6	6.2
17/01/2018	6.5	6.5	6.5
24/01/2018	6.3	6.5	6.6
31/01/2018	6.2	6.1	6.6
07/02/2018	6.5	6.4	5.8

Nutrient Uptake Efficiency

Each of the three green manure treatments increased soil fertility components compared to the initial soil analysis (Table 4). Before the green manure treatment, soil organic carbon was very low at 0.82%. It increased to 1.52% for the sesbania-cattle manure treatment. Both, the azolla-cattle manure and the pure cattle-manure was even higher at 1.82% SOC after harvest. SOC strongly influences the growth and development of rice plants. The function of the organic compound is to improve the biological, physical, and chemical properties of the soil (Espe et al. 2015). Liu et al. (2014) point to the five main functions of the soil organic carbon, which are: (1) a granulator-forming soil structure; (2) a source of nutrients of N, P,

Table 4 Results of analysis of soil tests in the laboratory

Parameter	Initial soil analysis	Treatments		
		P1	P2	P3
Organic carbon (%)	0.82	1.82	1.82	1.52
Nitrogen (%)	0.10	0.28	0.19	0.15
C/N ratio	8.00	6.50	9.58	10.13

and S; (3) increase the water holding capacity of the soil; (4) increase the ability of the soil to hold nutrients; and (5) an energy source for soil microorganisms.

Consistent with the C-organic parameter, there is an increase of nitrogen (N) after the application of green manure. In the soil analysis before the treatment, the total nitrogen was only 0.10%. It increased to 0.28% in P1, 0.19% in P2, and 0.15% in P3 (Table 4). Nitrogen (N) is the most important nutrient for plant growth. Lack of N causes a suboptimal plant growth (Chhogyel et al. 2017).

In this study, Azolla and Sesbania green fertilizers were given in the fresh form. This is a possible reason that the total N-content in P2 and P3 is lower than in P1. This interpretation is supported by the C/N ratio, as P2 and P3 have a higher C/N ration than P1. The C/N ratio is the ratio of carbon and nitrogen contained in the inorganic material. The higher C/N ratio indicates that the organic material has not yet completely decomposed while the lower ratio indicates that organic material has further decomposed to humus (Wahyuni et al. 2017).

To evaluate the extent of nutrient uptake by plants, observations were made on several characteristics of rice plants, namely leaf chlorophyll content, the 100-seeds-weight, and grain yield. The results of the analysis show that the coefficient of variation (CV) for each character ranges from 4.79 to 7.78% (Table 5). The CV value indicates the level of accuracy of the treatment. A CV of less than 20% is considered as good (Brown 1998; Prayoga et al. 2016). It indicates that a trial error for the observed characters is relatively small. The greater the CV value, the greater the uncertainty of a study. With a CV of less than 10%, trial errors and uncertainty levels of this study were relatively small.

The leaf chlorophyll content trait reveals significant differences between the three varieties as shown in Table 6. Inpari 34 had with 21.8% the highest leaf chlorophyll content, however, this did not significantly differ from the local variety Bangir with 20.59%. Chlorophyll is a magnesium-porphyrin attached to proteins. It serves as

Table 5 Analysis of variance for each character

Character	Varieties	Treatments	CV (%)
Leaf chlorophyll content	**	*	7.78
Weight of 100 seeds	**	*	4.79
Grain yield	ns	ns	6.99

*Significant in $P < 0.05$; **Significant in $P < 0.01$; ns = non significant

Table 6 Comparisons of leaf chlorophyll content (%)

Varieties	Treatments			Average of varieties
	P1	P2	P3	
Bangir	18.15	20.61	23.01	20.59^{a,b}
Ciherang	15.37	20.60	20.44	18.80^{b}
Inpari 34	19.67	21.88	23.83	21.80^{a}
Average of treatments	17.73^{b}	21.03^{a,b}	22.43^{a}	

Note Numbers followed by the same letter are non-significant ($P < 0.05$)

catalyst for photosynthesis and is found in the thylakoid membranes as green pigments in photosynthetic plant tissues. They are loosely bound to proteins and easily extracted into lipid solvents such as acetone and ether. One component of chlorophyll is Nitrogen. Therefore, leaf chlorophyll content is closely related to the ability of plants to absorb nitrogen from the soil (Sange et al. 2014).

The ability of plants to absorb nutrients has an effect on the grain density of rice plants. When forming grains or grain, nutrients are required for metabolic processes and for starch translocation when filling grains. A lack or excess of nutrients limits or even disrupts the metabolic process of plants and has a negative impact on the filling of grains, while an optimum absorption of nutrients potentially increases the grain density (Putri et al. 2017).

The results of the analysis show that there is a highly significant difference between varieties with significant differences between treatments. Further tests using Tukey's HSD revealed that Bangir had the smallest grain size compared to the Ciherang and Inpari 34 varieties. Generally, the grain size character largely depends on genetic traits and less on the environment. Therefore, grain size characteristics considerably vary between varieties, which has a very influential impact on grain weight and yield (Lee et al. 2015) (Table 7).

With respect to the 100-seeds-weight, the Bangir variety has with 2.71 grams a significantly smaller weight than Ciherang and Inpari 34 varieties. However, the yield per hectare was highest for Bangir with 7.12 tons/ha compared to Ciherang yielded only 6.96 tons/ha, and Inpari 34 reached 6.70 tons/ha. The ANOVA results

Table 7 Comparisons of the weight of 100 seeds (gram)

Varieties	Treatments			Average of Varieties
	P1	P2	P3	
Bangir	2.46	3.03	2.65	2.71^{b}
Ciherang	2.80	3.31	2.88	3.00^{a}
Inpari 34	2.85	3.36	2.92	3.04^{a}
Average of treatments	2.70^{b}	3.23^{a}	2.82^{a,b}	

Note Numbers followed by the same letter are non-significant ($P < 0.05$)

Table 8 Comparisons of grain yield (tons/ha)

Varieties	Treatments			Average of Varieties
	P1	P2	P3	
Bangir	6.34	7.11	7.90	7.12[a]
Ciherang	6.67	7.37	6.84	6.96[a]
Inpari 34	6.51	6.80	6.78	6.70[a]
Average of treatments	6.51[a]	7.10[a]	7.17[a]	

Note Numbers followed by the same letter are non-significant ($P < 0.05$)

reveal no significant differences between varieties and treatments. Thus, in general, yield character per hectare of each variety in each treatment is the same.

Overall, the rice grain yield is a quantitative character that is influenced by the genetic base and at the same time by the environment (Assanga et al. 2017). The yield per hectare indicator is used to calculate the nutrient uptake efficiency. Table 8 demonstrates that by adding green fertilizers such as Azolla or Sesbania the yield for each variety can be increased. Figure 2 illustrates that the highest increase in yield was recorded for the Bangir variety in treatment P3 (10 tons/ha of cattle compost + 2 tons/ha Sesbania rostrata). Adding two tons/ha of sesbania can increase the grain yield of Bangir to 1.56 tons/ha. The lowest increase occurred in the Ciherang variety on treatment P3 (0.17 ton/ha).

The local Bangir variety in the azolla-compost (P2) treatment had the highest nutrient uptake efficiency reaching 78.00%. The concept of nutrient uptake efficiency generally describes how well plants use the nutrients available in the soil and are able to turn it into productive growth (Razie et al. 2013). The greater the value of the nutrient uptake efficiency of a variety, the better this variety has the ability to utilize the nutrients from the soil. In this study, it can be concluded accordingly, the greater

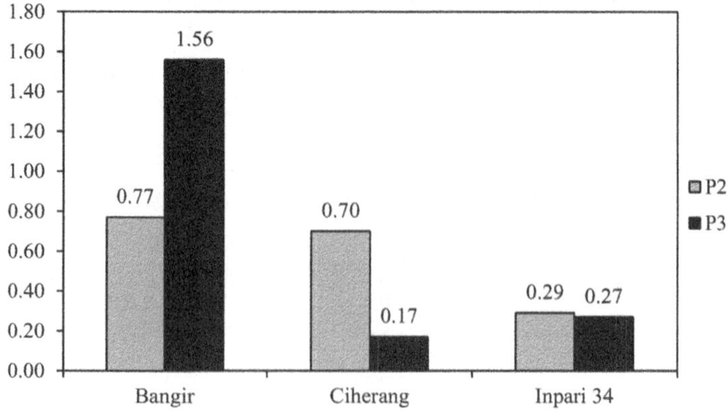

Fig. 2 Increased grain yield of three varieties at each treatment

Table 9 Nutrient uptake efficiency of three varieties in each treatment of green manure

Varieties	Nutrient uptake efficiency (%)		
	P2	P3	Average
Bangir	7.70	78.00	42.85[a]
Ciherang	7.00	8.50	7.75[b]
Inpari 34	2.90	13.50	8.20[b]
Average	5.87[b]	33.33[a]	

Note Numbers followed by the same letter are non-significant ($P < 0.05$)

the nutrient uptake efficiency, the better or easier the green fertilizer is used by the rice plants.

Table 9 shows that the Sesbania enriched green fertilizer has a generally higher nutrient uptake efficiency compared to Azolla green manure. For example, the Bangir variety had a nutrient uptake efficiency of 7.7% in P2 and 78% in P3. The Ciherang variety had a 7% nutrient uptake efficiency in P2 and 8.5% in P3. Inpari 34 reached on average 2.9% in P2 and 13.5% in P3. Referring to these results, the addition of 2 tons/ha of Sesbania rostrata to 10 tons/ha of cattle compost was relatively efficient in all three varieties.

Sesbania is the best green manure plant because it is easily biodegradable, able to restore and maintain soil moisture, produce organic matter and binds nitrogen in the soil (Wabi et al. 2016). Sesbania enriched manure can be highly absorbed by rice plants. 2.5 tons of biomass of Sesbania rostrata per hectare can increase the rice grain weight up to 15%. The addition of 2 tons Sesbania rostrata per hectare increases the grain yield equivalent to the provision of 60 kg N per hectare (Latt et al. 2009; Rauf et al. 1989; Soil Research Center 2009). The potential of Sesbania rostrata as green fertilizer is supported by its morphology, as it contains nodules on the roots and stems. Sesbania enriched green manure can increase the soil capacity to better absorb nutrients, improve soil structure, and increase soil microbial activity (Wabi et al. 2016).

Participatory research conducted jointly by the research farmers and the university increased the ability of farmers to carry out simple research. Through climate field labs research farmers learned to design the research, to take notes, and to conduct a first analysis. Climate Field Labs (CFL) opened the space for farmers to choose, create, develop, and practice new technologies. Herewith, farmers broaden their problem-solving capacities to develop new climate-resilient farm management strategies. Farmers learn through the process of seeing, analyzing, and evidencing in such CFL activities. According to Falsafi et al. (2015) learning-by-doing in farmer-led field experiments are more easily accepted and can accelerate the process of dissemination of innovations. Research farmers that were involved in the climate field lab activities are expected to be a role model for other farmers around and become a new source of peer-to-peer information and farmer exchange.

Conclusion

The use of Azolla and Sesbania green manure has increased the soil organic carbon, nitrogen, and grain yield of rice. The highest nutrient uptake efficiency was obtained by Bangir variety fertilized with 10 tons/ha cattle compost + 2 tons/ha Sesbania. Consequently, Bangir variety produced with the cattle manure plus Sesbania supplement is one suitable technology to promote climate-resilient rice farming.

The application of Azolla and Sesbania as green manure is an innovative technology for farmers, and many farmers do not know how to cultivate and propagate it. The Climate field lab is a powerful extension strategy to increase the adaptive capacity of farmers by testing ecologically sustainable adaptation strategies on their farms. In a further step, farmers capacities need to be built on cultivation techniques of Azolla and Sesbania and their prospective challenges in propagation and application are to be addressed.

Acknowledgements This study is part of the Climate-resilient Investigation and Innovation Project (CRAIIP) funded by the German Non-Governmental Organization Bread for the World (first phase: 2017–2018), which is implemented by JAMTANI. The authors are equally grateful to the farmers from the Taruna Tani Mekar Bayu Farmer Group, Ciganjeng village, Padaherang subdistrict, Pangandaran district, for their active participation in the design, implementation, and discussion of results and for the close collaboration throughout the experimental period.

References

Assanga SO, Fuentealba M, Zhang G, Tan CT, Dhakal S, Rudd JC, Ibrahim AMH, Xue Q, Haley S, Chen J, Chao S, Baker J, Jessup K, Liu S (2017) Mapping of quantitative trait loci for grain yield and its components in a US popular winter wheat TAM 111 using 90K SNPs. PLoS ONE 12(12):1–21

BPS (2017) Statistics on Indonesian agricultural production. Central Bureau of Statistics. Jakarta. Indonesia

Brown CE (1998) Coefficient of variation. In: Applied Multivariate Statistics in Geohydrology and Related Sciences CE, Brown (eds) Berlin, Heidelberg: Springer Berlin Heidelberg. pp 155–157

Carlson K, Gerber MJS, Mueller ND, Herrero M, MacDonald GK, Brauman KA, Havlik P, O'Connell CS, Johnson JA, Saatchi S, West PC (2017) Greenhouse gas emissions intensity of global croplands. Nat Clim Chang 7:63–68

Ehsan S, Niaz A, Saleem I, Mehmood K (2014) Substitution of major nutrient requirement of rice-wheat cropping system through Sesbania green manuring. Sci Agric 8:99–102

Espe MB, Kirk E, van Kessel C, Horwath WH, Linquist BA (2015) Indigenous nitrogen supply of rice is predicted by soil organic carbon. Soil Sci Soc Am J 79:569–576

Falsafi P, Kerdari M, Kashani SJ (2015) Acceleration of transfer of findings as a new strategy the iranian national innovation system lessons from agricultural extension in Iran. J Agric Sustain 7(2):150–170

Fehr W (1987) Principles of cultivar development, vol I, theory and technique. Macmillan Publishing Co., New York

Filho WL, Manke F (2014) Managing the impacts of climate change in Latin America: the need for technology transfer. Springer International Publishing, Switzerland

Ilagan LA, Tablizo RP, Roberto BB Jr, Marquez NAA (2014) Soil fertility evaluation for rice production in Catanduanes Province, Philippines. Int J Sci Technol Res 3:81–87

IPCC (2013) Contribution of working group I to the fifth assessment report of the intergovernmental panel on climate change. Cambridge University Press

IPCC (2015) Revised 2006 IPCC guidelines for national greenhouse gas inventories: reference manual. In: Eggleton HS, Buendia L, Miwa K, Ngara T, Tanabe K (eds) Prepared by The national greenhouse gas inventories programme. IGES Japan

Latt YK, Myint AK, Yamakawa T, Ogata K (2009) The effects of green manure (*Sesbania rostrata*) on the growth and yield of rice. J Fac Agr Kyushu Univ 54(2):313–319

Lee C, Park J, Kim B, Seo J, Lee G, Jang S, Koh H (2015) Influence of multi-gene allele combinations on grain size of rice and development of a regression equation model to predict grain parameters. Rice 33(8):1–10

Linquist B, van Groenigen KJ, Adviento Borbe MA, Pittelkow C, van Kessel C (2012) An agronomic assessment of greenhouse gas emissions from major cereal crops. Glob Chang Biol 18:194–209

Liu SL, Huang DY, Chen AL, Wei WX, Brookes PC, Li Y, Wu JS (2014) Differential responses of crop yields and soil organic carbon stock to fertilization and rice straw incorporation in three cropping systems in the subtropics. Agr Ecosyst Environ 184:51–58

Minamikawa K, Fumoto T, Itoh M, Hayano M, Sudo S, Yagi K (2014) Potential of prolonged midseason drainage for reducing methane emission from rice paddies in Japan: a long-term simulation using the DNDC-rice model. Biol Fertil Soils 50:879–889

Ogbodo EN (2013) Impact of the use of inorganic fertilizers to the soils of the ebonyi state agro-ecology, South-Eastern Nigeria. J Environ Earth Sci 3(7):33–38

Poehlman S (1995) Breeding field crops, 4th edn. Iowa State University Press

Prashar P, Shah S (2016) Impact of fertilizers and pesticides on soil microflora in agriculture. Springer International Publishing, Switzerland

Prayoga MK, Rachmady M, Wicaksana N (2016) Yield characters of 15 black soybean genotypes (Glycine soja (L.) Merr) in intercropping pattern of 2:1 with Corn. J Agrikultura 27(2):89–93

Putri FM, Suedy SWA, Darmanti S (2017) The effect nanosilica fertilizer on numbers of stomata, chlorophyll content, and growth of black rice (*Oryza sativa* L. cv. Japonica). Bull Anat Physiol 2:72–79

Rathnayake WMUK, De Silva RP, Dayawansa NDK (2016) Assessment of the suitability of temperature and relative humidity for rice cultivation in rainfed lowland paddy fields in Kurunegala district. Tropic Agric Res 27(4):370–388

Rauf M, Usman AM, Djamaluddin S Saenong, Subandi (1989) Effect of seed quantity and method of planting on the N, P, and K content of *Sesbania rostrata* plants. Agrikam 4(1):33–41

Razie F, Anas I, Sutandi A, Sutandi, Gunarto L (2013) Efficiency of nutrient uptake and rice yield with SRI cultivation on tidal land with enriched compost application. J Agron Indonesia 41(2):89–97

Sange MO, Ryan AA, Letchford KA, Macgowan SA, Mielke T (2014) Chlorophylls, symmetry, chirality, and photosynthesis. Symmetry 6:781–843

Setiawati MR, Damayani M, Herdiantoro D, Suryatmana P, Anggraini D, Khumairah FH (2017) The application dosage of *Azolla pinnata* in fresh and powder form as organic fertilizer on soil chemical properties, growth, and yield of rice plant. In: AIP conference proceedings

Simarmata T, Joya B, Turmuktini T (2011) Water saving and organic fertilizers based technology to remediate the health of paddy soils and to increase rice productivity in Indonesia. In: Proceeding of conference on international research on food security. Natural Resource Management and Rural Development, http://www.tropentag.de/2011/abstracts/full/694.pdf

Simarmata T, Citraresmini A, Sujana B, Setiawati MR (2015) Organic based rice intensification innovation technology (IPAT-BO) for increasing the soil health and rice productivity toward the food sovereignty in Indonesia. In: Conference proceedings. Bappenas international conference on best development practices and policies

Soil Research Center (2009) *Sesbania rostrata* and *Azolla pinnata* can partially substitute nitrogen fertilizers. News Agric Res Dev 31(3):11–12

Wabi MA, Verzelen K, Houngnandan P, Vanhove W, Lucas GKR, Van Damme P (2016) *Sesbania rostrata* (Bremek and Oberm.) as biological nitrogen fixator for sustainable lowland rice production. Tropentag. University of Natural Resources and Life Sciences, BOKU Vienna, Austria, pp 18–21

Wahyuni TH, Kusnadi H, Honorita B (2017) The nutrient status of organic carbon and nitrogen of paddy fields in three districts in Bengkulu province. In: Proceedings of the suboptimal land national seminar

Xiao Y, Ma W, Lu N, Wang Z, Wang N, Zhai W, Kong L, Qu G, Wang Q, Wang J (2019) Genetic variation of growth traits and genotype-by-environment interactions in clones of Catalpa bungei and Catalpa fargesii f. duclouxii. Forests 10:1–19

An Assessment of the Impacts of Climate Change on African Catfish Fingerling (*Clarias gariepinus* Burchell, 1822)

Jude Awoke and Johnny Ogunji

Abstract Climate change standard forecasts have heightened incidence of severe weather actions like floods and protracted famine. This in effect increases periodic volume of water level fluctuation in lentic water bodies thereby altering their natural composition. Too much water level variation is projected to have more serious effect on the morphology, role and biodiversity of marshy localities than effects due to increase in temperature due to change in climate pattern. This study was undertaken, to unravel specific areas and extent of water level effects on fishes. It is specifically targeted at assessing the effect of water levels on *Clarias gariepinus* fingerlings. The results will be useful for aquaculture as adaptation strategies are needed to contain the effect of water level fluctuation in the wild. A total of 180 *Clarias gariepinus* fingerlings were randomly distributed into 18-plastic tanks at 10-fish per tank for a 56-day growth trial. Growth performance, haematology and carcass composition were assessed at the end of the experiment. Results showed a significant difference ($P > 0.05$) in mean weight gain of fishes reared at varying water levels. The packed cell volume, red blood cell; mean corpuscular haemoglobin and corpuscular volume also showed significant differences ($P > 0.05$). Results indicate that water level variation has significant effect on haematology, growth and survival of *Clarias gariepinus* fingerlings. These results highlight the vulnerability of fish, especially *Clarias gariepinus* in the face of climate change.

Introduction

Global weather conditions are fluctuating. Climate change is being countered all over the world and also portraying its physical impacts everywhere (Shakoor et al. 2015).

J. Awoke (✉)
Department of Fisheries and Aquaculture, Ebonyi State University (EBSU), Abakaliki, Nigeria
e-mail: judeawoke786@yahoo.com

J. Ogunji
Department of Fisheries and Aquaculture, Alex Ekwueme Federal University, Ndufu Alike (AE-FUNAI), Abakaliki, Ebonyi, Nigeria
e-mail: ogunjijo@yahoo.com

W. Leal Filho et al. (eds.), *Climate Change, Hazards and Adaptation Options*, Climate Change Management, https://doi.org/10.1007/978-3-030-37425-9_47

The changing weather patterns affect both the amount and quality of water resources available for drinking, irrigation, fish farming, and power generation (EPA 2010). Unfortunately it is also bringing about increasing temperatures, giving rise to both negative as well as positive impacts on fishing and fish culture systems according to the region and latitude (Magawata and Ipinjolu 2014). Due to rising demands and universal climatic change, the water level structure in most freshwater bodies will probably adjust (IPCC 2007) as a result of drought or flood effect.

Arnell and Gosling (2016) have stated that flood-related impacts are expected to worsen due to global environmental change with flood risk increasing by 187% from increasing temperature in the HadCM3 climate model. Alfieri et al. (2017) reported that flood magnitude will increase due to intensified water cycling resulting from as little as a 1.5 °C global average temperature increase. Flooding according to Talbot et al. (2018) is a major disturbance that impacts aquatic ecosystems and the ecosystem services that they provide. It is however important to note that the effects of flooding on aquatic ecosystems can be both negative and positive.

Intense or premature water level fluctuation cause unfavourable impacts on the environment and on living organisms (Bond et al. 2008). Adding to water losses in underground sources and other types of water usage, climate change forecasts a heightened incidence of weather-related natural disasters like floods and protracted famine (IPCC 2013). This may in turn increase the periodic volume of water level fluctuation in lentic water bodies thereby altering their natural composition (Magnuson et al. 1997). Too much water level variation is projected to have more serious effect on the morphology, role and biodiversity of marshy localities than an increase in temperature due to change in climate pattern (Hulme 2005; Abrahams 2008). Diminished amount of normal water level variation is equally thought to be unfavourable as was experienced in the low lying man-made water bodies in Holland (Coops and Hosper 2002).

Therefore, an investigation on the effect of water levels on the growth performance, survival and haematology of *Clarias gariepinus* fingerlings is valuable in understanding the interrelationship between water level variation and modification to the nature of aquatic animals. No recorded study about the effect of water level variations on the haematology and of *Clarias gariepinus* is available. Considering the serious effect of climate change on fish and fish culture, this study will provide information on the effect of slight water level changes on *Clarias gariepinus.* It will illuminate the sensitivity of *Clarias gariepinus* to water level alterations and climate change effect. The results will be useful for aquaculture in developing adaptation strategies to contain the effect of water level fluctuation in the wild.

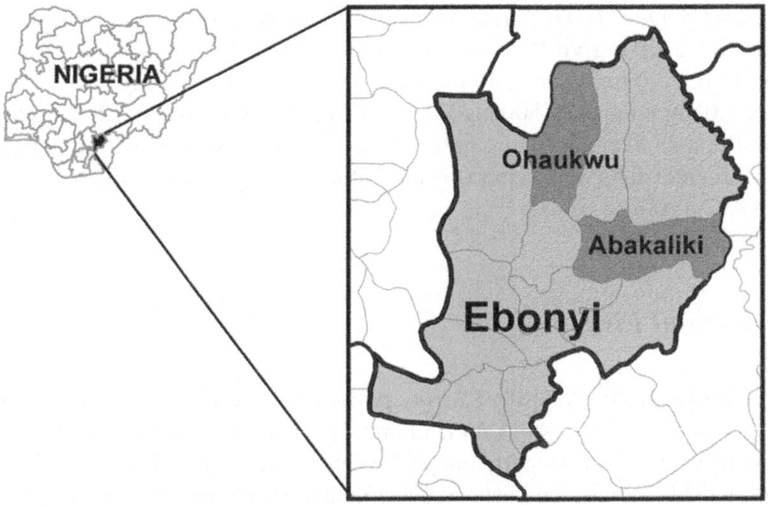

Fig. 1 Map of Ebonyi State showing the study area. *Source* Ebonyigov

Materials and Methods

Study Area

The study was carried out at the Department of Fisheries and Aquaculture, Faculty of Agriculture and Natural Resource Management, Ebonyi State University Abakaliki, Nigeria. Ebonyi State is located between Latitude: 6° 15′ 18″ N, Longitude: 8° 05′ 55″ E (Fig. 1) and it has Land Area—5533 km^2 on the South-Eastern part of Nigeria (EbonyiGov 2019). The mean temperature range is 27–31 °C minimum and maximum, respectively. The annual rainfall range is 1700–2000 mm with a mean of about 1700 mm. The relative humidity at dry weather and rainy season is between 60 and 90% (Ofomata 1995).

Preparation of Experimental Diets

One experimental diet, with protein content of 41.97% dry matter, was prepared. Housefly maggot meal (50%) was used as the main dietary protein source in the diet. Other ingredients that were used in the formulation include soya bean meal (34%), maize (14.5%), fish oil (0.25%), groundnut oil (0.25%) and vitamin premix (1%). The diets were pelleted using pelleting machine and the pellets were sundried. Maggot meal was produced according to the description of Ogunji et al. (2008). Vitamin premix supplying for each kilogram of food at 5 kg/tone addition: 20,000 i.u, 30 mg Manganese, 4 mg Copper, 40 mg Zinc, 0.2 mg Selenium, 100 mg Lysine,

100 mg Methionine, 100 mg Anti-oxidant, Vitamin A, 2000 i.u,Vit. D3, 200 mg Vit E, 8 mg Vit K3, 20 mg Vit B1, 30 mg Vit B2, 12 mg Vit B6, 50 mg Pantothenic acid, 0.8 mg Biotin, 2.0 mg Cobalt, 40 mg Iron, 5.0 mg lodine, 150 mg Niacin, 0.05 mg Vit B12, 4.0 mg Folic acid, 500 mg Vit C, 600 mg Choline chloride, 200 mg Inositol, 200 mg Betaine. Pearson square method was used in feed formulation. Samples of the experimental diets were sent to the laboratory for proximate analysis as described by A.O.A.C. (2000).

Experimental Fish

One hundred and eighty (180) *Clarias gariepinus* juveniles were procured from Regina Pacis fish farm in Abakaliki, Ebonyi State and transported in a 50 L gallon filled with water to the Department of Fisheries and Aquaculture wet laboratory, Ebonyi State University. The fish were acclimated in a tarpaulin tank (4 × 2 × 1 m) for two weeks and were fed commercial fish feed (Coppens) throughout the 2 weeks acclimation period.

Experimental Design

A total of one hundred and eighty (180) *Clarias gariepinus* fingerlings (initial mean weight 4.33 ± 0.03 g) were randomly distributed into 18 aquarium plastic tanks at ten fish per tank for growth trials. The experimental tanks were organized in three locations namely: laboratory, outdoor and greenhouse. Test diets were randomly assigned using completely randomized design to triplicate tanks located in laboratory, outdoor and greenhouse. Each treatment location had two volumes of water (20 L and 10 L). The fish were fed a restricted ration of 5% body weight per day in two portions (by 9.00 and 15.00 h) for 56 days in static water. Quantity of feed was adjusted forth nightly after batch-weighing of experimental fish. The aquaria were cleaned and water partly replaced by siphoning every three days to avoid fouling resulting from faeces and uneaten food. The experiment was conducted between the months of June and August, 2017.

Water Quality Analysis

Water temperature, dissolved oxygen, pH, nitrate, nitrite and ammonia were checked daily using mercury in glass bulb thermometer, Fresh Innovative Multitech, NIFFRI and water testing kit (Nice chemicals India) respectively in the six treatment tanks. Water temperature in the experimental tanks was measured three times daily in the morning (6.00 h), afternoon (14.00 h) and evening (18.00 h).

Growth Analysis

At the end of the experiment, all the fish was weighed and data obtained from triplicate tanks were used to calculate weight gains, specific growth rate (SGR), feed conversion ratio (FCR) and percentage body weight. Weight gain = final weight – initial weight,

$$\mathrm{SGR} = (\mathrm{LnW2} - \mathrm{LnW1})/(\mathrm{T2} - \mathrm{T1})100$$

where, W1 and W2 = initial and final weight of fish and T1 and T2 = time in days.

$$\mathrm{FCR} = \frac{Feed\ fed}{live\ weight\ gain}$$

Protein efficiency ratio (PER) = live weight gain (g)/protein fed (g)

where

F1 = number of fish at the end of experiment,
F2 = number of fish at the beginning of experiment.

Biochemical and Heamatological Analysis

For the purpose of blood analysis, samples were collected at the beginning and at the ending of the experiment. Blood was drawn from the caudal peduncle of the fish using a disposable needle and syringe into an EDTA sample bottle to avoid clotting. Packed cell volume (PCV) was analyzed with micro-haematocrit by means of heparinized 25 mm capillary tubes. Red and white blood cell counts were analyzed as described by Blaxhal and Diasley (1973). While, haemoglobin concentration (Hb) was estimated using the method outlined by Wedemeyer and Yasutake (1977). Other haematological indices like mean cell haemoglobin (MCH), mean cell volume (MCV) and mean cell haemoglobin concentration (MCHC) were determined using the formula put forward by Dacie and Lewis (2001) thus:

$$\mathrm{MCH\ (pg)} = \left[\mathrm{Hb}\left(\mathrm{g\ dl^{-1}}\right) \times 10\right]/\mathrm{RBC}\left(10^6\ \mu\mathrm{L}^{-1}\right)$$

$$\mathrm{MCV\ (fl)} = \mathrm{Hct/RBC}\left(10^6\ \mu\mathrm{L}^{-1}\right)$$

$$\mathrm{MCHC}\left(\mathrm{g\ l^{-1}}\right) = \left(\mathrm{Hb}\left(\mathrm{g\ dl^{-1}}\right) \times 10\right]/\mathrm{Hct} \times 100$$

Protein (N × 6.25) was determined by the Kjeltec System (Tecator) and crude fat by Soxtec.

System HT (Tecator) using petroleum ether. Ash was determined by burning in a muffle furnace at 550 °C for 10 h. Gross energy was calculated using the following factors: crude protein = 23.9 kJ/g, crude lipids = 39.8 kJ/g and NFE = 17.6 kJ/g (Schulz et al. 2005).

Statistical Analysis

Data resulting from the experiment were subjected to one-way analysis of variance (ANOVA). The significance of difference between means was determined by Duncan's Multiple Range Tests (DMRT) using the SPSS computer statistic package for windows 7 (version 21). Values were expressed as means ± SE.

Results

Fish death was observed in the greenhouse treatment when water temperature reached 40 °C. All the experimental fish in the greenhouse treatment died on the 8th day of exposure. Consequently, there was no fish in this particular treatment for subsequent and final sample collection at the end of the experiment. Nonetheless, there was no significant ($P > 0.05$) mortality in the remaining treatments. Mean values and standard error (SE) of all water quality parameters for each treatment group is presented (Table 1). Temperature values differed significantly ($P < 0.05$) in the three locations but did not differ significantly in the varying water levels (volumes). Highest temperature value was recorded in greenhouse tanks. That explains why experimental fishes died after one week of the experiment. Dissolved oxygen, pH, nitrate, nitrite and ammonia were observed not to have significant difference ($P < 0.05$) in all treatment groups.

Proximate nutrient compositions of feedstuffs and experimental diet used in this trial are presented (Tables 2 and 3). The dietary crude protein composition was 41.97% while dietary crude fat was 3.56%. The Food Conversion Ratio (FCR), Standard Growth Rate (SGR) and other growth performance parameters of *Clarias gariepinus* fingerlings fed experimental diet are presented (Table 4). At the ending of the experimentation, the fingerlings increased from an initial mean body weight of 4.33 g to represent a final mean body weight 13.10 g. Highest mean weight gain MWG and SGR were observed in *Clarias gariepinus* reared in (LB20), followed by fish in (OD20) and (LB10) while (OD10) showed least growth. FCR presented no significant ($P > 0.05$) difference among the treatment groups. However, mean weight gain was highest in (LB20). PER showed no significant difference between (LB20) and (OD20) but there was significant difference ($P < 0.05$) between (LB20) and (OD10).

The body composition proximate of *Clarias gariepinus* fed experimental diet is presented (Table 5). At all phases of the experiment, it was noticed that the dietary protein increased in all the treatments when compared with the initial status. However, fish reared in 20 L of water (OD20) accumulated more body protein and was

Table 1 Water quality parameters in experimental tanks*

Parameter	LB20[1]	LB10[2]	OD20[3]	OD10[4]	GH20[5]	GH10[6]
Temperature (°C)	26.53 ± 0.01^{b}	26.55 ± 0.01^{b}	26.06 ± 0.01^{a}	26.19 ± 0.01^{a}	31.52 ± 0.00^{c}	30.96 ± 0.25^{c}
Dissolved oxygen (mg/L)	5.29 ± 0.00^{c}	5.15 ± 0.00^{b}	5.20 ± 0.00^{b}	5.20 ± 0.00^{c}	0.67 ± 0.00^{a}	0.66 ± 0.00^{a}
pH	6.78 ± 0.00	6.79 ± 0.00	6.79 ± 0.00	6.80 ± 0.00	1.02 ± 0.00	1.05 ± 0.00
Nitrate (mg/L)	1.83 ± 0.00	1.83 ± 0.00	1.83 ± 0.00	1.83 ± 0.00	0.12 ± 0.00	0.12 ± 0.00
Nitrite (mg/L)	0.01 ± 0.00^{a}	0.01 ± 0.00^{a}	0.00 ± 0.00^{a}	0.00 ± 0.00^{a}	0.00 ± 0.00^{a}	0.00 ± 0.00^{a}
Ammonia (mg/L)	2.38 ± 0.00	2.38 ± 0.00	2.08 ± 0.00	2.08 ± 0.00	0.21 ± 0.00	0.21 ± 0.00

*All the values are means of triplicate feeding groups and values in the same row with different superscripts are significantly different ($P < 0.05$)

[1]LB20 (Laboratory tank 20 L treatment)

[2]LB10 (Laboratory tank 10 L treatment)

[3]OD20 (Outdoor tank 20 L treatment)

[4]OD10 (Outdoor tank 10 L treatment)

[5]GH20 (Greenhouse tank 20 L treatment)

[6]GH10 (Greenhouse tank 10 L treatment)

Table 2 Proximate composition of maggot meal and soybean meal used for diet formulation

Proximate components	Maggot meal (MGM)	Soybean meal (SBM)
Crude protein (%)	44.87	43.78
Crude fat (%)	7.38	3.67
Crude fibre (%)	6.88	6.48
Crude ash (%)	7.95	5.96
Moisture content (%)	7.35	5.63
NFE (%)*	25.57	34.48

*Nitrogen free extract + fibre, (NFE) = 100 − (% protein + % fat + % ash)

Table 3 Proximate composition (%) of experimental diet (maggot meal)

Parameter	Diet
Crude protein (%)	41.97
Crude fat (%)	3.56
Crude fibre (%)	2.54
Crude ash (%)	7.69
Moisture content (%)	8.05
NFE (%)*	36.19
Total	100

*Nitrogen free extract + fibre, (NFE) = 100 − (% protein + % fat + % ash)

Table 4 Growth performance of *Clarias gariepinus* fingerlings fed experimental diet*

Parameters	LB20	LB10	OD20	OD10
Initial weight (g)	4.34 ± 0.26^{e}	4.22 ± 0.12^{e}	4.33 ± 0.03^{e}	4.31 ± 0.05^{e}
Final weight (g)	13.10 ± 1.47^{g}	$10.01 \pm 0.59^{f,g}$	11.17 ± 1.93^{g}	5.61 ± 2.94^{f}
Weight gain (g)	8.76 ± 1.21^{g}	$5.80 \pm 0.50^{f,g}$	6.83 ± 1.94^{g}	1.30 ± 2.98^{f}
FCR[1]	1.61 ± 0.10^{f}	1.87 ± 0.09^{f}	1.67 ± 0.73^{f}	1.97 ± 0.94^{f}
SGR[2]	1.96 ± 0.10^{g}	1.54 ± 0.07^{g}	1.64 ± 0.30^{g}	1.79 ± 0.43^{g}
PER[3]	0.21 ± 0.03^{g}	$0.14 \pm 0.01^{f,g}$	0.16 ± 0.05^{g}	$0.03 \pm 0.07^{e,f}$

*All the values are means of triplicate feeding groups and values in the same row with different superscripts are significantly different ($P < 0.05$)

[1]Food conversion ratio = food fed (g)/live weight gain (g)

[2]Specific growth rate (%/d) = (InW2 − InW1/T2 − T1) × 100

[3]Protein efficiency ratio = live weight gain (g)/protein fed (g)

Table 5 Initial and final carcass composition of *Clarias gariepinus* fingerlings fed experimental diet (%)[a]

	Initial status	LB20	LB10	OD20	OD10
Crude protein (%)	58.97 ± 0.00^{c}	63.97 ± 0.06^{d}	63.97 ± 0.06^{d}	71.28 ± 0.00^{h}	69.56 ± 0.00^{g}
Crude fat (%)	8.25 ± 0.01^{g}	5.87 ± 0.06^{c}	5.94 ± 0.00^{c}	6.89 ± 0.00^{f}	6.78 ± 0.00^{e}
Crude ash (%)	12.49 ± 0.01^{g}	7.58 ± 0.06^{c}	7.59 ± 0.00^{c}	9.07 ± 0.00^{d}	9.25 ± 0.00^{f}
Moisture (%)	8.4 ± 0.01^{h}	6.68 ± 0.06^{f}	6.89 ± 0.00^{g}	5.25 ± 0.00^{c}	5.37 ± 0.00^{d}
NFE (%)[b]	11.71 ± 0.01^{f}	15.90 ± 0.14^{g}	16.6 ± 0.00^{h}	7.51 ± 0.00^{c}	9.07 ± 0.00^{d}

[*]All the values are means of triplicate feeding groups and values in the same row with different superscripts are significantly different ($P < 0.05$)
[1]Nitrogen free extract + fibre, (NFE) = 100 − (% protein + % fat + % ash)

significantly different ($P < 0.05$) from other treatments. Crude ash, crude fibre, crude fat and moisture were significantly decreased ($P < 0.05$) from the initial status. Data from (GH20) and (GH10) were not shown because experimental fish in the treatments died before the end of the experiment.

Haematological indices of *Clarias gariepinus* fingerlings fed experimental diet and exposed to varying water levels are presented (Table 6). The packed cell volume results showed that the initial PCV count significantly decreased ($P < 0.05$) in all treatments except OD10. The results obtained for haemoglobin Hb showed that the initial status was not significantly different from OD10. However, other treatments

Table 6 Haematological indices of *Clarias Gariepinus* fingerlings fed experimental diet and exposed to varying water levels[*]

Parameter	Initial	LB20	LB10	OD20	OD10
PCV (%)	26.00 ± 0.58^{c}	20.00 ± 1.54^{b}	15.83 ± 0.44^{b}	20.00 ± 0.58^{b}	21.00 ± 1.15^{c}
Hb (g/100 mL)	8.70 ± 0.58^{c}	6.60 ± 0.23^{b}	5.30 ± 0.23^{b}	6.60 ± 0.17^{b}	7.13 ± 0.15^{c}
WBC (× $10^{9/L}$)	$78.00 \pm 57.74^{c,d}$	57.00 ± 230.94^{b}	51.00 ± 57.74^{b}	58.00 ± 173.20^{b}	64.33 ± 88.19^{c}
RBC (× $10^{12/L}$)	6.40 ± 0.12^{c}	3.80 ± 0.17^{b}	3.20 ± 0.07^{b}	6.90 ± 0.12^{d}	4.23 ± 0.12^{c}
MCHC (%)	30.56 ± 3.13^{b}	33.09 ± 0.76^{b}	33.45 ± 0.57^{b}	33.01 ± 0.09^{b}	34.10 ± 1.20^{b}
MCH (pg)	7.36 ± 0.18^{b}	17.93 ± 0.71^{d}	16.54 ± 0.37^{c}	9.58 ± 0.41^{c}	16.86 ± 0.14^{c}
MCV (fl)	40.62 ± 0.17^{c}	52.57 ± 0.65^{d}	49.47 ± 0.39^{c}	29.00 ± 1.34^{b}	49.53 ± 1.33^{c}

[*]All the values are means of triplicate feeding groups and values in the same row with different alphabet superscripts are significantly different ($P < 0.05$). *PCV* packed cell volume; *WBC* white blood cell; *RBC* red blood cell; *Hb* haemoglobin; *MCHC* mean corpuscular haemoglobin concentration; *MCH* mean corpuscular haemoglobin; *MCV* mean corpuscular volume

were significantly decreased ($P < 0.05$) from the initial status. With regards to WBC, all the treatment groups did not differ significantly except (OD10). RBC values in all treatment groups significantly decreased ($P < 0.05$) from the initial status except (OD10) which had no significant difference.

Discussion

In this experiment all the experimental fish in the greenhouse died after 8 days of exposure. This happened around 1400 h when the temperature in the water reached 40 °C. According to Cnaani (2006) temperature outside optimum limits negatively impacts the health of fish as a result of metabolic strain which eventually hinders immunity levels, reproductive capability and growth. The other water quality parameters obtained during the experiment were in the range allowed for *Clarias gariepinus* growth.

The totality of NH_3 observed in the rearing containers in this experiment may be as a result of excretory activities of the fish. This is in agreement with the studies carried out by Obasa et al. (2013) where the occurrence of NH_3 in water was ascribed to excretory actions of juvenile fish. Dissolved oxygen (5.15 ± 0.00 to 5.29 ± 0.00) and pH (6.78 ± 0.00 to 6.80 ± 0.00) values for the treatment groups were within the range suitable for rearing fish (Akinrotimi et al. 2010; Dienye and Olumuji 2014). Research has proved that fish feeding, growth and reproduction are improved at DO level of 5 mg/L and above. Fish will feed inadequately and will be malnourished at low concentration of DO that is below 3 mg/L (NAERLS 1996). NAERLS (1996) further posits that pH is a key chemical factor to consider because it influences metabolic activities and other body processes of aquacultural animals. A suitable pH range (6.8–8.7) ought to be sustained for satisfactory growth efficiency and production capacity.

Fish growth performance was good as evidenced by an increase in body weight and growth of the experimental fish. At the end of the experiment, the fish developed from an original body weight of 4.33 g to a finishing average body mass of 13.10 g. Growth performance of *Clarias gariepinus* was highest in (LB20) having final mean weight of 13.10 ± 1.47 g (Table 4) whereas (LB10) was 10.01 ± 0.59. Likewise (OD20) was 11.17 ± 1.93 and (OD10) 5.61 ± 2.94 (Table 5). Feed conversion ratio (FCR) is a significant indicator of the excellence of fish feed. A lower FCR indicate improved use of the fish feed (Mungo-Bundi et al. 2013). In this experiment, FCR presented no significant difference in all the treatments. However, the best FCR and SGR were observed in fish reared in high volumes of water (LB20 and OD20). It is known that FCR values of less than 1 have been documented even though 1.2–1.5 is the normal range for fish reared with properly compounded diet (De Silva and Anderson 1995; Ogunji et al. 2008). The low FCR of 1.61–1.98 exhibited by fish in all the treatment groups is an indicator that the fish utilized the feed well.

Results of the proximate composition of carcass in this study showed the different protein composition of the treatment groups. There was significant difference ($P <$

0.05) (Table 5) between the treatment groups. All the treatment groups showed higher levels of crude protein than the initial value. This shows that the experimental diet (Maggot meal) was well accepted and utilised by *Clarias gariepinus* (Fasakin et al. 2003; Ogunji et al. 2007; Idowu and Afolayan 2013). This result showed good fish performance because of an increment in the body mass of the experimental fishes. However, the crude protein composition in was highest in OD20 (71.25 ± 0.00). This may be ascribed to the reality that the water level of the tank was conducive for the fish and the temperature of the outdoor environment was at range acceptable for fish growth. This aligns with the work of Nlewadim et al. (2011) when he reported that higher water levels favoured the growth of *Clarias gariepinus* fingerlings at both early and later stages of growth.

All the blood characteristics determined in this experiment fell within the authorized body limits documented for *Clarias gariepinus*. Packed cell volume (PCV) limits 20–26% recorded in this investigation are inside the limits of 20–50% documented by Zarejabad et al. (2009) and Dienye et al. (2014). However, values above 50% are rarely reported (Etim et al. 1999). Though initial PCV differed significantly ($P < 0.05$) in all the treatment groups except OD10 treatment. PCV is used to find out the ratio of plasma to corpuscles in the blood and the oxygen-carrying capacity of the blood (Audu et al. 2014).

Haemoglobin is a very important constituent of blood. It works by binding to oxygen and then transported to the rest of the body. When haemoglobin is low, it means that the body does not have the ability to carry oxygen as efficiently as it should. The haemoglobin result showed a decrease from the initial value. The haemoglobin (Hb) range 5.30–8.70 g/100 mL (Table 6) recorded in this study was high and fell within the range 5.6–15.8 g/100 mL documented for *Esox lucius* (Mulcahy 1970). It equally matched appropriately with 8.7 g/100 mL for *Clarias gariepinus* (Sowunmi 2003). These records were also more than 4.46 g/100 mL as reported by Fagbenro et al. (2000) for *Heterotis niloticus*. In this study however, the decrease in haemoglobin from the initial value does not constitute any problem since the values still fell within the required standards.

Lymphocytes are very abundant cells which consist of leucocytes. Their function is to produce antibodies and other chemically oriented materials that protect the body from infections. The basic implication of detected alterations in the quantity of leucocytes in fish subjected to stressful conditions is the repression of the immunity condition and a high vulnerability to infections (Omitoyin 2006). In the present investigation the WBC count reduced in all the treatment groups when compared with the initial value. White blood cells (WBC) and lymphocytes are the defence cells of the body and the quantity has consequence on immunity response and the capability of the organism to ward off infections. An elevated WBC count is typically linked with bacterial infections or the circulating system (Oyawoye and Ogunkunle 1998). Thus, the low WBC recorded in this investigation is a sign that the fish had no microbial contamination and this is advantageous to the health condition of the fish.

Also, a decrease in the erythrocytes (RBC) was detected in the treatment groups except OD20. RBC for OD 20 (outdoor tank 20) (Table 6) was higher than the initial value. It is acknowledged that a low quality and small amount of red blood cells as well as a low level of haemoglobin leads, to a decline in the availability of oxygen in the body. Apart from transportation of oxygen, red blood cells perform other important functions in the body and inadequate amounts and quality of red blood cells would have a chain of consequences on metabolic activities other than just the provision of oxygen for respiration (Gross et al. 1996).

The mean corpuscular haemoglobin concentration (MCHC) did not differ between treatment groups. The MCHC range (30.56–34.10%) reported in this experiment for fish fed mainly on maggot meal diet measured favourably well at 30.7% as documented for catfish reared in Asejire dam in Oyo state (Adedeji and Adegbite 2011). Mean corpuscle haemoglobin (MCH) and mean cell volume (MCV) results showed a significant increase from the initial status. The increased values of MCH and MCV are signs of the extent of shrinking cell size of erythrocytes, strained due to fluctuating temperatures (Akinrotimi et al. 2010). The reduction in haematological characteristics observed in this study implies that the varying water levels in the rearing containers may have affected the normal functioning of the fish physiology.

Conclusion

Result from this study implies that water level variations affected the growth and body composition of *Clarias gariepinus*. The mean weight gain MWG, FCR, SGR and PER showed significant difference between the treatment groups. Fish reared in different water levels showed significant difference in growth. Fish reared in higher water level (LB20) and (OD20) performed better than fish reared in lower water level (LB10) and (OD10). However, the experimental fish showed good results in growth performance and body composition. In the same vein, fish reared in (OD20) and (OD10) showed significant difference in carcass composition. Furthermore, fish reared in lower water level especially (LB10) showed lower haematological indices than fish reared in higher water level. It is also recommended to rear fish in higher water volumes since this study has demonstrated that *Clarias gariepinus* perform better in higher water levels.

In the bid to device adaptation strategies for cushioning climate change effect, findings of this study will facilitate increase in fish production which will lessen the consequences of global warming on man and food security. The following issues need to be put in their right perspective in line with the conclusions drawn from this study. Climate change and global warming are altering rainfall patterns, making heavy rain more frequent in many areas and this gives rise to increase in water volume due to flooding. However this can be seasonal. This increase in water volume through flooding amplifies the dynamics of the watercourse and affect energy expenditure by fish. The increase in dynamics of the watercourse can bring about destruction of breeding areas and sweeping away of eggs thus inimically affecting reproduction. On the other hand predatory species may also be introduced into habitats which will destabilize and undermine fish production.

All these are aspects that were not evaluated in this study. More close study should be designed bearing in mind the aforementioned. Additional meticulous research is required to corroborate this result.

References

Abrahams C (2008) Climate change and lakeshore conservation: a model and review of management techniques. Hydrobiologia 613:33–43. https://doi.org/10.1007/s10750-008-9470-5

Adedeji OB, Adegbite AF (2011) Comparative haematological parameters of the Bagrid catfish (*Chrysichthys nigrodigitatus*) and African catfish (*Clarias gariepinus*) from Asejire Dam Southwestern Nigeria. J Appl Sci Res 7(7):1042–1046

Akinrotimi OA, Uedeme-Naa B, Agokei EO (2010) Effects of acclimation on haematological parameters of *Tilapia guineensis* (Bleeker, 1862). Sci World J 5 (4):1–4

Alfieri L, Bisselink B, Dottori F, Naumann G, Wyser K, Feyen L, de Roo A (2017) Global projections of river flood risk in a warmer world. Earth's Future 5:171–182

Arnell NW, Gosling SN (2016) The impacts of climate change on river flood risk at the global scale. Clim Change 134:387–401

Association of Official Analytical Chemists International (AOAC) (2000) Official methods of analysis, 17th edn. AOAC International, Gaithersburg, MD, USA

Audu BS, Adamu KM, Nonyelu ON (2014) Changes in haematological parameters of *Clarias gariepinus* exposed to century plant (*Agave americana*) leaf dust. Int J App Bio Res 6(1):54–65

Blaxhall J, Daibley S (1973) Routine hematology methods for use with fish blood. J fish Bio 5:771–781. https://doi.org/10.1111/j.1095-8649.1973.tb04510.x

Bond NR, Lake PS Arthington AH (2008) The impacts of drought on freshwater ecosystems: an Australian perspective. Hydrobiologia 600:3–16

Cnaani A (2006) Genetic perspective on stress and disease resistance in Aquaculture. Isr J Aqua Bamidgeh 58:375–383

Coops H, Hosper SH (2002) Water-level management as a tool for the restoration of shallow lakes in the Netherlands. Lake Res Man 18(4):293–298

Dacie JV, Lewis SM (2001) Practical Haematology 9th edn. Churchill Livingstone, London. p 633

De Silva SS, Anderson TA (1995) Fish nutrition in aquaculture. Chapman and Hall, London, p 31

Dienye HE, Olumuji OK (2014) Growth performance and haematological responses of African mud catfish *Clarias gariepinus* fed dietary levels of *Moringa oleifera* leaf meal. Net J Agri Sci 2(2):79–88

Ebonyi Government (2019) Ebonyi state profile. http://www.ebonyistate.gov.ng/profile.aspx. Retrieved 6/3/19

EPA—United States Environmental Protection Agency (2010) Climate change and society. https://digital.library.unt.edu/ark:/67531/metadc949667/m2/1/high_res_d/Climate_Change_Society.pdf. Retrieved 6/3/19

Etim L, Lebo PE, King RP (1999) The dynamics of an exploited population of Siluroid Catfish (*Schilbe Intermedius Ruppell*, 1832) in the Cross River. Nig Fish Res 40:295–307

Fagbenro O, Adedire CO, Ayotunde EO, Faminu EO (2000) Haematological profile, food consumption and enzyme assay in the gut of the African bony–tongue fish *Heterotis niloticus* (clupisudis) (Cuvier 1829) (Osteoglossidae). Trop Zoo 13:1–9

Fasakin EA, Balogun AM, Ajayi OO (2003) Nutrition implication of processed maggot meals; hydrolysed, defatted, full-fat, sun-dried and oven dried in the diets of *Clarias gariepinus* fingerlings. Aqua Res 9(34):733–738

Gross R, Gliwitzki M, Gross P, Klaus F (1996) Food and nutrition. Bulletin 17(1). http://archive.unu.edu/unupress/food/8F171e/8F171E00.htm. Retrieved 6/1/19

Hulme PE (2005) Adapting to climate change: is there scope for ecological management in the face of a global threat? J App Ecol 42(5):784–794

Idowu EO, Afolayan EB (2013) The effects of supplementing of fish meal with maggots at varying levels in the diet of Clarias gariepinus. Int Arc App Sci Tech 4(4):41–47

Intergovernmental Panel on Climate Change (IPCC) (2007) Climate change 2007: synthesis report. In: Pauchauri RK, Reisinger A (eds) Contribution of working groups I, II, and III to the fourth intergovernmental panel on climate change. IPCC, Geneva, Switzerland. https://www.ipcc.ch/site/assets/uploads/2018/02/ar4_syr_full_report.pdf Retrieved 02/7/2019

Intergovernmental Panel on Climate Change (IPCC) (2013) Climate chance 2013. The physical science basis. In: Stocker TF, Qin D, Plattner GK, Tignor M, Allen SK, Boschung J, Nuells A, Xia Y, Bex V, Midgley PM (eds) Contribution of working group I to the fifth assessment report of the intergovernmental panel on climate change. Cambridge University Press, New York, pp 255–317

Magawata I, Ipinjolu JK (2014) Climate change: mitigation and adaptation strategies in fisheries and aquaculture in Nigeria. J Aquatic Sci 9(4):257–261

Magnuson JJ, Webster KE, Assel RA, Bowser CJ, Dillon PJ, Eaton JG, Evans HE, Fee EJ, Hall RI, Mortsch LR (1997) Potential effects of climate changes on aquatic systems: Laurentian Great Lakes and precambrian shield region. Hydrol Process 11(8):825–871

Mugo-Bundi J, Oyoo-Okoth E, Ngugi CC, Manguya-Lusega D, Rasowo J, Chepkirui-Boit V (2013) Utilization of *Caridina nilotica* (Roux) meal as a protein ingredient in feeds for Nile tilapia (*Oreochromis niloticus*) Aqua Res 2:1–12

Mulcahy MF (1970) Blood values in the pike, *Esox lucius* (L.). J Fish Bio 2:203–209

National Agricultural Extension and Research Liaison Services (NAERLS) (1996) Water quality management in fish culture. Extension bulletin no. 98 Fisheries series no. 3

Nlewadim AA, Udoh JP, Aniekan JO (2011) Growth response and survival of *Heterobranchus longifilis* cultured at different water levels in outdoor concrete tanks. AACL Bioflux 4(3):404–411. http://www.bioflux.com.ro/aacl

Obasa SO, Alegbeleye WO, Akinyemi AA, Idowu AA, Bamidele NA, Adesanya AN (2013) Replacement of maize meal by toasted African breadfruit (*Treculia africana*) seed meal in the diet of *Clarias gariepinus* (Burchell 1822) fingerlings. Livestock Res Rural Dev 25(108). Retrieved 21 Nov 2015, from http://www.lrrd.org/lrrd25/6/obas25108.htm

Ofomata GEK (1995) Nigeria in maps: Eastern states. In: Ofomata GEK (ed) Ethiopic Publishing Cooperation, Benin, City, pp 14

Ogunji JO, Nimptsch J, Wiegand C, Schulz C (2007) Evaluation of the influence of Housefly maggot meal (Magmeal) diets on Catalase, Glutahione, S-Transferase and Glycogen concentration of the liver of *Oreochromis niloticus fingerling*. Comp Biochem Physiol Part A 20(147):942–947

Ogunji JO, Kloas W, Wirth M, Schulz C, Rennert B (2008) Housefly maggot meal (Magmeal) as a protein source for *Oreocromis niloticus* (Linn). Asian Fish Aqua Sci 8:141–147

Omitoyin BO (2006) Haematological changes in the blood of *Clarias gariepinus* (Burchell 1822) juveniles fed poultry litter. Livestock Res Rural Dev 18(11):23–34

Oyawoye EO, Ogunkunle M (1998) Physiological and biochemical effects of raw Jack Beans on broilers. Pro Ann Conf Nig Soc Ani Prod 23:141–142

Schulz C, Knaus U, Wirth M, Rennert B (2005) Effects of varying dietary fatty acid profile on growth performance, fatty acid, body and tissue composition of juvenile pike perch (*Sander lucioperca*) Aqua Nutr 11:1–11

Shakoor U, Saboor A, Baig I, Afzal A, Rahman A (2015) Climate variability impacts on rice crop production in Pakistan. Pak J Agri Res 28:19–27

Sowunmi AA (2003) Haematology of the African catfish, *Clarias gariepinus* (Burchell 1822) from eleyele reservoir, Ibadan, Southwest Nigeria. The Zoologist 2(1):40–44

Talbot CJ, Bennett EM, Cassell K, Hanes DM, Minor EC, Paerl H, Raymond PA, Vargas R, Vidon PG, Wollheim W, Xenopoulos MA (2018) The impact of flooding on aquatic ecosystem services. Biogeochemistry 141:439–461

Wedemeyer GA, Yasutake WT (1977) Clinical methods for the assessment of the effects of environmental stress on fish health. Technical papers of the US fish and wide life service no. 89. Washington, DC, US. Dept of the interior, fish and wild life service, p 180

Zarejabad AM, Sudagar M, Pouralimotlagh S, Bastami KD (2009) Effects of rearing temperature on hematological and biochemical parameters of great sturgeon (*Huso huso* Linnaeus, 1758) juvenile. Comp Clin Path Clin Pathol 19(4):367–371. https://doi.org/10.1007/s00580-009-0880-1

A Novel Transdisciplinary Methodology and Experience to Guide Climate Change Health Adaptation Plans and Measures

Marilyn Aparicio-Effen, James Aparicio, Cinthya Ramallo, Mauricio Ocampo and Gustavo J. Nagy

Abstract Climate change and socio-economic and environmental determinants of health (SDH) are the first acknowledged root-causes of infectious diseases. Controlling for SDH would reduce disease burden and promote adaptation. How can we determine which non-health sectors contribute the most and how to health vulnerability? No interdisciplinary and participative methodologies have yet been devised to address, from a complex systems perspective, the degree of responsibility non-health sectors and climate change have in disease occurrence, as a basis for adaptation. This study aims to identify climate change adaptation options for Dengue fever based on Eco-health and watershed approaches to influence public policies in the Bolivian Chaco Ecosystem. We carried out a transdisciplinary "Methodology for climate change Health Vulnerability Assessment considering Eco-health and Watershed Approaches" (MHVA), in selected areas of Pilcomayo watershed. Results established the level and type of current and future Dengue vulnerability, and the degree of responsibility of health and non-health sectors. Then, adaptation

M. Aparicio-Effen (✉) · C. Ramallo
Unidad de Cambio Climático, Ambiente y Salud, Instituto Boliviano de Biología de la Altura (IBBA), Facultad de Medicina, Universidad Mayor de San Andrés, La Paz, Bolivia
e-mail: marilyneffen@gmail.com

C. Ramallo
e-mail: cynthia.ramallo@gmail.com

J. Aparicio
Investigador Asociado, Unidad de Cambio Climático, Ambiente y Salud. Instituto Boliviano de Biología de Altura (IBBA) Facultad de Medicina. Universidad Mayor de San Andrés, La Paz, Bolivia
e-mail: james.k.aparicio@gmail.com

M. Ocampo
Investigador Asociado, Unidad de Cambio de Climático, Ambiente y Salud. Instituto Boliviano de Biología de Altura (IBBA) Facultad de Medicina. Universidad Mayor de San Andrés, La Paz, Bolivia
e-mail: mauiocampo@gmail.com

G. J. Nagy
Instituto de Ciencias Ambientales y Ecología, Facultad de Ciencias (IECA-FC), Universidad de La República (UdelaR), Iguá 4225, POB 11400, Montevideo, Uruguay
e-mail: gnagy@fcien.edu.uy

W. Leal Filho et al. (eds.), *Climate Change, Hazards and Adaptation Options*, Climate Change Management, https://doi.org/10.1007/978-3-030-37425-9_48

options were participatory prioritised, designing a *Climate Change Health Strategic Adaptation Plan,* which implementation has already begun. The use of MHVA and current adaptation experience would help in identifying the most vulnerable locations and target adaptation actions.

Introduction

Climate change influences the environmental, economic and social determinants of health such as air quality, access to safe water, food availability, family income, education, housing, and essential services. Social determinants of health (SDH) refer to the conditions in which people are born, live and work (Irwin and Scali 2010). Climate change affects health in three ways (IPCC 2012, 2014):

1. Directly, such as the mortality and morbidity due to extreme weather events in which climate change may play a role.
2. Indirect impacts from environmental and ecosystem changes, such as shifts in patterns of disease-carrying mosquitoes and ticks, or increases in waterborne diseases due to warmer conditions and increased precipitation and runoff.
3. Indirect impacts mediated through societal systems (e.g. undernutrition from altered agricultural production, stress caused by population displacement, or other environmental stressors, and damage to health care systems by extreme weather events).

Diseases have a multifactorial origin, considering the many determinants that influence them to which has been added climate change and variability, and extreme events. Thus, "*Climatic factors, coupled with multiple human, biological and ecological determinants, influence the emergence and re-emergence of infectious diseases such as arbovirosis, malaria, leishmaniasis, Chagas disease*" (Amarakoon et al. 2008). With the global expansion of vectors that are able to transmit Dengue, Zika, and Chikungunya virus (Kraemer et al. 2015), neglected tropical diseases that are associated with climate dynamics reiterate the need for intensified political attention and global health action (Watts et al. 2017), particularly in Latin America (Leal Filho et al. 2018).

Operationally, the management of health determinants is a responsibility of health sector as well as of other sectors, for example an optimal quality and availability of water will favour or avoid, the occurrence of cases of diarrhoeas, or the number of hours of water supply contributes to occurrence or not, of cases of dengue fever. Such conditions exist outside the domain where the health sector has direct influence since in most countries water management agencies are not part of health sector. So the policies, strategies, and performance of other sectors will have visible results on occurrence of different diseases and health outcomes.

As far as the stakeholders are concerned, there are some requirements which also need to be met, if their vulnerability is to be reduced, such as to foster more awareness on the means via which vector-borne diseases are transmitted, which may substantiate and add a greater degree of reliability to prevention efforts (Leal Filho et al. 2018).

However, no interdisciplinary and participatory methodologies have yet been devised to address, from a complex systems perspective, the degree of responsibility non-health sectors, and climate change and variability have in disease occurrence, as a basis for adaptation. The project "Climate Change, vulnerability and health-Colombia and Bolivia", funded by IDRC, carried out a transdisciplinary vulnerability assessment "Methodology for climate change Health Vulnerability Assessment considering Eco-health and Watershed Approaches" (MHVA) (Aparicio-Effen et al. 2016a), aimed at developing adaptation decisions, to reduce the deleterious impacts of climate change on health, which include vulnerability dimensions, climate and non-climate factors. We applied MHVA to assessing the climate vulnerability related to Dengue fever in Bolivian Pilcomayo watershed. After that, adaptation options were participatory prioritised, designing a Climate Change Health Strategic Adaptation Plan, which implementation has already begun.

The overall goal of this project research is: To identify climate change adaptation options for Dengue fever from Eco-health and integrated watershed approaches to influence public policies and reduce vulnerability to Dengue and associated diseases like the Chikungunya and Zika in Bolivian Chaco Ecosystem.

To achieve the goal as mentioned above we have worked with (i) decision-makers at the local, regional and national level; (ii) the health networks of project area; (iii) urban planners of the municipal governments and (iv) representatives of social movements and ethnicities (e.g. indigenous groups and urban, peri-urban and rural populations), of the Pilcomayo basin. We focused on strengthening climate adaptation capacities, which included the evaluation of the current and future vulnerability initially, the rise of adaptation options and their implementation.

The use of MHVA and current adaptation experience would help in identifying the most vulnerable locations and target adaptation actions.

Conceptual and Methodological Framework About Health Vulnerability Assessment

Methodology

The methodology followed in this project to assess climate-related health vulnerability to Dengue (V_{CCD}) considered Watersheed approach (IWRM) and Ecosystem approach to health (Eco-health) (Charron 2012). Eco-health is an emerging field of transdisciplinary research that brings together public health, environmental health people, veterinarians, ecologists, social scientists, policy makers, local authorities and experts from other fields and community members to explore how ecosystem changes can have adverse impacts on human health and implement practical solutions to address these health challenges (Hung and Tran 2016). Six principles of this approach include transdisciplinarity, participation, gender, and social equity, system-thinking, sustainability and research-to-action (Charron 2012). Eco-health research integrates different disciplines and world-views to tackle difficult health issues (Waage et al. 2010).

The Integrated Water Resources Management (IWRM) (Rahaman and Varis 2005) of river basins is a systematic process for the sustainable development, allocation, and monitoring of water resources used in the context of socio-economic and environmental objectives. When several agencies share responsibilities for drinking, irrigation, and environmental water, the lack of cross-sectoral linkages leads to uncoordinated water resource development and management, resulting in conflict, waste and unsustainable systems.

Climate vulnerability is "*a condition of a natural or human system with a propensity to be adversely affected by vulnerability factors such as exposure, the nature, and magnitude of the change in weather patterns, sensitivity, and adaptability*" (IPCC 2007). It should also be considered that all dimensions of human activity: physical, socio-economic and environmental act together (inter and transdisciplinary) to build vulnerability, serving as a basis to study the direct and indirect effects of climate on specific diseases (e.g. V_{CCC} for Chagas (Nagy et al. 2017).

We followed the "Methodology for climate change Health Vulnerability Assessment (MHVA) considering Eco-health and IWRM Approaches" to address, from a complex systems perspective, the degree of health and non-health sectors responsibilities and climate change effects on disease occurrence, as a basis for adaptation options (Aparicio-Effen et al. 2016a).

Vulnerability Factors

The adaptive capacity of health sector depends on vulnerability factors and dimensions, the types of human and socio-economic development, social organisation, and relationship with access, coverage to health services and social participation of community. The adaptive capacity of health sector and other health-related sectors is a fundamental part of the vulnerability. The climate vulnerability (V_{CC}) includes the following factors (IPCC 2007; Aparicio-Effen et al. 2016a):

$$V_{CC} = f\left(Exp, CMC, Sens, CR_{-Adap}\right)$$

where

Exp: Exposition, variation to which a system is exposed
Sen: The sensitivity
CMC: The character, magnitude, and rapidity of change in weather patterns

C_(R-Adap): The ability to respond and or adapt.

(a) Exposure: The exposure is determined by geographical location of an area of analysis, the physiographic and climatic characteristics, which can determine the condition of fragility or sensitivity that determines the greater or lesser impact of climate change.
(b) Sensitivity: characteristic of system that is being evaluated, which can be altered positively or negatively (resilient or fragile). The epidemiological and environmental dimensions are sensitive to climate variability, gradual global climate change, and changes in land use.
(c) Nature and magnitude of change: Changes generated in climate behaviour, due to natural variability, or anthropogenic climate change, which affects or alters the condition of the system under study. Climate Change Vulnerability (VCC) is Character and Magnitude of change = Differences (Δ) between current climate and historical climate and future climate and current climate = CMC: f (Δ1) (Nagy et al. 2017).

where: Δ1: ((current climate (CA) − Historical Climate (CH))
CMC: f − Δ 2
Δ 2: ((future climate (CF) − Current Climate (CA))
Future climate scenarios (Aparicio-Effen et al. 2016a; Nagy et al. 2016)

CMC represents the influence of climate in series that are available in the reference area which includes the thermal amplitude between the minimum and maximum average temperatures recorded and the precipitation patterns. These values are associated with altitude ranges in which people live and influence both temperature and precipitation. Thus the climatic index is obtained (IC) being its expression the following equation:

$$IC_i = IPP\ (Tmax - Tmin)/\Delta h$$

where:

ICi: Climate Index
IPP: Precipitation index that is the ratio between the observed precipitation during the studied period and the average precipitation over the analysed periods.
Tmax: Maximum average temperature of the studied period.
Tmin: Minimum average temperature of the studied period.
Δh: Temperature variation range due to the change in altitude determined by Ti: *i*th (observation of the average temperature), Hi: (*i*th observation of the altitude where the temperature was measured) and N (Number of observations of the ten years series of observation).

The valuation of the index mentioned above applies to each of the diseases to be analysed (e.g. Chagas, Dengue, Zika) and allows us to read how the climatic conditions are modified according to the precipitation rate and the average thermal amplitude.

Vulnerability Dimensions

The representation of vulnerability factors and interdisciplinary dimensions of the MHVA is shown in Fig. 1.

The vulnerability corresponds to a historical and temporal context, so the evaluation of climate system should be done considering baseline, the current and future climate, and spatial area of analysis (river basin, location, altitude, and topography) where evaluated human system is located (Aparicio-Effen et al. 2016a).

The IWRM approach seeks to balance the use of natural resources with its conservation, taking into account climate change within the river basin system, with subsystems (dimensions of vulnerability). The "integrated" concept includes goals of production and social development, with the consequent impact on human health. For this process, key actors for development and adaptation measures must be same inhabitants and users of watershed. In this context, the eco-health approach in watersheds (spatial scope), also incorporates the temporal scope of change (climate) and allows the improvement of productive and social interactions with current and future human and ecosystem needs, which potentially improves the health and equity of gender and race.

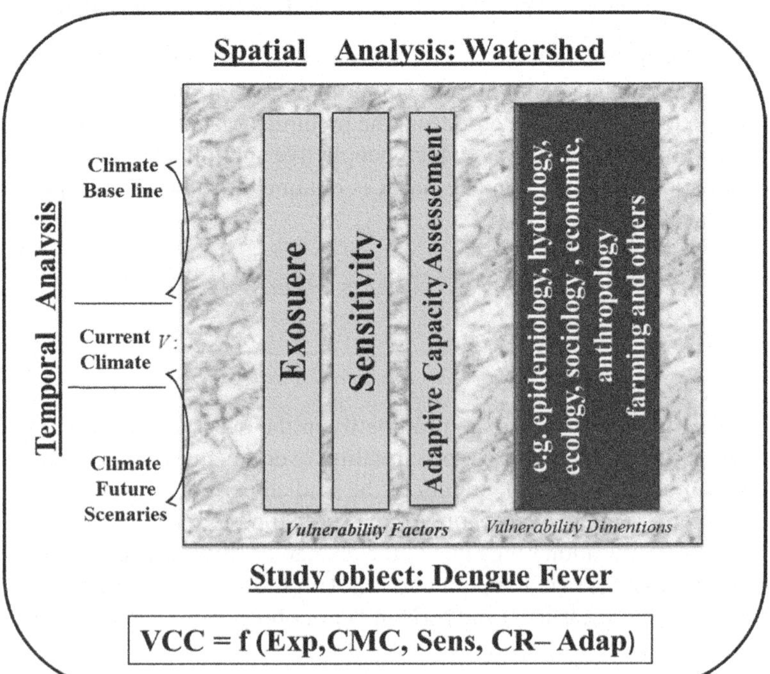

Fig. 1 The methodological framework for climate-related health vulnerability assessment considering eco-health and watershed approaches. Modified from Aparicio-Effen et al. (2016a) and Nagy et al. (2017)

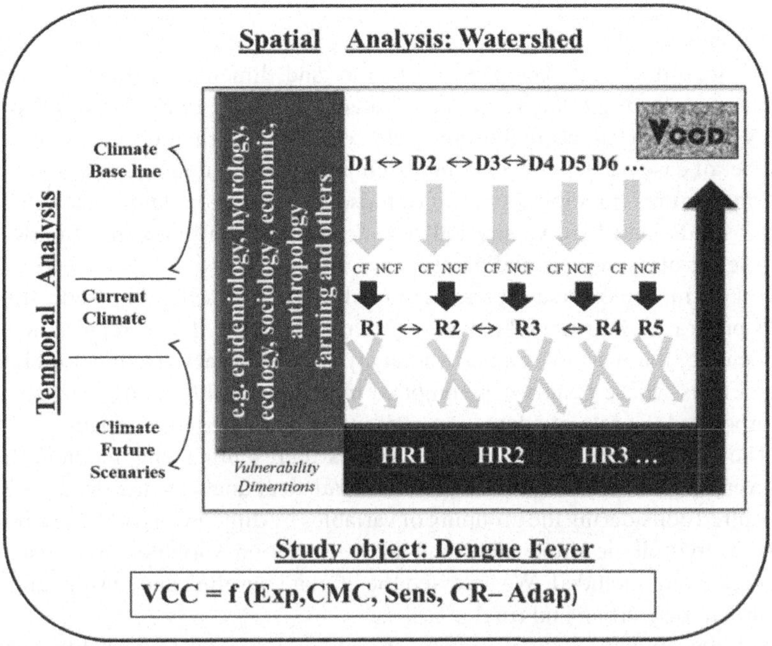

Fig. 2 Climate-related health vulnerability (V_{CCD}) constructivism process

The MHVA refers to a study object (or problem) related to discipline (e.g. epidemiology, hydrology, ecology, sociology, and climatology), science or sector, which analyses the exposure, sensitivity and adaptive capacity to the climatic and non-climatic pressures that exert influence on the object as mentioned earlier. The MHVA initially follows a deconstructivist process of the discipline favouring a health central element and leaving aside other no-health elements considered peripheral. This opposition, centre/periphery, can be inverted and deconstruct the discipline, by virtue of the "difference" and later these health central elements obtained from differents disciplines will follow a constructivist process for obtein climate change health vulnerability for specific disease, in this case Dengue Fever (VCCD) (Aparicio-Effen et al. 2016a; Nagy et al. 2017) (Fig. 2).

The MHVA allows the assessment of the V_{CC}. VCC results from the action of climate and non-climate factors on vulnerability dimensions of studied object with following time space variations and magnitude.

Vulnerability Results

The project determined the level and type of current and future V_{CCD} in the five prioritised municipalities of the Pilcomayo watersheed (Box 1).

Box 1 Health vulnerability to climate change and variability for dengue (V_{CCD})

We identified and compiled the factors and dimensions of vulnerability for dengue, from all disciplines and dimensions under study. We applied the MHVA and then proceeded to integrate collected information (aedic indices, number of cases of dengue, maximum temperature, minimum temperature, Δ1 of temperatures, precipitation, use of mosquito nets, storage of water, habits, waste solids, and biodiversity indices such as Shanon index and Simpson's inverse for birds, insects and plants).

Then a multiple linear regression model was developed, processed with the SAS program (Statistical Analysis System), obtaining the correlation matrix for dengue, that was interpreted and analysed to parameterise the correlation coefficients whose results were emptied in a vulnerability matrix, which also incorporated the rate of adaptation capacity of evaluated municipality.

The results were presented with colours representing low, medium, high and very high levels of vulnerability. For statistical analysis, the database was structured considering the grouping of variables by dimensions of vulnerability. For the hydrologic dimension, annual precipitation variables and observed decrease were included. We assessed the climate baseline and current climate for urban, peri-urban and rural areas.

For the epidemiological dimension, epidemiological indices for dengue were considered: housing infestation indexes (IIV), Breteau Index (IB) and vessel Infestation Index (IIR) associated with some confirmed cases and case numbers Suspects added by urban neighbourhoods and peri-urban neighbourhoods. In each category, the capacities of health centres (infrastructure, equipment, laboratory and health personnel were assessed).

Also, we assessed the confidence of people in health services by applying qualitative methods to evaluated the adaptive capacity of health system.

The application of the MHVA to human health made it possible to recognise the systemic relationships between health and different biophysical, socioeconomic and environmental factors. The observed differences in exposure and vulnerability are related to climate and non-climate factors and they are the result of multidimensional inequalities produced by a disparate and inequitable development process, as discussed in previous works by the authors (e.g. Aparicio-Effen et al. 2016b; Nagy et al. 2018) evident in the following levels.

Political-Administrative

The V_{CCD} in the municipalities of Department of Tarija shows that in Yacuiba climatic factors prevail over non-climatic factors. The meteorological variables are optimal for vector development, so seasonally Aedes *aegypti* finds favourable conditions for

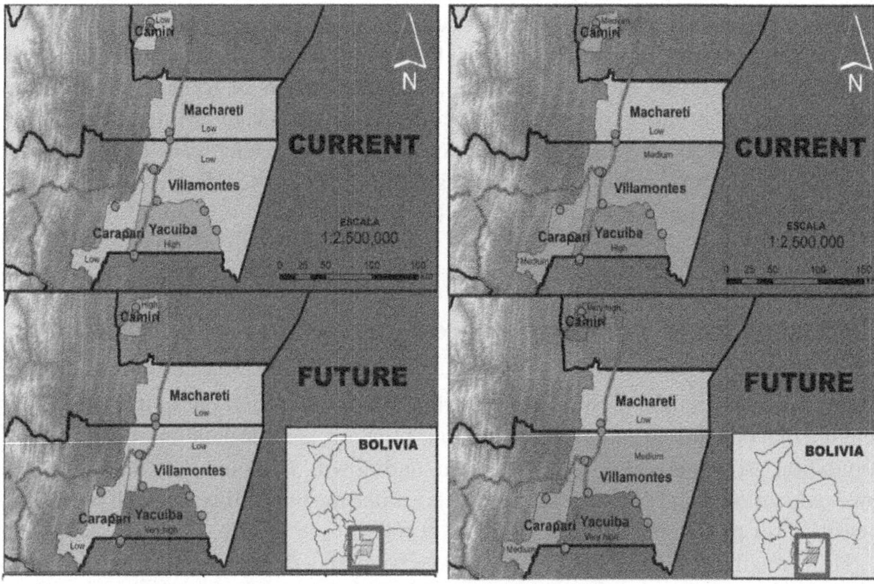

Fig. 3 Current and future climate-related health vulnerability for Dengue (V_{CCD}) for urban (left) and peri-urban (right) areas evaluated with MHVA methodology at the five studied municipalities

its reproduction, particularly in the rainy season. Future climate scenarios will accentuate the current changes in maximum and minimum temperature increases, associated with a 12% decrease in precipitation, which will generate water accumulation by population, and other human and municipal management factors. In Villamontes and Carapari, non-climatic factors prevail, but greater climatic pressures are expected in the future.

The current V_{CCD} of urban area is high in Yacuiba and low in urban Camiri. The V_{CCD} is high in the peri-urban area of Yacuiba and average in Villamontes and Camiri. Under the 2030 future climate scenario, the urban V_{CCD} will become very high for Yacuiba and high for Camiri. In the evaluated peri-urban neighbourhoods, the V_{CCD} will become very high for Yacuiba and Camiri and will remain average in Villamontes. The future V_{CCD} due to the effect of the changing climate 2020–2030, added to population growth in urban and peri-urban area, increases it from high to very high, so it is urgent to implement adaptation measures that reduce the degree of vulnerability (Fig. 3).

Watershed, Ecosystem, and Production

The impacts of extreme events (e.g. droughts, forest fires) highlight the exposure and vulnerability of Chaco ecosystems and their human systems to the current climate

(Aparicio-Effen et al. 2016b). Extreme events produce alteration of ecosystems, loss of biodiversity (with impacts on health, for example, the almost extinction of the guild of insectivorous birds in Yacuiba), disorganisation of food production and water supply (continuity of services and availability), damage to infrastructure and settlements, impacts on agricultural production, favouring Dengue. Both climatic and non-climate factors, the former predominantly indirect on health, have a summative effect on health vulnerability.

The Site of Residence and Culture

The project has studied urban and peri-urban (indigenous and peasant) groups, evidencing that peri-urban areas with precarious housing, service deficits, and unplanned population growth are exposed to Dengue, and now to Chikungunya and Zika. The lower educational and socioeconomic levels reduce the possibility of using preventive measures for dengue and access to health services.

These results evidence that the Pilcomayo watersheed, in particular at the five prioritised municipalities, which show high climate-related health vulnerability, V_{CCD} will be very high by 2030, agreeing with the assertions made by the IPCC (IPCC 2007, 2014). The methodology implemented allowed to identify climate and non-climate factors, with different effects and graduations for studied municipalities and different degree of vulnerability for Dengue.

Methodological Contributions for Adaptation Prioritisation

> Adaptation can contribute to well-being of populations, the security of assets and the maintenance of ecosystem goods, functions and services now and in the future. Adaptation is place- and context-specific. The first step towards adaptation to future climate change is reducing vulnerability and exposure to present climate variability. Integration of adaptation into planning, including policy design, and decision-making can promote synergies with development and disaster risk reduction. Building adaptive capacity is crucial for effective selection and implementation of adaptation options. (IPCC 2014)

Climate change will affect species distribution, demography and life histories, with consequences for human livelihoods including changing patterns of human disease distribution (McNeely and Mainka 2009). Until the middle of the century, the impact of projected climate change will affect human health mainly due to the aggravation of existing health problems. Throughout the 21st century, climate change is expected to cause an increase in poor health in many regions and especially in low-income developing countries, compared to the reference level without climate change (IPCC 2014).

Adaptation can reduce the risks of climate change impacts, but there are limits to its effectiveness, especially with greater magnitudes and rates of climate change. Taking a longer-term perspective, in the context of sustainable development, increases the likelihood that more immediate adaptation actions will also enhance future options and preparedness (IPPC 2014).

Climate change and social, economic and environmental determinants of health (SDH) are the first acknowledged root-causes of infectious diseases. Controlling for SDH would reduce disease burden and promote adaptation. The most effective short-term health vulnerability reduction measures are programs that implement and improve necessary public health measures such as providing clean water and sanitation, ensuring essential health care that includes immunisation and child health services, higher capacity for preparedness and response to disasters, and poverty alleviation (IPCC 2014).

These and other measures to reduce vulnerability and to adapt are developed within a temporal and spatial context, which are addressed through the MHVA methodology. Once the current and future V_{CCD} was estimated for the Project area, the approach was continued focused on transcending the health sector, quantifying the degree of responsibility of non-health sectors, and transferring the scientific findings in sustainable public policies and actions.

The results were presented to the different stakeholders of the Pilcomayo basin, to prioritise participatory adaptation measures, trying to avoid the implementation of reactive adaptation measures, which in most cases mean a loss of economic and human resources, pollution of waters, soils, and loss of biodiversity.

The Methodology Followed for the Participatory Prioritisation of Adaptation Measures in the Bolivian Chaco

The implementation of knowledge management tools facilitated the assessment of progress of direct, key and strategic actors concerning:

- Changes promoted by the Project (Diaries and Scope Reports).
- Map of Actors and Map of Actors of Trust.
- Communication of results (e.g. triptychs, comic strips, radio series, and press activities).

For participatory prioritisation of local climate adaptation measures, we developed workshops in each municipality, as follows.

Before the Workshop

Step 1 Identify and summon the actors.

The scope diary progressively allowed knowing the actors and their progress on study problem (object), their knowledge and behaviour. The updating of actors map and the construction of trustworthy actors map oriented the identification of those that should be summoned to participatory process of prioritisation. The workshops involved decision makers, municipal, local, departmental, health network personnel, hospitals, civil society, two indigenous ethnic groups, and peasants. The development of various communication products supported the change process, contributed to raising awareness, levelling information and knowledge, and motivating the participation of actors in the prioritisation process.

Step 2 Preparation of the workshop

Before the workshop, the research team prepared two types of materials: PowerPoint presentations with the objectives and results of research, explaining the workshop dynamics and investigation results, and participatory instruments for prioritisation of adaptation measures.

The participatory instruments included the preparation of a prioritisation matrix and cards with adaptation measures list. Previously, the research team, together with health authorities and networks, defined the most critical adaptation measures, which would be consulted in the workshop, based on research results. Once the measures were defined, we identified the most relevant analysis dimensions for measures set, including, for example, changes in behaviour (in people), political will (in the authorities), institutional changes and new legal frameworks.

At the same time, different types of activity were defined, corresponding to main change involved in the implementation of each type of activity, and each one received a different colour. All activities were transcribed into colour cards (one activity per card) according to their type and for all measures. Finally, all participant stakeholders designed a large prioritisation matrix that included as columns: the measure, the activities, who leads each action, who will participate in its execution, who will finance the actions, the scale of intervention, the time scope and the score (for prioritisation).

All the researchers assumed their role in the workshop: facilitate the groups, help identify the cards in the matrix, take notes and clarify doubts.

During the Workshop

Step 3 Presentation of research results: levelling knowledge and promoting collaborative spaces

At the beginning of each workshop, the research team announced the key findings to stakeholders, reporting on objectives and expected results of workshop, levelling the information and knowledge product of research, and preparing for participation in the activity.

The systemic approach of process allowed to socialise research results related to levels of water pollution, type, and distribution of the solid waste, diseases to which they face, the role of different sectors in diseases prevention, the possible reasons that originate them, deficiencies of infrastructure and management. A round of questions after each presentation allowed to exchange points of view and to satisfy doubts of participants.

Box 2 Prioritised adaptation measures for Dengue selected by the stakeholders in the Project area (Table 1).

1. Safe water
2. Water for hygiene
3. Integrated solid waste management
4. Early Warning System for Dengue
5. Improvement of the management of protected areas
6. Sectoral and intersectoral capacity that encompassed: Organising awareness workshops and strengthening health networks.

The measures mentioned above were followed by the implementation of a massive campaign and the implementation of a multisectoral platform to address health issues. The discussions held with the stakeholders allowed knowing which would be the activities that would have more significant support, besides the type of measure, showing the vision that actors have about problem and solutions. The construction of dialogue and consensus and the organisation of awareness campaigns were essential actions to generate behavioural changes. Regarding safe water, water boiling was the activity with the highest consensus, followed by water filtering, monitoring, and maintenance of the distribution network and tank construction. This result shows that prioritised activities highlight those that demand behavioural changes in people and families, and those that require infrastructure works in charge of municipalities and families themselves.

The prioritised adaptation measures respond to research results that show the population contribution, health sector, and other health-related sectors, and how the deficiencies or well-performed sectoral work are made explicit in environmental and social determinants, that will favour or prevent the presentation of Dengue cases.

Step 4 Prioritisation of adaptation measures: incorporation of all perspectives

Due to the socio-cultural characteristics of participants, we decided to implement a methodology based on debate and consensus, and not on numerical weighting of measures. Although numerical weighting could contribute to objectivity of decision, there was a risk of transforming the process into a technical exercise of validation of opinions instead of being a moment of knowledge dialogue that would allow expressing the participants opinion in their complexity (idiomatic, worldview, educational, social position and perception).

Table 1 A matrix of participatory prioritisation of the adaptation measures

Measure	What is?	Leaders	Participants	Funding	Scale of application	Initiation	Prioritisation
Safe water	Access to safe water at homes & schools. Monitoring and alliances	Municipal government	Government users	Government	Quarters community	Short-term	1
Water for hygiene	Water at homes and schools; hand washing	Families	Representatives of civil society (ROSC)	Government corporate social responsibility	Family	Short-term	2
Solid waste integrated management	Garbage and sanitary landfill management	Municipal government	Municipal government ROSC families	Municipal government	Municipality	Short-term	3
Early warning system (EWS) for dengue	Observation network strengthening and EWS normative	National government	Municipal and departmental governments ROSC families	Municipal government	National	Medium-term	4
Improvement of natural protected areas management	Installation of municipal dumps; recycling; ordinances	Managers	Departmental Governor and Deputy Governor	Departmental government	Regional	Short-term	5
Sectoral and intersectoral capacity	Building capacity for change; citizens for health	Managers	Community peasants unions of health professionals	Municipal Government	Departmental Government	Short-term	6

Once results were presented, the different stakeholders were organised in groups by type and sectors of activity (NGO, urban organisations, indigenous peoples, health sector and local governments), where the facilitators presented the set of selected adaptation measures, explaining their meaning and scope. Each measure was placed in the first column of the matrix.

Identification of Strategic Activities

The cards with activities were organised in groups and were placed in a visible place for all the participants. Participants were asked to select two activities for each adaptation measure. The possibility was left open for each group to incorporate new activities by consensus. Then, with the help of the facilitator, each group agreed on the main activities that should be implemented for each adaptation measure. The cards prioritised by each group are placed in the second column along with each adaptation measure.

Identification of Stakeholders

> Adaptation planning and implementation can be enhanced through complementary actions across levels, from individuals to governments. National governments can coordinate adaptation efforts of local and sub-national governments, for example by protecting vulnerable groups, by supporting economic diversification and by providing information, policy and legal frameworks and financial support. (IPCC 2014)

Each group was given cards in which the name of different institutions was written by type: Municipal government, national government, university Health networks, which were requested to identify, by consensus, the institutions that should or might

- Lead the implementation of activity.
- Be summoned to design activity.
- Follow-up of activities.
- Have potential or interest in funding them.

The cards with the names of the institutions identified by each group were placed in the third, fourth and fifth column of matrix together with each activity or group of prioritised activities with which they were related.

Subsequently, each group discussed the scale of intervention (family, neighbourhood, municipality, department, Country) and initiation of activities (short, medium and long term). The selected cards were placed in the sixth and seventh column.

Prioritisation of Adaptation Measures

The work in groups allowed building different matrices of prioritisation of measures and activities, which included the contributions of each type actor, placing together with each one, a number from 1 to 6 (there were six measures of adaptation) according to the assigned importance. Subsequently, a single matrix synthesis was consolidated.

After the Workshop

Step 5 Analysis of results

> Adaptation experience is accumulating across regions in the public and private sectors and within communities. There is increasing recognition of the value of social (including local and indigenous), institutional, and ecosystem-based measures and the extent of constraints to adaptation. Adaptation is becoming embedded in some planning processes, with a more limited implementation of responses. (IPCC 2014)

The research team and health sector counterparts, processed the information jointly constructed, emptying the prioritisation matrices in an analysis matrix of measures and activities prioritised by actor type. The analysis matrix allowed identifying the number of actors that prioritised each measure guiding those that are of greater consensus or importance at the local level.

As a final result of the project, climate adaptation measures for dengue were prioritised following Eco-health and watershed approaches. Then, adaptation options were participatory prioritised, designing a "Climate Change Health Strategic adaptation Plan", which implementation has already begun. The Plan initially designed for Dengue serves as the basis of adaptation for Chikungunya and Zika that share the same vector and that are new diseases in the Pilcomayo Watersheed.

Many adaptations and mitigation options can help address climate change, but no single option is sufficient by itself. Effective implementation depends on policies and cooperation at all scales and can be enhanced through integrated responses that link mitigation and adaptation with other societal objectives (IPCC 2014). In this sense, the use of MHVA and current adaptation experience would help in identifying the most vulnerable locations and target adaptation actions.

Conclusions

Based on current and future climate and non-climate scenarios, and participatory consultation process developed within the Methodology for climate change Health Vulnerability Assessment considering Eco-health and Watershed Approaches" (MHVA) framework, we highlight the following findings.

The climate-related health vulnerability resulted from the action of climate and non-climate factors and vulnerability dimensions related to the studied object (Dengue Fever: V_{CCD}) with time-space and magnitude variations.

The vulnerabilities are the result of multidimensional inequalities produced by a disparate and inequitable development process.

Climate adaptation options for Dengue were prioritised based on Eco-health and Watershed approaches through a participatory workshop.

The primary outcome of the MHVA and participatory process is the "Climate Change Health Strategic adaptation Plan" initially designed for Dengue Fever, that serves as an adaptation basis for Chikungunya and Zika, two new diseases in Pilcomayo Watersheed that share the same vector.

As stated by the Intergovernmental Panel on Climate Change (IPCC), the effective implementation of adaptation measures depends on policies and cooperation at all scales and can be enhanced through integrated responses that link management options with other societal objectives. In this sense, the use of MHVA and current adaptation experience would help in identifying the most vulnerable locations and target adaptation actions for health and other health-related sectors.

References

Amarakoon D, Chen A, Rawlins S, Chadee DD, Taylor M, Stennett R (2008) Dengue epidemics in the Caribbean-temperature indices to gauge the potential for onset of dengue. Mitig Adapt Strat Glob Chan 13:341–357

Aparicio-Effen M, Arana I, Aparicio J, Ramallo C, Bernal N, Nagy GJ, Ocampo (2016a). Chapter 14: Climate change and variability health vulnerability in Bolivian chaco ecosystems. In: Leal Filho W, Azeiteiro Alvez F (eds) Climate change and health: improving resilience and reducing risks, 231–259. Climate change management series. Springer, Heidelberg, Germany. eBook ISBN 978-3-319-244660-4_14, Hardcover ISBN 978-3-319-24658-1, Series ISSN 160-2010. https://doi.org/10.1007/978-3-319-24660-4_14, http://www.springer.com/us/book/9783319246581, 480 pp

Aparicio-Effen M, Arana I, Aparicio J, Cortez P, Coronel G, Pastén M, Nagy GJ, Galeano Rojas A, Flores L, Bidegain M (2016b). Chapter 26: Introducing hydro-climatic extremes and human impacts in Bolivia, Paraguay and Uruguay. In: Leal Filho W, Azeiteiro Alvez F (eds) Climate change and health: improving resilience and reducing risks, pp 449–473. Climate change management series. Springer, Heidelberg, Germany. ISBN 978-3-319-24658-1, https://doi.org/10.1007/978-3-319-24660-4_26, 480 pp

Aparicio-Effen M, Arana-Pardo I, Ocampo M, Aparicio J, Roque S, Nagy GJ (2018). A successful early warning system for hydroclimatic extreme events: the case of La Paz City mega landslide. In: Climate change adaptation in latin America: managing vulnerability, fostering resilience. Climate change management series. Springer, Heidelberg, Germany, pp 241–265. https://doi.org/10.1007/978-3-319-56946-8

Charron DF (2012) Ecosystem approaches to health for a global sustainability agenda. EcoHealth 9:256–266 PMID: 22961374 https://doi.org/10.1007/s10393-012-0791-5, https://www.ncbi.nlm.nih.gov/pubmed/22961374

Hung N-V, Thi THT (eds) (2016) Ecosystem approach to health (Ecohealth): theory and practices in environmental health research in Vietnam. Medical Publishing House, Hanoi, Vietnam. https://cgspace.cgiar.org/handle/10568/77089

IPCC (2007) Climate change 2007: impacts, adaptation and vulnerability. In: Parry M, Canziani O, Palutikoff J, van der Linden P, Hanson C (eds) Contribution of working group II to the fourth assessment report of the intergovernmental panel on climate change. University Press, Cambridge, Cambridge, p 976

IPCC (2012) Managing the risks of extreme events and disasters to advance climate change adaptation. In: Field CB, Barros V, Stocker TF, Qin D, Dokken DJ, Ebi KL, Mastrandrea MD, Mach KJ, Plattner G-K, Allen SK, Tignor M, Midgley PM (eds) A special report of working groups I and II of the intergovernmental panel on climate change. Cambridge University Press, Cambridge, UK, and New York, NY, USA, 582 pp

IPCC (2014) Climate change 2014 impacts, adaptation, and vulnerability: working group II contribution to the fifth assessment report of the intergovernmental panel on climate change. Cambridge University Press, Cambridge. https://doi.org/10.1017/CBO9781107415379

Irwin A, Scali E (2010) Action on the social determinants of health: learning from previous experiences. Social determinants of health discussion paper 1 (Debates). World Health Organization, Geneva. https://www.who.int/social_determinants/corner/SDHDP1.pdf

Kraemer MU, Sinka ME, Duda KA, Mylne AQ, Shearer FM, Barker CM, Moore CG, Carvalho RG, Coelho GE, Van Bortel W, Hendrickx G, Schaffner F, Elyazar IR, Teng H-J, Brady OJ, Messina JP, Pigott DM, Scott TW, Smith DL, Wint GW, Golding N, Hay SI (2015) The global distribution of the arbovirus vectors Aedes aegypti and Ae. albopictus. eLife 4. https://doi.org/10.7554/eLife.08347

Leal Filho W, Boenecke J, Spielmann H, Azeiteiro UM, Alves F, Lopes de Carvalho M, Nagy GJ (2018) Climate change and health: an analysis of causal relations on the spread of vector-borne diseases in Brazil. J Cleaner Prod 177:589–596. https://doi.org/10.1016/j.jclepro.2017.12.144

McNeely JA, Mainka SA (2009) Conservation in a New Era. IUCN, Gland, Switzerland. https://portals.iucn.org/library/efiles/documents/2009-026.pdf

Nagy GJ, Coronel G, Pastén M, Báez J, Monte Domecq R, Galeano Rojas A, Flores, Ciganda C, Bidegain M, Aparicio-Effen M, Arana I, Aparicio J (2016) Impacts on well-being and health of excessive rainfall and floods in Paraguay, Uruguay and Bolivia. In: Leal Filho W, Azeiteiro U, Alvez F (eds) Climate change and health: improving resilience and reducing risks. Climate change management series. Springer, Heidelberg. https://doi.org/10.1007/978-3-319-24660-4

Nagy GJ, Cabrera C, Coronel G, Aparicio-Effen M, Arana I, Lairet R, Villamizar A (2017) Addressing climate adaptation in education, research and practice: the CliVIA-network. Int J Clim Change Strategy Manag 4(9). https://doi.org/10.1108/ijccsm-04-2016-0056

Nagy GJ, Leal Filho W, Azeiteiro UM, Heimfarth J, Verocai JE, Li C (2018) An assessment of the relationships between extreme weather events, health vulnerability and the impacts on human wellbeing in latin American countries. Int J Environ Res Public Health 15(9):1802. https://doi.org/10.3390/ijerph15091802

Rahaman MR, Varis O (2005) Integrated water resources management: evolution, prospects and future challenges. Sustain Sci Pract Policy 1(1):15–21. https://doi.org/10.1080/15487733.2005.11907961

Waage J, Banerji R, Campbell O, Chirwa E, Collender G, Dieltiens V, Dorward A, Godfrey-Faussett P, Hanvoravongchai P, Kingdon G, Little A, Mills A, Mulholland K, Mwinga A, North A, Patcharanarumol W, Poulton C, Tangcharoensathien V, Unterhalter E (2010) The millennium development goals: a cross-sectoral analysis and principles for goal setting after 2015. Lancet and London international development centre commission. Lancet 376(9745):991–1023

Watts N, Amann M, Ayeb-Karlsson S, Belesova K, Bouley T, Boykoff M, Byass P, Cai W, Campbell-Lendrum D, Chambers J, Cox PM, Daly M, Dasandi N, Davies M, Depledge M, Depoux A, Dominguez-Salas P, Drummond P, Ekins P, Flahault A, Frumkin H, Georgeson L, Ghanei M, Grace D, Graham H, Grojsman R, Haines A, Hamilton I, Hartinger S, Johnson A, Kelman I, Kiesewetter G, Kniveton D, Liang L, Lott M, Lowe R, Mace G, Odhiambo Sewe M, Maslin M, Mikhaylov S, Milner J, Latifi AM, Moradi-Lakeh M, Morrissey K, Murray K, Neville T, Nilsson M, Oreszczyn T, Owfi F, Pencheon D, Pye S, Rabbaniha M, Robinson E, Rocklöv J, Schütte S, Shumake-Guillemot J, Steinbach R, Tabatabaei M, Wheeler N, Wilkinson P, Gong P, Montgomery H, Costello A (2017) The lancet countdown on health and climate change: from 25 years of inaction to a global transformation for public health. The Lancet. https://doi.org/10.1016/S0140-6736(17)32464-9

Potential Biophysical Climate Change Impacts at World Natural Heritage Sites in the Brazilian Atlantic Forest

Felipe Bittencourt, Melina Amoni, Augusto Schmidt and Cecília Loureiro

Abstract There are 209 UNESCO World Natural Heritage sites around the world that provide an outstanding amount of universal value. Among them, 37 are in Latin America, and seven in Brazil. There are two natural sites in the Brazilian Atlantic Forest of significant concern: (i) Atlantic Forest South-East Reserves and (ii) Discovery Coast Atlantic Forest Reserves. Climate change poses a very high threat to these forest reserves. The fragmented forests are believed to be particularly vulnerable to the expected impacts of climate change such as erosion and water availability. This paper presents an analysis about the potential biophysical climate change impacts in these two World Natural Heritage sites. An impact model was built using climate variables and climate extremes from two regional climate models available for Brazil: Eta-HadGEM2-ES and Eta-MIROC5. Simulations were made using two different greenhouse gases emissions scenarios (RCP4.5 and RCP8.5) and two timeframe windows (1961–2005/2071–2100). Erosion, soil water availability and occurrence of phytophysiognomies were analyzed in this research. Critical areas in the reserves were identified and should now be set by government as priority in the national climate adaptation strategy.

Introduction

The rapid growth of human population over the last 200 years has led to significant global environmental change, especially concerning biodiversity loss (Lanz et al. 2018; Newbold et al. 2016). Worldwide biodiversity decline and ecosystem degradation have long been considered primary consequences of habitat loss and forest landscape fragmentation, which are driven by the growth of agriculture and industry, and lead to significant land-use change (Phillips et al. 2017). The rapid expansion of

F. Bittencourt (✉) · M. Amoni
WayCarbon, Rua Paraíba 1000, Belo Horizonte, MG 30130-141, Brazil
e-mail: felipe.bittencourt@waycarbon.com

F. Bittencourt · A. Schmidt · C. Loureiro
Universidade Federal de Minas Gerais, UFMG, Avenida Presidente Antônio Carlos, 6627, Belo Horizonte, MG 31270-901, Brazil

W. Leal Filho et al. (eds.), *Climate Change, Hazards and Adaptation Options*, Climate Change Management, https://doi.org/10.1007/978-3-030-37425-9_49

road networks, mining activities and urban sprawl (Haddad et al. 2015) has caused deforestation in many habitats, inducing ecosystem simplification (Reich et al. 2012), and the loss of both genetic traits and ecosystem services, with a direct impact on society (Thom and Seidl 2016).

Climate change and forest conservation and management are fundamentally related (Bugmann et al. 2017). Forests play an important role as carbon sinks and thus help mitigate climate change. Moreover, forest degradation, lack of proper forest management and land-use change result in increased carbon dioxide emissions. The Food and Agriculture Organization of the United Nations (FAO) points out that abiotic (e.g., drought, heat, acidification) and biotic (e.g., disease and alien species) factors have great influence on the health and vitality of forests, and can provoke significant economic and environmental losses. Vegetation growth and survival, quality of wood and non-wood forest products, wildlife habitat, recreation and scenic and cultural values are all directly affected by these disturbances. Global climate change is increasing the magnitude of these impacts (IPCC 2014) and the most vulnerable forests are the fragmented ones (Santo-Silva et al. 2016).

Currently, the main focus of climate change policies (e.g., Kyoto Protocol and Paris Agreement) is the reduction of greenhouse gases (GHG) emissions (Shishlov et al. 2016), while less attention is given to forest conservation and management. Although scattered, recent efforts have been set out to mitigate deforestation and forest degradation, the second leading cause of global warming and responsible for about 15% of global GHG emissions (World Bank 2018). Discussed since 2005 at UNFCCC and already in place in the Voluntary Carbon Market, Reducing Emissions from Deforestation and Forest Degradation (REDD) mechanism is one of these efforts. Despite its ambitious aims, REDD struggles with political and economic interests favouring deforestation and degradation.

Considering the natural environment is of major importance for all of humanity and its degradation produces significant economic, social and cultural impacts, it is critical to safeguard this unique, irreplaceable and invaluable resource. The General Conference of the United Nations Educational, Scientific and Cultural Organization meeting, held in Paris in 1972, recognized the importance of ensuring the identification, protection, conservation, presentation and transmission to future generations of the World Cultural and Natural Heritage. The latter is defined by UNESCO as the natural sites, features, and geological and physiographical formations precisely delineated areas of outstanding universal value from the point of view of science and conservation (UNESCO 1972).

One of the 35 world conservation hotspots is the Atlantic Forest in Brazil, where only 15.2% of the natural vegetation cover still remains and in an intensely fragmented state (SOS Mata Atlântica and INPE 2017), which results in high vulnerability of this biome to climate change. Atlantic Forest South-East and Discovery Coast Atlantic Forest Reserves are both considered World Natural Heritage sites by UNESCO and are located in the most strategic region in Brazil from the demographic point of view (e.g., they are home to more than 120 million inhabitants) and, from the economic point of view, they account for more than 70% of Brazil's

GDP. The Atlantic Forest is in fact extremely heterogeneous, with several phytophysiognomies. The Atlantic Forest consists mainly of an evergreen to semi-deciduous forest, however, deciduous forests, mangroves, swamps, *restingas* (coastal white-sand woodlands), inselbergs, high-altitude grasslands (*campo rupestre* and *campo de altitude*), and mixed Araucaria pine forests (Scarano 2002) are also part of this biome. This mosaic of habitats is currently home to nearly 18,000 species of plants (REFLORA 2017), 263 mammals, 936 birds, 306 reptiles, and 475 amphibians (Mittermeier et al. 2005). Moreover, outstanding levels of endemism make the Atlantic Forest one of the most distinctive biogeographic units and one of the 35 biodiversity hotspots for conservation prioritization (Myers et al. 2000).

In this scenario, it is important to develop and implement policies and practices to scale down negative impacts or even adapt forests to new conditions posed by climate change (IPCC 2007). Climate projections are especially important to understand how priority areas for conservation might respond to environmental and anthropic pressures, and guide to an adaptive approach (Perry 2015). Therefore, the adaptation of forest management to climate change requires the projection of potential impacts to be based on solid principles. This study aims to perform an analysis of climate change vulnerability in the two Atlantic Forest hotspots in Brazil, based on three biophysical potential impact models.

Data and Methodology

Study Area

Two remaining areas of the Atlantic Forest listed as World Natural Heritages have been selected for this study:

- Atlantic Forest South-East Reserves, located in the states of Paraná and São Paulo, with 25 protected areas (about 470,000 ha in total); and
- Discovery Coast Atlantic Forest Reserves, located in the states of Bahia and Espírito Santo, with eight protected areas (112,000 ha in total).

Methodological Approach

Three potential impacts were analyzed for each study area: hydric erosion, soil water availability and occurrence of phytophysiognomies. Climate exposure related to biophysical sensitivity has been considered for modelling each one of these impacts. The input data for evaluating the sensitivity of the impact model were represented by a stable biophysical component because of the lack of reliable projections, (as in the case of soil use and occupation) or because of the absence of a dynamic component, e.g., topography. Model impact input data for climate exposure, on the other hand, varies over time, once they result from climate models.

All data was georeferenced (raster format) and standardized, ensuring spatial comparability (pixel-to-pixel) and the possibility of aggregating them in an weighted composite index. The standardization was done with the Max–Min[2] methodology, subtracting the minimum value and dividing by the interval of each variable (Naumann et al. 2014). This methodology guarantees that all factors are standardized between 0 and 1, removing units of measurement. Two distinct simulations were run, loaded with temperature and precipitation (mean and extreme values) from Eta-HadGEM2-ES and Eta-MIROC5 for two timeframe windows: 1961–2005 and 2071–2100, calculated for two different GHG emission scenarios, referred to as Representative Concentration Pathways (RCPs). RCP8.5 (high emissions) is considered the pessimistic scenario, while RCP4.5 (low-intermediate emissions) is considered the optimistic one. In order to establish a benchmark for comparison, the 1961–2005 period was considered the baseline. For hydric erosion and soil water availability, impacts were evaluated both for rainy (December, January and February) and dry seasons (June, July and August). Therefore, results from impact models differ from each other because of different climate projections from Eta-HadGEM2-ES and Eta-MIROC5.

The 1961–2005 period was used as the baseline, while the 2071–2100 period was chosen with the intent of analysing the impact in the long term, focusing on the end of the century, which is the reference for the Paris Agreement goals. The study was performed between 2016 and 2018.

Potential Impact Models

The models used in this study were run based on an attentive literature review, in search of up-to-date references on equations, explanatory factors and parameters that are usable for comprehension and mapping of the investigated impacts. Each of the following subsections briefly presents the theoretical basis and the environmental data list, thus representing the biophysical attributes of the environmental systems and processes under analysis.

Hydric Erosion Model

The impact of raindrops and surface runoff are the active hydric erosion agents. The splash is the initial stage of the erosive process, caused by the raindrop impact on the soil. In order to understand and evaluate the precipitation impact on erosion, Wischmeier (1959) proposed a rainfall erosion index (MJ ha^{-1} mm h^{-1} y^{-1}). Silva (2004) organized a set of equations to determine the spatial variation of erosivity in Brazil, based on monthly and annual mean precipitation data. The equations that describe the indicator performance within the study area boundaries were used.

Soil Water Availability Model

Climate change, mainly in terms of precipitation frequency and volume, is responsible for changing groundwater availability dynamics. Even with unchanged total precipitation, increased intensity of rainfall events, concurrently with a reduction of their frequency—i.e. longer droughts—will affect soil humidity and, consequently, plant dynamics and conditions (Porporato et al. 2004). It is worth highlighting the importance of temperature increase in soil water availability final balance, due to evapotranspiration losses.

The calculation of climate change impact on groundwater availability is undertaken with the application of a differential equation, through a numerical model implemented for the operational estimate of soil water availability (Porporato et al. 2004). This estimate considers trends in precipitation and evapotranspiration variables only (Allen et al. 1998), besides current data on land use and soil texture to describe the ecosystem biophysical and edaphic properties.

Occurrence of Phytophysiognomies Model

The Atlantic Forest tree species spatial distribution is intrinsically linked to climate characteristics, especially precipitation and temperature (Salis et al. 1995; Rolim et al. 2005; Colombo 2007). A change in the distribution of plant classes may occur in the medium and long term, given the high plant versatility and plasticity (Knapp et al. 2002). Therefore, especially for the occurrence of phytophysiognomies model, the analysis considered the 2071–2100 period in the RCP 8.5 scenario. A conservative analysis was adopted, considering the most drastic scenario and the longest term, once the vegetation component generally presents great climate resilience. The following vegetation classes have been analysed: (i) ecotone; (ii) deciduous seasonal forest; (iii) semideciduous seasonal forest; (iv) open ombrophilous forest; (v) dense ombrophilous forest; (vi) mixed ombrophilous forest; (vi) pioneer formation.

The projections have been performed with the Maximum Entropy model, or MAXENT (Phillips et al. 2004). System entropy is the measurement of its 'disorder'. In this study, entropy (Shannon 1949) refers to a probability distribution uncertainty. MAXENT is an algorithm that aims to calculate and minimize this uncertainty, obtaining the future distribution closest to the actual uniform distribution, based on available environmental variable restrictions. Information on current phytophysiognomies, as well as historical environmental variables, are input data for the model to calibrate and define the set of restrictions. Subsequently, these restrictions are used to define the phytophysiognomy potential occurrence with respect to the restrictions identified in future climate variable spatial distributions.

Input Data

Climate Data: Eta-HadGEM2-ES e Eta-MIROC5

The climate scenarios were produced from the results of regionalization of global climate models HadGEM2-ES (*Met Office Hadley Centre*) and MIROC5 (*Japan Agency for Marine-Earth Science and Technology, Atmosphere and Ocean Research Institute*) conducted by INPE in South America for the RCP 4.5 and RCP 8.5 scenarios. Most data resolution used for this research was of 400 km^2 (20 km × 20 km). A detailed description of the global and regional models (Eta) can be found in Chou et al. (2014a, b).

Biophysical Data

For this analysis six biophysical variables were considered for the two Atlantic Forest Reserves: soil types (Embrapa 2011), vegetation cover (SOS Mata Atlântica and INPE 2017), phytophysiognomies distribution (Atlantic Forest Law nº 11.428/2006), water resource (Agência Nacional de Águas 2000), land use (IBGE 2002) and topography (TOPODATA 2011). The information was selected based on availability and the best spatial resolution of public data. Additionally, only data within the limits of both areas analysed were selected.

Results and Discussion

Hydric Erosion Model

For the hydric erosion evolution in the Atlantic Forest South-East Reserves, during the rainy season (Fig. 1a), the Eta-HadGEM2-ES model predicts considerable erosion reduction in the RCP 4.5 scenario, ranging from −37 to −58%. The RCP 8.5 scenario presents even further reduction, ranging from −57 to −78%. On the other hand, the Eta-MIROC5 model presents a completely different evolution, with a considerable positive variation in both scenarios, ranging from +46 to +86% in the RCP 4.5 scenario and +88 to +135% in the RCP 8.5 scenario.

The models also show opposite behavior of erosion evolution for the dry season (Fig. 1b). Eta-HadGEM2-ES predicts an increase ranging from +18 to +44% in the RCP 4.5 scenario and +46 to +101% in the RCP 8.5 scenario. On the other hand, Eta-MIROC5 presents a reduction between −34 and −51% in the RCP 4.5 scenario and −32 and −52% in the RCP 8.5 scenario (Table 1).

In the Discovery Coast Atlantic Forest Reserves, during the rainy season (Fig. 2a), the Eta-HadGEM2-ES model presents very distinct results in the two scenarios: while RCP 4.5 ranges from a decrease of −24% to an increase of +6%, RCP 8.5 shows

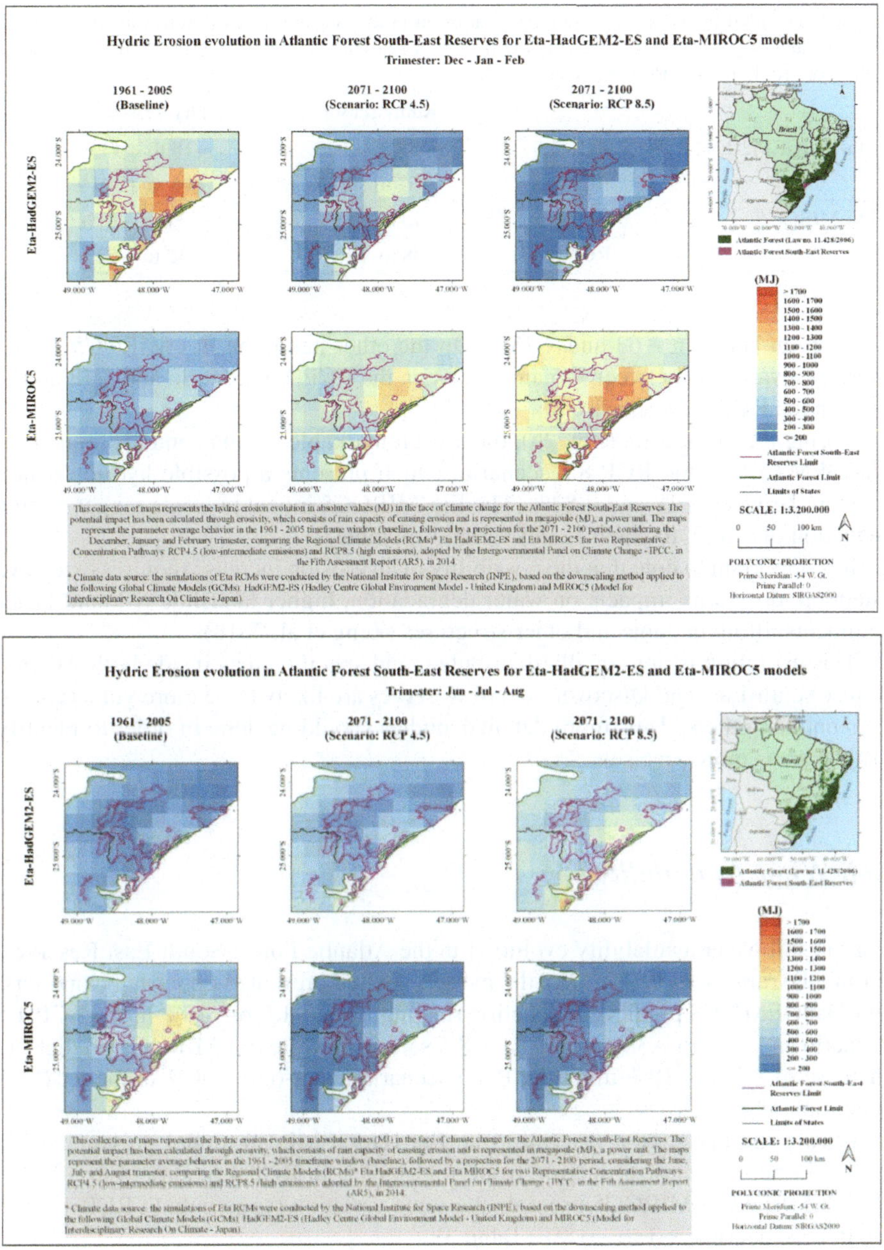

Fig. 1 Maps of hydric erosion evolution in the Atlantic forest South-East reserves for both Eta-HadGEM2-ES and Eta-MIROC5 climatic models, in two RCP scenarios (4.5 and 8.5) during **a** December, January and February (rainy season) and **b** June, July and August (dry season)

Table 1 Detailed hydric erosion range of variation in the Atlantic forest South-East reserves for the rainy and dry seasons, according to the results of Eta-HadGEM2-ES and Eta-MIROC5 models using both RCP 4.5 and RCP 8.5 scenarios

		Rainy season	Dry season
Eta-HadGEM2-ES	RCP 4.5	−37 to −58%	+18 to +44%
	RCP 8.5	−57 to −78%	+46 to +101%
Eta-MIROC5	RCP 4.5	+46 to +86%	−34 to −51%
	RCP 8.5	+88 to +135%	−32 to −52%

a reduction between −63 and −75%. On the other hand, the Eta-MIROC5 model predicts a variation between −5 and +45% in the RCP 4.5 scenario and −18 and + 36% in the RCP 8.5 scenario.

During the dry season (Fig. 2b), the most remarkable results come from the Eta-HadGEM2-ES in the RCP 8.5 scenario, which presents a possible hydric erosion increase between +18 and +82%. The Eta-MIROC5 model presents ranges from a reduction of −32% to stability (no variation) (Table 2).

It is important to note that the possibility of an erosion increase requires extensive attention, due to its impacts on water degradation, higher landslide risk and faster sediments filling in dams and other structures (Zeng et al. 2018).

This evolution is not equally distributed, and specific areas inside both Atlantic Forest South-East and Discovery Coast Reserves are likely to be more vulnerable to erosion than others. Thus more detailed studies should be done in order to identify critical areas more precisely.

Soil Water Availability

For the soil water availability evolution in the Atlantic Forest South-East Reserves, during the rainy season (Fig. 3a), the models applied presented opposite behaviours: Eta-HadGEM2-ES predicts a reduction ranging from −13 to −26% in the RCP 4.5 scenario and −15 to −38% in the RCP 8.5 scenario, while Eta-MIROC5 predicts an increase of +5 to +18% in the RCP 4.5 scenario and +6 to +30% in the RCP 8.5 scenario.

During the dry season (Fig. 3b), the Eta-HadGEM2-ES model presents stable results in both scenarios, ranging from −3 to +5% in RCP 4.5 and −4 to +7% in RCP 8.5. The Eta-MIROC5 presents the same variation range in both RCP 4.5 and RCP 8.5 scenarios: −1 to −18% (Table 3).

In the Discovery Coast Atlantic Forest Reserves, during the rainy season (Fig. 4a), the Eta-HadGEM2-ES points to a soil water availability reduction between −6 and −22% in the RCP 4.5 scenario and a more critical decrease in the RCP 8.5 scenario, ranging from −21 to −41%. On the other hand, the Eta-MIROC5 model is more

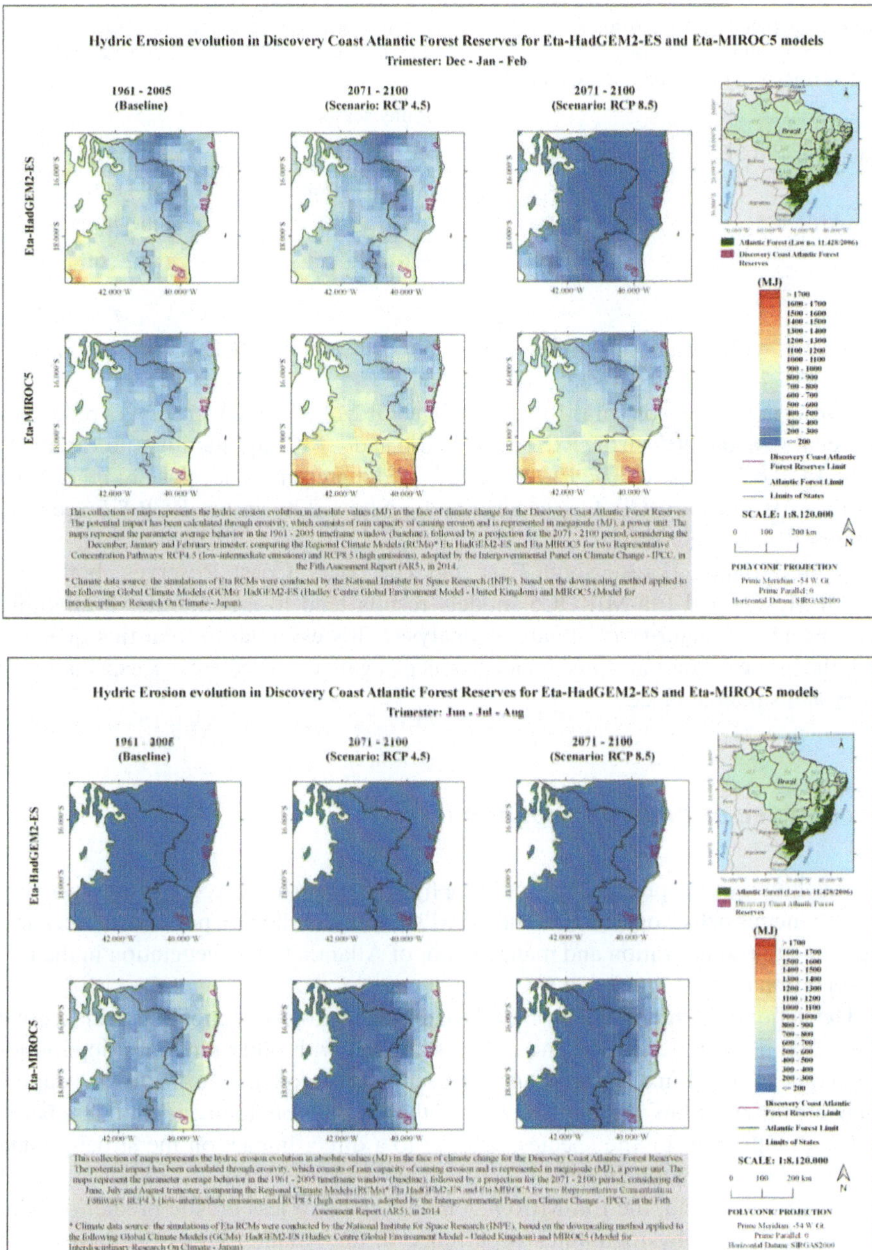

Fig. 2 Maps of hydric erosion evolution in the discovery coast Atlantic forest reserves for both Eta-HadGEM2 -ES and Eta-MIROC5 climatic models, in two RCP scenarios (4.5 and 8.5) during **a** December, January and February (rainy season) and **b** June, July and August (dry season)

Table 2 Detailed hydric erosion range of variation in the Discovery coast Atlantic forest reserves for the rainy and dry seasons, according to the results of Eta-HadGEM2-ES and Eta-MIROC5 models using both RCP 4.5 and RCP 8.5 scenarios

		Rainy season	Dry season
Eta-HadGEM2-ES	RCP 4.5	−24 to +6%	−32 to 0%
	RCP 8.5	−63 to −75%	+18 to +82%
Eta-MIROC5	RCP 4.5	−5 to +45%	−16 to −47%
	RCP 8.5	−18 to +36%	−22 to −56%

optimistic, with variations ranging from −6 to +3% in the RCP 4.5 scenario and −3 to −16% in the RCP 8.5 scenario.

During the dry season (Fig. 4b), the Eta-HadGEM2-ES model presents a reduction between −5 and −20% in the RCP 4.5 scenario and −14% and stability (no variation) in the RCP 8.5 scenario. The Eta-MIROC5 model presents a decrease ranging from −3 to −11% in the RCP 4.5 scenario and −5 to −22% in the RCP 8.5 scenario (Table 4).

In the Atlantic Forest South-East and Discovery Coast Reserves, both Eta-HadGEM2-ES and Eta-MIROC5 models results tend to a soil water availability decrease in the majority of situations analysed. It's essential to treat this question with the proper attention, since it can produce negative consequences, especially for ecosystems maintenance.

Occurrence of Phytophysiognomies

The evolution of the probability of phytophysiognomies occurrence presents a very heterogeneous behavior. The Eta-HadGEM2-ES model shows a pessimistic scenario regarding the conservation and management of Atlantic Forest vegetation in the two preservation areas (Figs. 5 and 6).

The major distribution increase of the ecotone phytophysiognomy reveals a great vegetation decharacterization and a direct conflict with other phytophysiognomies and biomes. The climatic conditions foreseen in this model, along with the edaphic changes, do not allow the permanence of the original phytophysiognomies, being replaced with other biome species, leading to a direct impact on the structure and associated fauna composition.

On the other hand, the Eta-MIROC5 model shows a less drastic prediction concerning the phytophysiognomies occurrence in the Atlantic Forest Reserves. The ecotone distribution decrease shown would lead to a higher confrontation between different vegetation types. At the same time, this more abrupt change may cause greater vegetation fragmentation, with forest border reduction and consequential environmental structure instability (Magnago et al. 2015; Scarano and Ceotto 2015).

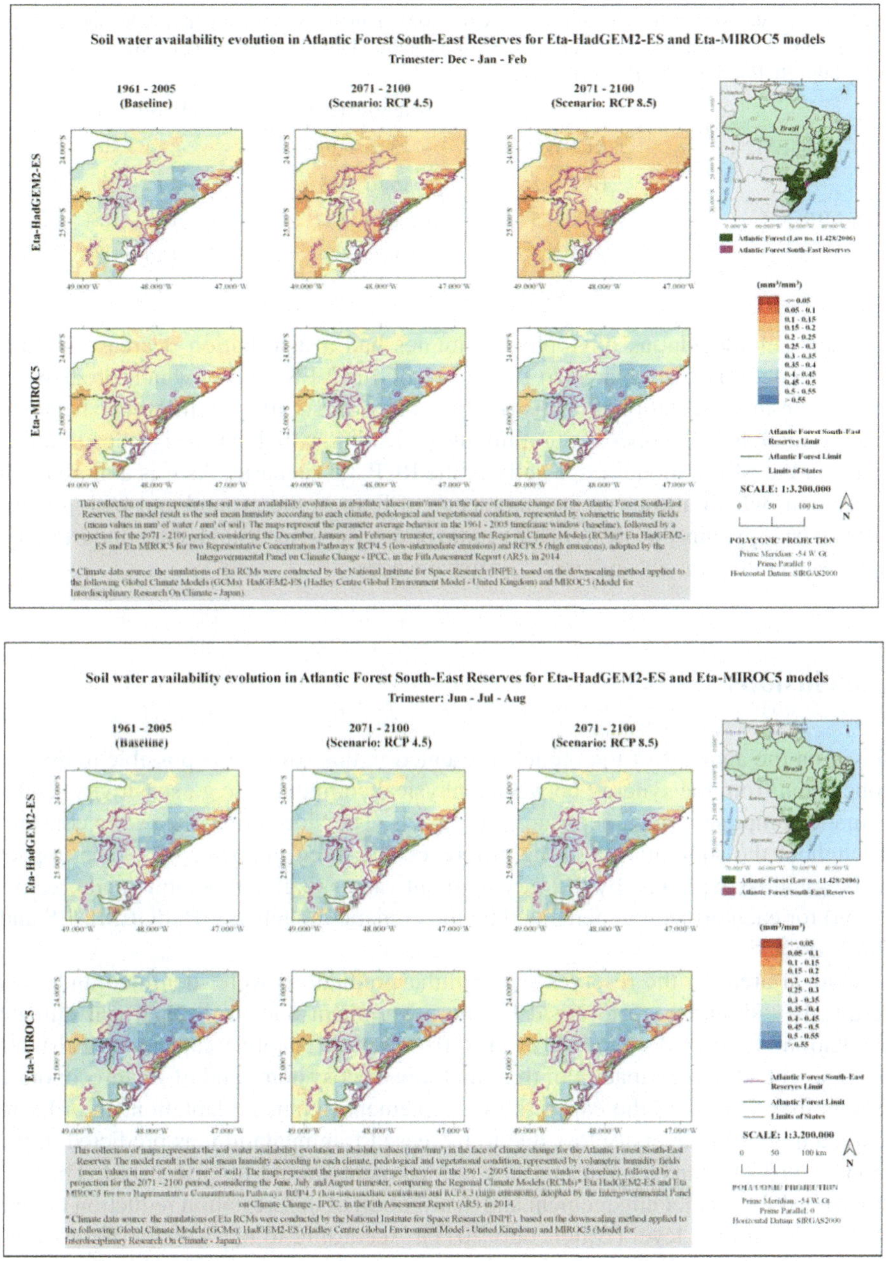

Fig. 3 Maps of soil water availability evolution in the Atlantic forest South-East reserves for both Eta-HadGEM2 -ES and Eta-MIROC5 climatic models, in two RCP scenarios (4.5 and 8.5) during **a** December, January and February (rainy season) and **b** June, July and August (dry season)

Table 3 Detailed soil water availability range of variation in the Atlantic forest South-East reserves for the rainy and dry seasons, according to the results of Eta-HadGEM2-ES and Eta-MIROC5 models using both RCP 4.5 and RCP 8.5 scenarios

		Rainy season	Dry season
Eta-HadGEM2-ES	RCP 4.5	−13 to −26%	−3 to +5%
	RCP 8.5	−15 to −38%	−4 to +7%
Eta-MIROC5	RCP 4.5	+5 to +18%	−1 to −18%
	RCP 8.5	+6 to +30%	−1 to −18%

Seasonal semideciduous phytophysiognomies had a distribution decrease in Eta-Miroc model when compared to the Eta-HadGEM2-ES model, which predicts an increase in this vegetation type distribution. Another significant difference between the two models is the open ombrophilous vegetation, which has a significant dominance in the Miroc model, especially in the RCP 8.5 scenario. As it is a more open vegetation, which allows greater incidence of light in the ground and with no significant canopy, this physical factor can favor other species establishment during the time, especially grasses and exotic invasive weeds.

Conclusions

Based on the analysis of the models presented in this study, it is possible to draw a conclusion from the synergy between a physical component (e.g., sensitivity) and a climatic component (e.g., exposure) of impacts. The biophysical data were constant throughout the simulations, so the differences between the behavior of the impact models can be explained by the varying input values (e.g., temperature and precipitation) for each projection generated by the climatic models Eta-HadGEM2-ES and Eta-MIROC5.

Critical areas in the reserves with similar predictions were identified and, as a result, they should be prioritized by the government and in the national climate adaptation strategy. As Lindner et al. (2010) states, adaptive capacity consists of the inherent adaptive capacity of trees and forest ecosystems and of socio-economic factors that determine the capability to implement planned adaptation. In order to avoid a higher biodiversity loss due to increased fragmentation, as predicted in the models, proper management and law enforcement should be prioritized. There is still a need for greater engagement and distribution of deforestation reduction targets at the sub-national level (Bittencourt 2011). According to the Atlantic Forest Law (Law nº 11.428) settled in 2006 in order to protect this environment, this biome was classified as public interest, yielding partially to the constant pressure of logging and endangering the remaining vegetation. Beyond the lack of enough protected areas, the fragility of its infrastructure, insufficient investments and proper qualified

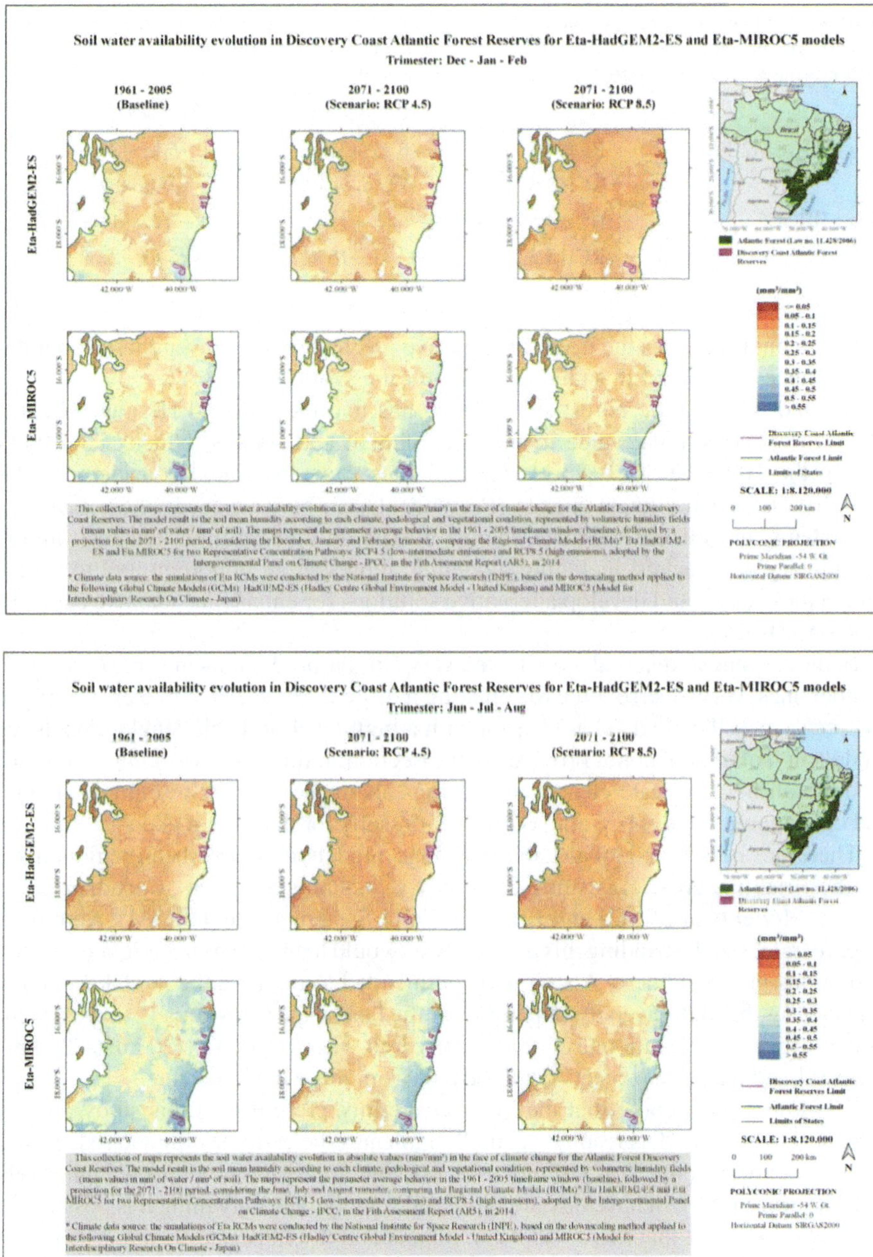

Fig. 4 Maps of soil water availability evolution in the discovery coast Atlantic forest reserves for both Eta-HadGEM2 -ES and Eta-MIROC5 climatic models, in two RCP scenarios (4.5 and 8.5) during **a** December, January and February (rainy season) and **b** June, July and August (dry season)

Table 4 Detailed soil water availability range of variation in the discovery coast Atlantic forest reserves for the rainy and dry seasons, according to the results of Eta-HadGEM2-ES and Eta-MIROC5 models using both RCP 4.5 and RCP 8.5 scenarios

		Rainy season	Dry season
Eta-HadGEM2-ES	RCP 4.5	−6 to −22%	−5 to −20%
	RCP 8.5	−21 to −41%	−14 to 0%
Eta-MIROC5	RCP 4.5	−6 to +3%	−3 to −11%
	RCP 8.5	−3 to −16%	−5 to −22%

management reinforces the need for incentive mechanisms and awareness about the significance and role of this unique biome for the World Heritage.

Beyond the Atlantic Forest sites, Brazil has five more World Heritage Natural sites, all of exceptional value to the environment and society. Nineteen hotspots among the 209 UNESCO World Heritage sites were listed and considered to be at higher risk from climate change (Perry 2011). To better manage these complex ecosystems and mitigate climate change effects on conservation hotspots, multi-disciplinary approaches, including modeling, represent an important tool to assess data in these cases and develop alternative solutions for enhancing adaptive capacity and conservation.

In the present study, models achieved very similar predictions in some cases and, in other ones, very diverse forecasts, making it difficult for decision makers to choose one result over the other. That is, even with a high level of detail, single models, or models in isolation, can not provide clear direction in this case. Working with more models can help indicate clearer conclusions about the impacts of climate change and endangered areas, providing convergent results for decision makers.

There are now four global climate change models that have been regionalized for the Brazilian area. This study used only two, since the others, Canadian Earth System Model (CanESM2) and Brazilian Earth System Model (BESM), were only recently released. Expanding this data analysis would help overcome the conclusion's limitations presented here. Also it is recommended to improve the model accuracy, considering the high diversity of patterns in the landscape in an already limited area. The information resolution used for this research was of 400 km^2 (20 km × 20 km) per pixel and adaptive projects could benefit from more detailed data.

It is important to note that the high uncertainty presented in this study—when considering both models results—actually indicates the extreme climatic risk we are currently experiencing. It is still not even possible to assure the impacts high levels of GHG emissions will have on our nature and society. The only thing it is possible to highlight is the fact that there will be an impact and it will be significant. Therefore, divergence of climate change models results should not be used to justify not to act. On the contrary, risks are so high that mitigation and adaptation actions must be planned and carried out as soon as possible.

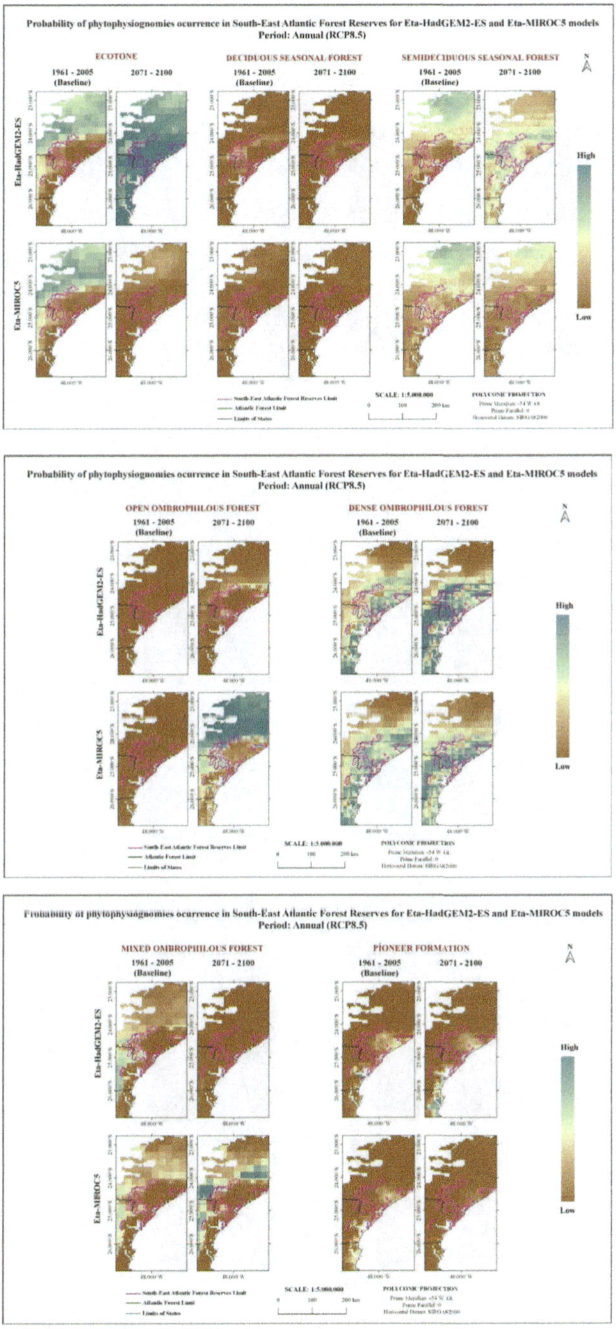

Fig. 5 Maps of phytophysiognomies occurrence probability in the Atlantic forest South-East reserves for both Eta-HadGEM2-ES and Eta-MIROC5 climatic models, in the 8.5 RCP scenario for seven phytophysiognomies. **a** Ecotone, deciduous seasonal forest and semideciduous seasonal forest; **b** open ombrophilous forests and dense ombrophilous forests; and **c** mixed ombrophilous forests and Pioneer formation

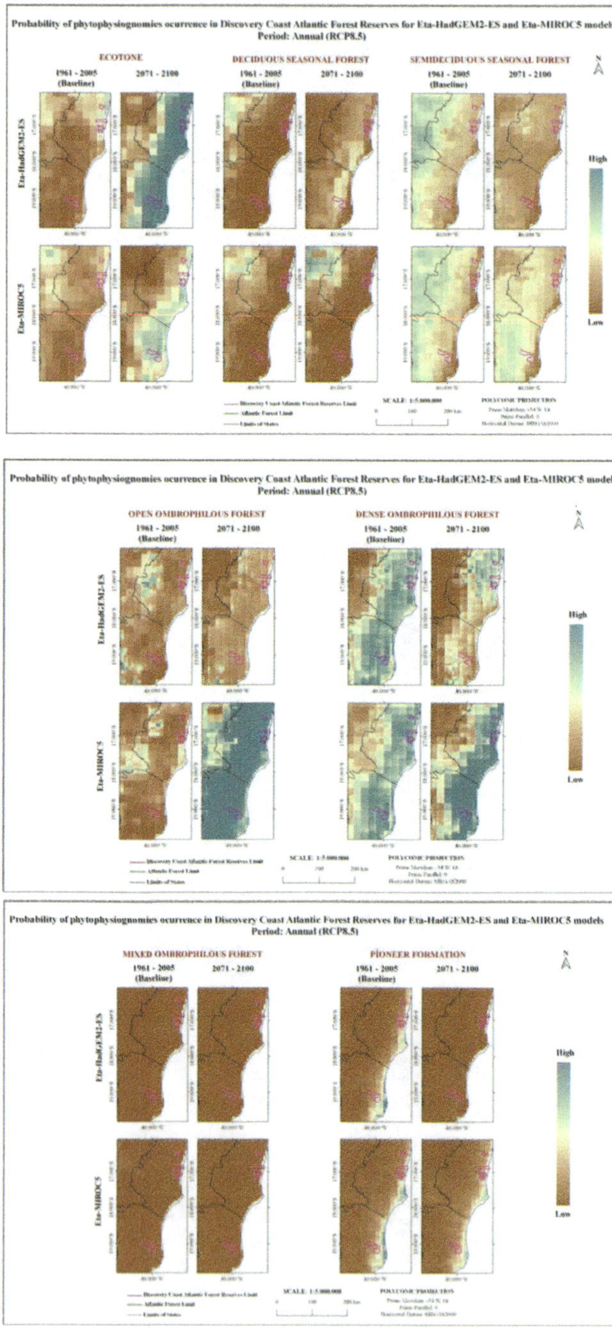

Fig. 6 Maps of phytophysiognomies occurrence probability in the Discovery coast Atlantic forest reserves for both Eta-HadGEM2-ES and Eta-MIROC5 climatic models, in the 8.5 RCP scenario for seven phytophysiognomies. **a** Ecotone, deciduous seasonal forest and semideciduous seasonal forest; **b** open ombrophilous forests and dense ombrophilous forests; and **c** mixed ombrophilous forests and Pioneer formation

References

Agência Nacional de Águas (2000) http://mapas.sosma.org.br/dados/. Accessed on 27 Nov 2018

Allen RG, Pereira LS, Raes D, Smith M (1998) Crop evapotranspiration-Guidelines for computing crop water requirements-FAO irrigation and drainage paper 56. Fao, Rome 300(9):D05109

Atlantic Forest Law nº 11.428 (2006) http://www.mma.gov.br/biomas.html. Accessed on 14 Sept 2018

Bittencourt FR (2011) Modelagem dinâmica de distribuição de metas de redução de desmatamento, estimativas de potencial de emissão de gases de efeito estufa e de receitas financeiras via Redd+ em níveis sub-nacionais: um estudo de caso para a Amazônia Brasileira. Belo Horizonte, Brazil, UFMG, p 325

Bugmann H, Cordonnier T, Truhetz H, Lexer MJ (2017) Impacts of business-as-usual management on ecosystem services in European mountain ranges under climate change. Reg Environ Change 17(1):3–16

Chou SC, Lyra A, Mourão C, Dereczynski C, Pilotto I, Gomes J, Bustamante J, Tavares P, Silva A, Rodrigues D, Campos D, Chagas D, Sueiro G, Siqueira G, Nobre P, Marengo J (2014a) Evaluation of the Eta-simulations nested in three global climate models. Am J Clim Change 3:438–454

Chou SC, Lyra A, Mourão C, Dereczynski C, Pilotto I, Gomes J, Bustamante J, Tavares P, Silva A, Rodrigues D, Campos D, Chagas D, Sueiro G, Siqueira G, Nobre P, Marengo J (2014b) Assessment of climate change over South America under RCP 4.5 and 8.5 downscaling scenarios. Am J Clim Change 3:512–525

Colombo AF (2007) Conseqüências potenciais das mudanças climáticas globais para espécies arbóreas da Mata Atlântica. UNICAMP, Campinas, Brazil, p 87p

Embrapa (2011) http://mapoteca.cnps.embrapa.br/. Accessed on 27 Nov 2018

Haddad NM et al (2015) Habitat fragmentation and its lasting impact on Earth's ecosystems. Sci Adv 1(2):e1500052

IBGE (2002) http://www.mma.gov.br/biomas/mata-atlantica/mapa-da-area-de-aplicacao/. Accessed on 27 Nov 2018

IPCC (2007) The physical science basis. Contribution of working group I to the fourth assessment report of the intergovernmental panel on climate change. In: Solomon SD, Qin M, Manning Z, Chen M, Marquis KB, Averyt M, Tignor and HL, Miller (eds) Cambridge University Press, Cambridge, United Kingdom and New York, NY, USA, 996 pp

IPCC (2014) Summary for policymakers. In: Field CB, Barros VR, Dokken DJ, Mach KJ, Mastrandrea MD, Bilir TE, Chatterjee M, Ebi KL, Estrada YO, Genova RC, Girma B, Kissel ES, Levy AN, MacCracken S, Mastrandrea PR, White LL (ed) Climate change 2014: impacts, adaptation, and vulnerability. Part A: global and sectoral aspects. Contribution of working group II to the fifth assessment report of the intergovernmental panel on climate change. Cambridge University Press, Cambridge (in press)

Knapp AK, Fay PA, Blair JM, Collins SL, Smith MD, Carlisle JD, Harper CW, Danner BT, Lett MS, McCarron JK (2002) Rainfall variability, carbon cycling, and plant species diversity in a mesic grassland. Science 298(5601):2202–2205

Lanz B, Dietz S, Swanson T (2018) The expansion of modern agriculture and global biodiversity decline: an integrated assessment. Ecol Econ 144:260–277

Lindner M, Maroschek M, Netherer S, Kremer A, Barbati A, Garcia-Gonzalo J, Rupert S, Delzond S, Corona P, Kolström M, Lexer MJ, Marchettie M (2010) Climate change impacts, adaptive capacity, and vulnerability of European forest ecosystems. For Ecol Manage 259(4):698–709

Magnago LFS, Rocha MF, Meyer L, Martins SV, Meira-Neto JAA (2015) Microclimatic conditions at forest edges have significant impacts on vegetation structure in large Atlantic forest fragments. Biodivers Conserv 24(9):2305–2318

Mittermeier RA, Fonseca GD, Rylands AB, Brandon K (2005) Uma breve história da conservação da biodiversidade no Brasil. Megadiversidade 1(1):14–21

Myers N, Mittermeier RA, Mittermeier CG, Fonseca GAB, Kent J (2000) Biodiversity hotspots for conservation priorities. Nature 403:853–858

Naumann G, Barbosa P, Iglesias A, Vogt J (2014) Exploring drought vulnerability in Africa: an indicator based analysis to be used in early warning systems. Hydrol Earth Syst Sci 18:1591–1604

Newbold T, Hudson LN, Arnell AP, Contu S, De Palma A, Ferrier S (2016) Has land use pushed terrestrial biodiversity beyond the planetary boundary? A global assessment. Science 353:288–291

Perry Jim (2011) World Heritage hot spots: a global model identifies the 16 natural heritage properties on the World Heritage list most at risk from climate change. Int J Heritage Stud 17(5):426–441

Perry J (2015) Climate change adaptation in the world's best places: a wicked problem in need of immediate attention. Landscape Urban Plan 133:1–11

Phillips HR, Newbold T, Purvis A (2017) Land-use effects on local biodiversity in tropical forests vary between continents. Biodivers Conserv 26(9):2251–2270

Phillips SJ, Dudík M, Schapire RE (2004) A maximum entropy approach to species distribution modeling. In: Proceedings of the twenty-first international conference on machine learning. ACM, p 83

Porporato A, Daly E, Rodriguez-Iturbe I (2004) Soil water balance and ecosystem response to climate change. Am Nat 164(5):625–632

REFLORA (2017) Lista de espécies da flora do Brasil. Rio de Janeiro, Brazil: Jardim Botânico do Rio de Janeiro. Retrieved from: http://floradobrasil.jbrj.gov.br/. Accessed on 27 Sept 2018

Reich PB, Tilman D, Isbell F, Mueller K, Hobbie SE, Flynn DF, Eisenhauer N (2012) Impacts of biodiversity loss escalate through time as redundancy fades. Science 336(6081):589–592

Rolim SG, Jesus RM, Nascimento HE, Do Couto HT, Chambers JQ (2005) Biomass change in an Atlantic tropical moist forest: the ENSO effect in permanent sample plots over a 22-year period. Oecologia 142(2):238–246

Salis SM, Shepherd GJ, Joly CA (1995) Floristic comparison between mesophytic forests of the interior of the state of São Paulo, S.E. Brazil. Vegetation 119:55–164

Santo-Silva EE, Almeida WR, Tabarelli M, Peres CA (2016) Habitat fragmentation and the future structure of tree assemblages in a fragmented Atlantic forest landscape. Plant Ecol 217(9):1129–1140

Scarano FR, Ceotto P (2015) Brazilian Atlantic forest: impact, vulnerability, and adaptation to climate change. Biodivers Conserv 24(9):2319–2331

Scarano FR (2002) Structure, function and floristic relationships of plant communities in stressful habitats marginal to the Brazilian Atlantic rainforest. Ann Bot 90(4):517–524

Shannon CE (1949) Communication theory of secrecy systems. Bell Syst Tech J 28(4):656–715

Shishlov I, Morel R, Bellassen V (2016) Compliance of the parties to the Kyoto Protocol in the first commitment period. Clim Policy 16(6):768–782

Silva AM (2004) Rainfall erosivity map for Brazil. CATENA 57:251–259

SOS MATA ATLÂNTICA (2017) Atlas dos remanescentes florestais da Mata Atlântica: Período 2015–2016. https://www.sosma.org.br/link/Atlas_Mata_Atlantica_2016-2017_relatorio_tecnico_2018_final.pdf. Accessed on 27 Nov 2018

Thom D, Seidl R (2016) Natural disturbance impacts on ecosystem services and biodiversity in temperate and boreal forests. Biol Rev 91:760–781

TOPODATA (2011) http://www.dsr.INPE.br/topodata/acesso.php. Accessed on 27 Nov 2018

UNESCO (1972) Convention concerning the protection of the world cultural and natural heritage: adopted by the general conference at its seventeenth session. Paris

Wischmeier WH (1959) A rainfall erosion index for a universal soil-loss equation. Soil Sci Soc Am Proc 23:246–249

World Bank (2018) Forest carbon partnership facility. https://www.forestcarbonpartnership.org. Accessed on 27 Nov 2018

Zeng Z, Estes L, Ziegler AD, Chen A, Searchinger T, Hua F, Guan K, Jintrawet A, Wood EF (2018) Highland cropland expansion and forest loss in Southeast Asia in the twenty-first century. Nat Geosci 11(8):556–562

Towards a Prevention-Driven Adaptation Strategy, as Applied to Indigenous Peoples' Internal Climate Migration: Some Inputs Based on a Rights-Based Approach

Armelle Gouritin

Abstract Climate change-induced migration as a phenomenon is frequently reported and ever more evident. This phenomenon is expected to exacerbate as climate change impacts are susceptible to amplification. Against this background, some scholars contend that climate change induced migration is the result of a successful adaptation strategy. In this chapter, this claim is challenged for not taking into account climate migration diversity, such as the case of Indigenous Peoples being forced to leave their traditional lands or territories and resettle internally. The argument put forward is that a rights-based approach calls for preventative adaptation measures. This chapter builds on the Inter-American Court of Human Rights' case law and the Mexican case to illustrate the argument. Insights gained from engaging in this research are as follows. First, a rights-based approach calls for preventive adaptation measures based on Indigenous Peoples' relationship with their lands and territories which goes beyond lands and territories' mere geographical dimension but encompasses spiritual, cultural and historical dimensions. Second, a rights-based approach provides Guiding Principles that apply to preventative adaptation processes and the climate migration scenarios identified by Prof. Walter Kälin where a prevention approach applies (sudden-onset disasters, slow-onset environmental degradation, "sinking" small islands, and "unrest seriously disturbing public order, violence or even armed conflict"). Those Guiding Principles stem from the positive obligations and horizontal effect doctrines. Concretely, they range from environmental protection measures to natural resources management to avoid socio-environmental conflicts.

Introduction

Climate induced migration is one of the major types of climate change impacts, one that calls for urgent adaptation measures. The recent mass migration of Central American citizens which formed a caravan with a view to entering the territory of the

A. Gouritin (✉)
Cátedra CONACYT, Facultad Latinoamericana de Ciencias Sociales, Sede México, Carretera al Ajusco 377, Colonia Héroes de Padierna, Tlalpan, 14200 Ciudad de México, Mexico
e-mail: armellegouritin@gmail.com

W. Leal Filho et al. (eds.), *Climate Change, Hazards and Adaptation Options*, Climate Change Management, https://doi.org/10.1007/978-3-030-37425-9_50

USA passing by Mexico illustrates but one of the clearest and evident examples of a climate induced migration phenomenon (Miller 2018) and has been widely covered by mass media. A very recent study estimates that by 2050 and if no "concrete climate and development action" was taken, over 143 million people from Sub-Saharan Africa, South Asia, and Latin America would be forced to migrate internally (i.e. within their own country). The estimation for Latin America amounts to over 17 million, representing up to 2.6% of Latin America's population (Rigaud et al. 2018). Attention to this issue has grown exponentially as scholars and institutions become more and more interested in the forced displacement induced by climate change phenomenon echoes the greatest awareness on the phenomenon's current and projected amplified scale.

Among those scholars, some portray climate migration as a clear example of successful adaptation to climate change. They argue that (*ex post*) climate induced migration can increase resilience to climate change (for example if those who migrated send money or goods back to home communities, which would enable the home communities to raise resilience) (Black et al. 2011). Others argue that "migration is a form of adaptation, not a failure to adapt": "migration is (…) a form of adaptation at household level in response to threats to the sustainability of livelihoods" (Geddes and Jordan 2012). In other words, migration is "part of the solution rather than the problem" (Tacoli 2009). Those scholars make a very good point in trying to escape the "securisation" of climate related migration (Nishimura 2015).

Yet, their claim might be nuanced and some exceptions might be relevant. Adding to scholars who question this approach on the ground that it eliminates the space of the political and omits to present climate change as a social problem that could be tackled (Methmann and Oels 2015), this chapter aims at putting forward and testing another claim: the internal climate migrants phenomenon rather requires *ex ante* preventive actions when the populations at stake have a particularly strong relationship with their lands and territories. The foundations for this claim are Indigenous Peoples' identity, and a rights-based approach grounded on the Inter-American human rights law.

Mexico is a good case to illustrate the claim within the Latin American context. Not only is Mexico particularly vulnerable to Climate Change. Mexico and Guatemala alone are expected to "reach an average of 1.4–2.1 million" people forced to migrate internally by 2050, and the phenomenon is expected to increase by a factor of two between 2020 and 2050" (Rigaud et al. 2018). Also, 17 millions indigenous people lived in Mexico by 2010, representing 15.1% of Mexican's population (ECLAC 2014). Eventually, Mexican authorities have also addressed the climate migration phenomenon within the scope of adaptation: "Regarding the vulnerability and adaptation issues, it is necessary to keep strengthening research and actions with regards to (…) climate induced migration" (SEMARNAT and INECC 2012, translation of the author). In addition, the Mexican Climate Law was modified in 2016 and recognizes the "internal displacement of persons induced by climate change related events" (Fig. 1).

The chapter begins with elements of definition of the climate induced forced displacement phenomenon. The claim is subsequently tested and verified: the rights

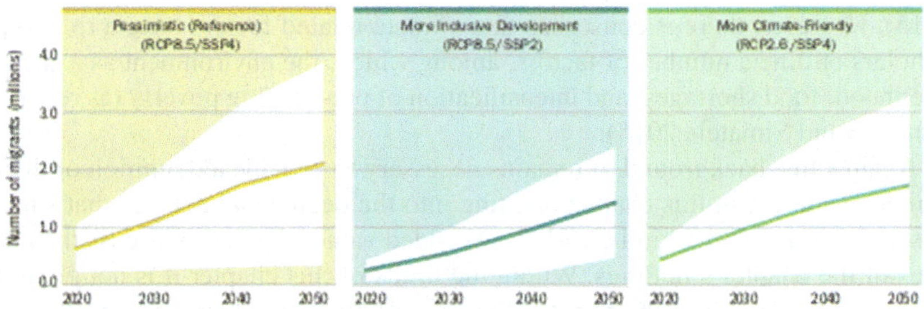

Climate migrants as a percentage of the total population

Year	2020	2030	2040	2050	2020	2030	2040	2050	2020	2030	2040	2050
%	0.37	0.60	0.87	1.03	0.14	0.29	0.48	0.68	0.25	0.51	0.72	0.85

Note: Dark lines represent the average runs for each scenario. Unshaded white areas represent the 95th percentile confidence intervals. The wide intervals are in part a reflection of the fact that climate and sectoral models were selected to represent the widest possible range of outcomes.

Fig. 1 Projected number of internal climate migrants in Mexico and Central America under three scenarios. *Source* Rigaud et al. (2018), p. 101

based approach applied to Indigenous Peoples calls for a preventive driven adaptation approach and provides some tools that must guide the preventive driven approach. Concrete requirements and examples illustrate the findings.

"Climate Induced Forced Displacement": Elements of Definition

The current and expected scale of the phenomenon demonstrates the social relevance of the phenomenon, the urgent need to address the phenomenon and, more specifically, the necessity to protect climate migrants. To begin with, we ought to define the phenomenon and describe it at more length.

There is a growing body of literature on the phenomenon, consensus on the existence of an environmental and climate trigger to migration, and endorsement of climate migration in the political agenda of various institutions at the international (for example during the UNFCCC COP24 in Katowice, Poland), regional and national levels. However, the definition of climate migration and identification of its constituting elements remains a challenge and highly debated.

It can be explained by the multiple and linked drivers and factors of the migration patterns: economic, social, political, and environmental/climate drivers (Geddes and Jordan 2012). It is hard to isolate the environmental/climate drivers within this web of interlinked factors with a view to identifying "purely or mainly environmental/climate migrants." Identifying underlying reasons for a decision or obligation to migrate for climate related reasons is even more complex (and accurate) if the analysis is done through a micro, macro and meso perspective (Simatele and Simatele

2015). However, there is consensus that climate related factors have a role to play. Scholars outline a number of factors, among which: the environment as a source of livelihood, food shortages, and intensification of pre-existing poverty (as reported in Simatele and Simatele 2015).

Against this background, it remains necessary to provide a definition of the phenomenon studied in this chapter. Getting into the details and debates that surround this definition matter would clearly be of added value in order to discuss the boundaries of the chapter's findings. Within the limits of this chapter it is not possible to get into details on the debates regarding the definition of climate migrants (or "environmental/climate change refugee", "environmentally/climate displaced persons" …). Therefore, the definition put forward is by no way meant to engage into current debates, but is rather meant to provide the reader with elements of understanding and delimitate the specific object of the chapter.

The International Organization for Migration's working definition of environmental migrants reads: "Environmental migrants are persons or groups of persons who, for reasons of sudden or progressive changes in the environment that adversely affect their lives or living conditions, are obliged to have to leave their habitual homes, or choose to do so, either temporarily or permanently, and who move within their territory or abroad" (IOM 2007).

The International Organization for Migration's working definition is a good starting point to further distinguish and specify the internal climate migration phenomenon addressed in the chapter. First, the phenomenon covers climate change effects that are both slow onset (e.g. biodiversity loss, soil degradation) and rapid onset (e.g. natural disasters such as hurricanes). Second, the phenomenon addressed in this chapter is exclusively internal (climate migration within states' borders). Third, it covers both short and long term migration. Fourth, the migration phenomenon is forced (as opposite to migration as a matter of choice or voluntary). Eventually, the scope of application of the concept is limited to Indigenous Peoples who are forced to leave their traditional lands or territories.

The first element refers to the slow and rapid onset impacts of climate change and deserves more explanation with a view to distinguishing "environmental" from "climate" migration. Concretely, a variety of climate-related drivers influence the decision or constraint to migrate. Prof. Kälin (former Representative of the United Nations Secretary-General on the Human Rights of Internally Displaced Persons) offers a typology that distinguishes five scenarios: (1) "Sudden-onset disasters, such as flooding, windstorms (hurricanes/typhoons/cyclones) or mudslides", (2) "Slow-onset environmental degradation caused, inter alia, by rising sea levels, increased salinisation of groundwater and soil, long-term effects of recurrent flooding, thawing of permafrost, as well as droughts and desertification or other forms of reduced water availability", (3) "So-called 'sinking' small island states present a special case of slow-onset disasters. As a consequence of rising sea levels and their low-lying topology, such areas may become uninhabitable", (4) "Governments may designate areas as high-risk zones too dangerous for human habitation on account of environmental dangers", and (5) "Finally, unrest seriously disturbing public order, violence or even armed conflict may be triggered, at least partially, by a decrease in essential resources due to climate change (such as water, arable land or grazing grounds)" (Kälin 2010). Except the fourth scenario that clashes with the hypothesis put forward that focuses on preventive measures (i.e. cases where displacement is not yet a necessity), they are all addressed in the subsequent sections. The four types of climate drivers are addressed under a rights based approach and illustrated by the Mexican case.

The subsequent section focuses on Indigenous Peoples' relationship to their traditional lands and territories and grounds theoretically and within human rights law the claim for a preventive driven adaptation process applied to Indigenous Peoples as internal climate migrants with a view to reducing the risks generated by climate change.

Human Rights Require a Preventive Driven Adaptation Approach

The notion of territory is crucial in the context of the claim put forward in this chapter. A merely "administrative" or narrow understanding of the notion of territory applied to Indigenous Peoples traditional lands and territories and the recognition of Indigenous Peoples' property over those implies a risk of dispossession (Surrallés and Hierro 2004).

The narrow notion of territory is limited to its spatial or geographical scale. This narrow definition is criticized and obsolete according to the large majority of scholars from a broad range of disciplines (including geography). The wider notion of territory embraces territory's whole dimensions, namely its environmental, symbolic, economic, social, organizing, cultural and historical dimensions (Valle 2012). More specifically, the territory is both the scene for social relations and a social construct (Gómez and Mahecha 1998).

Turning to Indigenous Peoples' relationship with their lands and territories, not only do Indigenous Peoples mobilize them for social reproduction purposes. They are also crucial with regards to their mere identity: "the territory is the space where a community's culture is embedded, the base for collective memory and ancestral past (…) for indigenous peoples the territory is the center of the cosmovision, the core of its identity's referencing, and has a sacred value" (Gallardo Robles 2016, translation of the author). It results from this definition that ensuring the continuity of Indigenous Peoples' relationship with their lands and territories is crucial. If they define themselves according to their lands and territories and vice versa, breaking this link should be prevented with a view to guaranteeing Indigenous Peoples' identity survival.

The theoretical necessity to guarantee Indigenous Peoples' relationship with their lands and territories is supported by Inter-American human rights law. The Inter-American Court of Human Rights (Organization of American States) endorses a notion of territory that resonates with the theoretical cultural and symbolical dimensions mentioned above and articulates the state's obligation to preserve the relationship between Indigenous Peoples and their lands and territories. The Court's case law on Indigenous Peoples' right to consultation and consent, forced displacement in the context of development projects, and land claims illustrates this obligation and grounds it on identity survival (Gouritin and Aguilar 2017).

Adding to subsistence, the Court emphasizes Indigenous Peoples territories' various dimensions. In its settled case law the Court acknowledges Indigenous Peoples' own worldview and cultural and spiritual identity in connection with their territories (IACtHR 2012, p. 155). The Court takes due account of the territory's cultural and symbolic dimensions in a historical context. To do so, it considers testimonies from Indigenous Peoples' members and experts (usually sociologists and anthropologists). The *Mayagna (Sumo) Awas Tingni Community versus Nicaragua* case illustrates the Court's settled case law very clearly:

> Indigenous groups, by the fact of their very existence, have the right to live freely in their own territory; the close ties of indigenous people with the land must be recognized and understood as the fundamental basis of their cultures, their spiritual life, their integrity, and their economic survival. For indigenous communities, relations to the land are not merely a matter of possession and production but a material and spiritual element which they must fully enjoy, even to preserve their cultural legacy and transmit it to future generations. (IACtHR 2001, p. 149)

Expressed even more clearly: "[T]his Court has underlined that the close relationship of indigenous peoples with the land must be acknowledged and understood as the fundamental basis for their culture, spiritual life, wholeness, economic survival, and preservation and transmission to future generations" (IACtHR 2005, p.131). The recognition of the territories' cultural and symbolic dimensions stems on communal property as a condition to maintain their cultural identity and "very survival" (IACtHR 2005, p. 147).

The Court identifies what is encompassed by the term "identity" mentioning a series of elements, such as oral expressions and traditions, customs and languages, arts and rituals, "knowledge and practices in connection with nature, culinary art, customary law, dress, philosophy, and values. In connection with their milieu, their integration with nature and their history, the members of the indigenous communities transmit this non-material cultural heritage from one generation to the next, and it is constantly recreated by the members of the indigenous groups and communities" (IACtHR 2005, p. 154).

By contrast, the Mexican Constitution the Constitution does not recognize as such the various cultural and symbolic dimensions of Indigenous Peoples territories, even though it recognizes the link between Indigenous Peoples and their territory, including as an element of Indigenous Peoples' definition (Article 2). Inter-American human rights requirements nevertheless bind Mexico (Mexican Constitution, Article 1). Accordingly, the obligation to preserve Indigenous Peoples' territories in all its dimensions applies in the context of resource extraction activities (e.g. logging, mining concessions) or development projects (e.g. hydro-electric plants).

The obligation to prevent forced displacement is settled case law (IACtHR 2013, p. 220). Similarly, the obligation to prevent any rupture between Indigenous Peoples and their lands and territories is settled case law (IACtHR 2007). The Court's case law applies the positive obligations doctrine to human rights protection. According to this doctrine, states not only have to refrain from interfering with human rights. They also have to adopt the practical measures that guarantee human rights' actual enjoyment: human rights must be "effective in reality and actual practice" (IACtHR 2005, p. 141). Those measures can be administrative or legal (IACtHR 2015a, p. 104; Gouritin 2016).

The obligation to preserve Indigenous Peoples' relationship with their lands also applies in the context of forced climate induced displacement. The analogy lies on the rupture between Indigenous Peoples and their lands. In the subsequent section the attention is directed at the practical measures that should be undertaken for the state to fulfill its obligation to prevent Indigenous Peoples from being displaced from their lands and territories because of climate change triggers.

Human Rights Law Provides Guiding Principles for Preventive Driven Adaptation Public Policies

The rights based approach does not only ground the state's obligation to prevent Indigenous Peoples from being forced to leave their lands and territories because of climate change triggers. The rights based approach also provides some tools and safety nets that must guide states' public policy and legal frameworks. There are four such tools: the positive obligations doctrine applied to risk management, the horizontal effect doctrine, the standstill effect, and the necessity/proportionality test (Gouritin 2016). A full analysis of the four tools would have provided a greater analysis and a full picture that might have nuanced the outcome of the chapter. However, within the limits of this chapter it is not possible to perform such analysis. In the subsequent lines two of those tools are applied to four climate displacement scenarios distinguished by Prof. Kälin, and more specifically to Indigenous Peoples' potential loss of their lands and territories. The tools are the positive obligations doctrine and the horizontal effect doctrine.

The positive obligations doctrine mentioned in the former section also applies to environmental protection obligations: "the right to use and enjoy the territory would be meaningless for indigenous and tribal communities if that right were not connected to the protection of natural resources in the territory" (IACtHR 2012, p. 146). Accordingly, states bear the obligation to prevent or halt an environmental degradation that could generate Indigenous Peoples' constraint to forcefully leave their lands and territories.

On the one hand, environmental protection can help prevent or attenuate slow onset and rapid onset events. Environmental protection of Indigenous Peoples' lands and territories can prevent the forced departure of Indigenous Peoples from their lands and territories. It is now commonly acknowledged that slow onset environmental degradation is amplified by climate change. It is also true for Mexico.

Concretely, environmental degradation could hamper Indigenous Peoples from having access to fish stocks, harvesting, or performing their rituals. Therefore, states must protect and preserve the environment, so as to prevent a degradation of such an extent that would force Indigenous Peoples to leave their lands and territories. Also, environmental protection may not only slow down the environmental degradation amplified by climate change. Environmental protection can also attenuate the effects of sudden onset events. For example, halting down deforestation or reforesting can help attenuate flooding effects. This function has been acknowledged by the Mexican National Commission of Human Rights (CNDH 2017b).

Consequently, nation states should adopt and ensure compliance with environmental protection measures. The situation in Mexico can illustrate this first aspect. The environment in Mexico is largely under threat (Sarukhán et al. 2017). Even more relevant, the third scenario of Prof. Kälin is clearly identified in Mexico: the Mexican territory encompasses islands that are threatened by the sea level rise phenomenon, with the difference that those islands are not states but part of the Mexican territorial landscape (they are, among others: Islas Marías, Mayas, Pericúes, Cochimíes and Seris) (Fig. 2).

A great wealth of environmental protection laws virtually apply in Mexico and they reach the higher standards, but they suffer from a very bad implementation (Ugalde 2010). Accordingly, Mexico's obligation to duly implement environmental protection law and solve the structural problems in that respect is triggered by the obligation to prevent Indigenous Peoples' forced climate displacement.

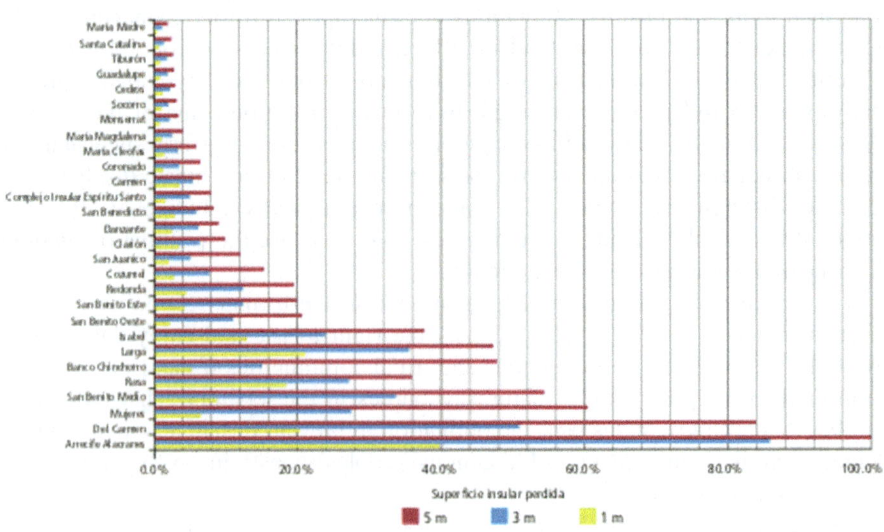

Fig. 2 Lost superficies in various Mexican islands in case of sea level rise of 1, 3, and 5 m. *Source* SEMARNAT and INECC (2018), p. 474

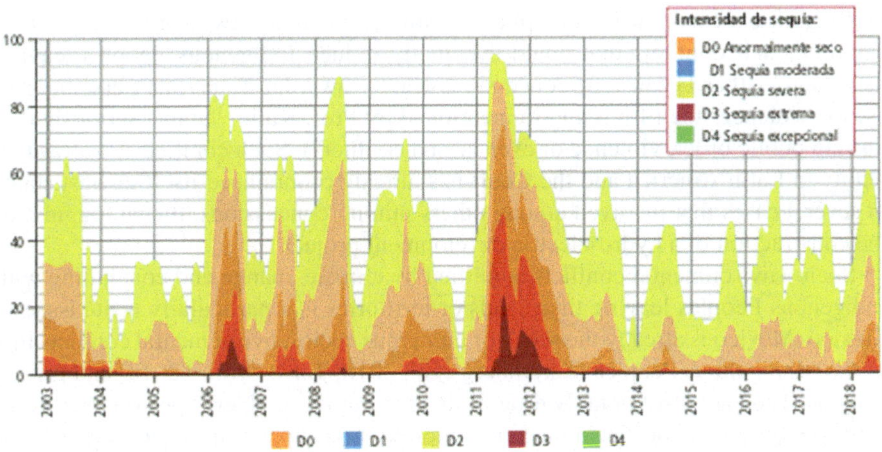

Fig. 3 Percentage of the Mexican territory impacted by one or more draught. *Source* SEMARNAT and INECC (2018), p. 451

On the other hand, climate change exacerbates natural resources' scarcity. For example, water scarcity is already a very contentious issue in Mexico (SEMARNAT and INECC 2018) (Fig. 3).

Against this background, Mexico is bound to adopt and enforce a water management policy that would ensure Indigenous Peoples' access to waters. An equilibrium should be reached between productive sectors' demand in waters (for example agriculture) and Indigenous Peoples' needs. Once again, the benchmark is not the mere adoption of public policy measures. The benchmark is human rights' effective and actual enjoyment.

The horizontal effect doctrine is the second human rights tool applied in this chapter with a view to unveiling some guiding principles in the context of a preventive driven adaptation approach. It refers to the actors who can interfere with human rights. Under the traditional vertical effect doctrine states and their emanations can interfere with human rights, and the state is hold responsible for those if the interference amounts to a human rights violation. According to the horizontal effect doctrine, when private persons interfere with human rights the states are held responsible on the ground that they did not prevent a human rights violation (through the adoption of administrative or legal measures, because of deficient measures or a deficient enforcement of those measures). In its settled case law the Court applies the horizontal effect (e.g. IACtHR 2015b that concerns mining concessions on Indigenous Peoples lands). The horizontal effect is crucial in the context of this chapter.

Climate change amplifies natural resources scarcity (as mentioned above), which could in turn amplify existing socio-environmental conflicts or generate new ones. Against this background, the horizontal effect doctrine applies to potential or existing conflicts and requires states to adopt administrative or legal measures, modify existing measures, or ensure compliance with them with a view to preventing socio-environmental conflicts that could eventually induce Indigenous Peoples leaving their lands either because of concessions legally awarded to private persons or to escape violence (the fifth scenario mentioned by Prof. Kälin: "unrest seriously disturbing public order, violence or even armed conflict"). Such conflicts are already a reality in Latin America and the Court has already pronounced itself against states in several cases that involved among others mining concessions, illegal logging, or the construction of a hydro-electric development project.

Socio-environmental conflicts amplified by climate change and concluding with Indigenous Peoples leaving their lands or territories is a particularly acute issue in Mexico. Mexico is already the scene of numerous socio-environmental conflicts that involve Indigenous Peoples (Zaremberg et al. 2018; Torres-Wong 2018) and private persons. They are most notably drug trafficking entities and legal persons (i.e. companies) whose legal or illegal behaviors more or less supported by the state induces very violent conflicts in cases where Indigenous Peoples traditional lands or territories are under threat. The Mexican National Human Rights Commission has already had various opportunities to sanction the state in the context of socio-environmental conflicts involving Indigenous Peoples, among others in the context of water scarcity. The cases concern competing uses of waters: water pollution generated by unproperly planned growing industrial activities and population density (CNDH 2017a), construction of tourism facilities that destroyed protected humid areas ("manglars") and jeopardized water quality and quantity (CNDH 2017b), and the construction of an aqueduct with a view to providing water to an urban area, at the prejudice of rural areas where Indigenous Peoples live (CNDH 2012). In the latter case, very violent conflicts occurred. The state was among others found to violate human rights because it had been an active actor in generating the conflict and not having prevented it. Exactly the same scenario can already occur in the specific case of concern for this chapter (Indigenous Peoples forced to leave their lands and territories because of competing uses of water in a scarcity context caused by or amplified by climate change).

Adding to and linked with socio-environmental conflicts, Indigenous Peoples social vulnerability is also to be taken into account when considering private persons' interference with Indigenous Peoples lands and territories within a climate change and climate induced migration context. Indigenous Peoples pre-existing vulnerability or climate induced vulnerability is a factor that amplifies the risk to be forcefully displaced. The Court's settled case law in this context is very enlightening. It mobilizes a broad range of rights (right to health, food, education to the right to life among others) and calls upon the state to take due account of Indigenous Peoples' vulnerability (IACtHR 2005, p. 163):

> the Court must establish whether the State generated conditions that worsened the difficulties of access to a decent life for the members of the Yakye Axa Community and whether, in that context, it took appropriate positive measures to fulfill that obligation, taking into account the especially vulnerable situation in which they were placed, given their different manner of life (different worldview systems than those of Western culture, including their close relationship with the land) and their life aspirations, both individual and collective.

This requirement is particularly relevant in Mexico, where many Indigenous Peoples are already in a situation of vulnerability (and more particularly women and children), and climate change is expected to amplify this vulnerability by (Gouritin 2018).

In sum, states are to adopt measures, modify existing measures, and duly enforce them with a view to preventing private persons from degrading Indigenous Peoples lands and territories, occupying them, or unduly exploiting the natural resources therein (legally or illegally). The trigger for such obligation is private persons' actions combined with climate change induced degradation and/or its effect on natural resources' scarcity caused by or amplified by climate change, which could eventually result in Indigenous Peoples being forced to leave their lands or territories. When doing so, the state also has to take due account of Indigenous Peoples vulnerability.

Conclusions

The claim that climate displacement is a successful adaptation to climate change should be nuanced and take due account of the displaced persons' diversity. The rights based approach provides the foundations for such nuance. In the case of Indigenous Peoples who have a relationship a their lands and territories that goes beyond possession and a mere spatial dimension, adaptation first calls for measures aimed at reducing the risks of forced displacement induced by climate change.

The rights based approach endorsed by the Inter-American Court of Human Rights supports this claim. It also gives flesh to the preventive adaptation approach. Among other human rights tools, the positive obligation and horizontal effect doctrines applied by the Court provide guiding principles that should trigger adaptation public policy and legal frameworks. Concretely, the rights based approach requires the adoption of environmental protection measures and enforcement of those, scarce natural resources management, and the prevention of socio-environmental conflicts, all of them taking due account of Indigenous Peoples' vulnerability.

The rights based approach applied in the chapter and illustrated by the Mexican case relies on the Inter-American Court of Human Rights' case law. Accordingly, the findings extend beyond the Mexican case as an illustration: they apply to Latin American states and provide guidance for attending the climate induced forced migration of Indigenous Peoples. The phenomenon urgently calls for public policy and legal answers, as it is constantly reported as an ever growing phenomenon.

The findings of this chapter call for further research regarding the definition of Indigenous Peoples' identity, autonomy, and the identification of Indigenous Peoples' strong relationship with their lands and territories in the context of climate migration as a window of opportunity. The purpose of this research and corresponding legal frameworks and public policy would be enhancing Indigenous Peoples' emancipation through the critical climate justice and self-defined autonomy lens rather than using this window of opportunity to deepen Indigenous Peoples' oppression through a discriminatory and exclusion approach.

References

Black R, Bennett SR, Thomas SM, Beddington JR (2011) Climate change: migration as adaptation. Nature 478(7370):447–449

Comisión Nacional de los Derechos Humanos (CNDH) (2012) Recomendación no. 37/2012 Sobre el caso del Proyecto "Acueducto Independencia", en el estado de Sonora. Mexico

Comisión Nacional de los Derechos Humanos (CNDH) (2017a) Recomendación no. 10/2017 Sobre la violación a los derechos humanos a un medio ambiente sano, saneamiento del agua y acceso a la información, en relación con la contaminación de los ríos Atoyac, Xochiac y sus afluentes; en agravio de quienes habitan y transitan en los municipios de San Martín Texmelucan y Huejotzingo, en el estado de Puebla; y en los municipios de Tepetitla de Lardizábal, Nativitas e Ixtacuixtla de Mariano Matamoros, en el estado de Tlaxcala. Mexico

Comisión Nacional de los Derechos Humanos (CNDH) (2017b) Recomendación no. 67/2017 Sobre el caso de la violación a los derechos humanos a un medio ambiente sano, a la seguridad jurídica y al principio de legalidad, por la remoción de manglar para la realización del desarrollo "Malecón Cancún" (Proyecto Tajamar) en Quintana Roo. Mexico

Economic Commission for Latin America and the Caribbean (ECLAC) (2014) Guaranteeing indigenous people's rights in Latin America: progress in the past decade and remaining challenges. Santiago

Gallardo Robles LJ (2016) La defensa del Chuchutsipi: la lucha de las comunidades Totonocas por la vidad. Master thesis in social anthropology, Centro de Investigaciones y Estudios Superiores en Antropología Social. Mexico

Geddes A, Jordan A (2012) Migration as adaptation? Exploring the scope for coordinating environmental and migration policies in the European Union. Environ Plan C: Gov Policy 30(6):1029–1044

Gómez GM, Mahecha OD (1998) Espacio, territorio y región: conceptos básicos para un proyecto nacional. Cuadernos de Geografía: Revista Colombiana de Geografía 7(1–2):120–134

Gouritin A (2016) EU Environmental Law, International Environmental Law, and Human Rights Law: confirmation, complement and conflicts. The Case of Environmental Responsibility. BRILL, International Environmental Law Series, Leiden

Gouritin A, Aguilar A (2017) The adoption of the American declaration on the rights of indigenous peoples: a critical analysis from the perspective of environmental rights. Anuario Mexicano de Derecho Internacional, XVII, pp 291–327

Gouritin A (2018) Human rights violations in the context of socio-environmental conflicts: wind farms in San Dionisio del Mar, Oaxaca (Mexico). In: Primova R, Gouritin A (eds) The road from Paris to sustainable development: integrating human rights and gender equality into EU external climate policies. Heinrich-Böll-Stiftung European Union, Brussels

Inter-American Court of Human Rights (IACtHR) (2001) Mayagna (Sumo) Awas Tingni Community vs. Nicaragua. Serie C No. 79

Inter-American Court of Human Rights (IACtHR) (2005) Yakye Axa Indigenous Community Vs. Paraguay. Serie C No. 125

Inter-American Court of Human Rights (IACtHR) (2007) Pueblo Saramaka vs. Surinam. Serie C No. 172

Inter-American Court of Human Rights (IACtHR) (2012) Pueblo Indígena Kichwa de Sarayaku vs. Ecuador. Serie C No. 245

Inter-American Court of Human Rights (IACtHR) (2013) Comunidades Afrodescendientes desplazadas de la Cuenca del río Cacarica (operación Génesis) vs. Colombia. Serie C No. 270

Inter-American Court of Human Rights (IACtHR) (2015a) Comunidad Garífuna Triunfo de la Cruz y sus miembros vs. Honduras. Serie C No. 305

Inter-American Court of Human Rights (IACtHR) (2015b) Comunidad Garífuna de Punta Piedra y sus miembros vs. Honduras. Serie C No. 304

International Organization for Migration (IOM) (2007) Discussion note: migration and the environment, MC/INF/288, 94th session of the council. Available at https://governingbodies.iom.int/94th-session-council-2007 on 13 Dec 2018. Geneva

Kälin W (2010) Conceptualising climate-induced displacement. In: McAdam J (ed) Climate change and displacement: multidisciplinary perspectives. Bloomsbury Publishing, London

Miller T (2018) Why the migrant caravan story is a climate change story. Drought, crop failure, storms, and land disputes pit the rich against the poor, and Central America is ground zero for climate change. Yes!Magazine, Journalism for People Building a Better World, 27 Nov 2018. Available at https://www.yesmagazine.org/peace-justice/why-the-migrant-caravan-story-is-a-climate-change-story-20181127 on 15 Jan 2019

Methmann C, Oels A (2015) From 'fearing' to 'empowering' climate refugees: governing climate-induced migration in the name of resilience. Secur Dialogue 46(1):51–68

Nishimura L (2015) 'Climate change migrants': impediments to a protection framework and the need to incorporate migration into climate change adaptation strategies. Int J Refugee Law 27(1):107–134

Rigaud KK, de Sherbinin A, Jones B, Bergmann J, Clement V, Ober K, Schewe J, Adamo S, McCusker B, Heuser S, Midgley A (2018) Groundswell: preparing for internal climate migration. The World Bank, Washington, DC

Sarukhán J et al (2017) Capital natural de México. Síntesis: evaluación del conocimiento y tendencias de cambio, perspectivas de sustentabilidad, capacidades humanas e institucionales. Comisión Nacional para el Conocimiento y Uso de la Biodiversidad. Mexico

Secretaría de Medio Ambiente y Recursos Naturales (SEMARNAT), Instituto Nacional de Ecología y Cambio Climático INECC) (2012) México, Quinta Comunicación Nacional ante la Convención Marco de las Naciones Unidas sobre el Cambio Climático. Mexico, Mexico

Secretaría de Medio Ambiente y Recursos Naturales (SEMARNAT), Instituto Nacional de Ecología y Cambio Climático INECC) (2018) México, Sexta Comunicación Nacional y Segundo Informe Bienal de Actualización ante la Convención Marco de las Naciones Unidas sobre el Cambio Climático. Mexico, Mexico

Simatele D, Simatele M (2015) Migration as an adaptive strategy to climate variability: a study of the Tonga-speaking people of Southern Zambia. Disasters 39(4):762–781

Surrallés A, Hierro PG (eds) (2004) Tierra adentro: Territorio indígena y percepción del entorno (No. 39). IWGIA, Copenhague

Tacoli C (2009) Crisis or adaptation? Migration and climate change in a context of high mobility. Environ Urbanization 21(2):513–525

Torres-Wong M (2018) Natural resources, extraction and indigenous rights in Latin America: exploring the boundaries of state corporate crime in Bolivia, Peru and Mexico. Routledge, Abingdon, New York

Ugalde V (2010) La aplicación del derecho ambiental. In: Graizbord B, Lezama JL (eds) Los grandes problemas de México. El Colegio de Mexico, Mexico

Valle LM (2012) Apuntes para pensar el territorio desde una dimensión social. Ciências Sociais Unisinos 48(1):12–18

Zaremberg G, Wong MT, Guarneros-meza V (2018) Descifrando el desorden: instituciones participativas y conflictos en torno a megaproyectos en México. América Latina Hoy 79:103–124

Carbon Inventories Implementation as Competitive Strategy in Mexican Industry

Yolanda Mendoza Cavazos, Yesenia Sánchez Tovar and Mariana Zerón Felix

Abstract Nowadays, Mexico occupies the 14th position in the Global Carbon Atlas for GHG emissions. Geographic features and unfavorable social conditions bring the Country in a very vulnerable place for the effects of global climate change. For this reason, Mexico committed to reducing by the 30% its Green House Gases (GHG) emissions by 2013, and 50% by 2050. Sustainability management is a strategy triggering enormous potential to businesses and public bodies, being an overarching driver for innovation and competitiveness. Proactive firms look at environmental protection duties as an opportunity and an additional business input able to breed economic benefits in its region. The purpose of this paper is to present the need for carbon inventories as the base for a sustainable strategy in the competitiveness of Mexican companies. A particular emphasis is given to the assessment of each organization as a potential contributor to climate change, as well as their level of awareness about existing mitigation and adaptations strategies or national commitments. This study reveals the crucial role of federal government and regional context in their cultural and social humus, necessary for the implementation of climate change mitigation strategies in pursuit of a local and global sustainable development.

Introduction

All the diverse ecological crises that the planet is going through, such as climate alterations, scarcity of natural resources, as well as pollution and deterioration of natural capital, urge the establishment of a new model for economic development,

Y. Mendoza Cavazos (✉) · Y. Sánchez Tovar · M. Zerón Felix
Facultad de Comercio y Administración Victoria, Universidad Autónoma de Tamaulipas, Tamaulipas, Mexico
e-mail: ymendoza@docentes.uat.edu.mx

Y. Sánchez Tovar
e-mail: yesanchez@docentes.uat.edu.mx

M. Zerón Felix
e-mail: zfmariana@docentes.uat.edu.mx

W. Leal Filho et al. (eds.), *Climate Change, Hazards and Adaptation Options*,
Climate Change Management, https://doi.org/10.1007/978-3-030-37425-9_51

different from the one followed and taught so far (Mazzucato and Penna 2015). Climate change refers to the gradual change in Earth's climate and physical geography that supports an increase in the Earth's temperature; it is one of the most significant challenges facing life on earth (Chan 2018).

Climate change generates global risks and several concerns, one of the most critical issues addressing it is the excessive increase of greenhouse gases (GHG). This phenomenon demands long-term planning and immediate actions to adapt to potentially adverse impacts, as well as reducing the GHG emissions.

According to Silva (2018) only a few regions in the world are prepared to deal with climate change because of their adaptation facilities; however, Mexico as a developing country has still certain limitations in respect to financial resources and approach to technologies. The country is facing warmer climatic periods over more extended periods as well as greater droughts, heavier rainfalls, more intense storms, and regular floods even though the country contributes only with the 1.4% of the global CO_2 emissions.

Dafermos et al. (2018) states that there are two main financial risks related to climate change, those including the re-valuation of carbon assets as a result of shocks compared to the transition to a low carbon economy and the physical risks that are linked to the economic damages caused by climate-related events. Dealing with climate change should be an immediate matter for Mexico if it intends it to be prevented from hurting the economy (Silva 2018).

Global warming, in this respect, does not change the nature of natural disasters, but the escalation of intensity and the likelihood of such events have significant implications for organizations. Climate change generates global risks and uncertainties due to the extreme weather events that have occurred in recent years, having negative impacts not only for companies but also for industrial operations. Despite the interdependence between companies and the environment and the growing interest in corporate sustainability, the management literature has only recently dedicated attention to the pressing issues of climate change. In the same matter, companies have also committed limited attention to climate change This is because, among other factors, management decisions are rarely based on their adaptability to the natural environment (Orsato et al. 2017).

Companies have a profound influence on society, generating jobs, investing capital, acquiring goods and doing daily business. However, they also have a great responsibility to contribute to a prosperous economy that would represent their social responsibility. Business activity contributes significantly to economic growth, but it is also one of the major causes of environmental deterioration due to excessive consumption of resources and emissions of greenhouse gases for production and extraction of raw materials (Labandeira et al. 2007). This led some companies to introduce strategies contributing to the sustainable development of the region, and not only to individual economic goals (Carrillo and Gonzalez 2006).

Although some companies consider sustainability as an added value to their product or service, most of them still intend it as an immediate expense, and not as an investment that can generate more significant benefits to their surroundings.

Eco-innovations and corporate social responsibility (CSR) define a significant step toward an industrial sustainability agenda, presenting competitive advantages and innovative approaches (Bocken et al. 2014). During the last two decades, carbon footprint has become a highly valued asset for society to display. Kyoto Protocol has established carbon inventories as one of the most crucial CO_2 reduction mechanisms capturing worldwide attention, increasing the interest in carbon inventories allowing the integration of environmental aspects in financial issues.

The primary objective of those companies which develop environmental efforts is to reduce greenhouse gas emissions. Although they exist in nature as the result of earth´s natural processes, emissions mainly generated by fossil fuels combustion and industrial processes have increased the detention of heat in the atmosphere, providing 78% of GHG emissions globally. Carbon footprint is a popular methodology in companies' reporting initiatives (Chang et al. 2015). It refers to the total accounting of direct and indirect greenhouse gases linked to specific activities of any company, attributed to several phases of production. In their corporate responsibility report of 2010/2011, Ernst and Young declare integrated carbon management as a primary corporate strategy. This includes the evaluation of all the resources, actors and processes within the corporate ecosystem, as well as its supply chain, from materials extraction, transport, production and operations, distribution, use, reuse, and recycling to its final disposal.

This paper aims to present carbon inventories cases as an instrument for the Mexican industry to become a competitive market. Firstly, it analyses the climate change problem in the Mexican context; then, it describes carbon footprint initiatives in the Mexican Industry Context, followed by an evaluation of sustainability implementations. Finally, it presents the case of the construction industry and its contribution to the regional and national strategy toward sustainable development.

The Climate Change in the Mexican Context

Mexico is the southernmost portion of North America bordered to the north to the United States of America; to the southeast with Guatemala, Belize, and the Caribbean Sea; to the east by the Gulf of Mexico and to the south and west by the Pacific Ocean; it is the 3rd American country with greater coasts length. With a surface of 1,964,375 km^2 it is the 14th largest country in the world and the 3rd in Latin America.

It is the 11th populated country in the world with an estimated 124 million people by 2017, just after Brazil in Latin America. According to its GDP, Mexico is the 15th world economy, with around $1.149 tn USD.

It is an ecologically megadiverse country, home of 10–12% of world biodiversity, ranking 4th in the world. Although Spanish is the primary language in the nation, there are 67 mother indigenous languages. Due to its rich culture and history, Mexico ranks first in America and 7th in the world UNESCO World Heritage Sites. In 2014 it became the first Latin American member of the OECD.

Mexico is a member of the United Nations, the World Trade Organization, the G8+5, the G20, the Uniting for Consensus group of the UN and the Pacific Alliance Trade bloc and signer of the Kyoto protocol and the Paris Agreement.

Due to its geographical characteristics and its socio-environmental, economic and health conditions, the problem of climate change may be felt more intensively in Mexico. According to the 2017 Risk Index by Germanwatch, Mexico occupied the 7th place in economic losses from natural disasters from 1997 to 2016, with USD 2957 million. (Eckstein et al. 2017). Mexico is especially vulnerable to the effects of climate change. In this perspective, the INFORM Global Risk Index 2018 ranks Mexico in seventh place for Hazard and Exposure Risk. In this aspect, Mexico became one of the few developing countries to create a National General Climate Change Law (NGCCL) in 2012. Mexico's NGLCC is the main climate policy instrument in the country; it provides the mandate to craft comprehensive long-term policies including the National Climate Change Strategy (NCCS).

The NCCS sets a roadmap to incorporate climate change in Mexico's development plans, impacting critical strategic decisions to establish priorities for national and regional development. It pursues a socially equitable nation, integrating a green economy, including climate change resilient ecosystems and communities with sustainable cities.

To mitigate the vulnerability risks and increase the populations' resilience, the NCCS establishes critical cross-cutting issues that set the foundation of climate policy in the country both in adaptation and mitigation (see Fig. 1).

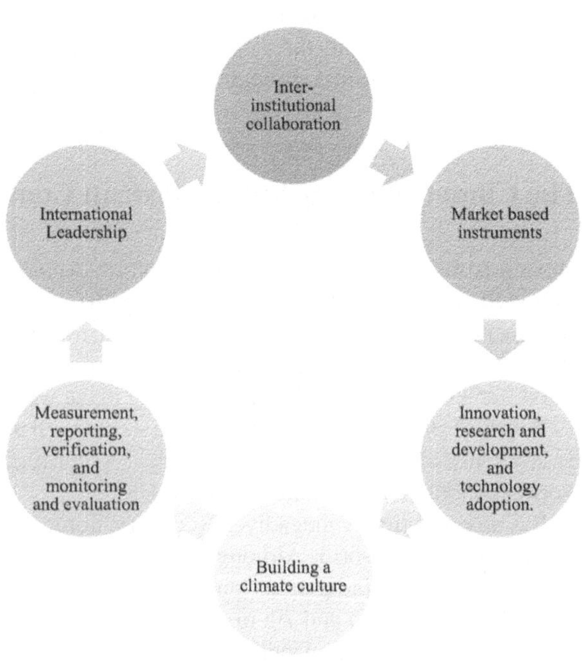

Fig. 1 Cross-cutting issues from Mexican National climate change strategy (From Mexico's climate change mid-century strategy 2016)

The climate adaptation strategy, based on Mexico's vulnerability assessment, identifies three main areas of action:

- to reduce vulnerability factors and building social resilience,
- to promote ecosystems-based adaptation
- to protect strategic infrastructure and production systems.

The NCCS translates the latter list into five actions, starting from those having the most significant potential for reducing emissions at the lowest cost and with more environmental, economic and social benefits in order:

- to accelerate the transition to clean energy.
- to reduce energy intensity through efficiency and responsible consumption schemes
- to transit to sustainable city models with mobility systems, integral waste management, and low carbon footprint buildings
- to promote better agricultural and forestry practices preserving natural carbon sinks.

Carbon Footprint in Mexican Industry

Nowadays, Mexico occupies the 14th position in the Global Carbon Atlas for GHG emissions. Despite being a small emitter, Mexico is exposed to the severe alterations caused by these gases; as a Paris Agreement signer, Mexico has committed through the NGCCL to reduce by the 30% its GHG emissions by 2030 and 50% by 2050. Indeed, Mexican companies are increasingly interested in Carbon Footprint issues and Carbon Disclosure Projects (CDP).

According to the Mexican National Inventory of short-life climate pollutants and greenhouse gases, 63% of the total GHG emissions that comes from industry in Mexico correspond to CO_2, 26% to CH_4 and 8% to N_2O.

To be able to understand the complexity of reducing GHG emissions, it is relevant to observe the industrial behavior in the country. Table 1 shows the leading industries in Mexico and their GHG emissions.

The previous table (Table 1) shows only the leading industries based on burning fuel activities; these emissions represent 82% of the total GHG emissions in the country. Energy Industries, Manufacture and Building industries and Transport Industries represent 75% of the total GHG from fuel consuming industries in Mexico; as seen in the following chart (Fig. 2).

Evidently, from Fig. 2, the transport sector is the industry with higher impact with the 32% of the emissions from direct fuels combustion, followed by electrical generation with the 31% and manufacture and building industries with the 12%.

To compare the impact of these industries in the Mexican economy, Table 2 shows the contribution of each sector to the national GDP.

Table 1 Leading industries in Mexico with GHG emissions

Leading industrial activities in Mexico	Total CO_2e emissions Gg CO_2e
1A Burning fuel activities	436,459.96
1A1 Energy industries	164,636.82
1A1A Main activities for heat and electricity production	125,124.43
1A1B Oil refinery	11,815.40
1A1C Solid fuels manufacture and other energy industries	27,696.99
1A2 Manufacture and building industries	63,490.20
1A2A Iron and steel	4335.50
1A2C Chemical substances	8886.47
1A2D Pulp, paper, and printing	2396.00
1A2E Food, beverages and tobacco processing	1636.06
1A2G Transport equipment	445.45
1A2K Building	874.42
1A3 Transport	171,354.84
1A3A Civil aviation	6284.89
1A3B Motor transport	159,944.08

From Mexico's National Greenhouse Gases Inventory (INECC 2016)

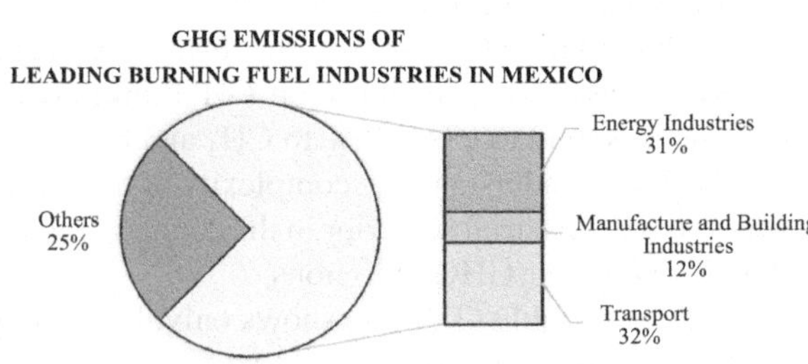

Fig. 2 GHG emissions distribution from burning fuels activities in Mexican Industry. *Source* INECC (2016)

Analyzing Table 2, it is possible to conclude that the emitters of the 81% of the total GHG emissions in the country provide 21% of the GDP showing the critical role of these activities in Mexico´s CO2 global contribution. It is also relevant to observe that the Manufacture and Building industries emissions represent the higher GDP contribution; above energy generation and transport. Evidencing the need for a paradigm shift in the traditional productive activities, reducing the use of fossil fuels and using energy more efficiently (Ruiz 2011).

Table 2 GHG emissions from leading burning fuels industries in Mexico with their corresponding GDP proportion

Leading industrial activities in Mexico	CO_2e emissions proportion	GDP proportion
Total	100	100
1A Burning fuel activities	81.64	21.9
1A1 Energy industries	30.796	4.6
1A1A Main activities for heat and electricity production	23.405	1.3
1A1B Oil refinery	2.21	2.9
1A1C Solid fuels manufacture and other energy industries	5.181	
1A2 Manufacture and building industries	11.876	11.1
1A2A Iron and steel	0.811	0.7
1A2C Chemical substances	1.662	1.5
1A2D Pulp, paper, and printing	0.448	0.3
1A2E Food, beverages and tobacco processing	0.306	4.8
1A2G Transport equipment	0.083	3.4
1A2K Building	0.164	3.9
1A3 Transport	32.052	6.2
1A3A Civil aviation	1.176	0.2
1A3B Motor transport	29.918	5

Source Mexico's National Greenhouse Gases Inventory (INECC 2016)

The 34th article of the NGCCL establishes regulations to reduce GHG emissions in the following areas:

- generation and use of energy
- transport sector
- agriculture sector and other land use and preservation of ecosystems and biodiversity
- waste sector
- industrial process sector
- Editing and changes in behavior patterns, consumption, and production.

Accordingly, the Climate Change Special Program in Mexico (CCSP) aims to detonate cost-effective actions, with environmental benefits and significant impacts of GHG mitigation in growing sectors. As it might be expected, CCSP's strategies are focused on the implementation of energy efficiency actions, cogeneration, clean energy use, and sustainable mobility schemes. All of them related to CO_2 inventories to promote and facilitate the collaboration of the private sector in the development of appropriate national mitigation actions.

Sustainability as a Competitive Strategy

The term sustainable development at a corporate level is increasingly taken into account, to the extent that today is considered as a precondition for doing business. As a result, in the '90s the World Business Council for Sustainable Development created a group of leading companies in the subject to facilitate the dialogue between the public and private sector looking for sustainable solutions in the cities. Currently, about 200 international companies create sustainability departments and deliver sustainability reports (Dyllick and Hockerts 2002). Financial institutions and investors are becoming increasingly interested in social policies and environmental corporations. As a result, many companies have integrated environmental responsibility into their planning and operations as remediation and preservation efforts (Gibson 2012; Snider et al. 2003), bonding internal and external activities to respond to environmental sustainability.

Environmental protection activities are interrelated with the rest of the strategic aspects of any company and can alter their contribution to the economics result since they have the potential to influence both incomes and cost overruns. Consequently, the adoption of an environmental strategy is not innocuous: while a company with a lagged ecological approach sees the consideration of the environment as a negative aspect that brings a loss in competitiveness due to its costs, other companies with proactive positions see environmental protection as an opportunity.

In this sense, sustainability could be seen as a smart strategy that represents additional business opportunities with the vast potential to streamline efficient processes; producing greater profitability within companies by reducing the impact on their environment (Instituto Global para la Sostenibilidad 2011).

Also, Porter states that environmental improvement can benefit industry products resources: "from the productivity point of view, environmental improvement and competitively belong together" (Porter and Van der Linde 1995). Environmental regulations imposed by a nation incite companies to seek new production forms that lead to reducing pollution. This approach has been recognized in the literature as the "Porter's Hypothesis," and indicates a win-win situation: by seeking better environmental results, it delves in innovative processes, products, and systems. This approach allows it to access to competitive advantages in its sector and that in turn reflects lower production costs and higher business profits.

Since Porter (1991), several authors have tried to determinate how the sustainable strategy can generate a competitive advantage for the company. Harts' work (1995) is a crucial starting point for many others (Russo and Fouts 1997; Aragón-Correa 1998; Sharma and Vredenburg 1998; Klassen and Whybark 1999; Christmann 2000) by relating the Resource-Based Vision with environmental strategies. Hart (1995) proposes the development of three environmental capabilities (pollution prevention, life cycle assessment, and sustainable development) as a potential source of competitive advantages for the company. In this context, other forms of competitively measures appear in Table 3.

Table 3 Competitiveness variables in the most relevant researches on CSR

Author	Competitiveness variable
Akben-Selcuk (2016)	Return on Assets (ROA), Gross Profit Margin, Tobin Q, Sales growth, Financial leverage level, export company
Kovács (2012)	The difference between the final value and the initial value in cash, active capital, assets account growth, market share
Yang et al. (2013)	Corporate Image value, Quality of service improvements, customer satisfaction, productivity levels, financial benefits
Man et al. (2002)	Product image, sales increases, market share, corporate image improvements, higher short-term profits, higher long-term profits
Rao (2002)	Product Quality, productivity improvements, cost reductions
Bodnaruk et al. (2016)	Returns of assets (ROA), Return on equity (ROE), market share, sales growth
Yang et al. (2015)	Quality of the product or service, ability to respond to the changing market, market share

In addition to Hart (1995), Kolk et al. (2010), integrates climate change as a variable to evaluate Corporate Social Responsibility Disclosure, including, climate change issues, direct GHG emissions, voluntary agreements and targets, emissions trading schemes and, carbon-reducing projects.

The Tamaulipas Case Study

The building sector; considered worldwide as the origin of 65% of the pollutants is close related with several industries such as energy, iron and steel, and transport among others (Centro Mario Molina 2012). The growth of this sector is notorious: in 2004, 4.7 million jobs in the country were generated, 5.9 million in 2014 and 6.1 million in 2015, all of which were directly related it to the building sector, considered as one of the main engines of the country's economy (INEGI 2017).

With the purpose to understand the implementation of sustainable strategies in the industry including carbon inventories, several building companies from Tamaulipas Mexico were studied based on their internal and external practices impacting the three pillars of sustainability. Tamaulipas is the 6th largest state in Mexico, bordered to the north with the United States of America and to the east to the Gulf of Mexico.

The database was obtained from the building companies associated with the Mexican Chamber of Building Industry in Victoria City (capital of the state of Tamaulipas). To calculate the probabilistic sample for the application of the instrument, the total number of MCBI associates from 2016 was considered. As a result, 88 construction companies were studied.

The instrument was elaborated from literature research using sustainability models and sustainability certifications (Table 4). The physical and electronic distribution of the questionnaire was applied during February, March, and April of 2017.

Table 4 Literature references considered for the elaboration of the sustainability measurement instrument

Instrument	Author	Description
Sustainable small business designation program checklist	Pittsburgh Sustainability Department	Voluntary program from Pennsylvania USA. Evaluation of Sustainability actions implemented in business
Environmental sustainability check-list	WWFN World Wide Fund for Nature	Evaluates sustainable development for projects. It includes the impact on the social and environmental aspects
Business sustainability check-list	Michigan Commerce Chamber	An instrument for measuring sustainable practices implemented in local business
A checklist for assessing sustainability performance of construction projects	Li-Yin Shen, Jian Li Hao, Vivian Wing-Yan Tam y Hong Yao	Evaluates all building processes according to social, economic and environmental aspects
Sustainable enterprises penta-dimensional model	Luis Portales, Consuelo García de la Torre, Gloria Camacho Ruelas, Osmar Arandia Pérez	Presents a model for business sustainability practices

Within the theoretical framework of this study, the development of sustainable aspects in building companies is seen as mainly focused on the economy and natural environment issues, leaving aside social factors. Also, the economic factor has a direct connection with reducing the carbon footprint including variables such as the generation of policies for green procurement, renewable energy strategies implementation and, the use of local products.

For instance, only 13% of the companies which answered always consider their projects as contributors to reducing GHG emissions as can be seen in Fig. 3.

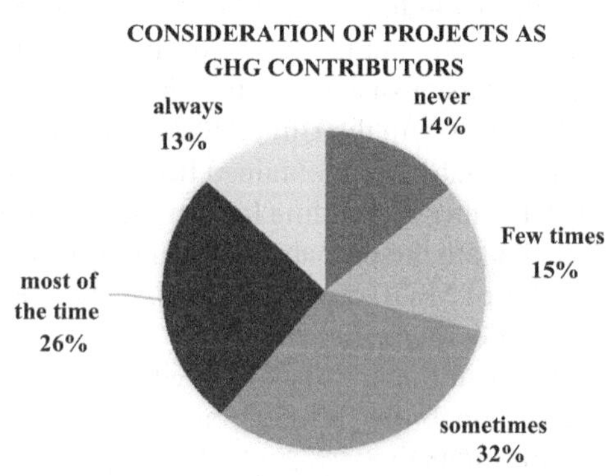

Fig. 3 Contribution of projects to GHG reduction

Although the study presents the implementation of diverse sustainability strategies in the sector, carbon inventories and their relation to global impact are still weak. Only 6% of the companies have a sustainable policy implemented. Leaving carbon footprint concepts, carbon inventories, and national GHG commitments behind, as well as its impact on their competitivity. None of them has an active disclosure program for their emissions neither an inventory for internal control.

Yet, there are at least four important limitations for this study. Firstly, most of the literature considered from Mexican strategies and policies was sourced from official documents that have not been subject of academic peer review to guarantee that they are based on reliable scientific knowledge. Also, existing research in the field in Mexican context is very shallow. Lastly, due to social concerns and security problems in the region, some of the data is considered risky to publish; all these and the lack of interest of the building companies in the subject made the data collection difficult.

In contrast, the interest in the implementation of sustainability strategies in Mexican Global companies in the market is observed with the generation of the Sustainable IPC in 2011 at the Mexican Stock Exchange.

Similarly to the Dow Jones Sustainability in the United States, FTSE4Good in England and Bovespa-IFC Sustainability Ltd in Brazil, the S & P/"BMV IPC Sustentable" index seeks to measure the performance of leading companies in Mexico concerning economic, environmental and social criteria. The objective of this index is to position Mexico as a country which market is committed to CSR and Sustainability; urging all companies to adopt policies and measurements systems in these areas.

With this also called "green index" the 30 top organizations with sustainable efforts in the market are identified (Table 5); 40% of them related to the industrial sector and materials. CEMEX is one of the most important companies in the building sector, producing building materials which contribute significantly to GHG emissions. According to Andrew (2018) the third largest source of anthropogenic emissions of carbon dioxide, after fossil fuels and land-use change.

As an incentive to adopt Carbon footprint strategies, during February 2018, the Mexican Stock Exchange launched their official first carbon bond simulation (MEXICO2 "Mexican Carbon Platform") due to the international growth of this market. Till this date, it has only been voluntary. Therefore, there are no evidence that this strategy has an impact on their competitivity.

Conclusions

Nowadays, Mexico occupies the 14th position in the Global Carbon Atlas for GHG emissions. Geographic features and unfavorable social conditions bring the Country in a very vulnerable position for the effects of global climate change. For this reason, Mexico committed to reduce by the 30% its Green House Gases (GHG) emissions by 2013, and 50% by 2050. Mexico is implementing several strategies that evidently

Table 5 2017 companies at "BMV IPC Sustentable"

Mexican sustainable index Mexican stock exchange		
No.	Emission	%
1	AC	1.80
2	ALFA	3.63
3	ALSEA	1.07
4	AMX	12.67
5	ASUR	2.46
6	AXTEL	0.05
7	BIMBO	2.10
8	BOLSA	0.45
9	CEMEX	10.30
10	ELEKTRA	0.70
11	FEMSA	13.47
12	GENTERA	1.11
13	GFNORTE	9.64
14	GMEXICO	8.66
15	HCITY	0.24
16	HERDEZ	0.30
17	IENOVA	1.07
18	KIMBER	2.01
19	KOF	2.53
20	LAB	0.51
21	LALA	0.71
22	MEXCHEM	2.04
23	OMA	1.02
24	PE&OLES	1.92
25	RASSINI	0.15
26	SANMEX	2.07
27	TLEVISA	8.78
28	VESTA	0.49
29	VOLAR	0.96
30	WALMEX	7.07

contribute to the incorporation of sustainable strategies in their development plans. However, these are not getting the right impact.

Sustainability management is a strategy triggering enormous potential to businesses and public bodies, being an overarching driver for innovation and competitiveness. Proactive firms look at environmental protection duties as an opportunity and an additional business input able to breed economic benefits in its region.

The purpose of this paper is to present the need for carbon inventories as the base for a sustainable strategy in the competitiveness of Mexican companies. A particular emphasis is given to the assessment of each organization as potential contributor to climate change, as well as their level of awareness about existing mitigation and adaptations strategies or national commitments. Nowadays, if a company's vision is to maintain and transcend in a competitive global market, the need to operate under a competitive concept is evident. They should integrate changing innovative process permanently, consider social and environmental aspects in their business activities; and therefore, maintain and lead in a global market.

As a result of this research it is evident that local level industries are not familiarized with sustainability strategies neither GHG emissions and their contribution to global impact. Therefore, this study calls for the crucial role of national government and regional context in their cultural and social humus, necessary for the implementation of climate change mitigation strategies in pursuit of a local and global sustainable development. It is evident that the country needs profound support for a change from the sector level, to the company level to adopt new perspectives on sustainability.

In conclusion, in order to confront climate change successfully, transform all production and consumption patterns is imperative. To achieve this, Mexico requires an informed, conscious, committed and participative society. Therefore, educational programs and effective dissemination programs are crucial. The implementation of strategic communication campaigns and transversal collaboration between universities, government, and private sector should provide favorable results shortly.

Further research should be done to find the appropriate tools and channels for transferring the knowledge about carbon inventories as a sustainable strategy and their impact on their competitivity. Not only for global market industries but also for smaller and local companies which also contributes significantly to regional development.

Acknowledgements This work is the result of the NPTC (511-6/18-9249) project research process for New full-time faculty member from the School of Commerce and Management, at Universidad Autónoma de Tamaulipas. Including the collaboration of the academic body "UAT-CA-103 *Regional Development Strategies*" and the support of the faculty support department PRODEP.

References

Akben-Selcuk E (2016) Does firm age affect profitability? Evidence from Turkey. Int J Econ Sci 5(3):1–9. https://doi.org/10.20472/ES.2016.5.3.001

Aragon-Correa J (1998) Strategic proactivity and firm approach to the natural environment. Acad Manag J 41:556–567. https://doi.org/10.2307/256942

Bocken Nancy, Short Samuel, Rana P, Evans Steve (2014) A literature and practice review to develop sustainable business model archetypes. J Clean Prod 65:42–56. https://doi.org/10.1016/j.jclepro.2013.11.039

Bodnaruk A, O'Brien W, Simonov A (2016) Captive finance and firm's competitiveness. J Corp Finan 37(C):210–228. https://EconPapers.repec.org/RePEc:eee:corfin:v:37:y:2016:i:c:p:210-228

Carrillo Gonzalez G, Gonzalez Parodi G (2006) Gestión de proyectos empresariales: el ecodiseño, un intrumento para la sostenibilidad. Administración y tecnologías para el diseño. Anuario 2006, administración para el diseño. UAM Azcapotzalco, CyAD, Departamento de Procesos y Técnicas de Realización, pp 15–31. Mexico http://hdl.handle.net/11191/275

Centro Mario Molina para Estudios Sobre Energia y Medio Ambiente A.C. (2012) Evaluación de la sustentabilidad de la vivienda en Mexico. Mexico D.F. http://centromariomolina.org/wp-content/uploads/2012/09/14.-Evaluaci%C3%B3nSustetabilidadViviendaM%C3%A9xico_fin.pdf. Accessed on 15 Nov 2018

Chan EY (2018) Climate change is the world's greatest threat—in celsius or fahrenheit? J Environ Psychol 60:21–26. https://doi.org/10.1016/j.jenvp.2018.09.002

Chang, D-S, Yeh L-T, & Liu W (2015) Incorporating the carbon footprint to measure industry context and energy consumption effect on environmental performance of business operations. Clean Technol Environ Policy 17(2):359–371. https://doi.org/10.1007/s10098-014-0785-9

Christmann P (2000) Effects of "best practices" of environmental management on cost advantage: the role of complementary assets. Acad Manag J 43(4):663–6670. https://doi.org/10.2307/1556360

Dafermos Y, Nikolaidi M, Galanis G (2018) Climate change, financial stability and monetary policy. Ecol Econ 152:219–234. https://doi.org/10.1016/j.ecolecon.2018.05.011

Dyllick T, Hockerts K (2002) Beyond the business case for corporate sustainability. Bus Strategy Environ 11(2):130–141. https://doi.org/10.1002/bse.323

Eckstein D, Künzel V, Schäfer L (2017) Global climate risk index 2018. Germanwatch, Berlin. https://germanwatch.org/sites/germanwatch.org/files/publication/20432.pdf. Accessed on 27 Nov 2018

Gibson RB (2012) Why sustainability assessment? In: Bond A, Morrison-Saunders A, Howitt R (eds) Sustainability assessment: pluralism, practice and progress, chap 1, 1st edn. Taylor & Francis, London

Hart S (1995) A natural-resource-based view of the firm. Acad Manag Rev 20(4):986–1014. https://doi.org/10.2307/258963

INEGI (2017) Instituto Nacional de Estadística y Geografía (México). Anuario estadístico y geográfico de los Estados Unidos Mexicanos 2017 National Institute of Geography and Stadistics. México, D.F. http://internet.contenidos.inegi.org.mx/contenidos/Productos/prod_serv/contenidos/espanol/bvinegi/productos/nueva_estruc/AEGEUM_2017/702825097912.pdf. Accessed on 15 Nov 2018

INFORM (2018) Index for risk management 2018. http://www.inform-index.org/Portals/0/InfoRM/2018/INFORM%20Annual%20Report%202018%20Web%20Spreads.pdf?ver=2017-11-29-171105-863. Accessed on 15 Nov 2018

Instituto Global para la Sostenibilidad (2011) Acuerdo por un México sostenible, cumbre negocios verdes. México D. F. https://static1.squarespace.com/static/56228080e4b04bb42a1523da/t/567876a5c647ad2a4bc58c9b/1450735269478/Reporte%2BCGNV.pdf. Accessed on 15 Nov 2018

Klassen R, Whybark D (1999) The impact of environmental technologies on manufacturing performance. Acad Manag J 40(6):599–615. https://doi.org/10.2307/256982

Kolk A, Pinkse J (2010) The integration of corporate governance in corporate social responsibility disclosures. Corporate Soc Responsib Environ Manag 17(1):15–26. Available at https://ssrn.com/abstract=1350939UK

Kovács Z (2012) The key to competitiveness: forecast. Acta Oeconomica 62(4):505–518. https://doi.org/10.1556/AOecon.62.2012.4.5

Labandeira X, León CJ, Vázquez MX (2007). Economía Ambiental. Pearson, Madrid, España

Man TWY, Lau T, Chan KF (2002) The competitiveness of small and medium enterprises: a conceptualization with focus on entrepreneurial competencies. J Bus Ventur 17(2):123–142. https://EconPapers.repec.org/RePEc:eee:jbvent:v:17:y:2002:i:2:p:123-142

Mazzucato M, Penna C (2015) Mission-oriented finance for innovation: new ideas for investment-led growth. Policy Network and Rowman & Littlefield International, UK

Orsato RJ, Barakat SRF, de Campos, JG (2017) Organizational adaptation to climate change: learning to anticipate energy disruptions. Int J Clim Change Strat Manage 9(5):645–665. https://doi.org/10.1108/IJCCSM-09-2016-0146

Porter M, Van der Linde C (1995) Toward a new conception of the environment-competitiveness relationship. J Econ Perspect 9(4):97–118. Retrieved from http://www.jstor.org/stable/2138392

Rao Purba (2002) Greening the supply chain: a new initiative in South East Asia. Int J Oper Prod Manag 22(6):632–655. https://doi.org/10.1108/01443570210427668

Ruiz NP (2011) Estimación de los costos relativos de las emisiones de gases de efecto invernadero en las ramas de la economía mexicana. El trimestre económico 78(309):173–191

Russo M, Fouts PA (1997) A resource-based perspective on corporate environmental performance and profitability. Acad Manag J 40(3):534–559. https://doi.org/10.5465/257052

SEMARNAT-INECC (2016) Mexico's climate change mid- century strategy. Ministry of Environment and Natural Resources (SEMARNAT) and National Institute of Ecology and Climate Change (INECC), Mexico https://unfccc.int/files/focus/long-term_strategies/application/pdf/mexico_mcs_final_cop22nov16_red.pdf. Accessed on 15 Nov 2018

Sharma S, Vredenburg H (1998) Proactive corporate environmental strategy and the development of competitively valuable organizational capabilities. Strateg Manag J 19(8):729–753. https://onlinelibrary.wiley.com/doi/abs/10.1002/(SICI)1097-0266(199808)19:8%3C729::AID-SMJ967%3E3.0.CO;2-4

Silva Rodríguez de San Miguel JA (2018) Climate change initiatives in Mexico: a review. Manag Environ Qual Int J 29(6):1042–1058. https://doi.org/10.1108/MEQ-03-2018-0066

Snider J, Hill RP, Martin D (2003) Corporate social responsibility in the 21st century: a view from the world's most successful firms. J Bus Ethics 48(2):175–187. https://doi.org/10.1023/b:busi.0000004606.29523.db

Yang C-S, Lu C-S, Haider JJ, Marlow PB (2013) The effect of green supply chain management on green performance and firm competitiveness in the context of container shipping in Taiwan. Transp Res Part E: Logistics Transp Rev 55(C):55–73. https://EconPapers.repec.org/RePEc:eee:transe:v:55:y:2013:i:c:p:55–73. Accessed on 15 Nov 2018

Yang J, Zhang F, Jiang X, Sun W (2015) Strategic flexibility, green management, and firm competitiveness in an emerging economy. Technol Forecast Soc Chang 101:347–356. https://doi.org/10.1016/j.techfore.2015.09.016

How to Govern Climate Change Without Being Able to Govern: Adaptation Governance in Colombia

Kerstin Mohr

Abstract The coordination of climate change adaptation is not merely a technical issue but, above all, requires effective governance. Solid governance structures provide the framework for adequate knowledge, funding, and planning security for practical measures. This is particularly challenging in Colombia, where a difficult topography and weak infrastructures result in high demands for technical solutions. Moreover, the country has experienced over five decades of rural violence leading to political, social and economic disparities as well as to an outspoken mistrust between civil society and political leaders. The hence stretched relations between authorities and citizens have negative impacts on the country's ability to create solid governance structures and adapt successfully—independent of how advanced technical solutions may be. This twofold analysis first elaborates the current state of the art of adaptation governance, outlining the specific challenges and depicting basic elements which have to be addressed when designing an adaptation governance framework. As one of the few papers currently addressing this issue, this analysis pays special attention to the impact of power politics. Second, against the backdrop of this theoretical outline, the wicked case of Colombia is analyzed. Through a qualitative case study design using textual analysis, this chapter focuses on hindering and enabling factors under the conditions of limited state resources. In order to create a holistic understanding, it draws on an interdisciplinary perspective. The analysis concludes with a discussion of the findings. Inferences show that climate change adaptation is highly complex and that technical and societal aspects must both be considered in order to design sustainable solutions. Besides the well-known challenges, a strong research-policy linkage and power dynamics are decisive challenges for the creation of an adaptation governance framework. The case study confirms the importance of these factors.

K. Mohr (✉)
Faculty of Economics and Social Sciences, University of Potsdam, August-Bebel-Str. 89, 14482 Potsdam, Germany
e-mail: kmohr@uni-potsdam.de

W. Leal Filho et al. (eds.), *Climate Change, Hazards and Adaptation Options*, Climate Change Management, https://doi.org/10.1007/978-3-030-37425-9_52

Introduction

Scientific findings on anthropogenic climate change are increasing constantly, providing more and more evidence and gradually lessening some of the many uncertainties related to it (Caesar et al. 2018; Steffen et al. 2018). The impacts of climate change are expected to be severe on human security and to result in high costs for society (Stern 2007). In order to reduce the risks and impacts of climate change, the Paris Agreement stipulated in 2015 the long-term goal of holding "the increase in the global average temperature to well below 2 °C above pre-industrial levels" and to pursue "efforts to limit the temperature increase to 1.5 °C above pre-industrial levels" (UN 2015, p. 2). However, despite the outcome of COP21 being a great success, the Paris Agreement also left much to be determined in the following COP negotiations (cf. Falkner 2016). Three years later, results are mixed: Countries such as the USA and Brazil have announced their intentions to withdraw from the agreement, COP24 in Katowice faces the mammoth task of stipulating the rule book, and the current nationally determined contributions (NDCs) allow for global temperature to rise to a minimum of 3 °C (Climate Action Tracker 2018; Escobar 2018). In line with these developments, the IPCC Special Report of 2018 finds that human activities have already caused a temperature rise of approximately 1.0 °C (±0.2 °C) and that "[i]mpacts on natural and human systems from global warming have already been observed" (IPCC 2018, p. 6f), thus underlining once more that comprehensive actions to tackle climate change have to be taken rapidly.

The primary focus of tackling climate change has been mitigation. However, it has become increasingly clear that "failure to mitigate climate change will require society to learn how best to respond to the challenges of living in a climatically altered world" (Hill 2013, p. 32). Accordingly, the focus is slowly shifting towards adaptation, which is most commonly defined as "adjustment in natural or human systems in response to actual or expected climatic stimuli or their effects, which moderates harm or exploits beneficial opportunities" (IPCC 2001, p. 982). For a long time, these adjustments were understood as technological solutions. However, it has become clear that the coordination of climate change adaptation is not merely a technical issue but, above all, requires effective governance. Solid governance structures provide the framework for adequate knowledge, funding, and planning security for practical measures. Creating a governance framework for adaptation is of particular importance in countries most vulnerable to climate change. If and how a country's population is affected depends on its socio-economic and socio-ecologic structure. Therefore, lacking or inadequate infrastructure and administrative capacities render inhabitants of developing countries particularly vulnerable (Abeling and Leopold 2012, p. 4). This paper examines the case of Colombia as an example of a country which is highly vulnerable, characterized by weak state structures and which therefore faces the challenge that "[a]daptation is most urgently needed where it is most difficult to implement" (Houghton 2012, p. 24).

The twofold analysis begins with an analytical examination of theories and concepts of adaptation governance, outlining its development and current state of the art.

By elaborating basic elements and specific challenges which ought to be addressed when designing adaptation governance, an analytical framework is created. Special attention is given to the impact of power politics. In the second part, the analytical framework is deployed to understand the case of Colombia by means of a qualitative case study design using content analysis. In order to gain a holistic understanding, relevant sources supplementing the body of scientific literature include, i.a., key government and ministerial documents (e.g. communications, agreements, press releases, speeches) and NGO publications. The scope of the literature comprises publications in Spanish, English and German. The analysis was conducted throughout 2016–2018 and focuses on the 2010–2018 Santos administration. Besides the depicted methods, it is informed by an underlying regional expertise derived from a ten year engagement with Colombia and a three week research stay in 2016 which included participant observation and numerous informal conservations within civil society. These aspects have particularly added to building an understanding of political dynamics which are "often hidden, half spoken and deeply sensitive for most people" (Nightingale 2017, p. 14). Moreover, the study is enriched by its interdisciplinary perspective. According to common agreement, it takes multidisciplinarity to find sustainable solutions for climate change. Yet, interdisciplinary research on the matter is still scarce and social sciences in particular have been slow to respond (Javeline 2014; Keohane 2015). Drawing on aspects of political science, sociology, history, economics and public law to explore impacts of institutions, societal and governmental structures and laws enacted to regulate climate change adaptation, this paper creates a holistic picture. In so doing, it contributes to the groundwork for an urgently needed comprehensive understanding of adaptation under the condition of limited state resources.

The introduction is followed by an analysis of the conceptual frameworks of adaptation governance. After an analysis of the Colombian case, findings are discussed. Results are summarized in the final conclusion.

From a Technical to a Political Understanding: Developing Adaptation Governance

The most common definition of adaptation in use today was published by the IPCC in 2001. Despite the common agreement on this general definition, the underlying understanding of adaptation has changed from a merely technical issue to a more complex notion and the question of how to govern adaptation processes.

In the third and fourth IPCC assessment reports, the understanding of adaptation was a mere technical one with an emphasis on biophysical vulnerability and technical solutions, such as infrastructural adjustments in general (IPCC 2001), and the introduction of adaptive capacity into infrastructure design as well as technical development (IPCC 2007, p. 20). Only by 2007 had awareness for parameters other than biological, technological and economic ones begun to increase. Institutional and political constraints were identified as possible limitations (IPCC 2007, p. 19), and an understanding began to emerge that limits were not only exogenous but also endogenous to society. Adaptation no longer meant providing only technological solutions, but also to tackling the "underlying causes of vulnerability" (IPCC 2014, p. 836). The understanding increased that technological solutions for "infrastructure, agriculture, public health, coastal protection, conservation, and other fields all depend on political variables, for their implementation and effectiveness" (Javeline 2014, p. 420). As climate change not only requires adaptation to new hazards but also to new forms of knowledge and changes in access to and power over resources, adaptation is highly political and contested (cf. Eriksen et al. 2015). Calls for governance were raised to manage the complexity arising from these factors (cf. Adger et al. 2009). Governance[1] comprises the coordination and management of inter-relationships between different state and non-state stakeholders (cf. Benz and Dose 2004; cf. Fröhlich and Knieling 2013), hence governing refers to

> the totality of interactions, in which public as well as private actors participate, aimed at solving societal problems or creating societal opportunities; attending to the institutions as contexts for these governing actions; and establishing a normative foundation for all those activities. (Kooiman 2003, p. 4)

Over time, global developments demanded a new understanding of processes needing to be governed and governance systems. Awareness began to emerge that developments such as technological innovation, the spread of diseases and especially climate change challenged the steering capacity of conventional governance systems. As it became clear that shocks and disturbances required a shift in governing, governance was no longer understood as managing linear, predictable, social, economic and ecological processes:

> There is a move away from command-and-control management performed by Weberian bureaucrats within centralized national bureaucracies toward a plethora of different schemes of self-government, public-private partnerships, collaboration efforts, policy entrepreneurs, and participatory initiatives usually gathered under the umbrella term of "governance". (Duit and Galaz 2008, p. 328f.)

While mitigation governance has become a well-established field of governance, adaptation governance is a rather recent topic on governance agendas throughout the world and yet to be developed (cf. Termeer et al. 2017). Adaptation governance is a wicked problem[2] par excellence (O'Brien et al. 2008; Davoudi et al. 2009) and can be defined as

> the patterns that emerge from the governing activities of social, political, and administrative actors in the realm of climate change adaptation. The governing activities refer to their combined efforts to adapt to climate change, their efforts to found institutions that deal with

[1]The concept of governance is highly complex due to which there is no single definition. As the focus of this paper is on adaptation governance, the concept of governance itself is not further elaborated. For further reading on the concept of governance see Benz and Dose (2004), Benz et al. (2007) and Kooiman (2003).

[2]For further reading on the concept of wicked problems, see Rittel and Webber (1973).

> the issues, and finally they refer to their ideas about appropriate normative underpinnings for the way climate change adaptation should be governed, taking into account wider social and political beliefs and systems. (Huitema et al. 2016, p. 37)

Adaptation governance needs to address the specific characteristics of climate change in order to ensure adequacy. The key characteristics are uncertainty, longevity and the features of an emerging policy domain:

Uncertainty exists because despite considerable sound knowledge on the sensitivity of the climate system and impacts of changes within this system, knowledge remains incomplete due to a lack of information and disagreement on what is known and knowable (cf. Fröhlich and Knieling 2013). The high number of stakeholders and their different perspectives add to further controversy around assured and unassured knowledge. The difficult task of linking most different types of knowledge ranging from scientific to indigenous knowledge (cf. Makondo and Thomas 2018) and transferring it through the multi-level system can add to uncertainty. This has direct impacts on adaptation governance, as knowing what to adapt to is the prerequisite for knowing how to adapt to it (Termeer et al. 2017).

The **long-term nature (longevity)** of climate change impacts requires long-term, often even intergenerational thinking. These time frames exceed policy cycles and decision-making periods. Projections predict impacts decades ahead and imply uncertainty, hence creating a dilemma: Long-term investments to combat climate change impacts compete with short-term economic investments for the allocation of (limited) resources. The alignment of different timeframes of physical, economic and social impacts further complicates the task. As expensive long-term investments are highly unattractive from a political viewpoint, this dilemma often results in a constant postponing of action (cf. Fröhlich and Knieling 2013; Termeer et al. 2017, p. 7).

Adaptation governance is an **emerging policy field**, as depicted in its evolution. Unlike mitigation governance, it is a comparatively young and not yet well-structured policy domain which involves numerous stakeholders as well as sectors. The fact that objectives, responsibilities, procedures and evaluation methods are still only weakly defined, creates opportunities as well as barriers at the same time (Termeer et al. 2017, p. 5).

Figure 1 displays the features of climate change adaptation and their resulting governing challenges. The three key features of uncertainty, longevity, and the characteristics of an emerging policy field create particular challenges that adaptation governance needs to address. These challenges comprise cross-sectoral alignment, framing, multi-actor and multi-level coordination, timing and choice of policy instruments. As the figure illustrates, "power" resembles an additional element which can be found in several of the six challenges, yet it has not been well researched. Also, while the challenges are well identified, the body of literature on how to handle these challenges is not yet very comprehensive.

The following part elaborates the six key challenges as well as possible solutions depicted in the recent literature on adaptation governance.

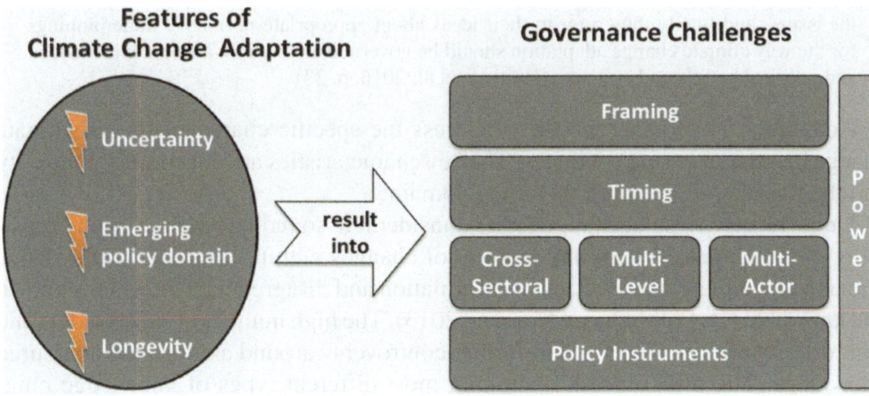

Fig. 1 Features of climate change adaptation and resulting governance challenges. *Source* Original depiction

Challenges to Adaptation Governance

Framing and prioritization are among the first steps when governing adaptation: Climate change impacts require numerous and comprehensive adaptation measures. However, as knowledge about impacts is still uncertain, knowledge about the most urgently needed adaptation measures remains uncertain too. Since climate change adaptation is not a single problem but instead highly interconnected, competing framings, not only among skeptical and convinced stakeholders may exist (cf. Hoffman 2011). A strong research policy-linkage ensures adequate framing and can prevent single framing by the most powerful stakeholder. Morton et al. (2011) showed that aversion to uncertainty in combination with a negative framing, i.e. highlighting possible losses, decreases the intention to act. Framing uncertainty positively, however, i.e. not materializing losses, can increase the willingness to act. Grant et al. (2015) suggest that framing disastrous events as caused by climate change ("climatization") can help create a sense of urgency.

Timing decisions can be influenced by current events such as floodings, extreme heat periods or droughts, as they facilitate raising climate change on agendas (cf. Grant et al. 2015). Such windows of opportunity must be used wisely as they only open for limited time periods. Timing is also a challenge regarding investments: Long-term adaptation measures compete with short-term needs for restrained resources. In combination with uncertainty about ideal measures, this easily results in postponing, waiting and responsibility negation, as taking action is unattractive on all levels of actions. A close policy-research linkage is important to ensure informed decisions (Ampaire et al. 2017) as well as flexibility as a design principle. Flexible policies allow for adjustment, i.e. they can be intensified or enhanced when knowledge or circumstances change over time. Flexibility can therefore help to translate long-term measures into short-term policies.

Cross-sectoral alignment is necessary due to the high number of affected sectors. Among the most climate-sensitive sectors are agriculture and forestry, health, energy, water management and flood protection, as well as infrastructure and housing. The depicted interconnectivity demands an alignment across sectors. The cross-sectoral nature, however, results in a lack of a clear policy domain, which in turn results in the need for strong leadership for its coordination. Hill (2013) and Madzwamuse (2010) suggest that this role be filled by government entities as—assuming that most decisions regarding adaptation are made on national level—other actors lack the necessary authority to make policy decisions on this level while at the same time factoring in multi-scale dimensions (Madzwamuse 2010, p. 81f.; Hill 2013, p. 34). Kreft et al. (2011) confirm that national ministries are in the position to fulfill this role as they make decisions and coordinate resources. Therefore, key ministries (such as the ones for finance or economics) as well as the prime minister's office should play key roles in the development of adaptation strategies (Kreft et al. 2011, p. 8). Ostrom (2009) states that a single governmental unit in charge is insufficient as it would cause free-rider problems. However, drawing on their empirical findings in a recent case study on Uganda where central government agencies are in charge, Ampaire et al. (2017) find that the inclusion of actors across levels remains insufficient. Moreover, it has to be taken into consideration that the allocation of climate change to a certain ministry implies aspects of power as some ministries are more influential than others (Nightingale 2017).

The **level of action** is a particular challenge. While mitigation policies require action mainly on national and international level, adaptation requires multilevel governance. Ostrom (2009) finds that single actions at a global scale only are insufficient to induce the collective action needed and instead proposes a polycentric approach. Within the field of adaptation, many policies are embedded in the United Nations Framework Convention on Climate Change (UNFCCC) or strategies of the European Union, and main decisions are made on national level. Yet, most implementation measures take place on local and regional levels, the levels on which there is also the most precise knowledge about what form of adaptation is most needed. Knowledge transfer and a coherent multilevel policy alignment require coordination and cooperation in order to avoid mismatches between the level of decision-making and the level of implementation. Special attention is to be given to the inclusion of the local level, which tends to be overseen in decision-making processes led by the most powerful actors (Ampaire et al. 2017).

Involved **actors** and stakeholders are numerous and diverse, too. They comprise the public and private sectors as well as civil society, i.e. the government with several ministries, international organizations, NGOs, companies, and research institutes (public universities as well as private think tanks). The multitude of actors, perspectives, understandings of adaptation and priorities results in high diversity and adds to competing framings. Moreover, it indicates that there is not only one ideal adaptation governance strategy. Ostrom (2009) finds that a polycentric approach requires oversight over local, regional and national stakeholders. An insufficient inclusion of actors in policy formulation and implementation leads to a lack of awareness of existing measures and hence to sectoral segregation (Ampaire et al. 2017, p. 87). Furthermore, the high number of actors can result in a lack of ownership. Special effort should be given to strong policy-research linkages (e.g. through engaging policy makers with evidence and including them in research processes) to ensure informed policy making (Ampaire et al. 2017, p. 88). The fact that compromises and priorities have to be found holds both opportunities and barriers, since besides knowledge and technological feasibility, the influence and power of actors are crucial. The literature on managing climate change stakeholders suggests that the identification of all relevant stakeholders is crucial, followed by their engagement in dialogues, taskforces, working groups and consultations (Carman 2013). Most publications suggest that government agencies fill the role of managing stakeholders. Yet, recent findings from a case study in Iran show that stakeholder coordination remains one of the main challenges in climate change adaptation[3] (Amiraslani and Caiserman 2018).

Regarding the type of **policy instruments**,

> the main choice here is between creating and imposing a set of enforceable social norms hierarchically, i.e., through the instrument of regulation, allowing them to emerge and disseminate via market-based instruments in markets, or relying on flatter and more network-based modes in which trust-based incentives play a more important role. (Huitema et al. 2016, p. 3)

The state has direct and indirect policy instruments at its disposal. Direct instruments are of a regulatory nature, i.e. laws, regulations and directives. Indirect instruments include financial ones (taxes, subsidies, etc.), structural ones (provisioning of infrastructure and institutions) and "soft" ones which aim at argumentation and promotion and rely on communication and cooperation through conferences, round tables, etc. (cf. Braun and Giraud 2008). While the number of actors and stakeholders might call for a hierarchical regulation, the fact that impacts mainly occur on the local level might favor bottom-up solutions. Given the high interconnectivity and its cross-sectoral nature, the integration of adaptation into already existing policies across sectors is also discussed (cf. Mubaya and Mafongoya 2017). Institution building has been one of the most frequently applied instruments over recent years.[4] The idea of institution building is to "bring stakeholders at different levels into cooperative arrangements (institutions) to govern resources that cross jurisdictional boundaries" (Nightingale 2017, p. 11). Institutions, formal as well as informal ones, can be of a public, private and civic nature with their distinction sometimes being fuzzy (cf. Mubaya and Mafongoya 2017). While institutions play a key role in governing adaptation, the strong focus on institutional design principles, however, is too limited in scope. A recent strand of literature shows that power struggles within institutions also affect whether their output is positive or not: Nightingale (2017) and Eriksen et al. (2015) stress that "political realities" have to be taken into consideration as questions of governing (scarce) resources and questions of access to and

[3]The gap between these two findings represents a gap between theories and empirical findings often found in adaptation governance literature.

[4]Institution building is a common measure especially in developing countries, which is stipulated for example within National Adaptation Plans (NAPAs) which are funded by the UNFCCC (Nightingale 2017).

control over such resources are questions of power. The authors stress that adaptation projects—even more so in politically contentious contexts—are not solely based on objective examinations of climatic changes and feasible technological solutions, but also always imply the question of whose needs are prioritized (cf. Eriksen et al. 2015; Nightingale 2017, p. 14). Hence, the allocation of climate change to a certain ministry implies aspects of power, since some ministries are more influential than others. Even apparently objective decisions, such as regarding the first best technical solutions, may imply power aspects when, e.g., choices are linked to party politics (Nightingale 2017, p. 15).

The results of this analysis are structured in Table 1, which lists the identified six challenges with a short description and their respective hindering and enabling factors. The table demonstrates once more the importance of research-policy linkages and power and serves as the analytical framework for the case study of Colombia.

Climate Change Adaptation Governance in Colombia

Governing climate change adaptation in Colombia is particularly wicked due to the country's features: Characterized by the Andes and located on both the Caribbean and the Pacific Ocean, Colombia presents a difficult topography, a high risk of being affected by extreme climatic events and weak infrastructures, which result in high demands for technical solutions. In line with these characteristics, Colombia is categorized as highly vulnerable to climate change (Ardila et al. 2013, p. 39f.; Leal Filho and Mannke 2014, p. 99). Politically and culturally, the country is characterized by a history of an internal conflict which has lasted over 50 years and resulted in political, social, and economic disparities. The isolation of regions due to a lack of transportation and communication has led to self-contained economies and strong regional identities. The outspoken mistrust between the state and its society is a particular challenge since "where governance is already weak and the relations between citizens and the authorities are stretched, inappropriate responses to climate change have the potential to do harm inadvertently" (Smith and Vivekananda 2012, p. 77). This view is supported by Scheffran et al. (2012, p. 105) who find that "rapid or drastic climate change could overwhelm the adaptive capacity of social systems and lead to periods of instability". Since the ratification of a peace agreement between the government and the Revolutionary Armed Forces of Colombia (FARC) in 2016, the country has begun to transition into a post-conflict era. This phase brings about further challenges and opportunities such as the reallocation of power and access to resources (i.a. gold, coal, oil, gas, rare earths, copper, water).

Not least due to its history of conflict, Colombia's environmental, and even more so its climate and adaptation policies are not very comprehensive and partially contradictory. With the focus of its political leaders on prestige, the country has been part of important international conventions and agreements, such as the Kyoto Protocol and the Paris Agreement, has acted as leader at the UNFCCC which it joined in 1994, and has advocated climate protection policies (Bustos 2018, p. 34). Yet,

Table 1 Features of adaptation governance challenges

Challenges	Description	Hindering and enabling factors
Framing	Clear communication of problem despite high complexity	• Adequate framing of uncertainty to increase willingness to act • Cautious use of climatization of disasters • Ensure research-policy linkage • Consider power dynamics
Timing	Raise (long-term) adaptation measures on political agendas	• Use windows of opportunities • Flexibility as a design principle to translate long-term measures into short-term policies • Ensure research-policy linkage
Cross-sectoral	Inclusion of and policy alignment across all climate-sensitive sectors	• Allocation of the main responsibility to powerful enough key (governmental) actor(s) • Ensure research-policy linkage • Consider power dynamics
Multi-level	Inclusion of all governance levels	• Adequate inclusion of international, national, regional, local levels to ensure polycentric approach and avoid mismatches between decision-making level and level of implementation • Coordination and cooperation to ensure knowledge transfer and policy alignment • Special focus on inclusion of the local level • Consider power dynamics
Multi-actor	Inclusion of multitude of actors	• Identification and engagement of public, private, national, international stakeholders, civil society, academia • Special focus on inclusion of local level stakeholders • Ensure ownership • Ensure research-policy linkage • Consider power dynamics
Policy instruments	Use of direct and indirect instruments to create governance framework	• Balance between top-down and bottom-up approaches • Balance between direct and indirect instruments • Focus on institutional design • Consider power dynamics

Source Original depiction. Due to high interconnectivity, a strict division into single challenges and hindering and enabling factors is hardly possible and of an exemplary nature on a domestic level, it is characterized by decades of economic planning without an environmental focus (Semana 2011) and by "traditional patterns of unsustainable development" (Bustos 2018, p. 29).

Regarding climate change, the Colombian focus on mitigation has been stronger than on adaptation. The focus on adaptation was effectuated in 2010–2011: During this time, Colombia experienced an unusually strong El Niño Southern Oscillation (ENSO), known as La Niña. The multi-hazard weather event led to floods, landslides and windstorms, causing losses of several billions US dollars[5] through the destruction of infrastructure, agricultural lands and crops and a resulting need for government subsidies (Hoyos et al. 2013; Bustos 2018, p. 40). Besides monetary losses, La Niña caused several hundred fatalities and left several thousands of people injured and/or homeless. This weather event marked a clear turning point in Colombian climate policies, as not only the public became aware of climate change but especially government officials recognized the need for adaptation (Ardila et al. 2013, p. 34). President Santos[6] declared the weather event a national catastrophe which had been caused by climate change and publicly put climate change adaptation on the agenda (Semana 2011). Shortly after, a comprehensive institutional reform was implemented: The country's main body addressing climate change, the Climate Change Mitigation Group (CCMG), was moved from the Ministry of Environment (MoE) to the National Planning Department (NPD). This move from one authority to another equaled a gain in decision-making power: The NPD receives instructions from the President and thus has the authority and political power to coordinate the ministries involved (Comstock et al. 2012, pp. 4–14). This reform led to the NPD coordinating a high-level commission comprising the ministers of Environment, Agriculture, Finance, Energy and Mines, Transport, Foreign Relations and Social Development. Below the Commission, a lower-level inter-agency Financial Committee, an Executive Secretary located in the MoE, consulting groups and four sub-committees also comprising adaptation were put into place (Comstock et al. 2012, pp. 9–17). The Ministry of Environment underwent a reform in the same year and was renamed to Ministry of Environment and Sustainable Development. Within the framework of the National Development Plan (Plan Nacional de Desarrollo) 2010–2014, four strategies to tackle climate change were stipulated. These four strategies aimed at reducing vulnerability and fostering adaptation, comprising the formulation and implementation of a National Climate Change Adaptation Plan (Plan Nacional de Adaptación al Cambio Climático—PNACC) (PNACC 2012, pp. 10–19). The initiative to create the PNACC was led by the NPD and supported by the MoE as well as the Institute of Hydrology, Meteorology and Environmental Studies (Instituto de Hidrología, Meteorología y Estudios Ambientales—IDEAM), a government agency specialized on generating scientific knowledge for the MoE. IDEAM also took part in a cooperation between the NPD and the Inter-American Development

[5]Over US$ 7.8 billion according to Hoyos (2013); 11.2 trillion COP, i.e. the equivalent to a 2.2 loss of GDP according to Bustos (2018).

[6]The Santos administration was in office from August 2010 to August 2018. It was followed by the Duque administration after two electoral terms.

Bank (IADB) who commissioned a comprehensive study on the economic impacts of climate change in 2014. Despite the results on adaptation remaining very generalized in this study (cf. NPD 2014), the allover collaboration between ministries and science in the field of climate change has been intense and successful (Pinto et al. 2018). The PNACC comprises the four objectives of generating knowledge, making use of possible benefits, incorporating risk management in the planning of sectoral development and identifying and implementing adaptation measures. It also stresses the need for stakeholder inclusion (PNACC 2012, p. 11). Moreover, the creation of a National System of Climate Change (Sistema Nacional de Cambio Climático—SISCLIMA) was stipulated in the PNACC (PNACC 2012, p. 11). SISCLIMA is an intersectoral commission for coordinating climate change in the multi-level system with a focus on funding and international affairs. Consulting with different stakeholders and sectors is part of SISCLIMA's mandate. Moreover, the commission is in charge of policy formulation and mainstreaming climate into existing projects. SISCLIMA was created and operational in 2011, yet it was not formalized through a decree until 2016. Gaining ministries to support the cross-sectoral approach, which cuts across departmental agendas, has been one of the main challenges (IDEAM; Jaramillo 2014; Rojas-Laserna 2014; Bustos 2018, p. 44). The following 2014–18 National Development Plan continues the prominent features of environmental aspects with a particular focus on green growth. Making green growth a central element had been a particular demand during Colombia's OECD accession process, which had begun in 2013 and was completed in 2018 (OECD 2014).

All of these actions were taken on national level. Special attention was given to international cooperation. In 2012, international cooperation partners included (but were not limited to) the USA, Germany, the UK, and the EU; non-profit organizations such as the World Resources Institute (WRI) and the Center for Clean Air Policy (CCAP); and multilateral institutions such as IADB, UNDP and the World Bank. Colombia received financial and technical support through these cooperation and was considered a "supporter's darling" (Comstock et al. 2012, pp. 9–17). The inclusion of the private sector is considered important and public-private sector cooperation is sought (Comstock et al. 2012, p. 9; Green Climate Fund 2018). The inclusion of the civil society has been very limited and local environmental governance has been rather weak (Jaramillo 2014; García and Slunge 2015, p. 2; Bustos 2018). The reason for this is an alleged lack of technical expertise which results in a lack of interest in consulting NGOs (Bustos 2018, p. 35). Despite a clear institutional focus on adaptation and the promotion of both adaptation and mitigation policies on international level, contradictions remain on the national level as can be seen in the case of mining: While the government undertook strong measures to end illegal mining, a major source of pollution, in 2010, it also makes strong efforts to attract multinationals to start operations in the country to exploit hydro-carbons and coal (Semana 2011).

Discussion of Findings: Achievements and Shortcomings

The depicted developments in adaptation governance in Colombia show that after the 2010–2011 La Niña hazard, the Colombian government used timing and climatization to raise adaptation upon the agenda. In line with the theoretical findings of the first part of this analysis, institutional reform has been used extensively and can be categorized as the primary instrument. The analysis of the institutional reform shows that the government has assumed a leading role in decision-making and the coordination of stakeholders. This top-down strategy was enhanced by a change in the way adaptation was framed: Adaptation was no longer framed as only an environmental problem, but also as an economic and social problem requiring cross-sectoral solutions. Moreover, possible benefits were emphasized. The need for multi-actor, multi-level and multi-sector inclusion was attended to through the creation of a special institution. Policy alignment, especially the integration of climate into existing policies was prioritized. Further achievements include the policy-science linkage for informed decision-making processes.

Regarding shortcomings, the analysis shows that despite a multi-actor and multi-level understanding of adaptation, the main focus has been on the national level, neglecting the regional and especially the local level. The inclusion of stakeholders other than international actors has been insufficient and civil society in particular has been neglected.

Building on Table 1, Table 2 summarizes the features of adaptation governance in Colombia and thus provides an illustration of the finding that four out of six challenges have been addressed while two have been neglected.

Yet, despite the number of achievements apparently outweighing the number of shortcomings, the analysis also shows that despite years of strong promotion, adaptation governance remains weak and outcomes in terms of concepts and actions are fuzzy. The stronger focus on mitigation rather than on adaptation is one explanation for this apparent contradiction. Generating most of its electricity via hydro-power, "[…] Colombia can call for ambitious climate-mitigation measures without necessarily having to implement a massive transformation of the electricity sector at home" (Bustos 2018, p. 41). Mitigation is hence the "low-hanging fruit" compared to adaptation.

Table 2 Features of adaptation governance in Colombia

Governance challenges	Colombia
Framing	✓
Timing	✓
Cross-sectoral	✓
Multi-level	∘
Multi-actor	∘
Policy-instruments	✓

However, mitigation as the only explanatory factor seems to fall short. A closer look at the factors of international cooperation and economic aspects help to gain a deeper understanding: It is common for emerging countries to seek international cooperation. Steinberg (2003) finds that environmental policies in developing countries are driven by those "who combine international scientific and financial resources with the domestic political resources needed to usher through major reforms" (Steinberg 2003, p. 27). This holds true for the Colombian case, yet there are further reasons for the focus on international cooperation: As Bustos (2018) states, the focus of Colombian decision-makers has traditionally been on international prestige. In the case of climate change governance, progressive international positions can help cleanse the image of Colombia as a "failing state". Moreover, the benefits for politicians on domestic level are twofold: The cooperation with powerful international actors (and powerful national actors for that matter, as can be seen in the focus on public-private partnerships) helps to retain power and brings funding opportunities—not only for adaptation but possibly for other aims, too. Bustos (2018, p. 38f.) finds that the Santos administration promoted the idea to use international funding for climate change for projects that would also enhance the peace process at municipal level. Likewise, the Santos administration stated that the exploitation of hydro-carbons and coal by multinationals had to be promoted intensely, as the country's monetary profits were crucial for the funding of the peace process (González Espinosa 2015; Strambo et al. 2018, p. 13). Moreover, the fact that while illegal mining was strongly combated, efforts were made to attract multinationals shows that preservation of power and economic interest prevail over climate change considerations. This view is supported by the OECD country report on Colombia which criticized the lack of coherence between environmental goals and economic sector planning as the main shortcoming (OECD 2014, p. 14). These findings illustrate the balancing act between economic and environmental interests with which many countries struggle. However, they also illustrate the strong impact of power politics which is characteristic for countries with weak state structures. In the case of Colombia this leads to inconsistencies and contradictions in adaptation governance. The impact of the power factor is also reflected in the focus on the national level: The strong focus on the national level indicates that, in a country where the government did not have complete sovereignty over state territory, state power and its people for several decades, a constant focus is on regaining and maintaining power. Assigning the decision-making authority to a department responding to the president assured the control on the part of the latter. Hence this assignment enforces the president's claim of leadership within the state institutions. Power within the field of adaptation is of importance since it includes questions of access to and power over resources.

Regarding the insufficient inclusion of civil society, it has to be taken into account that interest groups in developing countries are less organized and less well funded than interest groups in developed countries. Their influence is therefore not as strong (Bustos 2018, p. 45). Especially in a country with a 50-year history of conflict, it can be assumed that civil society participation is not well developed and that the focus of NGOs' work has been on the peace process rather than on climate change. Yet, since the government has not identified a lack of technical expertise on part of NGOs, and since such organizations would be helpful in the civil-social arena regardless of their level of technical expertise, the argument that NGO inclusion is not necessary due to their lack of technical expertise does not seem valid. Whether traditional power structures are an explanatory factor is yet to be researched. Generally speaking, it seems that the weak cooperation between the government and civil society reflects their stretched relationship. In order to achieve effective adaptation, this challenge has to be tackled as the analysis has shown that the inclusion of civil society and the local level is one of the main shortcomings.

Further developments against the backdrop of the 2016 peace agreement and especially the presidential elections in 2018, which were won by Iván Duque, remain to be seen.

Besides this temporal restriction, it must be pointed out that this analysis focused on one country only and that the emphasis was on the national level, for which findings cannot be generalized. The single country focus was necessary, however, in order to do justice to the exploratory character of this qualitative study. This paper therefore provides a basis for further research which should include in-depth case studies of the Colombian case focusing on regional and local level as well as further in-depth studies of similar countries in order to provide sufficient samples to draw general conclusions.

The study was designed as a desktop research and therefore relied on access to sources. It cannot be assumed, however, that the entire universe of documents was accessed as some may have been confidential or inaccessible due to other reasons. Thorough research and comprehensive coverage of available sources through a trilingual approach were applied in order to restrict this inevitable limitation to a minimum. Further research based on different types of data collection, such as interviews, can contribute to scrutinize and strengthen the results of the analysis at hand and hence to closing the identified knowledge gaps.

Conclusion

The analysis of the theoretical concepts of adaptation governance first explained the shift from a technical to political understanding of adaptation and elaborated on the need for a governance framework. It then identified the factors cross-sectoral alignment, framing, multi-actor and multi-level coordination, timing and policy instruments as the six main challenges for the creation of an adaptation governance framework. It confirmed the particular importance of a research-policy linkage and elaborated on the impacts of power politics which is a recent topic on research agendas.

The findings show that despite awareness and common agreement on the main challenges, the body of literature on possible solutions is not very comprehensive yet. Moreover, several theoretical findings on solutions have been contradicted by findings from recent case studies, hence indicating a research gap. This served as motivation for the second part of this chapter analyzing the case of Colombia.

Colombia, similar to other countries with weak state structures, is vulnerable to climate change and requires adaptation governance. In line with the findings of the theoretical analysis, the case study proves that the identified six factors are of high importance. It also confirms the importance of the additional factor power dynamics, in particular with regards to the ongoing peace process and resulting resource allocations. Moreover, the case study shows that the focus of several key actors on the factor power reduces the effects of measures in the other six areas, resulting in insufficient adaptation governance.

These findings call for further research in order to generate knowledge on the depicted research gaps. Regarding the Colombian case, further in-depth studies based on different types of data collection, are needed in order to complement the findings at hand. Further analyses of similar country case studies can help to generalize findings. Both approaches supplement each other and contribute to the urgently needed creation of adaptation.

Acknowledgements I would like to express my appreciation to Prof. Dr. Harald Fuhr, the Potsdam Graduate School, and the reviewers and editors of this book.

References

Abeling T, Leopold A (2012) Climate change and security: adapting the discussion to the evidence. Global Governance Institute (GGI), Brussels, Belgium

Adger WN, Dessai S, Goulden M et al (2009) Are there social limits to adaptation to climate change? Clim Change 93:335–354. https://doi.org/10.1007/s10584-008-9520-z

Amiraslani F, Caiserman A (2018) Multi-stakeholder and multi-level interventions to tackle climate change and land degradation: the case of Iran. Sustainability 10:2000. https://doi.org/10.3390/su10062000

Ampaire EL, Jassogne L, Providence H et al (2017) Institutional challenges to climate change adaptation: a case study on policy action gaps in Uganda. Environ Sci Policy 75:81–90. https://doi.org/10.1016/j.envsci.2017.05.013

Ardila G, Andrade G, Benavides J et al (2013) Desarrollo económico y adaptación al cambio climático. Friedrich Ebert Stiftung en Colombia (fescol), Bogotá, Colombia

Benz A, Dose N (eds) (2004) Governance—Regieren in komplexen Regelsystemen: Eine Einführung, 1st edn. VS Verlag für Sozialwissenschaften, Wiesbaden, Germany

Benz A, Luetz S, Schimank U, Simonis G (eds) (2007) Handbuch Governance: Theoretische Grundlagen und empirische Anwendungsfelder. VS Verlag für Sozialwissenschaften, Wiesbaden, Germany

Braun D, Giraud O (2008) Politikinstrumente im Kontext von Staat, Markt und Governance. In: Schubert K, Bandelow NC (eds) Lehrbuch der Politikfeldanalyse 2.0. De Gruyter Oldenbourg, Munich, Germany

Bustos MC (2018) What shapes Colombia's foreign position on climate change? Colombia Int 27–51. https://doi.org/10.7440/colombiaint94.2018.02

Caesar L, Rahmstorf S, Robinson A et al (2018) Observed fingerprint of a weakening Atlantic Ocean overturning circulation. Nature 556:191. https://doi.org/10.1038/s41586-018-0006-5

Carman R (2013) Multi stakeholder decision making for NAMAs and LEDs. Low emission capacity building programme. UNDP, Maseru, Lesotho

Climate Action Tracker (2018) For the Talanoa dialogue: input from the Climate Action Tracker. New Climate Institute, Ecofys, Climate Analytics, Germany
Comstock M, Santelices I, Vanamali A (2012) Colombia's national climate change process. Center for Clean Air Policy (CCAP), Washington DC, USA
Davoudi S, Crawford J, Mehmood A (eds) (2009) Planning for climate change: strategies for mitigation and adaptation for spatial planners, 1st edn. Routledge, London, Sterling, VA
Duit A, Galaz V (2008) Governance and complexity—emerging issues for governance theory. Governance 21:311–335. https://doi.org/10.1111/j.1468-0491.2008.00402.x
Eriksen SH, Nightingale AJ, Eakin H (2015) Reframing adaptation: the political nature of climate change adaptation. Glob Environ Change 35:523–533. https://doi.org/10.1016/j.gloenvcha.2015.09.014
Escobar H (2018) Scientists, environmentalists brace for Brazil's right turn. Science 362:273–274. https://doi.org/10.1126/science.362.6412.273
Falkner R (2016) The Paris Agreement and the new logic of international climate politics. Int Aff 92:1107–1125. https://doi.org/10.1111/1468-2346.12708
Fröhlich J, Knieling J (2013) Conceptualising climate change governance. In: Knieling J, Leal Filho W (eds) Climate change governance. Springer Berlin Heidelberg, Berlin, Heidelberg, Germany, pp 9–26
García JH, Slunge D (2015) Environment and climate change management: perspectives for post-conflict Colombia. Sida's helpdesk for environment and climate change. Göteborgs Universitet, Chalmers, SLU, Sweden
González Espinosa A (2015) La paradoja del sector minero-energético en el Plan Nacional de Desarrollo 2014–2018: financiador de la paz y generador de tensiones en el territorio. Universidad Externado de Colombia, Bogotá, Colombia
Grant S, Tamason CC, Jensen PKM (2015) Climatization: a critical perspective of framing disasters as climate change events. Clim Risk Manag 10:27–34. https://doi.org/10.1016/j.crm.2015.09.003
Green Climate Fund (2018) Colombian President calls for enhanced public-private climate action. https://www.greenclimate.fund/-/colombian-president-calls-for-enhanced-public-private-climate-action
Hill M (2013) Climate change and water governance: adaptive capacity in Chile and Switzerland. Springer Netherlands, Netherlands
Hoffman AJ (2011) Talking past each other? Cultural framing of skeptical and convinced logics in the climate change debate. Organ Environ 24:3–33. https://doi.org/10.1177/1086026611404336
Houghton K (2012) Climate change in fragile states: adaptation as reinforcement of the fabric of the state. In: Climate change and fragile states: rethinking adaptation. United Nations University, Institute for Environment and Human Security, Bonn, Germany, pp 24–43
Hoyos N, Escobar J, Restrepo JC et al (2013) Impact of the 2010–2011 La Niña phenomenon in Colombia, South America: the human toll of an extreme weather event. Appl Geogr 39:16–25. https://doi.org/10.1016/j.apgeog.2012.11.018
Huitema D, Adger WN, Berkhout F et al (2016) The governance of adaptation: choices, reasons, and effects. Introduction to the special feature. Ecol Soc 21. https://doi.org/10.5751/es-08797-210337
IDEAM Política Nacional de Cambio Climático. Instituto de Hidrología, Meteorología y Estudios Ambientales, Bogotá, Colombia. https://www.cambioclimatico.gov.co/de/directorio-del-cambio-climatico. Accessed on 9 October 2018
IPCC (2001) Climate change 2001: impacts, adaptation, and vulnerability: contribution of working group II to the third assessment report of the intergovernmental panel on climate change. Cambridge University Press, Cambridge, UK, New York
IPCC (2007) Climate change 2007: impacts, adaptation and vulnerability. Contribution of working group II to the fourth assessment report of the intergovernmental panel on climate change. Cambridge University Press, Cambridge, UK, New York

IPCC (2014) Climate change 2014: impacts, adaptation, and vulnerability. Part A: global and sectoral aspects. Contribution of working group II to the fifth assessment report of the intergovernmental panel on climate change. Cambridge University Press, Cambridge, UK

IPCC (2018) Global warming of 1.5°C. Intergovernmental Panel on Climate Change, Switzerland

Jaramillo M (2014) The coordination of climate finance in Colombia. E3G/GIZ. Overseas Development Institute, London, UK

Javeline D (2014) The most important topic political scientists are not studying: adapting to climate change. Perspect Politics 12(2):420–434

Keohane RO (2015) The global politics of climate change: challenge for political science. PS Political Sci Politics 48:19–26. https://doi.org/10.1017/s1049096514001541

Kooiman J (2003) Governing as governance. Sage, London, UK

Kreft S, Oumar AK, Harmeling S (2011) National adaptation plans. Towards effective guidelines and modalities. Germanwatch and WWF International, Germany, Switzerland

Leal Filho W, Mannke F (2014) Managing the impacts of climate change in Latin America: the need for technology transfer. In: International perspectives on climate change. Springer International, Switzerland, pp 95–106

Madzwamuse M (2010) Climate governance in Africa—adaptation strategies and institutions. Heinrich Böll Stiftung, Unity Press, Cape Town, South Africa

Makondo CC, Thomas DSG (2018) Climate change adaptation: linking indigenous knowledge with western science for effective adaptation. Environ Sci Policy 88:83–91. https://doi.org/10.1016/j.envsci.2018.06.014

Morton TA, Rabinovich A, Marshall D, Bretschneider P (2011) The future that may (or may not) come: how framing changes responses to uncertainty in climate change communications. Glob Environ Change 21:103–109. https://doi.org/10.1016/j.gloenvcha.2010.09.013

Mubaya CP, Mafongoya P (2017) The role of institutions in managing local level climate change adaptation in semi-arid Zimbabwe. Clim Risk Manag 16:93–105. https://doi.org/10.1016/j.crm.2017.03.003

Nightingale AJ (2017) Power and politics in climate change adaptation efforts: struggles over authority and recognition in the context of political instability. Geoforum 84:11–20. https://doi.org/10.1016/j.geoforum.2017.05.011

NPD (2014) Impactos Económicos del Cambio Climático en Colombia. Departamento Nacional de Planeación (National Planning Department), CEPAL, BID, IDEAM, Bogotá, Colombia

O'Brien G, O'Keefe P, Meena H et al (2008) Climate adaptation from a poverty perspective. Clim Policy 8:194–201. https://doi.org/10.3763/cpol.2007.0430

OECD (2014) OECD environmental performance reviews: Colombia 2014. OECD Publishing, United Nations Economic Commission for Latin America and the Caribbean (ECLAC), Paris, France

Ostrom E (2009) A polycentric approach for coping with climate change. The World Bank, policy research working papers

Pinto AD, Loboguerrero AM, Londoño M et al (2018) Informing climate policy through institutional collaboration: reflections on the preparation of Colombia's nationally determined contribution. Clim Policy 18:612–626. https://doi.org/10.1080/14693062.2017.1321521

PNACC (2012) Plan Nacional de Adaptación al Cambio Climático. Departamento Nacional de Planeación (National Planning Department), Colombia

Rittel HWJ, Webber MM (1973) Dilemmas in a general theory of planning. Policy Sci 4:155–169

Rojas-Laserna M (2014) Adaptation to climate change in Colombia. Effective institutional arrangements for NAP formulation and implementation. Ministry of the Environment and Sustainable Development Colombia, Bonn/Germany

Scheffran J, Brzoska M, Brauch HG et al (2012) Climate change, human security and violent conflict: challenges for societal stability. Springer Science & Business Media, Berlin, Heidelberg, Germany

Semana (2011) Climate change and Colombia, Santos' new deal. https://www.semana.com/international/print-edition/articulo/climate-change-and-colombia-santos-new-deal/238322-3. Accessed 27 Oct 2018
Smith D, Vivekananda J (2012) Climate change, conflict, and fragility: getting the institutions right. In: Scheffran J, Brzoska M, Brauch HG et al (eds) Climate change, human security and violent conflict: challenges for societal stability. Springer Berlin Heidelberg, Berlin, Heidelberg, Germany, pp 77–90
Steffen W, Rockström J, Richardson K et al (2018) Trajectories of the earth system in the anthropocene. PNAS 115:8252–8259. https://doi.org/10.1073/pnas.1810141115
Steinberg PF (2003) Understanding policy change in developing countries: the spheres of influence framework. Glob Environ Politics 3:11–31
Stern NH (2007) The economics of climate change: Stern review on the economics of climate change. Cambridge University Press, New York, USA
Strambo C, Espinosa ACG, Velasco AJP, Atteridge A (2018) Privileged coal: the politics of subsidies for coal production in Colombia. Stockholm Environment Institute (SEI), Sweden
Termeer CJAM, van Buuren A, Dewulf A et al (2017) Governance arrangements for the adaptation to climate change. In: Storch H von (ed) Oxford research encyclopedia of climate science
UN (2015) Adoption of the Paris Agreement. Conference of the Parties. Twenty-first session, Paris, France

The Andean Farmers of Peru: Farm-Household System Vulnerability to Climate-Related Hazards

Mariana Vidal Merino, Diana Sietz, Francois Jost and Uta Berger

Abstract Agriculture in the Andes is subject to multiple climate-related risks, typical of complex mountain ecosystems. Most of the strategies used to confront or reduce these risks are based on the adaptive capital of the farm households, such as the availability of labor, extension, and distribution of agricultural land, access to markets, among others. In order to increase the adaptive capacity of farm households, it is first necessary to understand the heterogeneity of the factors that explain their vulnerability. This article presents an analysis of archetypes (patterns) of climate vulnerability based on empirical data of farm household systems in the central Andes of Peru. The study uses mixed methods, combining qualitative tools and quantitative techniques, including cluster analysis. The results demonstrate the suitability of the methodology for explaining the vulnerability of farm household systems to climate-related hazards. For the case study, seven factors explain differences in vulnerability between five archetypes of agricultural households, including agricultural area, availability of irrigation, use of different agro-ecological zones, and access to non-agricultural employment.

M. Vidal Merino (✉) · U. Berger
Institute of Forest Growth and Forest Computer Sciences, Technische Universitaet Dresden (TU Dresden), Pienner Straße 8, 01737 Tharandt, Germany
e-mail: MarianaVidal@forst.tu-dresden.de

U. Berger
e-mail: uta.berger@tu-dresden.de

D. Sietz
Earth System Analysis, Potsdam Institute for Climate Impact Research, Research Domain 1, Telegrafenberg A 31, D-14473 Potsdam, Germany
e-mail: sietz@pik-potsdam.de

F. Jost
School of Environment and Sustainability, Royal Roads University, 2005 Sooke Road, Victoria, BC V9B 5Y2, Canada
e-mail: francoisjost@gmail.com

W. Leal Filho et al. (eds.), *Climate Change, Hazards and Adaptation Options*, Climate Change Management, https://doi.org/10.1007/978-3-030-37425-9_53

Introduction

The Tropical Andes of Peru is categorized as one of the geographical areas that is most vulnerable to—and will be most affected by—climate change (Vuille et al. 2008; PNUD 2013). The persistence of poverty and heavy dependence on agriculture as a source of food and income exacerbates this risk (PNUD 2013). The Peruvian government has developed a number of policies, strategies, and action plans designated to address climate change issues and support a paradigm shift towards a climate change-resilient development in this region.[1] The agricultural sector is prioritized for the implementation of main measures, and the emphasis is to be given to vulnerable populations and those living in poverty (FAO and MINAGRI 2012; MINAM 2015). However, the implementation of concrete measures to reduce the vulnerability of Andean populations remains a complex challenge (MINAM 2015).

This complexity derives from the Andes' remarkable heterogeneity in terms of environmental conditions, ecosystems, land-use patterns, and the framework of factors influencing decision making—including livelihoods, policies, market prices, among others (IPCC 2014). Furthermore, climate change impacts are manifold and multidimensional. Changes in one aspect, such as water availability (amount and distribution), can have adverse effects on different sectors such as drinking water supply, agriculture, or power generation (Barnett et al. 2005; Vergara et al. 2007; Salzmann et al. 2009). The vulnerability in one sector is often linked with vulnerability in others.

On the other hand, distinct processes reappear within the multitude of socio-ecological contexts. Recurrent features can be generalized to define archetypical patterns so that the underlying processes can be better understood and insights into generic strategies that foster sustainable development can be provided (Sietz et al. 2017; Vidal Merino et al. 2018). As an example, different archetypes of farming systems' vulnerability in the drylands of sub-Saharan Africa were differentiated according to erosion sensitivity, governance, and degree of urbanization (Sietz et al. 2017).

We present here a mixed-method approach to analyzing archetypes of the climate vulnerability of farm household systems[2] (FHSs) using the Central Andes of Peru as a case study to illustrate this approach. First, we characterize the FHSs and their adaptive capital and surrounding socio-ecological context. In the second step, according to the similarities and differences in their adaptive portfolios, we classify FHSs in archetypes. Then, we explain the climate vulnerability outcomes of farmers using the identified archetypes. Finally, we discuss the main findings and the proposed methodology as well as its potential contribution to developing adaptation pathways.

[1] See Pramova et al. (2015) for an overview of adaptation and mitigation to climate change-related policies in Peru.

[2] The FHS is composed of three closely linked subsystems: the farm, the household, and the off-farm employment (including temporary migration) by household members. For more information on FHSs, see McDonnell and Dillon (1997) and Jost (2016).

FHS Vulnerability Framework

Vulnerability is "the propensity or predisposition to be adversely affected" (IPCC 2014) and is a function of exposure, sensitivity, and adaptive capacity (Adger 2006; Easterling et al. 2007). Complementary, contextual vulnerability is defined as the system's current ability to cope with external pressures such as changing climate conditions (O'Brien et al. 2007). It is an essential characteristic of socio-ecological systems, generated by multiple factors and processes (O'Brien et al. 2007; IPCC 2014).

We take this multi-dimensional view of biophysical and socio-economic interaction, and select the FHS level as the unit of analysis (see Fig. 1) to target factors leading to differential vulnerability to climate-related hazards.

The vulnerability, that is, the level of exposure, sensitivity, and adaptation capacity of a specific FHS at a given time is determined by the interaction of external and internal determinants. External determinants are changing biophysical conditions as well as dynamic social, economic, institutional, political, and technological structures and processes (Adger 2006; Fellmann 2012) that are outside an FHS's control. On

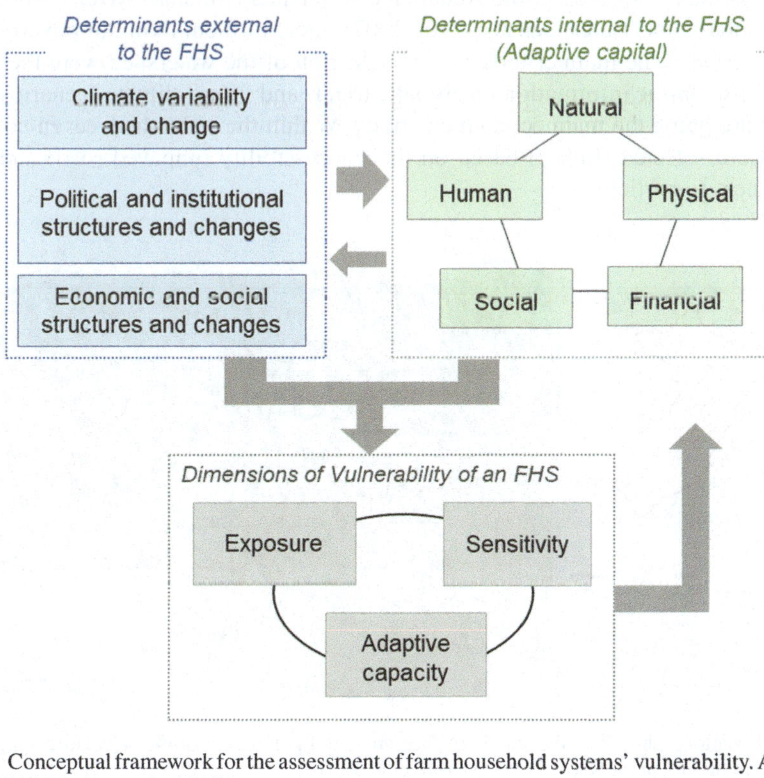

Fig. 1 Conceptual framework for the assessment of farm household systems' vulnerability. Adapted from Vidal Merino et al. (2018)

the other hand, internal determinants define the ability to cope with external stressors and are characterized by the set of adaptive capitals available to the FHS.

Based on the Sustainable Livelihood Approach (see Chambers and Conway 1992; Carney 1998), the adaptive capital portfolio includes natural, financial, social, human, and physical capitals. These capitals define the livelihood strategies of a given FHS, e.g. crop and livestock selection and diversification of on-farm and off-farm activities (Valdivia and Gilles 2001; Heltberg et al. 2009). Often, maintaining a pool of different forms of capital including secure and liquid assets to cope with short-term stresses affects the expected returns and limits long-term adaptation, potentially entrapping FHSs further into an asset-based poverty cycle (Heltberg et al. 2009). Thus, an FHS's adaptive portfolio is crucial for its coping capacity, adaptation to climate change, and risk management since it determines manifold FHS livelihood strategies used to minimize vulnerability.

Case Study: The Central Andes of Peru

The case study is located in the Achamayo watershed (Mantaro River basin), with 15,183 inhabitants (latest census: INEI 2007), most of them living in poverty (Vilcapoma 2008). The main criteria for the selection of the watershed were the availability of scientific information on climatic trends and future climate scenarios, and; agriculture being the main economic activity. Within the watershed, seven communities were selected (Fig. 2) based on their accessibility (unpaved road) and their willingness to participate.

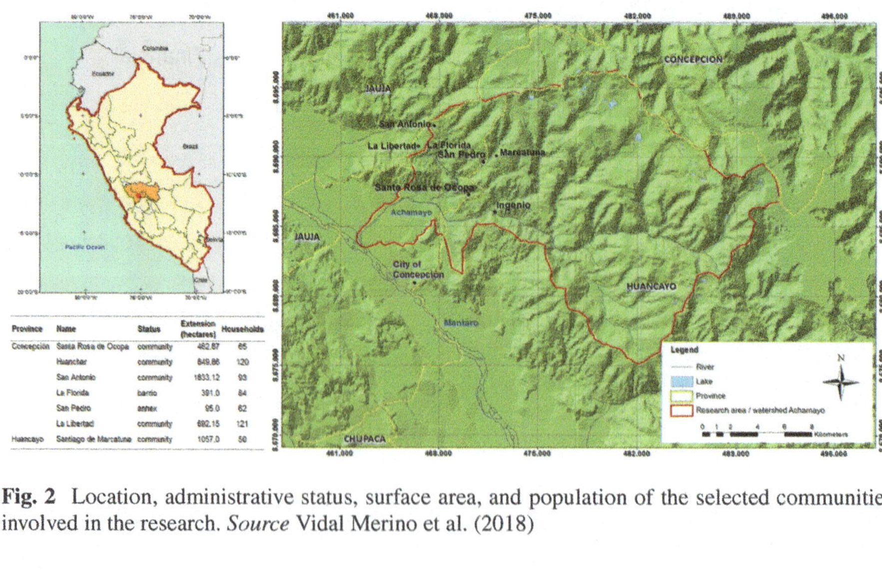

Province	Name	Status	Extension (hectares)	Households
Concepción	Santa Rosa de Ocopa	community	462.87	65
	Huanchar	community	849.86	120
	San Antonio	community	1833.12	93
	La Florida	barrio	391.0	84
	San Pedro	annex	95.0	62
	La Libertad	community	692.15	121
Huancayo	Santiago de Marcatuna	community	1057.0	50

Fig. 2 Location, administrative status, surface area, and population of the selected communities involved in the research. *Source* Vidal Merino et al. (2018)

Weather is temperate and dry, with temperatures ranging from 21 to 0 °C, a rainy season with 83% of annual precipitation occurring between October and April, and a dry season with the minima (20 mm/month) occurring during the coldest months of June and July (IGP 2005a). Future climate scenarios for the year 2050 (IGP 2005b) predict an increase in average temperatures of 1.3 °C during the dry season and a decrease of 19% in precipitation in the central part of the Mantaro River Basin, where the study area is located. Also expected is an increase in the diurnal amplitude of temperature of approximately 1 °C, as well as an increase in the number of days of frost during the dry season to 40 days.

Methodology

The mixed-method approach to determining archetypes of climate vulnerability outlined in this study offers a structured way of revealing the heterogeneity in farm household systems' vulnerability to climate-related hazards. It presents a comprehensive sequence of steps to structure a participatory assessment of vulnerability, starting from the single farm household and upscaling the results to farm household typologies. Due to the limitation of space, we did not include in this paper the detailed description of the methodology used. However, a complete description is available in Vidal Merino et al. (2018).

Field Laboratories

The fieldwork had as its main objectives: (1) to gather qualitative data to gain an understanding of the complexity of the socio-ecological environment in which the FHS is embedded, i.e. a comprehensive analysis of the external determinants contributing to the vulnerability of the particular system under study; (2) to develop indicators of adaptive capital as FHS internal determinants; (3) to select an indicator of agricultural losses as a vulnerability outcome; and (4) to collect data on the identified indicators.

The first three objectives were achieved by means of socio-economic field laboratories. "Field laboratories" is an umbrella term for a set of participatory and flexible methods that belong to the Participatory action research (PAR) and use some elements of "social learning processes" (Rist et al. 2006). Participants of field laboratories are usually from many different walks of life and include small farmers and their representatives, public authorities, members of development organizations, researchers, and academics. The aim of field laboratories is the joint production of knowledge based on a collectively constructed systemic view of farming and forestry systems and livelihood strategies in the different areas of intervention (Lindner et al. 2017). Participatory rural assessment (PRA) tools used as part of the field laboratories included four participatory rural workshops, key informant interviews, field observations, and secondary data sources. The field research was conducted between June 2011 and January 2012. The specifics of the application of these tools have been reported in Lindner et al. (2017) and Vidal Merino et al. (2018).

The data collection on potential adaptive capital indicators, as well as on agricultural losses was done by means of structured interviews at the FHS level. The number of interviews conducted for the case study was 137, which, taking into account a population size of 595 households, at a 95% confidence interval, provides a 7% error margin and a response distribution (prevalence) of 50%.

Adaptive Capital and Vulnerability Indicators

Poor FHSs depend on small-scale agriculture and animal husbandry as their main income-generating activities, having little adaptive capital to draw on and difficulty in adjusting their income sources to changing conditions (Sietz et al. 2012). By contrast, better economic conditions are reflected not only in higher financial capital but also in increased off-farm employment.

Based on the outcomes of the PRA and the household surveys, 18 indicators of adaptive capital were pre-selected, including gender, age and educational level, number of household members, access to credit, number of farm plots, trees, livestock and other assets, among others. After carrying out a principal component analysis (PCA) to exclude the indicators that did not contribute significantly to the variance, the number of indicators of adaptive capital was reduced to seven (Table 1).

Natural capital is expressed as total farm area and the agro-ecological zones in which these are located. The maximization of the vertical control over agro-ecological zones, also known as verticality, is a household- and community-level climate risk reduction strategy. The cultivation of multiple plots under different agro-ecological conditions may result in the cultivation of a greater diversity of crops according to the FHS's food requirements, market prices, expected weather conditions, and soil properties.

The irrigation ratio is an indicator for physical capital applied in the study area as a strategy against frost and droughts, thus having a higher capacity to adapt to climate fluctuations (Deressa et al. 2008). Additionally, irrigated farm plots hold a higher productive potential and are therefore used for cash crops.

Financial capital is determined by livestock and asset ownership. Added to the production of food, wool, and plowing force, livestock can be used as a liquid asset to cover short-term needs (Sietz et al. 2012; Jost 2016). FHSs may build up assets when possible, to be used in bad years to recover from shocks, as a self-insurance mechanism (Lindner et al. 2017).

Off-farm employment, an indicator for human capital, includes the different non-farm related activities along with the wage work. Off-farm employment is reported to be used as an ex-post strategy for crop losses as a way of earning extra income when hardship occurs (Wittmann 2013; Lindner et al. 2017). As it enhances adaptive

Table 1 Adaptive capital indicators for farm household systems (FHS)

Adaptive capital	Indicators	Explanation	Descriptive statistics	Measure
Natural	Total farm area	Total area (owned or rented) used by the household for agriculture	Range (111–34,323) Mean (9759.4) SD (9521.2)	Square meters
	Agro-ecological zones	Spatial distribution of the farm plots across agro-ecological zones	Range (1–3) Mean (1.5) SD (0.6)	Low (3000–3500); intermediate (3500–4000); high (>4000)
Physical	Irrigation ratio	Total irrigated land divided by total agricultural land	Range (0–1) Mean (0.6) SD (0.4)	
Financial	Animal husbandry	Animals owned by the household	Range (0–13) Mean (4.4) SD (3.5)	Tropical livestock units (TLU)
	Number of assets	Assets owned by the household: cell phone, radio, TV	Range (0–3) Mean (2.2) SD (0.8)	No assets = 0; one asset = 1; two assets = 2; three assets = 3
Human	Off-farm employment	Any income-generating activity outside own farm	Range (0–2) Mean (0.9) SD (0.9)	No member = 0; at least one member (agricultural labor) = 1; at least one member (non agric. labor) = 2
Social	Visits to the city	Refers to the adult that most frequently visits the nearest city	Range (0–3) Mean (2.6) SD (0.7)	<Once a month = 0; once a month = 1; twice a month = 2; weekly = 3

Source Vidal Merino et al. (2018)

capacity, FHSs specialized in non-agricultural off-farm employment are expected to be less vulnerable to climate-related hazards (Sietz et al. 2012).

As an indicator for social capital the frequency of visits to the nearest city, a proxy for accessibility and connectivity, was used. FHSs located in remote rural areas have more restricted access to services like health, education, and credit, and also limited market access and off-farm employment opportunities.

In this study, we selected "money lost" related to extreme weather events as a quantitative indicator for an important vulnerability outcome. Considering specific weather-related hazards, crop damage and economic losses during the past five years were estimated by farmers which were then transformed to total investment loss to compose this indicator. It is important to take into consideration that the magnitude of the damage is partially determined by the sensitivity and adaptive capacity of the FHS.

Archetype Analysis

The archetype analysis comprised: (i) correlation and variance analyses to explore the data structure and select input indicators of adaptive capitals for the cluster analysis; (ii) a cluster analysis to identify archetypical patterns of climate vulnerability revealing recurrent indicator combinations; and (iii) a vulnerability index to link results from the cluster analysis to a vulnerability outcome, thus enabling a ranking of archetypes of climate vulnerability.

Correlation and principal component analyses (PCA) are applied to reveal the strength of the relationships between indicators and to reduce the dimensionality of the data set prior to clustering. Data preparation for these analyses includes handling outliers and missing data as well as data normalization. After data preparation, we identified correlations within the data set. We ran a correlation matrix of all variables applying standard Pearson correlations to examine significant relations. For example, the indicator "total farm area" was positively correlated with "money lost" (0.3), which might explain a higher exposure to climate-related risks by larger-scale farmers. Thereafter, the PCA was applied to reveal the variance structure in the data set (Jolliffe 2002). To heighten the relevance of indicators in the archetype analysis, the indicators with low loadings were excluded. Finally, a second PCA was carried out with the remaining indicators, which explained 71% of the total variance.

For the cluster analysis a two-step cluster procedure, previously used to identify archetypes of farmers' vulnerability in the southern Peruvian Andes (Sietz et al. 2012), was used (as described by Janssen et al. 2012) to identify archetypes of FHS vulnerability to climate-related hazards. Despite potential variations on the relative importance of indicators, all indicators were given an equal weighting as a first approximation in the present analysis. To demonstrate a clear differentiation among households according to the available indicators, consistency measures were determined. Such differentiation is useful when interpreting FHS vulnerability. Outcomes from the consistency measure indicated a maximum for partitions with five clusters (used as base for our analysis).

The vulnerability outcome was then used to validate the results of the cluster analysis and assess the severity of vulnerability. We calculated the vulnerability index as a ratio of the sum of indicators per cluster and the total money lost due to extreme weather events per cluster (Eq. 1).

$$V_i = \frac{\sum_1^j (1 - X_{ij})}{(1 - Y_i)} \tag{1}$$

V_i: Vulnerability index of cluster i
X_{ij}: Value at cluster center for indicator j in cluster i
Y_i: Value at cluster center for indicator “money lost”.

Results and Discussion

Climate-Related Hazards in Agriculture

Within the myriad of external determinants of climate vulnerability in agriculture, the study looked specifically at climate-related hazards. Rather than relying on national or subnational statistics, results were derived from the joint reflections of local farmers, authorities, technicians, and researchers, following the methods already presented in the previous section. This bottom-up approach allows looking into the specifics of the agro-ecological system under study.

The results on the most frequent extreme weather events which reportedly cause damage in agriculture are presented in Fig. 3. Most frequent extreme weather events are frosts (47%), excess rainfall (11%), agricultural droughts (9%), and hail (9%). The impacts of these climatic hazards on the FHS were specified as dependent on the season of the year, the physical characteristics of the farm plot, the crop type and its stage of development, the agricultural technologies available, and the intensity and duration of the climatic hazard.

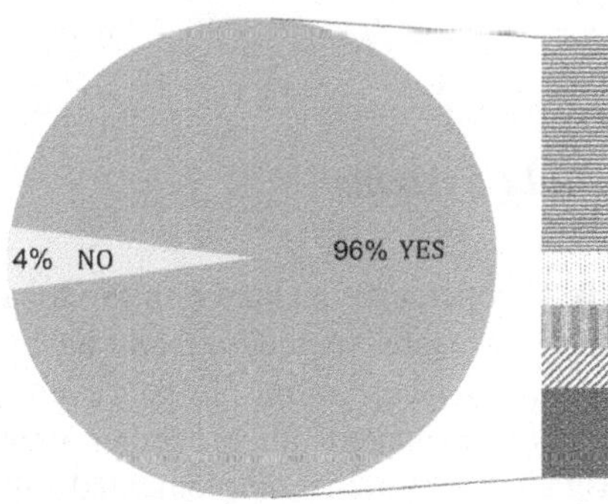

Extreme weather event	N° respon-dents	N° reported episodes	Average monetary loss (nuevos soles)	SD
Frost	103	139 (47%)	751	837
Excess rainfall	31	34 (11%)	828	961
Agric. Drought	24	27 (9%)	600	673
Hail	21	26 (9%)	601	790
Others	56	58 (20%)	534	547

Fig. 3 Agricultural damage caused by climate-related hazards. 96% of all respondents reported having agricultural losses caused by climatic hazards in the past 5 years. The table shows the type of hazard, the number of respondents having experienced an event, the number of reported events, average monetary loss per event and standard deviation of the monetary losses. *Source* Vidal Merino et al. (2018)

The highest average monetary losses per event were reported for excess rainfall (828 nuevos soles), followed by frost (751 nuevos soles), with great variability within the sample. Despite the latter having less average monetary loss per episode, the higher frequency of events makes frost responsible for overall higher economic losses.

Other types of hazards, mainly pests and diseases, were repeatedly reported as a consequence of, or closely related with, the occurrence of one or more climatic factors. For instance, potato plantations were reported to suffer widespread attack by fungi after the occurrence of consecutive episodes of excess rainfall followed by intensely hot days. The main pests and diseases reported were late blight (*Phytophthora infestans*), early blight (*Alternaria spp.*), Andean potato weevil (*Premnotrypes spp.*), and potato rust (*Puccinia pittieriana*).

In addition, the research aimed to investigate the farmers' perceptions of changes in local climatic conditions. It can be said unequivocally that farmers recognize a change in the climate. Although they acknowledge the fact that there is normal and predictable variability, they also perceive that in recent years this variability has become abnormal (i.e. there are more extreme events) and unpredictable (e.g. indicators are not working anymore).

In this sense, there is no perception of steady changes in climatic conditions but a strong perception of a future increase in climate extremes, i.e. changes in the probability distribution of climatic hazards: excess rainfall, frosts and agricultural droughts are likely to increase in frequency and intensity in the future. Indeed, trends and future climate scenarios for the study area show that the frequency of frosts and droughts has been increasing and that this trend is likely to continue in future decades (IGP 2005b, c). As stated by Tucker et al. (2010), perceptions are critical in shaping farmers' response capacities. In effect, extreme changes in climate can go beyond farmers' capacities (Smit et al. 2000) and motivate adaptive changes (Tucker et al. 2010).

Adaptation Strategies of Farm Household Systems

Results show that, even if farmers of the study area perceive a change in the local climate, they are not taking additional adaptation measures as they already have a broad set of techniques and strategies that allow them to face extreme climatic events. These include the use of different crop varieties that are planted either together in the same plot (intercropping) or separate; under irrigation or rainfed; complemented with trees (agroforestry); planted at the same time or having different sowing times. As a rule, crop rotations and fallow periods are practiced, varying in length according to the agro-ecological zone where agriculture takes place. Additional adaptation strategies practiced include the accumulation of savings in the form of farm animals and trees. Further, communal practices of reciprocity and exchange in agriculture are common in the study area, with 77% of the survey respondents practicing one or more of these. They include labor exchange in agriculture (62%), borrowing farm

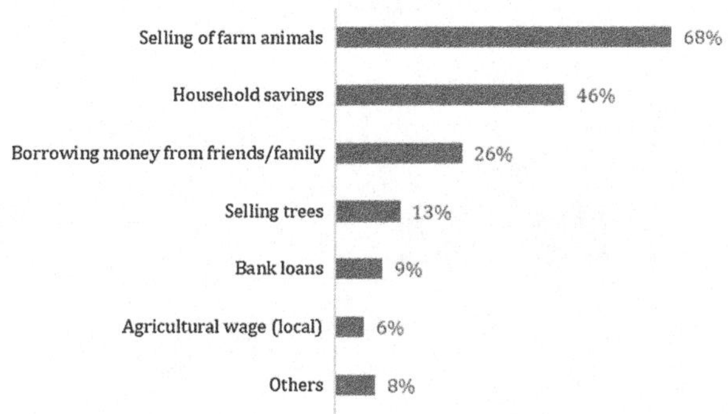

Fig. 4 Ex-post risk adaptation strategies reported to be practiced by FHSs in the study area. *Data Source* INCA field survey (2011)

areas from the community for a crop season (11%), and—as an alternative to renting land—"al partir," i.e. instead of money, the peasant who cultivates the farm plot gives the owner a portion of the final harvest (28%).

By contrast, communal ex-ante communal strategies to reduce or mitigate the effects of climatic hazards are known by the population but no longer used, or practiced only by a few farmers in an individual manner.

Ex-post risk adaptation strategies, i.e. actions taken to make up for losses after the occurrence of a climatic shock, are diverse (see Fig. 4). At the FHS level, the most common ones are selling off farm animals (68% of survey respondents), household savings (46%), borrowing money from family or friends (26%), selling trees (13%), taking out a bank loan (10%), and earning an agricultural wage (6%). As "other" ex-post strategies, temporary migration and remittances were mentioned.

Archetypes of Climate Vulnerability

The vulnerability of an FHS arises from the interaction of the analyzed external and internal determinants. We categorized FHSs based on these factors, making visible different archetypes of climate vulnerability. To achieve this, we first characterized external and internal determinants.

As shown in Fig. 5 (top-right), FHSs were categorized in five clusters according to the similarity in their adaptive capitals, i.e. internal determinants of vulnerability. The identified clusters were interpreted based on the indicator values at the cluster centers, which are representative of all FHSs categorized in a given cluster. At first glance, a clear division between clusters becomes apparent: Cluster 1 contains particularly resource-constrained FHSs; Clusters 2 and 3 represent predominantly agriculture-oriented systems (with values for off-farm employment of ≤ 0.1); Clusters 4 and 5 depict off-farm oriented systems (with values for off-farm employment of 1 and 0.8 respectively).

In the next step "money lost" related to extreme weather events was selected as a quantitative indicator for an important vulnerability outcome. This outcome indicator

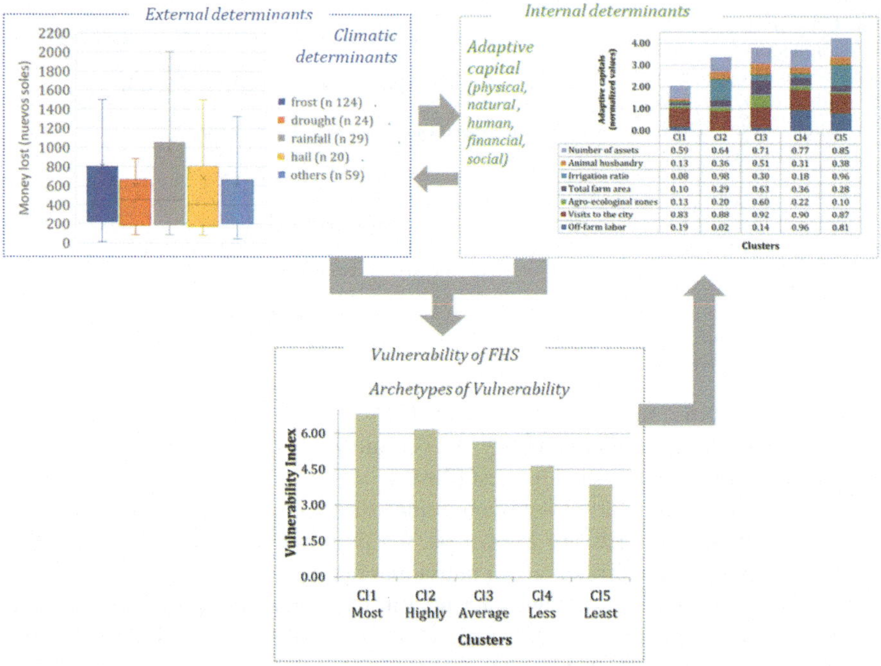

Fig. 5 Implementation of the vulnerability framework and resulting FHS archetypes. Top-left: Main climatic factors (external determinants) contributing to the vulnerability of FHS in agriculture and related outcome expressed as economic losses. Information on economic losses and crop damage were collected during the field survey. Farmers were asked to estimate their crop losses during the past five years, considering specific weather events. In particular, they were asked to recall those losses in terms of: (1) investment lost, (2) seed lost, and/or (3) estimated yield lost. As part of the data processing, all estimates were transformed to total investment loss using current local market prices and local estimates of crop productivity. The magnitude of the damage depends on the sensitivity and adaptive capacity of the household. Top-right: Profiles of FHS-internal determinants (adaptive capital) with indicator values specifying the importance of each form of capital averaged at cluster centers. High indicator values contribute to higher adaptation capacity. Bottom: Archetypes of climate vulnerability

depicts economic losses related to climatic hazards and their effects on crop yields. The money losses disaggregated per specific extreme-weather event are presented in Fig. 5 (top-left).

Last, the identified FHS typologies were linked to the outcome indicator "money loss" in a vulnerability index, revealing a clear distinction between the five archetypes of vulnerability. The "most vulnerable" cluster (Cluster 1) represents 19% of the population sample, and is formed by agriculture-oriented FHSs with particularly constrained resources; besides having limited access to off-farm employment, they also lack capital for on-farm activities, suggesting that FHSs belonging to this cluster are exposed to extreme poverty. This leaves them with little to no possibility of coping with climatic stressors, and they are therefore considered the most vulnerable group.

Here, strategies for managing and coping with risks are likely to be insufficient. The "highly vulnerable" cluster (Cluster 2), which represents 22% of the population sample, is also formed by agriculture-oriented FHSs with constrained financial, natural, and human capital; this cluster stands out because of its reduced agricultural area and restricted use of agro-ecological zones. Used commonly as a measure against seasonality issues and to cope with loss i.e. from climatic hazards, migration is a key source for income generation in highly vulnerable areas where options are more limited such as in the highland communities with little access and more detrimental weather conditions (Dixon et al. 2001).

The "averagely vulnerable" cluster (Cluster 3) is again agriculturally oriented, but agents in this cluster have greater adaptive capital, with larger agricultural areas spread across a higher variety of agro-ecological zones. The "less vulnerable" cluster (Cluster 4), is formed by off-farm oriented FHSs with high values for cumulative adaptive capital, mainly due to the indicators of off-farm employment, number of assets and frequent visits to the nearest city; yet agents have a rather limited availability of physical (irrigation) and natural (total farm area and agro-ecological zones) forms of capital. Off-farm activities reduce the covariance among the sources of vulnerability, and so diminish the avoidable excess of exposure in their FHS. In addition, farmers in this cluster can profit more from their advantage in accessibility. Finally, the "least vulnerable" cluster (Cluster 5) represents 26% of the population sample and is formed by off-farm oriented FHSs; complementing their off-farm activities, they have relatively small agricultural areas, but these are highly irrigated. The access to irrigation is externally restricted by the supply and distribution which are dependent on the local water users' rights. Taking advantage from the better access to irrigated land, a livelihood strategy identified for these farmers is the intensification of the FHS's production, to be able to accumulate resources and so create a buffer stock to use for coping with future hazards. Intensification in these areas can be achieved with improved breed, crops, and varieties increasing their productivity. However, as described by Valdivia et al. (2003), a setback of this strategy emerges whenever capital accumulation during the good years does not suffice to cover the losses of the bad years.

The main study limitations include: (i) the cross-sectional nature of the FHS data set, which does not reflect the dynamics of vulnerability directly but rather in a specific point in time and; (ii) the deliberate exclusion of communal strategies, e.g. traditional social-coping mechanisms such as trust, reciprocity, and collective action. By focusing only on the FHS, possible equally important strategies to reduce vulnerability and enhance resilience among community members may not be considered. Therefore, the interpretation of the results needs to take into account that the scope of the defined archetypes of vulnerability is the FHS level.

Conclusions

Andean farmers recognize a change in the climate, and while climate perception is in its essence subjective, they have clearly identified that this variability has increased and become unpredictable in recent years. Perceiving change in the climate is a crucial precursor of encouraging farmers to undergo adaptive changes.

In spite of the preceding, most farmers lack the motivation to consciously engage in taking additional preventive adaptation measures or ex-ante strategies. Even if the latter tend to be more satisfactory in preventing negative effects on the FHS, most farmers rely on ex-post responses.

The close correlation between the archetypes and the vulnerability index proves the suitability of the methodology used to inform decision making. The relative ease of use and applicability of the indicators throughout the archetype analysis facilitates their replicability and comparison with farming systems in other regions. Moreover, the use of archetypes allowed outcomes to be applied beyond individual households, with findings offering insights into the vulnerability of smallholder farmer groups. Determining the vulnerability at this level informs decision-makers on the characteristics of specific groups and helps them design and put into effect more suitable adaptation measures.

According to their adaptive capitals, FHSs were divided into five archetypes, ranging from "most vulnerable" (resource-constrained and agriculture-oriented) to "least vulnerable" (off-farm oriented) FHSs.

A number of strategies were revealed among the different archetypes. Scenarios including some of these strategies, such as increasing the proportion of off-farm labor activities, thus allowing FHSs to become less climate dependent, are expected to enhance the response capacity of farmers while reducing their sensitivity to climate change, thus making the system more resilient.

Finally, the loss of the Andean verticality principle, limiting the agro-ecological diversity within the FHSs, and the lack of consolidation and participation in organizations have reduced farmers' livelihood options and affected their response capacity (Jost 2017). Therefore, additional forms of adaptive capital such as reciprocity and collective action, which were not included in the present analysis, are worth exploring.

Acknowledgements This research was conducted as part of the International Network on Climate Change in the Andean Region (INCA project), implemented by the Institute of International Forestry and Forest Products of the Technische Universitaet Dresden (TUD) and funded by the Deutscher Akademischer Austauschdienst (DAAD) and by the Graduate Academy of the TUD.

Disclosure statement No potential conflict of interest was reported by the authors.

Funding This work was supported by the Deutscher Akademischer Austauschdienst (DAAD).

References

Adger WN (2006) Vulnerability. Glob Environ Change 16:268–281. https://doi.org/10.1016/j.gloenvcha.2006.02.006

Barnett TP, Adam JC, Lettenmaier DP (2005) Potential impacts of a warming climate on water availability in snow dominated regions. Nature 303–309. https://doi.org/10.1038/nature04141

Carney D (1998) Implementing the sustainable rural livelihoods approach. In: Carney D (ed) Sustainable rural livelihoods: what contribution can we make? London, pp 3–26

Chambers R, Conway G (1992) Sustainable rural livelihoods: practical concepts for the 21st century. Institute of Development Studies

Deressa T, Hassan RM, Ringler C (2008) Measuring Ethiopian farmers' vulnerability to climate change across regional states. Food Policy 806:32

Dixon J, Gulliver A, Gibbon D (2001) Farming systems and poverty. Improving farmers' livelihoods in a changing world. Food and Agriculture Organization of the United Nations, Rome and Washington, DC

Easterling WE, Aggarwal PK, Batima P et al (2007) Food, fibre and forest products. In: Climate change 2007: impacts, adaptation and vulnerability: Contribution of working group II to the fourth assessment report of the intergovernmental panel on climate change, pp 273–313

FAO, MINAGRI (2012) Plan De Gestión De Riesgo Y Adaptación Al Cambio Climático En El Sector Agrario, Período 2012–2021 (PLANGRACC-A). Lima

Fellmann T (2012) The assessment of climate change related vulnerability in the agricultural sector: reviewing conceptual frameworks. Seville

Heltberg R, Siegel PB, Jorgensen SL (2009) Addressing human vulnerability to climate change: toward a "no-regrets" approach. Glob Environ Change 19:89–99. https://doi.org/10.1016/j.gloenvcha.2008.11.003

IGP (2005a) Atlas Climático de precipitación y temperatura del aire de la Cuenca del Río Mantaro

IGP (2005b) Vulnerabilidad actual y futura ante el cambio climático y medidas de adaptación en la cuenca del río Mantaro

IGP (2005c) Diagnóstico de la Cuenca del Mantaro bajo la visión del cambio climático

INEI (2007) Censos Nacionals de XI Población y VI Vivienda. Sistema de consulta de resultado censales. http://censos.inei.gob.pe/cpv2007/tabulados/. Accessed 20 Apr 2001

IPCC (2014) Climate change 2014: impacts, adaptation, and vulnerability. In: Working group II contribution to the intergovernmental panel on climate change's fifth assessment report (WGII AR5), 31 Mar 2014

Janssen P, Walther C, Lüdeke M (2012) Cluster analysis to understand socio-ecological systems: a guideline. Potsdam

Jolliffe IT (2002) Principal component analysis, 2nd edn. Springer-Verlag New York Inc., New York, NY, United States

Jost FP (2016) Trees in the Andes: sustainable livelihood strategies for risk reduction. Technische Universität Dresden

Jost F (2017) Modeling agricultural and forestry production systems in the Andes: Identifying livelihood strategies for risk reduction through the optimization of resource allocation. ISBN 9783844031829. Shaker Verlag, Germany

Lindner A, Jost F, Vidal Merino M et al (2017) Aligning Socio economic Field Laboratories and Agent Based Models assessing local climate change adaptation measures of Andean farmers. J For Landsc Res 2:7–17. https://doi.org/10.13141/jflr.v2i1.77

McDonnell DJ, Dillon JL (1997) Farm management for Asia: a systems approach. Rome, Italy

MINAM (2015) Estrategia Nacional ante el Cambio Climático. 85

O'Brien K, Eriksen S, Nygaard LP, Schjolden ANE (2007) Why different interpretations of vulnerability matter in climate change discourses. Clim Policy 7:73–88. https://doi.org/10.1080/14693062.2007.9685639

PNUD (2013) Informe sobre Desarrollo Humano Perú 2013. Lima

Pramova E, Monica Di G, Locatelli B (2015) Integrating adaptation and mitigation in climate change and land-use policies in Peru. Bogor, Indonesia

Rist S, Chiddambaranathan M, Escobar C, Wiesmann U (2006) "It was hard to come to mutual understanding …"—the multidimensionality of social learning processes concerned with sustainable natural resource use in India, Africa and Latin America. Syst Pract Action Res 19:219–237. https://doi.org/10.1007/s11213-006-9014-8

Salzmann N, Huggel C, Calanca P et al (2009) Integrated assessment and adaptation to climate change impacts in the Peruvian Andes. Adv Geosci 22:35–39. https://doi.org/10.5194/adgeo-22-35-2009

Sietz D, Choque SEM, Lüdeke MKB (2012) Typical patterns of smallholder vulnerability to weather extremes with regard to food security in the Peruvian Altiplano. Reg Environ Change 12:489–505. https://doi.org/10.1007/s10113-011-0246-5

Sietz D, Ordoñez JC, Kok MTJ et al (2017) Nested archetypes of vulnerability in African drylands: where lies potential for sustainable agricultural intensification? Environ Res Lett 12:095006. https://doi.org/10.1088/1748-9326/aa768b

Smit B, Burton I, Klein RJT, Wandel J (2000) An anatomy of adaptation to climate change and variability. In: Societal adaptation to climate variability and change. Springer Netherlands, Dordrecht, pp 223–251

Tucker CM, Eakin H, Castellanos EJ (2010) Perceptions of risk and adaptation: coffee producers, market shocks, and extreme weather in Central America and Mexico. Glob Environ Change 20:23–32. https://doi.org/10.1016/j.gloenvcha.2009.07.006

Valdivia C, Gilles J (2001) Gender and resource management: households and groups, strategies and transitions. Agr Hum Val 18:5–9. https://doi.org/10.1023/a:1007608717996

Valdivia C, Gilles JL, Jette C et al (2003) Coping and adapting to climate variability: the role of assets, networks, knowledge and institutions. In: Programs NO of G (ed) Insights and tools for adaptation: learning from climate variability. Climate and Societal Interactions, Washington, DC, pp 189–199

Vergara W, Deeb AM, Valencia AM et al (2007) Economic impacts of rapid glacier retreat in the Andes. EOS Trans Am Geophys Union 88:261–264. https://doi.org/10.1029/2007EO250001

Vidal Merino M, Sietz D, Jost F, Berger U (2018) Archetypes of climate vulnerability: a mixed-method approach applied in the Peruvian Andes. Clim Dev 1–17. https://doi.org/10.1080/17565529.2018.1442804

Vilcapoma (2008) Plan de Desarrollo Local Concertado 2009–2015 del Distrito Heroínas Toledo

Vuille M, Francou B, Wagnon P et al (2008) Climate change and tropical Andean glaciers: past, present and future. Earth-Sci Rev 89:79–96. https://doi.org/10.1016/j.earscirev.2008.04.002

Wittmann N (2013) Non-farm and agricultural wage income in the Achamayo Region, Peru. Technische Universität Dresden

The Importance of Climate Change Education in Urban Planning: A Review of Planning Courses at UK Universities

Alice Preston-Jones

Abstract Climate change is affecting communities and environments across the world. If immediate, concerted effort is not made to reduce levels of greenhouse gas emissions, the effects could be catastrophic. Urban areas are the primary contributors of greenhouse gases, yet they are also likely to be most vulnerable to the effects of climate change. The need to adapt is extremely important. Urban planners have a unique responsibility in shaping the built environment and as such create solutions to the climate change challenge. A lack of integration of climate change knowledge and planning has been highlighted by academics and global policy makers, and the sufficiency of climate change-related teaching in planning education has been questioned. This paper evaluates the prevalence of climate change-related teaching in urban planning courses in universities across the UK. Topic by topic analysis is carried out to highlight key teaching gaps. It is argued that, due to planners' central responsibility in addressing the climate change challenge, it is paramount that climate change education is integrated into all courses. Discourses around interdisciplinary collaboration and globally-focused planning education are considered as potential solutions to the lack of climate change-related teaching. The results of this paper emphasise a need for immediate reform of University curricula to ensure future planning graduates have adequate knowledge to address future urban climate change challenges.

Introduction

Climate change is currently affecting communities and environments across the world. If immediate and concerted effort is not made to drastically reduce levels of greenhouse gas (GHG) emissions, the effects could be catastrophic (IPCC 2014a). As the primary contributors of GHG emissions, policy makers across the world are looking to urban areas to lead the way in progressive response to climate change (UN-Habitat 2017). However, urban areas are also increasingly vulnerable to the effects of climate change. Effective adaptation will be crucial, considering that the impacts of

A. Preston-Jones (✉)
Sheffield, UK

W. Leal Filho et al. (eds.), *Climate Change, Hazards and Adaptation Options*,
Climate Change Management, https://doi.org/10.1007/978-3-030-37425-9_54

climate change will still be felt even if global temperatures are stabilised to less than two degrees above pre-industrial levels (UN-Habitat 2011a). It is essential that all built environment professionals come together to create solutions to mitigate against, and adapt to, climate change (Susskind 2010; Walsh et al. 2011; Hebbert and Jankovic 2013; While and Whitehead 2013; IPCC 2014b).

Urban planners play a significant role in shaping the development of urban areas and will be at the forefront of ensuring cities adapt to withstand the impacts of climate change. It is vital that urban planners have sufficient knowledge to address issues relating to climate change across the world. The 2014 IPCC Report (2014b) emphasises the importance of planning and design in regard to building climate resilience and responding to climate risks. Whilst there has been a substantial amount of theoretical literature written on the integration of climate change and urban planning practice, there is concern that many planning professionals are not sufficiently prepared to respond to the challenges of climate change (Viguié and Hallegatte 2012; Hebbert and Mackillop 2013; IPCC 2014b). This insufficiency has been attributed to inadequate climate change education in urban planning curricula (Hamin and Marcucci 2013; Hess and Collins 2018).

The integration of climate change education into all disciplines and levels of education has become a recent priority, despite the fact that it was previously regarded as a subject for scientists (Leal Filho 2010). The importance of climate change education has been highlighted within several recent policy documents; Article 12 of the Paris Agreement states that *"Parties shall cooperate in taking measures, as appropriate, to enhance climate change education" (UN* 2015*b, p. 16)* and Goal 13 of the SDGs, Target 13.3 is to *"improve education, awareness-raising and human and institutional capacity on climate change mitigation, adaptation, impact reduction and early warning" (UN* 2015*a, p. 27)*. The UN has developed several programs to address climate change education in primary and secondary schools and for university education (Whitehouse 2017). Whilst it is important that climate change is integrated into all levels of education, it is argued that climate change-related teaching is even more important for those in higher education due to their imminent progress into sectors that will be responding to climate change challenges (Leal Filho 2010). However, there is concern that the current integration of climate change teaching into schools and universities is insufficient (Ibid.). A world survey of climate change at universities (of 1250 students from 166 universities) found that 43% of students thought that climate change was not as well covered as they would like, despite 60% regarding climate change as either essential or very important to their studies (Ibid.). A separate study of 100 American universities echoed these findings; that climate change is insufficiently integrated into the general education curriculum (Hess and Collins 2018). It was discovered that it is easy for a student to graduate with a bachelor's degree with no educational exposure to climate science or related climate change topics through core curriculum requirements. This has been attributed to the slow pace of change in education, and the fact that curriculum is often dictated by the priorities of the relevant professors and administrators (Ibid.).

Whilst it is clear that many policy makers and academics are aware of the importance of climate change education, there is obviously a gap between theory and implementation. It is important that such gaps are highlighted so that relevant curriculum changes can be made to ensure that future students will graduate with the ability to be able to address issues relating to climate change in practice. For several decades there has been global concern regarding the integration of climate change education in urban planning (Frommes 1980). In 1980, climate experts were concerned that architects and town planners were not adequately educated in meteorological factors, and consequently a publication of *Fundamental Knowledge in Urban and Building Climatology* was produced and distributed to Europe's architecture and planning schools (Hebbert and Jankovic 2013). Whilst there has been a growing focus on sustainable cities and climate change mitigation and adaptation since this time (Frank et al. 2014; Adams et al. 2015), many authors still express concern regarding planners' lack of climate science knowledge (Crane and Landis 2010; Norman 2017).

Whilst there have been several studies on climate change and education, and urban planning and climate change, there are relatively few that analyse the interface between climate change, education and urban planning. In a study by Wamsler et al. (2013), integrating urban planning climate adaptation into education was highlighted as a key solution for addressing the lack of climate adaptation implementation in planning practice. A 2009 report by UN-Habitat presented an unpublished world survey conducted by the Global Planning Education Association Network (GPEAN), in which they found 550 universities that offer urban planning courses and found that only "one third of planning schools teach climate change" (UN-Habitat 2009, p. 191). Whilst this shows an alarming deficiency in climate change education within planning schools, it is important to acknowledge that the study was conducted ten years ago and that there are limitations in the methodology.[1] In an Australian study conducted in 2009, 50 Australian planning practitioners were surveyed regarding their perception of planning graduates' knowledge (Hurlimann 2009). It was found that practitioners deemed planning graduates were least prepared in 'environmental issues'. 'Climate change implications of adaptation and mitigation' was perceived as the most inadequate area of environmental knowledge for planning graduates, and 'climate change: mitigation; adaptation; social impacts; sea level rise' was believed to be the environmental knowledge area that needed to be most imminently addressed in education. In a North American study conducted in 2013, the prevalence of climate change-related teaching was assessed in 120 planning schools accredited by the Association of Collegiate Schools of Planning (Hamin and Marcucci 2013). It was found that only 22 universities (18%) offered climate change-related modules. Through thematic analysis they found a varying coverage of climate change-related topics, and a particular deficiency was found in teaching students about the role of planners in implementing responses and adaptation measures to climate change.

The relative paucity of literature regarding the intersection of climate change, education and urban planning highlights a gap in research that needs to be addressed. Though different parts of the world may face varying scales of issues related to climate change, there is a universal responsibility to mitigate against climate change, put in

[1] See UN-Habitat (2009), footnote 65, p. 198 for further information on the limitations of the study.

place sufficient adaptation measures, and prepare future practitioners for the effects that will be felt globally (Susskind 2010). Transnational collaborations between universities has been proposed as a solution to sustainability problems such as climate change (Caniglia et al. 2017). Whilst there are identified challenges in implementing such a solution, Caniglia et al. (2017) argue that connecting universities across the world would enable leaning, sharing and adapting of the lessons learnt in different contexts. The need for adequate climate change education for built environment professionals has prompted the development of climate change curricula by global organisations such as the United Nations (UN). In 2015 UN-Habitat released a climate change curriculum aimed at students undertaking courses related to the built environment (UN-Habitat 2015). Whilst this illustrates that there is a global effort taking place to create sufficient climate change-related educational materials for built environment professionals, it is clear from two in-depth studies on planning education in Australia and North America that climate change education is yet to permeate planning curricula (Hurlimann 2009; Hamin and Marcucci 2013). Further in-depth studies into the prevalence of climate change-related teaching in planning courses in other countries is needed in order to determine whether the deficiencies noted in these studies are universal.

Research Design

Research Focus

Through analysis of curriculum documents, this study investigated the teaching of issues related to climate change in RTPI-accredited undergraduate or combined masters planning courses in the UK.[2] Two stages of analysis were utilised to answer the following research questions:

RQ.1 How prevalent is the teaching of issues related to climate change in RTPI-accredited planning courses in the UK?
RQ.2 What topics are present in climate change-related modules in RTPI-accredited planning courses in the UK?
RQ.3 Based on the prevalence of climate change-related teaching, and the climate change-related topics present, is the level of climate change-related teaching in planning courses in the UK sufficient to equip students with adequate climate change knowledge for future practice?

[2] 'RTPI-accredited undergraduate or combined masters planning courses in the UK' will henceforth be shortened to 'RTPI-accredited courses'.

Research Methods

Methodology, Data Sources and Collection Methods

The research design was based on qualitative research principles using an embedded case study methodology. The study was undertaken for 9 months from January 2018 to September 2018. Though the RTPI accredit several courses outside of the UK, the analysis focused on UK universities only to ensure institutional consistency and enable valid comparison. Whilst there are a range of RTPI-accredited master's degrees in the UK, it is important to assess the knowledge of climate change-related issues that all graduates will possess. Since not all students will pursue a master's course, this study focused solely on undergraduate or combined master's courses. A desk-based study found that, at the time of writing, there were 28 undergraduate or combined master's courses accredited by the RTPI at 17 universities in the United Kingdom (see Table 1).

It was found that thirteen universities provided easily accessible module information online, whilst four had insufficient data. To obtain the required information, email correspondence was carried out with these four universities. Two replied with attachments containing the required information. However, Leeds Beckett (2 courses) and University of Dundee (1 course) did not respond and had insufficient information online to be included in the analysis. Therefore, of the 17 universities and 28 courses in the UK accredited by the RTPI, 15 (88%) universities and 25 (89%) courses formed the basis of data collection and analysis.

Data Analysis

The present study used a qualitative approach to analyse these data in two stages. Utilising a method implemented by Hamin and Marcucci (2013) and Hess and Collins (2018), the first stage undertook a key-word search to identify the prevalence of climate change-related content. The key-words chosen were 'climate change', 'climate science', 'global warming' and 'greenhouse'. If any of these key-words were present in either the module title or module/syllabus content the course was deemed a climate change-related module. Optional modules were discounted from further analysis as there is a possibility that students may not choose to study climate change-related optional modules, and this study was concerned with the level of climate change knowledge that <u>all</u> planning students will graduate with. Through assessing weekly content, core modules where climate change-related teaching was only a minor part of the course (due lack of week-by-week climate change-related teaching) were discounted from further analysis, as they were deemed to have insufficient climate change-related teaching for content analysis. For the remaining modules, the proportion of climate change-related content for each course was analysed to determine the amount of time courses dedicated to climate change-related teaching. This was

Table 1 List of RTPI accredited undergraduate or combined master's courses

University	Course title	Level
Birmingham City	Property development and planning	MPlan
Cardiff	Urban planning and development	BSc
	Human geography and planning	BSc
Heriot Watt	Urban planning and property development	BSc
Leeds Beckett	Human geography and planning	BA
	Master of planning	MPlan
London South Bank	Urban and environment planning with diploma in town planning	BA
Newcastle	Urban planning	BA
	Urban planning and green infrastructure and landscape	MPlan
	Urban planning and urban conversation	MPlan
	Urban planning and regeneration	MPlan
Oxford Brookes	Urban design, planning and development	BA
	Planning and property development	BA
Queens University Belfast	Planning, environment and development plus diploma	BSc
	European planning	MPlan
Ulster University	Integrated master's in planning, regeneration and development	MSci
University College London	Urban planning, design and management	BSc
	Planning and real estate	BSc
University of Dundee	Town and regional planning	MA
University of Liverpool	Master's in planning	MPlan
University of Manchester	Master of planning	MPlan
	Planning and real estate	BSc
	Master of planning and real estate	MPRE
University of Reading	Real estate with diploma in urban planning and development	MSc
University of Sheffield	Urban studies and planning	MPlan
University of the West of England	Urban planning	BSc
	Architecture and planning	BA
University of Westminster	Designing cities: planning and architecture	BA

calculated using the credit rating of the climate change-related module in comparison to the total number of credits attributed to all core modules in each course.

The second stage utilised thematic analysis to investigate topic-by-topic coverage of climate change teaching. A coding frame was established to examine the data and consisted of two levels of categories. The main overarching category was 'Climate

Change'. The sub-categories were formed using concept-driven data; where categories are based on previous knowledge rather than data-driven (Schreier 2012). To form these sub-categories, a 'Baseline Climate Change Knowledge' was established using the seven climate change modules produced by UN-Habitat in 2015. Descriptions and indicators for each sub category were established to ensure consistency in subsequent data analysis. Using a method informed by Krizek and Levinson (2005), the module content of each course, using all information available; module descriptions, syllabus content, learning outcomes, etc., were analysed using the established coding framework, in order to determine which sub-topics were covered. The results of the coding highlighted the prevalence of each climate change-related topic in the 25 UK planning courses. This was followed by an in-depth critique of each topic in relation to its coverage. The two-stage approach to data analysis ensured a rigorous investigation of the state of climate change-education in UK planning courses, and enabled identification of knowledge gaps and implications for future research.

Limitations

The main limitations of this research were in relation to the data analysed. As highlighted in similar research by Pezzoli and Howe (2016) and Krizek and Levinson (2005), the varying level of detail provided to describe module content is a key challenge in analysing curriculum data. Hess and Collins (2018) highlight a limitation in using the key-word search approach; some modules may cover climate change without explicitly stating so in the module title or description and, as such, will not be captured in the research. However, they argue that their approach still provides a good indication of climate change-related teaching in the courses which deem it significant enough to depict in the module descriptions.

Whilst every attempt was made to secure as much information as possible on the module content for the thematic analysis, the length of module outlines ranged from 100 to 3000 words. Further, there was some level of inconsistency in the data sourced. For example, some module outlines offered module outlines, the learning objectives, reading lists and weekly timetables, whilst some simply offered a short description of the module. Whilst the flexibility of qualitative content analysis meant that, despite the varying amounts of data, themes could still be drawn for all of the 15 modules, the lack of data for some modules represented a substantial limitation. Nonetheless, due to the importance of climate change teaching in planning education outlined in the literature review, arguably the teaching of climate change-related topics should be clear even in short descriptions of module content. There are also limitations in using 'concept-driven' categories for the thematic analysis. Using pre-defined categories may leave some material unaccounted for (Schreier 2014). For example, there may be climate change-related topics covered in RTPI accredited courses that were not identified because they did not match the BCCK topics. Nonetheless, the comprehensive nature of the modules created by UN-Habitat provided a strong base for comparative analysis.

Results and Discussion

Stage 1: Assessing the Prevalence of Climate Change Education in Urban Planning Courses

Key-Word Search

The results of the key-word search yielded 25 climate change-related core modules across 15 courses and 33 climate change-related optional modules across 12 courses. Of all 25 RTPI-accredited planning courses, 8% (n = 2) did not provide any climate change-related modules, 32% (n = 8) only offered optional climate change-related modules and 60% (n = 15) offered climate change-related teaching in core modules (and in some instances in further optional modules). A breakdown of the results of this word search, by university, is shown in Fig. 1.

The mean number of climate change-related core and optional modules offered per course was 1.7 and 2.1 respectively. Courses varied from containing no optional or core climate change-related modules to up to 4 core climate change-related modules. Some courses offered a strong mix of core and optional modules. It is clear from this analysis that the number of climate change-related modules differed widely between UK universities. In view of the importance that planning graduates are educated in issues relating to climate change, it was anticipated that all RTPI accredited courses would have climate change-related teaching embedded into the core module teaching. This was found not to be the case. However, the prevalence of climate change-related teaching in UK planning courses (60%) was shown to be higher than in the 2013 North American study (18%) and the 2009 UN-Habitat study (33%). It should be noted that the North American and UN-Habitat studies were conducted 6 and 10 years ago respectively.

For students interested in climate change-related teaching, it is encouraging that 92% of RTPI-accredited planning courses offer at least one climate change-related core or optional module. However, it is concerning that 32% of courses only offer climate change-related teaching as an optional module. It is even more concerning that two universities do not offer any climate change-related modules; optional or core. A potential explanation for this could be differing research interests of the academics teaching at different institutions, or the availability of qualified academics for teaching climate change-related modules (Hess and Collins 2018). Further investigation is needed to identify why climate change teaching is not included within core curricula of 40% of RTPI-accredited courses and to explore how to ensure better integration.

Proportion of Climate Change-Related Teaching

The proportion of climate change-related teaching within each RTPI-accredited course was established through evaluation of module credit ratings. Within each

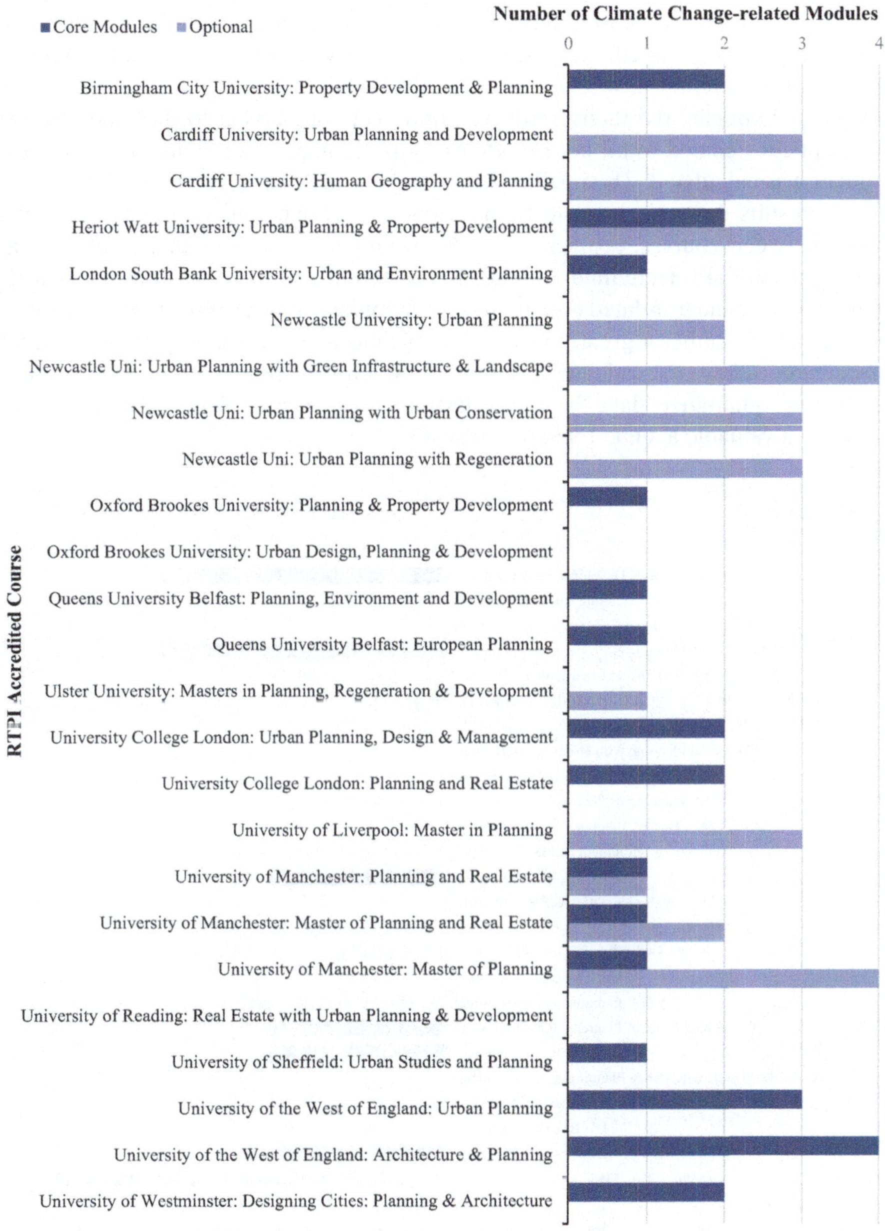

Fig. 1 Number of climate change-related modules offered on RTPI courses

module the credit weighting, and consequently duration of study hours, differed; eight modules were worth 15 credits, six modules were worth 20 credits, and one was worth 10 credits. Using the number of climate change-related core modules offered per course, and their credit weighting (in comparison to the total number of credits designated to all core modules), a percentage of climate change-related content was calculated. The results of this analysis are displayed in Fig. 2.

The results demonstrate that the proportion of climate change-related teaching varies between courses. A mean of 3.7% of climate change-related teaching was found for all RTPI-accredited courses. It was found that there was not necessarily more climate change-related content in 4-year courses. Further research is necessary to determine whether a greater proportion of time spent on climate change-related teaching correlates to graduates obtaining more climate change-related knowledge. Collecting qualitative data through questionnaires once students have graduated could be a valuable avenue for future research.

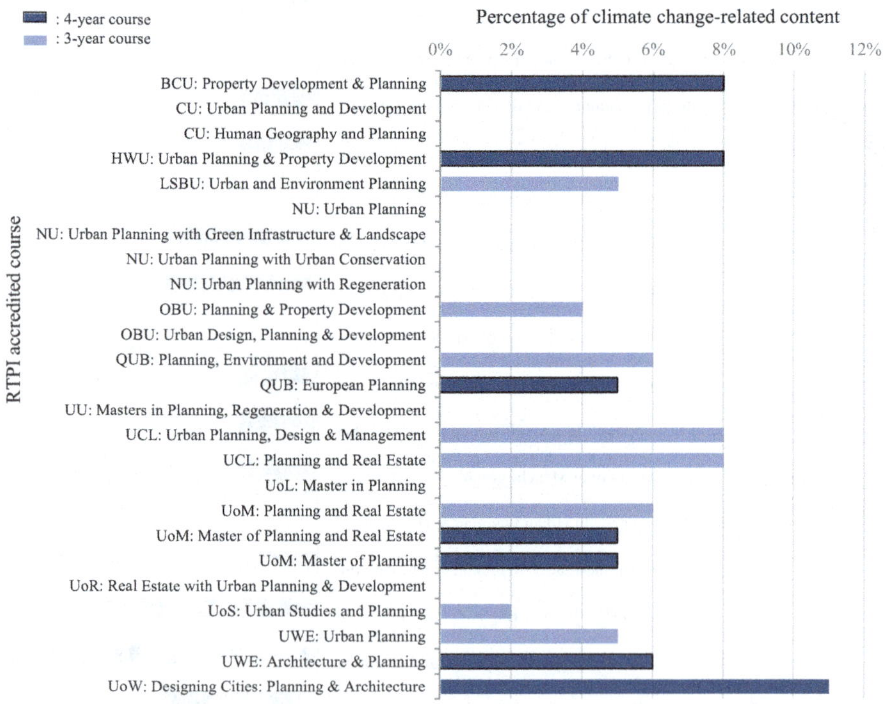

Fig. 2 Proportion of climate change-related teaching in each RTPI accredited course (university names have been abbreviated—see Fig. 1 for the full titles)

Stage 2: Assessing the Content of Climate Change Education in Urban Planning

Establishing a Baseline Climate Change Knowledge

UN-Habitat's seven climate change modules for students studying courses related to the built environment (UN-Habitat 2015) were used to establish a baseline climate change knowledge. UN-Habitat (2015) state that the primary objective of the production of the curricula is to "provide a comprehensive body of knowledge on climate change in the context of cities and to provide universities…with ready-to-use materials and case studies to be utilised in their…educational programs". The global collaboration to produce these modules is demonstrated in the range of academics that contributed; from Australia, the USA, Russia, South Africa, Mongolia, Algeria, Mexico, Germany and the UK. For each module UN-Habitat provides a course syllabus, lecture slides, lecture notes, reading lists and, where applicable, further material such as information on case studies. These modules, intended by the UN to address the lack of a clear baseline for global climate change education, provide a comprehensive outline of the climate change-related topics deemed to be of key importance for urban environment professionals (see Table 2 for a breakdown of the topics covered in each module). The modules will henceforth be referred to as 'Baseline Climate Change Knowledge' (BCCK) and used to determine the coverage of climate change-related topics in RTPI-accredited UK planning courses.

In order to make comparisons regarding climate change-related teaching in UK planning schools, it is important to consider the amount of study hours UN-Habitat dedicate to the modules (see Table 3). In UK universities, the credit weighting of a module relates to the number of study hours; 1 credit surmounts to approximately 10 study hours (teaching and independent study), as such a 15-credit module warrants 150 hours of studying. The combined hours of the UN-Habitat modules total approximately 36 credits, which is approximately two and a half 15-credit UK modules. Therefore, it is important to note that all BCCK topics cannot be feasibly covered within one UK module.

Topic-by-Topic Analysis

Using thematic analysis, teaching content within core climate change-related modules on RTPI-accredited courses was evaluated against the BCCK. A graphical representation of topic coverage in comparison to all RTPI-accredited courses is presented in Fig. 3, while a more detailed representation of the results is depicted in Table 4. It is clear that the coverage of BCCK topics in RTPI-accredited courses varies; from only seven courses teaching Climate Change Shelter and Housing, to 14 courses teaching Planning for Climate Change. To assess in detail the coverage of the BCCK topics across RTPI-accredited courses, the following section discusses each of the seven topics individually. The content of these topics is briefly presented, followed

Table 2 An overview of UN-Habitat climate change modules (UN-Habitat 2015)

Module	Topics covered
Theory and concepts of climate change	*Understanding climate and climate change—the science* *Climate change mitigation* Greenhouse gas emissions and emitters Cities and climate change mitigation Acting on mitigation *Climate change adaptation* Climate change impacts Adaptation planning
The practice of urban climate change adaptation and mitigation	*How cities are affected by climate change* *How cities contribute to climate change*—assessing greenhouse gas (GHG) contributions *Introduction to climate change risk and vulnerability assessments* *Climate change mitigation and adaptation*—rationale, objective and measures *Climate change mitigation and adaptation for sustainable development*
Planning for climate change	*Planning frameworks* *Approaches to climate change planning* Ad hoc versus strategic (stand-alone) plans versus mainstreaming *Climate change planning process* General process approaches Public participation Scenarios *Policy decision rules*
Climate change and urban water cycle management	*Urban water resources*—the global and urban water cycles *Urban water infrastructure* *Impacts of climate change on water resources and infrastructure* *Climate change impacts on urban water infrastructure and urban sanitation*
Climate change and urban energy	*The link between climate change, cities and energy* *Energy and cities*: some fundamental concepts *Energy*: related climate action *Energy policy and planning for climate change*
Climate change and urban mobility	*Effects of climate change on mobility and effects of mobility on climate change* *Adaptation response strategies* *Mitigation*: technology and logistical responses *Comprehensive response strategies* (adaptation and mitigation combined)

(continued)

Table 2 (continued)

Module	Topics covered
Climate change and shelter and housing	*Planning* *Housing design* *Renewable energy substitutes* *Carbon offset through photosynthesis and carbon sinks* *Behaviour change*

Table 3 Study hours (h) given for each UN-Habitat module (UN-Habitat 2015)

Modules	Class (h)	Student-led (h)	Individual study (h)	Total (h)
Theory and concepts of climate change	8	24	8	40
Climate change adaptation and mitigation	8	24	8	40
Planning for climate change	15	12	40	67
Climate change and urban water	6	46	15	67
Climate change and urban energy	16	12	40	67
Climate change and urban mobility	4	n/a	n/a	Approx. 40
Climate change and shelter and housing	4	n/a	n/a	Approx. 40
Total				Approx. 360

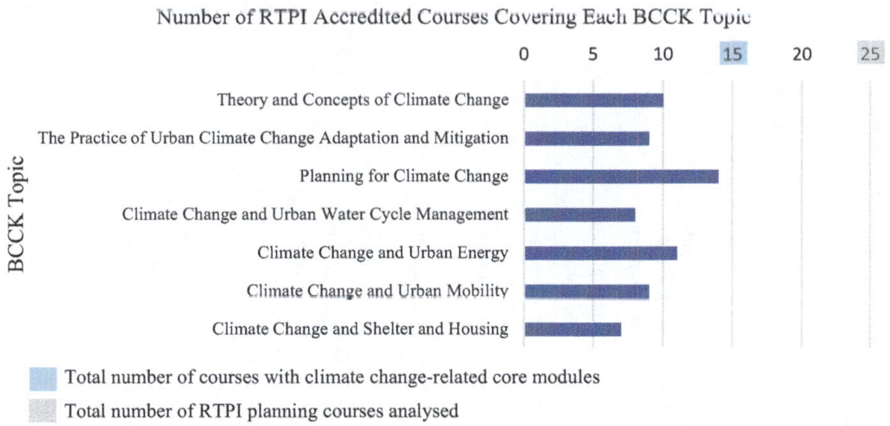

Fig. 3 BCCK topic coverage in RTPI accredited courses

Table 4 BCCK topic coverage in RTPI accredited courses

Degree course	Module	Baseline climate change knowledge (BCCK)						
		CC theory and concepts	CC adaptation and mitigation	Planning for CC (frameworks and policy)	CC and urban water cycle	CC and urban energy	CC and urban mobility	CC and shelter/housing
Property development and planning MPlan *Birmingham City University*	Introduction to the built environment			I				
	Smart policies and plans			I		I	I	I
Urban planning and property development BSc *Heriot Watt University*	Living planet and environmental change	I						
	Introduction to the environment	I	I	I	I	I		I
Urban and environmental planning BA *London South Bank University*	Environmental management and CC	I	I	I	I	I	I	
Planning and property development BA *Oxford Brookes University*	Introduction to spatial planning		I	I		I		

(continued)

Table 4 (continued)

Degree course	Module	Baseline climate change knowledge (BCCK)						
		CC theory and concepts	CC adaptation and mitigation	Planning for CC (frameworks and policy)	CC and urban water cycle	CC and urban energy	CC and urban mobility	CC and shelter/housing
Planning, environment and development BSc/European planning MPlan *Queens University Belfast*	Planning and CC	I	I	I		I	I	I
Urban planning design and management BSc/Planning and real estate BSc *University College London*	Green futures	I	I	I		I	I	I
	Urban and environmental politics		I		I	I		
Planning and real estate BSc/MPRE/Master of planning MPlan *University of Manchester*	Introduction to urban and environmental economics			I				

(continued)

Table 4 (continued)

Degree course	Module	Baseline climate change knowledge (BCCK)						
		CC theory and concepts	CC adaptation and mitigation	Planning for CC (frameworks and policy)	CC and urban water cycle	CC and urban energy	CC and urban mobility	CC and shelter/housing
Urban studies and planning MPlan *University of Sheffield*	The environmental challenge	I	I	I	I		I	
Urban planning BSc/Architecture and planning BA *University of the West of England*	Healthy sustainable communities	I	I		I	I		
Architecture and planning BA *University of the West of England*	Future places		I	I			I	
Planning and architecture BA *University of Westminster*	Contemporary issues in designing cities	I				I	I	
	Sustainability and environment			I	I	I		I
Topic total (of 15 courses)		10	9	14	8	11	9	7

Note Climate change is abbreviated to CC

by analysis of the number of courses that cover each topic and discussion of the implications of these results in relation to the literature.

Theory and Concepts of Climate Change

The teaching of *Theory and Concepts of Climate Change* is described by UN-Habitat as integral for urban planners (UN-Habitat 2015). The topic is split into four parts; a basic overview of the science of climate change, theory of climate change mitigation, theory of climate change adaptation and three global case studies. Whilst the topic does not cover climate science extensively, it offers a basic theoretical outline which is viewed as fundamental by many (Crane and Landis 2010; Norman 2017). This topic was covered in 40% of RTPI-accredited courses. In a study conducted in North American planning schools, teaching of the basic science of climate change was present in 50% of modules that had a primary focus on climate change-related teaching (Hamin and Marcucci 2013). This highlights that this topic was more effectively integrated into climate change-related modules in planning schools in North America in 2013 in comparison to the UK in 2018. A deficiency in the teaching of climate science in planning education was identified almost 40 years ago (Frommes 1980). It is clear from the result of this paper that this deficiency is still present within RTPI-accredited planning courses in the UK. An immediate solution could be to further the debate on the interdisciplinarity nature of planning education, so that available climate change modules delivered by other subjects, such as geography or the natural sciences, can be used to enhance knowledge (Crane and Landis 2010; Davison et al. 2013; Norman 2017).

The Practice of Urban Climate Change Adaptation and Mitigation

This topic further explores the concepts of climate change adaptation and mitigation and looks at practical implications and solutions. The topic is divided into roughly five parts; how cities are affected by climate change, how they contribute to climate change, the risk and vulnerabilities associated with climate change, the rationale and measurement of adaptation and mitigation, and how adaptation and mitigation relate to sustainable development. Climate Change Adaptation and Mitigation was covered in 36% of RTPI-accredited courses. This same insufficiency was highlighted by Hurlimann (2009) in a study on planning education in Australia. Thus, deficiencies in this topic of climate change-related teaching is not isolated to the UK. Walsh et al. (2011) believe that adaptation to, and mitigation of, climate change is one of the biggest challenges facing urban planners today. A study by Wamsler et al. (2013) argues for educational mainstreaming of urban planning climate change adaptation as a solution to the weak integration of adaptation planning practice. However, despite a five-year advance since the study, over two thirds of UK planning courses are yet to achieve this. The importance of this topic, stressed in both academia and policy, highlights the urgency at integrating this topic into all planning courses (OECD 2010; Hebbert and Jankovic 2013; IPCC 2014b; UN 2015).

Planning for Climate Change

This topic is divided into four parts; planning frameworks, approaches to climate change planning, climate change planning process and policy. *Planning for Climate Change* had the highest coverage of any BCCK topic and was found in 56% of RTPI-accredited courses. This is encouraging as the literature highlights the importance of spatial planning in regard to building climate resilience and responding to climate risks (IPCC 2014b; Walsh et al. 2011). Indeed, *Planning for Climate Change* is perhaps one of the most important topics to be covered in planning courses as it has been argued that some key principles of traditional planning practices are questionable in the context of climate change (Wamsler et al. 2013). However, in Hamin and Marcucci's (2013) study of North American planning schools they found that the *discussion of the role of planners in* climate change was lacking in modules that had a primary focus on climate change-related teaching. Whilst this topic had the highest coverage, the fact that there are still 44% courses in the UK without it, and that similar findings were found in North America in 2013, effort must be made to ensure its full integration to planning courses globally. Indeed, the need to bridge the gap between education and practice has been highlighted to ensure that students are gaining the most relevant knowledge and that practice benefits from the broad coverage and complexity that research and education can provide (Hamin and Marcucci 2013).

Climate Change and Urban Water Cycle Management

This topic is divided into four parts; an overview of the global and urban water cycles, urban water infrastructure, impacts of climate change on water resources and infrastructure and climate change impacts on water infrastructure and sanitation. Water Management was covered in 32% of all RTPI-accredited courses. In the study of North American planning schools, water management was only covered in 20% of modules that had a primary focus on climate change-related teaching. These findings illustrate a deficiency in the teaching of climate change and water in both the UK and North America. The low coverage reflects the findings of the literature review; that water management is a topic that is often overlooked in practice (OECD 2010). Water management is critical in terms of adapting to climate change, particularly in areas where droughts are increasing in frequency and intensity. It is possible that water management is overlooked in UK education, and other countries in the global North such as the US, because problems relating to Water Management are more common in countries of the Global South (UN-Habitat 2011b). Nonetheless, as the literature highlights the global nature of climate change responses, it is critical that this topic is incorporated into planning education globally. A possible solution could be to look at transnational collaborations between universities to allow for comparison, transfer and adaptation of teaching across different geographical and cultural contexts (Wiek et al. 2013; Withycombe Keeler et al. 2016; Caniglia et al. 2017; Ratnayake and Butt 2018). Indeed, Caniglia et al. (2017) argue that a vital element to addressing global sustainability problems, including climate change, is connecting universities from all over the world.

Climate Change and Urban Energy

This topic is divided into four parts; the link between climate change, cities and energy, the fundamental concepts of energy and cities, related climate change action and energy policy and planning for climate change. Reducing energy use and finding alternatives is highlighted as one of the key solutions to climate change (Crane and Landis 2010; OECD 2010; UN-Habitat 2011b; Emmanuel and Baker 2012). This topic was covered in 44% of RTPI-accredited courses. In a study of North American planning schools, energy was only covered in 30% of modules that had a primary focus on climate change-related teaching (Hamin and Marcucci 2013). Whilst *Climate Change and Urban Energy* is the second most well covered topic, it is problematic that over half of planning students from RTPI-accredited undergraduate or combined master's courses may graduate without knowledge of this topic. In view of the fact that spatial planning is considered the key solution to managing urban energy (Emmanuel and Baker 2012), it is vital that this topic is better integrated into UK planning curricula.

Climate Change and Urban Mobility

This topic is divided into four parts; effects of climate change on mobility and the effects of mobility on climate change, adaptation response strategies, mitigation technology and logistical responses and combined response strategies. Urban Mobility was found in 36% of RTPI-accredited courses. In Hamin and Marcucci's (2013) study, transport was covered in 40% of modules that had a primary focus on climate change-related teaching. This highlights that there is similar integration of this topic in both the UK and North America. It is clear from the literature that an important aspect to successful mitigation against climate change is effective mobility and transport planning (Booth 2012; Emmanuel and Baker 2012; Droege et al. 2010; Girardet 2008). It is important to note that topics surrounding mobility and transport may be embedded in other modules not identified through the key-word search on climate change.

Climate Change and Shelter and Housing

This topic is the only one of the seven UN-Habitat modules that places an emphasis on design. This BCCK topic was most poorly covered; present in only 28% of RTPI-accredited courses. Hamin and Marcucci's (2013) study of North American planning schools did not present any findings for the topics *shelter* or *housing* through the thematic analysis undertaken for modules which contained climate change-related teaching as a primary focus. This highlights a severe deficiency in this topic both in the UK and North America. The lack of design curricula in planning courses is an ongoing debate (Manley and Parnaby 2000; Higgins and Forsyth 2006). This is attributed to the fact that many planning schools have adopted a more social-science methods approach to teaching, in contrast to the technical, architecture studio-type approach that was taught in planning schools in the early 20th century (Frank 2006). Whilst the role of the planner is not necessarily to be a designer, the literature highlights the importance of urban planners' knowledge of design, as they often work closely with architects and engineers (Manley and Parnaby 2000). It is clear from this UN-Habitat module that an understanding of design in relation to climate change is important. Adaptation of shelter and housing will be critical in places increasingly affected by climate change to protect from floods, droughts and extreme weather events. Stronger integration of design elements into planning curricular would increase planners' capacity to create effective adaptation measures.

Conclusion

Urban planners play a crucial role in shaping and adapting the built environment in response to climate change issues. Therefore, it is vital that they have the knowledge and tools to do this successfully. The importance of climate change-related teaching in urban planning education has been highlighted by many authors. Specifically, integrating urban planning climate adaptation into education has been highlighted as critical to ensure planning practice implements effective adaptation measures. There is concern that climate change-related teaching has not been sufficiently integrated into urban planning courses. Despite this, there are few studies addressing the topic. As such, the present study looked to address the prevalence of climate-related teaching in planning schools in the UK context.

The first stage of analysis determined the prevalence of climate change-related teaching in RTPI-accredited UK planning courses. It was found that 40% of courses did not contain any climate change-related teaching as part of the core curriculum. The second stage of analysis looked at the content covered in the 15 courses that contained climate change-related modules. A baseline climate change knowledge consisting of seven topics was defined and used as a reference to analyse the content within each course. It was found that the coverage of BCCK topics in all RTPI-accredited UK planning courses was low; ranging from topic coverage of 28–56%.

Deficiencies in the teaching of these crucial topics reflects the findings of recent literature; that many aspects of climate change-related knowledge are yet to sufficiently permeate urban planning education.

Further research is required in order to understand why these deficiencies exist and explore optimal methods for capacity building. Educational reforms often take time to implement, and as such immediate solutions must be proposed. Three possible solutions have been identified; encouraging interdisciplinary learning across disciplines, transnational collaborations across universities, and strengthening the links between practice and education. This study calls for all those with the power to shape planning curriculums in the UK, and globally, to begin to normalise climate change-related teaching in planning education and consequently increase climate change knowledge in planning practice. With climate change impacts already effecting the lives of many urban dwellers, the planning world has a responsibility to ensure that practitioners have relevant climate change knowledge to create positive and enduring solutions.

References

Adams D, Larkham P, Pain K (2015) Viewpoint re-evaluating the place of urban planning history. Town Plan Rev 86(4):373–380

Booth CA (2012) Solutions to climate change challenges of the built environment. Wiley-Blackwell, Chichester

Caniglia G, Luederitz C, Groß M, Muhr M, John B, Keeler L, von Wehrden H, Laubichler M, Wiek A, Lang D (2017) Transnational collaboration for sustainability in higher education: lessons from a systematic review. J Clean Prod 168:764–779

Crane R, Landis J (2010) Introduction to the special issue: planning for climate change: assessing progress and challenges. J Am Plan Assoc 76(4):389–401

Davison A, Brown P, Pharo E, Warr K, McGregor H, Terkes S, Boyd D, Abuodha P (2013) Distributed leadership: building capacity for interdisciplinary climate change teaching at four universities. Int J Sustain High Educ 15(1):98–110. https://doi.org/10.1108/IJSHE-10-2012-0091

Droege P, Rosenwax J, Asselin S, AECOM (eds) (2010) Climate design: design and planning for the age of climate change [a collection of works from academics and AECOM's thought leaders]. ORO Ed, Pt. Reyes Station, CA

Emmanuel MR, Baker K (2012) Carbon management in the built environment. Routledge, London, New York

Frank AI (2006) Three decades of thought on planning education. J Plan Lit 21(1):15–67

Frank AI, Mironowicz I, Lourenço J, Franchini T, Ache P, Finka M, Scholl B, Grams A (2014) Educating planners in Europe: a review of 21st century study programmes. Prog Plan 91:30–94

Frommes B (1980) Fundamental knowledge in urban and building climatology. International Federation of Housing and Planning, The Hague

Girardet H (2008) Cities people planet: urban development and climate change. Wiley, Chichester, New York

Hamin E, Marcucci D (2013) Mainstreaming climate in the classroom: teaching climate change planning. Plan Pract Res 28(4):470–488

Hebbert M, Jankovic V (2013) Cities and climate change: the precedents and why they matter. Urban Stud 50(7):1332–1347

Hebbert M, Mackillop F (2013) Urban climatology applied to urban planning: a postwar knowledge circulation failure: urban climatology and knowledge circulation failure. Int J Urban Reg Res 37(5):1542–1558

Hess DJ, Collins BM (2018) Climate change and higher education: assessing factors that affect curriculum requirements. J Clean Prod 170:1451–1458

Higgins M, Forsyth L (2006) Raising the standard: design skills for practising planners. Town Plan Rev 77(5):605–624

Hurlimann AC (2009) Responding to environmental challenges: an initial assessment of higher education curricula needs by Australian planning professionals. Environ Educ Res 15(6):643–659

IPCC (2014a) Climate change 2014: impacts, adaptation, and vulnerability: working group II contribution to the fifth assessment report of the intergovernmental panel on climate change. Cambridge University Press, New York, NY

IPCC (2014b) Chapter 8: Urban areas. In: Climate change 2014: impacts, adaptation, and vulnerability: working group II contribution to the fifth assessment report of the intergovernmental panel on climate change. Cambridge University Press, New York, NY

Krizek K, Levinson D (2005) Teaching integrated land use-transportation planning: topics, readings, and strategies. J Plan Educ Res 24(3):304–316

Leal Filho W (ed) (2010) Climate change at universities: results of a world survey. In: Universities and climate change. Climate change management. Springer, Heidelberg [Germany], New York, pp 1–21

Manley S, Parnaby R (2000) The return of the architect-planner? Plan Pract Res 15(3):269–277

Norman B (2017) Sustainable pathways for our cities and regions in the future planning within planetary boundaries. Taylor & Francis. Retrieved 21 Feb 2018, from http://search.ebscohost.com/login.aspx?direct=true&scope=site&db=nlebk&db=nlabk&AN=1691989

OECD (2010) Cities and climate change. OECD Publishing, Paris. https://doi.org/10.1787/9789264091375-en

Pezzoli K, Howe D (2016) Planning pedagogy and globalization. J Plann Educ Res 20(3):365–375

Ratnayake R, Butt A (2018) Encounters with the unfamiliar: international planning education. Int Plan Stud 23(1):51–64

Schreier M (2012) Qualitative content analysis in practice. SAGE, Los Angeles

Schreier M (2014) The SAGE handbook of qualitative data analysis (U. Flick, Ed.). SAGE Publications Ltd, United Kingdom. Retrieved from http://methods.sagepub.com/book/the-sage-handbook-of-qualitative-data-analysis

Susskind L (2010) Responding to the risks posed by climate change: cities have no choice but to adapt. Town Plan Rev 81(3):217–235

UN (2015) Transforming our world: the 2030 agenda for sustainable development. Retrieved from https://sustainabledevelopment.un.org/content/documents/21252030%20Agenda%20for%20Sustainable%20Development%20web.pdf

UN-Habitat (2009) Planning sustainable cities: global report on human settlements 2009. Earthscan, London, UK

UN-Habitat (ed) (2011a) Cities and climate change: global report on human settlements, 2011. Earthscan, Washington, DC

UN-Habitat (ed) (2011b) Cities and climate change. Earthscan, London

UN-Habitat (2015) Cities and climate change academy. UN-Habitat. Retrieved 7 Aug 2018, from https://unhabitat.org/urban-initiatives/initiatives-programmes/cities-and-climate-change-academy/

UN-Habitat (2017) New urban agenda. Retrieved 15 Apr 2018, from http://habitat3.org/wp-content/uploads/NUA-English.pdf

United Nations (2015a) Transforming our world: the 2030 Agenda for Sustainable Development [online] Retrieved from: https://sustainabledevelopment.un.org/content/documents/21252030%20Agenda%20for%20Sustainable%20Development%20web.pdf

United Nations Framework Convention on Climate Change (2015b) Adoption of the Paris Agreement FCCC/CP/2015/L.9/Rev.1

Viguié V, Hallegatte S (2012) Trade-offs and synergies in urban climate policies. Nat Clim Change 2(5):334–337

Walsh CL, Dawson RJ, Hall JW, Barr SL, Batty M, Bristow AL, Carney S, Dagoumas AS, Ford AC, Harpham C, Tight MR, Watters H, Zanni AM (2011) Assessment of climate change mitigation and adaptation in cities. Proc Inst Civ Eng Urban Des Plan 164(2):75–84

Wamsler C, Brink E, Rivera C (2013) Planning for climate change in urban areas: from theory to practice. J Clean Prod 50:68–81

While A, Whitehead M (2013) Cities, urbanisation and climate change. Urban Stud 50(7):1325–1331

Whitehouse H (2017) Point and counterpoint: climate change education. Curric Perspect 37(1):63–65

Wiek A, Bernstein MJ, Laubichler M, Caniglia G, Minteer B, Lang DJ (2013) A global classroom for international sustainability education. Creat Educ 4(4):19–28. https://doi.org/10.4236/ce.2013.44A004

Withycombe Keeler L, Wiek A, Lang DJ, Yokohari M, van Breda J, Olsson L, Evans J (2016) Utilizing international networks for accelerating research and learning in transformational sustainability science. Sustain Sci 11(5):749–762. https://doi.org/10.1007/s11625-016-0364-6

When Fighting Climate Change Leads to Better Cities: A Study of Actions Implemented by 100 Cities in Spain

Xira Ruiz-Campillo

Abstract This study presents an empirical analysis of how cities change as a result of the fight against climate change. It analyses the measures adopted by 100 cities and towns participating in the Covenant of Mayors in Spain in order to reduce their emissions in the areas of public lighting, municipal buildings, transport, energy as well as other measures such as the increase in green spaces. Although reducing greenhouse gases has unintended consequences, such as the so-called rebound effect, in general we can maintain that the fight against climate change is leading to greener and healthier cities for citizens.

Background

Cities and urban areas of all kinds are of utmost importance in the fight against climate change because of the agglomeration of population within them and the impact that this has on climate change. In 2018, for instance, 55% of the world's population resided in urban areas and this figure will continue to increase, reaching 68% by 2050 (UN Population 2018). It is also in cities that more than 70% of the world's greenhouse gases emissions related to energy are generated, and, with the increase in population, two thirds of the world's population—with Asia and Africa containing the fastest-growing urban areas—will live in cities by 2050 (UNDESA 2014; UN Population 2018). All this makes the commitment of cities and towns of all sizes indispensable to advance towards a less polluted, more sustainable, more environmentally friendly and more livable world. Ensuring that urban areas grow in a sustainable way can have the double benefit of reducing climate change and increasing the wellbeing of citizens. This study analyses the type of transformations that can be found in cities adopting policies to fight climate change within the framework of the Covenant of Mayors (CoM) and how they can contribute to the wellbeing of citizens, as evidence shows that environmental quality is closely linked to people's wellbeing (Streimikiene 2015).

X. Ruiz-Campillo (✉)
University Complutense of Madrid, Madrid, Spain
e-mail: xiraruiz@ucm.es

W. Leal Filho et al. (eds.), *Climate Change, Hazards and Adaptation Options*,
Climate Change Management, https://doi.org/10.1007/978-3-030-37425-9_55

The European Union Covenant of Mayors was launched in 2008 by the European Commission as a way to enhance the action of cities and towns (from now on all will be referred to as cities) in the fight against climate change. The initiative was launched just after the adoption of the 2020 climate and energy package, which established a number of targets for 2020: to reduce 20% of greenhouse gas (GHG) emissions with respect to the levels of 1990 and to increase the share of renewable energies and energy efficiency by 20% (European Council 2007). Even though all these efforts are binding for Member States, the idea of the Covenant of Mayors was to mobilize other actors that were already seen as necessary to reach these targets: cities.

In 2015, the Covenant of Mayors joined the Mayors Initiative on Adaptation (launched in 2014 to urge municipalities into adopting measures in order to anticipate and adapt to the effects of climate change), with the resulting emergence of the Covenant of Mayors for Climate and Energy. This new covenant adopts a more holistic approach in which the aim is not only that cities reduce their emissions of greenhouse gases, but also that they implement measures to adapt to climate change by 2030. This new initiative was more ambitious as it took on the objectives that the European Union had already established for the year 2030: reducing greenhouse gases by 40% and increasing the use of renewable energy and energy efficiency by 27% (European Council 2014). By joining the Covenant, cities committed to the reduction of their emissions through a flexible array of instruments that go from increasing access to sustainable transport and greener areas to investing in low consumption lighting or renewable energy. All these strategies have an impact on urban GHG emissions by shaping individual and collective behaviors and by positively affecting the efficiency of energy, water and waste infrastructures (Croci et al. 2016).

Description of the Study Areas and Methodology

The working hypothesis adopted in this study is that cities that fight climate change and adopt reduction or adaptation strategies experience positive transformations that will improve the wellbeing of their citizens. To test this hypothesis, the study analyses the measures implemented by 100 cities in Spain in order to reduce their emissions in the framework of the Covenant of Mayors (CoM). The aim in this regard is to highlight both the opportunities that the fight against climate change creates for citizens and the possible setbacks that can already be identified, in order to prevent or minimize their impact. In addition, existing literature on the relation between the environment and its impact on health has been examined to understand the links between the reduction of greenhouse gas emissions and the improvement of the wellbeing of citizens.

This study adopts the perspective that cities are necessary actors in the fight against climate change, as evidenced by a strand of the literature on global environmental governance (Bulkeley and Schroeder 2011; Bulkeley and Castán Broto 2012; Alber and Kern 2008; Kern and Bulkeley 2009). With more than half of the world's population living in urban areas, cities are indispensable actors in the fight against climate change. At the same time, cities are in a privileged position to promote and influence GHG emissions reduction because of their proximity to businesses, companies and citizen's daily lives, as well as their lifestyle and consumption patterns.

As of February 2019 there were 7755 signatories (most of them in the European Union) to the Covenant of Mayors for Climate and Energy, of which almost 24% were Spanish. A total of 1817 Spanish municipalities were part of the Covenant of Mayors at the beginning of 2019, of which 1393 were of less than 10,000 inhabitants; 330 had between 10,000 and 50,000 inhabitants; 79 between 50,000 and 250,000 inhabitants; 9 between 250,000 and 500,000 and 6 more than 500,000 inhabitants. The bulk of the signatories were, therefore, municipalities with less than 10,000 inhabitants.

As seen in Table 1, Spain is the country with the highest number of signatories of the Covenant only after Italy. Both countries, Italy and Spain, add up to 66% of the total signatories of the Covenant. Lagging behind are Belgium, Greece and Portugal, and the rest of countries contributed to the Covenant with a very limited number of cities. Although the transformations experienced in Spanish cities are not necessarily the most common in other countries, many of them do represent a pattern that can be seen in other Spanish cities that are also part of the Covenant, and others are just pointed out here as they could serve as inspiration for other cities.

After signing the Covenant, signatories can transit through different phases. Signing the Covenant is the first step, after which, cities have up to two years to submit their Sustainable Energy and Climate Plan (SECAP) (step 2), and after two years

Table 1 Number of signatories to the CoM as of February 2019

Country	Nº of signatories	Country	Nº of signatories
Italy	3319	Slovenia	29
Spain	1817	Bulgaria	25
Belgium	337	Cyprus	24
Greece	156	Malta	24
Portugal	119	Netherlands	24
France	89	Latvia	21
Romania	73	Austria	13
Croatia	70	Slovakia	13
Germany	64	Finland	12
Sweden	56	Ireland	12
Poland	40	Lithuania	15
Hungary	40	Czechia	11
Denmark	37	Estonia	5
UK	36	Luxembourg	2

Author's elaboration

Table 2 N° of Spanish cities in the Covenant, in the monitoring phase and in this study

Population	Total n° cities in the Covenant	Total n° cities in monitoring phase	N° cities used in study	% of the sample
>500,000	6	3	2	66.6
250–500,000	9	3	3	100
50–250,000	79	36	29	80
10–50,000	330	123	30	24
<10,000	1393	371	36	9.7
Total	1817	536	100	18

Author's elaboration

still they will undergo a monitoring process in which the extent to which cities have reached their goals will be evaluated (step 3), with the aim of checking the compliance of the interim results with the foreseen goals cities have adopted.

In the case of Spain, as of February 2019 there were 984 cities that had submitted their SECAP, out of which 536 were already in the monitoring phase. This research will focus on the latter in order to the initiatives they have implemented in order to reach the 2020 or 2030 targets. Concretely, a sample of 100 cities that were in the monitoring phase by the end of 2017 was analyzed, in order to check the degree of implementation of each of the following areas: municipal buildings, equipment and facilities; tertiary buildings, equipment and facilities; residential buildings, public lighting; industry; transport; local electricity production; local heat/cold production, and others. These domains are established by the EU Covenant of Mayors as prefixed areas and cities have to communicate their progress two years after submitting their action plans. In addition to figures and data taken from each city's SECAPs, data has been collected through a survey done between October and November 2017 and answered by 27 municipalities in the monitoring phase. In total, 18% of the cities that are currently in the monitoring phase were analyzed for this study (Table 2).

Cities in Motion: Areas Transformed by Climate Change

Data gathered from the CoM clearly shows the areas where cities have been more active to reduce their emissions. These include changes in public lighting (33% of all the cities analyzed had completed this action by the end of 2017), municipal buildings (represent 20% of the implemented actions), followed by actions in other areas (18%) as well as tertiary buildings (17%). The areas where fewer actions were implemented by cities were those related to industry (4%) and to the local production of heat/cold (2%) (see Fig. 1).

In their SECAPs, cities are free to choose different actions for each of the areas mentioned above. For instance, in the area prefixed by the CoM as "transport", cities can decide to renew all the public transport fleet in favor of a more efficient and cleaner

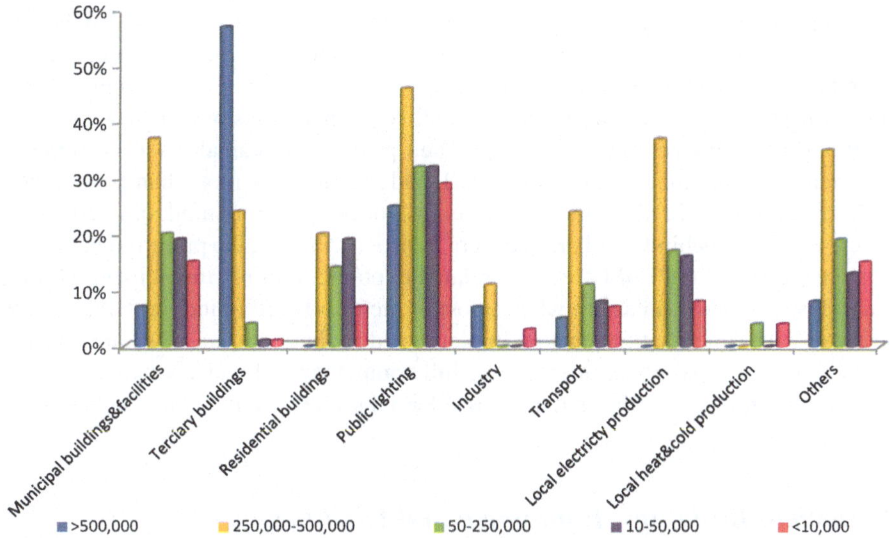

Fig. 1 Percentage of completed initiatives in cities. Author's elaboration with data retrieved from CoM between September and December 2017

one or to give incentives for the acquisition of electric cars by citizens (or both). In the "others" section cities can include actions that go from planting more trees in the city to the construction of bicycle paths or the contracting of renewable energy. Although changing public lighting or public transport fleets may have an immediate impact on the reduction of GHG emissions, actions included in the "others" sections show whether city councils are willing to implement actions that can transform cities and have an additional positive impact on the health and lives of citizens. It is probable that with more financial and technical support from national and regional governments more initiatives would be adopted, as for 74% of the respondents to the survey submitted for this study, the main problem for cities regarding the monitoring process of the CoM was the lack of economic resources to follow up on the action plans. This is in line with the setbacks found in many of the SECAPs consulted for this research. Ten of the cities analyzed pointed out in their reports that financial constrains were the main problems found in the implementation of their action plans. The lack of technical experience was also highlighted in some SECAPs as a medium problem, together with the lack of support to cities more generally.

The review of the percentage of planned actions completed by cities reveals that except for actions implemented in tertiary buildings, where cities of more than 500,000 inhabitants are clearly ahead, in the rest of areas the leading cities are those of a large size (between 250 and 500,000 inhabitants). For instance, such cities have implemented 37% of their foreseen actions in municipal buildings (against the 20% implemented by the following group of cities, those between 50 and 250,000 inhabitants); they also lead actions in public lighting, industry, transport and electricity production. Almost in all cases (and with the exception of XL cities, that only stand out in the case of tertiary buildings) the level of actions completed depends on the size of the city. The larger the city, the higher the percentage of actions implemented. For example, in the case of transport, large cities have completed 24% of their actions, medium size cities 11%, small cities 8% and very small cities only 7%.

When analyzing actions in relation to their progress we can also see a difference between large and small cities. While XL, L and M cities have more than 30% of their actions in progress in almost all areas, this is not the case for small and very small cities, which are behind in all areas except transport. Clearly, the preferred actions to attain the goals of the CoM are those related to public lighting, municipal buildings and transport. These areas are where most of the efforts are found, probably due to the ease with which actions can be implemented in these areas as they are areas in which the municipal governments have full competences. Precisely because of the lack of power, reaching the tertiary or residential sectors seems more challenging.

Municipal Buildings, Equipment and Facilities

This area includes buildings and facilities managed by local authorities the energy costs of which are paid by the city council. These can include municipal and public buildings such as town halls, libraries, schools, sports or cultural centers amongst others.

This is the second area in which most actions implemented by cities can be found (Fig. 1). Such actions consist mainly in the improvement of energy efficiency in public and municipal buildings. For instance, it is a constant to find that cities change their air conditioners, heaters and light bulbs to others that are more efficient. They also invest in the improvement of energy efficiency with the installation of motion detectors and the improvement of isolation (e.g. the installation of double glaze windows, double doors, awnings or a better roof isolation). Many cities also invest in the centralization and computerization of energy consumption and costs in all municipal buildings, which enables having a better idea of overall energy costs (as seen in the city of Basauri). One example of how these actions can reduce municipal energy consumption is found in Los Montesinos, a small town in Alicante that has managed to reduce its energy consumption with the additional result of a 14% of reduced emissions in the reported period. In towns like Castelldefels, Castellar del Vallés and Argentona there has been a reduction of 25%, 4% and 10% respectively in the energy consumption of municipal buildings as reflected in their SECAPs.

However, there are other cities in which there has been an increase in municipal energy consumption despite a decrease in their population (i.e. Badalona, Cerdanyola). This can be the result of the so called rebound effect. It has happened to many citizens and households that after moving to low energy consumption they experienced an increase in energy consumption, induced most probably by the lower cost of light. The so-called rebound effect can imply that energy efficiency improvement leads to a decrease in the cost of energy services and thus eventually to an increase in the demand for such good or service (Sorrell and Dimitropoulos 2008; Sorrell 2010;

Jin and Kim 2019). The rebound effect can be triggered by many factors going from economics to consumer information, economic resources and physical space, as well as socio-psychological factors, time or skills (Font Vivanco and Van der Voet 2014; Santos et al. 2018). Although the rebound effect may be larger than conventionally assumed (Sorrell 2007) and should be taken into account when implementing policies to reduce GHG, other sources prefer to point out the benefits of improved energy efficiency, such as its health benefits, or its impact on poverty alleviation and improved productivity, which balance the outcomes of the rebound effect (IEA 2014). Strategies to reduce this unwanted effect can go from consuming more efficiently, consuming differently or consuming less (Sorrell 2010; Font Vivanco et al. 2016) all of which can be promoted through policies that have been proposed in the literature, such as giving information to serve as moral suasion, regulating prices, improving efficiency standards and eco-taxes, promoting environmental and sustainable communication with business and citizens as well as the performance of policy instruments (Van den Berg 2011; Santarius 2012; Font Vivanco et al. 2016).

Another common action found in different SECAPs is the training of staff in charge of the maintenance of buildings and equipment in cities to make a better use of energy, water and municipal resources, as in the case of Badalona. Some town councils have even created the figure of the "energy city manager", in charge of informing and auditing the energy use in municipal and public buildings on a constant basis (i.e. Bigastro and Benissa). In some cases, all staff working for the town councils have received information about energy efficiency, recycling and responsible consumption of goods and services in their offices. In other cases, town councils have demanded electric companies to supply 100% of the energy from renewable sources, which undoubtedly can set an example of good practices for citizens as is the case of Cádiz, Barcelona, Alicante, Madrid and hundreds of cities in Catalonia, all of which have contracted the purchase of green energy from providers certified by the National Commission of Markets and Competition.

Others have incorporated additional or other environmental criteria in the purchasing of municipal goods and services, as is already being promoted by the European Union, through the green public procurement (GPP). Although the latter is (still) a voluntary set of criteria, the Council of the European Union and the European Parliament adopted in 2014 two directives to simplify public procurement and ensure the greater inclusion of goals related fro example to environmental protection, social responsibility, combating climate change, public health or environmental considerations, amongst others (OJEU 2014a, b). This set of criteria is already being used by many public administrations when contracting a wide arrange of goods and services, from cleaning services and computers to electricity, road construction or gardening services. The inclusion of GPP criteria in tender competitions is a powerful tool with a great impact on the consumption of products and services that meet certain environmental criteria, thus promoting among providers and consumers sound environmental and green practices and fostering the creation of growth and jobs at enterprises which develop green technology (Milios 2018; Palmujoki et al. 2010; European Commission 2016).

Through all the actions mentioned above, cities are reducing their emissions by increasing their energy efficiency, which results in a reduction of the council's energy bills. These savings can have a positive impact on the council budget which could then be used to improve services and infrastructures for citizens.

Residential Buildings

Energy consumption in this sector comes mainly from private residences: air conditioning, heating, lighting, hot water, electric or gas cookers, electric appliances (TV, refrigerator, computers, etc.) and elevators in the buildings.

In the residential sector we can distinguish two big areas of action: the promotion of subsidies to increase (1) the use of solar thermal energy and the use of biomass, and (2) to purchase more efficient electric appliances or light bulbs. Another area in which almost all cities coincide is in the development of awareness campaigns, the most preferred ones being those related to energy efficiency and sustainable energy (like in Benimassot or Bolulla) followed by campaigns on water use and recycling (in Benalúa de Guadix). This area also includes the creation of communication channels through which information on grants and subsidies is rendered available for citizens interested in making investments in energy saving, energy efficiency, the installment of renewable energy in households, or, in the more exceptional cases, the creation of a consumer unions to deal with electric companies (i.e. Carmona).

The residential sector is the area with the highest demand of energy and therefore a key sector if GHG emissions are to be reduced. Obviously, it is easier for cities to implement decisions in the areas in which they have competences—like renewing their transport fleet or changing the public lighting, than forcing citizens to use renewable energy or to purchase more efficient appliances. This is a key sector, however, for cities that want to decrease their emissions. In Spain, in 2016, the residential sector consumed 15% of the total energy in the country while the tertiary and public services sectors together consumed 10.6% (IDAE 2016). Although most actions to reduce GHG coming from cities focus on increasing efficiency in the public sector, more funds should be allocated to the residential sector.

Programs for the renewal of households' electrical appliances and air conditioning units such as the "Plan Renove" launched in autonomous communities in Spain like Andalusia, the Basque Country, Aragón or Castilla La Mancha are useful and have had an impact on households' energy bills. For instance, in Andalusia in 2012 more than 17,000 citizens changed their air conditioning units to more efficient ones (class A or above). Through this program, the community of Andalusia spent 3.9 million euros (Junta de Andalucía 2012), managed by the Andalusian Energy Agency and the cities included in their action plans. These kinds of plans have also targeted the renewal of windows and boilers, although the amount allocated to programs of this kind is clearly insufficient for such a vast number of citizens. For example, the Community of Madrid had a budget of 17.2 million euros in 2016 (Comunidad de Madrid 2016) but only 800,000€ were allocated to the renewal of windows that year

(Es por Madrid 2016) decreasing to 745,000€ in 2017 (Fenercom 2017). This seems insufficient for a population of more than 6 million people.

In many cases, the residential sector has decreased its consumption in the period in which cities have participated in the Covenant of Mayors. The towns of Benimantell and Benillup, both of them in Alicante (Community of Valencia), published their energy consumption by sector in their monitoring reports. Benimantell, a city of 499 inhabitants, has reduced its consumption in the residential sector by 3.6% between 2013 and 2015 while it increased its population by 2.87%. The same happened with Benillup, a small town of only 104 inhabitants that managed to decrease its residential consumption by 3.7% while increasing the number of citizens by 7% in the period between 2011 and 2014. The case of Argentona (Catalonia) shows a similar pattern: while there was an increase of 9.6% in the population, there was a decrease in energy consumption of 5.7%, although this is not always the case (i.e. Santander), probably due to the already mentioned rebound effect.

There are evidences that promoting energy efficiency in the residential sector helps advance both economic growth and climate action, as well as having particularly high economic and social returns. For instance, better insulated households keep air conditioning inside during warm seasons and heat during cold ones making homes more comfortable and enjoyable for dwellers. According to the IEA (2007), each additional dollar invested in more efficient electrical equipment, appliances, and buildings avoids more than two dollars in energy supply costs, which could mean that reducing energy consumption in households could prove to be beneficial for lower-income residents who use a higher share of their rent for energy than wealthier residents Bill de Blasio, mayor of New York, pointed out in 2014 when he committed to decrease GHG emissions by 80% by 2050 with regards to the levels of 2005 (De Blasio 2014). Energy efficiency not only has an impact on disadvantage population: there are also considerable gains for the rest of the citizenry through reduced local pollution, improved health and well-being, job creation, energy savings both at the public and private level and better energy prices that benefits all (IEA 2014).

Public Lighting

This area includes all the electric consumption of public street lighting as well as all the electricity consumed in municipal buildings, equipment and facilities in the city. Without any doubt, public lighting is the area where most actions have been implemented by cities signatories of the Covenant of Mayors. In the two very big cities analyzed here (Barcelona and Málaga), actions foreseen in this area have already been implemented or where in process by the time of the monitoring report. In cities between 250,000 and 500,000 inhabitants, actions in public lighting reached 97%, and medium cities reached 89% of actions in this area, followed by 73% in small cities and 67% in very small cities. In all cases this is the area in which the highest percentage of actions were implemented.

The most common actions adopted by cities in this area include the substitution of street lamp lights and traffic lights with LED technology so there is a reduction in GHG emissions, and the use of remote management systems or lighting flow dimmer-stabilizers (i.e. Logroño). All these actions are relatively easy to implement by cities and have an immediate impact on the reduction of emissions.

Reducing the energy bill of public lighting as a result of the increase of energy efficiency has led to another unwanted consequence besides the rebound effect mentioned above. Although energy efficiency has been increased, lower costs have led to an increased demand and greater use of outdoor lighting in cities, resulting in more LED lighting being installed outdoors (Kyba et al. 2017). This is leading to an increasing light pollution that threatens biodiversity by changing the night habits (related for example to reproduction, foraging, vocal communication or migration) of fish, birds, bats, insects and other animals (Hölker et al. 2010; Van Doren et al. 2017). To reduce this unwanted effect of the increased amount of light in cities, Van Doren et al. (2017) recommend selective removal of light pollution during nights with substantial bird migration. Not only animals and plants seem to be impacted by greater outdoor lighting, however, since there are suggestions that it could also represent a novel risk factor for breast cancer (James et al. 2017).

Tertiary Buildings and Industry

Tertiary buildings and facilities in the tertiary sector include those that are not property nor managed by the city council. It includes private offices, banks, small and medium enterprises, commercial businesses, restaurants, etc. The public and municipal sector, together with the tertiary sector where responsible for 12.5% of the final energy consumed in Spain in 2015 (IDAE 2015). Therefore, actions in both sectors are of great importance for reaching the targets of the Covenant of Mayors.

Together with the tertiary sector, industry is beyond any doubt the area where fewer actions to implement the Covenant of Mayors have been taken in cities. The reasons are partially the same: it is a sector where public action is pretty limited since it depends very much on the goodwill of companies. Besides, regulations in the industrial sector in Spain are very much made at the national level, so the ability of cities to influence the industrial sector is clearly limited. However, in the cases studied, we find that a small number of cities have launched awareness campaigns for saving energy and increasing energy efficiency in the industrial sector. Apart from this, a number of cities have included amongst its actions in this area the promotion of energy audits in the sector.

Local Electricity Production

Local electricity production is improved by cities in two main ways: one is the installation of photovoltaic energy stations or biomass boilers in municipal and public buildings and the other is the promotion of renewable energy in households and the industrial sector either by granting permits for the construction of photovoltaic, wind or biomass energy stations, or by promoting the use of renewable energy in households. Other minor initiatives include the collection of organic material from public or private sources (mainly from gardens, green areas or public forests) to produce either energy or compost for private or public areas (i.e. Viso del Alcor). All the cities that responded to the survey for this study indicated that being part of the CoM had contributed to the development of sustainable energy in their cities, with only 18.5% of the respondents stating that they appreciated little connection, and 66.6% who considered that the Covenant had considerably contributed to the promotion of sustainable energy at the local level.

Although this is one of the areas least developed by cities to reduce their GHG, the potential for developing renewable energies at the local level is high. Apart from being an example to follow for citizens, developing renewable energies can reduce the energy dependence of cities and contribute to the creation of jobs locally (IEA 2009). At the consumers level, local administrations can use financial incentive schemes to encourage the use of renewables, and can make the installation of renewables easier at the domestic level, through taxes, through the creation of standards or through awareness raising campaigns (IEA 2009).

Transport

This area refers to both public transport (buses, trains, trams, cars/motorbikes/trucks used by council staff or staff of companies working for the council) and private transport. Together with "Other areas", this is where more interesting and inspiring initiatives capable of leading the transformation of life in cities are found. Three main groups of actions implemented by town councils can be identified in the SECAPs examined: the renewal of the municipal transport fleet, the improvement of public transport services and the promotion of pedestrian and cyclist mobility.

Cities tend to opt for their fleet renewal with more efficient and environmentally friendly vehicles, such as electric, hybrid or biodiesel buses and cars used by council staff. In this area there are also actions related to the control, monitoring and use of programs destined to improve the understanding of the use of fuel in municipal fleets. Other actions focus on improving the quality of public transport as a way to discourage the use of private vehicles. Ways for doing this can include increasing the number of buses within a city, buses and/or train lines, as well as the frequencies of trips (i.e. Ronda). Other initiatives try to make public transport more attractive for citizens by improving shelters for urban bus stops and the accessibility to public transport, for instance, through buses with access ramps for disabled people.

Another main area of action in transport is the improvement of pedestrian and cyclist mobility. Many cities have demarcated areas where speed is limited to 30 km/h making them safer for pedestrians; others have created parking-meter areas in city centers to discourage citizens from accessing cities by car; and others have opted for setting up deterrence car parks as a way of discouraging the use of private cars in city centers (i.e. Almensilla). These kinds of actions together with an improvement in public transportation can decrease the use of private vehicles and render urban centers more citizen-friendly.

Other examples of the improvement of pedestrian and cyclist mobility can be found in the improvement of pavements, the pedestrianization of historical centers and commercial streets, the removal of architectural barriers to make walking more pleasant, the purchase of bikes, the constructions of more kilometers of cycle lanes and the installation of car parks for bicycles (i.e. Irún). All these actions have a positive impact on the wellbeing of citizens as evidenced in numerous studies. For instance, there is a relation between the building of bike lanes and improved health. When there are bike lanes, cycling (as well as other forms of exercise) tends to increase since it becomes safer through designated areas for cycling (Gu et al. 2017; Che et al. 2012). The construction of bike lanes can also be seen as a way of addressing multiple public health problems as it improves air quality in cities at the same time as it increases physical activity, impacting therefore on citizens' health (Gu et al. 2017; Pucher and Dijkstra 2003).

These are the most popular actions found in the cities analyzed, but there are other initiatives to reduce GHG emissions in the transport sector, like the promotion of car pooling and car sharing, the promotion of electric vehicles (with for example the reduction of registration or circulation taxes or the installation of recharging points), the training in efficient and safe driving for drivers of municipal fleets and for farmers in small towns, or the promotion of safe school paths that helps improving pedestrian mobility in the surroundings of schools.

Other Areas

In this section there are three main areas were cities act. The first one focuses on waste management, with proposals that go from the reduction in the use of inorganic fertilizers (through the recovery of biomass to be later used as either fertilizers or for the production of energy), or the promotion of the selective collection of waste, to the inclusion of containers for the collection of used oil or organic waste.

Another area is related to water management in cities. Some cities opt for improving their irrigation network and their water supply systems in order to reduce leakages. Cities like Logroño have promoted water-efficient irrigation techniques in the town, and others have approved and implemented ordinances related to the saving and efficient use of water in their cities. The creation of an environmental information point to promote saving in water and energy and the use of organic compost is another initiative seen in one of the cities analyzed (Carmona).

Finally, there are other actions implemented by cities that include the increase of green areas (with more parks or reforestation plans), the promotion of ecological crops in the municipality or the development of a municipal green label as a sign of good environmental practices in shops and restaurants (i.e. Castelldefels). The health benefits of moving to green areas have also been studied, showing that living next to green areas can improve the overall health of citizens, reduce stress, depression, increase longevity and can be especially beneficial for people with a lower socioeconomic status (Alcock et al. 2014; De Vries and Verheij 2003; Maas and Verhij 2009; Mitchel and Popham 2008; Van den Berg et al. 2010). All this suggest that cities can be important agents in the promotion of healthier environments and better welfare among citizens by increasing green areas in cities, while at the same time fighting climate change. A specific question asked to cities in the survey was to what extent they considered that their participation in the Covenant had contributed to the improvement of the quality of life of citizens. All respondents agreed that their participation had contributed to such improvement, although to different degrees. For 63% the participation had contributed somehow; for 33.3% it had a considerable impact, and only for 3.7% it had contributed a lot.

Conclusions

This study has showed a wide array of initiatives and actions implemented by Spanish cities to reduce greenhouse gas emissions at the local level and fulfill their commitments within the Covenant of Mayors, going from the contract of renewable energy in city councils to the increase of green areas. Although it is difficult to attribute these initiatives only to the signing of the CoM, there is no doubt that a good part of the actions implemented have helped to reconfigured cities making them more sustainable and livable for citizens. This is in line with other studies that highlight how a shift to more compact urban growth and better connected cities with strong mass transit systems renders them more productive, socially inclusive, resilient, cleaner, quieter and safer as in addition to contributing to boost economic prosperity and tackle climate change (New Climate Economy 2014). According to more than 80% of the respondents to the survey done for this study, the CoM has contributed to develop a local strategy to reduce GHG and to improve their public image. This shows how the adoption of specific targets in cities to reduce greenhouse gas emissions can be efficient to reducing pollution but also to earn prestige at the regional, national and even international level.

Research done for this study has provided evidence of a link between the reduction of pollution and the increase in the wellbeing of citizens, as a result of the savings in energy efficiency and in the higher availability of green spaces or safer cycling options. The connection between the fight against climate change and the wellbeing of citizens should be used in a more efficient way by authorities when communicating the adoption of policies that may be seen by citizens as counterproductive in the beginning. If there is scientific evidence that reducing GHG emissions is good for our wellbeing, then public administrators should work in the promotion of more initiatives that help reducing GHG and improve our cities.

The results of this research yielded some interesting findings: actions implemented to reduce GHG by cities can have a positive effect on citizens, but also unwanted consequences that policymakers should take note of, as is the case of the rebound effect in energy efficiency and the impact of more outdoor lighting on biodiversity. It is clear that when finding solutions to climate change, some problems go away but new ones arise and should be faced.

References

Alber G, Kern K (2008) Governing climate change in cities: modes of urban climate governance in multi-level systems. In: Competitive cities and climate change. OECD conference proceedings, Milan, Italy, 9–10 Oct 2008

Alcock I, White M, Wheeler B, Fleming L, Depledge M (2014) Longitudinal effects of mental health of moving to greener and less green urban areas. Environ Sci Technol 48:1247–1255

Bulkeley H, Schroeder H (2011) Beyond state/non-state divides: global cities and the governing of climate change. Eur J Int Relat 18(4)

Bulkeley H, Castán Broto V (2012) Government by experiment? Global cities and the governing of climate change. Trans Inst Br Geogr 38(3)

Che L, Chen C, Srinivasan R, McKnight C, Ewing R, Roe M (2012) Evaluating the safety effects of bicycle lanes in New York City. Am J Public Health 102(6):1120–1127

Comunidad de Madrid (2016) Ley 6/2015, de 23 de diciembre, de Presupuestos Generales de la Comunidad de Madrid para el año 2016

Croci E, Lucchitta B, Janssens-Maenhout G, Martelli S, Molteni T (2016) Urban CO_2 mitigation strategies under the Covenant of Mayors: an assessment of 124 European cities. J Clean Prod 169:161–177

De Blasio (2014) De Blasio orders a greener city, setting goals for energy efficiency of buildings. The New York Times, 20 Sept 2014

De Vries S, Verheij R (2003) Natural environments—healthy environments? An exploratory analysis of the relationship between greenspace and health. Environ Plan 35(10)

Es por Madrid (2016) Plan Renove 2016 de ventanas de la Comunidad de Madrid, 27 Sept 2016. Consulted on Mar 2019. https://www.espormadrid.es/2016/09/plan-renove-2016-de-ventanas-de-la.html

European Commission (2016) Buying green! A handbook on green public procurement, 3rd edn. European Union

European Council (2007) Presidency conclusions 8/9 March 2007. Brussels, 2 May 2007

European Council (2014) European Council conclusions. Brussels, 24 Oct 2014

FENERCOM (2017) Plan Renove de Ventanas. Información general

Font Vivanco D, Van der Voet E (2014) The rebound effect through industrial ecology's eyes: a review of LCA-based studies. Int J Life Cycle Assess 19:1933–1947

Font Vivanco D, Kemp R, Van der Voet E (2016) How to deal with the rebound effect? A policy-oriented approach. Energy Policy 94:114–125

Gu J, Mohit B, Muenning P (2017) The cost-effectiveness of bike lanes in New York City. Inj Prev 23:239–243

Hölker F, Wolter C, Perkin EK, Tockner K (2010) Light pollution as a biodiversity threat. Trends Ecol Evol 25

IDAE (2015) Consumo de Energía Final 2015. Consulted on Mar 2019. http://sieeweb.idae.es/consumofinal/bal.asp?txt=2015&tipbal=t
IDAE (2016) Balance del consumo de energía final. 2016 (avance). Consulted on Mar 2019. http://sieeweb.idae.es/consumofinal/bal.asp?txt=2015&tipbal=t
IEA (2007) World energy outlook 2006. OECD/IEA
IEA (2009) Cities, towns and renewable energy. Yes in my front yard. OECD/IEA
IEA (2014) Capturing the multiple benefits of energy efficiency. OECD/IEA
James P, Bertrand K, Hart J et al (2017) Outdoor light at night and breast cancer incidence in the nurse's health study II. Environ Health Perspect 125(8)
Jin T, Kim J (2019) A new approach for assessing the macroeconomic growth energy rebound effect. Appl Energy 239
Junta de Andalucía (2012) Más de 17.000 andaluces cambian su aire acondicionado por otro más eficiente con el plan renove de la consejería de economía. Consejería de Empleo, Formación y Trabajo Autónomo. Consulted on Mar 2019
Kern K, Bulkeley H (2009) Cities, Europeanization and multi-level governance: governing climate change through transnational municipal networks. JCMS 47(2)
Kyba C, Kuester T, Sánchez de Miguel A et al (2017) Artificially lit surface of earth at night increasing in radiance and extent. Sci Adv 3
Maas J, Verhij R (2009) Morbidity is elated to a green living environment. Epidemiol Community Health 63(12)
Milios L (2018) Advancing to a circular economy: three essential ingredients for a comprehensive policy mix. Sustain Sci 13:861–878
Mitchel R, Popham F (2008) Effect of exposure to natural environment on health inequalities: an observational population study. Lancet 372
New Climate Economy (2014) Better growth, better climate. The New Climate Economy Report. Global Commission on the Economy and Climate
Official Journal of the European Union (2014a) Directive 2014/24/EU of the European Parliament and of the Council of 26 February 2014 on public procurement and repealing Directive 2004/18/EC
Official Journal of the European Union (2014b) Directive 2014/25/EU of the European Parliament and of the Council of 26 February 2014 on procurement by entities operating in the water, energy, transport and postal services sectors and repealing Directive 2004/18/EC
Palmujoki A, Parikka-Alhola K, Ekroos A (2010) Green public procurement: analysis on the use of environmental criteria in contracts. Rev Eur Community Int Environ Law 19(2)
Pucher J, Dijkstra L (2003) Promoting safe walking and cycling to improve public health: lessons from the Netherlands and Germany. Am J Public Health 93(9)
Santarius T (2012) Green growth unravelled. How rebound effects baffle sustainability targets when the economy keeps growing. Heinrich Böll Foundation. The Green Political Foundation
Santos R, Matias J, Abreu A, Reis F (2018) Evolutionary algorithms on reducing energy consumption in buildings: an approach to provide smart and efficiency choices, considering the rebound effect. Comput Ind Eng 126:729–755
Sorrell S (2007) The rebound effect: an assessment of the evidence for economy-wide energy savings from improved energy efficiency. Project report
Sorrell S (2010) Energy, economic growth and environmental sustainability: five propositions. Sustainability 2
Sorrell S, Dimitropoulos J (2008) The rebound effect: microeconomic definitions, limitations and extensions. Ecol Econ 65(3)
Streimikiene D (2015) Environmental indicators for the assessment of quality of life. Intellect Econ 9
UN Department of Economic and Social Affairs (2014) World urbanization prospects, 2014 revision. UN
UN Population (2018) World urbanization prospects: the 2018 revision. UN

Van den Berg J (2011) Energy conservation more effective with rebound policy. Environ Resour Econ 48:43–58
Van den Berg A, Maas J, Verjrok T, Groenewegen P (2010) Green space as a buffer between stressful life events and health. Soc Sci Med 70(8)
Van Doren B, Horton K, Dokter A et al (2017) High-intensity urban light installation dramatically alters nocturnal bird migration. PNAS 114(42)

The manufacturer's authorised representative in the EU is Springer Nature Customer Service Centre GmbH, Europaplatz 3, 69115 Heidelberg, Germany. If you have any concerns regarding our products, please contact ProductSafety@springernature.com

Printed and bound by CPI Group (UK) Ltd, Croydon, CR0 4YY
26/09/2026
02222889-0001